Modern Livestock and Poultry Production

9th Edition

Frank B. Flanders and James R. Gillespie

CENGAGE Learning

Australia • Brazil • Mexico • Singapore • United Kingdom • United States

Modern Livestock and Poultry Production, 9th Edition
Frank B. Flanders and James R. Gillespie

SVP, GM Skills & Global Product Management: Dawn Gerrain

Product Director: Matthew Seeley

Product Team Manager: Erin Brennan

Product Manager: Nicole Sgueglia

Senior Director, Development: Marah Bellegarde

Senior Product Development Manager: Larry Main

Content Developer: Richard Hall

Product Assistant: Jason D. Koumourdas

Vice President, Marketing Services: Jennifer Ann Baker

Marketing Manager: Scott Chrysler

Senior Production Director: Wendy Troeger

Production Director: Andrew Crouth

Production Service: Prashant Kumar Das, MPS Limited

Senior Content Project Manager: Elizabeth C. Hough

Senior Art Director: Benjamin Gleeksman

Digital Project Manager: Christina Brown

Composition: MPS Limited

Cover images: Hay field background image: ©Mikael Goransson/Shutterstock.com; hereford cow: ©iStockphoto/John Nielsen; two piglets: ©iStockphoto/Jeff Fullerton; feeding goats: ©iStockphoto/BORSEV; Horse: ©iStockphoto/Cathleen Abers-Kimball; hens: ©iStockphoto/johnnyscriv; dairy cow:©iStockphoto/Jason Lugo

Design images: Hay field background image: ©Mikael Goransson/Shutterstock.com; dairy cow: ©iStockphoto/Jason Lugo; hens: ©iStockphoto/johnnyscriv; two piglets: ©iStockphoto/Jeff Fullerton; horse: ©iStockphoto/Cathleen Abers-Kimball; Hereford cow: ©iStockphoto/John Nielsen; feeding goats: ©iStockphoto/BORSEV

Library of Congress Control Number: 2014939094

ISBN: 978-1-133-28350-8

Cengage Learning
20 Channel Center Street
Boston, MA 02210
USA

Cengage Learning is a leading provider of customized learning solutions with office locations around the globe, including Singapore, the United Kingdom, Australia, Mexico, Brazil, and Japan. Locate your local office at **www.cengage.com/global**

Cengage Learning products are represented in Canada by Nelson Education, Ltd.

To learn more about Cengage Learning, visit **www.cengage.com**

Purchase any of our products at your local college store or at our preferred online store **www.cengagebrain.com**

Notice to the Reader

Publisher does not warrant or guarantee any of the products described herein or perform any independent analysis in connection with any of the product information contained herein. Publisher does not assume, and expressly disclaims, any obligation to obtain and include information other than that provided to it by the manufacturer. The reader is expressly warned to consider and adopt all safety precautions that might be indicated by the activities described herein and to avoid all potential hazards. By following the instructions contained herein, the reader willingly assumes all risks in connection with such instructions. The publisher makes no representations or warranties of any kind, including but not limited to, the warranties of fitness for particular purpose or merchantability, nor are any such representations implied with respect to the material set forth herein, and the publisher takes no responsibility with respect to such material. The publisher shall not be liable for any special, consequential, or exemplary damages resulting, in whole or part, from the readers' use of, or reliance upon, this material.

Printed in the United States of America
Print Number: 08 Print Year: 2023

Dedication

Dr. Frank Flanders dedicates his work on this book to his father, Ivy Artis Flanders (1923–2006). "To him I owe more than I could ever repay. He provided me the opportunity to succeed."

Contents

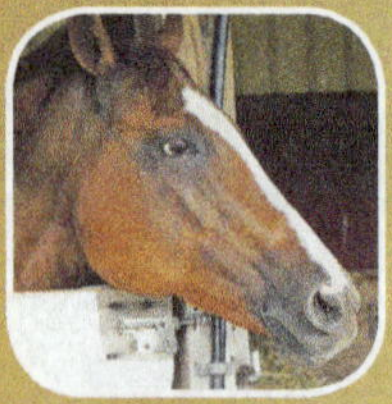

Preface

Modern Livestock and Poultry Production, Ninth Edition, is designed for agriculture education students who require competency in all phases and types of livestock production. Its comprehensive, balanced development emphasizes readability, organization, and hands-on activities. The text is based on the most up-to-date information available and is applicable to all areas of the United States.

How This Text Is Organized

Section 1 includes a general introduction to the livestock industry—its history, the careers available, and the importance of safety and environmental considerations. Section 2 introduces the student to the topics of anatomy, physiology, feeding, and nutrition, while Section 3 provides a sound basis for the understanding and practice of animal breeding. Sections 4 through 10 detail the production of beef cattle, swine, sheep and goats, horses, poultry, dairy cattle, and alternative animals. Each section presents information on selection of stock, feeding, management, housing, diseases and parasites, and marketing. The section on horses also includes a chapter on training and horsemanship.

Features of This Text

This text is designed in such a way as to enhance the teaching and learning experience. A bulleted format for important points, numerous subheads, and chapter summaries facilitate the learning process. A list of suggested Student Learning Activities at the end of each chapter provides the student and teacher with ideas for the hands-on experiences that are essential for success in the livestock production field. The end-of-chapter questions provide a comprehensive review of the material presented in each chapter and allow the teacher to measure student understanding.

Other outstanding features include the text's many color illustrations and photographs, tables and charts, an appendix, and an expansive glossary.

New to This Edition

Every effort has been made to thoroughly update and revise this very successful textbook to reflect the most current information available. The ninth edition of *Modern Livestock and Poultry Production* is designed with the teacher and student in mind:

- A new chapter format includes learning objectives and a list of key terms at the beginning of each chapter.
- The learning objectives focus attention on the information and skills to be learned, which are then tested in the review section at the end of the chapter.
- In addition, having the vocabulary presented at the beginning of each chapter allows teachers and students to obtain a general idea of some of the concepts to be covered in the chapter. Teachers may use these terms for the classroom word wall, and students can build their vocabulary as they develop skills in livestock and poultry production.
- Connection boxes are placed throughout the book to integrate concepts in livestock and poultry production with other curricula and life applications. Each Connection provides interesting information related to a specific animal science topic. These newly added elements are designed to spark student interest and encourage students to think beyond the scope of the classroom.
- A major revision of Section 6, Sheep and Goats, includes one of the most comprehensive, in-depth references available on judging goats. The production of goats, and especially meat goats, has become a major agricultural enterprise in recent years. This revised chapter prepares students for the production of goats as well as showing and judging goats.

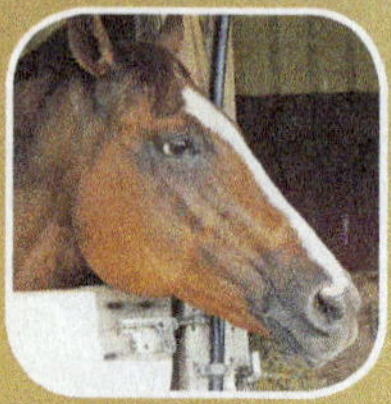

- This revision also contains updated and easier to understand graphics and tables.
- The statistical facts and figures presented were verified or revised based on the most up-to-date information available.
- The example problems in Chapter 8, Balancing Rations, were reformatted in a cleaner and easier to understand design.

Supplement to This Text

Modern Livestock and Poultry Production, ClassMaster CD-ROM

ISBN: 978-1-1332-8352-2

A complete instructor's teaching package on one CD-ROM! The ClassMaster is an innovative set of teaching tools designed to correlate directly to the text materials. They work together, reinforcing each other, but may also be used individually.

The ClassMaster contains the following educational tools:

- Instructor's Resource Guide to the text (in PDF format). The Instructor's Resource Guide includes: lesson plans; answers to the end of chapter questions; an objective test with answer key, lab manual answer key, and additional instructor information.
- Instructor support slide presentations that can be customized in PowerPoint® format focusing on key points for each chapter. Approximately 1,000 slides (about 15 to 20 slides per chapter) are available to accompany the textbook.
- A computerized test bank created in ExamView® makes generating tests and quizzes a snap. With more than 1,000 questions and different question formats from which to choose, you can create customized assessments for your students with the click of a button. Add your own unique questions and print rationales for easy class preparation.
- New! An Image Library containing all the figures illustrated in the text. This new library can be used in slide presentations or as part of classroom discussion.
- Various Correlation and Conversion Guides help you map content to National AFNR Standards and move from earlier editions to the new ninth edition.

Instructor Companion Website

New! The instructor companion website provides online access to many of the instructor support materials provided on the ClassMaster CD-ROM, including the *Instructor's Resource Guide* computerized test bank files, correlation guides, and support slides. To access the available materials, sign up for a faculty account at login.cengage.com. Add the core textbook to your bookshelf using the 13-digit ISBN that appears on the back cover of the textbook.

Lab Manual

New to this edition is a lab manual, complete with exercises and activities to accompany the critical content areas covered throughout the book. This resource reinforces text content through the practical application of real-world examples. It is recommended that students complete corresponding lab components to further develop understanding of essential skills and concepts discussed in the text.

MindTap for Modern Livestock and Poultry Production, 9E

New! MindTap is a personalized teaching experience with relevant assignments that guide students to analyze, apply, and improve thinking, allowing you to measure skills and outcomes with ease.

- Personalized Teaching: Becomes yours with a Learning Path that is built with key student objectives. Control what students see and when they see it. Use it as-is or match to your syllabus exactly—hide, rearrange, add and create your own content.
- Guide Students: A unique learning path of relevant readings, multimedia and activities that move students up the learning taxonomy from basic knowledge and comprehension to analysis and application.
- Promote Better Outcomes: Empower instructors and motivate students with analytics and reports that provide a snapshot of class progress, time in course, engagement and completion rates.

The MindTap for *Modern Livestock and Poultry Production*, 9E features a complete integrated course combining additional quizzing and assignments, interactivities and video clips along with the enhanced ebook to further facilitate learning.

About the Authors

Dr. Frank B. Flanders is an Assistant Professor of Agricultural Education at the University of Georgia. He has served agriculture education for more than 35 years. Dr. Flanders taught high school agriculture education for six years and was a Special Instructor in Agricultural Education at the University of Georgia for 15 years. Dr. Flanders has extensive experience in developing agriculture education teaching materials. He served as the Agriculture Education Curriculum Coordinator in Georgia, where he developed and maintained the Agricultural Education website, produced instructional DVDs and CDs for teachers, managed curriculum projects, and taught numerous workshops for teachers. He served three years as State Director of the Career, Technical, and Agricultural Education Resource Network Curriculum Unit for Georgia. Dr. Flanders served on the writing team for the National Standards in Agriculture Education and served as lead writer for the National Standards in Animal Science. He is the ninth generation of his family to be actively involved in American agriculture. In 2003, Dr. Flanders was inducted into the Georgia Agriculture Education Hall of Fame and received the National Outstanding Faculty Advisor Award from the Professional Fraternity Association. He has also received the Outstanding Faculty Member in Agriculture Education Award and the Outstanding Service and Faculty Member Award from the University of Georgia Student Government Association.

James R. Gillespie, the original author of this text, had extensive training and experience in the field of livestock production and agricultural education. He received his BS and MS degrees in Agricultural Education from Iowa State University and an Education Specialist degree in School Administration from Western Illinois University. Mr. Gillespie taught agricultural education at the high school and adult education levels. In addition to other agriculture-related positions, he was self-employed in farming and employed by the Illinois State Board of Education, Department of Adult, Vocational, and Technical Education, Program Approval and Evaluation as a Regional Vocational Administrator, Region II. Mr. Gillespie was a member of Phi Delta Kappa and the Honor Society of Phi Kappa Phi. James R. Gillespie passed away in 2006. His previous editions of this book set the gold standard for instruction in livestock and poultry production. The eighth and ninth editions, revised by Dr. Frank Flanders, carry on the excellence in textbook publications on livestock and poultry production established by Mr. Gillespie.

Acknowledgments

The author expresses appreciation to the many people who made contributions to the revision of *Modern Livestock and Poultry Production*, Ninth Edition. Special thanks are extended to Alyssa K. Elrod, Bailey M. Taylor, and Shannon R. Lawrence for their expert editing, research assistance, technical expertise, and photographs. Thanks especially to Alyssa and Bailey for their dedication to this project and for keeping it on track. Also, thanks to Dr. Ray V. Herren, Tommy Waldrop, and Christa Steincamp for their technical advice and encouragement. Thanks to the many University of Georgia students for their work in researching and assembling materials. These include Ashley Sapp, Samantha Meeks, Rikki Griffith, James Hale, and Lauren Ball. Lastly, thank you to everyone on the editorial and publishing staff of Cengage Learning and others for their hard work and patience throughout this revision; special thanks to Laura Stewart, Betsy Hough, Richard Hall, Nicole Sgueglia, and Prashant Kumar Das.

Frank B. Flanders

The authors and Cengage Learning would like to thank those individuals who reviewed the manuscript and offered suggestions, feedback, and assistance. Their work is greatly appreciated.

Frederick Aherns—Hernando High School, Brooksville, FL
Gary Blankenship—Department Chair, Hinckley-Big Rock High School, Hinckley, IL
Jonathan Merriam—Denair High School, Denair, CA
Ken Moncrief—Central Valley High School, Ceres, CA
Dusty Spencer—Agriculture Teacher, Honey Grove High School, Honey Grove, TX

How to Use This Textbook

Modern Livestock and Poultry Production, Ninth Edition, offers students and teachers features and benefits that enhance learning, many of which are highlighted here.

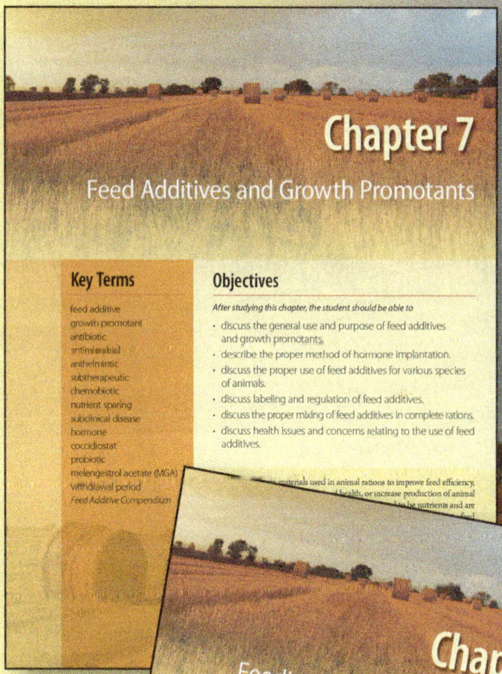

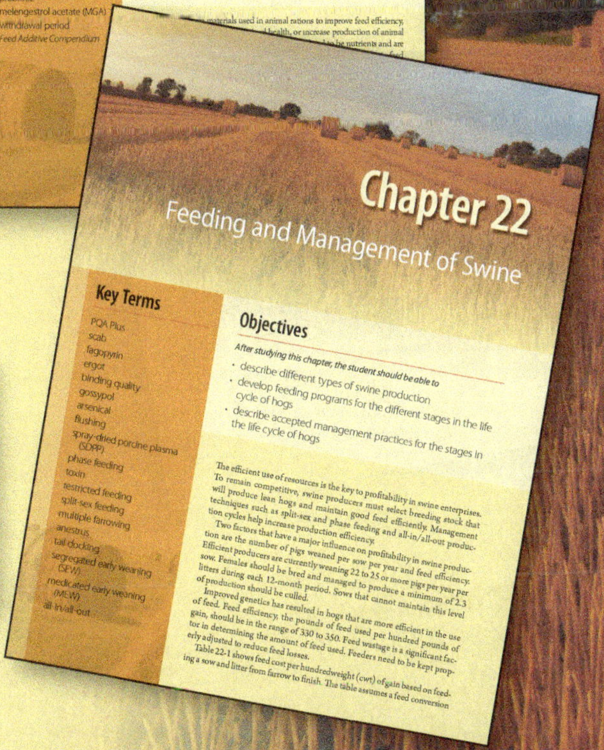

Chapter Objectives

Objectives at the start of each chapter focus attention on the information and skills to be learned and then tested in the review section at the end of the chapter.

New! Key Terms Lists

Having the vocabulary presented at the beginning of each chapter allows teachers and students to obtain a general idea of some of the concepts to be covered in the chapter. Key terms are also highlighted in the readings for cross reference.

New! Connections

New Connections features are placed throughout the book to integrate concepts in livestock and poultry production with other curricula and life applications.

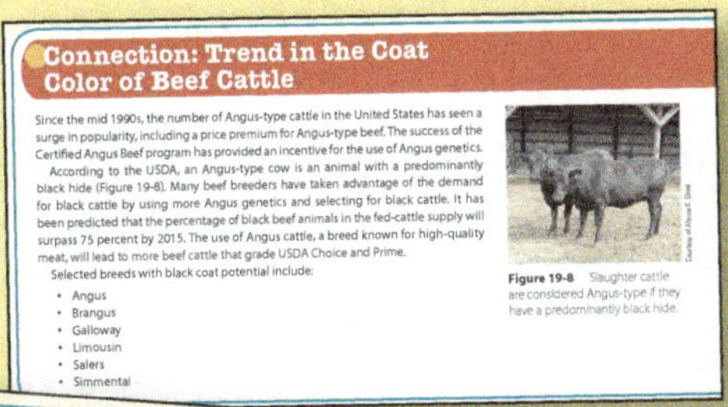

Connection: Trend in the Coat Color of Beef Cattle

Since the mid 1990s, the number of Angus-type cattle in the United States has seen a surge in popularity, including a price premium for Angus-type beef. The success of the Certified Angus Beef program has provided an incentive for the use of Angus genetics.

According to the USDA, an Angus-type cow is an animal with a predominantly black hide (Figure 19-8). Many beef breeders have taken advantage of the demand for black cattle by using more Angus genetics and selecting for black cattle. It has been predicted that the percentage of black beef animals in the fed-cattle supply will surpass 75 percent by 2015. The use of Angus cattle, a breed known for high-quality meat, will lead to more beef cattle that grade USDA Choice and Prime.

Selected breeds with black coat potential include:

- Angus
- Brangus
- Galloway
- Limousin
- Salers
- Simmental

Figure 19-8 Slaughter cattle are considered Angus-type if they have a predominantly black hide.

Connection: Kobe Beef

Kobe beef is becoming more popular in the United States (Figure 19-4). Many producers of Wagyu cattle in the United States refer to their beef as "American Style Kobe beef" or "Wagyu beef." While many Wagyu cattle in the United States are crossbred, the beef is still noted for having exceptionally high marbling and taste. Kobe beef is available at specialty stores and can cost more than $100 per pound.

Price Comparison of Traditional Beef vs. Kobe Beef (Restaurant Pricing)

Filet Mignon **12 oz. Traditional American Beef**	**$45.00**
Filet Mignon **12 oz. Wagyu (Kobe) Beef**	**$107.00**

Figure 19-4 Kobe beef is generally considered a delicacy, renowned for its flavor, tenderness, and fatty, well-marbled texture.

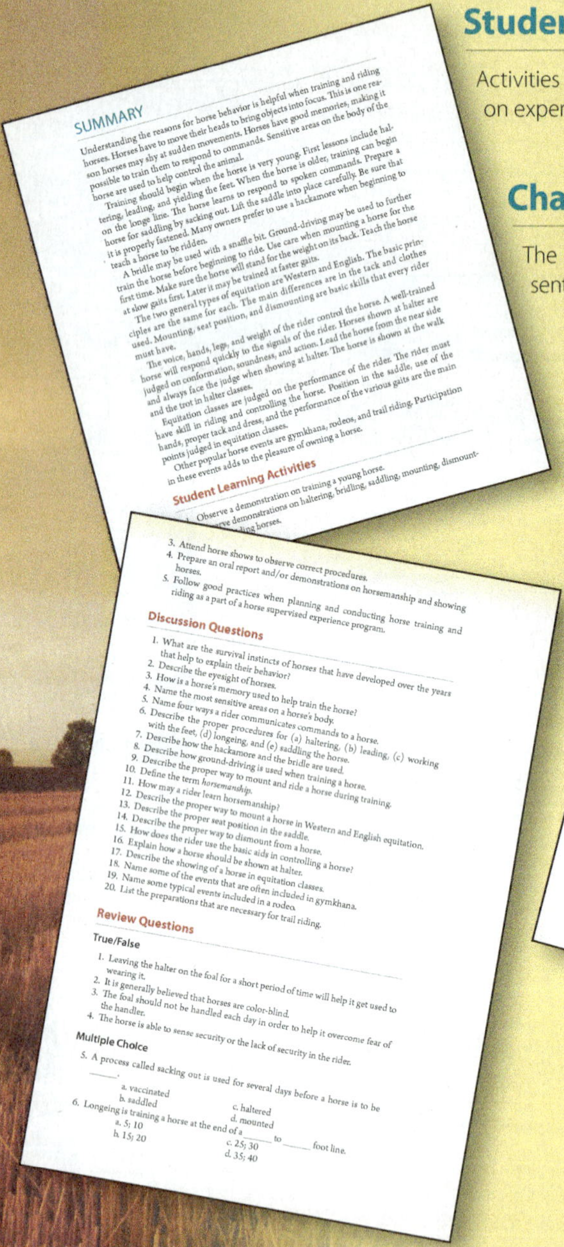

Student Learning Activities

Activities at the end of each chapter provide the student and teacher with ideas for hands-on experiences essential for success in livestock production.

Chapter Review

The end-of-chapter questions provide a comprehensive review of the material presented in each chapter and allow the teacher to assess student understanding.

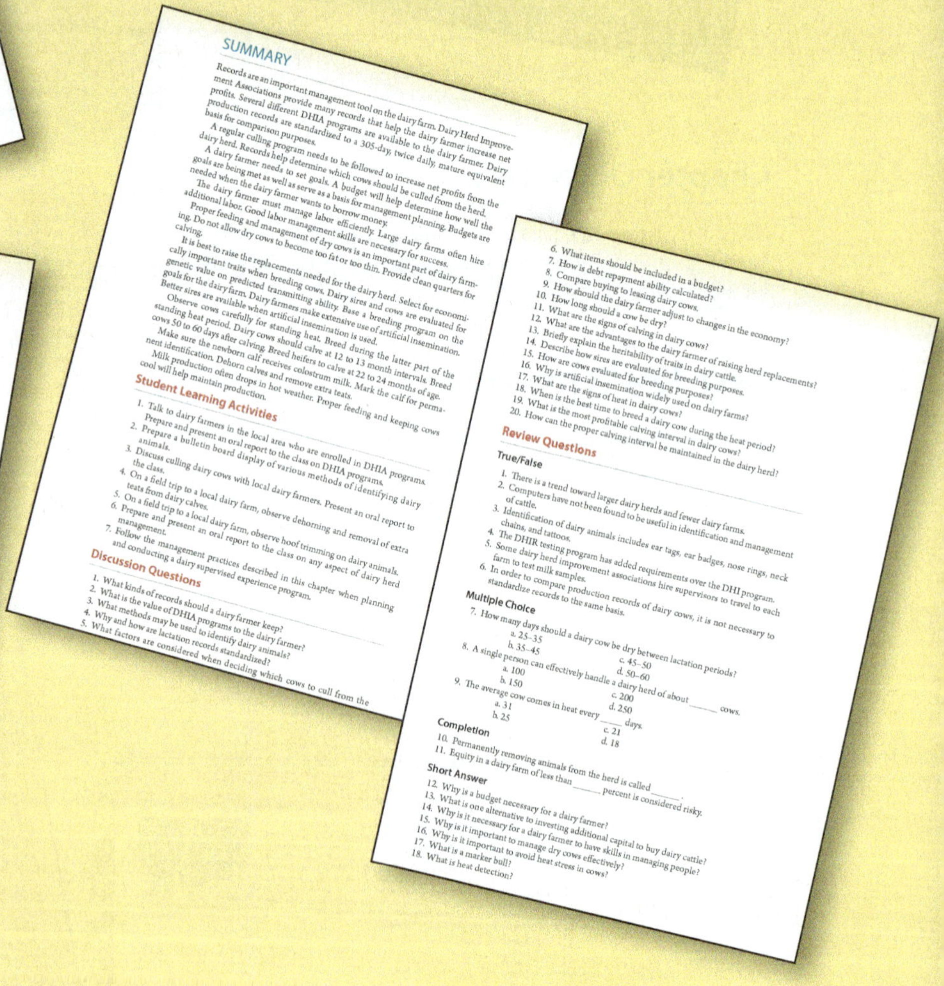

SECTION 1

The Livestock Industry

Chapter 1

Domestication and Importance of Livestock

Key Terms

domesticate
Bos taurus
Bos indicus
selection
crossbreeding
Sus scrofa
Sus vittatus
Mouflons
Asiatic urial
Eohippus
draft animal
Gallus gallus
Anas boschas
homeothermic
non-ruminant
ruminant
rumen
by-product
insulin
cortisone
thrombin
heparin

Objectives

After studying this chapter, the student should be able to

- discuss briefly the history of the domestication of farm animals.
- list and explain the functions of livestock.
- describe the size of the livestock industry in the United States.

Livestock production is an important part of farming in the United States. People are dependent upon livestock production for food, clothing, and many other products. The production of livestock involves selection, breeding, feeding, care, and marketing of animals.

Success in raising livestock depends on many factors. Farmers must have knowledge, skill, and patience. They must use the results of research by animal scientists. Research in animal science is carried on by many state universities and the U.S. Department of Agriculture (USDA). Many of the commercial firms that supply and service the animal industry also do research in animal science.

DOMESTICATION OF ANIMALS

Our present civilization has its roots in the domestication of animals. To **domesticate** means to adapt the behavior of an animal to fit the needs of people. The domestication of animals began when early humans had contact with wild animals, which they hunted for food and skins. After a period of time, these early humans began to confine and breed animals to ensure a steadier supply of food and clothing. Humans later learned to select and breed animals with certain desirable characteristics. As a result of selective breeding, identifiable breeds began to be developed that would breed true for those characteristics that were determined to be desirable. Before the human race learned to tame and raise animals, it was dependent on hunting and wild plants for food and clothing. With the domestication of animals came the beginnings of a more settled way of life. Domesticated animals supplied a surer source of food and clothing. A better food supply meant an increase in population. More people made it possible to divide the labor within the tribe. Some historians believe that the human race would never have become civilized without the domestication of animals.

Key Terms *continued*

epinephrine
rennet
corticotropin
soring
Hazard Analysis and Critical Control Points (HACCP)
critical control points (CCPs)
food irradiation
radura

Cattle

Modern cattle are descendants of ***Bos taurus*** and ***Bos indicus***. *Bos taurus* are domestic cattle that came from either the Aurochs or the Celtic Shorthorn. The Aurochs were common in Europe. The Celtic Shorthorn were found in the British Isles. *Bos indicus* are the humped cattle originating in tropical countries and are common in modern cattle production throughout the United States. They are more resistant to some diseases, parasites, and heat than cattle that came from *Bos taurus*.

Cattle were probably tamed early in the Neolithic (New Stone) Age, which occurred about 18,000 years ago. Early humans used cattle for pulling loads, meat, and milk. Cattle were also a measure of wealth. They are mentioned in records that are at least 4,000 years old. Various types of cattle were known at that time.

Selection and crossbreeding of cattle for different purposes began early in the history of agriculture. **Selection** means to identify and use for breeding purposes those animals with traits that are considered by the breeder to be desirable. **Crossbreeding** is the mating of animals of different breeds. Selection and crossbreeding resulted in improvement of the animal and the development of different breeds. Most of the improvement has taken place since the middle of the eighteenth century.

Cattle are not native to the United States. Christopher Columbus brought cattle to the New World on his second voyage in 1493. More cattle were brought by Portuguese traders in 1553. The English were the first to bring large numbers of cattle to the United States when they founded the Jamestown colony in 1611. The Spanish type of longhorn cattle arrived in Mexico in 1521. They later spread throughout the western United States as they were brought to Christian missions built by the Spanish.

The number of cattle soon began to grow. Early pioneers took cattle with them as they moved westward. The major growth of the large cattle herds took place in the Great Plains states. This happened because plentiful grazing land was available in this area. The North Central states, which have good supplies of grain, became the main region for finishing cattle (feeding the cattle to market weights) (Figure 1-1).

Courtesy of USDA

Figure 1-1 Many cattle are shipped to be finished in feedlots, which are cost efficient due to abundance of corn in the region.

The early pioneers took dairy cattle as well as beef cattle with them as they moved westward. Before the 1850s, most farms had at least one or two dairy cows that provided milk and butter for the family. During the last half of the nineteenth century, however, dairy herds began to increase in size as a market for dairy products began to develop in the growing cities of the United States.

The numbers of dairy cattle in the United States continued to increase until they reached a peak in the middle 1940s. Since that time, the dairy cattle population of the United States has been declining steadily. Milk is still produced in all of the 50 states, but dairying as a major enterprise on the farm tends to be concentrated in several states, as shown in Table 1-4.

Swine

American breeds of swine come from two wild stocks: the European wild boar (***Sus scrofa***) and the East Indian pig (***Sus vittatus***). Some wild types of piglike animals, which have never been tamed, still exist in certain parts of the world.

The first use of swine for food probably occurred in the Neolithic Age. The first people to tame swine were the Chinese. Written records show that this took place about 4900 B.C. Biblical references to swine occur as early as 1500 B.C. The keeping of swine in Great Britain is mentioned as early as 800 B.C.

Courtesy of USDA. Photo by Scott Bauer

Figure 1-2 The Corn Belt states are the nation's leading hog-producing states. Much of the feed used in raising hogs is produced in the Corn Belt.

Swine were brought to the New World by Columbus on his second voyage in 1493. More swine were brought later by Spanish explorers. The first major increase in the number of swine in the United States occurred in 1539 when the Spanish explorer Hernando DeSoto brought 13 head of hogs to Florida. As DeSoto moved westward during his exploration, the hogs were taken along. Three years later, by the time the Spanish had reached the upper Mississippi valley, the number of swine had increased to 700 head.

English settlers also brought swine with them to America. The herds grew rapidly in size. Production was soon greater than the local need. Pork and lard were then exported in trade for other products. The main expansion of the swine industry occurred in the Corn Belt states, an area where feed for finishing hogs for market was available (Figure 1-2).

Sheep

Sheep were among the first animals tamed by the human race. They were first tamed during the early Neolithic Age. Sheep are shown on an early Egyptian sculpture dated about 4000 B.C. and are mentioned in the earlier passages of the Bible. Sheep bones have been found in caves and lake dwellings used by early people of Europe.

Wool fabrics have been found in the ruins of Swiss Lake villages. These date back to between 10,000 and 20,000 years ago. The Babylonians used wool for clothing around 4000 B.C. Sheep flocks were important in both Spain and England by 1000 A.D. and by 1500 A.D, both countries were major sheep-producing areas.

The ancestry of sheep is not as well known as that of other domestic animals. There are more than 200 breeds of sheep in the world. All of the breeds are timid, defenseless, and the least intelligent of the tamed animals. These traits are the result of selection for herding in large bands.

Most present-day sheep probably came from the wild sheep called ***Mouflons*** and the ***Asiatic urial***. The wild, big-horned sheep of Asia are also the ancestors of some of the present-day breeds.

The only sheep native to North America are the Big Horn or Rocky Mountain sheep. The present-day domestic breeds came from sheep that were imported. Columbus brought sheep to the New World on his second voyage in 1493. Merino sheep were brought to Mexico by Cortez in 1519. British breeds of sheep were imported to Virginia in 1609. Sheep were used by the early colonists mainly for wool production. Importation of breeding stock continued and flock owners selected certain animals to breed that would improve the wool produced.

Courtesy of USDA. Photo by Scott Bauer

Figure 1-3 Sheep grazing on western range land convert roughage into wool and meat.

By 1810, the northeastern part of the United States was the sheep-producing center of the country. However, sheep production gradually moved westward. The number of sheep increased, and by 1840, there were 19 million head of sheep in the United States. The center of the sheep-producing industry moved west when inexpensive rangeland became available (Figure 1-3).

Goats

Goats were first tamed in the Neolithic Age. They may have been the first tamed animals in Western Asia. The goat is believed to be descended from the Pasang or Grecian Ibex. These are species of wild goats found in Asia Minor, Persia, and other countries. Other wild species, such as the Markhors and Tahrs, may be ancestors of some of today's domestic breeds of goats.

Goats are closely related to sheep. The following are the major differences between the two species.

Sheep	Goats
Stockier body	Shorter tails
Spirally twisted horns	Horns that are long and grow upward, backward, and outward
Males do not have beards	Males have beards
Males do not give off a strong odor in the rutting (breeding) season	Males give off a strong odor in the rutting (breeding) season
Males have scent glands in the face and feet that secret pheromones that attract females	Males do not have scent glands in the face and feet that secret pheromones that attract females
Less intelligent animal with less ability to fight and fend for itself	More intelligent animal with a greater ability to fight and fend for itself
Unable to easily return to the wild state	Able to easily return to the wild state

Goat remains have been found in the Swiss Lake villages of the Neolithic Age. The Bible refers to the use of mohair from goats.

Early goat importations into the United States came from Switzerland. Records show that milk goats were brought to Virginia and the New England states by Captain John Smith and Lord Delaware. Angora goat flocks in the United States came from Turkey. Most of the increase in milk goat numbers has occurred since 1900. Milk goats are found all over the United States. Many are kept on small farms, and there are few large herds of milk goats. Production of mohair from Angora goats has also increased mainly in the twentieth century. Most Angora goats are found in the western states. Texas has the largest number of Angora goats in the United States.

Horses

The horse evolved from a tiny four-toed ancestor called **Eohippus** (dawn horse). Eohippus (Figure 1-4) was about a foot high and lived in swamps about 58 million years ago. The descendants of Eohippus gradually grew in size. Changes in the feet and skeleton occurred. Eventually, the animal was better adapted to the prairie than to the swamp. Eohippus was native to the North American continent. However, it had disappeared entirely before the Europeans discovered the New World. There were no horses present in the New World when Columbus made his voyages.

The first domestication of the horse seems to have been in Central Asia or Persia before 3000 B.C. The use of the horse spread from there into Europe. Babylonian records show the use of the horse by 2000 B.C. Egyptians were using horses by 1600 B.C. Horses were also used by the ancient Greeks and Romans. The Arabs did not begin to use horses until after about 600 A.D. Columbus brought horses to the New World on his second voyage in 1493. In 1519, Cortez imported horses into Mexico. The first importation of horses into what is now the United States was by DeSoto in 1539. During DeSoto's explorations throughout the Southeast, many of these horses were left behind. Other Spanish explorers also brought horses with them.

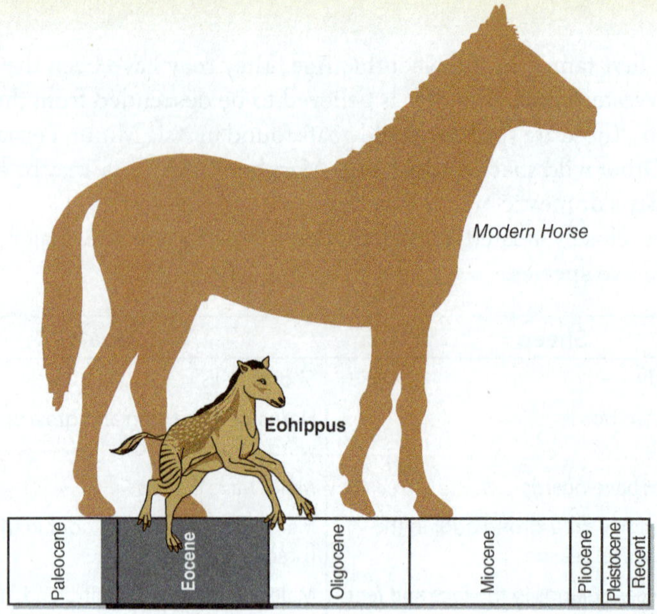

Figure 1-4 Eohippus, the horse ancestor, was native to North America and lived about 58 million years ago.

Courtesy of USDA

Figure 1-5 Horseback riding is a major recreational activity in the United States.

These horses spread throughout the western part of what is now the United States. Wild horses found on western ranges are the descendants of these tamed horses.

Saddle horses and draft horses were brought to the United States by the early colonists. For many years, oxen were used as the main **draft animal**, or an animal used for pulling loads, of the colonists. Horses served mainly as pack animals and for riding. Heavier draft horses were brought into the colonies by Dutch, Puritan, and Quaker settlers. These horses were used mainly for work. The early development of the horse in the United States was primarily associated with riding on plantations. Horse racing also developed into a sport during the 1700s and early 1800s. With the development of other sources of power, the use of draft horses on farms declined. Most horses in the United States are now used for riding and racing (Figure 1-5).

Poultry

Chickens were domesticated in India and were being raised by the Chinese and Egyptians about 1400 B.C. Drawings of chickens have been found in Egyptian tombs dating from around 1400 B.C.

Although poultry and eggs were used for food early in history, poultry raising has only recently become a major commercial enterprise. In the past, most poultry was raised on small family flocks.

The wild jungle fowl of India (***Gallus gallus***) may have been the early ancestor of most tame chickens (Figure 1-6). Some other wild species may also have been involved in the development of chickens.

Turkeys are believed to be descended from two wild species. One is found in Mexico and the other in the United States. The turkey was probably tamed by the people originally living in America. Most of the American varieties were probably developed from the species found in the United States.

The wild mallard duck (***Anas boschas***) is thought to be the ancestor of all domestic breeds of ducks. Ducks were tamed at an early date. The Romans referred to ducks more than 2,000 years ago. China has probably raised ducks on a commercial basis longer than other parts of the world.

Courtesy of Randy Glance

Figure 1-6 The wild jungle fowl of India (*Gallus gallus*) is the ancestor to most domestic chickens today.

The goose was probably tamed shortly after the chicken. It was regarded as a sacred bird in Egypt 4,000 years ago. Geese were well distributed over all of Europe 2,000 years ago.

Poultry were brought to the New World by early explorers and colonists. There were poultry in the Jamestown settlement in 1607. Poultry shows, which began to be held around 1850, were important in the development of recognized breeds. In the second half of the nineteenth century, older breeds were perfected and new breeds were developed. The American Poultry Association was formed in 1873. The *American Standard of Perfection* was first published in 1874. The purpose of the association and the *Standard of Perfection* was to standardize breeds for purposes of showing.

The American poultry industry grew out of the small home flocks raised by early settlers. For a long time, poultry raising was mainly a small enterprise on the farm. However, as the population grew, the demand for poultry products increased. Poultry production began to be more specialized in the late 1800s and today is a highly specialized industry. Much of the poultry industry is concentrated in the southern part of the United States (Figure 1-7). Few small farm flocks remain; however, this type of flock is growing again in popularity because of consumer demand for organic or free range products. Most poultry is raised in large confinement flocks.

© Enrico Jose/www.Shutterstock.com

Figure 1-7 Today most eggs are produced by caged layers in large confinement flocks.

CLASSIFICATION OF COMMON FARM ANIMALS

A great many species of plants and animals exist on the earth. It is necessary to have a logical system to classify these different species in order to study and describe them. In the past, all living things were divided into two large groups called kingdoms—Plantae (plants) and Animalia (animal). Some forms of life (bacteria, fungi, and protozoa) do not fit easily into either of these kingdoms, however, so other systems of grouping have been developed. Some systems use three large kingdoms, some use four, and some use five.

A series of terms is used for the scientific name of a member of a kingdom. This is called a hierarchy of categories. Living things are classified as follows: kingdom, phylum (or division), class, order, family, genus, species. Each division of the hierarchy contains one or more groups from the next lower division. For example, a genus contains a group of related species, a family contains a group of related genera, an order contains a group of related families, and so on. This method of grouping continues up the hierarchy so that a kingdom contains all the groups listed below it. Major groups may be divided into subphyla, subclasses, suborders, subfamilies, or subgenera if necessary. Species may also be subdivided into varieties, breeds, or strains for further grouping.

A species is a group of animals with many common traits. When members of this group mate within the group, they will produce fertile young.

All of the domestic animals found on the farm are included in the Animalia kingdom. Domestic farm animals are further classified as phylum Chordata, subphylum Vertebrata, and either class Mammalia (for livestock) or class Aves (for poultry).

Members of class Mammalia, or mammals, share the following important characteristics:

- four-chambered hearts
- diaphragm that separates the thoracic (chest) and abdominal cavities; enables more efficient breathing
- mother secretes milk for the young through the mammary glands
- warm-blooded
- bodies are covered with hair, which provides insulation
- homeothermic
- embryo develops in the mother's uterus, and the young are born alive

There are other traits of the class Mammalia, but these are some of the more important ones shared by farm animals. Common domesticated animals that belong to this class include cattle, sheep, goats, horses, donkeys, swine, bison, alpacas, llamas, and rabbits.

Members of class Aves also have four-chambered hearts and are **homeothermic**, meaning they can maintain their own body temperature. Instead of hair, their bodies are covered with feathers, which provide insulation. Instead of a diaphragm, they have hollow bones and an air-sac system attached to the lungs. They lay eggs in which their young develop. They do not secrete milk to nourish the young. These animals have beaks and gizzards instead of teeth. Chickens, ducks, emus, geese, ostrich, rheas, and turkeys all belong to this class.

The scientific name of an animal includes its genus and species names. The genus name is capitalized; the species name is not. In written material, the genus and species names are underlined or italicized.

The classifications of some common farm animals are listed in Table 1-1.

TABLE 1-1 Classification of Common Farm Animals

Common Name	Phylum	Subphylum	Class	Order	Family	Genus	Species
Alpaca	Chordata	Vertebrata	Mammalia	Artiodactyla	Camelidae	*Llama*	*pacos*
Bison	Chordata	Vertebrata	Mammalia	Artiodactyla	Bovidae	*Bison*	*bison*
Cattle	Chordata	Vertebrata	Mammalia	Artiodactyla	Bovidae	*Bos*	*taurus*
Cattle	Chordata	Vertebrata	Mammalia	Artiodactyla	Bovidae	*Bos*	*indicus*
Chicken	Chordata	Vertebrata	Aves	Galliformes	Phasianidae	*Gallus*	*domesticus*
Donkey	Chordata	Vertebrata	Mammalia	Perissodactyla	Equidae	*Equus*	*asinus*
Duck	Chordata	Vertebrata	Aves	Anseriformes	Anatidae	*Anas*	*platyrhyncha*
Emu	Chordata	Vertebrata	Aves	Casuariiformes	Dromiceidae	*Dromiceius*	*novaehollandiae*
Goat	Chordata	Vertebrata	Mammalia	Artiodactyla	Bovidae	*Capra*	*hircus*
Goose	Chordata	Vertebrata	Aves	Anseriformes	Anatidae	*Anser*	*anser*
Horse	Chordata	Vertebrata	Mammalia	Perissodactyla	Equidae	*Equus*	*caballus*
Llama	Chordata	Vertebrata	Mammalia	Artiodactyla	Camelidae	*Llama*	*glama*
Ostrich	Chordata	Vertebrata	Aves	Struthioniformes	Struthionidae	*Struthia*	*camelus*
Rabbit	Chordata	Vertebrata	Mammalia	Lagomorpha	Leporidae	*Oryctolagus*	*cuniculus*
Rhea	Chordata	Vertebrata	Aves	Rheiformes	Rheidae	*Rhea*	*americana*
Sheep	Chordata	Vertebrata	Mammalia	Artiodactyla	Bovidae	*Ovis*	*aries*
Swine	Chordata	Vertebrata	Mammalia	Artiodactyla	Suidae	*Sus*	*scrofa*
Swine	Chordata	Vertebrata	Mammalia	Artiodactyla	Suidae	*Sus*	*vittatus*
Turkey	Chordata	Vertebrata	Aves	Galliformes	Meleagrididae	*Meleagris*	*gallopavo*

FUNCTIONS OF ANIMALS

Some functions of livestock benefit all of society; however, other functions are important mainly to individual farms. Taken together, the functions of livestock are a vital part of the total agriculture of a nation. It is useful to understand the functions of livestock when selecting enterprises for a farm. Selecting enterprises involves the choice of what kind of livestock is to be raised. Some functions of livestock, such as the conversion of roughage into food, are factors in this management decision.

Converting Feed into Food

Livestock convert feed grains and roughages into food for human consumption. There is some controversy over the use of feed grains as livestock feed. In the face of world food shortages, it has been suggested that this is not the most efficient use of limited resources.

Non-ruminant animals such as swine and poultry are fed large amounts of grain because they cannot use much roughage in their diet. However, about 30 percent of the feed fed to swine and poultry in the United States consists of fish meal, meat and bone meal, milling and fermentation by-products, and tankage. These are feeds that generally cannot be used directly by humans for food.

Ruminants are animals of the suborder Ruminantia of the order Artiodactyla that have a stomach that is divided into several compartments. These animals regurgitate and masticate their feed after they swallow it. Animals in the subdivision Tylopoda have a three-compartment stomach. Typical animals found in this subdivision include camels, llamas, and alpacas. Animals in the subdivision Pecora have a four-compartment stomach. The Pecora are referred to as true ruminants. Typical animals found in this subdivision include cattle, bison, sheep, and goats. The largest of the compartments in true ruminants, the **rumen**, contains microorganisms that allow ruminants to digest many kinds of feed that non-ruminant animals cannot use effectively. Antelope, deer, gazelles, and giraffes are examples of other animals in the subdivision Pecora.

Ruminants are important because they have the ability to convert large quantities of materials that humans cannot use directly into edible food. Almost half of the chemical energy in the major cereal crops such as corn, wheat, and rice is found in parts of the plant, such as the stems, which are not used by humans for food. These crop residues can be converted to human food by ruminants. Waste products from a number of agricultural industries can be used as feed for ruminants. Examples include waste products from fruit and vegetable farming, citrus processing, sugar manufacturing, milling, and cotton ginning. Wood chips, sawdust, and shredded newspaper can also be used as feed for ruminants.

Hay, silage, and pasture are produced on about one-third of the cropland in the United States. Land used for all agricultural purposes accounts for approximately 52 percent of total U.S. land, while total grazing area (grassland pasture and range, cropland pasture, and grazed forests) makes up 35 percent of the total and two-thirds of all agricultural land. Forests and rangeland cover a large percentage of the land masses of the planet. Only about 11 percent of the world's land area is suitable for the production of foods that can be used directly by humans. About 75 percent of the total energy intake of beef and dairy cattle in the United States consists of roughages and other waste materials that cannot be used directly for human food. The ability of ruminants to convert much of what would not otherwise be available to humans as food adds significantly to the world's total food supply.

In nations with limited grain supplies, ruminants are fed largely on roughages. In the United States and some other nations of the world where grain is plentiful and relatively inexpensive, ruminants are fed rations high in grain during the finishing period. However, during the total lifetime of a beef animal, about 80 percent of the total feed used comes from roughages. Generally, these roughages cannot be used by humans directly for food.

About 80 percent of the human population of the world gets most of its protein, fats (lipids), iron, niacin, and some vitamins (including vitamin B12) from the meat produced by ruminant animals. Because of the kind of land and/or climate where they live, about 14 percent of these people have no other practical source of these nutrients. Food products from ruminants provide about 45 percent of the protein, 32 percent of the fat, 50 percent of the phosphorus, and 77 percent of the calcium found in diets of people living in the United States. About one-third of the total amount of food eaten by people in the United States comes from ruminants.

Animal products (ruminant and non-ruminant combined) are important sources of nutrients in the average diet in the United States. They provide 35 percent of the energy, 68 percent of the protein, 78 percent of the calcium, 39 percent of the iron, 42 percent of vitamin A, and 37 to 98 percent of various B vitamins consumed by people in the United States.

The most important livestock sources of protein and energy for human consumption are swine, beef, poultry, and sheep. The efficiency of the major farm animals in converting feed into protein and energy for human use is shown in Figure 1-8.

Clothing

Livestock provide fiber and skins for the production of clothing. The demand for animal fibers for clothing is lower now because of the increased use of synthetic

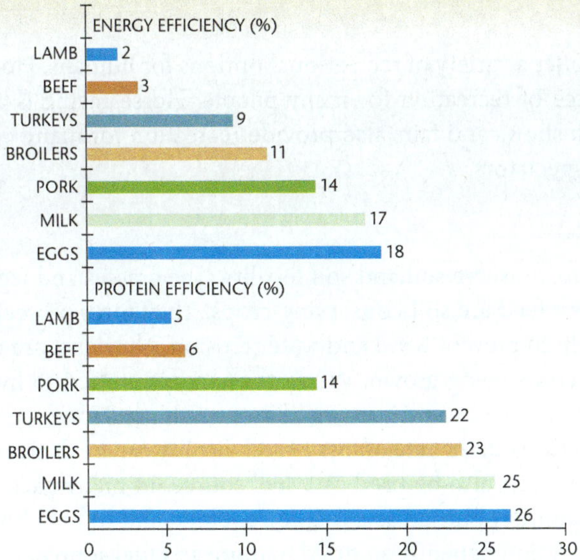

Figure 1-8 A comparison of the efficiency of major farm livestock in converting feed calorie intake to food calories output (energy efficiency) and converting crude protein in feed into edible protein in the form of meat, milk, and eggs (protein efficiency).

Sources: Vocational Agricultural Service, University of Illinois, and National Academy of Sciences.

fibers for clothing. The use of wool produced in the United States has decreased from 123.7 to 25 million pounds from 1980 to 2013 (Figure 1-9).

Most synthetic fibers are oil based, and the price of oil has risen dramatically in recent years; this means that animal fibers continue to be an important resource in human society. Leather is used for shoes, belts, gloves, and clothing, as well as for other products used by humans. From 5 to 10 percent of the market value of animals comes from the sale of hides. Leather has some characteristics that make it superior to synthetics for the production of clothing. It can allow air to pass through, is more durable, and is warmer than clothing made from synthetics.

Power

At one time, animals provided much of the power used by the human race. With the development of other sources of power, this use has declined. Very little animal power is now used in the United States. However, in some parts of the world, animals still provide much of the power used by humans.

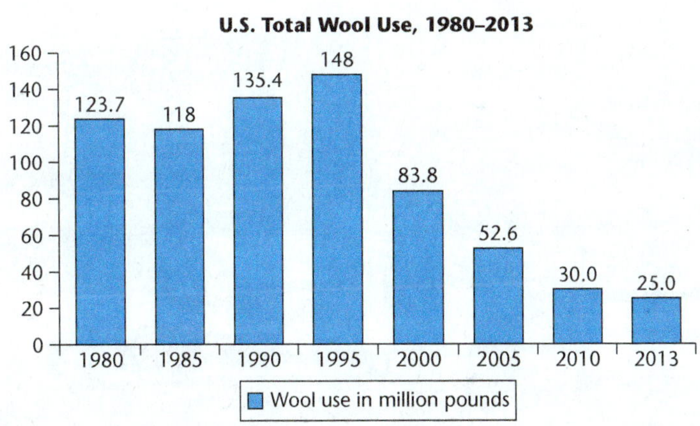

Figure 1-9 Total U.S. use of wool in millions of pounds.

Source: U.S. Wool Supply and Use, 1976–2013, *Cotton and Wool Yearbook 2014*, Economic Research Service, USDA.

Recreation

Livestock also offer a variety of recreational options for humans. Horseback riding is a major source of recreation for many people. Horse racing is also a popular sport. Livestock shows and fairs also provide recreation for many people, both as exhibitors and spectators.

Conservation

Livestock help to conserve soil and soil fertility. The grasses and legumes that are used for livestock feed are soil-conserving crops. They form protective covers on the land and help to prevent wind and water erosion. Nutrients are removed from the soil by the crops being grown. When these crops are fed to livestock, about 80 percent of the nutrient value is excreted in the manure. However, by putting the manure back on the soil, the rate of loss of soil fertility can be decreased.

Animal manure can also be used as a fuel source. In many parts of the world, dried animal manure is burned as a fuel for cooking and to heat homes. Some of the world's population depends on dried manure as a fuel source.

Animal manures can be used as a raw material in methane gas digesters. The use of methane gas converters has increased worldwide as a result of the energy crisis brought about by higher oil prices. Fuel for the electricity, cooking, and heating needs of an average U.S. farm could be supplied by the manure from about 40 cows. Some large farms and feedlots in the United States have built biogas plants to utilize the animal manure produced.

Higher feed costs have led to an increase in research on using animal manures as a supplement in feeds. Chicken litter, for example, is sometimes used in cattle feed.

Stabilize Farm Economy

Livestock help bring stability to the farm business. Raising livestock makes good use of the resources already available to the farmer such as land, labor, capital, and management ability, and can also increase farm income. Including livestock in a farm business helps to spread the risks involved in farming over more enterprises (Figure 1-10). Thus, the farmer is not dependent on only one or two sources of income. In addition, both labor and income are spread more evenly throughout the year.

Courtesy of Harveststore-Engineered Storage Products Co

Figure 1-10 Diverse farm enterprises which include livestock and crops help spread the risk and distribute labor needs evenly throughout the year.

Connection: Collagen and Jell-O

You may have noticed the jelly-like substance in the pot after meat has been cooked. Any idea what it is? It is Jell-O (Figure 1-11)! Well, more correctly, it's gelatin. Jell-O is a brand name synonymous with gelatin in the United States. Gelatin manufacturers cook the skin, bones, hooves, and other animal parts and then collect and purify the gelatin. Gelatin has many uses, from creating various dishes to making skin more youthful and firm. In agriculture, gelatin is employed as an ingredient of fertilizers and animal feeds.

Gelatin has a consistency unlike most products. It is formed by extracting collagen from the tissues, bones, and skin of cattle and pigs. Collagen is the main protein of the connective tissue found in mammals. It gives elastic qualities to skin and has many beneficial traits. It has several proteins and amino acids, which promote the feeling of fullness, help regulate metabolism, reduce joint pain, and help improve bones, teeth, and nails.

Hospitals often serve gelatin to patients. It has qualities that make the dessert easy to digest, and it melts in the body as a liquid. This allows the body to take in calories without losing them.

Gelatin is highly nutritious and easily digested and absorbed. It is colorless, transparent, odorless, and tasteless in a purified form. Many gelatin products contain fruit or fruit juices to add flavor. Gelatin dissolves in hot water and forms a gel, or jelly, upon cooling.

Courtesy of iStockPhoto/Juan Moyano

Figure 1-11 Gelatin is used in a variety of everyday products.

Concentrate Bulky Feeds

It costs a lot of money to transport bulky feeds such as hay. Livestock convert these bulky feeds into a more concentrated form. This reduces transportation costs to market, a great advantage for farms that produce large amounts of bulky feeds.

By-Products

Meat, wool, and leather are not the only products that come from animals. Any product from the animal carcass, other than meat, is called a **by-product**. Thus, wool and leather are by-products from the slaughter of animals. Many other products come from the animal carcass. These include fat, bone, intestine, brain, stomach, blood, and various glands. These by-products are used in the manufacture of many products.

Edible by-products that come from animals include variety meats such as brains, tongue, kidneys, and heart. Oleo stearine, which comes from the fat in the animal carcass, is used in the manufacture of candy and gum. Hooves, horns, bones, and hides produce gelatin, which is used in the production of gelatin desserts, marshmallows, canned meats, and ice cream.

Hides used in the manufacture of leather goods are the most important of the inedible by-products that come from animals. Examples of leather goods made from animal hides include clothing, belts, shoes, purses, furniture, drumheads, and sports equipment.

The inedible fats are used in the production of cosmetics, waxes, soap, lubricants, and printing ink. Bones, horns, and hooves are also used in the production of glue, buttons, bone china, camera film, sandpaper, dice, piano keys, wallpaper, and toothbrushes. Hair from animal hides is used in making brushes, rug padding, house insulation, and upholstering materials for furniture. Artists' paintbrushes can be made from the fine hair found in the ears of animals.

Feeds for livestock are made from animal by-products. These include blood meal and meat and bone scraps.

More than 100 drugs used by humans for medical purposes are made from animal by-products. Insulin, which is used in the treatment of diabetes, is extracted from the pancreas of animals and is now produced in the lab. **Cortisone** comes from the adrenal glands and is used for the treatment of rheumatoid arthritis, adrenal insufficiency, some allergies, diseases of the connective tissues, and gout. **Thrombin** comes from the blood of animals and is a coagulant used in surgery to help make blood clot. It is also used in skin-graft operations and for the treatment of ulcers. **Heparin** comes from the lungs and is used to prevent blood clotting during operations. It also helps prevent heart attacks. **Epinephrine** comes from the adrenal glands. It is used for the treatment of some kinds of allergies and to help relieve the symptoms of hay fever and asthma. **Rennet** comes from the stomachs of cattle and is used in cheese making. It also helps babies digest milk. **Corticotropin** (ACTH) comes from the pituitary glands in the brain. It is used for the treatment of some breathing problems, severe allergies, mononucleosis, and leukemia.

By-products are also used in the manufacture of perfumes, fertilizers, candles, lanolin, and glycerine, as well as many other products. Animal by-products make valuable contributions to society through their use in medicine, clothing, household materials, and a variety of other areas of life.

CONSUMPTION OF LIVESTOCK PRODUCTS

The per capita consumption of major livestock products is shown in Figures 1-12A through F. Producer organizations have conducted marketing campaigns financed by check-off programs based on producer sales of animals and milk to attempt to increase market share for their particular product. An increased emphasis on producing products that are perceived by the prospective consumer as being safe and healthy has contributed to the slower decline in demand for livestock products. These promotional programs are discussed in more detail in the units related to the marketing of specific livestock products.

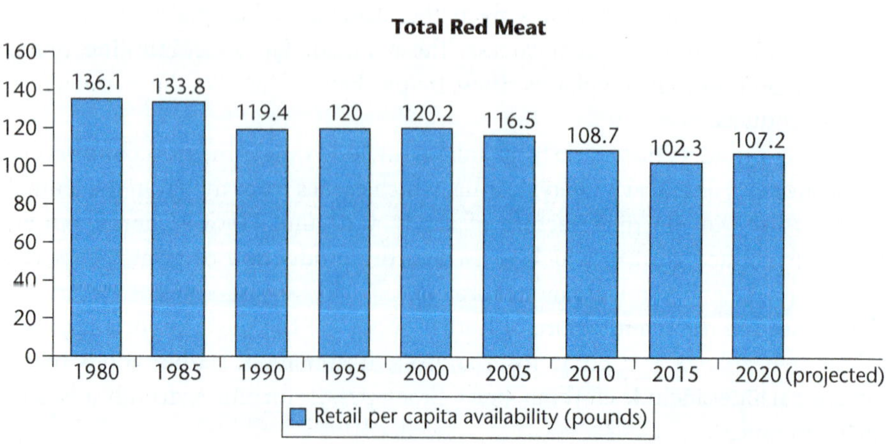

Figure 1-12A Total red meat (beef, veal, lamb, pork) retail per capita availability in pounds in the United States.

Source: USDA/Economic Research Service, Total red meat: Supply and disappearance, February 1, 2014.

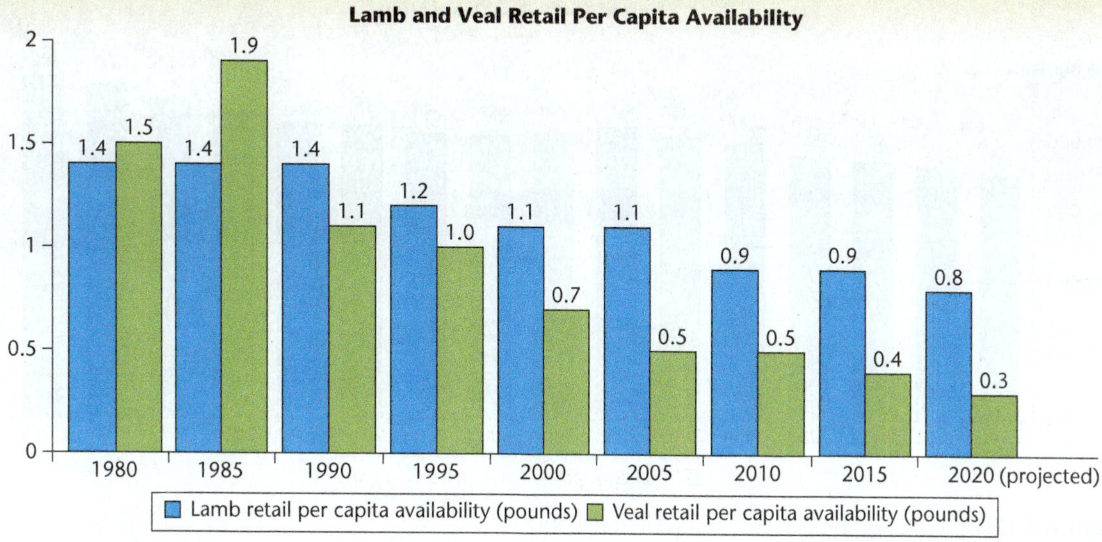

Figure 1-12B Veal and lamb retail per capita availability in pounds in the United States.
Source: USDA/Economic Research Service, Veal and lamb: Supply and disappearance, February 1, 2014.

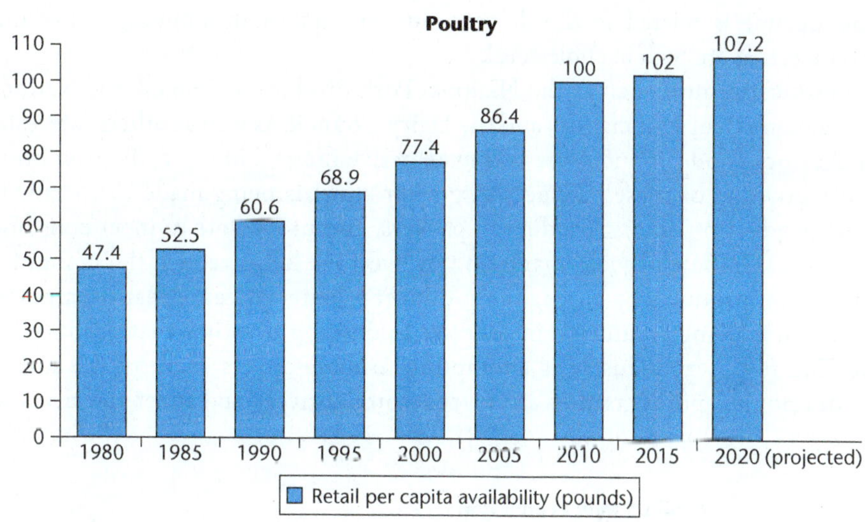

Figure 1-12C Total poultry retail per capita availability in pounds in the United States.
Source: USDA/Economic Research Service, Poultry supply and disappearance, February 1, 2014.

Eating habits have also been influenced by more food consumption away from home and an increase in demand for meals that are quick and easy to prepare at home. Marketing methods and price competition have also influenced consumer demand for various livestock products.

Consumers are concerned about food safety and nutrition. Food safety concerns are discussed later in this chapter. Nutritional concerns about food content include:

- cholesterol level
- fat content
- salt content
- food additives
- sugar content
- artificial coloring

Milk and animal fats are a source of cholesterol in the diet. The use of animal fats has declined significantly since 1967. During this same period, the use of vegetable

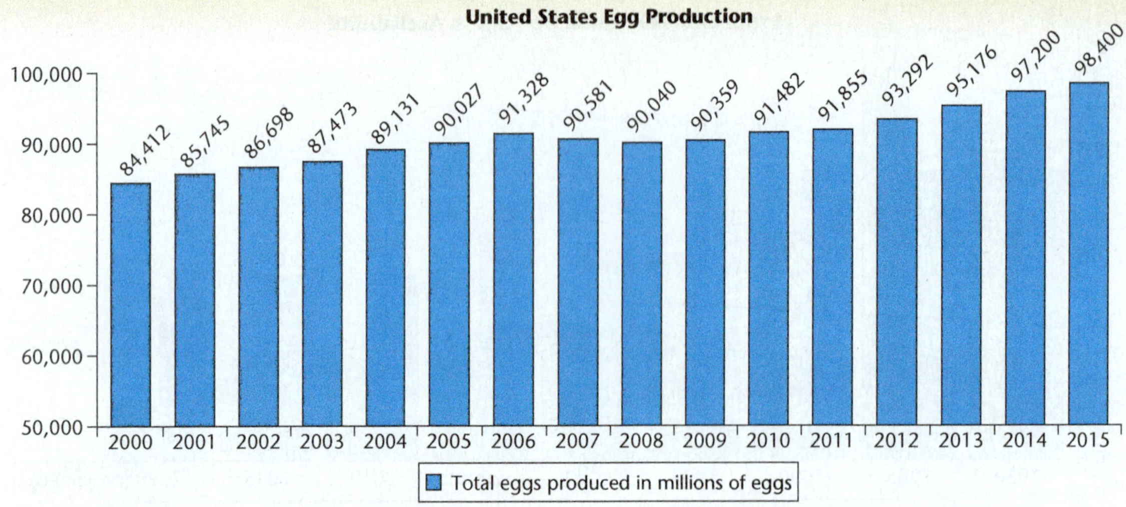

Figure 1-12D Total egg production in the United States.

Source: USDA.

fats, which are seen as having less cholesterol, has risen rapidly. The per capita consumption of cholesterol in the diet of people in the United States is declining. This decline is related to the decrease in per capita consumption of livestock products that are high in cholesterol.

Producer groups such as the National Pork Producers Council, the National Cattlemen's Beef Association, and the Dairy Council, as well as others interested in livestock food products, are taking an active interest in research, promotion, and marketing of these products. A greater effort is being made to coordinate meat production among producers, packers, and processors to meet consumer concerns. Grade and yield marketing of livestock helps reduce the amount of fat in meat products by paying the producer a better price for leaner carcasses. Research is being conducted in genetics to develop livestock that produce less fat. The red meat industry is attempting to maintain its share of the market by developing products that are priced competitively and meet the needs of

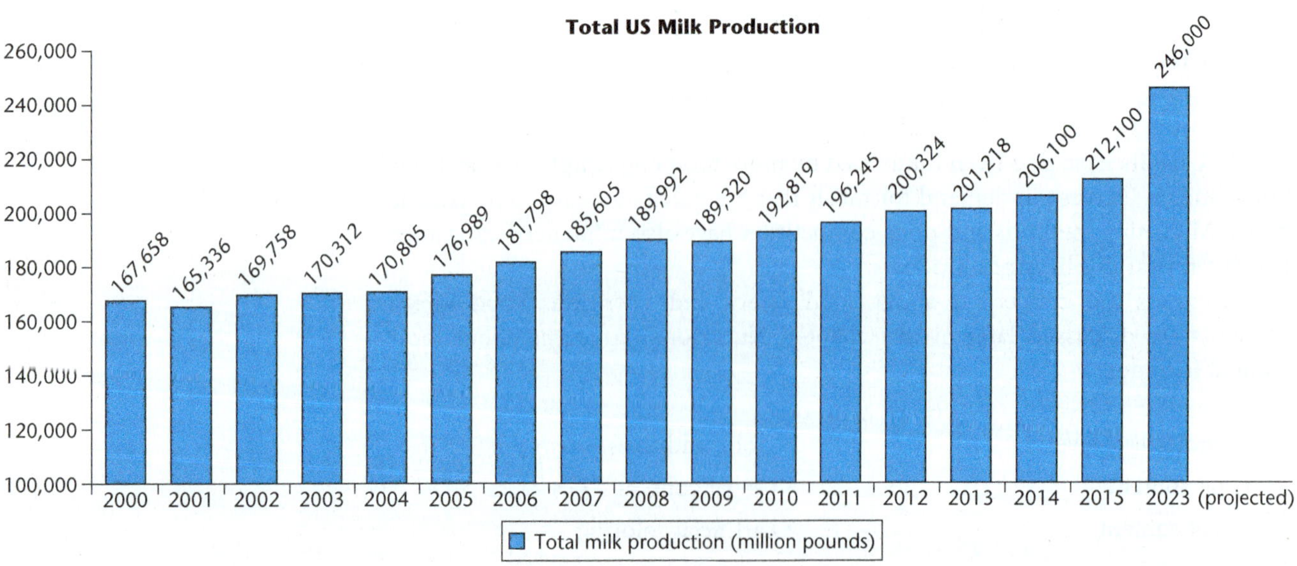

Figure 1-12E Total milk production in millions of pounds in the United States.

Source: USDA.

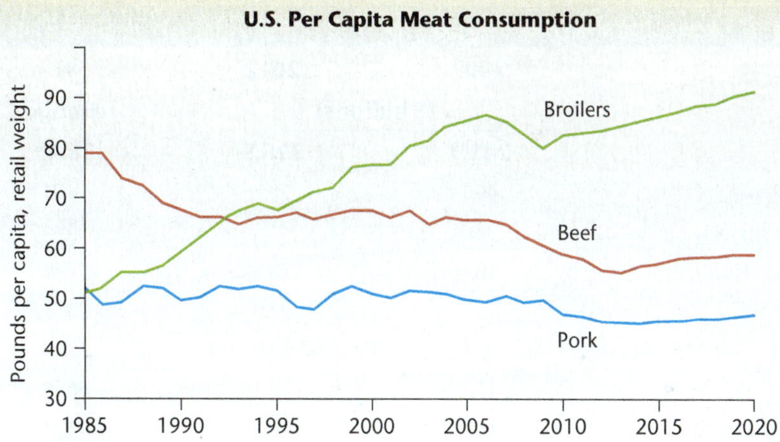

Figure 1-12F U.S. per capita meat consumption for broilers, beef, and pork.
Source: USDA Long-Term Projections, February 2012.

the consumer. The poultry industry is also actively promoting and marketing products both to meet consumer needs and to address concerns related to health and the safety of food products.

SIZE AND SCOPE OF THE LIVESTOCK INDUSTRY IN THE UNITED STATES

Income and Costs

The total value of agricultural products sold from farms in the United States in 2012 was nearly $395 billion.

In 2012, farmers spent more than $328 billion for total production expenses. Table 1-2 shows the value of agricultural products sold from 2007 to 2012. Table 1-3 shows a breakdown of production expenses for U.S. farmers in 2007 and 2012.

Leading States in Livestock Production

Table 1-4 shows the 10 leading states in several categories of livestock and livestock products production. The ranking of the states shown in the table may change slightly from year to year. Consult the USDA website for current rankings. However, the states listed tend to be those that are the major producers year after year of the livestock or livestock product category shown.

TABLE 1-2 Total Value of Agricultural Products Sold in the United States, 1982–2012 (in $1,000)

Year	Value
1982	$131,900,223
1987	$136,048,516
1992	$162,608,334
1997	$196,864,649
2002	$200,646,355
2007	$297,220,491
2012	$394,644,481

Source: USDA Census of Agriculture, 2007 and 2012.

TABLE 1-3 Agricultural Production Expenses, 2007 and 2012

	2007	2012	%
	($ billions)		change
Total	**241.1**	**328.9**	**36.4***
Feed	49.1	75.7	54.2*
Livestock & poultry purchases	38.0	41.6	9.4
Fertilizer	18.1	28.5	57.6*
Hired labor	21.9	27.0	23.4*
Cash rent	13.3	21.0	58.2*
Seeds	11.7	19.5	66.0*
Supplies and repairs	15.9	18.9	18.7*
Gasoline, fuels, and oils	12.9	16.6	28.4*
Chemicals	10.1	16.5	63.4*
Other	50.1	63.7	27.1*

source: USDA NASS, 2012 Census of Agriculture.
**statistically significant change.*

TABLE 1-4 Leading States in Livestock and Livestock Product Production

Ranking	Beef Cattle and Calves	Eggs	Broilers	Dairy Products	Hogs	Turkeys	Leading States in Cash Receipts from Livestock
1	Texas	Iowa	Georgia	California	Iowa	North Carolina	California
2	Nebraska	Georgia	Arkansas	Wisconsin	Minnesota	Minnesota	Iowa
3	Kansas	Pennsylvania	North Carolina	New York	North Carolina	Indiana	Nebraska
4	Iowa	Ohio	Alabama	Idaho	Illinois	Arkansas	Texas
5	Colorado	Indiana	Mississippi	Pennsylvania	Indiana	Missouri	Minnesota
6	Oklahoma	Texas	Texas	Texas	Missouri	South Carolina	Illinois
7	California	Arkansas	Kentucky	Minnesota	Oklahoma	Virginia	Kansas
8	South Dakota	California	Maryland	Michigan	Nebraska	California	Wisconsin
9	Missouri	North Carolina	South Carolina	New Mexico	Ohio	Ohio	Indiana
10	New Mexico	Alabama	Delaware	Washington	Kansas	South Dakota	North Carolina

Source: USDA, Farm Income: Cash Receipts, 2012.

Number of Livestock on Farms

Over a period of years, livestock and poultry numbers respond to economic conditions and also reflect changes in consumer demand. Table 1-5 shows the trend in beef cattle, dairy, and hog numbers in the United States, and Table 1-6 shows the trend in turkey and poultry numbers in the United States. Figures 1-13A and B illustrate the trend in sheep, lamb, and goat numbers in the United States. All of the data shown in Tables 1-5 and 1-6 and Figures 1-13A and B were obtained from the USDA National Agricultural Statistics Service. These data are based on previously collected data and future projections.

TABLE 1-5 Trend in Beef Cattle, Dairy Cattle, and Hog Numbers in the United States

	1995	2000	2005	2010	2015	2020
Beef Cattle	35,156	33,569	33,055	31,376	32,877	34,130
Dairy Cattle	9,487	9,190	9,005	9,115	9,025	8,940
Hogs	58,264	59,138	61,449	64,887	66,118	70,565

Note: Inventory numbers shown are in thousands.
Source: USDA Long-Term Agricultural Projection Tables, February 2011.

TABLE 1-6 Trend in Poultry and Turkey Numbers in the United States

	2009	2011	2013	2015	2017	2020
Chicken	35,961	37,543	38,399	39,829	41,211	43,496
Turkey	5,997	5,709	5,873	6,095	6,244	6,498

Note: Total supply numbers shown are in million pounds.
Source: USDA Long-Term Agricultural Projection Tables, February 2011.

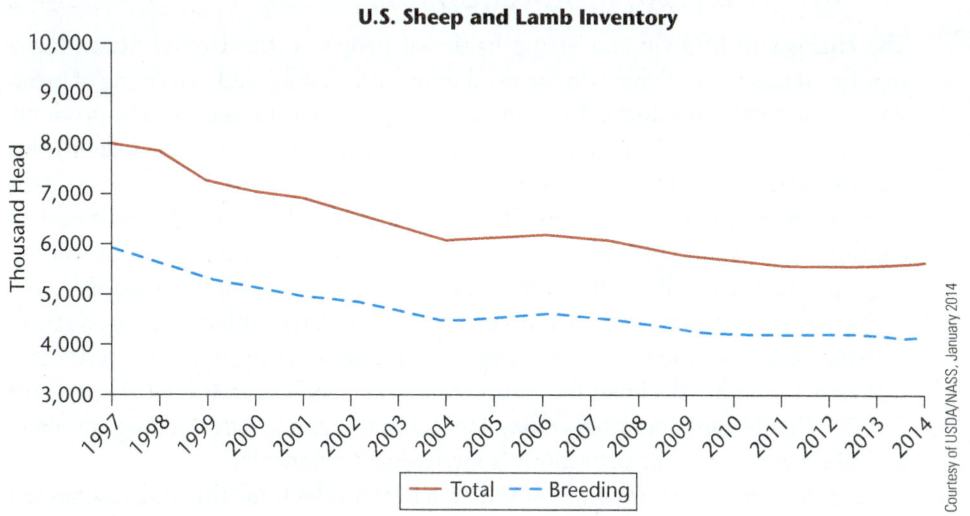

Figure 1-13A U.S. sheep and lamb inventory.

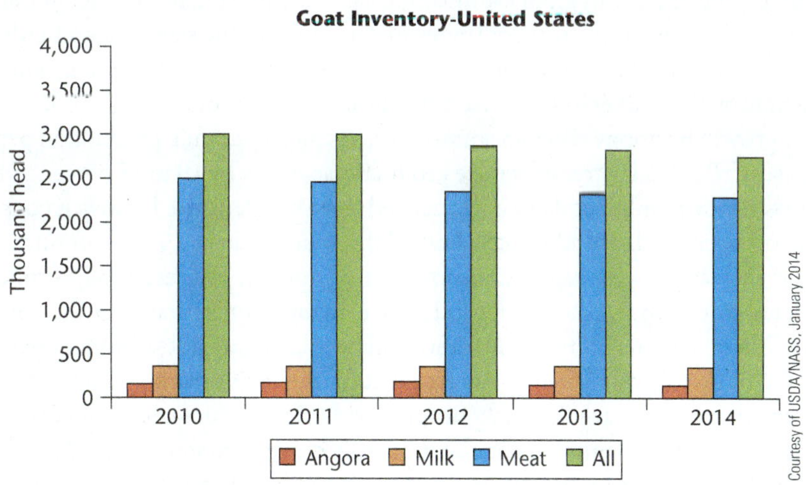

Figure 1-13B U.S. goat inventory.

ANIMAL HEALTH PRODUCTS

The development and sale of animal health products is a large and growing industry. The three categories of products included are feed additives, biologicals, and pharmaceuticals. Feed additives are those products used in livestock and poultry production to control or prevent disease, enhance growth, or improve feed efficiency. Biologicals include vaccines, bacterins, and antitoxins. Pharmaceuticals include medicines used in disease control and prevention.

The research and development of animal health products is expensive. In addition, companies have the expense of defending the continued use of products already approved. These costs are passed along to producers, and eventually to consumers, in the form of higher food prices.

TRENDS IN ANIMAL AGRICULTURE

Consumption and Production

The changes in lifestyle and eating habits of people in the United States has a significant impact on changes in the production, processing, and marketing of meat, poultry, and dairy products. The concern over cholesterol levels in food products in recent years has had a major influence. Consumers are demanding leaner meat and less fat in meat and dairy products. Competition in the marketplace continues to drive increased efficiency in livestock production to lower the per unit cost to consumers. It is anticipated that the need for increased efficiency and resulting lower product cost will result in larger operations that use more automation.

The swine industry has moved rapidly toward large, integrated production systems, much like the broiler industry. The number of farms in the United States that raise swine has declined from approximately 1 million in the 1970s to about 71,000. The declining number of hog farms has not greatly affected hog numbers because the size of swine operations is increasing dramatically.

Large feedlot operations are prominent in beef production. The Midwest has fed large numbers of cattle because of the availability of corn; however, a number of feedlot operations are now found outside of the Midwest. Because feeding practices tend to put more emphasis on the use of roughages and other crop by-products for feeding cattle, the use of grain for finishing cattle is expected to decline moderately.

The number of dairy farms continues to decrease, but the size of dairy herds is increasing. The total number of dairy cows in the United States is decreasing. However, total milk on the market has increased due the increase in production per cow. The use of the growth hormone rBST, in some cases, has increased milk production per cow. The use of rBST can increase average production per cow by 10 to 15 percent, further decreasing the number of dairy cows needed to supply demand. However, consumer concerns over the use of rBST have limited the widespread use of this product.

The increase in per capita consumption of poultry, and especially broilers, is expected to continue. The broiler industry is mainly concentrated in the southern states. There is a possibility of some movement of large integrated broiler operations into the Midwest, where feed grains are plentiful and major markets are located. This movement would reduce the transportation costs of feedgrains and poultry products.

The trend in sheep numbers has been sharply downward since 1945. The number of sheep on U.S. farms peaked in 1945 at 56 million head. In 2014, there were 5.2 million sheep in the United States. While this general trend is expected to continue into the near future, in recent years there have been some encouraging signs for the sheep industry. Some of the leading grocery chains have started promoting American lamb, and nontraditional markets—which include on-farm sales, farmers markets, and small processors serving ethnic communities—have grown rapidly. In addition, lamb and wool prices have generally increased in recent years. The sheep and wool industry organizations continue promotional campaigns to expand consumption and production of sheep products.

The goat industry is growing, and projections are for continued growth into the near future. Worldwide, more people eat goat meat and dairy products than the meat and dairy products of any other animal. Goats have proven to be an enduring source of healthy, low-fat meat and high-calorie milk for people in some of the globe's harshest climates. A similarly sized serving of goat meat has a third fewer calories than beef and a quarter fewer than chicken, and it has much less fat: up to two-thirds less than a similar portion of pork and lamb and less than half as much as chicken. However, in the United States, goat meat and milk consumption has traditionally been low and is just beginning to increase. Meat goat production and consumption has been gaining in popularity in the United States in recent years thanks to several factors, including growing populations of ethnic groups that favor goat meat and faith-based consumers who prefer it. Goats fall into three categories: meat, dairy, and Angora. The largest segment of goat production in the United States is the meat goat segment. National estimates based on import data indicate that the United States' supply of goats is deficient—providing an opportunity for the expanding goat industry.

Research in livestock production is continuing to improve production efficiency. Areas of research that are having and will continue to have a significant impact on the livestock industry include:

- higher embryonic survival rate in swine.
- shortening of calving interval for dairy cattle.
- improved methods of artificial insemination.
- biotechnology.

Another trend in livestock production is the increasing emphasis on disease prevention rather than treatment. Livestock producers can lose approximately 20 percent of their income from disease, parasites, and toxins. By placing increased emphasis on disease prevention rather than treatment, livestock producers can reduce costs and increase efficiency.

Biotechnology

Biotechnology is the use of biology or biological processes to develop helpful products and services. It is leading to many new products and strategies that can benefit agriculture, human health, and the environment. Research in genetics and livestock production is continuing to improve production efficiency. Areas of research that have had a significant impact on the livestock industry include cloning, superovulation, sex determination, *in-vitro* fertilization, and embryo transfer.

ANIMAL WELFARE AND ANIMAL RIGHTS

The domestication and use of animals for the benefit of humans began many thousands of years ago. There are varying degrees of beliefs regarding the use of animals in agriculture. Often, people fall in or between the two categories of animal welfare and animal rights. There is a significant difference between those people who are concerned about animal welfare and those who believe in animal rights. There are many different groups within these two views, each with their own agendas.

Animal Welfare

Animal welfare supporters emphasize the humane treatment of animals, both in research and production agriculture. They believe that animals can be used to benefit humans. Some animal welfare advocates support essential uses of animals (for example, for food and medical research), and do not support nonessential uses of animals (such as entertainment).

Livestock producers generally support proper feeding and housing, veterinary care, and good management practices because these activities result in more efficient production of meat, milk, eggs, and wool.

Animal Rights

The views of animal rights activists vary. Some advocate the total elimination of all animal use by humans (animal liberation), while others recognize that animal use probably is not going to be eliminated and, therefore, work to eliminate animal suffering to the greatest extent possible. If all animal use by humans were to be eliminated, it would have a major impact on society. Those who take a more moderate approach try to achieve their goals by influencing legislation and through public education campaigns.

Some of the livestock production practices that are under attack by animal rights activists include:

- use of hormones, antibiotics, and additives in animal feeding.
- caging laying hens.
- production of veal calves in crates.
- raising swine in confinement and using farrowing crates for sows.
- management practices such as castration, docking, beak trimming, and dehorning.
- having them as household pets.
- using animals in medical and scientific research (including biotechnology).
- consuming animals' flesh.
- using animal skins (leather and fur) for clothing or other products.
- using animal products (including milk and eggs) for food.
- making animals the target of hunting and trapping.
- featuring animals in entertainment activities such as horse and dog racing.

Some militant animal rights activists have been willing to break laws by such actions as stealing research animals and damaging property to draw attention to their cause. Major targets have included biomedical research facilities, food production and food retail facilities, and retail fur facilities. Some activists have targeted meetings of farm groups and exhibitors at livestock shows with demonstrations protesting the use of animals and animal products. Investigations by the U.S. Department of Justice and the U.S. Department of Agriculture show that there is an increasing tendency for animal rights violence to be directed at food production facilities and individuals involved in animal production.

Animal rights activists often distribute their message by providing free materials to schools under the label of educational material. They also maintain sites on the Internet to distribute their message. This material should be reviewed and compared to the information in other factual and reliable resources regarding animal welfare and animal rights.

Legislation

Societies for the prevention of cruelty to animals were organized in both England and the United States in the 1800s. Today, there are federal, state, and local laws that address the humane treatment and care of animals. The Animal and Plant Health Inspection Service (APHIS) of the USDA has the responsibility of enforcing federal legislation regarding animal welfare in the United States.

In the United States, the first federal law dealing with the humane treatment of animals was passed in 1873; this legislation mandated that feed and water be provided for farm animals being transported by barge or railroad. Other federal legislation includes the following:

- The Federal Humane Slaughter Act of 1958 requires federally inspected slaughter plants to comply with humane methods of slaughter.

- The Animal Welfare Act of 1966 addresses the sale, transportation, and handling of dogs and cats used in research institutions.
- The Animal Welfare Act amendments of 1970 expand coverage of the 1966 act to most other warm-blooded animals used in research, animals in zoos and circuses, marine mammals in sea life shows and exhibits, and animals sold in the wholesale pet trade. Retail pet shops, game ranches, livestock shows, rodeos, state and county fairs, and dog and cat shows are not covered by this act.
- The Animal Welfare Act amendment of 1976 extends the 1970 act to include care and treatment while animals are being transported by common carriers and outlaws animal fighting exhibits (dog or cock fights) unless specifically permitted by state law.
- The Horse Protection Act of 1970 (amended in 1976) prohibits the soring of horses, as well as the transport of sored horses across state lines to compete in shows. **Soring** is the practice of using chemical or mechanical irritants on the forelegs of the horse. A sored horse lifts its front legs more quickly to relieve the pain. The practice is strictly prohibited by most horse industry organizations and associations.
- The Animal Welfare Act amendment of 1985 provides for the establishment of special committees at all research facilities to oversee animal use; it also requires exercise for dogs and provides for the psychological well-being of nonhuman primates at such facilities.
- The Animal Enterprise Protection Act of 1992 adds a section to the federal criminal code to deal with vandalism and theft at animal research facilities and threats to research workers.
- The Animal Welfare Act amendment of 1993 was passed to help prevent the use of lost or stolen pets in research. It establishes requirements for more documentation from dealers selling animals to research facilities; it specifies that dogs and cats must be held by pounds and animal shelters for at least 5 days, including a Saturday, before releasing them to dealers.

The Animal Welfare Act, with its amendments, provides only a minimum level of standards for the adequate care and treatment of animals covered by its provisions. Areas addressed in the standards include housing, handling, sanitation, nutrition, water, veterinary care, and protection from extreme weather and temperatures. Farm animals used for food, fiber, or other agricultural purposes are not covered by the Animal Welfare Act.

Livestock producers and others interested in the production and use of animals need to understand that societal concern about the welfare of animals will continue to influence legislation and regulations. Decisions about the treatment and use of animals will increasingly be made by society as a whole rather than by individuals. More research is needed on issues that affect the welfare of animals. It is important that decisions about animal welfare and use be made on the basis of scientific information and not emotions.

People who produce or work with animals need to do the best possible job of providing humane treatment for their animals. Animals that are well managed are more efficient producers of meat, milk, eggs, wool, and other products. Livestock producers need to let people know that they care about the welfare of the animals they raise. Students who show livestock need to be aware of the image they are projecting when they are competing. Concern for the welfare of their animals should always be a priority when students are participating in livestock shows.

ANIMAL IDENTIFICATION

A variety of methods are currently used to identify animals. These include ear tags, ear notching, tattoos, electronic collars, electronic ear tags, ear buttons, implants, microchips, and rumen boluses with microchips installed. There are many commercial companies that manufacture animal identification systems and the software packages used to record the data in a computer database.

The livestock industry has developed a system of unique identifying numbers that can be utilized with various livestock identification methods. This system can be used to establish a database of information about individual animals. Computer technology exists that allows embedded microchips to be read, and the information secured can then be transferred to a database. Standards for security of information and rules for access to database information must be considered. Additional information about the animal, including place of origin, health records, breeding records, and production records, can be added to the database.

A unique animal identification coupled with a database of information has been developed by the Holstein Association. The system is known as National Farm Animal Identification and Records (FAIR). The system is voluntary, but many dairy producers are using it as a means of securing better information for managing their dairy herds. The USDA also administers a voluntary system known as the National Animal Identification System (NAIS). This system is endorsed by some livestock producers; however, there are many controversial concerns associated with this program.

The advantages of developing a system of unique animal identification numbers coupled with a database of information include:

- providing data for disease control and eradication programs.
- providing a record of vaccinations and treatments for diseases.
- providing a record of good management practices used by the producer.
- monitoring emerging diseases.
- improving quality assurance programs and food safety.
- improving genetics in breeding programs.
- identifying animals so they may be returned in case of theft or loss.
- improving access to international markets.

FOOD SAFETY

Despite the fact that the United States has one of the safest food supplies in the world, food safety is becoming a major area of concern among consumers. Much of this concern is fueled by special-interest groups, consumer groups, and the news media. Major issues regarding food safety include:

- bacterial contamination
- pesticides in food (generally related to crops)
- drug residues in food
- irradiation of food
- genetic engineering
- contamination of food by processors

Consumers are looking for zero health risk in relation to their food. Food safety specialists know there is no such thing as zero risk; they define food safety in terms of risks and benefits. A procedure designed to improve food safety may pose some degree of risk for some consumers. However, the benefits to most consumers may be many times greater than the risk. For example, many additives are used in foods

Connection: What Is Salmonella?

Salmonella are rod-shaped bacteria found in warm- and cold-blooded animals and in the environment. These bacteria can cause a foodborne illness known as *Salmonellosis*, a zoonotic infection. Zoonotic infections can be transferred from animals to humans and vice versa. People can contract salmonella by drinking polluted water, consuming undercooked or raw meat, and eating food contaminated with the excretions from infected animals and people.

Healthy adults must ingest a relatively large number of salmonella bacteria to contract an illness. Small numbers of salmonella are destroyed by the acidity of the stomach and, therefore, are usually not harmful to adults. Infants, toddlers, and the elderly are much more susceptible to salmonella because they have more sensitive or weak digestive tracts. People who become infected can experience symptoms such as nausea, diarrhea, fever, and abdominal pain. The symptoms may continue for up to 7 days and may not require treatment. Antibiotics may be necessary in some places.

The number-one way to prevent the spread of salmonella is cleanliness. Hand washing with soap and hot water after touching raw meat and eggs is a good way to reduce the chances of contamination (Figure 1-14). Some other ways to help stop the spread of disease-causing bacteria include cooking red meats, eggs, and poultry thoroughly and storing foods at the appropriate temperature.

Figure 1-14 Salmonella is one type of bacteria that can cause foodborne illnesses.

for a variety of purposes, including enhancing flavor, maintaining freshness, and preservation. A small number of people may have an adverse reaction to the additive; most people will not be affected. If the benefits to the majority of consumers far outweigh the potential health hazard to a small number of people, the risk–benefit ratio of using these additives is considered to be acceptable.

Most of the problems with foodborne illness are caused by bacteria (66 percent). Other sources of foodborne illness include chemical (25 percent), viral (5 percent), and parasitic (4 percent). The most common bacterial causes of foodborne illness are:

- *Salmonella* species—often found in eggs, milk, chickens, beef, and turkey.
- *Campylobacter*—often found in poultry, raw milk, and drinking water.

- *Clostridium botulinum*—lives in the soil, grows in many meats and vegetables; multiplies in improperly processed canned or smoked foods
- *Staphylococcus aureus*—found on human skin; may enter food supply from improper handling of food by workers.
- *Shigella*—normally found in the intestinal tract of humans; may be transmitted to food by improper sanitation procedures by food handlers.
- *Escherichia coli (E. coli)* O157:H7—a more deadly form of the common *E. coli* bacteria found in undercooked ground beef; may also be found in other foods such as lettuce, salami, unpasteurized apple cider, and unpasteurized milk.
- *Listeria monocytogenes*—this bacteria is more resistant to acidity, salt, nitrite, and heat than many other microorganisms; often found in soft cheese, unpasteurized milk, imported seafood products, frozen cooked crab meat, cooked shrimp, and cooked surimi [imitation shellfish]; these bacteria can survive and grow at low temperatures.
- *Clostridium perfringens*—most often found in meat and meat products; bacteria that survive cooking multiply when food is not kept hot enough [above 140°F]; do not leave prepared foods at room temperature for cooling before storage—refrigerate immediately.

According to the Centers for Disease Control and Prevention (CDC), the four major bacterial pathogens that contaminate meat and poultry products are *Salmonella, Campylobacter, E. coli* O157:H7, and *Listeria monocytogenes.* Deaths and illnesses occur each year in the United States due to people consuming contaminated meat or poultry products.

These bacteria can cause illness ranging from diarrhea, nausea, vomiting, breathing difficulty, and fever to death. Children and the elderly are generally more susceptible and often have more severe reactions to these bacteria. Various species of *Salmonella* are the most common cause of foodborne illness. *Clostridium botulinum* grows in improperly canned foods and produces a toxin that can cause botulism, an illness that often results in paralysis and death. *Escherichia coli (E. coli)* is a common cause of diarrhea among travelers.

All fresh-food products contain some bacteria. Proper handling practices such as cooking and holding temperature, personal hygiene of food handlers, and kitchen sanitation should prevent most bacteria problems in food.

Regulation of food additives is the responsibility of the Food and Drug Administration (FDA), with additional review provided by the U.S. Department of Agriculture. In 1958, the Delaney Clause was added to the Food, Drug, and Cosmetic Act of 1938. The Delaney Clause prohibits the use of any food additive that causes cancer in humans or animals at any dosage. However, there are many naturally occurring chemicals in foods that can cause cancer at high enough levels or if eaten over a long enough period of time.

The Food Quality Protection Act of 1996 revised the Delaney Clause to set a new standard for carcinogens in food specified as "a reasonable certainty of no harm." This removed the zero tolerance policy relating to food additives and carcinogens that was imposed by the original Delaney Clause.

The USDA's Food Safety and Inspection Service (FSIS) is responsible for verifying that meat and poultry processing plants meet regulatory requirements and taking enforcement action when a plant fails to meet these requirements. Historically, the inspection of slaughter plants was based on visually checking for signs of sick animals or birds before slaughter and inspecting the carcass for foreign matter, abscesses, or feces contamination. There was no routine microbiological testing done on raw meat or poultry.

Connection: Foodborne Illness

The last time you came down with a "stomach bug," it very easily could have been a mild case of a foodborne illness. The Centers for Disease Control and Prevention (CDC) reports an estimated 76 million cases of foodborne illness in the United States every year. Even though most of these cases are minor, some can be much more serious. Foodborne illness lands an estimated 325,000 people in the hospital each year and leaves 5,000 people dead.

What exactly is a foodborne illness? A foodborne illness is any illness caused by consuming contaminated food or drink. The contamination mainly occurs from bacteria, viruses, or parasites, but it can also come from traces of chemicals left in the food product. The CDC estimates there are more than 250 foodborne illnesses, but the most well known include *Campylobacter, Salmonella, E. coli* O157:H7, Norwalk-like viruses (or *Calicivirus*), and botulism.

Most foodborne illnesses are caused by bacteria, but some are caused by viruses. They are rarely diagnosed and are believed to be caused by the virus spreading from one infected person to another. This means that if the person preparing the food has the virus and does not use proper hygiene measures, the virus can be spread to other people through the handling of the food.

While the symptoms of foodborne illnesses can be serious, these illnesses can usually be avoided by taking simple precautions. By simply maintaining good hygiene (that is, washing hands, cleaning up after cooking, etc.) and ensuring meat is cooked to the recommended temperature (Figure 1-15), the chances of contracting a foodborne illness can be greatly reduced.

Figure 1-15 Foodborne illness can be caused by improperly cooked meats.

Outbreaks of foodborne illnesses showed that the original system of plant inspection did not correct the problem of bacteria that are a major cause of foodborne illness. The USDA determined that the meat and poultry inspection system needed to be revised.

The Sanitation Standard Operating Procedures (SSOPs), which addresses sanitation problems in plants, along with other pathogen reduction regulations and concepts for enforcement were introduced in 1997 by the FSIS. The focus is on making sure that the plants detect potential sanitation problems such as unclean equipment or poor worker hygiene that may cause contamination of food products by harmful bacteria, and then prevent those problems from occurring. These regulations and concepts place the responsibility for food safety on the plants; it also gives plants flexibility in the development and implementation of innovative measures for the production of safe foods. The role of FSIS inspectors and compliance officers is primarily one of verifying that the plant is meeting industry standards and taking enforcement action when necessary. The basis for the current regulations is the belief that if plants properly maintain sanitation and process controls, food products produced by the plant will not be contaminated by dangerous pathogens. The FSIS will not approve the product of the plant if the control systems fail to prevent contamination by pathogens. It is the responsibility of the plants to address any deficiencies found by FSIS inspectors. Failure to correct the deficiencies may result in a suspension of all or part of the plant's operation. Plants have the right to receive notice of alleged violations and file appeals of actions by FSIS.

Plants are required to keep accurate records to verify that their control measures work properly and the product that they produce is safe for human consumption. Criminal prosecution may result if a plant is found to be maintaining false or deceptive records.

In 1971, a system called **Hazard Analysis and Critical Control Points (HACCP)** was developed for the National Aeronautic and Space Administration (NASA) to monitor the production of food for the space program. After that, some plants used HACCP on a voluntary basis. A phase-in of mandatory use of HACCP in all slaughter plants began in January 1998, with all large plants (500 or more employees) required to implement a HACCP plan. Small plants (10 or more employees but fewer than 500) were required to implement a HACCP plan by January 1999. All remaining slaughter plants were required to have a HACCP plan in place by January 2000.

The National Advisory Committee on Microbiological Criteria for Foods (NACMCF) defined seven principles for ensuring safety in the food supply using HACCP. The HACCP systems used in slaughter plants must be based on those seven principles:

1. The plant conducts a hazard analysis to determine potential food safety hazards and identify preventive measures the plant can use to control those hazards.
2. The plant identifies **critical control points (CCPs)**. This is any point, step, or procedure in the process where control can be applied to prevent, eliminate, or reduce a food safety hazard to an acceptable level. Food safety hazards are defined as any biological, chemical, or physical property that may cause a food to be unsafe for human consumption.
3. Critical limits are established for each CCP. A critical limit is defined as the maximum or minimum value to which a physical, biological, or chemical hazard must be controlled at a critical point to prevent, eliminate, or reduce it to an acceptable level.
4. CCP monitoring requirements must be established. These activities are necessary to ensure that the process is under control at each CCP. The HACCP plan must list each monitoring procedure and its frequency.
5. Corrective actions must be established whenever monitoring indicates a deviation from an established critical limit. The HACCP plan must identify the corrective actions to be taken if a critical limit is not met. The purpose of corrective actions is to prevent unsafe food from reaching the public.
6. Record-keeping procedures must be established by all plants. These records must include its hazard analysis, written HACCP plan, records documenting the monitoring of CCPs, critical limits, verification activities, and the handling of processing deviations.
7. Procedures must be established to verify that the HACCP system is working as planned. Plants are required to validate their own HACCP plans. FSIS does not

Connection: Who is FAT TOM?

FAT TOM is a mnemonic device, meaning it is a verbal learning aid using the first letter of each word to form a phrase. It is used in the food service industry (Figure 1-16) to describe six favorable conditions required for the growth of foodborne pathogens. It is an acronym for food, acidity, time, temperature, oxygen, and moisture. These are factors that help us understand how to keep foods safe. Each of the six conditions that foster the growth of food-borne pathogens is described here:

Food	Some foods promote microorganism growth more than others. Protein-rich foods such as meat, milk, eggs, and fish are most susceptible.
Acidity	Foodborne pathogens require a slightly acidic pH level of 4.6–7.5, while they thrive in conditions with a pH of 6.6–7.5. The FDA regulations for acid/acidified foods require that the food be brought to a pH of 4.5 or below
Time	Food should be removed from "the danger zone" (see below) within 2 hours, either by cooling or heating. While most guidelines cite 2 hours, a few indicate that 4 hours is still safe.
Temperature	Foodborne pathogens grow best in temperatures between 41°F to 135°F, a range referred to as the temperature danger zone (TDZ). They thrive in temperatures that are between 70°F to 120°F.
Oxygen	The presence of oxygen can be both helpful and harmful to the growth of pathogens. Aerobic pathogens need oxygen to grow, whereas anaerobic pathogens do not.
Moisture	Water is essential for the growth of foodborne pathogens. Water activity (wa) is a measure of the water available for use and is measured on a scale of 0 to 1.0. Foodborne pathogens grow best in foods that have a wa between 0.86 and 1.0. FDA regulations for canned foods require a wa of 0.85 or below.

Figure 1-16 Foodborne pathogens grow at specific conditions.

approve HACCP plans in advance but does review them for conformance with the rules. Verification procedures include review of HACCP plans, CCP records, critical limits, and microbial sampling and analysis. The HACCP plan must include verification tasks to be performed by plant personnel. FSIS inspectors will also perform some verification tasks. Microbial testing to be conducted by both FSIS and the plant is included as one of the verification activities.

Slaughter plants must regularly test animal carcasses for the generic form of *E. coli* to show that their control system for preventing fecal contamination is working. The frequency of sampling is determined by the plant's production volume. Test results must be recorded and be made available to inspectors.

Testing for *Salmonella* contamination is required in slaughter plants and any plants that produce raw ground products. The selection of *Salmonella* as a performance standard was based on its prevalence in raw meat and poultry products and the fact that reliable laboratory tests are available. Also, steps taken to prevent *Salmonella* contamination may also reduce contamination from other bacteria such as *E. coli* O157:H7 and *Campylobacter*. The FSIS has developed a baseline standard based on the prevalence of *Salmonella* in raw products that all plants must meet. The FSIS will continue to collect data and revise its performance standards as necessary. The best way for plants to meet the standards is to develop and implement process controls to prevent product contamination.

Food irradiation is the treatment of food with radioactive isotopes to kill bacteria, insects, and molds that are present in the food. Cobalt-60 is the most widely used source of the short wavelength radiation used for food irradiation. Cesium-137 is sometimes used as the radiation source. The process is carried out in an irradiation facility that confines and directs the energy. The facility has an irradiation chamber and thick concrete walls that confine the energy and protect the workers from stray radiation. The food is moved through the irradiation chamber on a conveyor.

Irradiation does not raise the temperature of the food as it is treated. The process can be used on fresh or frozen foods. The energy waves are not retained by the food; the food does not become radioactive. No significant difference in nutritional quality has been found when irradiated foods are compared to foods processed by other methods. There are a few foods that show undesirable changes when irradiated. An undesirable flavor change occurs in dairy products that are irradiated, and some fruits, such as peaches and nectarines, show a softening of tissues.

Irradiation of food as a method of food preservation began in 1950. The irradiation of food was implemented to help improve the safety of the food supply. There are several benefits that result from the irradiation of food, including those listed here:

Preservation of perishable foods	Shelf life of foods is extended by destroying organisms that cause spoiling and decomposition
Sterilization	Makes it possible to store foods for long periods of time without refrigeration
Inhibition of sprouting	In potatoes, carrots, onions, garlic, and ginger
Delay in ripening	In bananas, mangos, avocados, and other noncitrus fruits
Control of insect damage	In grains, fruits, vegetables, dehydrated fruits, spices, and seasonings
Control of foodborne illness in meat, poultry, and fish	Eliminates microorganisms such as *Salmonella* species, *Campylobacter*, *Clostridium botulinum*, *Staphylococcus aureus*, *Shigella*, and *Escherichia coli*

Courtesy of the U.S. Food and Drug Administration

Figure 1-17 The international radura symbol must be printed in green.

Because irradiation is classified as a food additive, the FDA must approve its use in the United States. Inspection and monitoring of irradiated meat and poultry and the enforcement of FDA regulations concerning these products are the responsibility of the USDA. All irradiated foods must carry the international symbol called a **radura**, along with a statement that they have been treated by irradiation (Figure 1-17).

Slaughter plants are not required to irradiate their products. Some of the problems involved in the irradiation of meat include uncertainty about consumer acceptance, the volume of product produced each year, and the cost of irradiation compared to other methods of reducing bacteria in foods.

As is the case with the advent of many new technologies, food irradiation has its opponents. Some people are confused because the word "irradiation" sounds like radioactive, thus raising fears about the possibility of cancer or other illnesses. Another concern is the possibility of toxic radiation products forming in the food. Research over many years has not shown any health problems resulting from the use of irradiated foods. The use of food irradiation has been significantly delayed in the United States by the activities of those opposed to this technology.

Improper handling of meat and poultry in the home is a major cause of foodborne illness. Everything that touches food must be kept clean. In order to help prevent foodborne illness:

- Wash hands thoroughly in hot soapy water before preparing food and after handling raw meat or poultry.
- Do not allow juices from raw meat or poultry to come into contact with other foods.
- Use different plates for cooked foods and raw meat or poultry.
- Wash all utensils that are used on raw meat or poultry in hot soapy water before they are used for cooked foods.
- Wash and sanitize counters, cutting boards, and other surfaces where raw meat or poultry was placed before placing cooked foods on them.

Bacteria are destroyed by high temperatures (165° to 212°F) when meat and poultry are cooked. The internal temperature of meat or poultry should be checked in several spots with a meat thermometer, especially if cooking is done in a microwave oven. Internal temperatures of cooked foods vary more when they are prepared in a microwave oven instead of in a conventional oven. Make sure the meat or poultry is thoroughly cooked; there should be no pink meat left internally in the prepared food. Frozen foods usually need about 1.5 times the normal cooking time used for thawed foods. Reheat leftover foods at 165°F to kill any bacteria.

Many food retailers now offer food safety pamphlets with specific instructions for safely cooking and storing food. By following the advice in these guides, consumers can prevent many cases of foodborne illnesses. Safe handling instruction labels are now attached to many meats and poultry. A typical label is shown in Figure 1-18.

Many farm animals are believed to be carriers of the common bacteria that cause foodborne illnesses. It is usually not possible to determine by visual inspection which animals are carrying these bacteria and which are not. Doing simple serum antibody tests or examining cultures from blood or feces samples does not always produce definite proof that the bacteria are present. If the bacteria are present in low numbers, a serum test will not show its presence. Some bacteria live in specific tissues in the animal's body and are not detected by culture tests. Research is needed to develop an accurate and simple test for the presence of bacteria that cause foodborne illnesses that can be used on the farm.

Because of the difficulty in determining which animals are carriers, the possibility of contaminating many uncontaminated carcasses in the packing plant is always present. Ground meat products produced at the plant normally contain meat from several animals. If one of these animals is contaminated, then the entire product becomes contaminated.

More research is needed to determine exactly how these bacteria infect animals and thus get into the food chain. Following good management practices on the farm is the best recommendation that can be given. Maintain clean housing, keep animals healthy, and do not use contaminated feed or water.

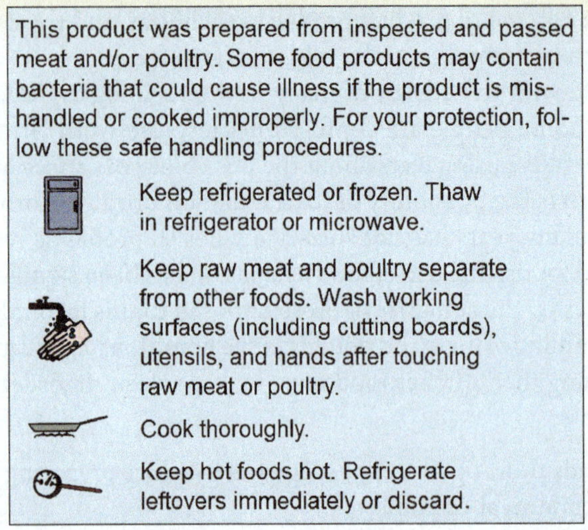

Courtesy of USDA

Figure 1-18 A typical food label showing instructions for safe handling.

Closer inspection and regulation of livestock production on the farm by the USDA can be expected as part of an overall program to improve food safety. Tracking of infected animals from the packing plant back to the farm of origin can be expected as a part of this effort. The determination of normal levels of bacteria in livestock is under study and will help establish baselines for inspection and regulation. Livestock producers have a vital interest in helping to ensure the safety of the food supply. If consumers lose confidence in that safety, livestock producers lose money.

ENDANGERED SPECIES

Changes in livestock production have resulted in a threat of extinction for many species that were once commonly found on the farm. The American Livestock Breeds Conservancy, founded in 1977, is a nonprofit organization working to prevent the extinction of many breeds of livestock once common in agriculture. Livestock breeders increasingly select for those factors that increase production, such as faster rate of gain, increased feed efficiency, larger litters, or more milk production. More uniformity within a breed is developed by the use of artificial insemination, embryo transfer, and cloning. There is an increasing emphasis on crossbreeding to produce animals that are considered better able to meet the market demand for their products.

As a result of these changes, many purebred lines of traditional breeds of livestock have almost disappeared. Selection for uniformity, crossbreeding, and the decline in the numbers of many purebred lines has resulted in a loss of genetic diversity within a species. Genetic diversity is important to maintain within a species because it permits adaptation to changing conditions. When genetic diversity is lost, specific breeds within the species are more likely to become extinct.

Modern livestock production is highly dependent upon the use of antimicrobial compounds and anthelmintics to control diseases and parasites. Natural resistance to diseases and parasites is not usually selected when breeding animals. The result is a loss of natural resistance to infestations of diseases and parasites. Some disease organisms and parasites develop immunity to the drugs used for control. Most swine breeds used in the United States are not resistant to scours caused by *Escherichia coli*. Some breeds of Chinese hogs, however, are resistant to this form of scours. These breeds may be of importance in developing a genetic resistance to scours in swine breeds in the United States. Gulf Coast Native sheep, Caribbean hair sheep, and Florida Cracker cattle show a genetic resistance to parasites. These breeds may be of importance in developing this resistance in other sheep and cattle breeds that are currently not resistant to parasite infestations. The lack of breeding stock with a natural resistance to diseases or parasites could result in a serious threat to the food supply.

The preservation of genetic diversity within a species makes it possible to develop new breeds with characteristics that meet changing market needs. Cattle breeds such as the Santa Gertrudis and Senepol were developed in response to a need for cattle that can be raised in hot climates. The Finn sheep has the ability to produce multiple lambs instead of singles or twins. This ability is being developed in the sheep industry in crossbreeding programs.

Maintaining genetic diversity within species also contributes to scientific knowledge and research. Minor breeds often have unusual characteristics that allow study of adaptation, feed use, disease and parasite resistance, and reproductive abilities under a variety of conditions. This knowledge may be essential in the future for the maintenance of an adequate food supply for the world's population.

SUMMARY

The domestication of animals played a vital role in the development of civilization. Domesticated animals provided a more dependable source of food and clothing. Most of the ancestors of present-day farm animals were first tamed in the Neolithic (New Stone) Age. Written records show that animals were being used more than 4,000 years ago. Early explorers brought the various species of farm animals to the United States. These animals spread across the United States as early explorers and colonists moved westward.

Animals have many useful functions. They convert feed into food, are a source of materials for clothing, and provide power and also recreation. They help conserve natural resources and contribute to good farm management by increasing farm profits. Animals concentrate bulky feeds, making them easier to market. Many by-products of animals are also important to society. Consumer concerns affect the trends in the livestock industries.

Biotechnology research is greatly affecting livestock production practices. Animals may be identified by a variety of different methods, including ear tags. Concern regarding the use of animals in agriculture is usually categorized as animal rights or animal welfare. Many laws and regulations have been passed relating to the safety of both the livestock industry and the food supply. The threat of extinction for some species is also a problem in the livestock industry.

Student Learning Activities

1. Survey the farms in a community, collecting data on size of farm, acres of various crops grown, and kind and size of livestock enterprises on the farms. Prepare a bulletin board display showing a map of the community with the data summarized.
2. Prepare and present an oral report on one of the following topics:
   a. the domestication of animals
   b. the spread of livestock in the United States
   c. the functions of animals
   d. trends in animal agriculture

e. animal rights issues
f. food safety
3. Prepare graphs that show trends in livestock numbers in the United States.
4. When planning and conducting a supervised experience program, take into consideration the most common agricultural enterprises found in the local area.

Discussion Questions

1. Why was the domestication of livestock important to the development of civilization?
2. Why did the raising of large cattle herds develop in the Great Plains states whereas the finishing of cattle developed in the North Central states?
3. From what two wild stocks of swine were the American breeds of swine developed?
4. What was the main use of sheep by the early colonists in the United States?
5. Describe the major differences between sheep and goats.
6. What was the main use of the horses brought to the United States by the early colonists?
7. What changes have taken place in the poultry enterprise from colonial times to the present?
8. Name and briefly describe the eight functions of animals in our society.
9. How important economically is the livestock industry as compared to the total farm industry in the United States?
10. Briefly explain the trends in consumption of livestock products per person in the United States.
11. What is the trend in consumption of wool in the United States?
12. Briefly explain current trends in animal agriculture.
13. Briefly explain at least one issue associated with animal rights.
14. Why are animal by-products important to human society?

Review Questions

True/False

1. Domestication means to adapt the behavior of an animal to fit the needs of people.
2. The wild mallard duck has no relationship to the modern breeds of domesticated ducks.
3. Livestock producers generally support proper feeding, housing, and other good management practices because these activities lead to more efficient production.

Multiple Choice

4. Adapting the behavior of animals to fit the needs of humans is called ____________________.

a. training
b. domestication
c. breeding
d. culling

5. The trend in the consumption of poultry has continually shown a(n) ________________.

   a. increase
   b. decrease
   c. tendency to remain the same
   d. drastic variation in demand

6. One of the earliest animals tamed by humans was ________________.

   a. the horse
   b. sheep
   c. the goat
   d. swine

Completion

7. Modern cattle are descendants of the ____________________.
8. The three categories of animal health products are __________, __________, and ________________.
9. Chickens were being raised by the Chinese in about ______________ B.C.

Short Answer

10. Why was domesticating and confining animals so important to the food supply of early civilizations?
11. List and discuss three functions of animals.
12. What is the basic difference between the concerns of animal welfare and animal rights activists?

Chapter 2

Career Opportunities in Animal Science

Key Terms

associates degree
bachelors degree
masters degree
doctoral degree
talent
interest
citizenship

Objectives

After studying this chapter, the student should be able to

- discuss employment trends and opportunities in agriculture and animal science.
- describe the process of choosing an occupation.
- identify employer expectations and good work habits.
- identify good citizenship skills.

EMPLOYMENT IN AGRICULTURE

Agriculture is a major industry in the United States. The number of jobs most closely related to production agriculture has been decreasing, while the number of jobs that are in industries related to agriculture has been increasing. More than 300 different careers are available to persons with an interest in agriculture. Many of those careers require a minimum of 2 years of education beyond high school and many are in the field of animal science.

Agriculture and agriculture-related industries provide employment for about 17 percent of the work force in the United States. There are five major categories of employment in agriculture-related jobs:

1. *Farm production and agricultural services.* Typical jobs include farmers, hired farm workers, farm managers, and veterinarians—about 2 percent of the total work force.
2. *Input suppliers.* Typical jobs include employment in wholesale and retail sales of farm equipment; seed, feed and feed ingredients, fertilizer,

Figure 2-1 Feed and feed ingredients salespeople must have a broad knowledge of animal science.

and chemical suppliers; farm machinery manufacturing; and manufacturing and sale of farm chemicals (Figure 2-1).

3. *Processing and marketing.* Typical jobs include employment in apparel and textile manufacturing and processing meats, dairy goods, fruits, vegetables, bakery products, and beverages.
4. *Agricultural wholesale and retail trade establishments.* Typical jobs include employment in grocery stores, restaurants, convenience, and carry-out stores.
5. *Indirect agricultural businesses.* Typical jobs include employment in chemical and fertilizer mining and manufacturing food-processing machinery.

The number of full-time farms in the United States has been declining since 1936. This trend is expected to continue in the future. The average size of full-time farms has grown steadily larger over the years. However, there has also been an increase in the number of small, part-time farms because of an increased consumer demand for locally grown and organic products. Approximately 97 percent of the farms in the United States are individually owned, and the other 3 percent are owned by partnerships or corporations.

An expanding population in the United States and the world has resulted in an increased demand for food. This has created more jobs in the businesses that process, package, and market food.

Many job opportunities related to animal science are also available in the fields of agribusiness, communications, science, government, education, and sales. The emergence of new technologies provides additional job opportunities for people with an interest in animal science. The right combination of skills, education, training, and experience can lead to exciting careers in helping to feed and clothe the people of the United States and the world.

People develop an interest in animal science careers in various ways. Middle and high school agriculture programs are one source of information about animal

Connection: Locavores

The local food movement is a growing movement in the United States and elsewhere that spawned as interest in sustainability and eco-consciousness. Locavores are people who are interested in eating food that is locally produced and not moved long distances to market. The food may be grown in home gardens or grown by local commercial groups interested in keeping the environment as clean as possible and selling food close to where it is grown. Some people say that local food tastes better than food that is shipped long distances because it is fresher, riper when harvested, and may not depend as heavily on chemicals or irradiation to increase shelf life. On the other hand, local food is less regulated, so freshness, chemical use, and quality are variable and depend largely on the producer.

Local food networks play a role in efforts to eat what is local. These include community gardens, food cooperatives, community-supported agriculture, farmers' markets (Figure 2-2), and seed-saver groups. Preserving food for those seasons when it is not available fresh from a local source is one approach some locavores include in their strategies.

Those in the locavore movement generally seek to keep the use of fossil fuels to a minimum, thereby releasing less carbon dioxide into the air. Other reasons for locavorism may include a desire for fresher, less processed foods or to support the local economy and small and family farms. Many approaches can be developed, and they vary by locale. Many advocates of the local food movement encourage only a partial dependence on local food and suggest taking into account food miles, production methods, and availability in your decision.

Critics of the local food movement point out that transport is only one component of the total environmental impact of food production and consumption (only 4 percent of greenhouse gases emitted by the food industry come from transportation). In fact, any environmental assessment of food bought by consumers needs to take into account how the food has been produced and what energy is used in its production.

©Aigars Reinholds/Shutterstock.com

Figure 2-2 Locally grown foods can be found at farmers' markets.

science occupations (Figure 2-3). Courses in animal science in middle and high school can help students learn more about agriculture and develop basic skills (Figure 2-4). Many colleges also prepare students for various agriculture careers. There are many jobs for men and women with experience and training in animal science.

There are many jobs available that relate to animal science and increasingly these jobs require some type of education beyond high school. Jobs may require an **associates degree**, a **bachelors degree, masters degree**, or even a **doctoral degree.** Table 2-1 lists some of the career opportunities available in animal science.

USDA Agricultural Research Service; photo by Stephen Ausmus

Figure 2-3 High school agriculture programs provide students with a hands-on approach to learning about animal production and animal science occupations.

Courtesy of Frank B. Flanders

Figure 2-4 In agriculture education classes, middle school and high school students learn skills in the care and handling of animals.

TABLE 2-1 Sample Career Opportunities in Animal Science for College Graduates

Veterinary Medicine
- Practice
- Research
- Product development
- Teaching
- Inspection

Meat or Dairy Foods
- Product development
- Quality control
- Distribution and marketing

Livestock Promotion and Marketing
- Breed organizations
- Livestock publications
- Livestock sales/market reporting

Sales
- Agricultural Sales
- Feed
- Pharmaceuticals
- Agricultural chemicals
- Livestock supplies
- Distribution

Management
- Sales/marketing companies
- Food production/distribution

Financial Institutions
- Banks
- Lending agencies

Service Organizations
- Extension service
- Agriculture agents
- 4-H Agents
- Teaching

Computer Specialists
- Computer programmers
- Software distributors
- Modeling

Biotechnology
- Laboratory technical support
- Research scientists
- Genetics and animal breeding
- Population genetics
- Molecular genetics
- Genetic engineering
- Reproductive management
- Endocrinology
- Cloning
- Embryo technology
- Nutrition
- Feeding programs
- Nutrition/reproduction interactions
- Nutrition/health/immunity interactions
- Food science
- Product development
- Food processing fermentation

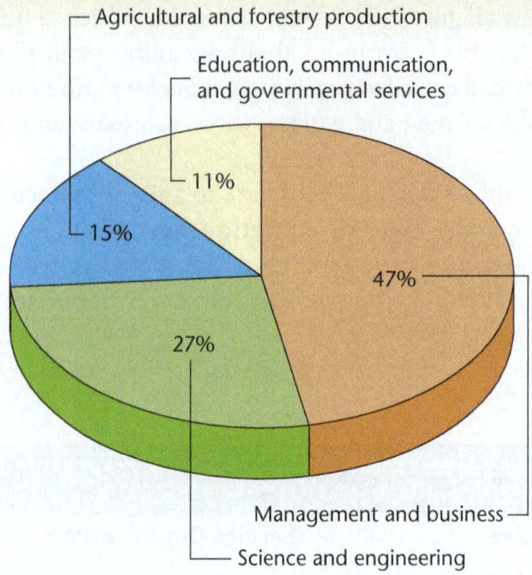

Figure 2-5 Employment opportunities for college graduates in the field of agriculture in the United States.

Source: Goecker, Allan D., Smith, P. Gregory, Smith, E., and Goetz, R. (2010). Employment Opportunities for College Graduates in Food, Renewable Energy, and the Environment, United States, 2010–2015.

EMPLOYMENT OPPORTUNITIES FOR COLLEGE GRADUATES IN AGRICULTURE

There is a growing concern among agribusinesses about a future shortage of qualified college graduates with training in the field of agriculture. Employment opportunities for college graduates with expertise in the food, agricultural, and natural resources are expected to remain strong. For example, a shortage of agricultural education teachers has been an ongoing problem for many years. Figure 2-5 shows an annual percentage of employment opportunities for college graduates in various fields of agriculture. Some examples of careers for college graduates in differing agricultural fields are included in Table 2-2.

Jim Pickerell Stock Connection Wo ldwide/Newscom

Figure 2-6 One possible job in the field of animal science is that of meat inspector.

TABLE 2-2 Example Careers for College Graduates in Agriculture

Scientists, Engineers, and Related Specialists

- Agricultural engineer
- Animal scientist
- Biochemist
- Entomologist
- Environmental engineer
- Food engineer
- Food scientist
- Geneticist
- Microbiologist
- Nutritionist
- Veterinarian

Social Services Professionals

- Dietitian
- Food inspector
- Nutrition counselor

Managers and Financial Specialists

- Business manager
- Credit analyst
- Economist
- Financial analyst
- Food service manager
- Retail manager
- Wholesale manager

Marketing, Merchandising, and Sales Representatives

- Food broker
- Grain merchandiser
- Livestock buyer
- Market analyst
- Marketing manager
- Sales representative
- Technical service representative

Agricultural Production Specialists

- Farmer
- Feedlot manager
- Professional farm manager
- Rancher

Education Communication and Information Specialists

- Agriculture education teacher
- College faculty member
- Cooperative extension agent
- Editor
- Public relations specialist
- Reporter

Courtesy of Frank B. Flanders

Figure 2-7 A veterinarian may specialize in large animal or small animal medicine or both.

OCCUPATIONS IN ANIMAL SCIENCE

Many occupations are possible in the field of animal science (see Figures 2-6, 2-7, and 2-8). Table 2-3 contains some of these occupations listed by type of animal. Some general occupations that deal with more than one phase of animal science are also listed.

CHOOSING AN OCCUPATION

Selecting an occupation involves three basic steps. The first step is to evaluate individual academic abilities, skills, characteristics, and career goals. The second step is to obtain as much information as possible about potential careers and occupations. The third step is to take into consideration individual abilities and interests along with career requirements to determine the most desirable career path.

Self-Analysis

When choosing an occupation, it is important to keep in mind natural abilities. Grades in school and testing done by guidance counselors can help to determine a person's abilities. It is important to be as realistic as possible when matching abilities to occupations.

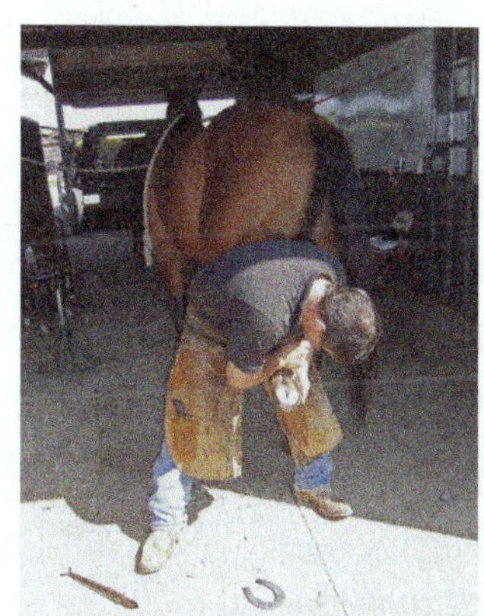

Courtesy of Wayne Nunn Farrier Services; photo taken by Frank B. Flanders

Figure 2-8 A farrier is a person who shoes horses. Farriers are a vital part of the horse industry.

TABLE 2-3 Occupations in Animal Science

General Occupations

- Agricultural consultant
- Agricultural economist
- Agricultural educator/FFA advisor
- Agricultural engineer
- Agricultural equipment dealer
- Agricultural equipment designer
- Agricultural extension agent
  - Extension specialist
- Agricultural journalist
- Agricultural loan officer
- Agricultural news director
- Animal behaviorist
- Animal breeder
- Animal breeding scientist
- Animal caretaker
- Animal cytologist
- Animal ecologist
- Animal groomer
- Animal health products representative
- Animal inspector
- Animal keeper
- Animal nursery worker
- Animal nutritionist
- Animal physiologist
- Animal science teacher
- Animal scientist
- Animal taxonomist
- Animal technician
- Animal trainer
- Animal treatment investigator
  - Animal control officer
- Aquaculturist
- Artificial breeding distributor
- Artificial breeding technician
- Artificial insemination herd sire evaluator
- Artificial insemination herd sire manager
- Artificial inseminator
- Bacteriologist
- Biological aide
- Biotechnologist
- Breed association field worker
- Breeding researcher
- Butcher
- Cell biologist
- Central test station manager
- Central test station worker
- College teacher
- County agricultural agent
  - Agricultural agent
  - County adviser
  - County agent
  - Extension agent
  - Extension-service agent
  - Extension worker
  - Farm adviser
  - Farm agent
- Custom-feed-mill operator
- Drug company representative
- Embryologist
- Farm broadcaster
- Farmhand
- Farm investment manager
- Farm loan officer
- Farm manager
- Farmworker, general
- Farmer, general
- Feed and farm management adviser
- Feedlot manager
- Feed mill operator
- Feed ration developer and analyst
- Feed research aide
- Field sales representative
  - Animals and feed
  - Chemicals, pharmaceutical
- Food and drug inspector
- Food processing supervisor
- Food scientist
- 4-H club agent
- Geneticist
- Inspector, grain mill products
- Laboratory technician
- Land bank branch manager
- Livestock agent
- Livestock auctioneer
- Livestock building dealer
- Livestock building salesperson
- Livestock buyer
- Livestock caretaker
- Livestock commission agent
- Livestock equipment dealer
- Livestock equipment salesperson
- Livestock farmhand
- Livestock herder
- Livestock geneticist
- Livestock inspector
- Livestock producer
- Livestock rancher
- Livestock sales representative
- Livestock seller
- Livestock yard attendant
- Livestock yard supervisor
- Magazine writer
- Mammalogist
- Manager of retail feed and supply store
- Market research analyst, agriculture
- Marketing analyst
- Marketing manager
- Market news analyst
- Meat buyer
- Meat cutter
- Meat grader
- Meat inspector
- Meat science researcher
- Microbiologist
- Muscle biology researcher
- Nutritionist
  - Researcher
  - Physiologist
- Ova transplant specialist
- Parasitologist
- Pathologist
- Pharmaceutical chemist
- Pharmaceutical company representative
- Production researcher
- Reproduction researcher
- Reproductive physiologist
- Research technician
- Salesperson, animal feed products
- Sales manager
- Sales representative, agricultural supply
- Show cattle caretaker, supervisor
- Superintendent, grain elevator
- Supervisor, animal cruelty investigation
  - Animal humane agent supervisor
- Supervisor, artificial breeding ranch
- Supervisor, feed mill
- Supervisor, livestock yard
- Supervisor, stock ranch
- Technical service representative
- Toxicologist
- Training manager
- Vertebrate zoologist
- Veterinarian
- Veterinarian aide

TABLE 2-3 Occupations in Animal Science (*continued*)

- Veterinarian, laboratory animal care
- Veterinarian, avian
- Veterinarian's assistant
- Veterinarian technician
- Veterinary anatomist
- Veterinary epidemiologist
- Veterinary livestock inspector
- Veterinary meat inspector
- Veterinary parasitologist
- Veterinary pathologist
- Veterinary pharmacologist
- Veterinary physiologist
- Veterinary virologist
- Veterinary virus-serum inspector
- Zoologist

Beef Occupations

- Beef breeder
- Beef farmer
- Beef farmhand
- Beef herder
- Cattle buyer
- Cattle feeder
- Cattle rancher
- Feedlot maintenance worker
- Feedlot supervisor
- Stock ranch supervisor

Dairy Occupations

- Cheese maker
- Dairy cattle herder
- Dairy farm worker
- Dairy farmer
- Dairy helper
- Dairy herdsman
- Dairy herd supervisor (DHIA)
- Dairy management specialist
- Dairy nutrition specialist
- Dairy sanitarian
- Dairy scientist
- Dairy technologist
- Dairy tester
- Field contact technician, dairy
- Manager, dairy farm
- Milk plant supervisor
- Milk sampler
- Milker, machine
  - Milking machine operator
  - Sampler
- Supervisor, dairy processing
  - Butter production supervisor
  - Cheese production supervisor
  - Instant-powder supervisor
  - Pasteurizing supervisor
- Supervisor, dairy farm

Swine Occupations

- Hog buyer
- Swine breeder
- Swine farmer
- Swine farmhand
- Swine herder
- Swine rancher

Sheep and Goat Occupations

- Fleece tier
- Lamber
- Breeder
- Farmer
- Farmhand
- Herder
- Sheep shearer
- Supervisor, wool-shearing
- Wool buyer
- Wool-fleece grader
- Wool-fleece sorter

Horse Occupations

- Equine dentist
- Farrier
- Hoof and shoe inspector
- Horse breeder
- Horse exerciser
- Horse farmhand
- Horse herder
- Horse rancher
- Horse stable attendant
- Horse trainer
- Paddock judge
- Race horse trainer
- Racing secretary and handicapper
- Stable attendant

Poultry Occupations

- Blood tester, fowl
- Caponizer
- Chick grader
  - Poultry culler
- Chick sexer
- Chicken breeder
- Egg candler
- Farm worker, poultry
- Field service technician, poultry
- Grader, dressed poultry
- Laborer, poultry farm
  - Laborer, brooder farm
  - Laborer, chicken farm
  - Laborer, egg-producing farm
  - Laborer, fryer farm
  - Laborer, pullet farm
  - Laborer, turkey farm
  - Laborer, poultry hatcher
  - Hatchery helper
  - Incubator helper
- Manager, poultry farm
- Manager, poultry hatchery
  - Manager, chicken hatchery
  - Manager, duck hatchery
  - Manager, turkey hatchery
- Poultry breeder
  - Chicken fancier
- Poultry breeding researcher
- Poultry debeaker
- Poultry farmer
  - Duck farmer
  - Poultry farmer, egg
  - Poultry farmer, meat
  - Turkey farmer
- Poultry farmhand
- Poultry field service technician
- Poultry geneticist
- Poultry grader
- Poultry inseminator
  - Artificial insemination technician
- Poultry nutrition researcher
- Poultry products sales manager
- Poultry products technologist
- Poultry scientist
- Poultry tender
- Poultry vaccinator
- Supervisor, poultry farm
  - Supervisor, brooder farm
  - Supervisor, egg-producing farm
  - Supervisor, fryer farm
  - Supervisor, pullet farm
  - Supervisor, turkey farm
- Supervisor, poultry hatchery
  - Supervisor, chicken hatchery
  - Supervisor, turkey hatchery
- Turkey breeder
- Turkey producer

Occupations are sometimes chosen on the basis of talents. A **talent** is a natural aptitude a person possesses for performing an activity particularly well. Hobbies may be an indication of talents. Being able to handle animals is an important talent for those considering a job in animal science.

Interests are those things that hold a person's attention. People do best when they work at an occupation that interests them. Individuals must consider whether or not an interest in a particular occupation is temporary or lasting. Other factors that may influence occupational choices include the following

- health condition
- work experience
- educational aspiration
- personality
- attitudes
- values
- self-image
- flexibility

Studying an Occupation

When studying an occupation, it helps to have an outline of key points to follow. The nature of the work, or what a person has to do on the job, should be considered. The following points are examples of things to examine when seeking information on a specific job.

- educational requirements
- special skills required
- working hours
- salary
- geographical location
- indoor vs. outdoor work
- physical activity required
- travel requirements
- tools or machines used
- possible physical hazards
- variety of tasks involved
- certification or licensing
- employment availability
- job security
- insurance options
- retirement plans

Making a Decision

After self-analysis and occupational study, an occupation can be selected. It is a good idea to have several occupations in mind. Individuals should try to match themselves to the best possible occupation, but they should also leave room for a change of career plans in the future.

SOURCES OF INFORMATION ABOUT OCCUPATIONS

Information about occupations is available from a variety of sources. School guidance counselors and a variety of reliable Internet sites can provide a great deal of information. Teachers, those employed in a specific field, parents, and state employment services are also good sources of information about occupations. Obtain information from as many different sources as possible. The more information a person has, the better the chances of selecting a suitable occupation.

ON THE JOB: EMPLOYER EXPECTATIONS AND GOOD WORK HABITS

Finding a suitable job is an important step in career planning and advancement. After being hired, it is important to understand that employers have certain expectations that must be met to be successful in a career. Desirable and undesirable work habits can have a strong influence on career success. Good citizenship skills are important, and both employees and employers have a right to expect certain things from one another. Specific expectations and work habits will vary from job to job.

Attitude

One of the most important traits that employers expect in their employees is a positive attitude. Workers with poor attitudes often have difficulty holding a job. Some indications of a positive attitude include:

- being friendly, with a ready smile; looking people in the eye when interacting with them but not being overbearing in manner.
- being willing to accept new ideas when appropriate; recognizing the right of other workers to hold different points of view.
- rarely complaining about the job or other workers; not criticizing others on the job.
- being willing to accept responsibility for one's own actions, including mistakes; not making excuses for one's own behavior.
- being willing to dress in a manner appropriate for the job.
- giving consideration to what is good for other workers and being willing to help others succeed on the job.
- showing enthusiasm for the job.
- being loyal to the employer.

Honesty and Dependability

Employers generally place a high value on honesty and dependability in their employees. A number of worker actions demonstrate these traits. These include:

- being on time for work, not taking excessive breaks during the workday, and putting in a full day's work.
- using money that belongs to the company only for legitimate company business and keeping an accurate account of company money.
- maintaining a high level of ethical behavior in relationships with fellow workers, customers, suppliers, government officials, and any others one comes into contact with during the course of one's work.
- not disclosing proprietary or confidential company information to unauthorized persons.
- not using inside information if buying or selling securities that relate to one's employer; to do so is a violation of the law and subjects one to severe penalties.
- not engaging in behavior in personal relationships or financial activities that creates even the appearance of a conflict of interest.

Working Well with Others

The ability to work with other employees is an important job skill that employers find desirable. Maintaining a pleasant personality, having patience, and working together to achieve common goals are qualities often expected of employees. Good social skills, such as the ability to meet and talk with people, may be important in some jobs (Figure 2-9). It is important to understand appropriate relationships on the job both worker-to-peer and worker-to-supervisor. Show respect for authority in the workplace and be able to reach a compromise when necessary to resolve differences of opinion.

The workplace of today differs from what it was just a few years ago. One must be careful to not create a hostile or intimidating work environment or to harass other workers. Jokes that may be considered offensive, vulgar and profane language, and involvement in disputes between other workers should be avoided. Making sexual advances to another worker, subordinate, or supervisor may be construed as harassment and can result in legal action against both the worker behaving in this manner and the company that allows this type of behavior. Most companies have very specific policies that deal with these types of behaviors. Failure to learn and obey these policies may result in job loss.

Figure 2-9 Meeting and communicating with others is an important skill.

Time Management

Workers who learn to manage their time on the job in a productive manner are usually considered to be more valuable employees. Workers should be punctual arriving and leaving the job. Some jobs have set hours; however, some jobs may have more flexible time requirements.

It is not enough to just come to work; one must also be a productive worker. Planning how to use time and producing quality work are good skills to develop. Taking pride in one's work, not wasting materials, and taking care of the equipment being used are also important considerations. Learn to listen carefully, follow directions, complete assigned tasks in an appropriate manner, and be able to work under pressure. Sometimes jobs need to be completed in a shorter time than normal; a good worker is able and willing to adjust to this need and still produce quality work. It is also important to show initiative at work; do not wait to be told what to do next. Learn the requirements of the job and be willing to move ahead to the next task without waiting for direction. The worker who cannot see what needs to be done next may not have a job for long.

Safety on the Job

Most workplaces have safety rules that are designed to protect the workers and customers. Workers are expected to learn, follow, and practice safe work habits on the job. Failure to do so may result in injury to oneself or to others in the workplace, or loss of employment.

Good Communication Skills

The ability to communicate with others on the job is a desirable employability skill. Learn to use correct grammar, spelling, punctuation, and sentence construction in written communication. Many jobs require good speaking skills. Develop a pleasant speaking voice and enunciate words clearly. Use correct grammar and avoid slang and colloquial expressions. Think about what is to be said before speaking, ask questions when instructions are not clear, and be willing to contribute ideas at company meetings.

Remember that one communicates with more than just speaking or writing. Body language and facial expressions may reveal more about what one is thinking than the words being spoken. Be aware of the signals being sent to others by body language and facial expressions. Practice maintaining a pleasant demeanor at all times. This will become a habit that will serve one well in working with others.

Appropriate Use of Computers and the Internet

Workers often use computers on their jobs and have access to the Internet through these computers (Figure 2-10). The ability to use this technology is increasingly important to employers. Workers must become familiar with company policy as it relates to the use of the Internet. Companies may not permit use of their e-mail systems for personal use. Use of social networking sites at work may also be considered inappropriate. Employers may also view the social networking profiles of job applicants and use this as a hiring factor. Websites that contain material such as sexually explicit images should never be accessed from a computer at work. Be aware of any written policies about appropriate Internet use, and be sure to follow those policies because failure to do so may result in job loss. Remember that computers retain history data files that employers may access to determine where a user has been on the Internet. Some employers are adding software to company computers that records keystrokes and permits a reconstruction of activity on that computer. The computer at work belongs to the employer and courts have upheld the right of the employer to have absolute control over its use. Employees do not have an expectation of privacy regarding their use of company computers.

Courtesy of Frank B. Flanders

Figure 2-10 Computer skills and skills using other technology are increasingly important to employers.

If the job entails working with company data that are stored or transmitted electronically, the worker is obliged to protect those data from misuse, unauthorized access, and loss. Use good backup systems to build redundancy into the data system and a good firewall to isolate the computer from access by outsiders on the Internet. This is especially important where the computer is part of a local area network and is connected to the Internet all the time.

Appropriate Behavior

Discrimination on the basis of age, race, ethnicity, religion, gender, or any other protected class under applicable law is illegal. Do not discriminate or show prejudice on the job. Federal and state laws prohibit discrimination and provide penalties for engaging in it. Many companies also have policies that govern worker relationships. Employers expect employees to learn and follow the applicable policies.

If you work at a job where workers and supervisors occasionally gather in social settings, such as an annual holiday party, be careful of your behavior in such situations. Dress in an appropriately conservative manner and do not indulge in excessive alcohol consumption. When in doubt, it is wise to be more conservative in your behavior.

Be cautious about romances with fellow employees. Some companies have policies that specifically prohibit such relationships; others do not. It is wise to proceed cautiously if you are tempted to pursue romance on the job. Remember that you may have to work with the individual even after the romance has ended. Be willing to accept the consequences when getting involved with a fellow worker. If you are in a supervisory capacity, never become involved with a subordinate. To do so may be perceived as harassment on the one hand or as blatant opportunism by the subordinate on the other.

In a survey on the impact of office romances, approximately half of respondents indicated that office romances result in favoritism or at least the perception of favoritism; almost one-third of respondents believed that the careers of those who were involved were damaged. The best path to follow is to exercise caution and proceed slowly. Always be aware of and do not violate company policy.

Health and Grooming

To the extent that it is possible, the worker should strive to maintain the level of physical fitness necessary to perform the job. The maintenance of good health is closely related to proper diet and exercise. Getting enough sleep so as to be rested when coming to work contributes to better job performance. Do not abuse drugs of any kind (prescription or illegal) or alcohol. It is difficult to perform well on the job when impaired by drug or alcohol use.

Develop good grooming habits such as cleanliness. Proper grooming is an indication of respect for those you work with. Grooming habits may influence how your job performance is evaluated by supervisors and affect the opportunities for promotion or increased pay.

Different jobs require different kinds of clothing. Sometimes a uniform is required; other jobs may leave the choice of clothing to the worker. Always dress in a manner appropriate for the job you are doing. If in doubt, ask the supervisor what is considered appropriate dress for the job you are assigned to do.

Many companies permit business casual dress if you are working in an office setting. Do not confuse casual dress with sloppy dress. Clothing that is neat, clean, and conservative is generally appropriate in an office setting. Many jobs do not permit workers to wear T-shirts, jeans, or sneakers.

GOOD CITIZENSHIP SKILLS

Good **citizenship** is an important part of one's life and a part of a successful career. Workers who practice good citizenship are generally viewed as valuable employees. Many different aspects go into being a good citizen.

A good citizen understands his or her constitutional rights and exercises those rights without violating the rights of others. For example, one has the right to freedom of speech but not the right to slander others. Understanding and following the laws that protect the rights of workers and children in the workplace are also part of being a good citizen. For example, child labor laws prohibit companies from hiring children under a certain age to do jobs that are considered too dangerous for young people. Minimum wage laws protect workers from being exploited by employers who do not want to pay proper wages.

A good citizen understands the need for taxation to support the needs of society and complies with applicable tax laws with honesty and integrity. Taxes help pay for things such as public education, police protection, and highways. It is also important to understand the consequences of failing to follow the laws of society. Breaking laws may result in heavy monetary fines or imprisonment.

Another mark of good citizenship is active participation in the political and economic systems of one's country. One should be an informed voter and know the issues and positions taken by the various candidates for public office (Figure 2-11). Exercise the right to vote in all elections because local issues are as important and worthy of one's attention as are state and national issues. Respect the rights of other members of society. Disagreements should be resolved in a respectful manner.

Be loyal to one's country; however, remember that loyalty does not mean blind obedience to those in positions of authority. It is important to exercise one's right to dissent and to raise public awareness of problems in society. Good citizens contribute to the progress and growth of the society in which they live and are willing to take part in civic and community activities that contribute to the general good. For example, serve on the local board of education, city council, or fire commission. Taking an active role by giving one's time and talent to worthy civic causes is an indication of good citizenship.

Being a good steward of the environment is important for future generations. In work and personal life, know, understand, and follow environmental standards that protect the soil, air, and water. Be aware of the regulations of state and federal environmental agencies and do not violate those regulations. Many companies have policies that require workers to follow all applicable regulations that protect the environment. Failure to follow such company policies can result in loss of the job.

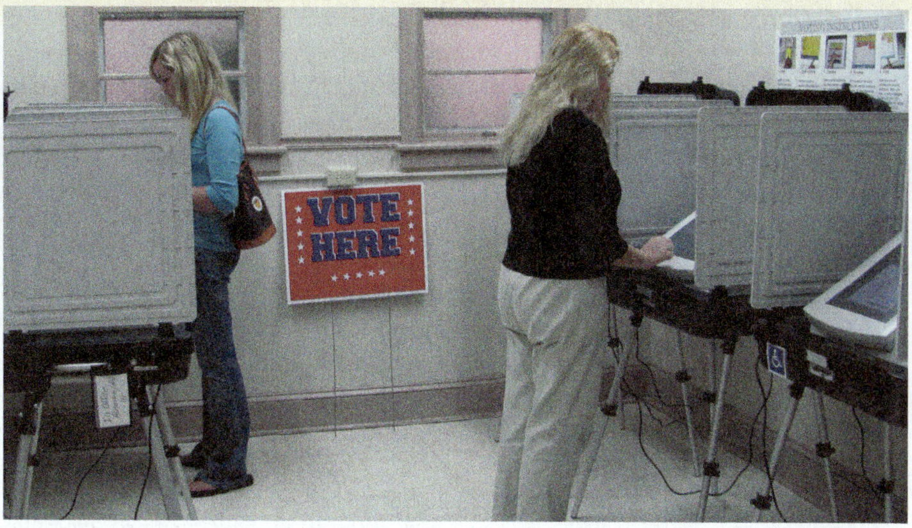

Courtesy of Frank B. Flanders

Figure 2-11 Voting is a civic duty.

SUMMARY

Agriculture offers many different kinds of jobs and occupations. Both agriculture and agriculture-related industries provide many opportunities for employment. Livestock production and its related services and supplies, processing, distributing, and marketing are all a part of the agriculture industry. There are also many possible careers for those who are interested in animal science.

Three basic steps should be followed in choosing an occupation: (1) self-analysis, (2) studying the occupation(s) of interest, and (3) matching personal traits and skills to the occupation(s). Many sources of information about occupations are available. School guidance counselors, teachers, people who work in the occupation, printed material, and various reliable Internet sources all provide information about various occupations.

To be successful on the job, a worker must have good work habits that meet the expectations of employers. These include positive attitudes, honesty, dependability, working with others, good time management, communication skills, appropriate use of new technologies, considerate behavior, good health, and proper grooming. Workers who lack some of these skills will find it more difficult to hold a job and have a successful career.

Another important aspect of career success is possessing good citizenship skills. Some important aspects of good citizenship include knowing one's rights and respecting the rights of others, contributing one's time and talents for the good of society, taking an active role in the political and economic system, obeying the law, and being a good steward of the environment.

Student Learning Activities

1. Prepare a bulletin board display of occupations in animal science.
2. Survey the local community to determine employment opportunities in animal science.
3. Interview a person working in an animal science occupation and present an oral report to the class based on the interview.
4. Invite people in animal science occupations to speak to the class.
5. Practice good work habits and citizenship skills in school to help prepare you for planning and conducting a supervised experience program.

Discussion Questions

1. About what percent of the total labor force in the United States is employed in agriculture and agriculture-related jobs?
2. What are two ways in which a young person can learn more about a career in animal science?
3. Name five general occupations in animal science that might require a college degree.
4. Name three occupations in each of the following areas: (a) beef, (b) swine, (c) sheep, (d) horses, (e) poultry, (f) dairy.
5. List the three basic steps in choosing an occupation.
6. Name five factors that play a part in self-analysis.
7. What types of information should be known about an occupation in order to make a wise career decision?
8. What are three good sources of information about occupations?
9. Name and briefly describe some of the skills that employers expect of employees.
10. Name and briefly describe some of the good citizenship skills that one should have to be a productive member of society.

Review Questions

True/False

1. An expanding population in the United States and the world has caused an increased demand for food.
2. Selecting an occupation involves three basic steps: self-analysis, obtaining information about the occupation, and making a decision based on steps 1 and 2.
3. The number of full-time farms in the United States has been increasing since 1936.

Multiple Choice

4. The number of careers available to those interested in agriculture number more than ______________.

   a. 60 c. 500
   b. 300 d. 150

5. Approximately what percentage of the farms in the United States are individually owned?

   a. 18 c. 97
   b. 3 d. 75

6. When studying an occupation, the ______________, or what a person has to do on the job, should be considered.

   a. job flexibility c. nature of the work
   b. educational requirements d. job security

Completion

7. Employers generally place a high value on ________ and ________ in their employees.

Short Answer

8. List the five major categories of employment in agriculture-related jobs.
9. List five occupations that relate to animal science.
10. List four types of degrees that jobs may require.
11. How might people develop an interest in agriculture or agriculture-related careers?

Chapter 3

Safety in Livestock Production

Key Terms

zoonosis
biosecurity
bioterrorism
agroterrorism
agrosecurity

Objectives

After studying this chapter, the student should be able to

- explain the importance of safety procedures when working with livestock.
- discuss four types of hazards related to livestock production.
- list the safety practices to be followed when working with livestock and poultry.
- develop a livestock safety checklist.
- distinguish between agroterrorism and biosecurity.
- discuss the dangers of bioterrorism and agroterrorism.
- discuss the procedures to be followed in providing agrosecurity.

Agriculture is one of the most dangerous occupations in the United States (Figure 3-1). Recent data show a death rate of 26 workers per 100,000. All industries combined show a death rate of 3.5 workers per 100,000. There are approximately 150,000 disabling accidents involving farm workers. Most fatal and nonfatal farm injuries are caused by machinery. Although livestock cause relatively few deaths each year, they are the second highest cause of farm injuries. Other major causes of nonfatal injuries on farms, in descending order of importance, are machinery (except tractors), hand tools, slips and falls, and tractors. Unfortunately, many of the deaths that occur in the agricultural industry are among young workers under the age of 20. Most of these deaths result from accidents involving farm machinery. The annual cost of farm accidents in the United States is estimated to be between $4 and $5 billion. Figure 3-2 shows recent safety statistical data.

Figure 3-1 Agriculture is one of the most dangerous occupations in the United States.

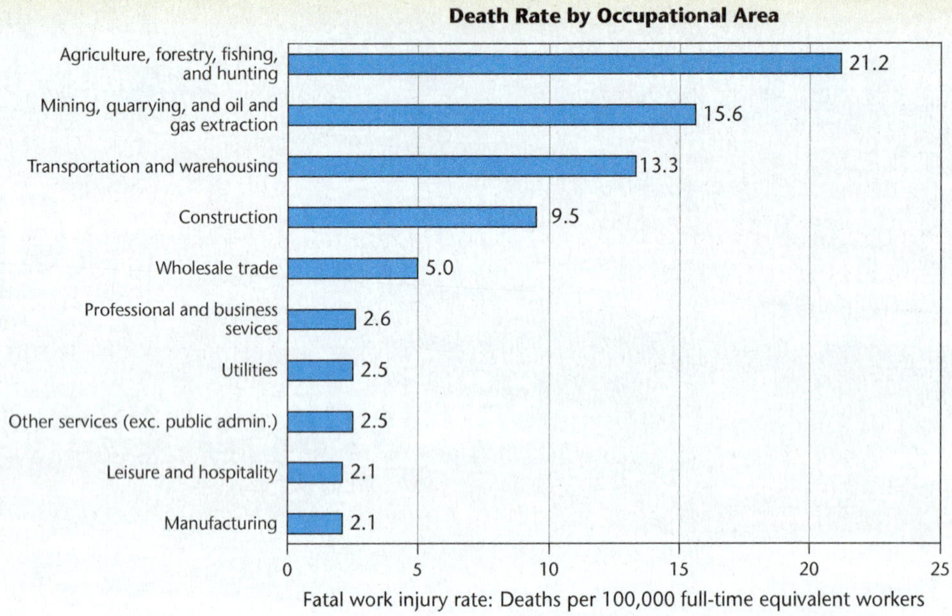

Figure 3-2 National Safety Statistics: Industry-Related Work Injuries

Source: US Bureau of Labor Statistics, US Department of Labor, 2011.

Farmers who hire labor are required by law to provide safe and healthy workplaces for their employees. They must inform workers about safety practices. Employees must be told about their rights and responsibilities under the regulations of the Occupational Safety and Health Act (OSHA). Posters have been designed by OSHA for this purpose. OSHA requires farm employers to keep records of work-related injuries and illnesses if 11 or more employees work on the farm. Farmers must permit OSHA inspectors to check their farms to see that the law is being followed.

TYPES AND KINDS OF INJURIES

Most of the people injured by animals on the farm are farm family members. Fewer than 10 percent of the injuries are to hired help or visitors to the farm. The most serious accidents often occur with horses and bulls; however, accidents also happen when working with cows and hogs.

The National Safety Council reports that people in the 45 to 64 age group are most often hurt in accidents with cattle, and injuries caused by hogs occur most commonly to people in the 25 to 64 age group. Surveys reported by the National Safety Council show that most of the people hurt by cattle and hogs are men and boys. Hogs may bite, step on people, or knock them down. Injuries from cattle and hogs usually occur in farm buildings or in lots close to the buildings. People are usually injured by cattle when they kick, step on, or push someone against a hard surface, such as the side of a pen; however, people may also be injured by falling when working with cattle.

The National Safety Council reports that children in the 5 to 14 age group and young adults in the 15 to 25 age group are most often hurt by horses. Surveys show that the number of women and girls hurt by horses is almost the same as the number of men and boys. Accidents involving horses are more common in barnyards, fields, lanes, woods, and along public roads.

HUMAN AND ENVIRONMENTAL FACTORS RELATING TO SAFETY

Human error is usually a major factor in the cause of accidents. Being tired, not paying attention, and using poor judgment are frequent causes of accidents that involve animals.

People 15 years old and younger, and those older than 65, have more accidents on the farm than people between the ages of 16 and 64. Farming is an occupation in which children are likely to be in the work area. Their curiosity and lack of experience can easily lead them into situations where they may get hurt. As people grow older, they tend to lose some of their strength and agility. They may have poorer balance and failing vision, which may cause them to have more accidents around animals.

Sometimes, workers are not properly instructed in handling animals. This can also result in accidents involving livestock.

A worker who does not feel well may be more likely to have an accident. Sometimes people are in a hurry to get the job done. This can lead to mistakes in judgment that cause accidents. Long hours of work during a day are often common in farm operations. Being tired increases the chance of having an accident.

Workers who fail to use personal protective equipment in dangerous environments are more likely to be injured than those who dress correctly. There are many dangerous environments involved in livestock operations. These include slippery floors, manure pits, corrals, dusty feed areas, silos, automatic feeding equipment, and confinement livestock and poultry buildings (Figure 3-3).

Photo by Sarah Kane, Courtesy of Cabe Brothers Farm.

Figure 3-3 Poultry house with tunnel ventilation.

Many confinement livestock buildings have a manure storage pit that is cleaned out only a few times each year. If the building is not properly ventilated, the pit gases can kill workers and livestock. Vent pipes must be installed properly. Improperly installed vent pipes may allow gas fumes to be recycled into the building, which can cause illness and possibly death. Figure 3-4 shows a tunnel ventilation system for poultry.

A standby source of electrical power is recommended for modern livestock farms. This is especially important for farms with confinement livestock buildings.

Photo by Sarah Kane, Courtesy of Cabe Brothers Farm.

Figure 3-4 Tunnel vents in chicken houses help to circulate and cool the air and reduce the danger from the buildup of toxic gases.

If pit fans do not operate because of an electrical power failure, a buildup of toxic gases can result. An emergency source of power can be a life-saving measure should this situation occur.

CHEMICAL SAFETY

The U.S. Environmental Protection Agency (EPA) issued regulations under the Worker Protection Standard for farm chemicals that became effective January 1, 1995. These standards are designed to reduce the health risks associated with pesticide use. They apply to farm employees who work with farm chemicals or related equipment. Farmers and their immediate families are exempt from the regulations. The standards apply to the use of pesticides for plant production on farms, forests, nurseries, and greenhouses. There are exceptions for the use of pesticides for livestock, pasture, rangeland, structures, gardens, lawns, rights-of-way, and post-harvest applications.

The major provisions of the regulations include the following:

- Personal protective equipment must be provided to pesticide applicators or handlers.
- Restrict entry to fields treated with:
  a. highly toxic material for 48 hours.
  b. moderately toxic material for 24 hours.
  c. less toxic material for 12 hours.
- Workers must be told, either in person or by posted signs, that a field has been treated with a pesticide and when it is safe to enter the field after treatment.
- Notices must be posted in easily seen locations showing the date, time, and location of treated fields. The notices must include the brand name, active ingredients, and EPA registration number of the pesticide used.
- Workers must be trained in safe pesticide handling methods at least once every 5 years.
- Written safety information must be given to workers, and pesticide safety posters must be displayed.
- Water, soap, and towels for washing and decontamination must be provided and located where they are readily accessible to the workers.
- Emergency transportation to a medical facility must be readily available.
- If the worker is applying a Class I (highly toxic) pesticide, he or she must be seen or talked to by the farmer or supervisor every 2 hours while the work is being done.

The EPA requires that labels on pesticides must provide information that employers need to properly notify their employees about the safe use of the pesticide. Much of the information required under the Worker Protection Standard must be included on the pesticide label. Although the Worker Protection Standard applies only to employees, farmers and their families who work with chemicals should be careful to protect their own health. Following the Worker Protection Standard that protects workers can also protect farmers and family members.

Many chemicals that are used in growing crops or raising livestock are dangerous to people. All workers must be instructed in safe practices relating to the use of these chemicals. Make sure workers who handle farm chemicals read and understand the label instructions on the chemical containers, including the instructions for first aid treatment in case of an accidental spill or ingestion of the chemicals.

Connection: LD50: It's a Killer!

Have you ever wondered how chemical companies determine which products are toxic enough to have the words CAUTION, WARNING, or DANGER on their labels (Figure 3-5)? In toxicology, the definition of *acute toxicity, lethal dosage value, median lethal dose, or LD50* (lethal dose, 50 percent) is simply the amount of a toxic substance needed to kill at least 50 percent of the population being tested. LD50 is measured in milligrams (mg) of the chemical per kilogram (kg) of body weight. Ultimately, the LD50 of a substance is used to predict how much of the particular toxic material it would take to kill a human. By figuring out a product's LD50, companies can determine how hazardous the chemical is to humans. The EPA categorizes LD50 by three methods of entry: oral, dermal, or inhalation. Oral LD50 values are obtained when test subjects are fed pesticide-treated feed or water. Dermal LD50 values are obtained when the pesticide is applied to the skin of the test subjects. Inhalation LD50 values are obtained when the test subject breathes in the pesticide. Often the inhalation LD50 is lower (more toxic) than the oral LD50, which is in turn lower (more toxic) than the dermal LD50. In addition to pesticides, commonly used household chemicals, drugs, and many food substances are assigned LD50 values. The following list gives the LD50 values for some substances arranged from the most toxic (lowest LD50) to the least toxic (highest LD50):

- rattlesnake venom 1 mg/kg
- ivermectin (an antiparasitic drug) 50 mg/kg
- caffeine 192 mg/kg
- aspirin 200 mg/kg
- Roundup™ 5,600 mg/kg

Figure 3-5 All chemicals must include one of the three signal words based on their toxicity.

OSHA requires that a Safety Data Sheet (SDS), once known as a Material Safety Data Sheet, be available for all chemicals in the workplace. Any retail outlet that sells hazardous chemicals is required to provide the SDS for that chemical if a consumer requests it. Anyone using a hazardous chemical should carefully read and understand the SDS for that chemical before handling it. Be sure the SDS is current. Check the date it was prepared; it may be wise to contact the manufacturer to make sure it contains the most recent information available.

Information typically found on an SDS includes the:

- identity of the chemical, the manufacturer's name and address, and an emergency contact number; a nonemergency number for more information may also be included.
- hazardous ingredients found in the chemical.
- physical and chemical characteristics of the chemical.
- fire and explosion potential of the chemical.
- health hazards posed by the chemical.
- precautions for safe handling and use of the chemical.
- procedures for controlling spills of the chemical.
- control measures for the use of the chemical.
- LD50 values for the chemical.

In some parts of the United States, a special permit is required to buy and use certain farm chemicals. Information regarding these requirements may be secured from the local Agricultural Extension Service office.

CAUTION

Always check label requirements and local and state disposal regulations before discarding any unused pesticides or empty containers.

Information about the proper disposal of unused pesticides and empty pesticide containers is found on the label of the container. Always follow these instructions to reduce health hazards for workers. Empty containers must be kept in a safe storage area until they can be disposed of properly. Metal, plastic, and glass containers should be rinsed three times and disposed of in a landfill designated for pesticide disposal. Recycling is a disposal option that is gaining in popularity, especially for metal and plastic containers. Do not pour unused pesticides down a drain or put them in the trash. Generally, commercial producers who may accumulate excess pesticides or diluents are limited to three options: continue to store them, remove them to a landfill specifically designated for pesticide disposal by the state or the EPA, or use the pesticides according to label directions.

Information about chemical regulations may be obtained by contacting the EPA headquarters, a regional EPA office, or a regional poison control center.

Chemicals, such as pesticides, must be handled with care, and clothing that has been contaminated must be properly washed or disposed of. Some chemicals can enter the body through the skin; this is the most common way dangerous chemicals get into a person's body. Other dangers include inhaling or swallowing the chemical.

A person who has handled chemicals should always wash his or her hands and face with soap and water. After completing a job that requires the use of chemicals, a worker should shower to remove all the chemicals from all parts of the body. If an accidental, massive contamination of the body occurs, the worker should immediately take a shower to avoid absorbing any of the chemical into the body.

Care of Clothing Worn While Using Farm Chemicals

The proper care or disposal of clothing worn while using farm chemicals depends upon the level of toxicity of the chemical used. If the label says CAUTION, the chemical is slightly toxic and the clothes may be cleaned with one to three machine washings. If the label says WARNING, the chemical is moderately toxic and the clothes will need more than three machine washings. If the label says DANGER POISON, the chemical is highly toxic and the clothes must be disposed of according to the directions on the label. Clothing that is contaminated with a concentrated chemical should also be disposed of following the label directions.

Guidelines to follow for washing clothes contaminated with farm chemicals include the following:

- Before washing, store the clothes in a plastic bag, separate from other clothes.
- Handle the contaminated clothes with neoprene or rubber gloves; do not use these gloves for any other purpose.
- Wash the clothes within 8 hours of use.
- Before washing, rinse the clothes by soaking in water in a tub, hang them on a line outside and hose them down with water, or run them through a prewash cycle with agitation in the washing machine.
- Be careful when disposing of the rinse water. If it contains herbicides, do not dispose of it on a garden or lawn area.
- Do not mix contaminated clothes with other clothes in the washing machine.
- Put only a few items of contaminated clothing in the machine at a time, grouping together those that are contaminated by the same chemical, and set the machine for a full load.
- Use hot water (140°F to 150°F) for washing.

- Do not use a suds-saver cycle on the washing machine.
- Use a heavy-duty liquid detergent to remove oily chemical residues from emulsifiable concentrate or use a powdered detergent with a phosphate base to remove chemical residues from wettable powders.
- Use a normal 12- to 14-minute wash cycle with two rinses.
- Do not use bleach if the clothing was contaminated with ammonia fertilizers. The ammonia and bleach can combine to form a deadly chlorine gas.
- Increase the amount of detergent by 1.25 times on clothes treated with soil- or water-repellent spray.
- Dry the clothes on a clothesline to avoid contaminating the clothes dryer.
- To remove pesticides from the machine after washing contaminated clothes, run the machine empty through a complete cycle using hot water and detergent.
- After cleaning, wipe the tub with isopropyl alcohol; this will help remove all traces of the chemical.

If any articles of clothing, including shoes, have become badly contaminated with toxic chemicals, they should be burned or buried. It is more difficult to remove oil-based chemicals from synthetic fibers than from clothing with natural fibers such as cotton. It may, therefore, be more likely to be necessary to destroy contaminated synthetic fiber clothing than clothing made from natural fibers. Care must be taken when burning or burying articles of clothing that are contaminated to avoid further contamination of the air or the worker.

Do not wash and reuse disposable chemical respirators. Washing will not remove chemical contamination from respirators. They are designed to be used once and then thrown away.

Closely woven fabrics are more likely than other types of fabric to become contaminated with chemicals. The close-knit fabric acts like a wick to absorb chemicals. It is very difficult to remove chemicals from this type of clothing. Leather gloves and boots also readily absorb chemicals and are almost impossible to decontaminate.

Protective clothing can be made from a wide variety of materials such as paper-like fiber, treated wool and cotton, leather, neoprene, rubber, plastics, and olefin. Each material is effective against a particular hazard; therefore, it is important to choose appropriate protective equipment based upon the work environment. Examples of protective clothing include lab coats, coveralls, jackets, aprons, and full body suits (Figure 3-6).

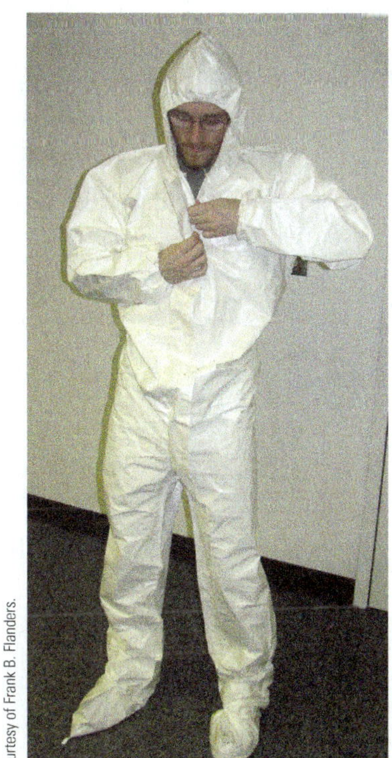

Figure 3-6 Personal protective equipment (PPE) includes contamination suits.

Storing Chemicals

Farm chemicals, such as pesticides, should be stored in a safe place where children or others cannot accidentally get into them. Do not use the wrong types of containers, such as soft drink bottles, for chemical storage and be sure the container is properly labeled. The label should contain a warning that the contents are dangerous. Many injuries and deaths related to improper chemical storage and labeling occur each year. For example, a child may not realize the contents of a container are dangerous chemicals and accidentally drink the contents.

Chemicals should be kept locked in a cabinet located in an area that does not freeze. Many chemicals used on the farm are damaged or destroyed by freezing. Keep only chemicals in the cabinet. The area around the cabinet should be kept free of other objects and should provide drainage in case of an accidental leak or spill from any of the containers. The area should drain into a safe collection area and not into regular sewer or drainage lines.

FIRST AID KITS

First aid kits containing the proper medical supplies should be kept in the home, in livestock buildings, on all major pieces of equipment, and in all vehicles. Several sizes of first aid kits are available commercially.

Many commercially available first aid kits may not contain all of the recommended items for a farm first aid kit. Therefore, combing a commercially available first aid kit with additional items may be helpful.

HEAT AND HUMIDITY FACTORS RELATING TO SAFETY

High temperature combined with high humidity can be a health hazard for farm workers. Some general guidelines on the relationship between temperature and humidity relating to safety when engaging in physical labor are shown in Figure 3-7. Danger from heat-related problems exists both with lower temperatures and higher humidity and higher temperatures with lower humidity. Under these conditions, workers may be subject to heat emergencies classified from least severe to most severe as heat cramps, heat exhaustion (sometimes called heat prostration), and heatstroke.

Heatstroke can result from prolonged exposure to high temperatures and occurs when the body cannot get rid of excess heat fast enough by sweating. Some people have a greater risk of heat-related problems. Young children and older people are at greater risk because their bodies do not cope with changes in body temperature as well. People who use alcohol, are taking some kinds of medications, have certain health conditions, or engage in heavy physical work outside during hot weather are also at higher risk from heat-related problems.

The symptoms of heat cramps include excessive sweating, fatigue, thirst, and muscle cramps. Symptoms of heat exhaustion include headache, dizziness, weakness, nausea and vomiting, and cool, moist skin. The symptoms of heatstroke include fever

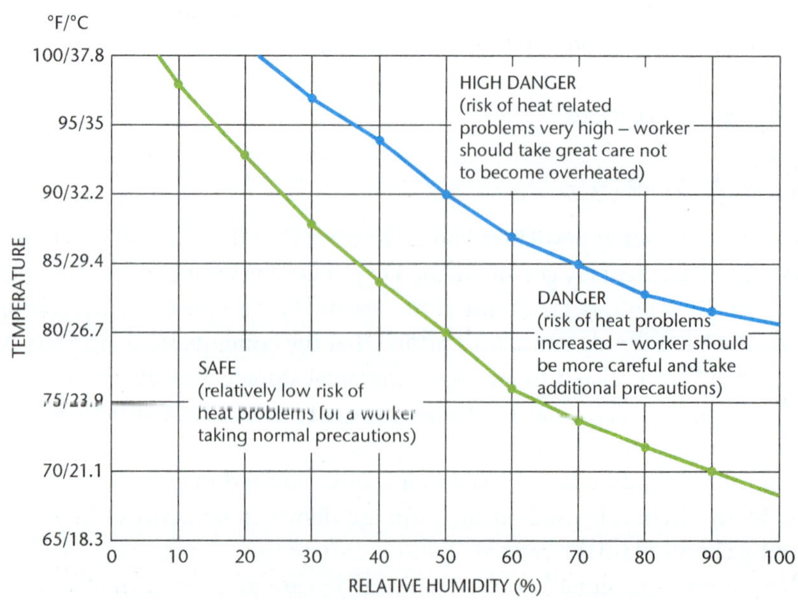

Figure 3-7 The relationship of temperature and humidity to safety when engaging in physical labor.

(above 104°F), confusion, irrational behaviors, flushed skin, rapid and shallow breathing, rapid and weak pulse, unconsciousness, and seizures. Heatstroke can result in serious complications and even death when untreated. If someone is demonstrating the symptoms of heatstroke, emergency medical help should be contacted immediately. To treat heat cramps or heat exhaustion, or to aid with heatstroke before medical care arrives, the following may help: moving to a shady or air-conditioned place, cooling off with damp cloths and a fan, taking a cool shower or bath, and rehydrating with water or sports drinks.

To help prevent heat-related problems, people—especially those who are at an increased risk for a heat related problem—can take some precautions. Select clothing that is lightweight, loose fitting, and light colored. A cotton or cotton blend fabric is preferred because it allows air to pass through. Drink plenty of fluids. If possible, avoid strenuous activity outside during the hottest parts of the day. Taking breaks from working in the heat or allowing your body time to adjust to hot weather can also help.

HAZARDS IN HANDLING LIVESTOCK

Whenever livestock are handled, there is the possibility of injury to the worker. Loading and unloading operations are particularly dangerous because the animals are excited and confused by what is happening. It is important to have solid facilities for handling livestock. Temporary or makeshift gates, pens, and chutes increase the chances of a worker being injured. Squeeze chutes and headgates should have solid latches to prevent accidental opening and possible injury to the worker (Figure 3-8).

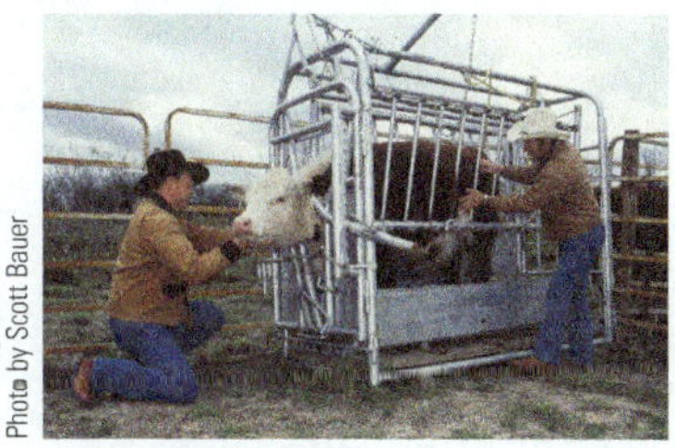

USDA Agricultural Research Service; Photo by Scott Bauer

Figure 3-8 The use of well-constructed headgates and squeeze chutes when treating livestock reduces danger to the workers.

Facilities should be designed so that the worker does not have to enter a small or enclosed area with animals. An escape gate (mangate) or other means of quickly getting out of pens should be provided. Catwalks should be a part of chutes and alleys so that the workers do not have to get into the area with the livestock. A guardrail should be placed on all catwalks that are more than 18 inches off the ground.

Floors must not be slippery or cluttered with things that might trip the worker. Sharp corners, pinch points, and protrusions should be eliminated from livestock handling facilities.

Lighting should be adequate so that workers can see what they are doing. The National Safety Council recommends at least 10 foot-candles of light in squeeze and loading chute areas. Lighting should be diffused and even. There should be no bright spots that might confuse cattle and cause them to balk.

Cattle will seldom attack a person. However, sudden noises may startle them. The cattle may then injure a person by crowding him or her against a hard surface. Never approach cattle from the side or rear. Approach the animal from the front while talking to it. This alerts the animal to the presence of the worker. Always wear boots or hard shoes when working around cattle. Tennis shoes should never be worn around livestock. Cattle kick forward and then to the rear. This is a hazard that must be watched for when working with an individual animal. Many of the safety practices outlined for horses also apply to cattle.

An understanding of why cattle behave in certain ways helps reduce stress for both the animal and the worker and increases safety. Reducing stress also increases cattle productivity.

The eyes of cattle are located on the sides of their head, which gives them a panoramic view of their surroundings. They have a wider range of peripheral vision than humans. Cattle have limited depth perception and see things in various shades of black and white. As a result, they are sensitive to movement in their field of vision and also react strongly to contrasting patterns of objects around them. The eyes of sheep, goats, horses, and chickens are also located on the sides of their heads. The visual reaction of these animals is very similar to that of cattle.

Curved chutes that have solid sides and are a uniform color make it easier and safer to move cattle. Anything that creates shadows and contrasting patterns in cattle-working facilities should be avoided because cattle will slow or stop when they see these things. Cattle sometimes refuse to move across a shadow or bright contrasting patterns on the floor in front of them. Articles of clothing that can move in the wind should not be left hanging on the sides of fences or chutes because cattle may balk when confronted with these sudden movements.

Cattle react negatively to sudden, loud noises, as well as to high-pitched sounds. When moving cattle, do not yell or make other sudden loud noises. This only tends to confuse the animals and will not get them to move in the desired direction. Cattle should be handled calmly and quietly at all times.

Cattle behavior is patterned as a result of previous experiences. When cattle are handled roughly in working pens, they remember the experience and it becomes very difficult to get them to re-enter the area for further treatment. Avoid the indiscriminate use of electric prods, yelling, punching, and arm waving to move cattle.

To increase safety and make it easier to work with cattle, get them used to being around people when they are on pasture. Don't try to push cattle too hard, and allow them to follow the natural herd leaders. When catching cattle in a headgate, make sure to catch them the first time. It is very difficult to catch them in the headgate after they have been missed on the first attempt.

Cattle hesitate to enter what appears to them to be a dead end. That is one of the main reasons for using curved chutes. Cattle will move more easily into headgates and chutes that are open ahead of them. Also, cattle hesitate to enter a darkened building when it is light outside. Lighting the interior of a truck, especially at night, makes it easier to get cattle to move into the truck.

After treating cattle in a headgate or working pen, allow them to move out of the area at their own speed. Yelling at them to get them to move will make it more difficult to get them to enter the area again.

People who work with cattle need to be aware of the "flight zone" concept (Figure 3-9). The flight zone is an imaginary circle around the animal or the herd. The radius of the circle is fairly small for animals that are used to being around people and much bigger for animals that have not been handled much in the past. The flight zone for cattle generally ranges from 5 to 20 feet. However, cattle that are used to being handled, such as dairy cows, may allow a worker to walk up and touch them without moving away. When the worker enters the animal's or herd's flight zone, the animal or the herd will move away. If the worker stays on the edge of the flight zone, the animals will move in a calm, steady manner.

When moving cattle, the worker should stay in a position where the animal can see the worker. If the worker is directly behind the animal, a sudden, unpredictable movement to one side or the other may result. When the worker is positioned at the edge of the flight zone and about 30 degrees to one side, the animal will move straight ahead; when the worker moves to a position at the edge of the flight zone about 45 degrees to one side of the animal, the animal will turn. Cattle tend to turn in a circle to keep the worker in view. When moving a herd of cattle, the worker should adjust his or her position at the edge of the herd flight zone relative to the lead animal in the herd.

Do not try to head off cattle when they turn and try to go back into an alleyway. A more effective way to handle this situation is to move out of the animal's flight

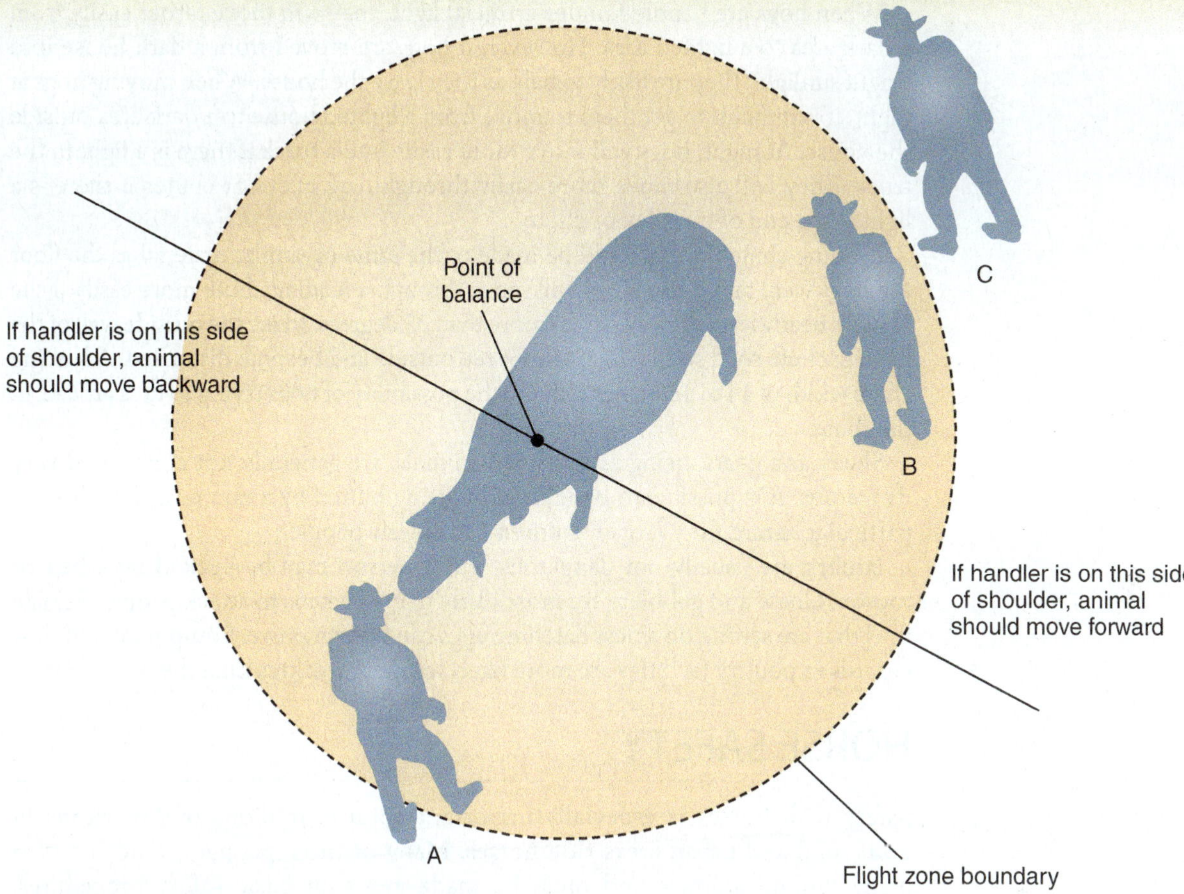

Figure 3-9 Flight zone and point of balance. If the handler moves to position A, in front of the point of balance, the animal will move backwards. If the handler is behind the point of balance position B, the animal will move forward. When the handler moves inside the flight zone, the animal will move. If the handler moves outside the flight zone, position C, the animal will stop or slow movement.

zone when it starts to turn back. This will generally allow the worker to reestablish control over the animal's movements. This is also a safer way to handle the animal because there is a danger of being knocked down by a charging animal when trying to head it off.

A sow may attack a person if she thinks her piglets are hurt or threatened. Never work with small pigs in the same pen as the sow. Hogs may bite or knock a worker down, so care must be taken when moving these animals. A hurdle or solid panel should be used when handling hogs. It is possible to move a hog backward by placing a basket over its head. It will try to back out and can be guided fairly easily to where it is wanted. Small children and visitors should be kept out of hog pens. Do not let children pet hogs through the fence.

An understanding of how hogs behave makes it easier and safer to handle them when they are being moved from one area to another. Unlike cattle, sheep, goats, horses, and chickens, the eyes of hogs are located further forward on their heads. This gives hogs better binocular vision and greater depth perception.

Hogs have a tendency to want to stay in or to return to an area with which they are familiar. When an attempt is made to separate a hog from the herd, that hog will try to return to the herd. However, hogs in a group will follow a leader, trying to maintain visual and physical contact with each other. This makes it easier to separate a group of hogs, rather than one hog, from the herd.

When hogs are handled under artificial light, they will move rather easily from a dark area to a lighted area. However, if hogs are moved from a dark house into bright sunlight, they are likely to balk as they leave the house. When moving hogs at night, it is difficult to get them to move from a lighted house to a dark area outside the house. At night, hogs will move more easily into a truck if there is a light in the truck. They will also move more easily through dark alleys or chutes if there is a light at the end of the alley or chute.

Loading chute floors should be made of the same or similar material as the floor the hogs were raised on. Hogs will move through a loading chute more easily if the chute is nearly level or slopes at no more than 25 degrees. Make the outside wall of the loading chute solid so the hogs cannot see outside and become distracted. A loading chute width of 14 to 16 inches reduces the possibility of hogs trying to turn around in the chute.

Sheep and goats, being rather small animals, are generally not considered very dangerous. It is possible to be injured by being butted by a ram or buck. This is a particular hazard for younger children and elderly people.

Poultry are usually not dangerous, but a person may be pecked by a hen or rooster. Geese and gobblers are more likely than chickens to attack people. Female fowl that are setting on a nest hatching eggs can be aggressive. Equipment and dust hazards in poultry facilities are more likely to injure workers than the poultry are.

HORSE SAFETY

Safety with horses is especially important because millions of Americans in both rural and urban areas ride horses. Many of these people are not used to being around animals and must be made aware of basic safety procedures. Serious injury can result from failure to follow safety rules. Horses are variable in temperament. Some are timid and will react violently when frightened. If safety rules are followed when riding or working with horses, these activities can be enjoyable experiences.

Horses have good hearing, but they do not see well directly in front or to the rear. Always speak to the horse as you approach. Approach the horse at a 45-degree angle from the shoulder, never directly from behind. Pet the horse by rubbing its neck or shoulder. Do not reach for the end of its nose. Stay out of kicking range when walking behind the horse. Do not step over the tie rope or walk under it.

Handling

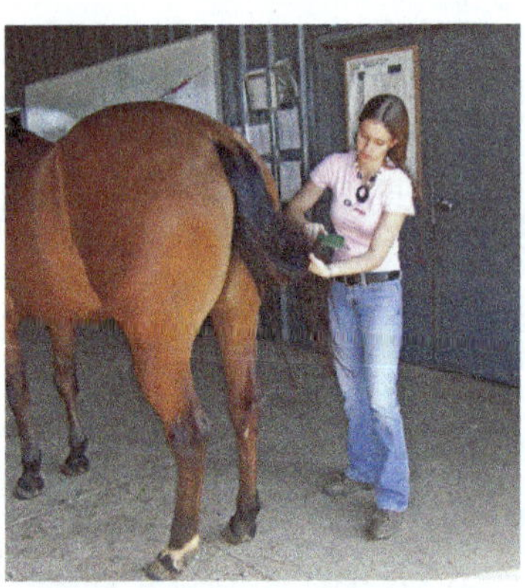

Courtesy of Frank B. Flanders.

Figure 3-10 Stand to one side and face the rear when working with the horse's tail.

The most important rule to remember is to stay out of kicking range of the horse whenever possible. Another safety measure is to stand close to the horse when working with it. Then, if it kicks, the full force of the kick will not be felt. Work as much as possible from a position near the shoulder of the horse. When working with the horse's tail, take a position near the point of the buttock. Stand to the side and face the rear (Figure 3-10). Do not stand directly behind the horse.

A nervous handler makes the horse nervous. A calm, confident manner that is kind but firm should be used. Let the horse know what is happening. Move slowly when doing things with the horse, such as lifting his feet. A person should learn the peculiarities of his or her horse and tell others who may be working or riding the animal about it.

Use simple methods of restraint. Tying or holding the head is safest when working with the horse. Horses should be tied with about 2 or 3 feet of rope. This should be at the height where the lead rope attaches to the halter. Do not leave a halter on a loose horse. The horse might catch the halter on a post or other object. Break-away halters provide a measure of safety when the halter must be left on a horse for a short period.

Never tease a horse. If it is necessary to punish the horse, do so at the moment of its disobedience. Never strike a horse around the head.

Protective footwear should always be worn around horses. A horse may step on a person's foot, or there may be nails around the barn that could cause injury. Boots or hard-toed shoes are better footwear than tennis shoes. Never go barefooted around horses or barns.

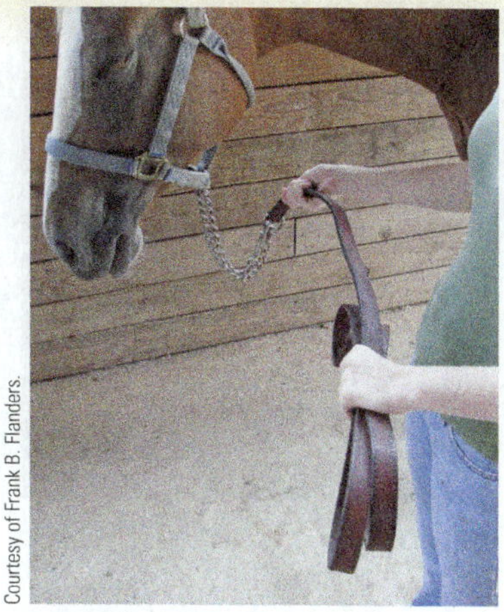

Courtesy of Frank B. Flanders.

Figure 3-11 The lead strap is folded accordion style when leading a horse.

Leading

When leading a horse, walk beside it rather than ahead or behind. Turn the horse to the right, walk around it and keep it on the inside. Horses are usually stronger than people, so it is unwise to try to out pull them. The horse will usually respond to a quick snap on the lead strap if properly halter-broken.

The lead strap, halter shank, or reins should not be wrapped around the hand, wrist, or body. Fold the lead strap accordion style in the left hand when leading the horse (Figure 3-11). The right hand should be extended slightly toward the horse. The horse's shoulder will make contact with the elbow first and move a person out of the way.

Be especially careful when leading a horse into a box stall or pasture. Always turn the horse to face the door or gate before releasing it. Otherwise, the horse may bolt forward when released.

Tying

Horses should be tied with about 2 feet of rope. The rope should be tied with a quick-release knot. Horses should not be staked out. Be sure the horse is tied far enough away from other horses so they cannot fight. Long lines and leads should be kept off the ground. This will keep the horse's or rider's feet from being tangled in the lines. The lead shank should be untied before taking the halter off the horse.

Bridling

Do not try to bridle a nervous horse in close quarters. When bridling, stand close to the left side of the horse, just behind the head. In this position, the horse's neck action will push you clear if it throws its head to avoid the bridle.

Saddling

When saddling a horse, a person should stand well back in the clear and reach forward. Do not let the cinch ring strike the off (right) knee of the horse. The front cinch of a Western double-rigged saddle should be fastened first. Fasten the rear cinch last. When unsaddling the horse, unfasten the rear cinch first, then the front cinch. Be sure the rear cinch is not so loose that the horse could get its foot caught in it. Make sure the strap connecting the front and rear cinch is secure.

The saddle should be swung into position easily. Do not drop it. Adjust the saddle carefully and check the cinch after walking the horse a few steps. The cinch must be checked again after riding a short distance.

Mounting

Do not mount the horse in a barn or near fences, trees, or other overhanging projections. Horses should be trained to stand perfectly still while the rider is mounting.

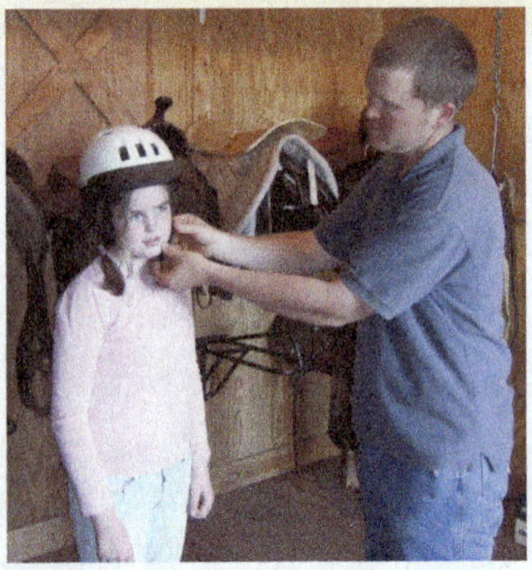

Figure 3-12 When riding a horse, it is advisable to wear a helmet to prevent head injury.

Riding

Horses may be frightened by unusual objects or noises. If the horse is frightened, steady it and give it time to calm down. After the horse is calm, ride or lead it past the obstacle. If the horse attempts to run, turn it in a circle and tighten the circle until it stops.

Walk horses up or down hills and go slowly on rough ground or in sand, mud, ice, or snow. Always allow the horse to pick its own way. Try to stay off paved roads. If it is necessary to cross or be on a paved road, slow to a walk or lead the horse.

Do not ride away from another rider who is mounting a horse. Leave a safe distance between riders and do not rush past other riders who are going slower. If it is necessary to pass others, approach slowly and pass on the left side.

Night riding is more dangerous than riding during the day. If the ride is on a road, follow the same rules as for pedestrians. Light-colored clothing should be worn. Riders should carry flashlights and reflectors. Allow the horse more freedom to pick its way. The horse's senses are much keener than the rider's. Riding a horse may be dangerous, so it is best to wear a helmet (Figure 3-12).

Equipment and Clothing

All equipment should be kept in the best possible condition. Replace any strap that is too worn to be safe. Make sure all tack fits the horse. Do not wear spurs on the ground because of the danger of tripping. Clothing should be neat and well fitted so that it will not snag on equipment. Boots and shoes should have heels so the foot will not slip through the stirrup. The horse's feet must be properly trimmed and shod.

Hauling Horses

It is safer for two people to load a horse on a trailer than for one person to attempt it alone. Stand to one side and never directly behind a horse when loading or unloading it from a trailer. The horse should be trained so that it can be sent into the trailer. Do not lead a horse into a trailer unless there is a front exit from the trailer. Be sure the ground around the trailer gives firm footing. Remove all equipment from the horse before loading. Use the halter for leading the horse. Always untie the horse before opening the trailer gate or door when unloading. Make sure the trailer is in good condition.

HAZARDS OF ANIMAL DISEASES

Diseases and parasites that may be transmitted between humans and animals are called **zoonoses**. Some of the common zoonoses are rabies, brucellosis, bovine tuberculosis, trichinosis, salmonella, leptospirosis, swine erysipelas, ringworm, tapeworm, and spotted fever. Some of these diseases and parasites are very dangerous to humans.

Cleanliness, vaccination, quarantine of sick animals, and avoiding exposure are some of the ways to prevent these diseases. Wear rubber gloves when treating sick animals. A doctor should be called if a person becomes sick after contact with animals.

Cases of human rabies in the United States are extremely rare, with only an average of about two or three human deaths per year reported. However, the disease is nearly always fatal once symptoms appear. Because of the seriousness of rabies if contracted, people who work around livestock need to take precautions against the possibility of becoming infected by the virus that causes rabies.

The virus is normally transmitted in the saliva of the infected animal. Domestic animals may contract rabies by being bitten by infected wild animals such as skunks,

Connection: Zoonotic Diseases

Sneezing and coughing are ways to spread colds, influenza, and other diseases, but did you know approximately 60 to 75 percent of all diseases can be passed from animals to humans or humans to animals? These are called zoonotic diseases (Figure 3-13).

Humans who come into contact with infected animals, their environments, and their feces become susceptible.

Rabies is a well-known zoonotic disease. It is caused by a virus. Rabies may be spread through bites or scratches.

Avian influenza, also known as bird flu; the H1N1 virus, also known as swine flu; and the West Nile virus are zoonotic diseases that have appeared in recent years. Avian influenza and H1N1 are strains of the flu virus that have evolved so they may be passed from human to animal or animal to human. West Nile virus is spread by infected mosquitoes and can cause disease in humans, horses, cats, bats, squirrels, rabbits, and other mammals.

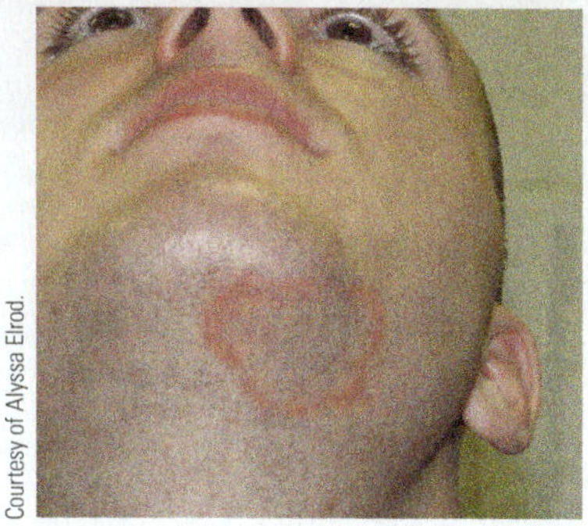

Figure 3-13 Ringworm is one type of zoonotic disease.

raccoons, or bats. Domestic dogs rarely become infected because of widespread immunization against rabies. Vaccinations must be kept up-to-date for farm pets such as dogs and cats to reduce the danger of rabies infection.

Workers who treat sick animals should wear rubber gloves to help prevent infection through open cuts on the hands. If the sick animal dies and there is suspicion that rabies might be involved, the animal must be sent to a testing laboratory to determine whether it had rabies. When handling the animal, keep your hands away from its mouth. Notify local health officials of a suspected case of rabies.

Be careful around any wild animal that appears to be acting strangely. Do not pick up or play with wild animals. If there is a suspicion that the animal has rabies, call your local animal control or police department. Local health officials should be notified immediately when a wild animal suspected of having rabies is killed.

Rabies in humans is preventable with medical care. If you are bitten by any animal, quickly wash the wound with soap and water and have a doctor check the wound. People who come into contact with animals with confirmed cases of rabies must be vaccinated immediately. People at high risk of exposure to rabies, such as animal handlers, may want to consider getting the vaccine as a preventive measure.

PERSONAL PROTECTIVE EQUIPMENT

Several kinds of personal protective equipment should be used around livestock. Bump caps or hard hats protect the head around livestock facilities. Respirators should be used around dusty or moldy hay. They should also be used in silos, manure storage areas, and for the use of some pest-control chemicals. Eyes should be protected by goggles from dust, chaff, and chemicals. Glasses should have impact lenses. Protective gloves should be used for certain jobs. Cotton or canvas gloves may be used as hand protection for light work. Leather gloves may be used for heavier work or when working with barbed wire fencing to increase gripping power and protect the hands when handling rough or abrasive materials. Rubber gloves should be worn around sick animals or when assisting a birth. Safety shoes should always be worn on the farm.

Workers in livestock confinement buildings suffer a high rate of respiratory and other problems. Typical problems include coughing, shortness of breath, scratchy throat, headaches, and watering eyes. Permanent lung damage may result from continued exposure to a contaminated atmosphere in confinement livestock buildings. Some deaths have resulted from toxic gases produced in liquid manure storage pits.

Livestock workers are often exposed to three types of atmosphere contamination. Dust and particulate matter come from feed, animal hair, and fecal matter. Pesticides used in treating livestock may cause health problems when inhaled by workers. Toxic and asphyxiating gases are produced in liquid manure storage pits. The four main kinds of dangerous gases produced are ammonia, hydrogen sulfide, methane, and carbon dioxide.

There are two kinds of respiratory protection equipment available for use by workers. Air-purifying respirators use filters to remove the contaminants from the air before it is inhaled. One type of purifying respirator removes particles from the air. The other type removes vapor and gas from the air. A combination filter is available for removing both kinds of contaminants from the air before it is inhaled.

Atmosphere-supplying respirators supply air from a source independent from the surrounding air. One type supplies air through a compressed air line, and the other type uses a compressed air cylinder.

Selection of respirator protection is based on the type and concentration of contaminant found in the air. Use only respirators that have been tested and meet the minimum standards of the National Institute for Occupational Safety and Health (NIOSH). Such respirators can be identified by a NIOSH number and label that describes the kind of hazard it protects against.

A coding system has been in use since July 1998 to identify three types of nonpowered particulate filter respirators. The "N" code means the respirator is not oil resistant, the "R" code means it is oil resistant, and the "P" code means it is oil proof. Efficiency levels of 99.7 percent, 99 percent, and 95 percent are available in all three classes of filters. For example, a filter coded R99 means that it is at least 99 percent efficient and is resistant to oil.

FACILITIES

There are many hazards in the facilities used for livestock. Slippery steps or floors may cause falls. Electric shock is a possible danger in damp areas. Electrical cords or appliances that may come into contact with water are hazards. Waterers that overflow or faucets that do not turn off completely may cause slippery conditions or electrical shock hazards. Strains can result from lifting heavy loads. Other areas that are hazardous include manure pits, lagoons, livestock confinement buildings, and grain storage areas.

Courtesy of Harveststore-Engineered Storage Products Co.

Figure 3-14 Silos pose a hazard for farm workers. Workers may fall from the silo or be harmed by the dangerous gases that build up inside. Always use caution when entering a silo.

Silo Hazards

As noted earlier, silos are dangerous for several reasons. People can fall from the silo or may be injured by equipment used to fill the silo (Figure 3-14). Nitrogen gases that form from ensiling green material can be deadly. Fermentation of the ensiled material produces mainly carbon dioxide and nitrogen dioxide; hydrogen sulfide and ammonia are produced in smaller quantities. Carbon dioxide is odorless and colorless. Nitrogen dioxide is yellow-brownish to reddish-brown in color and smells like bleach. It is heavier than air and can sink down the unloading chute into the feed room at the base of the silo. The most danger exists during the first 12 to 60 hours after filling the silo. Silo gas can be a danger in an unvented silo for several weeks after the silo is filled. Nitrogen dioxide at low concentration can result in eye irritation and coughing or lung injury. At high concentration it can lead to unconsciousness and even cause death.

Workers should not enter the silo for at least 3 weeks after it has been filled. Warning signs of the presence of deadly gas include yellow or brown stains on the sides of the silo chute or feed room wall and dead birds or small animals lying on the floor below the unloading chute. If it is necessary to enter the silo during this period, mechanically vent the silo by running the silo blower for at least 30 minutes before entering. Keep the blower running the entire time the worker is in the silo. Do not depend on respiratory equipment for protection unless it carries its own oxygen supply. Never enter the silo alone during the danger period. Always wear a body harness with a safety lifeline and have another person present to observe the entire time the worker is in the silo.

The silo room should be ventilated for 2 weeks after the silo is filled. Keep any doors between the silo room and the barn closed. Do not let children or visitors enter the silo room or the silo during the danger period. Get treatment immediately if exposed to silo gas.

Grain Handling and Storage Hazards

Unloading grain from a storage bin can pose several dangers for workers. Unloading equipment sometimes becomes plugged and a worker must enter the storage facility to correct the situation. The worker may become trapped in the flowing grain if proper precautions are not taken. It only takes 4 or 5 seconds for a worker to become submerged to the point of helplessness. Within 20 seconds a worker can become completely covered by the grain.

Another hazard is created when feed and grain bridge over in a storage facility and a hollow space forms near the bottom. If a person falls into this space, he or she may be buried under the feed or grain. This could result in suffocation.

Dust and molds in grain storage areas can cause workers to become sick. Toxic organic dust syndrome can cause symptoms much like those of the flu—coughing, fever, chills, headaches, muscle aches, shortness of breath, and fatigue. Symptoms can last from 1 to 7 days. Farmer's lung disease produces similar symptoms but is more serious because it may cause permanent damage. Breathing in large amounts of dust can cause toxic organic dust syndrome. An allergic reaction to mold spores in the air from moldy grain, hay, or straw causes farmer's lung disease. Symptoms of either illness will become apparent within 4 to 12 hours after exposure to dust and molds in the air. A person with symptoms of these illnesses should see a doctor for treatment. A blood test can determine which illness is present.

Safety precautions to be followed when unloading grain from a storage facility include:

- making sure the control circuit on automatic unloading equipment is locked in the off position before entering the storage bin.
- not entering the storage bin when the unloading equipment is operating.
- not entering a storage bin from the bottom when the material is caked to the sides of the bin or is bridged overhead.
- using a long pole made of nonconductive material to break up bridged or caked material. Work from the outside of the bin through a door or hatch in the roof. Poles made of conductive material may contact a power line, posing a danger of electrocution.
- always using safety harness or safety belts that are secured firmly to lines in such a manner that a person will be kept above the material in case he or she

Figure 3-15 Silo explosions.

falls. Never work alone. Always have at least one other person, who knows what to do in case of an accident, outside the bin and also equipped with a safety harness or belt. (The use of a safety harness or safety belt is mandated by OSHA in facilities subject to OSHA regulations.)

- being aware that an accumulation of dust in a grain bin can be highly explosive. Do not do anything that might cause an accidental spark that could result in an explosion (Figure 3-15).
- being sure the enclosed area is well ventilated before entering and keeping it ventilated while working in the area.
- using a respirator designed for toxic dust/mist that is approved by the National Institute for Occupational Safety and Health. It will have the words "dust/mist" stamped on the mask. Thin paper filters should not be used because they will not filter out the fine dust particles that can penetrate into the lung.

Livestock Confinement Building Hazards

Livestock confinement buildings can have toxic concentrations of gases such as ammonia, methane, hydrogen sulfide, and carbon monoxide. These gases, with the exception of ammonia, are heavier than air and sink to the bottom of manure pits. This forces oxygen out of the area and creates a dangerously lethal atmosphere. Hydrogen sulfide is poisonous to humans and smells like rotten eggs. It only takes one to three breaths of hydrogen sulfide to kill a person if the gas is present in a high concentration. At a lower concentration it can cause nausea, dizziness, sudden collapse, and severe breathing difficulty.

Workers in livestock confinement buildings must exercise caution because of the danger from toxic gases. Proper ventilation of these buildings is especially important. The greatest danger from lack of ventilation is during the winter when ventilation rates may be reduced below the minimum recommended level in order to conserve heat. The danger from toxic gases is higher when the manure in the storage pit is being agitated.

It is safest to remove animals and workers from the building when agitating the manure in the pit. If this is not possible, observe from a safe distance when starting agitation and be prepared to stop the pump if any sign of trouble becomes apparent. Workers should never enter a manure pit during or just after agitation because the dangerous gas levels are at their highest at that time.

A power failure that stops the ventilating fans also poses a hazard both to workers and to the livestock in the building. An alarm system should be installed to warn of power failures and auxiliary power should be available to keep the ventilating system working. Commercial gas monitors are available to measure the level of toxic gases in the building.

Workers should not enter manure pits unless it is absolutely necessary. Before entering the pit, test for oxygen with an oxygen meter and test for toxic gas levels, especially hydrogen sulfide, using an appropriate detector. Continue to monitor the oxygen and gas levels while the worker is in the pit. Make sure the pit is well ventilated, and use self-contained breathing equipment and a safety line attached to the person entering the pit. Always have another worker, also equipped with self-contained breathing equipment and a safety line, available outside the pit to rescue the person going into the pit if necessary. Never enter the manure pit without being properly equipped, even to rescue another worker; to do so may result in the death of the person attempting the rescue as well as the person already in the pit. Have a third person outside the confinement area to provide additional help if needed. This person should also have self-contained breathing equipment available for use. At least one of the workers present to provide assistance should be trained in cardiopulmonary resuscitation (CPR) and first aid.

The "Five Die on Farm from Deadly Gas in Manure Pit" account is based on a true story of a tragic farm accident involving methane gas buildup in a manure pit. Unfortunately, this type of accident is not uncommon.

FIVE DIE ON FARM FROM DEADLY GAS IN MANURE PIT

USA, 2008—Deadly methane gas emanating from a dairy farm's manure pit killed five people, including four members of one family, authorities said.

Emergency workers speculate that after the first victim was overcome by the gas, the others climbed into the pit in a frantic rescue attempt. "It was a domino effect with one person going in and becoming unconscious, and the others going in one at a time to try to save them," said local rescue personnel. As each person entered the pit he or she was immediately overcome by methane gas.

The victims were a dairyman, his wife, their two children, and a farmhand. Another farmhand said he tried to save some of the victims by hooking them from above and pulling them up. The second farmhand, visibly shaken, said, "I tried to hook them but I couldn't." He said he recognized the futility of going into the pit himself, as he would certainly have been the sixth victim without the proper breathing apparatus, including a separate oxygen supply.

The accident began when the dairyman tried to transfer manure from one pit to another. The pipe that was transferring the manure became clogged, so he climbed into the pit to unclog the pipe. It was probably something he had done many times. There was gas in the pit and he immediately succumbed.

Methane gas is an odorless and colorless by-product of liquefied manure. The pit was nearly enclosed and poorly vented.

Carbon monoxide gas may accumulate to dangerous levels in poorly ventilated confinement buildings. Malfunctioning space heaters and unvented radiant heaters in the building are often the source of this gas. Carbon monoxide can cause abortions in sows and humans. It may also cause mental retardation, an outcome that human fetuses are more susceptible to than are pig fetuses. Pregnant women should be careful if they are working in livestock confinement buildings. The use of personal protective equipment to filter air for breathing is recommended for all workers who spend 2 or more hours per day in these buildings. Minimum protection is provided by disposable paper mask filters; better protection is provided with quarter-face or half-face masks with a screw-on cartridge. An air stream helmet provides the greatest protection.

Even with proper ventilation, dust can be a major problem in livestock confinement buildings. Workers should use a high-quality dust filter that provides protection from 0.3-micron-diameter particles. Inexpensive filters generally do not provide this level of protection. The addition of soybean oil or tallow to feed has been shown to reduce the level of dust in confinement buildings. Another danger from dust is the possibility of an explosion and fire. Keeping the building and equipment clean can help reduce the danger of high concentrations of dust.

Provide tight covers for ground-level and below-ground manure storage pits. Keep lagoons and manure storage basins fenced off and posted with signs warning people to keep out.

Any open flames, including cigarettes, should be prohibited in enclosed manure storage areas. If repairs, such as welding, are necessary in the area, be sure it is well ventilated to reduce the risk of explosion and fire. Use only explosion-proof electric motors in livestock confinement buildings. Make sure all lights and electrical wiring are maintained to prevent accidental sparking. Methane gas, which is produced in manure pits, is highly explosive.

Locate first aid supplies and rescue equipment close by the manure storage area. Train all workers in proper first aid procedures for toxic gas problems. Emphasize to workers that they are not to enter manure pits without the proper equipment, even to attempt the rescue of another worker.

Keep the telephone number of the local fire department or rescue squad readily available. When phoning for help, be sure to let the fire department or rescue squad know the nature of the problem so they will bring the appropriate equipment. Emergency medical help may also be necessary.

FIRE SAFETY

Fire can be one of the most serious hazards on livestock farms. Fires are usually caused by carelessness. Most fires start from electrical equipment, heaters, or careless smoking. Other causes of fires are lightning, arson, and spontaneous combustion.

About 70 percent of all farm fires are caused by some problem in electrical wiring or electrical equipment. Livestock confinement buildings often have high levels of moisture and ammonia gas that can corrode electrical wiring and equipment. In high concentrations, dust can explode if an ignition source such as a spark from electrical wiring is present. Electrical control panels should not be located within the confinement area; put them in an adjacent office or outside the building. Sources of information about the electrical requirements for confinement buildings include the *Agricultural Wiring Handbook* and the

National Electrical Code, which provide ratings of materials suitable for use on the farm. Code requirements for home use are not adequate for livestock confinement buildings.

Many things around a farm burn easily. Most buildings are of wood construction, and hay, bedding, and feed are easily ignited. Once a fire starts in a livestock building, it burns and spreads very rapidly.

Safety practices to prevent fires include the following:

- protecting buildings from lightning
- storing fuels properly
- practicing good housekeeping by keeping all areas in and around buildings clean and free of debris
- not allowing people to smoke in and around buildings
- making sure all electrical wiring and equipment is in good condition and meets code requirements for use in livestock confinement buildings
- being careful when using heaters and brooders in livestock buildings
- avoiding conditions that could cause spontaneous combustion in stored feed
- when planning new buildings, spacing them at least 50 feet from other buildings to help slow the spread of fire (spacing of 75 feet will allow better access by fire trucks)
- using fire-resistant exterior materials (metal siding, asphalt shingles) on new construction
- locating fire extinguishers near doorways in all buildings
- maintaining fire extinguishers properly, with periodic inspection to make sure they are fully charged
- providing water outlets and hoses in and around the buildings
- having a pond located near the buildings can provide an emergency supply of water for fire fighting
- storing combustible chemicals in a safe place
- instructing all workers and family members in methods of fire prevention and what to do in case of fire

People should also know what to do in case a fire starts. When a fire is seen, the first thing to do is call the fire department. Fire extinguishers, water hoses, wet gunnysacks, and shovels and dirt should be used to fight the fire until the fire department arrives. If it is possible, animals should be removed from the buildings. They should be led to a place that is out of the way of the fire fighters.

Classification of Fires

Fires are classified as Class A, Class B, Class C, or Class D. Class A fires are those in which the burning material is wood, paper, textiles, grass, trash, and other similar materials. Water can be used to extinguish Class A fires.

Class B fires are those in which the burning material is grease, gasoline, oils, paints, kerosene, and solvents. Do not use water on a Class B fire because the water can spread the fire. Class B fires must be smothered to extinguish them. Blanketing agents such as carbon dioxide, water-based foam, or a wet blanket may be used to smother a small Class B fire.

Class C fires are those involving burning electrical equipment. Do not use water on a Class C fire unless all power to the area has been cut off. The fire fighter can suffer severe electrical shock if water is used on a Class C fire. High-pressure water fogs can be used on Class C fires.

Checklist of Farm Safety Practices

- ✓ Establish good sanitation, vaccination, and inoculation programs.
- ✓ Plan ahead when working with animals in an enclosed space to provide a way out; have at least two exits from the area.
- ✓ Use proper equipment when handling livestock; make sure all pens, gates, loading chutes, and fences are strong enough for the job and in good repair.
- ✓ Be sure livestock handling is done only by those with enough strength and experience for the job.
- ✓ Use caution when approaching animals to avoid startling them.
- ✓ Teach workers the correct safety measures for handling livestock.
- ✓ Know the animals.
- ✓ Be patient with animals.
- ✓ Do not work with animals when you are tired.
- ✓ Have enough help available to do the job.
- ✓ Be careful when leading animals and handle lead lines properly.
- ✓ Do not allow horseplay around animals.
- ✓ Keep children and visitors away from animals.
- ✓ Dehorn dangerous animals.
- ✓ Check equipment carefully when riding horses.
- ✓ Do not allow smoking in and around farm buildings and fuel storage and refueling areas; post no smoking signs in these areas.
- ✓ Have working, fully charged ABC-type fire extinguishers in barns and other major buildings.
- ✓ Remove all trash and junk in and around buildings to prevent fires and falls.
- ✓ Keep all buildings in good repair.
- ✓ Keep electrical wiring in good condition; check insulation, connections, outlets, and electrical equipment.
- ✓ Use adequate lighting in all buildings.
- ✓ Use proper ventilation in buildings and silos; make sure vents are clear and fans operate properly in all confinement buildings.
- ✓ Keep floors and ramps clean and free of broken concrete and slippery spots to ensure good footing.
- ✓ Keep a well-maintained first aid kit in all major buildings.
- ✓ Keep emergency telephone numbers posted.
- ✓ Have telephones or radios in vehicles and major buildings.
- ✓ Keep entrances to grain, feed, and silage storage areas closed and locked to keep children out.
- ✓ Post warning signs in grain and feed storage areas to warn of the hazard of becoming trapped in flowing grain or feed.
- ✓ Maintain silo and bin ladders in good condition.
- ✓ Shield auger inlets to prevent contact with the auger.
- ✓ Cover loading troughs on augers, elevators, and conveyors with grating.
- ✓ Use caution when moving augers and elevators; check for overhead power lines in the area.
- ✓ Check that the proper shields are in place on all feeding, grinding, and other equipment; DO NOT REMOVE SHIELDS.
- ✓ Use protective equipment such as bump caps, respirators, goggles, and gloves when needed.
- ✓ Store chemicals, fertilizers, medicines, and hardware away from animals and in a room or building that can be locked.
- ✓ Post warning signs at the entrance to areas where chemicals are stored that identify the hazards inside and provide information to fire fighters in case of fire.
- ✓ Mix all chemicals outside or in an open, well-ventilated area in the building.
- ✓ Have first aid equipment and plenty of water available in the area where chemicals are handled.
- ✓ Properly dispose of all chemical containers, following directions on the label.
- ✓ Store only chemicals in the chemical storage area.
- ✓ Carry an ABC-type fire extinguisher (minimum size 10 lb) in the combine.
- ✓ Carry an ABC-type fire extinguisher (minimum size 5 lb) in the tractor.
- ✓ Maintain ladders and steps on tractors, combines, and other equipment in good repair and free of mud and grease.
- ✓ Keep the operator's platform on combines free of mud, grease, and tools.
- ✓ Keep cab windows and mirrors clean for maximum visibility.
- ✓ Check mufflers and other parts of the exhaust system on tractors, combines, trucks, and other powered equipment to make sure there are no leaks of exhaust fumes.
- ✓ Maintain tires in good condition and properly inflated on all equipment.
- ✓ Make sure all fuel, oil, and hydraulic systems on all equipment is in proper condition.
- ✓ Use reflectors and Slow Moving Vehicle emblems on equipment; make sure they are clean and positioned where they can be easily seen.
- ✓ Make sure all tractors are equipped with rollover protection cabs or roll bars.
- ✓ Keep all farm ponds fenced to keep children out.
- ✓ Do not carry loaded guns in vehicles, tractors, combines, or other equipment.
- ✓ Keep guns unloaded and locked in a cabinet or gun rack in the home; keep ammunition stored in a location separate from the guns.
- ✓ Make sure all tractors are equipped with roll-over protection cabs or roll bars.
- ✓ Keep all farm ponds fenced to keep children out.
- ✓ Do not carry loaded guns in vehicles, tractors, combines, or other equipment.
- ✓ Keep guns unloaded and locked in a cabinet or gun rack in the home, keep ammunition stored in a location separate from the guns.

Class D fires are those involving combustible metals like sodium, potassium, titanium, and magnesium. These fires must be controlled by removing air with a blanket of nonreactive powder like sodium chloride or graphite. Water and carbon dioxide will not control Class D fires because these materials provide a source of oxygen for the burning metal.

Fire Extinguishers

The proper type of fire extinguisher must be used for the different classes of fires. Fire extinguishers are marked with a combination of letters and colors for the class of fire on which they can be used.

- Class A extinguishers are marked with an A in a green triangle.
- Class B extinguishers are marked with a B in a red square.
- Class C extinguishers are marked with a C in a blue circle.
- Class D extinguishers are marked with a D in a yellow five-pointed star.

Class A and Class B fire extinguishers use numbers with the letters. The number with the Class A extinguisher indicates its relative effectiveness for putting out a fire. For example, a 5 would indicate that the extinguisher is five times as effective as an extinguisher marked with a 1. The number on the Class B extinguisher gives an indication of the maximum square foot area of a liquid fire that can be extinguished by the fire extinguisher.

Class C and Class D extinguishers do not use numbers to rate their effectiveness. Select a Class C extinguisher on the basis of the type of construction surrounding the electrical equipment. The nameplate of the Class D extinguisher lists its effectiveness for various metals.

No one type of extinguisher is effective on all types of fires. Some are designated as multipurpose extinguishers and are effective for the classes of fires listed on their labels. Typical multipurpose extinguishers are for Class A, B, and C fires. Providing multipurpose fire extinguishers in buildings is recommended to help reduce confusion when a fire occurs.

BIOSECURITY AND AGROTERRORISM

A safe and healthy food supply has been taken for granted during most of the history of the United States. Terrorism and the threat of worldwide disease epidemics have changed the way we must think about our food supply.

The terrorists who struck the United States on September 11, 2001, changed forever the way Americans think about security—including the security of our food supply. Every aspect of the American economy and society must now be continuously evaluated for vulnerabilities to attack. Many have expressed surprise that our nation's food supply has not already been targeted.

In addition, livestock producers have been concerned in recent times about diseases in confinement livestock operations. These systems have large numbers of animals in close contact where diseases can spread quickly. Movement of people and animals around the world with modern transportation methods has increased the problem with the greater probability of the introduction of new diseases. Most confinement, high-capacity livestock enterprises severely limit visitors and take many precautions to prevent the introduction of diseases (Figure 3-16).

Many terms are used to describe the deliberate or incidental harm to the food production system and the precautions taken for prevention.

- **biosecurity**: protection from biological harm to living things from diseases, pests, and bioterrorism.

Courtesy of Frank B. Flanders.

Figure 3-16 Biosecurity at facilities such as this grandparent broiler stock farm is of vital importance. Entry is strictly controlled.

- **bioterrorism**: the deliberate use of biological or chemical weapons. In agriculture it is referred to as agroterrorism.
- **agroterrorism**: the deliberate use of biological or chemical weapons to bring harm to agricultural enterprises.
- **agrosecurity**: the use of all possible means and procedures to guard against deliberate or incidental harm to the food production system.

Ensuring security of the nation's food supply is a major responsibility and of major importance to every citizen, and especially to individual producers. One incidence of a major disease such as porcine reproductive and respiratory syndrome (PRRS) could be disastrous economically for a producer. Other diseases such as foot-and-mouth disease could spell disaster for an entire industry. Almost 6 million cattle had to be destroyed in the 2001 outbreak of foot-and-mouth disease in the United Kingdom. Avian influenza (bird flu) could negatively impact the poultry industry as well as result in the death of many people. humans. Because there are no vaccines for some diseases, biosecurity is the only line of defense (Figures 3-17 and 3-18).

USDA. Photo by Ken Hammond.

Figure 3-17 A member of the Beagle Brigade inspects the luggage of returning international travelers. Beagles are specially trained to sniff-out food products, soil, etc., that may harbor agriculture animal disease of major concern such as foot-and-mouth disease.

Damage to the agriculture industry could come from pests, diseases, and chemicals, among other threats. These agents of destruction could be spread naturally though animals, wind, accidentally through the movement of people, vehicles, and equipment, or deliberately as an act of terrorism. Anything a producer does to prevent such destruction of the food industry can be considered agrosecurity.

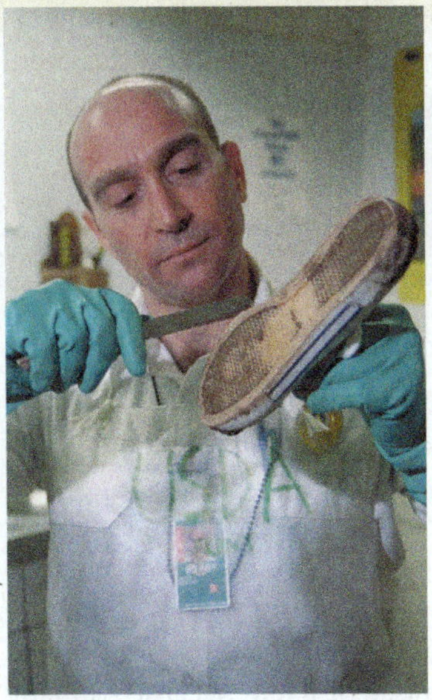

USDA. Photo by Ken Hammond.

Figure 3-18 A federal inspector cleans and disinfects the shoes of returning international travelers who had visited a farm in another country.

Courtesy of Frank B. Flanders.

Figure 3-19 At many sites workers are required to wash and disinfect their shoes before entering. Visitors may be required to shower-in and shower-out and may wear farm provided clothing.

Agroterrorism is the intentional use of any weapon such as chemicals, biological agents, and explosives against the nation's agricultural and food industries. Attacks of agroterrorism are intended to cause destruction of agricultural resources and serious economic harm.

While biosecurity measures are taken at both the national and state levels to keep diseases from entering the country or individual states, it is most important to start at individual herd levels (Figure 3-19). It is necessary that herd owners or management teams implement programs that prevent diseases from even entering their herds and spreading diseases to animals in the herds. To prevent diseases from entering the herd, livestock producers buy healthy animals, and, once at the farm, they keep new animals away from their other herds for 10 to 30 days. This period allows for the livestock producer or manager to detect any diseases that the animal may have and to follow up on vaccinations and immunizations. Here are some helpful tips that livestock producers may use to help prevent disease.

- Limit the number of people who visit the farm and have access to the herd or facility.
- Keep a distance between the herd or facility and other operations.
- Keep the farm as clean as possible. Proper disposal of manure is essential. For poultry and swine facilities, only enter with clean, disinfected attire, and keep all cages and pens clean.
- Do not bring diseases home. Disinfect all outside equipment brought to the facility before using it on the animals.
- Limit use of borrowed equipment. If equipment is borrowed, disinfectit before it comes to your property.
- Know the warning signs and symptoms of diseases.
- Report sick animals to a local veterinarian and/or a state veterinary department.

- Livestock keepers should reevaluate the security on their premises, keeping in mind the risk of intruders entering the farm at night. Animal houses, feed storage facilities, fertilizer storage areas, water tanks, and milk houses all should be tamper-free.
- Closed herds or flocks should be maintained, and all purchased animals should be kept away from other animals for a minimum of 10 days.

SUMMARY

Farming is a dangerous occupation. Many people are injured and some are killed each year when working with livestock. Children and older people are more likely to be injured by livestock.

Many environmental factors play a part in farm accidents. Facilities, especially silos and confinement livestock buildings, should be checked for safety. Protective equipment should be worn when necessary.

It is important that all workers be familiar with the correct handling procedures for various types of livestock. Horse safety is a unique problem because so many people who have little other experience with animals are around horses.

Many accidents can be avoided by preventing hazardous situations and knowing safety rules. Good housekeeping on the farm prevents many accidents.

Everything a producer does to reduce the risk of deliberate or incidental spread of disease or other destruction or disruption of the food production system is considered agrosecurity or biosecurity.

Student Learning Activities

1. Give an oral report on safety practices with livestock.
2. Prepare a bulletin board display of newspaper and magazine stories related to livestock safety.
3. Prepare posters about livestock safety.
4. Survey farms in the community using a livestock safety checklist and formulate recommendations for improvement.
5. Present livestock safety programs to local agricultural groups.
6. Prepare a livestock safety exhibit for display in the community.
7. Follow good safety practices when planning and conducting a livestock supervised experience program.

Discussion Questions

1. What is the death rate per 100,000 for workers in agriculture?
2. Why do people under 25 and those over 64 have more accidents on farms than people between 25 and 64 years of age?
3. List five environmental dangers to people working with livestock.
4. Describe the proper procedure for washing clothes that have been contaminated by farm chemicals.
5. Describe safety practices for the proper storage of farm chemicals.
6. List the recommended contents of a first aid kit for use on the farm.

7. What are the symptoms of heat exhaustion?
8. How should heat exhaustion be treated?
9. What are the symptoms of heat stroke?
10. How should heat stroke be treated?
11. What circumstances might cause some people to be at greater risk from heat-related problems than other people?
12. Why do cattle sometimes refuse to move across a shadow or bright contrasting pattern on the floor in front of them?
13. Based on an understanding of how cattle react to noise and sudden movement, list several good practices to follow when moving or working with them.
14. Describe the "flight zone" concept and how it may be used to control the movement of cattle.
15. Describe several practices that make it easier to move hogs from one place to another.
16. Describe how livestock facilities can be designed to prevent accidents.
17. What safety precautions should be followed when approaching a horse?
18. Describe three safety procedures a person should follow when handling a horse.
19. What is the safest way to hold the horse's lead strap?
20. List three safety rules to follow when mounting and riding a horse.
21. Why is it dangerous to enter a silo for a period of time after it has been filled?
22. Describe several safety practices that should be followed to protect workers from the danger of rabies.
23. List three types of atmosphere contamination that might affect people working with livestock.
24. List the four main kinds of dangerous gases found in livestock confinement buildings.
25. Describe the kinds of respiratory protection equipment available for use by workers on farms.
26. List three safety practices that should be followed to reduce the danger from toxic gases in livestock confinement buildings.
27. What safety equipment must a person use when entering the manure pit in a livestock confinement building?
28. List 10 safety practices to follow to prevent farm fires.
29. List the four types of fire extinguishers and tell what kinds of fires each may be used on.
30. Name the three most common causes of fires on farms.
31. What should be done if a fire starts on the farm?
32. Describe two ways in which the dangers of animal diseases to humans can be reduced.
33. Name three kinds of personal protective equipment a worker may need when working around livestock.

Review Questions

True/False

1. Agriculture is not considered to be among the more dangerous occupations in the United States.
2. The Environmental Protection Agency requires that Material Safety Data Sheets be available for all chemicals in the workplace.
3. After using chemicals, the best option is to only rinse the clothing worn in cold water.

Multiple Choice

4. Farm chemicals, such as pesticides, should be stored in places where:

   a. children cannot get to them c. there are no sewer or drainage lines
   b. they will not freeze d. all of the above

5. Which of the following is considered a heat emergency?

   a. heat exhaustion c. heatstroke
   b. heat prostration d. all of the above

6. When unloading grain from a storage bin, a worker can be in a dangerous situation within:

   a. 10 to15 seconds c. 60 to 90 seconds
   b. 4 to 5 seconds d. 20 to 30 seconds

Completion

7. If the chemical is moderately toxic, the label says ______ .
8. The ______ issues regulations designed to reduce health risks associated with chemicals.
9. ______ are diseases and parasites that can be passed between humans and animals.
10. ______ is the deliberate use of biological or chemical weapons to bring harm to agricultural enterprises.

Short Answer

11. List some precautions for the safe handling of cattle.
12. List examples of toxic gases found in livestock confinement buildings.
13. List and describe the types of fire extinguishers.
14. Discuss ways to avoid heatstroke in workers.

Chapter 4

Livestock and the Environment

Key Terms

point source
agronomic nitrogen rate
diversion
diversion terrace
drainage channel
debris basin
holding pond
disposal
masking agent
counteractants
deodorant
digestive deodorant
effective ambient temperature (EAT)
comfort zone
lower critical temperature (LCT)
upper critical temperature (UCT)
wind chill index
estray

Objectives

After studying this chapter, the student should be able to

- describe livestock production problems relating to the environment.
- describe methods of handling livestock wastes that reduce environmental pollution and are within the guidelines of current laws and regulations.
- describe the proper way to dispose of dead animals from livestock production operations.
- explain farmer liability under animal trespass laws.

Farmers have become more vulnerable to environmental lawsuits than at any time in the past. Changes in federal and state environmental laws make it easier to take action against farmers and ranchers who knowingly or accidentally damage the environment with chemicals or animal wastes. Lawsuits may be filed by property owners who have suffered damage or sometimes by individuals or groups that have not been directly damaged. Civil and criminal penalties, including heavy fines and imprisonment, may be applied. Those convicted may also be required to pay for cleanup costs, attorney fees, and costs of the prosecution. Farmers need to be aware of changes in environmental laws that affect their current operations or restrict their ability to develop new enterprises or expand existing ones. Insurance carried by farmers may not cover environmental liability. It is wise to develop an environmental compliance plan and keep good written records.

Connection: Right-to-Farm Legislation

The following paragraph shows the easy-to-understand wording used in right-to-farm legislation.

> It is the policy of this state and this community to conserve, protect and encourage the development and improvement of agricultural land for the production of food, and other products and also for its natural and ecological value. This notice is to inform prospective residents that the property they are about to acquire lies partially or wholly within an agricultural district and that farming activities occur within the district. Such farming activities may include, but not be limited to, activities that cause noise, dust and odors.

This notice shall be provided to prospective purchasers of property within an agricultural district or of property that has boundaries within 500 feet of a farm operation located in an agricultural district.

Another area of growing concern for farmers and ranchers is the Endangered Species Act. When a species is declared to be endangered and a particular area is considered to be a critical habitat for that species, then all human activity in that area must be stopped. This includes all farming or ranching activities.

In response to people moving out of cities into rural areas during the late 1970s and early 1980s, all 50 states passed "right-to-farm" laws. These laws were designed to protect farmers from nuisance lawsuits based on subjective perceptions of people who moved into the area from the city and were unfamiliar with the reality of the sounds and smells of livestock production on farms. The laws protected farmers against nuisance suits as long as the complaints were not based on violations of federal or state laws, negligence in operating the farm, water pollution, or excessive soil erosion.

Property rights of individuals are coming under increasing attack by environmental groups that are urging more government control over land and water use. Some farm groups are working to secure legislation that will help protect private property rights.

Some states are more closely regulating the disposal of medical waste such as needles, syringes, scalpel blades, and blood vials that are used in the treatment of livestock. Before these materials can be disposed of, they must be treated to eliminate the possibility that they might transmit infections. Special containers that are puncture and leak resistant are available from veterinarians, waste haulers, hospitals, and local health departments. The medical waste is placed in the special container and then shipped to an approved infectious medical waste facility. State environmental protection agencies may be contacted to determine the location of approved facilities.

Livestock producers must deal with animal wastes, odors, and dead animals in ways that do not harm the environment. They are also legally liable for any damage their livestock may do to other people or their property. Many of these problems require costly solutions. Society must decide whether the benefits are worth the investment.

ENVIRONMENTAL PROBLEMS WITH LIVESTOCK PRODUCTION

Changes in Livestock Production

The trend toward larger livestock operations has caused an increase in the concentration of animal wastes on individual farms. The use of large livestock confinement buildings presents special problems related to the disposal of livestock wastes in a manner that is not harmful to the environment or objectionable to others living in the area. Confining cattle in large feedlots results in greater problems in the disposal of livestock waste (Figure 4-1). Many operators of large feedlots do not have the land on which to spread the manure, which increases the potential for pollution problems.

USDA Agricultural Research Service. Photo by John Neinaber

Figure 4-1 Holding cattle in confined spaces creates problems with manure buildup.

The Changing Environment of Agriculture

Many people are moving into farm areas to get away from the problems of large cities. Large recreational developments also attract city dwellers to rural areas. Those who live in the cities often find farm odors offensive. They may not realize that odors and livestock wastes are a natural part of livestock production. Farmers must deal with this attitude while still maintaining production.

Handling farm waste poses different problems than handling waste from cities. The cost for handling these wastes falls on individuals rather than on a whole community. Cities solve their waste problems by building waste disposal plants that may cost several million dollars. However, the cost for each person living in the city may be only $100 to $200. A poultry farm of 200,000 hens, a beef feedlot of 1,200 head, or a 10,500-head hog operation may produce as much waste as a city of 20,000 people.

Although several farms in a community may have large feeding operations, they are usually too far apart to use a single disposal plant. Thus, each farm has to bear the cost of taking care of its own waste. Farmers must develop systems of waste control that are acceptable to others in the changing rural environment. These systems must be a part of their total management plan and must be affordable, as well as meet the expectations of nonfarming people who live nearby.

Federal and State Laws

The Federal Water Quality Act of 1965

The Federal Water Quality Act of 1965 requires states to have water quality standards. Public hearings must be held before the standards are set up. These standards apply to waters that move from one state to another. They also affect any portion of these waters that are within the borders of a state. The U.S. Environmental Protection Agency (EPA) must approve the standards that are set by the state.

Changes in federal law relating to water pollution control are contained in the Federal Water Pollution Control Act of 1972, also known as the Clean Water Act. This law lists national goals for water quality and establishes cooperation between the federal and state governments to reach these goals. The law prohibits the discharge of any pollutants from a **point source** into a river or stream without a permit. A large feedlot is an example of a point source of pollution. Thus, waste from a large feedlot cannot be allowed to run into a river or stream. Nonpoint sources of pollution were added to the law in the 1987 amendments to the Federal Water Pollution Control Act. A field that has manure spread on it is an example of a nonpoint source of pollution. The main goal of the 1987 amendments is the development and implementation of programs to control both nonpoint and point sources of pollution. The act gives the states the main responsibility for controlling water pollution.

The Refuse Act of 1899

The Refuse Act of 1899 gives the U.S. Army Corps of Engineers control over some animal waste pollution problems. The Army Corps of Engineers can approve or deny an application for permits to let waste run into navigable waters or their tributaries. A permit is required if a confinement feedlot feeds more than 1,000 animal units per year and if there is a direct discharge of waste to the waters. Runoff from natural causes is not considered to be a discharge of waste.

The Solid Waste Disposal Act of 1965

The Solid Waste Disposal Act of 1965 was amended in 1970. This act sets up federal guidelines for solid waste management, which includes the management of animal manure.

The Federal Clean Air Act

The Federal Clean Air Act establishes national air quality standards. The act makes the states primarily responsible for making sure that these standards are met, maintained, and enforced. State laws deal with how the standards are to be met. These standards deal with things like dust, grit, organic matter, and open burning as sources of air pollution.

State Laws

Most states have set up some type of environmental protection agency. Different states have different names for these agencies, but they all have the job of seeing that the federal and state laws that affect the environment are enforced. Livestock producers are affected by many of the regulations of these agencies.

Many states have laws that deal with nuisances. Nuisances may include odors, dust, chemicals, water pollution, and animal noises. These laws also often control the disposal of dead animals.

USDA Agricultural Research Service

Figure 4-2 Manure dropped or washed into streams causes environmental damage as well as health risks.

Water Pollutants

Water pollution is a major concern when determining the best way to dispose of animal wastes. The EPA monitors six water pollutants that can be measured directly. These are biochemical oxygen demand (BOD), fecal coliform, fecal streptococcus, suspended solids, phosphorus, and ammonia.

Oxygen is required to digest the organic matter in animal manure that is discharged into a river, stream, pond, or lake. Biochemical oxygen demand is a measure of the amount of oxygen-demanding organic matter in the water. Fish and other organisms living in the water need oxygen to live. If the BOD level is too high, there may not be enough oxygen left in the water to support life.

There are many types of bacteria found in the intestinal tracts of humans and animals. Two types of these bacteria are fecal coliform and fecal streptococcus. Water that is contaminated by human or animal waste contains measurable amounts of one or both of these bacteria (Figure 4-2). Contamination of water by coliform bacteria is undesirable because they can transmit disease to humans and animals.

Some materials like oil, grease, organic matter, and minerals may not dissolve in water. They may also float on the surface of a river, stream, pond, or lake. These materials are called suspended solids. They are undesirable because of the odors they may cause, and they also give a river, stream, pond, or lake an unsightly appearance.

Algae problems in water may be caused by contamination with phosphorus and ammonia contained in animal waste. These materials provide nutrients that cause the algae in the water to grow excessively.

HANDLING LIVESTOCK WASTES

Objectives of Waste Management

Animal manure must be handled so that odors, dust, flies, rodents, and other nuisances are controlled. Nitrate problems in water supplies caused by nitrogen in the manure must also be prevented. The system of waste handling must not allow the waste to be dumped into streams, rivers, lakes, or reservoirs. The waste must be in a form that is easily handled and disposed of without causing health and safety hazards to people or animals.

The main objective of manure handling is to prevent surface and groundwater pollution. Biological and chemical treatments of animal wastes are too expensive for farmers to use. Generally, the wastes must be held in some way until they can be properly disposed of on the land.

Amount of Waste Produced

The amount of raw manure that an animal produces depends on many factors, including the ration fed and the age of the animal. The data presented in Table 4-1 are based on American Society of Agricultural Engineers (ASAE) research, except where otherwise indicated in the table. Data in Table 4-1 that are not based on ASAE research are based on common assumptions and are not proven. The nutrient content data represent guidelines to the approximate fertilizer value of the manure and are not precise. Nutrient content of specific samples of manure may vary above or below the data given. Some factors that affect the nutrient content of manure include (1) length of time in storage, (2) methods of treatment, (3) amount and type of bedding used, and (4) amount of dilution by water entering the system.

About 2 billion tons of manure are produced each year on livestock farms in the United States. Some of this manure is deposited on pastures and rangeland by the animals. A much larger volume of the total is in feedlots, barnyards, and stockpiles. All of this manure has to be taken care of in some way. This is one of the major problems that the livestock producer must solve.

Selecting a System of Manure Handling

An animal feeding operation refers to facilities that house livestock for production purposes. The EPA specifies the conditions that must exist to define an animal feeding operation as one subject to regulation. Livestock must be housed at the location for at least 45 days out of each 12 months, and neither crops nor any other plants should be normally grown in the area. The 45-day figure is simply the total number of days, not necessarily continuous, during the 12 months. A livestock pasture that is sodded over is not considered to be an animal feeding operation under these regulations. If the farm has two or more kinds of livestock operations that share the same waste disposal system, it is defined as one animal feeding operation for the purposes of EPA regulation.

The choice of a system of manure handling depends primarily on the kind of animal that is being raised. There are many different kinds of facilities that can be used. The farmer must decide how to collect, handle, treat, and dispose of the manure. The ration fed to the animals influences the characteristics of the manure produced. The type of housing and management also affects the kind of system selected (Figure 4-3).

Photo by Bob Nichols, USDA Natural Resources Conservation Service

Figure 4-3 Animals in confinement operations create a manure disposal problem.

TABLE 4-1 Manure Production and Nutrient Content of Manure Produce by Various Species of Animals

A. 1 Type and Production Grouping	Weight	Total Solids	Moisture Content	N	P	K	Total Manure
	Lbs	Lbs/day	%	Lbs/day	Lbs/day	Lbs/day	Lbs/day
Beef-Cow	N/A	15	88	0.42	0.097	0.30	N/A
Beef-Growing Calf	N/A	6.0	88	0.29	0.055	0.19	50
Beef-Finishing Cattle	N/A	780	92	55	7.3	38	9,800
Dairy-Calf	330	3.2	83	0.14	N/A	N/A	19
Dairy-Heifer	970	8.2	83	0.26	0.044	N/A	48
Dairy-Veal	260	0.27	96	0.033	0.0099	0.044	7.8
Dairy-Lactating Cow	N/A	20	87	0.99	0.17	0.23	150
Dairy-Dry Cow	N/A	11	87	0.50	0.066	0.33	83
Swine-Nursery Pig	27.5	10	90	0.91	0.15	0.35	87
Swine-Grow/ Finish	154	120	90	10	1.7	4.4	1,200
Swine-Gestating Sow	440	1.1	90	0.071	0.020	0.048	11
Swine-Lactating Sow	423	2.5	90	0.19	0.055	0.12	25
Swine-Boar	440	0.84	90	0.061	0.021	0.039	8.4
Poultry-Laying Hen	N/A	0.049	75	0.0035	0.0011	0.0013	0.19
Poultry-Broiler	N/A	2.8	74	0.12	0.035	0.068	11
Poultry-Turkey (Males)	N/A	20	74	1.2	0.36	0.57	78
Poultry-Turkey (Females)	N/A	9.8	74	0.57	0.16	0.25	38
Poultry-Ducks	N/A	3.7	74	0.14	0.048	0.068	14
Horse	1100	8.6	85	0.34	0.073	0.21	57

Based on data from the American Society of Agricultural Engineers (ASAE) unless otherwise indicated.

Different sizes and types of feeding operations require different systems. A large operation needs more equipment and automation. A small operation requires less money and more labor. The amount of money the farmer can invest also affects the selection of a system, as does personal preference.

Climate is important when selecting a system. The amount of rainfall and when it occurs and the amount of evaporation are other factors. Temperature also makes a difference in the kind of system used. The usual direction of the local wind may also influence system selection.

Photo by Jeff Vanuga, USDA Natural Resources Conservation Service

Figure 4-4 Manure handling is not much of a problem when cattle are on pasture.

Each farm has characteristics that affect the system selected for that farm. How big the farm is, the soil type present, whether the land is sloping or level, and the kind of crops the farmer grows all affect the decision.

Regulations must be taken into account when planning waste handling systems. Federal, state, and local laws must be followed. Zoning may affect the kind of system permitted in an area. Nearby neighbors must also be taken into consideration.

Facilities for producing livestock are classified as either unconfined or confined. Unconfined facilities usually make use of pasture or range. In this case, most of the manure is left on the pasture or range by the animal (Figure 4-4). Only small amounts of manure will be deposited in barns and lots. This is a low-cost system from the manure-handling standpoint. Confined facilities may be open lot, lot and shelter, or a totally enclosed shelter. These types of facilities mean greater costs for handling the manure produced.

Animal manure may be collected and handled as a solid. Other systems collect and handle the manure as a liquid. If the manure is handled as a solid, then bedding may also be handled with the manure. Liquid systems generally cannot handle bedding; however, newspaper bedding usually breaks down enough to be usable with these systems. Flushing systems add to the amount of water in the raw manure.

Photo by Jeff Vanuga, USDA Natural Resources Conservation Service

Figure 4-5 Some swine facilities use slotted flooring for waste handling.

Several types of floors are used, depending on the way in which the manure is to be handled. Housing that uses a solid system may have concrete floors, dirt floors, slotted floors, or solid floors with gutters. Those buildings with liquid systems may use solid concrete floors that are flushed with water. Slotted floors are also used with liquid systems (Figure 4-5).

Liquid systems use pits, lagoons, or storage basins for storing and handling manure. Pits are pumped out and the manure is spread on the land in liquid form. Storage basins may be above or below ground. Above-ground systems are more expensive to build than underground basins. In both systems, a liquid manure pump is used to unload the basin so that the manure can be hauled to the field (Figure 4-6).

Manure pits may be recharged by draining the pit through pipes into a lagoon and then pumping fresh water from the lagoon into the pit. Benefits from recharging the manure pit about once a week include less buildup of dangerous gases and odors, improved feed efficiency, lower death rate of animals in the confinement building, and less medication needed for the animals. Recharging systems are used mainly in swine confinement buildings.

USDA. Photo by Tim McCabe

Figure 4-6 Liquid manure applied to fields serves as a fertilizer.

To determine the storage space needed for manure in a holding pit or tank, multiply the number of days in the holding period by the daily production of manure. This result is then multiplied by the number of animals that are producing the manure. Water must be added to the manure for proper storage and handling. The storage capacity requirement is increased by about 150 percent to allow for this added water.

The manure in lagoon systems (Figure 4-7) is not unloaded and hauled to the field. Instead, the waste material is broken down by bacteria. Lagoons may be either aerobic or anaerobic. Aerobic systems must have oxygen for the bacteria to work.

Photo by Jeff Vanuga, USDA Natural Resources Conservation Service

Figure 4-7 A state of the art lagoon waste management system. This facility is highly automated and is temperature controlled.

These systems are also shallower and produce less odor but require more area than anaerobic lagoons. They usually require mechanical aerators to control sludge and reduce odors.

Anaerobic systems make use of bacteria that work without oxygen. They can handle a larger volume of manure with less cost, labor, and maintenance than aerobic lagoons.

Lagoons must be designed to hold the total amount of manure produced by the livestock, plus any extra water that may be added. Extra water may result from rainfall, feedlot runoff, building wash water, and overflow from livestock waterers. The total design volume needed is about twice the volume needed for just the livestock waste.

Determining the Amount of Livestock Waste to Apply on the Land

The amount of available nitrogen per unit of yield necessary to produce a given crop is called the **agronomic nitrogen rate**. If the recommended agronomic nitrogen rate is followed when applying manure to the land, the crop will be provided with all the nitrogen it needs. If manure is applied at a higher rate than recommended, the excess nitrogen may pollute a water source. The recommended agronomic nitrogen rates for various crops are given in Table 4-2. The agronomic nitrogen rate is shown in pounds of nitrogen per bushel or weight of crop yield.

The nitrogen content of animal wastes varies with the species of animal and the waste storage method used on the farm. The approximate values for nitrogen content of livestock wastes from various sources utilizing different storage methods are given in Table 4-3. These values are only guidelines because there can be a wide variation in the nutrient content of animal wastes. The actual nutrient content of any given sample of livestock waste is affected by the kind of ration fed and the method of collecting and storing the waste. Actual nutrient content can be determined by laboratory analysis.

For a sample calculation, assume that a corn yield of 150 bushels per acre is expected. The livestock waste to be applied comes from a beef feeding operation using pit storage. Table 4-2 shows that 1.3 pounds of nitrogen per bushel of expected yield is required. Therefore, multiply 150 times 1.3 to arrive at 195 pounds of nitrogen required per acre ($150 \times 1.3 = 195$). Table 4-3 shows that waste from a beef feeding operation using pit storage contains from 25 to 50 pounds of nitrogen per 1,000 gallons of waste. For this example, assume the nitrogen content to be 30 pounds per acre divided by 30 pounds per 1,000 gallons of waste, which results in 6,500 gallons of livestock waste per acre needed $[(195 \div 30) \times 1{,}000 = 6{,}500]$.

Not all of the nitrogen is available to the plant in the year in which the waste is applied to the soil. It may take from 3 to 5 years for all the nitrogen from a given application to become available. More animal waste will need to be applied per year during this period if the amount needed for the expected crop yield is to come from this source. After a 3- to 5-year period, the amount available each year from the calculated application rate plus that available from applications in previous years will about equal the needs of the crop for the expected yield. An alternative to applying more animal waste during this period is to make up the difference with the application of commercial fertilizer.

Some farmers prefer to use the phosphorus requirements rather than the nitrogen requirements for determining the application rate for animal wastes. It takes less animal waste per acre to meet the phosphorus requirements. The additional nutrients needed are then supplied by adding commercial fertilizer.

TABLE 4-2 Agronomic Nitrogen Rates

Crop	Pounds of Available Nitrogen
Corn (grain)	1.3/bushel
Corn (silage)	7.5/ton
Barley (straw removed)	1.5/bushel
Grain sorghum (grain)	2.0/100 pounds
Grain sorghum (silage)	7.5/ton
Oats (straw removed)	1.1/bushel
Reed cana rygrass	55.0/ton
Rye (straw removed)	2.2/bushel
Sorghum-Sudan grass	40.0/ton
Tall fescue	30.0/ton
Wheat (straw removed)	2.3/bushel

TABLE 4-3 Nitrogen Content of Livestock Waste

Species	Nitrogen	
	lb/1000 gal	lb/ton[1]
Beef		
Pit storage	25–50	
Open lot—runoff	0.5–5.0	
Open lot—solids		10–12
Bedded confinement solids		10–15
Anaerobic lagoon	10–15	
Oxidation ditch	10–25	
Dairy		
Pit storage	20–40	
Open lot—runoff	0.5–5.0	
Open lot—solids		7–10
Bedded confinement solids		10–15
Anaerobic lagoon	10–15	
Poultry		25
Swine		
Pit storage	30–55	
Open lot—runoff	0.5–5.0	
Open lot—solids		10–12
Bedded confinement solids		10–15
Anaerobic lagoon	10–15	
Oxidation ditch	10–25	

[1]At approximately 50% moisture content.

Disposing of Manure

Most animal waste is eventually spread on the land. All solid handling systems work in this way. Liquid systems, except lagoons, also involve moving the manure to a field to dispose of it.

Manure is valuable as a fertilizer. When commercial fertilizers were inexpensive, the value of animal manure was low. With the increasing cost of chemical fertilizers, animal manure has become more valuable as a fertilizer. The amount of fertilizer value in manure is shown in Table 4-1.

Farmers who have fields to which manure can be hauled are less likely to use lagoons. However, large confinement feeding operations may not have fields available and often use lagoons to handle the manure.

When livestock wastes are applied to the land, care must be taken to avoid polluting the environment. The following points reflect EPA regulations and should be considered when applying animal waste to the land.

1. Animal waste must be incorporated or injected into the soil if the:
    a. percent of slope of the land on which it is being applied is greater than 5 percent or more than 5 tons of soil per acre is lost per year from erosion.
    b. field is in a 10-year flood plain (the field floods at least 1 year in 10).
2. Injection or incorporation of the livestock waste will reduce odors.

3. Generally, do not apply livestock waste on frozen or snow-covered ground. Such application may be done if the land slopes less than 5 percent or adequate soil erosion control is practiced.
4. Do not apply livestock waste immediately before or during a rainstorm or to soil saturated with water.
5. Do not apply livestock waste to grass waterways.
6. Do not apply livestock waste within 200 feet of surface waters or within 150 feet of a well.
7. Reduce the amount of livestock waste applied to the soil if there is a high water table present or the soil is highly permeable. This will reduce the chances of polluting water supplies.

Photo by Tim McCabe, USDA Natural Resources Conservation Service

Figure 4-8 This hog operation utilizes a filtering system that consists of a series of hillside terraces that form constructed wetlands that also use bacteria to purify the wastewater.

Feedlot Runoff Control

In an open feedlot, runoff is caused by rainfall or snowmelt. A great deal of manure is carried off the feedlot by the running water. If the runoff gets into a stream or river, it may cause fish kill. Many states have laws to prevent runoff from being channeled into a stream or river.

Five ways to prevent runoff are diversion, drainage, debris basins, detention ponds, and disposal. A feedlot operator must plan each of these control measures carefully.

Diversion is preventing surface water from outside the feedlot from getting onto the feedlot. Some feedlots are located at the top of a slope. This prevents much of the water from rain or snowmelt from getting onto the feedlot. If the feedlot is located at the bottom of a slope, a **diversion terrace** must be built (Figure 4-8). A diversion terrace forces the water to go around the feedlot.

The surface of the feedlot must be properly shaped to divert the runoff water to a **drainage channel** or pipe in the shortest distance possible. If the surface is graded correctly, water will not collect in ponds in the feedlot. Mounds are sometimes made in feedlots (Figure 4-9). They are constructed with clay soil and packed down. They should be 4 to 5 feet high and large enough at the top to let all the animals onto dry ground in wet weather. Feed bunks and waterers should be located so that the animals can get to them on dry ground. There should be no high piles of manure within the pen.

Photo courtesy of Iowa State University

Figure 4-9 Dirt mounds in feedlots allow cattle to stay on drier ground in wet weather.

Debris basins are used to catch runoff from the pens. The solids settle to the bottom and the liquids are drained into holding ponds. Debris basins keep about 50 to 85 percent of the solids from reaching the holding ponds. This helps to reduce odor from holding ponds. Solid manure should be removed regularly from the debris basin. The manure should never be more than 1 foot deep. A wide, flat channel works best for collecting the debris. It should be about 10 feet wide and not more than 3 feet deep.

A **holding pond** is a temporary storage area for runoff. It is not designed for waste treatment. It should be big enough to hold the runoff from the maximum 24-hour rainfall expected once in 10 years. In drier parts of the United States, the holding pond can be an evaporation area.

Disposal is the final step in controlling runoff from feedlots. The collected water can be used for irrigation of the land, or it may be allowed to evaporate. Holding ponds need to be pumped out fairly often. There must be enough space to hold the runoff from future rains or snowmelts. Local weather bureau records are used to determine how often the pond needs to be pumped.

Gases and Odors from Livestock Wastes

Gases and odors are given off by animal manure. This is caused by anaerobic bacteria breaking down the organic part of the manure. Anaerobic bacteria work when no oxygen is present. The gases produced can be dangerous to people and animals in a confinement facility. Odors may cause people who live close to the farm to take legal action against the farmer.

Gases and odors can be reduced by mixing air with the manure. In liquid manure systems, this is done by installing equipment to force air through the liquid. In solid manure systems, it is more difficult to prevent the gases and odors from forming. In a feedlot, the cattle keep the surface stirred, which allows air to mix with the manure.

It is hard to control odors when hauling manure onto the land. The best way to do this is to mix the manure in the soil as soon as possible after hauling. This can be done by plowing it under or disking it into the soil.

There are four general types of chemical and bacterial culture odor-control products that may be used to control odors from animal manure. These products include: (1) masking agents, (2) counteractants, (3) deodorants, and (4) digestive deodorants. **Masking agents** cover up the odor of wastes with the introduction of another odor. They are considered the most effective of the four types of odor control. However, the masking odor is not always pleasant, and a different odor problem may be created. **Counteractants** attempt to neutralize the odor so no odor remains. These substances are the second most effective type of control. **Deodorants** are chemicals that kill the bacteria that cause the odor. They are not as effective as the first two types. **Digestive deodorants** are bacteria that create a digestive process that eliminates the odor. They are the least effective of the four types. All of these control methods are expensive.

ENVIRONMENT AND NUTRITION

Environmental stress affects the nutrient requirements and intake of animals.

The environment of animals that are raised in confinement is usually carefully controlled to minimize stress caused by temperature extremes and humidity. Animals that are raised in less confined environments may be subject to more environmental stress; this needs to be considered when determining their nutritional needs.

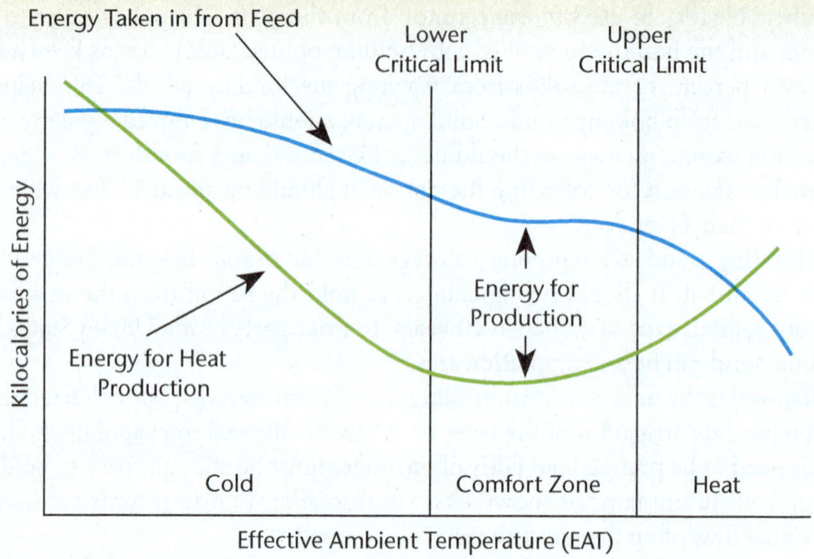

Figure 4-10 The general relationship of temperature to energy intake in the feed, energy used for maintenance, and energy available for production.

The temperature of the air is the primary factor that affects the efficiency of energy use by farm animals; secondary influences include humidity, precipitation, wind, and heat radiation. The combined effect of these factors is referred to as **effective ambient temperature (EAT)**. Within limits, animals attempt to compensate for changes in the EAT by altering feed intake, metabolism, and heat dissipation (Figure 4-10).

The **comfort zone**, or *thermoneutral zone*, is the range of effective ambient temperatures within which an animal does not have to increase normal metabolic heat production to offset heat loss to the environment. The thermoneutral zone varies with livestock species and may shift up or down as an animal becomes acclimatized to warmer or colder temperatures. For example, as cattle become accustomed to the winter season, their thermoneutral zone may shift downward as much as 27°F.

The **lower critical temperature (LCT)** is the temperature at which animals will show symptoms of cold stress; feed intake increases, as does metabolic heat production. The **upper critical temperature (UCT)** is the temperature at which animals will show symptoms of heat stress; feed intake is generally lower as animals attempt to reduce the rate of metabolic heat production when the upper critical temperature is reached.

Mature ruminants at high feeding and production levels produce more metabolic heat, have small surface areas relative to total body mass, and have a large amount of insulative tissue. For these reasons, they have significantly lower critical temperatures than do smaller animals such as swine, poultry, or young animals.

Because of their higher metabolic rate, mature ruminants have more difficulty adjusting to high temperatures than to lower temperatures; they become heat stressed more easily than they become cold stressed. The primary method of heat loss in hot environments is through evaporation from the surface of the skin or from the respiratory tract. Providing shade for animals during periods of high temperature will help reduce heat stress.

During periods of high humidity, it is more difficult for animals to lose heat by evaporation. Animals such as cattle (which depend more on sweating to lose excess heat) are more affected by high humidity than are those such as swine (which do not sweat but lose heat through the respiratory system).

TABLE 4-4 Wind Chill Index

NWS WindChill Chart

Temperature (ºF)

Wind (mph)	Calm 40	35	30	25	20	15	10	5	0	-5	-10	-15	-20	-25	-30	-35	40	45
5	36	31	25	19	13	7	1	-5	-11	-16	-22	-28	-34	-40	-46	-52	-57	-63
10	34	27	21	15	9	3	-4	-10	-16	-22	-28	-35	-41	-47	-53	-59	-66	-72
15	32	25	19	13	6	0	-7	-13	-19	-26	-32	-39	-45	-51	-58	-64	-71	-77
20	30	24	17	11	4	-2	-9	-15	-22	-29	-35	-42	-48	-55	-61	-68	-74	-81
25	29	23	16	9	3	-4	-11	-17	-24	-31	-37	-44	-51	-58	-64	-71	-78	-84
30	28	22	15	8	1	-5	-12	-19	-26	-33	-39	-46	-53	-60	-67	-73	-80	-87
35	28	21	14	7	0	-7	-14	-21	-27	-34	-41	-48	-55	-62	-69	-76	-82	-89
40	27	20	13	6	-1	-8	-15	-22	-29	-36	-43	-50	-57	-64	-71	-78	-84	-91
45	26	19	12	5	-2	-9	-16	-23	-30	-37	-44	-51	-58	-65	-72	-79	-86	-93
50	26	19	12	4	-3	-10	-17	-24	-31	-38	-45	-52	-60	-67	-74	-81	-88	-95
55	25	18	11	4	-3	-11	-18	-25	-32	-39	-46	-54	-61	-68	-75	-82	-89	-97
60	25	17	10	3	-4	-11	-19	-26	-33	-40	-48	-55	-62	-69	-76	-84	-91	-98

Frostbite Times: 30 minutes, 10 minutes, 5 minutes

$$\text{Wind Chill (ºF)} = 35.74 + 0.6215T - 35.75(V^{0.16}) + 0.4275T(V^{0.16})$$

Where, T= Air Temperature (ºF) V = Wind Speed (mph) *Effective 11/01/01*

Source: National Weather Service.

Heat loss from the body by convection and evaporation is affected by the movement of air surrounding the animal. The rate of change in heat transfer is greatest at lower air velocities. **Wind chill index** is a measure of the combined effect of air temperature and speed of air movement. Cold air in motion has a greater adverse impact on animals than cold air that is still. Providing a windbreak for animals during cold weather helps reduce cold stress. Table 4-4 shows the wind chill index for a variety of wind speeds and temperatures.

Precipitation (in the form of rain or wet snow) combined with low temperature and wind can cause animals to lose heat at a rapid rate. The insulation value of an animal's hair, wool, or fur coat is reduced when it is wet or becomes matted by rain or snow, and the animal loses heat more rapidly by conduction. When the hair, wool, or fur coat dries, the animal loses heat by evaporation.

Water intake generally increases as the temperature rises and decreases in colder weather. Cattle will drink more water during cold weather if it is heated; heating water for sheep during cold weather does not appear to increase water intake. Water intake of cattle and sheep tends to decrease as the relative humidity increases. Pregnant and lactating animals have a higher water requirement than nonpregnant and nonlactating animals. Animals fed feeds with a high dry-matter content tend to have a higher water requirement than those fed feeds with a higher moisture content.

Feed efficiency is reduced when the temperature is outside the animals' comfort zone. This results in an economic loss for the livestock producer. The benefit of controlling environmental conditions when raising livestock must be weighed against this economic loss when making management decisions concerning the expenditure of capital assets.

Some nutritional adjustments can be made during cold or hot weather. The energy requirements of the animal during cold weather are higher; however, the protein requirement is about the same as it is when the temperature is in the comfort zone. The percentage of protein in the ration can be reduced during cold weather because feed intake is higher to meet the energy requirements of the animal. Because feed intake is reduced during hot weather, increasing the amount of fat in the ration will help to maintain caloric intake. Fat has a lower heat increment than carbohydrate or protein. The percentage of protein in the ration may need to be increased in hot weather to meet the needs of the animal.

DISPOSAL OF DEAD ANIMALS

Most states have laws that require the disposal of dead animals within a given period of time after death—usually 24 to 48 hours. It is generally the responsibility of the owner of the animal to dispose of it. This must be done in such a way that no health hazard is created. Approved methods of disposal vary from state to state. Some of the methods often approved include a licensed disposal plant, burying, disposal pits, burning, and composting.

Diseases may spread from dead animals to people or to other animals. Always treat dead animals as though they were diseased. Trucks or other equipment used to haul dead animals should be disinfected after use. Sometimes there is doubt about the cause of death. If this is the case, a diagnostic laboratory should check the animal for possible diseases.

Dead animals should be hauled in a covered, metal, leak-proof vehicle. Skinning an animal makes it more difficult to haul the animal and also increases the chance of spreading diseases.

The best place to dispose of dead animals is a rendering or disposal plant. There are fewer of these plants today than in previous years. The operating costs have gone up, and the prices for products they make have gone down. They also have trouble with liability for air and water pollution.

Another method used to dispose of animal remains is to have them burned at a commercial incinerator facility. However, this can sometimes be difficult and costly and can produce unpleasant odors. In addition, it is only practical for smaller animals. Burning animal remains on old brush piles or other open burning methods is not recommended and is illegal in many places.

When a local disposal plant or sanitary landfill is not available, farmers may bury dead animals. A site should be selected that does not require moving the animal over a public road or someone else's property. The burial site should be far away from other buildings, public roads, and property boundaries and it should be high enough so that it is above the water table. The site should have good drainage away from water sources such as wells, springs, or streams.

Burying animals requires the use of excavation equipment to dig a large enough pit. Some farmers prepare an excavation in advance. This prevents delay in burying the dead animal.

The carcass should be placed at least 4 feet below the surface and then covered with 4 feet of earth. Covering must be done so that water does not pond in the excavation. The excavation is dug like a trench. The dead animal should be placed at the high end of the trench.

A disposal pit should be rectangular in shape and have vertical sidewalls. If the soil caves in easily, it is necessary to wall the pit with concrete blocks or pressure-treated wood. This increases the cost of the pit. The pit cover may be of concrete, pressure-treated wood, or steel. The cover must be:

- watertight.
- able to keep flies and predators out.
- able to keep most of the odor in.
- able to provide an easy way to drop the carcass in.
- constructed in such a way that it can be skidded from an old pit to a new one.

There are several chemicals that help speed the breakdown of the carcass. They also help disinfect and control the spread of diseases. It is not practical to use chemicals on large animals because of the high cost. Always follow the directions and precautions on the label of any chemicals used.

In some areas it is possible to dispose of animal carcasses in sanitary landfills. Check with the operator of the landfill ahead of time to see if it receives dead animals. If it does, the advance notice gives the landfill operator time to prepare for disposing of the animal. In this way, it can be disposed of as soon as it arrives at the landfill.

Some states permit the composting of dead animals, especially poultry and hogs. The poultry industry began the practice of composting dead carcasses in the 1980s. Composting is one of the most commonly used methods of dead carcass disposal throughout the poultry industry in the United States. Poultry carcasses will typically decompose in 4 to 6 weeks, depending on the size of the bird. Composting is beginning to gain acceptance as a method for disposal of dead animals in the swine industry. The facility needs to be larger when used for swine and the decomposition time is longer than for poultry.

Composting bins are usually constructed on a concrete floor and should have a roof to protect the material in the bin. Sufficient size should be allowed to permit the use of end loaders or a tractor with a loader to handle the compost material. The moisture content of the composting material should be in a range of 40 to 60 percent. A variety of materials may be used to cover the carcasses in the bin, including sawdust, wood chips, ground corn cobs, wood shavings, or poultry barn litter (Figure 4-11). The temperature in the compost pile should reach 120 to 150°F; this temperature level hastens the decomposition process and helps kill harmful microorganisms. A properly constructed compost pile does not give off odors or attract flies and rodents. The animal carcasses are placed on a layer of composting material, then more material is added on top. This layering process continues until the bin is full (typically about 5 feet). After an initial decomposition period, the pile is moved to a secondary bin where the composting process is completed. The finished product is safe to spread on fields as fertilizer.

Photo by Gary Farmer

Figure 4-11 Poultry litter stack houses are used to compost dead chickens and litter before they are applied to pastures and crops.

Regulations regarding the disposal of dead animals vary from state to state. Farmers should check state and local regulations to determine which methods are legal in their area. Local and state health departments, the Cooperative Extension Service, and reputable governmental and educational websites are other good sources of information regarding legal methods for dead animal disposal.

LIVESTOCK LAWS

Animal Trespass

In many states, the owner of an animal is liable for damage an animal does if it strays onto another person's property. Animals that are trespassing may usually be held by the person whose property is damaged until the owner makes good on the damage the animal has done. The person holding the animal must notify the owner. If the owner of the animal is not known, public notice must be posted that the animal is being held.

Estray means a domestic animal of unknown ownership that is running at large. Although laws vary from state to state, the following generally applies to an estray. A person may take possession of, and use, an estray found on his or her property. However, public notice of possession must be made. If the owner appears, the person holding the animal can collect the cost of keeping, feeding, and advertising the animal. If the estray dies or gets away, the person holding it is not held responsible.

The owner of an estray horse, mule, ass, or a head of cattle has 1 year in which to claim the animal, and the owner must pay any charges against the animal. The estray becomes the property of the person holding it if the owner does not claim it within 1 year. The owner of an estray hog, sheep, or goat must appear within 3 months to claim the animal. Estray laws generally do not apply to poultry.

Animals on Highways

Animals sometimes stray onto public roads. A driver running into such an animal may try to collect damages. Whether or not damages may be collected depends on each individual case. Usually, negligence by the owner of the animal must be proven before damages can be collected.

SUMMARY

Society's growing concern for the environment is affecting livestock production. Farmers must now plan for handling animal wastes within laws and regulations that protect the environment. Larger feeding operations, with more confinement feeding areas, have increased the problems of handling animal wastes. In addition, more people are moving into rural areas and may complain when livestock feeding operations cause odors.

Farmers must select systems of manure handling based on the kind of animal that is being raised. These systems should not pollute the environment. The major types of systems use either solid waste management or liquid waste management. Equipment costs can be low if more manual labor is used to handle the manure. On the other hand, high-cost systems reduce the amount of labor required.

Manure is valuable as fertilizer. Thus, the best disposal method is often just to apply the wastes to the land. However, some systems, such as lagoons, do not use land for disposal of animal manure.

Feedlots must be constructed carefully to prevent runoff of the waste into streams and rivers. Proper construction and management of the feedlot also helps to prevent odors from feeding operations.

Gases and odors from livestock wastes can be dangerous and unpleasant. Managing the facility so that the manure is aerated helps to minimize the amount of odor given off. Chemical treatments can be used, but these are expensive.

Dead animals must be disposed of properly. Most states have laws that control the method of disposal. Care must be taken to prevent the spread of diseases.

Farmers are liable for damage their animals do to other people's property. Many states have laws that cover taking possession of stray animals. If an animal strays onto a public road and is involved in an accident, the owner may have to pay for damages if negligence can be proved.

Student Learning Activities

1. Present oral reports on various phases of livestock production as they relate to the environment.
2. Survey the local community to determine what problems farmers have experienced relating to livestock and the environment, and how these problems were solved.
3. Interview farmers to learn their views about current laws and regulations relating to livestock and the environment.

4. Attend local meetings concerning laws and regulations as they affect livestock production.
5. Follow all applicable environmental laws when planning and conducting an animal supervised experience program.

Discussion Questions

1. What changes have occurred in livestock production in recent years that have increased environmental problems?
2. What changes have occurred in the environment of agriculture in recent years that have made the handling of livestock wastes more difficult?
3. List and briefly describe the federal and state laws that affect livestock production as it relates to the environment.
4. From an environmental standpoint, what is the main objective of manure handling?
5. Approximately how many tons of manure are produced each year by livestock on farms in the United States?
6. How does the size of the livestock enterprise affect the system of manure handling used?
7. Describe the operation of a liquid manure-handling system.
8. How valuable is livestock manure as a fertilizer for crop production?
9. List and describe five ways to control runoff from feedlots.
10. Describe how gases and odors from livestock wastes may be controlled.
11. How does the method of livestock raising influence the effect of the environment on nutrition?
12. Define *effective ambient temperature, comfort zone, lower critical temperature,* and *upper critical temperature,* and explain their effect on animal nutrition.
13. What is *wind chill index,* and how does it affect stress in livestock?
14. How does the temperature affect water intake of animals?
15. What nutritional adjustments may need to be made when the weather is very cold or very hot?
16. List five disposal methods for dead animals.
17. What precautions should be taken when disposing of dead animals?
18. What is the responsibility of the owner of an animal that strays onto another person's property?
19. What is an estray, and how should it be handled?
20. Name the six water pollutants that the EPA monitors and briefly discuss each.
21. Name four factors that affect the nutrient content of animal manure.
22. What conditions must exist for the EPA to define an animal feeding operation as subject to regulation?
23. Define *agronomic nitrogen rate.*
24. Why should the rate of application of livestock waste not exceed the agronomic nitrogen rate?
25. How much livestock waste from a swine-feeding operation that uses pit storage should be applied per acre for an expected 130-bushel corn crop?

26. Under what conditions should animal wastes be injected or incorporated into the soil?
27. List four points that should be considered when applying livestock wastes to the land.

Review Questions

True/False

1. In some states, individuals or groups who do not directly suffer damages may file lawsuits against farmers.
2. When an animal strays onto public roads, a driver does not have to prove negligence in order to collect damages.
3. The Endangered Species Act is not a concern among farmers and agricultural interests because it does not affect farming operations.
4. States do not regulate the disposal of dead animals; it is the responsibility of each individual farmer to dispose of dead livestock in a safe and proper manner.
5. The Environmental Protection Agency determines whether an animal feeding operation, based on the conditions surrounding it, is subject to regulations.

Multiple Choice

6. The Environmental Protection Agency monitors all of the following water pollutants relative to animal wastes except:
   a. fecal coliform   c. chlorine
   b. BOD   d. ammonia
7. Bacteria that work without oxygen are called:
   a. protozo   c. one cell
   b. anaerobic   d. aerobic

Completion

8. ______ with low temperature and wind can cause animals to lose heat at a rapid rate.
9. ______ is reduced when the temperature is outside the animal's comfort zone.

Short Answer

10. How does temperature affect feed efficiency?
11. What types of risks do dead animals present to humans?
12. What are the "right-to-farm" laws?

SECTION 2

Anatomy, Physiology, Feeding, and Nutrition

Chapter 5

Anatomy, Physiology, and Absorption of Nutrients

Key Terms

cell
organ
system
quadruped
ossein
osteocyte
endoskeleton
exoskeleton
cancellous bone material
compact bone material
fibrous joint
amphiarthroses joints
diarthroses
synovial membrane
striated voluntary muscle
smooth involuntary muscle
cardiac muscle
pharynx
larynx
trachea
bronchi
bronchioles
alveoli

Objectives

After studying this chapter, the student should be able to

- describe the functions of the 11 body systems.
- describe the functions of the parts of the digestive systems of ruminant and non-ruminant animals.
- classify farm animals as ruminant or non-ruminant.
- explain the relationship of types of digestive systems to the ability of ruminants and non-ruminants to digest and absorb different classes of feed.

A knowledge of animal anatomy and physiology is helpful in order to be successful in the feeding and care of domesticated animals. This text provides a brief overview of general anatomy and physiology common to most of the domesticated species of farm animals.

CELLS AND TISSUES

All higher organisms are built from **cells**, beginning as a single cell (the fertilized egg or ovum) and developing into multicellular organisms. As cells divide and grow they differentiate into tissues with a variety of functions. The tissues normally found in an animal's body are:

- *muscle:* contractile tissues that allow the animal to move.
- *connective:* tissues that hold other tissues together; for example, muscle tissues are covered by a sheath of connective tissue. Bones are also classified as connective tissue.
- *nerve:* bundles of tissues that transmit information throughout the body.
- *epithelial:* tissues that form the covering for most of the internal and external surfaces of the body and its organs; skin is an example.

TABLE 5-1 Body Systems

1. Skeletal
2. Muscular
3. Respiratory
4. Circulatory
5. Integumentary
6. Immune
7. Nervous
8. Endocrine
9. Excretory
10. Reproductive
11. Digestive

The tissues of an animal's body are grouped together to form the **organs** that perform complex functions in the body. Any given organ consists of two or more types of tissue. Examples of bodily organs include skin, heart, blood vessels, stomach, small intestine, large intestine, liver, bones, brain, kidney, and bladder. A group of organs that carries out a major function is called a **system**. There are 11 body systems (Table 5-1). Although any given system is similar in structure and function among all species, there are differences in the detail of bodily systems among species. The greatest difference in a given system is between mammals and poultry.

SKELETAL SYSTEM

Components of the Skeleton

The skeletal system of class Mammalia (mammals) is composed of bones, cartilage, teeth, and joints. Figure 5-1 shows the skeletal system of the cow, which is typical of **quadrupeds**, mammals that have four legs. The skeletal system of a chicken (class Aves) is shown in Figure 5-2. Although the skeletal systems of these two classes are similar, there are important differences between the two.

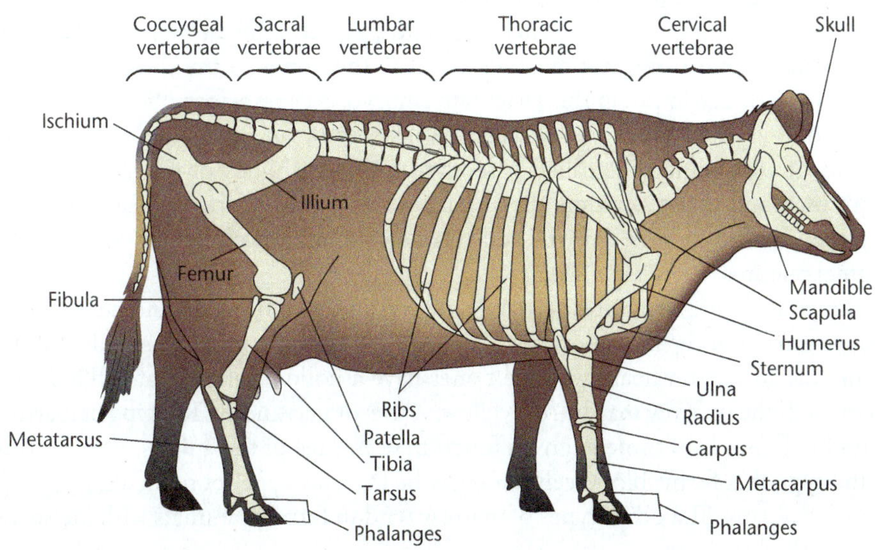

Figure 5-1 The skeletal system of a cow.

Key Terms *continued*

syrinx
arterioles
pulmonary circulation system
systemic circulation system
plasma
hemoglobin
digestion
absorption
concentrate
roughage
esophagus
cloaca
vent
cud
enzyme
rumination
bacteria
protozoa
amino acid
omasum
abomasum
gastric juice
chyme
bile
villi
cecum
crop
epithelium
gizzard
metabolism
anabolism
catabolism
oxidation

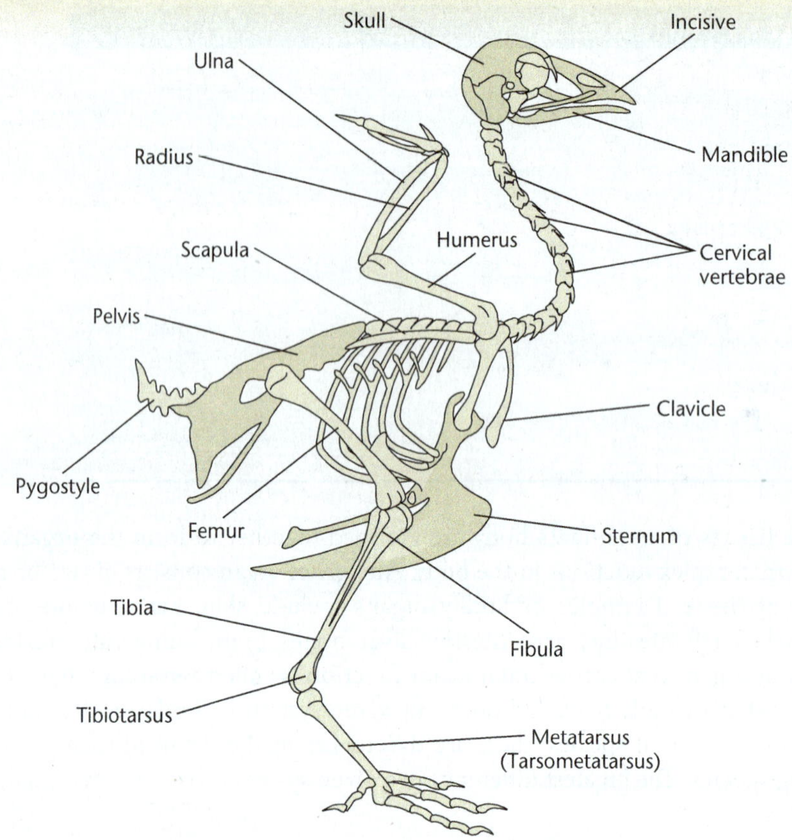

Figure 5-2 The skeletal system of a chicken.

Formation of Bone

Cartilage is a tough connective tissue that is flexible and elastic. It forms the early skeletal structure of the vertebrate embryo. As the bones develop, they follow the form of the cartilage skeletal structure and gradually replace much of the cartilage. The adult vertebrate skeleton retains some cartilage in the joints of bones and in other specialized structures, such as the trachea, nose, larynx, and external ear. Cartilage is also found between the bones of the vertebral column in the form of discs that separate and cushion the bones.

Bones are composed of calcium compounds, a gelatin-like protein called **ossein,** and small amounts of other minerals. Most of the bones of the skeleton develop from the cartilage found in the vertebrate embryo. This transformation of cartilage into bone is carried out by specialized cells that break down the cartilage and replace it with bone cells called **osteocytes**. The flat bones of the skull develop from membranes. By the time the animal is born, most of the cartilage has been replaced by bone. Bone growth and secondary bone development occur as the animal matures.

Bones are living structures containing blood and lymph vessels and nerve fibers. Bones grow and repair themselves if damaged. Nutrients must be supplied if the bones are to remain healthy. Some bones have a hollow center that is filled with marrow. Bone marrow may be red, yellow, white, or gelatinous. The type of marrow found in bones is to some degree a function of the age of the animal. Red marrow is the type that forms blood cells and may be the only type found in animals with short life spans. The other types of marrow tend to form in animals with life spans greater than 10 years.

Functions of the Skeletal System

The main functions of the skeletal system are to provide form, protection, support, and strength for the body. With the muscles attached, the bones of the skeleton act as levers permitting the animal to move. The bones also store minerals and the bone marrow produces blood cells.

Types of Skeletal Systems

Skeletal systems may be rigid or hydraulic. Animals with rigid skeletal systems are able to move because muscles are attached to the rigid skeleton and apply force that results in movement. There are two types of rigid systems: endoskeletal and exoskeletal. The **endoskeleton** is found on the inside of the body; the **exoskeleton** is typically a hard shell on the outside of the body. Vertebrates, such as farm animals, have endoskeletons. Arthropods, such as insects and crustaceans, generally have exoskeletons.

Some animals have a hydraulic skeletal system. They achieve movement by the application of force to a fluid confined in a small space and surrounded by muscle tissue. Examples of animals with hydraulic skeletal systems include worms, the octopus, and starfish.

Kinds of Bones

The two kinds of bone material, cancellous and compact, are based on structure. **Cancellous bone material** is spongy and generally found in the ends of long bones. **Compact bone material** is usually found on the outside of bones and surrounding the bone marrow. Figure 5-3 is a section of a long bone. This type of bone structure is not found in small mammals such as mice or birds.

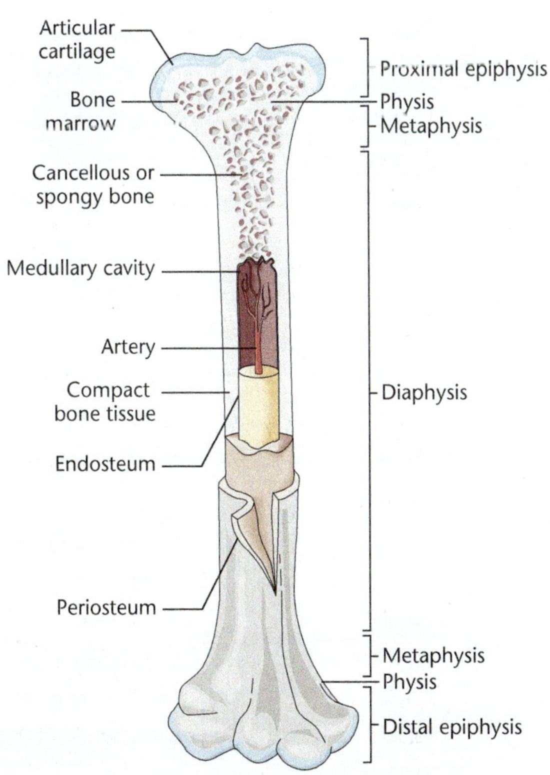

Figure 5-3 Section of a long bone.

Bones may be divided into four classifications based on shape: long, short, flat, and irregular. Classification of the major bones of an animal's body is shown in Table 5-2. Long bones have more length than width and are generally cylindrical in shape. Some long bones have modifications based on their function in the body. Short bones come in a variety of shapes. Their length and width are nearly the same compared to long bones. Flat bones are found mostly in the head, where they protect the brain. These bones are composed of a spongy layer between two layers of compact bone. Flat bones are usually thinner than other bones. Irregular bones have a variety of sizes and shapes.

Parts of the Skeletal System

The three components of the skeletal system are the axial skeleton, pectoral limb, and pelvic limb. The skull, vertebral column, and rib cage are the main parts of the axial skeleton. The pectoral limbs are the front legs of the animal and the pelvic limbs are the hind legs.

Joints of the Skeleton

Three types of joints are found in the skeleton of an animal: synarthroses or fibrous joints, amphiarthroses or cartilaginous joints, and diarthroses or synovial joints. These classifications are based on the structure of the joint and the movement it is capable of performing.

TABLE 5-2 Classification of Bones in Vertebrates

Skeletal Component	Name of Bone	Classification
Axial	Skull	Group of flat and irregular bones
bones	Vertebra	Irregular bones
	Ribs	Modified long bones
Pectoral limb	Scapula	Modified long bone
	Humerus	Long bone
	Radius	Long bone
	Ulna	Long bone
	Carpus	Group of short bones
	Metacarpals	Long bones
	First and second phalanx	Long bones
	Third phalanx	Modified short bone
	Digital sesamoid bones	Short bones
Pelvic limb	Illium	Modified long bone
	Ischium	Modified long bone
	Pubis	Modified long bone
	Femur	Long bone
	Patella	Short bone
	Tibia	Long bone
	Fibula	Modified long bone
	Tarsus	Short bone
	Metatarsus	Long bone

Synarthroses or **fibrous joints** are joined by fibrous tissue, or in some cases, cartilage tissue. These joints generally do not permit any type of movement. The joints in the bones of the skull are an example of this type of joint.

The vertebrae are joined by **amphiarthroses joints**. These joints consist of discs of a fibrous cartilage that separate and cushion the vertebrae, allowing very limited movement. If one of these joints becomes damaged, it can cause serious injury or paralysis of the animal. Limiting the movement of the vertebrae helps to protect the spinal cord from injury.

Joints that allow free movement and have a fluid-filled cavity are called **diarthroses** or *synovial joints*. A typical joint is surrounded by ligaments that help hold the joint together. A **synovial membrane** is usually found inside the ligaments. The joint is lubricated by the fluid produced by the synovial membrane. A cartilage disc may be found between the surfaces of the joint. These joints provide a variety of movements depending upon the functions required. Classifications include ball-and-socket, gliding, pivot, and hinge.

The joint between the scapula and the humerus is an example of a ball-and-socket joint. The remaining joints in the pectoral limb are hinge joints. In the pelvic limb, the femur connects to the pelvis with a ball-and-socket joint. The mandibles are connected to the skull by both hinge and gliding joints. The first vertebra attaches to the skull by means of a hinge joint and attaches to the second vertebra with a pivotal joint. This allows the animal to move its head from side to side as the neck is twisted on this pivotal joint.

Major Differences Between Mammalian and Avian Skeletal Systems

Avian bones tend to be thinner, harder, and more brittle than mammalian bones. Generally, more of the bones of the avian skeleton contain air spaces than do the bones of mammals. Avian bones develop almost exclusively from cartilage over a period of several months after hatching. The avian rib cage does not move as much as the rib cage in mammals and has little involvement in breathing. The avian skull does not contain any teeth.

MUSCLE SYSTEM

Types of Muscle

The three kinds of muscle found in the body are skeletal, smooth, and cardiac. These muscle types are differentiated based on their structure, location, and method of control. All three muscle types have the ability to contract; generally they do not stretch. Figure 5-4 shows these three muscle types.

Skeletal (Striated Voluntary)

Much of the flesh referred to as meat in farm animals is composed of skeletal muscle (also called **striated voluntary muscle**). It is called striated because it has dark bands that cross each muscle fiber (Figure 5-4). These bands can be seen when the muscle is examined under a microscope. The cell nucleus is located near the periphery of the cell. Skeletal muscle may be subdivided into two types based on color. Muscle that is involved with sustained work is generally red in color, while muscle that is activated on a more intermittent basis is white or pale in color.

shape and the muscle fibers appear to branch. Contraction of the heart muscle normally begins in the sinoatrial node that is located in the upper right atrium. Nerve stimulation is not required for this contraction to occur. An important feature of the contraction mechanism in cardiac muscle is that when contraction begins in the sinoatrial node it rapidly spreads to the entire muscle. Another feature is that a contraction is followed by a relaxation period during which it cannot be stimulated to contract again. These properties result in the rhythmic beating of the heart that is essential to the circulation of blood in the body.

Cardiac muscle contains two components not found in other types of muscle: intercalated discs and Purkinje fibers. The intercalated discs are dark thickenings that

Connection: Muscle Types and Goose Bumps

There are three types of muscles in the body—skeletal muscle, smooth muscle, and cardiac muscle.

Skeletal muscles are voluntary muscles used for walking, lifting, and other bodily functions. Smooth muscles are involuntary muscles found within the walls of organs and structures like the esophagus, stomach, intestines, bladder, blood vessels, and skin. Cardiac muscle is involuntary muscle found in the heart.

As seen in Figure 5-6, the arrector pili is a smooth muscle, found at the base of hair follicles, that causes piloerection (goose bumps). Goose bumps are produced by an involuntary muscle response to a stimulus (i.e. temperature, fear, strong emotions). The term "goose bumps" is believed to have originated from the bumps' resemblance to the skin of a goose whose feathers have been plucked.

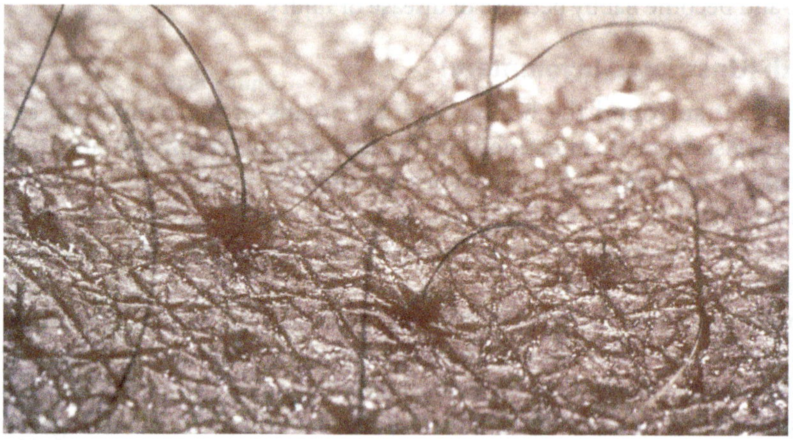

Figure 5-6 The arrector pill is a smooth muscle, found at the base of hair follicles, that causes goose bumps.

The contraction of the arrector pili muscle causes the hair or feathers to stand up. In response to cold, goose bumps are used to raise hair or feathers to trap an insulating layer of air to retain body heat. Although hair of modern humans is too thin to retain much body heat, birds and other mammals use this mechanism to maintain body heat. On cold days, many birds will fluff their feathers and mammals will fluff their hair to keep warmer. In addition goose bumps, which make hair and feathers stand on end, make an animal appear larger and more intimidating when it is threatened.

cross the muscle fibers and separate the cardiac cells. Purkinje fibers are specialized muscle fibers that are found in the lateral ventricles of the heart. They are a part of the contractile system, carrying the contraction impulses to the ventricle muscles. The autonomic nervous system can speed up or slow down the rate of heart muscle contraction but it does not start the contractions.

Muscle Functions

In conjunction with the skeletal system, the muscle system provides form, support, and movement for the body. It also generates body heat in the chemical processes that provide energy for the movement of the muscles. The skeletal muscle system is the primary one involved in movement. The smooth muscle system is involved in digestion and other activities of the internal organs of the animal. The cardiac muscle system maintains a rhythmic heartbeat that keeps the blood circulating throughout the body.

RESPIRATORY SYSTEM

Animal Respiration

Animals use oxygen to release energy in their cells by oxidation of molecules that contain carbon. A by-product of this oxidation is carbon dioxide. In its simplest form, respiration is the process by which oxygen is brought into the body and carbon dioxide is removed from the body. Water and other gases that the body does not need are also expelled during respiration. The method of exchanging gases in the respiratory system depends entirely on the diffusion of gases from an area of higher concentration to one of lower concentration through the cell wall membrane.

The oxygen is secured from the earth's atmosphere, which contains oxygen (20.95 percent), carbon dioxide (0.03 percent), nitrogen (78.09 percent), and some other gases (0.93 percent). Nitrogen and the other gases are inert from the standpoint of vertebrate animals. The amount of moisture in the atmosphere varies considerably and does not affect the composition just listed.

Mammalian Respiratory System Structures

The structures of the respiratory system in vertebrates are as follows:

1. *Nostrils.* Air is drawn into the system through the nostrils; air may also be drawn into the system through the mouth.
2. *Nasal cavity.* Here the air is warmed and moistened and dust particles are filtered out. Smelling also occurs here. The nasal cavity is separated from the mouth by the hard and soft palates.
3. Pharynx. This is where the passages from the nostrils and the mouth are joined. The air and food passages cross in the pharynx. The esophagus is the food passage. The epiglottis is a flap of tissue that closes when food is swallowed, thus preventing the food from entering the passage to the lungs. The epiglottis opens when a breath is drawn in, allowing the air to pass to the lungs.
4. Larynx. Air passes from the pharynx through the glottis to the larynx, an area composed of cartilage structures. The larynx contains vocal cords that vibrate when air passes across them, producing sound.
5. Trachea. The larynx opens into the trachea, a tube that leads to the bronchi. The wall of the trachea is lined with a series of C-shaped rings of cartilage. These help to maintain the shape of the passage during breathing.

6. **Bronchi**. At the lower end of the trachea, the tube divides into two branches called the bronchi (singular bronchus). These branches lead to the two lungs of the animal. The lungs of mammals are located in the thoracic cavity.
7. **Bronchioles**. Each bronchus keeps subdividing into smaller and smaller branches, called bronchioles.
8. **Alveoli**. The bronchioles terminate in the alveoli (singular alveolus). The walls of the alveoli are very thin, about one cell thick, and are covered by a film of fluid that acts as a surfactant. This fluid is a complex mix of protein, polysaccharides, and phospholipids. The alveoli are surrounded by tiny blood capillaries. This is where the actual exchange of gases occurs, which is the function of the respiratory system. Because the air contains a higher concentration of oxygen than the blood, oxygen passes through the cell wall membrane into the bloodstream. Conversely, there is a higher concentration of carbon dioxide in the blood than in the air. Therefore, the carbon dioxide passes from the bloodstream through the cell wall membrane into the air. Carbon dioxide is then flushed out of the system when the animal exhales.

Physiology of Respiration

An animal breathes by utilizing muscles to enlarge the chest cavity, thus forcing air in, and to reduce the chest cavity, thus forcing air out. Air enters the body through the nostrils or mouth, passes through the nasal passage to the pharynx, then goes through the trachea to the lungs. The lungs contain air sacs called alveoli, where oxygen from the air is exchanged for carbon dioxide from the body. When the air is expelled from the lungs, the carbon dioxide is removed from the body. The respiratory system also helps to control the temperature of the animal's body and to produce sounds by utilizing the larynx.

Avian Respiratory System

The avian respiratory system is different in some ways from the mammalian system. Birds do not have nostrils; they do have a nasal chamber in the upper mandible that opens into the mouth. The trachea leads down from the mouth to the lungs and divides into the bronchi. The **syrinx**, a structure that allows the bird to make sounds, is found at the lower end of the trachea. Birds have a system of air sacs that are extensions of the bronchi. These air sacs extend into the bones of the bird. Instead of alveoli, the avian respiratory system has small air capillaries in the lung tissue through which air circulates and gas exchange takes place. Birds do not have a diaphragm; breathing is accomplished by muscle action in the thoracic and abdominal regions and is additionally helped by the movement of the wings.

CIRCULATORY SYSTEM

The heart, arteries, capillaries, and veins make up the major parts of the circulatory system. The heart is a muscular organ that pumps blood throughout the body. Blood leaves the heart through the aorta, which branches into smaller arteries and eventually into the capillaries that reach tissues in all parts of the body. In the capillaries an exchange takes place: nutrients and oxygen go into the cells while carbon dioxide, water, and waste products enter into the blood. From the capillaries the blood enters small veins that converge into larger veins, eventually leading back to the heart. Figure 5-7 shows a diagram of the circulatory system.

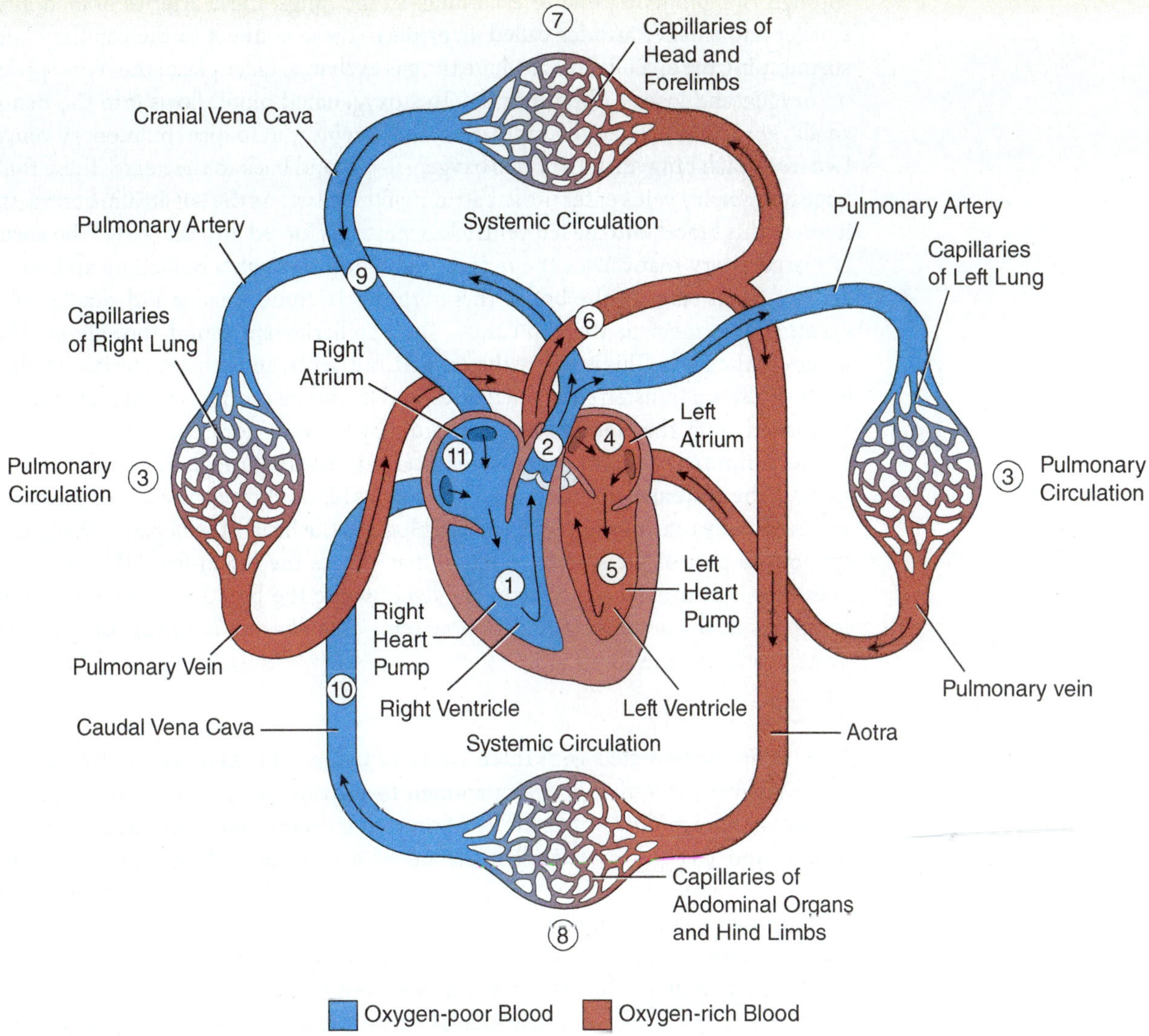

Figure 5-7 The circulatory system carries nutrients from food and oxygen that is dissolved in the blood to all of the cells in the body.

Heart

The heart is the organ that pumps the blood through the circulatory system. In mammals, it has four chambers. The left and right ventricles are found in the lower part of the heart, and the left and right atria (singular atrium) are found in the upper part. Contraction of the heart muscle begins with the ventricles and proceeds to the atria. When the ventricles are contracting, the atria are relaxing; as the atria contract, the ventricles relax. This rhythmic cycle of contraction and relaxation forces the blood to move through the circulatory system, which is a closed system in vertebrates.

Valves located between the atria and ventricles function in a manner that keeps the blood flow moving in one direction. Blood enters the right atrium of the heart through two large veins. One vein, the anterior vena cava, comes from the head and upper part of the body while the other, the posterior vena cava, comes from the lower part of the body. This blood, which is low in oxygen and high in carbon dioxide levels, is referred to as deoxygenated blood. The contraction of the right ventricle forces the blood into the pulmonary artery that carries it to the lungs through two branches, one to each lung. In the lungs, these arteries branch into smaller and smaller arteries called **arterioles**. These connect to the capillary bed surrounding the alveoli. This is where the gas exchange takes place; the blood picks up oxygen and loses carbon dioxide. This oxygenated blood flows into the many small veins connected to the capillaries. These veins join to form pulmonary veins, two from each lung, that carry this oxygen-rich blood back to the heart. These four large pulmonary veins enter the left atrium of the heart. As the left atrium contracts, it forces this blood into the left ventricle, where it is forced into the aorta. The aorta is a large artery that carries the oxygenated blood through a branching system of arteries to all parts of the body. These arteries become smaller and smaller and eventually connect to the capillaries. The capillaries are found throughout the tissues of the body. This is where the oxygen, nutrients, and other materials needed by the body are transferred to the tissues, and the waste products from metabolism are picked up by the blood to be transported by the veins for elimination.

The **pulmonary circulation system** carries blood through the lungs of the animal. The **systemic circulation system** carries blood through the rest of the body and eventually returns the deoxygenated blood to the heart. The hepatic circulation system is a part of the systemic system that carries the blood from the stomach, intestines, spleen, and pancreas to the liver, where the blood is filtered before it continues in the systemic circulation system. This filtering removes some of the waste products from the blood.

Blood Vessels

Arteries are constructed with three layers of tissue: (1) the outer wall made up of connective tissue that is strong enough to provide protection from cutting or tearing; (2) the middle wall made up of smooth muscle cells and elastic connective tissues; and (3) the inner wall made up of a layer of endothelium cells over connective tissue. Arteries are elastic enough to smooth out the pulsation of the blood caused by the pumping action of the heart, which helps to maintain a steady blood pressure. The inner wall generally does not allow blood or other material to pass through; that is, it forms an impervious membrane.

As arteries branch, they become smaller, eventually reaching a size less than 0.2 mm in diameter. These tiny arteries are called arterioles; their walls are constructed almost entirely of smooth muscle cells. The arterioles branch into capillaries that are about 8 microns in diameter. Blood flow and pressure is generally smooth and steady in the capillary bed. This is important to allow the proper exchange of nutrients and waste products to occur. Capillary walls are composed of cells that permit the passage of gases and some fluids to and from body tissues. It is in the capillaries that oxygen and nutrients exchanged between the blood and the body tissues and waste products are picked up from the body tissues. All body tissues except hair, cartilage, hooves, horns, nails, part of the teeth, and the cornea of the eye contain capillary beds.

The capillaries begin to come together again in the tissues, forming small veins called venules. The venules continue to recombine to form larger vessels called veins and eventually these return the blood to the heart. The walls of the veins are also composed of three layers containing elastic tissue and smooth muscle. The walls of veins are thinner than the walls of arteries; therefore, veins have a greater capacity than arteries.

The rate of blood flow in the blood vessels is inversely proportional to the diameter of the blood vessel. This means that blood flows faster in those parts of the system with the smallest diameter or cross section and slowest in those parts with the greatest diameter. It is the total cross section, however, of all the branches of the system in a particular area taken together that determines the diameter. As the arteries branch into smaller sections, the overall cross-section diameter in that region increases and the rate of blood flow is slower. As the capillaries combine to form venules, the total cross section of the vessels in the region decreases and the rate of blood flow increases.

The rate of blood flow in the arteries also varies directly with the rate and strength of the heart contractions. The pressure is highest (systolic pressure) when the heart contracts and is lowest (diastolic pressure) when the heart relaxes. This measurement is taken relatively close to the heart (in humans at the upper arm). The farther away from the heart the lower the systolic and diastolic pressures become, until in the capillary beds they are essentially nonexistent. Blood pressure is necessary to force the blood through the system against the force of gravity, through the smallest blood vessels, to overcome the resistance to blood flow caused by friction in the system, and to help blood flow back through the veins to the heart.

Blood pressure is reduced as it enters the capillary beds and is at its lowest in the veins. Blood does not move as rapidly or pulse in the veins as it does in the arteries. Veins are generally found closer to the surface of the skin than are arteries. Veins also have valves to prevent the backflow of blood in the system and keep it moving toward the heart. Most of the valves in the veins are found in the parts of the body farthest from the heart and where two veins join.

Blood

Blood is the material that circulates through the circulatory system. Blood serves a number of major functions in the body: (1) to transport nutrients from the digestive system to the body tissues, (2) to transport oxygen from the lungs to the tissues and carbon dioxide from the tissues to the lungs, (3) to transport waste products so they can be eliminated from the body, (4) to help regulate body temperature by transferring heat from the internal organs toward the surface of the body, (5) to transport hormones as needed in the body, (6) clotting to reduce blood loss when injuries occur, (7) to maintain the electrolyte balance and pH level in the body, and (8) to help protect the body against disease organisms.

The fluid portion of whole blood is called **plasma** and makes up about 50 to 60 percent of the total volume. The other 40 to 50 percent of the volume of the whole blood consists of three components: red blood cells (erythrocytes), white blood cells (leukocytes), and platelets (thrombocytes). Approximately 90 percent of the plasma consists of water. The remaining 10 percent, dissolved in the water, consists of proteins, organic nutrients, inorganic ions, hormones, gases, and waste products.

Red blood cells are formed in the red bone marrow of adult vertebrates. They are present in larger numbers than the white cells or the platelets. In mammals, the red blood cells do not have nuclei, although they originate from cells that do contain nuclei. As the cells mature, they lose their nuclei and acquire hemoglobin. **Hemoglobin** consists of iron and globulin (a protein); it helps in transporting the oxygen and carbon dioxide in the blood throughout the body and gives blood its characteristic red color. Red blood cells also help maintain a normal pH level in the body. In birds and many other vertebrates, the red blood cells retain their nuclei. On average, these cells live about 120 days in domestic animals; they are continually being replaced by new cells produced by the red bone marrow.

White blood cells are not as numerous in the body as the red blood cells; they are larger and have nuclei but do not contain hemoglobin. There are five types of leukocytes, all of which function as part of the body's immune system. The granulocytes are formed in the red bone marrow and the agranulocytes are formed in lymph nodes. White blood cells are not confined to the blood but may be found in other tissues of the body.

Platelets are small particles that occur in large numbers in the blood. They live about 10 days and are being continually replaced from the bone marrow. Platelets attach themselves to injuries in blood vessels and release a chemical necessary for blood clotting. They are sticky and form a plug at the site of an injury.

INTEGUMENTARY SYSTEM

The integumentary system is the organ system that protects the body from damage. This system includes the skin, hair, scales, feathers, and nails of animals. The integumentary system is the largest organ system of the body. Some of the most important functions of the integumentary system include the following: defending the body against infectious organisms, protecting the body from dehydration, excreting waste materials through perspiration, and producing vitamin D through ultraviolet exposure.

IMMUNE SYSTEM

The immune system is a group of organs and cells that defend the body against infection and disease. A major component of the immune system is the lymphatic system. The lymphatic system is auxiliary to the circulatory system, carrying lymph (a tissue fluid) into the capillaries of the circulatory system. A small amount of lymph is carried back to the blood through a system of fine capillaries of the lymphatic system that are located adjacent to the blood capillaries. The lymphatic capillaries merge into larger trunks that eventually feed into two large ducts that merge into the circulatory system. Both the lymphatic system and the circulatory system absorb nutrients from the digestive system (primarily the small intestine) for transport to the tissues of the body.

NERVOUS SYSTEM

The nervous system transmits information to and from the various parts of the body. The two major parts of the nervous system are the central nervous system and the peripheral nervous system. The central nervous system is located in the skull and vertebral column and is the master control system for the entire body; it consists of the brain and the spinal cord. The nerves that radiate from the central nervous system to all other parts of the body make up the peripheral nervous system. The sensory nerve fibers of the peripheral nervous system carry information to the central nervous system where it is analyzed and appropriate responses are transmitted to the body. The autonomic nervous system is a part of the peripheral nervous system that controls those activities of the body under automatic control.

ENDOCRINE SYSTEM

Hormones are chemical messengers that influence the growth and development of the body; they are secreted by endocrine glands. These glands do not have ducts to transport the hormones in the body. Blood passing through the gland absorbs the hormone and transports it to its target organ or tissue. Many of the endocrine glands and the hormones that they secrete are shown in Table 5-3. The pituitary gland secretes hormones that regulate hormone production in many other endocrine glands. Interactions between the endocrine glands, the level of hormone in the blood, and the actions of the target organs maintain the appropriate level of hormones in the body by utilizing a feedback mechanism.

TABLE 5-3 Major Hormones

Gland	Hormone	Action of Hormone
Adrenal (cortex)	Aldosterone	A steroid hormone that regulates the salt and water balance in the body.
	Glucocorticoids (group of steroid hormones)	Metabolism of carbohydrate, protein, and fat; also have anti-inflammatory properties. Examples: cortisone, cortisol, and corticosterone.
	Mineral corticoids (group of steroid hormones)	Regulate the balance of water and electrolytes in the body. Example: aldosterone.
Adrenal (medulla)	Epinephrine, also called Adrenaline	Released into the bloodstream in response to stress such as fear or injury; increases heart rate and blood pressure, muscle strength, and sugar metabolism increasing the concentration of glucose in the blood. Associated with the "fight-or-flight" response; prepares the body for strenuous activity.
	Norepinephrine, also called Noradrenaline	Similar in action to epinephrine.
Hypothalamus	Gonadotropin-Releasing Hormone (GNRH), also called Luteinizing Hormone-Releasing Hormone (LHRH)	Causes the anterior pituitary gland to begin secreting luteinizing hormone and follicle-stimulating hormone.
	Thyrotropin-Releasing Hormone	Stimulates release of thyrotropin.
Ovary (corpus luteum); placenta	Progesterone	Prepares the uterus for implantation of the fertilized ovum; helps to maintain pregnancy; development of the alveoli in the mammary glands.
	Relaxin	Softens the cervix and relaxes the pelvic ligaments during parturition.
Ovary (follicle); placenta	Estrogen	Induces estrus; development of female reproductive organs and secondary sexual characteristics.
Pancreas (in the islets of Langerhans)	Glucagon	Increases blood sugar level; opposes the action of insulin.
	Insulin	Helps to lower the glucose (sugar) level in the blood.
Parathyroid	Parathyroid (PTH), also called Parathormone	Regulates calcium metabolism in the body; reduces the concentration of phosphorus ions that normally combine with calcium to form a relatively insoluble salt
Pituitary (anterior lobe)	Adrenocorticotropic (ACTH)	Stimulates the adrenal cortex to secrete cortisone and other steroid hormones.
	Follicle-Stimulating Hormone (FSH)	Stimulates the growth of follicles in the ovary; stimulates estrogen production by the follicles; induces the formation of sperm in the testis.
	Gonadotropin (Gonadotrophin)	Any of several pituitary hormones, such as luteinizing hormone and follicle-stimulating hormone, that stimulate the function of the ovaries and testis.
	Luteinizing Hormone (LH)	Stimulates ovulation and the development of the corpus luteum and production of progesterone in the female; stimulates production of testosterone by the interstitial cells of the testis in the male.
	Prolactin	Stimulates and maintains the secretion of milk.
	Somatotropin (Somatotrophin), also called Somatotropic Hormone (commonly called Growth Hormone)	Stimulates body growth; affects metabolism of proteins, carbohydrates, and lipids.
	Thyrotropin (Thyrotrophin), also called Thyroid-Stimulating Hormone (TSH) or Thyrotropic Hormone	Stimulates production of thyroid hormones; regulates the activity of the thyroid gland.
Pituitary (posterior lobe)	Antidiuretic (ADH), also called Vasopressin	Controls water retention in the kidneys, reduces excretion of urine, and raises blood pressure.
	Oxytocin	Stimulates the contraction of the smooth muscle of the uterus during parturition and facilitates milk letdown from the mammary gland.

(continues)

TABLE 5-3 Major Hormones (continued)

Gland	Hormone	Action of Hormone
Testes	Testosterone	Development of the external genitals in the male fetus; development and maintenance of male secondary sex characteristics; sperm production; stimulates male sex drive (libido). Also produced in small amounts in the ovaries and placenta of the female.
Thyroid	Thyroxine (Thyroxin)	Increases the rate of cell metabolism; regulates growth. Mineral iodine is required for this hormone to function properly.

Sources: D. Acker and M. Cunningham, *Animal Science and Industry*, 4th ed. (Englewood Cliffs, NJ: Prentice-Hall, 1990); C. E. Stufflebeam, *Principles of Animal Agriculture* (Englewood Cliffs, NJ: Prentice-Hall, 1983); M. J. Swenson, *Dukes' Physiology of Domestic Animals*, 9th ed. (Ithaca, NY: Comstock Publishing Associates, 1979); R. E. Taylor, *Scientific Farm Animal Production*, 5th ed. (Englewood Cliffs, NJ: Prentice-Hall, 1995).

EXCRETORY SYSTEM

The major components of the excretory system are the kidneys, ureters, bladder, and urethra. Blood passes through the kidneys, where waste products and some water are filtered out. This liquid, urine, is then passed through the ureters to the bladder, where it is stored until the bladder is full before being voided from the body through the urethra. Poultry do not have a bladder or urethra; the ureters lead to the cloaca, where the urine is voided along with the feces.

REPRODUCTIVE SYSTEM

The reproductive system is important to the survival of a species, and is the basis of economic animal production systems. Mammals have a different reproductive system than birds. The reproductive system is discussed in detail in the chapter on animal reproduction (Chapter 10).

DIGESTIVE SYSTEMS

Knowledge of the different kinds of digestive systems helps in selecting the proper livestock feeds. Understanding the chemical and physical changes that take place after the feed is eaten leads to more efficient livestock feeding.

Digestion is the process of breaking feed down into simple substances that can be absorbed by the body. **Absorption** refers to taking the digested parts of the feed into the bloodstream. Ruminants are cud-chewing animals that have a stomach that is divided into several parts (Figure 5-8). Cattle, sheep, and goats are ruminants. Non-ruminants are animals that have simple, one-compartment stomachs (Figure 5-9). Pigs, horses, and poultry are non-ruminants.

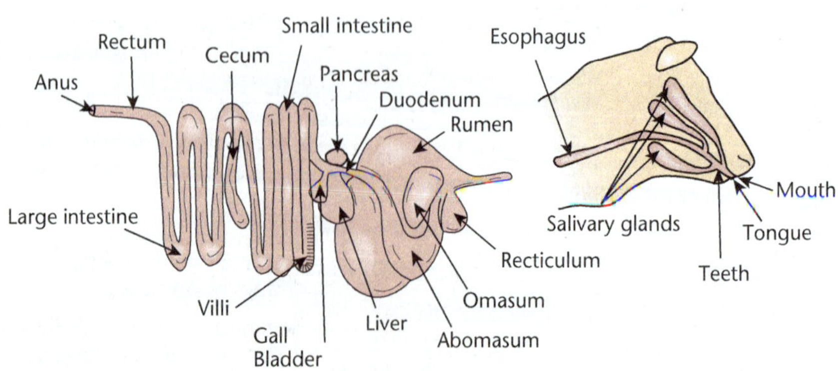

Figure 5-8 The digestive tract of the cow, a ruminant animal.

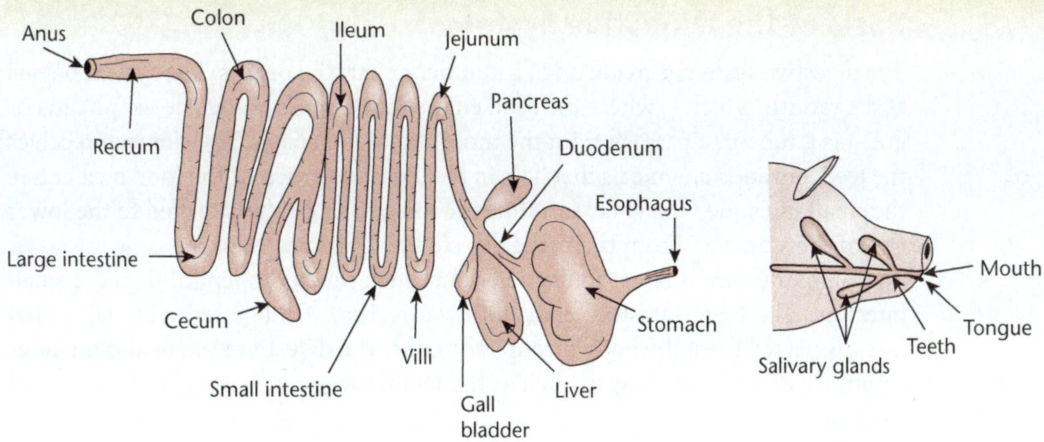

Figure 5-9 The digestive tract of the pig, a simple-stomached (non-ruminant) animal.

The digestive system consists of the parts of the body involved in chewing and digesting feed. This system also moves the digested feed through the animal's body and absorbs the products of digestion.

The capacities of digestive systems vary greatly among different species of animals (Table 5-4). Within species, the age, breed, and size of animals also affect the capacity of their digestive systems. The digestive systems of ruminant animals are generally larger than those of non-ruminants.

There is a great deal of difference among animals in their ability to digest various kinds of feed. This difference is mainly the result of differences in their digestive systems. Ruminant animals can digest large quantities of fibrous feeds, such as hay and pasture. Non-ruminant animals need a high-energy, low-fiber ration, such as grain. Grains and protein supplements are called **concentrates**.

Cattle and sheep can digest about 44 percent of the roughage they eat. **Roughage** refers to high-fiber feeds, such as hay, silage, and pasture. Horses are able to digest about 39 percent of the roughage in their ration. Swine can digest only about 22 percent of the roughage they eat.

Ruminants can digest large quantities of roughage because of the bacteria present in their digestive systems. These bacteria can produce proteins, B-complex vitamins, and vitamin K. Because ruminant animals produce certain nutrients themselves, the job of balancing their ration is somewhat easier than it is for non-ruminants.

TABLE 5-4 Capacities of the Digestive System of Selected Species

Organ	Swine (qts)	Horse (qts)	Cattle (qts)	Sheep/Goat (qts)
Rumen			80–192	25
Reticulum			4–12	2
Omasum			8–20	1
Abomasum			8–24	4
Stomach in non-ruminants	8	8–19		
Small intestine	10	27–67	65–69	10
Cecum	1–1.5	14–35	10	1
Large intestine	9–11	41–100	25–40	5–6
Total	28–30.5	90–221	200–367	48–49

Parts of the Digestive System

The digestive system is made up of a number of parts or organs. The system begins at the mouth, which is where the food enters the animal's body. The **esophagus** or gullet is a tube-like passage from the mouth to the stomach. The stomach receives the feed and adds chemicals that help in the digestive process. The food next enters the small intestine. The small intestine is a long, folded tube attached to the lower end of the stomach. From there, the digested material passes to the large intestine. The large intestine is larger in diameter but much shorter in length than the small intestine. The large intestine ends with the rectum. Undigested material, called feces, is passed from the body through the anus. The digestive system also includes a number of accessory organs such as the teeth, tongue, salivary glands, liver, and pancreas.

The teeth, tongue, and salivary glands are located in the mouth. The liver is the largest gland in the body. It is located along the small intestine just past the stomach. The pancreas is located along the upper part of the small intestine.

The digestive systems of most livestock are very similar in terms of the parts they contain. However, there are some differences in the poultry digestive system. Poultry have no teeth, and have a crop and gizzard. They also have two blind pouches, called ceca, which are attached to the small intestine. The **cloaca** is an enlarged part connected to the large intestine. Elimination (passing of feces) in poultry is through the **vent**.

THE DIGESTIVE PROCESS

Mouth and Esophagus

The chewing action of the mouth and teeth breaks, cuts, and tears up the feed. This increases the surface area of the feed particles, which, in turn, helps the chewing and swallowing process. Saliva also stimulates the taste. In ruminants, saliva is important in the chewing of the **cud**.

In most animals, saliva contains the enzymes salivary amylase and salivary maltase. **Enzymes** are substances called organic catalysts that speed up the digestive process. Salivary amylase changes some starch to maltose or malt sugar. Salivary maltase changes maltose to glucose.

Ruminants do not chew their food completely when they eat. Roughages, such as hay and silage, and coarse feeds, such as unbroken kernels of corn, are re-chewed later. These feeds form ball-like masses in the stomach. The material is then forced back up the esophagus to be chewed again. This is called **rumination** or chewing the cud.

The tongue helps direct the feed to the throat for swallowing. The chewed material enters the esophagus. Food is carried down the esophagus by a series of muscle contractions. The lower esophageal sphincter prevents food in the stomach from coming back into the esophagus.

The Ruminant Stomach

The four compartments of the ruminant stomach are the rumen, reticulum, omasum, and abomasum (Figure 5-10). Because of this four-compartment stomach, digestion in ruminants differs from that in non-ruminants.

Ruminants eat rapidly. They do not chew much of their feed before they swallow it. The solid part of the feed goes into the rumen. The liquid part goes into the reticulum, then to the omasum, and on into the abomasum. In the rumen, the feed is mixed and partially broken down by bacteria. A slow churning and mixing action takes place.

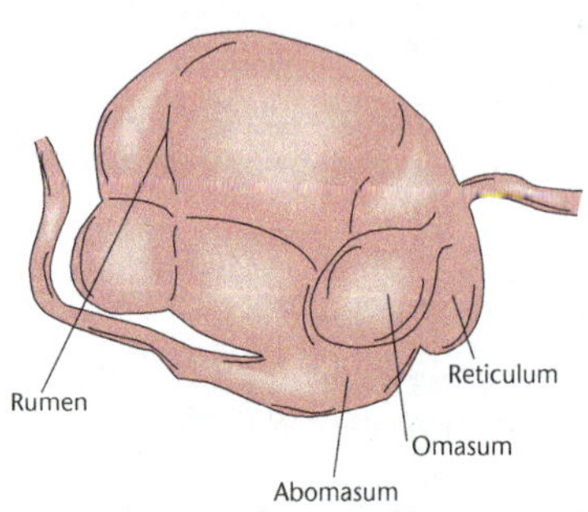

Figure 5-10 The four divisions of the ruminant stomach.

When the rumen is full, the animal lies down. The feed is then forced back into the mouth and rumination occurs. Cattle chew their cud about six to eight times per day. A total of 5 to 7 hours each day is spent in rumination.

There is no division between the rumen and the reticulum. Together, they make up about 85 percent of the stomach. They are located on the left side of the middle of the animal. The rumen and reticulum contain millions of microorganisms called **bacteria** and **protozoa**. Muscles in the rumen and reticulum help break the food into smaller particles. This makes it easier for the bacteria to act in the digestive process. Saliva and water are added to the feed to aid in digestion.

The bacterial action in the rumen allows ruminants to use large amounts of roughage. These bacteria can change low-quality protein into the amino acids needed by the animal. **Amino acids** are compounds that contain carbon, hydrogen, oxygen, and nitrogen. Some amino acids also contain sulfur, phosphorus, and iron. They are essential for growth and maintenance of cells. Bacteria also produce many of the vitamins needed by the animal. As the bacteria die, the animal digests them. This process makes protein and vitamins available to the animal.

The kind of ration that is fed affects the growth of the microorganisms in the rumen. The number of microbes in the rumen is reduced when the ration is made up of a large amount of fine-particle material.

The effect of fine-particle material on the rumen is reduced when animals are fed several times per day. Also, increasing the level of forage in the ration improves the growth rate of microbes in the rumen.

The pH level in the rumen also affects the growth of microbes. Microbes grow best when the pH of the rumen is in the range of 5.5 to 8.0 (optimum pH level is 6.5). A pH level outside of this range reduces the growth rate of the rumen microbes.

Rumen microbes use ammonia and amino acids, energy, and minerals for growth. A shortage of any of these can cause reduced microbe growth in the rumen. Ammonia deficiency may result in reduced efficiency of microbe growth and thus a reduction in the rate and extent of digestion of organic matter in the rumen. This may lead to reduced feed intake by the animal. Providing additional protein sources may stimulate microbe growth in the rumen. The cheapest source of ammonia for rumen microbes is generally some form of nonprotein nitrogen, such as urea.

Animals sometimes swallow foreign objects, such as wire and nails. These are held in the reticulum and damage organs depending on positioning. A small magnet is sometimes inserted down the throat of the cow and into the stomach. When the magnet enters the reticulum, it attracts the wire and nails and holds them so that they do not injure the animal.

A large amount of carbon dioxide and methane gas is released by bacterial action in the rumen. These gases must be disposed of through the digestive system. Sometimes, the gases form faster than the animal can eliminate them. This may happen when the animal eats large amounts of fresh grass or legumes, and it can cause the animal to bloat.

The **omasum** is the third part of the ruminant stomach. It makes up about 8 percent of the stomach. The omasum has strong muscles in its walls. The purpose of the omasum is not exactly known. It grinds up a certain amount of feed and may also squeeze some of the water out of the feed.

The **abomasum** is called the true stomach of the ruminant. It makes up about 7 percent of the stomach. Feed is mixed with gastric juice in the abomasum. Digestion is carried on here the same as in non-ruminant animals.

The Non-ruminant Stomach

When feed enters either the stomach of the non-ruminant or the abomasum of the ruminant, **gastric juice** begins to flow. This fluid comes from glands in the wall

Connection: Hardware Disease

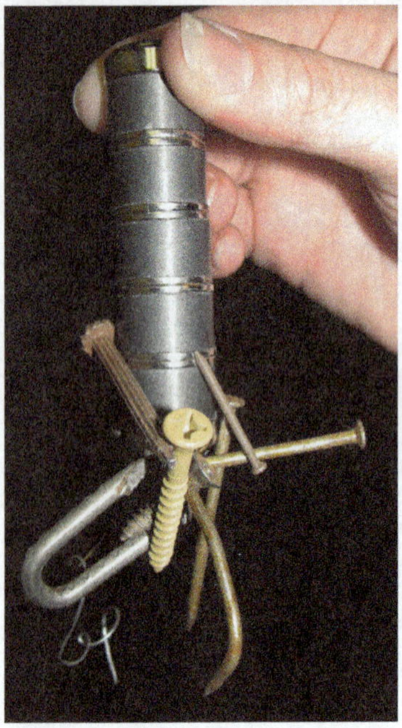

Courtesy of Rikki Griffith, Georgia Agricultural Education

Figure 5-11 A cow magnet attracts metal objects.

Cattle often pick up and swallow pieces of metal such as nails, wire, and staples while grazing or eating harvested forages. These pieces of metal often lodge in the honeycombed walls of the reticulum causing hardware disease. The metal may cause damage to surrounding organs, irritation, and inflammation. Cattle with hardware disease may lose their appetite, produce less milk, and lose weight. One veterinary tool, commonly used by dairy farmers, to help prevent hardware disease is a cow magnet (Figure 5-11). This magnet is inserted into the stomach of a cow (Figure 5-12) and helps prevent hardware disease by attracting stray metal from the folds and crevices of the reticulum. The magnet remains in the cow's stomach for the life of the cow.

www.TheWifeOfADairyman.blogspot.com

Figure 5-12 Cow magnets are often inserted into cattle, especially dairy cows, to help prevent hardware disease.

of the stomach. The gastric juice contains 0.2 to 0.5 percent hydrochloric acid. It stops all action of amylase when it mixes with the feed. The gastric juice contains additional enzymes called pepsin, rennin, and gastric lipase. These enzymes act on the feed in the following ways. Pepsin breaks the proteins in the feed into proteoses and peptones. The casein of the milk is curdled by the rennin. Emulsified fats are split by the gastric lipase into glycerol and fatty acids. However, most of the fat entering the stomach is not emulsified.

The muscular walls of the stomach churn and squeeze the feed. Liquids are pushed on into the small intestine. Gastric juice then acts on the solids that remain in the stomach.

Small Intestine

When the partly digested feed leaves the stomach, it is an acid, semifluid, gray, pulpy mass. This material is called chyme. In the small intestine, the chyme is mixed with three digestive juices: pancreatic juice, bile, and intestinal juice.

Pancreatic Juice

Pancreatic juice is secreted by the pancreas. It contains the enzymes trypsin, pancreatic amylase, pancreatic lipase, and maltase.

Trypsin Trypsin breaks down proteins not broken down by pepsin. Some of the proteoses and peptones are broken down by trypsin to peptides. Proteoses, peptones, and peptides are combinations of amino acids. Proteoses are the most complex compounds, with peptides being the simplest.

Pancreatic Amylase Pancreatic amylase changes starch in the feed into maltose. Pancreatic amylase is more active in this process than salivary amylase because it is found in greater quantities and has a longer time to work on the feed. Sugar and maltose are broken down into even simpler substances. When acted on by maltase, they are changed into a simple sugar called glucose.

Lipase Lipase works on fats in the feed. It changes them into fatty acids and glycerol.

Bile

The liver produces a yellowish-green, alkaline, bitter liquid called **bile**. Bile is stored in the gallbladder in all animals except the horse, which does not have a gallbladder. Bile aids in the digestion of fats and fatty acids. It also helps in the action of the enzyme lipase. In a final step, fatty acids combine with bile to form soluble bile salts.

Intestinal Juice

Glands in the walls of the small intestine produce **intestinal juice**. This fluid contains peptidase, sucrase, maltase, and lactase, which help in the digestion process. Proteoses and peptones are broken down by peptidase into amino acids. Starches and sugars are broken down by sucrase, maltase, and lactase into the simple sugars glucose, fructose, and galactose.

Absorption

The wall of the small intestine is lined with many small fingerlike projections called **villi**. These increase the absorption area of the small intestine. Most food nutrients used by the animal are absorbed from the small intestine.

The Cecum

The cecum or "blind gut" is found where the small intestine joins the large intestine. The **cecum** is a small organ and has little function in most animals, except the horse. In the horse, roughage feeds are digested by bacterial action in the cecum, which explains why the horse can eat large amounts of roughage. The digestive system of the horse is shown in Figure 5-13.

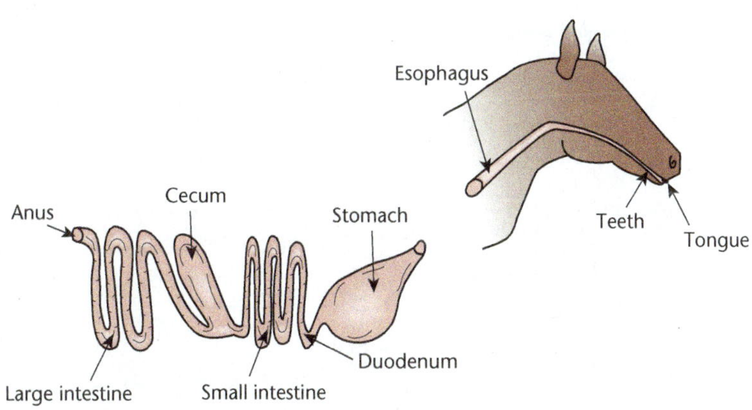

Figure 5-13 The digestive tract of the horse, a non-ruminant animal.

Large Intestine

The large intestine is larger in diameter but much shorter in length than the small intestine. The main function of the large intestine is to absorb water. Material that is not digested and absorbed in the small intestine passes into the large intestine. The enzymes of the small intestine continue to work on the material even after it has passed into the large intestine. Some bacteria also work on the material. The large intestine adds mucus to enable the material to pass through more easily.

Feed materials that are not digested or absorbed are called feces. This material is moved through the large intestine by muscles in the intestinal walls. The undigested part of the feed is passed out of the animal's body through the anus, the opening at the end of the large intestine.

DEVELOPMENT OF THE RUMINANT STOMACH

The abomasum is the only part of the stomach of young ruminant animals that functions. Therefore, young ruminants cannot use roughages in their diet. Milk fed to the animal goes directly to the abomasum. This action continues until the other parts of the stomach have developed.

When the animal is born, the rumen is a very small organ found in the upper left part of the abdomen. After about 2 months, the rumen moves to its normal position in the mature animal. The reticulum and the omasum grow and develop rapidly during the first 2 months of the animal's life. By the time the animal is 3 months old, the rumen has grown large enough to begin to function. The animal can then begin to use more solid feeds and roughage in its diet.

DIGESTION IN POULTRY

Poultry possess certain special digestive organs that are not found in other animals (Figure 5-14). Feed taken in by poultry first goes to the **crop** for storage. Here, it is softened by saliva and secretions from the crop wall.

The feed moves from the crop, through the glandular stomach (the proventriculus), into the muscular stomach (the gizzard). The large red walls of the muscular stomach are thick, powerful muscles. The muscular stomach is lined with a thick, horny membrane-like material called the **epithelium**. Feed particles are crushed and mixed with digestive juices by the **gizzard**.

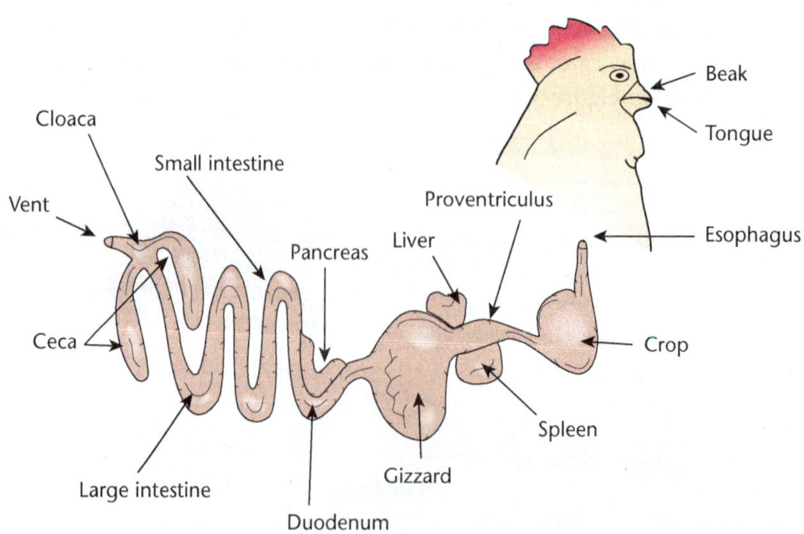

Figure 5-14 The digestive tract of the chicken, a non-ruminant.

Two blind pouches, or ceca, are found where the small intestine joins the large intestine. These are about 6 to 7 inches long in adults. The functions of the ceca include the absorption of water from fecal material, fermentation of coarse materials, and the production of some fatty acids and B vitamins. Few of these nutrients are absorbed though because the ceca are located near the end of the digestive system.

The cloaca is an enlarged part found where the large intestine joins the vent. Feces from the large intestine are passed out of the body through the vent. Eggs from the oviduct and urine from the kidneys also pass out through the vent.

ABSORPTION OF FEED

Most absorption of digested feed takes place in the small intestine. Some feed is also absorbed from the large intestine. Feed is not broken down enough to be absorbed in the mouth or esophagus. In ruminants, some fatty acids are absorbed from the rumen. The stomach tissue of non-ruminants is not suited for absorption.

Villi, the millions of small finger-shaped projections on the wall of the intestine, are the key to absorption because they greatly increase the surface area of the intestine. Each villus (singular of villi) has a network of blood capillaries and a lymph vessel through which nutrients enter the bloodstream.

Digested protein is absorbed in the form of amino acids; some sodium must be present for absorption to occur properly. Digested carbohydrates (starches and sugars) are present as the monosaccharides (simple sugars) glucose, fructose, and galactose. ATP supplies energy for the absorption of monosaccharides and fatty acids. Sodium must also be present for monosaccharide absorption. As the substances are absorbed by the blood capillaries in the villi, they pass through the liver and then into the blood.

Digested fats, in the form of soaps and glycerol, undergo a slightly different process. They are formed again into fats and absorbed by the lymph vessel in the villi. They are then carried through the lymphatic system. They pass through the thoracic duct in the neck into the circulatory system.

Villi also absorb water and dissolved minerals into the bloodstream. Some minerals are present in the form of compounds that are not readily absorbed. Phosphorus from grains, held in the compound phytin, is only partially absorbed by swine and none of it is absorbed in poultry. Ruminants, however, can use this phosphorus because the bacteria in the rumen release enzymes that break phytin down so the phosphorus can be absorbed.

The large intestine also absorbs water and some nutrients directly into the bloodstream. This is done through capillaries in the wall of the intestine. Nutrient absorption in the large intestine is especially important in the horse because much of the microbial digestion of roughage occurs in the cecum.

Most nutrients end their journey in the muscle cells. Some are deposited in the liver. The animal uses the nutrients to replace worn-out cells and to build new ones. Some nutrients are used for energy; others are stored in the form of fat for later use.

METABOLISM

Metabolism is the sum of the processes, both chemical and physical, that are used by living organisms and cells to handle nutrients after they have been absorbed from the digestive system. The metabolic processes include anabolism, catabolism, and oxidation of nutrients. **Anabolism** is the formation and repair of body tissues. **Catabolism** is the breakdown of body tissues into simpler substances and waste products. **Oxidation** of nutrients provides energy for the animal.

SUMMARY

An animal's body is composed of a number of different kinds of tissues. These tissues are grouped to form organs and these organs are organized into systems that carry out some major function in the body. The 11 major body systems include the skeletal, muscle, respiratory, circulatory, nervous, endocrine, excretory, reproductive, immune, integumentary, and digestive system.

The skeletal system gives form, protection, support, and strength to the body. The muscle system helps the body to move and performs other vital functions, such as maintaining the beating of the heart. The heart is a part of the circulatory system that acts as a pump to keep blood moving throughout the body. Blood carries oxygen and nutrients to the cells where they are needed and removes waste products from those cells. The respiratory system draws air into the body, where the oxygen is extracted to oxidize molecules to provide energy for body activities. Air carrying waste products such as carbon dioxide is then expelled out of the body.

The nervous system provides a method for cells to transmit signals from one part of the body to another as needed. The endocrine system secretes hormones that are needed for growth and development of the body. The excretory system carries some waste products out of the body. The integumentary system protects the body from damage. The immune system defends the body against infection and disease. The reproductive system is a group of organs that allow for reproduction.

Digestion is the breaking down of feeds into simple substances that can be absorbed into the bloodstream and used by the body cells. Enzymes do most of the job of digestion.

Ruminant animals can use a lot of roughage in their rations. They have a four-compartment stomach in which bacteria break down the roughage. Non-ruminants must have more concentrates, such as grain, in their ration because they have a simple, one-part stomach.

Most digested feed is absorbed from the small intestine of the animal. The small intestine has millions of tiny projections called villi through which nutrients are absorbed and enter the bloodstream.

Student Learning Activities

1. Secure digestive systems of various species of animals from a packing plant. Dissect the systems, identify their parts, and explain their functions.
2. Prepare visuals and give oral reports on various digestive systems, identifying the parts and explaining the functions.
3. Prepare a bulletin board display with pictures of various species classified as ruminant and non-ruminant animals. Using pictures or samples of feeds, show typical feeds that are fed to each kind of animal.
4. Using an understanding of animal body systems, plan and conduct a supervised animal experience program in order to learn how to more effectively feed and manage animals.

Discussion Questions

1. List and give examples of the tissues normally found in an animal's body.
2. What are the main functions of the skeletal system?
3. What are the three components and the major parts of each that comprise the skeletal system?
4. Describe the kinds of bones normally found in the mammalian skeletal system.
5. Name several functions of bones.
6. Name and describe the three types of muscle normally found in animal bodies.
7. Describe the functions of each of the three types of muscles found in animal bodies.
8. Describe how a typical muscle works.
9. How does the cardiac muscle differ from other muscles found in the body?
10. Name and describe the functions of the major parts of the mammalian respiratory system.
11. How does the avian respiratory system differ from the mammalian system?
12. Describe the physiology of respiration.
13. What are the major components of the circulatory system?
14. Describe how the heart functions to keep blood circulating.
15. Name the two kinds of blood vessels and describe how they function.
16. Where does the exchange of gas and waste products take place in the circulatory system?
17. Name and describe the major components of blood.
18. What is the function of blood in the circulatory system?
19. What is the function of the immune system?
20. What is the function of and what are the major parts of the nervous system?
21. What is the importance of the endocrine system?
22. Describe the function of the excretory system.
23. Why can ruminants digest large quantities of roughage?
24. What are the special digestive organs found in poultry that are not found in other animals? Describe their function.
25. What digestive action occurs in the mouth, small intestine, and large intestine of an animal?
26. What is meant by "chewing the cud"?
27. What is the function of each part of the ruminant stomach?
28. Why are young ruminants unable to digest large quantities of roughage?
29. How is feed absorbed in an animal's body?
30. What use does an animal make of absorbed nutrients?

Review Questions

True/False

1. The kind of ration that is fed affects the growth of microorganisms in the rumen.
2. Non-ruminants do not require high levels of grain but can also survive on roughage.

Multiple Choice

3. Feed taken in by poultry first goes to the______where it is softened by saliva and secretions from the walls.
   a. crop
   b. gizzard
   c. esophagus
   d. stomach
4. The key to absorption is the small fingerlike projections on the walls of the intestine called________.
   a. sugars
   b. hairs
   c. villi
   d. fatty acids

Completion

5. ______ are a group of animals that can digest large quantities of roughage in their digestive systems.
6. The ______ is called the true stomach of a ruminant.
7. ______ refers to high-fiber feeds such as hay, silage, and pasture.
8. Feed particles in poultry are crushed and mixed with digestive juices by the ______.

Short Answer

9. What causes an animal to bloat?
10. What is the importance of the bacterial action in the rumen?

Chapter 6

Feed Nutrients

Key Terms

nutrient
carbohydrate
nitrogen-free extract (NFE)
fiber
dry weight
commercial feed tag
fat
oil
protein
urea
crude protein
digestible protein
vitamin
trace organic compounds
conception rate
mineral
deficiency

Objectives

After studying this chapter, the student should be able to

- describe the major functions of the basic nutrient groups, and identify feeds that are sources of each.
- describe the characteristics of nutrient sources for each basic nutrient group.

A **nutrient** is defined as a chemical element or compound that aids in the support of life. A nutrient becomes a part of the cells of the body. Nutrients are necessary for cells to live, grow, and function properly. They are divided into five groups: (1) energy nutrients (carbohydrates, fats, and oils), (2) proteins, (3) vitamins, (4) minerals, and (5) water.

Many different kinds of nutrients are needed by animals. In addition, they must have the right nutrients in the proper balance. Too much of one nutrient and not enough of another may result in unhealthy stock and high feed costs. A lack of one or more nutrients may slow normal growth or production.

Animals differ in the kinds and amounts of nutrients they need. It is the job of animal nutritionists to determine which nutrients animals need. They make recommendations for feeding each kind of animal based on the results of feeding experiments. Proper feeding is necessary for the production of healthy livestock.

ENERGY NUTRIENTS—CARBOHYDRATES

The main energy nutrients found in animal rations are carbohydrates. **Carbohydrates** are made up of sugars, starches, cellulose, and lignin. Carbohydrates are chemically composed of carbon, hydrogen, and oxygen.

In a chemical reaction very much like burning, carbohydrates provide energy for the body cells. This energy powers muscular movement such as the heartbeat, walking, breathing, and digestive contractions. Carbohydrates also produce the body heat that helps to keep the animal warm. Extra carbohydrates are converted to fat and stored in the body.

Simple Carbohydrates

The simple carbohydrates are sugars and starches. They are easily digested. Simple carbohydrates are referred to as **nitrogen-free extract (NFE)** and come from the cereal grains, such as corn, wheat, oats, barley, rye, and sorghum.

Complex Carbohydrates

The complex carbohydrates, called **fiber**, are cellulose and lignin. These substances are more difficult to digest than simple carbohydrates. Fiber is found mainly in roughages such as hay and pasture plants. Examples are alfalfa, bromegrass, orchard grass, and bluegrass.

Figure 6-1 Ruminant animals can digest large amounts of fiber found in roughages such as hay and pasture.

Carbohydrate and Fiber Content of Feeds

The **dry weight** of most grains and roughages ranges from 65 to 80 percent carbohydrates. The dry weight of a feed refers to its weight with the moisture content removed. Mature roughages contain more fiber than those harvested when less mature; however, the mature plant is not easily digested. Hay plants harvested at an early stage of maturity have a higher feed value because they are more easily digested.

Ruminant animals, such as cows, sheep, and goats, can digest large amounts of fiber. A high percent of their ration is roughage (Figure 6-1). Simple-stomached animals, such as swine, cannot digest large amounts of fiber. Their ration must be mostly cereal grains that are more easily digested (Figure 6-2).

A **commercial feed tag** is the label attached to a bag of feed purchased at a grain elevator or feed store. The maximum amount of fiber found in that feed is shown as a percent of the total weight of the bag of feed.

The sources of carbohydrates used for that feed are found in the list of ingredients printed on the tag. Some examples of carbohydrate sources found on a feed tag are cane molasses, ground corn, wheat middlings, and oats. There are many other sources of carbohydrates used in commercial feeds. The ones used depend on the kind of animal for which the feed is made.

Figure 6-2 Simple-stomached animals require primarily grain in their ration; they cannot digest large amounts of roughage.

ENERGY NUTRIENTS—FATS AND OILS

Fats and oils are also energy nutrients. They are chemically composed of carbon, hydrogen, and oxygen. They contain more carbon and hydrogen atoms than carbohydrates do. For this reason, the energy value of fats is higher than that of carbohydrates. In fact, fats have 2.25 times the energy value of carbohydrates. **Fats** are solid at body temperature, and **oils** are liquid at body temperature. In animal nutrition, both fats and oils are commonly referred to as fats.

Fats are easily digested by animals. They provide energy and body heat. They also carry the fat-soluble vitamins that are in the feed.

Fats come from both vegetable and animal sources. Cereal grains such as corn, oats, and wheat range from 1.8 to 4.4 percent fat. Cereal products used in rations, such as brewer's dried grains, corn gluten meal, distiller's dried grains, and wheat middlings, range from 1.1 to 8.2 percent fat. Animal and vegetable protein concentrates range from 1 to 10.6 percent fat.

The amount of fat in a commercial feed is shown on the feed tag as a percent of the total weight of feed in the bag. The fat content in the feed is expressed as a guaranteed minimum level.

PROTEINS

Proteins are organic compounds made up of amino acids. They supply material to build body tissues. The ligaments, hair, hooves, horns, skin, internal organs, and muscles of the animal body are all partially formed from protein. Protein is essential for fetal development in pregnant animals.

If an animal takes in more protein than it needs, the nitrogen is separated and given off in the urine. The material that is left is converted into energy or body fat by the animal.

Ten amino acids are considered to be essential for swine, and 14 are considered to be essential for poultry (Table 6-1). Nonessential amino acids are needed by animals, but are synthesized in the body from other amino acids and therefore do not have to be provided in the ration for either ruminant or non-ruminant animals. Non-ruminant animals cannot synthesize the essential amino acids fast enough to meet their needs; therefore, those amino acids must be provided in their rations. Ruminants can generally synthesize the essential amino acids by rumen bacterial action at a rate sufficient to meet their needs.

Research suggests that fermentation and bacterial action in the rumen may reduce the quality of protein in the feed. This may result in a deficiency of one or more essential amino acids in ruminant rations. Encapsulating the protein feed reduces exposure to fermentation in the rumen. Heat or chemical treatment with tannic acid or formaldehyde reduces the tendency of the protein to degrade in quality in the rumen. When the high-quality protein is protected, the rumen bacteria are more likely to use nonprotein nitrogen sources in fermentation processes. Research is continuing in this phase of animal nutrition.

TABLE 6-1 Essential and Nonessential Amino Acids for Swine and Poultry

Essential	Nonessential
Arginine	Alanine
Histidine	Aspartic acid
Isoleucine	Citrulline
Leucine	Cysteine
Lysine	Cystine
Methionine	Glutamic acid
Phenylalanine	Glycine
Threonine	Hydroxyglutamic acid
Tryptophan	Hydroxyproline
Valine	Norleucine
Additional Essential for Poultry:	Proline Serine Tyrosine
Alanine	
Aspartic acid	
Glycine	
Serine	

Sources of Protein

Animal protein sources are considered to be good-quality proteins since they usually contain a good balance of the essential amino acids. Plant protein sources are usually thought of as poor-quality proteins because they often lack some of the essential amino acids.

The amino acid needs of ruminants can be met by feeding proteins from vegetable sources. They can also be met by including urea in the ration. **Urea** is a synthetic nitrogen source that is mixed in the ration to provide nitrogen for the making of amino acids in the rumen.

It is necessary to feed simple-stomached animals protein from sources that provide a balance of essential amino acids. This is done to make sure they get the essential amino acids in their feed. If cereal grains are combined in the right combination, they will provide a balanced ration. Soybean oil meal, for example, is one of the most commonly used protein sources for feeding hogs.

A commercial feed tag shows the guaranteed minimum of crude protein contained in that feed, shown as a percent of the weight of the bag of feed. The amount of **crude protein** (total protein) is the amount of ammoniacal nitrogen in the feed multiplied by 6.25. It may contain materials that are not true protein. Plant protein sources often used in commercial feeds include linseed meal, soybean meal, cottonseed meal, and dehydrated alfalfa meal. Animal protein sources often used are meat meal, fish meal, condensed fish solubles, dried whey, casein, dried milk albumin, and dried skim milk. The synthetic nitrogen source, urea, is also used for ruminant feeds. The list of ingredients on the feed tag shows the protein sources used in the mix.

Not all of the crude protein in a feed is digestible. Thus, the animal cannot use all of it. Some may come from hair, hooves, feathers, or other sources that have a low digestibility. About 60 percent of the crude protein in a roughage ration is considered to be digestible. About 75 percent of the crude protein in a high-concentrate ration is considered to be digestible. **Digestible protein** is approximately the amount of true protein in a feed.

VITAMINS

Vitamins are **trace organic compounds**. In other words, they are needed only in very small amounts by animals. All vitamins contain carbon, but they are not alike chemically.

Vitamins are divided into two groups: fat-soluble and water-soluble. Fat-soluble vitamins can be dissolved in fat. Water-soluble vitamins can be dissolved in water.

The fat-soluble vitamins are A, D, E, and K. Vitamin A is associated with healthy eyes, good **conception rate**, and disease resistance. Vitamin D is associated with good bone development and the mineral balance of the blood. Vitamin E is necessary for normal reproduction and muscle development. Research suggests that increasing the level of vitamin E alone or in combination with selenium in the diet may help strengthen the immune system of several animal species. Vitamin K helps the blood to clot and prevents excessive bleeding from injuries.

Some sources of the fat-soluble vitamins are green leafy hay, yellow corn, cod liver and other fish oils, wheat germ oil, and green pastures. Vitamin D is produced in the animal's body if the animal is in sunlight part of the day.

The water-soluble vitamins include vitamin C and the B-complex vitamins. The B-complex includes many different vitamins. These are B1 (thiamine), riboflavin, niacin, pyridoxine, pantothenic acid, biotin, folic acid, benzoic acid, choline, and vitamin B12.

Vitamin C helps in teeth and bone formation and in the prevention of infections. The B-complex vitamins are necessary for chemical reactions in the animal's body. They help improve appetite, growth, and reproduction.

Sources of Vitamins

Some sources of vitamin C are green pastures and hay. All farm animals seem to produce enough vitamin C in their bodies. Therefore, it does not need to be included as a specific nutrient in their rations. Some sources of the B-complex vitamins are green pastures, cereal grains, green leafy hay, milk, fish solubles, and certain animal proteins.

Commercial feeds usually include necessary vitamins in the mixture. Vitamins often included in commercial feeds for ruminants are A, D, and E. Vitamins often included in commercial feeds for swine are A, D, E, K, riboflavin, niacin, *d*-pantothenic acid, B12, and choline chloride.

MINERALS

Minerals are inorganic materials needed in various amounts by animals. Minerals contain no carbon. Thus, if a feed were completely burned, the ash that was left would be the mineral content of the feed.

Connection: Avast, Ye Scurvy Blige Rats!

The life of a pirate or other seaman exploring the New World was not as glamorous as the movies sometimes portray. Hundreds of years ago, seamen had a 50 percent survival rate. A major cause of death was a nutritional deficiency known as scurvy.

Scurvy caused symptoms including swollen legs, putrid gums, fatigue, ulcers, and loss of teeth. Old scars would break open just as if the wound had recently been inflicted.

A British doctor observed that sailors who ate citrus fruits (Figure 6-3) did not develop scurvy. Using this information, sailors were assigned a daily ration of lime juice to prevent the disease. Hence, the sailors became known as "limeys."

Figure 6-3 Citrus fruits, such as oranges, can help prevent scurvy.

Pirates thought scurvy was a curse because their animals did not get the disease. It was later determined that scurvy was caused by a lack of vitamin C, which the human body requires. Animals, unlike humans, make their own vitamin C. Humans have to get vitamin C from food. Because marine travel was slow, fruits and vegetables high in vitamin C were usually in short supply.

Vitamin C (ascorbic acid) is needed by many body organs and body systems, including the adrenal glands, blood, bones, capillary walls, heart, nervous system, and teeth. It helps prevent infection and enhances the immune system.

Minerals are important in animal feeding. They provide material for the growth of bones, teeth, and tissue. Minerals also regulate many of the vital chemical processes of the body. They aid in muscular activities, reproduction, digestion of feed, repair of body tissues, formation of new tissue, and release of energy for body heat. If there is a lack of a certain mineral in an animal's ration, this is called a **deficiency**.

For example, without iron in the blood, oxygen could not be carried to the body cells. A deficiency of iron and copper in baby pig rations will cause anemia. The bones and teeth will not form properly if the animal lacks calcium and phosphorus in its ration.

Research suggests that increasing the level of trace minerals in livestock rations above the currently recommended level may improve an animal's immunity to disease. For example, increasing the level of selenium in the diet of calves and swine has increased the antibody response of the immune system. A small increase in the zinc level in the diet has increased the production of white blood cells to fight infections. More research is needed before this practice can be recommended for use on the farm.

Minerals are divided into two groups. Major minerals are those needed in large amounts. Trace minerals are those needed in small amounts.

Major minerals that are often lacking in animal rations are salt (sodium and chlorine), calcium, and phosphorus (Figure 6-4). It is important to make sure that these minerals are included in the animal's ration. The ratio of calcium to phosphorus in swine rations should not be greater than 1.5 to 1. The ratio of calcium to phosphorus for ruminants can be as high as 7 to 1.

Trace minerals that are necessary for animals include potassium, sulfur, magnesium, iron, iodine, copper, cobalt, zinc, manganese, boron, molybdenum, fluorine, and selenium. Most of these trace minerals are found in common animal feeds. Usually there are adequate amounts of these minerals in the feed.

In some areas and under certain conditions there may be deficiencies of some trace minerals in the feed available to livestock. These deficiencies may be widespread in a given geographic area or they may occur only in parts of selected fields. Table 6-2 lists trace minerals and where problem areas of deficiencies may occur in the United States. If a trace mineral deficiency in the feed is suspected, it may be wise to have the feed analyzed. When a deficiency is identified, the ration must be properly supplemented with the missing trace minerals.

Some trace minerals, when present in excess quantities, are toxic to livestock. Toxicity may occur when the minerals accumulate in plants grown in the area. Some toxicity problems are discussed in Table 6-3.

Courtesy of Christy Bryan

Figure 6-4 A good vitamin and mineral mix is imperative for livestock health.

TABLE 6-2 Trace Mineral Deficient Areas of the United States

Element	Deficiency Occurrence in the United States
Boron	Throughout the United States on sandy light-textured soils, acid soils, calcareous soils, soils with low amounts of organic matter. Most serious with alfalfa.
Cobalt	Sandy soils in New England states and South Atlantic Coastal Plain, especially in Florida. Legume plants in these areas; grasses and cereal grains in all parts of the United States.
Copper	Most often found in organic soils or very sandy soils in Central wheat belt, New England states, lower Atlantic Coastal Plain. Generally in legumes grown in these areas.
Iodine	Great Lakes states, Dakotas, Montana, Idaho, Washington, Oregon, Nevada, Utah, Colorado, Wyoming, western Nebraska, southeastern state of the Appalachian Range.
Iron	Alkaline soils of western United States and very sandy soils. Intermountain region in western United States including northern Nebraska to Kansas, Colorado, western Oklahoma, eastern New Mexico, northwest Texas; also Iowa and California. Often highly localized in specific fields.
Magnesium	Sandy and loamy soils with a high level of available potassium.
Manganese	Great Lakes region and Atlantic Coastal Plain states. Problem is greatest on calcareous soils, peats, mucks, coarse-textured soils, and poorly drained soils.
Molybdenum	Poorly drained acid soils in the intermountain valleys of western United States and poorly drained acid soils in South Central Florida.
Selenium	Great Lakes states, New England states, upper Appalachian area, Atlantic Coastal Plain, Florida. Also, Washington, Oregon, northern two-thirds of California, Nevada, Idaho, western Montana, western Utah. Deficiency may be highly localized in some areas.
Sulfur	Pacific Northwest and some parts of the Great Lakes states.
Zinc	Western United States (generally highly localized on irrigated land), southeastern United States on sandy, well-drained, acid soils, or soils from phosphatic rock parent material. Some localized areas of deficiency in other areas of the United States.

Major and trace minerals are usually supplied in commercial feeds. They may be included in a protein supplement or added by the use of a mineral premix. Salt and mineral blocks are often used to provide the additional minerals needed in the ration. Monocalcium phosphate, dicalcium phosphate, ground limestone, steamed bonemeal, and calcium carbonate are usually included in commercial feeds and mineral supplements to provide calcium and phosphorus.

Commercial feed tags show a guaranteed minimum and maximum percent of calcium. The minimum percent of phosphorus is shown on the tag. If salt is added, the minimum and maximum percent of salt is also shown on the feed tag. Major minerals, except salt, are guaranteed in terms of the individual element rather than as compounds. Salt is guaranteed as the compound sodium chloride (NaCl). Trace minerals are guaranteed on the feed tag as a minimum percent of the weight of the feed in the bag.

TABLE 6-3 Trace Mineral Toxicity Areas of the United State

Element	Toxicity Problem Areas in the United States
Boron	San Joaquin Valley of California. Semiarid regions with alkaline soils in other areas of the western United States.
Manganese	On poorly drained, acid soils. More likely to affect crop growth than livestock production.
Molybdenum	Forage crops on some alkaline soils in western United States and poorly drained organic soils in South Central Florida. Cattle and sheep most likely to be affected.
Selenium	Some areas of the Plains and Rocky Mountain states; most often on well-drained alkaline soils. Some shrubs and weeds native to semiarid and desert rangelands have an unusual ability to extract selenium from the soil. If grazed, these plants can cause selenium toxicity in livestock. Range grasses and field crops growing in the same areas do not accumulate excessive amounts of selenium in their tissues.

Figure 6-5 A fresh and clean supply of water is required for all animals.

WATER

Water is so common that its importance as a nutrient is often forgotten (Figure 6-5). However, water makes up the largest part of most living things. The amount of water in an animal's body varies with the kind of animal, its age, and its condition. In general, the amount of water in animal bodies ranges from 40 to 80 percent. Younger animals have a higher percent of water in their bodies than older animals.

Water has many important functions. It helps to dissolve the nutrients the animal eats. It also helps to control the temperature of the animal's body. Water in the blood acts as a carrier of the nutrients to different parts of the animal's body. Water is necessary for many of the chemical reactions that take place in the body.

A fresh, clean supply of water is necessary for animals to grow and produce efficiently. A continuous supply is best for rapid growth and efficient production. If animals do not have a good water supply, they will not make good use of the other nutrients supplied in the ration. Animals can survive longer without food than they can without water.

SUMMARY

Nutrients are chemical elements or compounds that aid in the support of life. Animals must have five different groups of nutrients to grow and produce efficiently. Energy nutrients provide the energy necessary for movement and production of body heat. Proteins supply material to build body tissues, hooves, horns, hair, and skin. Vitamins help to regulate many of the body's functions. Minerals provide material for bones, teeth, and tissues and help to regulate chemical activity in the body. Water dissolves and carries nutrients, regulates temperature, and is necessary for chemical reactions in the body.

Nutrients are supplied by the grains and forages fed to the animal. Additional nutrients needed by the animal are supplied by commercial feed mixes.

Student Learning Activities

1. Collect samples of nutrient sources from farms and feed stores. Develop an exhibit of these samples showing the approximate percent of each nutrient that comes from each source and include information about the function of each nutrient.
2. Present a short oral report to the class about new developments in feed nutrients that have been reported in farm magazines and newspapers.
3. Develop a bulletin board exhibit of commercial feed tags. Include information about the functions of the nutrients listed on the tags.
4. Take a field trip to a local grain elevator, feed store, or commercial feed manufacturer. Observe the sources of feed nutrients used in feed mixing, and the method of mixing complete feeds.
5. Take a field trip to a local farm that has facilities for mixing complete feeds. Observe the nutrient sources and methods used to obtain a complete feed mix.
6. Select and use appropriate feed sources containing the necessary feed nutrients when planning and conducting a supervised experience program in animal production.

Discussion Questions

1. What is a nutrient?
2. List the five groups of nutrients.
3. What is nitrogen-free extract?
4. What are the four substances that make up carbohydrates?
5. What is the function of carbohydrates?
6. What is the function of fats and oils?
7. List five feed sources of fats and oils.
8. What are proteins?
9. What is the function of proteins?
10. Name the 10 essential amino acids for swine.
11. What is the difference in value between animal protein sources and plant protein sources?
12. What is urea?
13. List three plant protein sources and three animal protein sources.
14. What is the difference between crude protein and digestible protein?
15. What are vitamins?
16. What is the function of vitamins?
17. Which vitamins are fat soluble and which are water soluble?
18. What are minerals?
19. What is the function of minerals?
20. List the major and trace minerals often needed in animal rations.
21. What is the function of water in an animal's body?

Review Questions

True/False

1. Some sources of vitamin C are green pastures and hay.
2. Major and trace minerals are usually supplied in commercial feeds.
3. Salt is usually guaranteed as the compound calcium nitrate.

Multiple Choice

4. The main energy nutrient found in animal rations is:
   a. proteins c. fats
   b. carbohydrates d. oils
5. Which have the highest energy value?
   a. carbohydrates c. proteins
   b. fats d. vitamins
6. Water does the following in the ration:
   a. dissolves nutrients c. enhances chemical reactions
   b. controls temperature d. all of the above

Completion

7. A chemical element or compound that aids in the support of life is a ______.
8. The amount of ammoniacal nitrogen in the feed multiplied by 6.25 is known as ______ ______.
9. When there is a lack of a certain mineral in an animal's ration, this is called a ______.

Short Answer

10. List three sources of B-complex vitamins.
11. Because simple-stomached animals such as swine cannot digest large amounts of fiber, what must be included in their rations?
12. List some sources of carbohydrates in animal feeds.

Chapter 7

Feed Additives and Growth Promotants

Key Terms

feed additive
growth promotant
antibiotic
antimicrobial
anthelmintic
subtherapeutic
chemobiotic
nutrient sparing
subclinical disease
hormone
coccidiostat
probiotic
melengestrol acetate (MGA)
withdrawal period
Feed Additive Compendium

Objectives

After studying this chapter, the student should be able to

- discuss the general use and purpose of feed additives and growth promotants.
- describe the proper method of hormone implantation.
- discuss the proper use of feed additives for various species of animals.
- discuss labeling and regulation of feed additives.
- discuss the proper mixing of feed additives in complete rations.
- discuss health issues and concerns relating to the use of feed additives.

Feed additives are materials used in animal rations to improve feed efficiency, promote faster gains, improve animal health, or increase production of animal products. These materials are not generally considered to be nutrients and are used in small amounts in the ration. They are often added to the basic feed mix and require careful handling and mixing. Hormone implants are pelleted synthetic or natural hormones or hormone-like compounds placed under the skin or in the muscle of the animal. Implants are used to lower production costs by improving both rate and efficiency of gain. Hormone implants are sometimes called **growth promotants**.

Feed additives came into common use in livestock feeding in the early 1950s. Since then, a wide variety of feed additives and hormone compounds have been developed, tested, and approved for use with livestock. The use of some of these materials has been discontinued because of toxicity, high cost, lack of proven benefit, or excessive residues in meat and livestock products.

However, many feed additives and hormones are still widely used in livestock production. Hormone implants are used mainly in beef cattle operations. The kinds of materials used in livestock production include antimicrobial compounds (such as **antibiotics** and chemoantibacterials), hormones and hormone-like substances, **anthelmintics** (dewormers), buffering agents, feed flavors, and bloat preventatives.

KINDS OF FEED ADDITIVES AND HORMONES

Antimicrobial Drugs

Antibiotics and chemoantibacterial (chemotherapeutic) compounds are called **antimicrobial** drugs because they kill or slow down the growth of some kinds of microorganisms. These compounds are often used as feed additives for livestock and poultry rations.

There are many different kinds of microorganisms (microbes) that live in the bodies of animals. Some of these microbes are beneficial to animals and some are harmful. When livestock are raised in confinement, there is a greater potential for the spread of harmful microbes among the animals because they are crowded more closely together; the use of antimicrobial drugs helps to keep these harmful microbes under control. The use of these drugs at a lower level in the feed to aid in the prevention of illness is referred to as **subtherapeutic**. The use of antibiotics at subtherapeutic levels in large cattle feedlots has been declining in recent years. Cattle feeders have become increasingly concerned about the development of resistant strains of bacteria when drugs are used at the subtherapeutic level.

The major difference between antibiotics and chemoantibacterial compounds is the way in which they are produced. Antibiotics are produced by living microorganisms. Chemoantibacterial compounds are made from chemicals. Sometimes an antibiotic and an antibacterial are combined into one compound, called a **chemobiotic**, to combat a problem that is not susceptible to either one individually. In this text, the term *antibiotic* is used generically to refer to all of these compounds.

There are hundreds of antibiotics. Some most commonly used in livestock production include chlortetracycline (Aureomycin), neomycin, oxytetracycline (Terramycin), penicillin, streptomycin, and tylosin. Polyether antibiotics are called ionophores and are usually used in the production of ruminant animals. Monensin (Rumensin) and lasalocid sodium (Bovatec) are two commonly used ionophore antibiotics used in beef production. Some chemoantibacterial compounds used in livestock feeding include furazolidone and sulfamethazine.

There is a variation in response to the use of antibiotics among different species and under various feeding and housing conditions. However, in general, the use of antibiotics in livestock nutrition results in an increase in rate of gain, improved feed efficiency, and general health improvement. In modern livestock facilities, where strict biosecurity measures are practiced, there may be little or no benefit gained from the feeding of antibiotics at a subtherapeutic level.

When a substance is referred to as **nutrient sparing**, it means that the substance allows animals to use available nutrients more effectively. Antibiotics act in several ways to accomplish a nutrient-sparing effect in livestock. Some antibiotics stimulate microbes that are present in the digestive tract to produce more nutrients than they would without the presence of the antibiotic. Some microbes in the digestive tract compete for essential nutrients without improving the performance of the animal; some antibiotics slow down this competition for essential nutrients. An animal absorbs nutrients through the intestinal wall in order to use them in metabolism; some antibiotics help in the development of a thin, healthy intestinal wall, which allows for easier absorption of nutrients.

The rate of metabolism in young, growing animals changes when they are fed antibiotics. When antibiotics are included at low levels in the ration, the daily feed intake is greater and the conversion of feed to meat becomes faster and more efficient as compared with animals that do not have antibiotics included in the ration.

Subclinical diseases are those that are present in the animal's body at levels too low to cause visible effects. Under many conditions of farm feeding, subclinical diseases are present in the animals being fed. When antibiotics are included in the ration, these subclinical diseases are continuously controlled. As a result, the animal is healthier and more vigorous. Because subclinical diseases are controlled, the rate of gain is more uniform among groups of animals that are being fed antibiotics.

Different types of antibiotics vary in the range of microorganisms that they control. Some control many different microorganisms and are thus referred to as broad-spectrum antibiotics; those that control only a few microorganisms are called narrow-spectrum antibiotics. Broad-spectrum antibiotics are preferred for use as feed additives. They generally give better results in terms of rate of gain, feed efficiency, and improved animal health. Narrow-spectrum antibiotics are more often used to control a specific disease problem that may be present in the group of animals being fed.

Hormones and Hormone-Like Compounds

Hormones are substances produced in the animal's body. Natural hormones are secreted into the body fluids, such as the bloodstream, by various glands in the body. The adrenal cortex, pancreas, pituitary, ovaries, and testes all secrete hormones. Hormones regulate many body functions, such as growth, metabolism, and the reproductive cycle. Hormone-like compounds are synthetic substances that act like hormones in the body. Hormones and hormone-like compounds are produced commercially to be used as feed additives, primarily in beef nutrition.

Androgens, progestogens, and estrogens are hormones produced by the sex glands of animals. These hormones increase the rate of protein synthesis and muscle development. They are used in rations to improve feed efficiency and increase the rate of growth. Beef cattle have shown the greatest response to their use either as feed additives or implants. The first synthetic hormone developed for use in animal feeding was stilbestrol (diethylstilbestrol [DES]). DES is a synthetic estrogen that was approved by the U.S. Food and Drug Administration (FDA) in 1954 as an additive for use in beef cattle finishing rations. However, in 1979, the FDA banned the use of DES as a feed additive due to research indicating a link between DES and cancer development.

The use of hormones and hormone-like compounds in livestock production continues to be controversial. Surveys have shown that a majority of consumers believe that hormone residues in meat are a health hazard. This perception has greatly affected the livestock industry.

The debate over the use of hormones has affected the international markets for meat produced in the United States. Since January 1, 1989, the European Union (EU) has banned the importation of any meat for human consumption that has been treated with anabolic agents. Meat produced with the use of hormones or hormone-like compounds cannot be sold in the EU. Individual meat producers who do not use these compounds in livestock production can still export their products to the EU under individual commercial agreements. The EU has established criteria to monitor these products. However, a problem still exists in monitoring this meat because it is not possible to differentiate between hormones that the animal produces naturally and those that have been implanted.

The EU ban is based on consumer concern for human health and safety. However, research by the FDA has shown that when these compounds are properly used, there are no adverse effects on human health. Researchers at the World Health Organization (WHO) have reached the same conclusion. The amount of these hormones that the human body produces naturally is far greater than any residues found in meat.

Some of the consumer concern is related to the possible carcinogenic effects of these compounds. Research shows that the hormones approved for use in the United States for livestock production would have to be consumed in extremely high dosages to act as carcinogens. When they are used in the proper manner at the approved levels, they pose no danger as carcinogens to humans.

The continuing controversy over the use of these compounds in meat production creates a greater need for livestock producers to exercise care in their use. Careless use provides ammunition for those who would ban these compounds in the United States. If such a ban were imposed, it would result in an estimated 10 percent or more drop in the amount of beef produced, reduced profits for beef producers, and higher beef prices for the consumer. The controversy over the use of these compounds has been marked with a great deal of sensationalism and misinformation. Beef producers face a major problem in providing accurate information to the consuming public regarding this issue.

Anthelmintics

Anthelmintics (dewormers) are compounds used to control various species of worms that may infest animals. They may be provided to the animal in either feed or water.

Some of the more common species of worms that can infest animals at various stages of production include large roundworms, nodular worms, and whipworms. The presence of worms in the animal's system reduces feed efficiency and rate of gain. The level of performance improvement gained from the use of anthelmintics depends on the level of worm infestation found in the animals being treated. Where good management practices (such as rotating pastures or keeping manure cleaned from pens) are followed, there will be less response from the use of anthelmintics. However, if there is an indication of worms in any of the animals in a group, it is safe to assume that all animals in the group have some level of worm infestation. This condition indicates the need to treat the entire group of animals for worm infestation.

Other Feed Additives

A number of substances may be used for specific purposes in livestock feeding. In some cases, these are designed to improve growth, rate of gain, or feed efficiency. In other cases, they are used for specific purposes.

Coccidiostats are sometimes added to cattle, sheep, swine, and poultry rations to prevent the disease coccidiosis. Coccidiosis is an intestinal parasitic infection that results in diarrhea and decreased performance. Coccidiostats work to prevent growth of the coccidia protozoa that cause coccidiosis.

The pH level (the acid-base balance) of the fluids in the digestive tract influences the proper digestion of feeds. Sodium bicarbonate and ground limestone are sometimes used in rations of ruminant animals to regulate the pH level and thus improve digestion.

Ruminant animals may bloat when they eat too much lush, green alfalfa, or too much grain. Foaming may occur in the rumen, or a slime layer may build up over the liquid in the rumen that prevents the gases in the rumen from escaping. This causes a noticeable swelling of the animal's midsection. A bloat preventive may be added to ruminant feeds when bloat is a problem. It may be used as a precautionary measure when putting animals on pasture in the spring or when starting them on feed in the feedlot. Bloat preventives work by breaking up the foam or the slime layer and allowing the gases to escape.

Probiotics are compounds such as yeasts and lactobacilli that change the bacterial population in the digestive tract to a more desirable type. In some cases the use of probiotics has improved animal performance, but the improvement has not been as great as that obtained by the use of antibiotics. Probiotics may be used in conjunction with antibiotics in the feeding program.

BEEF CATTLE

The use of antibiotics and hormones as feed additives is one of the most effective management tools available to beef cattle producers. The use of these products can significantly increase feed efficiency and rate of gain.

Antibiotics

Cattle being fed a high-energy ration will show about a 3 to 5 percent improvement in rate of gain and feed efficiency when a continuous low level (35 to 100 mg per head per day) of antibiotic is included in the ration. Daily gain and feed efficiency is even higher for growing cattle being fed low-energy rations.

The improved performance that results from the use of antibiotics generally results from the action of the antibiotics against harmful microorganisms. These microorganisms cause such feedlot disorders as foot rot, liver abscess, respiratory diseases, and shipping fever. The greatest response from the use of antibiotics usually occurs when cattle are under stress or just starting on feed for finishing in the feedlot. However, there is a continued response throughout the feeding period when antibiotics are fed continuously at low levels.

A high percentage of rations for cattle on feed contain either monensin sodium (Rumensin) or lasalocid sodium (Bovatec). Both are ionophore antibiotics. Monensin sodium generally improves feed efficiency and the rate of gain. Addition of monensin sodium to the ration will typically decrease feed intake. Lasalocid sodium is added to the ration to improve feed efficiency and also increase the rate of gain, but addition of lasalocid sodium to the ration does not affect feed intake.

These compounds affect the fermentation of feed in the rumen, decreasing the proportion of methane gas and increasing the proportion of propionic acid produced. As a result, the conversion of feed energy to growth is improved by a factor of about 5 to 10 percent. Feedlot bloat and acidosis are also reduced by these compounds.

Cattle on a high-energy (high-grain) ration show an initial reduction in feed intake of about 20 percent when monensin is first added to the diet. This reduction in feed intake lasts about 3 to 4 days. Feed intake then gradually rises to a level of about 10 percent less than the feed intake of cattle that are not being fed monensin. While cattle being fed monensin eat about 2 to 4 percent less than those not being fed monensin, the rate of gain is about the same. The efficiency of feed use is increased by the addition of monensin to the diet.

Monensin can also be used as a feed additive for cattle on pasture. Cattle on low-energy (pasture or high-roughage) diets do not respond in the same manner as those on high-energy diets. Feed intake is not reduced, but the cattle make better use of the energy in the feed. This reduces the daily energy maintenance requirements of the cattle and results in faster gains. Cattle on pasture with monensin in their diet gain about 16 percent faster than cattle that are not being fed monensin.

Lasalocid sodium (Bovatec) acts in a manner similar to monensin. Research shows that feed efficiency is improved by about 8 percent, while rate of gain is increased by about 5 percent. A major difference between lasalocid sodium and monensin is lasalocid is less toxic to horses and swine than is monensin.

In all cases, directions for use and claims on labels should be followed closely. Current approvals for feed additive use are subject to change and the user is cautioned to follow current regulations when using any feed additive.

Courtesy of Frank B. Flanders

Figure 7-1 An implant instrument is commonly used to insert growth hormone pellets in the ear of market beef animals.

Hormones and Hormone-Like Compounds

Many hormones or hormone-like compounds have been developed for use with beef cattle. Some are naturally occurring hormones and others are synthetic substances. In the following discussion, the generic term hormone is used to refer to both natural hormones and synthetic hormone-like compounds. While some of these compounds may be added to the feed, it is a common practice to use others as implants in the ear of the animal (Figure 7-1).

The practice of using hormones in beef production is recognized as one of the most effective management tools available to increase feed efficiency and improve the rate of gain of cattle. Research shows that rate of gain is improved an average of 10 percent, while feed efficiency is improved from about 6 to 8 percent through the use of hormones with feedlot cattle.

Hormone Feed Additive

Melengestrol acetate (MGA) is a synthetic hormone similar to progesterone. It suppresses estrus (prevents the heifer from coming into heat), which reduces the continual mounting seen when heifers are coming into heat in the feedlot. MGA also increases the rate of gain and improves feed efficiency in fattening heifers. Research shows that average daily gain is improved by about 5 to 11 percent and feed efficiency is improved about 5 percent when MGA is included in the diets of heifers in the feedlot. MGA gives no response in feed efficiency or rate of gain when fed to steers.

Hormone Implants

Implants improve both rate of gain and feed efficiency when they are properly implanted. Combining implants with feed additives improves performance more than using either alone.

Using growth hormone implants with nursing calves is one of the most economically justifiable practices in the beef industry. Implants have been shown to increase weaning weights of nursing calves in hundreds of research trials. Stocker and feedlot calves exhibit even greater responses than nursing calves. Implanting returns more revenue per dollar invested than any other management practice. However, only about one-third of cow-calf producers nationwide use growth-promoting implants. Unless calves are to be marketed as hormone-free, nursing calves intended for sale should be implanted prior to weaning.

Implants stimulate growth hormones in the animal's pituitary gland and change the hormone balance in the animal's body. Implants produce a slightly greater response in steers as compared to heifers; bulls show less response than either steers or heifers. Weight gain response also differs among different ages of cattle. Implants increase weight gains by about 8 to 15 percent in growing and finishing cattle and by about 8 percent in suckling calves. Feed efficiency in growing and finishing cattle is improved by about 6 to 10 percent. Rate of gain is higher in implanted suckling heifer calves compared with suckling steer calves.

Research studies have shown that using implants with large-frame steers increased the protein requirements of the ration. Rate of gain increased by about 30 percent and feed efficiency improved by about 19 percent when the ration contained 12.5 to 14 percent crude protein compared to a ration containing 9.5 to 11 percent crude protein.

Variation in response to hormone implants is greatest among cattle on pasture and high-roughage diets. Research indicates that cattle should be gaining about 1 pound per day before there is significant response to the use of implants. Some studies have shown that implanting cull cows on a high-concentrate ration will improve rate of gain and feed efficiency. The best response to hormone implants is obtained when cattle are on a high-concentrate finishing diet. Reimplanting before the effective period of the original implant expires will produce improved performance.

Because implants contain an active ingredient that is hormone-like, they can interfere with reproduction or cause complete sterility. Implants should not be used on animals that are to be kept for breeding purposes. The development of the testicles is so severely impaired when implants are used with young bull calves that they cannot be used for reproductive purposes. Research indicates that there is less effect on the sexual development of heifers, but it is still recommended that they not be implanted if they are to be kept for breeding purposes. However, if heifers are not identified as replacements until later in life, they can be implanted once if label directions are carefully followed. Heifers that have been implanted may show a slightly lower conception rate when they are first bred.

There have been reports of undesirable side effects from implants, such as "buller" steers, high tailheads, and udder development. Buller steers show an unusual amount of sexual activity in the feedlot, which reduces gain and increases labor costs. It is believed that these side effects are probably the result of improper techniques used in implanting.

The proportion of lean meat deposition in the carcass is higher when implants are used on feeder cattle. This may result in fewer carcasses grading choice. Therefore, cattle may need to be fed to slightly higher weights to achieve the same level of marbling reached by cattle that are not implanted.

Implanting Procedure

The length of time implants are effective appears to be related to proper technique in applying the implant. If any of the pellets are crushed during implantation, the effective period of the implant is decreased. Some of the undesirable side effects that have been observed with implants appear also to be traceable to improper implantation technique. Improper implanting may also result in some of the animals losing the implant.

Rushing to do the implanting can result in serious economic loss from improperly placed implants. It has been estimated that improper implanting procedures may cause losses of as much as 20 percent if compared with the loss rate when proper procedures are followed.

The following guidelines should be followed to secure the proper placement of the implant.

1. Restrain the animal securely in a headgate or squeeze chute. Use a halter to secure the animal's head to prevent up-and-down head movement.
2. Do not use an instrument with a dull needle as that will make it difficult to penetrate the skin. Make sure the needle is not bent and that it has neither burrs nor rough edges. These conditions often result in crushed pellets.

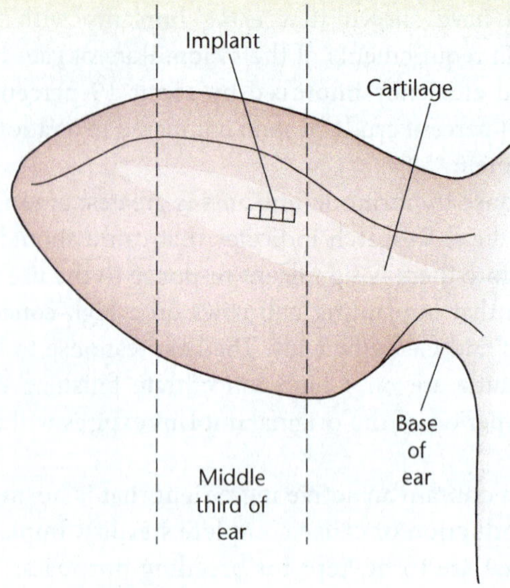

Figure 7-2 Proper location for implant.

3. Be sure the needle and the implantation site are both clean. Use a cotton swab or sponge soaked in alcohol to clean dirt and manure from the implantation site on the ear. Use a different cotton swab or sponge soaked in alcohol to clean the needle before injecting the animal. Do not use the same cotton swab or sponge used to clean the needle. Failure to clean the needle and the implantation site on the ear properly can result in infection. The infection may encapsulate the implant in connective tissue, rendering it ineffective.
4. Select the proper implanting site on the back surface of the ear. The recommendation is to place all implants in the middle one-third of the ear—that is, about 1.5 to 2 inches from the base of the ear (Figure 7-2).
5. With the implanting instrument pointed toward the head and parallel to the ear, lift the loose skin with the point of the needle. Push the needle in, being careful not to go into or through the ear cartilage or hit a vein. Placing the implant in the ear cartilage may cause encapsulation of the implant and result in slower-than-normal absorption of the implant material. Hitting a vein can cause excessive bleeding and possibly result in the loss of the implant.
6. Withdraw the needle slightly before starting the implant. Push the plunger or pull the trigger while slowly withdrawing the needle. Feel for the implant with the opposite hand as the needle is being withdrawn to make sure the pellets have been properly deposited. If the needle is withdrawn too rapidly, the pellets may be crushed or balled up in one pocket. The pellets should be deposited in a line between the skin and the ear. Crushing the pellets may result in too rapid absorption and undesirable side effects.

Using Combinations of Feed Additives and Hormones

Antibiotics are sometimes used in combinations in beef production. These combinations are claimed to promote growth and be effective against diseases. Use of antibiotics in combination must follow current regulations and label directions for mixing must be carefully followed.

The ionophore antibiotics (monensin sodium and lasalocid sodium) may be used in combination with hormone implants. Melengestrol acetate (MGA) can also be used in combination with hormone implants. The use of these combinations results in significant increases in rate of gain and feed efficiency. Rate of gain is increased by about 18 percent and feed efficiency is increased by about 12 percent when combinations are used.

Special-Purpose Additives

Many beef producers treat all newly purchased cattle for worms, as well as cattle that have just come off of pasture into the feedlot. Feed additives are also available for bloat prevention and for controlling flies. The use of a feed additive for fly control works by controlling growth of fly larvae in the manure and is therefore more effective when all cattle in the area consume the additive.

The materials discussed here are not intended to be inclusive of all the kinds of special-purpose additives that may be used in beef production. Other products are available for a variety of purposes. Label directions and current regulations must be followed when using any drug in livestock production.

DAIRY CATTLE

The use of feed additives is not as widespread among dairy farmers as it is with beef producers (Figure 7-3). However, there are a number of additives that may be used with the dairy herd that can increase profits. The returns from the use of these additives vary with the environment and the management practices of individual producers.

Antibiotics

Antibiotics may be used with replacement heifers that are under 16 weeks of age and are not growing properly. The goal of using of antibiotics with these young heifers is to improve their growth rate and reduce stress from the environment. Antibiotics may be used as a feed additive for calves to promote growth, improve feed efficiency, and treat diarrhea or for non-lactating dairy cows for growth

Figure 7-3 Feed additives are used more in beef cattle feed than dairy cattle rations.

promotion, feed efficiency, and treatment or prevention of diarrhea, liver abscesses, foot rot, respiratory diseases, anaplasmosis, and bloat. Antibiotics may also be used for lactating dairy cows for treatment or prevention of diarrhea, foot rot, and respiratory diseases and for improved milk production.

Monensin may be used to improve feed efficiency and rate of gain in replacement dairy heifers. Label directions and current regulations must be followed when using antibiotics in dairy cows.

Other Additives

Anhydrous ammonia (NH_3) may be used as a forage preservative for corn silage, low-quality forages, or wet hay. It improves the digestibility of the fiber in the forage and also provides a nonprotein nitrogen source for ruminants.

Buffers such as sodium bicarbonate (baking soda) and sodium sequecarbonate may be added to feed to help preserve a pH of 6.5 to 6.8 in the rumen. This pH level is desirable because it improves feed intake and feed digestibility. Buffers may also be used to advantage when cattle are heat stressed, off-feed, or when finely ground forages or grain are being fed.

Isoacids may be fed to lactating dairy cows to improve feed efficiency and the digestion of fiber in the rumen. The use of isoacids in the ration may result in from 2 to 6 pounds more milk per day. It may take as long as 4 weeks from the time feeding of this additive is started until milk production begins to increase.

Propionic acid may be used as a preservative for hay or high-moisture corn. It prevents mold and heating of the hay or corn. It is recommended for use when there is danger of spoilage when the hay or corn is too wet. When used at the recommended level, milk fat tests are not lowered.

Propylene glycol may be used to treat ketosis. It is recommended for use when symptoms of ketosis appear in urine tests or milk ketone tests. Ketosis is a nutritional disorder, especially in dairy cattle. Blood sugar drops to a low level.

SHEEP AND GOATS

There are fewer feed additives available for use with sheep and goats than cattle or swine. The size of the sheep and goat industries is smaller compared to other livestock industries; therefore, commercial companies that produce feed additives have a smaller financial incentive to conduct research on sheep and goat feed additives. Because feed additives do help to prevent disease problems, the Minor Use and Minor Species Animal Health Act of 2004 was signed into law. This law is designed to encourage development of new drugs for minor species such as sheep and goats. Local veterinarians, extension specialists, nutritionists, or other qualified professionals should be able to provide more information on approved feed additives for sheep and goats.

Antibiotics

Some producers include a broad-spectrum antibiotic in feeder lamb rations. Most feeder lambs will respond with improved feed efficiency and rate of gain when an antibiotic is used. Response is often better when the antibiotic is included in the ration during the first month of the feeding period or when high-energy rations are fed.

Some producers will also use a broad-spectrum antibiotic in the ration for pregnant ewes to prevent abortion per industry terminology. If abortions have been a problem in the past, antibiotics may be added to the ration for the last 6 weeks of gestation. If an outbreak of abortions begins to occur, many producers will immediately add a broad-spectrum antibiotic to the ration.

A coccidiostat may be used if a coccidiosis problem exists with feeder lambs. Lasalocid sodium may also be used to improve rate of gain and feed efficiency in lambs.

Special-Purpose Additives

Some producers will choose to use a feed additive to control gastrointestinal worms in sheep and goats. Some anthelmintic products can be used as an additive to the feed or salt. When using these products, the label directions must be closely followed and all withdrawal times observed.

Hormone Implants

Growth-promoting hormone implants are available for sheep. Although these implants do enhance the lambs' rate of gain and feed efficiency, most producers do not use implants in lambs. The implants can result in premature hardening of the lamb's bones, and at slaughter, the lamb could be classified as mutton and not lamb. Most producers find the advantage in rate of gain and feed efficiency is not sufficient to compensate for the potential reduced value of the lamb if it is classified as mutton at slaughter.

SWINE

Feed additives are widely used in swine rations. It is important for the swine producer to determine the specific purpose for using any particular feed additive in swine feeding. The economic returns expected from the use of the additive should be greater than the cost of the additive.

Young swine give a greater economic return from the use of antibiotics than any other age group. The use of antibiotics during the breeding, farrowing, and weaning phases of production gives greater returns than at other stages of production. Other factors that affect the economic returns from the use of feed additives include the level of management, the environment in which the hogs are raised, and the history of disease problems on the individual farm.

Antibiotics

Antibiotics may be used in swine rations for improving rate of gain and feed. Antibiotics should either not be used at all or used only at low levels when growing gilts unless disease problems are clearly present.

The use of high levels of antibiotics in the ration for 10 days before or 10 days after breeding may increase conception rate and litter size. There are reasons other than bacterial infections that may cause low conception rates or small litters in sows. A veterinarian should be consulted if serious problems related to conception rates or litter sizes exist over a period of time.

Antibiotics are generally not needed during the gestation period unless there is a high level of disease or environmental stress present. Sows and gilts should be treated with an anthelmintic 3 to 4 weeks before farrowing.

The survival rate and the performance of pigs are improved when antibiotics are used in farrowing rations for a period of 7 days before to 14 days after farrowing. Because this is normally a high-stress period, the use of antibiotics to prevent infections during this period can produce significant economic returns.

The use of antibiotics may help prevent lactation problems when sows are nursing their litters. However, there are reasons other than bacterial infections that may cause lactation problems in sows. If problems with poor lactation persist, a veterinarian should be consulted.

Stress and disease risks are high with pigs that have been weaned early or orphaned. Antibiotics should be included in prestarter rations. Starter rations used for pigs weighing up to 30 pounds should also include antibiotics because stress and disease risks are high during this period. Scours is often a problem during the weaning period. It is recommended that a veterinarian should determine the specific bacteria that are causing a scours problem so the proper antibiotic can be selected. This will reduce the cost of treating for scours.

During the growing period when pigs weigh from 25 to 60 pounds , low levels of antibiotics might be used unless a disease problem exists on the farm. Higher levels of antibiotics should be used for purchased feeder pigs of this weight because of the increased stress of shipping.

During the 60- to 125-pound stage of development, the use of antibiotics will result in more economical gains. Higher levels should be used if a disease problem exists on the farm.

The lowest level of response to the use of antibiotics in swine feeding is during the finishing stage of 125 pounds to marketing. If an antibiotic is used, preference should be given to one that does not require a **withdrawal period** before slaughter.

Anthelmintics

Swine should be treated for worms any time there appears to be an infestation. Pigs housed outdoors are more likely to have worm problems than pigs housed in confinement. A veterinarian should examine a sample of the manure to determine the exact nature of the worm infestation. The recommendation of the veterinarian regarding the anthelmintic to use should be followed.

Other Additives

Various compounds may be added to swine diets to promote growth and improve feed efficiency. These additives must be in compliance with the label and withdrawn from the feed before slaughter. Additional feed additives are available for fly control in pigs. Some additives are used in swine production for medicinal purposes. All feed additives in swine diets must be in compliance with FDA regulations. More information can be obtained from the local veterinarian, extension specialist, nutritionist, or other qualified professional.

POULTRY

Feed additives in poultry are used primarily to improve feed efficiency and growth in broiler chick and market turkey rations. Some feed additives are also used to improve egg production in laying flocks or for the prevention or control of diseases.

Good management practices need to be followed when raising poultry in confinement to reduce disease problems and lessen the dependence on antibiotics for disease control. Access to poultry houses should be restricted and strict cleanup procedures should be followed by all persons who enter the houses. All poultry houses should be thoroughly cleaned and disinfected after the birds are moved out and before bringing new birds into the houses. Disease-carrying predators should

Connection: Hormones in Chickens?

Despite the popularity of chicken, some consumers still express concerns about eating chicken because they fear growth hormones have been fed or administered to the chickens (Figure 7-4).

©iStockphoto/fotokon

Figure 7-4 Hormone use in poultry production is illegal in the United States.

Like all living things, chickens contain natural hormones; however, no added hormones are given to the chickens. This is due to three important reasons:

1. Genetic selection has created chickens that grow without the need for added hormones.
2. Hormones would have to be injected into the animal multiple times, which is costly and not feasible.
3. Hormones are illegal to use in poultry production in the United States.

be controlled. Avoid overcrowding the birds and make sure the litter is kept dry. Proper ventilation and temperature control in poultry houses also helps to reduce disease problems. If an outbreak of disease does occur, it should be treated promptly and with the proper drugs.

HORSES

Antibiotics are generally not used for low-level feeding over a period of time in horse rations (Figure 7-5). Feed additives may be used for fly control and for gastrointestinal worm control. As with other species of animals, it is important to consult the local veterinarian, extension specialist, nutritionist, or other qualified professional about the use of feed additives.

Courtesy of Frank B. Flanders

Figure 7-5 Antibiotics are not used on a regular basis in horse feeds.

REGULATION OF THE USE OF ADDITIVES

The use of feed additives and hormone implants is strictly regulated in the United States by the FDA. The FDA publishes information concerning approved antibiotics and other animal drugs in the United States. Detailed information on feed additive use and regulations may be found in the ***Feed Additive Compendium***, published by the Miller Publishing Company. The compendium is updated during the year on a regular basis. In Canada, the Animal Health and Production Branch of the Canada Department of Agriculture is responsible for animal drug regulation. Detailed information is available in the compendium of Medicating Ingredient Brochures.

Government regulations regarding antibiotics and other animal drugs change from time to time as new information on effects and dangers becomes available. Many additives may be used only within certain specified levels and for certain species or types of animals. Many additives must also be withdrawn from use in the feed within specified times of marketing the animal. Local veterinarians, extension specialists, nutritionists, or other qualified professionals should be able to provide current information on the use of feed additives for livestock.

Feed Label Requirements

Any feed that contains any level of one or more drugs is defined as a medicated feed. Medicated feeds may be a complete feed mix or may be a premix that is mixed with other feeds to make a complete feed. Premixes contain higher levels of the drug ingredients than do complete mixed medicated feeds.

The FDA requires all manufacturers of medicated feeds to provide information on the label relating to the use of the feed. The word "medicated" must appear under the name of the feed. The specific purpose of the drug or drugs included in the feed must be stated on the label. The name and amounts of all active drug ingredients must be listed. The required withdrawal period prior to slaughter of the animals must be given along with the cautions against misuse and the directions for the use of the feed.

Medicated feeds that are custom mixed at a local mill must meet the same label requirements as commercial medicated feeds. All labels on medicated feeds are assigned a lot or control number so the batch can be identified at a later date if necessary.

MIXING AND RESIDUE AVOIDANCE

The proper mixing of medicated feeds is important for their safe use. Be sure the drug ingredients are added in the correct proportions and are uniformly mixed in the batch. Failure to do so can have undesirable results; for example, some animals may get too much of the drug and others too little. This reduces the effectiveness of the medicated feed and may result in higher production costs or undesirable side effects.

Drug residues in livestock and livestock products can cause financial losses for farmers. Carcasses of animals may be condemned at slaughter if illegal drug residues are found. Public concern over drug residues in meat and livestock products may result in additional regulation of or restriction on the use of feed additives, which could result in higher production costs for farmers.

The U.S. Department of Agriculture conducts targeted and routine tissue testing in all federally inspected slaughter plants. If evidence of unapproved drugs is found in the tissues of a producer's animals, the producer is notified and the animals are held until they are free of the illegal substance. It usually takes 2 weeks to 30 days to get test results back. This delay causes an economic loss for the farmer. Additional animals cannot be marketed until the tests show them to be free of the unapproved drug.

Both horizontal- and vertical-type mixers may be used on the farm for mixing medicated feeds. Regardless of which type of mixer is used, care must be taken to ensure a complete mix of the drug ingredients in the complete feed. Proper cleaning of the mixing equipment must be done after each batch of medicated feed is mixed to avoid residue problems in later batches of feed.

A regular procedure should be used when mixing medicated feeds to avoid problems with drug carryover. The following recommendations will help avoid problems.

- Know the labeled uses, mixing instructions, and withdrawal times for all medications used in the feed.
- Clean the mixer before use to avoid carryover of drugs from one batch to the next.
- Do not try to mix more feed in the batch than the mixer can hold.
- Premix the drugs in a large enough quantity to allow accurate weighing and mixing in the complete feed.
- Follow the manufacturer's instructions for premixing and adding medications.
- Use an established order for mixing the medication with the feed.
- Mix all feeds containing medication first, clean the equipment, and then mix the unmedicated feeds. Mix any withdrawal feeds before mixing other nonmedicated feeds.
- Clean the mixer.
- Keep feed additives stored in a clean, orderly area in their original packages. Control rodents and insects in the storage area. Do not store other chemicals in this area.
- Read all labels and observe withdrawal times.
- Keep labeled samples of purchased premixes, supplements, and mixed medicated feeds for at least 3 months after the livestock to which they were fed have been marketed.
- Make sure all augers, holding bins, feed wagons, and feeders are thoroughly cleaned before withdrawal feeds are used. Do not mix withdrawal feeds with medicated feeds in bins or feeders.

KEEPING RECORDS

A good set of up-to-date records on the use of medicated feeds can help the farmer avoid problems with feed contamination and drug residues in livestock and livestock products. The following are recommended records to be kept.

- The date the batch was mixed.
- The mixing order and the amount of medication added.
- The mixing time for the batch.
- The location where the feed is stored.
- The number, age, and weight of the animals fed from the batch and the amount given per head.
- The medication that was used, the amount, and the concentration in the batch of feed.
- The date of cleaning mixers, bins, conveyors, and feeders.

In addition to keeping records, the farmer should have a long-term plan for the use of feed additives. Because microorganisms develop resistance to antibiotics, the farmer should avoid feeding some types of antibiotics so that they might be used in case of an outbreak of disease. Additionally, it is recommended that the farmer periodically change the type of antibiotic being used to reduce the chances of microorganisms building immunity to the additive. If good records are kept and a long-term plan for the use of feed additives is followed, it is easier for a veterinarian to properly treat animals if a disease outbreak should occur.

HEALTH CONCERNS

There has been a growing concern in recent years that the continued use of antibiotics in animal agriculture may have an adverse effect on human as well as animal health. This concern centers around the use of antibiotics at subtherapeutic levels in livestock feeding. It is feared that this continuous low-level use of antibiotics might result in the development of resistant strains of microorganisms that could not be effectively treated with antibiotics. Bacterial resistance to drugs has been observed almost from the time antibiotics were first used in animal feeding.

The impact of a ban on the use of antibiotics in animal feeding would vary from species to species, but the overall effect would be to raise the cost of animal products to the consumer. There would be less meat and livestock products produced and the costs of production would be higher. More feed would be required per animal raised and the rate of gain of animals on feed would decrease. There would be an increase in the death loss among animals being raised.

The Food and Drug Administration has also expressed concern about the possible carcinogenic (cancer-causing) effects of some feed additives. The FDA prohibits adding to food any substance that is a known carcinogen. There is ongoing research into the possible carcinogenic effects of feed additives. If these effects are shown to be present in any additive, it will be withdrawn from use in animal feeding.

SUMMARY

Feed additives are materials that are not considered to be nutrients and are used in small amounts in the ration to improve feed efficiency, promote faster gains, improve animal health, or increase production of animal products. Hormone implants are pelleted synthetic or natural hormones or hormone-like compounds that are used to improve rate and efficiency of gain.

Antibiotics and chemoantibacterial compounds are used to kill or slow the growth of some kinds of microorganisms. The use of these compounds at low levels in the ration over a period of time is a common practice in livestock feeding.

A number of other feed additives may be used in livestock production. Anthelmintics (dewormers) are often used to control various species of worms that may infest animals. Other specialized feed additives are used to prevent or treat various diseases that may affect animals.

Beef producers are major users of feed additives and hormone implants. Feed efficiency and rate of gain are significantly increased by the use of these products. A high percentage of beef cattle rations use either monensin sodium (Rumensin) or lasalocid sodium (Bovatec) alone or in combination with other feed additives.

Hormone implants improve both rate of gain and feed efficiency when they are properly used. The recommendation is to place all implants in the middle one-third of the ear, about 1.5 to 2 inches from the base of the ear. Implants should not be used with breeding stock.

The use of feed additives is not as common with dairy animals as it is with beef cattle. Most of the use of feed additives with dairy cattle is with younger animals.

There are few feed additives approved for use with sheep and goats. The Minor Use and Minor Species Animal Health Act of 2004 may lead to more feed additives being available for sheep and goats.

Feed additives are widely used in swine rations. The greatest economic returns from the use of antibiotics occur when they are used for younger animals.

Feed additives in poultry are used especially with broilers and market turkeys. These additives can improve feed efficiency and growth in broiler chick and market turkey rations, improve egg production in laying flocks, or aid in the prevention or control of diseases. Hormones are not used in poultry.

Feed additives are generally not used with horses. Some antibiotics are approved for use with younger horses to promote growth, improve feed efficiency, and reduce stress.

The use of feed additives and hormone implants is regulated in the United States by the Food and Drug Administration. Their use in Canada is regulated by the Animal Health and Production Branch of the Canada Department of Agriculture. Regulations on the use of additives and hormones change from time to time; therefore, care should be taken to follow current regulations on use of these products.

Any feed that contains any level of one or more drugs is considered a medicated feed. There are strict requirements relating to the labeling of medicated feeds.

It is important that all medicated feeds be properly mixed. Failure to do so can have a negative economic impact on livestock producers. A regular procedure should be followed when mixing medicated feeds and careful records should be kept.

There is concern about possible health hazards resulting from the use of feed additives in livestock production. The concern focuses on the possible development of resistant strains of microorganisms that cannot be treated with antibiotics. There is no persuasive evidence of animal or human health problems resulting from the use of antibiotics in animal feeding.

Student Learning Activities

1. Present a short oral report to the class about new developments with feed additives and/or hormone implants as reported in farm magazines and newspapers.
2. Prepare a bulletin board display of medicated feed tags.
3. Take a field trip to a local farm that has facilities for mixing complete feeds and observe the proper technique for mixing feed additives in the ration.
4. Give a demonstration on the proper technique for hormone implants in beef cattle.
5. Select and use the appropriate feed additives and hormone implants as necessary when planning and conducting a supervised experience program in animal production.

Discussion Questions

1. Why are feed additives used in animal rations?
2. Why are hormone implants used in livestock production?
3. What is the major difference between antibiotics and chemoantibacterial compounds?
4. List four ways in which antibiotics produce the results they do when used in livestock production.

5. What is the difference between broad-spectrum and narrow-spectrum antibiotics? Which type is preferred for use as a feed additive?
6. What is the function of anthelmintics?
7. When feeding beef cattle a high-energy ration, what is the range of percent improvement in rate of gain and feed efficiency that can be expected from the continuous feeding of a low level of antibiotic?
8. What results can be expected from feeding monensin sodium to feedlot cattle?
9. What results can be expected from feeding lasalocid sodium to feedlot cattle?
10. What is the effect of using melengestrol acetate (MGA) as a feed additive in beef heifer rations?
11. List and describe the use of hormone implants in beef cattle.
12. Describe the proper procedures for implanting hormone pellets in beef cattle.
13. Briefly explain the use of feed additives with dairy cattle.
14. Why are there few feed additives available for use with sheep and goats?
15. Why is a broad-spectrum antibiotic sometimes used in feeder lamb rations?
16. Which age group of swine gives the greatest economic return from the use of antibiotics in the ration?
17. For what two kinds of poultry are feed additives most commonly used?
18. What agency regulates the use of feed additives and hormone implants in the United States? In Canada?
19. What publications are good sources of current regulations regarding the use of feed additives and hormone implants in the United States? In Canada?
20. List six things that must be on the label of all medicated feeds.
21. Why is the proper mixing of medicated feeds important to livestock producers?
22. How might drug residues in livestock or livestock products cause economic losses for farmers?
23. List the recommendations for the proper mixing of medicated feeds.
24. List the records that should be kept when using medicated feeds.
25. What is the major concern relating to possible health hazards from the use of feed additives in livestock production?
26. What would be some of the effects of a total ban on the use of antibiotics in livestock production?

Review Questions

True/False

1. Probiotics are sometimes added to cattle, sheep, swine, and poultry rations to prevent the disease coccidiosis.
2. The rate of metabolism changes in young animals when they are fed antibiotics.

Multiple Choice

3. The use of drugs at a lower level in feeds than would be used to treat a sick animal is referred to as:

a. chemobiotic	c. antibiotic
b. subtherapeutic	d. none of the above

Concentrates

Livestock feeds that contain less than 18 percent crude fiber when dry are called concentrates. There are two classes of concentrates: protein supplements and energy feeds.

Protein supplements are livestock feeds that contain 20 percent or more protein. They are divided into two groups based on their source. Those that come from animals or animal by-products are called **animal proteins**. Those that come from plants are called **vegetable proteins**.

Some common animal proteins are tankage, meat scraps, meat and bonemeal, fish meal, dried skimmed milk, dried whole milk, blood meal, and feather meal. (**Tankage** is animal tissues and bones from animal slaughterhouses and rendering plants that are cooked, dried, and ground.)

Most animal proteins contain more than 47 percent crude protein. The protein is more variable in quality than protein from vegetable sources. Animal proteins contain a more balanced amount of the essential amino acids than do plant proteins. Thus, animal proteins are sometimes used for balancing rations for swine and poultry.

Some common vegetable proteins are soybean meal, cottonseed meal, linseed oil meal, peanut oil meal, corn gluten feed, brewer's dried grains, and dried distiller's grains. Most vegetable proteins contain less than 47 percent crude protein. Soybean meal is used more than any of the other protein supplements for livestock rations. Soybean meal can supply the necessary amino acids to balance a swine or poultry ration with cereal grains. Vegetable proteins can be used as the only protein supplement for ruminants. Non-ruminants, however, may need some animal protein in their ration. Animal proteins give the amino acid balance needed in non-ruminant rations when plant protein sources other than soybean oil meal are used.

Commercial protein supplements are made by commercial feed companies. They are mixes of animal and vegetable protein feeds. Each commercial supplement is usually made for one class of animal. Feed companies often mix minerals, vitamins, and antibiotics in their protein supplements. The feed tag on the supplement tells the class of animal for which it is designed. The tag also gives feeding directions and lists the contents of the feed. Feeding directions must always be followed carefully. If feed supplements contain antibiotics, they usually must be taken away from the animal for a period of time (withdrawal period) before the animal is sent to market. This practice is required by law. The antibiotic must not be present in the meat when humans eat it.

Livestock feeds with less than 20 percent crude protein are called **energy feeds**. Most of the grains are energy feeds. Some common energy feeds are corn, sorghum grain, oats, barley, rye, wheat, ground ear corn, wheat bran, wheat middlings, dried citrus pulp, dried beet pulp, and dried whey. Corn is the most widely used energy feed. Sorghum grain, oats, and barley are the other commonly used energy feeds.

RATION CHARACTERISTICS

An animal must receive the proper amounts of nutrients in the right proportion to efficiently produce meat, milk, eggs, wool, work, etc. A ration is said to be balanced when it provides the nutrient needs of the animal in the proper proportions. Strictly speaking, a **ration** is the amount of feed given to an animal to meet its needs during a 24-hour period; however, in common practice, the term may refer to feed provided without reference to a time period. A **balanced ration** is one that has all the nutrients the animal needs in the right proportions and amounts. The term **diet** refers to the ration without reference to a specific time period.

Key Terms *continued*

- total digestible nutrients
- neutral detergent fiber (NDF)
- acid detergent fiber (ADF)
- relative feed value (RFV)
- one hundred percent (100%) dry-matter basis
- as-fed (air dry) basis
- free choice
- limit-fed
- self-fed
- feeding standards

TABLE 8-1 Poisonous Plants[1]

Common Name(s)	Animals Affected
Arrow grass	Cattle; sheep
Aster	Sheep
Azalea, western	Sheep
Baccharis	Cattle
Baccharis, eastern; silverling; groundsel tree; consumption weed	Cattle; sheep; poultry
Bitterweed; sneezeweed	Sheep; cattle; horses
Bracken	Horses; cattle, sometimes sheep
Buttercup	All livestock; most commonly cattle
Cherry, wild	Sheep; cattle; horses
Chokecherry	Sheep; cattle
Cocklebur	All livestock; especially hogs; chickens, if seeds ground in feed
Copperweed	Cattle; sheep
Death camas; black snakeroot; crow poison; pink death camas; poison sage; swamp grass; alkali grass; poison onion	Sheep; cattle; horses
Drymary	Cattle
Dutchman's breeches	Cattle
Goldenrod	Cattle; sheep; horses
Halogeton	Sheep; occasionally cattle
Hemp; marijuana	Cattle; horses
Henbane	Cattle; sheep; horses
Horsebrush; spring rabbit brush; coal-oil brush	Sheep
Horsetail	Horses
Indian hemp; dogbane; Indian physic	Cattle; horses; sheep
Japanese yew	All livestock
Jimmyweed, rayless goldenrod	All livestock
Jimpson weed; thornapple	All livestock
Larkspur; staggerweed (rocket, azure, tall, dwarf)	Cattle
Laurels (Black sheep and Mountain)	Sheep; goats; cattle; other animals to lesser degree
Locoweed	Cattle; horses; sheep; goats
Lupine; bluebonnet; wild bean; blue pea	Sheep; goats; cattle; hogs; horses
Milkweed (several species)	Cattle; sheep; goats; horses
Nightshade (Black; Deadly), Other species: Horsenettle; bullnettle	All livestock
Oaks	Cattle; sheep; goats; horses; occasionally hogs by the acorns
Oleander	All livestock
Paperflower, greenstem	Sheep
Peganum	Sheep; cattle
Poison bean	Cattle; sheep; goats

(continues)

TABLE 8-1 Poisonous Plants (*continued*)

Common Name(s)	Animals Affected
Poisonvetch	Cattle; sheep
Ragwort; groundsel	Cattle; sheep; horses
Rubberweed (Bitter and Colorado)	All livestock; especially sheep
St. John's wort; goatweed	Animals with white skin and hair
Snakeroot, white	All livestock
Snakeweed	Cattle; sheep; goats
Spring parsley; wild carrot	Cattle; sheep
Tarweed	Horses; cattle; hogs
Timber milkvetch	Cattle; sheep; goats; horses
Water hemlock	All livestock

[1]Many other plants may be poisonous to livestock under certain conditions. This list is not intended to be all-inclusive but only presents some of the more common poisonous plants that affect livestock.

A ration must be **palatable**, or taste good, in order for the animal to eat it. Moldy feed is often not palatable. Insect and weather damage also lower the palatability of feed. Feed is of no value if the animal will not eat it.

Feed accounts for approximately 75 percent of the total cost of raising livestock. To feed livestock profitably it is necessary to develop rations that are as economical as possible. The ration must be palatable and meet the nutritional requirements of the animals. Homegrown feeds are used as much as possible because they are generally less expensive than purchased feeds. Commercial feeds are used when homegrown feeds are not available and also to supply nutrients not provided by homegrown feeds.

Feeds used in rations must not be harmful to the animal's health or lower the quality of the product. Poisonous plants should not be included in diets for livestock as these plants may make animals sick. Poisonous plants sometimes grow in hay and pasture fields (Table 8-1). Eradicate these plants before harvesting the hay or allowing animals to graze the pasture. Usually, animals will not eat poisonous plants, but if they are in the hay the animals may not sort them out. If the pasture is sparse, animals may eat poisonous plants that are growing there.

It is necessary to balance the intake of roughage and concentrates for the particular species and age of livestock being fed. Ruminants can use more roughage in their diets than non-ruminants. Also, younger animals cannot use as much roughage in their diets as can more mature animals. The purpose for which the animal is being fed must also be considered when including roughage in the diet. For example, fattening animals generally should be fed less roughage than breeding animals.

Micronutrients and feed additives are used in small quantities in the diet. Care must be taken to thoroughly mix these materials to ensure uniform distribution in the feed. Failure to do so may result in one animal getting too much of the micronutrient or additive while another animal may get too little. Excessive amounts of some additives may be harmful to an animal. Feed only the recommended amounts of these materials and make sure they are well mixed with the rest of the feed ingredients.

The functions of a ration must be considered when determining the nutrient requirements of livestock. These functions include maintenance, growth, fattening, production, reproduction, and work.

RATION FUNCTIONS

Maintenance

The primary use of the nutrients in a ration is for maintaining the life of the animal. The animal must have energy for the functioning of the heart, for breathing, and for other vital body processes. These activities make up what is called the **basal metabolism** of the animal. Maintenance also includes the use of energy supplied by the ration to keep the animal's body temperature normal. Protein in the body tissues breaks down. Protein from the ration is used to repair these body tissues. Minerals and vitamins are continually being lost from the body and are replaced by those in the ration. Certain fatty acids are needed for good health, and must be supplied by the animal's ration. Water is required for all bodily activities.

If the animal is not being fed enough feed, it may need to use its entire ration for maintenance. Thus, it will have none left for other activities, such as growth. Normally, about one-half of the ration that is fed to an animal is used for maintenance. An animal on full feed will use about one-third of its ration for maintenance. **Full feed** means to give an animal all it wants to eat.

Growth

Nutrients in the ration are used for growth only after the maintenance requirements of the animal are met. Animals grow by increases in the size of muscles, bones, organs, and connective tissues. If they do not grow properly, they will not be productive when they are mature. Animals grow fastest when they are young and the growth rate slows down as they get older. The larger species of animals usually mature slower than the smaller animals, but the growth rate of larger animals is faster than that of smaller animals.

Finishing

Feed nutrients that are not used for maintenance or growth may be used for fattening. Fat is stored in the tissues of the body. Fat stored within the muscles is called **marbling**. Marbling helps make meat juicy and good tasting. The consumer does not want too much fat, however. The object of fattening is to obtain the right amount of fat in the muscle without getting too much fat. Feeds that are high in carbohydrates and fats are used for fattening. They are less expensive than protein feeds.

Production

Animals produce many products that require specific nutrients. Cows, swine, horses, sheep, and goats produce milk to feed their young, while dairy goats and cows also produce milk for human use. Chickens produce eggs and sheep and goats produce wool and mohair.

Reproduction

Proper nutrition is required for reproduction. An animal may become sterile if it does not get an adequate level of nutrients. A sterile animal cannot be bred. Nutrition is extremely important for pregnant animals. Most of the growth of the **fetus**, the unborn animal when it is still in the mother's womb, takes place during the last third of the pregnancy. Animals need additional amounts of nutrients during pregnancy.

Connection: Fat Equals Flavor

Flavorful meat is a valuable source of nutrients, but many people shy away from eating fatty meats because of misconceptions about fat content. Most consumers want a low-fat meat, but are disappointed with the lean meat taste and lack of tenderness. There are several different types of fat. Meat with marbling (Figure 8-1), or intramuscular fat, grades higher and is generally better. Marbling, especially in beef, adds flavor and increases the perception of the juiciness of meat. This causes more salivation and makes the meat seem tenderer. Marbling does not transfer heat as well as protein, so the more marbling there is in a cut of meat, the less likely the meat is to get overcooked.

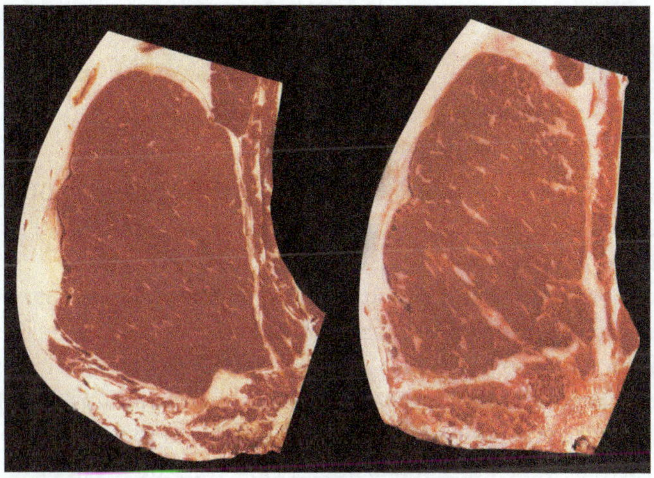

Courtesy of Frank B. Flanders

Figure 8-1 The steak on the left has slight marbling, making it less flavorful and easier to overcook than the steak on the right, which has moderately abundant marbling.

Work

Livestock may be used to perform work. For example, horses do work when they are ridden. The energy needed for work comes from carbohydrates, fats, and extra protein in the ration. The other needs of the body are met before nutrients are available for work. The animal may use fat stored in the body for work if the ration does not provide enough nutrients. Animals sweat more when they work. This creates a need for extra salt to make up for that lost by sweating.

BALANCING RATIONS

General Principles

The livestock ration must meet the nutritional needs of the animal. The nutrient allowance figured in the balanced ration should not be more than 3 percent below the animal's requirement, which should be met as closely as possible.

An animal must have a certain amount of dry matter in its ration in order to satisfy hunger. Its digestive system will not function properly if it does not receive enough dry matter. There is also an upper limit on the total amount of dry matter that an animal can eat. This varies with the kind of animal being fed and its size.

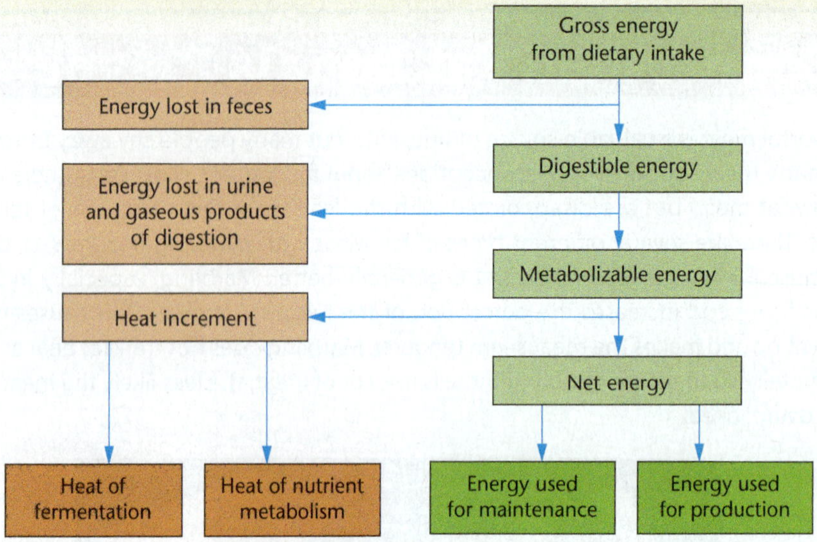

Figure 8-2 Utilization of dietary energy by animals.

The total dry matter in the ration of a full-fed animal should not be more than 3 percent above its need. Total dry matter for animals not on full feed can be considerably above or below their listed needs.

The amount of protein in the ration may be measured by the total protein (TP) need of the animal. Digestible protein (DP) may also be used as the measure to balance the ration. The essential amino acids must be included when balancing rations for non-ruminants. Some feeds being used may be below the average protein content listed in feed composition tables. It is acceptable to allow 5 to 10 percent more protein in the ration than the animal needs. However, too much protein above the animal's needs will raise the cost of the ration.

Four methods are commonly used to measure the energy provided by a ration. Energy may be calculated as digestible energy (DE), total digestible nutrients (TDN), metabolizable energy (ME), and net energy (NE) as illustrated by Figure 8-2. The **gross energy** of a feed is measured in a laboratory with a device called a **bomb calorimeter**. The feed is completely burned (oxidized) in the bomb calorimeter, which contains 25 to 30 atmospheres of oxygen. The gross energy is the total amount of heat released by burning in the bomb calorimeter. **Digestible energy** is the gross energy of a feed minus the energy remaining in the feces of the animal after the feed is digested. **Metabolizable energy**, for ruminants, is the gross energy in the feed eaten minus the energy found in the feces, the energy in the gaseous products of digestion, and the energy in the urine. The energy in the gaseous products of digestion is not considered when determining metabolizable energy for birds and simple-stomached animals.

Net energy is the metabolizable energy minus the heat increment. It is energy used either for maintenance or production, or both. Net energy for maintenance only is written NE_m; for production only it is written NE_p; and for maintenance plus production it is written NE_{m+p}. Net energy for maintenance is the amount of energy used to keep the animal in energy equilibrium; that is, there is no net gain or loss of energy in the animal's body tissues. The net energy for production is that amount of energy needed by the animal above the amount used for maintenance. The net energy for production is used for work, tissue growth, fat production, fetus growth, or milk, egg, or wool production. The different kinds of net energy are indicated as follows: for work, NE_{work}; for egg production, NE_{egg}; for gain, NE_{gain}; for fetus growth, NE_{preg}; for wool production, NE_{wool}; and so on.

The **total digestible nutrients (TDN)** in a feed are the total of the digestible protein, digestible nitrogen-free extract, digestible crude fiber, and 2.25 times the digestible fat. It gives a measure of the total energy value of a feed when it is fed to an animal. The TDN value of a feed varies with the class of animal to which it is fed. The energy in the ration should not be more than about 5 percent greater than the animal's needs. Animals are limited in the total amount of energy they can use.

Calcium and phosphorus are two important minerals in balancing rations. The ratio of calcium to phosphorus should be between 1:1 and 2:1 for most animals. For sheep, the calcium to phosphorus ratio should be 2:1 or higher. In fact, ruminants can tolerate calcium to phosphorus ratios as high as 7:1. This ratio is more important than the total amount being fed. The total calcium and phosphorus available in the ration is often more than the animal needs when the other requirements are met. The other mineral needs of the animal are usually not considered when balancing rations. Trace-mineralized salt will usually meet these needs.

Vitamin A is taken into account when balancing rations. The other vitamins needed are added to the ration without calculating the vitamin content of the feed. The amount needed to meet the minimum daily requirement for the animal is added with a vitamin supplement. The amount of vitamin A in feed will often be more than the animal needs, but this is not harmful. Vitamin deficiencies may occur in cattle and sheep during pregnancy if low-quality legume hay is fed. A vitamin supplement should always be added to a pregnancy ration.

Some feeds are cheaper sources of nutrients than other feeds. Energy feeds should be compared on the basis of price per pound of energy (TDN, DE, ME, or NE). Protein feeds should be compared in terms of the price per pound of total protein or digestible protein. The least expensive sources of nutrients should be used as much as possible.

Sampling and Analyzing Feeds

A person must know the nutrient composition of the feeds used to properly balance a ration. Average nutrient composition values can be found in nutrient composition tables. However, some feed, especially forages, can vary considerably from these average values. To get the maximum benefit from the ration, it is wise to have the feeds used analyzed for nutrient content.

Milk cows, beef cows that are nursing calves, and young growing animals require high nutrient levels. Forages fed to these animals should be analyzed for nutrient content. Haylage, silage, and high-moisture grains may vary greatly in moisture and nutrient content. Better rations can be developed if these feeds are also analyzed for moisture and nutrient content.

Growing conditions, such as excessive rainfall or drought, can affect the nutrient content of feeds, especially forages. When unusual growing conditions occur, it is a good idea to have these feeds analyzed. Other indications that a feed should be analyzed include musty odor, unusual amounts of foreign material or pests present in the feed, and a high level of leaf shattering on forages.

The most important tests to be done on feed grains are moisture, protein, and energy content. Forages should be tested for moisture, protein, acid detergent fiber, and neutral detergent fiber. It seldom pays to test for major or trace mineral content. Under some growing conditions, it is wise to check forages for nitrate and prussic acid content. Excessive amounts of these substances can be harmful to animals.

Take representative samples of the feed to be analyzed. Take random samples of hay from at least 20 bales. Insert the sampling tube into the center of the bale.

About 15 samples from silage and 5 from grain will usually give enough to be representative of the entire lot. Samples of silage or total mixed rations should be taken from the silage feeder or feed mixer as it is being fed. Avoid the top of the feed because it is drier than the entire batch.

Mix the samples from one type of feed and take a subsample from the mixture for analysis. Seal the samples in polyethylene freezer bags. Store dry samples in a cool area and freeze samples that contain more than 15 percent moisture. Send the samples to a testing laboratory as soon as possible. Use a laboratory that is certified by the National Forage Testing Association to ensure the highest accuracy of the test.

A feed analysis report will generally include the following measures:

- *dry matter (DM)*—the mass of a material when the moisture content is removed
- *crude protein (CP)*—the total of both true protein and nonprotein nitrogen
- *insoluble crude protein (ICP)*—the amount of indigestible crude protein in the feed resulting from overheating
- *adjusted crude protein (ACP)*—calculated value, adjusted for insoluble crude protein. If the ICP/CP ratio exceeds 0.1, use this value instead of crude protein when balancing a ration
- **neutral detergent fiber (NDF)**—relatively insoluble material found in the cell wall of plants, which may be used to predict feed intake. A low NDF is desirable
- **acid detergent fiber (ADF)**—measures the least digestible part of the feed; includes cellulose, lignin, silica, insoluble crude protein, and ash. A low ADF is desirable
- *digestible dry matter (DDM)*—percent of forage that is digestible
- *net energy (NE)*—an indicator of the true value of a feed. It is the energy left after determining the energy lost through the feces, urine, gas, and heat generated by metabolism
- *total digestible nutrients (TDN)*—the total of the digestible parts of crude fiber, protein, fat, and nitrogen-free extract
- *dry-matter intake (DMI)*—estimated maximum consumption of forage dry matter by the animal. It is shown as a percentage of body weight
- **relative feed value (RFV)**—an evaluation of the quality of hay and haylage by combining into one number digestibility and feed intake

The energy value of forage can also be estimated by plant maturity at harvest and the amount of weather damage to the feed. Late-cut, mature plants are lower in energy than early-cut, immature plants. Weather damage lowers the energy content of forages regardless of the stage of maturity when cut.

Relationship between One Hundred Percent Dry-Matter Basis and As-Fed Basis

Many publications list nutrient requirements and feed composition on a **one hundred percent (100%) dry-matter basis.** All feeds contain some moisture. The amount varies with the feed, the form of the feed, the stage of growth at which it was harvested, the length of time it was stored, and the conditions under which it was stored.

One hundred percent dry-matter basis means that the data presented are calculated on the basis of all the moisture being removed from the feed. The term **as-fed (air dry) basis** means the data are calculated on the basis of the average amount of moisture found in the feed as used on the farm.

Using the 100 percent dry-matter basis makes it easier to compare feeds that have different moisture contents on an as-fed basis. The values, when given on a 100 percent dry-matter basis, must be changed to the as-fed basis to find the amounts of feed to actually use. If the feed being used is analyzed, the actual dry-matter content is used for this conversion. If no analysis is available, the average dry-matter content given in the feed composition table is used. These are averages and actual feeds being used may vary from these figures.

The method of converting from one basis to the other is as follows.

Let a = pounds of feed on a 100 percent dry-matter basis

b = pounds of feed on an as-fed basis

c = percent of dry matter in the feed

To convert from an as-fed basis to a 100 percent dry-matter basis:

$$a = b \times c$$

That is, the pounds of feed on a 100 percent dry-matter basis equals the pounds of feed on an as-fed basis multiplied by the percent of dry matter in the feed.

Example 1: A ration calls for 5.6 pounds of #2 dent corn on an as-fed basis. The feed composition table shows that corn has 89 percent dry matter. Therefore: $5.6 \times 0.89 = 5.0$ pounds on a 100 percent dry-matter basis.

Example 2: A ration calls for 22.4 pounds of alfalfa hay on an as-fed basis. The alfalfa has been cut in the mid-bloom stage. The feed composition table shows that mid-bloom alfalfa hay has 89.2 percent dry matter. Therefore: $22.4 \times 0.892 = 20$ pounds on a 100 percent dry matter basis.

Example 3: A ration calls for 21.8 pounds of corn silage on an as fed basis. The feed composition table shows that mature corn silage has a dry-matter content of 55 percent. Therefore: $21.8 \times 0.55 = 12$ pounds on a 100 percent dry-matter basis.

To convert from a 100 percent dry-matter basis to an as-fed basis:

$$b = a \div c$$

That is, the pounds of feed on an as-fed (air-dry) basis equals the pounds of feed on a 100 percent dry-matter basis divided by the percent of dry matter in the feed.

Example 1: A ration calls for 5 pounds of #2 dent corn on a 100 percent dry-matter basis. The feed composition table shows that #2 dent corn has 89 percent dry matter. Therefore, on an as-fed basis:

$$5 \div 0.89 = 5.6 \text{ pounds}$$

Example 2: A ration calls for 20 pounds of alfalfa hay on a 100 percent dry-matter basis. The alfalfa has been cut in the mid-bloom stage. The feed composition table shows that mid-bloom alfalfa hay has 89.2 percent dry matter. Therefore, on an as-fed basis:

$$20 \div 0.892 = 22.4 \text{ pounds}$$

Example 3: A ration calls for 12 pounds of corn silage on a 100 percent dry-matter basis. The feed composition table shows that mature corn silage has a dry matter content of 55 percent. Therefore, on an as-fed basis:

$$12 \div 0.55 = 21.8 \text{ pounds}$$

When using nutrient requirement and feed composition tables given on a 100 percent dry-matter basis, it is easier to work out the ration on the dry-matter basis and then convert the final figures to an as-fed (air-dry) basis.

Rules of Thumb for Balancing Rations

Beef

A maintenance ration for beef cows is primarily roughage. Some supplement may be required, depending on the quality of the roughage fed. The amount of air-dry roughage to feed should equal about 2 percent of the body weight of the animal. For example, if the beef cow weighs 1,213 pounds the amount of air-dry roughage to feed for maintenance would be about 24 pounds If the ration is being calculated on a 100 percent dry-matter basis, 1.8 percent of the animal's body weight is used as the rule of thumb. For a 1,213-pound cow, this would be about 22 pounds of 100 percent dry-matter roughage.

Cows nursing calves in drylot should be fed about 50 percent more than dry cows. Silage is substituted at the rate of three parts silage for each one part of dry roughage. Vitamin A supplement may be needed if the roughage is of poor quality.

Fattening rations for beef should be about 2 to 2.5 percent of the animal's body weight, fed as air-dry grain and protein supplement. If the ration is calculated on a 100 percent dry-matter basis, use 1.8 to 2.25 percent of the animal's body weight. About 0.5 to 1.0 percent of body weight should be fed as air-dry roughage. On a 100 percent dry-matter basis, use 0.45 to 0.9 percent of the animal's body weight. About one part of protein supplement should be fed to each 8 to 12 parts of grain. The total ration should have about 10 to 15 percent air-dry roughage. On a 100 percent dry-matter basis, about 9 to 13.5 percent of the total ration should be roughage. Rations with a high grain content give the fastest and most efficient weight gains.

Fattening cattle must be fed a mineral supplement. If a high-concentrate (grain) ration is fed, a mineral supplement consisting of two parts dicalcium phosphate, two parts limestone, and six parts trace mineralized salt should be fed free choice. **Free choice** means that the supplement is available at all times to the animal. When feeding a high roughage ration, a mineral supplement consisting of two parts trace mineralized salt and one part dicalcium phosphate should be fed free choice.

Swine

Bred sows and gilts that are limit-fed should receive about 3.5 to 4.5 pounds of air-dry feed in the ration. Feed about 3.15 to 4 pounds on a 100 percent dry-matter basis.

Limit-fed means that the amount of feed given the animal is controlled or limited to less than the animal would eat if given free access to the feed. **Self-fed** means that the animal is given free access to all the feed it will eat.

The ration for bred sows and gilts should contain about 14 percent total (crude) protein. Ground roughage may be added to a self-fed ration to limit the energy intake.

Sows nursing litters should receive about 10 to 15 pounds of air-dry feed in the ration. Feed 9 to 13.5 pounds when calculating the ration on a 100 percent dry-matter basis. The ration should contain about 15 percent total protein.

Growing-finishing pigs are fed according to their size. Fifty-pound pigs should be fed about 6.5 percent of their body weight as air-dry feed (5.8 percent on a 100 percent dry-matter basis). The total protein in the ration should be about

16 percent. Pigs weighing 100 pounds should be fed about 5.5 percent of their body weight on an air-dry basis (5 percent on a 100 percent dry-matter basis). The ration should be about 14 percent total protein. Pigs from about 170 pounds to market weight should be fed about 4.5 to 3.5 percent of their body weight as air-dry feed (4 to 3 percent on a 100 percent dry-matter basis). The ration should have about 13 percent total protein. As the pig becomes larger, the percent of its body weight to be fed as feed in the ration should decrease.

In all cases, the balance of amino acids in the ration is as important as the amount of protein.

Sheep

Sheep maintenance rations should have about 3 percent of body weight as air-dry roughage (2.7 percent on a 100 percent dry-matter basis). Supplement may be needed to balance the ration. In fattening rations for sheep, about 0.5 to 1 percent of body weight should be fed as air-dry roughage (0.45 to 0.9 percent on a 100 percent dry-matter basis). The ration should have about 2 to 3 percent of body weight fed as air-dry grain and protein supplement (1.8 to 2.7 percent on a 100 percent dry-matter basis).

Goats

Rules of thumb for goats are similar to those for sheep. Milk goats should receive about 0.5 pound of air-dry grain (0.45 pound on a 100 percent dry-matter basis) for each pound of milk produced. This is fed in addition to the needs of the animal for maintenance, growth, fetal development, and mohair production.

Horses

Horse rations are based on the amount of work the horse is doing. Table 8-2 gives some guidelines for use in balancing rations for horses. Amounts given in Table 8-2 are on an air-dry (as-fed) basis. Use 90 percent of the amounts given to calculate a ration on a 100 percent dry-matter basis.

Poultry

Poultry rations are made up almost entirely of grain and protein supplements. Laying hens need a great deal of calcium for forming the egg shell. A ration for poultry is about 10 percent of body weight, fed as air-dry feed (9 percent if ration is calculated on a 100 percent dry-matter basis).

TABLE 8-2 Rules of Thumb for Feeding Light Horses

Class	Hours Use per Day	Pounds/100 lb Body Weight	
		Roughage	Grain
Idle	0	1.5–2.0	0
Light work	1–3	1.25–1.5	0.5–0.75
Medium work	3–5	1.0–1.5	0.66–1.0
Heavy work	5–8	1.5	1.0–1.4
Idle mares nursing foals	0	1.5–2.0	0.33
Growing colts after weaning	0	0.75–1.5	1.0–1.5
Stallions (breeding season)		0.5–1.5	0.75–1.5
Pregnant mares		0.5–0.75	0.5–1.5
Foals before weaning			0.5–0.75

Source: "Feeding Light Horses," Clemson Cooperative Extension.

General Steps in Balancing a Ration

Step 1: Identify the kind, age, weight, and function of the animal(s) for which the ration is being formulated. In this text, suggested rations and feeding programs are found in the units referring to specific species of animals. These may be used as general guides in formulating rations.

Step 2: Consult a table of nutrient requirements to determine the nutrient needs of the animal(s). These requirements are called **feeding standards**. Feeding standards are based on average requirements and may not meet the needs under specific feeding conditions. If unusual conditions such as weather stress are present, adjustments in the diet may be needed.

Step 3: Choose the feeds to be used in the ration and consult a feed composition table to determine the nutrient content of the selected feeds. Note that the nutrient content of a feed may be different for different species. Values given in a feed composition table are average values and may not represent the actual composition values of the feeds being used. An analysis of feeds being used is a more accurate method of determining feed composition.

Step 4: Calculate the amounts of each feed to use in the ration. Several methods may be used to do this. The Pearson Square or algebraic equations may be used to balance a ration using two or more feeds. Computer programs may also be used to balance rations.

Step 5: Check the ration formulated against the needs of the animal(s). Be sure it meets the requirements for minerals and vitamins. If there is an excessive amount of a nutrient present, it may be necessary to recalculate the ration to bring it more closely in line with the requirements.

Determining Ration Costs

Check the cost of the nutrients in the ration to determine if this is the most economical ration that is practical to feed. Calculate the cost of the ration per pound or ton. The daily cost of feeding the animal may also be calculated if a daily consumption rate is known or assumed. In some cases, it may be necessary to feed certain nutrients, such as salt or other minerals, on a free-choice basis in addition to the amounts provided in the formulated ration.

The most commonly purchased feeds in rations are the protein supplements. Compare these on the basis of the cost per pound of the nutrient content. For example, several protein supplements may be compared as follows:

Percent Protein	Price per Ton	Pounds Protein per Ton	Price per Pound Protein
14%	$186	280	$0.66
16%	$200	320	$0.625
18%	$213	360	$0.59
20%	$219	400	$0.5475

In this example, the 20% protein feed has the lowest cost per pound of protein content although its cost per ton is higher than the other examples.

Other factors to consider when comparing prices include the transportation costs of the feed from the supplier to the farm and the suitability of the particular feed for the class of animal being fed. Feed prices vary over a period of time. Sometimes, it is profitable to change feeds used in the diet as prices change. However, the availability of the feed and the effect of a diet change on the performance of the animals being fed must be considered before making changes. Some species of animals react unfavorably when major changes are made in their diet. Changes in diet usually need to be made gradually to avoid a reduction in feed intake, which may result in a reduction in rate of gain or production.

If the livestock feeder knows about how much feed will be needed for a given period of time, it may be profitable to purchase that feed in larger quantities and when prices are lower. The amount of money available and alternate uses of the money must be considered before making this kind of an investment in a feed supply.

Evaluating Diets Using the Computer

The National Research Council publishes *Nutrient Requirements of Beef Cattle and Nutrient Requirements of Swine* that utilize computer programs to evaluate diets. Both of these programs generate tables of nutrient requirements based on much more detailed information about management, breed (beef program), and environmental conditions than was previously provided in earlier nutrient requirement publications. Both programs evaluate diets based on detailed user input. It is possible for the user to add feeds and to add different feed analyses to the feed library used in the programs. The programs take into account the many variables that interact in livestock feeding and thus are much more accurate in evaluating diets than other methods described in this book. Because both programs require the user to select feeds and enter estimated amounts to be fed, it is useful to use the ration-balancing techniques and the rules of thumb for feeding described in this text to help provide a starting point in these computer programs. It is then possible to quickly refine the inputs to develop a diet to closely match the nutrient requirements of the specific animals being fed.

The methods for balancing rations as outlined in this chapter may still be used to gain a general understanding of livestock nutrition and formulating diets. Balancing a ration using two feeds or groups of feeds and balancing for two desired nutrients can be completed using simultaneous algebraic equations; however, computer programs have made these calculations much simpler. The use of the computer programs provided in these revised nutrient requirement publications will also allow the user to do a much better job of evaluating diets for a wider range of conditions.

Use of the Pearson Square

It is difficult to balance a ration by trial and error. The Pearson Square is a useful tool for simplifying the balancing of rations. It shows the proportions or percentages of two feeds to be mixed together to give a percent of the needed nutrient. Figure 8-3 shows an example problem using the Pearson Square method.

The Pearson Square can be used to find out how much of two grains should be mixed with a supplement. The example shown in Figure 8-4 is calculated on an as-fed basis. The proportions of each grain to be used must be known or decided upon first. The same method is also used to mix two protein supplements with one grain and to mix two grains and two protein supplements. In each case, the proportions of like feeds to each other (such as the two grains) must be decided upon in advance. The weighted average percent of protein is then found. Finally, the Pearson Square is used to balance the mix. Any of the measures of nutrients in the feed may be used. To balance on energy needs, use TDN, NE, ME, or DE. To balance on protein needs, use total (crude) protein or digestible protein.

Figure 8-5 shows an example of balancing a swine ration. This ration utilizes two feedstuffs. Figure 8-6A and Figure 8-6B shows an example of balancing a beef ration.

Check the mix to make sure the protein need is met. Multiply the pounds of corn by the percent of protein in the corn ($1,724 \times 0.089 = 153$; or $782 \times 0.089 = 69.6$). Multiply the pounds of soybean meal by the percent of protein in the soybean meal ($276 \times 0.458 = 126$; or $125 \times 0.458 = 57.2$) Add the pounds of protein together. Divide by the total weight of the mix.

$$153 + 126 = 279$$
$$(279 \div 2{,}000) \times 100 = 14\%$$
$$69.6 + 57.2 = 126.8$$
$$(126.8 \div 907) \times 100 = 14\%$$

The mix is balance for crude protein content.

Pearson Square Method

Example: 2,000 pounds of feed are needed to feed a 100-pound growing hog. A 14% crude protein ration is needed. Corn and soybean meal are selected as feeds. The corn has 8.9% and soybean meal has 45.8% crude protein. How much corn and soybean meal need to be mixed together for 2,000 pounds of feed?

Step 1: Draw a square with lines connecting the opposite corners. Write the percent of crude protein needed (14) in the center of the square where the lines cross. Write the feeds to be used and their crude protein percentages at the left-hand corners of the square.

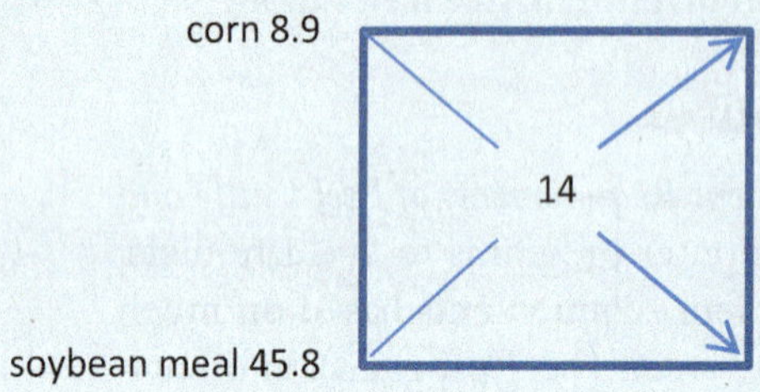

Step 2: Subtract the smaller number from the larger along the diagonal lines. Write the difference at the opposite end of the diagonals. (14 – 8.9 = 5.1 and 45.8 – 14 = 31.8).

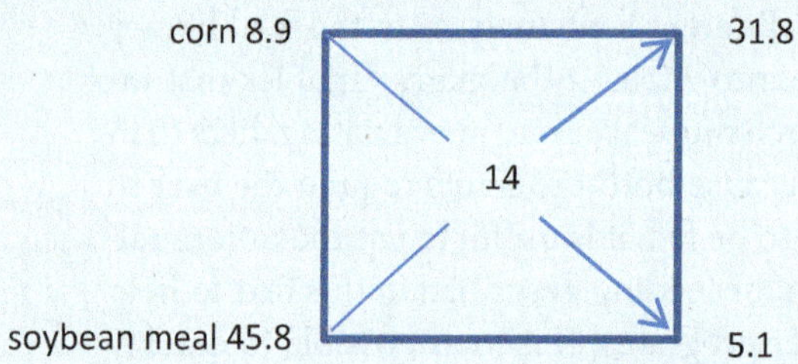

The difference between the percent protein in the soybean meal (45.8) and the needed percent protein in the ration (14) is the parts of corn needed (31.8). The difference between the percent protein in the corn (8.9) and the percent protein needed in the ration (14) is the parts of soybean meal needed (5.1).

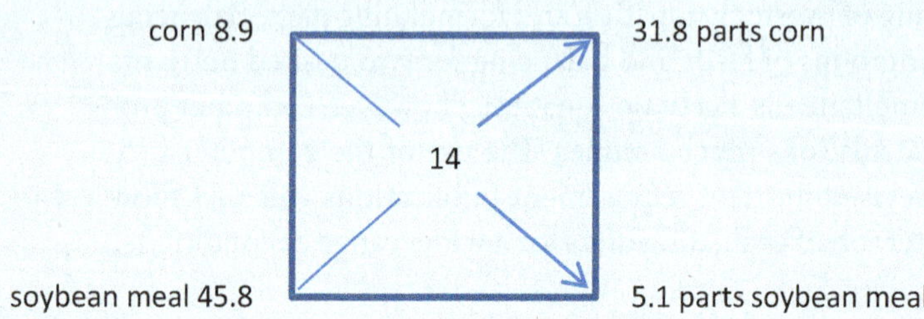

Step 3: Add the parts corn to the parts soybean meal to determine the total parts ration (36.9). Divide the parts of each feed by the total parts to find the percent of each feed in the ration.

(31.8 ÷ 36.9) x 100 = 86.2% corn
(5.1 ÷ 36.9) x 100 = 13.8% soybean meal

Step 4: It is known that 2,000 pounds of the mixture is needed. Multiplying the percent of corn in the mix by the total pounds of the mix.

2,000 x 0.862 = 1,724 pounds of corn

Multiplying the percent of soybean meal in the mix by the total pounds of the mix.

2,000 x 0.138 = 276 pounds of soybean meal

Courtesy of Alyssa K. Elrod

Figure 8-3 An example problem using the Pearson Square Method.

Pearson Square Method: Mixing Two Grains with a Supplement

Example: Assume that a 2,000 pound mix of corn, oats, and soybean meal is needed. The mix is to contain 16% digestible protein. A decision is made to use .75 corn and .25 oats in the mix. Thus, the proportion of corn to oats is 3 to 1. How many pounds of corn, oats, and soybean meal are needed?

Step 1: The weighted average percent of protein in the corn and oats is found first. Multiply the proportion of corn (3) by the percent digestible protein in corn (7.1). Do the same for oats (digestible protein 9.9). Add the two answers together and divide by the total parts (4). The answer is the weighted average percent of digestible protein in the corn-oats mix.

$$\begin{aligned} 3 \times 7.1 &= 21.3 \\ 1 \times 9.9 &= \underline{\ 9.9} \\ & \quad 31.2 \end{aligned}$$

31.2 ÷ 4 = 7.8% digestible protein in the corn-oats mix.

Step 2: Draw the Pearson Square and use it to find the pounds of the corn-oats mix and the soybean meal needed.

corn-oats mix 7.8 | 25.7 parts corn-oats mix

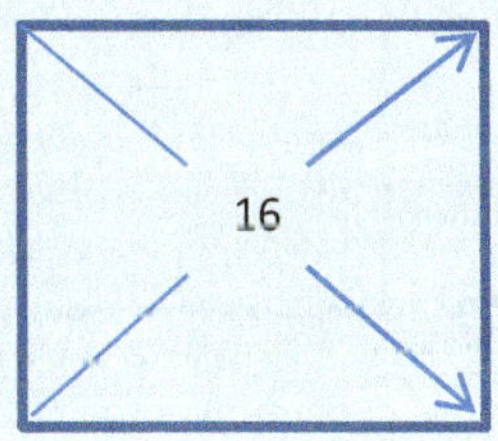

soybean meal 41. 7 | 8.2 parts soybean meal

Step 3: Add the parts corn-oats and the parts soybean meal to determine the total parts ration.

25.7 + 8.2 = 33.9 total parts

Step 4: Divide the parts of each feed by the total parts to find the percent of each feed in the ration.

(25.7 ÷ 33.9) x 100 = 75.8% corn-oats mix
(8.2 ÷ 33.9) x 100 = 24.2% soybean meal

Step 5: It is known that 2,000 pounds of the mixture is needed. Multiply the percent of each component of the mixture by 2,000 to find the number of pounds needed.

0.758 x 2,000 = 1,516 pounds corn-oats mix
1,516 x .75 = 1,137 pounds corn needed
1,516 x .25 = 379 pounds oats needed

0.242 x 2,000 = 484 pounds of soybean meal needed

Courtesy of Alyssa K. Elrod

Figure 8-4 An example problem using the Pearson Square Method to mix two grains with a supplement.

Balancing a Swine Ration

Example: A ration is needed for a 45 kilogram (99 lbs) growing hog.

Step 1: The daily requirements are found in a feed standards table. They appear as follows:

Feed intake (kg)	ME (kcal)	Lysine (g)	Ca (g)	P (g)
1.9	6,200	14.3	11.4	9.5

Step 2: Feeds to be used are selected. Their compositions are as follows:

Feed	ME (kcal/kg)	Lysine (%)	Ca (%)	P (%)
Corn	3,420	0.25	0.034	0.33
Soybean oil meal	3,220	2.9	0.38	0.78

Step 3: Use the Pearson Square to find out how much corn and soybean meal to mix together to make 1.9 kilograms of feed. A shortage of the amino acid lysine often slows the weight gain of the hog. Therefore, in this example, the ration is balanced on the lysine needs. The other needs are then checked. The ration must provide 14.3 grams of lysine. This is 0.0143 kilograms. Divide 0.0143 by 1.9 (feed intake) and multiply by 100, which gives 0.75 as the percent of lysine in the ration. The Pearson Square is set up using the percent of lysine in corn and soybean meal.

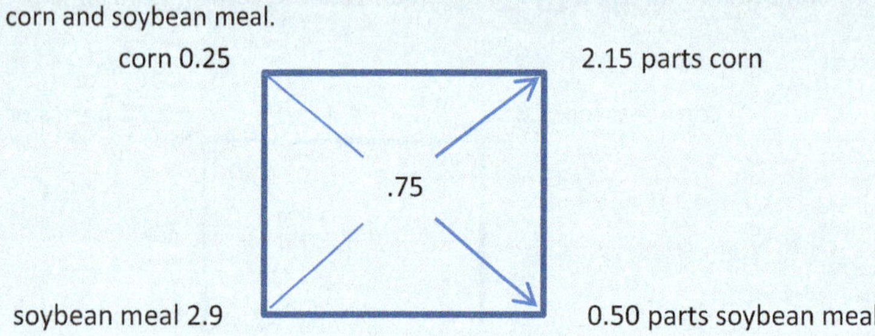

Step 4: Add the parts corn and the parts soybean meal to determine the total parts ration.

2.15 + 0.50 = 2.65 total parts

Step 5: Determine the amount of corn and soybean meal needed by first finding the percentage of parts corn and parts soybean meal desired. Use this percentage to find how many kilograms of each part are needed for 1.9 kg of feed.

(2.15 ÷ 2.65) x 100 = 81.1% corn
0.811 x 1.9 = 1.54 kg corn needed

(0.50 ÷ 2.65) x 100 = 18.9% soybean oil meal
0.189 x 1.9 = 0.36 kg soybean oil meal needed

Step 6: The amount of each nutrient in the ration is checked against the needs of the animal. The ration meets or exceeds the needs of the animal for lysine and metabolizable energy. The extra amounts that this ration gives for these nutrients will not harm the animal because they are within the allowable limits for a ration. However, the ration is short on calcium and phosphorous. This need can be met by feeding a mineral supplement. The amounts needed can be added to the mix to meet the needs of the animal.

Feed	Lysine (kg)	ME (kg)
Corn	1.54 x .0025 = 0.0039	1.54 x 3,420 = 5,266.8
Soybean oil meal	0.36 x .029 = 0.0104	0.36 x 3,220 = 1,159.2
Total provided	0.0143	6,426
Needed by animal	0.0143	6,200
Excess or deficiency	0.0	+226
Feed	**Ca (g)**	**P (g)**
Corn	1,540 x 0.00034 = 0.5236	1,540 x 0.0033 = 5.082
Soybean oil meal	360 x 0.0038 = 1.368	360 x 0.0078 = 2.808
Total provided	1.8916	7.89
Needed by animal	11.4	9.5
Excess or deficiency	−9.5084	−1.61

Courtesy of Alyssa K. Elrod

Figure 8-5 Balancing a swine ration example.

Balancing a Beef Ration

Example: A ration is needed for an 800 pound medium-frame steer with an expected daily gain of 2.0 pounds.

Step 1: Find the requirements for the animal in a feeding standards table. For this example, determine the energy need in metabolizable energy. Measure the protein need as total protein. Calculate the ration on a 100% dry matter basis and convert the final figures to an as-fed basis. The steer in this example has the requirements shown below. Find the megacalories/day and the pounds/day requirements by multiplying the dry matter intake/day by the megacalories/pound and the percents for the appropriate nutrients.

Dry Matter Intake (lb) /day	ME (Mcal/lb)	Protein (%)	Ca (%)	P (%)
18.6	1.11	9.2	0.31	0.20
	Total Mcal /day	**Pounds Protein/day**	**Pounds Ca/day**	**Pounds P/day**
	20.64	1.71	0.057	0.037

Step 2: Select the feeds to be used. Find their composition in a feed composition table. The following feeds are used in this example:

Feed	Dry Matter (%)	ME (Mcal/lb)	Protein (%)	Ca (%)	P (%)
Bromegrass hay	89	0.9	10.0	0.30	0.35
Corn	87	1.36	9.0	0.07	0.27
Soybean meal	90	1.38	49.9	0.33	0.71

Step 3: Use the rules of thumb to determine the amount of each feed to use. The rule of thumb for the amount of 100% dry matter roughage to feed to fattening beef is 0.45 to 0.9 percent of body weight. For this example, 0.9% is used:

0.009 x 800 = 7.2 pounds of hay to feed

Calculate the amount of ME and protein the hay will provide.

7.2 x 0.90 = 6.48 Mcal of ME from hay
7.2 x 0.10 = 0.72 pounds protein from hay

Subtract the amount of ME and protein that comes from the hay from the amounts the animal needs:

20.64 – 6.48 = 14.16 Mcal of ME from concentrate mix
1.71 – 0.72 = 0.99 pounds protein from concentrate mix

Divide the deficit amount of Mcal in ME by the Mcal in 1 lb of corn. Use the Mcal in corn because the largest portion of the concentrate mix is corn. This gives the pounds of concentrate mix needed.

14.16 ÷ 1.36 = 10.4 pounds of concentrate mix needed

The pounds of deficit protein is divided by the pounds of concentrate needed and multiplied by 100, which gives the percentage of protein needed in the concentrate mix.

(0.99 ÷ 10.4) x 100 = 9.5 % protein in concentrate mix

The Pearson Square is used to determine the amount of corn and soybean meal needed. The parts corn and parts soybean meal are added together to give the total parts ration which is 40.9

(continued)

Courtesy of Alyssa K. Elrod

Figure 8-6A Balancing a beef ration example.

Balancing a Beef Ration (continued)

Step 3 (continued):

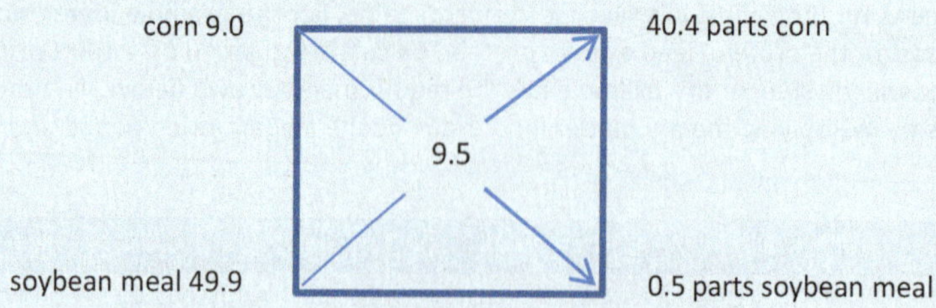

(40.4 ÷ 40.9) x 100 = 98.8 % corn in concentrate mix
0.988 x 10.4 = 10.28 pounds of corn needed

(0.5 ÷40.9) x 100 = 1.22% soybean meal in concentrate mix
0.0122 x 10.4 = 0.13 pounds of soybean meal needed

Step 4: Check the amount of each nutrient in the ration against the needs of the animal. This ration is balanced for the metabolizable energy and protein needs of the animal. It is slightly deficient in calcium and has a slight excess of phosphorus. Although these variations are within the allowable limits for a ration, it might be wise to add a little calcium with a mineral supplement. The ration was calculated on a 100% dry matter basis. The calculated amounts are converted to an as-fed basis by dividing by the percent of dry matter in each feed:

7.2 ÷ 0.89 = 8.09 pounds bromegrass hay
10.28 ÷ 0.87 = 11.82 pounds corn
0.13 ÷ 0.9 = 0.14 pounds soybean meal

The total pounds of feed to be fed is found by adding the pounds of each feed:

8.09 hay + 11.82 corn + 0.14 soybean meal = 20.05 pounds

The animal will eat about 2.5 to 3 percent of its body weight in feed each day. This ration provides 2.50% of the body weight in the total ration.

(20.05 ÷ 800) x 100 = 2.50%

Feed	ME (Mcal)	Protein (lb)
Bromegrass hay	7.2 x 0.9 = 6.48	7.2 x 0.1 = 0.72
Corn	10.28 x 1.36 = 13.98	10.28 x 0.09 = 0.93
Soybean oil meal	0.13 x 1.38 = 0.18	0.13 x 0.499 = 0.06
Total provided	20.64	1.71
Needed by animal	20.64	1.71
Excess or deficiency	0.0	0.0

Feed	Ca (lb)	P (lb)
Bromegrass hay	7.2 x 0.003 = 0.0216	7.2 x 0.0035 = 0.0252
Corn	10.28 x 0.0007 = 0.0072	10.28 x 0.0027 = 0.0278
Soybean oil meal	0.13 x 0.0033 = 0.0004	0.13 x 0.0071 = 0.0009
Total provided	0.0292	0.0539
Needed by animal	0.057	0.037
Excess or deficiency	−0.0278	+0.0169

Courtesy of Alyssa K. Elrod

Figure 8-6B Balancing a beef ration example.

Using Algebraic Equations to Balance Rations

Algebraic equations may be used instead of the Pearson Square to balance rations (Figure 8-7). Letters are used as a means of representation in mathematical statements. For example, let "*x*" represent corn and "*y*" represent soybean meal. A mathematical statement describing the relationship between *x* and *y* can then be set up and used to solve for an unknown.

Using Algebraic Equations to Balance Rations

Example: 2,000 pounds of a 14% protein feed is needed using corn and soybean meal. This ration is to be balanced for protein using two feeds. The basic equations are:

X = pounds corn needed
Y = pounds soybean meal needed

Equation 1:

X + Y = total pounds of mix needed

Equation 2:

(% protein in corn) x (X) + (% protein in soybean meal) x (Y) = pounds of protein desired in mix

Step 1: Find the pounds of protein desired in mix by multiplying the quantity of feed by the percent (or amount/lb) of the nutrient desired.

$$2{,}000 \times 0.14 = 280$$

Step 2: Place the desired values in equation 2. Express all percents as decimals. Protein in corn is 8.9% and protein in soybean meal is 45.8%.

$$0.089X + 0.458Y = 280$$

Step 3: Either X or Y must be canceled by the multiplication of equation 1 by the percentage of nutrient for either X or Y, and the resulting equation 3 is subtracted from equation 2. This example uses the percentage crude protein for corn (0.089). The value 178 is found by multiplying 0.089 times 2,000 lb. Equation 3 is:

$$0.089X + 0.089Y = 178$$

Step 4: Subtract equation 3 from equation 2:

$$\begin{array}{r} 0.089X + 0.458Y = 280 \\ \underline{-0.089X - 0.089Y = -178} \\ 0.369Y = 102 \\ Y = 102/0.369 \end{array}$$

Y = 276 pounds of soybean meal

Step 5: The value of X may be found by substituting the value of Y in equation 1 and solving for X:

$$X + 276 = 2{,}000$$
$$X = 2{,}000 - 276$$

X = 1,724 pounds of corn

Courtesy of Alyssa K. Elrod

Figure 8-7 Algebraic equations may be used to balance rations.

Using Fixed Ingredients When Formulating Diets

Example: A 2,000 pound (1 ton) mix is needed to feed finishing hogs weighing 125 pounds. However, 55 pounds of the mix are to be fixed ingredients that provide additional minerals and vitamins and do not add energy or protein to the ration. The amount of the fixed ingredients is not to be included in the energy/protein calculations. The major ingredients selected are corn (IFN 4-02-935) and soybean meal (IFN 5-01-600). The ration is to be balanced for daily requirements of lysine and metabolizable energy (ME).

Step 1: Set up the requirements and composition of the feeds:

	Lysine	ME
Daily requirement (kg)	0.0122	6,320 kcal
Corn	0.0025	3,300 kcal/kg
Soybean meal (SBM)	0.0279	2,972 kcal/kg

Step 2: Set up the algebraic equations and solve.

X = amount of corn needed per day
Y = amount of soybean meal needed per day

Equation for lysine: $0.0025X + 0.0279Y = 0.0122$
Equation for energy: $3{,}300X + 2{,}972Y = 6{,}320$

Divide 3,300 by 0.0025 to get a factor that is multiplied times the equation for lysine. The resulting equation is then subtracted from the equation for energy to eliminate the X unknown and solve for Y.

$$3{,}300 \div 0.0025 = 1{,}320{,}000$$

$$\begin{aligned} 3{,}300X + 2{,}972Y &= 6{,}320 \\ -3{,}300X - 36{,}828Y &= -16{,}104 \\ \hline -33{,}856Y &= -9{,}784 \\ Y &= 0.289 \end{aligned}$$

Substitute the value of Y in the equation for lysine and solve for X.

$$\begin{aligned} 0.0025X + (0.0279 \times 0.289) &= 0.0122 \\ 0.0025X + 0.008 &= 0.0122 \\ 0.0025X &= 0.0122 - 0.008 \\ 0.0025X &= 0.0042 \\ X &= 1.68 \end{aligned}$$

Step 3: Determine the amount of corn and soybean meal to mix together to make 1,945 pounds of mix. The amounts of corn and soybean meal needed daily are added together and each amount is divided by the total to determine the percent of each ingredient in the ration. This percentage is then multiplied times 1,945 pounds to determine how many pounds of corn and soybean meal are necessary in the total mix. The balance of the 2,000 pounds is composed of the fixed ingredients previously determined to provide the added minerals and vitamins needed in the ration.

	kg/day	% diet	lb/ton
Corn	1.68	85.3	1,659
SBM	+ 0.289	+ 14.7	+ 286
Total	1.969	100	1,945

Figure 8-8 An example of using fixed ingredients when formulating diets.

USING FIXED INGREDIENTS WHEN FORMULATING DIETS

Feed mixes formulated to provide a complete diet for the animal normally have small amounts of minerals, vitamins, and antibiotics added. Generally, these make up less than 10 percent of the total mix and provide little of the protein or energy needed in the diet. However, these fixed ingredients must be taken into account when formulating diets if the final computed protein and energy needs of the animals are to be met.

The first step in formulating a diet using fixed ingredients is to determine what these ingredients are and how much of each is to be in the final mix. An International Feed Number (IFN) is a six-digit number used to classify feedstuffs on a worldwide basis. Next, determine if any of these fixed ingredients provide any of the nutrients for which the ration is being balanced. If they do, then these amounts must be calculated and subtracted from the amount to be provided by the major ingredients in the diet. After this is done, then the procedures outlined earlier may be followed to balance the major ingredients for the mix. An example of using fixed ingredients when formulating a diet is shown in Figure 8-8.

SUBSTITUTING SILAGE FOR HAY

Silage replaces hay in a ration at the rate of three parts silage to one part hay. In the beef ration example shown earlier in Figures 8-6A and 8-6B, assume that one-half of the roughage is from silage. The amount of hay and silage used is calculated as follows:

$$7.2 \div 2 = 3.6 \text{ pounds of hay}$$

$$3.6 \times 3 = 10.8 \text{ pounds of silage}$$

The ration is then calculated using the same method as in the example shown in Figures 8-6A and 8-6B. The ME and protein content of the silage and hay is balanced by adding grain and supplement. The pounds of ME in the hay and silage are added together when finding the pounds of concentrate mix needed. The pounds of protein in the hay and silage are added together when finding the percent of protein needed in the concentrate mix.

SUMMARY

Livestock feeds are classified as roughages and concentrates. Roughages have a crude fiber content of more than 18 percent. Concentrates have less than 18 percent crude fiber. Roughages are either legume or nonlegume. Legume roughages can use nitrogen from the air. They are higher in protein content than nonlegume roughages. Concentrates are either energy feeds or protein supplements. Energy feeds are usually grains such as corn or oats. Protein supplements have more than 20 percent protein content. They come from either animal or vegetable sources.

A ration is the amount of feed an animal is given during a 24-hour period. It is balanced if it provides all the nutrients the animal needs for good growth, gain, or production. In addition, the ration must taste good to the animal and it must be economical. The right balance of roughages and concentrates must be in the ration. No harmful materials or excessive amounts of additives should be fed. Micronutrients and additives must be carefully mixed in the right amounts in the ration.

A ration is fed for several purposes. It must provide the nutrients needed for maintenance, growth, pregnancy and, sometimes, work. Many animals are also fed to fatten for market or to produce milk, eggs, or wool. Proper nutrition is also essential for reproduction.

Rations are balanced for the protein and energy needs of the animal. Feeding standards and tables of feed composition are used in balancing rations. The mineral and vitamin needs of the animal are also considered. The price of the feed used is important when calculating least-cost rations. Certain rules of thumb, which give general guides to use for rations, may be used.

There are five general steps in balancing a ration: (1) The kind of animal to be fed is identified. (2) The needs of the animal are found. (3) Feeds are selected and the composition of the feed is found. (4) The amount of each feed to use is calculated. (5) The ration is checked against the needs of the animal to make sure it is balanced. The Pearson Square is a helpful tool to use in balancing rations. Computers may also be used when balancing rations.

Student Learning Activities

1. Secure prices of feeds from local sources and calculate least-cost, balanced rations for various classes of animals.
2. Prepare an exhibit showing feeds classified as roughages and concentrates.
3. Present an oral report on balancing a ration for livestock on your home farm.
4. Take a field trip to a feed company that uses a computer for balancing rations and observe the method being used.
5. Calculate and use appropriate balanced rations when planning and conducting an animal supervised experience program in animal production.

Discussion Questions

1. What are roughages?
2. Name the two general classes of roughages.
3. What are the sources of each of these two classes of roughages?
4. What are concentrates?
5. Define the term *ration.*
6. List and briefly explain the six functions of a ration.
7. What are the two common ways to measure the amount of protein in a ration?
8. What are the four common ways to measure the amount of energy provided by a ration?
9. What should be the ratio of calcium to phosphorus in a ration?
10. What are the rules of thumb for balancing rations for (a) beef cattle, (b) swine, (c) sheep and goats, (d) horses, (e) poultry?
11. List and briefly explain the five general steps to follow when balancing a ration.
12. Give an example showing how to use the Pearson Square when balancing a ration.
13. Select an animal from the home farm or in the local area and balance a ration for that animal using feeds available locally.
14. How much silage can be substituted for hay in a ration for beef?
15. Show how to convert the amounts in a ration calculated on a dry-matter basis to an as-fed basis.

Review Questions

True/False

1. Legume roughages can use nitrogen from the air.
2. Rations are balanced for the protein and energy needs of the animal.
3. When balancing rations, the term *air-dry* means the same as the term *as-fed*.
4. Peas, alfalfa, and soybeans are all legumes.
5. Roughages are feeds that have more than 10 percent crude fiber when dry.

Multiple Choice

6. Livestock feeds are classified as ______:

    a. roughages c. crude fiber
    b. concentrates d. a and b

7. A maintenance ration for beef cattle is primarily ______:

    a. carbohydrates c. roughages
    b. vitamins d. fats

8. The net energy for production is that amount of energy needed by the animal above the amount used for ______ .

    a. growth c. work
    b. fat production d. maintenance

9. When determining a ration, the ______ must be considered when determining the nutrient requirements.

    a. weight of the ration c. size of the ration
    b. type of the ration d. functions of the ration

Completion

10. Breathing, function of the heart, and other vital body processes are called ______ ______.
11. Livestock feeds that contain 20 percent or more protein are called ______ ______.
12. ______ is animal tissues and bones from animal slaughterhouses and rendering plants that are cooked, dried, and ground.

Short Answer

13. How does proper nutrition affect reproduction?
14. What is the primary use of nutrients in a ration?
15. What is tankage?
16. What is meant by withdrawal period?
17. What is the difference in a ration and a diet?
18. Define *palatable*.
19. Name three poisonous plants.
20. What is the difference in limit-fed and self-fed?

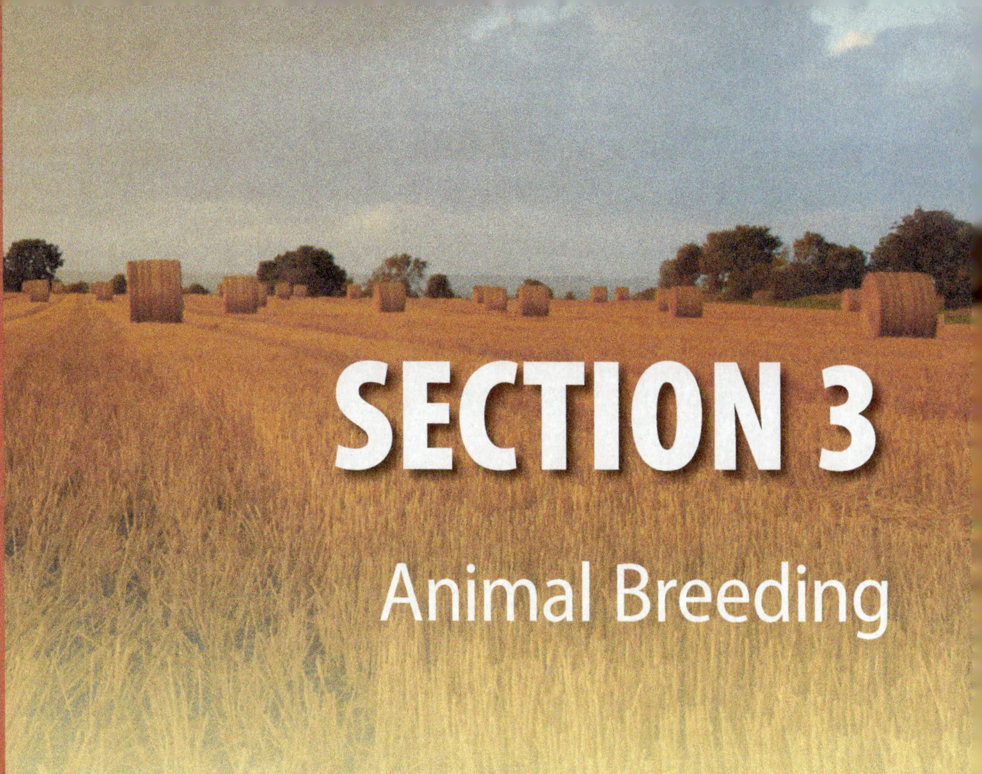

SECTION 3

Animal Breeding

Chapter 9

Genetics of Animal Breeding

Key Terms

genetics
environment
phenotype
genotype
additive gene effect
nonadditive gene effect
heterosis
heritability
heritability estimate
breeding value
protoplasm
nucleus
chromosome
cytoplasm
lysosome
ribosome
cell membrane
mitosis
diploid
prophase
metaphase
anaphase
telophase
meiosis
gamete
sperm

Objectives

After studying this chapter, the student should be able to

- explain how genetics relates to improvement in livestock production.
- describe how cell division occurs.
- diagram and explain how animal characteristics are transmitted.
- diagram and explain sex determination, linkage, crossover, and mutation.

THE IMPORTANCE OF GENETICS

Farm animals today are more efficient than they were 100 years ago. They produce more meat, milk, eggs, and wool on less feed. Much of this progress in livestock efficiency is the result of the use of genetics. **Genetics** is the study of heredity, or the way in which traits of parents are passed on to offspring. Good breeding programs are based on an application of the principles of genetics (Figure 9-1).

An Austrian monk named Gregor Johann Mendel is considered to be the founder of the science of genetics. In a period from 1857 to 1865, Mendel did many experiments with garden peas. He proved that certain characteristics, such as color and height, are passed from parent to offspring. Livestock breeders use this fact to select animals for breeding that will produce offspring with desirable characteristics.

Not all differences in animals are caused by genetics. Some are caused by the **environment**, or the conditions under which the animals are raised. This makes the job of selection more difficult. However, methods have been developed that enable farmers to select parent animals with traits that are related to genetics rather than the environment.

USDA Agricultural Research Service; Photo by Peggy Greb

Figure 9-1 Good breeding programs are based on the application of genetics.

SELECTION BASED ON GENETICS

Additive and Nonadditive Gene Effects

Observation of any population of farm animals reveals variation in **phenotype** and, by inference, variation in **genotype**. *Phenotype* refers to the outward appearance of an animal without reference to its genetic makeup. *Genotype* refers to the combination of genes that an individual possesses. Two factors are responsible for the genetic variation in animals: additive gene effects and nonadditive gene effects.

When many different genes are involved in the expression of a trait, that expression is said to be controlled by **additive gene effects**. Individual genes have relatively little effect on the trait; the effect of each gene is cumulative with very little or no dominance between pairs of alleles. Each member of the gene pair has an equal opportunity to be expressed. Most of the economically important traits of livestock are controlled by additive gene effects. Carcass traits, weight gain, and milk production are examples of traits that have moderate to high heritability and are considered to be greatly influenced by additive gene effects.

Traits that result from additive gene effects are considered to be quantitative. There may be hundreds or even thousands of gene pairs, located on different chromosome pairs, that are involved in the expression of the trait. The environment the animal is raised in often influences the expression of the trait. It is difficult to classify the phenotypes of the animals into distinct categories because they usually follow a continuous distribution. It is hard to identify animals with superior genotypes for quantitative traits.

Nonadditive gene effects control traits by determining how gene pairs act in different combinations with one another. Generally, these traits are readily observable and are controlled by only one or a few pairs of genes. Typically, one of the genes in the pair will be dominant if the animal is heterozygous for the trait being expressed. When combinations of gene pairs give good effects, the offspring will be better than either of its parents. This is sometimes called hybrid vigor or **heterosis**.

Traits that result from nonadditive gene effects are considered to be qualitative. The phenotype of these traits can usually be identified easily, there is relatively little environmental effect on these traits, and the genotype can usually be easily determined.

Heritability Estimates

Heritability is the proportion of the total variation (genetic and environmental) that is due to additive gene effects. A **heritability estimate** expresses the likelihood of a trait being passed on from parent to offspring.

Key Terms *continued*

ovum
haploid
homologous
spermatogenesis
oogenesis
spermatogonia
spermatocytes
spermatid
oocyte
ootid
fertilization
zygote
gene
allele
deoxyribonucleic acid (DNA)
deoxyribose
adenine
guanine
thymine
cytosine
nucleotide
polynucleotide
semiconservative replication
ribonucleic acid (RNA)
uracil
codon
dominant gene
recessive gene
polled
albino
homozygous gene pair
heterozygous gene pair
incomplete dominance
locus
codominance
sex-limited genes
sex-influenced genes
sex-linked genes
linkage
crossover
mutation

TABLE 9-1 Heritability Estimates for Beef Cattle

Heritability Trait	Percentage
Number born	5
Calving interval (fertility)	10
Percent calf crop	10
Services per conception	10
Conformation score at weaning	25
Cancer eye susceptibility	30
Gain on pasture	30
Weaning weight	30
Yield grade	30
Carcass grade	35
Age at puberty	40
Birth weight	40
Body condition score	40
Carcass—percent lean cuts	40
Conformation score at slaughter	40
Cow maternal ability	40
Efficiency of gain	40
Preweaning gain	40
Yearling frame size	40
Yearling weight	40
Fat thickness	45
Feedlot gain	45
Dressing percent	46
Marbling score	50
Mature weight	50
Scrotal circumference	50
Tenderness	50
Final feedlot weight	60
Retail yield	60
Rib eye area	70

If a trait has a high heritability, the improvement in the animals' characteristics will be rapid. The improvement is slow for traits with a low heritability, requiring several generations of animals for the desirable trait to become strong. Tables 9-1, 9-2, and 9-3 list heritability estimates for several species of livestock.

Selecting Breeding Stock

There are computer programs and databases, developed by universities and breed associations, available that can provide information about the **breeding value** of animals. The use of estimated breeding value and expected progeny difference helps the producer make faster genetic improvement in livestock.

There are three types of systems that might be used to select breeding animals.

1. Tandem
   - Selection is for one trait at a time; selection for another trait begins when a desired level of performance is achieved in the first
   - An animal with one desirable trait but other undesirable traits may be kept for breeding purposes
   - For the most profitable production, emphasis needs to be placed on several traits when selecting breeding stock; tandem selection does not do this
   - Simple to use but not recommended; it is the least effective of the selection methods
2. Independent culling levels
   - Establishes a performance level for each trait in the selection program that an animal must achieve to be kept for breeding purposes
   - Selection for the breeding program is based on more than one trait
   - A disadvantage of this type of selection is that superior performance in one trait cannot offset a trait that does not meet the criteria for selection
   - Most effective when only a small number of traits are being selected for in the breeding program
   - This is the second most effective method of selection and is the one most widely used in the livestock industry
3. Selection index
   - An index of net merit is established that gives weight to traits based on their economic importance, heritability, and genetic correlations that may exist between the traits
   - Does not discriminate against a trait with only slightly substandard performance when it is offset by high performance in another trait
   - Provides more rapid improvement in overall genetic improvement in the breeding group
   - Extensive records are required to establish the index
   - Most effective method of achieving improvement in genetic merit

From a practical standpoint, it may be wise for a livestock breeder to use a combination of selection methods in the breeding program. A combination of the independent culling level method and the selection index method may work best for many producers. One or two traits that are particularly critical for the producer may be selected for using the independent culling level method. The selection index might then be used for other traits that are important for that producer.

THE CELL AND CELL DIVISION

An animal's body is made up of millions of cells. Cells are the basic and generally the smallest parts of the body that are capable of sustaining the processes of life (metabolism and reproduction). Figure 9-2 shows the parts of an animal cell. Most of the cell is made of a material called **protoplasm**. The **nucleus** contains the hereditary material of the cell, that is, the **chromosomes** that contain the genes. The nucleus also controls the cell's metabolism, growth, and reproduction. The nucleus is surrounded by the **cytoplasm**. The cytoplasm contains the smooth and rough endoplasmic reticulum, mitochondria, **lysosomes**, vacuoles, the Golgi apparatus, and **ribosomes**. The nucleus and cytoplasm are surrounded by the semipermeable **cell membrane**.

Mitosis

Each animal begins as a single cell. This cell divides to make two cells. The cells continue to divide, and groups of cells form specialized tissues and organs in the animal's body. This division of body cells in an animal is called **mitosis**. Mitosis increases the number of body cells, which causes the animal to grow. Old body cells that die are replaced by mitosis. Chromosomes occur in pairs in the nucleus of all body cells except the sperm and ovum. Each parent contributes one-half of the pair. The number of pairs of chromosomes is called the **diploid** number. The diploid number varies from species to species but is constant for each species of animal:

cattle	30	goat	30	chicken	39
swine	19	horse	32	rabbit	22
sheep	27	donkey	31		

During mitosis, the chromosome pairs are duplicated in each daughter cell, so they are exactly like the old cell. Figure 9-3 shows the steps in mitosis. A cell that is not dividing is in the interphase stage. During mitosis there are four typical stages in the division of the cell nucleus: **prophase**, **metaphase**, **anaphase**, and **telophase**.

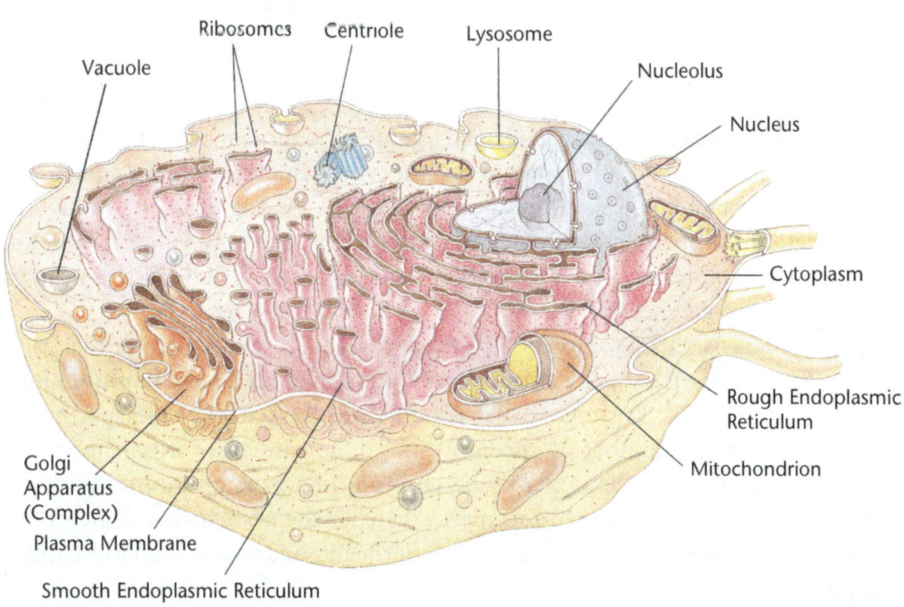

Figure 9-2 Parts of an animal cell.

TABLE 9-2 Heritability Estimates for Swine

Heritability Trait	Percentage
Litter survival to weaning	5
Litter size	10
Number farrowed	10
Number pigs weaned	12
Weaning weight (3 weeks)	15
Birth weight	20
Five-month weight	25
Number of nipples	25
Conformation	30
Feed efficiency	30
Age at puberty	35
Percent lean cuts	45
Probe back fat (live at 200 lb)	45
Carcass length	50
Loin muscle area	50
Percent of shoulder	50
Percentage carcass muscle	50
Percent ham	55
Percent fat cuts	60

TABLE 9-3 Heritability Estimates for Sheep

Heritability Trait	Percentage
Number born	13
Conformation score	15
Feed efficiency	20
Fat thickness	25
Milking ability	25
Birth weight	30
Weaning weight	30
Carcass—percent lean cuts	35
Fleece weight	40
Post-weaning daily gain	40
Skin folds	40
Weight of retail cuts	40
Yearling weight	40
Rib eye area	45
Face covering	50
Mature weight	50
Staple length	50

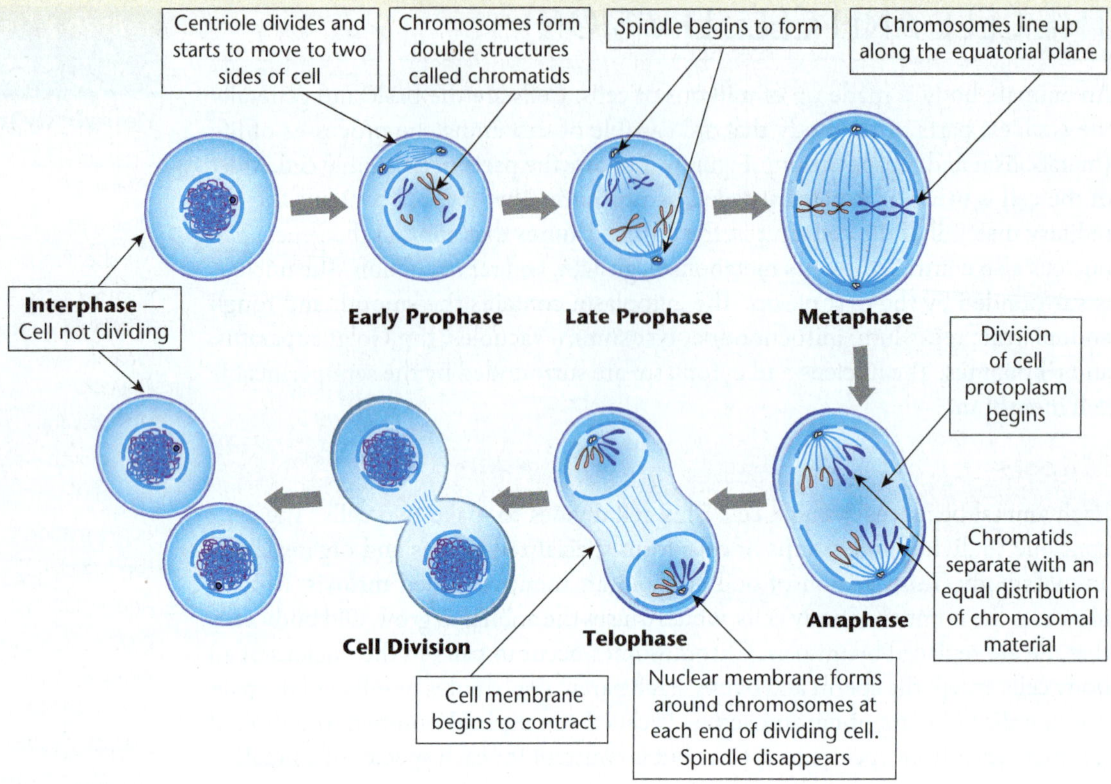

Figure 9-3 Cell dividing by mitosis.

The ability of body cells to continue to divide throughout the life of the animal is limited. At the end of each chromosome in the nucleus of the cell there is a specific repeating DNA sequence called a telomere. The presence of these telomeres is critical for successful cell division. Each time a cell divides, some of the telomere is lost from the end of the chromosome. As the animal ages, the telomeres become shorter and eventually the cell stops dividing.

Meiosis

Meiosis is a type of cell division that creates **gametes** (sperm or egg cells). Gametes are the reproductive cells of an organism. The male gamete is called a **sperm** cell and the female gamete is called an **ovum**, or egg, cell. During sexual reproduction two gametes (one sperm and one ovum) unite to form the zygote. In meiosis, the number of chromosomes in the gametes is reduced by one-half. Figure 9-4 shows the steps involved in meiosis.

During meiosis, the chromosome pairs are divided in such a manner that each gamete has one of each type of chromosome; the gamete cell has a **haploid** number of chromosomes. The zygote that results from the union of the sperm and ovum has a diploid number of chromosomes. One set comes from the sperm and one set comes from the ovum. The chromosome pairs are **homologous**—that is, the first chromosome in the sperm matches the first chromosome in the ovum; the rest of the chromosomes in the sperm and ovum match up in a similar manner.

Although both the spermatozoa and the ova are produced by meiosis, there are some differences in the production of each. The production of spermatozoa is called **spermatogenesis**; the production of an ovum is called **oogenesis** (Figure 9-5).

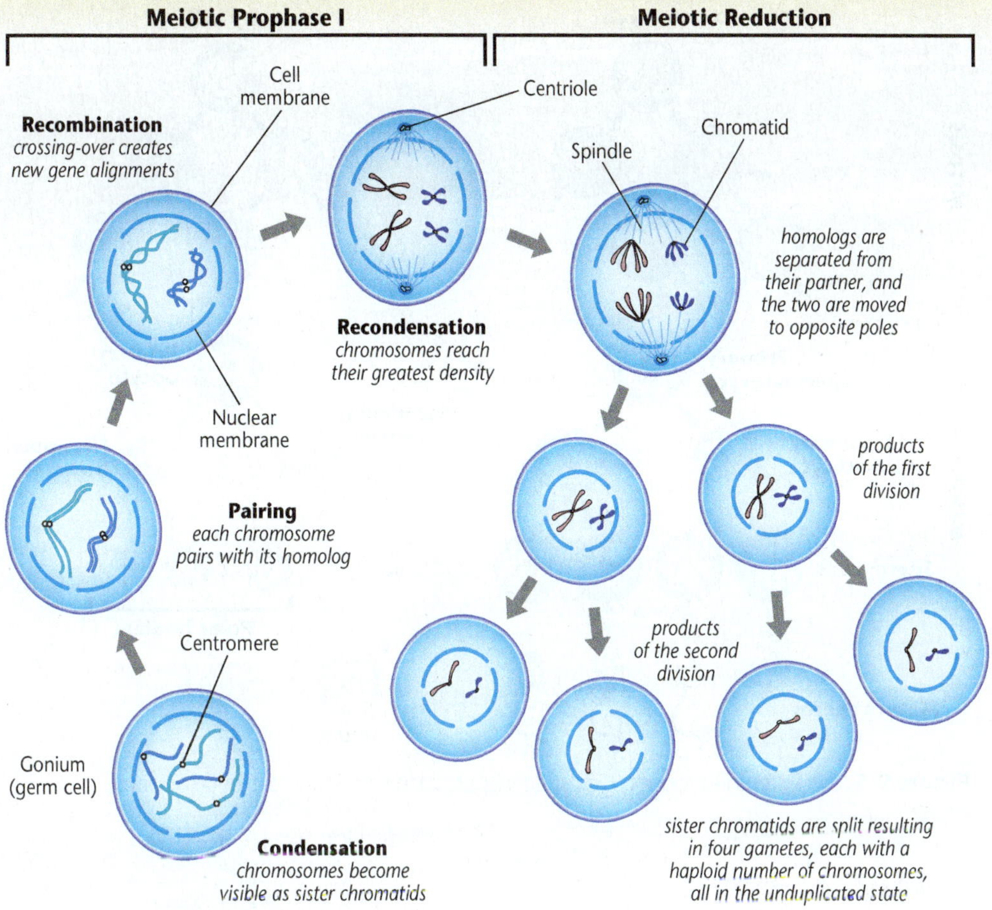

Figure 9-4 Cell dividing by meiosis.

When they reach sexual maturity, male animals begin producing spermatozoa (sperm cells) from spermatogonia in the seminiferous tubules in the testes. **Spermatogonia** are the progenitors of **spermatocytes** (diploid cells that divide by meiosis to produce four spermatids). A **spermatid** is any of the four haploid cells produced by meiosis that develop into spermatozoa. The first meiotic division acts on a primary spermatocyte to produce two secondary spermatocytes that then divide in the second meiotic division to produce four spermatids. Spermatozoa are small, with only a small amount of cytoplasm in the head that is, primarily, the nucleus; they develop a long flagellum or tail that gives them a high degree of motility.

At sexual maturity, female animals produce ova (egg cells) in the ovaries. The first meiotic division acts on a primary **oocyte** to produce two cells; one is the secondary oocyte and the other is the first polar body. The secondary oocyte is a relatively large body, while the first polar body is quite small. The second meiotic division acting on the secondary oocyte produces one large cell (**ootid**) and one small cell (second polar body). The ootid develops into the ovum. The first polar body may or may not divide during the second meiotic division; if it does it produces two second polar bodies. The second polar bodies produced are not functional and are reabsorbed. The single ovum that is produced is large and contains a lot of cytoplasm and stored food; this provides the initial nourishment for the zygote and embryo.

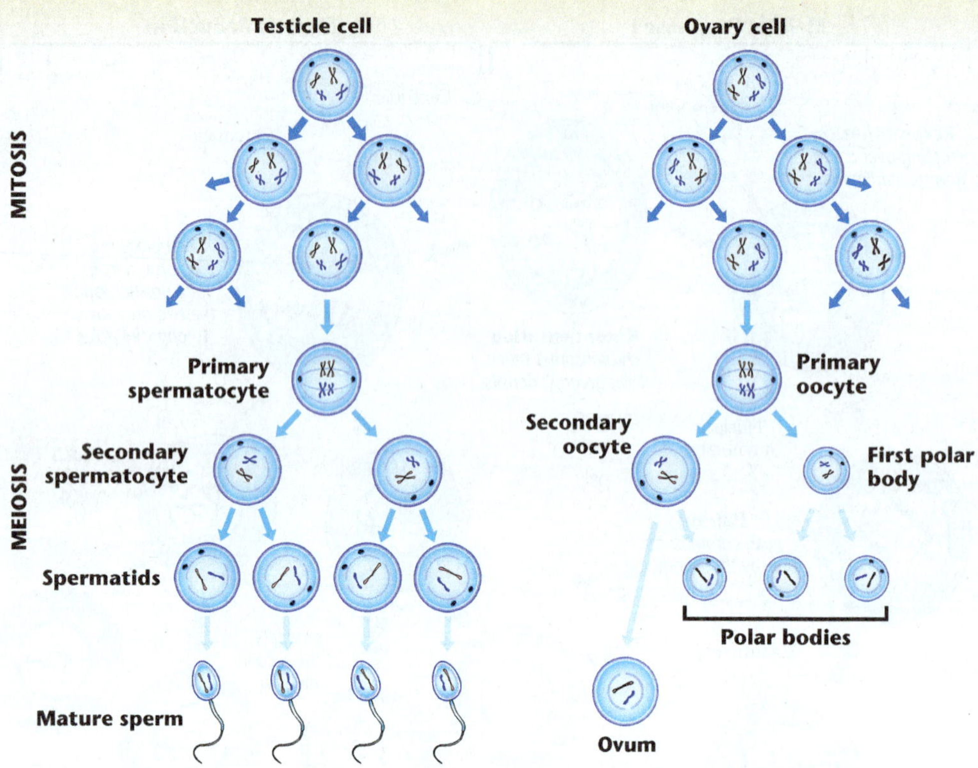

Figure 9-5 The steps in production of sperm (spermatogenesis) and the ovum (oogenesis).

Fertilization

When a sperm cell from the male reaches an egg cell from the female, **fertilization** takes place. The two haploid cells unite to form one complete cell called a **zygote**. The zygote is diploid; that is, it has a full set of chromosome pairs. This process results in many different possible combinations of traits in the offspring.

TRANSMISSION OF CHARACTERISTICS

Genes

The characteristics of an animal that are inheritable are passed from one generation to the next by genes. **Genes** are located on chromosomes and are composed of DNA. Because chromosomes occur in homologous pairs, the genes that they carry are also paired. The sequence of the locations of the gene pairs is the same in a homologous pair of chromosomes. If a gene pair at a given location is identical, they control a trait in the same way. For example, both may code for the color black. In this case, the gene pair is said to be homozygous. In some gene pairs, each gene codes for a different expression of the same trait. In this case, the genes are called **alleles**, and the gene pair is said to be heterozygous. For example, one allele may code for black color, and the other may code for red color. The same trait is being affected, but the alleles are coding for different effects.

The genes provide the code for the synthesis of enzymes and other proteins that control the chemical reactions in the body. These chemical reactions ultimately determine the physical characteristics of the animal. The physical appearance of an animal is determined by its genotype and is referred to as its phenotype.

Environmental conditions may also influence the physical appearance of an animal. For example, the genotype of a beef animal for rate of gain determines a range for that characteristic in which it will fall, but the ration it receives will determine where it actually is in that range.

Some traits are controlled by a single pair of genes; however, most of the traits are controlled by many pairs of genes. For instance, carcass traits, growth rate, and feed efficiency are controlled by many pairs of genes.

The Coding of Genetic Information

The **deoxyribonucleic acid (DNA)** molecule is shaped like a double helix (Figure 9-6). The sides of the molecule are composed of two strands of **deoxyribose** (a five-carbon sugar molecule). The deoxyribose molecules in the side strands are linked together by phosphoric acid. The side strands are linked together by nitrogen-containing bases. Two are purines (**adenine** [A] and **guanine** [G]) and two are pyrimidines (**thymine** [T] and **cytosine** [C]). A **nucleotide** is a combination of one of the nitrogenous bases, one phosphate, and one deoxyribose. The nitrogenous base is attached to one side of the deoxyribose and the phosphate to the other. The phosphate is also attached to the next deoxyribose in the chain, forming a **polynucleotide** chain. Two of these polynucleotide chains twist into a coil, forming the double helix of the DNA molecule.

The nitrogen bases combine in specific combinations as they link across to the other strand in the DNA molecule. Adenine connects only to thymine by two hydrogen bonds; cytosine connects only to guanine by three hydrogen bonds. The possible sequences are AT, TA, CG, and GC. With these sequences, four types of DNA nucleotides can be formed.

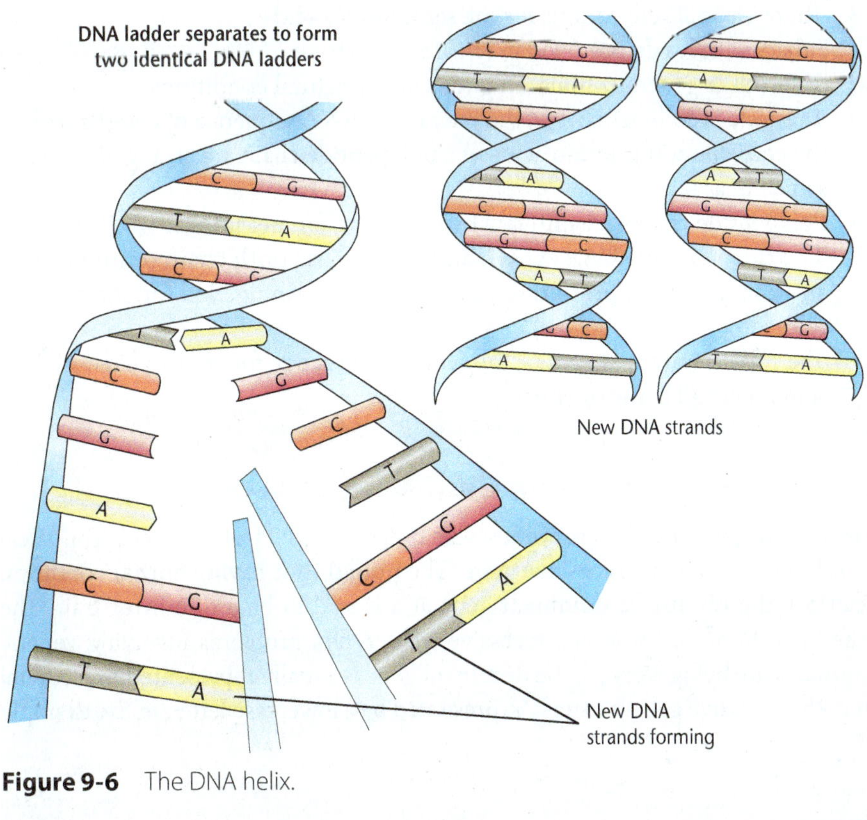

Figure 9-6 The DNA helix.

The factors that distinguish the DNA of one species from that of another are as follows:

- the number of AT and CG pairs
- the sequence of these pairs
- whether the connections are AT, TA, CG, or GC
- the number of these base pairs that are present (the length of the DNA molecule)

The DNA is duplicated when cells divide. The two strands of the double helix separate, and the nitrogenous base attracts its corresponding base. At the same time, the phosphate of one nucleotide is bound to the deoxyribose of the next by an enzyme (DNA polymerase), thus forming a new polynucleotide chain. This results in a new DNA double helix molecule that consists of one old strand and one new strand. The process is called **semiconservative replication**.

Ribonucleic acid (RNA) regulates protein synthesis in animals. The basic structure of RNA is similar to that of DNA except the nitrogenous base **uracil** (U) takes the place of thymine, the sugar is ribose instead of deoxyribose, and the binding enzyme during replication is RNA replicase instead of DNA polymerase. The primary function of RNA is carrying the genetic message from DNA for the building of the polypeptide chains that begin the process of protein synthesis.

The basic unit of the genetic code is the **codon**. The sequence of the four nucleotide bases in the DNA or the RNA is the key to the genetic code. Sixty-four different combinations are possible when three bases in specific sequences are used ($4 \times 4 \times 4 = 64$). This provides more than enough codons to identify the amino acids found in protein.

The characteristics of the genetic code are as follows:

- Codons exist as three bases in specific sequences
- There are multiple codons for the same amino acid
- Adjacent bases do not overlap to form codons
- The code is generally nonambiguous under natural conditions
- The code is colinear, that is, the messenger RNA sequence of codons and the corresponding amino acids of a polypeptide chain are arranged in the same linear sequence
- Polypeptide chains are initiated by the codon for methionine (AUG)
- The three codons UAA, UAG, and UGA cause both chain termination and chain release; protein release factors are also needed for chain release
- The code is universal, that is, the same codons code for amino acids similarly in all living organisms

Dominant and Recessive Genes

A **dominant gene** in a heterozygous pair hides the effect of its allele. The allele that is hidden is called a **recessive gene**. The **polled** (not having horns) condition in cattle is the result of a dominant gene and is said to be a dominant trait. The horned condition in cattle is a recessive trait. When problems involving genetic inheritance are being worked, the dominant gene is usually represented by a capital letter. The recessive gene is usually represented by a lowercase letter. In the example of polled cattle, the dominant gene is written as P. The recessive gene is written as *p* (P = polled condition; *p* = horned condition).

Some other examples of dominant and recessive traits are listed here:

Dominant	Recessive
• Black coat in cattle	• Red coat in cattle
• White face in cattle	• Colored face in cattle
• Black coat in horses	• Brown coat in horses
• Color in animals	• Albinism in animals
• Rose comb in chickens	• Single comb in chickens
• Pea comb in chickens	• Non barred feather pattern in chickens
• Barred feather pattern in chickens (also sex-linked)	• "Snorter" dwarfism in cattle
• Normal size in cattle	

Homozygous and Heterozygous Gene Pairs

A **homozygous gene pair** is one that carries two genes for a trait. For example, a polled cow might carry the gene pair PP. A horned cow must carry the gene pair *pp*. For a cow to have horns, it must carry two recessive genes for the horned trait.

A **heterozygous gene pair** is one that carries two different genes (called alleles) that affect a trait. For example, a polled cow might carry the gene pair P*p*. This cow is polled (because the P gene is dominant), but carries a recessive gene for the horned trait. If this cow is mated to a bull with a gene pair P*p*, some of the calves will be polled and some will have horns.

Six Basic Crosses

There are six basic types of genetic combinations possible when a single gene pair is considered:

- homozygous × homozygous (PP × PP) (both dominant)
- heterozygous × heterozygous (P*p* × P*p*)
- homozygous (dominant) × heterozygous (PP × P*p*)
- homozygous (dominant) × homozygous (recessive) (PP × *pp*)
- heterozygous × homozygous (recessive) (P*p* × *pp*)
- homozygous (recessive) × homozygous (recessive) (*pp* × *pp*)

It is possible to predict the results of crossing animals with varying genotypes, the kinds of gene pairs possessed by the animal. A Punnett square may be used to predict the results of crossing animals with various kinds of genotype. The male gametes are usually shown across the top of the Punnett square and the female gametes along the left side.

Results of crossing are often referred to in the following manner in genetics:

parental generation	P
first filial generation	F_1
second filial generation	F_2

Homozygous × Homozygous (PP × PP) (Both Dominant) A cross between two polled cattle that are homozygous for the polled trait would be set up as follows:

Male Gametes

Female Gametes

	P	P
P	PP	PP
P	PP	PP

- All the F_1 are homozygous for the polled trait with the genotype PP
- All the F_1 are polled
- If the parents are homozygous dominant, all the F_1 must be homozygous dominant

Heterozygous × Heterozygous (P*p* × P*p*) A cross between two cattle that are heterozygous for the polled trait is set up as follows:

Male Gametes

Female Gametes

	P	*p*
P	PP	P*p*
p	P*p*	*pp*

- The F_1 genotypic ratio is 1:2:1 (1 PP, 2 P*p*, 1 *pp*)
- The F_1 phenotypic ratio is 3:1 (3 polled, 1 horned)

Homozygous (Dominant) × Heterozygous (PP × P*p*) A cross between two cattle, one homozygous and one heterozygous for the polled trait, is set up as follows:

Male Gametes

Female Gametes

	P	P
P	PP	PP
p	P*p*	P*p*

- The F_1 genotypic ratio is 1:1 (2 PP, 2 P*p*)
- All the F_1 are polled

Homozygous (Dominant) × Homozygous (Recessive) (PP × *pp*) A cross between two cattle, one homozygous (dominant) and one homozygous (recessive) for the polled trait, is set up as follows:

Male Gametes

Female Gametes	P	P
p	P*p*	P*p*
p	P*p*	P*p*

- All the F_1 are heterozygous, with the genotype P*p*
- All the F_1 are polled

Heterozygous × Homozygous (Recessive) (P*p* × *pp*) A cross between two cattle, one heterozygous and one homozygous (recessive) for the polled trait, is set up as follows:

Male Gametes

Female Gametes	P	*p*
p	P*p*	*pp*
p	P*p*	*pp*

- The F_1 genotypic ratio is 1:1 (2 P*p*, 2 *pp*)
- The F_1 phenotypic ratio is 1:1 (2 polled, 2 horned)

Homozygous × Homozygous (*pp* × *pp*) (Both Recessive) A cross between two horned cattle, both homozygous (recessive) for the polled trait, is set up as follows:

Male Gametes

Female Gametes	*p*	*p*
p	*pp*	*pp*
p	*pp*	*pp*

- All the F_1 are homozygous for the horned trait, with the genotype *pp*
- All the F_1 are horned
- If the parents are homozygous recessive, all the F_1 must be homozygous recessive

Multiple Gene Pairs

When more than one trait is considered, the possible genotypes and phenotypes increase. For example, if a polled, black cow (P*p*B*b*) is crossed with a polled, black bull (P*p*B*b*), both animals are heterozygous for the two traits. The Punnett square is set up as follows:

Male Gametes

Female Gametes	PB	P*b*	*p*B	*pb*
PB	PPBB	PPB*b*	P*p*BB	P*p*B*b*
P*b*	PPB*b*	PP*bb*	P*p*B*b*	P*pbb*
*p*B	P*p*BB	P*p*B*b*	*pp*BB	*pp*B*b*
pb	P*p*B*b*	P*pbb*	*pp*B*b*	*ppbb*

- The genotypic ratio of the F_1 is 1:2:2:4:1:1:2:2:1 (1 PPBB, 2 PPB*b*, 2 P*p*BB, 4 P*p*B*b*, 1 PP*bb*, 1 *pp*BB, 2 P*pbb*, 2 *pp*B*b*, and 1 *ppbb*)
- The phenotypic ratio of the F_1 is 9:3:3:1 (9 polled, black; 3 polled, red; 3 horned, black; and 1 horned, red)

It is not possible to predict which calf will result from any given mating; the genetic combinations occur by chance. The predicted ratios occur only when a large number of matings occur. The number of offspring necessary in heterozygous matings to approximate the predicated ratios increases rapidly as the number of gene pairs being considered increases (Table 9-4).

Incomplete Dominance

When a heterozygous condition exists for a given trait, one allele is not always dominant over the other. **Incomplete dominance** occurs when the alleles at a gene **locus** are only partially expressed. This usually produces a phenotype in the offspring that is intermediate between the phenotypes that either of the alleles would express. Some references refer to this phenomenon as blending inheritance or codominance.

TABLE 9-4 The Number of Offspring Required in Heterozygous Matings to Get the Predicted Ratios and the Number of Genotypes and Phenotypes That Result (where *n* equals the number of pairs of genes involved)

Number of Pairs of Heterozygous Genes (*n*)	Number of F_1 Required to Get Predicted Ratio[a] $(4)^n$	Number of F_1 Genotypes $(3)^n$	Number of F_1 Phenotypes $(2)^n$
1	4	3	2
2	16	9	4
3	64	27	8
4	256	81	16
5	1,024	243	32
6	4,096	729	64
7	16,384	2,187	128
8	65,536	6,561	256
9	262,144	1,9,683	512
10	1,048,576	5,9,049	1,024
11	4,194,304	17,7,147	2,048
12	16,777,216	53,1,441	4,096
13	67,108,864	1,594,323	8,192
14	268,435,456	4,782,969	16,384
15	1,073,741,824	14,348,907	32,768

[a] This is also the total number of possible combinations of gametes.

Codominance

Codominance occurs when neither allele in a heterozygous condition dominates the other and both are fully expressed. Note how this definition differs from the definition for incomplete dominance.

The roan color of Shorthorn cattle is an example of codominance. The roan color appears to be an intermediate color between red and white. When the hair coat is examined closely, it is actually a mixture of red hairs and white hairs. The individual hairs are not a blend of red and white. To illustrate this trait, let R = red and W = white. If a red animal that is homozygous for red (RR) is mated with a white animal that is homozygous for white (WW), all the F_1 are roan:

Male Gametes

Female Gametes	R	R
W	RW	RW
W	RW	RW

In a mating of the F_1, the F_2 has a genotype ratio of 1:2:1 (1 RR, 2 RW, 1 WW). The phenotype ratio of the F_2 is also 1:2:1 (1 red, 2 roan, 1 white):

Male Gametes

Female Gametes	R	W
R	RR	RW
W	RW	WW

Sex-Limited Genes

The phenotypic expression of some genes is determined by the presence or absence of one of the sex hormones; its expression is limited to one sex. These are known as **sex-limited genes**. An example of this is the male and female plumage patterns in chickens. The neck and tail feathers are long, pointed, and curving in male chickens that have the cock-feathering plumage pattern. The hen-feathering and cock-feathering plumage patterns are controlled by a pair of autosomal genes (Table 9-5). The dominant gene, H, produces the hen-feathering plumage pattern if either sex hormone is present and the recessive gene, *h*, produces the cock-feathering plumage pattern if the female sex hormone is absent and hen-feathering if it is present.

TABLE 9-5 Effect of a Sex-Limited Gene on Plumage Patterns in Chickens

Genotype	Female Phenotype	Male Phenotype
HH	Hen-feathering pattern	Hen-feathering pattern
H*h*	Hen-feathering pattern	Hen-feathering pattern
hh	Hen-feathering pattern	Cock-feathering pattern

Sex-Influenced Genes

Some traits are expressed as dominant in one sex but recessive in the other sex; this action is called **sex-influenced genes**. In humans, male pattern baldness is an example of sex-influenced genes. Among farm animals, examples of sex-influenced genes include horns in sheep and color spotting in cattle. In sheep, the allele, H, for horns is dominant in the male; the allele, *h,* for polled is dominant in the female. The homozygote HH produces horns in both males and females; the homozygote *hh* produces the polled trait in both males and females. In cattle, mahogany and white spotting is dominant in males and recessive in females; red and white spotting is dominant in females and recessive in males.

Sex Determination

Mammals

The sex of the offspring is determined at the moment of fertilization. In addition to the regular chromosomes, the female mammal has two sex chromosomes shown as XX. Male mammals have sex chromosomes shown as XY. After meiosis, all the egg cells will have an X chromosome, but only one-half the sperm cells will have an X chromosome. The other half of the sperm cells will have a Y chromosome. Thus, the sex of the offspring is determined by the male parent. This can be shown by the use of the Punnett square:

Male Gametes

Female Gametes	X	Y
X	XX	XY
X	XX	XY

One-half of the offspring are females (XX). One-half of the offspring are males (XY).

Birds

In poultry, the female determines the sex of the offspring. The male carries two sex chromosomes shown as ZZ. The female sex chromosomes are shown as ZW.

After meiosis, all the sperm cells carry a Z chromosome. Only one-half of the egg cells carry a Z chromosome; the other half carry a W chromosome. The determination of the sex of the offspring by the female can be shown by use of the Punnett square:

Male Gametes

Female Gametes	Z	Z
Z	ZZ	ZZ
W	ZW	ZW

As can be seen in the Punnett square, one-half of the offspring are males (ZZ), and one-half of the offspring are females (ZW).

Sex-Linked Characteristics

Genes that are carried only on the sex chromosomes are called sex-linked genes. An example of a sex-linked trait is the barred color in chickens. Barred color (B) is dominant to black color. The gene for barred color is carried only on the sex chromosome. The color gene is indicated as a superscript on the sex chromosome. The results of crossing a barred female (Z^BW) with a black male (Z^bZ^b) are shown as follows.

Male Gametes (columns) / **Female Gametes** (rows)

	Z^b	Z^b
Z^B	Z^BZ^b	Z^BZ^b
W	Z^bW	Z^bW

All of the male offspring will be barred and carry a recessive gene for black. All of the females will be black and carry no gene for barred color.

More proof that the barred color is sex-linked is found by crossing a black female (Z^bW) with a barred male (Z^BZ^B). All of the offspring are barred.

Male Gametes (columns) / **Female Gametes** (rows)

	Z^B	Z^B
Z^b	Z^BZ^b	Z^BZ^b
W	Z^BW	Z^BW

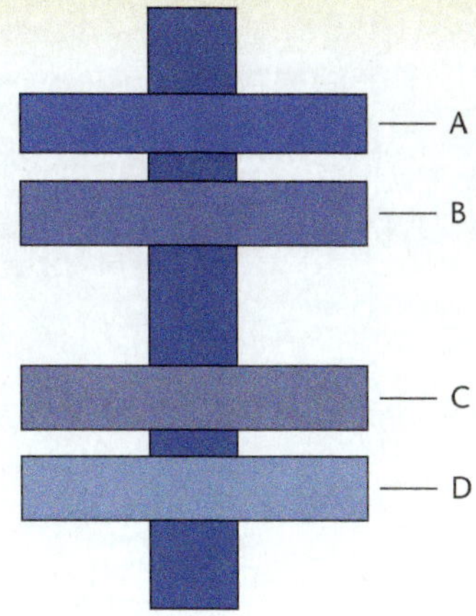

Figure 9-7 Gene linkage. Genes A and B will tend to stay together when the chromosomes divide, as will C and D. A and D are not as likely to stay together because they are farther apart.

Linkage

Early studies in genetics were based on the idea that all genes are redistributed in each mating. It was seen, however, that some groups of traits seemed to stay together in the offspring. More work showed that there is a tendency for certain traits to appear in groups in the offspring. This is called linkage. Traits can be sex-linked, but sometimes genes that are not on the sex chromosomes also seem to be linked together. This is because they are found on the same chromosome. The closer the genes are located together on a chromosome, the more likely they are to stay together (Figure 9-7).

Crossover

The fact that the predicted results of a mating do not always happen is sometimes the result of crossover. During one stage of meiosis, the chromosomes line up together very closely. Sometimes the chromosomes cross over one another and split (Figure 9-8). This forms new chromosomes with different combinations of genes. The farther apart two genes are on a chromosome, the more likely they are to end up in a new combination.

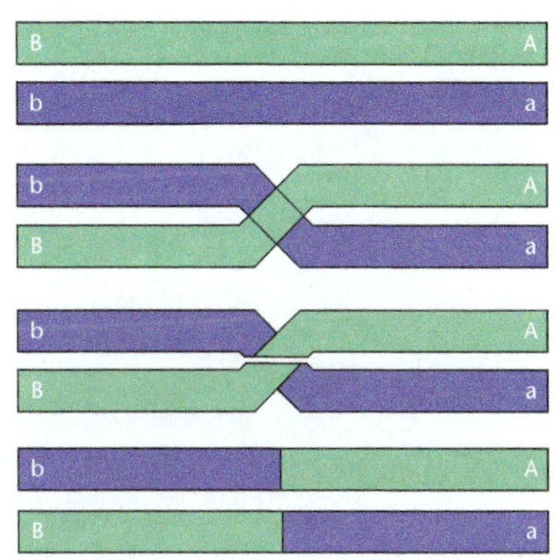

Figure 9-8 New combinations of genes are formed when chromosomes cross over and split.

Mutation

Usually genes are not changed from parent to offspring. However, sometimes something happens that causes a gene to change or mutate. A new trait shown that did not exist in either parent is called a mutation. Mutations may be beneficial, harmful, or not make much difference to the animal. Animal breeders do not depend on mutations to improve animals.

An example of a trait that is believed to be the result of a mutation is the polled Hereford. A cross between horned Herefords result in a polled animal. The polled condition is dominant. For the calf to be polled, at least one of its parents would have to be polled. Therefore, the polled calf from horned parents can be explained as being a mutation.

SUMMARY

Much of the improvement in livestock is the result of using the principles of genetics. The work of Gregor Johann Mendel proved that parents pass their traits to their offspring. The amount of difference between parents and offspring is caused by genetics and the environment. Heritability estimates are used to show how much of the difference in some traits might come from genetics.

An animal's body is made up of millions of cells. Animals grow by cell division. The nucleus of the cell contains chromosomes, which are found in pairs. One chromosome of the pair comes from the father, and one comes from the mother.

Ordinary cell division is called mitosis. In mitosis, each new cell is exactly like the old cell. The reproductive cells are called gametes. Gametes divide by a process called meiosis. In meiosis, the chromosome pairs split, and each chromosome of a pair goes to a different gamete.

The male gamete is called a sperm. The female gamete is called an egg. Fertilization occurs when the sperm cell penetrates an egg cell. The chromosome pairs are formed again when fertilization takes place.

Genes control an animal's traits. Genes are complex molecules found on the chromosomes in pairs. Some traits are controlled by a single pair of genes. Other traits are controlled by combinations of genes. Genes control traits by either additive gene effects or nonadditive gene effects. Traits that are highly heritable are controlled by additive gene effects. Traits that are improved by crossbreeding are controlled by nonadditive gene effects.

Some genes are dominant; others are recessive. Dominant genes hide or mask the effect of recessive genes. Some genes are neither dominant nor recessive, and result in incomplete dominance, or a mixture of the two gene effects.

An animal may carry two dominant or two recessive genes for a trait. These are called homozygous pairs. Some animals have a dominant and a recessive gene, forming a heterozygous pair. It is possible to predict the results of mating two animals by using a Punnett square.

The sex of mammals is determined by the male. The sex of poultry is determined by the female. Some characteristics are sex-linked. The genes for those traits are on the sex chromosomes. Some traits tend to always appear together in the offspring. This is because the genes for these traits are on the same chromosomes. However, chromosomes may exchange genes in a process called crossover. This results in new combinations of traits in the offspring. Genes are sometimes changed by mutation. Animal breeders do not depend on mutations to improve animals.

Student Learning Activities

1. Take a field trip to a local livestock producer. Observe the animals and make a list of traits that you think are (a) due to genetics and (b) due to environment. Suggest traits that should be selected for improving the offspring of the observed animals.

2. Prepare an oral report, including visual aids, on any aspect of genetics studied in this chapter.
3. Prepare a bulletin board display showing mitosis and meiosis or transmission of traits by mating individuals with a given genetic makeup.
4. When planning and conducting a supervised experience program in animal production, use the principles of genetics for selecting and using breeding stock.

Discussion Questions

1. What is a heritability estimate, and how is it used to improve livestock through breeding?
2. Name the parts of a cell.
3. How many pairs of chromosomes do each of the following animals have: (a) cattle, (b) swine, (c) sheep, (d) goats, (e) horses, and (f) chickens?
4. Describe mitosis.
5. Describe meiosis.
6. What is fertilization?
7. Why are genes important in animal breeding?
8. Name and briefly describe the two ways in which genes control inherited traits.
9. Define *dominant gene* and *recessive gene.*
10. Define *homozygous* and *heterozygous* gene pairs.
11. Demonstrate the use of the Punnett square to predict the traits of the offspring when the male and female carry heterozygous gene pairs of a given trait.
12. Define and give an example of *incomplete dominance.*
13. How is sex of the offspring determined in mammals?
14. How is sex of the offspring determined in poultry?
15. Define and give an example of a *sex-linked characteristic.*
16. Define *linkage, crossover,* and *mutation.*

Review Questions

True/False

1. Genetics accounts for all differences in animals.
2. The female determines the sex of mammals.

Multiple Choice

3. In poultry, the sex of the offspring is determined by ______.

   a. male
   b. female
   c. both a and b
   d. neither a nor b

4. Genotype refers to the ______.

   a. number of genes
   b. placing of genes
   c. splicing of genes
   d. combination of genes

5. When chromosome pairs are divided in such a manner that each gamete has one of each type of chromosome, the process is called ______.

   a. mitosis
   b. cell division
   c. sequencing
   d. meiosis

6. The diploid number of chromosomes in cattle is ______.

   a. 19
   b. 27
   c. 30
   d. 32

Completion

7. When cells divide to form specialized tissues and organs, this process in animals is called ______.
8. The ______ ______ molecule is shaped like a double helix.

Short Answer

9. What are the two factors responsible for genetic variation in animals?
10. When some traits are expressed as dominant in one sex but recessive in the other sex, what is this action called?
11. Define *mutation*.

Chapter 10

Animal Reproduction

Key Terms

reproduction
sexual reproduction
asexual reproduction
copulation
embryo
parturition
scrotum
sterile
ridgeling
cryptorchidism
testicles
testosterone
epididymis
vas deferens
spermatic cord
urethra
urine
seminal vesicles
seminal fluid
prostate gland
Cowper's gland
semen

Objectives

After studying this chapter, the student should be able to

- identify and describe the male and female reproductive organs.
- describe the function of the endocrine glands and hormones in reproduction.
- describe reproductive failures that may occur.
- define fertilization, gestation, parturition, and estrus cycle.

REPRODUCTION

When organisms multiply, or produce offspring, it is called **reproduction**. Reproduction may be sexual or asexual. **Sexual reproduction** involves the union of a male and a female gamete. **Asexual reproduction** does not involve the gametes. Simple cell division in bacteria is an example of asexual reproduction. All of the common farm animals reproduce by sexual reproduction.

Sexual reproduction begins with the mating of the male and female. This is called **copulation**. The male gamete (sperm) is placed in the reproductive tract of the female. The sperm moves toward the egg cell. Fertilization occurs when the sperm penetrates the egg cell. The new animal, called the **embryo**, begins to grow. It is fed and protected in the female reproductive tract until it is born. **Parturition**, the act of giving birth, is the final step in reproduction.

Key Terms *continued*

penis
sigmoid flexure
retractor muscle
sheath
alimentary canal
papilla
ovaries
estrogen
progesterone
corpora lutea
oviducts
infundibulum
uterus
cervix
vagina
bladder
vulva
clitoris
funnel
magnum
isthmus
estrus
ovulation
atrophy
fraternally related
identical
gestation
umbilical cord
placenta
diffusion
dystocia
colostrum
antibodies
afterbirth
incubation
cyst

MALE REPRODUCTIVE SYSTEM

Mammals

The male has special organs for reproduction. The reproductive organs of the bull are shown in Figure 10-1. The reproductive organs of other male mammals are similar to those of the bull.

The **scrotum** is the saclike part of the male reproductive system found outside the body cavity that contains the testicles and the epididymis. The testicles are held in the abdominal cavity of the fetus. After the animal is born, the testicles descend into the scrotum. Muscle tissue in the scrotum raises or lowers the testicles in response to the ambient temperature. During cold weather this muscle contracts, raising the testicles closer to the body, and during hot weather it relaxes, allowing the testicles to hang further from the animal's body. The temperature in the scrotum is slightly below the body temperature of the animal, allowing spermatogenesis to occur. Spermatogenesis cannot occur at body temperature. It is reduced if the ambient temperature is too high, producing a temporary reduction in fertility.

If the testicles of an animal are held in the body cavity, the animal is **sterile**, or cannot produce live sperm. A **ridgeling**, or ridgel, is a male in which one or both testicles are held in the body cavity. This is also called **cryptorchidism**, and it is an inherited trait. The animal is usually sterile if both testicles are in the body cavity. If one testicle is retained in the body cavity and the other descends into the scrotum, the animal will be fertile. An animal with cryptorchidism should not be used for breeding.

The **testicles** produce the sperm cells and the male hormone, **testosterone**. The presence of testosterone maintains the masculine appearance of the animal.

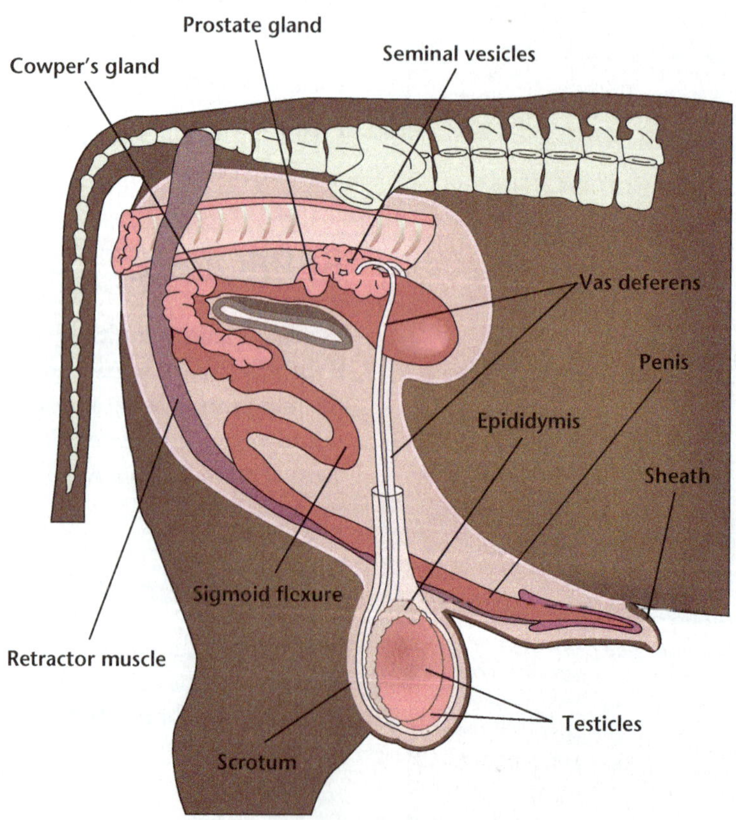

Figure 10-1 Reproductive organs of a bull.

A male that is castrated at an early age does not develop the typical masculine appearance of the species, and the reproductive organs do not continue to develop. When an adult male is castrated, the reproductive organs diminish in size and lose some of their function.

The **epididymis** is a long, coiled tube that is connected to each testicle. Sperm cells are stored in the epididymis while they mature. Sperm cells that are not moved out of the epididymis by ejaculation during copulation are eventually reabsorbed by the body.

The **vas deferens** is a tube that connects the epididymis with the urethra. Sperm cells move through the vas deferens to the urethra. The vas deferens is inside a protective sheath called the **spermatic cord**.

The **urethra** is the tube that carries urine from the bladder. This tube is found in both male and female mammals. In the male animal, both semen and urine move through the urethra to the end of the penis. The **urine** is the liquid waste that is collected in the bladder. The semen contains the sperm and other fluids that come from accessory glands. The three accessory glands are the seminal vesicles, the prostate gland, and Cowper's gland.

The **seminal vesicles** open into the urethra. They produce **seminal fluid**, which protects and transports the sperm.

The **prostate gland** is near the urethra and the bladder. It produces a fluid that is mixed with the seminal fluid.

The **Cowper's gland** produces a fluid that moves down the urethra ahead of the seminal fluid. This fluid cleans and neutralizes the urethra. This helps protect the sperm as they move through the urethra. The mixture of the seminal and prostate fluid and the sperm is called **semen**.

The **penis** deposits the semen within the female reproductive system. The urethra in the penis is surrounded by spongy tissue that fills with blood when the male is sexually aroused. This causes an erection that is necessary for copulation to occur. The **sigmoid flexure** (found in bulls, rams, and boars) and the **retractor muscle** extend the penis from the **sheath**, a tubular fold of skin. Horses and other mammals do not have a sigmoid flexure. After copulation, the blood pressure in the penis subsides, and the retractor muscle helps draw the penis back into the sheath.

Poultry

The reproductive system of the male chicken is shown in Figure 10-2. The testicles (which are held within the body cavity) produce the sperm and seminal fluid. The vas deferens carries the seminal fluid and sperm cells to the cloaca. The cloaca is the enlarged part where the large intestine joins the end of the alimentary canal. The **alimentary canal** is the food-carrying passage that begins at the mouth and ends at the vent. The **papilla** is the organ in the wall of the cloaca that puts the sperm cells into the hen's reproductive tract.

FEMALE REPRODUCTIVE SYSTEM

Mammals

The female has special organs for reproduction that are very different from the male reproductive system. The female produces sex cells in the form of eggs or ova (singular, ovum). The female must also provide the place for the fetus to grow. The fetus is the unborn animal in the later stages of its development. In the early stages of development it is called the embryo. A side view of the reproductive system of the female cow is shown in Figure 10-3. A top view of the female reproductive system

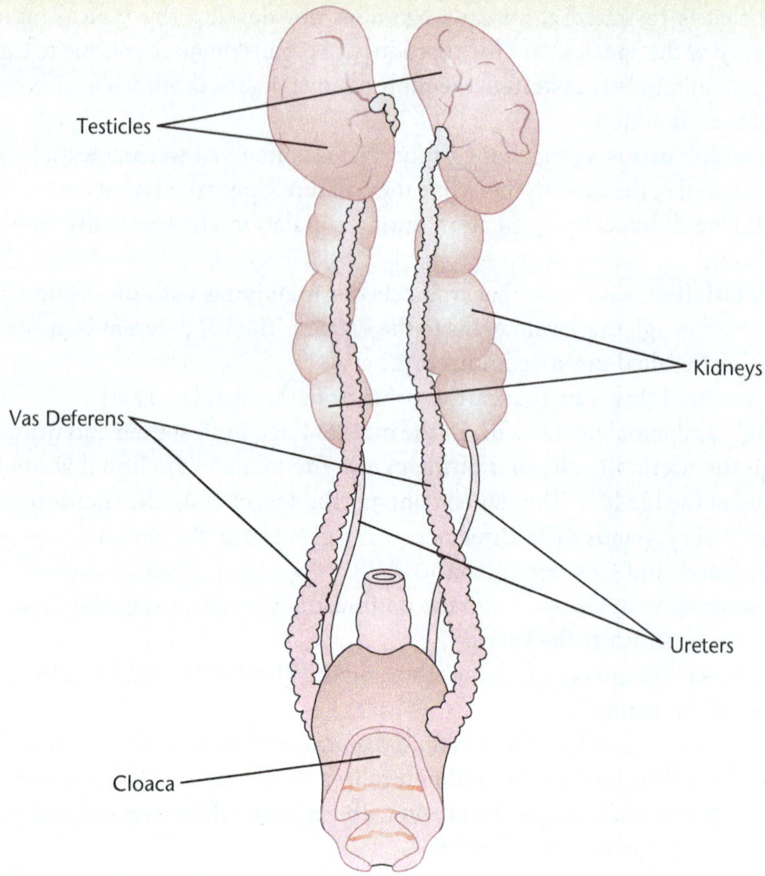

Figure 10-2 Reproductive system of the male chicken.

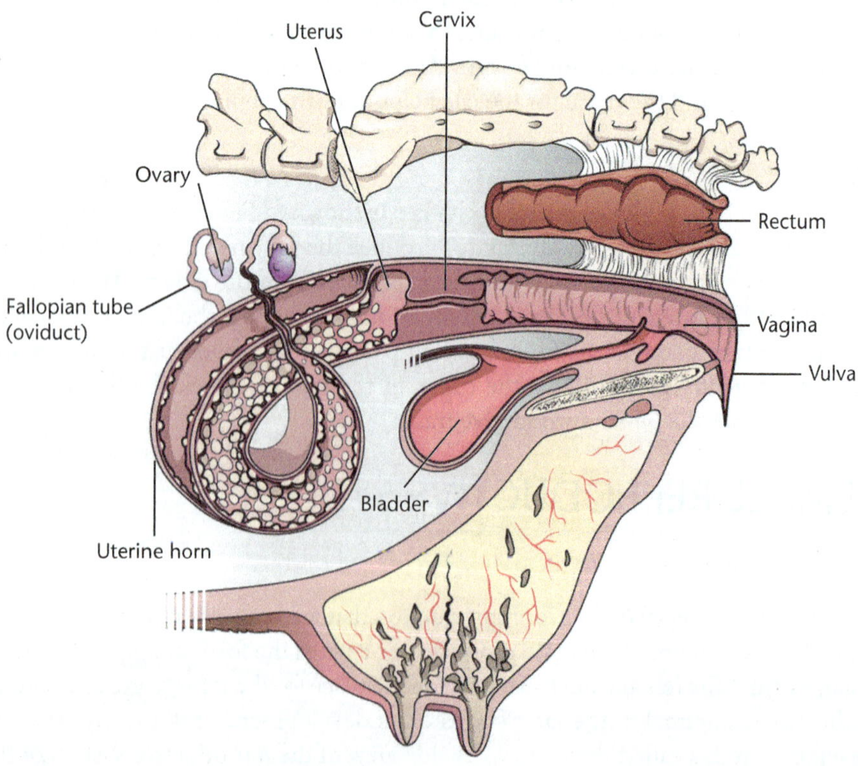

Figure 10-3 The female reproductive system of a cow, side view.

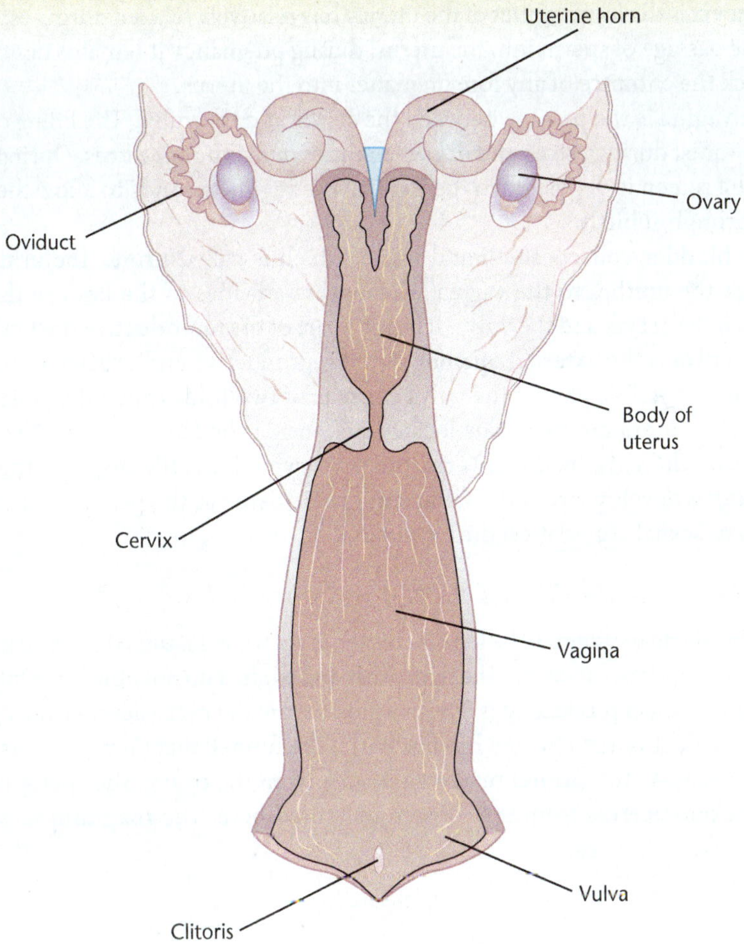

Figure 10-4 Reproductive system of a female cow, dorsal view.

is shown in Figure 10-4. The reproductive systems of other female mammals are similar to that of the cow.

Female farm mammals have two **ovaries** that produce the ova and two female sex hormones (**estrogen** and **progesterone**). There are hundreds of tiny follicles on each ovary. In cows, follicles are about the size of the head of a pin. The ova are produced in the follicles. The ovaries also form the **corpora lutea** (singular corpus luteum).

The **oviducts** are two tubes that carry the ova from the ovaries to the uterus. The oviducts are also called the Fallopian tubes. The oviducts are close, but not attached, to the ovaries. The funnel-shaped end of each oviduct that is close to the ovary is called the **infundibulum**. At ovulation the follicle ruptures, releasing an ovum that is caught by the infundibulum. After copulation, sperm move through the uterus to the oviduct. Fertilization of the ovum occurs in the upper end of the oviduct. The zygote (fertilized egg cell) moves to the uterus about 2 to 4 days after fertilization.

The **uterus** of mammals is a Y-shaped structure consisting of the body, two uterine horns, and the cervix. The size and shape of the uterus varies among the various species of farm animals. The upper part of the uterus consists of the two uterine horns that develop into the oviducts or Fallopian tubes. Mammals that normally produce litters, such as swine, have relatively large horns and a small body; those that normally produce single offspring or twins, such as cattle and sheep, have smaller horns and a larger body. In all of these species, pregnancy normally occurs in the uterine horns; in horses, pregnancy normally occurs in the body of the uterus. In all species of farm animals, the fetus grows within the uterus, where it remains until parturition.

The **cervix** is the lower outlet of the uterus. It is relatively relaxed during estrus to allow the passage of sperm into the uterus; during pregnancy it remains tightly closed to block the entrance of any foreign matter into the uterus.

The **vagina** is the passage between the cervix and the vulva. The lining of the vagina is moist during estrus and dry when the animal is not in estrus. During copulation, the semen is deposited in the vagina. The vagina expands to allow the fetus to pass through at birth.

The **bladder** collects the liquid waste, which is called urine. The urine passes through the urethra to the vagina. The urethra attaches to the floor of the vagina between the cervix and the vulva. It is not a part of the reproductive tract in females.

The **vulva** is the external opening of the reproductive and urinary systems. The exterior and visible part of the vulva consists of two folds called the labia majora. The labia minora are two folds located just inside the labia majora. Also located just inside the vulva is the **clitoris**, the sensory and erectile organ of the female. The clitoris develops from the same embryonic tissue as the penis in the male and produces sexual stimulation during copulation.

Poultry

The reproductive system of the female chicken is shown in Figure 10-5. The chicken has two ovaries and two oviducts. The right ovary and oviduct do not function. Only the left ovary and oviduct produce eggs. The ova produced in the ovary develop into egg yolks.

The oviduct of the chicken has five parts: the funnel, magnum, isthmus, uterus, and the vagina. The **funnel** receives the yolk from the ovary. The sperm cells that the chicken receives from the rooster are stored here. The **magnum** secretes the thick white of the egg.

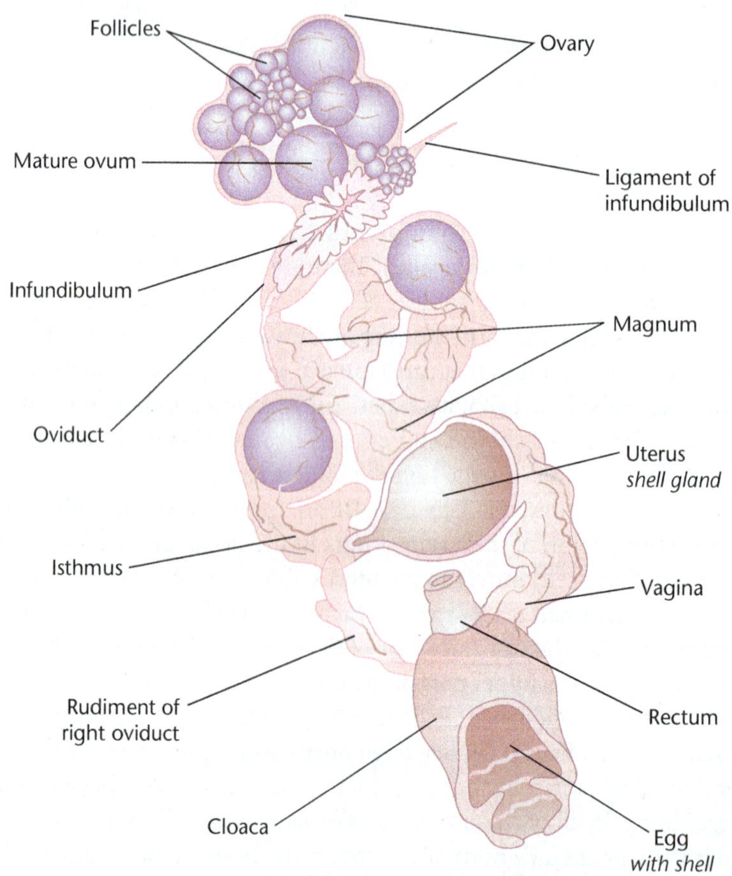

Figure 10-5 Reproductive system of the female chicken.

It takes about 3 hours for the thick white to be placed around the yolk in the magnum. The yolk and the thick white next moves to the **isthmus**, where two shell membranes are added. It takes about 1¼ hours for the shell membranes to be placed around the yolk and thick white.

The thin white and the outer shell are added to the egg in the uterus. The egg remains in the uterus about 20 hours. After the egg is completed, it moves to the vagina, where it stays for a short time, and then is laid. It takes about 25 to 27 hours for a chicken to produce one egg.

ESTRUS CYCLE

The **estrus**, or heat, period is the time during which the female will accept the male for copulation or breeding. The female mammal begins to have estrus periods when it is old enough to be bred. The estrus cycle begins when a follicle on the ovary begins to develop. The hormone estrogen is produced, and causes the animal to show the signs of estrus.

The signs of estrus in cattle include:

- standing when mounted by another cow (best indicator for time to breed, see Figure 10-6)
- nervousness
- swelling of the vulva
- inflamed appearance around the lips of the vulva
- frequent urination
- mucus discharge from the vulva
- trying to mount other cattle (cattle not in estrus may do this)

The signs of estrus in swine include:

- frequent mounting of other sows
- restless activity
- swelling of the vulva
- discharge from the vulva
- frequent urination
- occasional loud grunting

Image courtesy of James Lawhead, DVM

Figure 10-6 Standing estrus, or standing heat, may be determined by changes in animal behavior such as standing while mounted by a bull or another female. The lower cow is in standing heat.

Sheep do not show any visible signs of estrus. The only way to tell if the ewe is in estrus is if she accepts the ram. A teaser ram with an apron to prevent breeding is sometimes used to see if the ewe is in estrus. The apron prevents the ram from completing the act of copulation when mounting the ewe. Most sheep have seasonal estrus periods. They come into estrus only in the fall. Dorset and Tunis breeds, however, come into estrus the year around. Seasonal estrus in sheep seems to be the result of the shorter hours of daylight and the cooler temperatures in the fall.

The signs of estrus in goats include:

- nervousness
- riding other animals and standing when ridden
- shaking the tail
- frequent urination
- bleating
- swelling, red appearance of the vulva
- mucus discharge from the vulva

Dairy and Angora goats are seasonal breeders just as sheep are, but Spanish goats will breed year-round.

The signs of estrus in horses include:

- raised tail
- relaxation of the vulva
- frequent urination
- teasing of other mares
- apparent desire for the company of other horses
- slight mucus discharge from the vulva

Horses are partially seasonal in their breeding habits. The breeding season for horses is generally March through May. They have an irregular estrus cycle in the fall and early winter.

Table 10-1 shows the estrus cycles and other breeding information for the common species of farm animals.

TABLE 10-1 Estrus Cycles and Reproductive Traits[1]

Species	Age at Puberty (months)	Length of Cycle (days)		Length of Estrus		Time of Ovulation	Best Time to Breed	Length of Gestation (days)	
		Range	Average	Range	Average			Range	Average
Cattle	4–12 (6–8 more common)	16–24	21	6–35 hours	16–18 hours	20–40 hr from beginning of estrus 10–14 hr after end of estrus	1st, early in estrus; 2nd, 12–20 hr after start of estrus	278–289	283
Swine	4–7	18–24	21	1–5 days Sows: 40–60 hr Gilts: 24–28 hr	3 days	18–60 hr after start of estrus (24 hr before end of estrus)	Sows: last ½ of estrus; Gilts: 2nd day	111–115	114
Sheep	4–8 (16–20 Merino)	14–20	16	1 3 days	30 hr	Near end		144–152	148
Goats	1st autumn	12–30	22	2–3 days	2 ½ days	Near end	Last ½ of estrus	140–160	151
Horses	10–12	10–37	22	1–37 days	6 days	1 to 2 days before end of estrus	3rd day; 2nd time; if still in heat, 3 days later	310–370	336

[1]The literature reveals a great deal of variation in information concerning estrus cycles and reproductive traits. Information given here is based on most frequently mentioned data.

OVULATION

Ovulation is the release of the egg cell from the ovary. The number of young that an animal gives birth to at one time is an indication of the number of egg cells released. The time of ovulation is usually near the end of the estrus period. Some animals ovulate after the estrus period ends.

Before ovulation, the egg cell is contained in the follicle. The follicle breaks open, releasing the egg. The egg moves into one of the oviducts.

If live sperm are present, the egg may be fertilized. Shortly after ovulation, the corpus luteum forms on the ovary. It releases the hormone progesterone. This hormone causes four things to happen:

1. The fertilized egg (embryo) is implanted in the uterus.
2. Other eggs are stopped from forming.
3. The pregnant condition is maintained.
4. The mammary glands begin to develop. The mammary glands produce the milk to feed the young when they are born.

If the egg is not fertilized, the corpus luteum does not grow. It **atrophies**, or wastes away. This allows another follicle to grow and another estrus period to occur. The time between estrus periods is the amount of time it takes for a new follicle to grow.

Animals that have several young at one birth release more than one egg at ovulation. These offspring are said to be **fraternally related**. They each come from a different egg cell. Sometimes one egg cell divides to form two animals. These two animals are **identical**.

FERTILIZATION

Fertilization is the union of the sperm and the egg cells. During copulation, the male animal deposits sperm in the reproductive tract of the female. The sperm move through the reproductive tract of the female until they reach the infundibulum. If an egg cell is present, a sperm cell may penetrate it. Only one sperm cell can fertilize an egg cell. Many millions of sperm cells are present in the reproductive tract as a result of mating. This helps to make sure that at least one sperm will fertilize the egg cell.

Sperm cells cannot live very long in the female reproductive tract. In cattle, for example, sperm cells live only about 24 to 30 hours after mating occurs. The egg cell lives about 12 hours after it is released if it is not fertilized. For fertilization to occur, the animal must be bred to have the live sperm cells and the egg cell present at the same time. If fertilization does not occur, the egg and sperm cells are absorbed by the body. If the animal does not become pregnant, the estrus cycle will repeat itself.

GESTATION

The **gestation** period is the time during which the animal is pregnant. During pregnancy the fetus develops in the uterus. The fetus is surrounded by a watery fluid enclosed in membranes. Blood vessels in the umbilical cord supply nutrients and oxygen and carry off waste products. The **umbilical cord** connects from the navel of the fetus to the placenta. The **placenta** lies along the wall of the uterus. Food, oxygen, and wastes are exchanged with the mother through the placenta by a process called **diffusion**.

The fetus grows slowly. Most of its growth is in the last one-third of the gestation period. Early in the growth period, the head, nervous system, and blood vessels develop.

The bones and limbs are developed later. The position of the fetus shifts and changes during the gestation period.

PARTURITION

Parturition is the process of giving birth to the new animal. Near the end of the gestation period, the corpus luteum reduces the production of progesterone. There is an increase in the amount of estrogen in the body. This causes the uterine muscles to contract. The contraction of these muscles begins the process of birth.

The first water bag (part of the membranes that surrounded the fetus) soon appears. It gets larger and breaks open. Shortly thereafter, the second water bag, which contains the fetus, appears. The second water bag breaks open, and the presentation of the fetus begins. In cattle, normal presentation (position of the fetus at birth) is the front feet first. This is followed by the nose (Figure 10-7). Then the head, shoulders, middle, hips, rear legs, and feet appear.

Difficulty in birthing is called **dystocia**. Dystocia can be caused by a number of problems, including the fetus being in the wrong position for birth. But by far the leading cause of dystocia is a fetus too large for the birth canal. Selection of lower birth weight sires, especially for heifers and gilts, etc., and using pelvic dimensions as a selection criterion for females can help eliminate birthing problems. Sometimes, the young animal is still enclosed in the membranes when it is born. An attendant should be present at the time of birth to free the animal so that it does not suffocate if this occurs.

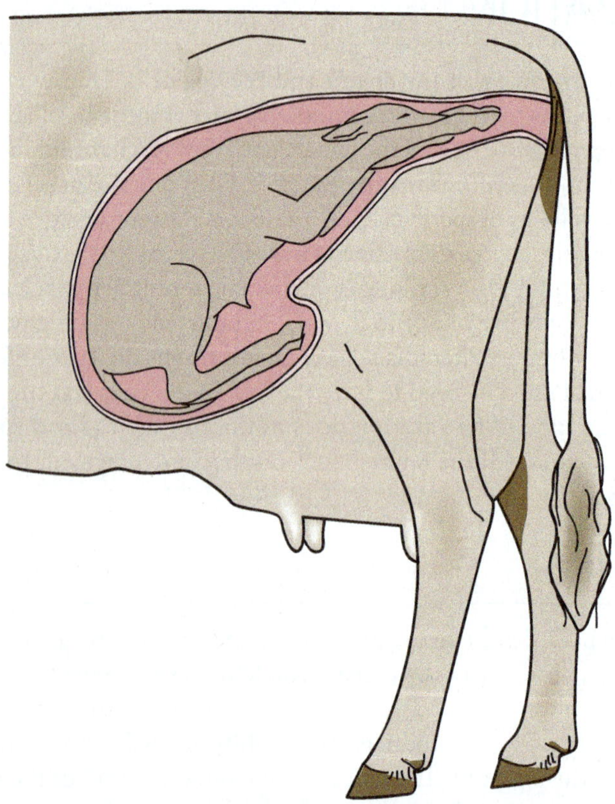

Figure 10-7 In cattle, the normal position of the calf when expelled from the uterus is the front feet first, followed by the nose, then the head, shoulders, middle, hips, rear legs, and feet.

Connection: Colostrum

Colostrum is a nutritional type of milk produced by mammals in the late stages of pregnancy. This thick, yellow, sticky substance is full of antibodies and immunoglobins, which help prevent disease in newborns.

Babies tend to thrive more when fed colostrum (Figure 10-8). It is important that newborn mammals nurse to receive colostrum within the first 12 hours after birth. After about 12 hours, a newborn's digestive system will not be able to utilize colostrum to its full benefit.

Colostrum helps protect newborns in the first 10 to 15 days of life. After this, the newborn's own immune system develops to protect the animal.

In some cases, cattle colostrum supplements are used for human consumption. Athletes sometimes take supplements containing colostrum to build muscle, prevent disease, and improve performance.

Figure 10-8 All female mammals, including humans, produce colostrum just before or after giving birth.

The umbilical cord is broken at birth. This causes the animal to begin breathing. Because the hormone progesterone has stimulated the mammary glands of the mother, she will normally have milk for the young animal to nurse. This first milk is called **colostrum**. It is rich in antibodies, vitamins, and minerals that the newborn animal needs. **Antibodies** are substances that protect the animal from infections and poisons. It is important that the newborn animal be given colostrum milk during the first 12 to 24 hours after birth. After the first day or so, the mother produces regular milk instead of colostrum.

After parturition, the afterbirth is expelled from the uterus. This may take several hours. The **afterbirth** consists of the placenta and other membranes that were not expelled earlier when the fetus was born. If the afterbirth is not expelled, it will decay inside the uterus, making the animal sick. If the afterbirth has not been expelled within a few hours, the help of a veterinarian may be needed. Care must be taken to prevent infection when removing the afterbirth from the uterus. The animal may become sterile if proper care is not taken when removing the afterbirth.

Do not allow the animal to eat the afterbirth. Remove it from the pen or area and bury it in lime or burn it. This helps prevent odors and the spread of disease.

REPRODUCTION IN POULTRY

Reproduction in poultry is different in some ways from reproduction in mammals. The young are not carried in the hen's body as they are in mammals. Instead, they develop in the fertilized egg outside of the hen's body.

The process begins with the placing of sperm into the oviduct of the female by the male. The papilla in the cloacal wall of the male deposits the sperm in the cloacal wall of the female. The sperm move up the oviduct to the funnel of the oviduct, where the egg is fertilized. Sperm cells will remain in the oviduct for 2 to 3 weeks after mating. They have full fertilizing ability for about 6 days. After that time, the ability of the sperm to fertilize the egg is decreased. By the tenth day after mating, the sperm have about 50 percent of their fertilizing ability. The fertilizing ability of the sperm is reduced to about 15 percent by the nineteenth day after mating.

After the egg yolk is fertilized, it moves through the reproductive tract of the female. The thick white, shell membranes, thin white, and the shell are added as it moves through the tract. After the egg is laid, the embryo grows inside the shell. It must have the right temperature and humidity to develop properly. During this time, the embryo is fed by the contents of the egg. After incubation, the chicken hatches, or breaks out of the shell.

Incubation of eggs is keeping them at the right temperature and humidity for hatching. An individual hen does this by sitting on the eggs in the nest. Commercial hatcheries use mechanical incubators to hatch chickens and other poultry.

The incubation period for chickens is 21 days. For ducks and turkeys, it is 28 days. For geese it is 29 to 31 days. Muscovy ducks have an incubation period of 33 to 35 days.

The temperature in the incubator should be 102°F to 103°F. The relative humidity in the incubator should be 60 percent for the first 18 days. It should be increased to 70 percent during the last 3 days of incubation. The eggs are laid horizontally in the tray. They should be turned twice daily for the first 15 days. For ducks, they should be turned for the first 22 days. This keeps the embryo from sticking to the shell. The incubator should have a small amount of air moving through it. This provides fresh oxygen and removes the carbon dioxide that collects in the bottom of the incubator.

REPRODUCTIVE FAILURES

The general physical condition of the animal has an effect on its ability to reproduce. An animal that is too fat or too thin may not become pregnant when bred. Proper nutrition and exercise will help prevent this problem. Animals in poor physical condition may have trouble giving birth.

There are several infections that affect the reproductive organs of the animal. Some may prevent pregnancy and others may cause abortion. If the animal does become pregnant, it may deliver weak young that may not live. Infection of the uterus or general poor health of the animal may cause difficulty in giving birth.

The sexual behavior of animals is affected by the secretion of hormones. When these hormones are not properly secreted, the animal may not be able to reproduce. Sometimes it is necessary to treat the animal with hormones to overcome this problem.

A **cyst** is a swelling containing a fluid or semisolid substance. Cysts may occur in the reproductive organs and cause breeding problems. Some cysts can be removed by surgery or treated with hormones. Not all cysts can be treated. Cryptorchidism may cause reproductive failure in the male. Not all female animals become pregnant when bred. The reasons for this reproductive failure are not always known.

SUMMARY

The mating of the male and female animal begins the process called reproduction. The female gamete is called the ovum or egg cell. The male gamete is called the sperm. The sperm cell is deposited in the female reproductive tract where it moves to the oviduct and fertilizes the egg cell.

The female has an estrus cycle during which she will accept the male for mating. The egg cell is released during the heat period.

If the egg cell is fertilized, it moves to the uterus, where it grows into the fetus. Hormones control the release of the egg cell and maintain pregnancy in the animal. When the fetus has reached the right time for birth, the uterine muscles contract, which forces the fetus through the birth canal.

Reproduction in poultry is similar in some ways to reproduction in mammals. The main difference is that the embryo develops in the eggshell outside of the mother's body. Eggs must be kept at the proper temperature and humidity to hatch. This is usually done in an incubator.

Student Learning Activities

1. Take a trip to a packing plant and observe the reproductive organs of male and female livestock.
2. Secure and dissect reproductive organs of male and female animals. Identify each organ by name and record its function.
3. Ask a veterinarian to talk to the class about problems associated with animal reproduction and birth.
4. Using visual aids, give an oral report on reproduction, gestation, or parturition.
5. Provide appropriate assistance in some phase of animal reproduction when planning and conducting a supervised experience program in animal production.

Discussion Questions

1. Name and briefly describe the parts of the reproductive systems of the bull and the cow.
2. Name and briefly describe the parts of the reproductive systems of the male and female chicken.
3. Describe the signs of heat in each of the following: (a) cattle, (b) swine, (c) sheep, (d) goats, and (e) horses.
4. Describe ovulation in the mammal.
5. Explain how the egg cell is fertilized in the mammal.
6. Describe what happens during gestation in the mammal.
7. Describe parturition in the mammal.
8. Briefly describe the reproduction process in poultry.

Review Questions

True/False

1. Asexual reproduction involves the union of a male and female gamete.
2. In poultry, the magnum secretes the thick white of the egg.
3. The main difference between reproduction in poultry and mammals is that in poultry the embryo develops outside the mother's body.
4. Cryptorchidism may cause reproductive failure in the male.

Multiple Choice

5. In which animal is the papilla a reproductive organ?

   a. horses c. poultry
   b. sheep d. cattle

6. The gland near the urethra and bladder that produces a fluid that mixes with the seminal fluid is called the ______.

   a. prostate c. penis
   b. vas deferens d. scrotum

7. What is the term for the period of time when a female will accept the male for breeding?

   a. gestation c. estrus
   b. conception d. parturition

Completion

8. ______ of eggs is keeping them at the right temperature and humidity for hatching.
9. ______ is the process of giving birth to the new animal.
10. In males, the ______ ______ open into the urethra.

Short Answer

11. Why are the testicles in males raised and lowered as the weather changes?
12. How long can sperm live in the female reproductive tract in cattle?

Chapter 11

Biotechnology in Livestock Production

Key Terms

agricultural biotechnology
biotechnology
transgenic
clone
genetic engineering
somatotropin
bovine somatotropin
recombinant bovine somatotropin (rbST)
superovulation

Objectives

After studying this chapter, the student should be able to

- define agricultural biotechnology.
- discuss the use of genetic engineering in animal science.
- discuss several current genetic engineering activities that have the potential to have a major impact on the livestock industry.
- discuss problems relating to the use of genetic engineering in animal science.
- list some current research projects in genetic engineering relating to plant science.
- discuss the use of embryo transfer in animal science.

BIOTECHNOLOGY

The human race has always used some form of biotechnology to change plants and animals or their products for commercial use. The science of altering genetic and reproductive processes in animals and plants is called **agricultural biotechnology**. This includes activities such as livestock breeding, crop improvement, and the production of food products.

In the 1970s, laboratory techniques were developed that allowed researchers to identify and manipulate the DNA that is found in the cells of all living organisms. Information contained in the DNA determines the characteristics of the organism. The term **biotechnology** generally refers to manipulating the DNA of living organisms at the molecular and cellular level to produce new commercial applications.

Connection: Transgenic Xenografts

People who need organ transplants often wait months or years for available donors, and many die before receiving a transplant. To alleviate the shortage, scientists have created xenografts, or xenotransplants. A *xenograft* is a procedure in which tissues or organs from one animal species are transplanted into a recipient animal of a different species.

Other than primates, the best match for human organs is the pig (Figure 11-1). The pig is used extensively in xenographic research. Their organs are similar in size and function to those of humans.

Xenografts have been attempted since 1906, but rejection of the organ by the recipient's immune system has always been a problem. Rejection may eventually result in organ failure and death.

Transgenic pigs are produced by inserting human genes into pig embryos. The first transgenic pig was born in 1985. Researchers are now pioneering research to create "designer pigs," modified with the future organ recipient's DNA. The rejection response of the recipient's immune system is expected to be weaker—or may be eliminated. Once fully grown, transgenic pigs could be harvested for their organs.

Figure 11-1 Other than primates, the best match for human organs is the pig.

Biotechnology is being used in a number of ways.

- genetic engineering
- identification of an individual organism by its DNA sequence (DNA fingerprinting)
- embryo transfer
- semen sexing
- cloning of animals
- rapid diagnosis of infectious diseases
- diagnosis of genetic disorders
- development of vaccines
- identifying genes that control specific traits

There is a need to increase food production to feed an expanding world population. The use of biotechnology has the potential to significantly increase food production and reduce the cost of production and yet cause less damage to the environment. The United States has led the world in the development and use of biotechnology in agriculture. There is resistance to the use of genetically modified organisms (GMOs) both in the United States and other parts of the world. This resistance may have an impact on the export of food products from the United States.

REGULATION AND SAFETY IN THE USE OF BIOTECHNOLOGY

In the United States, there are three federal agencies involved in the regulation of biotechnology in agriculture and testing for the safety of its products. These are the United States Department of Agriculture (USDA), the Food and Drug Administration (FDA), and the Environmental Protection Agency (EPA). Most state governments also monitor biotechnology.

The USDA is responsible for meat and poultry products, the FDA for all other domestic and imported foods and animal drugs, and the EPA for pesticides. The same standards of safety apply to foods and food ingredients produced by biotechnology that apply to all other foods and food ingredients. Permits must be secured for testing and evaluation of products produced by biotechnology. The product must be shown to be safe for human consumption and must not have a negative impact upon the environment. In the case of animal drugs, rigorous testing is required to show that the meat, milk, or eggs from animals treated with the drugs are safe for human consumption.

There is some controversy regarding the labeling of products produced by biotechnology. Some groups believe that all such products should be labeled as genetically modified. The FDA has yet to require the labeling of products just because biotechnology was involved in the production of those products. However, in situations where genes for proteins that some people are allergic to are involved, labeling may be used to let people know that a specific protein may be present. Labeling may also be required if the nutrient content of the food is altered.

Biotechnology is being used to develop resistance to insects in some plants. Because the EPA regulates the development and use of pesticides, it is involved in the regulation and safety assurances in the development of plants that are resistant to insects.

The three federal agencies involved in regulation and safety in the use of biotechnology cooperate with each other and with individual states as circumstances require. Typically, states become involved when field testing is done in their state or products are being moved in or out of the state. The level of state regulation of biotechnology varies from state to state.

CLONING

When cells or organisms are genetically identical to each other, they are said to be **clones**. Some organisms reproduce asexually, which results in new organisms that are genetically identical to the parent. Most organisms that reproduce asexually are single-celled microorganisms such as bacteria. Asexual reproduction usually does not occur in higher, multi-celled organisms. However, cloning does occur naturally in some invertebrates. For example, if an earthworm is cut into two pieces, each piece will grow into a new, complete earthworm that is a clone of the original.

Experimental work on frogs in the 1950s revealed that it was possible to clone vertebrate animals. The nucleus is removed from the egg cell (enucleation) and the nucleus of a body cell of an animal from the same species is placed in the egg cell. This results in a new animal that is genetically identical (a clone) to the animal that contributed the body cell nucleus. In these experiments, embryo body cells were used because the cells of the embryo are relatively unspecialized. As an embryo grows, its body cells specialize to form the various structures of the body. For many years, it was generally believed that it was not possible to produce clones from body cells that had matured into specialized structures.

The first successful cloning of a vertebrate organism from mature body cells occurred in Scotland in 1996. Researchers at the Roslin Institute near Edinburgh, Scotland, cloned a sheep, called Dolly, from a mature mammary gland cell of an adult sheep. The technique involved placing the mature mammary gland cell in a solution that stopped its growth by restricting its access to nutrients. Using electricity, the mammary cell was then fused with an enucleated egg cell. After embryo growth began, the egg cell was transplanted into a surrogate ewe to develop into a fetus. More than 200 attempts were needed before this technique resulted in the live birth of a healthy lamb. In addition to the problem of making the technique work, there is an indication that the body cells of the clone are the age of the original donor sheep rather than the chronological age of the clone.

Research is continuing in many laboratories around the world in an effort to develop practical methods of cloning animals (Figure 11-2). A variety of techniques for producing cloned animals are being explored, some of which involve the use of cells that have not yet begun to specialize and others using adult cells.

Interest in cloning animals is high because of the benefits that may result. These include:

- testing disease treatments on clones; because the clones are genetically identical, any differences in results would be due to the treatment.
- duplicating genetically modified animals to provide organs for human organ transplants.
- duplicating genetically modified animals for the production of pharmaceuticals.
- duplicating animals with desirable genes, such as those that result in faster growth, leaner meat production, or higher milk production.
- cloning animals of endangered species to increase their numbers and prevent the extinction of the species.

Courtesy of Dr. Steven Stice, University of Georgia, USA

Figure 11-2 Eight calves cloned from the same cow.

GENETIC ENGINEERING

Genetic engineering, also called recombinant DNA (rDNA) technology, is the process of identifying and transferring a gene or genes for a specific trait from one organism to another. Deoxyribonucleic acid (DNA) is a long strand of genes. Genes control traits of the organism, such as what it will look like and how its parts function. When doing genetic engineering, researchers must first identify the gene that controls the desired characteristic and locate the gene on the DNA strand. Using a restriction enzyme, one that recognizes a particular sequence of bases on the DNA strand, the DNA is cut at specific points to remove the desired gene. The cut piece of DNA carrying the desired gene is then spliced onto the DNA strand of the vector. The vector may be a virus or the plasmid of a bacterial cell. The vector is then placed in the organism where it will produce the desired action. Figure 11-3 shows George and Charlie, the first genetically altered cloned calves.

Courtesy of Dr. Steven Stice, University of Georgia, USA

Figure 11-3 George and Charlie were the first genetically altered cloned calves.

The use of genetic engineering can improve livestock performance in a number of ways (Figure 11-4). The benefits of genetic engineering in livestock production include:

- developing disease-resistant animals.
- developing growth regulators.
- developing new drugs and vaccines.
- specifying the sex of an animal before conception.
- developing animals that use feed more efficiently, grow faster, produce more lean meat, produce more milk, and have greater resistance to disease.
- making the production of pharmaceuticals easier and less expensive.

Genetic engineering is a complex technology. There are many thousands of genes in plants and animals. Often, a characteristic is controlled by a combination of genes rather than by a single gene. Many years of research are sometimes necessary just to identify and locate the gene or genes that control certain characteristics.

In some cases, such as in corn plants, the cell into which the DNA bit is spliced cannot regenerate itself to produce a new corn plant. A process involving tissue culture must be used with corn plants to generate new plants from cells that are used in recombinant DNA research.

BOVINE SOMATOTROPIN

The ability of dairy cows to increase the production of a bovine growth hormone called bovine somatotropin (bST) has been developed through genetic engineering. **Somatotropins** are proteins that affect the utilization of energy in the body. **Bovine somatotropin** causes energy derived from feed to be used for milk production rather than for weight gain, but it does not reduce the energy available for body maintenance. It also increases the amount of energy available by improving the breakdown of fat deposits and increasing appetite.

Small amounts of bST are produced naturally by the pituitary gland of the cow. The amount of bST produced by dairy cows has gradually increased over the years as a result of the selection of high-producing cows for breeding stock. However, the only source of bST for experimental work has been the pituitary glands of dead cows. Through the use of genetic engineering, large quantities of bST can be produced. The manufactured hormone is called **recombinant bovine somatotropin (rbST)**. The gene that controls bST production is spliced into the DNA of bacteria, which is then injected into the cow. Because rbST is a protein, it must be given by injection. If rbST were given orally, it would be broken down in the digestive system.

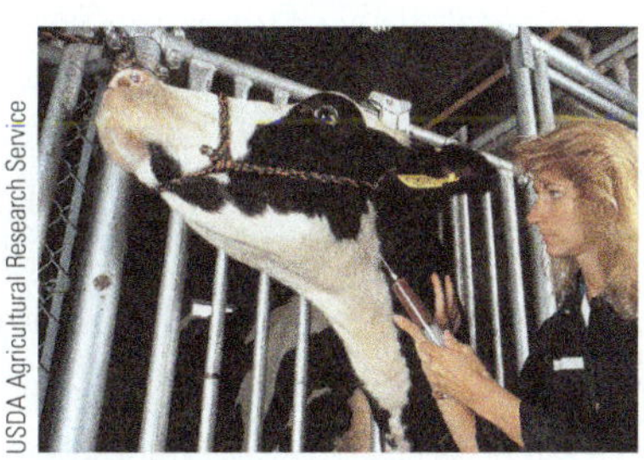
USDA Agricultural Research Service

Figure 11-4 Cattle performance can be improved with biotechnology.

Figure 11-5 Many dairy products are labeled rbST free.

The FDA approved the use of recombinant bovine somatotropin in 1993. Approval came after almost 10 years of testing that determined that milk and meat from cows treated with rbST are safe for human consumption. It became available for use in 1994.

Because bovine somatotropin is a naturally occurring hormone, trace amounts of it can be found in all milk. Therefore, the FDA ruled that milk from cows not treated with rbST cannot be labeled "bST free." Stores and dairies are permitted to use labeling that indicates the milk comes from cows not treated with rbST; however, they must be able to document such a claim (Figure 11-5). There is no requirement by the FDA for the mandatory labeling of milk from rbST-treated cows.

It takes about 2 or 3 days for milk production to increase after beginning treatment with rbST. Feed intake should increase 2 to 3 weeks after milk production increases. Greater attention to feeding and overall management is necessary when using rbST. Success with rbST depends upon feeding a balanced ration with high-quality forage in sufficient quantity to meet the dry matter intake requirements of the cow. It is not recommended for use when only low-quality forages are available. High feed intake is necessary if the use of rbST is to be effective in increasing milk production. Data show that the use of rbST may increase milk production by 10 to 15 percent.

The use of rbST does not overcome problems caused by poor health, poorly balanced rations, inadequate feeding, high somatic cell count, stress from the environment, or breeding problems. Its greatest potential for improvement of milk production is in herds that are already producing above average, have the genetic potential for high milk production, are in good health, and are well managed. There is no indication that the use of rbST causes any health problems or significant increases in mastitis in dairy cows. Because mastitis can be a problem in any high-producing cows, it is important to follow good management practices that reduce the chance for infection.

Consumer resistance to milk from rbST-treated cows may result from the fear of using animal products that contain any kind of additives. Most resistance to the use of rbST has come from consumer advocacy groups that generally oppose the use of all biotechnology in agriculture or the use of animals and animal products for human consumption.

Extensive studies and evaluations regarding the safety of milk and meat from animals treated with rbST have been done both in the United States and throughout the world. All of these studies have reached the conclusion that the use of rbST poses no health problems for humans. A National Institutes of Health study concluded that "composition and nutritional value of milk from rbST-treated cows is essentially the same as that of milk from untreated cows" and "meat and milk from rbST-treated cows are as safe as that from untreated cows." The American Medical Association reviewed the safety of rbST use and concluded that the milk is safe. The World Health Organization and the Food and Agriculture Organization of the United Nations concluded that rbST use presents no health concerns.

It is estimated that the use of rbST may increase overall milk production in the United States by about 2 percent. Some opponents of the use of rbST have expressed concern that its use will force small producers out of dairying and cause the loss of jobs in the dairy industry. Analysis by the USDA suggests that the use of rbST may not have as dramatic an effect on the long-term trend in the dairy industry toward fewer dairy farms, larger herds, and higher production per cow. The use of rbST is not expected to fundamentally change these trends.

PAYLEAN®

A common pig growth promoter is known as Paylean®. Paylean is the only orally active repartitioning agent. It directs nutrients from fat to lean. Paylean is often utilized by people who show hogs, but it is also sometimes used at low levels in commercial finishing operations. The active ingredient in Paylean is ractopamine hydrochloride, which increases feed efficiency, encourages growth in muscles, and lowers the fat content in hogs.

GENETIC ENGINEERING RESEARCH

Some additional areas in which research is being conducted or considered that involve genetic engineering are listed here:

- Identifying genetic markers for cattle and swine that make it easier to breed animals for greater resistance to disease and parasites, better quality meat, improved feed conversion, and higher milk production.
- Identifying the gene that determines the horned condition in cattle. This is the first step in developing naturally hornless cattle, eliminating the need for dehorning. While some cattle have developed the hornless condition through selection, most cattle carry the gene for horns.
- Reducing undesirable side effects of vaccines.
- Developing bacteria for use with non-ruminants to allow them to better use roughages in the ration.
- Modifying rumen microorganisms to better utilize low-quality feeds.
- Developing vaccines for controlling foot-and-mouth disease, bluetongue, and swine dysentery.
- Manufacturing bovine interferon to help control shipping fever and other diseases in cattle.
- Increasing the rate of twinning in cattle.
- Improving resistance to internal parasites.
- Accelerating growth rates in response to a specific feed additive.
- Developing a feed additive that activates the uterine gene, improving embryo survival.

- Developing the ability to screen progeny to determine growth potential, level of disease resistance, reproductive ability, metabolic efficiency, and nutritional capabilities.
- Producing the hormone relaxin for use as a therapeutic agent to help cattle during parturition and reduce calving losses.
- Making a commercially produced scours vaccine, utilizing cell fusion technology to help control calf scours.

OPPOSITION TO BIOTECHNOLOGY

There is some opposition to the development and use of genetically engineered products from people who fear the possibility of the release of some new type of uncontrollable disease or other long-term adverse effect on the environment from the use of these products. They claim that there is not enough knowledge regarding these long-term effects. Their solution to these concerns is to attempt to stop the development of genetically altered products in agriculture. Environmental groups have tried to block the testing of genetically engineered agricultural products by securing court orders prohibiting such testing. While these groups have not succeeded in permanently stopping testing, they have caused delays in the development and testing of genetically engineered products.

PATENTS AND GENETIC ENGINEERING

The U.S. Patent Office has ruled "non-naturally occurring nonhuman, multicellular living organisms, including animals, to be patentable subject matter." This ruling means that new animals developed through genetic engineering can be patented. The effect of this ruling is that anyone using an animal that has been patented must pay a royalty to the patent holder. A question has been raised concerning who owns the genes of the offspring of these animals. This question will probably have to be answered through the courts. Some people believe that the ruling of the Patent Office is desirable because it protects the investment of researchers in the development and testing of these animals. The impact of this ruling on the future of the purebred livestock industry is not clear. Some purebred producers believe that there will still be a demand for seedstock that is genetically desirable for use with other genetically engineered products such as growth hormones. It is also possible, however, that the development of genetically engineered animals with superior desirable characteristics will reduce the need for purebred producers. Superior animals may be supplied to farmers from just a few companies, much as hybrid seed for crops is currently produced and distributed.

EMBRYO TRANSFER

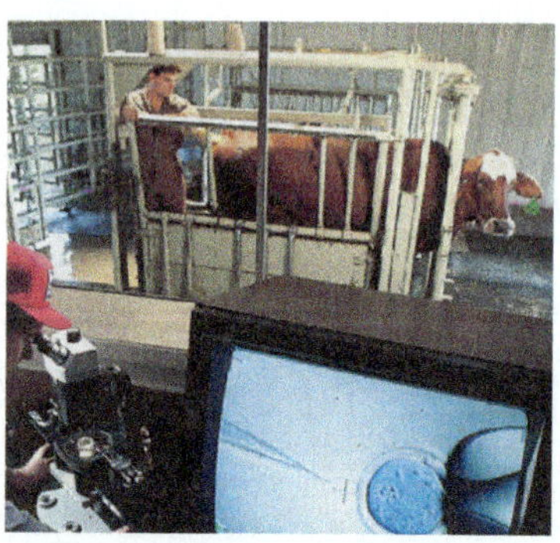

USDA

Figure 11-6 Computer-enhanced embryo transfer.

Embryo transfer, especially in the cattle industry, has become a fairly well-established technology (Figure 11-6). In cattle, the process involves the flushing of embryos from the reproductive tract of desirable donor cows and implanting them in other cows that may be of lower quality. The desirable cows can produce many embryos through a process called **superovulation**.

By utilizing this technology, it is possible to produce many more offspring from desirable cows than would be possible if the cows had to carry each embryo to term before producing another one. The transplanted embryos carry the desirable genetic traits of the donor cow even though they are carried to term in lower-quality host cows.

Other advantages of embryo transfer include:

- proving the productivity of dams and sires more quickly because of the increased number of offspring in a shorter period of time.
- extending the productive life of a female that has been injured and can no longer carry offspring to term.
- developing the ability to import and export desirable genetic traits with embryos when the animals cannot be imported or exported because of potential disease problems.
- developing faster genetic improvement in the herd.

Because of the relative high cost of embryo transfer, its use is generally limited to breeding stock. When the cost of the procedure is reduced, embryo transfer may become common for the average cattle raiser or dairy farmer. The success rate in embryo transfer is about 50 to 60 percent.

Superovulation refers to a process of inducing a cow to produce several oocytes during each estrus cycle. An oocyte is a cell that becomes an ovum that may be fertilized by a sperm cell to produce an embryo. Oocytes are contained in the ovaries of the female. Normally, a cow produces one oocyte per estrus cycle. Superovulation is induced by injecting the cow with a hormone that results in the release of several oocytes. Six to eight oocytes are normally produced by this procedure, although as many as 20 have been produced. Hormones that have been used to induce superovulation include gonadotropins, steroids, and prostaglandins. Progesterone and prostaglandin are two commonly used hormones in embryo transfer programs.

It is important to make sure that donor cows have proper nutrition and are carefully monitored during their estrus cycles. Most cows will come into heat 21 to 60 days after calving. A hormone injection to induce superovulation is generally done after day 60 following calving. Cows that come into heat shortly after day 21 could be injected for superovulation sooner. However, cows that are used more often in a donor program may need more time after calving before the injection is made. Frequent superovulation may result in the development of cystic ovaries. This is not a serious problem if the condition is properly diagnosed and treated.

After ovulation occurs, sperm is introduced into the reproductive tract by artificial insemination. The fertilized ova develop into embryos in 7 days. A saline solution is then used to flush the embryos from the reproductive tract. The embryos may be transferred immediately to another cow or they may be frozen for later use.

In the United States, the importation and exportation of embryos is regulated by the USDA's Animal and Plant Health Inspection Service (APHIS). A major concern is the possibility of transferring diseases through embryos. Research suggests that embryos that are properly handled do not transmit diseases; however, research in this area is continuing.

Some countries have very rigid testing requirements for the importation of embryos, while others are less rigid. APHIS is cooperating with the International Embryo Transfer Society to establish international standards for importing and exporting embryos. Contact the local USDA office for the location of the Animal and Plant Health Services, Veterinary Services division, for information on current regulations governing embryo exports.

SUMMARY

The science of altering genetic and reproductive processes in animals and plants is called agricultural biotechnology. Most of the work currently being done in agricultural biotechnology is with genetic engineering and embryo transfer.

Genetic engineering is based on a technology involving recombinant DNA. This involves taking a tiny bit of DNA containing the desired gene from one organism and splicing it onto the DNA strand in another organism. The recipient organism takes on the characteristic controlled by the transferred gene.

Genetic engineering has been used to increase the level of bovine somatotropin in dairy cows, which results in higher milk production. Paylean, or ractopamine hydrochloride, increases feed efficiency, encourages growth in muscles, and lowers the fat content in hogs. Many other areas of research are currently being conducted with genetic engineering in animal and plant science.

Embryo transfer has become an established technology in cattle production. The use of embryo transfer permits the production of many more offspring from genetically desirable animals. Embryo transfer is generally limited to breeding stock; however, it may become more common for the average cattle raiser.

Student Learning Activities

1. Prepare a bulletin board display of current information on the use of biotechnology in animal and/or plant science.
2. Ask a veterinarian who is familiar with or is using embryo transfer to speak to the class.
3. Interview a veterinarian who is familiar with or is using embryo transfer, and prepare a written report or give an oral report to the class.
4. Use one or more biotechnology methods as appropriate when planning and conducting a supervised experience program in animal production.

Discussion Questions

1. Define *agricultural biotechnology*.
2. Name the two major areas of research currently being conducted in agricultural biotechnology.
3. Define *recombinant DNA technology*.
4. Why is genetic engineering often a difficult process?
5. What are somatotropins?
6. What is the potential for increased milk production from the use of bovine somatotropin?
7. Briefly explain some things that must be considered if recombinant bovine somatotropin is used with a dairy herd.
8. Why is there some controversy concerning the use of bovine somatotropin?
9. List five other areas of research currently being conducted in genetic engineering in animal science.
10. Why are some people opposed to the use of biotechnology?
11. What are some of the possible effects of the U.S. Patent Office ruling concerning the patentability of genetically engineered animal and plant products?
12. List five advantages of embryo transfer.
13. What is the major current disadvantage of embryo transfer?
14. What is superovulation, and how is it induced?
15. What is the most common method of removing fertilized embryos from the reproductive tract of the cow?
16. What agency in the United States regulates the importation and exportation of embryos?

Review Questions

True/False

1. Genetic engineering is based on a technology involving recombinant DNA.
2. Transgenic refers to genetically engineered plants and animals.

Multiple Choice

3. Bovine somatotropin is a protein produced in which gland of the cow?
   a. lymph
   b. pituitary
   c. reproductive
   d. endocrine
4. Experimental work done on ______ in 1950 revealed that it was possible to clone vertebrate animals.
   a. frogs
   b. birds
   c. dogs
   d. rats
5. Genetic engineering has been used to increase the level of bovine somatotropin in cows, which results in higher ______ production.
   a. fat
   b. milk
   c. calf
   d. lean

Completion

6. ______ refers to a process of inducing a cow to produce several oocytes during each estrus cycle.
7. The science of altering genetic and reproductive processes in animals and plants is called ______ ______.

Short Answer

8. A common pig growth promoter is known as __________.
9. When cells or organisms are genetically identical to each other, they are said to be ______.

Chapter 12

Animal Breeding Systems

Key Terms

straightbreeding
purebred
inbreeding
closebreeding
linebreeding
outcrossing
linecrossing
grading up
grade
hybrid

Objectives

After studying this chapter, the student should be able to

- name and explain common breeding systems used in livestock production.
- explain the effects, advantages, and disadvantages of using various breeding systems.
- identify the factors involved in selecting a breeding system.
- calculate the percentage of parental stock in offspring, using various breeding systems.

SYSTEMS OF BREEDING

There are two basic systems of breeding in livestock production: (1) **straightbreeding** and (2) crossbreeding. Mating animals of the same breed is called straightbreeding. Mating animals of different breeds is called crossbreeding. Each system has a place in livestock production and is used for a particular purpose. Both systems have advantages and disadvantages, and there are several variations of each system. Straightbreeding includes purebred breeding, inbreeding, outcrossing, and grading up. The variations of crossbreeding include two-breed crosses, three-breed crosses, and rotation breeding.

The system of breeding to be used depends on the kind of livestock operation in which the animals are bred. Sometimes farmers use more than one kind of breeding system. The size of the herd, amount of money available, and goals of the farmer are other factors considered when selecting a system of breeding.

Purebred Breeding

A **purebred** is an animal of a particular breed that has the characteristics of the breed to which it belongs. Both parents of a purebred animal must have also been purebred. A purebred animal is eligible for registry in the purebred association of that breed if it has no disqualifications. Disqualifications are listed in the rules for registering animals in the breed association. A common disqualification is color markings that purebred breeders regard as undesirable. These are sometimes the result of recessive genes. To register purebred animals, it is necessary to become familiar with the rules for registering animals of that breed and obtain information from the breed association.

The ancestors of a purebred animal can be traced all the way back to the original animals accepted for registry in the herd book of the breed association. There is a tendency for purebred animals to be genetically homozygous. Usually, only a small number of animals were originally accepted in the herd book of the breed association, resulting in some **inbreeding** and **linebreeding** in the early history of the breed association. Inbreeding and linebreeding result in greater homozygosity of the genes in a given line of animals. Undesirable recessive characteristics may appear because of the homozygosity of the genes of the parent animals. Therefore, purebred animals are not necessarily better than nonpurebred animals.

The production of purebred animals is a specialized business. Purebred animals provide the foundation stock for crossbreeding to produce market animals. Purebred breeding requires a higher investment than raising market animals. A purebred breeder generally provides breeding animals for producers raising animals for market and sometimes for other purebred breeders. Purebred breeders often show their animals in purebred shows.

Closebreeding

(A represents the male; B the female)

1st mating: A X B
1st generation: ½ A and ½ B

2nd mating: A X AB
2nd generation: ¾ A and ¼ B

The offspring in the second generation have received 75% of their genetic inheritance from the sire A because he appears closer in the pedigree to the offspring than he does in linebreeding. They have received only 25% of their genetic inheritance from the female B.

Courtesy of Alyssa K. Elrod

Figure 12-1 Closebreeding of a sire to daughter.

Inbreeding

Inbreeding is the mating of related animals. Linebreeding and closebreeding refer to how closely related the animals are that are being mated. The most intensive form of inbreeding is **closebreeding**, in which the animals being mated are very closely related and can be traced back to more than one common ancestor. Examples of closebreeding include sire to daughter, son to dam, or brother to sister (Figure 12-1).

Linebreeding refers to mating of animals that are more distantly related and can be traced back to one common ancestor. Examples are cousins, grandparent to grandoffspring, or half-brother to half-sister (Figure 12-2).

Linebreeding

(A represents the male; B and C represent females)

1st mating: A X B A X C
1st generation: ½ A and ½ B ½ A and ½ C

2nd mating: ½ A and ½ B X ½ A and ½ C
2nd generation: ½ A and ¼ B and ¼ C

The offspring in the second generation have received 50% of their genetic inheritance from the sire A because he appears twice in their pedigree. They have received only 25% of their genetic inheritance form each of the females B and C.

Courtesy of Alyssa K. Elrod

Figure 12-2 Linebreeding of a half-brother to half-sister.

Inbreeding increases the genetic purity of the stock produced. The pairing of the same genes is increased, and the offspring become more genetically homozygous. The result of several generations of inbreeding is a high degree of genetic purity or homozygosity. Undesirable genes and desirable genes become grouped together in the offspring with greater frequency, which can make the undesirable and desirable traits more visible. The breeder can then eliminate animals with undesirable traits from the breeding program. It is possible to keep the desirable traits of an animal in the ancestry of the animals being produced. Inbred animals often transmit desirable genes to their offspring with great uniformity, which could help to improve the breed.

Inbreeding requires a carefully planned program of selection and culling to result in breeding stock with more desirable traits. It is expensive because all animals with undesirable traits must be removed from the breeding program. The average animal breeder generally does not find inbreeding a desirable system of breeding to use. It is often used by universities for experimental work and seedstock breeders that provide animals for crossbreeding in herds producing animals for market.

Outcrossing

Outcrossing is the mating of animals of different families within the same breed. The purpose of outcrossing is to bring traits that are desirable but not present in the original animal into the breeding program. Outcrossing is popular with purebred breeders because it reduces the chances of undesirable traits appearing in the offspring. The genes for those undesirable traits are still present; however, they are covered up by the outcrossing. Outcrossing is sometimes used in combination with inbreeding programs to bring in traits that are needed.

Linecrossing is mating animals from two different lines of breeding within a breed. The purpose is to bring together desirable traits from different lines of breeding. Some lines cross better than other lines because of different gene combinations.

Grading Up

Most of the animals on farms in the United States are not purebreds. **Grading up** is the mating of purebred sires to grade females. The mating of purebred sires with these grade animals is a good way to improve the quality of animals on the farm. A **grade** animal is any animal not eligible for registry. It does not require as much money because only the purebred sires, or their semen, must be purchased. How quickly the line of animals is improved depends on the species of animal. Animals with short generations, such as swine, are improved fairly rapidly. Those with longer generations, such as cattle or horses, take longer to produce improvement.

The greatest percent of improvement comes in the first cross because 50 percent of the genes of the offspring will be from the purebred sire. Second-generation offspring will be 75 percent purebred. The third generation will be 87.5 percent pure. If the use of a purebred sire continues long enough, the amount of grade breeding left in the offspring will be less than 1 percent (Figure 12-3).

The amount of improvement that results is dependent on the quality of sire selected for the breeding program. Most commercial producers get their purebred sires from purebred breeders. It is important to select the highest quality sire with performance records that the commercial breeder can afford. Offspring of grading up are generally not eligible for registry in the breed association because only one parent is registered. However, some breed associations do permit the offspring of grading up to be registered. A few associations are now requiring blood testing as a part of the registration process.

Grading Up

(A_1, A_2, A_3 represent purebred sires of a given breed; G represents a grade female)

1st mating: A_1 X G
1st generation: ½ A_1 and ½ G (50% purebred, 50% grade)

2nd mating: A_2 X ½ A_1 and ½ G
2nd generation: ½ A_2 and ¼ A_1 and ¼ G (75% purebred, 25% grade)

3rd mating: A_3 X ½ A_2 and ¼ A_1 and ¼ G
3rd generation: ½ A_3 and ¼ A_2 and ⅛ A_1 and ⅛ G (87.5% purebred, 12.5% grade)

Courtesy of Alyssa K. Elrod

Figure 12-3 Grading up with purebred sires on grade females.

Crossbreeding

Crossbreeding is the mating of two animals from different breeds and usually results in improved traits in the offspring. The resulting offspring is called a **hybrid**. Dominant genes tend to mask undesirable recessive genes.

Superior traits that result from crossbreeding are called hybrid vigor, or heterosis. Heterosis is measured by the average superiority of the hybrid offspring over the average of the parents. The kind and degree of superiority achieved by crossbreeding varies with different species. Traits with a high degree of heritability show little improvement from crossbreeding, but traits with low heritability usually show the greatest improvement as a result of crossbreeding.

Animals selected for use in a crossbreeding program must have the desired traits. There will be little or no improvement in the offspring over the parents if animals with undesirable traits are used in a crossbreeding program. Regardless of the crossbreeding system used, the producer must follow a good performance selection program, good management, good nutrition, and good herd health practices to achieve the desired results. Research has shown that well-planned crossbreeding programs can increase total productivity in beef herds by 20 to 25 percent.

Beef, swine, and sheep producers usually use crossbreeding for the production of market animals. It is rarely used by dairy producers because they are primarily interested in milk production. In addition, the Holstein breed, which is superior to other breeds in this trait, dominates the dairy industry. Poultry producers typically use strains that have been developed from crossing inbred lines.

Crossbreeding Systems for Beef

The use of crossbreeding in beef cow herds that produce animals for slaughter will generally result in higher profits. Some general considerations regarding a beef crossbreeding program include the following:

- Good records must be kept
- Calving difficulties may increase when crossing large-breed sires with small-breed dams
- Large-breed dams have fewer calving problems, but have higher maintenance costs
- Artificial insemination allows access to better bulls
- To avoid inbreeding, more than one breeding pasture may be required

Crossbreeding systems used with beef cattle range from those that are relatively simple to those that are complex. The following are some typical beef crossbreeding systems.

Terminal Sire Crossed with F_1 Females Replacement F_1 (crossbred) females in the herd are purchased and crossed with a terminal bull. All the offspring are marketed.

Rotate Herd Bull Every 3 or 4 Years The same breed of bull is used for several years then replaced with a bull of a different breed. Replacement females are selected from the herd.

Two-Breed Rotation Bulls from breed A are crossed with cows from breed B. The resulting heifers are bred to bulls from breed B for the duration of their productive life. Replacement heifers chosen from these matings are bred to bulls from breed A. Each succeeding generation of replacement heifers is bred to a bull from the opposite breed used to sire the replacement heifer.

Three-Breed Rotation The pattern of breeding is the same as in a two-breed system, except that a bull from a third breed is used in the rotation of sires.

Four- and Five-Breed Rotations In larger herds, bulls from a fourth or fifth breed may be used in the rotation of sires. This system requires a higher level of management than two- and three-breed systems.

Static Terminal Sire System Four breeding groups are needed for this system, as shown in Figure 12-4. The first group (25 percent of the herd) mates bulls (breed A) and cows (breed A) to produce replacement heifers (AA) for group one and group two. The second group (25 percent of the herd) breeds the AA heifers to a bull (breed B) of a different breed, producing crossbred heifers (breed AB). The third group (50 percent of the herd) breeds the AB heifers to a terminal (T) bull selected for ability to transmit a high rate of gain. A subgroup (group four, 10 percent of the herd) of the third group is composed of AB heifers being bred for the first time. These AB heifers are bred to a smaller breed (breed C) bull to reduce first-time calving problems. All the male offspring of groups one and two and all of the offspring of groups three and four are marketed. Any heifers from groups one and two that are not kept for breeding purposes are also marketed.

Rotational-Terminal Sire System Two breeding groups are needed for this system. Bulls from breeds A and B are used on a rotating basis on 50 percent of the herd, providing crossbred females for the entire herd. Mature cows in the herd are mated with a terminal bull to produce offspring, all of which are marketed.

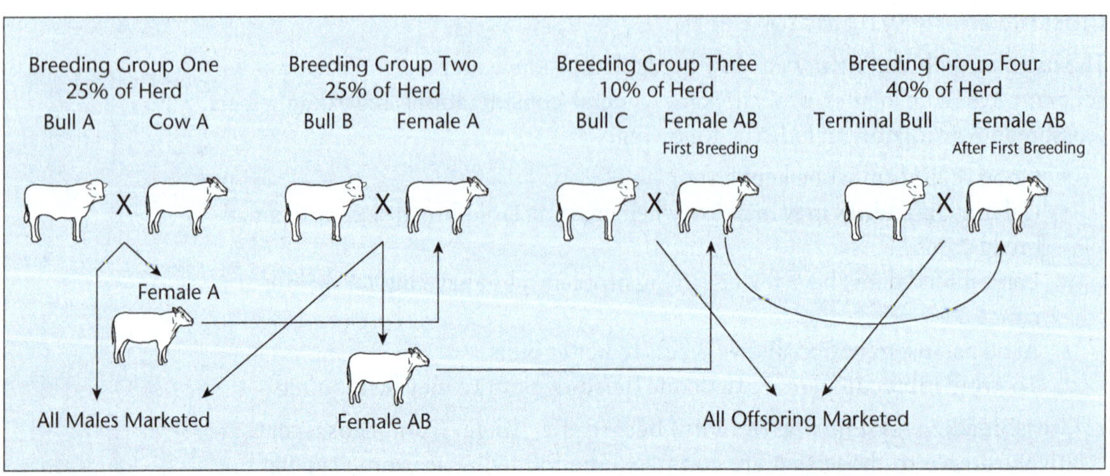

Figure 12-4 Static terminal sire system.

Replacement females generally come from the matings of bulls A and B with younger cows in the herd.

Composite Breeds The development of a new breed based on crossbreeding with four or more existing breeds of cattle to avoid inbreeding problems. After development, the composite breed is not crossbred with other breeds.

Crossbreeding Systems for Swine

Some general considerations related to a swine crossbreeding program include the following:

- Select breeds to use and replacement gilts and boars that meet the objectives of the breeding program.
- Select breeds that produce large litters and heavier weaning weights if they are to be used only on the female side of the matings. The white breeds are generally superior in these traits.
- Select breeds that have less back fat and higher rate of gain if they are used as terminal boars in the crossbreeding program. The Berkshire, Duroc, Hampshire, Poland China, and Spotted breeds are generally superior in these traits.
- If crossbred sows are used, select those that are at least 50 percent Chester White, Landrace, or Yorkshire. These crosses generally have the superior maternal traits that are desirable.
- Select boars that are from sows that rank in the top 25 percent of a herd, as measured by a Sow Productivity Index. Growth rate and back fat thickness are important considerations when selecting boars.
- Crossbred boars may be used in a crossbreeding program. Be sure their parents have desirable traits that fit the objectives of the program.

The following are some typical swine crossbreeding systems.

Rotational Crossbreeding In two-breed systems, a boar from breed A is mated with sows from breed B, producing offspring AB. Selected gilts (AB) are bred to a boar from breed B. Selected gilts from this mating are bred to a boar from breed A. The pattern is repeated, switching back and forth to the breed of the most distantly related boar. Three- or four-breed systems are more commonly used in swine production. The pattern is the same as that used in the two-breed system except that three or four breeds of boars are used in rotation, as shown in Figure 12-5.

Care must be taken to follow the planned order of breeds used in the rotation, or heterosis will be reduced. Because replacement gilts are selected from within the herd, the chance of bringing disease into the enterprise from purchased breeding stock is greatly reduced.

Terminal Crossing System Crossbred (F_1) females, with superior maternal traits, are bred to boars selected for desirable back fat and rate of gain. All of the offspring go to market. The breeder must either keep a separate herd to produce breeding stock or purchase replacement females. The costs involved in this breeding system are generally higher than in rotational breeding systems. There is some increased health risk if new breeding stock is brought into the herd. Terminal crossing does maintain the maximum advantage of heterosis and breed differences in the breeding system.

Rotaterminal System The rotational breeding system and the terminal breeding system are combined in this method of crossbreeding. Crossbred females are produced by breeding boars of different breeds in a rotating pattern to crossbred

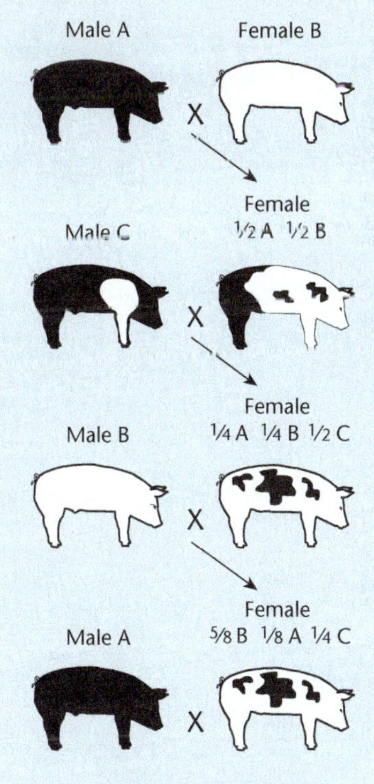

A, B, and C represent different breeds. Continue using males of three breeds in sequence. Using males of the three breeds in sequence is called rotating the males: hence, the inclusion of the term *rotation* in the name of this system.

Figure 12-5 Three-breed rotation cross.

females produced by previous matings in the system. Generally, breeds with good maternal traits are used to produce the crossbred females that are bred to terminal boars of other breeds. Terminal boars are selected for desirable back fat and rate of gain. All the offspring produced in the terminal breeding go to market. This system of crossbreeding maintains a high level of heterosis and allows the producer to select breeds with desirable traits. It does require the use of more boars of different breeds.

Crossbreeding Systems for Sheep

The use of crossbreeding generally increases profits from sheep flocks. Crossbred ewes are hardier, healthier, and produce more milk as compared to noncrossbred ewes. A study of crossbred sheep revealed that they produced higher grease (uncleaned), as well as cleaned, fleece weights compared to noncrossbred sheep.

The following are some typical sheep crossbreeding systems.

Rotational The same breeding pattern is used as in beef or swine rotational systems. The lambs are usually kept for flock replacements.

Static Replacement crossbred ewes for the flock are purchased and bred to a terminal ram. All the lambs are marketed.

Rotostatic A combination of rotational and static crossbreeding systems in which replacement ewes are produced from the flock. These ewes are bred to a terminal ram to produce market lambs. It takes about 25 to 30 percent of the flock to produce the replacement ewes. The best ewes should be kept for producing replacement ewes.

SUMMARY

The two basic systems of livestock breeding are straightbreeding and crossbreeding. The kind of system used depends on the size of the operation, the amount of money available, and the goals of the farmer.

Purebred animals are eligible for registry in the breed association. They tend to be genetically homozygous. Not all purebred animals are better than nonpurebreds. Purebreeding often requires more money and a more precise breeding system than commercial animal production.

Inbreeding increases the genetic purity of the livestock but generally reduces performance. Both desirable and undesirable traits become more visible. Therefore, inbreeding programs require careful selection and culling of breeding stock. Inbreeding is not commonly used by the average livestock producer. It is usually used by those who do experimental work to improve the breed.

Outcrossing brings genetic traits into the breeding program that tend to hide undesirable traits. Grading up is using purebred sires on grade females. Grading up is a good way for commercial producers to improve grade herds of livestock.

Crossbreeding is the mating of animals from two different breeds. Crossbreeding is used by many commercial producers. It usually results in hybrid vigor, which may improve some traits but have little effect on feed efficiency or carcass traits.

Student Learning Activities

1. Give an oral report on one type of breeding system, and explain how a farmer might use it and what results might be expected.
2. Prepare a bulletin board display illustrating matings in a given species and the percent of parental blood in the offspring.

3. Report on breeding systems used on your home farms and why they are used.
4. Interview local farmers and report on the breeding systems they use and the reasons for choosing these systems.
5. Use an appropriate breeding method to mate animals when planning and conducting a supervised experience program in animal production.

Discussion Questions

1. Define *straightbreeding* and *crossbreeding*.
2. What is purebred breeding?
3. Define and give two examples of *inbreeding*.
4. Why is inbreeding more commonly used by universities and seedstock producers than by the average livestock producer?
5. Define *outcrossing*, and tell why it might be used in a breeding program.
6. What is linecrossing, and why might it be used?
7. What is grading up, and why might it be used?
8. Why is crossbreeding used in breeding programs?
9. Name and briefly describe four systems of crossbreeding that might be used in breeding programs.

Review Questions

True/False

1. Inbreeding animals is a way of increasing performance.
2. Crossbreeding improves feed efficiency and carcass traits.

Multiple Choice

3. A(n) ______ animal is one that is of a particular breed.

    a. crossbred c. linebred
    b. purebred d. outcrossed

4. A person interested in registering animals should contact the ______.

    a. USDA c. appropriate breed association
    b. EPA d. cattle producers association

5. What is a common disqualification in rules for registering animals in a breed association?

    a. weight c. color markings
    b. size d. eyes

Completion

6. ______ and ______ are two basic systems for breeding livestock.
7. ______ is mating animals from two different lines within a breed.
8. In ______-______ sire system breeding, two breeding groups are needed for the system; bulls from A and B are rotated on 50 percent of the herd providing crossbred females for the entire herd.

Short Answer

9. Are purebred animals always better than nonpurebred animals?
10. Why is it important to select animals with desired traits when crossbreeding?
11. Why is crossbreeding rarely used with dairy animals?

SECTION 4

Beef Cattle

Chapter 13

Breeds of Beef Cattle

Key Terms

vealers
dewlap

Objectives

After studying this chapter, the student should be able to

- describe the characteristics of the beef industry.
- name and describe the breeds of beef cattle, giving their origin and breed characteristics.
- identify the various breeds of beef cattle by viewing pictures or live animals.

CHARACTERISTICS OF THE BEEF INDUSTRY

About 35 percent of the total income from all livestock and poultry marketing in the United States comes from the beef industry. Marketing of beef accounts for about 15 percent of the total income from all farm marketing (livestock, poultry, and crops) in the United States. The beef industry is the largest single segment of the U.S. agricultural economy. Raising and feeding beef cattle is an important source of income for many farmers.

On average, per capita consumption of beef has decreased since 2005. Consumption is linked to the supply of beef coming to market, the price of beef, and consumer perception that beef has a high cholesterol level. The concern about cholesterol as a health hazard has caused many people to increase their consumption of poultry and decrease their consumption of beef.

The demand for beef depends on (1) the number of people, (2) income per person, and (3) changes in people's meat preferences. When incomes are high, people tend to buy more beef instead of other kinds of meat. They

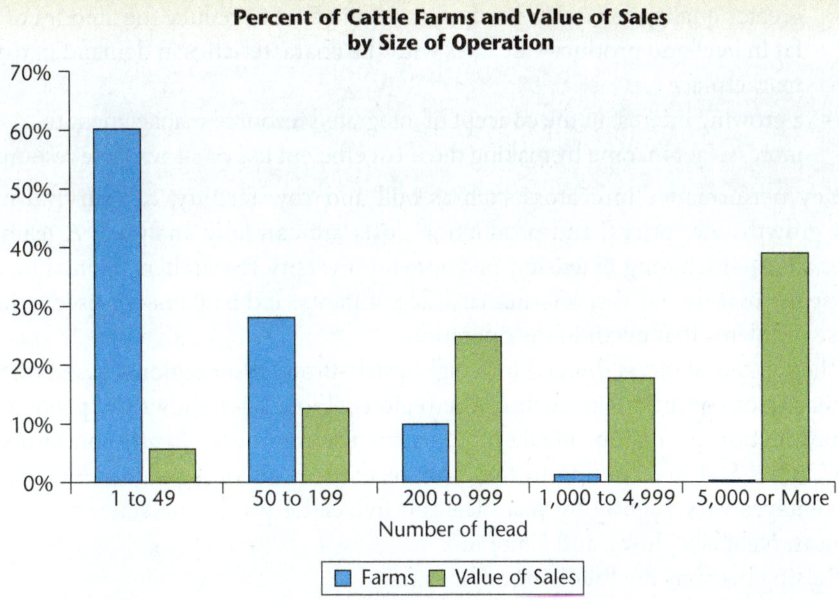

Figure 13-1 Percentage of cattle farms and value of sales by size of operation.
Source: USDA 2012 Census of Agriculture.

also want more grain-fed beef instead of **vealers** (calves grown for veal) and grass-fed cattle. Despite the recent drop in per capita consumption, the demand for beef is expected to continue to be strong in the future.

Most of the beef eaten in the United States comes from domestic production. About 9 percent of the beef supply in the United States comes from imports. About 7 percent of domestic beef production is exported to the world market.

Small-size herds are typical for beef cow-calf operations, as shown in Figure 13-1. About 60 percent of cattle farms have fewer than 50 head of cattle and account for 7 percent of sales. Cow-calf herds ranging in size from 50 to 199 head account for 28 percent of the total operations and make up about 12 percent of sales. About 10 percent of cattle farms have 100 to 499 head and account for 25 percent of sales. About 2 percent of cattle farms have 1,000 to 4,999 head and account for 18 percent of sales. Less that 1 percent of cattle farms have 5,000 or more head of cattle and account for 38 percent of sales.

There is a continuing trend toward fewer small (under 1,000-head capacity) cattle feedlots; however, there has not been any significant increase in the number of large-capacity feedlots. The total number of cattle feedlots in the United States has been declining in recent years. About 95 percent of the total feedlots are small; however, these feedlots feed only about 15 percent of the total cattle coming to market. About one-third of all cattle that are marketed come from a small number of feedlots with more than 32,000-head capacities.

The cattle-feeding industry has shifted in recent years from its formerly heavy concentration in the Corn Belt to the Central and Southern Plains states. Most of the high-capacity feedlots are found in Texas, Kansas, Nebraska, and Oklahoma. There is a trend toward custom feeding cattle in large feedlots, with cow-calf operators retaining ownership of the cattle.

Other changes occurring in the beef industry include:

- efforts to lower unit production cost
- development of closer coordination of breeding, growing, feeding, marketing, and processing operations, generally through contracts rather than through the vertical integration that is found in the poultry industry

TABLE 13-1 Percentage of Cattle Production by Region	
Southwest	23%
North Plains	19%
Southeast	16%
Corn Belt	12%
Mountain	10%
Pacific	8.5%
Lake States	7%
Northeast	4%

Source: USDA NASS 2012

- greater quality control in production in an effort to reduce the amount of fat in beef and produce carcasses with the characteristics in demand in the marketplace
- a growing interest in the concept of integrated resource management to increase net income by making the most efficient use of all available resources

Key performance indicators such as bull and cow fertility, calf survival rate, calf growth rate, prices, and production costs are carefully analyzed. A team of specialists—including extension personnel, university researchers, animal health company representatives, veterinarians, accountants, and bankers—is used to help solve problems that occur in the operation.

The United States is divided into eight cattle-production regions, based on the similarity of conditions in each of the regions. Table 13-1 shows the percentage of production by region. Figure 13-2 shows the numbers of cattle on farms in the United States according to the most recent Census of Agriculture; however, percentages vary slightly by year. The top five cattle producing states are Texas, Kansas, Nebraska, Iowa, and Colorado.

Raising beef has the following advantages:

- Beef use roughages that otherwise would be wasted for feed
- Labor requirements may be low
- Capital investment can be small
- Death losses are usually low
- Cattle are adapted for use in small operations as well as large ones
- There is high demand for meat

Disadvantages of raising beef include the following factors:

- Cattle feeding is a high-risk business
- Cattle are not efficient converters of concentrated feeds into meat
- It takes longer to develop a cattle herd and increase numbers than it does to develop hogs or sheep
- The capital investment in modern, efficient feeding operations can be high

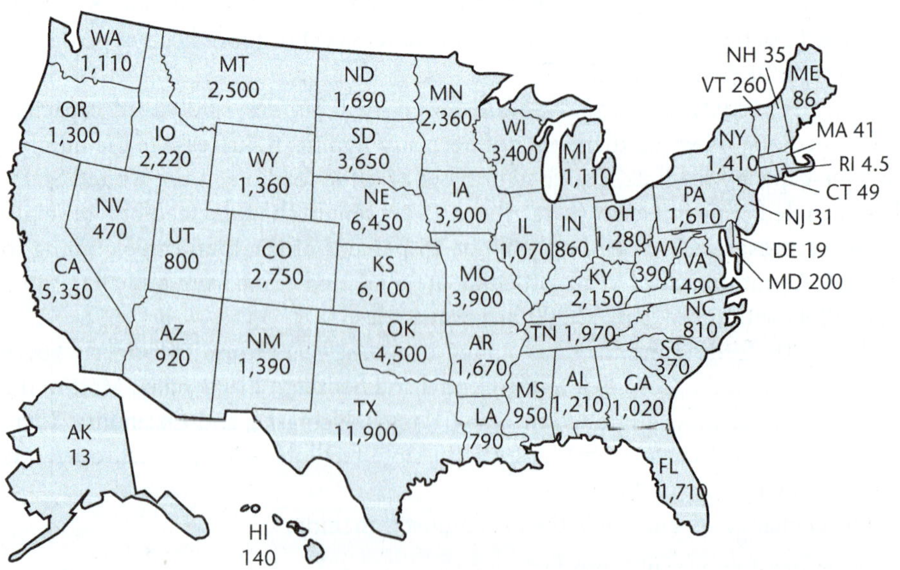

U.S. Total : 90,769

Figure 13-2 Number of cattle by state, in thousands.
Source: USDA NASS 2012

Connection: Freemartin Calves

What's the best way for cattle producers to double their calf crop? By getting their cows to produce twins, right? Twins are an unusual occurrence in cattle, and even though it sounds like a good idea, there are sometimes problems. First, cows are not built to nurse twin calves, and usually at least one of the calves is weak and does not grow well.

The biggest concern with twin calves, however, is the birth of a freemartin calf (Figure 13-3). A freemartin is an infertile female born twin to a fertile male. If both fetuses are the same sex, this is of no significance, but if they are different genders, male hormones pass through interconnected blood vessels from the male twin to the female twin. The male hormones render the female twin infertile—a freemartin. The male twin is relatively unaffected, although decreased testicle size may result in reduced fertility.

The freemartin is genetically and physically a female, but she is sterile and behaves and grows like a steer.

Figure 13-3 A freemartin is a sterile female calf born twin to a fertile male calf.

DEVELOPMENT OF BEEF BREEDS

Many modern beef breeds had their origin in Europe. Selection of cattle that later formed a breed was practiced mainly in the British Isles. Most of this selection began in the latter part of the 1700s.

Selection of animals was based on traits that local farmers considered best for their area. The most desirable animals were kept as breeding stock. The breeder culled (removed) from the breeding herd those animals that did not have the desired traits. This increased the gene frequency of the desired traits in the breeding herd. Outside breeding stock was rarely brought into the herds. The number of animals that had the same kind of traits increased. As these animals became more popular, breed societies were formed. A method of registering animals in the breed was established. Eventually, all animals registered in the breed traced their ancestry to the animals originally registered with the breed society.

Newer breeds of beef have been developed during the twentieth century in the United States. Some of these have been developed from crosses of existing breeds or strains of cattle. For example, the Brahman breed is the result of crossing several Zebu breeds from India. Other hybrid breeds have been the result of selecting crossbred animals that possessed desirable traits. These animals were then bred and the offspring selected for the desired traits.

During the late 1960s and 1970s, a number of so-called exotic breeds of cattle were introduced into the United States from Europe. Among these were the Simmental, Limousin, Blonde d'Acquitaine, Chianina, and Maine-Anjou. Other exotic breeds such as the Charolais, although introduced earlier into the United States, gained increased popularity during the 1970s.

There are more than 50 breeds of beef cattle available to producers in the United States. Many of these breeds are registered in small numbers with their respective purebred associations in any given year. Only a few beef breeds are widely used in commercial production. In rank order, Angus, Hereford, Polled Hereford, Simmental, Limousin, and Charolais have been the leading beef breeds in the United States based on total numbers registered since 1970. Table 13-2 shows the 10 breeds with the highest number of yearly registrations.

TABLE 13-2 Dominant Beef Breeds Based on Average Yearly Registrations
1. Angus
2. Hereford
3. Polled Hereford
4. Simmental
5. Limousin
6. Charolais
7. Red Angus
8. Gelbvieh
9. Shorthorn
10. Salers

Figure 13-4A Modern Angus bull.

Courtesy of American Angus Association

Figure 13-4B Modern Angus cow.

Courtesy of American Angus Association

SELECTION OF A BREED

The following are some points that should be considered when selecting a breed:

- All breeds have both strong and weak traits.
- No one breed is best for all traits.
- Every breed has a wide range of genetic variation.
- The selection of the best animals as breeding stock and the use of good breeding practices are more important than the particular breed selected.
- The breed selected should be one that seems to produce well in the area where it will be raised.
- The market demand for a given breed should be determined.
- Breeding stock should be available at a reasonable cost.

CHARACTERISTICS OF THE BREEDS

Angus

History

The official name of the Angus breed is Aberdeen-Angus, although the breed is often referred to as Black Angus. The breed originated in Scotland in the shires of Aberdeen and Angus. The earliest written records of Angus date to the early 1700s. Around 1800, farmers in Scotland began keeping records of pure breeding. In 1862, the first herdbook of the Angus breed was published.

George Grant, of Victoria, Kansas, imported four bulls into the United States in 1873. These were the first recorded importations of Angus into the United States. The American Angus Association was organized in 1883.

During the late 1800s and early 1900s, Angus were found mostly in the midwestern states. The breed has increased in popularity in the range areas. Today, Angus are found in every state in the United States. For many years, Angus led all other beef breeds in numbers registered.

Figure 13-5A Red Angus bull.

Red Angus Association of America

Description

Angus cattle are black in color (Figures 13-4A and B). They have a smooth hair coat and are polled. They are an alert and vigorous breed. Angus cattle perform well in the feedlot. They produce a desirable carcass of high-quality, well-marbled meat.

Nearly all Angus are pure for the dominant polled gene. When used in crossbreeding, nearly all the calves are polled. A few Angus carry a recessive gene for the red color. Sometimes, a red calf is born to black parents. The red calf is not eligible for registry in the American Angus Association.

Figure 13-5B Red Angus cow and calf.

Red Angus Association of America

Red Angus

History

Red Angus herds were begun in the United States about 1945. The herds developed from crossing red with red from the Black Angus. Because the red gene is recessive, the offspring of the red crosses are always red. Red Angus (Figures 13-5A and B) can be registered in the Angus Herdbook in Scotland. The American Angus Association registered Red Angus until 1917. The Red Angus Association of America was formed in 1954.

Description

Red Angus are similar to Black Angus except for their color. Because red absorbs less heat than black, the Red Angus can tolerate warmer temperatures somewhat better than Black Angus. In the future, as selection standards change, the Red Angus breed may become less like the parent Black Angus.

Photo compliments of Beefmaster Breeders United, San Antonio, Texas

Figure 13-6A Beef master bull.

Barzona

History

The Barzona breed was developed in Arizona beginning in 1942. The breed was begun by F. N. Bard for use in the desert areas of the southwestern United States. The first cross was between an Africander and a Hereford. Two herds were developed from the females from this cross. Santa Gertrudis bulls were used on one herd; Angus bulls were used on the other. Selection was made for fertility, mothering ability, and gain ability. The herd was closed in 1960; that is, no additional outside breeding stock was used. A cow that results from three topcrosses of Barzona registered bulls bred to a cow of a beef herd is registered as purebred. To be eligible for registration, a bull must be of fourth-generation Barzona breeding. The Barzona Breeders Association of America was formed in 1968.

Photo compliments of Beefmaster Breeders United, San Antonio, Texas

Figure 13-6B Beef master cow.

Description

Barzona cattle are nearly solid red with a little white on the underline and around the head. They are good mothers and are able to subsist on a grass-browse range.

Beefmaster

History

The Beefmaster breed (Figures 13-6A and B) began in 1931 in Texas. Further development of the breed took place in Colorado. The breed is the result of crosses among Herefords, Shorthorns, and Brahmans. The crossing was done with bulls from the three breeds under range conditions. The exact percent of blood of each breed is not known. It is estimated that the breed is about 25 percent Shorthorn, 25 percent Hereford, and 50 percent Brahman.

Beefmaster cattle are registered through Beefmaster Breeders United. Registered cattle are either descendants of the original herd or are from three consecutive topcrosses of Beefmaster breeding.

Each breeder uses a prefix, such as *Smith Beefmaster*, to identify cattle from his or her herd.

American Blonde d'Aquitaine Association

Figure 13-7A Blonde d'Aquitaine bull.

Description

The breed has a variety of colors. Reds and duns are more common than other colors. Some of the cattle are horned and some are polled. Selection has been mainly for good disposition, fertility, weight gain, conformation, hardiness, and milk production.

Blonde d'Aquitaine

History

The Blonde d'Aquitaine is a French breed of beef cattle (Figures 13-7A and B). The breed originated in 1961 when several French breeds were combined. In the 1960s, the French Herdbook was established. The cattle were originally used for draft, meat, and milk production. The first Blonde d'Aquitaine cattle were imported into the United States in 1972.

American Blonde d'Aquitaine Association

Figure 13-7B Blonde d'Aquitaine cow and calf.

Smith Bonsmara—Michael G. Smith

Figure 13-8A Bonsmara bull.

Description

The Blonde d'Aquitaine have heavily muscled bodies with deep chests. The hips are wide, and the hindquarters are well developed. Colors are yellow, brown, fawn, or wheat. Mature bulls weigh 2,500 pounds and cows weigh 1,500 pounds.

Bonsmara

History

The Bonsmara is of the Sanga breeds (Figures 13-8A and B). Sanga cattle are a mix of Zebu breeds that influence many indigenous African breeds. Developed in South Africa by Dr. Jan Bonsma, the Bonsmara breed is a composite breed that consists of 5/8 Africander, 3/16 Hereford, and 3/16 Shorthorn. Bonsmara have become increasingly popular in the United States. Bonsmara cattle were first introduced to the United States by Mr. G. R. Chapman, who imported embryos in 1997.

Smith Bonsmara—Michael G. Smith

Figure 13-8B Bonsmara cow.

Description

Bonsmara cattle are red and predominately horned. To be registered, Bonsmara cattle must be dehorned. They are known for their fertility, heat tolerance, and good temperament. They tolerate harsh conditions and less-than-desirable grazing conditions. Bonsmara beef is known for being of good quality.

Braford

History

Courtesy of United Braford Breeders

Figure 13-9A Braford bull.

Development of the Braford breed (Figures 13-9A and B) began in 1947 by crossing Hereford bulls on Brahman cows at the Adams Ranch near Fort Pierce, Florida. Because Hereford bulls were not well adapted to environmental conditions in southern Florida, crossbred Brahman-Hereford bulls that showed desirable characteristics were used to further develop the foundation herd for the Braford breed. Selection of breeding stock is generally made for those traits of highest economic value. Among these are high fertility, ease of calving, high calf survival rate, milking ability, high weaning rate, growth rate, efficient use of roughages in the diet, and adaptability to the environment.

Courtesy of United Braford Breeders

Figure 13-9B Braford cow and calf.

Description

The color of the Braford is red and shows a Hereford color pattern. The breed is about 5/8 Hereford and 3/8 Brahman. Calves grow rapidly and attain weaning weights of 500 to 800 pounds without supplemental feeding. Steers in the feedlot will produce 1,000-pound choice cattle in 12 to 15 months. Mature bulls weigh 1,500 to 2,000 pounds and mature cows weigh about 1,150 pounds. The breed is noted for its superior maternal ability.

Brahman

History

American Brahman Breeders Assn

Figure 13-10A Brahman bull.

The Brahman breed (Figures 13-10A and B) was developed in the southwestern part of the United States. Between 1854 and 1926, about 266 bulls and 22 females of the *Bos indicus* type of cattle were imported from India. *Bos indicus* cattle have a hump over the shoulders. These cattle are also called Zebu. Several strains of the Zebu cattle were bred to females from several British breeds of cattle. Early breeders selected for hardiness and ability to produce in the climate of the Southwest. Beef conformation and early maturity were also selected for in these early matings.

The major use of the Brahman in the United States is in crossing with other breeds. The resulting hybrids have proven to be desirable beef animals. The Brahman has been used in the development of a number of other newer breeds of beef cattle. The American Brahman Breeders Association was organized in 1924.

Figure 13-10B Brahman cow.

Description

The color of the Brahman is light gray or red to almost black. The most common color is light to medium gray. Red is becoming a popular color with a number of breeders. In addition to the characteristic hump over the shoulders, Brahmans have loose skin **(dewlap)** under the throat and large drooping ears.

Mature Brahman bulls weigh about 1,600 to 2,200 pounds. Mature cows weigh from 1,000 to 1,400 pounds. Brahman cattle have a very high heat tolerance. They are also resistant to cattle disease and insects. They are good mothers and have an excellent ability to forage on poor range. They gain rapidly and produce a quality carcass. However, they do tend to have an unpredictable disposition.

Brangus

History

The Brangus breed (Figure 13-11) was developed by crossing Brahman and Angus cattle. Early crossings of these breeds were done at the USDA Experiment Station at Jeanerette, Louisiana. Some of the first crosses were made as early as 1912. The American Brangus Breeders Association was formed in 1949. Since then, the name of the association has been changed to the International Brangus Breeders Association. The Brangus name is a registered trademark. Only animals registered with the breed association can be called Brangus. All present-day Brangus are descendants of foundation animals registered in 1949 or from registered Brahman and Angus cattle that have been enrolled since then. Brangus cattle are based on foundation stock that is 3/8 Brahman and 5/8 Angus. There are several recognized ways to produce Brangus cattle. A 1/4 Brahman, 3/4 Angus crossed with a 1/2 Brahman, 1/2 Angus is one way. Another is crossing an animal that is 3/4 Brahman, 1/4 Angus with a purebred Angus. The third method is mating of registered Brangus animals.

Figure 13-11 Brangus bull.

Description

Brangus cattle are solid black and polled. An inspection is necessary to determine conformation and breed character before the animal may be registered. Brangus are adaptable to different climates. They have good mothering ability, feed efficiency, and produce desirable carcasses.

Red Brangus

History

The Red Brangus breed was developed beginning in 1946 in Texas. Foundation stock came from purebred Angus and Brahman. No percent of blood of either breed is required. However, to be registered, the animal must show traits of both breeds.

Description

Red Brangus are red and are a polled breed (Figure 13-12). Animals that show mostly Brahman and Angus traits are listed as certified Red Brangus. Their offspring may be registered as Brangus if they meet the requirements of the association. In addition to being red and polled, they must meet size and conformation requirements. The

Figure 13-12 Red Brangus bull.

animals are registered with the American Red Brangus Association. The association was formed in 1956.

Charolais

History

Figure 13-13A Charolais bull.

Figure 13-13B Charolais cow.

The Charolais breed (Figures 13-13A and B) is one of the oldest of the French breeds of beef cattle. It was developed around Charolles in central France. In 1930, 2 bulls and 10 heifers were imported into Mexico. Two more importations were made in 1931 and 1937. A total of 37 animals were imported—8 bulls and 29 cows. In 1934, the King Ranch in Texas imported two bulls from Mexico. These were the first Charolais to be imported into the United States. The breed increased in numbers through breeding Charolais bulls to females of other breeds. New breeding stock was imported into Canada and the United States in 1966. Since then, there have been more importations from Canada, the Bahamas, England, Ireland, Japan, and France.

Description

Charolais cattle are white to light straw color with pink skin. They are a large, heavily muscled breed. Mature bulls weigh 2,000 to 2,500 pounds. Mature cows weigh 1,500 to 1,800 pounds. Most are naturally horned. Horns are white, slender, and tapered. Naturally polled Charolais may also be registered. Charolais have a high feed efficiency. They are heavily muscled in the round and loin because of generations of selection for this trait.

The Charolais are well adapted to many areas. They are used in many crossbreeding programs. Charolais are registered with the American-International Charolais Association. The association has an open herdbook. Animals may be registered after five generations of crossing with a Charolais bull.

Chianina

History

The Chianina (pronounced *Key-a-nee-na*) breed (Figure 13-14) originated in the Chiana Valley in Italy. It is one of the oldest breeds of cattle in Italy and probably one of the oldest in the world. The breed was in existence before the time of the Roman Empire. Chianina semen was first imported into the United States in 1971.

Chianina are used in many crossbreeding programs. The breed association is the American Chianina Association. The association does not establish weight or color standards for its members. There are large numbers of quarter-blood and half-blood Chianina in the United States. The color, type, and size vary considerably because of the different kinds of crosses used.

Description

Figure 13-14 Chianina bull.

The original Chianina cattle were white with a black switch. The skin pigment is black. They have a high heat tolerance and gentle disposition. Chianina are probably the largest breed of cattle. Mature bulls can grow to 6 feet at the withers. They can weigh as much as 4,000 pounds. Mature cows can grow to 5 feet at the withers. They can weigh as much as 2,400 pounds.

Chianina are popular in crosses for a number of reasons. They improve the growth rate of the offspring. In addition, they are good foragers and good mothers. They are well adapted to hot and cold climates, rough terrain, and have a high degree of tolerance to insects and diseases. Chianina have fine-textured meat.

Devon

History

Devon cattle originated in Devon and Somerset, which are counties in southwestern England. They are sometimes called North Devon to distinguish them from the closely related South Devon breed described later. It is an old breed of cattle. Some authorities believe they descended from *Bos longifrons*, a small type of aboriginal cattle in Britain.

Devon cattle were brought to the United States by colonists as early as 1623. They were used for milk, beef, ox teams, and leather. The American Devon Cattle Club was established in 1884. Devons are now registered with the Devon Cattle Association, Incorporated. The Devon Herdbook was established in 1850.

Description

The color of the Devon is rich red (sometimes called ruby red). They have yellow skin and creamy white, black-tipped horns. Polled Devons trace back to a mutation that occurred in 1915.

Most modern-day Devons are of the beef type. The body is long and moderately deep. The loin and hindquarters carry thick, natural fleshing. The breed is hardy and adaptable to many climates. It has a high heat tolerance.

Devon bulls may be crossed on any breed and the female offspring registered in the Devon Qualified Registry. This permits the breeder to upgrade the herd to registered purebred Devons in four generations. Performance testing and weight-for-age data on the offspring are encouraged by the breed association. The use of artificial insemination is also encouraged to promote superior bulls.

South Devon

History

The South Devon breed was developed in the southwest part of England (Figures 13-15A and B). The breed probably originated from large, red cattle that came from Normandy, France. They were brought to England at the time of the Norman invasions. South Devons are dual-purpose cattle (milk and meat). Devons are a single-purpose breed (meat). South Devon, however, are not related to the Devon cattle, which are also from England.

The South Devon was first imported into the United States in 1936 and 1947. Only five animals were imported at that time. In 1969 and 1970, 215 registered cattle were imported by Big Beef Hybrids of Stillwater, Minnesota, for crossbreeding purposes.

North American South Devon Association

Figure 13-15A South Devon bull.

Description

The color of the South Devon is a medium-red hair coat that is lighter in the twist. They have soft, curly hair and a very thick hide. They can also be black. The South Devon is a horned breed, with average size horns that curve forward and downward. Mature bulls weigh 2,000 to 2,800 pounds. Mature cows weigh about 1,500 to 1,600 pounds. It is a very docile breed.

North American South Devon Association

Figure 13-15B South Devon cow.

Galloway

History

The Galloway breed came to Canada from Scotland. The Vikings brought Galloways to modern-day southwestern Scotland in the ninth century, where they received recognition as a preferred breed as early as 1570. Galloways are the oldest known

British breed of record. The name derives from the District of Galloway, an area bordering northern England with a challenging climate. The first Galloway cattle were imported into the United States in 1866 from Canada with the North American Herdbook being started in 1882. Purebreds are registered with the American Galloway Breeders Association.

Description

Most Galloway cattle are black with soft, wavy hair and a thick undercoat (Figure 13-16A). Other colors include belted (Figure 13-16B), red, dun, and white. They are naturally polled. The Galloway is one of the smallest of the beef breeds.

The breed is noted for its hardiness, carcass quality, and foraging ability. Galloway cattle are capable of withstanding more severe weather conditions than many other breeds. Galloways need less feed to maintain good body condition in cold climates than most cattle. The thick hair coat of Galloway cattle protects the animals from harsh winters, requiring less digestible feed intake to maintain body weight. Rain hardly penetrates the hair coat, even in cold weather. Galloways thrive year round in the harshest of climates and require minimal shelter from winter cold to summer heat.

Figure 13-16A Black is the most common color of Galloway cattle. Black is a dominant trait preventing other colors within a black herd, except a rare red colored animal due to the red recessive gene.

Figure 13-16B The belt is a dominant trait and will exist in almost all calves sired by a belted bull.

Gelbvieh

History

The Gelbvieh breed was developed from four yellow breeds of cattle: Glan-Donnersburg, Yellow Franconian, Limburg, and Lahn. These breeds were developed around 1850. They were brought together in the Gelbvieh breed in 1920.

Description

Gelbvieh cattle are single-colored and vary from cream to reddish yellow (Figure 13-17). They are of medium weight and size, have good milking ability, and produce a very acceptable carcass.

Figure 13-17 Gelbvieh bull.

Hays Converter

History

The Hays Converter breed was developed in Canada by Harry Hays of Calgary, Alberta, Canada. The breed was developed from foundation stock from the Hereford, Brown Swiss, and Holstein cattle breeds. Breed development began in 1957. Animals from the foundation breeds were selected for rate of gain; large, strong frames; sound feet and legs; well-attached udders; good milking ability; high fertility; ease of calving; and winter hardiness. Selection for major traits continued for five generations, with animals showing undesirable traits being culled from the breeding herd.

Description

The animal is usually black with white face, feet, and tail. Some are red with white faces. Selection is not made on the basis of color. Bulls weigh about 2,200 pounds and cows weigh about 1,400 pounds. Calves reach market weight in 12 to 15 months.

Hereford

History

Figure 13-18 Hereford bull.

Herefords (Figure 13-18) originated in the county of Hereford in England. The early breeders selected for a high yield of beef and economical production. Native cattle were bred with white cattle from Flanders. A red bull with a white face was brought into the breeding from Yorkshire in 1750.

The first Herefords to come to the United States were imported by Henry Clay of Kentucky. These cattle were mixed with native cattle of the area. The first purebred breeding herd in the United States was established in 1840 in New York. During the 1870s, large numbers of Herefords were imported and the breed became popular in the United States.

Herefords have been registered by the American Hereford Association since 1881. More Herefords have been registered than cattle of any other breed.

Description

Hereford cattle have white faces and red bodies. They have white on the belly, legs, and switch. Herefords are a horned breed. They are docile in nature and easily handled.

The breed is well adapted to the western cattle-production regions of the United States. They have superior foraging ability, vigor, and hardiness. They produce more calves under adverse conditions than do many other breeds. When Herefords are used in crosses, the white color pattern tends to dominate.

Mature Hereford bulls weigh about 1,840 pounds. Mature cows weigh about 1,200 pounds. Herefords are popular for their general producing ability.

Polled Hereford

History

Polled Herefords (Figures 13-19A and B) originated in Iowa in 1901. Warren Gammon, an Iowa breeder, contacted all Hereford Association members asking if they had naturally polled animals in their herds. He located 4 bulls and 10 cows, which he purchased. These 14 animals became the foundation of the Polled Hereford breed. Later, other Polled Herefords were found and brought into the breeding. Polled offspring of crosses between Herefords and Polled Herefords have been used to improve the breed.

Photo courtesy of Inisfail Farm

Figure 13-19 Polled Hereford bull.

Description

Polled Herefords have the same traits as Herefords except for the horns. All are descended from purebred horned Herefords. Polled Herefords are eligible for registry in the American Hereford Association. The only distinction made on Hereford registration papers is a "P," indicating a polled animal.

Highland

History

The Highland breed (Figures 13-20A and B) originated in the Hebrides Islands near Scotland. This breed has been raised in northern Scotland for several centuries. A few of these cattle were imported into the United States in the early 1900s. More were imported in the 1930s. Interest in the breed increased and, in 1948, the American Scotch Highland Breeder's Association was formed. More cattle have been imported since that time. The breed is used in crossbreeding programs. The association is now called the American Highland Cattle Association.

Courtesy of American Highland Cattle Association, Denver, CO

Figure 13-20A Highland bull.

Description

Highland cattle have a long, coarse outer coat of hair with a soft, thick undercoat. The colors are black, brindle, red and light red, dun yellow, and silver. Animals of this breed are hardy and are excellent foragers. They are popular in crossbreeding because they give winter hardiness to the offspring.

Courtesy of American Highland Cattle Association, Denver, CO

Figure 13-20B Highland cow.

Courtesy of the North American Limousin Foundation

Figure 13-21 Limousin bulls.

Limousin

History

Limousin cattle were named after the province in west-central France where they originated about 7,000 years ago (Figure 13-21). In 1886, a breed association was formed and a herdbook was started. Limousin cattle entered the United States in 1968 when semen was imported from Canada, and registering began in the North American Limousin Foundation.

Description

Limousin cattle have light red hair with lighter circles around the eyes and muzzle. Purebred Limousin cattle can be red or black. Both fullbloods and purebreds can be polled, scurred, or horned. The Limousin head is small and short with a broad forehead. The neck is also short. Mature bulls weigh from 2,000 to 2,200 pounds. Mature cows weigh from 1,200 to 1,400 pounds. Limousin cattle are noted for their feed efficiency, carcass leanness, and large loin area.

Photo from American Maine-Anjou Association

Figure 13-22 Maine-Anjou bull.

Maine-Anjou

History

The Maine-Anjou breed (Figure 13-22) originated in France in the 1840s. The breed is the result of crossing English Shorthorns and French Mancelle cows. Maine-Anjou were originally work animals. Through selective breeding, milk producing and beef traits were developed. Today, the Maine-Anjou is considered an excellent beef producer.

Maine-Anjou semen was first imported into the United States from Canada in 1970. The first crossbreds born in the United States were registered in 1972. Animals are registered with the American Maine-Anjou Association.

Description

Maine-Anjou cattle are traditionally red and white. The natural red and white of the Maine-Anjou breed is a recessive trait. Due to crossbreeding Maine-Anjou can vary greatly in color. They have a lightly pigmented skin. They are a horned breed, with medium-size horns that curve forward. They are considered docile and are easily handled. Other traits include a fast growth rate and well-marbled carcass. Mature bulls weigh about 2,750 pounds.

Courtesy of Marky Cattle Association—Marchigiana Cattle

Figure 13-23A Marchigiana bull.

Marchigiana

History

The Marchigiana (pronounced *Mar-key-jahna*) breed (Figures 13-23A and B) originated in Italy near Rome during the fifth century, A.D. Foundation stock originated with crosses of native cattle, including the Chianina. The Italian Herdbook for the Marchigiana was established in 1930.

Courtesy of Marky Cattle Association—Marchigiana Cattle

Figure 13-23B Marchigiana cow.

Description

The Marchigiana is a horned breed. Cows are a grayish white. The bulls are somewhat darker. The skin, muzzle, and switch are dark. Bulls weigh from 2,650 to 3,100 pounds, while cows weigh 1,400 to 1,800 pounds. The Marchigiana are used in crossbreeding programs in the United States. Cattle are registered in the American International Marchigiana Society.

Murray Grey

History

The Murray Grey breed (Figures 13-24A and B) originated in Australia in 1905. It is the result of Shorthorn-Angus crosses. Semen was imported into the United States from Australia in 1969. The first live Murray Grey cattle were imported in 1972. Only small numbers of purebred bulls and cows have been imported into the United States since that time. The breed is used mainly in crossbreeding programs.

Description

The Murray Grey has a solid-color hair coat that is dark to silver gray. The breed is polled and is considered to be a docile animal. It produces a good-quality carcass on limited grain, and its females have good mothering ability.

The American Murray Grey Association was formed in 1970. All cows with not less than 7/8 Murray Grey breeding may be registered. Bulls with not less than 15/16 Murray Grey breeding may be registered. Both males and females with not less than 1/2 Murray Grey breeding may be recorded.

Courtesy American Murray Grey Asso, NV

Figure 13-24A Murray Grey bull.

Courtesy American Murray Grey Asso, NV

Figure 13-24B Murray Grey calf.

Pinzgauer

History

The Pinzgauer breed originated in the Alpine regions of Austria, Italy, and Germany (Figure 13-25A and B).

Description

They are a horned breed and brown. There is some white on the top and underlines. Bulls weigh 2,200 to 2,900 pounds and cows weigh 1,300 to 1,650 pounds. The Pinzgauer is a hardy breed and is used in crossbreeding programs in the United States. Cattle are registered in the American Pinzgauer Association.

American Pinzgauer Association

Figure 13-25A Pinzgauer bull.

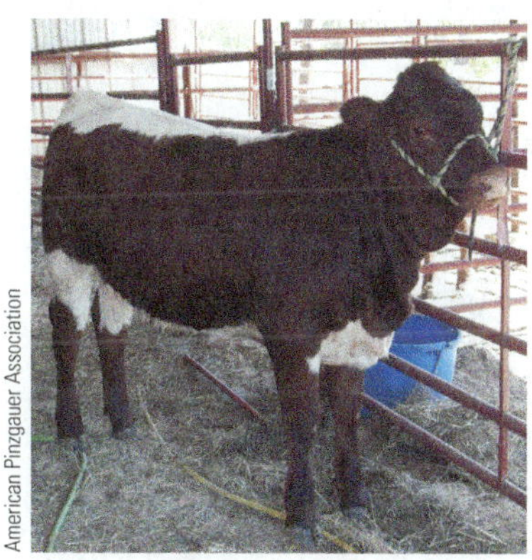

American Pinzgauer Association

Figure 13-25B Pinzgauer heifer.

Red Poll

History

The Red Poll breed originated in eastern England. Development of the breed began in the early 1800s. Native cattle of the shires of Norfolk and Suffolk were crossed to produce the breed. The first Red Polls were imported into the United States in 1873.

The Red Poll Cattle Club of America was formed in 1883. The Red Poll Cattle Society of Great Britain and Ireland was formed in 1885. The Red Poll Herdbook was established in 1873 in England. Registrations in the United States are in the Red Poll Herdbook, American Series, which is a continuation of the original herdbook. A Gain Register was established in 1960 to recognize preweaning calf gain records. Carcass merit and gain to slaughter is recorded in a Carcass Register. This was established in 1963. Cattle are registered in the American Red Poll Association.

Description

The color of the Red Poll is light red to very dark red. The tail switch contains some natural white. Some white is permitted on the underline. The skin is buff or flesh-colored. Solid black, bluish, or cloudy noses are not permitted for registration. The breed is polled and dual-purpose. They have been bred for both milk and meat production. The emphasis in recent years has been on beef type. The carcass has a high proportion of lean meat. There is little waste fat and the marbling is acceptable.

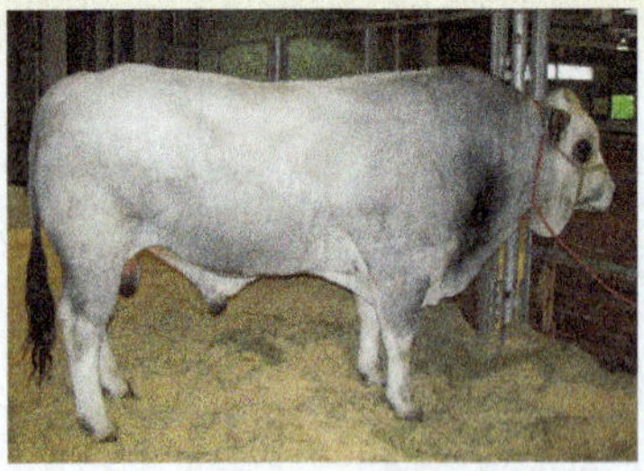

Courtesy of Alyssa K. Elrod

Figure 13-26 Romagnola bull.

History

The Romagnola (pronounced *Ro-ma-nola*) breed originated in northeastern Italy during the fourth century, A.D. The foundation stocks of this breed are the *Bos primigenius podolicus* and the *Bos primigenius nomadicus*. The *Bos primigenius podolicus* (forebears of the modern *Bos taurus*) lived in Italy and the *Bos primigenius nomadicus* (forbears of the modern *Bos indicus*) were brought to Italy in the fourth century by Gothic invaders. Several Italian breeds of cattle, all with similar characteristics, were developed from these ancestors. The Romagnola were originally used as draft animals in the fields, which resulted in the development of a muscular body type. After the mechanization of agriculture, the emphasis on breeding and development shifted to beef production. Romagnola cattle were first imported into the United States in the 1970s. Cattle are registered in the American Romagnola Association.

Description

The Romagnola are a horned breed—the females having lyre-shaped horns and the bulls have half-moon-shaped horns (Figure 13-26). The skin is black pigmented and the hair coat is ivory; bulls have gray hair around the shoulders and eyes. The black skin color is also found on the muzzle, horn tips, tail switch, hoofs, vulva, tip of sheath, and the base of the scrotum. Calves are a light reddish color when born and turn white at about 3 months of age. The average weight of adult bulls is 2,750 pounds, and the average weight of adult cows is 1,650 pounds. The breed is noted for economical feed conversion with rapid gains, a high dressing percentage, and early maturity.

Salers

History

The Salers (pronounced *Sa' lair*) breed of cattle are native to the Auvergne region of south-central France. Cave drawings, dating back about 7,000 years, found near Salers, France, show cattle believed to be the ancestors of the present breed. Historically, in France, Salers cattle were used for beef, milk, and as draft animals. The first Salers bull was imported into Canada in 1972. Semen from this bull was sold in both Canada and the United States. In 1975, one bull and four heifers were imported directly to the United States. Between 1975 and 1978, 52 heifers and 6 bulls were imported into the United States and another 100 head were imported into Canada. The present breed in the United States originated with these imports.

The American Salers Association was formed in 1974. The association accepts animals for registration from an upgrading breeding program. The continual use of registered fullblood or purebred Salers sires on herds will produce offspring that are eligible for registry with the association. Current requirements for registration are available from the association.

Description

Courtesy American Salers Association

Figure 13-27 American Salers bull.

The Salers breed is most commonly seen in modern herds as black and polled (Figure 13-27). However, originally, the breed was horned and dark mahogany red. This genetic diversity adds to their value in crossbreeding programs. Salers cows are noted for their ease of calving and good maternal ability. Other desirable characteristics of Salers cattle include good foraging ability on poor range, high weaning weights, and excellent carcass quality that meets current market demand for beef. Salers cattle are also used in crossbreeding programs.

Santa Gertrudis

History

The Santa Gertrudis breed (Figures 13-28A and B) was developed on the King Ranch in Texas. The breed is the result of crosses of Brahman bulls on Shorthorn cows. Crossbreeding began in 1910 using several different European breeds of beef cows. By 1918, the Brahman Shorthorn cross showed the most promise. In 1920, a bull named Monkey was born. This bull had 3/8 Brahman and 5/8 Shorthorn blood. Monkey showed outstanding traits and sired more than 150 useful sons. All present-day Santa Gertrudis cattle are descendants of this bull.

The breed association, formed in 1950, is the Santa Gertrudis Breeders International. The association has a system of compulsory classification. For registration, an animal must be inspected by an association classifier. A Standard of Excellence was established by the association. An animal that meets the standard is branded with an "S" and is certified as purebred.

Santa Gertrudis Breeders International

Figure 13-28A Santa Gertrudis bull.

Santa Gertrudis Breeders International

Figure 13-28B Santa Gertrudis cow and calf.

Description

The color of the Santa Gertrudis is cherry red. Most of the animals are horned. Some are polled, and these are eligible for registration. They have loose hides with folds of skin on the neck and a sheath or naval flap. The hair grows short and straight in warm climates. It is long in cold climates. Santa Gertrudis cattle are efficient in the feedlot. They produce desirable carcasses with little waste fat. They also resist diseases and insects.

Senepol

History

The Senepol is a Caribbean breed of beef cattle that was originally bred on the island of St. Croix in the U.S. Virgin Islands. The breed was developed in 1918 from a cross of the Red Poll and the N'Dama. The N'Dama is a humpless longhorn breed from West Africa. A few Senepol were imported into the United States, mainly to Florida and Tennessee, in the late 1970s. Since 1978, the USDA has been conducting crossbreeding experiments with the Senepol at the Subtropical Agricultural Research Station in Brooksville, Florida. The Senepol Cattle Breeders Association recognizes more than 500 Senepol breeders.

Senepol Cattle Breeders Association

Figure 13-29 Senepol bull.

Description

Senepol cattle range in color from tan to dark red, are polled, and do not have a hump (Figure 13-29). They are resistant to ticks and other insects, and tolerate hot weather and a wide range of rainfall conditions. Senepols are similar to the Brahman in adaptation to unfavorable environments but have a much gentler disposition. They are early maturing, have good mothering characteristics, and provide good milk production. They are seen as having value in the United States in crossbreeding programs to produce cattle with desirable characteristics for a subtropical environment. Senepol-Angus crosses have produced carcasses of a slightly higher quality grade than Brahman-Angus crosses.

Courtesy of American Shorthorn Association

Figure 13-30A Shorthorn bull.

Shorthorn and Polled Shorthorn

History

The Shorthorn breed (Figures 13-30A and B) originated around 1600 in the Tees River Valley of northern England. At that time, they were called Durhams. Major improvement of the breed began in the late 1700s. Shorthorn cattle were imported

Courtesy of American Shorthorn Association

Figure 13-30B Shorthorn cow.

into Virginia in 1783. The Coates Herdbook was established in 1822 to record Shorthorns. It was the first cattle herdbook and served as a model for other breed herdbooks that followed. Shorthorns were originally a dual-purpose breed. They were bred for both milk and meat production.

The breed was established in the United States on a permanent basis as the result of importations of cattle between 1820 and 1850. The American Shorthorn Herdbook was started in 1846. It was the first beef herdbook to be published in the United States. The American Shorthorn Association was organized in 1872.

Polled Shorthorns originated in Minnesota in 1881. They have the same traits as Shorthorns except for being naturally polled. Both horned and polled Shorthorns are registered with the American Shorthorn Association.

Description

Shorthorn cattle are red, white, or roan. They have short horns that curve inward. They are easily handled and have good dispositions. Mature bulls weigh up to 2,400 pounds. Mature cows weigh up to 1,500 pounds. Shorthorns are adaptable to many climates. They have excellent crossing ability with other breeds. Shorthorn cattle have been used in the bloodlines of more than 30 other recognized beef breeds. They are good mothers with excellent milking ability. Shorthorns produce a desirable carcass.

Simmental

History

Courtesy of American Simmental Association

Figure 13-31 Simmental bulls.

The Simmental breed (Figure 13-31) originated in the Simmen Valley of Switzerland. It is an old breed, dating back to the Middle Ages. The Simmental Herdbook was established in Switzerland in 1806. It required a performance pedigree for milk and conformation (physical appearance). Meat and carcass traits have since been added to the performance pedigree. About one-half of the cattle in Switzerland are Simmentals. It is the most popular breed of cattle in Europe. Simmentals were first brought into the United States from Canada in 1969. The American Simmental Association was formed in 1968. The herdbook is open to upgrading of beef and dairy stock. All animals must have a performance pedigree to be eligible for registration. The American Simmental Association has no color requirement for registration. Artificial insemination is encouraged. Heifers become purebreds in three topcrosses (7/8 Simmental blood). Bulls are registered as purebreds in four topcrosses (15/16 Simmental blood).

Description

Even though there is no color requirement for registration, black and red are the most common colors for Simmental cattle. They are a horned breed with medium-size horns. The Simmental is a large-bodied animal and is noted for being docile. Mature bulls weigh from 2,300 to 2,600 pounds. Mature cows weigh about 1,450 to 1,800 pounds. They will milk about 9,000 pounds of milk per lactation, and the milk will test about 4 percent butterfat.

Simmentals make extremely rapid growth, gaining about 3 pounds per day on roughage. They are thickly muscled and produce a carcass without excess fat. They are adaptable to a wide range of climates.

Texas Longhorn

History

Courtesy of Alyssa K. Elrod

Figure 13-32 Texas Longhorn.

The Texas Longhorn (Figure 13-32) originated from Spanish Andalusian cattle that were brought to Santa Domingo by Columbus on his second voyage in 1493. In the early 1500s, descendants of these cattle were taken to Mexico by Spanish explorers. As the Spanish continued their explorations, some of these cattle were taken with them into the region that became Texas. Many of the cattle escaped and adapted to the harsh environment of the Southwest. It is estimated that by 1860 about 4 million descendants of these cattle were running wild in Texas. After the Civil War, cattlemen in Texas began rounding up the Longhorns and moving them along trails to railheads, mainly in Kansas and Missouri. By the 1880s, the Longhorn began to be replaced by European breeds of beef cattle. The Texas Longhorn was almost extinct by 1900. The U.S. Congress appropriated money in 1927 to establish a federal herd of purebred Longhorns. At that time, 20 cows, 3 bulls, and 4 calves were located and established as seedstock at the Wichita Mountains Wildlife Refuge in Cache, Oklahoma. The Texas Longhorn Breeders Association of America (TLBAA) was established in 1964 by a small group of concerned cattle ranchers interested in preserving the unique existence of the Texas Longhorn cattle. The TLBAA was established to maintain the breed registry and to promote the magnificent breed. The association works to preserve the purity of the breed, and to promote Texas Longhorns as a distinct breed while encouraging its future through promotion, education, and research. Longhorn cattle may prove to be valuable in breeding programs aimed at producing cattle with meat that is leaner and has lower cholesterol levels.

Description

The Texas Longhorn has many shadings and combinations of colors. They have horns that curve upward and spread to 4 feet or more. Their legs are long and their shoulders are large and high. They have a large head with small ears and long hair between their horns. Their neck is short and stocky.

Texas Longhorns are slow maturing, have high fertility, are resistant to many diseases and parasites, and are well adapted to harsh environments. They have the ability to survive on sparse rangeland. They are noted for their easy calving ability, hardiness, and longevity.

Wagyu

History

The Wagyu is a *Bos taurus* breed of beef cattle that originated in Japan. Prior to 1910, crossbreeding was done with many British breeds. As a result, the Wagyu cattle of today are influenced by these breeds. The first Wagyu cattle were imported to the United States in 1976 and are becoming more popular.

Description

Wagyu cattle are black or red in color and are horned. They are known for their fertility and early maturity as well as their good temperament. Wagyu are most famous for beef quality, known as Kobe beef. Kobe beef is renowned for its flavor and tenderness and is generally considered to be the highest quality beef in the world. Beef from Wagyu cattle produced in the United States is known as Kobe-style beef.

SUMMARY

The beef industry is an important source of income for farmers. About 15 percent of cash income from crops and livestock comes from beef production. The use of beef in the United States has been decreasing in recent years. While many beef producers have small operations, there are also some large feedlots.

Most beef cattle breeds originated in Europe. Some of the newer breeds are the result of crossing breeds from India on European breeds.

Student Learning Activities

1. Present an oral report on the characteristics of the beef industry.
2. On field trips in the community, observe different beef breeds and their characteristics.
3. Conduct a survey of beef producers in the community to determine what beef breeds are used locally.
4. Prepare a bulletin board displaying pictures of the beef breeds.
5. Present an oral report on the history and characteristics of a beef breed to the class.
6. Write to beef breed associations for literature about their breed.
7. If you are planning and conducting a supervised experience program in beef production, select a breed or crossbred animal based on desirable characteristics of the breed.

Discussion Questions

1. How important is the beef industry as compared to the total livestock production industry?
2. In what part of the United States are most of the large beef feedlots found?
3. In what part of the United States are most of the smaller beef feedlots found?
4. List the advantages and disadvantages of raising or feeding beef cattle.
5. Describe briefly how modern beef breeds were developed.
6. List seven points that should be considered when selecting a beef breed.
7. Prepare a table that briefly describes the characteristics of each of the beef breeds.
8. Which beef breeds are most common in your area?
9. Which beef breeds were developed in the United States?

Review Questions

True/False

1. When incomes are high, the demand for beef increases while the demand for poultry decreases.
2. Most beef cattle breeds originated in Asia.

Multiple Choice

3. About ______ percent of the total income from all livestock and poultry marketing in the United States comes from the beef industry.

   a. 18
   b. 28
   c. 35
   d. 48

4. The United States is divided into ______ cattle-production regions.

   a. 4
   b. 8
   c. 10
   d. 12

5. Which breed of cattle was in existence before the time of the Roman Empire?

   a. Devon
   b. Beefmaster
   c. Chianina
   d. Brahman

6. Name the breed of cattle that the U.S. Congress appropriated money for in 1927 because it was almost extinct.

   a. Angus
   b. Beefmaster
   c. Hereford
   d. Texas Longhorn

Completion

7. The cattle breed that is red in color and originated in the United States in 1945 is the ______ ______.
8. The ______ ______ breed of cattle originated in Australia.

Short Answer

9. What is the origin of some of the newer breeds of cattle?
10. Name a breed of cattle developed in Canada.
11. What is a freemartin calf?
12. What is the official name of the Angus breed?
13. Which breed had the first herdbook and served as a model for other breeds to follow?
14. How was the Polled Hereford breed developed from the Hereford breed?
15. Define *dewlap* and give an example of a cattle breed with this characteristic.

Chapter 14

Selection and Judging of Beef Cattle

Key Terms

cow-calf system
feeder calf
yearling feeder
frame score
growthy
conformation
ultrasonics
performance testing
production testing
brood
progeny testing
pedigree
sire summary
maternal breeding value (MBV)
expected progeny difference (EPD)
yield grade

Objectives

After studying this chapter, the student should be able to

- describe the types of beef production systems.
- name the parts of a beef animal.
- select superior animals.

TYPES OF BEEF PRODUCTION

There are three main types of beef cattle production systems: (1) cow-calf producers, (2) purebred breeders, and (3) cattle feeders. A farmer may specialize in only one type of operation or combine several kinds of operations. For example, a farmer may produce calves from a cow herd and also feed the calves for slaughter.

Cow-Calf Producers

The **cow-calf system** of beef production involves keeping a herd of beef cows. These cows are bred each year to produce calves. The calves are then sold to cattle feeders who feed them to slaughter weights. Most of this type of beef production is done in the western range states and upper Great Plains. Land is used that is not suitable for growing crops. Beef cows are maintained mainly on roughage. Little or no grain is needed for this type of beef production. This type of operation requires less labor and a lower investment in equipment and facilities than other kinds of beef enterprises.

A larger investment in land is usually needed for this type of operation than is necessary for feeding cattle for slaughter. It is difficult to expand or reduce the size of operation quickly. The price received for calves tends to be more closely associated with the supply and demand for calves rather

Connection: Crossbreeding

Crossbreeding is one of the most powerful tools in the cattleman's tool chest. The benefits of a well-planned crossbreeding program are undeniable and are used widely in commercial cattle production. Well-planned crossbreeding programs produce healthier calves that grow faster, stay healthier, and perform better in the feedlot.

In traditional crossbreeding programs, English breeds were used, mostly Angus and Hereford. Improvement in the offspring was evident, a result of hybrid vigor, also known as heterosis. Breeders soon realized that hybrid vigor resulted from the crossing of unrelated breeds with diverse genetic lines (Figure 14-1). Crossbreeding systems were expanded as producers searched the world for new breeds and genetic lines that were distinctly different from breeds common in the United States. Beef cattle breeding systems were greatly expanded from the simple two-way crossbreeding programs. Sophisticated rotational breeding programs were established as the gene pool was expanded.

USDA Agricultural Research Service Photo by Scott Bauer

Figure 14-1 Crossbred cattle usually exhibit hybrid vigor or heterosis.

than with the cost of producing them. Therefore, a producer may not always recover production costs in this type of beef operation.

The cows are usually bred to calve in the spring. Most calves are weaned in the fall and sold as feeders. Sometimes the calves are fed roughage through the winter and sold the next year as yearlings. **Feeder calves** are weaned calves that are under 1 year of age and are sold to be fed for more growth. **Yearling feeders** are 1 to 2 years of age and are sold to be fed to finish for slaughter.

Purebred Breeders

Purebred breeders keep herds of purebred breeding stock. They provide replacement bulls for cow-calf operations. Cow-calf farmers sometimes buy cows or heifers from the purebred breeder to improve the commercial herd. Purebred breeders also sell to other purebred breeders. The purebred breeders are mainly responsible for the genetic improvements that have been made in beef breeds. Cow-calf producers use two or more purebred lines in order to capitalize on hybrid vigor.

A great deal of knowledge and skill is required to raise purebreds, and it should only be attempted by those with experience. The costs are usually higher in this type of cattle business. It takes many years to develop a high-quality herd and achieve success.

Cattle Feeders

The cattle feeder feeds animals for the slaughter market. The objective is to produce finished cattle in the shortest time possible. The operator usually buys feeders or yearlings and finishes them in the feedlot. Some producers feed cattle on pasture for a period of time and then finish them in the feedlot. There is a trend toward more confinement feedlot finishing of slaughter cattle.

While some roughage can be used in feeder cattle operations, this enterprise requires more grain than cow-calf or purebred production. It usually takes grain to get the quality of finish that is in demand in the marketplace. Feeder operations can easily adjust to changes in feed supplies, operating costs, labor supply, and economic outlook. The cattle feeder can expect a return on investment in 4 to 6 months.

The facilities required for confinement feeding of cattle are more expensive than those required for cow-calf operations. Feed costs, labor requirements, and

Connection: Black Baldy—A Classic

Legacy Herefords

Figure 14-2 Black Baldy calves are the classic symbol of crossbred cattle. They often inherit the best qualities of both Angus and Hereford cattle.

Nothing symbolizes classic crossbreeding in cattle like Black Baldies. Traditionally, Black Baldies were the result of crossing Hereford and Angus cattle. The calf is usually characterized by a white face similar to the Hereford but with the black body of an Angus. This is because white faces and black coat color are both genetically dominant in cattle.

At one time, cattle producers were generally loyal to a particular breed, and most cattle herds in the United States were either Angus or Hereford. However, the advantages of crossbreeding soon became evident, and crossbreeding of Angus and Herefords became common. The prevalence of Black Baldies increased significantly.

Black Baldy calves usually exhibit hybrid vigor, which means they inherit the best qualities of both Angus and Hereford (Figure 14-2). The calves grow faster, remain healthier, and perform better in the feedlot than purebred calves. Black Baldy cows are noted for their good mothering abilities and for remaining productive longer and healthier than purebred cows. When Angus bulls are also used on Hereford cows and heifers, the result is usually a smaller calf with reduced birthing problems (dystocia).

transportation costs are all higher in this type of enterprise. Cattle feeding is a high-risk enterprise because of the fairly large fluctuations in the price of finished cattle, and, in recent years, the fluctuations in the price of corn due to the use of corn to produce ethanol.

SELECTION OF BEEF ANIMALS

A beef herd is improved by selecting animals that have the desired traits. The producer must produce what is in demand in the marketplace. Selection is based primarily on conformation and performance records. Pedigree and show ring winnings are of less importance.

Cow-calf producers need to be aware of the relationship between cow size and weaning weights of calves that are produced. In general, for each 100-pound increase in cow size, there is a corresponding increase of 10 to 12 pounds in calf weaning weight. However, experimental work done in Texas has revealed that heavier cows (1,201 to 1,700 lb) require 18 percent more energy and 13 percent more protein per day than do medium-weight cows (1,000 to 1,200 lb). The average weaning weight of calves from the heavier cows was only 2.4 percent greater than the average weaning weight of calves from the medium-size cows. In this same experiment, age of the cow was shown to be a factor in calf weaning weights. Twelve-year-old cows produced calves with an average weaning weight that was 6 percent less than 2-year-old cows.

Producers with cow-calf herds need to consider the higher maintenance costs of heavier cows as well as the age of the cows when making selections of breeding stock. A valuable tool for selecting replacement heifers and herd bulls is the frame score. It is used by state and USDA cattle graders in beef performance testing programs. The **frame score** is a measurement based on observation and height measurements when calves are evaluated at 205 days of age. An estimate of the expected size of the animal when it reaches maturity can be made from its frame score. The use of the

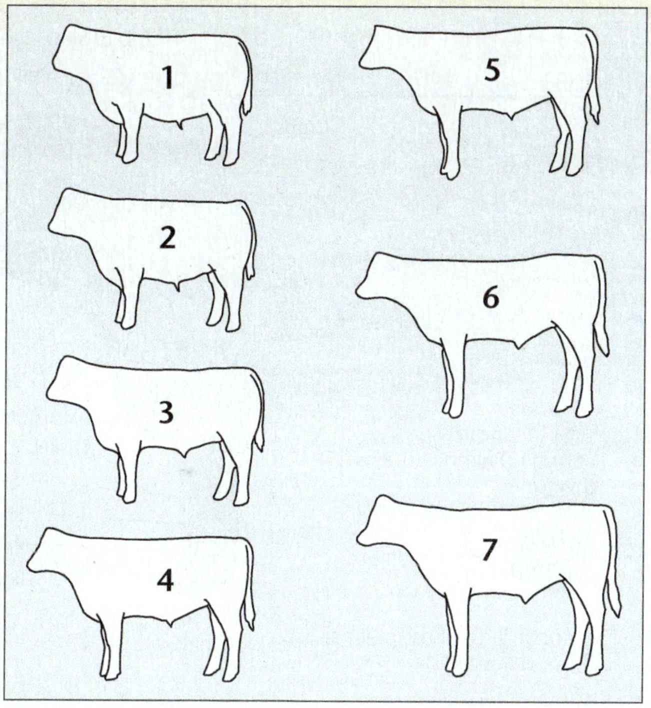

Figure 14-3 The seven frame scores used in many beef performance testing programs.

frame score in herd sire selection provides a more accurate prediction of expected genetic change in the herd.

Frame scores are made on a scale of 1 to 7 (Figure 14-3). The scores represent a range in body types of beef cattle and were developed at the University of Wisconsin. The English breeds of beef cattle are usually covered by body types 1 through 5. Charolais, Simmental, and similar-size cattle of other breeds usually require the use of body types 3 through 7. The age of the animal is a factor in determining its frame score and hip height. Many beef performance testing programs use the Missouri system of using a height measurement at the shoulder when determining frame scores.

Muscle conformation scores are also determined for the animal. These scores use a scale of 1 to 7, as follows:

1. An exceptionally thin calf
2. Very light muscled
3. Light muscled
4. Average muscled
5. Heavy muscled
6. Very heavy muscled
7. Double muscled

A conformation score on a scale of 1 to 17 is also determined for the animal. The low scores indicate inferior animals. Scores of 9 through 11 indicate animals that are below average for desirable characteristics. Scores of 12 through 14 indicate animals that are average or slightly below average for some of the desirable characteristics. Scores of 15 through 17 indicate superior animals that are **growthy**. *Growthy* is a term used to describe an animal that is large and well developed for its age, is well balanced with adequate muscle, and has sufficient frame and bone structure to indicate good growth potential.

Animals with conformation scores in the 14 to 17 range should be considered for herd sires. Replacement heifers should have conformation scores in the 13 to 17 range. Small-framed cattle generally should not be considered for herd replacements even if their conformation score is acceptable.

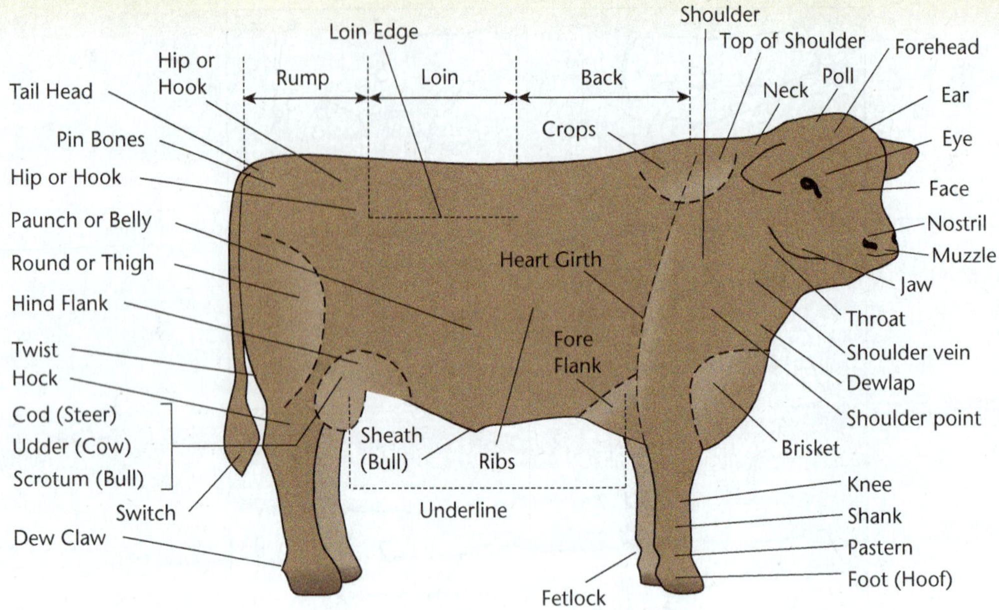

Figure 14-4 The body parts of the beef animal.

Conformation

Conformation refers to the appearance of the live animal. It includes the skeletal structure, muscling, fat balance, straightness of the animal's lines, and structural soundness. To describe the conformation of the animal, the parts of the animal must first be identified. The parts of the beef animal are identified in Figure 14-4.

Desirable conformation of the beef animal includes:

- long, trim, deep-sided body
- no excess fat on the brisket, foreflank, or hindflank
- no extra hide around the throat, dewlap, or sheath
- heavily muscled forearm
- proper height to the point of the shoulders
- correct muscling throughout the body
- maximum development of the round, rump, loin, and rib

An animal with the proper conformation will produce the maximum amount of high-value cuts. It will have a minimum of less-valuable bone and internal organs. It is necessary to learn the wholesale cuts of the beef animal to properly evaluate the live animal.

Figure 14-5 shows the location of the high- and low-value wholesale cuts of beef. The high-value wholesale cuts come from the round, rump, loin, and rib. The lower-value wholesale cuts come from the chuck, brisket, flank, plate or navel, and shank. Figure 14-6 illustrates the various cuts of beef that consumers are able to purchase and where each of the pieces come from.

Ultrasonics is the use of high-frequency sound waves to measure fat thickness and loin-eye area. It is a useful tool for selecting meaty animals for breeding purposes. The measurement is made on live animals. Therefore, meaty animals can be identified without killing them. The accuracy and value of the ultrasonic measurement depends on the skill of the operator. Increased use of ultrasonics may be expected as cattle breeders gain confidence in the results.

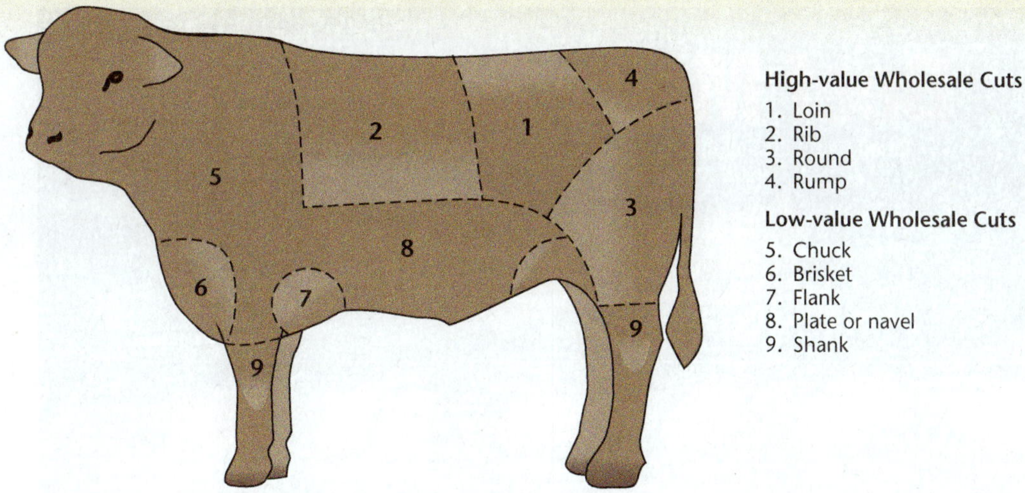

Figure 14-5 Location of high- and low-value wholesale cuts of beef.

Performance Records

One of the best ways to select beef animals is on the basis of performance records, of which there are several kinds. **Performance testing** refers to an animal's own performance in relation to important economic traits. It may be defined as a method of collecting records on beef cattle herds to be used for selecting the most productive animals. **Production testing** refers to measuring a brood cow's production by the performance of its offspring. (A **brood** animal is kept for breeding.) **Progeny testing** usually refers to the evaluation of a bull by the performance of a number of its offspring. Overall herd evaluation programs are often referred to as performance testing. This may include production and progeny testing.

Performance testing is used by both purebred and commercial beef producers. It helps the cattle producers improve the grade and weight of the calves in the herd more rapidly. Visual evaluation alone is not enough to select the best animals. Performance testing records are used to:

- cull low-producing cows
- check on percent calf crop (number of calves weaned divided by number of breeding females in the herd)
- select replacement heifers and bulls
- measure the productivity of each bull
- improve herd management
- improve the grade of calves produced
- increase the weaning weight of calves produced
- give buyers more information
- provide permanent records

Many states and breed associations have performance testing programs. A producer who desires to do performance testing should contact the local office of the state extension service or the proper breed association for information and forms to use. Setting up and conducting a testing program usually involves three main steps: (1) identifying each cow and calf, (2) recording birth dates of calves, and (3) weighing and grading the calves at weaning time.

Weaning weight is adjusted to a 205-day weight so that all calves in the herd can be compared on an equal basis. The weaning weight is adjusted for the age of the calf, the age of the dam, and the sex of the calf. Most states and breed associations have adopted the recommendations of the National Beef Improvement Federation

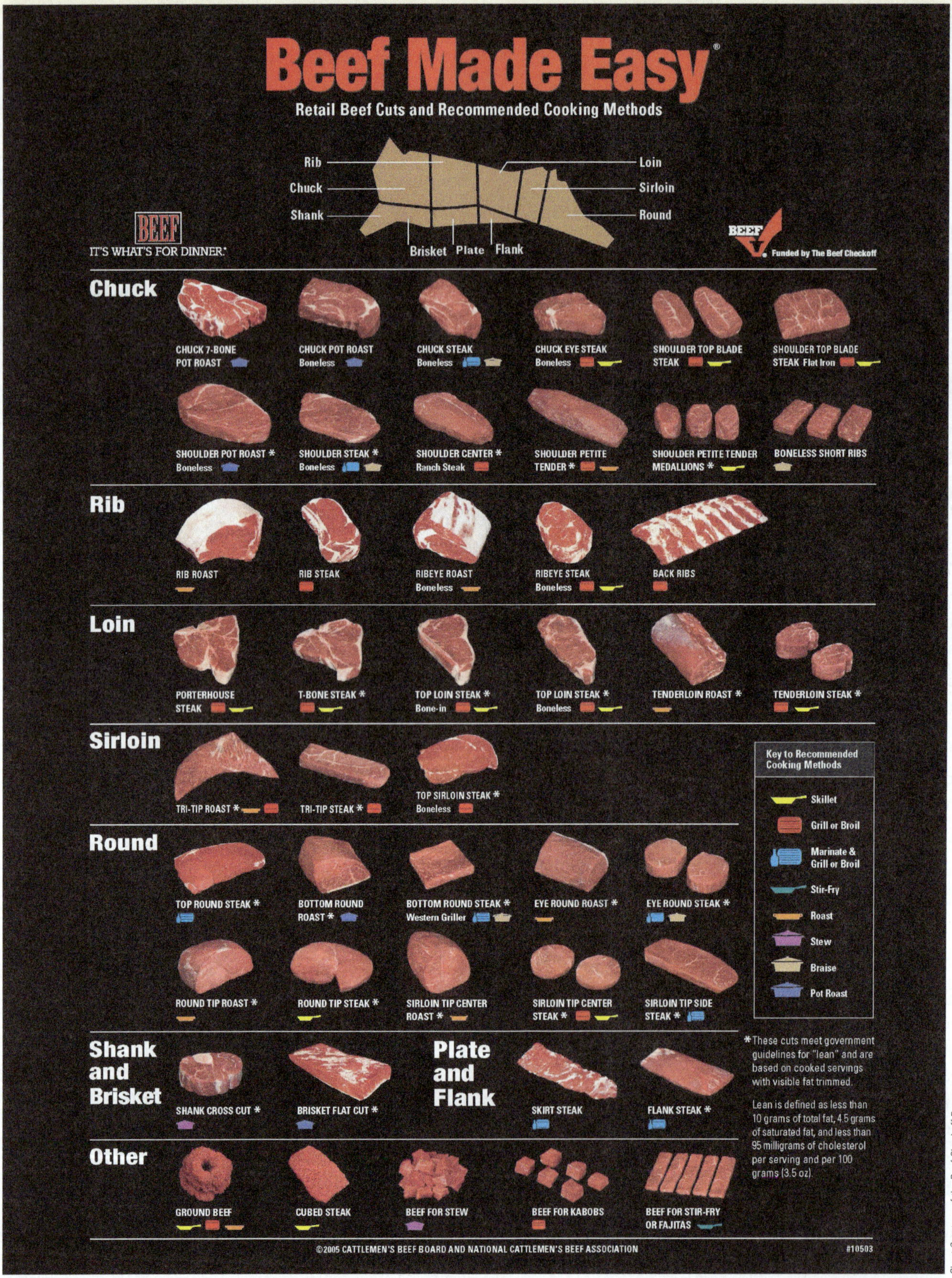

Chart Courtesy of the Beef Checkoff Program

Figure 14-6 Retail cuts of beef.

TABLE 14-1 Breed Standard Birth Weights Used in Performance Testing Programs

	Sex of Calf	
	Females	Males
Breed	Pounds	Pounds
Angus	65	75
Charolais	85	85
Chianina	80	80
Hereford	70	75
Polled Hereford	70	75
Limousin	75	80
Maine-Anjou	84	90
Shorthorn	70	70
Simmental	83	91

Data used with permission from the Beef Improvement Federation

for the 205-day age basis and the adjustment factors. The formula for the 205-day weight adjustment is:

$$\frac{\text{Actual weight} - \text{Birth weight}}{\text{Age in days}} \times 205 + \text{Birth weight}$$

If the calf was not weighed at birth, 70 pounds may be used as the average birth weight. Some breed associations have established their own standard average birth weight to be used if the calf was not weighed at birth. These standards are shown in Table 14-1.

Calves must be weighed between 160 and 250 days of age to be considered regular age and have 205-day weights and weight ratios calculated for them. If calves are weighed between 120 and 159 days or 251 and 290 days of age, the 205-day weight is calculated, but a weight ratio is not calculated for them. These calves are considered to be of irregular age. Performance data for irregular-age calves are not included in sex group summaries, sire summaries, or herd averages.

An adjustment in the 205-day weight of the calf is made for the age of the dam. These adjustments are as follows:

	Adjustment	
Age of Dam (years)	Male Calves (lb)	Female Calves (lb)
2	+60	+54
3	+40	+36
4	+20	+18
5 to 10	+ 0	+ 0
11+	+20	+18

The 205-day weight ratio within the sex group is calculated for each regular-age calf. The 205-day weight as adjusted for the age of the dam is divided by the average for the sex group and shown as a percentage. A weight ratio of 90 for a heifer would indicate a calf that is 10 percent below the average for all heifers in the herd. A heifer with a weight ratio of 110 would indicate a calf that is 10 percent above the average for all heifers in the herd.

An adjustment is also made for the sex of the calf. To adjust the 205-day weight to a steer-calf basis, add 5 percent to the weight of a heifer calf and subtract 5 percent from the weight of a bull calf. Adjusting the 205-day weights to a steer-calf basis makes it possible to directly compare heifers and bulls. The information gained from this process can be used for selection of breeding stock as well as in performance testing.

A 205-day adjusted weight ratio is determined for all calves in the herd that are classified as being of regular age. The 205-day weight (adjusted for age of dam and sex of calf) for each regular-age calf is divided by the herd average of 205-day adjusted weights for all regular-age calves. The calves can now be compared on an equal sex basis, and each cow's production can be compared to the herd average.

Individual cow productivity is compared by using the 205-day adjusted weight and weight ratio. Individual calf performance within its sex group is compared by using the 205-day weight adjusted for age of dam and the 205-day weight ratio within the sex group.

Performance testing programs also make use of adjusted yearling (365-day) weights. The formula for 365-day adjusted weight is:

205-day weight, adjusted for age of dam + (Average daily gain on test × 160)

Cattle must be at least 330 days of age and on test at least 140 days in order to calculate a 365-day adjusted weight.

Some producers use a 550-day adjusted weight (approximately 18 months of age) to help in the selection of replacement heifers. The final weight of the animal must be taken at 500 days or later but not before 500 days. The formula for calculating the 550-day adjusted weight is:

$$\frac{(\text{Actual final weight} - \text{Actual weaning weight}) \times 345}{\text{Number of days between weighings}} + \text{250–day weight, adjusted for the age of the dam}$$

Most states and breed associations with testing programs offer computer services to keep and analyze the records. Information provided is for use on the farm where it was gathered. It is not intended for use in comparing one herd with another.

The amount of improvement that can be expected from selecting superior animals depends on the heritability of the traits upon which the selection is based. The performance of an animal depends on genetics and environment. The heritability of a trait is the amount of the performance that comes from genetics. Heritability above 40 percent is high; from 20 to 40 percent is medium; below 20 percent is low.

Pedigree

A **pedigree** is the record of the ancestors of an animal. It usually contains only the names of the ancestors. In some cases, it may contain performance data. A person must be familiar with the performance of the individuals in the pedigree if it is to have much value in selection. Only the most recent ancestors are of importance in selection. Ancestors before the fourth generation contribute very little to the current generation. Sometimes the use of a bloodline becomes a fad. Cattle breeders must be careful not to keep poor animals just because they belong to a given bloodline.

Other Factors in Selection

Only healthy animals should be brought into a herd. The buyer should check the health of an animal carefully before buying it. Females should test negative for brucellosis (Bang's disease), vibriosis, and tuberculosis. It is best to buy females that have been calf-hood vaccinated. Check for mange, ringworm, and lumpy jaw.

The standards in the show ring are the result of what the consumer wants in the marketplace. Animals that do well in the show ring meet these standards. Therefore, show ring winners are often desirable animals to select for breeding purposes if they meet standards of conformation and performance records.

Selection of the Herd Bull

The selection of the herd bull is one of the most important decisions the cattle breeder makes. Each calf produced receives one-half of its genetic makeup from its sire. When replacement females are selected from the herd, 87.5 percent of the heifer's genetic makeup comes from the last three bulls used in her pedigree.

A major advancement in the improvement of the beef industry has been the development of sire summaries. Most breed associations provide sire summaries that are updated each year. These summaries can be secured directly from individual breed associations.

Most breed associations publish their sire summaries on the Internet. Entering a breed name in an Internet search engine will usually produce any existing websites for that breed. Some of these are sites maintained by the breed association, some are sites maintained by universities, and some are sites maintained by people with an interest in the breed (farmers, ranchers, owners, etc.). Always use caution when evaluating information provided by an Internet website. Consider the source of the information and any agenda the people maintaining the site might have.

Sire summaries provide information on traits that are economically important to cattle producers. Included is information regarding the ability of the bull to transmit growth rate to his offspring. This includes information about expected birth, weaning, and yearling weights. Several breed associations have added carcass information to the sire summaries. Included is information about rib-eye area, fat thickness, marbling scores, and carcass weight. It is expected that more breed associations will include carcass information in sire summaries in the future. Carcass data are important when selecting sires to produce animals to meet current market demand for leaner meat.

Many breed associations also include information regarding the daughters of the bulls in the sire summary. This information is referred to as **maternal breeding value (MBV)** and covers ease of calving and ability to wean heavy calves.

The ability of the sire to transmit genetic traits is defined in sire summaries as the **expected progeny difference (EPD)**. The EPD is a measure of the degree of difference between the progeny of the bull and the progeny of the average bull of the breed in the trait being measured. It is calculated from data collected on the progeny of the bull. Bulls in a sire summary for a given breed can be compared directly on the basis of their respective EPD values.

The EPD is usually given as a plus or minus value; however, some breed associations report EPDs as ratios. An EPD for yearling weight of +65 would show that the progeny of this bull should average 65 pounds more at 365 days of age than the progeny of the average bull of the breed. A minus value would indicate that the progeny of the bull would do poorer in the measured trait than the average bull of the breed.

The data from a sire summary can be used to select herd sires that are superior for traits that are desired. By using a succession of bulls that have high EPDs for selected traits, the producer can develop a superior herd. Care must be taken when selecting traits for a breeding program. There are some undesirable correlations between genetic traits. For example, yearling weight has a positive correlation with birth weight. This could lead to calving difficulties unless sires are selected with high EPDs for yearling weight and a moderate or low EPD for birth weight.

Bulls with high EPDs for the selected traits will produce some calves that are not better than the average for the breed. On the other hand, bulls with low EPDs for selected traits will sire some calves that are better than the breed average. However, the overall performance of the progeny of such bulls will be lower than the breed average. The use of sire summaries to select herd sires is a recommended practice for overall herd improvement.

The cow herd and the calves produced need to be evaluated before selecting the herd bull. The areas where the greatest improvement is needed must be found. After the weaknesses of the herd have been identified, a herd bull must be selected that will improve those areas.

Performance record information on the bull must be carefully considered. The bull selected should have desirable conformation, good health, and vigor. Breed traits are a matter of individual preference.

The adjusted weaning weight of the bull should be above the average of the herd. The bull must have plenty of size for age and be structurally correct. The adjusted 365-day weight should be at least 950 to 1,000 pounds. A bull that has gained rapidly from 8 to 12 months of age should be selected. Rate of gain is highly heritable.

The sire and dam of the bull should also have records as good producers. The dam should have been rated in the top half of the herd. The sire should have performed well in successfully impregnating cows and in the production of large-framed, fast-gaining calves.

There should be no record of dwarfism in the bull's pedigree. The semen should have been tested for fertility within 30 days of use for breeding. Testicle development should be normal. Both testicles should be present, fully descended, sound, and equal in size.

Measuring the circumference of the scrotum is a valuable method of determining the bull's fitness for breeding. Measurement is taken at the widest area of the scrotum using a tape with a slip cinch. The tape is pulled snug, but not so tight that the scrotum is wrinkled. Bulls with larger scrotums produce more sperm. Research has shown that they also sire heifers that reach puberty at an earlier age. Scrotum size has also been positively correlated with weight gain. English and crossbred bulls should have a minimum scrotum size of 11.8 inches at 12 to 14 months of age. A scrotum size of 13.4 inches at this age is considered very good. Research also reveals that it is possible to predict scrotum circumference in mature bulls by taking measurements when bulls are between 150 and 350 days of age. If the circumference of the scrotum at that age is less than 7 inches, it is probable that at maturity the bull will not have developed the minimum scrotum size needed for good breeding performance.

There should be no apparent physical or genetic defects. Bulls that are sickle-hocked (back feet are set too far forward), pigeon-toed (feet turn in), or splay-footed (feet turn out), or those with crooked ankles should not be selected. The bull should have a quiet disposition. Bulls that have unruly dispositions tend to produce calves of the same disposition, and calves of that kind do not gain rapidly in the feedlot.

JUDGING BEEF ANIMALS

Judging beef animals consists of comparing one animal to another based on the conformation of the animals. There are usually four animals in a class to be judged. The judge divides the animals into three pairs: a top pair, a middle pair, and a bottom pair. The four animals in the class are then examined and compared to the ideal animal. A good judge follows a definite procedure. The following is a summary of accepted judging procedures.

First, look at the animals from a distance of about 25 feet. This makes it possible to see all the animals at once. Look over the class for several minutes. Keep in mind the major differences among the animals. Placing should be made on major

Courtesy of Cabe Brothers Farm. Photo by Sarah Kane

Figure 14-7 Look at the animal from the side.

Courtesy of Cabe Brothers Farm. Photo by Sarah Kane

Figure 14-8 Look at the front view of the animal.

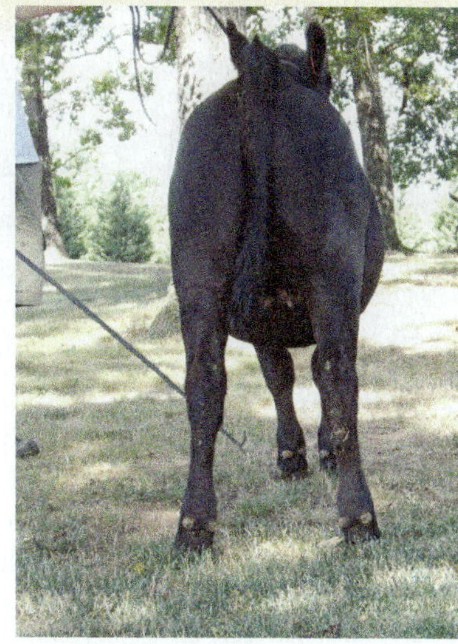

Courtesy of Cabe Brothers Farm. Photo by Sarah Kane

Figure 14-9 Look at the rear view of the animal.

differences. The judge's first impression of the class is usually best. Look at side, front, and rear views of the animals before making a placing (Figures 14-7, 14-8, and 14-9).

As the animals are walked, notice the soundness of feet and legs. Finally, study and handle the animals individually to check the placing. Handling the animals permits comparison of finish and natural fleshing. Feel the animals along the top of the shoulders, ribs, back, loin, rump, and round. Check for smoothness, firmness, and uniformity of finish (Figure 14-10).

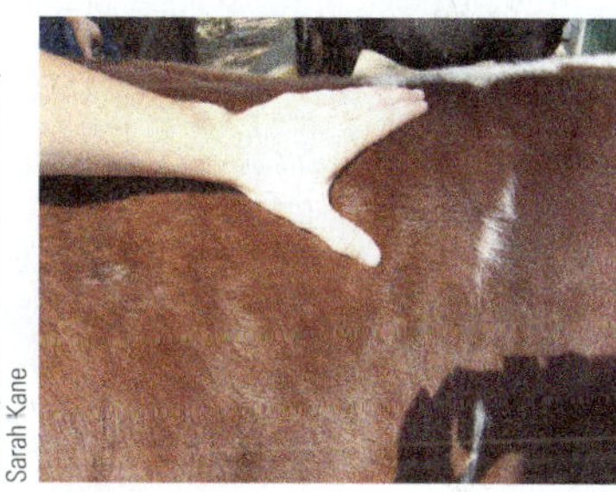

Courtesy of Cabe Brothers Farm. Photo by Sarah Kane

Figure 14-10 If rules allow, handle the animal to check the finish.

Each animal in the class is marked with a number, 1 through 4. The final placing is indicated by listing the numbers of the animals in the order of their placing. For example, if the number 3 animal was placed first, the number 1 animal placed second, the number 4 animal placed third, and the number 2 animal placed fourth, the final placing would be stated 3, 1, 4, 2.

When deciding the placing, begin with either the most or least favorable animal in the class, whichever is easier. Follow this by placing the other animals in the class.

Judging Market Classes

The main points to look for in judging market classes are (1) type, (2) muscling, (3) finish, (4) carcass merit, (5) yield, (6) quality, (7) balance, (8) style, and (9) smoothness.

Type (Beef Type)

Type refers to the general conformation of the animal and is best determined from the side view. The type most in demand is a thick, moderately deep-bodied animal. It should have medium length of leg and body, with straight lines and good balance. The current desirable type is one that is not too short-legged. A very low-set steer is just as wrong in type as an extremely tall animal.

Muscling

Muscling refers to the natural fleshing of the animal. It is an inherited trait. The animal's quarter should be thick, deep, and full. Good width of back, loin, and rump without too much fat also indicates muscling (Figure 14-11).

Finish

Finish is the amount of fat cover on the animal. The term *finish* is used only in market classes. A smooth, uniform finish that is not uneven is desirable. When the

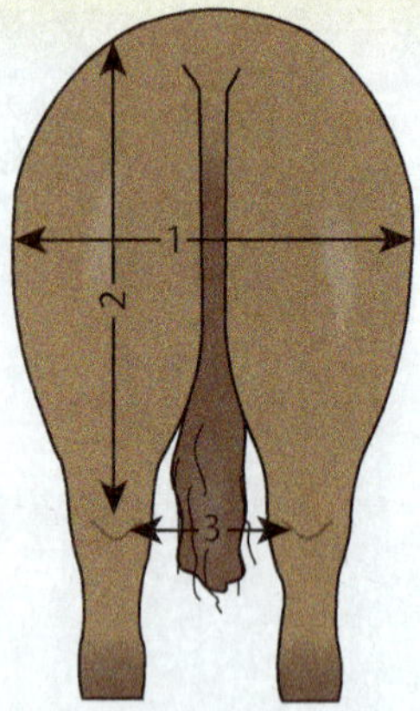

Things to Look For:
1. Width of round
2. Depth of round
3. Width between legs

Figure 14-11 Things to look for when checking muscling from the rear view of the animal.

animal is handled, the finish should be springy. A finish that is too hard or too soft is undesirable. The animal should not be too fat or too thin. A 1,000-pound steer should have from 0.3 to 0.6 inch of finish over the loin.

Carcass Merit

Carcass merit refers to the kind of carcass the animal will produce when slaughtered. A desirable carcass is thick, meaty, and correctly finished. There should be a lot of muscle in the high-priced cuts. Carcass merit must be estimated when judging live animals.

The carcass merit of cattle is expressed by yield grade and quality grade. There are five yield grades numbered 1 through 5. The fatter the animal, the smaller the percent of yield, since more fat must be trimmed from the retail cuts. **Yield grade** shows the percent of carcass weight in boneless, closely trimmed retail cuts from the round, loin, rib, and chuck that have been closely trimmed of fat. The percent of yield for each yield grade is:

- yield grade 1—over 52.4%
- yield grade 2—50.1%–52.3%
- yield grade 3—47.8%–50.0%
- yield grade 4—45.5%–47.7%
- yield grade 5—less than 45.5%

Lighter-weight, lean, heavily muscled cattle are yield grade 1. Heavier, fatter, light-muscled cattle are yield grade 4 or 5. The majority of cattle have a yield grade of 2 or 3.

The USDA uses a dual grading system that includes yield grade and quality grade. Quality grade (Prime, Choice, Select, Standard, Commercial, Utility, Cutter, and Canner) refers to the amount of marbling and other carcass grade factors.

Yield

Yield is the dressing percent (weight of the chilled carcass compared to the live weight) of the animal. It is found by dividing the chilled carcass weight by the live weight of the animal. This gives a percent of yield. A high yield is desirable. Conformation, finish, quality, and fill affect yield. Other factors affecting yield are the amount of refinement of the head, hide, and bone. An animal that is building fat through the middle will have a lower yield.

Quality

Quality refers to the refinement of the head, hide, bone, and hair. A steer should be medium boned. The head should be moderately refined and the hide should be thin and pliable. A soft, fine hair coat is desirable.

Balance

Balance refers to the general structure and proportion of the animal's body. A desirable animal is correctly proportioned.

Style

Style means the way the animal shows and the way it carries itself. It should be attractive and alert and should show to good advantage.

Smoothness

Smoothness refers to the lack of roughness in finish or bone structure. Rough finish can usually be seen along the top of the animal. Rough bone structure can be seen at the shoulders and hooks. A desirable animal will be smooth rather than rough.

Breeding Classes

Breeding classes are judged for several traits in addition to those listed for market classes. These are (1) condition, (2) size, (3) feet, legs, and bone, (4) breed character, and (5) sex character.

Condition

Condition refers to the amount of fat cover the animal has. This is referred to as finish in market classes. Finish depends on the amount of grain the animal has been fed. Always use the term *condition* rather than *finish* when judging breeding cattle. Muscling is more important than condition when judging breeding cattle because muscling is an inherited trait.

Size

Size in relation to age is important when judging an animal. Height, by itself, does not mean a larger size. A large animal is better than a small one when both are the same age. The overall size of the animals must be considered; however, size alone is not the only trait to look for. The other factors related to judging must be kept in mind.

Feet, Legs, and Bone

Legs should be strong and straight and have heavy bone. They should be set well out on the corners of the animal. The feet must be large enough to give a good foundation for the animal. Unsoundness in feet and legs makes the animal less valuable for breeding stock.

The Breed Character

The breed character is shown in the head and general appearance of the animal. Each breed has its own traits. A good judge must study the traits of the breeds in order to recognize breed character.

Sex Character

Sex character refers to the traits that distinguish the animal as a male or a female. The male should show a heavier development of the forequarters. The neck is thicker and supports a strong, bold head. The overall appearance should be massive and powerful. A female shows more refinement in appearance. The neck and shoulder are lighter. The head and bone show more refinement than those of the male.

Judging Terms

A good livestock judge uses the correct terms to describe the animal being judged. The following is a list of favorable terms used to describe beef cattle. When an animal is being criticized, the opposite of these terms is used. The terms are given in their comparative forms (for example, thicker rather than thick) as they would be used to compare one animal to another in an actual judging situation.

General Terms Used for Breeding and Slaughter Cattle

- growthier
- heavier muscled
- higher quality
- more style
- more correct set to feet and legs
- stronger top
- more structurally correct
- more even width (uniform width)
- more stretch
- tighter framed
- deeper rib
- wider loin
- deeper twist
- smoother handling
- freer, easier in movement
- typier
- thicker
- more nicely balanced
- deeper body
- smoother top (hooks, tail head, shoulders)
- straighter lined
- wider, stronger top
- bigger framed, more upstanding
- longer, leveler rump
- more spring of rib
- wider back
- fuller behind the shoulders
- thicker, deeper quarter
- thinner hide

Terms and Expressions Used for Slaughter Cattle

- more uniformly (desirably, correctly) finished
- smoother finished
- firmer finished
- trimness in brisket, twist, and rear flank
- carries more finish over the rump (loin, ribs, back)
- thicker fleshed throughout
- more finish all down the top
- beefier, heavier muscled
- deeper, plumper, more bulging round
- more natural width and muscling over the back and loin
- wider over back and loin
- fuller in the rib
- wider and fuller in the stifle
- thicker, more muscular round
- heavier muscled loin
- longer, meatier loin
- meatier further down the shank
- more width through the center of the round
- meatier, heavier quartered
- wider through the thigh
- longer and wider through the rump
- larger framed
- more size for age
- trimmer middled (fronted, brisket)
- more refinement of head (hide, bone)
- more uniform in width from front to rear
- would yield a meatier carcass (more of the high-priced cuts)
- would have a more desirable yield grade
- would hang up a meatier carcass would hang up a thicker, heavier muscled carcass

Terms and Expressions Used for Breeding Cattle

- typier heifer
- more weight for age
- larger framed heifer
- more size or scale and capacity
- larger, more rugged heifer
- growthier, beefier heifer
- growthier, thicker heifer
- more uniform in width from front to rear
- deeper, fuller spring of rib
- more refined in muscling (female)
- moderate thickness of muscling (female)
- longer muscling in round (female)
- more strength and thickness of muscling down the top (bull)
- more bulge and fullness of muscle in the stifle area (bull)
- stands more correctly on its legs
- stands and walks more correctly
- stronger fronted
- wider fronted (wider at the chest)
- easier walking (more correct on the move)
- walks with more style and balance
- more feminine head (female)
- strong head (bull)
- shows more masculinity and ruggedness (bull)
- more masculinity about head and neck (bull)
- better developed testes (bull)
- stronger udder attachment (female)
- better balanced udder (female)
- more correct udder development (female)
- no excessive fat deposits
- no excess fat along underline
- trimmer, less fatty brisket, rear flanks
- more desirable condition
- smoother blending of shoulders and rib
- smoother through the hooks
- neater about the tail head
- more style, symmetry, and balance
- smoother blending of parts

Oral Reasons

Oral reasons are given to explain the differences between animals that influenced their placing. The differences between two animals are described by using the terms noted earlier and by following the basic guidelines described here. Oral reasons are presented before a contest official after the contestant has judged one or more classes of animals.

Completeness

Mention the main differences; do not spend time on minor details.

Length

Do not speak more than 2 minutes. Be definite and concise.

Presentation and Delivery

When giving your reasons, stand 6 to 8 feet from the contest official. Look the official straight in the eye. Hands are kept at the side or behind the back. Stand straight, with feet apart. Reasons are given in a logical order and should be well organized. Speak clearly and loud enough to be heard, but not too loud and not too rapidly. Use complete sentences with correct grammar. Speak with confidence; do not hesitate. Make a concise final statement explaining why the animal at the bottom of the class was so placed.

Accuracy

Only make statements that are true. For example, describe strong points of the animal only if they exist. Grant or admit the good qualities of the other animal in the pair.

Terms

Use the correct terminology for the class the reasons were written for. A set of reasons is organized in the following way:

1. *Introduction*—Give the name of the class and how one placed it.
2. *Top pair*—Give the reasons for placing the top animal over the 2nd place animal, using comparative terms. Grant the advantages of the 2nd place animal over the 1st place animal, also using comparative terms. Finally, criticize the 2nd place animal using either descriptive or comparative terms.
3. *Middle pair*—Follow the same outline used for the top pair.
4. *Bottom pair*—Follow the same outline used for the top pair.
5. *Summary*—Give a final statement that explains why the animal at the bottom of the class was so placed. Mention the animal's strong points, if any, and describe its major faults.

Class #3: Market Steers		Placing: 3-1-2-4	
1 Black	2 Hereford	3 Black Baldy	4 Red
3/1	Heavy muscled More fat cover Apt to grade choice More lbs. of product Heavy boned	1/3 Grant Higher cutability Moderate frame 1 (Fault) Poor balance Over-finished	
1/2	Better off hind two legs Heavy muscled Market-ready look Wider chested	2/1 (Grant) Shapely, more expressive Clean lines Level hip 2 (Fault) Short coupled Narrow made	
2/4	Harvest with a more shapely carcass More even fat deposition	4/2 (Grant) Extended Structurally sound Deep bodied 4 (Fault) Light muscled Under finished Requires more days to grade choice Fine boned	

Courtesy of Alyssa K. Elrod

Figure 14-12 A sample set of notes for a class of market steers. Overall placing is shown at the top, with #3 being the most desirable animal in the class and #4 being the least desirable.

Taking Notes

Simple notes should be taken while evaluating each class. They are used to prepare oral reasons but are not read to the contest official. Notes should be put away before presenting reasons to the official. Notes should allow the contestant to recall each animal, form a mental picture, and prepare an oral presentation. The class number, name, and placing should be written as a heading. A description of each animal should be listed. Boxes are used to show the grants and faults of each animal. A comparison of the top, middle, and bottom pair should be included. Figure 14-12 shows a sample set of notes for a market class of crossbred steers.

SUMMARY

The three main kinds of beef cattle production systems are (1) cow-calf producers, (2) purebred breeders, and (3) cattle feeders. Cow-calf producers keep herds of cattle and produce calves for sale to slaughter cattle producers. Purebred producers furnish breeding stock to cow-calf producers. Cattle feeders buy calves or yearlings and feed them to market weights.

Beef animals are selected mainly on the basis of conformation and performance records. Desirable conformation refers to the finish and structure in demand in the marketplace. An animal should produce the maximum amount of the high-priced cuts. Performance records help the producer select animals based on records of previous performance. Some traits are more heritable than others. Cattle producers who use the knowledge of heritability of traits do a better job of improving their beef herds.

Beef animals are judged on the basis of their conformation. A good judge learns the parts of the animal and develops a logical system for evaluating the animal. Presentation of oral reasons, using the proper terms, provides the opportunity to explain why animals were placed in a given way.

Student Learning Activities

1. Using pictures or a live animal, name the parts of the animal.
2. Judge breeding and market classes of beef animals from pictures or live animals. Give oral reasons.
3. Take a field trip to a farm to observe the performance testing procedures being used.
4. Give an oral report to the class on various aspects of selecting beef cattle.
5. Evaluate pedigree information on several beef animals and select the most desirable animals.
6. Attend a beef consignment or dispersal sale and select animals you would buy based on conformation, pedigree, and performance records.
7. Observe the use of ultrasonics in beef selection.
8. Calculate 205-day adjusted weaning weights and adjusted yearling weights.
9. When planning and conducting a supervised experience program in beef production, use all the appropriate criteria learned in this chapter to select animals.

Discussion Questions

1. Name the three main types of beef cattle production systems.
2. Briefly describe each system named in question 1.
3. Which type of system is most common in your area? Why?

4. Describe the kind of conformation that is considered desirable in beef cattle.
5. Why is good conformation important when selecting beef breeding stock?
6. Name and briefly describe the kinds of performance records that might be used when selecting beef breeding stock.
7. List nine uses of beef performance testing records.
8. What is a pedigree?
9. How might a pedigree be used when selecting beef breeding stock?
10. Why is selection of the herd bull so important to success in beef production?
11. Briefly describe the traits that are desirable in a beef bull.
12. Describe the general procedure to be followed when judging a class of four animals.
13. List and briefly describe the nine main points to look for when judging a class of market beef animals.
14. Name and briefly describe the five additional points to look for when judging a class of breeding beef animals.
15. What are the five points to remember about procedure when giving a set of oral reasons?
16. Describe the organization of a set of oral reasons.
17. Describe a system for taking notes when preparing to give oral reasons in a judging contest.

Review Questions

True/False

1. The cow-calf system of beef production involves keeping a herd of beef cattle.
2. Beef animals are judged on the basis of their conformation.

Multiple Choice

3. Purebred breeders provide stock for:
   - a. feedlots
   - b. cattle feeders
   - c. replacement bulls
   - d. none of these
4. Bang's disease is another name for:
   - a. tuberculosis
   - b. vibriosis
   - c. brucellosis
   - d. mad cow disease

Completion

5. The use of high-frequency sound waves to measure fat thickness and loin-eye area is called ______.
6. Breed associations maintain performance records for the daughters of bulls in their sire summary. This information is referred to as ______ ______ ______.
7. ______ ______ are calves that are 1 to 2 years old and are sold to be fed to finish for slaughter.

Short Answer

8. What does EPD mean?
9. What is the value of performance testing?

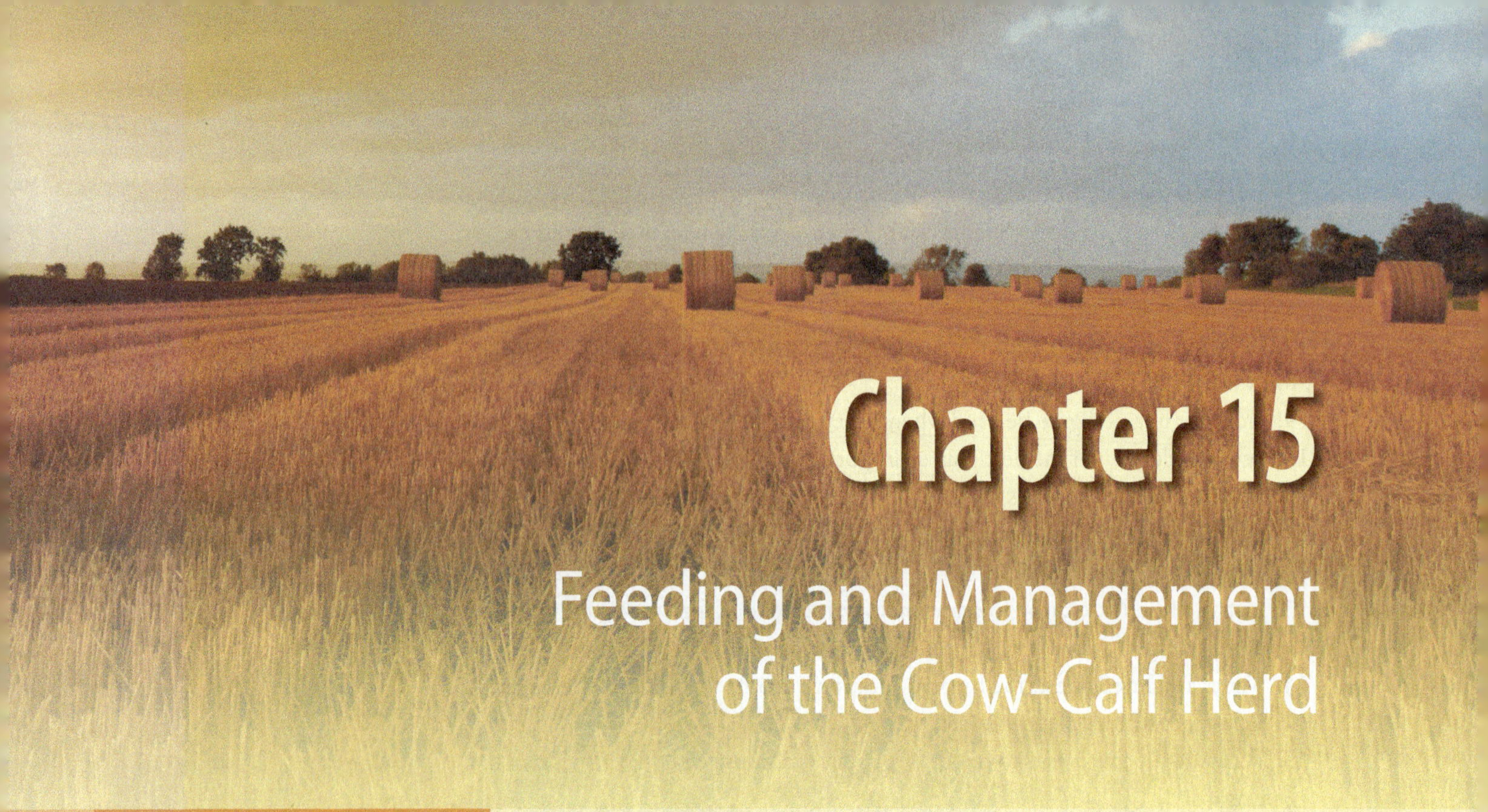

Chapter 15

Feeding and Management of the Cow-Calf Herd

Key Terms

integrated resource management (IRM)
feed composition table
rotation grazing
carrying capacity
husklage
creep feeding
puberty
artificial insemination
synchronization of estrus
castration
steer
staggy
burdizzo
elastrator bands
preconditioning
backgrounding

Objectives

After studying this chapter, the student should be able to

- plan a feeding program for a cow-calf herd.
- list and describe approved practices for managing a cow-calf herd.

INTEGRATED RESOURCE MANAGEMENT

Integrated resource management (IRM) is a management tool available to beef producers that uses a team approach to improve the competitiveness, efficiency, and profitability of their beef business. A typical team consists of accountants, bankers, extension personnel, veterinarians, university researchers, soil and water conservationists, and people from allied industries. Team members are all experts in their respective fields. The team analyzes the beef production practices of the farm or ranch to diagnose and solve inefficiencies and make sure the production activities complement each other in a positive manner.

Important components of IRM include setting production and financial goals and accurate record keeping. Both production and financial performance data are gathered and analyzed using the Standard Performance Analysis (SPA). Cattle producers keep accurate financial and production records to be used in the analysis. Typically, the operation is analyzed by its component enterprises. These may include the cow herd, heifer development, feedlot sales, bull sales, pasture use, and forage production. This permits the analysis to pinpoint areas of strength and weakness within the total operation.

The SPA is available as a computer software program to analyze the data. This program was developed by the U.S. National Cattlemen's Beef

Association Producer Committee and Extension specialists. The software has two components: SPA-F for financial information and SPA-P for production information. The SPA-F part of the program helps the operator analyze the economic and financial performance of the operation. It converts cash basis records and inventory changes to an accrual accounting basis. Information returned includes cost per pound of calf weaned, investment per cow, and return on assets. The financial data conform to the recommendations of the Farm Financial Standards Task Force. The SPA-P part of the program provides information on many production measures, including reproductive and growth performance and feed use. The data may be submitted for inclusion in a national database for comparison with other beef herds in the region. The SPA may also be completed as handwritten records.

A simpler form of the SPA, called SPA-EZ, was developed in an effort to get more cattle producers involved in the program. The simplified form runs in a Microsoft Excel format. It provides accurate performance information but in less detail than the full program. The program will determine the annual cost of a cow and the cost per hundredweight of producing a calf.

A full SPA analysis that spans several years provides management information that helps the farm or ranch manager:

- determine profitability of the enterprise
- identify components of the business that are meeting or exceeding goals and components that need improvement
- make better marketing, investment, and production decisions
- develop goals and track progress toward meeting those goals
- compare alternative investment opportunities
- develop employee incentive programs
- monitor and control costs
- determine how competitive the total business is, as well as the individual enterprise on the farm or ranch
- evaluate the use of resources and identify ways to improve resource use
- provide information to multiple owners, lenders, and advisors

Over a 5-year period, beef producers enrolled in the SPA program in Illinois were able to significantly lower their total costs per cow. This was done by improving the efficiency of the operations after SPA analysis identified areas where cost reduction was possible. Both commercial and purebred beef herds were enrolled in the SPA program, and larger herds had lower total costs per cow than did smaller herds. The management concepts used in IRM are being increasingly used in other segments of agriculture. Examples include integrated pest management, integrated crop management, and total quality management.

FEEDS

Feeding programs for beef cow-calf herds are based on the use of roughages. Different kinds of roughages are used, depending on where in the United States the beef herd is located. **Feed composition tables** give the analysis of feeds that can be used for rations for beef herds. Typical roughages used include pasture, hay, silage, straw, corncobs, and other crop residues. Corn silage is widely used as roughage feed for beef herds. Alfalfa is the most common roughage in the Midwest and West. Coastal Bermuda grass is more common in the southern coastal states. Native range grasses are utilized to a great extent in the western states.

Less commonly used roughages include oat straw, barley straw, and wheat straw. Cottonseed hulls, peanut hulls, oat hulls, and rice hulls are also used. In the Corn

Belt, husklage, corn stover (stalks), corn stover silage, and corncobs may be used. Sorghum and soybean residues also have value for feeding cows.

Grasses often used for pasture or hay include fescue, orchard grass, reed canary grass, smooth bromegrass, Kentucky bluegrass, perennial ryegrass, timothy, redtop, Bahia grass, dallis grass, and Sudan grass. Some of the legumes used include red clover, alsike clover, *Sericea lespedeza*, peanut hay, sweet clover, cowpeas, and soybean hay.

Roughages provide the cheapest source of energy for the cow and the calf. However, adjustments must be made so that the ration meets the needs of cows and calves. If the quality of the roughages is high, little or no supplement is needed. The use of high-quality roughages also increases the weaning weight of calves. Usually, some additional minerals are needed in the ration.

MANAGING FEED SOURCES

Forages

Forages should be handled in a way that will keep the labor requirement low. Grazing should be used as much as possible. In some areas, weather does not permit year-round grazing. In this case, harvesting and storing forages increases the amount available for the beef herd. Crop residues, such as cornstalks, can be used as feed for the beef herd. However, only about 15 to 30 percent of the amount produced is recovered when grazing is the only method of harvesting.

Pasture and Hay Land

The proper management of pasture and hay land increases the yield of forage. The soil should be tested and the correct type and amount of fertilizer added. If too many cows are fed on a pasture, its quality is reduced. When rotation grazing is used, the carrying capacity of the pasture is increased. **Rotation grazing** is the division of a field by the use of temporary fencing. The cow herd is allowed on only part of the field at a time.

Crop Residues

Grazing crop residues reduce feed costs. In northern areas, two acres of cornstalks will carry a pregnant beef cow for about 80 to 100 days. Heavy snow cover reduces the carrying capacity of cornstalk fields (Figure 15-1). The **carrying capacity** of a pasture refers to the number of animals that can be grazed on the pasture during the grazing season.

Crop residues may be harvested in several ways. Equipment such as balers, forage harvesters, and stackers may be used. Yields of 1 to 2 tons per acre may be expected. The moisture content may range from 20 to 55 percent. When stacked or baled, the residues should be below 30 percent moisture content at harvest to prevent spoiling. For corn stover silage, the moisture content should be above 50 percent at harvest, and 60 to 70 percent is better. Water should be added if necessary.

Husklage is the material that comes from the corn combine that can be used as a feed. Trailing units are made that will collect this material. It is then dumped in piles and can later be loaded and hauled to a storage site or can also be chopped for silage or baled. Sometimes husklage is left in piles in the field for the cattle to eat during the winter.

The digestibility and protein content of harvested crop residues can be increased by treating the residue with anhydrous ammonia (NH_3). Treat the crop residue with an amount of NH_3 equal to 2.5 to 3.0 percent of the weight of the residue. The

Photo by Jeff Vanuga, USDA Natural Resources Conservation Service

Figure 15-1 These Wyoming cattle have to be given hay in deep snow conditions when they cannot graze on pasture.

material to be treated is sealed under a 6- to 8-mil ultraviolet-resistant, clear or black plastic cover. The ammonia is applied through a hose or pipe, with the amount being controlled with a regulator or gauge. The NH_3 must be applied slowly (1 to 5 minutes per ton of ammonia) to prevent ballooning of the plastic.

The sealed environment must be maintained long enough to complete the treatment of the residue. At low temperatures (under 41°F), the environment must be kept sealed longer (more than 8 weeks) than at higher (over 86°F) temperatures, where the treatment can be completed in less than 1 week. The plastic cover can be left on the roughage until it is to be fed. It takes about 1 day for the free ammonia to dissipate after the plastic is removed.

Treatment with NH_3 can increase the protein content of low-quality roughages by 4 to 6 percentage units. Digestibility may be increased by as much as 10 percentage units. For example, crude protein content might be increased from 4 percent in untreated material to 9 percent in treated material, and digestibility might go from 46 percent in untreated material to 56 percent in treated material. Dry matter intake is also increased when the material is treated with ammonia.

Some problems have been observed when feeding with ammonia-treated roughages. Usually, these problems resulted when the roughage that was treated was high quality, with a high soluble sugar content; was above 20 percent moisture content when treated; and when an excessive amount of NH_3 (over 3 percent of forage dry matter) was used in the treatment. No problems have been observed from feeding low-quality roughages that have been treated with ammonia.

Use of Round Bales

The use of large round bales (Figure 15-2) to harvest some types of forage can cut labor requirements by up to 60 percent. However, there is about a 20 percent loss of dry matter, energy, and protein content when harvesting this way as compared to using rectangular bales. Research shows that about 20 to 32 percent more dry matter is needed in the ration when large round bales are fed. Weight gains and cow condition scores are lower when large round bales are fed outside, free choice.

Courtesy of Frank B. Flanders

Figure 15-2 Some types of forage can be harvested in large round bales.

Weight gains and cow condition scores are higher with limit-fed rectangular bales that are stored inside.

Large round bales should be baled at 16 to 20 percent moisture to prevent molding. Forage baled to wet or "green" may combust due to a buildup of heat within the center of the bale. Make the bale tight to prevent loss and store it in a well-drained area.

Losses can be reduced when large round bales are stored inside or under some type of cover. If the hay is not protected during storage, losses can be as high as 50 percent of the value of the hay. Research shows that losses of dry matter in large round bales is approximately 6 percent for inside storage, 15 percent when the bales are stored on wooden platforms and covered with plastic, and 66 percent when the bales are stored unprotected on the ground.

A pole-barn shed can be used for large bale storage. This is probably the best method of storage to prevent losses. Other methods of protection include covering the bales with plastic or completely enclosing them in plastic bags. Equipment is available that puts bales into long plastic tubes. Each tube is 200 feet long and will hold approximately 36 large round bales. If grass hay is enclosed in plastic bags, it is possible to treat it with ammonia to increase its protein content. To prevent spoilage on the bottom of the bales, put them on old tires, pallets, crossties, poles, gravel, or other objects that will keep them from making direct contact with the ground. Geotextile materials, such as spun or woven fabric, may be used to create a barrier between the ground and the rock or gravel spread on top. The material is rolled out on the ground and then 6 inches of rock or gravel is spread on top. This surface becomes very compact over a period of time and provides a good way to reduce waste when large round bales are stored on it.

Some research has been done on using a specially hydrogenated liquid animal fat (tallow) or plant fat (a soy protein isolate) to completely cover the outside of large round bales. After application, the fat hardens into a paraffin-like consistency. No mold or spoilage was found in the bales when the hay was baled at less than 18 percent moisture. Almost no loss in nutrients occurred in bales treated with this covering. Animals will eat the outer layer of hay covered by the material used in the treatment. A tightly wrapped bale, with 4-inch twine spacing, gives the best result when using this treatment. The treatment is less successful when bales are loosely wrapped.

The economic returns from protecting large bales in storage are higher in areas with high rainfall. However, some research indicates that there is some loss in feeding value when the bales are unprotected even in areas of low rainfall.

Bales should be placed in such a way that air can circulate around them. If the bales are stored outside, place the ends north to south so they will get more sunlight and dry out quicker after rain.

Access to bales, husklage piles, or stacks of forage is controlled to prevent waste. Feeding gates or electric fencing may be used to control access. A hay-feeding rack on skids can also be used. Any system of feeding hay will result in some loss, but the amount of loss can vary widely based on the methods used for feeding and controlling access to the feeding area. When cattle are given free access to the feeding area, losses may be as high as 60 percent. Poor management of forage feeding may increase the total cost of production by as much as 10 percent. Careful management can reduce losses to as little as 2 percent. Using feed racks, rings, or feeders for any type of forage bales will significantly reduce waste. The quality of the hay also influences the amount of waste; usually more low-quality hay than higher-quality hay is wasted. It may be necessary to limit the amount of hay fed at any one time to no more than a one-day supply in order to reduce waste.

FEEDING DRY, PREGNANT COWS AND HEIFERS

Dry, pregnant cows are fed enough to keep them in good flesh from fall to spring calving. Cows that are of normal weight in the fall should not lose more than 10 percent of their body weight. Thin cows should be fed enough to cause some weight gain during the winter. If cows do not receive enough feed, the calf crop percent is lowered.

Overfeeding should be avoided for several reasons. Results of overfeeding include:

- cows becoming too fat
- higher production costs
- breeding/calving difficulties
- increased calf losses
- decreased milk production

During the fall grazing season, fall pastures, permanent pastures, and crop residues from small grain fields or cornstalks may be used. During the winter, hay, silage, or harvested crop residues may be fed to the cow herd. If the climate permits, pasture may be used all winter.

Young cows and heifers are still growing and require more feed than mature animals. The amount of feed the cow receives is more important than the kind of feed. The amount of energy the cow needs varies according to size, condition, age, and weather. During cold weather, increase feed or energy intake by about 1 percent for each degree of cold stress.

During the last 30 to 45 days of pregnancy, cows generally need a 10 to 15 percent increase in the protein in the ration, especially if they are being fed stored roughage such as hay. This can be provided with an extra 2 pounds of high-quality hay per day. Additional protein supplement may be fed if the roughage is low in quality. Nutrient requirement tables may be used to develop rations for the beef breeding herd. Sample rations for wintering dry, pregnant beef cows are given in Table 15-1; sample wintering rations for bred heifers are given in Table 15-2.

Minerals should be fed free choice to the cow herd. Mineral mixes should include calcium, phosphorus, salt, and any trace minerals that are known to be deficient in the area. In areas where grass tetany is a problem, magnesium oxide may be included in the mix. A good mixture to use is one part trace mineral salt and one part dicalcium phosphate.

Protein supplement is often not needed if high-quality roughages are used. A ration that feeds low-quality roughage, such as range grasses, requires a protein supplement. As much as one-third of the protein in the ration may have to be provided as a supplement to cows on range.

Protein blocks are a convenient way to feed protein to the cow herd. The salt content and the hardness of the block control the amount the cow eats. Liquid protein supplements may be fed in lick tanks. Cubed or pelleted protein supplements are handy for hand feeding. However, plenty of bunk space must be provided if they are used. "Boss" cows will get more than their share if bunk space is limited.

Care must be taken to prevent overeating of protein. A hungry or thirsty cow will eat too much protein. Overeating of protein may be partially controlled by feeding plenty of roughage. Make sure there is plenty of water available because this will help flush out excess protein if the cow overeats. Self-feeding of urea is not recommended because urea may be toxic to the cow if too much is eaten. The intake of urea must be carefully controlled.

Vitamin A may be needed if poor-quality roughages are fed. Good quality green roughages usually provide plenty of vitamin A. If the cow has been on good summer pasture, enough vitamin A will be stored in the body to carry the animal for several months in the winter. If the ration is short on vitamin A, 30,000 international units (IU) should be fed during the last months of pregnancy.

TABLE 15-1 Rations for Wintering Dry, Pregnant Beef Cows (wt. 1,000 to 1,100 lb)

Ration	Amount per Day (lb)
1. Legume hay	16–25
2. Mixed legume-grass hay ($\frac{1}{3}$ legume)	18–22
3. Legume hay	5–10
Straw or low-quality grass hay	10–15
4. Legume-grass haylage	30
5. Corn or grain sorghum silage	35–50
Protein supplement (48% total protein)	0.5–1
6. Legume-grass hay	10
Straw or cobs	10
7. Corn or sorghum silage	30
Legume hay	5
Straw, low-quality grass hay, cottonseed hulls, ground corncobs, or other low-quality roughage	Unlimited
8. Prairie or grass hay	Unlimited
Protein supplement	0.5–1
9. Grass silage	30–40
Straw or low-quality grass hay	Unlimited
10. Grazing crop residue	Unlimited
11. Cornstalk silage	40
Legume hay or hay silage	4–5
12. Corn silage	30
Mixed hay	4
13. Grass silage	25–35
Grass or mixed hay	10
14. Mixed or grass hay	10
Pea-vine silage	25–35
15. Husklage	11
Alfalfa brome hay	7
16. Corn stover	12
Alfalfa brome hay	7
17. Corn stover	15
Shelled corn	4
Protein supplement (35% total protein)	1.1
18. Husklage	15
Shelled corn	3
Protein supplement (35% total protein)	1.1
19. Corn silage	43
20. Soybean stover	20
Alfalfa brome hay	7

Source: Data compiled from various university Cooperative Extension Service bulletins

LACTATION RATIONS

The ration needed for the cow that is nursing a calf depends on how much milk the cow produces. Heavier milk producers have higher requirements than average or low milk producers. The protein requirements for lactation are 160 to 268 percent greater than for dry cows and energy needs are 36 to 68 percent more. Calcium and phosphorus needs are 100 to 250 percent higher, while Vitamin A needs are 18 to 88 percent greater.

TABLE 15-2 Wintering Rations for Bred Heifers (wt. 800 to 900 lb)

Ration	Amount to Feed (lb)
1. Legume-grass hay	20
2. Corn silage	50
Soybean meal	1
3. Corn silage	25
Legume-grass silage	10
4. Corn silage	45
Protein supplement (48% total protein)	1.5
5. Legume-grass haylage	35
6. Legume-grass hay	20
Corn (shelled)	5.6
7. Alfalfa hay	22.8
Corn (shelled)	3.4

(Feed vitamin–mineral mix free choice.)

Source: Data compiled from various university Cooperative Extension Service bulletins

Pastures of high quality can usually meet the needs of the lactating cow. Salt and minerals should be provided free choice. When the roughage is of poor quality or limited in amount, some grain should be added to the ration. If there is only a limited amount of legume forage in the ration, a protein supplement should also be added.

Heifers calving for the first time require more feed because the heifers are still growing and developing. Weight that is lost from calving must be regained. Additional feed is necessary if the heifer is to produce enough milk for the calf. The heifer must be in good condition for rebreeding.

When pasture is not available, the lactating cow is fed in drylot. Table 15-3 gives some drylot rations that can be used for lactating cows. Table 15-4 shows sample rations for first-calf heifers in drylot.

TABLE 15-3 Lactating Rations for Cows in Drylot (wt. 1,100 lb)

Ration	Amount to Feed (lb)
1. Legume-grass hay	10
Corn silage	40
Vitamin A	40,000 IU
2. Legume-grass hay	30
3. Legume-grass hay	20
Corn	4
4. Legume-grass silage	74
5. Legume-grass haylage	50
6. Corn or grain sorghum silage	60
Protein supplement (48% total protein)	1.5
7. Mixed hay	20
Shelled corn	5
8. Corn stover	20
Shelled corn	5
Protein supplement (35–40% total protein)	2.8
9. Alfalfa haylage (55% DM)	29.8
Shelled corn	8
10. Corn silage	50–55
Legume hay	4–5

Source: Data compiled from various university Cooperative Extension Service bulletins

TABLE 15-4 First-Calf Heifers—Lactation Rations—Drylot

Ration	Amount to Feed (lb)
1. Legume-grass hay	25
Ground shelled corn	3
2. Corn silage	60
Soybean meal	1.5
3. Corn silage	30
Legume-grass hay	13
4. Legume-grass silage	65
Ground shelled corn	3

Source: Data compiled from various university Cooperative Extension Service bulletins

CREEP FEEDING OF CALVES

The Decision to Creep Feed

Creep feeding is a way of providing calves with extra feed, which may be grain, commercial creep feed mix, or roughage. It is fed in a feeder in an area cows cannot reach. Creep feeding may or may not be profitable. The kind of operation, production conditions, and marketing practices influence whether or not creep feeding is economical.

Creep feeding:

- produces heavier (30 to 70 lb) calves at weaning
- produces higher grade and more finish at weaning
- results in calves going on feedlot rations better at weaning time
- creates less feedlot stress
- allows cows and calves to stay on poorer-quality pasture for a longer time

Creep feeding is often used if:

- calves are to be sold at weaning
- calves are to be fed out on high-energy rations
- cows are milking poorly
- calves are from first-calf heifers
- calves were born late in the season
- calves have above-average inherited growth potential
- calves were born in the fall
- calves are to be weaned early (45 to 90 days)
- calf-feed price ratio is favorable
- pastures become dry in late summer
- cows and calves are kept in confinement

Creep feeding has the following disadvantages:

- If calves are well-fed after weaning, the weight advantage from creep feeding is lost
- When production testing, it is harder to detect differences in inherited gaining ability
- Replacement heifers may become too fat
- Calves that are not creep fed usually make faster and more economical gains after weaning as compared to calves that were creep fed before weaning

Creep feeding is generally not used if the:

- calves are to be fed through the winter on roughages
- cows are above-average milk producers

TABLE 15-5 Rations for Creep Feeding Calves—Self-Fed (100 lb Mix)

Ration	Amount to Feed (lb)
1. Shelled corn	100
2. Shelled corn	50
Oats	50
3. Ground ear corn	90
Soybean meal	10
4. Corn or barley (rolled, cracked, or coarsely ground)	50
Whole oats	30
Protein supplement (6.7 to 26% total protein)	10
Molasses (dried or liquid)	10
5. Shelled corn	90
Soybean meal	10

Source: Data compiled from various university Cooperative Extension Service bulletins

- calf-feed price ratio is poor
- calves are on good pasture
- heifers are to be kept for herd replacements
- milk production of the dam is to be measured

When and How to Creep Feed

Calves will start to eat grain at about 3 weeks of age. They eat only small amounts until about 6 to 8 weeks of age. It takes 6 to 9 pounds of feed for each one pound of gain. About 280 to 480 pounds of feed are required for 40 to 60 pounds of gain.

Grains alone will often meet the energy needs of the calf. Milk and pasture provide the protein, minerals, and vitamins the calf needs. Whole oats and cracked corn mixed 50:50 is a good, simple creep feed. If molasses is added to the mix, the feed will taste better, and the calf will eat more. Research has shown that calves prefer rolled shelled corn and linseed meal pellets over whole oats or whole shelled corn. Table 15-5 gives some suggested creep feed rations for self-feeding.

Commercial creep feeds are available. Pelleted mixes are well liked by calves. They are easy to handle and do not blow away in the feeder. The calves eat more and gain more rapidly. Some commercial mixes are medicated. Mixes usually include minerals and vitamins.

Locate the creep feeder near the area where the cows loaf, preferably in the shade. If salt for the cows is put near the creep feeder, they will gather in the area. This brings the calves to the area, and they will use the creep feeder. Waterers should be located nearby.

The amount of feeder space needed is 4 to 6 inches per calf. Self-feeders or bunks that are portable should be used. When feeding for limited energy intake, include a hay self-feeder in the creep. Cover the feeders to protect the feed from the weather. The opening into the creep must be small enough to keep the cows out. An opening 16 inches wide and 36 inches high is about right. An adjustable opening is best.

GROWING REPLACEMENT HEIFERS

At least 15 percent of the cows from the breeding herd are lost each year due to death, breeding failure, or aging. About 30 to 40 percent of the heifers are saved each year to replace the cows that leave the herd. Performance records are the best way to decide which heifers to keep. Only those in the top half in weaning weight should be kept.

Heifer conception rates are lower than cow conception rates. Check the heifers for pregnancy 60 to 90 days after breeding. About 60 to 70 percent of those bred are kept, based on pregnancy and adjusted yearling weight. Replacement heifers of the British breeds should gain about 1.0 to 1.25 pounds per day from weaning to breeding. Larger breeds should gain about 1.25 to 1.75 pounds per day.

Puberty is the age at which heifers come into heat. In an efficiently managed cow-calf herd, heifers should reach puberty at 12 to 14 months of age. This goal can be attained with proper selection and feeding of replacement heifers. Generally, heifers reach puberty when they have attained about 65 percent of their mature weight. Heifers of the English breeds should weigh about 550 to 625 pounds at puberty. Larger breed and crossbred heifers of larger breeds should weigh about 675 to 750 pounds at puberty. Heifers should be bred according to weight rather than age.

Heifers on good pasture will gain about 0.75 to 1.4 pounds per day. When the pastures are poor, it is good to feed about 3 to 5 pounds of grain per head per day. At calving time, heifers should weigh about 900 to 1,050 pounds

Feed for heifers must be palatable. Coarse, poor-quality feed is not as good in the ration as better-quality feeds. In regions of cold weather, the heifers need more feed for energy to maintain body heat. Nutrient needs increase 1 percent for each degree of temperature below freezing and when the heifers are in unprotected areas.

As much as 15 percent of the feed that is fed is wasted. Allowance for waste must be calculated when supplying feed. The amount of feed must be increased as the heifers become heavier. As weight increases, the need for energy is greater. Young heifers do not make good use of non-protein nitrogen sources such as urea.

Rations for growing replacement heifers that are weaned at 450 to 500 pounds are given in Table 15-6. These rations give a daily gain of about 1.0 to 1.25 pounds. Vitamins and minerals are to be fed free choice with these rations.

TABLE 15-6 Rations to Grow Replacement Heifers (450–500 lb) (Gain about 1–1.25 lb Daily)

Ration	Amount to Feed (lb)
1. Legume-grass hay	10
Oats	3
2. Legume-grass haylage	25
3. Legume-grass hay	10
Ground ear corn	4
4. Corn silage	30
Soybean meal	1.5
5. Corn silage	20
Legume-grass hay	6
6. Alfalfa hay	12.5
Shelled corn or ground ear corn	2.2
7. Sorghum silage	43.4
Protein supplement (35% total protein)	1.7
8. Alfalfa hay	4.6
Corn silage	25.7
9. Grass hay (brome, orchard grass, canary grass)	11.2
Shelled corn or ground ear corn	3.4
10. Alfalfa hay	12.5
Oats	2.6

(Feed vitamin–mineral mix free choice.)

Source: Data compiled from various university Cooperative Extension Service bulletins

GROWING, FEEDING, AND CARE OF BULLS

Wean bulls at 6 to 8 months of age. Feed high-energy rations for about 5 months after weaning to discover which bulls gain best. However, avoid fattening the bulls. The best-gaining bulls are used in the herd or kept for sale.

Bulls are full fed until spring. They are then put on pasture to complete growth. Well-grown bulls may be used for breeding at 15 to 18 months of age. Bulls continue to grow slowly until about 4 years of age.

Bulls fed corn silage are also fed grain at the rate of 1 percent of body weight. When hay or haylage is used, feed grain at the rate of 1.5 percent of the body weight of the bull. Poor-quality roughage must be supplemented with protein. Minerals are fed free choice. Vitamin A is fed at the rate of 30,000 to 50,000 IU per day if the ration is mostly corn silage or limited hay.

Bulls may be self-fed (Figure 15-3) or hand-fed. When bulls are self-feeding, use plenty of roughage so the bull will not become too fat or go off feed (fail to eat the proper amount of feed). Table 15-7 lists feed mixtures to use for bulls from weaning to about 700 pounds. Table 15-8 gives feed mixtures for bulls weighing more than 700 pounds. Trace mineralized salt is fed free choice with these rations.

USDA Agricultural Research Service; Photo by Peggy Greb

Figure 15-3 Bulls may be self-fed or hand-fed but should not be allowed to get too fat.

TABLE 15-7 Feed Mixtures for Weaned Bull Calves (Weaning to About 700 lb)

Ration	Amount to Feed (lb)
1. Corn	1,200
Alfalfa-grass hay (ground)	600
Soybean meal	200
2. Corn	1,200
Oats	600
Soybean meal	200
3. Ground ear corn	1,700
Soybean meal	300
4. Corn silage	1,860
Protein supplement (32–35% total protein) (Includes vitamin mix and trace minerals)	156
5. Shelled corn	460
Corn silage	1,431
Protein supplement (32–35% total protein)	156

(Includes vitamin mix and trace minerals)

Source: Data compiled from various university Cooperative Extension Service bulletins

TABLE 15-8 Feed Mixtures for Bulls Weighing More Than 700 lb

Ration	Amount to Feed (lb)
1. Shelled corn	1,200
Alfalfa brome hay	756
2. Corn silage	1,940
Urea	50
Dicalcium phosphate	10
3. Shelled corn	440
Corn silage	1,500
Urea	50
Dicalcium phosphate	10

(Feed trace mineralized salt free choice.)

Source: Data compiled from various university Cooperative Extension Service bulletins

Yearling bulls should gain 1.5 to 2 pounds per day. During the winter, grain may have to be added to the ration for these bulls. The amount of grain required is from 0.5 to 1.0 percent of body weight. If corn silage is included in the ration, grain need not be added.

Two- to 4-year-old bulls need more energy and protein than does the cow herd during the winter. These bulls should gain 1.0 to 1.5 pounds per day. A full feed of silage plus 2 pounds of 40 percent protein supplement may be used. A good quality legume hay at the rate of 16 to 20 pounds plus 10 pounds of grain per day will also meet the needs of these bulls. Feed a mineral supplement with the ration.

Mature bulls in good condition are fed the same as the cow herd. If the bulls are thin, feed 5 to 6 pounds of grain per day above the amount fed the cow herd. If at least one-half of the feed is legume hay, corn silage, sorghum silage, or grass silage, there will be enough Vitamin A. If there is no legume in the ration, feed 1 to 2 pounds of a high-protein supplement. Feed a mineral supplement with the ration.

Bulls lose weight during the breeding season. Enough feed must be fed so that they gain back the weight lost. Additional feed is often needed 6 to 8 weeks before the start of the next breeding season. Bulls that are too fat or too thin have poor fertility. They should be kept in medium flesh and given plenty of exercise. During the breeding season, the bull may need 1 pound of protein supplement and 5 pounds of grain per day. The amount of feed given is based on the condition of the bull.

Rotate the use of bulls in a large herd during the breeding season. Keep the bull separate from the cow herd when not breeding. On farms where there are no facilities to keep the bull separate, the bull can be run with steers or pregnant cows. Care must be taken when handling or working around bulls because they can be dangerous.

A lame or long-hooved bull will be a slow or reluctant breeder. If necessary, the hooves should be trimmed several weeks before the breeding season begins. In addition, the bull's semen should be checked for fertility before the breeding season begins. Ninety percent of all breeding problems can be eliminated by giving the bull a physical examination for reproduction soundness before the breeding season.

MANAGEMENT OF THE HERD DURING BREEDING SEASON

The goal of the cow herd owner is a 100 percent calf crop. Careful management helps in reaching this goal. The cow herd should be observed closely during the breeding period. Check for injured or diseased cows or bulls. Watch to see if the bull is servicing (mating with) the cows.

The increased use of sire performance records and scrotal circumference measurements in selecting herd sires has resulted in beef bulls that are more fertile than in the past. These bulls can be mated with more cows than was previously possible. Young bulls with a scrotal circumference of 13.4 inches can mate with 20 to 25 cows with no loss in pregnancy rate. Mature bulls can mate with 25 estrus-synchronized cows or 35 to 40 nonsynchronized cows without a decrease in the pregnancy rate. Under range conditions, many ranchers plan to use 4 bulls per 100 cows. Research suggests that the increased pregnancy rate when using 4 bulls per 100 cows is enough to justify the added cost of the bulls.

The bull should be replaced if a high percentage of the cows do not become pregnant after two matings. Check the semen for fertility if it is suspected that the bull is a poor breeder. Run the bull with the herd for no more than 60 days to maintain a short calving season (40 to 60 days). The majority of the cows should calve early in the season. Begin breeding 20 to 25 days after half of the calves are born. This allows a second and possibly a third heat period for those cows that do not settle (become pregnant) the first time. Older cows come in heat sooner after calving than do first-calf heifers. If yearling heifers are bred 20 days before older cows, they will stay on schedule for the second breeding season.

The conception rate is higher for cows that are gaining weight just before and during the breeding season. Cows that are too fat or too thin are poor breeders. Check the cows for pregnancy 60 to 90 days after the breeding season. Any cows that are still not bred at that time should be sold. Hot weather lowers the conception rate. Shade, water, and fly protection are necessary during breeding.

ARTIFICIAL INSEMINATION

Artificial insemination is the placing of sperm in the female reproductive tract by other than natural means. The breeder uses an inseminating tube to deposit sperm in the cervix and uterus of the cow. Figure 15-4 illustrates the proper insertion of the tube into the uterus. Sperm is collected from the bull by any one of several methods. The most common is the use of the artificial vagina. At each ejaculation, a bull produces about 5 cc of semen containing millions of sperm. Five hundred or more cows can be bred from the semen collected in one ejaculation.

An artificial vagina is a tube from 10 to 14 inches in length and about 2.5 inches in diameter. The outer tube is usually made of some type of heavy rubber, although metal or plastic is sometimes used. There is an inner lining of a thinner rubber. A collection cone and vial is attached to one end of the tube. The other end is open to permit the entrance of the bull's penis. The space between the inner lining and the outer tube is filled with warm water, usually about 140°F.

The bull is trained to mount either a dummy or a live animal. Cows in heat have been used, although it is becoming common practice to use a steer. The bull's penis is guided into the artificial vagina when he mounts the dummy or the live animal. The bull ejaculates into the artificial vagina and the semen is collected in the vial attached to the device. The technician must be trained and have experience to properly collect usable semen for use in artificial insemination.

Artificial Insemination

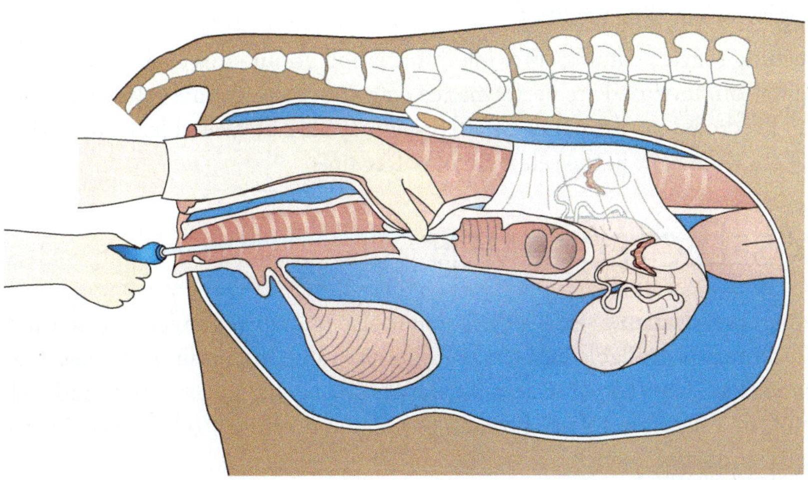

Figure 15-4 The instrument must place the semen through the cervix and into the uterus in cows.

Figure 15-5 Technician guides the straw of semen into the reproductive tract.

The semen is cooled slowly after collection because rapid cooling kills the semen. The semen can be stored for about 1 week at 41°F. Semen that is to be stored for longer periods of time is cooled to −320°F. Semen stored at this lower temperature will remain fertile for several months.

Semen is placed in the cow's reproductive tract by the rectovaginal method (Figure 15-5). A plastic glove is placed over the hand and arm of the inseminator. The hand is inserted into the rectum of the cow and the feces removed. The cervix is then grasped through the wall of the rectum. The inseminating tube is inserted through the vagina into the cervix. It is guided by the hand in the rectum. The semen is deposited anywhere from the middle of the cervix to just into the body of the uterus.

The use of artificial insemination with beef cow herds depends on the facilities available. There must be a system that allows close observation of the herd to detect cows in heat. The system must also provide for confining and holding the cow for breeding. It is difficult to use artificial insemination under range conditions and is more practical for smaller herds.

When using artificial insemination, the following practices should be followed.

1. Check the herd in the early morning and late evening to detect cows in heat.
2. Breed about 12 hours after the cow is first seen in heat.
3. Provide good handling facilities, including a breeding chute.
4. Handle the cows gently.
5. Use a good identification system for the cows.
6. Keep accurate records of date bred, date returned in heat, and calving date.
7. If breeding a purebred herd, check breed association rules for the use of artificial insemination.

Some of the advantages of using artificial insemination are that it:

- permits use of superior, performance-tested bulls, in any herd
- permits easier use of exotic breed bulls
- improves records for performance testing
- increases the number of cows that can be bred to superior bulls
- coordinates well with estrus synchronization programs
- reduces the spread of disease

Some of the disadvantages of artificial insemination are that it requires:

- a trained inseminator
- more time and supervision of the herd
- sterile equipment
- special handling facilities

Breeding service catalogs provide information about sires available for artificial insemination. Included are performance data about the animal's sire, dam, and offspring. Predicted performance of offspring may also be available. Careful study of these data will result in the selection of better sires.

SEX-SELECTED SEMEN

Commercially available sex-selected semen, also known as gender-selected and sex-sorted semen, is available and has commercial value in certain instances. Current techniques are expensive but feasible depending on the goals of the producer and value of the offspring. Use of sex-selected semen promises to be a valuable tool in the cattle industry.

Sexed semen is produced by sorting sperm cells based on the X or Y chromosome. The X-bearing sperm contains 4 percent more DNA than the Y sperm. A dye that attaches to the DNA makes the X sperm brighter when viewed by the proper electronic and computer equipment. Sperm cells pass a light beam one at a time to be sorted.

Sex-sorted semen consistently produces offspring of the desired sex with 90 percent accuracy. However, pregnancy rates are slightly lower than with unsexed semen. Conception rates are generally 70 to 90 percent of artificial insemination with unsexed semen. So, if the normal conception rate is 70 percent, then a conception rate of 49 to 63 percent might be expected with sexed sperm. Ten percent of pregnancies with sexed semen are expected to be of the "wrong" sex. In other words, if 100 animals conceive with sex-sorted semen, about 10 offspring would be expected to be the wrong sex.

Several factors affect the cost and use of sexed semen. The cost of sex-sorted semen is much higher than the cost of unsorted semen from the same bull. The sorting process, while expensive, also wastes 50 to 75 percent of the semen because only one gender is usually desired from a given bull. Sex-sorted semen is not available from the most popular bulls because owners can sell all of the semen they can produce. Because of the cost of semen and the reduction in pregnancy rate, the value of one gender over another must be substantial to justify the use of sexed semen. Improvements in technology and techniques may come over time allowing more common use of sex-sorted semen.

Calves born through the use of sexed semen are completely normal with the exception of gender ratio. Abortion rates, neonatal death rates, gestation length, birth weights, weaning weights, and incidence of abnormalities have been shown to be similar with sexed semen compared to animals bred with unsexed semen. Producers may desire either heifer or bull calves, depending on their goals and the market. Dairymen usually want heifer calves. Dairy heifers are valuable as replacements for cows culled from the milking herd or to sell to other producers. Cow-calf producers, if producing animals for the feedlot, may want bull calves because males grow larger, faster, and have a better feed conversion ratio than heifers. A purebred operator may want either heifers or bulls depending on the market. Sexed semen is a tool that can help producers meet their goals.

SYNCHRONIZATION OF ESTRUS

Synchronization of estrus is the use of various compounds (usually hormones) to cause all of the females in a herd to come into heat within a short period of time. The ultimate purpose of synchronization of estrus is to have all of the cows in the herd calve within a short period of time.

The advantages of estrus synchronization include the following:

- The labor involved in detecting estrus in the herd is reduced.
- The scheduling of artificial insemination is easier because most of the females are in heat during a short period of time.
- More uniform calf crops are produced.
- The breeding and calving seasons are reduced.

Some disadvantages of estrus synchronization include the following:

- Conception rates are sometimes low.
- The cost per head is fairly high.
- Labor for artificial insemination and calving is concentrated in a short period of time.
- Many cows are calving during the same 24 to 36 hours.
- Adequate handling facilities must be available.

An estrus synchronization program requires a high level of management. Cows in poor condition will not respond well to treatments used to synchronize estrus. It is important to maintain the proper level of nutrition for the breeding herd and make sure they are in good health. Good facilities are needed for handling the animals, and care must be taken to avoid causing the animals stress when handling them. Skilled labor is necessary to help during breeding and calving periods. It is wise to plan a synchronization program with a veterinarian.

WHEN TO BREED HEIFERS

Size is the most important consideration when breeding yearling heifers that are sexually mature. Heifers should weigh 550 to 750 pounds when bred. This weight should be from growth, not from fattening.

Age is the second consideration. The goal is to breed the heifers to calve at 2 years of age. When achieved, this will result in an average of one more calf produced during a cow's lifetime. Calving at 2 years of age brings cows into production at lower cost. It keeps a higher percent of cows in the herd in production. Fewer replacement heifers are needed each year to maintain a stable herd size.

The conception rate is lower for yearling heifers than for older cows. This lowers the percent of calf crop produced. A longer calving season normally results from breeding yearling heifers. Younger heifers need more help at calving than do older cows.

Good management reduces the calving problems of 2-year-old heifers. Heifers are bred earlier so that they will calve about 20 to 30 days before the older cows. This allows more time for heifers to come in heat after the first calving. Heifers require more feed, and should be kept separate from the older cow herd.

Breeding to a bull that sires smaller calves also results in less trouble at calving time. Give each heifer additional attention when they are calving. Breed heifers for 40 to 60 days, and then check for pregnancy 60 to 90 days later. Heifers that are not bred should be sold.

CALVING

Calving should occur during a 40- to 60-day period. This results in calves that are more uniform in weight and age. Also, it is easier to manage the herd when a shorter calving season is maintained.

Most cow herd operators calve in the spring of the year. Spring calving requires less housing, less feed during the winter, and less labor. Cows with calves do better on pasture. The calves are then ready for the feedlot in the fall when the feed supply is largest. Calving should start 6 weeks to 3 months before pasture season. The exact time depends on the local climate.

Observe the herd closely for signs that the cows are about ready to calve. Good breeding records help the producer plan for calving time. When the cow is close to calving, the udder and vulva swell. There is a loosening in the area of the tail head and pin bones.

Figure 15-6 Calving on clean ground helps prevent scours.

An effort should be made to save every calf that is born alive. Calving on clean pasture helps prevent scours (Figure 15-6). Clean, dry sheds or stalls should be used on late winter or early spring pasture in areas where the weather is bad.

Very few older cows need help at calving. About half of the 2-year-old first-calf heifers will need help at calving. Provide help at calving only if it is necessary. If the cow has been in labor for 2 hours or more and there is no sign of progress in giving birth, help is needed. Calving problems are also indicated if (1) only the tail or head of the calf is visible, (2) more than two feet are visible, (3) the feet are visible beyond the knees and the head is not visible, and (4) the head and only one foot are visible. The experienced cattle raiser can handle some of these problems if the right equipment is available. More serious problems require the help of a veterinarian.

After the calf is born, make sure that it breathes. It may be necessary to clean the mucus from the mouth and nostrils. If necessary, artificial respiration is given by alternate pressure and relaxation on the wall of the chest.

In cold weather, the calf must be kept warm and dry until it is on its feet. The navel cord is disinfected with a 2 percent iodine tincture solution. The calf needs to nurse shortly after birth (Figure 15-7). A weak calf must be helped to nurse. The first milk from the cow, called colostrum, contains nutrients (such as vitamins A and E) and antibodies that the calf requires. The placenta (afterbirth) is normally expelled within 12 to 24 hours after the calf is born. If it is not, call the veterinarian.

USDA Agricultural Research Service; Photo by Scott Bauer

Figure 15-7 Newborn calves need to nurse shortly after birth.

Separate the cows that have calved from those that have not. Provide plenty of clean, fresh water for the cows. Cows with calves need more feed to produce the milk to nurse the calf.

The calf is identified with an ear tag or tattoo. The date of birth should be recorded. Difficulties in calving and the birth weight of the calves are also recorded for performance records.

CASTRATION

Castration is the removal of the testicles of the bull calves. A castrated beef animal is called a **steer**. Calves can be castrated at birth. Bull calves that are going into the feedlot are castrated before they are 3 to 4 months of age. Bulls that are castrated after 8 months of age have a **staggy** appearance—that is, the looks of a mature male—and bring less profit when sold. There are more problems with bleeding and weight loss when older bulls are castrated. Bull calves gain slightly faster to weaning weight; however, this is not an advantage. The shrinkage that results from the shock of castration at the older age offsets the initial faster gain.

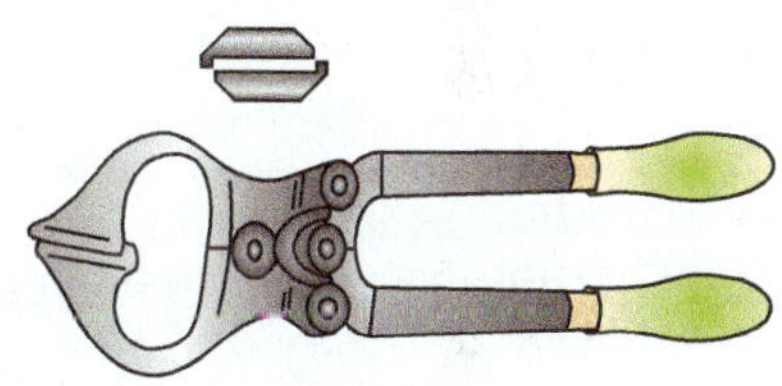

Figure 15-8 Burdizzo emasculator, a type of mechanical pincers used in bloodless castration.

Bull calves are castrated in several different ways. The most widely used method is the use of a knife. This method should be used only at times of the year when flies are not a problem. Calves should be no more than 3 to 4 months of age. Using the knife removes the testicles completely. Because an open wound results, there is some danger from bleeding or infection.

There are two common methods of bloodless castration. One method is the use of pincers called **burdizzo** (Figure 15-8). This tool crushes the cords above the testicles and does not cause an open wound. However, if the pincers are not applied correctly, the cords may not be completely crushed. This may result in the steer showing signs of stagginess at an older age. This method is a good choice in areas where screwworms are a problem.

Another method is the use of **elastrator bands** (Figure 15-9A and B). A special instrument is used to place a tight rubber band around the scrotum above the testicles. The blood supply to the testicles is cut off. The testicles waste away from the lack of blood. There is no open wound with this method.

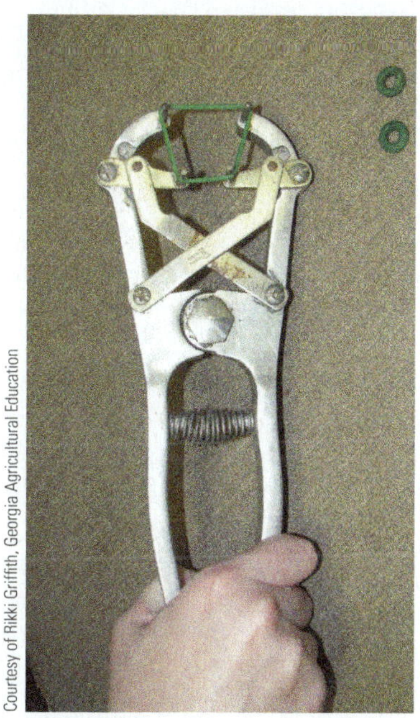

Courtesy of Rikki Griffith, Georgia Agricultural Education

Figure 15-9A One of the two common methods of bloodless castration is to use elastrator bands.

DEHORNING

There are several reasons for dehorning calves. First, horned calves often bring less money when sold. If calves are dehorned, less space is needed in feedlots and trucks, and there is less chance of cattle bruising one another. Dehorned cattle cause less damage to facilities.

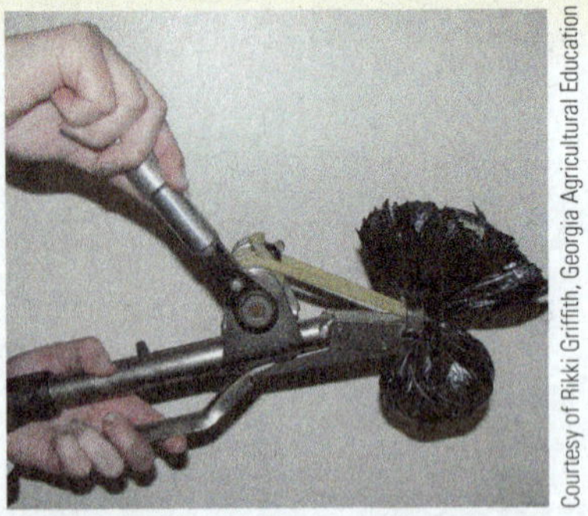

Figure 15-9B This photo simulates the placement of an elastrator band around the scrotum. The blood supply is cut off and the testicles waste away.

Calves should be dehorned when they are young. It is easier to handle a young calf, and there is less shock to the animal. If possible, do not dehorn during fly season. If flies are a problem, use a fly repellant to prevent maggots.

Several methods are used to dehorn calves and cattle. Chemical methods may be used when the calves are under 2 weeks of age. Liquids, pastes, or caustic sticks are used to apply a chemical to the horn button. Care must be taken to prevent the chemical from touching the skin. The hair is clipped from around the horn button and petroleum is put on the skin. The chemical is then applied to the horn button. The chemical must be dry before the calf is put back with the cow. Keep the calf out of the rain for several days after the chemical is applied. If a paste is used, the hair does not need to be clipped.

When the horns are past the button stage of growth, several other methods are used (Figure 15-10). Calves under 60 days of age may be dehorned with spoons, gouges, or tubes. Larger horns may be removed with Barnes-type dehorners. This type of dehorner has knives that cut off the horns. In range areas, hot irons are often used for dehorning. Hot irons may be used for calves up to 4 or 5 months of age. Electrically heated irons are also available, providing a fast and almost bloodless method of dehorning.

Older cattle are dehorned with dehorning clippers or saws. Bleeding is a major problem with older cattle. Use a forceps to pick out the main artery under the cut. The artery is pulled until it breaks, which usually prevents serious bleeding. Animals that have been dehorned by any cutting method should be watched closely. If bleeding continues, pull arteries that have not already been pulled. Another way to stop bleeding is to tie a heavy string around the poll below the dehorned area. Tight bandages may be put on the wound to stop bleeding. If bleeding still continues, call a veterinarian.

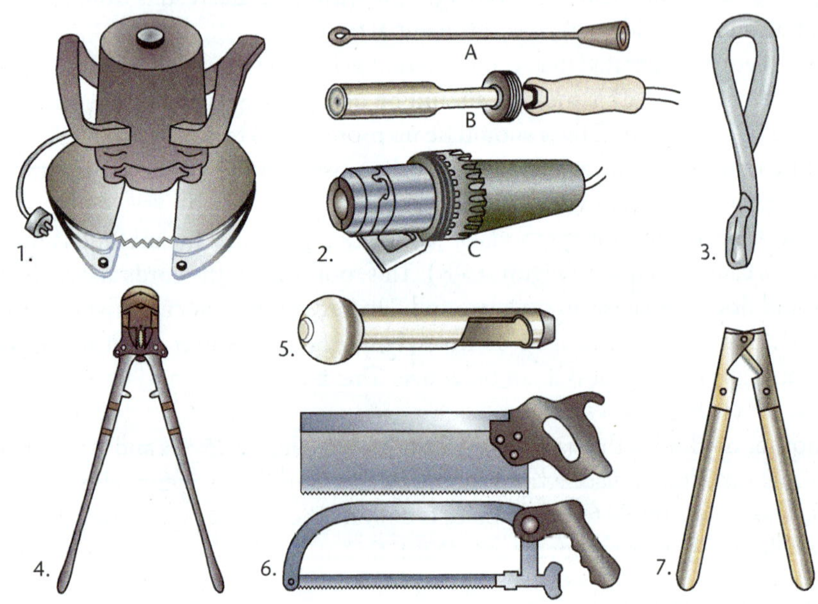

Figure 15-10 Instruments used for dehorning cattle.

BRANDING AND MARKING

Branding and marking cattle is a common practice when herds are large. It is required by law in some western states where cattle are run on rangeland. Most western states require that records of brands be kept on cattle that are slaughtered. Brands are often recorded by county or state governments.

Calves are usually branded before weaning. Hot irons, cold irons used with a commercial branding fluid, and freeze branding are three ways to brand cattle. The hot iron method is the oldest and most commonly used. The cold iron with branding fluid is not used very often. Freeze branding (Figure 15-11) is becoming more widely used.

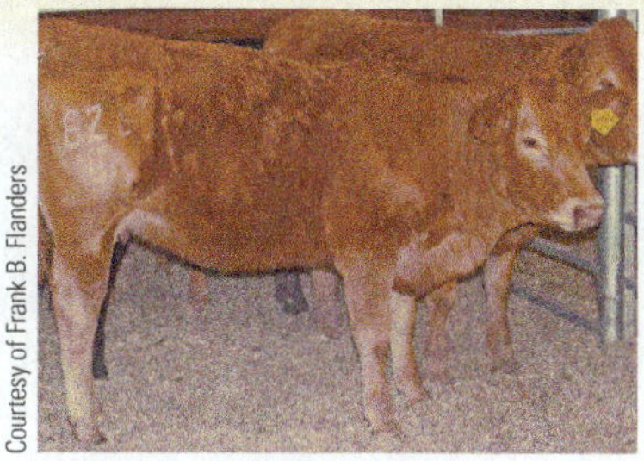

Courtesy of Frank B. Flanders

Figure 15-11 Freeze branding kills the hair pigment leaving a permanent brand.

Calves need to be restrained while being branded. It is easier to brand calves in a chute, and this method is becoming more widespread. The symbols in the brand should be about 4 inches in height. A width of 3/8 inch for the lines of the brand makes it easy to read.

Ear cuts, tattooing, ear tagging, and neck chains are other ways of identifying cattle. Cutting the ears is almost as common a method as branding. Earmarks are recorded in brand records and are protected by law. Either ear or both may be cut. Cutting is done in such a way that the mark can be seen from a direct front or rear view.

Ear tattooing is well adapted as a method of marking purebred cattle. It is a more permanent mark than ear cutting. A special instrument is used that has needlelike points (Figures 15-12A and B). The mark is made with indelible ink. No open wound is left.

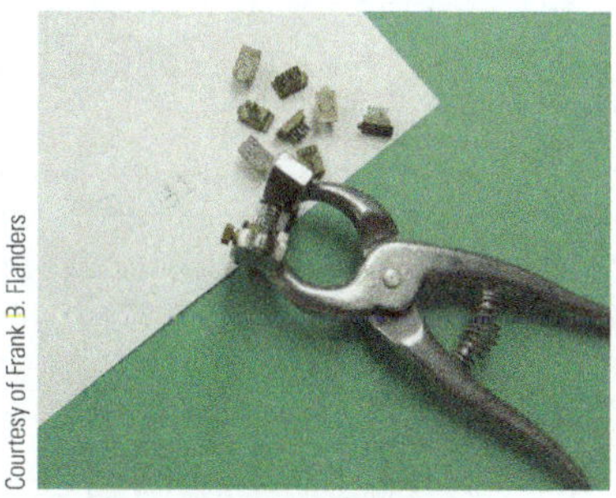

Courtesy of Frank B. Flanders

Figure 15-12A A simple type-tattooing instrument with interchangeable numbers and letters.

Courtesy of Frank B. Flanders

Figure 15-12B Test the tattooing tool on paper to be sure the desired mark will be obtained.

Ear tags (Figure 15-13) are also widely used on purebred herds. Special clamps are used to clamp tags or buttons in the ear. The identification number is on the tag or button, which may be made of metal or plastic

Neck chains may be used when the herd owner does not want to make permanent identification marks on the animals. Neck chains that carry tags with identification numbers on them are usually used with purebred herds. Neck chains are not the best choice when cows are on brushy range because the chains may become caught in the brush and be torn off.

USDA Agricultural Research Service; Photo by Scott Bauer

Figure 15-13 Calf identified with an ear tag.

MANAGING WEANED CALVES

The herd owner has several options in managing calves after they are weaned, including (1) selling feeder calves, (2) selling yearling feeders, and (3) growing and finishing beef animals. If calves are born in the spring, they weigh about 400 to 500 pounds in the fall. Heifer calves weigh about 5 percent less than steer calves at weaning. These calves are sold in the fall as feeder calves.

Calves that have grown to about 650 to 750 pounds are sold as yearlings for finishing to market weight. This system uses mostly roughage for feed. If the calves are born in the fall, they are weaned in the spring. They are then fed on pasture for sale as yearling feeders in the fall.

Some herd owners prefer to grow the calves on roughage and then finish them in the feedlot for 4 to 6 months. Corn silage or grain and roughage are used for the wintering ration. When the animals are on pasture, little or no grain is fed. The animals are then moved into the feedlot and grain feeding is begun.

PRECONDITIONING CALVES

Preconditioning is the process of preparing calves for the stress of being moved into the feedlot. Most of the procedures involved in preconditioning are accepted as good management practices. They are accomplished before the calves leave the producer's farm.

CERTIFICATE OF PRE-CONDITIONING

This certifies that the following pre-conditioning standards have been completed on the cattle herein described:

Identification: Number Steers ______________

Breed: Number Heifers ______________

Age: Total ______________

Practices, Treatments & Immunization	Date	Product Used
Weaned		
Castration (knife only)		
Dehorning		
Grub Control (if applicable)		
Internal Parasite Control		
Blackleg - Malignant Edema Vac.		
IBR - Vaccine		
Para Influenza - Vaccine		
Lepto - Vaccine		
Pasterurella Bacterin		
Other		

In addition to the above pre-conditioning practices, the heifers in this shipment are guaranteed open.

General Remarks:

Certified By

Information on this certificate is valuable to the purchaser and his or her veterinarian in making decisions on handling newly purchased cattle. It should reduce the cost of additional vaccinations and treatments.

Figure 15-14 A certificate of preconditioning.

Calves may be certified as preconditioned if a specific schedule of practices is followed. Preconditioning adds cost to the production of feeder calves. A sample certificate of preconditioning is shown in Figure 15-14.

Practices involved in preconditioning usually include:

- castration, dehorning, and identification by tattooing or branding
- maintaining an accurate health record
- vaccination for brucellosis, blackleg/malignant edema, infectious bovine rhinotracheitis (IBR), bovine virus diarrhea (BVD), para-influenza (PI_3), and leptospirosis
- weaning 4 to 6 weeks before sale
- training to eat solid feed from bunk and to drink water from a water tank
- worming and treatment for grubs, lice, and mange (if necessary)

Some research shows that feeding calves a preconditioning ration for 3 to 4 weeks after weaning and prior to shipping may not be a profitable practice for either the producer or the cattle feeder. Calves that were not weaned before being moved to the feedlot had higher feed efficiency in the feedlot compared to those that were on a preconditioning feeding program. A better feeding alternative appears to be not weaning the calves and limiting the creep feed to 1 to 3 pounds per head per day during the last 1 to 2 months before shipping the calves to the feedlot. A high-energy ration should be used for the calves.

BACKGROUNDING CALVES

Backgrounding is the growing and feeding of calves from weaning until they are ready to enter the feedlot. It may be done by the cow herd owner. Farmers who do not want to own cow herds may buy weaned calves and prepare them for the feedlot. They then sell the calves to others who finish them for market.

Backgrounding is done primarily with roughage rations. Calves are fed for 120 to 150 days. Expected daily gain is 1.5 to 2.0 pounds per day. Calves must not be allowed to become too fat because fat calves bring lower prices when going into the feedlot for finishing.

SUMMARY

Feeding programs for cow-calf beef herds are based on the use of roughages. Pasture in the summer and silage and hay in the winter are the common feeds used in the ration. The kinds of pasture, silages, and hay used depend on the part of the United States in which the cow-calf herd is located.

Hay can be harvested in rectangular bales or large round bales. Bales may be left in the field for self-feeding or they may be moved to a storage area. More feed is required when using large round bales because of increased losses from this method of feeding.

Dry pregnant cows are fed to prevent their becoming too fat or too thin during the winter. Younger cows and heifers require more feed than mature cows. Minerals and salt should be fed free choice to cow herds. Protein supplements are seldom needed if the hay quality is good. Vitamin A may be needed when low-quality roughages are fed.

Cows nursing calves require more feed than dry cows. High-quality pastures usually meet the needs of lactating cows.

Creep feeding calves may or may not be profitable. Creep feeding may pay when selling the calves at weaning. If the calves are being kept by the producer and fed roughages through the winter, then creep feeding is probably not economical. Grain is used in the creep feeder. Commercial pelleted mixes are also available for this use.

It is usually necessary to replace about 15 percent of the cows in the herd each year. The use of performance records to select replacement heifers is recommended. Replacement heifers should weigh about 550 to 750 pounds at breeding, depending on the breed. British breeds are bred at the lighter weights.

Bull calves to be kept for breeding purposes are weaned at 6 to 8 months of age. High-energy rations are fed for 5 months to determine which bulls have the best gaining ability. Bulls that gain best should be selected for breeding. The others should be marketed for slaughter.

Well-grown bulls may be used for breeding at 15 to 18 months of age. Bulls that are self-fed require large amounts of roughage so they will not become too fat. Winter rations for 2- to 4-year-old bulls should contain more energy and protein than cow rations contain. Mature bulls may be fed the same ration as the cow herd.

The beef producer desires a 100 percent calf crop each year. It is essential to make sure that the bull is fertile by testing the semen before the breeding season. Do not run too many cows with the bull when pasture breeding. A bull can breed more cows when pen breeding is used.

Artificial insemination is used in some herds. It requires good facilities and more work than natural breeding. A major advantage is the chance to use superior bulls on the cow herd.

Size is more important than age when determining when to breed heifers that are sexually mature. A good management objective is to have heifers calve at 2 years of age.

Farmers should aim to have calving occur in a 40- to 60-day period. If this procedure is followed, calves are more uniform in weight and age. Start calving 6 weeks to 3 months before pasture season, depending on local climate.

Observe cows closely for signs of calving. If the cow needs help, provide it. Most mature cows need no help in calving. A high percentage of heifers do require help.

Castration, dehorning, vaccination, and identification of calves are best done when they are young. There is less shock to the calf at that time. Preconditioning calves involves following a schedule of good management practices.

Backgrounding is the growing of calves on roughages from weaning until they are ready for the feedlot. The cow herd owner may want to background calves from the herd before selling them. Some farmers who do not own cow herds may buy calves at weaning and prepare them for the feedlot. These calves are then sold to others for finishing for slaughter.

Student Learning Activities

1. Visit a local beef cow herd, collect data on size of herd, system of production used, and available feeds. Plan a year-round feeding program for that herd.
2. Talk with local beef cow herd operators to find out how they care for the cows and calves at calving time and up to weaning. Report to the class on practices used in the area and recommend any improvements that you feel are needed.
3. Survey beef cow herd operators in the community to determine the system of production used locally. Find out why the farmers use these systems.
4. Observe and assist in the castration, dehorning, vaccination, and identification marking of calves.
5. When planning and conducting a supervised experience program in beef production, follow all the appropriate feeding and management practices as outlined in this chapter.

Discussion Questions

1. Why are cow-calf beef feeding programs based on the use of roughages?
2. What kinds of crop residues might be used for feeding cow-calf herds?
3. How can waste be reduced when feeding roughage to cow-calf herds?
4. Why is it important to control the weight gain of dry, pregnant beef cows?
5. What minerals and vitamins do dry, pregnant beef cows need, and how can they be supplied?
6. What are the protein, energy, mineral, and vitamin requirements of lactating cows as compared to dry cows?
7. State the reasons both for and against creep feeding calves.
8. How much gain per day should replacement heifers of (a) British breeds and (b) larger breeds have?
9. Describe how replacement bulls should be fed.
10. Describe the feeding of mature bulls (a) before the breeding season and (b) during the breeding season.
11. Briefly explain the number of bulls required for the cow herd during the breeding season.
12. When should breeding begin if a short calving season is desired?
13. Describe the process of artificial insemination.
14. List the advantages and disadvantages of using artificial insemination with a beef breeding herd.
15. When should the beef producer plan to have cows calve?
16. Describe the signs that indicate that a cow is about ready to calve.
17. Describe the methods that can be used to castrate bull calves.
18. Describe the methods that can be used to dehorn calves.

19. Describe some common methods for branding or marking beef cattle.
20. Describe three methods of managing weaned calves.
21. What is preconditioning of calves?
22. What is backgrounding of calves?

Review Questions

True/False

1. The standard performance analysis software has two components, one for financial information and one for production information.
2. As much as 15 percent of the feed that is fed is wasted and must be calculated when supplying feed.
3. Heifers on good pasture will gain about 0.75 to 1.4 pounds per day.

Multiple Choice

4. Wean bulls at approximately:
   a. 6 weeks
   b. 3–4 months
   c. 10–12 months
   d. 6–8 months of age
5. During cold weather, feed should be increased by what percent for each degree of cold stress?
   a. 1%
   b. 3%
   c. 5%
   d. 10%
6. Bulls are fed corn silage and grain at the rate of ______ percent of bodyweight.
   a. 3
   b. 4
   c. 2
   d. 1

Completion

7. The ______ ______ of a pasture refers to the number of animals that can be grazed on the pasture during the season.
8. ______ ______ is the placing of sperm in the female reproductive tract by other than natural means.
9. ______ is the process of preparing calves for the stress of being moved into the feedlot.

Short Answer

10. List three methods of marking and branding cattle.
11. List two reasons for dehorning calves.
12. Why is it necessary to adjust the rations of bulls during and after breeding season?

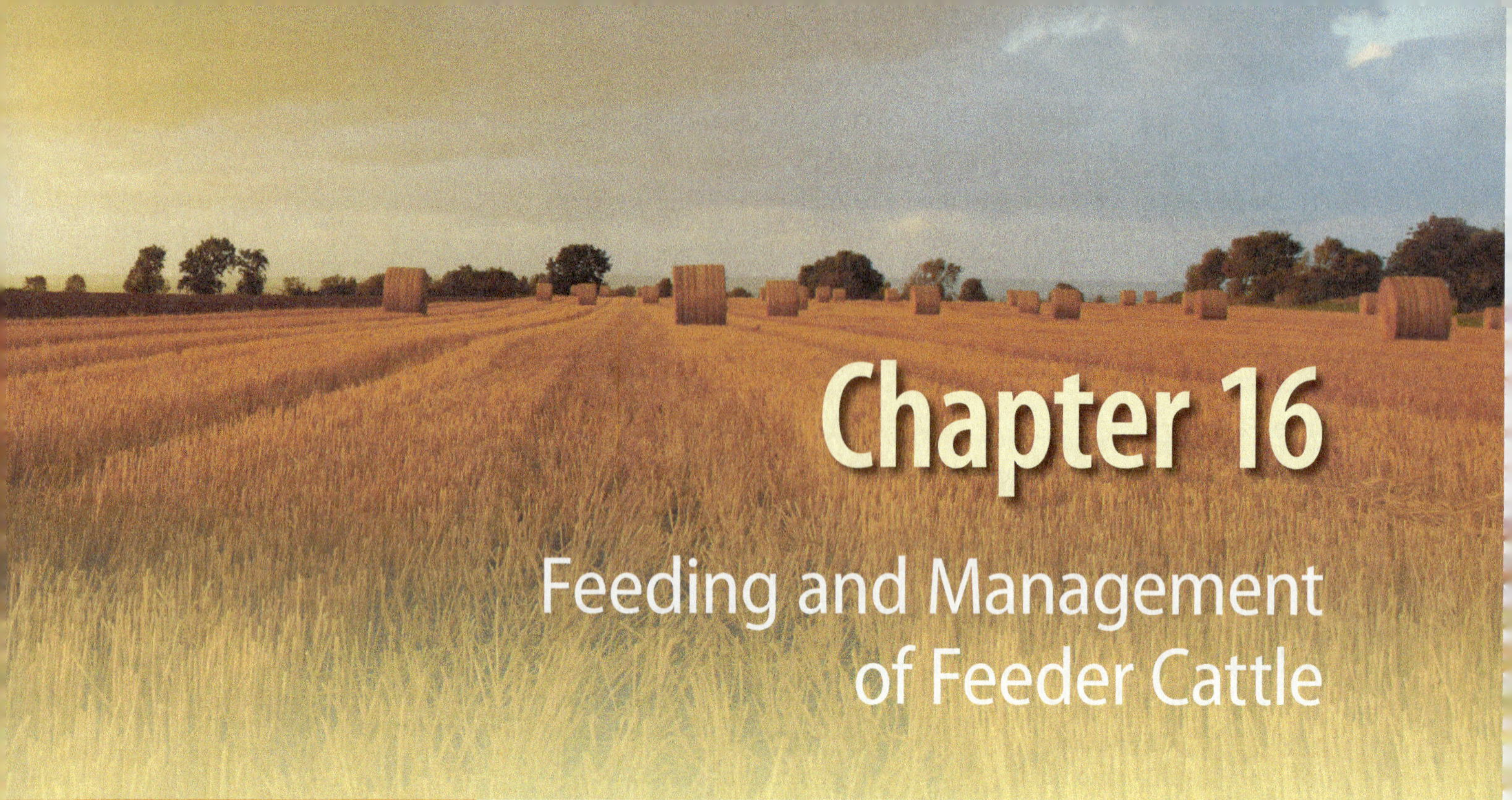

Chapter 16

Feeding and Management of Feeder Cattle

Key Terms

grass fed
efficiency in gain
haylage
boot stage
grinding
high-moisture storage
rolling
crimping
pelleting
cubing
middling
stillage
NEg
bacterin

Objectives

After studying this chapter, the student should be able to

- describe systems of cattle feeding and kinds of cattle that may be fed.
- identify grades of feeder cattle.
- describe the economics of buying feeder cattle.
- select feeds and develop feeding programs for various kinds of feeder cattle.
- list approved practices for managing various feeder cattle operations.

SYSTEMS OF CATTLE FEEDING

Cow-calf herds are found across the United States, especially in the western, southern, southwestern, and west north-central parts of the country. Finishing feeder cattle, on the other hand, is done mostly in the central and western parts of the United States, with Texas, Nebraska, Kansas, Colorado, and Iowa usually in the top five. Feeding cattle for slaughter requires more grain and protein supplement than that needed for cow-calf herds, which is why many cattle-feeding operations are found in the grain-producing states.

The majority of high-quality beef is produced in commercial feedlots where cattle are fed high rates of grains and protein supplement. However, in recent years the demand for **grass fed** beef has increased dramatically. The USDA grass fed standard in part states:

> **Grass (Forage) Fed Animals:** Grass and forage shall be the feed source consumed for the lifetime of the ruminant animal. Animals cannot be fed grain

or grain by products and must have continuous access to pasture during the growing season. Hay, haylage, baleage, silage, crop residue without grain, and other roughage sources may also be included as acceptable feed sources. Routine mineral and vitamin supplementation may also be included in the feeding regimen.

Grain-fed beef and grass-fed beef both have advantages and disadvantages. Grain-fed beef is grown faster, requiring less land and time. Grain-fed beef has more desirable meat qualities such as flavor, appearance, and tenderness, and it is less expensive in the supermarket. Grass-fed beef requires fewer resources and has less environmental impact than grain-fed beef. Claims that grass-fed beef contains more beneficial nutrients for humans have not been proven. Cattle on pasture generally have fewer health problems, including less stress.

Types of Feeding Operations

Commercial Cattle Feedlots

Large commercial cattle feedlots are generally defined as those with a capacity of 1,000 cattle or more. Often, many thousands of cattle are fed in these feedlots at one time. Most or all of the feed needed is purchased rather than grown by the cattle producers. The number of large commercial feedlots is increasing in areas where cow-calf operations are common. Most of the cattle fed in the Plains states (Iowa, Kansas, Minnesota, Missouri, Nebraska, North Dakota, and South Dakota), and in Colorado, Arizona, California, and Texas are in commercial feedlots. It is not necessary to move calves a great distance when feedlots are located close to the area where the calves are produced.

Farmer-Feeders

Farmer-feeders are farm operators who feed cattle mainly as a way of marketing feed raised on their own farms. Feedlot capacity of these operations is usually less than 1,000 cattle. The size of some of these feeding operations is large enough to make it necessary to purchase additional feed. The majority of the cattle-feeding operations in the central states, including Oklahoma and Texas, are of the farmer-feeder type.

Cattle feeders are in business to make a profit. Homegrown feed is fed with the hope of marketing it at higher prices through the sale of cattle. Feeders must produce the kind of product wanted by the buyers. The general demand is for steers that weigh between 1,000 and 1,250 pounds and heifers that weigh between 900 and 1,050 pounds. The goal of many producers is that 70 percent of the cattle fed should grade USDA Choice or Prime.

Types of Finishing

The two general ways to feed cattle for harvest are (1) finishing immediately and (2) deferred finishing. The finishing system used depends on the kind of cattle fed, how long they are fed, the feed used, and the market demand.

Finishing Immediately

Finishing immediately is a system in which the feeder cattle are brought to a full feed of grain (Figure 16-1). A full feed of grain means that the cattle are fed mostly grain and smaller amounts of roughage. They are then fed until ready for harvest. Steer calves are on feed for about 275 days and heifer calves for about 230 days. Yearling steers are on feed for about 175 days and yearling heifers for about 130 days. Older feeders are on feed for about 100 days. High-quality feeders are best suited to this system. Farms with limited amounts of roughage and plenty

Figure 16-1 Cattle in feedlot being finished on a feed with a high grain content.

of grain are well adapted to this system of cattle feeding. Small amounts of roughage and large amounts of grain are used in this feeding program. Older animals are fed more roughage than calves or yearlings. Heavier animals are particularly well adapted to this system.

Deferred Finishing Systems

Deferred finishing systems use more roughage and less grain. Calves are bought in the fall and wintered on roughage. Small amounts of grain are fed. Daily gain is about 1.25 to 1.5 pounds. The calves are pastured for 90 to 120 days the next summer. A small amount of grain or none at all may be fed on pasture. In the fall, the calves are put in the feedlot for 90 to 120 days. A full feed of grain is fed in the feedlot. Some feeders prefer to put the calves in the feedlot in the spring. The calves are then fed a full feed of grain for 120 to 150 days.

Yearling feeders may be used to clean up crop residues such as cornfields (Figure 16-2). The yearlings are then wintered on roughages. In the spring, the yearlings are finished with a high-grain ration. Finishing is done either on pasture or in drylot. Some feeders use little grain on pasture and finish the yearlings in the feedlot for later marketing. Wintering feeders on winter oat or winter wheat crops before putting them in the feedlot is common.

USDA Agricultural Research Service; Photo by Scott Bauer

Figure 16-2 Cattle can be used to clean up crop residues.

The deferred system is well suited for the farm that has roughage to market. Corn silage is one of the common feeds used. More beef per acre can be produced with corn silage than with any other crop.

KINDS OF CATTLE TO FEED

There are many choices for the cattle feeder when selecting cattle to feed. Selection is made on the basis of sex, age, weight, and grade of cattle.

Sex

Some heifers must be kept for herd replacements. Therefore, more steers are available for feeding. Steers gain about 10 percent faster than heifers if both are fed for the same length of time. Steers are 10 to 15 percent more efficient in gains. **Efficiency in gain** refers to the amount of feed needed for each pound of gain. The less feed required, the higher the efficiency. Heifers finish at lighter weights than do steers and are usually bought and sold for less money. Heifers are better fit for shorter feeding periods than steers are. Feeder heifers may be pregnant when purchased. This may result in discounts when they are sold for harvest. The developing fetus is of lower value to the packer.

Young bulls may be fed for market. At the same age and weight as steers, they gain faster and more efficiently. Young bulls produce lean carcasses of about the same quality as steers. However, they are not as well accepted in the marketplace. Generally, the feeding of young bulls for market is not recommended.

Age and Weight

Feeders are generally divided into three groups based on age and weight. These groups are calves, yearlings, and other feeders. The weight range is 350 to 1,000 pounds.

Calves

Calves are feeders that are less than 1 year old, usually weighing about 350 to 450 pounds. They are adapted to many different systems of cattle feeding. Gains are more efficient than they are in older cattle; however, it takes longer to feed calves to harvest weights. Calves need more grain and less roughage than older cattle when fed immediately for harvest. Calves are not well adapted to cleaning up crop residues, and they do not make good use of low-quality roughage. Death losses are usually higher with calves, and health problems are greater. Because calves are lighter in weight when purchased, most of the weight sold is gain. Success in feeding calves depends more on feeding skill than it does on ability to buy and sell.

Yearlings

Yearlings are feeders that are between 1 and 2 years old, usually weighing about 550 to 700 pounds. Yearlings are well adapted to feeding programs using more roughage than that given to calves. They are often used to clean up crop residues. Less time in the feedlot is necessary to finish yearlings for slaughter, and there are fewer health problems.

Older Feeders

Older feeders are those that are 2 years old or older. They weigh about 800 to 1,000 pounds. These feeders are fed for a short period of time, usually 90 to 100 days. Gains are fast but not as efficient as in younger feeders. Older feeders can make use of more roughage in the ration. Death losses are low. Much of the profit comes from reselling purchased weight. Therefore, more skill in buying and selling is needed for this type of feeder.

Grade

Feeder cattle grades are used by the USDA as the basis for reporting market prices of feeder cattle. They are also used as the basis for certifying the grade of feeder cattle that are delivered on future contracts. States use the grades for official feeder cattle grading programs.

The USDA standards for feeder cattle grades apply to cattle that are less than 36 months of age. They may also be used to describe stock cows for market reporting purposes. Three factors are used to determine the grade of feeder cattle: thriftiness, frame size, and thickness.

Thriftiness

Thriftiness refers to the apparent health of the animal and its ability to grow and fatten normally. An unthrifty animal is one that is not expected to grow and fatten normally in its current condition.

Frame Size

Frame size indicates the size of the animal's skeleton (height and body length) in relation to its age. When two animals are the same age, the large-framed animal is taller at the withers and hips and has a longer body than another animal with a smaller frame.

Thickness

Thickness means the development of the muscle system in relation to the size of the skeleton. When feeder cattle are the same age and frame size, differences in thickness are due to differences in bone structure, muscling, and degree of fatness. A standard degree of fatness, slightly thin, is used when evaluating thickness. Thicker feeder cattle have a higher ratio of muscle to bone when fed to the same degree of fatness as cattle that are not as thick. The thicker cattle will have a higher yield grade at the same degree of fatness as cattle that are not as thick.

There are 12 grades of feeder cattle that result from the use of the preceding standards (three frame sizes, including large, medium, and small, and four thickness groups, numbered 1 through 4): Large Frame, No. 1; Large Frame, No. 2; Large Frame, No. 3; Medium Frame, No. 1; Medium Frame, No. 2; Medium Frame,

No. 3; Small Frame, No. 1; Small Frame, No. 2; Small Frame, No. 3; and Inferior. The three frame sizes are shown in Figure 16-3. The four thickness standards are shown in Figure 16-4.

Large Frame

Large-frame feeder cattle are thrifty, have large frames, and are tall and long bodied for their age. To produce U.S. Choice carcasses (0.50 inch of fat at the 12th rib), steers would need to be more than 1,250 pounds live weight, and heifers would need to be more than 1,150 pounds live weight.

Medium Frame

Medium-frame feeder cattle are thrifty, have slightly large frames, and are slightly tall and slightly long bodied for their age. To produce a U.S. Choice carcass, a steer's live weight would need to be 1,000 to 1,250 pounds, and a heifer's live weight would need to be 1,000 to 1,150 pounds.

Small Frame

Small-frame feeder cattle are thrifty, have small frames, and are shorter bodied and not as tall as medium-frame cattle. Small-frame steers would produce U.S. Choice carcasses at live weights under 1,100 pounds, and heifers would produce U.S. Choice carcasses at live weights under 1,000 pounds.

No. 1

Feeder cattle that possess minimum qualifications for this grade usually display predominant beef breeding. They must be thrifty and moderately thick throughout. They are moderately thick and full in the forearm and gaskin, showing a rounded appearance through the back and loin with moderate width between the legs, both front and rear. Cattle show this thickness with a slightly thin covering of fat; however, cattle eligible for this grade may carry varying degrees of fat.

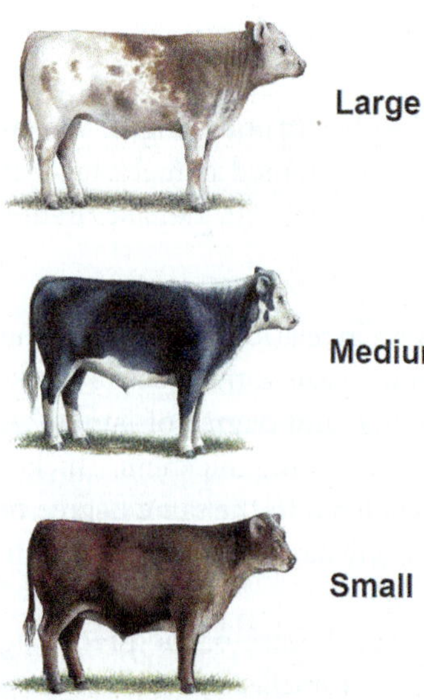

USDA

Figure 16-3 USDA feeder cattle frame sizes.

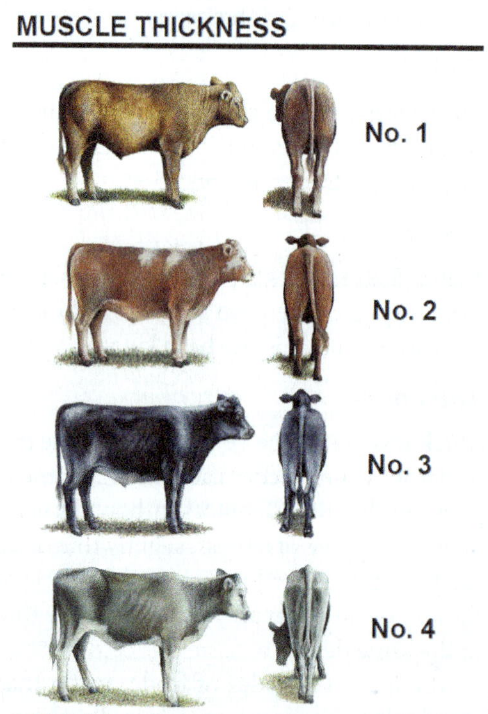

USDA

Figure 16-4 USDA muscle thickness grades for feeder cattle.

No. 2

Feeder cattle that possess minimum qualifications for this grade usually show a high proportion of beef breeding, and slight dairy breeding may be detected. They must be thrifty and tend to be slightly thick throughout. They tend to be slightly thick and full in the forearm and gaskin, showing a rounded appearance through the back and loin with slight width between the legs, both front and rear. Cattle show this thickness with a slightly thin covering of fat; however, cattle eligible for this grade may carry varying degrees of fat.

No. 3

Feeder cattle that possess minimum qualifications for this grade are thrifty and thin through the forequarter and the middle part of the rounds. The forearm and gaskin are thin, and the back and loin have a sunken appearance. The legs are set close together, both front and rear. Cattle show this narrowness with a slightly thin covering of fat; however, cattle eligible for this grade may carry varying degrees of fat.

No. 4

Feeder cattle included in this grade are thrifty animals that have less thickness than the minimum requirements specified for the No. 3 grade.

Inferior

Inferior feeder cattle are those that are unthrifty. They are not expected to grow or fatten normally. Unthriftiness may be caused by disease, parasites, extreme thinness, or other conditions. These conditions would have to be corrected before the animal could be expected to grow or fatten normally. Cattle that are double-muscled (muscular hypertrophy) are also graded Inferior. They cannot be expected to deposit intramuscular fat (marbling) normally. Feeder cattle in the Inferior grade may have any combination of thickness and frame size.

The higher grades of feeder cattle fit best into systems that use a longer feeding period. Less roughage and more grain are fed to the higher grades. These cattle finish to higher slaughter grades. Market prices for the better grades are generally higher in late summer and fall.

The lower grades of feeder cattle are the best choice for systems that utilize shorter feeding periods. More roughage can be profitably used in the ration. These cattle finish at lower slaughter grades. Market prices for the lower grades are generally higher in the spring and early summer.

BUYING FEEDER CATTLE

Auction markets are the main source for buying feeder cattle. Buying directly from farms or ranches is the second most important source. Only a small percentage of feeder cattle are bought through terminal markets.

Cattle for commercial feedlots are purchased mainly by salaried buyers or order buyers. Farmers who finish cattle raise their own feeders, buy cattle from other farmers or auction markets, or use an order buyer or cattle dealer to obtain feeders.

The supply of feeder cattle increases in the fall, with the peak movement of feeder cattle occurring in October. There is a trend toward keeping feedlots full year-round. This has created a demand for feeders at other times of the year. Cow-calf herd operators are adjusting their production to meet this demand.

Feeder and slaughter cattle prices tend to be lower in the fall and winter and higher in the spring and summer. Figure 16-5 shows the average seasonal variation in prices of feeder calves and slaughter steers and heifers in the United States. Prices of feeder and slaughter cattle vary greatly from year to year.

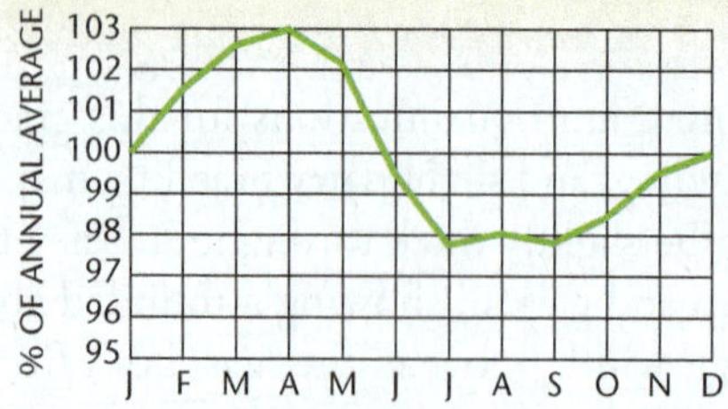

Average seasonal variation in prices of slaughter steers and heifers, United States.

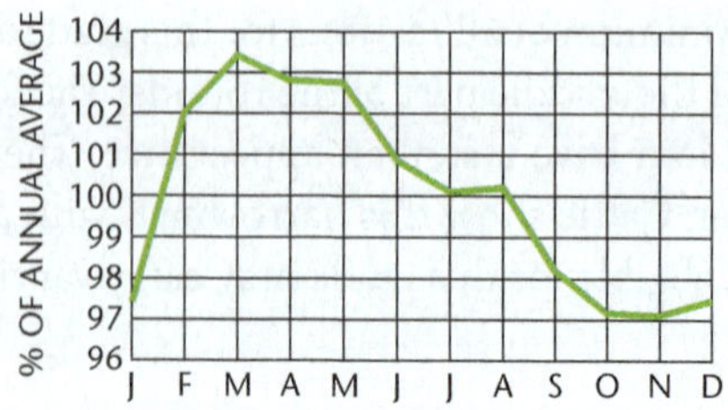

Average seasonal variation in prices of feeder calves, United States.

Figure 16-5 Average seasonal variation in the prices of slaughter steers and heifers and feeder calves in the United States.

Price trends are averages and do not always follow the same pattern every year. A cattle feeder may not be successful in trying to outguess the market. It is best for each cattle feeder to develop a feeding program that is adapted to his or her own feed supply. This program, when followed year after year, results in the greatest overall profit for the cattle feeder.

The profit from feeding cattle is equal to the value of the finished cattle minus the total costs. Costs include the price of the feeders, feed, labor, veterinary fees, fixed charges for buildings and equipment, interest on investments, and death losses. The cattle buyer must make some calculations before deciding on the price to pay for feeder cattle. The break-even feeder price is the price paid that just covers all costs without any profit. It may be found by the following formula:

$$\text{Break-even feeder price} = \left(\frac{\text{Fed weight}}{\text{Feeder weight}} \times \text{Expected price of fed cattle}\right) - \left(\frac{\text{Gained weight}}{\text{Feeder weight}} \times \text{Cost of gain}\right)$$

The gained weight can be determined by the following formula:

$$\text{Gained weight} = \text{Fed weight} - \text{Feeder weight}$$

Example: Given the following prices and weights, the break-even price is determined as shown.

700-pound feeders

finished cattle weight of 1,050 pounds

expected selling price of finished cattle, $45

cost per hundredweight (cwt) of gain, $35

$$\text{Break-even feeder price} = \left(\frac{1{,}050}{700} \times 45\right) - \left(\frac{350}{700} \times 35\right)$$
$$= (1.5 \times 45) - (0.5 \times 35)$$
$$= 67.50 - 17.50$$
$$= \$50 \text{ per cwt}$$

This calculation tells the producer that to make a profit, feeder cattle must be purchased at less than $50 per hundredweight.

The profit margin in feeding cattle may be calculated in the following way:

$$\text{Cost of feeder} = \frac{\text{Feeder weight}}{\text{Fed weight}} \times \text{Price of feeders per hundredweight}$$

$$\text{Cost of gain} = \frac{\text{Gained weight}}{\text{Fed weight}} \times \text{Cost per hundredweight of gain}$$

$$\text{Total cost} = \text{Cost of gain} + \text{Cost of feeders}$$

$$\text{Margin/hundredweight} = \text{Fed cattle price} - \text{Total cost}$$

$$\text{Profit per head} = \text{Margin/hundredweight} \times \frac{\text{Fed weight}}{100}$$

Example: Given the following prices and weights, the profit margin is determined as shown.

700-pound feeders

$46 per hundredweight of feeder fed weight of 1,050 pounds

feeding costs of $35 per hundredweight

fed cattle price of $47 per hundredweight

$$\text{Cost of feeders} = \frac{700}{1{,}050} \times 46 = \$30.67$$

$$\text{Cost of gain} = \frac{350}{1{,}050} \times 35 = \$11.67$$

$$\text{Total cost} = 11.67 + 30.67 = \$42.34$$

$$\text{Margin/hundredweight} = 47 - 42.34 = \$4.66$$

$$\text{Profit per head} = 4.66 \times \frac{1{,}050}{100} = \$48.93$$

Price is not the only factor to consider when buying feeder cattle. Some general rules to follow when selecting feeders are:

- If possible, buy from herds with performance records; cattle with good gaining ability are worth more
- Big-framed, rugged-boned, heavily muscled cattle usually are better gaining cattle
- Healthy, thin cattle make faster gains than fatter cattle
- More efficient gains are made by younger, lighter-weight cattle
- Steers are worth more as feeders than heifers
- Good crossbred cattle gain from 2 to 4 percent better than the average of the parent breeds

SELECTION OF FEEDS

Roughages

Only limited use is made of roughages in finishing rations. The amount of roughage in the ration varies from 5 to 30 percent. When less than 20 percent of the ration is roughage, the nutrient value of the roughage has little effect on gain.

Rate of gain is affected when the percent of roughage in the ration is higher than 20 percent. Better-quality roughages with higher energy content must be fed as the percent of roughage in the ration increases. As the amount of roughage in the ration increases, the rate of gain is slower. More roughages are used in the early part of the feeding period than are used later in the feeding period.

Roughage is an expensive component of feedlot diets based on the cost per unit of energy. Grains such as corn deliver more energy and increase weight faster than roughages. Reducing roughage and increasing the amount of grain in feedlot diets can improve feed efficiency. However, roughage cannot be removed from feedlot diets without negative effects on health and performance of the animal.

Hay

Hay is the most commonly used roughage in finishing rations in some areas of the United States. Legumes make the best hay for this purpose. Legume hays are higher in protein, calcium, and carotene than grass hays. Lower-quality hay requires more supplement to properly balance the ration.

Corn Silage

Figure 16-6 Corn silage is one of the best roughages to use for finishing beef cattle.

Corn silage is one of the best roughages to use for finishing beef cattle (Figure 16-6). More beef per acre is produced from corn silage than from any other roughage. It takes about 1 ton of corn silage to produce 100 pounds of beef. The corn silage must be supplemented with protein and minerals. Daily gains of 1.5 to 2 pounds can be expected when corn silage is used in the ration. More corn silage is used when corn prices are high to lower the cost of gain. Because cattle gain slower on corn silage, it takes more days in the feedlot to finish the cattle. This, in turn, increases non-feed costs. This is a greater problem for large commercial feeders than it is for smaller farmer-feeders.

Wet corn gluten feed may be substituted for corn silage in beef rations. Wet corn gluten is high in energy, protein, and phosphorus. It can be combined with crop residues in rations that require little additional supplement to meet the nutritional needs of the animals.

Sorghum Silages

Sorghum silages have from 60 to 90 percent of the feeding value of corn silage. The value varies with the amount of grain in the different varieties. Sorghum silage is harvested when the kernels are in the medium-dough stage. More mature grain is higher in feed value, but the kernels are harder. When harvested in the more mature stage, the kernels must be broken by grinding so that they are easier to digest. Sorghum silage is low in protein; therefore, protein and mineral supplements must be added to the ration when sorghum silage is fed.

Grass Silage

Grass silage is made from grass, legume, or grass-legume mixtures. Grass silages are lower in energy content than silages made from grain crops. Beef cattle make slower gains on grass silage. **Haylage** is low-moisture grass silage. It is more palatable than grass silage and contains more energy and protein. Beef cattle make slightly faster gains on haylage than they do on grass silage. A good-quality grass-legume haylage supplies most of the protein needed by the animal.

Small Grains

Small grains such as oats can be used for silage for finishing beef cattle. These are harvested just after reaching the boot stage. The boot is the upper sheath of the plant. The **boot stage** is that point in growth at which the inflorescence (the flowering part of the plant) expands the boot. From 100 to 200 pounds of carbohydrates are added at the time of ensiling. Corn, molasses, or citrus pulp may be added as the carbohydrate. The protein content is high and the silage is palatable. The quality and nutrient value is increased by the addition of the carbohydrate source.

Lower Quality By-product Roughages

Lower quality by-product roughages can also be used for finishing. Examples of these kinds of roughages are corncobs, cottonseed hulls, peanut hay, beet tops, oat hulls, and straw. They are usually low in protein, minerals, and vitamins. The ration must be carefully supplemented if they are used. The decision to use low-quality roughages is based on the relative prices of other feeds available.

Pasture

Good-quality pasture lowers the cost of gain for finishing cattle (Figure 16-7). When pastures are stocked at the best rates for the type of pasture, rotation grazing produces more beef per acre than continuous grazing. Daily gains per animal are about the same with either method.

Feeding concentrates at the rate of 1 percent of the animal's live weight produces a satisfactory gain, but cattle fed at this rate will generally finish to only USDA Standard or Select grades, rather than USDA Choice or Prime. Grains may be self-fed on pasture to reduce labor costs. However, feeding grain on pasture does increase the total cost of gain.

The protein content of pasture grasses is highest in the spring and lower in the fall. Therefore, more protein supplement is needed in the ration in the fall. The stocking rate must be such that the grasses are kept grazed down. This increases the palatability of the pasture.

Grazing pasture plants such as alfalfa, forage rape, wheat, and clover can cause bloat more than grasses or other forages. Bloating on pasture is prevented by feeding a bloat preventive such as poloxalene.

Grain intake must be limited when self-feeding cattle on pasture. Salt or fat may be added to the concentrate mix to limit feed intake on self-feeders. These additives reduce the palatability of the concentrate mix. A level of 10 percent salt in the ration limits the grain intake. Adding fat to the ration may be more profitable depending on the cost of the fat. Limiting the concentrate intake reduces the amount of grain needed by about 50 percent compared to the feed required in drylot.

Little or no protein supplement is needed for grass-legume pastures. Cattle on grasses fertilized with nitrogen or grass in the early stages of growth also need little protein supplement. Cattle on poor-quality pasture or grasses in more mature stages of growth do require protein and minerals added to the ration. Protein supplement

USDA Agricultural Research Service; Photo by Keith Weller

Figure 16-7 Good-quality pasture can lower the cost of feeding beef cattle.

is also needed under drought conditions. About 10 to 12 percent total protein in the concentrate ration is enough for most feeding conditions.

Concentrates

Grains usually used for finishing beef cattle are corn, milo (grain), barley, oats, and wheat. Corn and milo are the most commonly used grains. All other grains are compared to corn for feeding value. The relative prices of the different grains determine which is best to use at any given time. Grains vary in feeding value depending on the variety, how they are processed, and other factors.

No more than 50 percent of the grain mixture should be composed of wheat. Higher amounts of wheat in the ration tend to decrease feed intake and increase digestive problems and abscessed livers. Wheat should be coarse ground, rolled, or crushed when used in the ration, but it should never be finely ground. The use of a buffer such as sodium bicarbonate in cattle rations with a high wheat content is recommended. Wheat has a feeding value of 100 to 105 percent of the value of corn in beef rations.

Oats are limited to no more than 30 percent of the grain mixture in beef rations. Oats are a good feed for growth and development, but they are not as good for finishing cattle. Oats are lower in energy value and higher in crude fiber content compared to corn.

Milo has about 85 to 95 percent of the nutritional value of corn for beef cattle.

Dry corn gluten feed is a good source of energy and protein for feeder cattle. It is highly digestible and does not limit forage intake. Supplies of dry corn gluten feed may be limited in some areas.

Molasses is added to the ration to increase palatability. It also helps control dust. Molasses may replace up to 10 percent of the grain in the ration.

Methods of Processing Feeds

Grinding is processing a feed through a hammer mill. The feed is beaten with rotating hammers until it is broken up enough to pass through a screen. The diameter of the holes in the screen determines how finely the feed is ground. Different size holes are used for obtaining different degrees of fineness. The feed is dry when it is ground.

High-moisture storage refers to harvesting the grain at a high moisture content (22 to 30 percent for corn and milo) and storing it in a silo.

Rolling refers to processing the grain through a set of smooth rollers that are set close together. The grain is pressed into the form of a flake. Therefore, this process is sometimes called flaking. If the grain is heated with steam and then rolled, the process is called steam flaking. If the grain is dry, it is called dry-rolled grain. **Crimping** is the same as rolling, except the rollers used have corrugated surfaces.

Feeds may be cooked before use, although this practice is seldom profitable. Steam may be used to cook the grain. Roasting, popping, and grain exploding refer to dry heating of the grain.

The **pelleting** of feed refers to grinding the feed into small particles and then forming them into a small, hard form called a pellet. **Cubing** is the same process as pelleting, but the cubes are larger in size than the pellets. Individual feeds or complete rations may be pelleted or cubed (Figure 16-8).

Courtesy of Samantha Meeks

Figure 16-8 Individual feeds or complete rations may be pelleted or cubed.

Other methods of feed processing include wetting, soaking, fermenting, and germinating grains. None of these processing methods is widely used, although they may be used in special cases. However, they are not considered to be practical for general on-farm use.

Grinding

Grinding generally improves the value of grains fed to beef cattle. However, research conducted at several universities shows that grinding or rolling dry shelled corn has little effect on rate of gain or feed conversion for beef calves or yearlings.

Processing high-moisture shelled corn has shown varied results relating to rate of gain and feed efficiency. Some tests have shown that grinding high-moisture shelled corn before putting it in the silo has slightly reduced feed intake and rate of gain. It may be necessary to include some dry grain in the ration to maintain sufficient dry matter consumption. Animal performance is improved when some dry grain is included in the diet. It may not be profitable to process high-moisture shelled corn when diets containing less than 15 percent roughage are used.

Steam processing both barley and milo increases feed intake and rate of gain. The starch in milo is not as digestible as that in corn. Processing milo helps to make it more usable for the cattle. Heating and flaking milo can increase rate of gain by about 10 percent and improve feed efficiency by about 5 percent, as compared to using milo as a dry-rolled grain. Feed requirements are about 10 to 18 percent less for high-moisture milo compared to dry ground milo. High-moisture milo has not been shown to improve rate of gain.

High-moisture ground ear corn increases gains by about 3 percent. Feed efficiency is improved by about 10 percent as compared to dry ground ear corn. High-moisture ground ear corn is a palatable ration that the cattle like. It is easy to start cattle and keep them on feed with this kind of ration.

Pelleting grains gives an increase in feed efficiency of up to about 5 percent. Feed efficiency is increased because less feed is wasted when it is pelleted. Complete pelleted rations will slightly increase feed intake.

Grinding forages increases feed intake. Feed consumption is increased because the smaller particles of feed pass more quickly through the digestive system.

Protein Supplements

Natural protein sources often used in finishing rations for beef cattle include soybean, linseed, and cottonseed meals. Most commercial protein supplements for beef cattle are based on soybean meal with some urea included. Urea is a nonprotein nitrogen source that can be utilized by ruminant animals to meet some of their protein requirement.

Lighter-weight calves (under 600 lb) usually gain faster when being started on feed if a natural protein source is used instead of urea. This is because there is a 2- to 3-week adjustment period needed for cattle to become used to urea. If the calves are already eating several pounds of grain per day when started on feed, they will gain just as fast on urea as on a natural protein source.

When rations containing a high level of grain are used, urea may be used to provide half or more of the crude protein equivalent needed by finishing cattle. Yearling feeder cattle can meet their entire protein requirement from urea if the energy level of the diet is high enough. However, performance results have not been as consistent when all of the protein equivalent was from urea, as compared to diets that included natural protein sources.

Urea should be used to supply no more than one-third of the crude protein equivalent needed if a limited-grain ration is being fed. Microorganisms in the rumen must have enough energy available to properly utilize urea. If the diet is low in energy, then less urea should be used.

Urea Fermentation Potential

Younger cattle that are still growing require more protein per unit of gain than older cattle that are being fattened. The younger cattle are still developing muscle tissue, which requires more protein. The addition of fat for weight gain in older cattle requires less protein per unit of gain.

Some feeds are broken down more easily by microorganisms in the rumen than other feeds. Protein that is not broken down in the rumen moves on in the digestive tract and is called bypass protein. Metabolizable protein is the combined total of protein from bypass protein and degraded protein that is reformed by rumen microorganisms and is absorbed and used by the animal.

Urea fermentation potential measures the amount of fermentable energy in the ration that is needed to utilize additional urea. It is determined by the total digestible nutrient content of the feed and the amount of ammonia released when the protein is broken down in the rumen by fermentation. High-energy feeds with low protein content tend to have high urea fermentation potential. Low-energy feeds with high protein content tend to have lower urea fermentation potential.

After cattle reach about 700 pounds, the bypass protein concept is no longer used for ration formulation. The energy level of a corn/corn silage ration is high enough to allow the use of urea as the major protein source.

Feeding Raw Soybeans to Beef Cattle

When the price of soybeans is low enough, it may be profitable to feed them to beef cattle. Care must be taken not to feed soybeans that become moldy while in storage. These molds can be toxic to cattle. Beans must be dried to 13 percent moisture or below to prevent molding in storage.

Feeding the beans whole reduces feeding efficiency. The beans should be coarsely crushed when fed. Feed the crushed beans within 1 week of crushing in hot weather and 2 weeks in cold weather to avoid spoilage. Start adding the beans to the ration at a low level and gradually increase the amount fed till the desired feeding level is reached. Adding beans too rapidly to the ration can result in diarrhea and lower feeding efficiency. Beans should comprise no more than 8 percent of the ration for growing calves and no more than 6 percent for finishing rations. Do not feed raw beans to young calves with undeveloped rumens because this might lead to nutritional muscular dystrophy and abomasal torsion in the calves.

Feeding Wheat Middlings to Beef Cattle

Wheat **middlings** are a by-product of the flour milling industry. They provide protein and energy in the ration and contain a good supply of phosphorus and potassium. Some flour mills pellet or cube middlings for sale as cattle feed. Wheat middlings generally compare favorably to other supplements used for cattle feed.

Cattle need to become acclimated to wheat middlings, so they should be added gradually to the ration. There are some problems with transportation and storage of wheat middlings. Mold and spoilage can be problems in storage (Figure 16-9). Bulk middlings are more difficult to transport and store than pelleted middlings, but many farmers prefer to buy bulk middlings because they cost less than pelleted middlings.

Figure 16-9 Moldy feed should not be fed to animals. Moldy feeds do not always contain dangerous toxins, but the presence of mold may adversely affect production and health.

Distiller's Grains as a Protein Source

Distiller's grains may be used as a replacement for other protein sources in beef cattle rations. Distiller's grains are a by-product of grain alcohol fermentation and are called stillage. They are a good source of protein, energy, and phosphorus, but are low in calcium and potassium.

Wet distiller's corn grains, whether fed as the only protein source or in combination with urea, are equal to soybean meal or corn gluten meal in meeting the total supplementary protein requirement for growing and finishing beef cattle. However, sorghum stillage is not as good as soybean meal as a protein source for beef cattle.

Minerals

Salt, calcium, and phosphorus are the main minerals needed in rations for finishing cattle. Iodized salt is used in iodine-deficient areas. Salt provides the sodium and chlorine the animal needs. Calcium is supplied by limestone or oyster shell flour. Phosphorus is supplied by bonemeal or dicalcium phosphate. If the ration contains high-quality roughages and natural proteins, trace minerals do not give any additional gains. A ration of poor-quality roughages requires the addition of some trace minerals, such as copper and cobalt, to meet the needs of the animal. Urinary calculi (stones caused by precipitation of salts from the urine) may be controlled by adding 1 ounce of technical-grade ammonium chloride to the daily ration.

Vitamins

The vitamins most commonly added to beef cattle rations are A, D, and E. Cattle can produce vitamin K and the B-complex vitamins in the rumen, so these vitamins are usually not necessary in the ration.

Vitamin A is the most important vitamin to add to the ration. A daily intake of 20,000 to 30,000 IU is enough for cattle on feed. When cattle are under stress, the addition of 50,000 IU of vitamin A per head per day may be beneficial.

When the diet contains a high level of good-quality forage, no additional vitamin A supplementation is needed. Cattle convert carotene to vitamin A at the rate of 400 IU of vitamin A for each milligram of carotene. Good-quality legume forages can contain as much as 20 to 30 milligrams of carotene per pound. Good-quality grass hays can contain as much as 10 to 15 milligrams of carotene per pound.

Cattle that are in the sunlight part of each day do not require additional vitamin D. A shortage of vitamin D may occur in the winter when there is less sunlight. In that case, it should be added to the ration.

Rations that include leafy roughages usually have enough vitamin E. High-grain rations may need 2 to 5 IU of vitamin E added for each pound of feed. Injections of vitamin E for newly arrived cattle help to reduce sickness while starting them on feed.

Additives and Implants

A number of feed additives and hormone implants are available for finishing cattle. Growth-promoting hormone implants can result in significant improvement in both growth rate and feed efficiency.

FEEDING CATTLE

Feed Intake

Cattle with high feed intake generally have a higher feed efficiency than those with lower feed intake. The amount of feed that cattle will consume is related to (1) the energy level of the ration, (2) the weather, (3) feed palatability, (4) feed processing, and (5) degree of finish on the cattle (Figure 16-10).

Figure 16-10 Cattle with high feed intake generally have a higher feed efficiency than those with lower feed intake.

Research shows that cattle will continue to increase feed intake until the energy level of the ration reaches about 55 megacalories per hundredweight net energy for gain (**NEg**) on a dry matter basis. This is a level of about 78 percent total digestible nutrients (TDN) in the ration. Increasing the TDN above this level does not increase feed intake and reduces feed efficiency.

Cattle that are switched suddenly from a low-energy ration to a high-energy ration sometimes go off feed (reduce feed intake). The high-energy ration increases the availability of readily fermentable carbohydrates in the rumen; this may cause a low rumen pH level, reduced motility of the digestive tract, and acidosis. These problems can usually be avoided by gradually increasing the energy level in the ration.

Beef cattle generally decrease feed intake in hot weather and increase feed intake in cold weather. At temperatures above 93°F, cattle on full feed will reduce feed intake by 10 to 35 percent. When shade or cooling is available or a low-fiber diet is fed, the reduction in feed intake is less at these high temperatures. When the temperature is between 75°F and 93°F, feed intake may be reduced 3 to 10 percent. Feed intake is about normal when the temperature is between 59°F and 75°F. Feed intake may be increased by 2 to 5 percent when the temperature is between 40°F and 59°F. At temperatures of 23°F to 40°F, feed intake may be increased by 3 to 8 percent. At 4°F to 23°F, feed intake may be increased by 5 to 10 percent. When temperatures are below 4°F, feed intake may be increased by 8 to 25 percent. Extremely cold temperatures (under −13°F) or storms can cause a temporary decrease in feed intake.

Some management practices that a cattle feeder can follow to increase feed intake during hot weather include:

1. feeding cattle earlier in the morning.
2. feeding more of the ration in the evening.
3. feeding more frequently during the day to reduce spoilage in fermented feeds.
4. treating silage with anhydrous ammonia or other preservatives to increase bunk life.
5. increasing the concentrate level in low-energy diets.
6. feeding a drier ration to increase bunk life of the feed.

Rain can cause a temporary decrease in feed intake of 10 to 30 percent. Mud that is 4 to 8 inches deep may reduce feed intake by 5 to 15 percent. Deeper mud (12 to 24 inches) may reduce feed intake by 15 to 30 percent. Providing good access to feed and using suitable bedding helps overcome some of the reduction in feed intake caused by muddy conditions.

Palatability of the ration may be reduced by urea, dust, or mold in the feed. Adding molasses or other flavoring agents can help overcome palatability problems caused by the presence of these materials. Adding flavoring agents to the feed when there are no palatability problems present generally will not increase feed intake.

Cattle tend to increase feed intake until they reach about 85 to 90 percent of their market weight. Feed intake then levels off for a while and may decline as they get heavier. When feed intake of heavy cattle begins to decrease, it may be an indication that they are about ready for market.

Feeding Management

The time of year that cattle are placed on feed affects feed efficiency and rate of gain. A study revealed that feedlot performance for steers weighing 700 to 800 pounds was higher when they were placed on feed between March and May. The poorest performance occurred when they were placed on feed in October or November. There was less variation in feedlot performance when the steers weighed 800 to 900 pounds at the start of the feeding period.

While it is difficult to anticipate how much feed cattle will eat from one feeding to the next, the cattle feeder should try to deliver an amount of feed in the feed bunk equal to the amount the cattle will eat before the next feeding. Make sure the feed is fresh and palatable. This improves feed intake and results in fewer digestive problems, especially on high-energy diets. Using low-quality or spoiled feed lowers feed intake. If the feed intake is reduced by 5 percent, the rate of gain decreases by about 10 percent.

Cattle should be fed at least twice a day. Increasing the frequency of feeding to four or five times daily may increase feed intake. Increasing the frequency of daily feeding also makes the estimate of how much feed will be eaten between feedings less critical. Cattle should eat all the feed in the bunk each day; however, they should not be out of feed for an extended period of time.

Other good feeding management practices include calculating rations on a dry matter basis, cleaning the feed bunks at least once a week, and following proper mixing procedures to ensure that all ingredients, including additives, are thoroughly mixed into the ration. Proper mixing of the feed improves palatability and decreases the tendency of the cattle to sort through the feed, leaving some portions uneaten.

Starting Cattle on Feed

Many cattle coming into the feedlot have had little or no grain in the growing ration. Some cattle have had grain in a creep feeding ration or have been fed grain on pasture during the growing period. If cattle are put on a full feed of grain too quickly, the microorganisms in the rumen do not have time to adjust to the new ration. This will put the cattle off feed. Cattle should be brought up to a full feed of grain as rapidly as possible without their going off feed.

A good-quality grass hay or first-cutting alfalfa-brome hay makes good roughage for starting cattle on feed. Cattle will often eat grass hay better than legume hay when starting on feed. Second- or third-cutting alfalfa hay will often cause the cattle to scour. Grass hay helps to stop scours in cattle. Other roughages that may be used are oat hay, Sudan hay, or green chop.

Corn silage with a protein supplement may be used to start cattle on feed. Hay is provided separately in the ration.

A starting mixture of 80 percent concentrates and 20 percent roughage may be used for cattle that have been creep fed grain or have had grain on grass pasture. The roughage is decreased by 10 percent and the concentrate increased by 10 percent at the end of the first week in the drylot. The ration for the rest of the feeding period may be 90 percent concentrate and 10 percent roughage.

Cattle that have not had any grain before going into the feedlot are fed less concentrate and more roughage to start them on feed. Several methods may be used to bring them to a full feed of grain.

One method is to start the cattle on a mixture of 60 percent concentrates and 40 percent roughages. Over a 2-week period, the amount of concentrate is gradually increased while the roughage is decreased. On full feed, they receive 90 percent concentrate and 10 percent roughage in the ration.

Feeding may be started with 90 percent concentrate and 10 percent roughage in the ration. Using this method, the daily feed intake per head is limited to 1 percent of body weight. The amount of feed is gradually increased each day until, at the end of 2 weeks, the cattle are getting all they will eat.

Another method of starting cattle on feed is to use a ration of 60 to 70 percent roughage with 40 to 30 percent concentrate. The amount of concentrate in the ration is gradually increased over a 2-week period. On full feed, the cattle are eating grain at the rate of about 2 percent of body weight.

Cattle to be finished on pasture are started on a ration of 80 percent concentrates and 20 percent roughages fed free choice. The amount of concentrate is increased by 10 percent and the roughage is decreased by 10 percent each week, until the cattle are being fed a ration that is 100 percent concentrates.

The ration should be fortified with vitamin A, antibiotics, and minerals. For the first 3 weeks, add 50,000 IU of vitamin A per head to the ration on a daily basis. A broad-spectrum antibiotic at the rate of 350 milligrams per head daily for the first 3 weeks reduces sickness and improves the rate of gain. Minerals should be fed free choice. A mix of 60 percent dicalcium phosphate and 40 percent trace-mineralized salt meets the needs of the cattle.

Rations for Finishing Cattle

Table 16-1 gives some examples of rations for cattle on full feed. These examples are only general guidelines. The quality of available feeds and the relative prices of feeds will make it necessary for the producer to vary the rations.

Cattle feeders who have fast-growing cattle on high-energy rations may find that increasing the level of protein in the ration will improve feed efficiency and rate of gain. Feeding higher levels of protein may improve feed efficiency by as much as 7 percent and rate of gain by as much as 12 percent. The suggested protein levels vary with the current rate of gain:

- less than 3 pounds per day—11 to 11.75% protein
- 3 to 3.25 pounds per day—12 to 12.5% protein
- 3.5 pounds per day—13 to 13.5% protein

Response to increased levels of protein varies with the kind of cattle fed, implants used, and overall management. Research suggests that adjusting the protein level in the ration of feeder cattle during the feeding period may reduce feed costs. Research revealed that making an adjustment in the amount of protein in the ration about every 2 weeks resulted in reducing protein costs by 15 percent and reducing the amount of nitrogen in the manure by 37 percent.

Rations that are high in crude fiber have given satisfactory gains in experimental trials. These rations contain as much as 70 percent pelleted or cubed alfalfa hay. Gains on high-fiber rations are a little slower than on high-concentrate rations. Feed requirements are about 21 percent less. Fewer animals on high-fiber rations grade USDA Choice as compared to animals on high-concentrate rations.

TABLE 16-1 Sample Rations for Steers—Initial Weight 600–750 lbs.[1]

Example 1:	**lb/day**	**Example 2:**	**lb/day**
Grain	13–16	Grain	13–16
High-quality legume hay	6–8	Protein supplement	1.25–1.75
		Mixed or low-quality legume hay	5–7
Example 3:	**lb/day**	**Example 4:**	**lb/day**
Corn meal and cob meal	14–17	Grain	12–15
Protein supplement	1–2	Protein supplement	0.75–1.25
Mixed hay	4–6	High-quality legume hay	3–5
		Corn or sorghum silage	14–18

[1]Quantities of feed are averages for entire feeding period. Less feed is required in the early part of the feeding period and more in the later part of the feeding period. More roughage is used early in the feeding period and less in the later part of the feeding period.

Source: USDA

Sorting Cattle into Feeding Groups

Technology is available to determine the length of time necessary to feed an animal to a desired quality and yield grade. Cattle may then be sorted into pens for more efficient feeding because all the animals in a given pen are similar in their feed requirements.

Calculating Total Feed Needed

Some general guidelines may be used to calculate the total amount of feed needed to finish cattle. The exact amounts needed vary according to the age and condition of the cattle, kind of feed used, weather conditions, and management practices of the feeder.

Fattening cattle will eat about 2.5 to 3 percent of their body weight in feed each day. Multiplying the average weight of the animal by 2.5 to 3 percent and then multiplying by the number of days in the feeding period gives the amount of feed needed.

Example - Calculating Total Feed Needed:

Starting weight of steer	600 lb
Length of feeding period	210 days
Average daily gain	2.4 lb
Ending weight	1,104 lb
Daily ration:	
Corn	14.8 lb
Protein supplement	1.5 lb
Mixed hay	5 lb
Feed consumption	2.5% of average body weight

$$\frac{\text{Start weight} + \text{End weight}}{2} \times \left(\begin{array}{c}\text{Feed consumption as percent}\\ \text{of body weight}\end{array}\right) \times \text{Days on feed}$$

$$= \text{Total pounds of feed per head}$$

$$\frac{600 + 1{,}104}{2} \times 0.025 \times 210 = 4{,}473 \text{ Pounds of feed per head}$$

Consumption of each feed in ration:

Corn	14.8 lb × 210 days = 3,108 lb
Protein supplement	1.5 lb × 210 days = 315 lb
Mixed hay	5 lb × 210 days = 1,050 lb
Total feed consumption	4,473 lb

Care of New Feeder Cattle

The most critical time in feeding cattle is the first 2 or 3 weeks after they are put into the feedlot for finishing. Important management practices for new feeder cattle are feeding, immunization, parasite control, castration and dehorning, treatment of sick cattle, and minimizing stress.

Cattle must be handled carefully to avoid stress. Stressful conditions result when cattle are loaded and transported to the feedlot. Cattle should have a moderate fill of grass hay and water before loading for shipping. Loading should be done as quickly and quietly as possible to reduce stress. Observe the cattle carefully as they are unloaded to see if any of them show signs of sickness. Isolate sick cattle in separate pens to avoid spreading disease.

Cattle that have been through sale rings, auction markets, and holding yards are more likely to have been exposed to disease and to become stressed. Shipping fever symptoms usually appear within 7 to 10 days.

When cattle are first brought to the feedlot, they should be put into small lots with plenty of feeder and water space. This makes it easier for the producer to observe any cattle that do not go to the bunk to feed. Fresh, palatable feed should be provided several times per day. Dry, dust-free, well-drained lots are recommended.

Most cattle can be started on silage and soybean supplement upon arrival in the feedlot. Sick cattle do better if fed good-quality hay. Plenty of clean, fresh water helps the cattle to recover from shrinkage losses, helps to maintain health, and can be used to give medicines, if needed.

Cattle should be vaccinated as they come off the truck or as soon thereafter as possible. This decreases the amount of handling, thus reducing stress. No one vaccination program is best for all feedlots or groups of cattle. Knowledge of the past history of the cattle helps the operator to plan the vaccination program. About 3 weeks after arrival in the feedlot, the cattle can be given any vaccinations not given when they arrived.

Three vaccines that should be given to all cattle upon arrival are those for infectious bovine rhinotracheitis (IBR), para-influenza 3 (PI_3), and leptospirosis. Calves that are not castrated or dehorned probably have not been given any vaccines or bacterin. **Bacterin** is a liquid containing dead or weakened bacteria that causes the animal to build up antibodies that fight the disease caused by the bacteria. Vaccinations for blackleg and malignant edema should be included in the program for these cattle.

Other vaccines given depend on the history of the cattle and the problems that the feedlot operator has had in the past. Most of these vaccines may be given 3 weeks after the cattle arrive in the feedlot.

Vitamin A should be included in the ration. A level of 40,000 IU per pound of supplement is adequate. In addition, cattle can be injected with vitamin A as they arrive in the feedlot.

Parasites such as lice, mange, and grubs must be controlled on newly arrived cattle. To eliminate these parasites, cattle may be sprayed or dipped 3 weeks after they have been placed in the feedlot. Cattle may also need deworming. Dewormers should be used only after the cattle are eating well. Routine application of antibiotics to all cattle entering the feedlot is not recommended.

Castration is not done until bull calves have been in the feedlot for about 3 weeks. If they need dehorning, this can be done at the same time.

The feedlot operator must watch the cattle closely for signs of sickness. A check should be made at least three times a day. Right after feeding is a good time to spot sick animals.

Signs of sickness in cattle include:

- moving about slowly
- rising slowly
- drooping ears
- discharges from eyes or nose
- not eating

- thin appearance
- walking stiffly or dragging the hind feet
- carrying head in abnormal way or drooping head
- high temperature
- sunken or dull eyes
- difficult or rapid breathing, or coughing
- scours (diarrhea), especially if bloody

Sick cattle are isolated from the rest of the herd as soon as signs of sickness are observed. A veterinarian can give the most accurate diagnosis and prescribe the best treatment for sick cattle. A veterinarian can assist the producer in preventing diseases from spreading to the rest of the herd.

SUMMARY

Because of the high grain requirements in the finishing ration, many cattle are fed to harvest weight in the central states, near corn production areas, and in the western United States. Large commercial feedlots of 1,000 head or more are found in the plains states and in Colorado, Arizona, California, and Texas. Most of the smaller feedlots of less than 1,000 head are found in the Corn Belt.

Some feeders use more roughage in the ration under a system of deferred finishing. Cattle may be used to clean up crop residues before being put on a full feed. Steers make better gains than heifers in the feedlot. Calves make more efficient gains than older cattle but take longer to reach market weight. Crossbred feeder cattle gain faster than the parent breeds. More roughages can be used when feeding lower grades of feeder cattle.

Feeder cattle come from auctions or direct from the farm or ranch where they were produced. The peak movement of feeder cattle is in October. Profit from feeding cattle is equal to the value of the finished cattle minus the total cost.

The main feed for feeder cattle is corn. Corn silage is the best roughage and is the most common roughage used in feedlots. Common natural protein sources used in feeder cattle rations include soybean, linseed, and cottonseed meal. Urea may be used as a nonprotein source of nitrogen in feeder cattle rations. Cattle need salt, calcium, and phosphorus. It is sometimes necessary to add vitamins A, D, and E to finishing rations.

Cattle with higher feed intake generally have a higher feed efficiency than those with lower feed intake. Feed intake is influenced by energy level in the ration, weather, feed palatability, feed processing, and degree of finish on cattle. The feed intake level begins to decrease when cattle reach 85 to 90 percent of their market weight.

When cattle are started on feed, the proportion of roughage in the ration is higher at the start and then is gradually decreased. The proportion of concentrate is gradually increased until the cattle are on full concentrate feed.

New feeder cattle coming on the farm must be watched closely for signs of sickness. Put new cattle in small lots with fresh, palatable feed and water. Vaccination and parasite control programs must be carefully planned and carried out soon after the cattle arrive.

Student Learning Activities

1. Visit local cattle feeders who use different systems of feeding and feed various kinds of cattle. Ask why they follow these feeding programs and prepare an oral report for the class.

2. Prepare a bulletin board displaying pictures of the grades of feeder cattle.
3. Prepare a PowerPoint™ presentation and present it to the class on when to buy feeder cattle and how to calculate possible profit in cattle feeding.
4. Interview local cattle feeders to learn how they start newly purchased cattle on feed and what management practices are used with cattle on feed.
5. When planning and conducting a supervised experience program in beef production, follow approved feeding and management practices as discussed in this chapter.

Discussion Questions

1. Why is more cattle feeding done in the grain-producing states than in other parts of the United States?
2. What is the difference between commercial feedlots and farmer-feeders?
3. Describe (a) finishing immediately and (b) deferred finishing systems of feeding beef cattle.
4. List and briefly describe four factors that the cattle feeder must consider when selecting cattle for the feedlot.
5. What sources of feeder cattle are available to the feedlot operator?
6. Describe the seasonal variation in prices for feeder cattle and for slaughter cattle.
7. How is the break-even price for feeder cattle determined?
8. How is the profit margin in feeding cattle determined?
9. List six general rules a cattle feeder should follow when selecting feeder cattle.
10. Briefly explain the use of roughages in finishing cattle for the slaughter market.
11. Briefly explain the use of pasture when finishing cattle for the slaughter market.
12. What is the most commonly used grain for finishing cattle for the slaughter market?
13. What is the relative value of some other commonly used grains for finishing cattle?
14. Explain the processing of grain for feeding finishing cattle.
15. How does high-moisture ground ear corn compare to dry ground ear corn for finishing cattle?
16. Explain the use of urea in feeder cattle rations.
17. Name the factors that influence feed intake of feeder cattle, and briefly discuss each.
18. List four management practices that a cattle feeder might follow to increase feed intake during hot weather.
19. What minerals are commonly required for finishing cattle?
20. What vitamins should be included in the ration for finishing cattle?
21. Describe how cattle should be started on feed in the feedlot.
22. Give an example of a ration for cattle on full feed in the feedlot.
23. How can stress be minimized as new feeder cattle are started on feed?
24. List management practices that should be followed for cattle as they are brought into the feedlot.
25. List the signs of sickness that a producer should watch for in cattle in the feedlot.

Review Questions

True/False

1. Distiller's grain, as a replacement for other protein sources in beef cattle rations, is a poor choice.

2. Yearling feeders should never be used to clean up crop residues such as cornfields.
3. Urea should be used to supply no more than one-third the crude protein needed if a limited grain ration is being fed.
4. A starting mixture of 80 percent concentrates and 20 percent roughage may be used for cattle that have been creep fed grain or have had grain on grass pasture.

Multiple Choice

5. Yearlings are feeders between 1 and 2 years old weighing about ______ pounds.

   a. 100–250 c. 400–550

   b. 250–400 d. 550–700

6. Feed efficiency can be increased by ______ grain because less is wasted.

   a. wetting c. grinding

   b. pelleting d. powdering

7. A starting mixture of ______ percent concentrates and ______ percent roughage may be used for cattle that have been creep fed grain or have had grain on grass pastures.

   a. 80, 20 c. 50, 50

   b. 20, 80 d. 50, 50

8. Feeders are divided into these groups:

   a. calves c. older feeders

   b. yearlings d. all of the above

Completion

9. The price that covers all cost without any profit is the ______ feeder price.
10. Feedlot capacity for the farmer-feeder operation is usually fewer than ______ cattle.

Short Answer

11. How frequently should cattle be fed?
12. Contrast the weight gains of heifers and steers.
13. Why is it important for younger cattle to have more protein in the ration than older cattle when being fattened?

Chapter 17

Diseases and Parasites of Beef Cattle

Key Terms

bovine spongiform encephalopathy (BSE)
systemic insecticide

Objectives

After studying this chapter, the student should be able to

- explain the importance of maintaining healthy beef cattle.
- identify and recommend prevention and treatment for beef cattle diseases and parasites common to the local area.
- recognize and suggest controls for common nutritional health disorders of beef cattle in the local area.

HERD HEALTH PLAN

The beef producer needs to develop an overall plan for maintaining the health of the beef herd. The key to the success of the health plan is the prevention of problems. Being familiar with the diseases and parasites that affect beef cattle can help farmers plan preventive programs that reduce health problems and increase profits. An important part of the health plan is developing a good working relationship with a veterinarian. Scheduling routine visits by a veterinarian can save money by helping prevent health problems before they become serious.

Characteristics of a good herd health plan include the following practices:

- working with a veterinarian to develop a herd health program
- following good feeding practices that meet the nutritional needs of the animals
- keeping good records
- vaccinating at the correct time, following all label directions
- following proper procedures for handling and storing vaccines
- controlling parasites

- following good reproductive management procedures
- observing the animals to detect signs of disease, correctly diagnosing the disease, and treating with the appropriate drugs (Figure 17-1)

Photo courtesy B. K. Lilja, Fort Valley State University

Figure 17-1 Blood samples can be taken and analyzed to help diagnose a disease.

Observing the vital signs (temperature, pulse rate, and respiration rate) in an animal can help in the early detection of health problems. Vital signs will vary with activity and environmental conditions. Normal vital signs in beef cattle are:

- temperature: normal range is 100.4 to 102.8°F; average is 101.5°F. Usually temperature is higher in the morning than in the afternoon; younger animals will show a wider range of temperature than mature animals do
- pulse rate: normal range is 60 to 70 heartbeats per minute
- respiration rate: normal range is 10 to 30 breaths per minute

Body temperature is taken in the rectum using either a mercury thermometer or a battery-powered digital thermometer. Restrain the animal when inserting the thermometer into the rectum. The pulse rate is taken by finding the artery on the lower edge of the jaw. It may also be taken by finding the artery along the inside of the foreleg or the inside of the hind leg just above the hock. There is another artery that may be used located high on the underside of the tail. Respiration rate is determined by observing the number of times the animal breathes per minute.

It is better to prevent health problems than to try to cure them once they have occurred. Good sanitation programs are essential in preventing diseases and parasites. Feedlots and feed bunks must be kept clean. New additions to herds or to the feedlots should be handled carefully to prevent the spread of diseases or parasites to animals already on the farm or feedlot.

DISEASES

Anthrax

Anthrax is a disease caused by bacteria that may remain in the soil for 40 years or longer. Certain conditions cause the bacteria to become active. Anthrax affects mainly cattle and sheep. Infection may result from grazing on infected pastures. The bacteria usually enter the animal's body through the mouth but may enter through the nose or through open wounds. Biting insects, such as horseflies, may spread the disease from one animal to another.

Anthrax often results in sudden death of the infected animal. Less acute infections show symptoms of high fever, sudden staggering, difficulty breathing, trembling, and collapse. Death usually occurs within a few days after these symptoms appear (Figure 17-2).

The carcass of an animal that has died from anthrax should be burned or buried at least 6 feet deep and covered with quicklime. The carcass should not be buried near wells or streams. In some states, the carcass of an animal infected with anthrax may not be taken to a rendering plant. Care must be taken when handling the carcass of an animal suspected of having anthrax since the disease can be transmitted to people.

Vaccines may be used to control anthrax. In areas where anthrax is a problem, animals should be vaccinated on a yearly basis. Where it is not a common disease, vaccination should be done only on the advice of a veterinarian.

USDA

Figure 17-2 Death from anthrax usually occurs within a few days after symptoms appear.

Bovine Respiratory Syncytial Virus

Bovine respiratory syncytial virus (BRSV) affects the cells that line the respiratory system. As a result, the respiratory system is weakened and becomes more vulnerable to infection from other viruses and bacteria. Nursing and weaned calves are

Connection: The Cow That Stole Christmas

Bovine spongiform encephalopathy (BSE), commonly known as mad cow disease, was discovered in the United States on Christmas Eve in 2003. Due to the timing of the discovery, as well as the commotion it caused within the beef industry, the infected cow became known as "the cow that stole Christmas." When the official diagnosis was made, a recall was issued for all beef harvested in the same vicinity as the infected cow. The immediate response of the public was to stop buying beef, and several countries prohibited U.S. beef imports.

BSE is similar to other neurodegenerative disorders such as Alzheimer's, Parkinson's disease, and Creutzfeldt-Jakob disease in humans and scrapie in sheep and goats.

more likely to be affected than older cattle. Stress from moving or weaning calves increases the chances of infection from this disease. A combination vaccine has been developed for treatment of BRSV, IBR, BVD, and PI_3.

Bovine Spongiform Encephalopathy (BSE)

Bovine spongiform encephalopathy is a chronic degenerative disease that affects the central nervous system of cattle. This disease is a member of a class of brain diseases called transmissible spongiform encephalopathies (TSEs) that are relatively rare. Some TSEs affect animals and some affect humans.

TSEs affecting animals include:

- bovine spongiform encephalopathy (BSE)
- chronic wasting disease in deer and elk
- feline spongiform encephalopathy
- scrapie in sheep and goats
- transmissible mink encephalopathy

TSEs affecting humans include:

- Creutzfeldt-Jakob disease (CJD), first identified in the 1920s
- new variant CJD (nvCJD), first identified in 1995
- fatal familial insomnia
- Gerstmann-Straussler-Scheinker syndrome
- kuru

Bovine spongiform encephalopathy, sometimes called mad cow disease, is rare. It was first diagnosed in Great Britain in 1986 and has been found in a few other countries since then. The first case of BSE in the United States was found in 2003. The announcement had a dramatic economic impact on the cattle industry.

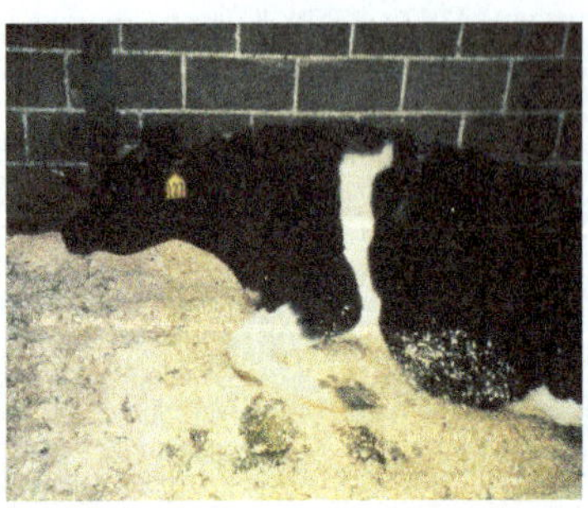

USDA, APHIS; Photo by Dr. Art Davis

Figure 17-3 A cow affected by BSE that is having difficulty trying to stand up.

Cattle that are affected by BSE show symptoms such as nervousness or aggression, muscle twitching, abnormal posture, loss of body weight, decrease in milk production, and difficulty in rising after lying down (Figure 17-3). There is no treatment or vaccine, and affected animals eventually die. Both beef and dairy cattle are susceptible to BSE.

BSE is caused by a microscopic piece of misfolded protein called a prion, which is neither a bacteria nor a virus. However, a prion is similar to a virus but lacks nucleic acid. Cattle may contract BSE by ingesting protein in feed that came from an animal protein source that was contaminated by the agent that causes the disease.

The incubation period for BSE ranges from 2 to 8 years. The animal usually dies within 2 weeks to 6 months after clinical symptoms appear. There is no test to determine if live cattle are affected; the presence of BSE can only be confirmed by postmortem microscopic examination of the brain. The brain tissue of infected animals has a spongy appearance when examined under a microscope.

There is no evidence that BSE can be transmitted to humans by direct contact with infected animals or by consuming their meat or dairy products. There also is no evidence that eating meat from BSE-infected cattle can cause Creutzfeldt-Jakob disease (CJD), a human brain disease.

In 1995, a new human neurological disease was found in England. It is called new variant Creutzfeldt-Jakob disease (nvCJD). Research indicates that the same agent that causes BSE in cattle may cause nvCJD in humans. The BSE causative agent has not been found in muscle meat or milk from cattle. It has been found in brain tissue, the spinal cord, corneal tissue, and some other central nervous system tissues of infected animals.

Bovine Virus Diarrhea

Bovine virus diarrhea (BVD) is a common disease throughout the United States. The disease may appear in mild, acute, or chronic forms. BVD spreads by contact, and it may be carried on a person's shoes from one herd to another. It can also result from fence-line contact with infected animals.

In the mild form, there are often no symptoms. If symptoms are present, they include fever, coughing, discharge from the nose, slow gains, rapid breathing, and mild diarrhea. Animals that have had the mild form of the disease are immune to further infection.

Animals infected with the acute form of BVD show symptoms of fever, difficulty breathing, discharges from the nose and mouth, and coughing. In addition, ulcers may develop on the mouth and the animal may become lame. Dehydration and weight loss also occur. Diarrhea begins 3 to 7 days after the animal becomes diseased. Pregnant animals may abort if the disease is contracted during the first 2 months of pregnancy. The fetus may mummify (absorb the fluids in the womb and become hardened) if the cow becomes infected with BVD from the 90th to the 120th day of pregnancy. In the later stages of pregnancy, BVD may cause the fetus to suffer brain damage, hairlessness, or underdeveloped lungs.

Chronic cases of BVD result in slow gaining. The hair coat is rough and the animal may become lame.

A modified live virus vaccination is used to prevent BVD. Calves are vaccinated between 1 day of age and 3 weeks before weaning. Feeder cattle may be vaccinated upon arrival in the feedlot or after they have been in the feedlot for 2 or 3 weeks. Calves vaccinated before weaning should be vaccinated again when placed in the feedlot.

Pregnant cattle should never be vaccinated. Vaccinate adult cattle only after calving and at least 3 weeks before breeding. Vaccinate replacement heifers between 9 and 12 months of age, but not during the last 3 weeks before breeding. A single vaccination of older cattle will give immunity for the productive life of the animal. Do not vaccinate animals during a stress period.

There is no cure for BVD. Treatments are given to control diarrhea and secondary infections.

Blackleg

Blackleg is a disease caused by bacteria that grow only in the absence of oxygen. The disease is most serious when the bacteria lodge in deep wounds. When the bacteria are exposed to air, they form a protective spore (covering), which allows them to live for many years in the soil. The spores enter the animal through the mouth or through open wounds. Young cattle are more commonly affected than older cattle. Blackleg may result in sudden death. Before death, the symptoms of blackleg are lameness, inability to stand, swollen muscles, severe depression, and in the early stages, high fever.

Black leg is a prevalent disease in the United States; however, it is not seen often because most cattle are vaccinated against it. Calves are vaccinated when young and again at weaning. Vaccination is vital because the spores that cause blackleg can live in the soil for at least 10 years, even if there has not been a recent outbreak. To prevent the spread of the disease, dead animals must be burned or buried.

The disease is treated with massive doses of antibiotics. Treatment is effective only if the disease is diagnosed early. Prevention is more effective and less costly than treatment.

Bovine Respiratory Disease (BRD)

Bovine Respiratory Disease (BRD) is a major problem for beef and dairy cattle producers. It continues to cause major losses in the cattle industry even after years of research into the cause and prevention of BRD as well as the development and use of improved vaccines, treatments, and modern management practices. BRD is the number one disease of stocker, backgrounder, and feedlot cattle. It accounts for up to 80 percent of all feedlot diseases and up to 50 percent of all deaths of feedlot cattle. Losses from BRD include:

- death loss
- decreased appetite
- increased labor costs
- increased medication costs
- weight loss

BRD is caused by several pathogenic organisms and brought on and complicated by multiple factors, including environmental factors, host factors, management factors, and stress. Many infectious agents have been associated with BRD, including various strains of viruses and bacteria. These pathogens are often found together, complicating diagnosis and treatment.

There is no single, simple cause of BRD, or simple solution. Several BRD related diseases are covered individually in this chapter. They include:

- Bovine Respiratory Syncytial Virus (BRSV)
- Bovine Viral Diarrhea (BVD)
- Infectious Bovine Rhinotracheitis (IBR)
- Parainfluenza 3 (PL3)
- Pasteurella
- Shipping fever

A wide range of stress conditions contribute to the onset and severity of BRD. Stress conditions include:

- Crowding
- Dehydration
- Dusty conditions
- Fatigue
- Hunger

- Injuries
- Medical treatments
- Muddy conditions
- Nutritional deficiencies
- Parasites
- Stress from handling and transportation
- Weaning anxiety
- Weather: hot, cold, windy, damp, etc.

Bovine Respiratory Diseases are divided into three main groups based on the parts of the respiratory system affected:

1. Upper respiratory tract infections of the nostrils, throat, and windpipe (trachea).
2. Diphtheria involving the larynx (voice box).
3. Pneumonia, a lower respiratory tract infection of the lungs.

Symptoms of BRD include nasal discharges, coughing, fever, varying degrees of breathing difficulty, rapid breathing, depression, loss of appetite, general unthrifty appearance including droopy ears, breathing from the mouth, discharge from the eyes, and death. Symptoms vary depending on the stage and extent of the disease process. A high percentage of cattle examined at harvest show signs of lesions on the lungs indicating that they had, or once had, BRD and were probably never diagnosed and treated. Many cattle carry one or more of the bacterial and viral agents in their upper respiratory system but show few, if any, symptoms unless stress factors become involved.

Brucellosis

Brucellosis, a disease caused by a microorganism, causes heavy economic losses in the cattle industry. Federal and state programs of eradication have resulted in the disease being less common. The germs that cause brucellosis are also dangerous to humans, causing undulant (Malta) fever.

Cattle with brucellosis often abort during the last half of pregnancy. Infected cows may retain the afterbirth (placenta). Other symptoms include sterility in cows and bulls, reduced milk flow in cows, and enlarged testicles in bulls. Calves born to infected cows may be weak.

Brucellosis is spread by infected cattle that are brought into the herd. It may also be picked up by fence-line contact with infected animals. An aborted fetus that carries the *Brucella* organisms may be brought from one farm to another by dogs or other carnivorous animals. Unborn calves may be infected by their mothers and become sources of infection after birth. Cattle can also contract brucellosis by eating feed or drinking water in which the organism is present. Sniffing or licking an aborted fetus or a calf from a cow with the disease can also spread the disease.

There is no cure for brucellosis. Prevention is accomplished by good management practices. Herd replacements should be purchased only from brucellosis-free herds. All animals new to the herd should be isolated and tested. The herd should be tested periodically (blood or milk tests) to determine whether the disease is present. Infected animals should be disposed of following state and federal guidelines. Calves should be vaccinated between 2 and 6 months of age to increase their resistance to the disease. A state that is nearing an eradication stage may not permit calfhood vaccination because calfhood vaccination does not give lifetime protection from brucellosis. Cooperation with state and federal eradication programs helps in controlling the disease.

Calf Enteritis (Scours)

Enteritis (scours) is a disease complex (group of diseases) that is most common in the fall, winter, and spring. It is a disease of young calves. Calves over 2 months of age seldom are affected by the disease.

Symptoms of scours vary. In the acute form, the calf is found in a state of shock. The nose, ears, and legs are cold and the animal may suffer from diarrhea. Acute scours results in sudden death. The chronic form shows symptoms, including diarrhea, for several days. The calf loses weight and dies after several days if not treated.

The most important factor in control is sanitation. Buildings in which calving occurs must be kept clean. Proper sanitation practices include keeping the bedding clean and dry and disinfecting buildings and pens where calves are kept. It is important that the calf be fed the first milk (colostrum) from the mother after birth because colostrum contains antibodies that help prevent scours. Vitamin A helps to control scours; therefore, supplementing the cow's diet with vitamin A just before calving may help prevent the disease.

The most common types of scours (*Escherichia coli*, reo virus, and corona virus) can be controlled by vaccines. The dam is vaccinated at least 30 days before calving; this causes the development of antibodies in the milk. The antibodies are passed on to the calf in the colostrum milk. A combination vaccine for scours and campylobacteriosis (vibriosis) is available and can be used 10 weeks before calving.

Extra care must be taken when calves are bucket-fed milk. If they receive milk from dirty buckets, they are more likely to become infected with scours. Since too much milk may also lead to scouring, it is good practice to keep the calves a little bit hungry for the first two weeks after birth. A gradual replacement of whole milk with milk replacer starting at about 2 weeks of age also helps to prevent scours.

Antibiotics and sulfa drugs may be used to treat scours after the symptoms appear. A veterinarian is the best source of information about which drugs are most effective for local conditions.

Campylobacteriosis

Campylobacteriosis is the new name for a reproductive disease formerly called *vibriosis*. Vaccine labels now carry the new name; however, some manufacturers are also including the trade name Vibrio on the label to reduce confusion among purchasers. Animals should be vaccinated 30 days before breeding, and vaccination must be repeated every year.

Campylobacteriosis in cattle is a disease that has both an intestinal and venereal form. It is a leading cause of infertility and abortion in the cattle industry, costing many millions of dollars per year.

The intestinal form of campylobacteriosis has little harmful effect on cattle. The venereal form of the disease is more serious. If the organism infects the uterus, there will be some abortion in the herd. However, the number of cows affected is normally small. Cows do not become sterile, and bulls are not affected.

Symptoms of the venereal form include infertility, abortion, and irregular heat periods. In herds that are newly infected, conception rates may drop below 40 percent. The calving season is longer, and there are more open cows in the fall. In herds that are chronically infected, the conception rate is lower than normal, usually about 60 to 70 percent. Heifers and new additions to the herd will require repeat breeding or will abort.

The disease is spread from infected bulls to clean cows during breeding. It is possible to treat valuable bulls with antibiotics, but the process is difficult. Infected cows may settle more easily if they are treated with antibiotics. Skipping two heat periods before attempting to breed the cow also improves the conception rate of infected cows. Cows with the disease usually develop immunity and eventually breed again.

The use of artificial insemination (AI) helps to prevent campylobacteriosis. The semen used for AI is treated with antibiotics to eliminate the disease organisms.

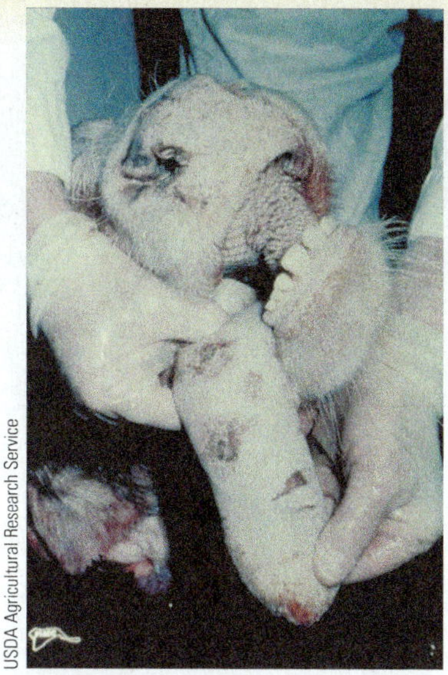

USDA Agricultural Research Service

Figure 17-4 The tongue of a cow with foot-and-mouth disease.

Foot-and-Mouth Disease

Foot-and-mouth disease (sometimes called hoof-and-mouth disease) is caused by a virus. The disease affects cloven-hoofed animals such as cattle, sheep, swine, deer, and other ruminants. The virus is easily spread by animal contact, through the air, by exposure to contaminated feed or contaminated meat products, and by mechanical means (such as on shoes or clothing). The main way it is spread is through the respiratory system, but it can also be spread through breaks in the skin. Figure 17-4 shows an example of the effects that are caused by foot-and-mouth disease. It is possible for humans to carry the virus in their nasal passages for up to 28 hours, and they could spread it to animals during that time.

The last outbreak of foot-and-mouth disease that occurred in the United States was in 1929. The disease is common in many other parts of the world and causes great economic loss to farmers. An outbreak in the United Kingdom in 2001 resulted in the loss of almost 6 million animals that were destroyed in the effort to contain the disease. This represented about 12 percent of the total farm population of animals in the United Kingdom. Another outbreak in the United Kingdom occurred in 2007. The USDA Animal and Plant Health Inspection Service (APHIS) has regulations in place to help prevent the re-entry of the disease into the United States from other countries.

The meat from animals infected with foot-and-mouth disease is safe to eat once it has been fully cooked. The virus that causes foot-and-mouth disease is killed during the cooking process.

People who have been on farms in areas where the virus exists need to take precautions to prevent spreading it to other areas. The virus can be killed by using an appropriate disinfecting agent. It is a good preventive measure to wear footwear and protective clothing provided by the facility that is being visited and to discard these items when leaving. Also, wash all clothing worn at the facility as soon as possible. These measures help to prevent the spread of the virus.

The incubation period for the virus is 1 to 7 days. Symptoms include a loss of appetite, fever, and the development of fluid-filled blisters in the mouth and on the feet. Animals may become lame, milk production may drop in lactating animals, and conception rates in breeding animals may be lowered. Some animals, especially younger ones, may die; the overall death rate is about 2 percent of infected animals. There is no cure for the disease. Most animals recover in 2 to 3 weeks but remain carriers of the virus for a period of time. Animals are slow to regain lost weight, and milk cows generally do not produce as well as they did prior to becoming infected.

In many parts of the world where foot-and-mouth disease is common, animals are generally vaccinated. While vaccination reduces the occurrence of the disease, it produces carrier animals that will continue to spread the disease. Destroying and burning the infected animals has been the method of control used in some countries, including the United States and the United Kingdom. Because the disease is so highly contagious and because of the tremendous losses that occur due to reduced production, these countries have decided to attempt to contain outbreaks by killing and burning infected animals rather than leaving a reservoir of the disease from vaccinated animals.

Foot Rot

Foot rot is caused by a variety of bacteria, fungi, and other organisms that are found in the feedlot. These organisms enter the animal when the skin of the foot is broken in some way. Sharp objects in the feedlot, such as stones, nails, or wire, can cause injuries that make it possible for the organisms to enter the animal's body. Feedlots that are muddy or contain large amounts of manure make the problem worse. Although not as likely, pasture animals can contract foot rot if there are any low-lying wetlands within the pasture.

The first noticeable symptom of foot rot is usually lameness. The animal loses its appetite, has a fever, and is depressed. It may not want to stand on its feet or move around. Death may eventually result.

There are no vaccines to prevent foot rot. Sanitation and paved lots help to prevent it. Good drainage and mounds in the feedlots are also helpful in eliminating conditions that encourage the disease. Spreading lime and 5 percent blue vitriol in the areas around water and feed bunks helps in its control. Penicillin, wide-spectrum antibiotics, or sulfa drugs may be used to treat foot rot. Always consult a veterinarian before beginning a treatment program.

Infectious Bovine Rhinotracheitis (IBR, Red Nose, IPV)

Infectious bovine rhinotracheitis is a disease caused by a virus. It occurs in several forms and is found in cattle throughout the United States and the rest of the world.

Respiratory IBR is the most common form of this disease. Symptoms are fever and a discharge from the nose. The animal may also foam at the mouth, breathe through the mouth, cough, and lose weight. The nose and muzzle become inflamed, which is why this form is sometimes called "red nose." Infection can involve anywhere from 15 to 100 percent of the animals in a herd. However, death losses from this disease are low.

Genital IBR is known as *infectious pustular vaginitis* (IPV). Symptoms are inflammation in the vagina and swelling of the vulva. Pustules (pus-filled blisters) form in the vagina, and there is a pustular discharge. In the bull, the disease is called *infectious pustular balanoposthitis* (IPB). Symptoms include inflammation of the sheath and penis. Lesions form on the penis and there is a pustular discharge.

Conjunctival IBR is similar to pinkeye. A clear discharge from the eye is the first symptom.

Abortion may be caused by IBR. Abortion may follow the symptoms of respiratory IBR or conjunctival IBR, or it may occur without any other symptoms being observed. Pregnant animals that are exposed to infected animals or vaccinated with modified live virus for IBR may abort.

Encephalitic IBR affects the brain and nervous system of the animal. Symptoms are inability to control voluntary muscular movement, depression, convulsions, and death. Encephalitic IBR affects younger cattle more frequently than older cattle. The symptoms are similar to listeriosis, which is discussed later in the chapter.

Because IBR is similar to some other diseases that affect cattle, positive diagnosis can be made only by laboratory tests. A veterinarian should be called if IBR is suspected.

A modified live virus is used to vaccinate for IBR. Calves are vaccinated after they are 6 months of age. Breeding animals are vaccinated at least 2 months before breeding if they have not been vaccinated as calves. Pregnant animals should never be vaccinated because vaccination may cause abortion.

Livestock producers must be sure that all new additions to the herd are disease free and have health certificates. Isolate new additions at least 30 days before they come into contact with the rest of the herd. Diseased animals must be isolated at once.

There is no treatment for IBR. Antibiotics and sulfonamides may be used to control other infections that may attack the weakened animal.

Figure 17-5 Johne's disease is caused by a bacterial infection leading to thickening of the intestinal wall. Infected animals lose weight and eventually die

Johne's Disease (Paratuberculosis)

Johne's disease (pronounced "yo-knees") is a contagious, chronic, and usually fatal infection of cattle that causes a thickening of the wall of the intestine (Figure 17-5). All ruminants are susceptible to Johne's disease. The disease is caused by a bacterium named *Mycobacterium avium.* Chronically infected animals rarely show signs of the disease until 2 or more years after the initial infection, which usually occurs shortly after birth. Infected animals have intermittent scours, but do not have a fever. Producers often treat infected animals for the diarrhea; however, the treatment has no effect because it is the damage to the intestine that causes the diarrhea. Infected animals lose weight and eventually die. Diagnosis is made with a *Johnin test,* which is done by a veterinarian. After diagnosis, the only way to solve the continual spreading of the disease is culling.

Some dairies have been put out of business because of Johne's disease. In a national study, researchers found that 22 percent of dairy farms have at least 10 percent of the herd infected by Johne's disease. Although the disease has traditionally been a dairy cattle disease, it is becoming prevalent in many beef cattle herds. The disease is controlled by checking the health history of animals being brought into the herd. They should not come from herds that have a history of Johne's disease. If the animal has had chronic diarrhea, suspect the presence of this disease and do not bring the animal into the herd.

Leptospirosis

Leptospirosis is caused by several strains of leptospira bacteria. Infected cattle may not show any symptoms. Acute cases may show a sudden rise in temperature, rapid breathing, loss of appetite, stiffness, bloody urine, jaundice, diarrhea, or abortion. In young cattle, the symptoms are very similar to shipping fever. Death rate ranges from 5 to 15 percent of the infected animals.

Infected animals can spread leptospirosis. Feed that is contaminated from the urine of infected animals can spread the disease. Leptospirosis may be transmitted both from cattle to hogs and from hogs to cattle. Keeping hogs and cattle separate will help control this disease. Good sanitation and isolation of new animals coming into the herd for a period of 1 month are additional control measures that may be used.

A regular program of vaccination helps control leptospirosis. There are five common strains of the disease. Antibodies from a vaccine for one of the strains do not give protection against other strains. A veterinarian should be consulted to determine the proper vaccine to use for local conditions. Cows should be vaccinated before the breeding season starts, and they should be given a booster shot 6 months later. Annual vaccination is required (Figure 17-6). Using high levels of antibiotics in the early stages of infection will stop the shedding of the organism. Animals that recover from leptospirosis develop immunity and the future breeding ability of the herd is usually normal.

Figure 17-6 A cow being injected with a vaccine for bovine leptospirosis.

Listeriosis

Listeriosis, a disease caused by a germ, occurs in three forms. It is spread from animal to animal by contaminated feed or water; by contaminated dust, manure, urine,

Image courtesy James Lawhead, DVM

Figure 17-7 A cow suffering from listeriosis. The cow is unable to rise. Note the obvious drooped ear.

or saliva; and, possibly, by breeding. It is more common in animals that are fed low-quality silage.

Encephalitic listeriosis affects the brain. Initial symptoms include fever, loss of appetite, and dullness. As the disease progresses, the animal may have difficulty standing and may begin to drool. The ears, eyelids, or lips droop. The head turns to one side and the animal wanders aimlessly in a circle. Eventually, it may not be able to stand. Cattle will show these symptoms for 4 to 14 days before death. If treated properly, some will recover. An animal suffering from listeriosis is shown in Figure 17-7.

Abortion is caused from one form of the disease. There may be premature birth of a live fetus or abortion of a dead fetus. Calves that are born live usually die shortly after birth. Most of the abortions occur in the last one-third of the gestation period. The cow usually shows no other symptoms of the disease.

Septicemic listeriosis is more common in older animals. However, it can occur in premature calves. The infection affects the entire body of the animal. Sudden death is a common result of this disease. The animal may become weak, depressed, and lie down much of the time. Hard breathing, slobbering, and a discharge from the nose are other symptoms, and the animal may die after several days.

There is no vaccine for control of listeriosis. Sanitation is the best control. Careful disposal of aborted fetuses and afterbirth is another control measure. Feed and water must be kept free from contamination.

Lumpy Jaw (Actinomycosis)

Lumpy jaw is a chronic disease and seldom causes the animal to die. It results in economic loss because the affected body parts are condemned when the animal is harvested.

The disease affects the jaw and surrounding bony part of the head. Sometimes it spreads to muscles and other internal organs. Symptoms are tumors or lumps on the jaw. These are filled with a yellow pus that has little odor. The teeth may become loose, making it difficult for the animal to chew. The jawbone becomes spongy and swells, which may result in breathing problems. In addition, the animal may lose weight because of its inability to eat.

Surgical treatment by a veterinarian may allow the animal to still be marketed. Complete recovery is usually not possible.

The organism that causes the disease enters the animal's body through wounds. To prevent lumpy jaw, be sure that there are no sharp objects such as wire and barley beards in the pasture or feedlot that could cause puncture wounds. Feed having sharp stickers may also cause injuries that allow the organism to enter the body of the animal. Isolate infected animals from the rest of the herd.

Image courtesy James Lawhead, DVM

Figure 17-8 A cow suffering from pinkeye. Pinkeye is an infectious disease that is commonly spread between animals by flies.

Malignant Edema

The symptoms, control, and treatment of malignant edema are much the same as those for blackleg.

Pinkeye (Infectious Keratitis, Keratoconjunctivitis)

Pinkeye is a disease carried by insects that affects the eyes of the animal (Figure 17-8). It has both mild and acute forms. In the mild form, the eyeball develops a pinkish color. The cornea (the transparent covering of the iris and pupil) becomes slightly clouded. The symptoms of the acute form are a flowing of tears from the eye and a cloudiness of the cornea. As the infection progresses, the cloudy condition becomes worse and ulcers may develop on the eye. The eye may be damaged so severely that blindness results. The condition may last 3 to 4 weeks. If not treated, it will spread to most of the herd.

White-faced cattle and cattle with white pigment around the eye are more likely to be infected with pinkeye. It occurs year-round and may affect any kind of cattle. It is most common during periods of maximum sunlight.

Pinkeye is spread by insects such as face flies, by direct contact with infected animals, by dust, and by tail switching. Cattle that are crowded at feed bunks or waterers may rub heads, which contributes to the spread of the disease. Controlling flies and other insects is one of the main ways to prevent pinkeye.

Vaccinations are now available to control *Moraxella bovis*, the bacteria that is considered to be a main cause of pinkeye. A veterinarian should be consulted for current information on the availability and use of vaccines for pinkeye.

Infected animals should be isolated in a dark place. Antibiotics and sulfa drugs may be applied to the infected eye at least twice a day. The medicine must be applied frequently because it is washed out of the eye by the flow of tears. A cloth patch can be used on the affected eye. A virus form of pinkeye is associated with IBR. The symptoms, treatment, and control of this form are discussed under the section on IBR.

Prolapse in Cattle

Prolapse is an abnormal repositioning of a body part from its normal anatomical position. There are two types of prolapse in cattle: vaginal and uterine (Figure 17-9). Vaginal prolapse is a result of pressure in the abdominal cavity during the latter stages of pregnancy. The prolapsed tissue usually appears as a pink mass comparable in size to a grapefruit or larger. Prolapsed tissue is subject to infection. Vaginal prolapses can be treated; however, the animal is likely to prolapse again prior to the next calving. Cows that have had a prolapse should be culled from the herd. Keeping replacement heifers from cows that have prolapsed is discouraged because vaginal prolapse can be an inherited trait. Vaginal prolapses are more likely to occur in older cows and Brahmans.

Figure 17-9 Uterine prolapse is common when a cow strains too much during labor. This cow's rectum in the upper position is intact; however, the huge red flesh at the lower location indicates uterine prolapse.

Uterine prolapses are usually larger and darker in color than vaginal prolapses. Typically, more blood loss is associated with this type of prolapse. It usually occurs just after calving and should be treated by a veterinarian immediately because the cow is at risk of dying from shock and blood loss. Uterine prolapses are not inherited. Factors that increase the chances of a uterine prolapse are calving difficulties, pulling the calf, and poor nutrition.

Ringworm

Ringworm is a contagious skin disease that may be spread to other animals and humans. Symptoms include round, scaly patches of skin that lack hair and that may appear on any part of the body (Figure 17-10). The affected area clears up, but the infection spreads to other parts nearby.

Sanitation helps control ringworm. Infected animals should be isolated from the rest of the animals. Iodine tincture or quaternary ammonium compounds may help in the treatment.

Figure 17-10 Heifer with ringworm. Ringworm can be spread to other animals and humans.

Shipping Fever (PI_3, Pasteurella, Bovine Respiratory Disease)

Shipping fever is a disease complex that affects the respiratory tract of the animal. It is most common in young cattle at times of stress. A combination of stress, viral infection, and bacterial infection is thought to be the cause.

The stress conditions that contribute to the start of the disease often occur when cattle are moved from the range to the feedlot. Extremes of heat or cold, dust, exhaust fumes, hunger, fright, and rough handling all create stress. This allows the virus and bacterial organisms, which are already present, to attack the respiratory tract of the animal.

The disease may vary from mild to acute. An early symptom is fever. The animal appears depressed, with the head down and the eyes closed. The ears are often drooping and a discharge from the nose develops. The eyes water and the animal may lose its appetite. Diarrhea and weight loss may occur. Difficulty breathing and coughing follow. Pneumonia may develop, and the animal may die. If it recovers, it will often be a slow gaining animal.

The disease is prevented by vaccination. Vaccination should be done after the animal is 4 months of age. The best time to vaccinate is 3 to 4 weeks before the animal is exposed to conditions that lead to the disease. Reduction of stress and exposure help to prevent the disease. Good feedlot management and careful handling of new cattle coming into the feedlot also help to reduce shipping fever.

Antibiotics and sulfa drugs are used to treat shipping fever. The treatment must begin as soon as possible after the symptoms are noticed. Treatment after an animal develops pneumonia, should this occur, is of little value.

Trichomoniasis

Trichomoniasis is a venereal disease in cattle caused by a protozoan, *Trichomona fetus*. The organism infects the genital tract of the bull and is transmitted to the cow during breeding. A clean bull can also be infected when breeding an infected cow. The disease can be transmitted through infected semen, even if artificial insemination is used.

Symptoms include abortion early in gestation, low fertility, irregular heat periods, and infection in the uterus. In females, there may be a discharge from the genital tract. An infected bull may not show any symptoms of the disease but is still capable of transmitting it to cows during breeding. The organism is identified by microscopic examination of material from an aborted fetus, the preputial cavity of the bull, or vaginal discharge from the cow.

There is no treatment available for bulls infected with trichomoniasis. Infected bulls should be eliminated from the herd. Prevent the disease by using a clean bull on cows that have not been exposed to an infected bull. Make sure bulls brought into the herd are free of the disease. When using artificial insemination, the spread of the disease can be prevented by using semen from clean bulls. No vaccination is available for trichomoniasis.

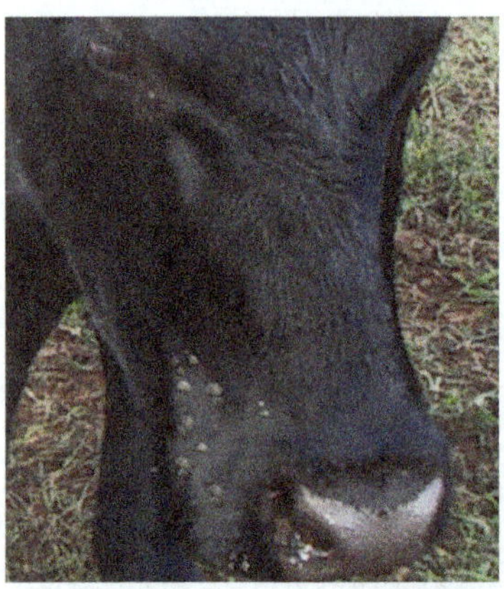

Alyssa K. Elrod

Figure 17-11 Warts are growths on the skin caused by a virus. They are highly contagious. Animals with warts are disqualified from shows. Infected animals should be separated from other animals, pens, rubbing posts, halters, and tattooing equipment should be disinfected to help prevent the spread of the wart virus.

Warts

Warts are growths on the skin caused by a virus. They are spread from one animal to another by contact, or by contact with posts, buildings, or other objects warty animals have touched.

Warts may appear singly or in clusters (Figure 17-11). Some are hard and others are soft. Calves are usually affected on the head, neck, and shoulders. On older cattle, warts often appear on the udder or teats.

Warts can be prevented by vaccination, and disinfecting pens and rubbing posts helps to prevent the spread of warts. Small warts may be clipped off with sterile scissors. Tying a sterile thread around the wart will cause it to slough off in a few days. Applications of glacial acetic acid, iodine tincture, silver nitrate, castor oil, or olive oil may be used to treat warts. Infected animals should be separated from other animals.

Wooden Tongue (Actinobacillosis)

Wooden tongue is a chronic but seldom fatal disease of cattle. It causes economic losses because affected body parts are condemned when the animal is harvested.

Symptoms include lesions on the soft tissues of the head and swelling of the lymph glands of the neck may also occur. These sometimes break open, discharging creamy pus. Lesions may develop on the tongue, which becomes hard and immobile and protrudes from the mouth. The animal will drool. As it becomes more difficult for the animal to eat, it takes in less feed and loses weight. Eventually, the animal may die.

The disease is thought to be spread through contamination of feed by infected animals. Infected animals should be isolated from other animals. Animals should not be fed stemmy hay or be allowed to graze on pastures with plants that might injure the mouth. Surgical treatment of animals with wooden tongue may be necessary. This should be done only by a veterinarian.

EXTERNAL PARASITES

External parasites of beef cattle include flies, lice, mange, mites, and ticks. Each year, there are high losses from these parasites in the United States. Some of these parasites irritate the animals and carry disease from one animal to another. Others are bloodsuckers; they slow down weight gains and, in some cases, damage the hides of the animals.

Chemical, biological, mechanical, and cultural control methods are used to reduce losses from parasites. Chemical control methods are the most economical. The use of chemical controls is regulated by state and federal laws. Insecticide registrations and recommendations for use change from year to year. The livestock producer should consult with county extension personnel, university insect control specialists, livestock supply stores, and veterinarians for recommended insecticides for local use on specific parasites.

Chemicals must be applied only to those animals for which they are intended and rates of application and withdrawal times must be observed. Carefully read and follow the directions on the labels. Keep a record of any chemical used, including the name, rate, dilution, percent of active ingredient, and dates of use.

A parasite control program includes the sanitation of pens, barns, and feedlots. Identification of insects and knowledge of their life cycle is important in the development of control programs. Manure that is thinly spread outdoors is less likely to provide a place for insect eggs and larvae to develop. Clean up wet litter and areas where seepage occurs. Good drainage in feedlots helps to control insects. Spraying barns with approved chemicals is a recommended practice for controlling insects.

The use of insecticide-impregnated ear tags has become popular in recent years for controlling some kinds of flies on beef cattle. However, not all fly species can be controlled with ear tags. There is also concern that some insects are developing resistance to the insecticides used in the ear tags. The removal of the tags at the end of the fly season appears to help reduce this problem. Some manufacturers are including more than one insecticide in the ear tags to reduce the buildup of resistant insects. The use of sprays or dusts for a quick kill of existing flies at the time of ear tag application is also suggested as a way to prevent a buildup of a resistant population.

Courtesy of Frank B. Flanders

Figure 17-12 Horn flies, a serious pest of cattle, pierce the skin to suck blood, often taking up to 20 blood meals per day.

Both internal and external parasites may be controlled by using a slow-release bolus placed in the rumen of the animal. The bolus is used primarily with cattle weighing 275 to 660 pounds that are on pasture. It is effective for up to 135 days and must be withdrawn 180 days before the animal is harvested. The company that developed the bolus claims that treated cattle gain 14 percent faster than untreated cattle.

Flies, Gnats, and Mosquitoes

Bloodsucking flies include the blackfly, horn fly, horsefly, mosquitoes and stable fly. Other flies that irritate livestock include the face fly, heel fly, housefly, and screw-worm fly.

Blackfly (Buffalo Gnats)

Blackflies are small, varying in length from $\frac{1}{25}$ to $\frac{1}{5}$ inch. Colors are orange, brown, and black. They have a humped prothorax.

Eggs are laid on an object in or near flowing water. Eggs hatch in 3 to 30 days, depending on the species. In northern areas, the winter is spent in the egg stage. Some species survive the winter as larvae. The larval stage varies from 10 days to 10 weeks, depending on the species. Larvae live underwater and spin a cocoon that attaches to underwater objects. The pupal stage is from 4 days to 5 weeks, depending on the species. Adults emerge from the pupal case, rise to the surface, and fly away.

Horn Fly

The horn fly is about one-half the size of the housefly. It is gray-black in color. Hundreds of horn flies may cluster on the face, back, horn, withers, and belly of the animal (Figure 17-12).

The horn fly lays its eggs in fresh manure. The eggs hatch in 1 to 2 days. Maggots become full grown in about 5 days. Mature maggots pupate (change from larva to pupa form) in manure or soil and emerge as flies in 5 to 7 days.

The horn fly is a problem mainly on pasture. Sanitation methods are of little value in control. Back rubbers containing recommended chemicals are a common control method.

Horsefly and Deerfly

Horseflies (Figure 17-13) and deerflies preputial are much larger than houseflies. They range from $\frac{1}{3}$ inch to $1\frac{1}{2}$ inches in length, depending on the species. They are usually gray, with preputial brown or black colors intermixed. Some have bright green heads or yellowish or reddish-brown bodies. The eyes are usually brilliantly colored.

The eggs are usually laid on vegetation growing in swamps, ponds, or other wet areas. Eggs hatch in 4 to 7 days. The larvae drop into the water or mud. They burrow into mud, moist earth, or decaying organic matter. Some species feed on organic matter, while others feed on insect larvae, snails, earthworms, or other small forms of life. Some species mature in 48 days or less, but others take almost a year to mature.

Nancy C. Hinkle

Figure 17-13 The horsefly is a pest and an irritant to cattle.

Mature larvae pupate in drier areas about 1 to 2 inches below the surface. The pupal period for most species is 2 to 3 weeks, but other species require 1 to 3 months. Adult flies then emerge and live 3 or 4 weeks. There is usually one generation per year, although some species may produce two.

Horseflies and deerflies irritate cattle, causing them to gain more slowly. They also feed heavily on blood. Twenty or 30 flies can take almost one-third of a pint of blood in 6 hours. They transmit a number of diseases, including anthrax and anaplasmosis.

Insecticides do not effectively control horseflies and deerflies although, if applied daily, they may provide some relief. Draining of swampy areas helps control these insects. Moving cattle from pastures in wooded or swampy areas when the flies are attacking them is one way to reduce the problem.

Mosquitoes

Mosquitoes are small flies with two wings. There are a number of different species and they vary in size. Eggs are laid on water or in low-lying areas that flood. The eggs of most species hatch in 2 or 3 days. Some species require a dry period before the eggs hatch. Eggs can remain dormant for many months and then hatch when flooded. Larvae live in the water, and most species must come to the surface to breathe. Larvae usually become pupae within 1 week. The pupal stage is about 2 days, after which the adults emerge.

Draining areas where mosquitoes lay their eggs is recommended. Insecticides are more effective when the insects are in the larval stage than when they are adults.

Stable Fly

The stable fly is about the same size as the housefly. It is grayish in color, with seven rounded dark spots on top of the abdomen. It stays on the animal long enough to feed and then spends the rest of the day on nearby fences, walls, or in barns.

The stable fly lays its eggs in moist straw, strawy manure, moist feed, or other decaying organic matter. The eggs hatch in 1 to 3 days. The maggots reach maturity in 11 to 30 days. The pupal stage takes 1 to 3 weeks, at which time adult flies emerge.

Stable flies are more of a problem in feedlots than on pasture. Sanitation is important in control. Insecticides do not give complete control. Spraying areas where the flies rest is also recommended.

Face Fly

The face fly is a little larger than the housefly. It is about the same color, being slightly darker than the housefly. The female feeds around the eyes and nose of cattle, but the male is usually not found on animals.

Eggs are laid in fresh cow manure. The eggs hatch in about 1 day and the larvae mature in 3 to 4 days. The mature maggots move to soil and change to the pupae stage. The adult fly emerges in 4 to 6 days. Face flies are difficult to control by sanitation. Insecticides are commonly used as a control measure.

Heel Fly (Cattle Grub)

The cattle grub is the larval stage of the heel fly. There are two species of cattle grubs in the United States. The common cattle grub is found in all states except Alaska. The northern cattle grub is found in Canada and the northern United States. The cattle grub causes greater losses in the beef industry than any other pest. The major losses come from lower rates of gain and damage to the meat and hides of cattle.

The adult heel fly is about the size of a honey bee, although in some areas it may be larger. The color is similar to that of the honey bee. The body has bands of yellow or orange hair. There are four longitudinal shiny bands on the thorax and black and orange hairs are found on the legs. The adult heel fly does not feed. It lives for only 2 or 3 days on food stored in the abdomen.

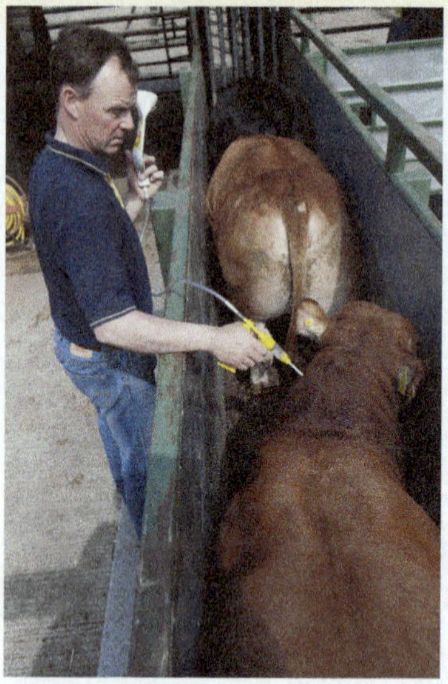

Figure 17-14 External parasites can be controlled in many ways, such as applying insecticides to the animal's skin.

The life cycles of the two species are similar. The adult fly lays eggs on the hair of cattle on the hind legs, flanks, or sides. The eggs hatch in 3 to 4 days. The larvae burrow into the skin. During the next 8 months, the larvae migrate through the body of the animal. When the larvae reach the back of the animal, they create a hole in the skin for breathing. Swellings, called *warbles*, appear on the back of the animal. The larvae, called *grubs*, stay in the back of the animal about 1 to 2 months. When they are mature they emerge through the hole and drop to the ground. The grub changes to the pupal form under trash, leaves, or other material on the ground. The pupal stage lasts from 20 to 60 days, depending on the temperature. The adult fly then emerges to mate and lay eggs. There is one generation per year.

Control of cattle grubs is accomplished most effectively with systemic insecticides. A **systemic insecticide** is one that spreads throughout the body of the animal. These insecticides may be applied as wet sprays, dips, back-line pour-ons, in mineral mixes, or as feed additives (Figure 17-14). Application is made as soon as possible after heel fly activity has stopped. Early applications are more effective than later ones. In southern states, the application time ranges from late spring to fall. In northern states, the application time ranges from early summer to late fall. Consult veterinarians, county agents, or university specialists for the exact time of application in a given area.

Housefly

The housefly is about $\frac{1}{4}$ inch in length. It is gray with a striped abdomen and brownish-yellow or red markings near the base. Eggs are laid in manure or other decaying organic matter and hatch in 8 to 20 hours. The larvae grow to maturity in 5 to 14 days. The pupal stage varies from 3 to 10 days, after which the adult emerges. Sanitation and insecticides are used to control the housefly.

Screwworm Fly

The adult screwworm is about twice the size of the common housefly, with a bluish-gray or gray body with three dark stripes down its back and has orange eyes. The larvae are pinkish in color. The adult female mates and then lays up to 400 eggs at one time in open wounds of animals. The female lives for about 31 days and can lay about 2,800 eggs during this time. The eggs hatch in 10 to 12 hours, and the larvae feed on the flesh of the living animal. They mature in 5 to 7 days, after which they drop off the animal and burrow into the soil to pupate. The pupal stage lasts 7 to 10 days, after which the adult flies emerge. The adults are ready to mate within 3 to 5 days after emergence.

Any open wound on an animal's skin may be a target for screwworm infestation. Multiple infestations may occur in the same wound. If the infestation is not treated, the animal may die in 7 to 14 days. Infested animals may go off feed, exhibit signs of discomfort, and isolate themselves from the rest of the herd. Treatment of an infestation is by topical application of an approved organophosphate insecticide. Continue treatment for 2 to 3 days and remove the larvae from the wound with tweezers.

The screwworm was eradicated in the United States by introducing sterile males into the native population. Pupae in a fly population in a production plant are exposed to gamma radiation. The adult flies are sterile. These sterile flies are released from an aircraft over areas where the screwworm exists. Over a period of time, this resulted in the complete eradication of the native screwworm fly population. Re-infestation is prevented by establishing a barrier zone of sterile flies between the screwworm-free area and the area where it has not yet been eradicated.

The screwworm remains a potential threat for livestock in the United States, particularly from animals imported from areas where it has not yet been eradicated. Any suspected infestation should be reported immediately to state or federal animal health authorities. A screwworm fly, marked during a research study, is shown in Figure 17-15.

USDA Agricultural Research Service; Photo by Peggy Greb

Figure 17-15 A screwworm fly marked so that fly dispersal, behavior, and longevity can be studied.

Lice

One species of biting lice and four species of bloodsucking lice attack beef cattle. Louse (singular of lice) eggs are laid on hairs on the animal's body. The eggs hatch in 1 to 2 weeks. Young nymphs emerge that mature to adults about 1 month after hatching.

The louse population is low during the summer and increases in the fall and winter. Cattle with lice have a rough appearance and do not gain at normal rates. Symptoms of lice include cattle rubbing against fences and feed bunks and hair balls left on fences. Bloodsucking lice weaken cattle so that they are more likely to be infected by diseases.

Lice are controlled by the use of insecticides that may be sprayed on, dusted on, used in a back rubber, or poured on. The best time for treatment is late fall or early winter.

Mites

Several species of mites attack cattle. They live on the skin or burrow into it and cause a condition known as *scab, mange,* or *itch*. All of the species that affect cattle are very small in body size. Mite populations are at their lowest in the summer and increase during the winter. Symptoms include the appearance of small pimply areas on the skin that lose hair. Infected animals rub, scratch, or lick at the affected areas. Cattle become restless and do not gain well. A heavy infestation may kill the animal. Treatments that control lice also control mites. Preventive treatment should take place in the fall or early winter.

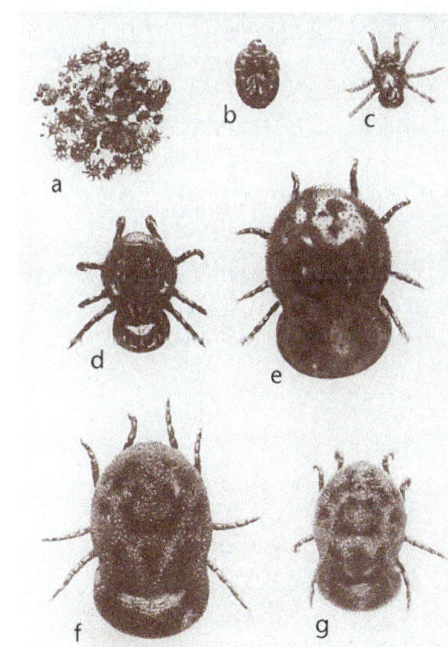

Figure 17-16 Ear tick. (a) Ticks and debris collected from ear; (b) engorged larva; (c) young nymph; (d) partially engorged nymph; (e) fully engorged nymph; (f) adult female; (g) adult male. Reprinted with permission from Price, Manning A., Hamman, Philip J., Newton, Weldon Harrison. (1969). *External parasites of cattle.* Volume 1080, Figure on page 8. Texas Agricultural Extension Service.

Ticks

Several species of ticks attack beef cattle. They are serious pests to cattle in some parts of the United States. Ticks are bloodsuckers and also transmit serious diseases among cattle. Cattle that are infested do not gain properly. The tick bites irritate the animals, causing them to rub and scratch at the affected area. This can result in a scabby skin condition or injury.

Ticks are flat, oval shaped, and dark brown or reddish in color. They are common in bushy pastures and wooded areas.

The *spinose ear tick* (Figure 17-16) attaches itself deep in the ear of the animal. This often causes severe irritation, wax buildup, and infection. Infested cattle rub their ears and shake their heads.

Ear ticks are controlled by dipping or treating each infested ear individually. Other ticks are controlled in the same way as lice.

INTERNAL PARASITES

There are a number of internal parasites that affect cattle. The most common are anaplasma, coccidia, flatworms, and roundworms.

Anaplasma

Anaplasmosis is a disease caused by protozoan parasites called anaplasma, which destroy red blood cells. It is spread by various biting insects.

Symptoms include anemia, weight loss, difficult breathing, abortion, and death. Older cattle are affected more frequently than young cattle.

Reducing insect populations helps to control the spread of anaplasmosis. Some feed additives are used to prevent infestation. Antibiotics are used in its treatment.

Coccidia

Coccidia are protozoan organisms that live in the cells of the intestinal lining. They cause irritation of the intestinal wall and bleeding. Symptoms include bloody diarrhea, weakness, and going off feed.

Prevention of *coccidiosis*, the disease caused by coccidia, is based on sanitation. Separate infected animals from the rest of the herd. Sulfa drugs and antibiotics are used to treat coccidiosis.

Flatworms

Flatworms found in cattle include tapeworms and the deer liver fluke. Diarrhea is a symptom of tapeworm infestation. Deer liver fluke symptoms include loss of weight, limping, and weakness in the hind quarters.

Tapeworms have an intermediate host. At one point in their life cycle, they are ingested by free-living soil mites. Feeding cattle on paved lots helps to break the cycle since the mites are not present under this condition.

Snails are the intermediate hosts of deer liver flukes. Pasturing cattle away from streams, ponds, swamps, or other wet areas helps to break the life cycle of the deer liver fluke.

Generally, treatment of flatworms has not been successful. Some new drugs are being tested that may provide effective treatments. Prevention of flatworms is the best control measure.

Roundworms

Roundworms found in cattle include stomach worms, nematodirus, threadworm, hookworm, cooperia, nodular worm, whipworm, and lungworm (Figure 17-17). The stomach worm is the most serious of these worms.

Symptoms of stomach worms include anemia, weakness, constipation, and diarrhea. A condition known as *bottle jaw*, which is a swelling under the jaw, sometimes develops. Worm infestations slow down gains. Symptoms of the other roundworms are similar to those of stomach worms.

Roundworms reproduce rapidly. Good sanitation helps to prevent infestation. Roundworms are treated with one of several chemicals, given as boluses (large pills), drenches, or feed additives.

Courtesy of the North Dakota Department of Education

Figure 17-17 Roundworms are major internal parasites of animals.

NUTRITIONAL HEALTH PROBLEMS

Bloat

Rapid fermentation (breakdown of carbohydrates by enzymes) in the rumen causes too much gas to be produced. The rumen swells and the animal cannot get rid of the gas. This condition is called bloat. The major cause of bloat is eating too much green legume too fast. Other feeds can also cause bloat. Some animals will bloat on dry feeds.

One of the main ways to avoid bloat is to prevent animals from overeating legumes in too short a period of time. Feeding grain, dry roughage, or silage before turning the animals onto legume pasture also helps in prevention. Free access to water should be provided at all times.

A stomach tube passed through the mouth helps the animal get rid of the gas. Other treatments include walking the animal on rough ground to make it belch, forcing the animal to drink mineral oil or poloxalene (trade name, Bloat Guard), or inserting a trocar and cannula (a sharp-pointed instrument attached to a tube) into the rumen through the side to allow the gas to escape. The use of the trocar and cannula should only be considered after other methods have failed.

Bovine Pulmonary Emphysema

Symptoms of pulmonary emphysema include panting, coughing, and difficulty in breathing. It occurs among cattle in feedlots. The cause is not known, although the disease may be due to some type of allergy. There is no known prevention or cure. Dust reduction, feeding less high-concentrate feed, and putting cattle on pasture may help relieve the condition.

Brisket Disease

Brisket disease is a heart condition of cattle that occurs at high altitudes. It may be caused by the enlargement of the heart in cattle that cannot adapt to high altitude. Diet may also be a factor.

Symptoms of brisket disease include swelling of the lower neck, brisket, and belly. Cattle are moved to lower altitudes when the symptoms are seen. Limiting water and salt intake and providing a well-balanced ration are also recommended as treatment of the symptoms. Cattle with this condition must be handled with care, or they will die of heart attacks.

Enterotoxemia (Overeating Disease)

Enterotoxemia usually affects cattle on high-concentrate rations. Symptoms include lameness, bloody diarrhea, and bloat. The animal may die in 1 to 24 hours. Vaccinating calves 2 weeks before putting them on full-feed, high-concentrate rations helps to prevent enterotoxemia. Treatment includes removing concentrates from the diet, feeding roughage, and vaccinating. The animal can gradually be put back on the high-concentrate ration after it is vaccinated.

Fescue Toxicity

There are three syndromes associated with tall fescue infected with the fungus *Acremonium coenophialum* growing within the plant (endophyte). Fescue foot, bovine fat necrosis, and fescue toxicity are all caused by the endophyte. Of these three, fescue toxicity is by far the most important (Figure 17-18).

USDA Agricultural Research Service; Photo by John Stuedemann

Figure 17-18 Two cows that suffer from fescue toxicosis.

Tall fescue is the most widely planted pasture grass in the central and eastern United States. Fescue toxicity causes more than $1 million in beef cattle losses each year from reduced calf weaning weights and reduced calving percentages. Beef cows with fescue toxicity are thin, feverish, and do not shed their winter hair coats in spring as cows normally do when it gets hot. The most serious effect is that conception rates can be reduced from a normal 95 percent down to 70 percent and even 50 percent, and growth rates of growing animals are often reduced by 50 percent.

The fungus is spread only through the seed. Removing the toxic fungus from the seed is not a solution because the fungus benefits the plant, making it more vigorous, more drought tolerant, and more competitive. Fungus-free tall fescue varieties do not cause toxicity, but they do not survive long when grazed hard, especially in summer.

Cattle toxicity can be reduced by diluting the pasture by planting clover or Bermuda grass, or even crabgrass, or feeding nontoxic hay in winter. However, these methods do not completely eliminate the problem. The best solution is to kill all the toxic fescue in a pasture, then replant in the dead sod with a variety that contains a nontoxic fungus, such as MaxQ tall fescue. MaxQ, a variety developed at the University of Georgia, results in tough plants that survive well and give excellent animal performance. Seed of this new variety has sold well as farmers have replanted pastures with MaxQ. It is one of the biggest forage research breakthroughs in recent years and is having a huge impact.

Fluorosis

Fluorosis occurs in parts of the United States where the fluorine content of the feed or water is too high. It is a poisoning effect that builds up over a period of time. Symptoms include abnormal teeth and bones, stiff joints, diarrhea, and damage to other organs of the body. The animal loses its appetite and becomes thin. Prevention and treatment involve using feeds that do not have high fluorine content.

Founder

Founder is a swelling of the tissue that attaches the hoof to the foot. It occurs among cattle in the feedlot. Founders can be caused by overeating concentrates, sudden changes in ration, drinking too much water, or standing in a stall for long periods. The animal becomes lame, shifts its weight from one foot to another, and has difficulty standing. Treatment includes taking the animal off high-concentrate rations and putting wet cold packs on the affected part. Antihistamines are sometimes used in treatment.

Grass Tetany

Grass tetany—also known as Hypomagnesemia, grass staggers, winter tetany, magnesium tetany, and wheat pasture poisoning—is found most often in cattle during the lactation period. It sometimes occurs in cattle that are not lactating. It occurs most often when cattle are grazing on grass pastures that are deficient in magnesium (Figure 17-19). Early symptoms of grass tetany include excitement, loss of coordination, and loss of appetite. The animal may show a trembling of the muscles, convulsions, and coma. Sometimes the animal cannot stand. Death may occur quickly, sometimes within 30 minutes. Animals seldom recover if not treated within 8 to 12 hours.

Prevention includes feeding magnesium in the ration in areas where the soil is deficient in this element. Including legumes in the pasture mix helps to prevent grass tetany. Treatment includes injecting a calcium and magnesium solution into the jugular vein of the animal. Cattle with this disease must be handled carefully because stress may kill the animal. A veterinarian should be called when grass tetany is suspected.

Figure 17-19 Grass tetany occurs most often during early lactation, on rapidly growing, lush spring forages with low magnesium levels and where fertility, especially nitrogen and potassium, is high.

Hardware Disease (Traumatic Gastritis)

Cattle sometimes pick up sharp metal objects, such as wire, nails, pins, and screws, with their feed. Metal objects collect in the reticulum. When they are sharp, they may puncture the wall of the reticulum, causing infection or damage to surrounding organs, such as the heart.

Symptoms of hardware disease include loss of appetite, arched back, fever, stiffness in moving, and less chewing of the cud. The animal may have pain when defecating and when lying down or getting up. The brisket is sometimes flabby and the animal may be bloated.

Hardware disease is prevented by making sure that metal objects do not accidentally become mixed with the feed. Checking for loose wire, nails, and other sharp metal objects in areas where cattle are kept also helps in prevention. A magnet may be placed in the cow's stomach to attract and hold the metal, thus preventing puncture of the wall. The disease is cured by surgically removing the metal.

Nitrate Poisoning

Too much nitrate in feed or water may cause nitrate poisoning in animals. The increase in the use of nitrogen fertilizers has resulted in more frequent occurrences of nitrate poisoning. Nitrate is converted to nitrite in the animal's body. Excess nitrite prevents oxygen from reaching the parts of the body where it is needed.

The affected animal has difficulty breathing and a faster pulse rate. The blood turns brown and the animal may froth at the mouth. Pregnant animals may abort.

Nitrate poisoning can be prevented by analyzing feeds to find the amount of nitrate they contain. If a feed has a high level of nitrate, it may be necessary to add non-nitrate sources of feed to the ration. Silo gases and drainage from silos also contain high nitrate amounts. Keep livestock away from these sources of nitrate poisoning.

Treatment includes the use of antidote tablets that can be secured from a veterinarian. In severe cases, call a veterinarian. Methylene blue is administered in extreme cases. Stop feeding suspect feeds until they have been tested for nitrate content.

Photosensitization

Photosensitization is a skin reaction that occurs in the presence of sunlight after the animal has eaten certain kinds of plants. Symptoms include scratching, rubbing, licking, and tail and head switching. Skin areas turn red and ooze a yellowish fluid. Crusting of the affected areas occurs. Animals lose their appetites and have a fever. Outbreaks occur most often in spring and summer.

There is no drug to prevent photosensitization although feeding dry roughages and rotation grazing help prevent against photosensitizations. Affected animals are moved to a shaded area and given dry feed. Paint or spray the affected body parts with methylene blue water solution. Sodium thiosulfate given orally or intravenously may be used as a treatment.

Poisonous Plants

Many kinds of plants can cause injury or poisoning when eaten by cattle (Figure 17-20). Symptoms vary with the kind of plant involved. Prevention consists of eliminating poisonous plants from pastures and rangeland and providing forage plants more desirable to cattle. Many common plants are poisonous to cattle but they are only eaten if normal grazing and supplemental feed are unavailable. Treatment is often of little value after the animal is affected.

Figure 17-20 Many plants are poisonous to cattle, such as wild black cherry, pokeweed, yew, and oak.

Urinary Calculi (Water Belly)

Minerals deposited in the urinary organs cause *urinary calculi*, or stones. Steers are affected with this condition more frequently than cows or heifers. Rations high in phosphorus are considered to be a partial cause. Symptoms include restlessness, lying down in a stretched position, and tail switching. The animal attempts to urinate but only dribbles. If the urinary tract becomes completely blocked, the bladder will burst. This causes death unless treated.

Prevention includes providing a ration with a calcium-phosphorus ratio of 2:1. Adding salt to the feed, which increases water intake, also helps in prevention. Surgery is the best treatment. Urinary tract relaxants may be given that help keep the urethra open.

Rumentitis (Liver Abscess Complex)

Cattle on high-concentrate rations are more likely to have liver abscesses. *Rumentitis* is caused by microorganisms that are in the soil and manure. Affected cattle do not gain well. No symptoms appear in cattle on feed; the first sign of the condition is usually seen when the animal is harvested and the abscessed liver appears. Infected livers are condemned at harvest. Feeding antibiotics may help prevent liver abscesses.

White Muscle (Selenium Deficiency)

Cattle fed in areas where there is a deficiency of the trace element selenium in the soil may be affected by *white muscle disease*. Muscle damage results from a shortage of selenium in the diet. The animal may have trouble walking, breathing, or may die of heart failure. Calves may be born dead or weak. Treatment and prevention consists of giving the animals selenium by injection or orally. Too much selenium in the ration can also be harmful to the animal.

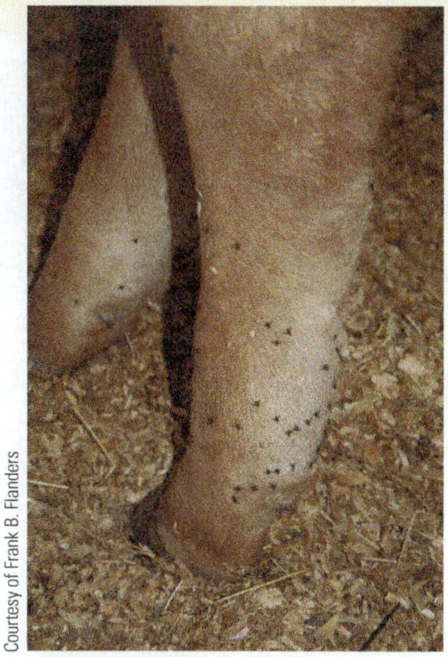

Courtesy of Frank B. Flanders

Figure 17-21 Parasites may pass diseases from a sick animal to a healthy one.

SUMMARY

Profits are reduced when diseases and parasites affect beef cattle. Good management and sanitation programs help to prevent health problems. Planning prevention programs with the help of a veterinarian is recommended.

Many diseases are prevented by vaccination. Antibiotics and sulfa drugs are used to treat some diseases. Buying animals from disease-free herds and isolating new animals help in control programs. Insects spread many diseases, and their control is important in preventing the spread of disease (Figure 17-21). The symptoms of many diseases are similar. Laboratory tests and the aid of a veterinarian are often needed to identify the disease that is present.

Flies, lice, mites, and ticks are the most common external parasites of cattle. Insecticides are used to control most of these parasites. Label directions must be followed carefully for safe use of insecticides.

Roundworms, flatworms, coccidia, and anaplasma are the most common internal parasites of beef cattle. The stomach worm is the most serious of the internal parasites. Sanitation is the most effective control for internal parasites.

A number of nutritional health problems affect beef cattle. Good management of feeding programs helps to prevent some of these problems. Metal and other foreign objects must be kept out of the feed and feedlot. Trace element deficiencies cause some health problems. The ration should be checked to be sure the necessary trace elements are included. Cattle must be kept away from poisonous plants.

Student Learning Activities

1. Prepare a bulletin board display showing pictures of unsanitary livestock equipment and facilities and contrasting pictures of good sanitary conditions.
2. Prepare and present an oral report on any phase of beef herd health problems.
3. Prepare a bulletin board display showing pictures of animals with diseases or parasites.
4. Practice herd health measures, such as giving injections.
5. Take field trips to observe herd health problems in the local area.
6. Ask a veterinarian to talk to the class about preventive measures for cattle health problems.
7. When planning and conducting a beef cattle supervised experience program, practice good herd health measures as described in this chapter.

Discussion Questions

1. Why is it important for the cattle producer or feeder to follow practices that help to maintain healthy cattle?
2. Prepare a table that briefly summarizes the symptoms, prevention, and treatment of the common diseases that affect beef cattle.
3. List and briefly describe each of the common external parasites that affect beef cattle.
4. Describe the major steps for control of parasites.

5. List and briefly describe the common internal parasites of beef cattle.
6. List and briefly describe the common nutritional health problems of beef cattle.

Review Questions

True/False

1. It is better to prevent health problems than to try to cure them once they occur.
2. The carcass of an animal that has died from anthrax should be sold for tankage.

Multiple Choice

3. Anthrax is a disease caused by bacteria that may remain in the soil for ______ years or longer.
   a. 20 c. 40
   b. 30 d. 50
4. Brisket disease affects the ______ in cattle.
   a. brain c. stomach
   b. lungs d. heart
5. ______ eggs laid in manure or organic matter hatch in 8 to 20 hours.
   a. Mosquito c. Tick
   b. Housefly d. Horsefly

Completion

6. Pinkeye in animals is carried by ______.
7. ______ is a disease caused by microorganisms in cattle that is also dangerous to humans causing undulant fever.
8. An organism that affects the genital tract of the bull and is transmitted to the cow during breeding is called ______ ______.

Short Answer

9. What are the most common external parasites of cattle?
10. What is the result of a deficiency of selenium in the ration of cattle?
11. What are the symptoms of hardware disease?
12. What is fluorosis?

Chapter 18

Beef Cattle Housing and Equipment

Key Terms

headgate
tilting table
dipping vat
cattle guard

Objectives

After studying this chapter, the student should be able to

- describe the steps in planning for facilities and equipment for beef enterprises.
- describe the facilities and equipment required for beef enterprises.

PLANNING FOR FACILITIES AND EQUIPMENT

Careful planning of facilities and equipment for beef cattle enterprises is important for the success of the operation. There are many different kinds of beef cattle operations. The kind of facilities and equipment vary with each individual farm.

Careful planning before building can make cattle handling easier. Facilities must be planned to be as safe as possible for the operator and the cattle. Wise planning also helps to save labor. Factors to consider when planning include the following:

- number of cattle in the enterprise
- space requirements per head
- kind of facilities
- location of the facilities
- environmental requirements
- feed storage and handling methods
- amount of land needed
- amount of money and labor that is available
- opportunity for expansion of the enterprise
- coordination of new facilities with existing facilities

Number of Cattle

Each operator must decide on the number of cattle to be handled in a given operation. This decision is based on several factors, including the availability of feed, labor, and housing and equipment. The major factor in determining the number of cattle that can be economically produced in breeding herds is the availability of grazing land.

Space Requirements

Facilities are planned according to the number of cattle that are to be handled in the enterprise. It is wise to plan for expansion of the facilities in the future. Build what is needed for the current operation, but allow room for enlarging the facilities if the initial number of cattle increases.

Kind of Facilities

The kind of facilities required depends on the kind of beef enterprise on the farm. Cow-calf operations can be run with limited facilities. The facilities for feedlot operations range from simple to complex. Facilities can be classified as one of the following types:

- confinement
- open feedlot
- open barn and feedlot
- feeding barn and lot

Location of Facilities/Environmental Factors

One of the most important decisions to be made is the location of the facilities. Easy access to good roads must be available, and it is important to plan for easy movement of cattle to and from pastures. Dust and odors from the facility may be objectionable to neighbors; therefore, the direction of prevailing winds must be considered. Runoff must be controlled. The facilities should not be located too close to streams, lakes, or ditches. A well-drained site and good drainage around feeding areas are necessary. The soil of the chosen location must be able to support large structures, such as silos.

Feed Storage and Handling

Feed storage and handling methods vary from simple, manual systems to complex, automated systems. The system selected must allow for future expansion. Plan so that large equipment will have easy access. An adequate power supply for the operation of feeding equipment must be available.

Amount of Land

The amount of land required varies with the size of the operation. For example, a feedlot for 500 head of cattle requires about five acres for the facilities. A 5,000 head facility requires about 35 acres of land. There must be enough land to provide space for lots, alleys, storage, and roadways. Some operators desire enough space for inclusion of an office and feed mill in the facilities. Always allow room for future expansion.

Money, Labor, and Opportunity for Expansion

The amount of money an operator has available affects the size and kind of facility that can be built. Certain types of facilities are more expensive than others. For example, warm confinement barns cost more than other kinds of facilities. The amount of automation and labor are closely related. Systems with little automation require more labor. If there is a limited labor supply, more highly automated systems should be considered. Careful planning and layout of the facilities permits future expansion, if desired. New construction should be coordinated with existing facilities.

COW HERD FACILITIES

Generally, beef cow-calf operations require the simplest facilities. Cows can calve on pasture in both spring and fall. Beef cows can be wintered outdoors with a minimum of shelter. Cows can be moved to a pasture or lot close to the farmstead just before calving. Open-front calving barns may be necessary in colder parts of the United States. Portable calf shelters can be used for calves on pasture. Separate feedlot areas should be available for different kinds of cattle, such as mature cows, heifers, bulls, and calves. A corral is needed for any kind of beef cattle operation. A holding pen, working chute, and headgate are essential features of a corral. A good water supply is also necessary.

FEEDLOTS

Confinement Barns

Cold confinement barns are open on one side, usually on the side away from the prevailing winds in the area. Table 18-1 shows general space requirements for cold confinement buildings. The inside temperature is about the same as the outside temperature. Pole-type buildings, ones in which the framework is attached to poles that are set in the ground, can be used. Many producers prefer open-span construction because it is easier to clean with mechanical equipment. Open-span construction has no poles or supports in the interior of the building.

Cold confinement barns provide certain advantages over open feedlots or other feeding facilities. First, they require less labor and land than other kinds of feeding facilities. It has also been found that cattle gain better when they have some protection from the weather. Cattle stay cleaner and yield about 1 to 2 percent more when slaughtered. The costs of building cold confinement barns are competitive with open lots when the size of the operation is greater than 300 head of cattle. It is usually easier to meet environmental regulations for confinement systems than for other kinds of feeding facilities. Manure handling costs are lower. If slotted floors are used, no bedding is needed. There are fewer problems with flies. It is easier to observe the cattle in a confinement system, especially in bad weather.

Warm confinement barns are closed buildings that are insulated and kept warmer than outside winter temperatures. They are the most expensive type of cattle-feeding facility to build. Research shows that there is no increase in rate of gain or feed efficiency when cattle are fed in warm confinement barns. This type of facility is, therefore, not recommended for cattle-feeding systems.

TABLE 18-1 General Space Requirements for Cold Confinement Buildings

Cold Confinement Buildings	Square Feet per Head
Solid floor—bedded	30
Solid floor—flushing	17–18
Totally or partially slotted	17–18
Calving pen	100
Calving space	1 pen/12 cows

Three types of floor systems are common in confinement barns: solid bedded, slotted, and solid flushing. Solid bedded floors may be either concrete or dirt. They are cheaper than other floor types but require more bedding and labor. Slotted floors may be either all slotted or partly slotted. Slats are 6 to 12 inches wide, and the slots between the slats are $1\frac{1}{4}$ to $1\frac{3}{4}$ inches. Partly slotted floors should be at least 40 percent slotted. Animals stay cleaner on slotted floors because manure drops through the slots and is stored in pits under the floor.

Solid flushing floors are usually made of concrete and are designed to slope toward gutters. Manure is flushed with water into the gutters, which empty into a lagoon. Animals do not stay as clean with this type of floor as they do with slotted floors. Larger animals tend to slip on this type of floor; therefore, the concrete should have a broom finish to reduce slipping. In addition, more moisture condensation occurs in cold weather with this type of floor than with other types.

Confinement barns have wall heights of 12 to 16 feet. High walls keep the building cooler in summer and reduce the amount of moisture condensation in the winter. Feed bunks and waterers are placed on opposite sides of the barn. This causes the animals to move around more, thus improving the movement of manure to lower areas on sloping floors or through the slots on slotted floors. Confinement barns must have handling facilities and sick pens, which are small pens for isolating sick animals.

Open Feedlots

An open feedlot has no buildings. General space requirements for feedlots are shown in Table 18-2. Protection for cattle is limited to a windbreak fence and sunshades. Several types of windbreak fences may be used to protect cattle from wind and reduce the amount of snow in the feedlot. The fence may be from 4 to 12 feet high and may be solid or semisolid. A solid fence piles the snow higher on the downwind side and is better for wind protection. A semisolid fence lets the snow spread out further on the downwind side and gives better snow protection. Local wind speeds and snow amounts need to be considered when determining how far to place windbreak fences from the feedlot. Fifty to 60 feet is about right for wind speeds of up to 40 miles per hour.

Exterior plywood, 1-inch boards, or 28-gauge corrugated metal may be used to construct the fence. A semisolid fence is usually constructed of 1-inch boards placed vertically and spaced 2 to 2.5 inches apart. Tree windbreaks may be used, but they allow the snow to spread further downwind. Therefore, they must be placed from 100 to 300 feet from the protected area. Open feedlots are usually not paved. Dirt

TABLE 18-2 General Space Requirements for Feedlots

Feedlot Space	Square Feet per Head
Unsurfaced (slope = 4–6%)	300–800
Partially surfaced	150
Surfaced with shelter	20 (lot); 30 (barn)
Surfaced no shelter	40–75
In sheds or under shade (by weight):	
400–800 lb	15–20
800–1200 lb	20–25
Over 1200 lb	25–30

TABLE 18-3 General Space Requirements for Dirt Mounds

Mounds (dirt in feedlot)	Square Feet
Per head (min) [Double if windbreak on top of mound—one-half each side of windbreak]	25 sq. ft.
Slope ratio	4:1

mounds are used to keep cattle out of the mud. General space requirements for dirt mounds are shown in Table 18-3. Concrete strips are sometimes installed along feed bunks and waterers. Open feedlots usually require more land area than other kinds of facilities. Good drainage and runoff control are important. Facilities for handling cattle and sick pens are included.

Open Barn and Feedlot

Protection for cattle housed in this kind of facility is provided by the use of an open-front barn. Feeding is done in an open lot, and mechanical or fence-line bunks are common. This kind of facility is common in the Midwest, where the climate is such that protection of the cattle gives better feed efficiency and rate of gain. This kind of facility is well adapted to smaller feedlots.

The floor of the barn and the lot are usually unpaved, although some pavement is often installed around the feed bunks and waterers. Some operators pave a strip along the open side of the barn extending inward about 4 to 6 feet to help control mud. Good drainage and runoff control are needed. Cattle-handling facilities are a part of this kind of facility.

Feeding Barn and Lot

The only major difference between this system and the open barn and feedlot system is that the feed bunks are located inside the barn. Placing the bunks inside has some advantages. The major advantage is that the feed and cattle are protected from the weather. Less feed is lost from the bunks by being windblown or damaged by the weather.

CORRALS

Cattle-handling facilities are an essential part of all beef operations. A corral system provides the following advantages:

- makes it easier to handle cattle
- reduces the amount of labor needed to handle cattle
- saves time when handling cattle
- reduces stress on the cattle when they are handled
- reduces injury and weight loss when handling cattle
- makes cattle handling safer for the workers
- makes it easier to treat diseased or injured animals
- makes fly and parasite control easier

Several pens are necessary in the corral for holding and working cattle. Water must be available in the holding pen and feed bunks may be needed if the cattle are to be held for any length of time. Access must be provided to the sorting and crowding pens and to the working chute.

Sorting pens are smaller than holding pens. A corral should have at least two sorting pens. The crowding pen is designed to narrow down to the working chute, thus moving the cattle along. Circular crowding pens make it easier to handle

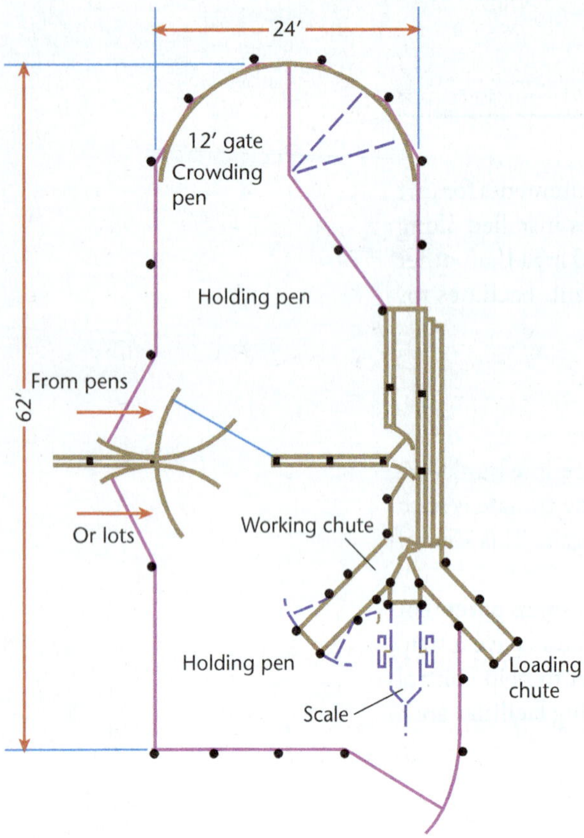

Figure 18-1 Plan for a corral for 300 to 1,000 head of cattle.

Sources: Reprinted with permission from *Beef Housing and Equipment Handbook*, 4th ed. (1987). Midwest Plan Service. Publication MWPS-6.

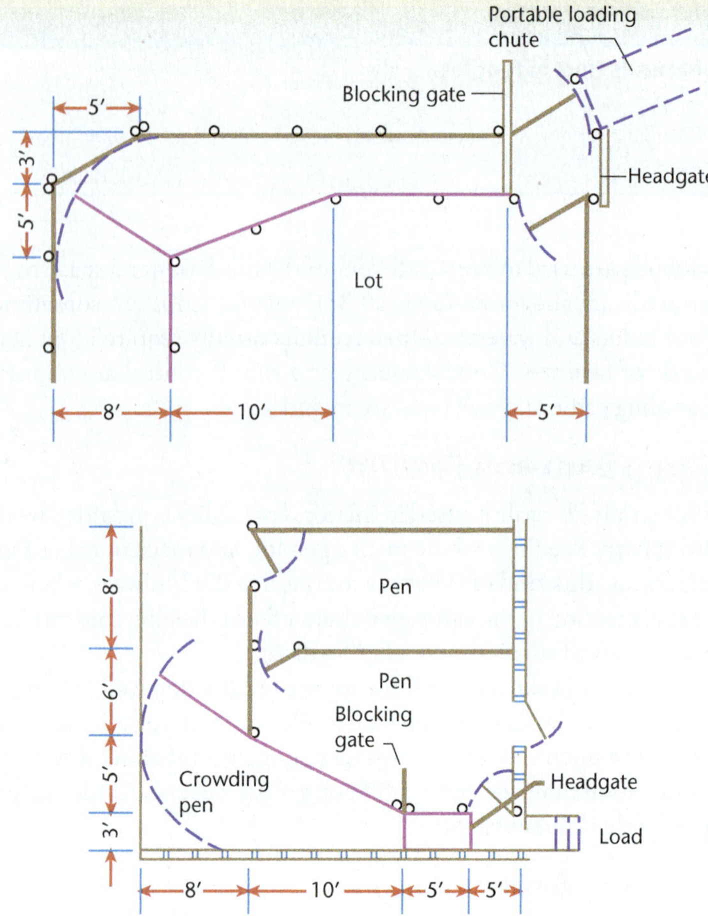

Figure 18-2 Two plans for small herds up to about 75 head.

Sources: Reprinted with permission from *Beef Housing and Equipment Handbook*, 4th ed. (1987). Midwest Plan Service. Publication MWPS-6.

cattle because it is easier to move cattle when they cannot see what lies ahead. Figures 18-1 and 18-2 show plans for two sizes of corrals.

The length of the working chute depends on the number of cattle to be worked. A minimum requirement is that the chute be large enough to hold at least three animals at one time. Some working chutes have slanted sides and are smaller at the bottom than at the top. The chute must be narrow enough so that the cattle cannot turn around. Both straight-line and curved working chutes are used. A walkway should be built along the side of the chute to make it easier and safer to work on the cattle. Many operations—such as parasite control, tattooing, branding, and veterinary treatments—are performed in the working chute. Larger cattle operators use working alleys to connect holding pens, the working chute, and feedlot. Figure 18-3 shows the entrance to a curved alley in a corral. General space requirements for a working chute are shown in Table 18-4. Table 18-5 shows general requirements for fence spacing outside of the working chute.

Courtesy of Cabe Brothers Farms. Photo by Sarah Kane

Figure 18-3 Half-circle chute. The entrance into a curved corral.

Squeeze chutes and **headgates** are used to hold the cattle while certain treatments are performed. Tilting tables are also used to hold calves for treatment. A **tilting table** (Figure 18-4) is a device for restraining the animal in a horizontal position while it is being treated. It may be a permanent part of a corral or portable for use in remote locations. The animal is restrained by a headgate or a squeeze gate.

TABLE 18-4 General Space Requirements for a Working Chute

Working Chute Space	Under 600 lb per Head	600–1200 lb per Head	Over 1200 lb per Head
Vertical side—width	18 in.	22 in.	26 in.
Sloping side—width			
bottom	15 in.	15 in.	16 in.
top	20 in.	24 in.	26 in.
Length (min)	20 ft		
Fence height (max)	5 ft		
Post depth (min)	3 ft		
Post spacing (max)	6 ft		

TABLE 18-5 General Space Requirements for Fence Spacing Outside of the Working Chute

Fence Space	Feet
Height (min)	5
Post depth (min)	2.5
Post spacing (max)	8

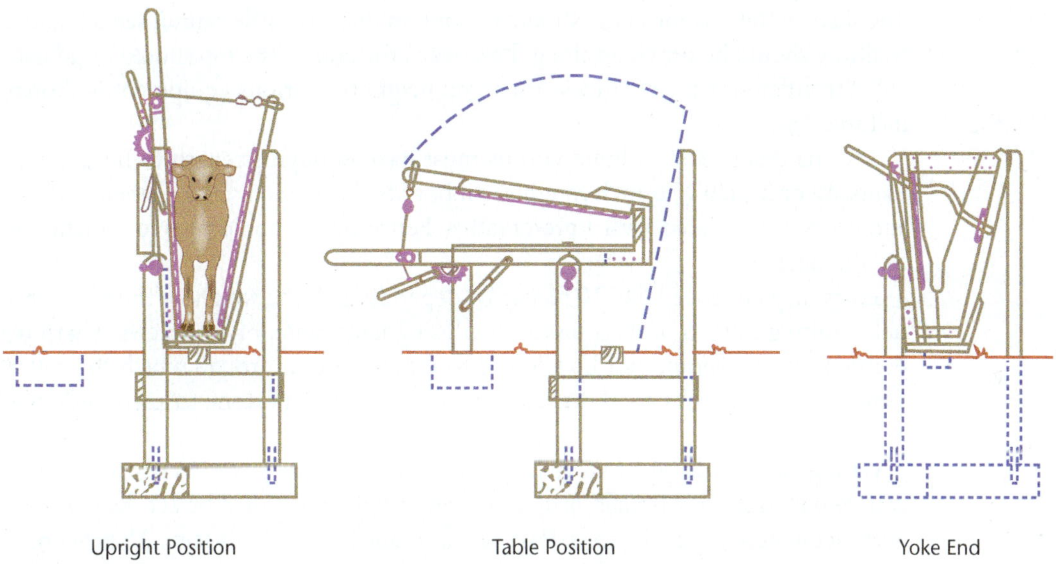

Figure 18-4 Construction and operation of a permanent, wood-frame tilting table for handling calves.

The entire device, with the animal restrained, is then rotated into a horizontal position. After the work on the animal is completed, the device is rotated back into a vertical position and the animal is released. The use of a headgate is shown in Figure 18-5. Some operators include dipping vats and scales in the corral.

A **dipping vat** is used to treat livestock for pest control. A metal dipping vat may be purchased or one may be built using concrete. An entrance chute guides the cattle into the vat, which is filled with the treatment solution. A typical vat for cattle will hold about 3,000 gallons of treatment solution. It is deep enough so that the animal is completely immersed when it is in the vat. The floor of the entrance end of the vat is a smooth slope that forces the animal into the treatment solution. A typical vat for cattle is about 28 to 32 feet in length. The exit end has steps so that the cattle can climb out of the vat into a drip pen. The drip pen provides a place for collecting the treatment solution that drips off the cattle as they come out of the vat. A cover should be placed over the vat when it is not in use to keep children out and prevent dilution or contamination of the treatment solution from rain, snow, dust, weeds, or small animals. Large-volume operations are more likely than smaller operations to use dipping vats.

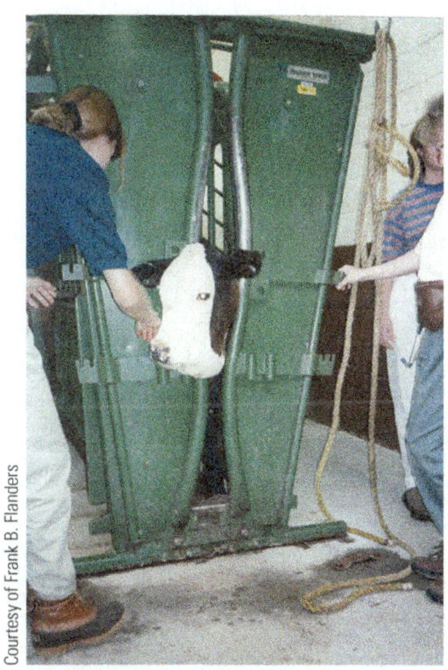

Figure 18-5 A headgate is used to hold cattle for treatment.

TABLE 18-6 General Space Requirements for Loading Chutes

Loading Chute Space	
Width	30–42 in.
Length (min)	12 ft
Slope (max)	
Sloped ramp	3.5 in./ft
Stepped ramp	4-in. riser/ 18-in. tread

TABLE 18-7 Ramp Height for Various Equipment

Ramp Height	Inches
Gooseneck trailer	15
Pickup truck	28
Stock truck	40
Tractor-trailer	48
Double-deck	100

A loading chute should be included in the corral design. Both stepped and sloping ramps are used in loading chutes. Table 18-6 shows general space requirements for loading chutes. Stepped ramps are safer and easier to use than sloping ramps. The sides of the loading chute should be solid so that the cattle cannot see out and a walkway should be provided along the sides of the chute. The top should be adjustable for different sizes of trucks. The ramp height for various equipment is shown in Table 18-7.

The materials used to build corrals must have enough strength to hold cattle. Planking or 2-inch × 6-inch rough-sawn boards should be used for fencing. Posts and boards are treated with a preservative. Fence posts should be 5 to 6 inches in top diameter.

Posts may be spaced 6 to 10 feet apart depending on the weight of the posts and rails. Bolting rails to posts is better than using nails, wire, or lag screws. Sharp or pointed objects should not be allowed to jut into areas through which the cattle move. If the area lacks good drainage, use concrete or limestone to keep cattle out of the mud.

Gate posts should be 6 to 8 inches in diameter. Set gate posts in crushed rock or concrete so that they remain firm. Posts set in light soils must be set deeper than posts set in heavy, gravelly soils. Post depths range from 2½ to 4 feet . All gate posts should be set at least 4 feet deep.

Gates must be heavy enough to restrain cattle. Metal gates last longer than wooden gates. Construct gates so they swing freely, do not sag, and can be opened easily. No more than 12 inches of clearance should be allowed under gates and gates are usually 10 to 12 feet long. Gates and fences should be the same height.

The corral should be in a convenient location. Good access to roads is important. The area where the corral is located should be large enough for trucks to move around it. Choose a well-drained area for the corral and take advantage of any windbreaks that are available. The corral should be located in such a way that winds do not carry odors to nearby houses. General requirements for corral space are shown in Table 18-8.

TABLE 18-8 General Requirements for Corral Space

Corral Space (square feet)	Under 600 lb per Head	600–1200 lb per Head	Over 1200 lb per Head
Holding pens	10–14	15–17	20
Sorting pens	8	12	15
Crowding pens	6	10	12

FEEDING FACILITIES

Feed Storage and Processing

Several kinds of silos are used to store feed. Upright silos include the gas-tight and conventional types. Upright silos (Figure 18-6) cost more but result in lower feed losses than horizontal silos. Gas-tight silos are the most expensive to build but do the best job of keeping feed and can be used for many different kinds of feed.

Horizontal silos include the trench, bunker, and stack types. Bunker silos are built above ground, while trench silos are built below ground level. Stack silos are used for temporary feed storage on the surface of the ground. Horizontal silos require good management to keep feed losses as low as possible.

Horizontal silos are usually the least expensive and best suited to the needs of small feeding operations. Upright silos work well with intermediate-size feeders, while large horizontal silos are the most economical for large feeding operations. The kind of ration used also affects silo choice.

Sealed silos can be used for grain or high-moisture corn. Metal bins can also be used for grain and protein storage. Feed grinding and mixing equipment may be located in a feeding center. Portable grinding and mixing equipment is used on some farms.

Hay may be stored as rectangular bales, small round bales, large round bales, or stacks. Large bales and stacks are handled with special equipment. Ground-level storage requires less labor than overhead storage. Well-drained areas are necessary for hay storage. Bales may be stored under some type of shelter to protect the hay from the weather. Large bales stored in the open should be separated by enough space so that the water runs off instead of soaking into the bales. Bales and stacks can be left scattered in the field for feeding. Hay may be self-fed on the ground or in feeders. If fed on the ground, temporary fences are used to reduce waste. A hay self-feeder is shown in Figure 18-7.

Augers, elevators, and other conveying equipment are used for easy handling of grain and protein. Figure 18-8 shows a feed center for a cattle-feeding facility.

Concentrates and roughages may be processed for feeding using a variety of equipment. Portable mixer-grinders can be used for processing both grain and roughage. Labor and operating costs are relatively high for portable mixer-grinders.

Courtesy of Harvestore–Engineered Storage Products Co.

Figure 18-6 Conventional silos used for storing grain and other feeds.

Courtesy of Cabe Brothers Farm. Photo by Sarah Kane

Figure 18-7 Portable feeder on wheels

Courtesy of Frank B. Flanders

Figure 18-8 Metal storage bins are used to store grain and protein.

Electric blender-grinder equipment is lower in operating cost and labor requirement. However, the capacity of this equipment is also relatively low. Concentrates can be measured, ground, and mixed on a continuous basis, but the system does not handle roughage. This type of equipment works well with completely automated systems of feeding concentrates.

Electric batch mill equipment provides accurate control over ration formulation. This type of equipment is relatively expensive and requires careful planning of the operation to ensure an orderly flow of feed through the system. A blender must be added to the system if more than one ingredient is used in the ration.

Hay mills and tub grinders may be used to process roughages for inclusion in the ration when automated feeding systems are used. This type of equipment is relatively expensive.

A well-planned feed center can handle receiving, drying, storage, unloading, elevating, conveying, and processing of feed. Some feed centers are designed to handle the feed in batches. Overhead bins, stationary grinding mills and mixers, and conveyors are included in this type of feed center. In continuous-flow feed centers, the ingredients are stored in overhead bins and fed to an automatic electric mill for processing. The feed is then blended and transported through a conveyor system for feeding.

Feed centers may be located in an area away from the feedlot. Feed is then loaded into a wagon or truck for delivery to the feedlot. If the center is not too far from the feedlot, augers may be used to deliver the feed to the bunks for feeding.

Feed Bunks

The kinds of feed bunks used for cattle feeding include:

- portable lot bunks
- mechanical bunks
- fence-line bunks
- covered bunks

Portable lot bunks are usually made of wood. They are less expensive than some of the other kinds of bunks. Lot bunks range in width from 3 to 5 feet and are usually about 14 feet in length. Both flat and V-shaped bottoms are used. Lot bunks may be moved with a forklift or a manure loader. General space requirements for feed bunks are shown in Tables 18-9 and 18-10.

TABLE 18-9 General Requirements for Feed Bunks (Lineal Space per Head)

Feed Bunks	Lineal Space per Head (inches)
All animals eat at once	
400–800 lb; one feeding/day	18–22
400–800 lb; two feedings/day	9–11
800–1200 lb; one feeding/day	22–26
800–1200 lb; two feedings/day	11–13
Over 1200 lb; one feeding/day	26–30
Over 1200 lb; two feedings/day	12–15
Feed available at all times:	
Hay or silage	9–14
Grain or supplement	3–6
Grain or silage	6
Creep or supplement	1 space/5 calves

TABLE 18-10 General Requirements for Feed Bunks (Feed Bunk Dimensions)

Feed Bunk Dimensions	Inches
Throat height	
400–500 lb	18
800–1200 lb	20
Mature cows/bulls	24
Width	
Fed from one side	18
Fed from both sides	48–60
	Feet
Concrete apron along bunks (0.75–1 in./ft slope away from bunk)	10–12

USDA Agricultural Research Service; Photo by Brian Prechtel

Figure 18-9 Fence-line feed bunks are filled from trucks or tractor-pulled equipment.

Mechanical bunks require a large capital investment; however, they save labor by automating cattle feeding. Two types of mechanical bunks are auger and chain-slat conveyors. The newer types are designed to reduce the separation of fine and coarse feed particles. Thus, all cattle get the same ration regardless of where they eat at the bunk. There are a number of different styles of each kind of mechanical bunk. Auger-type feed conveyors, when compared to chain-slat conveyors, do a better job of distributing the feed evenly when supplement is added into the complete ration. The chain-slat conveyor tends to leave the supplement behind. When grain-forage rations are fed, the chain-slat conveyor does a better job of distributing the feed evenly than does the auger type. All mechanical bunks require wiring, electric motors, and electrical controls. Mechanical bunks are wider than other types because the cattle eat from both sides.

Fence-line feed bunks (Figure 18-9) are most commonly used in feedlots in the western United States and are becoming more common in other areas. They are well adapted to large feedlots. The feed bunks are constructed along outside fences in the feedlot. They are filled from the outside by self-unloading wagons or trucks and the cattle eat from the other side of the bunk.

Fence-line feed bunks are made of wood, concrete, or a combination of the two materials. A concrete apron is usually built to keep the cattle out of the mud. The feed bunk is usually wider at the top than at the bottom. Typical widths are 30 inches at the top and 18 inches at the bottom. Feeding side height is 18 to 22 inches, and the outside height is 30 inches. Precast, tilt-up concrete fence-line feed bunks are commercially available.

Some feeders are designed with a roof to protect outdoor bunks from the weather. Covered bunks also protect mechanical feeding equipment and provide some shade for cattle. In northern areas, covered bunks should be placed so that the sun can reach the bunk for thawing.

OTHER EQUIPMENT

Cattle gain better when they have a good supply of fresh, clean water. Water should be at 68°F for maximum consumption by cattle. Automatic drinking fountains or waterers may be used. Concrete water troughs are also used. A low volume of water pumped continuously to the water trough supplies fresh water. The overflow must be to an area away from the trough. Water must be prevented from freezing

TABLE 18-11 General Water Needs of Cattle

Water Needs (per head/day)	Hot Weather (gallons)	Cold Weather (gallons)
Feeders	15–22	8–11
Cows	18–25	9–13

in the wintertime. Frost-free hydrants can be used to supply water to a trough. Table 18-11 shows the general water needs of cattle.

Commercial mineral feeders are available. Mineral feeders may be built out of a metal drum. A weather vane is used on swivel-type feeders to keep out the rain. The weather vane is on top of the feeder so that when the wind blows, the vane causes the swivel-mounted feeder to turn so that its opening is facing away from the wind. This protects the contents from rain that would otherwise blow into the opening. Stationary mineral feeders can be built of wood.

In areas of high temperatures, sunshades increase cattle gains. Cattle will use sunshades in any area, but the added cost of the shades may make them impractical in areas of lower temperatures. Permanent sunshades are built on posts set in the ground. Solid roofs of a material such as corrugated metal may be used. However, the area under a solid sunshade will not dry out as well and may create a fly problem. Snow fence or slatted fence makes a practical sunshade and can lower temperatures about 10 degrees.

Courtesy of Victoria Churchill

Figure 18-10 A cattle guard saves time in opening and closing gates.

Back rubbers provide treatment for parasite control and should be installed in places where cattle frequently gather, such as creep feeders and water troughs. Creep feeders, self-feeders, calf shelters, cattle-guard gates, and a cattle treatment stall are other kinds of equipment and facilities that the cattle raiser or feeder may want to include in the facilities. Figure 18-10 shows a cattle-guard gate in a pasture. A **cattle guard** allows equipment to be driven into an area without the operator having to stop and open a gate. However, cattle guards are not as popular as they once were. Cattle guards are more expensive than traditional gates. If vandalism and trespassing are problems in the area, cattle-guard gates may not be a good idea because they allow easy vehicle access to the property. Most cattle are afraid to cross a cattle guard, but some *Bos indicus* cattle and *Bos indicus* crosses (Indian breeds) have been reported to have less fear of the cattle guards than the *Bos taurus* (English breeds) and tend to jump the guards.

SUMMARY

Planning before building increases the value of cattle-raising facilities. Well-planned facilities save labor and make the operation more efficient. When planning a facility always plan for possible expansion in the future. Cow-calf herds have the simplest facility requirements. Feedlot operations range widely in their size, complexity, and use of automation.

Warm confinement barns for cattle feeding are not very practical. There is not enough additional gain in this type of housing to justify the added cost. Cold confinement barns provide adequate protection from the weather. Many newer facilities are being built using slotted floors to simplify manure handling. Open feedlots require some kind of windbreak protection for cattle. Dirt mounds are often used to keep cattle out of the mud.

Corrals are an essential part of all kinds of cattle operations. Corrals make it easier and safer to handle cattle. They provide a place to sort and work with the cattle when treatments are needed. A corral should include holding pens, sorting pens, alleyways, a working chute, squeeze chute, headgate, and loading chute. For large operations, a scale is useful. Corrals must be built with materials heavy enough to restrain cattle.

Feed is stored in silos and feed bins. Bunks of various kinds are used for feeding. Portable bunks are the least expensive. Mechanical feed bunks save labor but require greater initial investment.

Waterers are one of the most important kinds of equipment in the feedlot. Mineral feeders, back rubbers, sunshades, creep feeders, and self-feeders are other kinds of equipment often found in cattle operations.

Student Learning Activities

1. Take field trips to various beef-feeding or beef-herd facilities in the area. Observe and report on the kinds of facilities used.
2. Prepare visuals on housing facilities and/or feed-handling facilities and present an oral report to the class.
3. Prepare a bulletin board display of pictures showing beef facilities and feed-handling equipment.
4. If planning and conducting a beef cattle supervised experience program, plan and build appropriate facilities for the project.

Discussion Questions

1. List 10 factors that should be considered when planning facilities for beef cattle.
2. What are the four kinds of facilities commonly used for beef cattle?
3. What factors must be considered when choosing the location of beef facilities?
4. What kind of facilities might be needed for a cow-calf herd operation?
5. Name and briefly describe the two kinds of confinement facilities commonly used for beef operations.
6. Name and briefly describe the three kinds of floors found in beef confinement barns.
7. Describe an open feedlot for beef cattle.
8. List eight reasons for having a cattle corral in a beef operation.
9. Briefly describe the important parts of a cattle corral.
10. Describe the kinds of silos that are used for feed storage in a beef cattle operation.
11. Describe the methods of handling hay for a beef operation.
12. List and briefly describe four kinds of feed bunks used in beef operations.
13. Describe ways in which water can be provided for a beef operation.
14. Under what conditions should sunshades be provided for beef cattle?

Review Questions

True/False

1. In planning a beef housing facility, future expansion is not important.
2. Warm confinement barns for cattle feeding are economical and very practical.
3. A typical vat for cattle is about 15 to 20 feet long.
4. Animals stay cleaner on slotted floors because manure drops through the slots and into the pits under the floor.

Multiple Choice

5. A 5,000-herd facility requires about ______ acres of land.

a. 10	c. 35
b. 25	d. 50

6. What is one of the most important decisions to be made about cattle facilities?

   a. loss
   b. profit
   c. weather
   d. location

7. In confinement barns, cattle stay cleaner and yield ______ percent more when slaughtered.

   a. 1–2
   b. 2–3
   c. 3–4
   d. 4–5

Completion

8. In beef cattle operations, careful ______ is important for success.

Short Answers

9. What is an open feedlot?
10. What is a warm confinement barn?
11. List three types of floor systems that are common in confinement barns.

Chapter 19

Marketing Beef Cattle

Key Terms

supply
demand
stockyard
commission
yardage
auction market
fillback period
calf
cattle
slaughter calf
bullock
heifer
cow
bull
quality grade
hot carcass weight
dark cutting beef

Objectives

After studying this chapter, the student should be able to

- describe the supply and demand cycles for beef, and explain how they affect the market price.
- identify and describe the market classes and grades of beef cattle.
- describe the various methods of marketing beef cattle.
- describe how beef cattle should be handled to prevent losses when marketing.
- explain the use of the futures market with beef cattle.

BEEF CHECKOFF AND PROMOTION

The Beef Promotion and Research Act of 1985 established a $1 per head checkoff for every head of beef sold in the United States. Fifty cents of each dollar goes to the state beef council in the state where the sale occurs, and the other 50 cents goes to the Beef Promotion and Research Board for use at the national level. If there is no state council in the state where the sale occurred, all the money goes to the national level. The money collected is used to promote beef as a food, provide consumer information about beef, provide information to the beef industry, promote foreign marketing, and promote research in beef production and utilization.

The checkoff money is generally collected by auctions, packers, and other markets. Animals sold at private sales are also subject to the checkoff, but compliance enforcement is more difficult. Checkoff money is required to be forwarded to the state beef council or the beef board.

The activities of the Beef Promotion and Research Board are credited with helping beef maintain a higher share of the meat market than would have otherwise occurred. Research funded by the checkoff has addressed such issues as excess carcass fat, quality control, food safety, animal welfare, and environmental concerns. A recent analysis of the impact of spending in the beef checkoff program shows a net gain to cattle raisers of more than $5 for each $1 of checkoff money collected. The analysis revealed that consumers in the United States have responded favorably to beef promotion and education programs funded by the checkoff money. The programs have effectively increased the number of new beef consumers as well as the amount of beef purchased by current consumers.

Some beef producers have not agreed with the concept of the beef checkoff program and have challenged its legality. Federal courts in various states have ruled differently on this question. The U.S. Supreme Court may eventually settle the issue; however, the checkoff program continues to function until a final determination is made.

SUPPLY AND DEMAND

Supply is defined as the amount of a product that producers will offer for sale at a given price at a given time. As prices increase, producers are willing to offer more of the product for sale. As prices go down, less of the product will be offered for sale.

Demand is defined as the amount of a product that buyers will purchase at a given time for a given price. As prices go up, there is less demand for a product. As prices go down, there tends to be more demand for a product.

The combined effects of supply and demand govern the market price of beef cattle. Producers have little effect on the demand for beef; however, their decision to sell or not to sell beef affects the supply. Since cattle raising and feeding is a long-term project, the real influence on supply is made long before the cattle go to market (Figure 19-1). Management decisions to raise more cattle or fewer cattle eventually affect the price of cattle on the market. Beef cattle numbers tend to run in 9- to 13-year cycles.

Price patterns for feeder calves and slaughter steers and heifers vary by season and year. The price of feeder calves tends to be below the yearly average during the fall and winter months. During the spring and summer, the price of feeder calves tends to be above the yearly average, which reflects a stronger demand during this period. The price of slaughter steers and heifers follows a pattern similar to that of feeder calves. The peak prices tend to come in April, after which the prices taper off for the rest of the year.

In recent years, more cattle finishing has been concentrated in large commercial feedlots. This trend has tended to reduce the seasonal variation in fed-cattle prices. Large

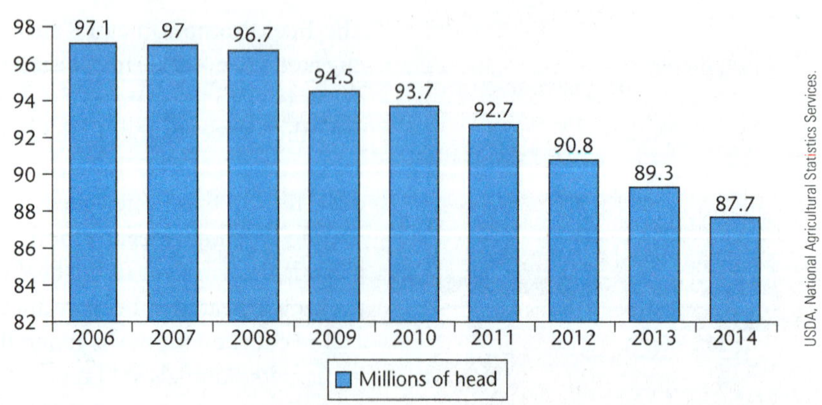

Figure 19-1 Total cattle inventory in the United States.

feedlots are more likely to follow a pattern of year-round placement and marketing of fed cattle. Another factor that has affected the seasonal variation in fed-cattle prices is the tendency to feed cattle for shorter periods of time than was common a few years ago.

The numbers of fed cattle marketed are generally lower during the last quarter of the year. However, an increase in marketing of cows and other non-grain-fed cattle and an increase in the pork supply tend to keep fed-cattle prices lower than the yearly average during this period. Beef marketing tends to be lower during the second quarter of the year; this causes an increase in seasonal prices during this period.

Knowledge of seasonal price changes can help in making management decisions about buying and selling cattle. However, price movements in any given year may be different than the long-term seasonal trends. Current information about supply and demand and other factors that may influence price movement must be carefully evaluated each year when making buying and selling decisions. Federal farm programs, general business conditions, feed supply, and price levels are all factors that may affect seasonal price trends for beef cattle.

METHODS OF MARKETING

Terminal Markets

Terminal markets are also called central markets or public **stockyards**. The facilities of the terminal market are owned by a stockyard company. The company charges for the use of the facilities and for feed fed to the cattle while they are in the stockyard. However, title to the livestock does not pass to the stockyard company. In other words, the stockyard never possesses ownership of the cattle.

Terminal markets have two or more commission firms. Cattle are consigned to a commission firm by the seller. When goods are consigned, they are not sold to another party, but merely given to another party who acts as a selling agent. The function of the commission firm, or selling agent, is to bargain with representatives of the purchaser to get the best price possible for the cattle. Commission firms charge a fee, called a **commission**, for their services in marketing the cattle. The commission firm does not take title of the cattle. Cattle that are consigned to the firm are sorted into uniform lots for sale.

Costs at a terminal market include charges for yardage, feed, insurance, and selling fees. **Yardage** is the fee charged for the use of stockyard facilities. These costs are deducted from the selling price of the cattle. The seller receives the net amount after the charges are deducted.

Terminal markets exist for both feeder cattle and slaughter cattle. Today, there are about 30 terminal markets compared, while in the 1920s and 1930s, there were about 80. Most terminal markets for feeder cattle are in the western and midwestern states. Slaughter-cattle terminal markets tend to be located near population centers and packing plants.

Commission firm sales of cattle are by private sales. Both buyers and sellers are represented by professionals. Some terminal markets are now using auction sales as a method of selling both feeder and slaughter cattle. However, the use of terminal markets as a method of selling cattle is decreasing. About 12 percent of the packer sources of slaughter cattle are from terminal markets as compared to more than 90 percent in 1925. Larger producers of cattle tend to use other methods of marketing their cattle.

Auction Markets

At **auction markets**, cattle are sold by public bidding, with the animals going to the highest bidder (Figure 19-2). Auction markets are also called local sale barns

Livestock Marketing Association

Figure 19-2 A public cattle auction where cattle are sold to the highest bidder.

or community auctions and can be found in local communities. Auction markets are popular because of their convenience for buyers and the open competition for cattle. Auction markets are of the most value to the smaller cattle producer.

Auction marketing costs include charges for yardage, feed, insurance, brand, and health inspection. Charges are based either on a percent of the selling price or a fixed fee. The costs are paid by the seller of the cattle.

Both feeder and slaughter cattle are sold through auction markets. In some states, cattlemen's associations sponsor consignment sales for feeder cattle. Cooperative auction markets are used in some areas to sell slaughter cattle. Increasing numbers of feeder cattle are being sold through auction markets (Figure 19-3). The number of slaughter cattle being marketed in this way is currently remaining steady. About 18.6 percent of the packers' sources of slaughter cattle come from auction markets, compared to 15.6 percent in 1960.

Direct Selling and Country Markets

Larger cattle producers tend to market their cattle by direct marketing methods. Typically, there are no commission firms or brokers involved in the marketing process. Contract sales are used to market cattle directly from feeder cattle producers

Livestock Marketing Association

Figure 19-3 Both feeder and slaughter cattle may be sold through auction markets.

to cattle feeders. The sale is made on the range or farm where the feeder cattle are produced. Large producers of feeder cattle often use contract sales as a method of marketing their cattle.

Cattle dealers buy feeder cattle for cattle feeders. Dealers are either paid on a commission basis or buy the cattle outright and then resell them at a higher price. Cattle dealers also buy slaughter cattle locally. Usually, they have facilities for handling cattle. These cattle are then shipped on to packing plants for slaughter.

Order buyers locate and purchase feeder cattle based on orders from companies and individuals who will feed the cattle to market weight. Cattle feeders use order buyers because the order buyer knows where the cattle are and is familiar with market conditions. Order buyers also buy slaughter cattle for packing plants. The cattle are bought on the farm and shipped directly to the packing plant. They are weighed at the packing plant and are paid for at prevailing prices for that day.

Increasing numbers of slaughter cattle are being bought on a grade and yield basis. Using this system, the value of the animal is determined after it is slaughtered. The carcass is graded, and the carcass weight or yield is determined. Good-quality cattle bring a higher price when marketed in this way. In recent years, about 23 percent of the cattle and 9 percent of the calves in the United States are marketed on a grade and yield basis. This compares to 14 percent of the cattle and 3 percent of the calves in 1967.

The number of slaughter cattle being marketed by direct marketing methods is increasing. About 69 percent of the packers' source of slaughter cattle is from direct marketing and country dealers as compared to 38 percent in 1960. Increasing numbers of feeder cattle are also being marketed through direct marketing methods.

Electronic Marketing

In some areas of the United States, livestock marketing associations have established a system of marketing utilizing computer technology. This method has been used mainly for marketing feeder cattle, but the potential exists for expansion into marketing other types of cattle.

The system works as a form of auction selling. Each bidder has a computer terminal online during the sale. Bids for each lot of cattle are entered by individual bidders on their terminals. When no more bids are received, the lot of cattle is sold and the bidding begins on the next lot.

Bidders can buy the cattle from their offices and do not have to travel to an auction barn. Accurate written descriptions of the cattle and terms of the sale are made available to the bidders several days prior to the beginning of the auction. This information is gathered by graders who work for the livestock marketing association. Information made available includes the number of head, breed or type, sex, estimated average weight, estimated grade, delivery date, weighing location, conditions of sale, flesh condition, feeding and pasture conditions under which the cattle were kept, and information concerning any preconditioning.

Video auctions are growing in popularity as a method of selling cattle. The cattle are videotaped on the ranch or farm where they are located. The videotape is sent to the transmitting station where the auction is located. The auction is telecast for viewing by prospective buyers. Typical information about the cattle includes breed, where they are located, number of head, frame size, estimated weight variance, vaccination and implant information, and current feeding program. Prospective buyers must register with the auction company in advance of the sale. A video auction allows the buyer to see the cattle without moving the cattle to the auction. The purchased cattle are moved directly from the seller's facility to the buyer's facility, which reduces the chances of the cattle having stress and health problems.

Some video sales cattle auction companies have now expanded their services to the Internet. Potential buyers can see the cattle and make bids online. Use the search term "cattle auctions" in an online search engine to find the URLs of these services. These URLs also provide links that allow trading in other agricultural supplies and products such as grain, hay, farmland, beef and dairy cattle embryos, chemicals, and livestock feeds. These commercial sites are designed to provide electronic commerce tools to the agricultural community. It is expected that the Internet will continue to increase in importance for livestock producers in terms of both commerce and information in the years ahead.

Packers and Stockyards Act

The Packers and Stockyards Act is a federal law that is administered by the USDA. All cattle that move across state lines are regulated by this act. The act sets the rules for fair business practices and competition. These rules apply to stockyards, auction markets, packers, market agencies, and dealers who engage in interstate livestock marketing. Individual farmers who buy or sell cattle as a part of their farming operation are not considered to be dealers under the provisions of the act.

Livestock dealers must register with the USDA and file a bond based on the volume and type of their business. Packer-buyers who buy only for slaughter do not have to file a bond. Records showing the nature of every transaction must be kept by dealers. Scales must be operated in a way that gives accurate weights. Serially numbered scale tickets are issued. Dealers are forbidden to use any unfair, discriminatory, or deceptive trade practices.

The Packers and Stockyards Act is enforced by representatives of the USDA. Violators of the act may be warned or ordered to stop violations. If the violation is serious, dealers may be suspended from registration and not allowed to continue to do business. Criminal violations are prosecuted by the Department of Justice.

Purebred Marketing

The marketing of purebred cattle is a specialized business. Purebred cattle are usually sold by private sales or by auction. Purebred producers advertise their herds through breed association magazines and other publications. Selling bulls that have been performance tested also provides advertising for the purebred herd owner. Exhibiting at fairs and shows is another way to advertise purebred cattle. Purebred associations sometimes sponsor consignment sales at which cattle are auctioned. The purebred producer should consign only cattle of the best quality to these sales.

Selecting a Market

The marketplace has two functions: to set the value of the cattle and to physically move the cattle from producer to consumer. The individual producer must determine the best market based on price, costs of marketing, and convenience. Many producers select markets mainly on the basis of convenience; however, price and costs of marketing should be given more importance when selecting a market.

Shrinkage

Shrinkage is the loss of weight that occurs as cattle are moved to market. The amount of loss varies from 1 to 5 percent of an animal's weight. The amount of shrinkage is affected by the following factors:

- the distance the cattle are moved; most weight loss occurs in the first few miles
- the amount and kind of feed and water the cattle receive just before shipping
- the weather

- the condition of the cattle; thin cattle shrink more than fat cattle
- how the cattle are handled
- the sex of the animal
- the amount of feed, water, and rest the cattle receive during shipping
- the length of the **fillback period** (time during which cattle are fed) after the cattle reach market

Losses from shrinkage are absorbed by the seller. A pencil shrink is sometimes used in certain kinds of marketing. In a pencil shrink, the amount of shrinkage is estimated and a deduction of weight is made based on the estimate. Tables can be used to determine the net bid when a pencil shrink is being used. For example, if the bid was $40.00/cwt with a pencil shrink of 3 percent, then the net bid is $38.80/cwt. If the producer decides to ship the cattle rather than taking the on-farm bid with a pencil shrink, then the cattle must bring a higher price to make up for the shrinkage while being shipped. In this example, if the cattle shrink 3 percent while being shipped, then the market price must be $41.20/cwt to be equal to the bid offered on the farm.

Price Information

If cattle are to be marketed profitably, the producer must be aware of current prices as well as price trends. Live cattle quotations can be obtained from websites, newspapers, radio, television, and by phone. The USDA publishes weekly livestock market news information to which producers can subscribe. Outlook information is available from the USDA website.

Labeling of Meat

The Agricultural Marketing Act of 1946 and recent Farm Bills require retailers to label covered commodities showing the country of origin. This includes ground and muscle cuts of beef, pork, lamb, chicken, and goat. Other commodities that do not come from the livestock industry are also included. The country of origin labeling program is generally referred to as the COOL program. The Agricultural Marketing Service of the USDA is responsible for making the rules for this requirement. The Farm Bill provides for enforcement and mandates fines for those who violate the regulations.

To use a "United States country of origin" label, the product must come from an animal that was born, raised, and slaughtered in the United States. The package may be labeled with a written label, stamp, mark, or other visible means that may be clearly seen by the consumer. Restaurants, bars, and food stands are exempt from the labeling requirement.

Record-keeping and tracking systems are being developed to meet the labeling regulations. COOL increases costs at all levels of the process, but eventually all costs are passed on to the final consumer. Proponents of country of origin labeling argue that it increases demand for domestic meat, thus raising prices for producers. Opponents cite the increased cost to the consumer and argue that there is little or no increase in demand for domestic meat and therefore no significant economic gain for producers. Data from the USDA indicates that imported meat amounts to about 1 to 2 percent of the total meat consumed in the United States. It is not clear that country of origin labeling of meat provides a significant economic benefit to producers. Opponents cite the active trade of cattle and cattle products with Canada. They ask, "What label should ground meat receive, which often contains meat from two or more countries, and how should cows that are born in Canada and finished in the United States be labeled?" It will probably be several years before any clear answers begin to emerge.

Connection: Kobe Beef

© iStockphoto/fi chaoshu

Figure 19-4 Kobe beef is generally considered a delicacy, renowned for its flavor, tenderness, and fatty, well-marbled texture.

Kobe beef is becoming more popular in the United States (Figure 19-4). Many producers of Wagyu cattle in the United States refer to their beef as "American Style Kobe beef" or "Wagyu beef." While many Wagyu cattle in the United States are crossbred, the beef is still noted for having exceptionally high marbling and taste. Kobe beef is available at specialty stores and can cost more than $100 per pound.

Price Comparison of Traditional Beef vs. Kobe Beef (Restaurant Pricing)

Filet Mignon	
12 oz. Traditional American Beef	**$45.00**
Filet Mignon	
12 oz. Wagyu (Kobe) Beef	**$107.00**

Use of Ultrasound to Determine Live Animal Quality

Research is being conducted to determine the value of ultrasound technology as a means of assessing the probable grade of live cattle. The use of the technology allows the producer to measure the percentage of marbling, the amount of fat cover, and the ribeye area in live cattle. Using ultrasound, cattle feeders can sort cattle into separate lots for feeding more uniform groups and determine when cattle are ready for market. Breeders can use the technology to help select breeding stock, especially bulls, with desirable carcass traits. As the cattle industry moves toward a marketing system based on carcass merit, the ability to accurately determine the quality of the carcass in live animals will become more important. Currently, the relatively high cost of the procedure has limited its use. Continued research in the use of ultrasound to determine carcass merit will improve the accuracy and value of this technology.

Effect of Mergers on Marketing

For a number of years, there has been an ongoing debate in agriculture regarding mergers of agribusiness corporations into larger corporations. Some observers believe that this trend is putting too much control of the market in the hands of too few people. Whenever the prices paid for livestock drop, the debate becomes more intense. Occasionally, there are suggestions that more legislation is needed to control such mergers. There are also those who believe the Justice Department should investigate possible antitrust violations in agriculture. The USDA estimates that more than 80 percent of all steer and heifer slaughter in the United States is controlled by the four largest packing companies. In 1980, the four largest packing companies controlled approximately 36 percent of the steer and heifer slaughter.

A statistical analysis by the USDA (U.S. Beef Industry: Cattle Cycles, Price Spreads, and Packer Concentration) concluded that there was no evidence that packer concentration had any major impact on the market price of live beef cattle. Some farmers, farm organizations, and agricultural economists did not agree with the conclusions of the USDA report. This is another issue that has yet to be resolved.

Photo by Bill Tarpenning. Courtesy of U.S. Department of Agriculture

Figure 19-5 Angus calf.

MARKET CLASSES AND GRADES OF BEEF

The purpose of this section is to provide a brief overview of the market classes and grades of beef cattle. The official U.S. standards for beef cattle may be secured from the USDA's Agricultural Marketing Service, Livestock and Seed Division website.

Calves

Beef animals younger than 1 year of age are called **calves** (Figure 19-5). After 1 year of age, beef animals are called **cattle**. Calves are further classed as veal calves (vealers) and slaughter calves.

Veal calves and slaughter calves are generally divided on the basis of the kind of carcass they will produce. The main difference between the two is the color of the lean meat in the carcass. The age and kind of feed the calf has had generally determines whether it is a vealer or a slaughter calf. Vealers usually have had only a milk diet and are under 3 months of age. **Slaughter calves** are usually between 3 and 8 months of age, but may be up to 1 year old. Slaughter calves are produced on feed other than milk for a period of time.

Light veal calves weigh less than 110 pounds, medium-weight veal calves weigh from 110 to 180 pounds, and heavy veal calves weigh more than 180 pounds. Weight classes for slaughter calves are light—less than 200 pounds; medium—200 to 300 pounds; and heavy—more than 300 pounds. The sex classes of vealers and slaughter calves are steers, heifers, and bulls. All three of these classes are graded on the same quality standards. The five grades of vealers and slaughter calves are (1) Prime, (2) Choice, (3) Good, (4) Standard, and (5) Utility. Standards for yield grades are not established for vealers or slaughter calves.

Slaughter Cattle

Cattle are first divided into feeder or slaughter cattle. This division is based only on their intended use. The class of cattle is based on sex definitions and the grade is based on the apparent carcass merit of the animals.

There are five sex classes of beef cattle. A *steer* is a male that was castrated before reaching sexual maturity and is not showing the secondary characteristics of a bull. A **bullock** is a male, usually under 24 months of age, that may be castrated or uncastrated and does show some of the characteristics of a bull. A **heifer** is an immature female that has not had a calf or has not matured as a cow. A **cow** is a female that has had one or more calves. An older female that has not had a calf but has matured is also called a cow. A **bull** is a male, usually over 24 months of age, that has not been castrated. When applying the official USDA standards, any castrated male bovine that shows, or is beginning to show, the mature characteristics of an uncastrated male is considered a bull.

The grades of live slaughter cattle are directly related to the grades their carcasses will produce. The quality grades for steers and heifers are Prime, Choice, Select, Standard, Commercial, Utility, Cutter, and Canner. The same quality grades apply to slaughter cows, with the exception that there is no Prime grade for cows. The quality grades for bullocks are Prime, Choice, Select, Standard, and Utility. The Prime, Choice, and Standard grades are generally used only for steers, heifers, and cows that are less than 42 months of age. The Select grade for steers, heifers, and cows is generally limited to animals that are no more than 30 months of age.

The Commercial grade is generally applied only to steers, heifers, and cows that are more than 42 months of age. The Utility, Cutter, and Canner grades may be applied to any age of steers, heifers, and cows. The quality grades for bullocks apply only to animals that are no more than 24 months of age.

Five yield grades are used to describe slaughter beef. These are numbered Yield Grade 1 through Yield Grade 5. Yield Grade 1 indicates the highest yield of lean meat. Yield Grade 5 indicates the lowest yield of lean meat. The grading of beef carcasses is directly related to live beef grading standards.

Live cattle **quality grades** are based on the amount and distribution of finish on the animal. The firmness and fullness of muscling and maturity of the animal are other factors involved in quality grading. Yield grades in live cattle are based on the muscling in the loin, round, and forearm. Fatness over the back, loin, rump, flank, cod, twist, and brisket is also used in determining yield grade. Fatness is more important than muscling in determining yield grade. U.S. slaughter quality grades are shown in Figure 19-6, and U.S. slaughter yield grades are shown in Figure 19-7.

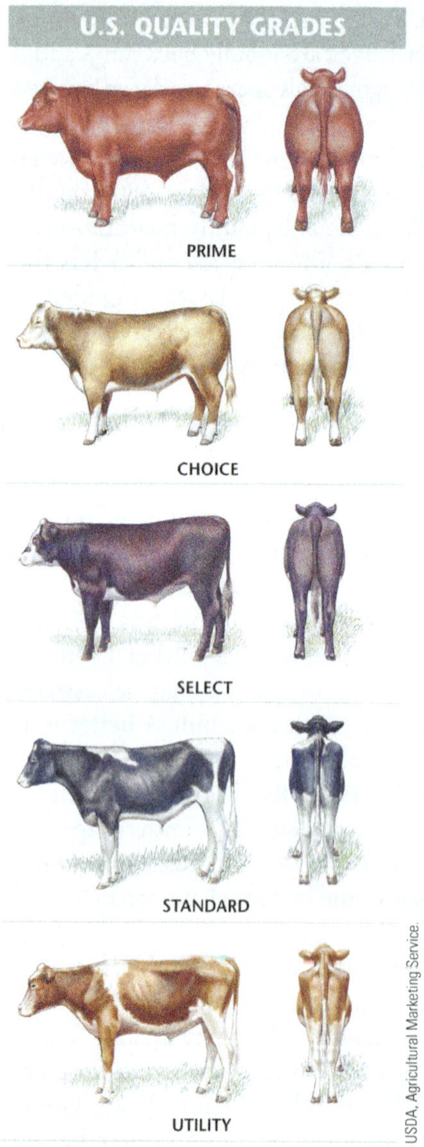

Figure 19-6 Slaughter steers—U.S. quality grades.

Figure 19-7 Slaughter steers—U.S. yield grades.

Connection: Trend in the Coat Color of Beef Cattle

Since the mid 1990s, the number of Angus-type cattle in the United States has seen a surge in popularity, including a price premium for Angus-type beef. The success of the Certified Angus Beef program has provided an incentive for the use of Angus genetics.

According to the USDA, an Angus-type cow is an animal with a predominantly black hide (Figure 19-8). Many beef breeders have taken advantage of the demand for black cattle by using more Angus genetics and selecting for black cattle. It has been predicted that the percentage of black beef animals in the fed-cattle supply will surpass 75 percent by 2015. The use of Angus cattle, a breed known for high-quality meat, will lead to more beef cattle that grade USDA Choice and Prime.

Selected breeds with black coat potential include:

- Angus
- Brangus
- Galloway
- Limousin
- Salers
- Simmental

Courtesy of Alyssa K. Elrod

Figure 19-8 Slaughter cattle are considered Angus-type if they have a predominantly black hide.

Carcass Beef

The age of the animal and the apparent sex designation when the animal is slaughtered determines the class of the beef carcass. The five beef carcass classes are (1) steer, (2) bullock, (3) bull, (4) heifer, and (5) cow. The quality and yield grades for carcass beef are the same as those for live slaughter cattle. The amount of marbling in the carcass affects the quality grade of the carcass. Marbling refers to the presence and distribution of fat and lean in a cut of meat. When grading beef carcasses, five maturity groups and seven degrees of marbling are used. Figure 19-9 shows the relationship between marbling, maturity, and quality grades of beef. Bullock carcasses are limited to maturity group A. The Select grade of beef is also restricted to maturity group A. Figure 19-10 shows the marbling grades used in the official U.S. standards for grades of carcass beef.

The yield grade of a beef carcass is influenced by **hot carcass weight**, which is the unchilled weight of the carcass after harvest and the removal of the head, hide, intestinal tract, and internal organs; the ribeye area; thickness of fat over the ribeye area; and the amount of kidney, pelvic, and heart (KPH) fat. A preliminary yield grade for a hot beef carcass is determined by the thickness of fat over the ribeye.

The base PYG is 2.0 for a carcass with no fat opposite to the ribeye. The greater the fat thickness, the higher the PYG will be. For every 0.1 inch of fat thickness above zero, add 0.25 of a grade to the PYG. Then, you will use the hot carcass weight to determine the ribeye area needed. During contests, the hot carcass weight will be given to you in order to determine the needed ribeye area. The base weight is 600 pounds with a needed ribeye area of 11.0 square inches. For every 25 pounds above the base weight, add 0.3 square inches to the needed ribeye area. For every 25 pounds below the base weight, subtract 0.3 square inches from the needed ribeye area. You will then determine the actual ribeye area. For each square inch of actual ribeye area that is greater than needed ribeye area, you will need to subtract 0.3 from the PYG. For each square inch of actual ribeye area less than the needed ribeye area,

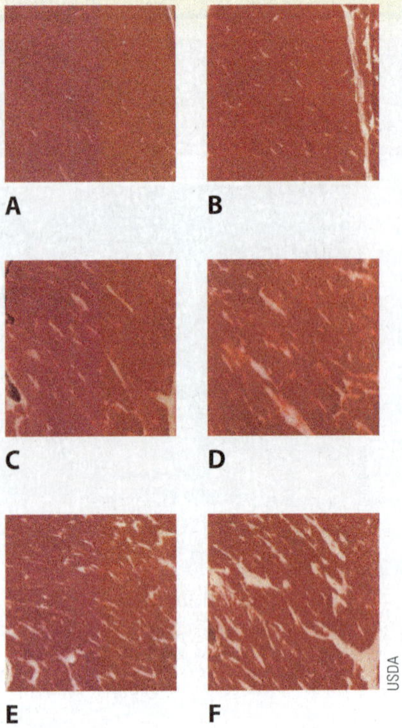

Figure 19-10 Marbling grades.
A = Slight
B = Small
C = Modest
D = Moderate
E = Slightly Abundant
F = Moderately Abundant

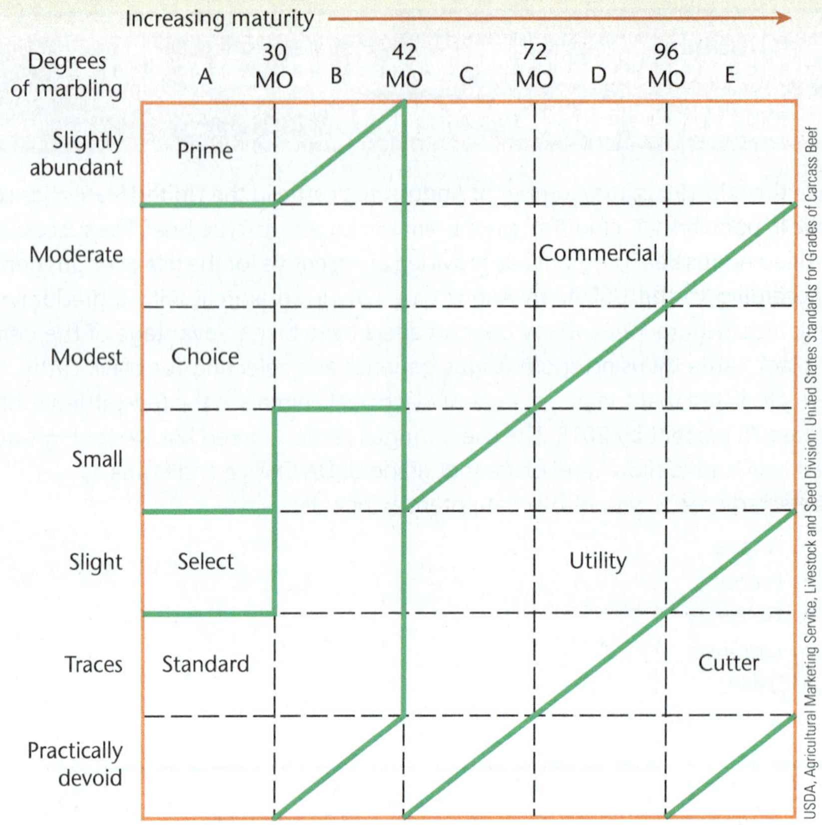

Figure 19-9 Relationship between marbling, maturity, and quality.

it will be necessary to add 0.3 to the PYG. Lastly, you will need to adjust for Kidney, Pelvic, and Heart (KPH) fat. An increase in the amount of KPH fats decreases the percent of retail cuts from the carcass. Each change of 1 percent of the carcass weight attributed to these fats causes a 0.2 change in the yield grade. The preliminary yield grade is adjusted for each percent that the kidney, pelvic, and heart fat is more or less than 3.5. For each percent more than 3.5, add 0.2 of a grade to the preliminary grade. For each percent less than 3.5, subtract 0.2 from the preliminary yield grade.

For example, if the thickness of fat over the ribeye is 0.2 inch, the preliminary yield grade will be 2.5. Assume the hot carcass weight is 600 pounds and the ribeye area is 12 square inches. The adjustment to the preliminary yield grade is then −0.3 because the actual ribeye area is one square inch above the needed ribeye area for 600 pound carcass. Assume further that the percentage of kidney, pelvic, and heart fat is 2.5. The adjustment to the preliminary yield grade is then −0.2 of a grade because the KPH is 1 percent lower than 3.5. The final yield grade would then be 2 $(2.5 - 0.3 - 0.2 = 2.0)$.

The higher grades of slaughter beef are usually grain-fed animals that have a high yield of lean cuts with the right amount of marbling. These animals bring higher prices in the market. About 99% of the graded beef in the United States grades U.S. Select, Choice, or Prime—with about 3% grading Prime, about 63% percent grading Choice, and about 33% grading Select. Approximately 77.5% of the commercial beef in the United States is quality graded.

Dark cutting beef is a condition in which the lean meat is darker than normal in color. It usually has a gummy or sticky texture. It is believed to be the result of reduced sugar content in the lean meat at the time of slaughter. The condition is sometimes found when cattle have been subjected to stress conditions just before slaughter.

The condition varies from being barely visible to nearly black in color (dark/black cutter). There is little evidence that the condition has any negative effect on the taste of the meat. It is considered in grading beef because it has an effect on consumer acceptance and therefore on the value of the meat.

Depending on how severe the condition is, the grade of carcasses that would otherwise be graded Prime, Choice, or Select may be reduced by one full grade when the condition is present. Beef carcasses that might have otherwise graded Standard or Commercial may be reduced in grade by up to one-half of a grade. Dark cutting beef is not considered when grading Utility, Cutter, and Canner grades.

HANDLING CATTLE PRIOR TO AND DURING MARKETING

Careful handling of cattle just before shipping and during transit to market can save the producer a great deal of money. Losses due to improper handling amount to millions of dollars per year in the United States. Feed additives and drugs must be removed from the feed for the proper length of time before marketing. Label directions on these additives and drugs must be followed closely.

Cattle can be conditioned before shipping to reduce shrinkage. If they have been on a ration of green feeds or silage, the amount of these feeds should be reduced and the amount of hay increased. Remove the protein from the ration 48 hours before cattle are to be shipped. Reduce the grain in the ration by one-half the day before shipping. Do not feed grain the last 12 hours before shipping and give the cattle free access to water up to the time of shipping.

If cattle are to be sold on a grade and yield basis, all feed can be taken away from them 48 hours before shipping. Allow them all the water they want up to the time of slaughter. Cattle that are to be weighed upon arrival at the plant can be fed the normal ration up to weighing time.

Cattle should be moved slowly and quietly when loading and handling. Do not use clubs, whips, or electric prods to force cattle to move. Clubs and whips can damage valuable cuts of meat on the carcass and electric prods overexcite the animals. Canvas slappers can be used with little injury to the cattle. Load only the proper number of cattle on trucks because overcrowding or underloading increases the chances that cattle will be injured during shipping. Large trucks should be divided into compartments, and sudden stops and starts with the truck should be avoided. In addition, the cattle need to be protected from the weather while being shipped.

CATTLE FUTURES MARKET

The cattle futures market is a system of trading in contracts for future delivery of cattle. The contract is a legal document calling for the future delivery of a given commodity. Trading in futures is common with many types of commodities. Trading in contracts means that the contract representing the commodity is bought or sold rather than the actual commodity. Trading in futures is done for both slaughter and feeder cattle. The trading is done at commodity exchanges. The futures market is supervised by the Commodity Exchange Authority of the USDA. The actual buying and selling is supervised by members of the commodity exchange. They charge a fee for their services.

A cattle feeder may use the futures market to hedge on the price of cattle. Hedging provides some protection against price changes. To hedge, or protect him- or herself from a future drop in the price of finished cattle, a cattle feeder might sell a contract on the futures market at the same time that he or she buys feeder cattle. Later, when the finished cattle are sold, the futures contract is bought back. This practice tends to lock in the profits from feeding the cattle.

Because the hedging operation involves opposite actions (buying and selling) in the cash and futures markets, the risk of loss is less if cattle prices go lower than expected. The sale of the futures contract at the start of the hedging period offsets the lower cash price for cattle at the end of the hedging period. However, if cattle prices go higher than expected at the end of the hedging period, the cattle feeder will not realize additional profits because he or she must buy back a contract at that time or deliver the live cattle to offset the sale of the futures contract at the start of the hedging period. Thus, while hedging reduces risk, it also reduces the opportunity for additional profit that might result from a higher than expected upward movement in the price of live cattle.

There is some risk in using the futures market. The cash price of cattle does not always closely follow the price of futures contracts. To use the futures market successfully, the cattle feeder must keep good records and understand the pricing system in the market. The futures market tends to bring price stability to beef cattle prices. It also tends to level out seasonal variations in the price of beef. Extensive use of the futures market by cattle feeders tends to take some of the risk out of the beef cattle business.

SUMMARY

Supply and demand govern the price of beef cattle. Beef producers have little effect on the demand for beef, but management decisions of producers do affect the supply of beef in the market. Beef cattle prices vary with the season of the year. There are also long-term trends or cycles in beef cattle prices.

Beef cattle are marketed through terminal markets, auction markets, and by direct selling. The importance of terminal markets has been declining in recent years. Auction markets are found in many local communities. The numbers of cattle marketed in auction markets have not changed much over the years. More cattle are being sold by direct marketing methods now than there were a few years ago. Larger cattle producers tend to use direct selling rather than other methods of marketing.

The sale of beef cattle that are transported across state lines is regulated by the Packers and Stockyards Act administered by the USDA. This act sets forth rules for fair business practices and competition in the marketplace.

Purebred cattle are sold by private sale and by auction. The reputation of the breeder and the methods used for advertising are important factors in the sale of purebred cattle.

Many producers select a marketing method based on convenience. However, price and costs of marketing are more important factors, and should be given first consideration when selecting a marketing method.

Cattle experience shrinkage when being shipped to market. Careful handling and management can reduce losses due to shrinkage and damage to carcasses that may occur during shipping.

Cattle are divided into classes based on use, age, weight, and sex. Quality grades are used to describe both feeder and slaughter cattle. Yield grades are also used in describing slaughter cattle. The age of the animal and the amount of marbling in the carcass affect the quality grade. The yield grade is determined by the amount of lean meat that can be cut from the carcass.

The cattle futures market can be used by the cattle producer to reduce price risk. The cattle producer must keep good records and have an understanding of the pricing system to successfully use the cattle futures market.

Student Learning Activities

1. Prepare a bulletin board of charts showing supply and demand cycles, price trends, and local price variations for beef cattle.
2. Take a field trip to various types of markets in the local area. Observe procedures; talk to sellers and buyers; and make a list of the advantages, disadvantages, and costs of each type of market.
3. Take a field trip to a feedlot or market and grade live cattle. If possible, arrange to see these same cattle after slaughter to observe actual grades.
4. Take a field trip to a packing plant. Grade carcasses of slaughtered animals, and observe damage to cattle carcasses from mishandling during shipping.
5. Ask a cattle feeder who uses the futures market to hedge cattle to explain the procedure to the class.
6. When planning and conducting a beef cattle supervised experience program, follow good marketing practices as described in this chapter.

Discussion Questions

1. Define the following terms: (a) supply; (b) demand.
2. How do producers affect the supply and demand for beef cattle in the market?
3. List and briefly describe three methods of marketing beef cattle.
4. How does the Packers and Stockyards Act affect the marketing of beef cattle?
5. How are purebred beef cattle marketed?
6. What are three factors that a producer should consider when selecting a market?
7. List eight factors that affect shrinkage when marketing beef cattle.
8. Who absorbs the losses from shrinkage?
9. What sources of price information for beef cattle are available to the producer?
10. List and briefly describe the market classes of beef cattle.
11. What are the factors involved in live cattle quality grades?
12. What are the factors involved in live cattle yield grades?
13. What is the most common grade of beef marketed in the United States?
14. How should cattle be handled during marketing to reduce losses?
15. What is hedging, and why might a cattle feeder use hedging?

Review Questions

True/False

1. Electronic marketing of cattle and beef is not expected to expand in the future.
2. The USDA reports that more than 80 percent of steers and heifers harvested in the United States are controlled by the four largest packing companies.
3. Beef animals younger than 1 year of age are called calves and after 1 year of age are called cattle.
4. It is not necessary for livestock dealers to register with the USDA.
5. The peak price for slaughter steers and heifers tends to come in April.

Multiple Choice

6. ______ govern(s) the price of beef cattle.

   a. Supply
   b. Demand
   c. Supply and demand
   d. USDA

7. In some cattle, a condition in which the lean meat is darker than normal is called:

   a. black cutter
   b. low cutter
   c. choice
   d. dark cutting beef

8. The number of fed cattle marketed is generally lower during the ______ quarter of the year.

   a. first
   b. second
   c. third
   d. fourth

Completion

9. The marketing of cattle directly from feeder cattle producers to cattle feeders uses ______ ______.
10. The cattle futures market is a system of trading ______ for the future delivery of cattle.
11. ______ refers to the presence and distribution of fat and lean in a cut of meat.
12. ______ is the amount of a product that buyers will purchase at a given time at a given price.

Short Answer

13. What is a video auction?
14. What is a bullock?
15. How can cattle be handled to reduce shrinkage?

SECTION 5

Swine

Chapter 20

Breeds of Swine

Key Terms

swirl
cryptorchidism
freckles

Objectives

After studying this chapter, the student should be able to

- list the main characteristics of the swine enterprise.
- identify the major breeds of swine by body characteristics.
- summarize the origin and development of the common breeds of swine.

OVERVIEW OF THE SWINE ENTERPRISE

The Corn Belt states remain the major swine-producing region of the United States. The major reason for the popularity of swine in this region is the availability of grain, mainly corn, which forms the basis for most swine diets. However, hogs are raised in every state. The leading states in hog production are Iowa, North Carolina, Minnesota, Illinois, Indiana, Nebraska, and Missouri (Figure 20-1A). About 75 percent of all swine in the United States are raised in these states. Figure 20-1B shows the dispersion of swine production in the United States.

Figure 20-2 shows the percent of operations and percent of swine inventories by the size groups of operations in the United States. Approximately 70.9 percent of the operations have fewer than 100 head of hogs; however, these operations account for less than 1 percent of the total swine inventory. About 4.5 percent of the total operations raise 1,000 or more head of hogs, accounting for 61 percent of the total swine inventory in the United States. This reflects an ongoing trend toward large, corporation-owned swine operations, which has raised concern among residents and state legislatures about the environmental impact of large swine operations. There has been an increase in the percentage of the smallest operations (1–99 head), most likely due to the increased consumer desire for more locally grown and organic products.

In the past, it was a common practice for farmers to raise hogs when the economic outlook was favorable and to go out of the hog business when returns were not profitable. This practice is not as common today. The investment required to raise hogs using the current technologies makes it more difficult to get into and out of the business over a period of time. When smaller producers quit raising hogs now, they generally do not return to the business. The producers who stay in the business are making larger investments in facilities, technology, and genetics to become more efficient and thus improve profitability.

The traditional model for swine production has been a farrow-to-finish enterprise on a farm. There is still a core of farmers in the United States who follow this traditional pattern, especially in the Midwest. A change to more specialization in swine production began in the 1990s. Three different sites, on different farms, are now a common pattern for the swine enterprise. The breeding, gestation, and farrowing phase is located at one site. When pigs are weaned, they are moved to a nursery site where they are fed for 8 to 10 weeks. When they reach 40 to 60 pounds, they are moved to a finishing facility where they are fed out to market weight.

Leading States in Swine Production ($ billions)

State	$ billions
Iowa	6.8
North Carolina	2.9
Minnesota	2.8
Illinois	1.5
Indiana	1.3
Nebraska	1.1
Missouri	0.9
Ohio	0.8
Kansas	0.7
Oklahoma	0.7

Figure 20-1A Leading Swine Production States, 2012.

Source: U.S. Department of Agriculture, National Agricultural Statistics Service

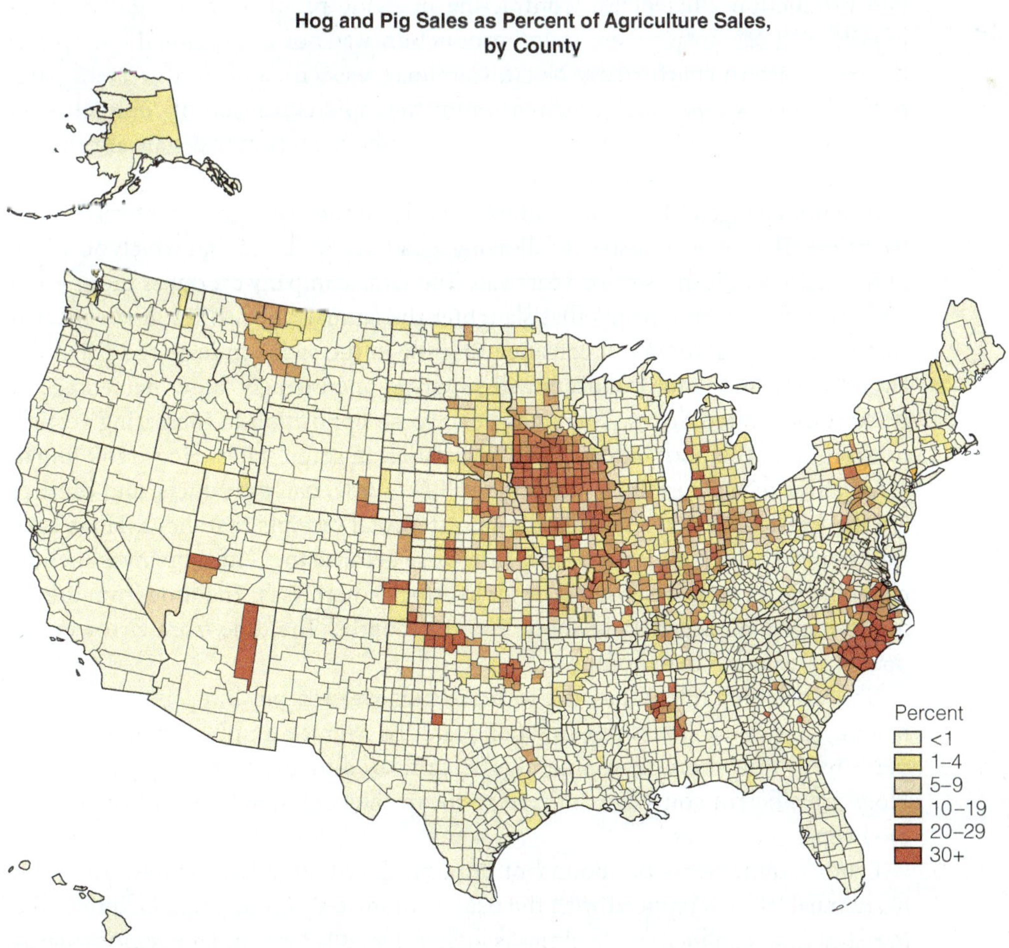

Figure 20-1B The dispersion of swine production in the United States.

Source: USDA, NASS, 2012 Ag Census Hog and Pig Farming; 07-M030

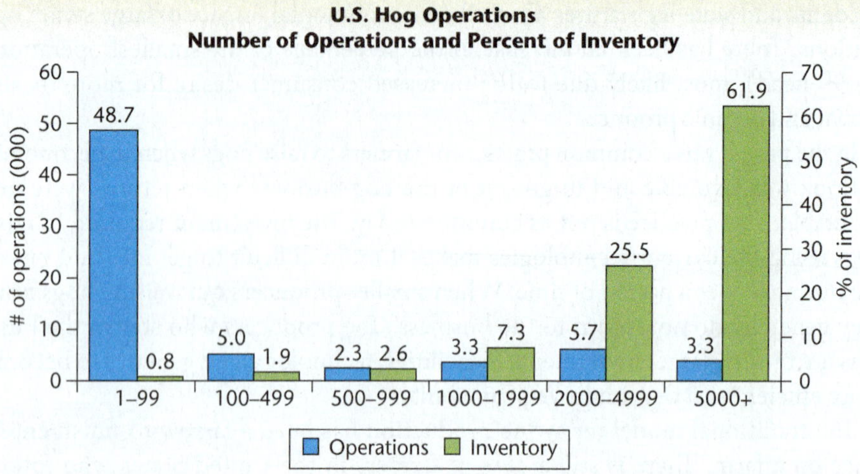

Figure 20-2 U.S. hog percent of operations and inventory, 2012.

Source: USDA, NASS

Production contracts are often used in one or more phases of this production model. The contractor (typically a large operator or corporation) provides the animals, feed, veterinary care, management services, and other necessary inputs. The contractee (typically a smaller farmer) provides labor and facilities in exchange for a guaranteed price per pig delivered. A bonus may be paid for reaching certain production efficiencies. Contracting in swine production has grown most rapidly in those areas where swine production was not a traditional enterprise: the southeastern (particularly North Carolina), western, and southwestern parts of the United States. Part of the reason for the rapid expansion of contracting in swine production in North Carolina is because the farmers in that state are already familiar with contracting in the poultry industry.

A major change in the swine industry has been the rapid growth of vertical integration. The swine industry is following a pattern similar to that which occurred in the poultry industry several years ago. The same company produces the animals and owns the packing plants that slaughter the animals for market. It is expected that vertical integration will continue to grow in the swine industry and that the largest companies and producers will supply an increasingly larger percentage of pork in the United States. There has been some opposition to increasing vertical integration in the pork industry with legislation in some states. Hogs rank fourth in livestock receipts in the United States. Beef cattle, dairy products, and broilers are the livestock enterprises that have greater total receipts than hogs. Hogs have traditionally been popular as a farm enterprise because the return on investment is generally faster than it is in most other livestock enterprises. The time from breeding to marketing is generally about 9 or 10 months. The time from farrowing to market is usually 5 to 6 months.

Improved technology has led to more pigs weaned per litter, better feed efficiency, and more litters per sow per year. The average number of pigs weaned per litter is now more than nine pigs as compared to about seven pigs in 1978. Hogs are efficient converters of feed to meat; they are more efficient than cattle or sheep.

Labor requirements per pound of pork produced are relatively low. The need for manual labor is reduced with the use of automated confinement facilities. The investment in confinement facilities is higher than it is for simpler systems such as raising hogs on pasture. However, few farmers today use pasture-raising systems for swine production.

SELECTION OF A BREED

It is the goal of the swine producer to raise breeding stock and market pigs that have rapid, efficient growth. These hogs should also yield a high percent of muscle when slaughtered.

There are differences among breeds in the traits that are considered economically important. When selecting breeds to use in crossbreeding programs, consider litter size, growth rate, feed efficiency, carcass length, leanness, and muscle (Table 20-1).

When selecting individual animals, performance test records should be examined. Performance records are sometimes not available for gilts and sows. Selection must often be made on the basis of the boar's performance test record.

Information on each breed is available from the individual breed associations. The location of breeding stock and information on the performance of different bloodlines can be secured from these associations.

Purebred herds provide breeding stock that can be used in crossbreeding programs to produce market hogs. Purebred animals may be registered in their respective associations if they meet the standards for registration.

Most swine purebred associations require the following information for the registration of individual hogs:

1. Date farrowed
2. Number of pigs farrowed
3. Number of pigs raised (males and females)
4. Ear notches (must be notched at birth)
5. Name and address of the breeder
6. Name and address of the owner at farrowing time
7. Name and registration number of the sire and dam
8. Litter number

Ear notches must follow the guidelines for the breed association with which the animal is being registered. Some breed associations require an ear tattoo that shows the year of the pig's birth and the herd it comes from. Litter numbers are assigned by the producer and must not be repeated during the same year.

TABLE 20-1 Comparison of Breed Performance Litter Size

Breed	Average Litter Size Ratio[a]
Yorkshire	100
Landrace	99
Duroc	91
Chester White	91
Spotted	83
Hampshire	83
Berkshire	77
Poland China	77

[a]Best breed performance is given 100 as compared to each breed.

Source: Personal communication, Charles J. Christians, Extension Animal Husbandman, University of Minnesota.

Leading Swine Breeds in the United States (in rank order)

1. Yorkshire
2. Duroc
3. Hampshire
4. Landrace

CHARACTERISTICS OF THE BREEDS

American Landrace

History

The Landrace breed originated in Denmark. In 1934, the first Landrace hogs were imported into the United States for experimental purposes by the U.S. Department of Agriculture. Additional imports of Landrace from the Scandinavian countries occurred in 1954 and in the 1970s.

Courtesy of the National Swine Registry

Figure 20-3 Landrace boar.

Description

Landrace hogs are white in color. They are long-bodied and their ears lop forward and down (Figure 20-3). Landrace sows are noted for their mothering ability and large litters (Figure 20-4).

The American Landrace Association Inc., the breed association, has its headquarters in West Lafayette, Indiana. The breed association was formed in 1950. Disqualifications for registry are black hair, erect ears, and less than six teats on a side.

Mapes Livestock Photos

Figure 20-4 Landrace gilt.

Berkshire

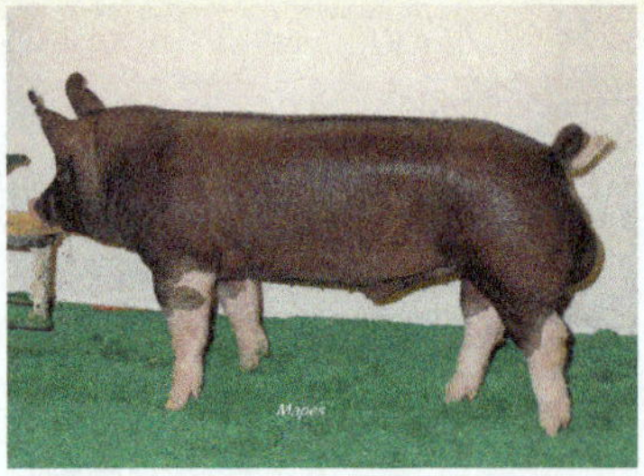

Figure 20-5 Berkshire boar.

Mapes Livestock Photos

History

The Berkshire originated in England in and around Berkshire and Wiltshire counties. The development of the breed began during the early and mid-1700s. The first importation of Berkshires into the United States occurred in 1823. Further importations occurred in 1857. All registered Berkshires in the United States trace their ancestry to the 1857 importations.

Description

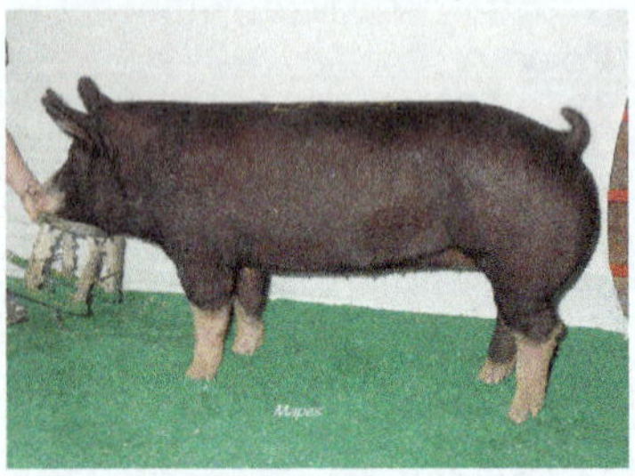

Figure 20-6 Berkshire gilt.

Mapes Livestock Photos

The Berkshire (Figures 20-5 and 20-6) is a medium-sized hog that produces an excellent carcass. The animal is black with six white points. Four white points are found on the feet. There is also some white on the face and the tail. The head is slightly dished and the ears are erect.

The American Berkshire Association was formed in 1875. Selection has placed emphasis on fast and efficient growth, meatiness, and good reproduction. Disqualifications for registration are swirls on the back or sides, large amounts of white hair on the body, or red hair. A **swirl** is hair growing in a circular pattern from the roots. It usually occurs along the top of the spine. In many breeds it is a disqualification for registry because of the undesirable appearance it gives the animal. However, it does not affect any of the economically important traits of market hogs.

Chester White

Figure 20-7 Chester White boar.

Mapes Livestock Photos

History

The Chester White (Figure 20-7) originated in Chester County, Pennsylvania. Additional development of the breed also occurred in Delaware County, Pennsylvania. The original name of the breed was Chester County White. Later the word *County* was dropped from the name.

Yorkshire, Lincolnshire, and Cheshire hogs, all of English origin, were mixed in breeding before 1815. An English white boar, which was called a Bedfordshire or Cumberland, was imported from England sometime between 1815 and 1818. This boar was used on the mixed breeding of the three English breeds mentioned earlier. From these matings, the Chester White breed originated.

Description

Figure 20-8 Chester White gilt.

Mapes Livestock Photos

To be eligible for registration, Chester Whites must be solid white in color, with no other color on the skin larger than a silver dollar and no colored hair (Figure 20-8). Pigs with any areas of skin pigmentation other than white that exceeds five in number are disqualified. The ears of the Chester White must be down, not erect, and of medium size. The breed is noted for its mothering ability.

Duroc

Figure 20-9 Duroc boar.

Courtesy of the National Swine Registry

History

The Duroc breed (Figure 20-9) originated from red hogs raised in the eastern United States before 1865. The New Jersey red hogs were called Jersey Reds. In New York, the red hogs were called Durocs. Some Red Berkshires from Connecticut are also thought to have been included in the early breeding. Intermingling of these breeds resulted in a breed called Duroc-Jersey. The name *Jersey* was later dropped and the breed became known simply as Duroc.

Description

The color of the Duroc is red (Figure 20-10). Shades vary from light to dark, with a medium cherry being the preferred shade. The Duroc has ears that droop forward. The breed has good mothering ability, growth rate, and feed conversion. It is one of the most popular breeds of hogs in the United States. Disqualifications for registry include having more than three black spots, black spots larger than 2 inches in diameter, **cryptorchidism** (retention of one or both testicles in the body cavity), fewer than six functional teats on each side of the underline, or white hair on the body (with the exception of white on the end of the nose).

The American Duroc-Jersey Record Association was formed in 1883. Several other associations were later formed to register red hogs. These associations merged to form the United Duroc Swine Registry in 1934.

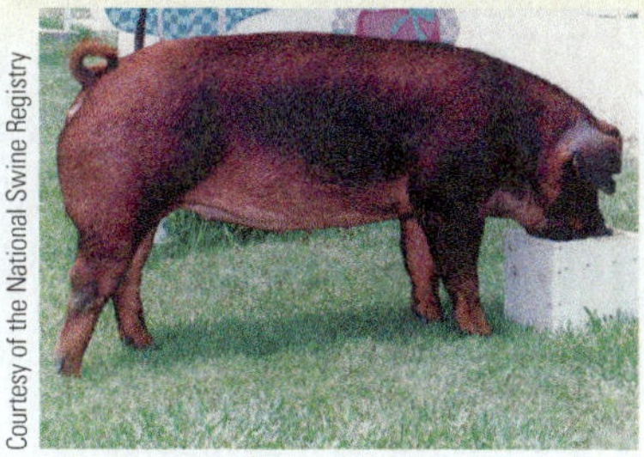

Courtesy of the National Swine Registry

Figure 20-10 Duroc gilt.

Hampshire

History

The Hampshire breed originated in England. The first Hampshires were imported into the United States between 1825 and 1835. Major development of the breed occurred in Kentucky, where the belted hogs were known as the Thin Rind.

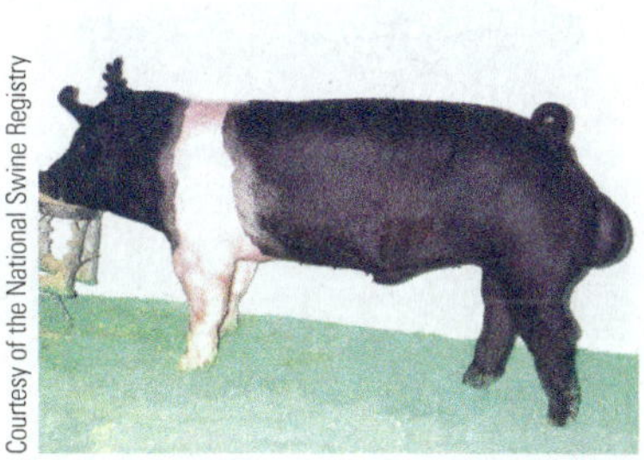

Courtesy of the National Swine Registry

Figure 20-11 Hampshire boar.

Description

The Hampshire is black with a white belt that encircles the forepart of the body (Figure 20-11). The forelegs are included in the white belt. To be eligible for registry, the white belt must include no more than two-thirds of the length of the body. White is permitted on the hind legs as long as it does not go above the bottom of the ham or touch the white belt (Figure 20-12). The Hampshire has erect ears. The breed is noted for its rustling, muscle, and carcass leanness. It is a popular breed and is used in many crossbreeding programs.

Disqualifications for registry are cryptorchidism, incomplete belt, or white belt more than two-thirds back on the body. Other disqualifications include white on the head (except on the front of the snout), black front legs, white going above the bottom of the ham, or white on the belly extending the full length of the body.

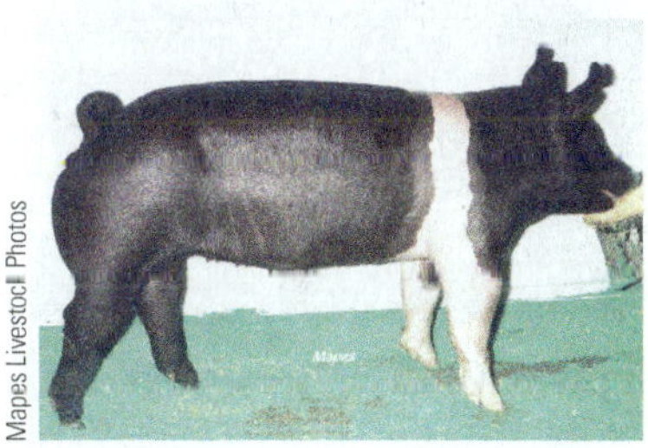

Mapes Livestock Photos

Figure 20-12 Hampshire gilt.

The belted hog was first registered in the American Thin Rind Association. The name of the breed was changed to Hampshire in 1904 and the association was renamed the American Hampshire Record Association. This association was discontinued in 1907, and a new association was formed in Illinois, called the American Hampshire Swine Record Association. The name Hampshire Swine Registry was adopted in 1939.

Mapes Livestock Photos

Figure 20-13 Hereford boar.

Hereford

History

The Hereford breed was developed in Missouri, Iowa, and Nebraska. Early development of the breed occurred from 1902 to 1925. Foundation stock used in the development of the breed included Duroc and Poland China. Chester White and Hampshire hogs may also have been included in the early breeding.

Description

Herefords are red with a white face (Figure 20-13). The ears are forward drooping. To be eligible for registry, Hereford hogs must be at least two-thirds red and have some white on the face. Herefords are prolific and good mothers (Figure 20-14). Disqualifications for registry include no white on the face, red color on less than two-thirds of the body, swirls on the body, or less than two white feet.

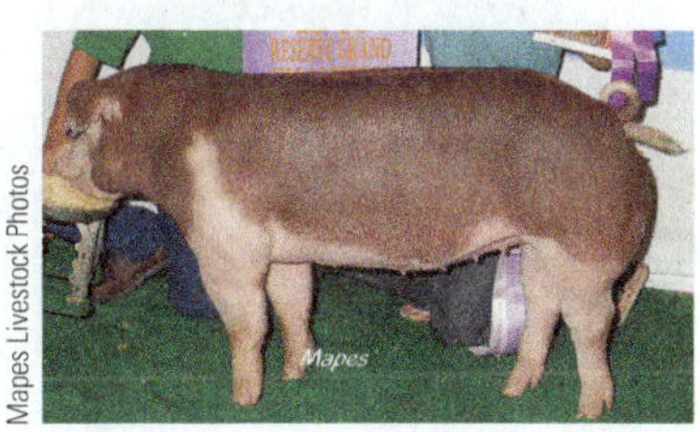

Mapes Livestock Photos

Figure 20-14 Hereford gilt.

Hypor, A Hendrix Genetics Company—Global Leader in Pig Genetics

Figure 20-15 Pietrain boar.

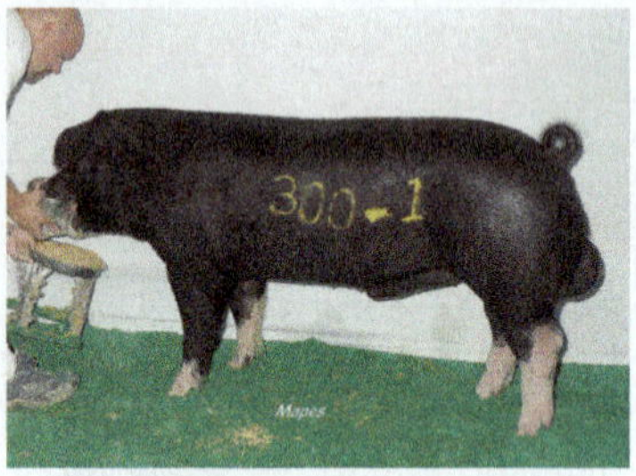
Mapes Livestock Photos

Figure 20-16 Poland China boar.

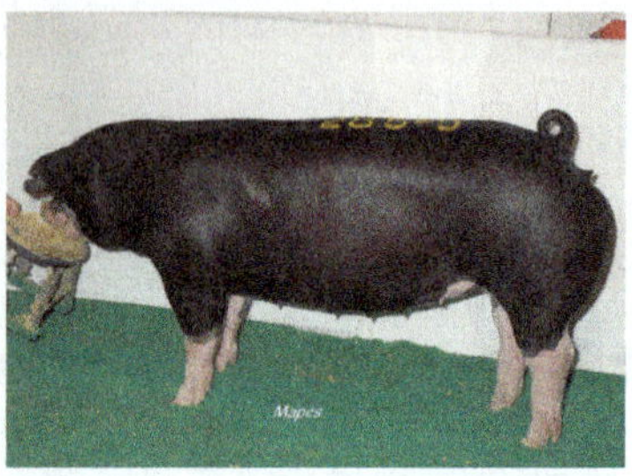
Mapes Livestock Photos

Figure 20-17 Poland China gilt.

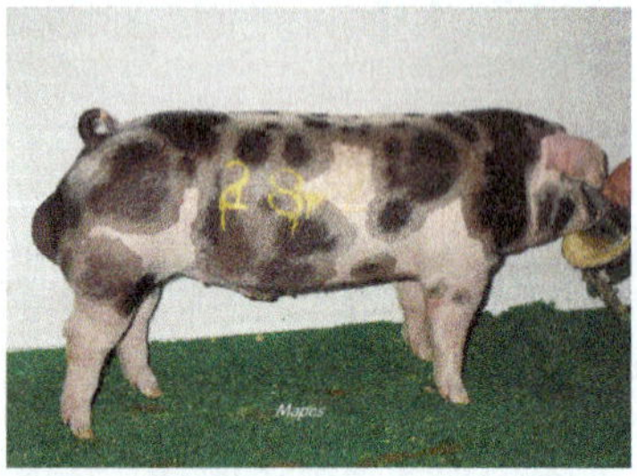
Mapes Livestock Photos

Figure 20-18 Spotted Swine boar.

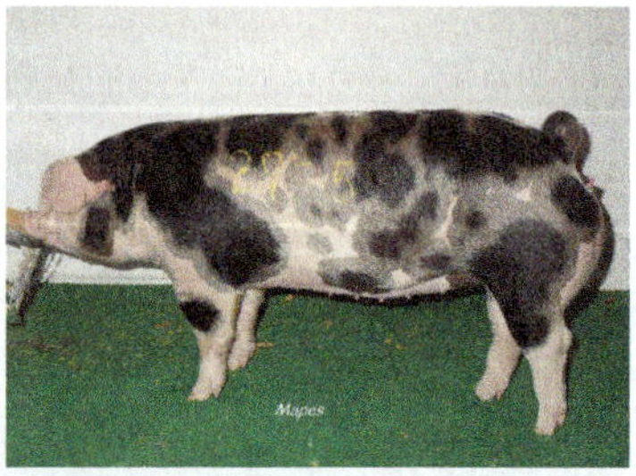
Mapes Livestock Photos

Figure 20-19 Spotted Swine gilt.

The National Hereford Hog Registry Association was organized in 1934. Hogs from Iowa and Nebraska were selected as foundation stock for original registry.

Pietrain

History

The Pietrain breed (Figure 20-15) originated near the village of Pietrain, Belgium, during the early 1950s. The Pietrain breed was brought to the United States to add muscle to breeding stock and improve the quality of market swine. Few, if any, purebred Pietrain swine remain in the United States, but the breed influence is evident in commercial swine. Most of the crossbred boars sold for breeding stock today have some Pietrain in their bloodline. Through crossbreeding programs, the Pietrain breed had a big impact on swine for show in the 1990s and early 2000s.

Description

The Pietrain breed is white with black spots. It is medium sized with a stocky build, erect ears, and short legs. The most distinctive feature of the Pietrain is the extremely muscular and bulging hams. The breed has an exceptionally high lean-to-fat ratio, which makes it desirable in genetic improvement programs. The Pietrain is known for its very high yield of lean meat, but this is often associated with the presence of the gene for porcine stress syndrome. For this reason, the use of purebred Pietrains in the United States is rare. It is most commonly found in crossbred lines.

Poland China

History

The Poland China breed (Figure 20-16) originated in the Ohio counties of Butler and Warren. The breed was developed between 1800 and 1850. Russian, Byfield, Big China, Berkshire, and Irish Grazer bloodlines were used in the development of the breed. It is generally believed that no new bloodlines were used in the breeding after 1846.

The breed was originally called the Warren County hog. The name Poland China was officially adopted at the National Swine Breeders Convention in 1872.

Description

The Poland China hog is black with six white points (Figure 20-17). The white points include the feet, face, and the tip of the tail. The Poland China has forward-drooping ears. Poland Chinas are one of the larger breeds of swine. These pigs produce carcasses with low back fat and large loin eyes. They are used in many crossbreeding programs.

Disqualifications for registry include less than six teats on each side of the underline, hernia, cryptorchidism, red or sandy hair and/or pigment, more than one solid black leg, or evidence of a belt formation. The absence of any of the white points is not objectionable nor is an occasional splash of white on the body.

Spotted Swine

History

The Spotted Swine breed (Figures 20-18 and 20-19) was developed in Indiana. It was created by crossing hogs of Poland China breeding with spotted hogs being grown in the area. Later crosses were made with hogs from England called Gloucester Old Spots.

Description

To be eligible for registration, Spotted Swine must be black and white and have forward-drooping ears. Erect ears, red tinted or brown spots, or a solid black head from ears forward will cause the swine to be ineligible for registration. The Spotted Swine breed is similar to the Poland China in body type. Breeders strive to produce a large-framed hog with efficient gains and good muscling.

The breed association was formed in 1914. At that time, only one recorded parent was necessary for registering an individual hog. The herdbook was closed in 1921, meaning that only animals whose parents are registered in the herdbook may be registered. The original name of the breed was the Spotted Poland China. The name was changed to Spotted Swine, or Spots, in 1960. The herdbook was opened to register purebred Poland Chinas in 1971. This was to provide a broader genetic base. The herdbook was closed again in 1975. The breed association is the National Spotted Swine Record.

Disqualifications for registry include brown or sandy spots and swirls on any part of the body. Cryptorchids (males with one or both testicles retained) are also disqualified.

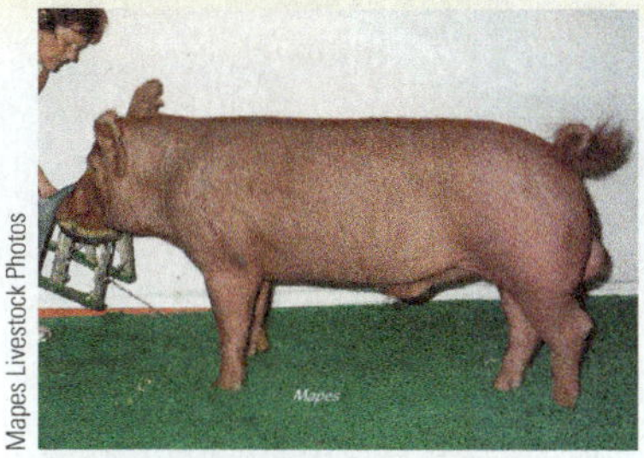

Mapes Livestock Photos

Figure 20-20 Tamworth boar.

Tamworth

History

The Tamworth hog originated in Ireland. Development of the breed took place in England in the counties of Stafford, Warwick, Leicester, and Northhampton. It is one of the oldest of the purebred breeds. Pure breeding began in the early 1800s. The first importations into the United States were in 1882.

Mapes Livestock Photos

Figure 20-21 Tamworth gilt.

Description

The Tamworth is red with shades varying from light to dark (Figure 20-20). The ears are erect, and it has a long head and snout. The sows are good mothers and have large litters (Figure 20-21). The breed is noted for its foraging ability. Swirls on the sides and back as well as inverted teats are disqualifications for registry. The breed association is the Tamworth Swine Association.

Courtesy of the National Swine Registry

Figure 20-22 Yorkshire boar.

Yorkshire

History

The Yorkshire hog (Figure 20-22) originated in England in the county of Yorkshire, where the breed was called Large White. Importations were made into the United States in the 1800s. At that time, many of the hogs were raised in Minnesota, where they became known as Yorkshires.

Description

Yorkshires are white (Figure 20-23) but the skin sometimes has black pigmented spots called **freckles**. Hogs with black spots can be registered, but this trait is considered undesirable. The ears are erect and the face is slightly dished. The Yorkshire was one of the early bacon-type breeds of hogs. Yorkshires have large litters, high feed efficiency, rapid growth, good mothering ability, and long carcasses (Figure 20-24). They are often used in crossbreeding programs.

Yorkshires are registered in the American Yorkshire Club. Disqualifications for registry include swirls on the upper third of the body, hair other than white, and blind or inverted teats. Other disqualifications include less than six teats on a side, hernia, and cryptorchidism.

Mapes Livestock Photos

Figure 20-23 Yorkshire gilt.

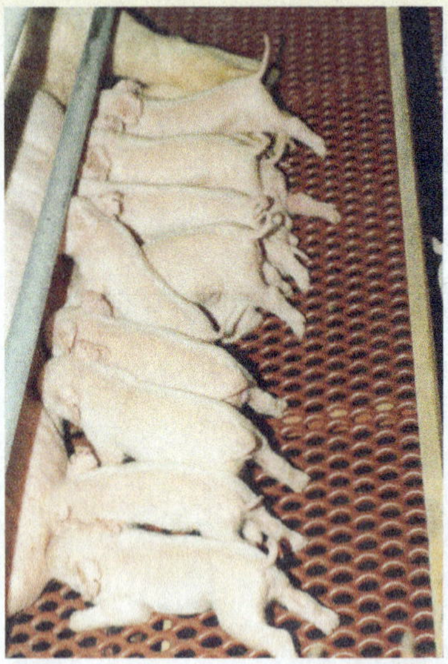

Figure 20-24 Yorkshires are known for large litters.

Figure 20-25 Meishan pigs.

Inbred Breeds

Beginning in 1935, the USDA and several state agricultural experiment stations conducted swine-breeding research by crossing purebred lines of hogs. As a result of this work, a number of inbred lines were developed. Later, some private individuals also developed inbred lines by crossing other breeds. Inbreeding programs were followed to fix the traits wanted in the inbred lines.

Chinese Pigs

A recent interest has developed in the use of several Chinese breeds of swine in crossbreeding programs. The main reason for this interest is the large litter sizes that are common for these breeds. The most prolific of the Chinese breeds include the Meishan (Figure 20-25), Fengjing, and Jiaxing Black. These breeds average 15 live pigs per litter at birth and wean 13 pigs per litter. They reach puberty at about 3 months of age and require 40 percent less feed intake per pound weaned as compared to breeds currently used in the United States.

Research to date indicates that these breeds have slower rates of gain, lower feed efficiency, and poorer carcass characteristics than western breeds. This would indicate that their usefulness in crossbreeding programs would be limited. Research is focusing on trying to determine why they are so prolific. It may be possible through genetic engineering to improve the prolificacy of U.S. breeds using DNA from Chinese breeds.

SUMMARY

Swine are produced in the areas of the United States where feed grains are raised. This is because the major part of the swine ration is grain. More than half of the hogs raised in the United States are raised in the midwestern states. Iowa leads the nation in the number of hogs raised. Hogs are a popular farm enterprise. They are efficient converters of feed to meat.

Selection of breeds to use in crossbreeding programs is based on data concerning litter size, growth rate, feed efficiency, carcass length, leanness, and muscle. Producers of breeding stock strive to produce the kind of carcass in demand on the market today.

Most swine breeds and lines of swine breeding stock produced in the United States were developed in the United States. Others were developed in Denmark and England. All purebred breeds have associations that register purebred animals. Standards for registration of the individual breeds are set by the breed associations.

Inbred lines of hogs were developed by the USDA and various state experiment stations, as well as by private breeders. Inbred lines have been used in crossbreeding programs to improve litter size and growth rate. Genetic engineering holds promise of improving swine breed characteristics.

Student Learning Activities

1. Prepare a bulletin board showing various breeds of swine.
2. Prepare a table that lists the common breeds of hogs, and briefly describes the traits of each.
3. Prepare a list of sources of purebred breeding stock for each of the major breeds in the local area or region.
4. Ask local swine breeders to speak to the class about their operations.
5. Use the breed information in this chapter to help select a breed of swine for a Supervised Agricultural Experience Program.

Discussion Questions

1. Why are more than half of the hogs produced in the United States raised in the midwestern Corn Belt states?
2. How important is swine production compared to all livestock enterprises in the United States?
3. Why are hogs popular as a livestock enterprise on the farm?
4. What type of hog carcass is in demand in today's market?
5. Explain how Chinese swine breeds may be used to improve U.S. swine breeds.

Review Questions

True/False

1. Some breed associations use ear notches for identification.
2. The price cycle for hogs is relatively long because the weight ranges for top butcher prices vary from year to year.
3. The main reason for interest in Chinese breeds of hogs is the large litter sizes common to many Chinese breeds.
4. Crossbred herds are used to provide breeding stock that can be used in purebred programs to produce market hogs.

Multiple Choice

5. ______ notches must follow the guidelines for the breed association with which the pig is being registered.

    a. Ear c. Hoof
    b. Nose d. Tail

6. The ______ hog is black with a white belt that encircles the fore part of the body.

    a. Duroc c. Chester White
    b. Berkshire d. Hampshire

Completion

7. The _________ is known as one of the oldest breeds of swine.
8. The __________ breed is known for extreme muscling.
9. The Hereford and ______ breeds are red.

Short Answer

10. Give an example of vertical integration.
11. What is the value of multiyear marketing contracts for hog producers?
12. Why is it so difficult for small producers to move in and out of the hog business based on the economic outlook at the time?

Chapter 21

Selection and Judging of Swine

Key Terms

contemporary group
estimated breeding value (EBV)
expected progeny difference (EPD)
porcine stress syndrome (PSS)

Objectives

After studying this chapter, the student should be able to

- identify the parts of the live hog and wholesale cuts of the carcass.
- select high-quality breeding stock using generally accepted criteria.
- correctly place and give oral reasons for placings of a ring of four market and four breeding hogs.

SELECTION OF BREEDING STOCK

There has been a major change in the type of market hog in demand today as compared to the typical market hog of the 1980s. Consumers want more lean meat and less fat in their pork. Producers can raise the kind of pork the market demands by using modern production technology and improved breeding programs. Purebred breeders pay careful attention to genetics when selecting bloodlines. Research shows that there are differences among various swine genotypes in their ability to efficiently produce the lean pork the market demands. The kind of environment and the health management techniques used by the producer also influence the rate and efficiency of lean growth in market hogs.

Practically all market hogs today are produced by some type of crossbreeding program. Rotational, rotaterminal, and terminal crossbreeding systems are used. Because a terminal crossbreeding system tends to produce leaner pork, it is now used to produce the majority of the market hogs in the United States. Good-quality, leaner, more heavily muscled hogs bring a premium price in the market (Figure 21-1).

Use of Ultrasound

Ultrasound can be used to measure the fat-free lean pork content of live hogs and carcasses. Measuring the amount of fat-free lean pork in live hogs can help producers make decisions about which animals to keep for breeding stock. Packers can use ultrasound measurements to help determine the value of the hog carcass. Two types of ultrasound machines may be used:

- A-mode machines use sound waves from a transducer and measure the time required for these waves to reflect back to the machine. This measurement is then converted into a distance measurement.
- Real-time (B-mode) machines use many sound waves at the same time to create a two-dimensional image that can be viewed on a monitor as it is created.

The A-mode machine is less expensive but is not as accurate in measuring the fat-free lean content in either live hogs or carcasses as is the real-time, B-mode, machine. The real-time machine is probably the better choice, despite its higher cost, for making genetic selections for lean content in breeding stock.

USDA

Figure 21-1 High-quality pork is in demand in today's market.

Parts of the Hog

Live Hog

A diagram of the parts of the live hog is shown in Figure 21-2. To describe, evaluate, or judge swine, it is necessary to be familiar with the parts of the hog.

Carcass

The four primal cuts of the hog carcass are ham, loin, Boston shoulder (Boston butt), and picnic shoulder. These four cuts represent the most valuable part of the hog carcass. In a typical 270-pound market hog, these cuts will make up about 44 percent of the live weight and represent about 75 percent of the total value of the animal. Assuming a carcass weight of about 197 pounds from this typical market hog, these cuts represent almost 60 percent of the retail cuts on a semi-boneless basis. These values will vary depending on the method of cutting the carcass and the quality of the carcass. Figure 21-3 shows the location of the wholesale cuts of the hog.

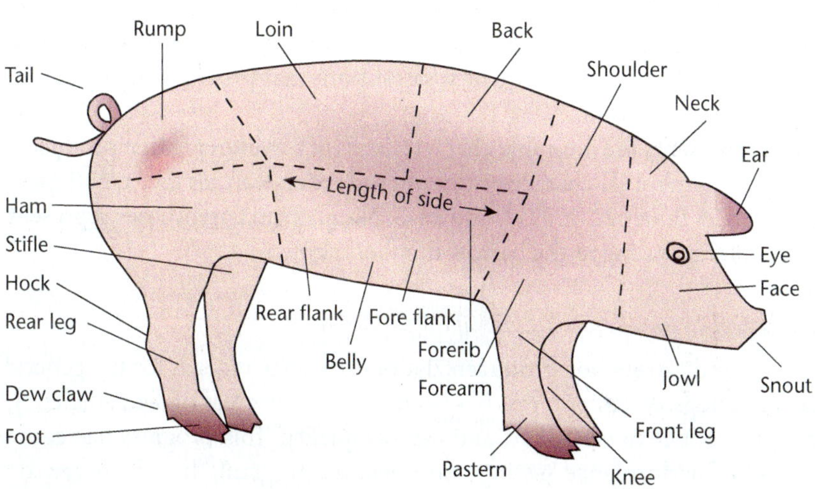

Figure 21-2 Parts of a hog.

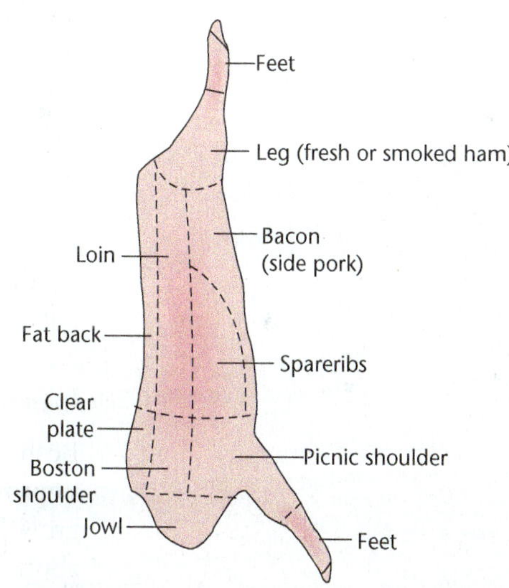

Figure 21-3 Wholesale cuts of the hog.

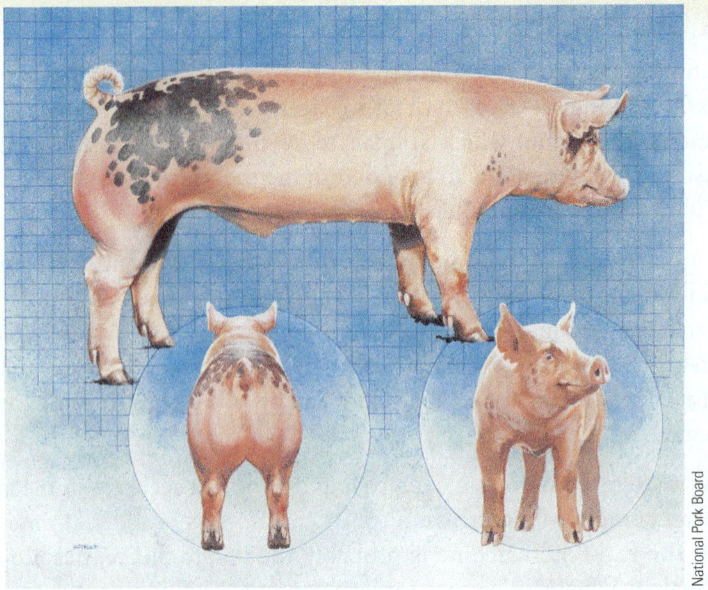

Figure 21-4 Symbol III is an ideal market hog. This ideal market hog has correctness of structure, production, performance, function, livability, attitude, health, and optimum lean yield.

Market Hog Description

Improvement in identifying and selecting desirable breeding lines through better genetics and the use of more technology in producing hogs has resulted in significant improvement in the quality of pork produced in the United States. Market hogs are more muscular, meatier, and have less fat as compared to market hogs produced in the past. The typical market hog today is marketed at about 270 pounds and produces a carcass weight of about 197 pounds. It has about 0.8 inch of back fat at the 10th rib and a loin eye area of about 7 square inches. The fat-free lean index is about 52 percent, and it produces about 105 pounds of lean meat.

The National Pork Producers Council (NPPC) originally described the ideal market pig named Symbol in 1983. Because of significant improvements in the quality of market hogs today, the NPPC has updated its description of the ideal market pig two times and now calls it Symbol III, adopted in 2005 (Figure 21-4). Symbol III is an ideal market hog that symbolizes profitability for every segment of the swine industry. The ideal market hog has correctness of structure, production, performance, function, livability, attitude, health, and optimum lean yield. Symbol III also produces the best quality, safest pork that provides the optimum nutrients for human nutrition.

Symbol III represents the desirable characteristics of a market pig that producers strive to achieve. Symbol III characteristics of production, carcass merit, and quality are described in Tables 21-1, 21-2, and 21-3. Numbers in parentheses represent gilt numbers corresponding to the barrow numbers shown.

Genetic Evaluation of Breeding Stock

Both seedstock and market hog producers use performance records for the genetic improvement of hogs. Much progress has been made in the areas of feed efficiency, growth rate, reproductive efficiency, and carcass quality. This progress has come about because of performance testing and selection programs based on genetic principles that result in the selection of genetically superior breeding stock.

TABLE 21-1 Production Characteristics

- live-weight feed efficiency of 2.4 (2.4)
- fat-free lean gain efficiency of 5.9 (5.8)
- fat-free lean gain of 0.95 pounds per day
- marketed at 156 (164) days of age
- weighing 270 pounds
- all achieved on a corn-soy equivalent diet from 60 pounds
- free of all internal and external parasites
- from a high-health production system
- immune to or free of all economically important swine diseases
- produced with Environmental Assurance
- produced under PQA™ and TQA™ guidelines
- produced in an operation that has been SWAP assessed (Swine Welfare Assurance Program)
- free of the stress gene (Halothane 1843 mutation) and all other genetic mutations that have a detrimental effect on pork quality
- result of a systematic crossbreeding system, emphasizing a maternal dam line and a terminal sire selected for growth, efficiency, and superior muscle quality
- from a maternal line weaning >25 pigs/year after multiple parities
- free of all abscesses, injection site blemishes, arthritis, bruises, and carcass trim
- structurally correct and sound, with proper angulation and cushion and a phenotypic design perfectly matched to the production environment
- produced in a production system that ensures the opportunity for stakeholder profitability from the producer to retailer while providing a cost competitive product retail price in all domestic and export markets
- produced from genetic lines that have utilized genomic technology to support maximum improvement in genetic profitability and efficiency

TABLE 21-2 Carcass Characteristics

- hot carcass weight of 205 pounds
- loin muscle area (LMA) of 6.5 (7.1) inches square
- belly thickness of 1.0 inch
- 10th rib back fat of 0.7 (0.6) inch
- Fat-Free Lean Index of 53.0 (54.7)

TABLE 21-3 Quality Characteristics

- muscle color score of 4.0
- 24-hour pH of 5.9
- maximum drip loss of 2.5%
- intramuscular fat level of 3.0%
- provides a safe, wholesome product free of all violative residues and produced and processed in a system that ensures elimination of all food-borne pathogens
- free of within-muscle color variation and coarse muscle texture
- free of ecchymosis (blood splash)
- provides an optimum balance of nutrients important for human nutrition and health

Economic Value of Improved Genetics

Market hog producers who want to improve their herds need to secure their breeding stock from seedstock producers who are consistently doing performance testing and following good selection principles. Seedstock producers using genetic improvement programs must keep good records, do performance testing, and use selection indexes that relate to important economic characteristics to select breeding stock for continued genetic improvement. Superior performance tested boars must be used by seedstock producers to achieve the kind of genetic improvement required to keep pork competitive with other sources of meat in the marketplace.

The major breed associations all have improvement programs in place to take advantage of genetic principles when producing seedstock. A number of companies produce hybrid seedstock using sound genetic principles. These types of breeding programs are expensive, and the producers of seedstock expect to make a fair return.

Buying better breeding stock will cost the commercial hog producer more money. However, using improved breeding stock can significantly improve feed and reproductive efficiency in the commercial herd. Carcass quality can also be improved, producing the kind of market hogs in demand by packers. Genetically improved breeding stock can potentially return more profit to the producer.

Use of Performance Records

There are many different performance records that might be collected on individual hogs or groups of hogs. When considering improvement programs, those records that have an impact on profitability are the most important. Such records include carcass traits such as loin eye area, length, muscling score, and meat quality. Other important traits include litter size, litter weight, milking ability, feed efficiency, and number of days to market.

Environmental effects can influence the performance of individual pigs. Typically, records are kept on contemporary groups to help reduce the effect of environmental factors. A **contemporary group** is a group of animals that are similar in a number of characteristics and have been raised under the same management practices. Age, sex, and breed or cross are often used to set up a contemporary group.

Performance records on close relatives of potential breeding stock are also important. It is possible to secure information about carcass traits after slaughter by keeping records on close relatives. In the past, some central boar testing stations used several animals from the same litter to secure information on traits such as average daily gain, feed efficiency, and carcass quality. The barrows that are littermates of the boar on test are slaughtered and carcass information is collected to help determine the breeding value of the boar. Today, most testing is done by the farms or companies that raise breeding stock. They keep detailed records on each genetic line they are working with.

Estimated Breeding Value

Estimated breeding values (EBVs) are determined by applying genetic principles to performance records. The EBV is a selection index because it combines information from a number of sources to determine the genetic merit of the individual. Important considerations are the number of records available, heritability of the measured traits, and pedigree information. The performance records of groups of close relatives are considered in determining the EBV of the individual. The performance records of close relatives are more important in determining the EBV than records of more distant relatives. The correlation among genetic traits is also taken into consideration. Traits with a high degree of heritability are expected to be passed on to the offspring of the breeding animal to a greater degree than those traits with a low degree of heritability.

Expected Progeny Difference

The EBV is the total estimated genetic value of an animal. Only one-half of the genes of each parent are passed on to the progeny. This is referred to as the **expected progeny difference (EPD)**, which is defined as one-half the EBV. In a given population, the average EPD is zero. Individual EPDs are expressed as deviations (+ or –) from the average. The EPD measures how much difference in performance on a given trait can be expected from the progeny of the breeding animal as compared to the average of all animals in the population. The EPDs of potential breeding animals can be compared across herds within the same breed. They cannot be compared between breeds.

Both purebred breeders and commercial companies producing seedstock collect performance records and compute the EPD for their breeding animals. Information from performance records, including records on relatives, may be used with a system called Best Linear Unbiased Predictor (BLUP) to determine the breeding animals with the highest EPDs. This procedure provides a highly accurate measure of the true breeding value of an individual. It also provides information regarding the correlation of multiple traits to determine breeding value.

Sire summaries and EPD values are made available by the various purebred associations, the National Swine Registry, and the National Pork Producers Council. Commercial seedstock companies vary in their policies on making this information available.

Performance Testing Programs

There are several genetic evaluation programs that may be used by swine breeders to help determine the breeding value of their hogs.

Computer programs called SWINE-EBV are available to analyze performance data collected on the producer's farm. The producer can run the software with an onsite computer so results are quickly available. Data, including data from relatives, are collected on reproduction (number born alive and 21-day litter weight) and post-weaning (days to 230 pounds and back-fat thickness) traits and are entered into the computer program. Maternal, general, and terminal sire indexes are computed from the data. Index numbers are ranked according to their economic importance.

The original Swine Testing and Genetic Evaluation System (STAGES) was developed by Purdue University, the USDA Agricultural Research Service, USDA Extension Service, National Association of Swine Records, several purebred associations, and the National Pork Producers Council. This performance and evaluation program was implemented in 1985. A new version of STAGES was implemented in 1998 to provide EPD information for the Yorkshire, Landrace, Duroc, and Hampshire breeds of swine.

Factors used to determine selection indexes for sows and boars include number of pigs born alive, number weaned, 21-day litter weight, growth rate to 250 pounds, back fat adjusted to the 250 pound weight, feed conversion, and carcass data. The Terminal Sire Index (TSI), Maternal Line Index (MLI), and Sow Productivity Index (SPI) are calculated as well as EPDs for most of the traits that are economically important. An EPD for pounds of lean adjusted to 250 pounds is also calculated.

Animals are evaluated within contemporary groups and breed-specific adjustments are made for the EPDs reported. Another feature of the revised STAGES is the use of a rolling base genetic year and the computation of new values every night. This method of computing EPDs provides a higher degree of consistency for the evaluation of each trait from year to year. A better evaluation of the true genetic breeding value for each animal also results from this change. EPDs cannot be compared across breeds.

The National Pork Producers Council has two national genetic evaluation programs for swine. The Terminal Sire Line program was begun in 1993 to evaluate boars from a number of different sire lines. The Maternal Sow Line program was implemented in 1997. These programs are designed to evaluate genetic lines for traits that are important for producing crossbred market hogs. These evaluation programs test more traits and more genetic lines than other evaluation programs.

The Acid Meat Gene Problem

There is some concern in the U.S. swine industry about a gene that is apparently found only in Hampshire pigs. The gene is also called the Rendement Napole (NP) or Napole gene. This gene has some negative effects on the quality of meat, causing it to be more acidic than normal. This reduces the water-holding capacity of the meat, which causes some loss for packers and processors. However, the gene has also been associated with a positive effect on growth and carcass traits and with improving the eating quality of the meat showing an increase in tenderness and juiciness.

The gene was first observed in Hampshire hogs in France. The frequency of the gene in Hampshire populations in the United States is not known. With DNA testing, work is being done by some breeders and companies to have the Napole gene genetically removed from some lines of Hampshires.

Other Considerations in Selecting Breeding Animals

The selection of breeding stock should always begin with a review of performance records and selection indexes. In addition to the genetic evaluation of breeding stock, there are other things the producer must consider when selecting breeding animals. These include:

- the type of breeding program used by the purchaser
- the production practices of the seedstock producer
- the soundness of the animal
- the health of the seedstock herd
- the minimum breeding age
- whether to buy or raise replacement gilts

Type of Breeding Program Used by the Purchaser

A producer using a rotational crossbreeding program needs to select boars that produce both desirable market hogs and desirable breeding gilts. Such boars should be selected from sows with good maternal characteristics. These boars also need to have good growth and leanness characteristics. Boars being selected for use in a terminal breeding program need to show outstanding growth and carcass traits. One of the reasons for the increased use of terminal crossbreeding programs in the production of market hogs is that it is difficult to make rapid genetic progress when selecting boars for both maternal and market hog characteristics.

Figure 21-5 Gilts selected for the breeding herd should have at least six functional teats on each side.

Production Practices of the Seedstock Producer

The production of seedstock for breeding programs is done by both individual purebred breeders and large breeding stock companies. The potential buyer needs to determine the reputation of the prospective seller by checking with other people who have purchased seedstock from the seller. Extension specialists and agribusiness people are other sources of information concerning a seedstock producer. The health program followed by the seedstock producer needs to be evaluated. Find out what vaccinations are used and if there have been any disease problems in the herd. Determine whether the producer routinely does performance testing to help assess the genetic value of the seedstock it produces.

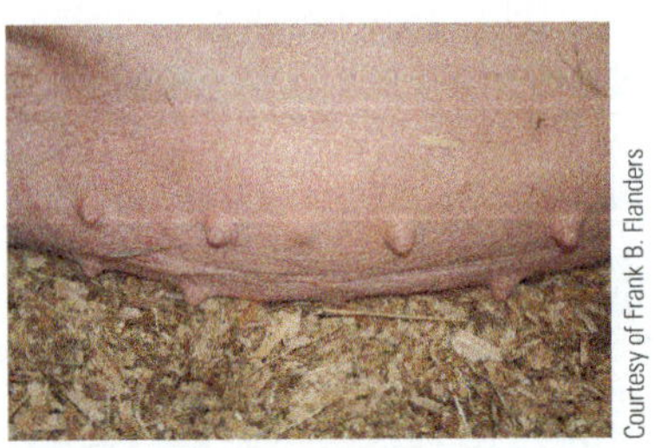

Figure 21-6 Example of a blind teat.

Soundness of the Animal

Select a boar that has visibly sound reproductive organs. The testicles should be well developed and of equal size. Do not select boars that have umbilical or scrotal hernias or other obvious structural problems. Select boars that are aggressive and show a desire to mate.

Sound underlines are important for both boars and gilts. Gilts should have six or more functional teats on each side (Figure 21-5). Check for proper spacing of the nipples and make sure there are no inverted, pin, or blind nipples (Figure 21-6).

Select gilts and sows that show normal development of the reproductive system. Gilts with small vulvas and vulvas that have the tip turned upward (tipped) should not be kept. This is often an indication of internal reproductive defects. Sows that have problems in farrowing should be culled from the herd. Select gilts with strong pasterns and sound feet and legs. Do not purchase breeding stock with obvious defects.

Health of the Seedstock Herd

The health of the herd from which breeding animals are purchased is important. Buy only healthy animals from healthy herds. Herds should be certified brucellosis and pseudorabies free. Breeding animals should be vaccinated for erysipelas and the six strains of leptospirosis. Purchase animals only if they are free of external and internal parasites. Herd health information should be available from the breeder.

Breeding stock should not be either homozygous (nn) or heterozygous (Nn) for the **porcine stress syndrome (PSS)** gene. This is an inherited neuromuscular disease. When stressed, animals carrying this recessive gene exhibit symptoms including heavy breathing, tail tremors, splotchy coloring, and occasionally death. When pigs with the PSS gene are harvested after being stressed, the meat is of lower quality. Two tests exist, either of which may be used to determine if an animal carries the PSS gene. The most accurate test is a check of the DNA of the animal. The DNA test will identify both homozygous and heterozygous carriers. The other test involves exposing the animal to the anesthetic Halothane. If the animal is homozygous for the PSS gene, it responds within 3 minutes by showing extreme muscle rigidity. The Halothane test cannot detect animals that are heterozygous for the PSS gene.

Minimum Breeding Age

Boars must be a minimum of 7½ months of age before they are used for breeding purposes. Breed gilts at the time of their second or third estrus after reaching puberty because they usually have a higher ovulation rate at this time. Replacement boars and gilts should be bought 30 to 60 days before they are to be used. This permits isolation for health checks. Also, they can adjust to the farm and boars can be test mated for breeding performance.

To Buy or Raise Replacement Gilts

The increasing use of terminal crossbreeding programs in market hog production means the producer has to make a decision regarding whether to buy or raise replacement gilts. It requires more capital investment to buy replacement gilts. This method may provide the producer with breeding stock of greater genetic value that may offset some of the additional cost. Raising replacement gilts requires additional facilities and a good record-keeping system. Producers need to evaluate their own swine enterprise to determine which method of securing replacement gilts works best for them.

SELECTING FEEDER PIGS

Common sources of feeder pigs are pig nurseries, farmers, auction barns, and dealers who buy and sell feeder pigs (Figure 21-7). Factors to consider when buying feeder pigs are: health, type, size, and uniformity.

Courtesy of Cole's Swine Farms

Figure 21-7 When purchasing feeder pigs, buyers should consider health, type, size, and uniformity of the pigs.

Health Only healthy pigs should be purchased. Pigs that have visible signs of sickness—such as coughing, infected eyes, rough hair coats, potbellies, gauntness, or listless appearance—should not be selected. Pigs should be wormed, tail docked, and castrated. Do not buy pigs that show signs of external parasites.

Type Meaty feeder pigs will produce the kind of carcass in demand on the market. Short, fat pigs will be over-finished when they reach market weight. The same type and conformation that is desirable in breeding hogs should be looked for in feeder pigs.

Size Feeder pigs usually range from 35 to 80 pounds in weight. Select pigs that have a good size for their age. Size for age is more important than condition or fatness when selecting feeder pigs.

Uniformity Uniformity in size, age, condition, and type is desirable in a group of pigs. When these traits are uniform, the pigs will feed out well together. All the pigs in the group will tend to reach market weight at about the same time.

JUDGING HOGS

A hog judge must know the parts of the live hog and the wholesale cuts of the carcass. Refer to Figures 21-2 and 21-3. To judge a class of hogs, look at them from a distance of about 15 feet. To identify them for judging, the hogs in the class have numbers marked on their bodies. Judge them as they move around the ring. Look at each hog and compare it with the ideal hog and the others in the class. Taking notes and giving reasons is done in the same manner as when judging cattle.

Using Performance Data in Judging Hogs

Traditionally, judging hogs has been done using only a visual appraisal of the animals in the class, with decisions being based on whether the class was market or breeding hogs. Because of the increased emphasis on genetic improvement in the swine industry, judging contests now often include performance data on the hogs in the class to be used in addition to the traditional visual appraisal in ranking the animals.

When performance data are used in a judging class, the contestants are given printed information about the class. This normally includes the breed and sex of the animals, the planned use of the animals, a description of the production environment, and the plan for marketing the animals or their progeny. Selection priorities may be given with the printed information or the contestants may have to develop their own priorities based on the information given them. Performance records are provided for each animal. These will include reproduction, growth, and composition records. Records may be provided as individual measurements, may be given as ratios or indexes, or may be expected progeny differences (EPDs). The most valuable information for judging purposes is the EPDs; the least valuable is individual measurements.

The EPDs are the best records because they combine more information; this makes possible a better evaluation of the true genetic merit of each animal. Individual records do not give enough information to fairly compare the animals in the class. Three indexes are typically used to provide information: (1) Sow Productivity Index, (2) Terminal Sire Index, and (3) Maternal Line Index. All of these indexes are calculated with EPDs. Most indexes use 100 as the average value; therefore, an index value above 100 is better than average and a value below 100 is below average. This provides a better comparison of the animals in the class than do individual measurements. When given a choice of performance data, use the EPDs to make preliminary rankings rather than individual measurements or indexes.

When performance records are given, make a preliminary ranking of the class based on those data. Then proceed to rank the class based on the visual traits as

Figure 21-8 At a 4-H/FFA livestock judging contest, contestants judge swine, focusing on body conformation, size, muscling, finish, and reproductive soundness.

described in the next section. The final ranking of the class must be made on the basis of both performance data and visual appraisal of the animals.

Visual Evaluation of the Class

A number of traits may be visually evaluated when judging hogs. These include body conformation, size, muscling, finish, and reproductive soundness (Figure 21-8).

Type refers to the conformation of the hog's body. It is judged on the basis of length of side and skeletal size (scale). Body length is related to growth rate and the productivity of the sow. The length of side may be estimated by looking at the distance from a point in the center of the ham to the forepart of the shoulder. A pig with desirable traits is shown in Figure 21-9.

A visual evaluation of the amount of muscling on a hog is best seen by looking at the rear view of the animal. In this view, the hog should show a wide back and loin and a deep rump. The ham should be deep, thick, and firm. The chest and shoulders should be wide. Be sure the observed width is due to muscling and not to excess fat. The correct shape across the back is an arc rather than a square There is greater width across the rump and ham than there is across the back. A narrow ham indicates poor muscling.

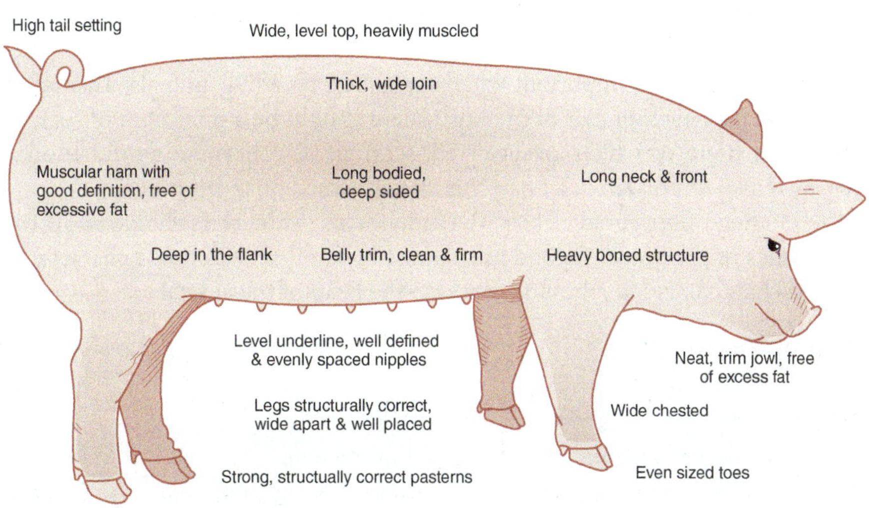

Figure 21-9 Desirable traits in swine.

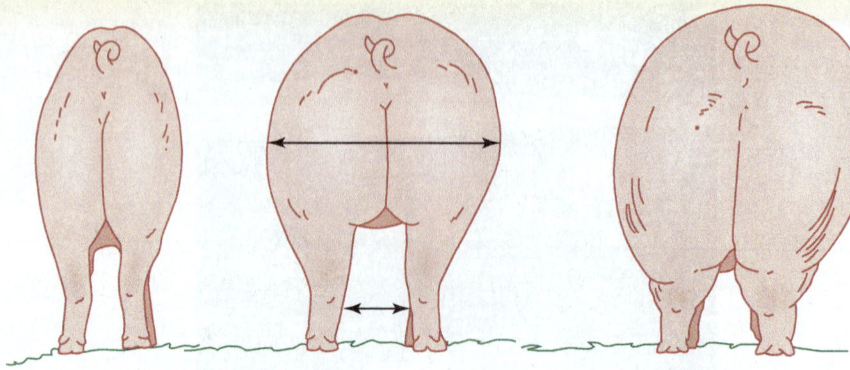

A B C

Figure 21-10 Pig A is too narrow across the hams and between the legs, indicating a lack of muscle. Pig C is an old-fashioned type of market hog with little muscling and a lot of fat. Pig B is wide across the middle of the hams and has a wide stance of the legs, indicating better muscling than Pigs A and C. Pig B is the best selection of this group.

Finish refers to the amount of fat on the hog. Some fat is desirable; excessive fat is undesirable (Figure 21-10). The amount of back fat is generally measured at the 10th rib and the last rib section of the loin. Performance data usually include information about the leanness and muscling of the animals in the class. This information needs to be combined with a visual evaluation. Indications of excessive fat include heavy, wasty jowl; shaky middle, square top; looseness in the ham and crotch; or a roll of fat over the shoulder. If the hog is too wide in the middle, it will not have a desirable yield of lean cuts. This lowers the market value of the hog.

Quality refers to the degree of refinement of head, hair, hide, and bone of the live hog. Moderate refinement is preferred over animals that are too coarse or too refined. A smooth hair coat and hide is desirable; the hide should be free of wrinkles. Pigs with rough skins or roughness over the shoulders are undesirable.

All parts of the hog should be properly proportioned. This is referred to as *balance*. A well-balanced hog is loose framed and moves well with the correct levelness from the point of the shoulder through the rump.

Soundness of feet and legs is important. Many hogs are raised on concrete floors and must have good feet and legs to do well under these conditions. Legs must come out of the shoulder and rump with correct slope and angle to have some give in the joints as the animal moves. The feet should be big and even-toed, and set out square on the animal's body.

Underline is especially important when evaluating breeding animals. This refers to the mammary development of the hog. There should be a minimum of six nipples per side, none of which are inverted, scarred, or otherwise nonfunctional. Underline is also important in a boar; this trait is passed on to his progeny.

In the judging of purebred classes, the individuals in the class should show the characteristics of the breed. A good judge must be familiar with breed characteristics. Breed character is often best observed in the head of the animal.

Judging Terms

General Terms for Market and Breeding Hogs

- frame size matches body weight
- better profile and heavy muscled
- longer and squarer in the rump
- squarer rump
- thicker through the rump
- meatier and wider at the loin

- more natural thickness down the top
- cleaner, trimmer along the loin edge
- cleaner fronted
- larger skeletal structure
- heavier muscled
- trimmer finished
- smoother
- sounder on front legs
- heavier bone
- has a better top shape
- longer, more correct muscle structure
- leveler over the top
- trimmer underline

Market Hog Terms

- longer, larger framed
- heavier muscled
- longer, stretchier side
- firmer finished barrow
- smoother side
- more uniform width
- longer rump
- trimmer middled
- heavier muscled barrow
- longer, deeper, fuller in the ham
- cleaner top
- more correctly finished
- more muscling over the top
- thicker loin
- trimmer in jowl and underline

Breeding Hog Terms

- sounder underline
- stands and walks more correctly
- more evenly spaced nipples
- shows more femininity
- wider fronted
- meatier gilt
- roomier-middled gilt
- shows more size and scale
- more desirable set to the legs
- broodier
- more breed character
- correct set to legs
- more rugged, heavier bone
- longer, deeper sided gilt
- growthier gilt
- deeper, wider sprung gilt

SUMMARY

Most market hogs produced in the United States today are the result of some type of crossbreeding program. Terminal crosses are most commonly used. Market hogs today are more muscular, produce a larger loin eye area, and have less back fat than those previously produced.

Performance records are widely used by seedstock producers and market hog producers to evaluate breeding stock. Genetics plays a large role in selection of breeding stock. Expected progeny difference (EPD) is a measure of how much difference in performance on a given trait can be expected from the progeny of a breeding animal as compared to the average of all animals in the population. EPDs can be compared within breeds and across herds, but not across breeds. Sire summary and EPD values are available from several sources on the Internet.

In addition to performance records, other criteria to use when selecting breeding stock include soundness of the animal and health of the breeding stock herd.

Feeder pigs must be healthy, meaty hogs to produce the best profits. Select feeder pigs that are healthy and have good size for their age. Uniformity within a group is also important.

In judging hogs, it is now common practice to include information from performance records combined with visual appraisal to place the class. The best performance records to use for judging hogs are EPDs. Visual appraisal should focus on body conformation, size, muscling, finish, and reproductive soundness. Final placing in a class is made on the basis of both performance data and visual appraisal of the animals.

Student Learning Activities

1. Using a live hog or photograph, name the parts of the hog.
2. Judge breeding and market classes of hogs and give oral reasons.
3. Practice good swine selection procedures when planning and conducting a swine Supervised Agricultural Experience Program.
4. Give an oral report to the class on an aspect of swine selection, testing, or judging.
5. Evaluate pedigree information on breeding swine and select the most desirable animals.
6. Attend a purebred swine producer's sale or locate pigs for sale online and select animals that you would buy based on type, pedigree, and production records.
7. Observe the use of ultrasonics in swine selection.

Discussion Questions

1. Describe the use of ultrasound to measure the fat-free lean content of live hogs and carcasses.
2. Name the four primal cuts of the hog carcass.
3. Describe the characteristics of the modern market hog.
4. Describe the goals for market hog characteristics suggested by the National Pork Producers Council.
5. What economic value can be realized by improving genetic lines of breeding stock in the swine industry?
6. What are the most important performance records to be measured in genetic improvement programs?
7. How are estimated breeding values determined?
8. What is expected progeny difference, and how is it used in genetic improvement programs?
9. List and briefly describe several performance testing programs currently being used in the swine industry.
10. What is the acid meat gene problem?
11. Name and briefly describe six factors other than performance records that should be considered when selecting breeding animals.
12. What is the minimum breeding age for boars and gilts?
13. How should performance data be used when judging hogs?
14. Name four factors that should be considered when buying feeder pigs.
15. Briefly describe the important visual traits that should be evaluated when judging hogs.

Review Questions

True/False

1. Underline or mammary development is not important in judging a boar.
2. Finish refers to the amount and condition of the hair on a hog.
3. When performance data is used in judging a class of pigs, the contestants are given printed information about the class.

Multiple Choice

4. A visual evaluation of the amount of muscle on a hog is best seen by looking at the ______ view of the hog.
   a. side
   b. rear
   c. front
   d. top
5. ______ refers to the degree of refinement of head, hair, hide, and bone of the live hog.
   a. Finish
   b. Quality
   c. Proportion
   d. Soundness

Completion

6. In hogs, the underline should have a minimum of ______ nipples per side.
7. ______ is the total estimated genetic value of an animal.

Short Answer

8. Contrast consumer demand today as compared with the typical market hog of the 1980s.
9. List important aspects of performance records.
10. What does soundness mean when selecting a boar?

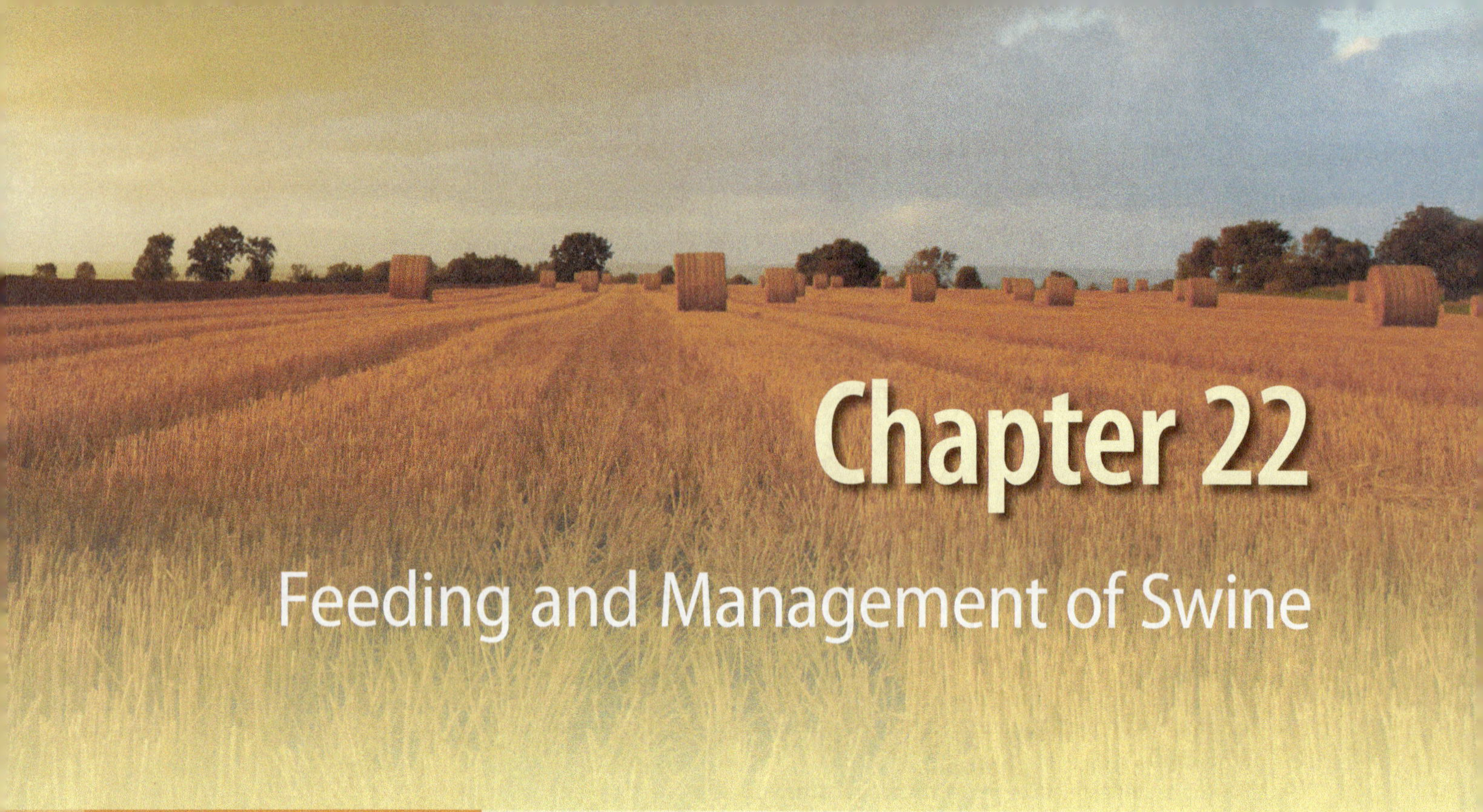

Chapter 22

Feeding and Management of Swine

Key Terms

PQA Plus
scab
fagopyrin
ergot
binding quality
gossypol
arsenical
flushing
spray-dried porcine plasma (SDPP)
phase feeding
free choice
complete mixed feed
toxin
restricted feeding
split-sex feeding
multiple farrowing
anestrus
tail docking
segregated early weaning (SEW)
medicated early weaning (MEW)
all-in/all-out

Objectives

After studying this chapter, the student should be able to

- describe different types of swine production.
- develop feeding programs for the different stages in the life cycle of hogs.
- describe accepted management practices for the stages in the life cycle of hogs.

The efficient use of resources is the key to profitability in swine enterprises. To remain competitive, swine producers must select breeding stock that will produce lean hogs and maintain good feed efficiency. Management techniques such as split-sex feeding, phase feeding, and all-in/all-out production cycles help increase production efficiency.

Two factors that have a major influence on profitability in swine production are the number of pigs weaned per sow per year and feed efficiency. Efficient producers are currently weaning 22 to 25 or more pigs per year per sow. Females should be bred and managed to produce a minimum of 2.3 litters during each 12-month period. Sows that cannot maintain this level of production should be culled.

Improved genetics has resulted in hogs that are more efficient in the use of feed. Feed efficiency, the pounds of feed used per hundred pounds of gain, should be in the range of 330 to 350. Feed wastage is a significant factor in determining the amount of feed used. Feeders need to be kept properly adjusted to reduce feed losses.

Table 22-1 shows feed cost per hundredweight (cwt) of gain based on feeding a sow and litter from farrow to finish. The table assumes a feed conversion

TABLE 22-1 Feed Cost per Hundredweight of Gain Based on Feeding a Sow and Litter from Farrow to Finish

Price/cwt of Supplement	Price of Corn per Bushel						
	$2.00	**$2.50**	**$3.00**	**$3.50**	**$4.00**	**$4.50**	**$5.00**
$12.00	20.28	22.80	25.32	27.84	30.36	32.88	35.40
$12.50	20.71	23.23	25.75	28.27	30.79	33.31	35.83
$13.00	21.13	23.65	26.17	28.69	31.21	33.73	36.25
$13.50	21.56	24.08	26.60	29.12	31.64	34.16	36.68
$14.00	21.98	24.50	27.02	29.54	32.06	34.58	37.10
$14.50	22.41	24.93	27.45	29.97	32.49	35.01	37.53
$15.00	22.83	25.35	27.87	30.39	32.91	35.43	37.95
$15.50	23.26	25.78	28.30	30.82	33.34	35.86	38.38
$16.00	23.68	26.20	28.72	31.24	33.76	36.28	38.80
$16.50	24.11	26.63	29.15	31.67	34.19	36.71	39.23
$17.00	24.53	27.05	29.57	32.09	34.61	37.13	39.65
$17.50	24.96	27.48	30.00	32.52	35.04	37.56	40.08
$18.00	25.38	27.90	30.42	32.94	35.46	37.98	40.50
$18.50	25.81	28.33	30.85	33.37	35.89	38.41	40.93
$19.00	26.23	28.75	31.27	33.79	36.31	38.83	41.35
$19.50	26.66	29.18	31.70	34.22	36.74	39.26	41.78
$20.00	27.08	29.60	32.12	34.64	37.16	39.68	42.20
$20.50	27.51	30.03	32.55	35.07	37.59	40.11	42.63
$21.00	27.93	30.45	32.97	35.49	38.01	40.53	43.05
$21.50	28.36	30.88	33.40	35.92	38.44	40.96	43.48
$22.00	28.78	31.30	33.82	36.34	38.86	41.38	43.90

of 5.04 bushels of corn and 85 pounds of supplement per hundred pounds of pork produced. These conversion rates may vary with individual producers.

TYPES OF SWINE PRODUCTION

Swine production can be divided into two types: purebred and commercial. Commercial production systems can be further divided into three systems: feeder pig production, buying and finishing feeder pigs, and farrow-to-finish.

The various types of swine production can also be classified according to the kind of housing used. Pasture systems, combinations of pasture and low-investment housing, and high-investment total confinement systems are three types of housing systems used in swine production. Swine on pasture with low-investment housing are shown in Figure 22-1.

The characteristics of a confinement management system include:

- high level of mechanization to reduce labor requirements
- high investment in buildings and equipment
- multiple farrowings per year, with a large number of hogs raised
- high level of management ability needed
- high degree of control over feeding operations
- better year-round working conditions
- stringent disease and parasite control programs
- a limited amount of land

Courtesy of Frank B. Flanders

Figure 22-1 One type of swine production utilizes a combination of low-investment housing and pasture.

Pasture Management

The rapid growth in confinement hog production in the 1970s brought about a massive shift in the commercial swine industry from pasture management to large-scale, indoor commercial facilities. By the late 1990s, larger producers had converted to indoor, confinement housing or had left the swine business. Pasture management in swine production was rarely used and was used almost exclusively by small producers.

Several factors have caused a slight resurgence in the small-scale, pasture system of swine operations. A niche market is developing in some areas based on concern for animal welfare and increased consumer interest in "all-natural" and "organic" pork. Locally raised organic or all-natural pork brings a premium in some markets. The high cost of confinement facilities provides a cost advantage for pasture production. The equipment for pasture production includes low-cost shelters or portable huts, watering systems, and feeders, —most of which can be easily moved when rotating pastures.

Properly managed pasture systems are friendly to the environment and provide better air quality than confinement operations. Producers do not have to worry about building air quality, and if the pastures are not overcrowded, manure disposal in pasture management is not a problem. Mud and erosion problems are avoided by raising hogs on sandy, well-drained soils and by rotating pastures.

Pasture management can be used with all breeds and colors of swine. However, it is best to use the darker colored breeds (Duroc, Hampshire, etc.) or a crossbreed of swine. The darker breeds tend to be better equipped for changes in the environment, especially the intense and prolonged exposure to sunlight in pasture management.

Purebred Production

The production of purebred hogs is a specialized business. Registered purebred hogs make up less than 1 percent of the total hogs raised in the United States. Purebred producers perform an important function in the swine industry by producing the breeding stock used in commercial hog production. Much of the improvement in hog type comes from the work of the purebred breeders.

Purebred producers must be excellent managers. They often have higher investments in labor and record keeping than commercial hog producers. The purebred producer must keep accurate records of the ancestry of the hogs produced. Careful recording of breeding and farrowing dates is essential for the purebred operation. Purebred producers spend a great deal of time advertising, showing, and promoting swine breeds.

In addition to purebred swine producers, breeding stock is also produced by specialized companies that use the latest techniques to produce superior animals. Purebred breeders are also using genetic principles to produce improved breeding stock that meets current consumer demand for lean pork.

Commercial Production

Most of the pork consumed in the United States is produced by large, commercial hog producers. Most commercial hog producers use some type of crossbreeding system to produce hogs for market. Purebred boars are often used on crossbred sows to produce market hogs. Good management is necessary for success in the commercial hog business.

Feeder pig production is an enterprise that produces pigs for sale to feeders, who then feed them to market weights. The feeder pig producer has a breeding herd of sows. The baby pigs are taken care of until they reach weaning weight.

A high-producing herd that raises large litters is required. It usually requires an average of 14 to 16 pigs marketed per sow per year to break even in feeder pig production. Health problems should be prevented and carefully treated when they occur. The goal of the producer is to raise uniform groups of feeder pigs for sale. Generally, only small investments are required for feeder pig production. A good manager tries to schedule farrowings so as to have a steady supply of feeder pigs for sale. Less total feed is needed for this type of production system than for the other types.

Buying and finishing feeder pigs is an enterprise in which the operator buys feeder pigs and raises them to market weight. This type of operation requires the least investment and managerial ability of any of the hog production systems. It is possible to feed pigs on pasture with very limited facilities. However, almost all large swine operations use confinement systems. The costs of this type of operation are greater, but hogs generally gain more efficiently in confinement systems. The high price of land makes its use as hog pasture questionable.

Buying and finishing feeder pigs does require a high investment to purchase the pigs to be fed. This system is well adapted to the producer who has large amounts of grain for feed. It requires less labor than other systems.

Buying and finishing also have certain disadvantages. The possibilities of health problems are greater in this operation because the purchased feeder pigs may bring diseases to the farm. Market prices of feeder pigs vary a great deal and there is a fairly high risk of not making a profit.

The farrow-to-finish system is the most common method of hog production. This operation involves having a breeding herd of sows, farrowing pigs, and caring for and feeding the pigs to market weights. Investment in facilities can be low for pasture systems. Confinement systems, on the other hand, can require very high investments in facilities. The trend is toward more confinement systems, with larger numbers of sows. This system permits spreading the production and, thus, the marketing of pigs, more evenly through the year.

Labor, management, and investment requirements vary considerably with the kind of system used. Pasture systems require more labor, less management, and lower investments. Confinement systems require less labor, more management, and much higher capital investment.

CONTRACTING IN THE SWINE INDUSTRY

In recent years, the practice of contracting with a packer to raise hogs under contract has become commonplace. A few large packers control a majority of the swine industry in the United States.

Under a typical contracting arrangement, the farmer provides the buildings and equipment. The company offering the contract provides the pigs and financing and makes most of the management decisions. These include the rations used, the type of breeding program, the animal health program and veterinary services used, and marketing decisions. Some contractors are offering financing for buildings and training in raising hogs for those not currently in the business. Typically, contracts may be offered for raising feeder pigs or for feeding feeder pigs to market weights.

Under the contracting arrangement, the farmer is generally paid a flat fee per pig raised or marketed. The fee may be adjusted based on the efficiency of the operation as measured by the feed conversion rate.

Contracting reduces the financial investment required by the farmer and reduces his risk. However, the farmer loses independence in making management decisions.

Control of the swine enterprise is placed in the hands of the company that offers the contract.

Contractors have the advantages of access to large amounts of capital, greater buying and selling power, greater ability to specialize, and the ability to adopt new technology more rapidly.

PORK QUALITY ASSURANCE PLUS PROGRAM

The **PQA Plus** is a management education program with major emphasis on swine herd health and animal well-being. In 1989, the National Pork Producers Council implemented the Pork Quality Assurance[SM] (PQA) Program. In 2007, the PQA program and Pork Checkoff's Swine Welfare Assurance Program (SWAP) joined to form the PQA Plus. Anyone who raises swine can take part in the program. Producers who are interested in participating in this program perform a review of their management practices, especially focusing on the use and handling of animal health products. A series of good management practices is reviewed and a plan is developed for any needed improvements. At the final step in the program, the producer's plans are reviewed and verified by a veterinarian, an extension specialist, or an agricultural education instructor.

REDUCING NITROGEN AND PHOSPHORUS EXCRETION

Two problems faced by swine producers are odor and pollution of the environment by excessive amounts of nitrogen and phosphorus in the manure. Some research suggests that changes in swine diets can help reduce these problems.

Much of the odor problem in confinement operations is related to the release of ammonia from the manure. This is caused by the nitrogen that is in the manure. It is possible to reduce the amount of nitrogen excreted by the pigs by substituting some synthetic lysine (amino acid) for soybean meal in the diet. This lowers the amount of nitrogen excreted and thus reduces the amount of ammonia generated, which in turn reduces the odor from the facility. Diets should be balanced to the first three to four limiting amino acids. Rations can be balanced for least-cost rations according to the feedstuffs that are available to the producer to help lower feed cost.

Another method of reducing nitrogen excretion is by using split-sex feeding. Barrows have a lower protein requirement than gilts. Feeding the barrows separately from the gilts means they can be fed less protein, thus reducing the amount of nitrogen excreted. Using four to five diets in phase feeding rather than two allows a reduction in the amount of protein fed, and this results in a corresponding reduction in the amount of nitrogen excreted in the manure.

Corn and soybean oil meal, which are the basis of most swine diets, have a fairly high phosphorus content. However, as much as 90 percent of the phosphorus content is in the form of phytic acid, which is not available to the pig. The diet must then be supplemented with additional phosphorus, which increases the amount of phosphorus that is released in the manure. This contributes to an excessive amount of phosphorus being released into the environment when the manure is applied to the land.

More phosphorus can be utilized by the animal when the enzyme phytase is added to the diet. Excretion of phosphorus in the manure has been reduced by as much as 30 percent when the enzyme is added to the diet. Research is also being done on using genetically modified corn that contains less phytate and thus allows the animal to digest more of the phosphorus in the corn. Experimental work with low phytase corn has shown that its use can result in as much as a 37 percent reduction in the amount of phosphorus excreted in the manure.

Phytic acid binds other minerals such as calcium, zinc, and manganese, reducing their availability to the animal. It appears to also have a negative effect on the digestibility of amino acids. Swine fed the low phytate corn in their ration developed a larger loin eye muscle, suggesting a better utilization of amino acids in the diet. More research is being conducted to determine the best levels of phytase to be added to the diet to secure optimal performance.

SELECTING FEEDS FOR SWINE

Feed costs range from 55 to 70 percent of the total cost of raising hogs. Combining the right kinds of feed in a well-balanced ration is one of the most important tasks of the hog producer. The nutrient needs of hogs include energy, protein, minerals, vitamins, and water.

Energy Feeds

Corn is the basic energy feed used in hog rations (Figure 22-2). It is high in digestible carbohydrates, low in fiber, and is palatable. Other feeds that are used as energy sources are compared to corn when determining their feeding value (Table 22-2).

USDA

Figure 22-2 Corn is the basic energy feed used in hog rations.

Number two (#2) dent corn contains 8.8 percent protein. However, corn lacks several of the amino acids that are essential for swine nutrition. Lysine and tryptophan are the two amino acids that are not found in corn in large enough amounts. Corn must be supplemented with protein, minerals, and vitamins when fed to hogs. The dry matter in high-moisture corn has the same feeding value as the dry matter in corn at normal moisture levels. High-lysine corn has been developed that may reduce the amount of protein that must be supplemented in swine rations.

Whole, shelled corn can be fed free choice. If a protein supplement is to be mixed in the ration, the corn should be ground. Ground ear corn has too much fiber for growing hogs. It can, however, be used in rations for pregnant sows.

Corn co-products coming from the corn-refining industry may be used in swine feeding. These products have been referred to as by-products in the past. Two products of interest to swine growers are corn gluten feed and corn germ meal.

Corn gluten feed is not the same product as corn gluten meal. They have different chemical compositions and are produced at different steps in the corn-refining process. Corn gluten feed has 75 to 80 percent of the energy value of corn grain and contains 1,120 kilocalories of metabolizable energy per pound and 22 percent crude protein. Corn gluten feed is available in three forms: wet corn gluten feed; dry, loose corn gluten feed; and dry, pelleted corn gluten feed. The dry, pelleted form is recommended for swine feeding. It is easier to handle and contains more available tryptophan than the other forms.

Dried corn gluten feed can be substituted in the ration for up to 30 percent of the corn in rations for growing-finishing hogs weighing more than 100 pounds. Because it is deficient in lysine and tryptophan, corn gluten feed must be properly supplemented with protein. Dried corn gluten feed may substitute for 40 to 50 percent of the corn in gestation rations. A general guideline to follow is that 100 pounds of corn gluten feed is equal to 88 pounds of corn and 12 pounds of soybean meal in a swine ration. Corn gluten feed is relatively low in phosphorus; swine rations in which it is used must therefore be supplemented with a minimum of 0.5 percent dicalcium phosphate or 0.1 percent inorganic phosphorus.

TABLE 22-2 The Relative Feeding Value of Various Energy and Protein Feed Sources for Swine

Feed (as-fed basis)	Relative Value % Compared to Corn	Relative Value % Compared to Soybean Meal	Maximum Recommended Percent of Complete Diet			
			Gestation	Lactation	Starter	Growing-Finishing
Alfalfa meal (dehydrated)	75–85	—	50	10	0	5
Alfalfa meal (sun-cured)	60–70	—	50	10	0	5
Bakery waste	95–110	—	40	40	20	40
Barley	90–100	—	80	80	25	85
Buckwheat	80–90	—	50	0	0	50
Buttermilk, dry	—	75–85	5	5	5	5
Corn (high lysine)	100–105	—	90	90	60	90
Corn (yellow)	100	—	80	80	60	85
Corn gluten meal	—	40–60	5	5	0	5
Corn silage	20–30	—	90	0	0	0
Cottonseed meal (solvent)	—	65–75	5	5	0	5
Fish meal, anchovy	—	140–165	5	5	5	5
Fish meal, menhaden	—	140–165	5	5	5	5
Grain sorghum (milo)	95–100	—	80	80	60	85
Linseed meal	—	55–65	5	5	5	5
Meat and bone meal	—	95–110	10	5	5	5
Molasses	55–65	—	5	5	5	5
Oats	85–95	—	80	10	0	20
Peanut meal, expeller	—	70–80	5	5	0	5
Potatoes	20–25	—	80	0	0	30
Rye	90	—	20	20	0	25
Skim milk, dried	—	95–100	0	0	20	0
Soybean meal, solvent	—	100	25	20	35	22
Soybean meal, solvent, dehulled	—	110–112	22	18	30	20
Soybeans, whole cooked	—	90–100	30	25	40	30
Tankage (meat meal)	—	115–130	10	5	0	5
Triticale	90–95	—	80	80	20	85
Wheat bran	60–65	—	30	5	0	0
Wheat, hard	100–105	—	80	80	60	85
Whey (dried)	135–145	—	5	5	20	5

Corn germ meal contains 1,360 kilocalories of metabolizable energy per pound and 20 percent crude protein. There is relatively little corn germ meal available for feeding because it is usually blended back into corn gluten feed in the corn-refining industry.

Wet corn gluten feed contains 40 percent dry matter and is similar to corn in protein content. It is low in available amino acids, especially lysine and tryptophan. Diets using wet corn gluten feed need to have their amino acid content balanced. Wet corn gluten feed contains only 32 percent of the metabolizable energy of corn.

Wet corn gluten feed must be properly balanced for calcium, trace minerals, and vitamins in sow diets. Additional amino acids and energy must also be provided in the diet. With a daily consumption of 6 pounds of wet corn gluten feed, sows should also receive 1.25 pounds of protein supplement and 1 pound of corn per day.

There is a danger of mycotoxins in wet corn gluten feed because the milling process does not remove molds from the corn. It should therefore be fed fresh because molds will grow easily in feed if stored damp.

Barley is a good substitute for corn. In some parts of the United States, more barley than corn is fed to hogs. Barley has a higher fiber content than corn and, therefore, slightly less digestible energy. The protein content of barley is higher than corn. However, like corn, it lacks some of the amino acids required by hogs. Barley must be supplemented with protein, minerals, and vitamins for hog feeding.

Barley is ground medium fine for hog feeding. It may also be rolled or pelleted for hog rations. Barley is not quite as palatable as corn and should be mixed with the protein supplement in the ration. Barley has a relative feeding value of 90 to 100 percent compared to corn. When up to one-third of the corn in the ration is replaced with barley, the hogs will gain as fast as they would on an all-corn ration. If barley is substituted 100 percent for corn, the hogs will show a slightly slower gain. Barley is sometimes infested with **scab**, a disease that attacks barley. This may make it poisonous to hogs; therefore, scabby barley should never be fed to hogs.

Buckwheat has 80 to 90 percent of the feeding value of corn in swine rations. It contains about 11 percent crude fiber and is not as palatable as corn. Generally, it should be mixed with other grains when fed to hogs.

It has 0.65 percent lysine, which is higher than the lysine content of corn. Less protein supplement may be needed when buckwheat is included in the ration. Buckwheat is not recommended for feeding lactating sows or small pigs. It can be used for gestating sows and in growing-finishing rations. While it can be used for up to 100 percent of the corn in growing-finishing rations, this level of feeding will result in a 5 to 10 percent drop in feed efficiency and growth rate. It is recommended that it not be used for more than 50 percent of the ration. Buckwheat contains **fagopyrin**, which is a photosensitizing agent. This can cause rashes and itching (buckwheat poisoning) when white pigs are exposed to sunlight.

Milo (grain sorghum) is grown widely in the southwestern part of the United States. Milo has a higher protein content than corn. It can replace all of the corn in hog rations. Milo must be supplemented with protein, minerals, and vitamins in hog rations. It has a relative feeding value of 90 to 95 percent compared to corn. Milo should be ground for feeding. Some varieties are unpalatable to hogs and should be mixed with protein for feeding.

Wheat (Figure 22-3) is equal to or slightly higher in feeding value than corn. Because of its higher protein, lysine, and phosphorus content, its relative feeding value ranges from 100 to 105 percent compared to corn. The energy value of wheat is slightly lower than corn. The relative price of wheat compared to other grains is a determining factor when considering its use in swine rations. Wheat must be coarsely ground when fed to livestock. If ground too fine, it forms a pasty mass in the animal's mouth and reduces feed intake. Feeding problems are reduced when wheat is processed through a roller mill rather than a hammermill.

USDA

Figure 22-3 Wheat has a high protein, lysine, and phosphorus content.

Oats have a higher protein content than corn, but the quality of the protein is poor. A protein supplement must be used when oats are fed in a hog ration. Oats also have a high fiber content. They have a relative feeding value of 85 to 90 percent compared to corn. Oats should not be substituted for more than 20 percent of the corn in the ration for growing-finishing hogs. If a greater amount of oats is used in the ration, the rate of gain will be slower. Oats for hog rations should be medium to

Figure 22-4 The rate of gain will be slower if more than 20 percent of the corn ration is substituted with oats for growing-finishing hogs.

finely ground. Hulled, rolled oats are an excellent feed for starter rations for baby pigs (Figure 22-4).

Rye is not a very good feed for hogs. Rye has a relative feeding value of 90 percent compared to corn and is less palatable than other grains. Not more than 25 percent of the grain in the ration should be rye. Rye is harder than corn and should be ground when fed to hogs. It is sometimes infested with a fungus called **ergot**. Rye containing ergot should never be fed to sows because it will cause abortions. Ergot-infested rye will slow down gains in growing-finishing hogs.

Triticale is a hybrid cereal grain that is the result of a cross between wheat and rye. Triticale has more lysine than corn but is not as palatable as corn in hog rations. No more than 50 percent of the ration should be made up of triticale. Some varieties of triticale may be infested with ergot. Ergot-infested triticale should not be fed to pregnant sows.

Potatoes may be fed to hogs. They contain mainly carbohydrates, and therefore must be fed with a protein supplement. Heavier hogs make better use of potatoes than younger hogs. It takes about 400 pounds of potatoes to equal the feed value of 100 pounds of corn. Potatoes should be fed at the rate of one part potatoes to three parts grain. They should be cooked before they are fed to hogs.

Bakery wastes may be fed as part of the hog ration. Bakery wastes include stale bread, breadcrumbs, cookies, and crackers. The average protein content of these foodstuffs is about 10 percent. Little is known about the amino acid content. A good protein supplement must be fed when bakery wastes are used.

Fats, tallow, and greases provide a high-energy source in hog rations. These substances usually make up less than 5 percent of the ration, depending on the price of fat. They are used to improve the binding qualities of pelleted feed. **Binding quality** refers to how well the feed particles stick together in the pellet. However, carcass quality is decreased if too much fat, tallow, or grease is added to the ration. These substances contain no protein, minerals, or vitamins. Proper nutrient supplements are essential when these substances are part of the ration.

Molasses provides carbohydrates in the ration. It can be substituted for part of the grain. However, molasses should not be more than 5 percent of the ration. Scours may result from overfeeding molasses.

Plant Proteins

Soybean oil meal is available with a 44 percent or 47.5 percent protein content. The 47.5 percent protein soybean oil meal is most often used for swine feeds. The two are about equal in value for growing-finishing hogs. The protein quality in soybean oil meal is excellent. Soybean oil meal is the most widely used protein source in hog rations.

The amino acid balance is generally good and soybean oil meal is very palatable to hogs. In fact, they will overeat soybean oil meal if it is fed free choice. It must be fed in a complete mixed ration or be mixed with less palatable proteins to prevent overeating. Soybean oil meal may be fed with corn as the only protein supplement. Minerals and vitamins must be added to the ration when soybean oil meal is not used as the protein source. Other feeds that are used as protein sources are compared to soybean meal when determining their feeding value (see Table 22-2).

Cottonseed meal is 40 to 45 percent protein. However, the protein quality of this material is poor. It is low in lysine and must be fed with other protein sources in hog rations. Cottonseed meal may be fed as 5 percent of the protein in the hog ration. Some cottonseed meal contains **gossypol**, which is toxic to hogs. If the gossypol has been removed, it may replace up to 50 percent of the soybean oil meal in the ration. Cottonseed meal is low in minerals and fair in vitamin B content. Hogs do not find it very palatable. Do not use cottonseed meal in starter rations.

Linseed meal is 35 to 36 percent protein. Because the protein is of poor quality, linseed meal must be fed with other protein sources in the hog ration. It usually makes up no more than 5 percent of the protein in the ration. Linseed meal contains more calcium than cottonseed meal or soybean oil meal. It has about the same amount of vitamin B as these protein sources. Linseed meal is best fed in combination with animal protein sources. When consumed in large amounts, linseed meal may act as a laxative.

Peanut meal is 47 percent protein. Because the protein is low in several amino acids, peanut meal must be fed with other protein sources. It becomes rancid if stored for more than a few weeks and is low in vitamins and minerals.

Whole soybeans contain about 37 percent protein and can be used to replace soybean oil meal in hog rations. They are higher in energy but lower in protein than the meal. Use 6 pounds of whole cooked soybeans to substitute for 5 pounds of soybean oil meal. The higher energy content of the whole soybeans may increase feed efficiency by about 5 percent. Do not use raw soybeans in swine growing-finishing rations; they contain an antitrypsin factor that prevents the action of the enzyme trypsin in non-ruminants such as swine and poultry. This reduces the availability of tryptophan, an essential amino acid. Heating the soybeans destroys the antitrypsin factor.

Animal Proteins

Tankage and meat scraps contain 50 to 60 percent protein. They have inadequate amounts of the amino acid tryptophan and, therefore, must be used with other protein sources in hog rations. The calcium and phosphorus content of tankage and meat scraps is high. The vitamin content is variable and the B vitamin pantothenic acid is sometimes low. Tankage is not as palatable to hogs as soybean meal. The maximum percent of tankage that should be included in various kinds of complete rations is 10 percent for gestation rations; 5 percent for lactation, grower, and finishing rations; and 0 for starter rations.

Meat and bone meal contains 50 percent protein. The amount of bone contained in the mix determines the feeding value. Meat and bone meal is low in lysine when compared with other protein sources. The maximum percent of meat and bone meal that should be included in various kinds of complete rations is 10 percent for gestation rations and 5 percent for lactation, starter, grower, and finishing rations.

Fish meal is 60 to 70 percent protein which is of excellent quality. Fish meal also is high in minerals and vitamins and very palatable to hogs. It is a good protein source but is usually too expensive to use except in creep rations. The maximum percent of fish meal that should be included in various kinds of complete rations is 5 percent.

Skim milk and buttermilk contain about 33 percent protein when dried. In liquid form, they are worth only about one-tenth as much as dried milk because the liquid form contains about 90 percent water. The protein quality of skim milk and buttermilk is good, and they are good sources of B vitamins. These milk products are often used in creep rations in dried form because young pigs cannot consume enough in liquid form to meet their protein needs. The maximum amount of dried skim milk that should be included in starter rations is 20 percent. Dried skim milk should not be included in gestation, lactation, grower, or finishing rations.

Whey in liquid form contains only about 1 percent protein, whereas dried whey contains 13 to 14 percent protein. The protein in whey is of excellent quality. However, liquid whey has a high water content, and hogs cannot consume enough to meet their protein needs. A starter ration may contain up to 20 percent dried whey. Limit gestation, lactation, grower, and finishing rations to no more than 5 percent dried whey.

Roughages

Alfalfa meal is 13 to 17 percent protein. It has large amounts of vitamins A and B, and is an excellent roughage for hogs. It is also a good source of minerals. Alfalfa meal should be limited to no more than 5 percent of the ration for growing-finishing hogs. It may make up as much as 50 percent of the ration for brood sows as it can help to prevent them from becoming too fat. A maximum of 10 percent of the lactation ration may be alfalfa meal. Do not use alfalfa meal in starter rations.

Alfalfa hay and other hays are generally not used in swine rations except for feeding the breeding herd. Hay must be ground and mixed in the ration for self-feeding sows and gilts. It can be used to make up as much as one-third of the ration.

Silage is most valuable in rations for the breeding herd. Ten to 12 pounds of corn or grass-legume silage can be fed per day to sows and gilts during pregnancy. This must be supplemented with protein and minerals. Moldy silage should never be fed to sows, gilts, or any other animal.

Pasture is also most valuable for feeding the breeding herd and is also of value for feeding boars. Good-quality pasture supplies the same nutrients as alfalfa meal and hay. Growing-finishing hogs on pasture will not gain as rapidly as those fed in the drylot. However, pregnant sows and gilts are given the exercise they need by being fed on pasture. A good-quality pasture supplies enough nutrients so that the amount of concentrates in the ration for the breeding herd may be reduced by up to 40 percent. A balanced ration must be fed when the sows and gilts are on pasture.

Minerals

Four major minerals and six trace minerals are frequently added to hog rations. The major minerals are calcium, phosphorus, sodium, and chlorine. The trace minerals are zinc, iron, copper, selenium, manganese, and iodine.

Salt adds sodium and chlorine to the ration. The most common calcium source is ground limestone. The use of dicalcium phosphate supplies both calcium and phosphorus in the ration. Other sources of calcium and phosphorus include steamed bone meal and defluorinated rock phosphate.

Feeding too much calcium or phosphorus may reduce the rate of gain for growing-finishing hogs. An excess of calcium will interact with zinc to cause a zinc deficiency. The ratio of calcium to phosphorus in swine diets should be 1.0 to 1.5 calcium to 1.0 total phosphorus in a grain-soybean meal diet.

Trace minerals are often found in commercial protein supplement mixes. Trace-mineralized salt is another source of trace minerals. Special trace mineral premixes are also available which can be added to the ration (Table 22-3).

Iron and copper, which help to prevent anemia, are especially important in baby pig rations. In addition to the iron supplied in the ration, baby pigs should always be given iron shots when they are 2 to 4 days old. Zinc is required to prevent parakeratosis. Early weaned pigs have a higher zinc requirement than older pigs.

Care must be taken when feeding minerals. Excess minerals in hog rations slow the rate of gain. Minerals should not be added to a ration containing commercial protein supplements unless the feed tag indicates that they are required. A mineral

Percentages of Minerals in Swine Rations

Mineral	Percentage
Salt	0.5%
Calcium	0.5–0.7%
Phosphorus	0.4–0.65%

Connection: Iron Naturally: Iron Deficiency Anemia

It is highly recommended that baby pigs be given an iron injection soon after birth. Pigs are born with a limited supply of iron in their livers so they must be supplemented to prevent anemia. Iron injections have been standard procedure for more than 40 years.

If supplemental iron is so important, how did pigs survive thousands of years without supplemental iron through injection? They obtained their necessary iron requirements from the soil. Pigs naturally root in the dirt (Figure 22-5). They accidentally ingest soil while rooting, or ingest it through food or from the exterior of their mother's udder. Soil provides a natural source of iron. Iron (Fe) is abundant in most soils. It is the fourth most common element on earth. Topsoil typically contains 1 to 5 percent iron in various forms. Iron oxides and hydroxides give soils and especially clay, their reddish and yellowing colors.

Pigs raised outdoors with access to soil do not usually develop iron deficiency. However, most pigs are now raised in confinement on concrete or other manufactured flooring and never touch the soil. So, iron supplementation is necessary for indoor pigs.

Figure 22-5 Pigs raised outside may obtain iron from soil.

TABLE 22-3 Suggested Vitamin-Trace Mineral Mix*

Nutrient	Amount per Pound of Premix**	Suggested Source
Vitamin A	900,000 IU	Vitamin A palmitate-gelatin coated
Vitamin D	100,000 IU	Vitamin D_3—stabilized
Vitamin E	5,000 IU	d1-tocopheryl acetate
Vitamin K (Menadione equivalent)	660 mgs	Menadione sodium bisulfite
Riboflavin	1,200 mgs	Riboflavin
Pantothenic acid	4,500 mgs	Calcium pantothenate
Niacin	7,000 mgs	Nicotinamide
Choline chloride	20,000 mgs	Choline chloride (60%)
Vitamin B_{12}	5 mgs	Vitamin B_{12} in mannitol (0.1%)
Folic acid	300 mgs	Folic acid
Biotin	40 mgs	D-Biotin
Copper	0.4%	$CuSo_4 \cdot 5H_2O$
Iodine	0.008%	KIO_4
Iron	4.0%	$FeSO_4 \cdot 2H_2O$
Manganese	0.8%	$MnSO_4 \cdot H_2O$
Zinc	4.0%	ZnO (80% Zn)
Selenium	0.012%	$NaSeO_3$ or $NaSeO_4$

*Vitamin and trace mineral mixes may be purchased separately. This is advisable if a combination vitamin-trace mineral premix is to be stored longer than 3 months. Vitamins may lose their potency in the presence of trace minerals if stored for a prolonged period.

**Premix is designed to be used at a rate of 5 lb per ton of complete feed for sows and baby pigs and 3 lb per ton of complete feed for growing-finishing swine.

Source: Luce, William G.; Hollis, Gilbert R.; Mahan, Donald C.; and Miller, Elwyn R. (1997). *Swine Diets, Pork Industry Handbook*, Urbana, IL: Cooperative Extension Service, University of Illinois mix can be fed free choice since hogs will not overeat minerals if they are receiving enough minerals in the ration.

Vitamins

Many of the vitamins required by hogs are already present in the feeds used in hog rations. Vitamins that must be added to the ration are A, D, E, K, riboflavin, niacin, pantothenic acid, choline, and vitamin B_{12}. Vitamins may be added to the ration as part of complete protein supplements, in mineral-vitamin premixes, or as vitamin premixes. The major differences among these sources are the amount of vitamins they contain and their cost. It is difficult to determine the exact amount of vitamins they contain because the feed tags usually do not list the amounts. Past experience with a particular mix is the best guide to follow in selecting a vitamin source.

Complete supplements and mineral-vitamin premixes usually cost more than vitamin premixes. However, if the producer does not have mixing equipment on the farm, it may be best to use complete mixes. Premixes are used in such small amounts per ton that it is difficult to mix them into the ration properly.

Water

Photo courtesy of Chore-Time Hog Production Systems, Milford, Indiana, USA

Figure 22-6 An automatic watering system provides hogs with fresh, clean water at all times.

Water is one of the most important nutrients in hog rations. Hogs should have plenty of water available at all times (Figure 22-6). The water should be fresh, clean, and no colder than 45°F. It should be checked periodically for nitrate content because too much nitrate or nitrite in the water is not good for hogs.

The estimated daily consumption of water by various classes of hogs that are not undergoing stress is as follows:

Type of Pig	Gallons/Day
25-lb pig	0.5
60-lb pig	1.5
100-lb pig	1.75
200-lb pig	2.5
300-lb pig	3.5
Gestating sows	4.5
Sow plus litter	6.0
Non-pregnant gilts	3.2
Pregnant gilts	5.5

Additives

Feed additives increase efficiency in hog production. Additives enable pigs to grow at a faster rate, improve feed conversion, and reduce disease and stress. The additives most commonly found in hog rations are anthelmintics, antibiotics, arsenicals, nitrofurans, and sulfa compounds.

Sources of additives include complete protein supplements, complete mixed feeds, and premixes. Premixes must be carefully mixed into the ration for even distribution. Factors to consider when evaluating additive sources include cost, which additives are to be included, and the amounts of additives in the source.

Feed tag instructions must be carefully followed in the use of additives. Withdrawal times must be observed when marketing hogs. These withdrawal times are always listed on the feed tag. Federal regulations govern the required withdrawal times. Penalties are enforced when these regulations are not followed.

FEEDING THE BREEDING HERD

When replacement gilts reach 150 to 200 pounds, they should be separated from market hogs. They may be fed a sow diet as shown in Table 22-4. Do not allow gilts to become too fat; they should gain about 1 pound per day before breeding. Breed gilts at about 7 to 8 months of age when they weigh about 250 to 300 pounds. Boars may be fed the rations shown in Table 22-4; follow the same general rules for weight gain as outlined above for gilts. Pasture may be used in the feeding program for both gilts and boars.

Flushing is a feeding program in which the amount of feed given is increased for a short period of time. A gilt may farrow a bigger litter, depending on her condition, if this practice is followed. When flushing, increase the ration to 6 to 8 pounds about 10 days before breeding. Sows that have been on restricted rations before breeding should also be flushed. Gilts and sows should be put back on limited feeding immediately after breeding so they do not get too fat. Limited feeding also helps to reduce fetal death during gestation.

TABLE 22-4 Some Representative Gestation and Lactation Diets for Sows and Replacement Gilts Using Several Different Grain Sources

	Diet Number						
Ingredient	**1 (lb)**	**2 (lb)**	**3 (lb)**	**4 (lb)**	**5 (lb)**	**6 (lb)**	**7 (lb)**
Corn, yellow	1,627	1,253	1,302	—	—	—	—
Barley	—	—	—	—	—	1,759	—
Oats	—	400	—	—	—	—	—
Sorghum grain	—	—	—	1,617	1,469	—	—
Wheat, hard winter	—	—	—	—	—	—	1,565
Wheat middlings	—	—	400	—	—	—	—
Soybean meal, 44%	295	270	225	306	260	128	165
Meat and bone meal, 50%	—	—	—	—	—	60	—
Dehydrated alfalfa meal, 17%	—	—	—	—	200	—	200
Calcium carbonate	19	19	25	20	13	16	12
Dicalcium phosphate	44	43	33	42	43	22	43
Salt	10	10	10	10	10	10	10
Vitamin-trace mineral mix*	5	5	5	5	5	5	5
Total	2,000	2,000	2,000	2,000	2,000	2,000	2,000
Calculated Analysis							
Protein, %	13.40	13.60	13.70	13.90	13.90	14.40	14.90
Lysine, %	0.62	0.62	0.62	0.62	0.62	0.62	0.62
Tryptophan, %	0.17	0.17	0.17	0.17	0.18	0.18	0.22
Threonine, %	0.51	0.51	0.51	0.48	0.49	0.44	0.50
Methionine + cystine, %	0.54	0.48	0.48	0.42	0.42	0.47	0.54
Calcium, %	0.91	0.90	0.91	0.90	0.91	0.90	0.90
Phosphorus, %	0.70	0.70	0.70	0.70	0.70	0.70	0.71
Metabolizable energy, kcal/lb	1,476	1,416	1,441	1,419	1,354	1,338	1,352

*See Table 22-3. It is also recommended that during the gestation period additional choline (550 grams per ton) be added to the diet. This can be provided by adding 2.5 lb of choline chloride premix containing 50% choline or 2.0 lb of a chloride premix containing 60% choline.

Source: Luce, William G.; Hollis, Gilbert R.; Mahan, Donald C.; and Miller, Elwyn R.; (1997). *Swine Diets, Pork Industry Handbook*, Urbana, IL: Cooperative Extension Service, University of Illinois

Gilts and sows should not be allowed to become too fat during gestation. A gain of 50 to 75 pounds is about right for sows. Gilts should gain 70 to 100 pounds during gestation. Four to five pounds of feed per day may be hand-fed during the first two-thirds of gestation. Increase this to 6 pounds during the last one-third of the gestation period. Rations shown in Table 22-4 may be used for sows and gilts.

Sows and gilts may be kept on pasture during gestation. Alfalfa, ladino, and red clover are good legume pastures to use. However, legume pasture may increase estrogen activity, which could impair reproduction. Orchard grass and Kentucky bluegrass are good non-legumes for hog pastures. Rations shown in Table 22-4 should be fed at the rate of 2 to 3 pounds per day on good-quality legume pasture.

Self-feeding may be used during the gestation period if a bulky ration is used. Adding a good-quality ground alfalfa hay to the ration provides the needed bulk. Ground ear corn and oats also add bulk to the ration. Sows and gilts may also be self-fed a high-energy ration. Access to the self-feeder must be limited if this practice is followed. Be sure adequate feeder space is available. Sows and gilts being self-fed high-energy feed, with limited access to feeders, do as well as those hand fed 4 pounds of the same feed each day.

Corn silage or grass silage may be used in the gestation ration. Silage substitutes for pasture or alfalfa meal in the ration. Feed only as much as the sows or gilts will completely finish eating in 2 to 3 hours. For sows, this is about 10 to 15 pounds. Gilts will eat about 8 to 12 pounds. Silage must be supplemented with 1.0 to 1.5 pounds of protein. Never feed moldy silage (or any other moldy feed) as it may cause abortion.

Modern sows have been bred to be heavy milkers. The main objective for nutrition of the lactating sow is to optimize milk production. Lactating sows produce 15 to 25 pounds of milk per day, resulting in daily nutrient requirements that are about three times higher than during gestation. Nutrient intake during lactation is directly related to the amount of milk produced and growth rate of nursing piglets. Fiber is not added to the ration but laxatives, more protein, and liquid fat are added to most lactation rations. A typical lactation ration will contain 18 percent protein and from 5 to10 percent fat. If a good ration is fed to sows during lactation, they will return to estrus faster after weaning and the conception rate will be higher.

During the first few days after farrowing, rations may be limit-fed. Gradually increase the ration until the sow or gilt is on full feed at 5 to 7 days. Some examples of lactation rations are shown in Table 22-4. Lactating sows and gilts will eat from 2.5 to 3 pounds of feed per day per 100 pounds of body weight. The amount of feed given may also be varied according to the number of pigs being nursed. A rule of thumb is 1 pound of feed for each pig nursed, plus 3 pounds per sow per day of a 15 percent protein ration.

Feed intake may be reduced by as much as 25 percent when the temperature in the farrowing house reaches 80°F as compared to feed intake at 60°F. Keeping sows cool will help maintain proper feed intake. Adding 10 percent supplemental fat to the diet during hot weather will help maintain the energy level of the diet on reduced feed intake. Wet feeding can also help maintain feed intake during hot weather.

Diets that are high in energy generally reduce feed intake by 5 to 11 percent compared to lower-energy diets. However, the intake of energy is about the same. High-energy diets must be properly supplemented to ensure adequate levels of other nutrients when the total feed intake is lower. Feed intake increases with high-fiber diets because of the lower amount of energy. Increasing the protein content of the lactation ration will increase feed intake about 1.1 pound for each 1 percent increase in protein content. Lactation and gestation rations should contain

0.8 percent calcium to 0.6 percent phosphorous. Feed intake is reduced about 9 percent for each variation of 0.1 percent (calcium) and 0.05 percent (phosphorus) above or below these levels. High levels (above 50 ppb aflatoxin or 4 ppm vomitoxin) of mold toxins also reduce feed intake. Reducing the dust content of feeds by pelleting, adding fat, or using a wet mash may increase feed intake up to 15 percent.

Raw soybeans may be fed to both pregnant and lactating sows. The raw soybeans can replace soybean meal on an equal-weight basis. The crude protein content of the ration is lower, but the lysine content is adequate to meet the requirements of the sows. The advantages of feeding raw soybeans to sows during gestation and lactation include (1) lower feed cost, (2) no processing cost for the beans, and (3) an increased fat content in the diet without adding liquid fat to the ration. Raw soybeans have an oil content of 18 percent. Research has shown that feeding raw soybeans to sows does not affect the number of live pigs farrowed or weaned. Pig birth weights are increased, but weaning weight is not affected. Feed intake during lactation is not affected, and the milk of the sows has a higher fat content.

The herd boar should be kept on limited rations during the breeding season so he will not become too fat. Fat boars are sluggish and make poor breeders. Young boars should be fed enough for moderate weight gains. Boars can be maintained on 4 pounds of feed when not in use. The same gestation diets recommended for sows may be used for boars (see Table 22-4).

Crude Protein Rations for Boars: Per Day Basis

Young boars	5–5.5 lb at 14% crude protein
Mature boars	5–6 lb at 14% crude protein
Breeding boars	6–8 lb at 16% crude protein during breeding season
Cold weather	for every degree below 68°F add 3.5–7 oz

FEEDING BABY PIGS

About one-fourth of the pigs lost before weaning are lost because of poor feeding. Make sure each pig nurses shortly after it is born. Pigs receive disease protection from the colostrum (first) milk of the sow. The sow reaches maximum milk production in 3 to 4 weeks. After that, milk production falls off. Pigs must be eating well by then in order to get the nutrients they need.

Baby pigs will start to nibble on a creep feed, if it is available, within a week after they are born. Small amounts can be provided in pans and replaced with fresh feed each day. Commercial creep feeds are available. Unless the producer has good mixing equipment, it is best to use a commercial creep feed. Baby pigs eat creep feeds better if they are sweetened. Use creep feeds that have the sugar mixed in the pellet rather than sugar-coated pellets. Creep and starter rations are shown in Table 22-5. Be sure that baby pigs are given plenty of clean, fresh water.

FEEDING GROWING-FINISHING PIGS

Scours (diarrhea) is sometimes a problem during the first 2 or 3 weeks after pigs are weaned. This is especially true with early weaned pigs. Replacing 10 to 15 percent of the corn in the ration with ground oats for 2 or 3 weeks may help prevent scouring.

When pigs are weaned early, they sometimes lose weight or do not gain weight for the first week or two (Figure 22-7). The digestive tract of the young pig is immature and only partially functioning. The stomach produces low levels of acid, and the pH level in the stomach and digestive tract is therefore higher (more alkaline). Enzyme activity and utilization of nutrients is more efficient and bacteria do not grow as readily when the pH level is lower. Research has shown that adding a small amount of fumaric acid to the diet may increase average daily gain and feed efficiency by about 8 percent.

Figure 22-7 If pigs are weaned too early they may not gain weight properly.

Research has also shown that adding dried whey to the diet for the first 2 weeks after weaning will help maintain weight gain. A feed containing 20 percent dried whey with soybean meal is recommended. The protein quality of dried whey varies;

TABLE 22-5 Representative Baby Pig Diets That May Be Used as Either Creep or Starter Diets. If post-weaning scours is a problem, substitute 200–400 lb of ground oats for corn or grain sorghum in diets 4, 5, or 7 for the first 2–3 weeks after weaning.

	Diet Number						
Ingredient	**1 (lb)**	**2 (lb)**	**3 (lb)**	**4 (lb)**	**5 (lb)**	**6 (lb)**	**7 (lb)**
Corn, yellow	990	1,060	1,211	1,396	625	1,159	1,279
Sorghum grain	—	—	—	—	625	—	—
Ground oats	—	—	—	—	—	200	—
Soybean meal, 44%	421	530	390	543	495	530	410
Fish meal, menhaden	100	—	100	—	—	—	—
Dried whey	400	—	200	—	200	—	200
Dried skim milk	—	200	—	—	—	—	—
Sugar*	—	100	—	—	—	—	—
Fat	50	50	50	—	—	50	50
Lysine, 78% L-Lysine	—	—	3	—	—	—	3
Calcium carbonate	7	15	10	15	13	15	13
Dicalcium phosphate	20	33	24	34	30	34	33
Salt	7	7	7	7	7	7	7
Vitamin-trace mineral mix**	5	5	5	5	5	5	5
Total	2,000	2,000	2,000	2,000	2,000	2,000	2,000
Calculated Analysis							
Protein, %	19.10	19.50	18.30	17.90	17.60	17.80	15.80
Lysine, %	1.15	1.15	1.15	0.95	0.95	0.95	0.95
Tryptophan, %	0.24	0.26	0.21	0.23	0.23	0.23	0.20
Threonine, %	0.80	0.78	0.76	0.68	0.69	0.67	0.63
Methionine + cystine, %	0.66	0.64	0.64	0.60	0.56	0.58	0.55
Calcium, %	0.85	0.86	0.85	0.75	0.75	0.76	0.77
Phosphorus, %	0.71	0.70	0.70	0.65	0.66	0.65	0.66
Metabolizable energy, kcal/lb	1,516	1,529	1,500	1,478	1,449	1,495	1,520

*Dextrose or hydrolyzed corn starch product.
**See Table 22-3.

Source: Luce, William G.; Hollis, Gilbert R.; Mahan, Donald C., and Miller, Elwyn R. (1997). *Swine Diets, Pork Industry Handbook*, Urbana, IL: Cooperative Extension Service, University of Illinois

care must therefore be taken to ensure proper protein content if it is to be included in the diet.

During the first 2 weeks after weaning, baby pigs have difficulty digesting soybean meal. Products containing processed soybean protein such as soy protein concentrate, soy flour, and isolated soy protein are more easily digested and do not cause as much intestinal damage in young pigs as does soybean meal. These products are very expensive and are recommended for use only during the first 2 weeks after weaning.

The use of **spray-dried porcine plasma (SDPP)** during the first 2 weeks after weaning will increase feed intake and rate of gain in young pigs. Spray-dried porcine plasma is also called plasma protein. It is a by-product of blood from pork slaughter plants. SDPP can replace some or all of the dried skim milk in the diet. The maximum recommended level of use is 8 to 10 percent of the diet. It may be necessary to add more methionine and lactose to the diet when SDPP is used.

Using antibiotics in the starter diet will generally improve feed efficiency by 5 to 10 percent and growth rate by 10 to 20 percent. Adding copper sulfate in combination with an antibiotic improves performance more than using either alone. The use of other feed additives does not generally result in significant improvement in rate of gain or feed efficiency for young pigs.

A **phase feeding** program is recommended when pigs are weaned at 3 weeks of age. Phase feeding is designed to meet the rapidly changing nutritional needs of pigs during the first weeks after early weaning. It helps reduce post-weaning growth lag and gets pigs started on a grain and soybean meal diet more quickly.

The phase I feeding period lasts for 7 to 10 days for pigs weaned at 3 weeks of age and 3 to 4 days for pigs weaned at 4 weeks of age. A pelleted diet containing 20 to 22 percent crude protein and 1.45 percent total lysine is used. The diet also contains 4 to 5 percent plasma protein, 20 percent food-grade whey, 10 percent food-grade dried skim milk, 4 to 6 percent cheese by-product, 2 to 3 percent egg protein, and 4 to 6 percent soy oil.

The phase II feeding period follows the phase I period and lasts for 1 to 2 weeks. The diet may be pelleted or meal and contains 18 to 22 percent crude protein and 1.35 percent total lysine. The diet also contains 10 to 15 percent food-grade whey, 2.5 to 5 percent menhaden fish meal, 2 to 3 percent blood meal, and a maximum of 8 percent soybean meal.

The phase III feeding period should start when the pigs weigh about 25 pounds and 3 to 5 weeks after they are weaned. A grain-soybean meal diet in pelleted or meal form with 18 to 20 percent crude protein and 1.1 percent total lysine is used. Five to 10 percent whey and 4 to 5 percent fish meal may be added to the diet. The phase III feeding period continues until the pigs reach 45 pounds.

The nutritional requirements of younger pigs are greater than those of older pigs. Protein content in the ration can be reduced as pigs get older. When pigs are fed on good legume pasture, the protein content of the ration can be reduced about 2 percent. Tables 22-6 and 22-7 show rations for growing-finishing pigs.

The concept of phase feeding has now been extended throughout the finishing period. The reason for dividing the finishing period into more phases is to more nearly match the changing protein needs of the animal. This practice also reduces feed costs and lowers the amount of protein fed. This in turn reduces the amount of nitrogen excreted in the manure, helping to reduce odor and pollution problems.

When the finishing period is divided into four or five feeding phases, the amount of protein in the diet is reduced in each progressive phase. The relationship between weight and the percent of crude protein in the diet is approximately:

Weight	Percent Crude Protein in Diet
40–65 lb	18–19%
65–95 lb	16–17%
95–140 lb	15–16%
140–195 lb	14–15%
195–255 lb	12–13%

Increasing the number of feeding phases means the producer must closely monitor the growth of the hogs and adjust the diet appropriately. Feed is approximately 60 to 65 percent of the total cost of swine production and about 75 to 80 percent of the variable cost. The time spent carefully monitoring the feeding program in a swine operation can result in significant savings and thus increase the profitability of the enterprise.

TABLE 22-6 Some Representative Diets for Growing Swine (40–125 lb) Using Several Different Grain Sources

	Diet Number							
	1	2	3	4	5	6	7	8
Ingredient	(lb)	(lb)	(lb)	(lb)	(lb)	(lb)	(lb)	(lb)
Corn, yellow	1,555	1,368	1,228	—	—	804	—	—
Barley	—	—	—	—	1,660	—	829	—
Oats	—	200	—	—	—	—	—	—
Sorghum grain	—	—	—	1,549	—	—	—	800
Wheat, hard winter	—	—	—	—	—	804	829	800
Wheat middlings	—	—	400	—	—	—	—	—
Soybean meal, 44%	395	383	327	400	293	342	293	350
Calcium carbonate	15	15	21	17	18	16	18	17
Dicalcium phosphate	25	24	14	24	19	24	21	23
Salt	7	7	7	7	7	7	7	7
Vitamin-trace mineral mix*	3	3	3	3	3	3	3	3
Total	2,000	2,000	2,000	2,000	2,000	2,000	2,000	2,000
Calculated Analysis								
Protein, %	15.30	15.40	15.60	15.70	16.00	15.90	16.30	16.10
Lysine, %	0.75	0.75	0.75	0.75	0.75	0.75	0.75	0.75
Tryptophan, %	0.20	0.20	0.20	0.20	0.22	0.21	0.22	0.21
Threonine, %	0.58	0.58	0.58	0.55	0.55	0.57	0.55	0.55
Methionine + cystine, %	0.54	0.54	0.52	0.46	0.48	0.56	0.53	0.52
Calcium, %	0.65	0.65	0.65	0.66	0.65	0.65	0.66	0.66
Phosphorus, %	0.55	0.55	0.55	0.55	0.55	0.55	0.55	0.55
Metabolizable energy, kcal/lb (kcal/kg)	1,494	1,464	1,461	1,438	1,360	1,465	1,397	1,437

*See Table 22-3.

Source: Luce, William G.; Hollis, Gilbert R.; Mahan, Donald C.; and Miller, Elwyn R. (1997). *Swine Diets, Pork Industry Handbook*, Urbana, IL: Cooperative Extension Service, University of Illinois

Rations may be fed free choice or as complete mixed feeds. **Free choice** means the supplement is fed separately from the grain. All the ingredients are mixed together in a **complete mixed feed**. The pigs receive about the same nutritional effect from either method of feeding and costs are about the same for each. More uniform growth results from a complete mixed feed. Pigs sometimes overeat protein if it is fed free choice. This is especially true if soybean oil meal is the protein used. Overfeeding increases the cost of gain and, therefore, should be avoided. On pasture, hogs gain a little faster when fed complete mixed feeds as compared to feeding free choice.

Growing-finishing pigs do not gain as efficiently on barley rations as they do on corn or milo rations. This is mainly because of the lower energy and higher fiber content of barley. Barley that weighs less than 48 pounds per bushel has a higher fiber content, which further reduces energy intake and rate of gain. Rate of gain may also be slightly lower on milo rations as compared to corn rations mainly because of the slightly lower energy level in milo.

Nutrient composition tables show the average protein content of corn. However, the crude protein content of corn can vary widely from sample to sample. Laboratory testing of corn for nutrient content can save the farmer money by allowing for more accurate balancing of the ration with protein supplement. A difference of 1 percent in the protein content of corn can mean a difference of 9 percent in the amount of

TABLE 22-7 Some Representative Diets for Growing Swine (125 lb to Market) Using Several Different Grain Sources

	Diet Number							
	1	**2**	**3**	**4**	**5**	**6**	**7**	**8**
Ingredient	**(lb)**	**(lb)**	**(lb)**	**(lb)**	**(lb)**	**(lb)**	**(lb)**	**(lb)**
Corn, yellow	1,662	1,473	1,329	—	—	902	—	—
Barley	—	—	—	1,770	—	—	—	882
Oats	—	200	—	—	—	—	—	—
Sorghum grain	—	—	—	—	1,649	—	852	—
Wheat, hard winter	—	—	—	—	—	800	851	883
Wheat middlings	—	—	400	—	—	—	—	—
Soybean meal, 44%	290	280	225	185	304	251	250	190
Calcium carbonate	16	16	19	19	17	16	17	18
Dicalcium phosphate	22	21	17	16	20	21	20	17
Salt	7	7	7	7	7	7	7	7
Vitamin-trace mineral mix*	3	3	3	3	3	3	3	3
Total	2,000	2,000	2,000	2,000	2,000	2,000	2,000	2,000
Calculated Analysis								
Protein, %	13.40	13.60	13.80	14.30	14.00	14.20	14.50	14.60
Lysine, %	0.62	0.62	0.62	0.62	0.62	0.62	0.62	0.62
Tryptophan, %	0.17	0.17	0.17	0.19	0.17	0.19	0.19	0.20
Threonine, %	0.51	0.51	0.51	0.48	0.48	0.51	0.48	0.48
Methionine + cystine, %	0.50	0.50	0.48	0.44	0.42	0.53	0.48	0.50
Calcium, %	0.61	0.60	0.62	0.61	0.61	0.60	0.61	0.60
Phosphorus, %	0.50	0.50	0.55	0.50	0.50	0.50	0.50	0.50
Metabolizable energy, kcal/lb	1,499	1,469	1,462	1,356	1,442	1,472	1,440	1,398

*See Table 22-3.

Source: Luce, William G., Hollis, Gilbert R.; Mahan, Donald C.; and Miller, Elwyn R (1997). *Swine Diets, Pork Industry Handbook*, Urbana, IL: Cooperative Extension Service, University of Illinois

soybean meal needed to balance the ration. Because the protein content of swine rations is critical for efficient growth and feed conversion, the small cost of laboratory testing for nutrient content of corn can be quickly recovered. Agricultural Cooperative Extension Service offices can provide information on the location of testing laboratories.

Molds that produce **toxins** (poisons) can grow in feed during hot weather. These toxins can reduce the rate of gain and affect animal health. Keeping feed bins and feeders clean and reducing the moisture content of all grains to 14 percent or less can reduce the danger of molds growing in the feed and producing toxins.

Some experimental work has been done on restricted feeding. **Restricted feeding**, also called limited feeding, means feeding 75 to 80 percent of full feed. Some improvement in feed efficiency results from restricted feeding. Daily gains are about 0.15 to 0.20 pound less for each 10 percent reduction in the amount of feed. This increases the time to market by 7 to 10 days for each 10 percent reduction in feed. If more than a 75 to 80 percent restriction is made, the amount of total feed needed to market increases. Carcass quality is improved by restricted feeding.

Under most farm conditions, restricted feeding is not yet a recommended practice. Full feeding from weaning to market seems to work better for most producers (Figure 22-8). Restricted feeding should never be used with pigs weighing less than 100 pounds. Extra time, labor, and equipment are needed for restricted feeding.

Figure 22-8 Pigs in a swine finishing unit.

Most swine producers feed growing-finishing barrows and gilts together. Research has shown that the nutritional needs of barrows and gilts are different and overall performance can be improved by feeding them in separate groups. This practice is called **split-sex feeding**.

Gilts need a higher concentration of dietary amino acids than do barrows to maximize lean growth rate. Barrows gain about 8 percent faster than gilts; however, gilts generally produce carcasses with less back fat and a larger average loin eye area. Barrows reach their maximum performance as measured by daily gain, feed efficiency, back-fat thickness, and loin eye area on a diet of 14 to 15 percent crude protein. Gilts reach their maximum performance as measured by the same traits on a diet of 16 percent crude protein and have a higher protein and lysine requirement than barrows. Gilts fed a higher protein level in the diet reach market at an earlier age. Split-sex feeding does require more facilities, such as separate feeding areas, separate feed-handling systems, and more feed storage bins. The producer must weigh the economic benefits against the increased requirements to determine the feasibility of using split-sex feeding.

PREPARATION OF FEEDS

Generally, grains used in hog feeding should be ground for most efficient use. Corn, barley, grain sorghum (milo), and oats should be finely ground. Wheat should be coarsely ground.

Pelleting complete feeds improves feed efficiency. Some of this improvement probably results from the lower feed waste that comes from pelleting. Higher-fiber rations are also improved by pelleting. Pelleting results in 4 to 8 percent improvement in rate of gain per ton of feed. Buying a complete pelleted feed may be less expensive than using meal. Most swine producers cannot justify the cost of owning equipment for feed pelleting. In addition, taking home-mixed feeds to a mill to be pelleted is usually not an economical practice.

Liquid or paste feeding reduces feed waste. Rate of gain may increase, but labor costs are generally higher with this method of feeding. A clear advantage for liquid or paste feeding hasn't been shown. There is an advantage in wetting complete mixed feeds when limit-feeding hogs. Mix 1.5 parts of water with 1 part of dry feed for best results.

There are commercial feeders available for use with wet-feeding programs. They are designed to mix the exact proportions of dry grain and water needed in the ration. Feeders made from different materials are available; stainless steel construction will last longer but is more expensive than some other types. Some feeders are made of plastic material, which makes them easier to keep clean. Feeders must be kept in areas that do not freeze, and they must be checked frequently. Better management is needed when following a wet-feeding program.

There is no advantage to cooking, soaking, or fermenting most feeds for hogs when they are full fed. The only exceptions are soybeans and potatoes, which are improved by cooking. Heating corn does not affect its nutritive value.

MANAGEMENT PRACTICES

Pre-Breeding Management

Producers must decide on the breeding system to be used. Crossbreeding hogs for slaughter is a recommended practice. Crossbred pigs generally grow faster and use feed more efficiently.

Multiple farrowing is arranging the breeding program so that groups of sows farrow at regular intervals throughout the year. Multiple farrowing usually results in a higher average price received for hogs on a yearly basis. The chances of selling at better prices are increased as the number of sales opportunities during the year is increased. Other advantages include spreading income more evenly through the year, making more efficient use of facilities, and reducing the investment per pig raised. Multiple farrowing requires better management than other systems. A year-round labor supply is necessary to handle the hogs.

Select replacement gilts at 4 to 5 months of age. Separate gilts from finishing hogs and feed separately. Worming of sows and gilts should take place before breeding. They should also be sprayed for external parasites at this time.

The boar should be purchased at least 45 to 60 days before use. Buy boars from healthy purebred herds that have good performance records. To prevent the spread of disease, isolate the boar from the rest of the herd when it is first brought to the farm. New boars should be treated for internal and external parasites. Semen test the boar or test breed it on a few market gilts before the breeding season begins to be sure it will breed.

The age of the boar is a factor in determining the number of times the boar can mate per day or week. Mating a boar to too many females in a short period of time will deplete the sperm reserve and reduce the boar's sex drive. Table 22-8 shows the current recommendations for the number of services per boar by age.

Conception rate and litter size can be increased by using more than one boar on each female. This is easier to do when using hand mating or artificial insemination. It can also be done with pen breeding by rotating the boars once a day between pens. Rotating boars from pen to pen also increases the boars' sex drive.

TABLE 22-8 Recommended Maximum Number of Services per Boar, by Age

	Individual Mating System Maximum Matings		Pen Mating System* Boar-to-Sow Ratio
	Daily	**Weekly**	**7–10 Day Breeding Period**
Young boar, 8–12 months of age	1	5	1:2 to 4
Mature boar, over 12 months of age	2	7	1:3 to 5

*Pigs weaned from all sows on the same day.

Artificial insemination has not been widely used in the swine industry in the past except for producers of purebred hogs. With improvements in the technology, artificial insemination has rapidly become a standard procedure in commercial swine herds, especially in larger operations. Some advantages of artificial insemination are that it:

- increases the ability to bring superior genetics to the herd
- makes use of the semen from a superior boar to inseminate many more sows than is possible with natural mating
- reduces risk of disease transmission
- makes it possible to bring new bloodlines into the herd

Breeding-Gestation Period

Gilts should be bred when they are 7 to 8 months of age and weigh 250 to 300 pounds. Gilts have larger litters if they are bred during their second heat period rather than during their first heat period. Gilts on pasture or in dirt lots begin having heat periods earlier than those confined to concrete feeding floors. Conception rates can be increased if gilts are moved to outside lots by the time they weigh 175 to 200 pounds. Gilts will begin cycling heat periods earlier if a boar is placed in an adjoining lot, allowing the gilt to see and smell the boar. Boars should be 7 ½ months of age before being used in a breeding program.

Allowing sows to have fence-line contact with boars stimulates estrus. When using hand mating or artificial insemination, sows and gilts should be checked for standing heat at least once a day. Checking for standing heat twice a day will increase the conception rate. Gilts should be bred at least twice at 12-hour intervals after standing heat is detected. Breed sows at least twice at 24-hour intervals after standing heat is observed. Breed the first time on the first day of standing heat. Breeding at these recommended intervals will increase the conception rate.

Gilts and sows should be kept separate during the gestation period. Boars of the same size and age can be run together during the off-breeding season. Do not allow boars of different ages to run together. The exercise area for a boar should be at least 0.25 acre.

Provide shade if animals are on pasture. Avoid exposing animals to excessive heat and be sure that plenty of fresh water is available. Separate the breeding herd from other hogs on the farm to avoid disease problems.

Common Reproductive Problems

Gilts normally reach puberty before they are 200 days of age. There are several reasons for gilts showing delayed puberty. Delay in reaching puberty is an inherited trait; crossbred gilts generally reach puberty at a younger age than non-crossbred gilts. Duroc and Yorkshire gilts reach puberty later than Landrace and Large White gilts. Age at puberty can be reduced by 20 to 30 days when gilts are exposed to boars beginning at 140 days of age. Gilts that mature during the period from July through September are more likely to show delayed puberty and anestrus. **Anestrus** is a period of sexual dormancy between two estrus periods. Gilts held in confinement, or in groups of more than 10, also show more delayed estrus and anestrus.

Estrus normally occurs 4 to 10 days after pigs are weaned from sows. There are several causes of delayed post-weaning estrus in sows. A lactation period of less than 21 days reduces the ability of the reproductive tract to recover from pregnancy; also, when young or high-producing sows nurse for more than 30 days, there may be a nutritional drain on the body that can delay post-weaning estrus. Increasing energy intake during lactation can help to overcome this problem. When weaning occurs during the period from July through September, sows take longer to come into heat and have more post-weaning anestrus. Sows put in large groups (more than six) when weaned do not come into heat as readily as those put in smaller groups.

Normally, about 5 percent of the sows and gilts bred will not become pregnant on the first breeding and will come into heat again in 18 to 25 days. There are several reasons for this, including failure to breed each female at least twice during estrus, low fertility of the boar, and poor sanitation in the breeding area. Low fertility in the boar may be caused by immaturity, illness or injury, or high temperature during the breeding season. Breeding areas should be cleaned regularly and kept as sanitary as possible.

From 2 to 3 percent of the females will normally come into heat again more than 25 days after breeding. There are several reasons a higher percentage may do this; matings during the July-through-September period result in more delayed returns to estrus, especially in young sows and gilts. Enteroviruses, porcine parvovirus (PPV), mycotoxins in the feed, and any disease that causes a fever in sows can also cause this condition.

Mycotoxins in feed can cause swelling of the vulva that is not associated with estrus. Feeds should be examined for the presence of mycotoxins if contamination is suspected. Contaminated feeds should not be fed.

An abortion rate of 1 to 2 percent in a swine-breeding herd is normal. There are a number of causes of higher abortion rates; among these are diseases including brucellosis, leptospirosis, pseudorabies, porcine parvovirus, or any other disease that causes a fever in sows. Other causes of higher abortion rates include mycotoxins, high levels of carbon monoxide from unvented heaters, and environmental stress. Sows that are bred during the July-to-September period also have more abortions.

A fetal mummification rate of 4 to 5 percent is normal in swine-breeding herds. Several diseases may cause an increase in this rate. These include enteroviruses, porcine parvovirus, and pseudorabies. Using older gilts, culling of females not immune to porcine parvovirus and vaccination are recommended solutions.

Unusually small litter sizes may result from a number of factors. Several diseases, including enteroviruses, porcine parvovirus, and pseudorabies, may be involved. Other factors causing unusually small litter sizes include low fertility of the boar, too few matings during estrus, the breed of the sow, breeding gilts too young, and rebreeding sows at the first post-weaning estrus. Follow good management practices to help overcome this problem.

About 6 to 8 percent of the pigs will normally be born dead (stillbirths). A higher rate of stillbirths may be caused by several factors. Larger litters normally produce more stillbirths. Older sows have more stillbirths. Overweight sows or gilts, high temperature (above 70°F to 75°F) in the farrowing house, or carbon monoxide toxicity can also increase the rate of stillbirths in the herd. Leptospirosis, eperythrozoonosis, or porcine parvovirus are also possible causes. Deficiencies of vitamin E or selenium in the diet can cause a higher rate of stillbirths. Reducing the average age of the sow herd, good nutrition during gestation, keeping sows cool in the farrowing house, vaccination, and good disease control can help reduce the incidence of stillbirths in the sow-breeding herd.

Farrowing Period

Closely watching the behavior of sows can help determine when they are about to farrow. A high percent of sows farrow within about 6 hours after they begin a period of intensive activity. "Intensive activity" is when the sow stands up and lies down more often than once per minute. Sows will start rooting and pawing at the pen floor when they are about ready to begin farrowing.

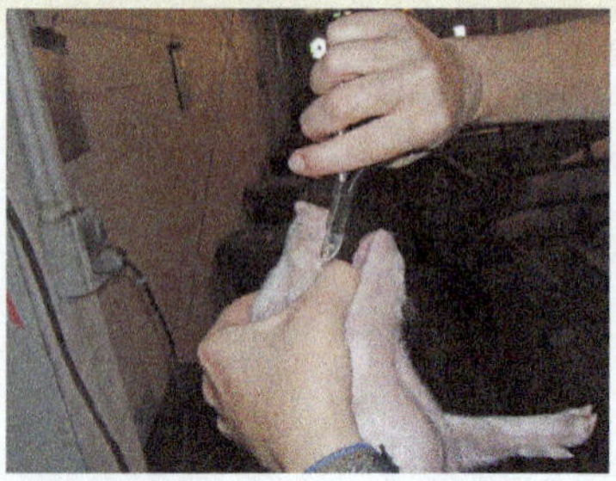

Courtesy of Frank B. Flanders

Figure 22-9 The needle teeth of piglets are clipped to prevent injury to the sow during nursing and to littermates.

Farrowing may also be induced by giving an injection of a hormone-based drug. The injection is given 111 to 113 days after breeding, and the sow usually farrows 18 to 36 hours later. There are several advantages to having a group of sows farrow within a short period of time: It is easier to even up litter sizes by cross-fostering pigs, labor is more efficiently utilized, and it is easier to keep a group of sows on a uniform rebreeding schedule. The breeding herd can be better managed because the farrowing time is more predictable.

Farrowing facilities must be cleaned and disinfected before the sows are placed in them. Traffic through the farrowing house should be kept at a minimum. Sows must be washed with soap and warm water before they are put in the clean stalls or pens. Sows should be moved into farrowing stalls or pens at least 1 day before they are due to farrow. Guardrails and artificial heat are used to protect and warm the baby pigs. For newborn pigs, temperature should be 85°F to 95°F under the heat lamp. Heat lamps are placed 18 inches above the pigs. After 4 or 5 days, the temperature is lowered to 70°F to 80°F by raising the heat lamp.

Many baby pigs can be saved by the operator being present at farrowing time. The sow may need assistance with a difficult farrowing. If pigs are trapped in the afterbirth, then the mucus must be wiped off and the pigs placed under the heat lamp. Baby pigs must be kept warm and dry.

Needle teeth should be clipped with disinfected clippers. The needle teeth of pigs less than 2 days old should be clipped at the gum line. Care must be taken not to cut the gum. Clip one-third to one-half of the tooth if the pigs are more than 2 days old. Figure 22-9 shows the method of holding the pig and clipping the needle teeth.

The navel cord should be clipped shortly after the birth of the pig. Cut the cord 1.0 to 1.5 inches from the body. Disinfect with tincture of iodine.

Pigs can be ear notched for identification. Identification is necessary for good record keeping. Accurate records help in selecting replacement animals. Problem litters can also be identified. Identification is useful in other management practices, such as determining rates of gain and feed efficiency.

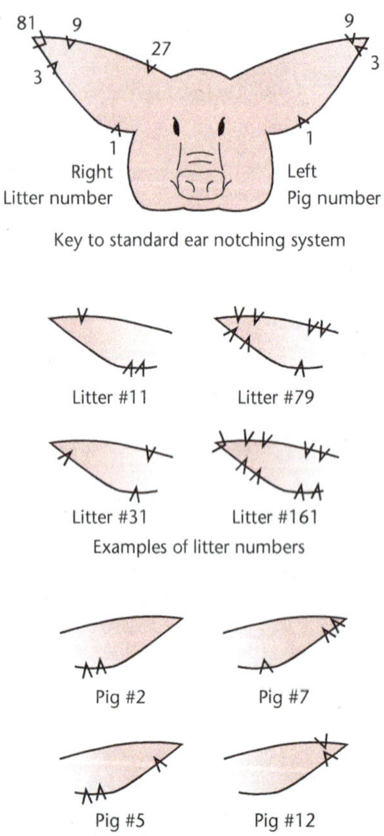

Figure 22-10 A standard ear notching system for pigs.

Purebred associations require ear notching for registration of pigs. However, many commercial swine producers do not ear notch their pigs or may only ear notch certain litters for tracking purposes. Some producers may notch only the litter number or may apply their own particular code. Ear notching pigs is labor intensive, and unless there is a good reason to do so, many commercial producers skip this procedure.

There are several systems of ear notching. Purebred associations specify which system they require for their records. A standard system is shown in Figure 22-10. The right ear shows the litter number. The left ear identifies the individual pig. Determination of right and left ear is made by viewing the pig from the rear.

Extra labor is needed to save runt (small) pigs. Using a commercial milk replacer increases their survival rate. A mixture of 1 quart milk, 1/2 pint half-and-half, and 1 raw egg may also be used. Feed 15 to 20 milliliters once or twice per day. Use a soft plastic tube with a syringe to give the milk replacer or mixture orally. This practice saves about one-half the pigs that otherwise might have died.

Litter sizes should be equalized. Move pigs from large litters to small litters in order to make the litter sizes about equal. This should be done during the first 3 days after the pigs are farrowed. Be sure that the pigs nurse colostrum milk before moving them. The largest pigs in the litter should be selected for moving. Make sure the sow has the nursing ability and the number of teats necessary for the number of pigs that are in the litter.

Farrowing to Weaning Period

Several important management practices are performed in the period between farrowing and weaning. **Tail docking** is cutting off the pig's tail 1/4 to 1/2 inch from the body. This should be done when the pigs are 1 to 3 days old. Side-cutting pliers, special cutting instruments, or a chicken debeaker may be used. The instrument in Figure 22-11 has an attachment that heat-treats the wound and encourages sealing on the blood vessels and promotes healing. Treat the tail stub with a disinfectant such as iodine spray. Disinfect the pliers between use on each pig. Tail docking helps to prevent tail biting among pigs in confinement. Producers of feeder pigs should always tail dock the pigs. Do not dock tails while pigs have scours. Tail docking of runts can be delayed until they are stronger.

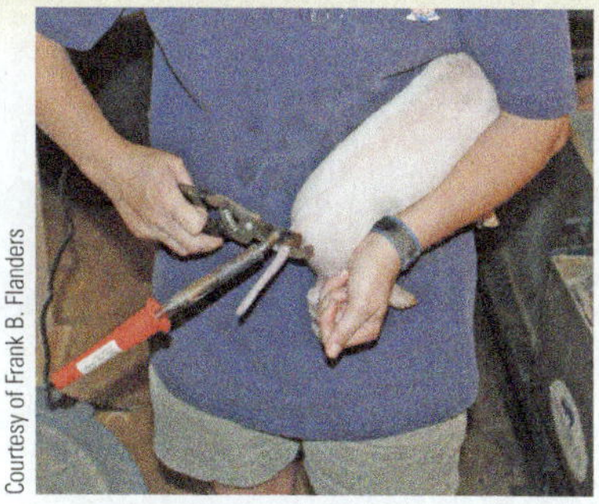

Courtesy of Frank B. Flanders

Figure 22-11 Tail docking is performed to prevent pigs from chewing on each other's tails.

Anemia is prevented by giving iron injections or oral doses of iron. This should be done when the pigs are 2 to 4 days old. Injections should be given in the neck area, rather than the ham or shoulder, to prevent staining of the meat (Figure 22-12). Iron-dextran shots are used at the rate of 100 to 150 milligrams per pig. Iron should be given carefully because overdoses may cause shock in the pigs. Repeat the dose at 2 weeks of age. Although not commonly practiced, iron can also be provided by placing fresh sod in the pen or it can be added to feed or water at the time of the second dose.

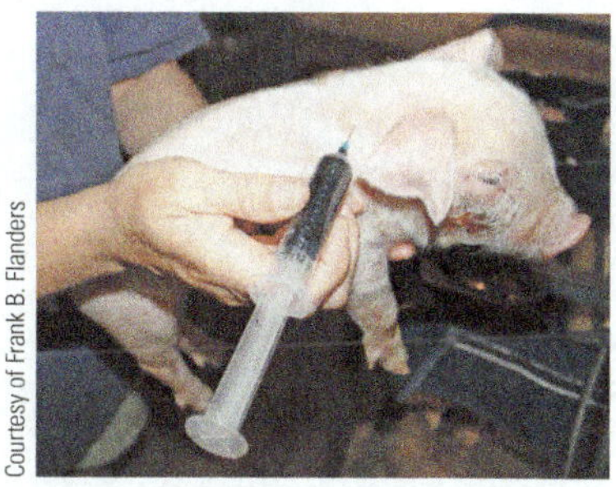

Courtesy of Frank B. Flanders

Figure 22-12 Iron injections are given in the neck to prevent staining the ham or other valuable retail cuts.

Watch the pigs closely for scours. Drugs given orally usually work better than injections in preventing scours. Medication dissolved in water can be used as the pigs become a little older. Sanitation helps in preventing scours. If scours becomes a serious problem, consult a veterinarian.

Male pigs that are raised for harvest must be castrated to avoid boar taint. Castration is best done when the pigs are young. There is less stress on the pigs and the job is easier. Males should be castrated before they are 2 weeks old. The knife and other instruments must be clean, sharp, and disinfected. Many producers castrate pigs at 3 days of age. There is very little stress on the pig and it is a one-person job. Pig holders are available that make the job easier for one person to accomplish. Castration, vaccination, and weaning should not all be done at the same time as this would place too much stress on the pigs.

Disease and parasite control is important in reducing losses. Vaccination and worming programs should be tailored to the problems of the individual farm. Consult a veterinarian for specific recommendations.

Pigs need to be started on feed as soon as possible. There is a trend toward earlier weaning of pigs, meaning weaning the pigs at an age of less than 21 days. Early weaning requires higher levels of management and nutrition. The average producer weans pigs at about 21 days of age.

Avoid drafts and changes in temperature when weaning pigs of any age. Three-week-old pigs require temperatures from 80°F to 85°F at weaning. Group pigs according to size at weaning. Groups should contain no more than 30 pigs, if possible.

Weaning to Market

From weaning to market, most management centers around feeding and facilities. During this stage, hogs are raised in confinement housing, but occasionally on pasture in small operations. Confinement requires more capital investment, but hogs gain a little faster under confinement conditions.

During finishing, hogs should be grouped in uniform size lots by weight. Groups should be no larger than 50 to 75 head. Weight range should be no more than 20 percent above or below the average weight of the group. Hogs are marketed at 230 to 280 pounds, depending on the market.

Feed accounts for 60 to 65 percent of the expense of raising hogs. Wasted feed reduces feed efficiency. Feed loss can be reduced by making sure feeders are properly adjusted. Self-feeders should be adjusted at least once or twice a week to reduce waste. Controlling rodents in feed storage and feeding areas can also reduce feed loss.

A management practice that can help reduce the incidence of disease in swine being fed for market is **segregated early weaning (SEW)**. In SEW systems, pigs are weaned from the sow at 9 to 20 days and placed in an isolated, clean facility. Research has shown that antibodies from the sow's colostrum protects pigs for about 20 days. After that, the pigs are more susceptible to being infected with disease from the sow. The average daily gain may be increased by 14 percent and feed efficiency may be increased by 9 percent when segregated early weaning is used. Death loss is also reduced using this system.

Some producers induce a hyper-immune response by inoculating sows with vaccines and feeding a broad-spectrum antibiotic 2 to 4 weeks before farrowing. Immunity is passed to the pigs at birth, providing protection beyond that provided by colostrum. Broad-spectrum antibiotics may be used in the sow's diet before farrowing and during lactation.

Segregated early weaning shows positive results when used with single-site production systems, but it is most effective when combined with all-in/all-out multiple-site systems. However, there are increased costs and other potential problems when multiple sites are used. More facilities are needed, and pigs and feed must be transported to the other sites. Another concern is the presence of other hog operations near each site used in the system. Depending on the disease problems present in the area, recommendations for distance from other hog-production facilities range from 2 to 10 miles. The all-in/all-out system is important in that all the pigs are placed together in a clean and isolated facility together and no other animals are added to the group to possibly infect the group. The entire group is sold together and the facility is cleaned prior to bringing in the next group.

While there are advantages to segregated early weaning, it has not proven popular with all producers. The current trend is to use a 21-day or older weaning age.

A term related to segregated early weaning, and occasionally used to describe a type of weaning system, is **medicated early weaning (MEW)**. In this system, pigs are weaned and isolated in a clean facility for finishing as in SEW, but the pigs are medicated as well with broad-spectrum antibiotics and vaccines.

Using the **all-in/all-out** method of raising hogs can improve rate of gain and feed efficiency. The incidence of disease in swine herds is reduced using this management method. The all-in/all-out system moves pigs as a group from nursery, through growing and finishing, and to market. Groups consist of pigs farrowed within a short period of time and of the same weight range.

Facilities are cleaned by power washing and are disinfected between groups of pigs. All bedding, manure, and feed are removed from the facility at the time it is cleaned. Use hot, soapy water and a 2 percent lye solution for cleaning, and rinse with clear water. Disinfectants should be used after cleaning. A number of commercial products containing approved disinfectants are available. Generally, these disinfectants are effective against a variety of bacteria, fungi, and viruses. Always follow label directions for the use of disinfectants. The facilities are left idle for a short period of time after cleaning and disinfecting before a new group of pigs is brought in.

Dust particles that adversely affect the health of workers in swine confinement buildings are a problem, especially in large operations where workers may spend the entire workday in the building. Research shows that using a light spraying of vegetable oil in the building can significantly reduce the dust and odor levels.

Figure 22-13 Weaned pigs at a swine unit.

FEEDER PIGS

Some producers prefer to sell pigs as feeder pigs. Other producers who do not want to invest in facilities and breeding herds prefer to buy feeder pigs to feed out to market weights. Feeder pigs generally are 8 to 9 weeks of age and average 35 to 50 pounds.

Feeder pig production provides a faster turnover in the volume of pigs handled. Less feed is required for each dollar's worth of pig sold. Labor must be available year-round for feeder pig production. Good sanitation and disease control programs are necessary. Large-volume operators have lower costs per pig produced than do small-volume operators. Net returns are higher for the large-volume producer.

Up to weaning, feeding and management practices are about the same for feeder pig production as for other hog enterprises (Figure 22-13). Good management and marketing practices are necessary if feeder pig production is to be profitable.

SUMMARY

The two types of swine production are purebred and commercial. Purebred production is a specialized business. Fewer than 1 percent of the hogs raised in the United States are registered purebreds. Purebred producers raise breeding stock used in commercial hog enterprises.

Commercial hog producers raise hogs to be sold for harvest. Most commercial producers use some system for crossbreeding. Commercial producers may merely produce feeder pigs; may buy and feed out feeder pigs; or may use a complete sow and litter system, also known as farrow-to-finish.

The most common feeds for hog production are corn and soybean oil meal. Other grains may be substituted for corn. However, most of them do not have the feeding value for hogs that corn does. Other sources of protein may also be used in place of soybean oil meal, although soybean oil meal has the best quality protein. Other sources must be fed in combinations to provide the amino acids required for proper swine nutrition.

Pasture is most valuable in feeding the swine-breeding herd. Market hogs raised on pasture gain a little slower than those raised in confinement. Almost all commercial swine are produced in confinement housing for all phases of production. Some pasture management is still used by smaller producers.

The most important minerals in hog rations are sodium, chlorine, calcium, and phosphorus. Salt, ground limestone, and dicalcium phosphate are the common sources of these minerals. Certain trace minerals are also needed in swine rations. These are usually added by the use of commercial mineral mixes.

Vitamins are added to swine rations by using commercial feeds and vitamin premixes. Water must be clean, fresh, and in adequate quantity for good hog production.

The breeding herd should be fed so that the sows, gilts, and boars do not become too fat. Limiting feed intake is a common way of avoiding over-fattening. Self-feeding a bulky ration with ground hay added will also help keep these animals from becoming too fat. Silage may be used in the ration for the breeding herd. Lactating sows should be fed a ration to maximize milk production.

Baby pigs should be started on feed as soon as they will eat. Starter and creep rations should be made available when the pigs are 1 week old.

Growing-finishing hogs may be fed free choice or on self-feeders. Limited feeding is not a recommended practice for the average hog producer. The protein level in the rations may be reduced as the hogs get older. Pelleted rations increase feed efficiency.

Crossbreeding and multiple farrowings are recommended practices. Gilts should be bred when they are at least 8 months old and weigh 250 pounds. Boars should be at least 7 1/2 months old before using them for breeding.

A producer can save more pigs by being present when sows are farrowing. Make sure pigs nurse to receive colostrum and that they are warm and dry. Clip needle teeth, ear mark pigs, disinfect and clip the navel cord, and dock tails during the first day or two after farrowing. Equalize litter sizes and give iron shots during the first few days. Castrate boars before they are 2 weeks old. Control scours in baby pigs and plan a disease prevention program to keep pigs healthy.

After pigs are weaned, feeding good rations is the most important management practice for successful swine production. After weaning, pigs should be grouped in uniform lots according to size.

Student Learning Activities

1. Give an oral report on feeding and management practices used by swine farms.
2. Use the practices outlined in this chapter when planning and conducting a feeding and management program for a swine Supervised Agriculture Experience Program.
3. Ask a swine producer in the area to talk to the class about feeding and management practices.
4. Observe and practice management practices such as clipping needle teeth, tail docking, and castration.
5. Do a community survey of swine producers on feeding and management practices and report to the class.

Discussion Questions

1. What is the function of the purebred hog business as it relates to commercial hog production?
2. What type of breeding system is commonly used to produce commercial hogs?
3. Describe the characteristics of a feeder pig production operation.
4. Describe the characteristics of an operation that buys and finishes feeder pigs.
5. Name and describe three energy feeds that may be used for feeding swine.
6. Describe the characteristics of soybean oil meal as a hog feed.

7. Explain why corn should be supplemented with protein.
8. Name and describe three protein feeds that may be used to feed hogs.
9. Describe the use of roughages to feed hogs.
10. Name four major minerals and six trace minerals, and explain how they may be supplied in the ration.
11. Describe the feeding of gilts selected for the breeding herd.
12. Describe the practice and purpose of flushing.
13. Describe a feeding program for gilts and sows during the gestation period.
14. What kind of ration should be fed at farrowing time?
15. Describe a feeding program for the herd boar.
16. Describe a supplemental feeding program for baby pigs.
17. Give a growing-finishing ration for hogs weighing 40 to 125 pounds.
18. Give a finishing ration for hogs from 125 pounds to market.
19. Compare feeding free choice to feeding complete mixed feeds.
20. What are the advantages of a multiple farrowing system?

Review Questions

True/False

1. Pasture is most valuable for feeding the breeding herd.
2. The nutrition requirements of younger pigs are greater than that of older pigs.
3. Baby pigs should be started on feed as soon as they will eat.
4. A fetal mummification rate of 4 to 5 percent is normal in swine breeding herds.

Multiple Choice

5. Many of the odor problems from confinement operations are related to the release of ______ from manure.

    a. carbon dioxide c. methane

    b. sulfur d. ammonia

6. Two factors that have a major influence on profitability in swine production are:

    a. number of pigs weaned per sow; feed efficiency

    b. breed; adaptability to environment

    c. number of pigs weaned per sow; adaptability to environment

    d. breed; feed efficiency

7. Pigs should be ear notched for ______, which is necessary for good record keeping.

    a. vaccination c. sexing

    b. identification d. grouping

Completion

8. The ______ ______ ______ ______ is a management education program with major emphasis on the swine herd health program.
9. ______ have a higher protein content than corn but the quality of protein is poorer.
10. About ______ of the pigs lost before weaning are lost because of poor feeding.

Short Answer

11. Why is whey in liquid form a problem for meeting the protein needs of hogs?
12. What term is used to describe swine operations when producers produce their own feeder pigs and raise them to market weight?
13. What practice helps prevent tail biting?

Chapter 23

Diseases and Parasites of Swine

Key Terms

abscesses
turbinate bone
specific pathogen free (SPF)
rotavirus
ampicillin

Objectives

After studying this chapter, the student should be able to

- explain the importance of maintaining good swine health.
- describe the causes, symptoms, prevention, and control of common swine diseases and parasites.
- list good management practices that help to prevent swine diseases and parasites.

The single most important management practice that affects profits from swine raising is maintaining a healthy herd. National averages show that 40 percent of the hogs that are farrowed do not reach market. One-third of the hogs farrowed do not reach weaning. Hog producers lose millions of dollars each year because of diseases and parasites.

Herd health problems have become more complex because of certain changes that have taken place in management in recent years. First, more hogs are being raised in confinement. In addition, hogs are being fed rations to push them to maximum growth rates. These two management changes create more stress on the pigs that may lead to an increase in herd health problems. Also, because of faster transportation, diseases may be spread farther in a shorter period of time.

DISEASE AND PARASITE PREVENTION

Herd health should be monitored on a continuous basis. Maintain accurate records of average rate of gain, feed intake, feed conversion, death rate, and treatments administered to the swine herd. These records form the basis for early detection of health problems. The use of a veterinary consultant will also help maintain good herd health. Veterinarians can provide diagnostic lab services to help prevent problems before they become serious. A thorough postmortem examination can accurately identify the cause of an animal's death and help to prevent spread of a disease through the herd. Blood testing is another health-monitoring technique that can measure exposure to diseases.

Observing the vital signs in an animal can help in the early detection of health problems (Figure 23-1). Vital signs will vary according to activity and environmental conditions. Normal vital signs in swine are:

- temperature: 102.0°F to 103.6°F; the average is 102.6°F. Temperature is usually higher in the morning than in the afternoon; younger animals will show a wider range of temperature than mature animals
- pulse rate: 60 to 80 heartbeats per minute
- respiration rate: 8 to 13 breaths per minute

Normal Vital Signs of Swine

Temperature	102.0°F to 103.6°F (Average is 102.6°F)
Pulse Rate	60 to 80 heartbeats per minute
Respiration Rate	8 to 13 breaths per minute

Figure 23-1 Vital signs of swine.

Body temperature is taken in the rectum using either a mercury thermometer or a battery-powered digital thermometer. Restrain the animal when inserting the thermometer into the rectum. There is no location on a hog's body where the pulse can be felt directly by finding an artery. The heart must be felt directly over the chest. Respiration rate is determined by observing the number of times the animal breathes per minute.

Many swine producers are participating in the pork industry's Pork Quality Assurance (PQA) Plus Program. This program emphasizes good management practices in the handling and use of animal health products and encourages producers to annually review herd health programs. This helps reduce drug residue problems in marketed swine, addresses consumer food safety concerns, and improves the export market for pork. The program is funded through a pork checkoff that collects money from producers based on hogs marketed.

A well-planned herd health program is aimed at prevention of diseases and parasites, rather than their treatment. Management practices such as sanitation, isolation of new stock, selecting healthy breeding stock, and proper care of sows and pigs are all part of good herd health maintenance.

The first step in a sanitation program is keeping the facilities clean. High-pressure water cleaning equipment and soapy water are useful in removing dirt and manure from the hog house. Use of a good disinfectant such as a hot lye solution, sodium carbonate, cresol, and iodine completes the process and kills most germs and viruses. Read and follow label directions carefully and always use caution when handling any chemical.

CAUTION

Lye solutions are caustic and must be handled carefully. Do not allow the solution to come into contact with the skin or eyes.

Pens for young pigs should receive the most attention when facilities are cleaned and disinfected because they are more susceptible to diseases. Clean pens are also important for older pigs; however, older pigs are not as susceptible to disease as young pigs. Proper ventilation and light improve sanitation in the hog house. Bacteria and viruses grow better in dark, damp, poorly ventilated areas. Concrete floors make the cleaning job easier.

Disease organisms may be carried into hog production areas by visitors. Restrict the entry of visitors into hog production facilities; if visitors are allowed to enter the facilities, make sure they are wearing disinfected footwear. A container of disinfectant and a brush for disinfecting shoes should be kept at the entrance for the use of those entering the facilities.

It is recommended that clean boots and coveralls be provided to be worn by any off-farm visitors entering hog production facilities. Provide facilities for loading and unloading trucks outside of the hog production area. Dogs, cats, and other

Figure 23-2 Animals may transmit diseases to swine farms.

(A) © iStockphoto/Oleg Kozlov; (B) Courtesy of Frank B. Flanders; (C) © HartmutMorgenthal/www.Shutterstock.com; (D) © iStockphoto/David Birkbeck

animals can carry diseases that may be transmitted to hogs; therefore, these animals need to be kept out of the hog production area (Figure 23-2). A good rat control program will also help reduce the transmission of diseases to hogs.

Other management practices that help control diseases and parasites include isolation of new breeding stock and proper care of sows and pigs at farrowing and during the lactation period. Vaccination and worming programs are adapted to the conditions of the individual farm. A program for a specific farm should be discussed with a local veterinarian. All buildings should be left unoccupied periodically to help prevent a buildup of diseases and parasites.

INFECTIOUS DISEASES

Abscesses

Abscesses, or swellings filled with pus, are the result of a bacterial infection that enters the body through the nose or mouth. They are also called *jowl abscesses, cervical abscesses, feeder boils,* or *hog strangles*. The swellings, which vary in size, may appear under the jaw and in the area of the neck. Not all infected hogs show symptoms of abscesses.

Hogs from weaning to market weight are more frequently affected than younger pigs. Affected hogs have a slower rate of gain and severe cases can cause death. Carcass value is reduced because affected parts are condemned at slaughter.

Abscesses can be prevented by vaccinating hogs with an antibiotic. Other practices that help in prevention are early weaning, buying replacement animals from herds free of abscesses, and good sanitation practices. Feeding antibiotics to young pigs also helps to control abscesses. Abscesses can be opened and allowed to drain, but this may spread the infection to other animals in the herd. Treating infected hogs with antibiotics offers some relief from the disease. The use and withdrawal times for drugs used in treatments are subject to change and current requirements should always be checked.

Actinobacillus Pleuropneumoniae

Actinobacillus pleuropneumoniae (APP) is caused by bacteria and may appear in chronic and acute forms. Symptoms of the chronic form include abdominal breathing, high fever, depression, and reluctance to move. Pigs with the chronic form may die, but most will survive usually with damaged lungs and an abnormal rate of gain. The acute form can be characterized by sudden death of apparently healthy pigs. The pigs may develop heavy breathing and die within a few minutes, possibly after a period of stress such as moving the pigs, mixing groups of pigs, weather changes, and poor ventilation in the housing facility. Death can occur in as short a time as 8 to 12 hours after exposure to the bacteria. The disease can occur in pigs of any age; however, it most commonly occurs in pigs from 40 pounds to market weight. Because the bacterium is spread through the air, 100 percent of the pigs in a group can be affected, with death losses reaching 20 to 40 percent if no treatment measures are taken. A definite diagnosis of the disease is made from cultures of lesions taken in a postmortem examination.

High levels of penicillin and tetracycline injections are the most effective immediate treatment. A specific antibiotic sensitivity test should be made to identify the proper antibiotic for continued treatment. All pigs in the same facility should be treated with an antibiotic injection to reduce the spread of the disease and death loss. Adding antibiotics to the feed or water may be effective as a preventive measure after the initial antibiotic injection is given.

Pigs that have recovered from the disease are carriers. Isolate any additions to the herd for 30 days; serum test at the beginning of the isolation period and again at the end to confirm that they are not carriers. Commercial vaccines are available to help prevent this disease; however, they do not provide complete immunity. Reducing overcrowding and providing good ventilation in the housing facility may help prevent this disease. These measures will also help reduce death losses and improve the chances for successful treatment when an outbreak does occur.

Atrophic Rhinitis

A mild form of atrophic rhinitis is caused by a toxin-producing bacterium, *Bordetella bronchiseptica*. A more severe form is caused by the toxin producing bacterium *Pasteurella multocida*. The turbinate bones may not grow properly or they may atrophy (shrink) as a result of the infection. The **turbinate bones** are located in the snout of the pig. They filter and warm the air before it reaches the lungs of the hog.

The milder form of the disease generally affects the nasal passages and the tonsils and has little effect on pigs after they are about 9 weeks of age. The more severe form generally affects the nasal passages, tonsils, and lungs and may damage the nasal cavities of pigs up to 16 weeks of age. It may also damage the liver, kidneys, ends of the long bones, and some components of the blood. Pigs infected with either form of the disease during the first week of life are more severely affected than those infected when they are older.

The disease spreads from one pig to another, and litters can be infected by the sow. A herd is usually infected by the introduction of infected breeding stock or feeder pigs. Pigs that appear to be healthy can still be carriers of the disease. It is thought that cats, dogs, and rats are also carriers of the bacteria.

Pigs are generally infected when young, usually when they are 2 to 3 weeks old. They can be infected after weaning. Symptoms—which include sneezing, tear flow, nasal discharge, nose bleeding, and twisting of the snout to one side—usually are seen at about 4 to 12 weeks of age. The atrophy of the turbinate bones leads to a secondary infection of pneumonia. The disease seldom causes death, but rate of gain is slowed, and feed efficiency is lowered.

Vaccines are available for atrophic rhinitis. Vaccinating sows helps reduce the prevalence of atrophic rhinitis in the pigs but does not eliminate the disease. Nursing pigs may be vaccinated, but it takes about 2 weeks for immunity to develop; by then, the pigs may already be infected.

Sulfonamides or oxytetracycline may be included in the sow diet during the last month of gestation to reduce the discharge of bacteria. The treatment of nursing pigs with injections of oxytetracycline or penicillin and streptomycin may reduce the severity of atrophic rhinitis. Medication may be used in the feed or drinking water of weaned pigs. Good sanitation and management practices must be used and followed with any vaccination and medication program. The following management practices are recommended:

- keep pigs of different ages separated
- reduce stress from cold, dampness, and drafts
- provide adequate ventilation in the housing facilities
- follow strict sanitation practices
- follow an all-in/all-out production system
- use a segregated early weaning program with multiple production sites
- provide proper care and nutrition at weaning
- use fewer gilts in relation to older sows in the breeding herd

Factors that can increase the chances of atrophic rhinitis are:

- large herds
- adding pigs to the herd from several sources
- large farrowing units
- large nurseries
- moving and mixing pigs frequently
- overcrowding
- poor ventilation in the housing facility
- unsanitary conditions
- continuous flow of pigs through the facilities
- excessive dust and gases in the facilities

- make sure new breeding animals are not infected
- use **specific pathogen free (SPF)** breeding stock. SPF hogs come from breeding stock that are surgically removed (generally by cesarean section) from the sows under antiseptic conditions. These hogs are raised in disease-free conditions

There is no way to eliminate the bacteria after a herd becomes infected. Following good management practices can help reduce the severity of the disease.

Atrophic rhinitis should not be confused with *bull nose*. Pigs with bull nose develop large abscesses on their snouts. These abscesses are filled with thick pus. Bull nose is associated with unsanitary conditions.

Avian Tuberculosis

Avian tuberculosis is caused by a bacterium. It is spread from chickens to hogs. Few symptoms of the disease appear in market hogs. Symptoms may be noticeable in older hogs that are kept for breeding purposes. The symptoms are a gradual loss in weight or an enlargement of the joints. A skin test is used to diagnose avian tuberculosis.

The main preventive measure is to keep chickens and hogs apart. Do not use poultry buildings for hog houses unless they have been disinfected, and do not allow hogs to eat poultry carcasses. There is no cure for avian tuberculosis.

Brucellosis

Brucellosis is caused by bacteria. Abortion early in the gestation period is a common symptom as well as failure to rebreed. The disease is spread mainly from infected boars. Boars with the disease show swelling of one or both testicles. Blood testing the breeding herd is the only way to determine if the disease is present.

Buying clean breeding stock is the main way to prevent brucellosis. The entire breeding herd should be tested at least once a year. There is no treatment for hogs infected with brucellosis.

Through the efforts of the Cooperative State-Federal Brucellosis Eradication Program, there are only a small number of swine herds left that are known to be infected with brucellosis. Under this program, a state is declared brucellosis free when no new cases are reported during a 24-month period and other requirements are met. The USDA is working toward the goal of complete eradication of brucellosis in swine herds in the United States. Most of the remaining known cases of brucellosis in swine herds are in the South and Southeast; the disease has not been a major problem in the Midwest for several years.

Cholera

At one time, cholera was one of the most serious of all hog diseases in terms of annual losses. A program of cholera eradication has been in effect in the United States for a number of years. The USDA announced in 1978 that hog cholera had been completely eradicated in the United States. Hog cholera is now classified as an exotic (foreign) disease by the USDA.

Clostridial Diarrhea

Clostridial diarrhea, a disease of baby pigs usually under 1 week of age, is caused by *Clostridium perfringens* bacteria—the same bacteria that cause necrotic enteritis in poultry. The bacteria produce a toxin or endotoxin that damages the cell linings in the intestine. Symptoms include listlessness and a yellow, bloodstained, watery diarrhea. The death loss of affected litters is often more than 25 percent. Death usually occurs within a day and a half after the symptoms are first seen.

A chronic, less severe, form of the disease is more difficult to diagnose. The chronic form usually occurs in pigs between 5 days of age and weaning. Symptoms include a gray-flecked diarrhea and emaciation. These symptoms may be confused with the symptoms of transmissible gastroenteritis, *Escherichia coli* scours, **rotavirus**, or coccidiosis. A positive diagnosis is made by laboratory analysis of smears taken from infected pigs. Pigs affected by the chronic form have lower feed efficiency and weight gain.

It is necessary to follow a good sanitation program in the farrowing facilities to help prevent this disease. Sanitize the farrowing crates or pens and wash the sows before putting them in the crates or pens.

Drug treatment of newborn pigs may offer limited protection but is generally not practical. Treatment after symptoms appear is generally not effective. Vaccinating sows and gilts at 5 weeks and 2 weeks before farrowing may provide some protection against the disease. Where the disease is a major problem, vaccinating at 7 weeks and 2 weeks before farrowing will allow the development of a higher level of antibodies in the colostrum milk.

The best way to control clostridial enteritis is to prevent the transmission of the disease from the sows to the baby pigs. Bacitracin methylene disalicylate may be fed to sows and gilts for 2 weeks before farrowing and continuing for 3 weeks after farrowing. Research has shown that using this treatment significantly reduces the death loss from clostridial enteritis. This treatment also improves the rate of gain in the baby pigs.

Swine Dysentery

Swine dysentery is also known as *vibrionic dysentery, bloody scours,* and *black scours*. It is caused by the bacterium *Serpulina hyodysenteriae*. Swine dysentery most commonly affects pigs from 8 to 14 weeks of age; however, any age may be affected. In adult swine, the disease is not as severe and may be difficult to diagnose.

Dysentery may be confused with other pig scouring problems. The first symptom is bloody diarrhea. After a time, the feces become watery and contain thick mucus. A positive diagnosis of swine dysentery is made by laboratory analysis of feces samples. Weight loss, fever, and refusal to eat are other symptoms. About 20 to 30 percent of the hogs infected die if no treatment is administered. Chronic infection results in slow rates of gain and poor feed efficiency.

The disease spreads from infected hogs to healthy ones. Sows may be carriers of the disease without showing outward symptoms. The bacteria live in the feces of the sow and may be transmitted to nursing pigs. The disease may be spread from farm to farm in feces carried on shoes, boots, or vehicle tires. Field mice are a major reservoir of the bacteria and can spread the disease from farm to farm. Dogs, birds, rats, and flies—even though they are not long-term reservoirs of the bacteria—can spread the disease from farm to farm.

Pigs that are infected with swine dysentery may be treated with drugs added to the drinking water. Because infected pigs do not eat much feed, treatments administered in the feed are of little value. Direct injection of drugs may be used in treatment, but this method is generally not practical. Many drugs are approved for treatment of swine dysentery. Check with a veterinarian or agriculture extension agent for recommendations.

Only healthy hogs should be brought into the herd (Figure 23-3). Animals should not be purchased from a herd that has had dysentery during the past 2 years. Isolate newly purchased stock. Avoid spreading the disease from one pen to another. The use of SPF hogs helps break the cycle of dysentery.

Courtesy of Frank B. Flanders

Figure 23-3 Only healthy hogs should be brought into the herd.

Other preventive measures that may be used include maintaining isolated herds, controlling rodents, controlling access to swine facilities, reducing stress, avoiding overcrowding in the facilities, maintaining strict sanitation, and maintaining a clean, dry environment.

Eliminating the disease from the herd may be more cost effective than trying to control it. One method is to totally depopulate the herd, clean and sanitize the facilities, and repopulate with breeding stock free of swine dysentery. A second option is to eradicate the disease without total depopulation of the herd. This involves using medications, cleaning and sanitizing the facilities, and eliminating the rodent population. A third option is using two- or three-isolated site production with partial depopulation. It is advisable to work closely with a veterinarian when developing a program for eliminating swine dysentery from the herd.

Edema

Swine edema is caused by toxins produced by a bacterial infection. Stress such as weaning, feed changes, and vaccination appear to make the problem worse. Anywhere from 10 to 35 percent of a herd may be affected. Death losses range from 20 to 100 percent of the pigs affected. Pigs between 3 and 14 weeks of age are most commonly affected.

Sudden death may be the first symptom seen. The disease may be confused with chemical poisoning. Other symptoms include refusal to eat, convulsions, staggering, and swollen eyelids. There is usually no fever.

Elimination of stress conditions helps to prevent edema. If symptoms appear, take feed away from pigs for 24 hours and then gradually increase to full feed. Sulfa and antibiotics in the feed or drinking water help in prevention and treatment. Observe withdrawal times on the label of the specific drug used.

Eperythrozoonosis

Eperythrozoonosis is caused by a small bacterium-like organism that attaches itself to the membrane of the red blood cell. This results in a destruction of red blood cells in the animal's body. The disease may appear in both chronic and acute forms. Because other infectious and noninfectious problems may produce similar symptoms, it is hard to diagnose this disease, especially in its chronic form. A blood test is available that is useful in identifying infected herds, but it is of limited value in identifying individually infected animals.

Symptoms of eperythrozoonosis include anemia, icterus (yellow discoloration of skin and mucous membranes), slow growth, and reproductive problems. The acute form of the disease is indicated by fever (104°F to 106°F), anemia, icterus, and an enlarged spleen. Hypoglycemia (low blood sugar) may occur in some hogs; this leads to coma, convulsion, and death. The acute form of the disease is more common in young pigs. The majority of the affected pigs recover but gain at a slower rate for the rest of their lives.

Sows may be affected by either the chronic or the acute form of the disease. The acute form occurs more often during periods of stress, such as farrowing, weaning, or breeding. At farrowing, infected sows may show udder edema (excessive accumulation of serous fluid in the tissue) and a lack of milk. The sow will usually begin to produce normal amounts of milk after a few days.

The chronic form of the disease in sows may be indicated only by persistent anemia. Other indications are decreased fertility, smaller litter size, occasional abortions, increased stillbirth rate, and smaller, weaker pigs at birth. Growing-finishing pigs may be affected by either the chronic or the acute form of the disease. The acute form is indicated by a yellow discoloration of the abdomen (yellow belly).

Eperythrozoonosis is not spread by direct contact between pigs. The most common method of transmission is by transfer of infected blood from a carrier to noninfected hogs. Infected blood may be transferred by biting insects, blood-contaminated needles, and castration instruments and other surgical equipment.

A veterinarian should be consulted for treatment of the disease. Tetracyclines and arsanilic acid have been effective in suppressing the symptoms, but they do not eliminate the carriers in the herd. The only way to control the disease is to avoid the transfer of infected blood from one animal to another. Control biting insects, lice, and mange to avoid transmission of the disease. Be careful when using needles and surgical equipment to avoid transferring contaminated blood from animal to animal. Be careful when introducing new breeding stock into the herd; new breeding stock should be selected from herds that have tested negative for eperythrozoonosis.

Erysipelas

Erysipelas is caused by a bacterium and occurs in acute, mild, and chronic forms. Hogs from weaning to market age are those most affected. Symptoms of the acute form include fever, withdrawal from the herd, lameness, depression, signs of chilling, and sudden death. Red skin lesions in diamond shapes may appear on the skin. The mild form shows the same symptoms as the acute form, but they are less severe. The chronic form may follow recovery from an acute or mild attack. Arthritis is the most common damage resulting from the chronic form. Prevention can be achieved by vaccinating hogs at 6 to 8 weeks of age. A combination of penicillin and erysipelas serum is used to treat infected hogs.

Exudative Epidermitis (Greasy Pig Disease)

Exudative epidermitis is caused mainly by a common bacterium that lives on the skin of the animal. It enters the animal's body when the skin is broken. Bite wounds, abrasions, lice and mange infections, dirty surgical or injection equipment, or viruses that cause blisters on the skin may create conditions that allow the bacterium to penetrate the skin. Pigs are usually affected during the first 5 to 6 weeks of life.

Symptoms of the disease are reddish areas that appear around the eyes, behind the ears, or under the legs. Fluid seeps from these areas, and dirt and dander from the pig accumulate in the fluid; this gives the pigs a dirty, greasy appearance. As fluid losses increase, dehydration, lethargy, and lack of appetite result. The average death rate for infected pigs is about 25 percent. In some cases, death may result within a few hours to 2 days after the initial signs of the disease appear. In other cases, death may be delayed for 6 to 10 days after the appearance of the initial signs of the disease. Pigs that recover may have a slower rate of gain and poorer feed efficiency because of kidney damage caused by the disease.

A veterinarian should be consulted about the appropriate drugs and control measures to use for this disease. In the early stages of the disease, injections of penicillin or tetracyclines may help. After the disease has had time to fully develop, these antibiotics are of little value. Washing the pigs with a mild soap and water or disinfectant solution may provide some help in treating the disease. Washing should be done daily for 3 to 4 days. Remove affected pigs from contact with those that are not affected. Treat unaffected pigs that have been in contact with affected pigs with penicillin or tetracyclines to reduce the chances of their becoming infected.

Disinfect equipment after treating each pig when giving injections, castrating, clipping needle teeth, or performing other surgical procedures to prevent spreading the bacterium with dirty equipment. Good sanitation procedures in the facilities will help prevent the spread of the disease. Provide a stress-free environment that is warm, dry, and free of drafts. Reduce conditions in the environment that may cause skin abrasions.

Influenza

Swine flu is a respiratory disease caused by a combination of a virus and bacteria. Symptoms include fever, difficulty breathing, coughing, going off feed, and weakness. Pigs become ill suddenly and usually recover in about 6 days. The death loss is low, but rate of gain is slowed. There is no drug or antibiotic that can be used to cure swine flu. Give sick pigs water and feed, reduce stress, and keep them warm.

If swine flu becomes a problem, it may be wise to use a vaccination program to help control the outbreak. Disease control can generally be achieved by vaccinating the sows two times per year. After the disease is brought under control, only replacement gilts need to be vaccinated. If vaccinating the sows does not control the disease, the pigs should be vaccinated when they are 7 to 8 weeks of age.

Leptospirosis

Leptospirosis is a disease caused by bacteria. The most common symptom of leptospirosis is abortion. Infected sows may farrow weak or stillborn pigs. Other symptoms that may appear in chronic cases include a slight fever, diarrhea, and loss of appetite. Convulsions and death may occur in more severe cases. The only sure way to identify leptospirosis is by blood testing the herd.

Feeding antibiotics in the ration helps to prevent the disease. Females should be vaccinated 2 to 3 weeks before breeding. A regular program of vaccination helps to prevent leptospirosis.

Mastitis-Metritis-Agalactia

The exact cause of mastitis-metritis-agalactia (MMA) is not known. It is believed that a bacterial infection, as well as a hormone imbalance, may be involved. Infected sows do not have any milk to nurse their baby pigs. As a result, the pigs scour and die of starvation. The main symptom of MMA is a lack of milk just before farrowing. The udder is hot and hard. The sow may have a fever, be depressed, and go off feed.

Reducing stress during pregnancy may help in prevention. Make sure the sows have plenty of water and the proper ration. Milk letdown is stimulated by washing the sow's udder with warm water and a mild disinfectant. Observe the sows carefully at farrowing to be sure that milk is being produced and that the pigs are nursing.

If MMA is observed, call a veterinarian. Treatment, which consists of antibiotics and hormones, must begin early. Feeding antibiotics before farrowing does not always prevent the problem.

When the sow has MMA, it is necessary to feed the baby pigs by hand to prevent starvation. Use cow's milk or a milk replacer, and make sure the milk or milk replacer is clean and at body temperature. Adding a tablespoon of corn syrup per pint of milk increases palatability. The baby pigs should be fed at least six times per day. Gradually reduce feeding to three times per day by the end of a week. Pigs should be put on a pig starter ration as soon as possible.

Mycoplasmal Arthritis (PPLO)

Two species of mycoplasma organisms cause arthritis in hogs. PPLO stands for pleuropneumonia-like organisms. Arthritis is a swelling or inflamation of a joint causing difficulty in moving and lameness. Animals with arthritis sometimes have a fever.

Prevention of this disease is difficult. Reducing stress may help in prevention. If the problem is chronic, it may be necessary to replace the entire herd. Treatment depends on the form that is diagnosed. Tylosin, lincomycin, penicillin, and corticosteroids may be used and must be withdrawn a certain number of days before harvest.

Mycoplasmal Pneumonia

Mycoplasmal pneumonia is also called swine enzootic pneumonia. It is caused by *Mycoplasma hyopneumoniae,* which are very small and pass through ordinary bacteria filters. It was once called virus pig pneumonia (VPP) because it was thought to be caused by a virus.

Mycoplasmal pneumonia is a chronic disease. Death losses are low. However, feed intake is reduced and rate of gain is slowed, resulting in economic losses. A high percentage (probably 99 percent) of all commercial swine herds in the United States are believed to be affected.

Symptoms of the disease usually appear in pigs that are between 6 and 10 weeks of age or older. Coughing, which may last for 1 to 2 months, is the most common symptom, while some symptoms are similar to other diseases that affect hogs. Exact diagnosis must be made by a laboratory test. The disease is often accompanied by secondary infections.

The disease is spread from one pig to another, generally in the growing-finishing facilities and usually by direct nose-to-nose contact. It may be spread from the sow to the baby pigs; however, this is not a major method of transmission.

Research has shown that antibiotics are of limited value in treating or controlling this disease. Vaccines are available commercially to prevent mycoplasmal pneumonia. Sanitation and isolation are two preventive measures. Selection of healthy breeding stock is important. The use of SPF pigs also helps to break the cycle. Use of segregated early weaning, an all-in/all-out production system, and a multiple-site system are methods that will help prevent or control the disease. Controlling internal parasites helps reduce the effects of the disease.

Necrotic Enteritis

Necrotic enteritis is a disease caused by bacteria. The disease is sometimes confused with swine dysentery. Symptoms include fever, loss of appetite, and diarrhea. Rate of gain is slowed, and infected animals may die.

Sanitation and good rations help in prevention. Diseased pigs should be isolated from healthy pigs to stop the spread of the disease. The use of bacitracin, chlortetracycline, furazolidone, neomycin sulfate, nitrafurazone, or oxytetracycline is recommended for prevention and treatment of necrotic enteritis. Follow all required withdrawal times before harvest.

Porcine Circovirus

In 1974, the porcine circovirus type 1 (PCV1) was identified and thought to be a non-disease-causing agent found in laboratory tissue cultures. In the early 1990s, veterinarians in Canada started having cases with clinical signs that included a progressive loss of body condition, pale skin, difficulty breathing, enlarged lymph nodes, diarrhea, and jaundice.

The name "post-weaning multisystemic wasting syndrome" (PMWS) was applied to the disease. In the late 1990s, a new circovirus (PCV2) was isolated. The disease has caused great economic loss to many producers. There is a vaccine available that seems to be very effective in preventing the disease.

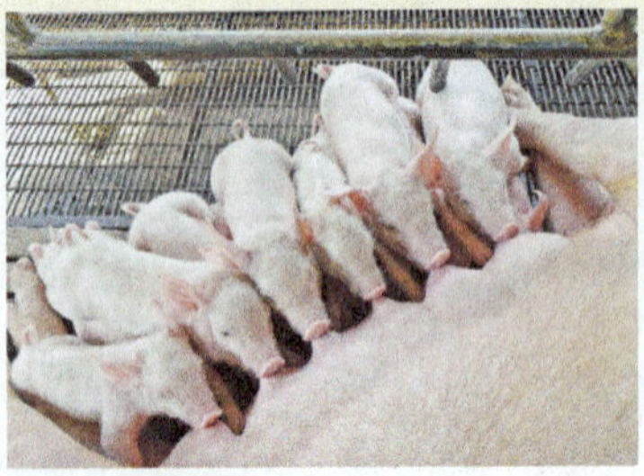

Figure 23-4 PEDv is most severe in baby pigs. Older pigs usually recover from the virus, but it often results in death for baby pigs.

Porcine Epidemic Diarrhea Virus

Porcine epidemic diarrhea virus (PEDv) is a viral disease of pigs that was first confirmed in the United States in 2013. PEDv is one of the most devastating diseases to affect the U.S. pork industry in many years. This highly contagious and deadly disease has resulted in extensive losses for swine farmers due to a high mortality rate of baby pigs. Although a relatively new disease to the United States, PEDv is common in China, Korea, and other Asian countries. It is unknown how the virus entered the United States. PEDv is not zoonotic, meaning that it does not affect people. The meat from infected swine is safe, posing no food safety concern. Other domestic animals are unaffected by the virus.

Symptoms of PEDv include severe diarrhea, fever, vomiting, and dehydration. The disease can affect pigs of any age but is typically a fatal disease for baby pigs under 3 weeks of age (Figure 23-4). The death rate in young pigs is often 100 percent. Older pigs are usually able to recover from the disease, but it often results in poor growth and feed conversion. Sows infected with PEDv often abort due to fever. The disease can be spread by pigs ingesting infected fecal matter and contaminated feed. The virus is often spread on farm equipment and clothing. The virus can survive for long periods of time in the environment, especially during cold, damp weather. A strict biosecurity program is the only way to prevent the introduction of the virus into the herd. Biosecurity recommendations include isolating new animals, cleaning and disinfecting anything coming onto the farm, limiting personnel and visitors to the farm, and showering and changing into clean boots and clothing before entering the swine facility.

Porcine Reproductive and Respiratory Syndrome

Porcine reproductive and respiratory syndrome (PRRS) is one of the most economically significant diseases facing the swine industry today. PRRS is a devastating disease of pigs worldwide. The symptoms of PRRS include late-term fetal death, abortion, weak pigs, and severe respiratory disease in young pigs. At one time, it was called mystery pig disease. It is now known to be caused by a virus. There is no treatment for PRRS; however, a vaccine is available. Outbreaks can occur even in herds that have been vaccinated. A genetic test has now been developed that can differentiate between the harmless strain of the virus found in the vaccine and the actual disease-causing virus. Following good management practices such as biosecurity will help reduce the incidence of PRRS in a producer's herd.

Porcine Respiratory Disease Complex

Several infectious agents combined with stress from the environment are responsible for porcine respiratory disease complex (PRDC). Included among the infectious agents are the PRRS virus, swine flu virus, *Mycoplasma hyopneumoniae, Salmonella cholerasuis,* and *Actinobacillus pleuropneumoniae.* This disease complex appears to be more common in multi-site operations.

PRDC usually affects pigs that are 14 to 20 weeks of age. Poor appetite, fever, and coughing are the major symptoms. Rate of gain slows significantly, with an accompanying decline in feed efficiency. Death rate is relatively low at about 4 to 6 percent of the affected population. Bringing pigs together from different herds appears to increase the incidence of this disease.

A modified live vaccine is available for use with unbred sows and replacement gilts. Killed PRRS vaccines may be effective when given to sows before farrowing. Antibiotics will help reduce the impact of the disease. Following good herd health management practices will help reduce the incidence of PRDC in producer herds.

Pseudorabies

Pseudorabies is caused by a virus. Outbreaks of this disease have increased in the swine-raising areas of the Midwest in recent years. Pseudorabies affects pigs of all ages, but the death losses are highest in baby pigs, often approaching 100 percent of the pigs on a farm.

Symptoms in young pigs include fever, vomiting, tremors, disorientation, incoordination, convulsions, and death. The pigs die within 36 hours after the symptoms appear.

Symptoms in pigs from 3 weeks to 5 months of age include fever, going off feed, depression, difficult breathing, vomiting, trembling, and incoordination. The death loss may be as high as 50 to 60 percent in pigs 3 weeks of age. In older pigs, the death loss may be as little as 5 percent. Pigs that recover have a slower rate of gain.

Symptoms in older hogs include fever, nasal discharge, coughing, and sneezing. Vomiting, diarrhea, or constipation may occur. Sows that are infected may abort their litters, or the pigs may be born dead. Infertility of the sows may also occur.

Pseudorabies may show symptoms much like transmissible gastroenteritis (TGE); leptospirosis; flu; and stillbirth, mummification, embryonic death, and infertility (SMEDI). Laboratory tests are necessary to identify the disease.

Antibiotics are not effective against pseudorabies. Vaccines currently in use do not cure the disease but do limit the transmission of the virus within the herd and prevent symptoms from appearing. The vaccines do not reduce susceptibility to other respiratory diseases that may affect the herd.

The prevention of pseudorabies involves management practices in two areas: (1) breeding and (2) herd security. Purchase breeding stock from a certified pseudorabies-free herd. New breeding stock should be isolated for a 60-day period. The isolation facility should be at least 100 feet from other animals. Blood test new breeding stock for pseudorabies before buying the animals. Retest the animals 2 to 3 weeks after purchase. Consider the use of artificial insemination or embryo transfer as a means of securing new genetic material in the breeding herd.

Herd security should begin with good fencing that does not allow other animals, including dogs, cats, and wildlife, into the hog production area. Restrict visitor traffic in the area, and provide clean boots and coveralls for those who are allowed to visit the facilities. A disinfectant boot wash should be used at all entry points. Screen open areas to keep birds out. Use an incinerator or deep burial to dispose of dead animals. If they are picked up by truck, locate the pickup area away from the production area. Place all truck loading and unloading areas away from the production area. Monitor the health of the herd on a continuous basis and keep good records.

Efforts are under way to eradicate swine pseudorabies. Eradication programs should be applied on an area-wide basis. Identify all infected herds in the area, and do herd cleanup simultaneously. The least expensive method of herd cleanup is to test the entire herd and market all infected hogs. This method works best when there is a low level of infection in the breeding herds and the virus has not spread. Another method is to remove all young pigs from their mothers and put them in a clean area to avoid any contact with infected hogs. The most expensive method of herd cleanup is to market all hogs in a herd that has some infected animals. Clean the facilities and leave them empty for 30 days before restocking with hogs from certified pseudorabies-free herds.

SMEDI

The term *SMEDI* is formed from the first letters of *stillbirth, mummification, embryonic death,* and *infertility*. SMEDI is caused by several viruses. It is a disease complex that shows one or more of the following traits:

- small litters of fewer than 4 pigs
- more than normal repeat breeding
- sows that come into heat after they were thought to be bred, but never farrow
- dead pigs (often mummified) included in small litters

The only control measures are managerial practices. The breeding herd should be maintained as a closed herd. All sows and gilts should be run together for at least 30 days before breeding. Fence-line contact is allowed with the boar. Do not expose breeding animals to other hogs. Do not allow people, birds, or rodents to enter the area where the breeding herd is kept. Be careful when moving feed trucks and other equipment into the area where the breeding herd is located.

There is no cure for SMEDI. Sows do develop an immunity to the disease. They will normally have good litters the next time they are bred if they have had the disease during the previous gestation.

Streptococcus Suis

Streptococcus suis is a bacterium that lives in the tonsils of pigs. It can cause disease in the brain (meningitis) and in other organs (septicemia). The disease can affect pigs of all ages; however, it is most commonly seen in pigs that have recently been weaned. Meningitis may appear as sudden death or convulsions and then death. More commonly, symptoms occur in this order: loss of appetite, reddening of skin, fever, depression, loss of balance, lameness, paralysis, paddling, shaking, and convulsions. Stress appears to be a factor in triggering an outbreak of meningitis.

Septicemia may appear as pneumonia, "fading piglet syndrome," arthritis, and abortion. Pneumonia caused by *S. suis* most often occurs when the pigs are 2 to 4 weeks of age. It can occur during the growing-finishing period. "Fading piglet syndrome" occurs when healthy newborn pigs stop nursing, become listless, are cold to the touch, and die in 12 to 24 hours after birth. *S. suis* infections are not common in breeding herds but can occur, resulting in a decrease in the farrowing rate.

Carrier pigs can appear healthy and yet have the bacteria in their tonsils or nasal passages. Introducing carriers into herds that are not infected usually results in the spread of the disease into the new herd. Dead hog carcasses can harbor the disease, and flies can carry it from farm to farm.

Observing the symptoms discussed above does not give a positive diagnosis of the disease. A laboratory analysis of cultured brain tissue from dead animals is the best way to definitely identify *Streptococcus suis.* Early treatment of affected pigs with injections of penicillin or **ampicillin** helps prevent death. Tests need to be done to determine the correct antibiotic to use for continued treatment. In some circumstances, it may be feasible to treat the entire group when some pigs are identified with the disease.

The key to controlling the disease is to reduce stress in the herd. Avoid overcrowding and drafts, and maintain good ventilation in the facilities. Make sure that new breeding stock comes from herds that are free of the disease. Artificial insemination and embryo transfer can be used to bring improved genetic stock from infected herds to uninfected herds. Using SPF pigs and segregated early weaning are other management techniques that will help reduce the incidence of *Streptococcus suis* in a swine herd.

Transmissible Gastroenteritis

Transmissible gastroenteritis (TGE) is caused by a virus and affects pigs of all ages. The most severe losses occur in pigs under 10 days of age. Death loss in baby pigs is almost 100 percent. Older pigs seldom die from the disease. The disease spreads rapidly, especially in farrowing houses. Almost all litters are affected at the same time.

Symptoms in young pigs include vomiting, diarrhea, and death. Death occurs within 2 to 7 days after the symptoms first appear. The feces are whitish, yellowish, or greenish in color. Dehydration and weight loss occur.

Symptoms in feeder pigs and older pigs include going off feed, vomiting, diarrhea, and weight loss. Death loss is low; however, the disease results in slower rates of gain in hogs that recover.

Symptoms in sows include going off feed and having mild diarrhea. Some sows may not show any symptoms. Sows that have the disease and recover have an immunity that lasts about 1 year. The immunity is passed on to the pigs in their litters. Because of this immunity, sows that recover from TGE should be used to breed another litter.

TGE is very contagious. Control measures consist mainly of good sanitation practices that prevent introduction of the disease. Keep people out of swine areas. Dogs and birds may carry the disease from one area to another and should also be kept out of swine areas. Disinfect equipment when moving it from one hog area to another. Isolate new stock, and do not bring new stock on the farm for 1 month before farrowing or 2 weeks after farrowing. Maintain a closed herd. Do not carry manure on shoes or clothing from one hog area to another. Breaking the farrowing cycle for about 1 month helps to control the disease.

Current vaccinations are not very effective. Some producers try to induce immunity in sows by exposing them to the disease about 1 month before farrowing. This is a practice that should be done only under the close supervision of a veterinarian.

There is no cure for TGE. Early diagnosis helps reduce losses. Proper disposal of dead pigs helps to prevent the spread of the disease. Dead pigs should be burned, buried, or sent to a rendering plant for disposal. Spread sows out for farrowing if the disease is diagnosed. Antibiotics can only help to reduce the effects of secondary infections.

White Scours (Colibacillosis: Bacterial Enteritis)

White scours is a highly contagious disease caused by several strains of bacteria. It is believed that most instances of baby pig diarrhea are caused by white scours. The bacteria are present in sows, which are the usual source of infection for baby pigs.

Baby pigs are most commonly affected. The disease has a high death rate. Symptoms include a watery, pale yellowish diarrhea. Pigs become listless and dehydrated, and lose weight. The death loss in older pigs is low, but many pigs that are affected gain at slower rates.

Good sanitation, proper nutrition, and reducing stress help to prevent the disease. Vaccination of the sows is helpful when a specific bacterium is identified as the cause.

Early treatment (within 24 hours) is important if the disease is identified. Tests to identify the bacterial strain are necessary. Treatment is made with the proper antibiotic or sulfa drug after the bacteria strain is identified.

Figure 23-5 Pigs raised with access to soil will generally obtain a sufficient amount of iron.

NUTRITIONAL HEALTH PROBLEMS

Anemia

Anemia is a condition mainly caused by a lack of iron or vitamin B12 in the diet. It affects mainly baby pigs when the sow's milk does not have enough iron for the needs of the nursing pigs. Symptoms appear in pigs from 1 to 2 weeks of age. Signs of anemia are poor growth, roughened hair coat, and difficulty in breathing. Sudden death of apparently healthy pigs is also a symptom. Anemia lowers the pig's resistance to other diseases.

Anemia is controlled by giving iron, either by intramuscular injection or mouth, to the baby pigs. Vitamin B12 is provided in the diet. Providing clean sod in the pen will also help to prevent anemia (Figure 23-5).

Hypoglycemia

Hypoglycemia is a condition caused by a lack of sugar in the diet. It has a high death rate. Symptoms include shivering, weakness, unsteady gait, dullness, and loss of appetite. The hair coat becomes rough. Diarrhea may develop in some pigs. Death occurs within a day and a half after the symptoms appear.

The main preventive measure is the reduction of stress factors, such as chilling. Be sure that the pigs nurse as soon as possible after they are born. Treat with dextrose solution or dark syrup. Dextrose may be injected or given by mouth. Syrup is given by mouth.

Parakeratosis

Parakeratosis is caused by a lack of zinc in the diet. High calcium content of the diet increases the need for zinc. The disease affects older hogs more frequently than younger ones.

Symptoms include rough, scaly skin, and slower than normal growth. The condition has the appearance of mange. Control and treatment consists of supplying adequate zinc in the ration.

Poisoning

Moldy feed is a common source of poisoning in hogs (Figure 23-6). Pitch, lead, mercury, pesticides, some plants, salt, and blue-green algae are other possible causes of poisoning.

Do not allow hogs to eat moldy feed or other possibly poisonous substances. Symptoms and treatments of poisoning vary with the type of poison involved. Consult a veterinarian if it is thought that hogs are suffering from poisoning.

Figure 23-6 Moldy feed is a common source of poisoning in hogs.

Rickets

Rickets is caused by a lack of calcium, phosphorus, or vitamin D in the diet. An improper ratio of calcium to phosphorus in the diet may also cause rickets. Symptoms include slower than normal growth and crooked legs. Control and treatment is accomplished by providing proper amounts of calcium and phosphorus in the correct ratio. Supply vitamin D in the diet or be sure that the pigs are exposed to sunshine.

EXTERNAL PARASITES

Lice

The hog louse is dull, gray-brown in color and about $\frac{1}{4}$ inch long. It is a bloodsucking pest. Lice tend to cluster in the ears, on the insides of the legs, and around the folds of skin on the neck of hogs. They cause an extreme irritation when they suck blood from the hog. They are most common in the winter, although they may be present year-round.

Economic losses occur due to reduced rates of gain and lowered carcass value. Infected hogs are more likely to be attacked by other parasites and diseases.

Lice are controlled by insecticides. Sprays and pour-on dusts are available. Observe label directions on limitations for use prior to slaughter.

All new additions to the herd should be treated with an insecticide. Sows should be treated in the fall or winter during the 6 weeks before farrowing. If an outbreak is detected, treat all the hogs on the farm.

Mange

Mange is caused by a tiny white or yellow mite that bores into the skin. The mite is so small it cannot be seen without magnification. Mites cause severe itching. The hogs are irritated and rub at the affected areas. The areas most often affected are around the eyes, ears, back, and neck. The area becomes inflamed and scabby. The problem is most severe in the fall, winter, and spring. Mites spread rapidly from hog to hog. Insecticides are used to control mange. Sprays or dusts may be used, although dusts are less effective than sprays. Observe label directions on limitations for use prior to slaughter.

INTERNAL PARASITES

Four groups of internal parasites affect hogs: (1) roundworms, (2) tapeworms, (3) flukes, and (4) protozoa. Roundworms cause more damage to hogs than any other internal parasite (Figure 23-7). The most serious of the roundworms is the large intestinal roundworm (ascarid). Other roundworms include stomach worms, intestinal threadworms, kidney worms, lungworms, nodular worms, whipworms, trichinae, and thorn-headed worms.

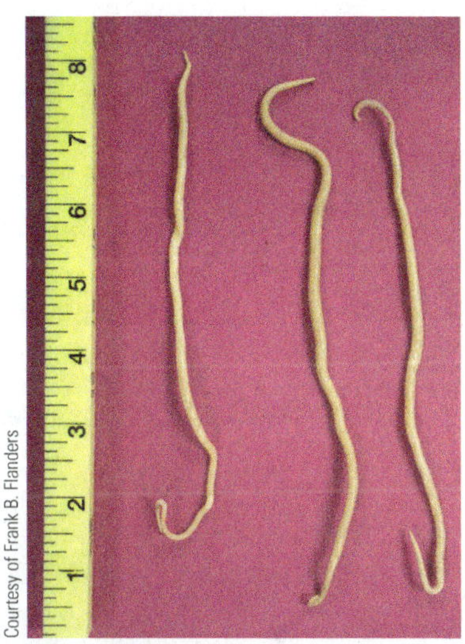

Courtesy of Frank B. Flanders

Figure 23-7 The large intestinal roundworm (ascarid) is the most damaging of the roundworms.

Bladder worms, which are incompletely developed tapeworms, are the only tapeworms that affect hogs. Flukes, which are soft leaf-like worms, are not very common in hogs. Protozoa are tiny parasites that can only be seen with a microscope. Coccidia and trichomonads are examples of protozoa that affect hogs. Neither is very common.

Hogs that are infested with internal parasites do not grow or gain weight as fast as other hogs. The infestation makes the hog more likely to have other health problems. A heavy infestation of internal parasites can cause thinness, rough hair coat, weakness, and diarrhea.

Internal parasites cause economic losses because infected hogs make slower gains. Other losses occur because of lower carcass value. Some parts of the carcass may have to be condemned at slaughter because of parasite infestation. Sometimes the entire carcass must be condemned.

Of all the internal parasites previously mentioned, the most common are large roundworms, lungworms, whipworms, nodular worms, and intestinal threadworms. Most worm treatments are aimed at these parasite problems.

Parasites are controlled by good sanitation practices and the proper use of drugs to treat the infestation. Make sure the sows and gilts are clean when put into farrowing crates or pens. A clean farrowing house is necessary. Rotate hog pastures to break the life cycle of parasites. Sows, gilts, and boars should be dewormed 1 week before breeding and 1 to 2 weeks before farrowing. Deworming breeding stock prevents the baby pigs from getting worms from the manure. Deworm young pigs when they are 5 to 6 weeks old and again 1 month later. Deworm feeder pigs when they arrive at the farm and repeat 3 to 4 weeks later.

Some dewormers are given in the feed and others in water. Check the label for information on the types of worms each drug controls, and instructions for its proper use. Observe label directions for withdrawal times before slaughter.

OTHER HEALTH PROBLEMS

Porcine Stress Syndrome

The sudden death of heavily muscled hogs is referred to as *porcine stress syndrome (PSS)*. Death is the result of a failure in the blood circulation system. It is most common in confinement housing. Genetic factors appear to be involved in the condition. It is believed to be carried on a recessive gene.

Courtesy of Frank B. Flanders

Figure 23-8 Pigs handled during extremely hot weather become stressed easily and may show signs of porcine stress syndrome.

Prevention is mainly achieved by selection of breeding stock. Boars and gilts that appear to be more than normally nervous should not be selected for the breeding herd. Signs of PSS are constant movement, tail twitching, and trembling ears. The skin tends to develop splotches when the hog is excited. The splotches are red on white pigs and purple on black pigs. A blood test is available to determine if the pig is stress prone. If a problem develops in a herd, replace the boar with one that is not a carrier of the recessive gene.

Care must be taken with herds in which the problem exists. Avoid stress when handling or moving hogs. Do not crowd hogs in pens, and provide plenty of feeder and watering space. Do not handle or move hogs in extremely hot weather. Provide plenty of bedding in cold weather. Some hogs will die even when these practices are followed, but in general, death losses will be less severe. Figure 23-8 shows a pig that is suffering from PSS.

SUMMARY

Disease and parasite problems cause millions of dollars in losses each year in the U.S. swine industry. Prevention of health problems is the key to reducing losses. Sanitation, selection of healthy breeding stock, and proper management are important factors in preventing losses.

There are many infectious diseases that affect hogs. Vaccinations, sulfa drugs, and antibiotics are useful in the prevention and treatment of many of these diseases. The use of SPF hogs and a segregated early weaning program are also helpful in controlling some diseases. The greatest losses from diseases occur among younger pigs.

Nutritional health problems are controlled by proper feeding. Minerals and vitamins are particularly important in the diet. Many health problems are related to a lack of these substances.

Lice and mites are the most common external parasites that affect hogs. Insecticide treatments are used to control these parasites.

The most serious internal parasite of hogs is the large roundworm. Other worms that are major problems are lungworms, nodular worms, whipworms, and intestinal threadworms. A regular program of sanitation and treatment with drugs helps to control these parasites.

Careful selection of breeding stock helps to reduce porcine stress syndrome. Careful handling of hogs in herds where PSS is a problem helps to reduce losses.

Student Learning Activities

1. Prepare a bulletin board display of pictures of unsanitary and sanitary facilities for swine.
2. Prepare and give an oral report on a particular phase of swine herd health management.
3. Prepare a bulletin board display of pictures of animals affected by diseases and parasites.
4. Plan and carry out good swine herd health practices as part of your swine supervised experience program.
5. Take field trips to observe swine herd health problems in the local area.
6. Ask a veterinarian to speak to the class about swine herd health management.
7. Conduct a community survey of swine herd health problems. Draw conclusions and present the results to the class.

Discussion Questions

1. Why have herd health problems increased in recent years in the hog industry?
2. List four management practices that will contribute to good swine herd health.
3. Describe a good sanitation program for the swine herd.
4. Prepare a table listing the infectious diseases of hogs, briefly describing the symptoms, prevention, and treatment.
5. Name the two common external parasites of swine and describe the symptoms, prevention, and treatment.
6. What are the four groups of internal parasites that affect swine?
7. Which is the most serious of the roundworms that affect swine?
8. Describe the kinds of losses caused by internal parasites of swine.
9. Describe how internal parasites of swine may be controlled.
10. Describe porcine stress syndrome (PSS), and tell how it may be prevented and treated.

Review Questions

True/False

1. Diseases cannot pass from sow to litter.
2. The use and withdrawal times for drugs used in treatments are not subject to change and current requirements should never be checked.

Multiple Choice

3. Early abortions during the gestation period can be caused by __________.

   a. cholera   c. brucellosis
   b. edema   d. anemia

4. __________ is a disease caused by bacteria, often confused with swine dysentery.

   a. Necrotic enteritis   c. Leptospirosis condition
   b. Pseudorabies   d. Influenza

5. Swine dysentery is also known as ________.
   a. black scours
   b. bloody scours
   c. vibrionic dysentery
   d. all of the above

Completion

6. ________ is caused by a virus. Symptoms in young pigs include fever, vomiting, tremors, lack of coordination, convulsions, and death.
7. ________, ________, ________, and ________ are four groups of internal parasites affecting hogs.
8. ________ is a condition caused by a lack of sugar in the diet.

Short Answers

9. What is the single most important management practice that affects profits when raising swine?
10. What is the best way to prevent avian tuberculosis?
11. List some of the most serious internal parasites of hogs.

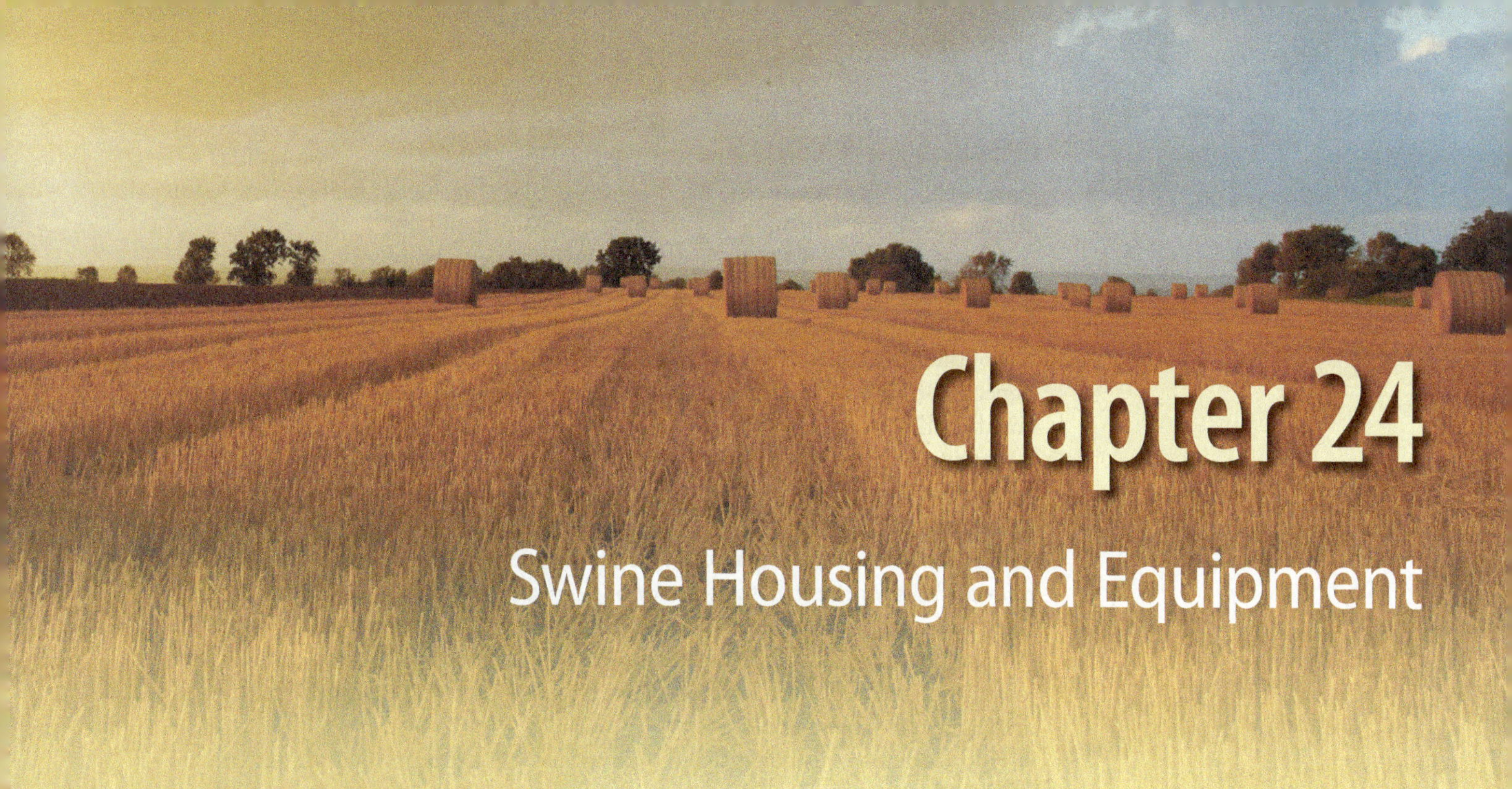

Chapter 24

Swine Housing and Equipment

Key Terms

zoning laws
warm confinement houses
cold confinement houses
growing period
breeding rack

Objectives

After studying this chapter, the student should be able to

- describe facilities required for swine production.
- describe equipment required for swine production.

The kinds of facilities used for swine production are much different today than they were 25 years ago. The trend in swine housing is moving away from pasture production systems and toward confinement systems. Today, there is more specialization in hog production. More hogs are being raised per farm, and there are fewer small producers. Electricity, mechanization, and slotted floors have also changed the kind of housing being used. These changes require greater investments in facilities but reduce labor requirements to produce each pig. The amount of net return from swine production depends more on good management than on the kinds of facilities used.

KINDS OF BUILDINGS

Several building systems are possible, based on stages of production. At one time, it was common to use one building from farrowing to finish. Since it does not use space efficiently, this is no longer a common practice. However, this type of system is still used with feeder pig production. About four farrowings per year can be done with this system when feeder pigs are sold.

The concept of using just two facilities for the swine enterprise is gaining increased interest. In this system, the pigs are weaned, usually at less than 3 weeks of age, and moved directly into the facility where they will remain until they reach market weight. Advantages appear to include less stress on the pigs and a saving in costs involved when moving them to a nursery and then to a finishing facility. The major disadvantage is the higher facility cost. A suitable environment for young pigs must be maintained initially.

Larger producers with six or more farrowings per year often use housing that is divided into three stages of production: a farrowing-to-weaning facility, a growing facility, and a finishing facility. The farrowing facility may contain a nursery. In any of the systems, pasture may be used in part of the program. The use of pasture depends on climate, land availability, the size of the operation, and the value of the land for other uses.

The use of hooped structures for housing hogs, a low-cost housing alternative for swine that originated in Canada, has been gaining popularity in the Midwest. Advantages claimed for this type of housing include lower costs for housing, lower energy costs, lower per pig production costs, better ventilation in the housing unit, multiple uses for the structure, less odor from solid manure, and ease of construction. Disadvantages include lower feed efficiency in winter because of a colder environment, lower average daily gains in hot weather because feed intake may be lower, the need for more bedding, difficulty in controlling parasites and disease, and higher labor requirement.

Hooped housing is constructed with arched metal frames that are fastened to posts in the ground. The entire structure is covered with a polyethylene tarp that is stretched over the frame and side walls. A typical size for this type of housing unit is 30 feet wide by 60 to 80 feet long. Capacity is considered to be about 12 square feet per finishing animal. The ends of the unit are movable and made from plywood and tarp. A concrete pad is usually constructed at one end of the unit. This pad is usually the width of the unit and extends 16 to 20 feet into the unit to hold feeders and waterers. The rest of the floor is dirt. Straw, corn stalks, ground corncobs, or other organic material may be used for bedding, which is allowed to build up into a deep pack. The use of hooped housing for hogs may be more attractive to a small producer than to a large growing-finishing operation.

SITE SELECTION AND BUILDING PLACEMENT

Facilities for hog production usually include units for handling hogs from farrowing to finish. This kind of facility requires a large capital investment. Careful planning is necessary to be sure the facility is built in the best location and provides for future expansion.

Odor problems always exist with hog production. Facilities should be located downwind from the residence on the farm. The location in relation to neighbors must also be considered. Protection from wind and snow are other factors to consider in selecting a site.

The facility must have access to an all-weather road. Power and water must be available. Access to these facilities must be considered. Good drainage and runoff control are also necessary. Do not locate buildings in low, wet areas where the drainage is poor. **Zoning laws** exist in some areas and should be checked. Zoning laws and regulations are used by counties, cities, and other local governments to define how property may be used. Governments adopt zoning plans to ensure that land is used for the common good. It may also be necessary to secure permits for building.

Methods of handling manure must be considered and planned as part of the total facility. Manure handling is usually accomplished by spreading it on a field or using a lagoon. If manure is to be spread on a field, it may be necessary to store manure during part of the year. Lagoon location must not interfere with future expansion.

VENTILATION, HEATING, AND INSULATION

Hog houses must be well ventilated. A ventilation system has several functions. In the summer, ventilation helps to control the temperature of the air. In confinement, hogs produce a great deal of moisture through body waste and respiration. Ventilation helps to remove the extra moisture that is present in large amounts, especially in the winter, from the house. Odors are also removed by ventilation.

Hogs are healthier when they have fresh air. A ventilation system helps to provide the necessary fresh air and also dilutes airborne disease organisms that are present in the hog house. The amount of ventilation needed for each of these functions varies. If enough ventilation is provided for fresh air and temperature control, the other functions are generally served.

The amount of ventilation required varies with the season. More air movement is needed in the summer than is needed in the winter. Winter ventilation is mainly for moisture and odor control. Summer ventilation is primarily for temperature control and to help keep the hogs cool. It is necessary to provide under-floor ventilation in buildings with slotted floors. This helps to remove the toxic gases and odors that come from the manure.

Warm Confinement Houses versus Cold Confinement Houses

There are two general types of hog houses. **Warm confinement houses** maintain desired temperatures regardless of the outside temperature. **Cold confinement houses** have temperatures that are only slightly warmer (3 to 10 degrees) than the outside temperature.

It is more important to have warm confinement houses for farrowing and nursery facilities than for growing-finishing facilities. Some producers also use warm confinement houses for growing-finishing hogs. However, hogs of this age do about as well in cold confinement buildings.

Warm confinement houses are more expensive to build and use mechanical ventilation systems. The design of the ventilation system depends on the area in which the house is located. Technical details of design and construction for this type of system can be obtained from agricultural engineers, university specialists, or commercial companies.

Cold confinement houses depend on natural ventilation for air movement. More moisture condensation, fogging, and frost occur in cold confinement houses; however, they are less expensive. Usually, the house is built so one side can be opened to take advantage of prevailing wind direction in the summer. Again, technical details of location and construction design should be obtained for the local area in which the house is to be built.

Baby pigs require temperatures of 80°F to 90°F. They must be kept dry and free of drafts. Heat lamps, brooders, and under-floor heating systems can be used to provide the conditions necessary for baby pigs. Heat lamps must be no more than 2 feet above the bedding in a pig brooder. Keep the lamp and cord out of the reach of the sow.

Sows are most comfortable with air temperatures of 50°F to 60°F. The body heat of the sows will provide enough heat if the building is well insulated. In cold areas, it may be necessary to provide some additional space heating. This is often done with vented, gas-fired space heaters. There is an increasing interest in the use of solar energy for heating hog houses. Consult with an agricultural engineer for the latest design details.

Weaned pigs are most comfortable at temperatures of 70°F to 75°F. Growing-finishing pigs are most comfortable at temperatures of 60°F to 70°F. High temperatures are more harmful than low temperatures for the breeding herd.

Energy costs are increasing. Proper insulation helps to reduce the cost of additional heat for hog houses. It keeps the interior surfaces warmer in the winter. This reduces the amount of moisture condensation on the walls and ceiling. Insulation also reduces the amount of animal body heat lost during the winter. This, in turn, helps to reduce heating bills. Fuel for uninsulated houses can cost as much as two to three times as much as that for insulated houses.

Insulation also reduces the amount of cooling necessary in hot climates during the summer. Proper insulation makes the building cooler in the summer. This increases the comfort of the hogs and their rate of gain. It also helps to reduce the cost of cooling the building. The amount of insulation required varies with the location of the building. Technical advice can be obtained from agricultural engineers and university specialists in the local area.

In hot weather, sows in farrowing houses require cooler temperatures than can be provided by the ventilation system. Some producers use a stream of cool air directed to the sow's nose in each farrowing crate to provide further cooling. Sprinkling systems are also used to help cool hogs. They work best when operated for 2 to 5 minutes each hour when the temperature is about 70°F to 75°F. A heavy spray should be used to cool the hogs. Fine mists or a fog are not recommended. Such systems are usually temperature and time controlled.

FLOORS

Three general types of floors are used in hog houses: solid, partially slotted, and totally slotted.

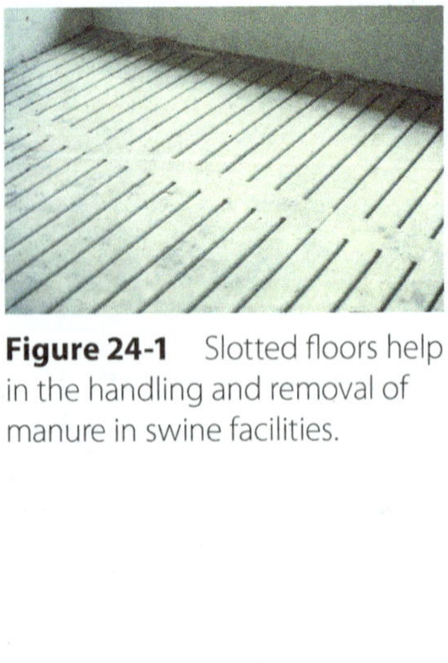

Photo courtesy of Dr. Robert Dove, University of Georgia

Figure 24-1 Slotted floors help in the handling and removal of manure in swine facilities.

Solid floors are usually made of concrete. The cost of solid floors is less than other types of floors. The major disadvantage of solid floors is the difficulty of handling the manure. Floors must be cleaned regularly, which requires a great deal of labor. Few confinement houses use totally solid floors.

Partially slotted floors are a combination of solid and slotted floor. Some hand cleaning of manure is required with partially slotted floors.

Totally slotted floors, like the floor pictured in Figure 24-1, almost eliminate the handling of manure. However, the cost of slotted floors is more than that of solid floors. Slotted floors are harder on the feet and legs of hogs, and it is harder to control the temperature in a slotted-floor building.

A number of different kinds of materials are used for slats. Commercial slats are available in concrete, metal, fiberglass, and plastic. Metals used include steel, stainless steel, and aluminum. Hardwoods, such as oak, can also be used for the slats. Soft woods should never be used for slats because they wear too fast, warp easily, and can be chewed through by the pigs. Concrete slats generally last the lifetime of the building. Metal slats may corrode if they are allowed to come into contact with wet manure for long periods of time and require more support to carry the weight of the hogs. Narrow slats, with spacing of ⅜ inch, should be used for baby pigs in farrowing stalls. Wider slats, with spacing of ¾ to 1 inch may be used in finishing houses.

In the farrowing pen, the slats should be placed parallel with the sow. This gives the sow better footing. Some producers use a partially slotted floor in the farrowing crate area. This provides the pigs with a warmer place to sleep.

Manure is handled as a liquid when slotted floors are used. Bedding is not used in liquid manure systems because it will interfere with the operation of some kinds of pumps used in liquid manure-handling systems. Centrifugal pumps without choppers may become clogged by bedding. The chopper-impeller-type pumps, however, can handle manure containing chopped bedding. Bedding decomposes slowly in lagoons and increases the odor problem.

BREEDING HERD FACILITIES

Facilities for sows, gilts, and boars may be portable or permanent buildings. Many producers are using total confinement buildings for their breeding herds. The cost for these facilities is higher than that for smaller, portable houses. However, better management of the breeding herd can be achieved in total confinement. Producers with limited capital or rented farms often choose to use more pasture for the breeding herd.

Houses with natural ventilation are adequate for the breeding herd. Boars must be separated from females. If pasture is to be used, locate the buildings so that it is easy to move the animals to pasture. Plan the arrangement of buildings so that it is easy to feed, clean, and water the animals. Boar housing and sow housing should be located close together to make breeding management easier.

Open-shelter housing may be used. Open-front buildings are described in the section on finishing facilities. If solid floors are used, provide about 15 square feet per sow. If individual pens are used for boars, provide a pen at least 6 × 8 feet per pen. Shade, water, and feeding facilities should be available. About 15 to 18 inches of feeder space is required per sow. If slotted floors are used, slats measuring 4 to 8 inches wide with 1- to 1¼-inch spacing are recommended.

When using a building with an outside apron, provide 11 to 12 square feet per sow inside and the same amount of space per sow on the apron. Provide 40 square feet per boar inside and the same amount per boar on the apron. In confinement systems, provide 40 square feet per breeding gilt on solid floors and 24 square feet on totally or partially slotted floors. For breeding sows, provide 48 square feet per sow on solid floors and 30 square feet on totally or partially slotted floors. Up to six gilts or sows can be put in each pen.

Mature boars need 60 square feet on solid floors and 40 square feet on totally or partially slotted floors. Put one boar per pen.

Gestating gilts should be provided 20 square feet per gilt on solid floors and 14 square feet per gilt on totally or partially slotted floors. Gestating sows should be provided 24 square feet per sow on solid floors and 16 square feet per sow on totally or partially slotted floors. Six to 12 gilts or sows can be put in each pen.

Individual breeding pens should be provided. Individual feeding stalls may be used to make it easier to limit-feed each sow.

GESTATION FACILITIES

Most large swine producers have used gestation crates (also know as gestation stalls) since the 1990s as the primary housing for gestating sows. In this system, sows stay in crates 24 hours a day from the time they are bred until they are moved into farrowing crates a few days before farrowing. Gestation crates have been a source of great controversy between animal rights activists and producers. Animal rights activists want to ban the use of gestation crates due to the lack of socialization, exercise, and limited mobility of sows in the crates. Oregon became the first state to pass legislation to ban gestation stalls. It is predicted that gestation crates will soon be a thing of the past. Most large pork production companies are phasing out gestation crates. Also, many of the largest pork buyers refuse to purchase pork from producers who use gestation crates.

Photo courtesy of Chore-Time Hog Production Systems, Milford, Indiana, USA

Figure 24-2 Free-access gestation stalls allow sows to gather in a socialization area and enter and exit individual stalls as desired.

Gestation housing to replace the crate system could include group housing in pens of a few sows each, outdoor housing, or free-access gestation stalls. Free-access gestation stalls is an innovation that originated in Europe. This system allows for a group of sows to be housed together in a community pen but provides a stall for each sow. The sows have free access to the stalls or community area. The community pen allows sows to exercise and socialize as they wish. The system allows sows to enter a stall to isolate themselves from other sows and allows sows to push open the stall door to an empty stall (Figure 24-2). The door locks behind the sow to prevent entry of other pigs. The door unlocks and opens when the sow backs out of the stall. Advantages of the free-access system include allowing sows to isolate themselves from dominant sows and providing a method of controlled feeding. The sows can be temporarily locked in the stall by the producer for breeding, pregnancy scanning, and treatment purposes. Research has shown that in a free-access gestation stall system, a sow will rest and eat in a particular stall about 90 percent of the time, and sows will stay in the gestation stalls more than 90 percent of the day. Most producers are moving away from gestation stalls—and may eventually be forced to do so by law—to use another system such as free-access gestation stalls that allow for free movement and socialization.

Photo courtesy of Chore-Time Hog Production Systems, Milford, Indiana, USA

Figure 24-3 Farrowing crate. Note the diagonal placement, a space-saving design that allows more crates to be placed in the facility.

FARROWING HOUSES

Portable farrowing houses cost less than confinement houses. Producers who farrow fewer than 20 sows per year often find it more economical to use portable houses. Other reasons for using portable housing for farrowing include (1) farrowing only once or twice per year, (2) limited capital, (3) using rotation pasture, or (4) renting the farm.

Confinement housing is more practical for producers who farrow more than 20 sows per year in multiple farrowing. Confinement housing might also be used when: (1) there is a desire to substitute capital for labor, (2) land is too valuable for use as pasture, or (3) a limited amount of space is available.

The main requirements for farrowing housing are that it should be kept warm, dry, and free of drafts. Farrowing usually is done in pens or crates. Pens require more labor in cleaning than do crates. Guardrails and pig brooders help protect the baby pigs in pens. Guardrails are 8 inches above the floor and project 6 inches from the wall. A 2 × 4 or ¾-inch pipe bolted to the wall can serve as a guardrail. Pens should be 4.5 to 5 feet wide and 12 to 14 feet long. Partitions between pens should be at least 3 feet high. Some producers use partially slotted floors in pens. In this case, about one-fourth of the floor area is slotted.

Photo courtesy of Dr. Robert Dove, University of Georgia

Figure 24-4 This swine facility is equipped with entirely slotted floors, farrowing crates, and automated feeders.

More pigs are saved using farrowing crates than open pens because the pigs are able to maintain a safe distance away from the sow and avoid crushing (Figure 24-3). In addition, less labor is required when farrowing in crates, and less bedding and floor space are needed. Partially or totally slotted floors are often used with crates. Sows may be fed in the crate or turned out each day for feeding. Crates are 4.5 to 5.0 feet wide and 6.5 to 8.0 feet long. About 2 feet of width is needed for the sow. About 1.5 feet are needed on each side for the baby pigs.

Automatic feeding and watering is often used in confinement houses. Careful adjustment of the feeders is necessary to prevent waste. The waterers should not leak. Figure 24-4 shows a farrowing facility equipped with farrowing crates, entirely slotted floors, and automatic waterers and feeders.

In large confinement houses, an area should be set aside for an office and equipment. Storage space for supplies is also needed. A refrigerator for cold storage and hot water for washing sows may be provided.

An area for washing the sows before putting them in the pens or farrowing crates should be included. Feed storage is commonly located in large storage bins just outside the confinement house. If feeding is not automated, a feed cart makes the job easier.

The nursery unit is often a part of the farrowing facility. When the baby pigs are a few days to 2 weeks old, the sow and litter may be moved from the farrowing house to a nursing unit. No more than three sows and litters should be grouped together in one pen. Provide 40 square feet per sow and litter. Partially slotted floors are recommended. A pig brooder should be provided at the upper end of the pen.

A pig brooder is a triangular area, usually in one corner of the pen, which is blocked off by boards so the pigs can enter but the sows cannot. It may be open on top with an infrared lamp hung above the area to provide heat. A hover-type brooder is closed on top, usually with a piece of plywood. An incandescent lamp with a reflector is placed on an opening in the cover to provide heat for the pigs. Lamps are more commonly used, but electric heating cables or hot water pipes can be used under the floor to provide heat in the pig brooder area in place of the lamps; however, these systems are more expensive.

GROWING FACILITIES

The **growing period** is from weaning to about 100 pounds. Some producers grow pigs on pasture. Portable or permanent shade structures or other shade must be provided for pigs on pasture (Figure 24-5). Portable feeders and waterers are also required. Permanent houses are used by some producers for this period. One kind of housing that may be used is similar to that used for finishing houses. Temperature and moisture control are important during this stage of production.

Weaned pigs may be grouped in pens of about 25 head per pen. Provide 3 to 4 square feet per pig. One drinking space should be allowed for each 20 to 25 head. Provide one feeding space for every four pigs. Feeding and watering may be automated to save labor. Pigs of this size tend to deposit dung over the entire floor of the house. Using partially or totally slotted floors reduces manure-handling problems.

Courtesy of Cabe Brothers Farm. Photo by Sarah Kane

Figure 24-5 A shade structure can be used in pasture systems with many animals, including swine.

FINISHING FACILITIES

Pigs weighing more than 100 pounds grow about as well in cold confinement houses as in warm confinement houses. Three general kinds of houses are used to finish pigs: (1) open front, (2) modified open front, and (3) totally enclosed. Pigs may also be finished on pasture. The same facilities are required for finishing on pasture as for growing. One acre of pasture will carry about 50 to 100 pigs. Figure 24-6 shows an enclosed finishing facility.

In confinement housing, temperature and moisture control are the most important considerations. The amount of space needed varies with the size of the pigs and the kind of floor used. Do not overcrowd pigs because problems such as tail-biting, cannibalism, and slower rates of gain result from overcrowding.

Open-Front Housing

Open-front buildings have about one-half of the pen area in front of the building. The open side is placed in such a direction that it protects the pigs in winter and takes advantage of summer wind directions for ventilation. Feeders are placed either under the roof area or in the open part of the pen. Slotted floors may be used in part of the open area for liquid manure handling. The solid part of the floor is sloped toward the front. Bedding or under-floor heat may be used in the building.

Photo courtesy of Chore-Time Hog Production Systems, Milford, Indiana, USA

Figure 24-6 A modern swine finishing facility.

Photo courtesy of Dr. Robert Dove, University of Georgia

Figure 24-7 An open swine finishing facility. Note the solid partitions that divide the pens. This helps control drafts in the winter.

About one-half of the rear of the building should be designed so that it may be opened in summer for ventilation. Drafts may be controlled in the winter by constructing a solid partition from floor to ceiling every 40 to 50 feet for the length of the building (Figure 24-7).

Modified Open-Front Housing

A modified open-front house is covered entirely by a roof (Figure 24-8). One or more sides are open. Provisions are often made to partially close the front during the winter.

Totally Enclosed Houses

Totally enclosed houses have no large openings to the outside. Insulation and ventilation must be provided. Floors may be solid, partially slotted, or totally slotted. When manure is stored in a pit under the building, provision must be made for ventilation in the storage pit. Some producers use a flush system to remove animal waste. This system uses a large holding tank (up to 1,000 gallons) to periodically dump this large volume of water onto the floor to wash the waste out of the building and into a lagoon. Recycled water from a secondary lagoon can be used to reduce the total amount of water used with the system.

Melissa Maher

Figure 24-8 A covered swine finishing facility with open sides.

In confinement buildings, pens are designed to hold from 20 to 30 pigs. Provide 8 to 10 square feet per pig. A drinking space for each 20 to 25 pigs is needed. Provide one self-feeder space for each four pigs. Service alleys are needed in the modified open-front and totally enclosed houses. The service alley should be 4 feet wide.

Feeding floors made of concrete, with shelters for the hogs, are still used on some farms. However, the trend in newer construction has moved away from this type of facility. The open-front system is an adaptation of this type of housing. In moderate and warm climate zones of the United States, a feeding floor with an extension of slotted floor over a lagoon may be used. Shade is provided on one side of the floor. Odor is a problem with this type of feeding floor. Care must be taken in locating the facility. Manure is disposed of in the lagoon and broken down by bacterial action.

FENCING

Pasture should be enclosed with strong, woven-wire fencing at least 3 feet high. Electric fencing can be used for temporary fences. The wire should be 6 to 8 inches off the ground.

Plank or concrete fencing is usually used with open-front or modified open-front houses. Windbreak fences may be made of wood or metal. They should be at least 6 feet high. Solid windbreak fences provide better wind protection. Fences that are about 80 percent solid are better for snow protection. Gates may be made of wood or metal. Sturdy construction and good hinges are important factors when selecting gates. Stock guards may be used instead of gates for convenience.

Courtesy of Frank B. Flanders

Figure 24-9 Modern adjustable chutes are used to load cattle, hogs, and other livestock. The platform can be lowered or raised depending on the height of the truck or trailer bed.

HANDLING EQUIPMENT

Holding and crowding pens with swinging gates make sorting and moving hogs easier. Alleyways are 20 inches wide. Cutting and blocking gates should be located in the alleyways to make sorting easier. The cutting gate swings in the alleyway and allows the operator to direct individual hogs into the desired pens. The blocking gate is also located in the alleyway just ahead of the cutting gate. It is used to stop the flow of hogs through the alleyway while individual hogs are directed into the desired pens by the cutting gate.

A **breeding rack** is used for hand mating. It should be 6 feet long, 3.5 feet high, and 32 inches wide. One end serves as a ramp that can be lowered to allow the sow and boar to enter. Shipping, weighing, and ringing crates are other types of equipment that hog producers find useful. Details of construction for these items are available from agricultural engineering specialists.

Loading chutes may be stationary or portable. Variable-height chutes permit loading into different kinds of trucks. Steps or cleats in the floor of the chute provide firmer footing for the hogs. Figure 24-9 shows an adjustable chute that can be used to load various types of livestock.

Exce Co-op Inc

Figure 24-10 Feed handling system for hog production.

FEEDING AND WATERING EQUIPMENT

Many different styles of feeders and waterers are available. Small portable feeders, larger walk-in feeders, fence-line feeders, and self-feeders may be used. Feeders may be constructed on the farm or bought commercially. Hogs are hard on this type of equipment, so the sturdiness of construction must be considered when buying any type of feeder. The ease and accuracy of feeder adjustment is important to consider. Feeders that are out of adjustment can release too much feed, which may be wasted. Feeders may also not release enough feed for pigs.

Many styles of waterers are available. A barrel or tank on a skid may be used on pasture or in the feedlot. Automatic waterers are popular with confinement systems. Electric heaters may be used to prevent the water from freezing in cold weather. Some waterers have provisions for adding medication to the drinking water.

Feed-handling methods range from the scoop shovel to fully automated equipment (Figure 24-10). Automation is more expensive initially but saves labor. Augers from holding bins are used to move the feed to the feeding areas. Design details are available from agricultural engineers or commercial companies.

Smaller Equipment Essential to Swine Production

- ear notchers
- vaccinating syringes
- needle teeth clippers
- ringing pliers
- castrating knives

SUMMARY

The trend in swine housing is toward more confinement housing. Housing is needed for farrowing, growing, and finishing hogs. Some producers still use pasture for all or part of the hog production cycle. Portable buildings and equipment are provided for hogs in pasture.

The location of permanent housing requires careful consideration. Odors are a major problem in locating hog houses. The direction of prevailing winds and the location of residences in the area must be considered.

Confinement buildings must have proper ventilation to control temperature and moisture. Additional heat may be required, especially in the farrowing house. Finishing pigs do about as well in open-front buildings as in warm confinement buildings.

The development of slotted floors has reduced the labor needed for manure handling. Most confinement systems use some type of slotted floor.

The plan for a hog enterprise must include fencing, handling equipment, and feeding and watering equipment. Careful planning saves labor. Increased automation also saves labor, but is more expensive.

Student Learning Activities

1. Take field trips to observe various kinds of swine facilities in the local area.
2. Survey the community to determine the types of facilities and equipment commonly used.
3. Construct portable swine equipment in the school shop or on your home farm.
4. Give an oral report on swine housing and equipment.
5. Prepare a bulletin board display of pictures of swine housing and equipment.
6. When planning and conducting a swine supervised experience program, use appropriate swine housing and equipment as described in this chapter.

Discussion Questions

1. What changes have occurred in the hog industry during the past 25 years that have affected the kinds of facilities used?
2. What factors must be considered when selecting the site and building location?
3. What is the function of the ventilation system in a hog house?
4. Describe the differences between warm and cold confinement houses for hogs.
5. What temperature is needed for baby pigs? How can that temperature be provided?
6. What temperatures are needed for (a) sows, (b) weaned pigs, and (c) growing-finishing pigs?
7. Why should insulation be used in hog houses?
8. Describe how hogs can be kept cool in hot weather.
9. Describe the three general types of floors used in hog houses.
10. What kinds of facilities are needed for the breeding herd?
11. Under what conditions are portable farrowing houses more practical than confinement housing?
12. Under what conditions are confinement farrowing houses more practical than portable farrowing houses?
13. Describe a farrowing pen.

14. Describe a farrowing crate.
15. What are the advantages of farrowing crates?
16. What kinds of facilities are needed for growing hogs?
17. Describe the three general kinds of houses used for finishing hogs.
18. Describe the kind of fencing needed for hogs.
19. What kinds of handling equipment are necessary for hog production?
20. Describe feeding and watering equipment for hogs.
21. What miscellaneous kinds of equipment are needed for hog production?

Review Questions

True/False

1. Swine production facilities are much the same as they were 25 years ago.
2. Sprinkling systems are of little value for cooling hogs because they have sweat glands.

Multiple Choice

3. Baby pigs require temperatures of ______ °F.
   a. 50–60 c. 70–80
   b. 60–70 d. 80–90
4. In confinement housing, ______ controls are the most important considerations.
   a. temperature c. light
   b. moisture d. a and b
5. Solid floors in hog houses are usually made of ______.
   a. wood c. concrete
   b. earth d. pea gravel

Completion

6. The trend in hog housing is more ______ housing.
7. Ventilation helps to remove ______ gases and odors that come from manure.

Short Answer

8. How is manure handled in today's hog operations?
9. What is the purpose of a breeding rack?

Chapter 25

Marketing Swine

Key Terms

boar taint
skatole
androsterone
shrinkage
futures contract
hedging

Objectives

After studying this chapter, the student should be able to

- explain three methods of marketing hogs.
- describe the grades of market hogs.
- describe grades of feeder pigs.

Making the right marketing decisions can mean the difference between profit and loss for the hog producer. Producers must carefully study the kinds of markets that are available. Factors such as the price offered, costs, convenience of the market, and the honesty and reliability of the hog buyer must be considered.

It is often hard to evaluate different markets on the basis of price. There are many factors that make up the price of hogs in a given market on a given day. The supply of hogs and the demand for hogs can vary from market to market on any given day. The costs at different kinds of markets also vary. The producer must calculate as closely as possible the highest net return that can be obtained for the hogs in each available market. Generally, the market giving the highest net return to the producer is the one to choose.

Producers of high-quality, meaty hogs will generally get better returns by selling on a grade and yield basis. An increasing number of hogs are being marketed on a grade and yield basis. Consumer demand for less fat in red meat has put pressure on the swine industry to produce leaner hogs. Grade and yield marketing provides a price mechanism to reward producers who raise the kind of hogs that are in demand.

PORK PROMOTION

The Pork Promotion, Research and Consumer Information Act of 1985 established a National Pork Board that collects funds through an assessment of 40 cents per $100 value of pork sold in the United States. These funds are used to promote the use of pork, develop foreign markets, provide consumer information, and conduct research and producer education programs.

Major advertising campaigns involving television and print media are used to influence consumer demand for pork as a safe, healthy, and nutritious food. Pork exports, especially to the Pacific Rim nations, have increased in recent years. Part of the checkoff funds is used to increase market share in the export market. Research and producer education programs include reviewing the pork production chain from producer to consumer; studying value-based marketing as a method of improving pork quality; and addressing issues of animal care, the environment, and food safety.

Checkoff funds are collected based on dollar sales at commercial markets on market hogs and feeder pigs. Seedstock producers and producers who sell farm to farm are responsible for making the proper payments. The National Pork Board regularly monitors swine sales nationwide to ensure compliance with the checkoff program. Producers who fail to make the required payments are reported to the USDA, which then takes steps to collect the money, and penalties may apply.

One analysis of the impact pork checkoff funds has had on returns to pork producers revealed that $4.79 was the overall net return for a period of years per invested checkoff dollar. Some of this return came from the benefits of research and technology development funded by checkoff dollars. Part of the net return can be attributed to an increased demand for pork and pork products resulting from checkoff programs. These returns are an average across the pork industry; for a variety of reasons, individual producers may not benefit to the same extent as the industry average.

BOAR MEAT

Research is being done to determine the feasibility of feeding boars to market weight instead of castrating them when young to produce barrows. Boars produce more lean meat and have a higher feed efficiency compared to barrows. Boar meat has gained some acceptance in the European market. Pork producers in the United States may need to develop methods of raising and marketing acceptable boar meat if they are to remain competitive in world markets.

The major objection to boar meat is the odor of the meat, called **boar taint**. Two compounds, **skatole** and **androsterone**, have been identified as the major causes of the odor problem with boar meat. European producers have overcome some of the odor and taste problem by marketing younger, lighter-weight boars. The amount of skatole and androsterone in the animal's body increases with age. More research is needed to determine the right combination of genetics and nutrition that can produce boar meat that is acceptable to the consumer. Castration of boars is a major objection of animal rights groups. Many European producers have had to face this opposition to castration.

KINDS OF MARKETS

Hogs may be sold through (1) direct marketing, (2) terminal markets, or (3) auction markets. Some hog producers participate in group marketing. Generally, the group markets the hogs through one of the three methods just listed.

Direct Marketing

Direct marketing involves selling to packing plants, order buyers, or country buying stations. There are several kinds of country buying stations. Some are owned by packing plants, others are independently owned, while still others are cooperatives.

Direct marketing accounts for the majority of the hogs sold in the United States. In this system, the producer deals directly with the buyer. When using direct marketing, the producer must possess selling skill and knowledge of the markets to be successful in obtaining the best price for the hogs. Generally, the animals are transported shorter distances when sold through direct marketing, and the shrinkage is less.

Terminal Markets

Hogs marketed through terminal markets are consigned to a commission firm. The commission firm deals with the buyer. Commission firms can help the producer select the best time to market hogs. They generally employ people who have a good knowledge of the market.

Less than 1 percent of the slaughter hogs sold in the United States are marketed through terminal markets. The trend in recent years has moved away from terminal markets and toward direct marketing. There are usually several buyers competing for hogs on the terminal market, which may produce a better price. Prices may vary more widely on the terminal market than in direct marketing. This variation is affected by the number of hogs coming on the market on any given day.

Courtesy of Frank B. Flanders

Figure 25-1 Premium prices are paid for hogs with more lean muscle and less fat.

Auction Markets

In some areas, auction markets are an important method of selling hogs. A small percent of the hogs sold in the United States are sold through auction markets. Auction markets are not widely used in the major hog-producing states. Some auction markets are developing new systems to obtain more buyers for the hogs offered. For example, telephone hookups and video auctions permit buyers who are not physically present to bid on the hogs.

Most auction markets have hog sale days once or twice a week. This limits the producer in choosing the day on which to market the hogs. Because auction markets are usually located in the local area, transportation costs and shrinkage are reduced with this method. Selling costs involved with terminal and auction markets include commissions, insurance, yardage, and feed costs.

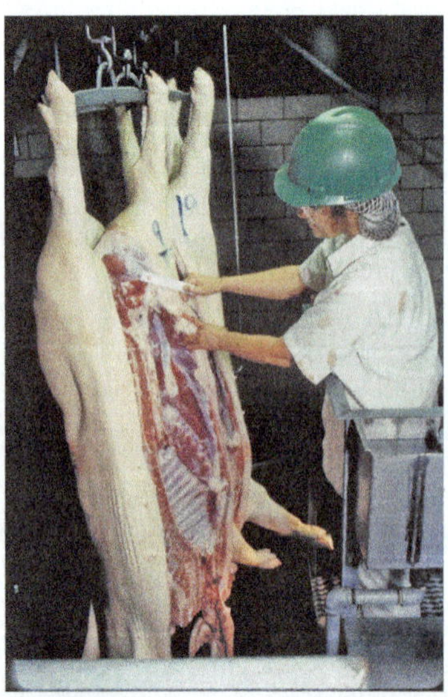

USDA

Figure 25-2 USDA inspectors check the wholesomeness of all meat produced in plants dealing in interstate commerce or foreign trade.

Group Marketing

Some producers use systems of group marketing. Group marketing systems have been established by some major farm organizations. Hog marketing cooperatives also exist in some areas. Some of these groups negotiate contracts with packers to supply hogs. The basic purpose is to obtain higher prices for hogs. The bargaining power of the producer may be increased by using a group marketing method.

PRICING METHODS

Approximately 70 percent of the market hogs sold in the United States are priced on the basis of carcass merit (Figures 25-1 and 25-2). Premium prices are paid for hogs with more lean muscle and less fat. This trend corresponds with the trend toward improved genetics in breeding, which results in the production of meatier hogs that meet consumer demand for leaner meat. Some hogs are still marketed on the basis of the weight of the animal without regard for carcass merit; however, this method of pricing is rapidly declining.

Another trend in setting hog prices in the market is the use of contracts with producers. Most hogs marketed are now sold under some form of contracting. The formula price contract is the most common type used. The price paid is based on an agreement between the packer and the producer, generally related to a quoted cash price and some formula that sets the final price. Formula pricing generally favors the packer more than the producer, especially when the cash price for hogs is low. Formula pricing does not change the variability found in hog prices because it moves up and down with the variation in the cash price for hogs. This method may increase the average price a producer receives for hogs, typically by $1.00 to $3.00 per hundredweight.

Several other methods of contract pricing are used for marketing hogs. A fixed cash price contract may be tied to the futures market margins. This type of contract may give some advantage to the producer when the cash price of hogs is above the break-even level.

Another type of fixed price contract may be tied to the price of feed or historical price data with an additional amount added for overhead. This type of contract may exist with or without a ledger being maintained. If a ledger is maintained, a debit is recorded to the producer when the cash market price is lower than the contract price; a credit is recorded to the producer when the cash market price is higher than the contract price. Eventually, debits and credits have to be reconciled so the actual cash price received over a period of time equals the market price over that same time period.

Window contracts provide for some sharing of the price risk between the packer and the producer. A window contract sets a floor for the price to be paid the producer. If the cash market price is lower than this floor, the producer is still paid the amount set by the floor. The packer assumes some of the loss when the price is below the price floor set by the contract and gets some of the gain when the price is above the price floor. Ledgers may or may not be maintained with this type of contract. However, if a ledger is maintained, then the producer may have to eventually repay the debt created by this type of contract when the cash price of hogs goes above the floor set in the window contract.

Some ledger contracts have sunset clauses that, in effect, forgive the debt accumulated by the producer after the contract expires. When a sunset clause exists, the producer may eventually realize an average price that is higher than the cash market price during the life of the contract. Generally, price risk is reduced only by cash contracts tied to futures market margins and contracts without ledgers.

Hogs that are not sold under some type of contract are sold on the cash market. This is referred to as the spot market. Prices on the spot market are determined by supply and demand for hogs on the day they are sold. The use of the spot market to set market hog price is decreasing and has almost no effect on the price of pork to the consumer (Figure 25-3).

USDA

Figure 25-3 Pork products displayed for sale. Consumer demand for pork is the number one factor in the price of hogs.

MARKET CLASSES AND GRADES

The use (slaughter or feeder), sex, and grade of swine determine their classification. Slaughter swine are those that are killed and sold as meat. Feeder swine are those that are sold to be fed to higher weights before harvest.

Four Lean Retail Cuts

1. Ham
2. Loin
3. Picnic shoulder
4. Boston butt

Classes of Swine

The five sex classes of slaughter and feeder swine are barrows, gilts, sows, boars, and stags. The definitions of each class are as follows:

- *barrow*—a male that was castrated while young. The physical traits of the boar have not developed
- *gilt*—a young female that has not farrowed and is not showing any signs of pregnancy
- *sow*—an older female that has farrowed or is showing signs of pregnancy
- *boar*—a male that has not been castrated
- *stag*—a male that was castrated after reaching maturity or is beginning to show the physical characteristics of a boar

USDA

Figure 25-4 USDA meat grade stamp.

Grades of Swine

The quality of the hogs and pigs is reflected in the grade standards established by the USDA (Figure 25-4). Grades of slaughter barrows and gilts are based on carcass quality and the yield of the four lean cuts.

The quality of the lean is referred to as *acceptable* or *unacceptable*. Acceptable carcasses have bellies that are at least slightly thick overall and no less than 0.6 inch thick at any point. Other factors used to determine quality are the amount and distribution of external finish and the firmness of fat and muscle. Carcasses with acceptable lean quality and firmness of fat are further classed in one of the four top grades, based on the expected yield of the four lean cuts. If the lean

IMPORTANT NOTE RELATED TO GRADES OF SWINE

It should be noted that although the official USDA market swine grades, which have been the basis of marketing swine since 1918, are discussed in this chapter, these grades are rarely used today in marketing swine. The USDA grades are still in place; however, the packing industry has moved away from this system. Almost all hogs are now marketed on the basis of fat-free lean percentage of the carcass. Fat-free lean percentage (FFL%) is an estimate of the amount of muscle tissue in a pork carcass. Higher prices are paid for animals that yield a greater percent of lean meat and less fat. The calculation for the FFL% is based on three factors:

- hot carcass weight (HCW)
- loin eye area (LEA)
- 10th rib back fat (TRBF)

Each packing plant may have its own formula for calculating FFL% based on HCW, LEA, and TRBF. Packing plants may also have their own price formulas dependent upon FFL%. In this system, the producer selling the hogs receives payment from the packing plant based on the FFL% calculated for the hogs delivered.

quality is unacceptable, the carcass is graded Utility. Animals that will produce carcasses with oily or less than slightly firm fat are also graded Utility.

The five official USDA grades for slaughter barrows and gilts are (1) *U.S. No. 1,* (2) *U.S. No. 2,* (3) *U.S. No. 3,* (4) *U.S. No. 4,* and (5) *U.S. Utility.* There are also grades of slaughter sows. Boars and stags are not graded.

The estimated back-fat thickness over the last rib and the muscling score are used to determine the official grade of slaughter barrows and gilts. These values are used in a mathematical equation to find the final grade. The equation is:

$$\text{Grade} = (4.0 \times \text{Back-fat thickness over last rib, in inches}) - (1.0 \times \text{Muscling score})$$

Muscling scores used in the equation are thin (inferior) = 1, average = 2, and thick (superior) = 3. If an animal has thin muscling, it cannot be graded U.S. No. 1.

For example, if an animal has an estimated back fat thickness over the last rib of 1.05 inches and thick (superior) muscling, the grade is determined as follows:

$$\text{Grade} = (4.0 \times 1.05) - (1.0 \times 3) = 1.2 \text{ (U.S. No. 1 grade)}$$

Another way to determine grade is to find a preliminary grade based on back-fat thickness over the last rib and then adjust the grade up or down by one grade based on thick (superior) or thin (inferior) muscling. The preliminary grade is based on the following back fat thickness ranges:

Less than 1.00 inch—U.S. No. 1	1.25 to 1.49 inches—U.S. No. 3
1.00 to 1.24 inches—U.S. No. 2	1.50 inches or over—U.S. No. 4

Using the same animal as in the previous example, the preliminary grade is U.S. No. 2 based on the back-fat thickness. This is improved one grade to U.S. No. 1 based on the thick (superior) muscling.

An estimate of the fatness and muscling is made by looking at the live animal. Fat deposits develop over the body at different rates, whereas muscle development is fairly uniform. Fat develops more rapidly over the back; at the loin edge; and on the rear flank, shoulder, jowl, and ham seam. As the animal gets fatter, these areas appear thicker and fuller than other parts of the body. The ham is least affected by increasing fatness; therefore, differences in the thickness and fullness of the hams are the best indicators of muscling.

Fatter animals are deeper bodied because of fat deposits in the flank and along the underline. Other signs of fatness are the fullness of the flanks and the thickness and fullness of the jowl.

The three degrees of muscling used when grading slaughter barrows and gilts are thick (superior), average, and thin (inferior). Thick-muscled animals with low fat have thicker hams than loins. The loins are full and well rounded. Animals with thick muscling and a high level of fat are slightly thicker through the hams than through the loins. The back is nearly flat with a slight break into the sides. Average-muscled animals with low fat are thicker through the hams than the loins. The loins are full and rounded. The hams and loins have about equal thickness in animals with average muscling and a high level of fat. A thin-muscled animal with low fat is slightly thicker through the shoulders and the center of the hams compared to the back. The loins are sloppy and fat. Animals with thin muscling and high fat are wider in the loins than the hams. There is a distinct break from over the loins into the sides. Figure 25-5 shows the USDA slaughter barrow and gilt grades.

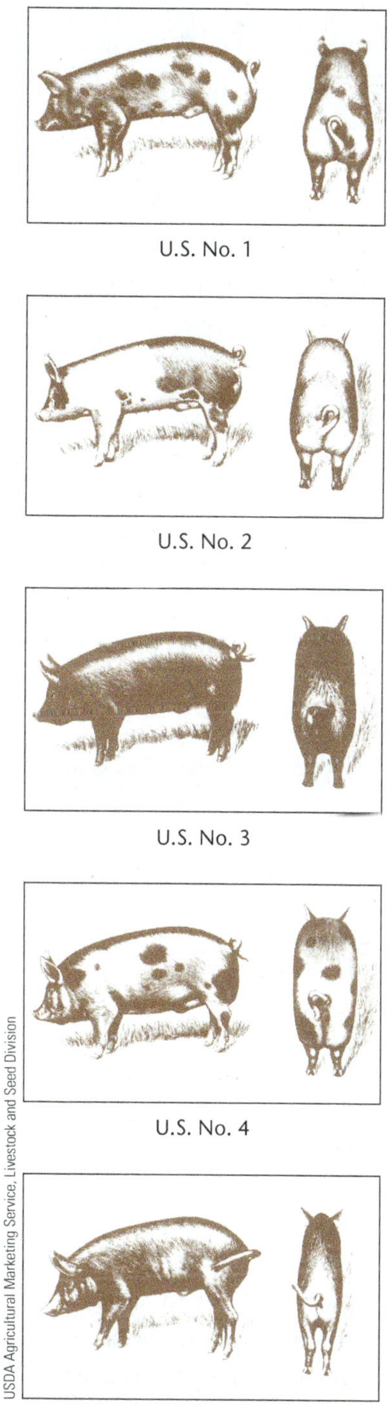

USDA Agricultural Marketing Service, Livestock and Seed Division

Figure 25-5 USDA slaughter grades of barrows and gilts.

Slaughter Barrows and Gilts Grade Descriptions

U.S. No. 1 Barrows and Gilts U.S. No. 1 barrows and gilts produce a chilled carcass yielding 60.4 percent, or more, of the four lean cuts. Muscling and back-fat combinations that allow hogs to be put into this grade are average muscling with less than 1.00 inch of back fat and thick muscling with less than 1.25 inches of back fat.

U.S. No. 2 Barrows and Gilts U.S. No. 2 barrows and gilts produce a chilled carcass yielding 57.4 to 60.3 percent of the four lean cuts. Muscling and back-fat combinations that allow hogs to be put into this grade are average muscling with 1.00 to 1.24 inches of back fat, thick muscling with 1.25 to 1.49 inches of back fat, or thin muscling with less than 1.00 inch of back fat.

U.S. No. 3 Barrows and Gilts U.S. No. 3 barrows and gilts produce a chilled carcass yielding 54.4 to 57.3 percent of the four lean cuts. Muscling and back-fat combinations that allow hogs to be put into this grade are average muscling with 1.25 to 1.49 inches of back fat, thick muscling with 1.50 to 1.74 inches of back fat, or thin muscling with 1.00 to 1.24 inches of back fat. Animals with 1.75 inches or more of back fat are not put into this grade.

U.S. No. 4 Barrows and Gilts U.S. No. 4 barrows and gilts produce a chilled carcass yielding less than 54.4 percent of the four lean cuts. Muscling and back-fat combinations that allow hogs to be put into this grade are average muscling with 1.50 inches or more of back fat, thick muscling with 1.75 inches or more of back fat, or thin muscling with 1.25 inches or more of back fat.

U.S. Utility Barrows and Gilts U.S. Utility barrows and gilts produce carcasses with unacceptable lean quality or unacceptable belly thickness. Also included are hogs with carcasses that are soft and oily. Muscling and back-fat thickness are not considered for hogs in this grade.

Slaughter Sow Grades

The five grades of slaughter sows are (1) *U.S. No. 1,* (2) *U.S. No. 2,* (3) *U.S. No. 3,* (4) *Medium,* and (5) *Cull.* These grades are based on differences in yield of lean and fat cuts and differences in quality of cuts. There is a close relationship between the grades of slaughter sows and the grades of sow carcasses. Average back-fat thickness is considered when grading sows. The average carcass back-fat thickness for each grade of slaughter sows is:

U.S. No. 1 — 1.5 to 1.9 inches
U.S. No. 2 — 1.9 to 2.3 inches
U.S. No. 3 — 2.3 or more inches
Medium — 1.1 to 1.5 inches
Cull — Less than 1.1 inches

The standard for each grade includes a description of the live animal and the minimum amount of finish for the grade. There are many combinations of characteristics that qualify an animal for a particular grade.

> Swine produce higher dressing percentages (approximately 70 to 75 percent on average) than cattle or sheep because they are monogastric and have a smaller digestive tract. The dressing percentage is calculated as follows:
>
> (Carcass weight/Live weight) × 100

U.S. No. 1 Slaughter Sows U.S. No. 1 slaughter sows have just enough finish to produce acceptable palatability in the cuts. These animals have moderate length and are slightly wide in relation to weight. They have a uniform width from top to bottom and front to rear. The back is moderately full and thick with a well-rounded appearance, blending smoothly into the sides. They have moderately long and slightly thick sides. Flanks are slightly thick and full. The fore flank may be slightly deeper than the rear flank. The hams have moderate thickness and fullness with a slightly thick covering of fat. The jowls are moderately thick and full with a trim appearance. Animals in this grade produce U.S. No. 1 carcasses.

U.S. No. 2 Slaughter Sows U.S. No. 2 slaughter sows have a little more than the minimum finish required to produce acceptable palatability in cuts. These animals are slightly short and are moderately wide in relation to weight. The body is usually wider over the top than at the underline. The shoulders tend to be slightly wider than the hams. The back is full and thick and is slightly flat with a noticeable break into the sides. They have slightly short and moderately thick sides. Flanks are moderately thick and full. The depth of the fore flank is about equal to the rear flank. The hams are thick and full with a moderately thick covering of fat. The jowls are generally thick and full. The neck is short. Animals in this grade produce U.S. No. 2 carcasses.

U.S. No. 3 Slaughter Sows U.S. No. 3 slaughter sows have much more than the minimum finish required to produce acceptable palatability in the cuts. These animals are short and wide in relation to weight. The body is wider over the top than at the underline. The shoulders tend to be wider than the hams. The back is very full and thick and is nearly flat with a definite break into the sides. They have short, thick sides. Flanks are thick and full. The depth of the fore flank is about equal to the rear flank. The hams are very thick and full with a thick covering of fat. The jowls are generally very thick and full. The neck is short. Animals in this grade produce U.S. No. 3 carcasses.

Medium-Grade Slaughter Sows Medium-grade slaughter sows have a little less than the minimum finish required to produce acceptable palatability in the cuts. These animals are long and moderately narrow in relation to weight. The body is usually narrower over the top than at the underline. The shoulders tend to be slightly narrower than the hams. The back is moderately thin and is slightly peaked with a distinct slope toward the sides. These animals have long and moderately thin sides. Flanks are thin. The depth of the fore flank is greater than the rear flank. The hams are moderately thin and flat. The jowls are generally slightly thin and flat. The neck is long. Animals in this grade produce Medium carcasses.

Cull-Grade Slaughter Sows Cull-grade slaughter sows have considerably less finish than that required to produce acceptable palatability in the cuts. These animals are long and narrow in relation to weight. The body is usually narrower over the top than at the underline. The shoulders tend to be narrower than the hams. The back is thin and lacks fullness and is peaked with a definite slope toward the sides. They have very long and thin sides. Flanks are very thin. The depth of the fore flank is considerably greater than the rear flank. The hams and jowls are generally thin and flat. The neck is long. Animals in this grade produce Cull-grade carcasses.

Feeder Pig Grades

Feeder pigs are classified in the same grades as slaughter hogs, with the addition of one lower grade, U.S Utility. U.S. Cull is the lowest grade of feeder pig. Feeder pig grades are used to indicate the expected grade of the pig when it reaches slaughter weight. Unthrifty pigs are classed in either U.S. Utility or U.S. Cull grades. These pigs have failed to grow and gain properly. Poor care or disease can cause a pig to be unthrifty.

The U.S. No. 1 Feeder Pig The U.S. No. 1 feeder pig has a large frame and thick muscling and is trim. The legs are set wide apart. The hams are wider than the loins. Feeder pigs in this grade should produce U.S. No. 1 grade carcasses when harvested.

The U.S. No. 2 Feeder Pig The U.S. No. 2 feeder pig has a moderately large frame with moderately thick muscling. The pig is a little fatter than the U.S. No. 1. The hams are slightly wider than the loin. The jowl and flank are a little fatter than the U.S. No. 1. The side shows less trim. Feeder pigs in this grade should produce U.S. No. 2 grade carcasses when harvested.

The U.S. No. 3 Feeder Pig The U.S. No. 3 feeder pig has a slightly smaller frame with slightly thin muscling. The hams and loin are about the same width.

The legs are set fairly close together. The jowl and flank show signs of too much fat. Feeder pigs in this grade should produce U.S. No. 3 grade carcasses when harvested.

The U.S. No. 4 Feeder Pig The U.S. No. 4 feeder pig has a small frame with thin muscling. The hams and loin are the same width. The back is flat. The legs are set close together. The jowl and flank are moderately full. The lower ham is beginning to show signs of too much fat. Feeder pigs in this grade should produce U.S. No. 4 grade carcasses when harvested.

The U.S. Utility Feeder Pig The U.S. Utility feeder pig shows unthriftiness because of disease or poor care. The skin is wrinkled and the head appears too large for the rest of the body. The pig is rough in appearance. Given good care, pigs in this grade can develop into a higher grade of slaughter hog. Feeder pigs in this grade will produce U.S. Utility grade carcasses when harvested unless the unthrifty condition is corrected.

The U.S. Cull Feeder Pig The U.S. Cull feeder pig has a poorer appearance than U.S. Utility. Improper care and disease cause a rough, unthrifty appearance. Pigs in this grade will gain at slower than normal rates and often will not be profitable. Figure 25-6 shows the USDA grades of feeder pigs.

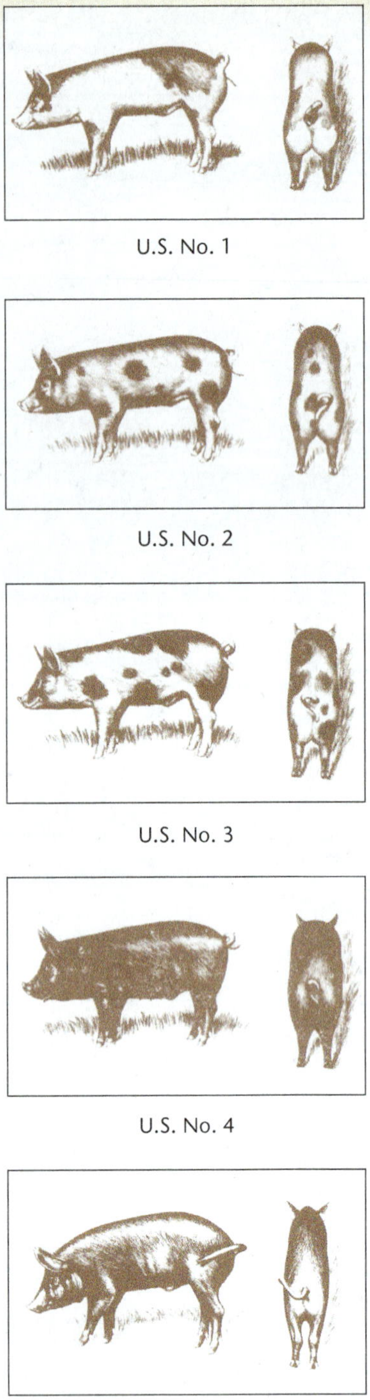

U.S. No. 1

U.S. No. 2

U.S. No. 3

U.S. No. 4

U.S. Utility

USDA Agricultural Marketing Service, Livestock and Seed Division

Figure 25-6 USDA grades of feeder pigs.

WEIGHT AND TIME TO SELL

The traditional recommended weight at which to sell slaughter hogs has been 200 to 220 pounds. This was based on two conditions related to weight gain: (1) research indicated that feed costs per pound of gain increased rapidly above 220 pounds, and (2) much of the weight added above 220 pounds was fat.

Research indicates that several factors should be considered when deciding on the best weight to market hogs: (1) type of hog, (2) hog-feed price ratio, (3) amount of discount for heavier hogs, and (4) the time of year when the hogs are marketed.

Research also shows that meaty hogs can be fed to heavier weights without a large increase in feed costs. The quality of the hog as measured by the ratio of lean to fat is not decreased appreciably with this type of hog. The experiments show that an increase in feed of 0.70 pound per 100 pounds of gain is necessary to go from lighter to heavier market weights. Two studies show a decrease in percent of lean cuts. One shows a decrease of 0.6 percent as hogs were fed from 200 to 250 pounds. The other study shows a decrease of 1.3 percent from 220 to 260 pounds.

As feed costs increase, the additional returns above feed costs decrease as hogs are fed to heavier weights. As the price of hogs increases, the returns above feed costs increase. It requires careful calculation to determine where the break-even point occurs as hog-feed price ratios change.

Packers typically discount the price of hogs below 220 to 230 pounds and above 270 to 280 pounds. This range varies with the number of hogs coming to market and the pricing practices of individual packers. If discounts are high for heavy hogs, it may not pay to feed to the heavier weights.

Part of the decision on feeding to heavier weights depends on the time of year. The price of slaughter hogs changes seasonally based on the supply of pork and consumer demand for pork. Figure 25-7 shows the seasonal variation in the price of slaughter hogs in the United States. Because there are more large commercial swine producers now than there were a few years ago, the supply of hogs coming to market is less variable than it used to be. Large producers tend to distribute farrowings more evenly during the year.

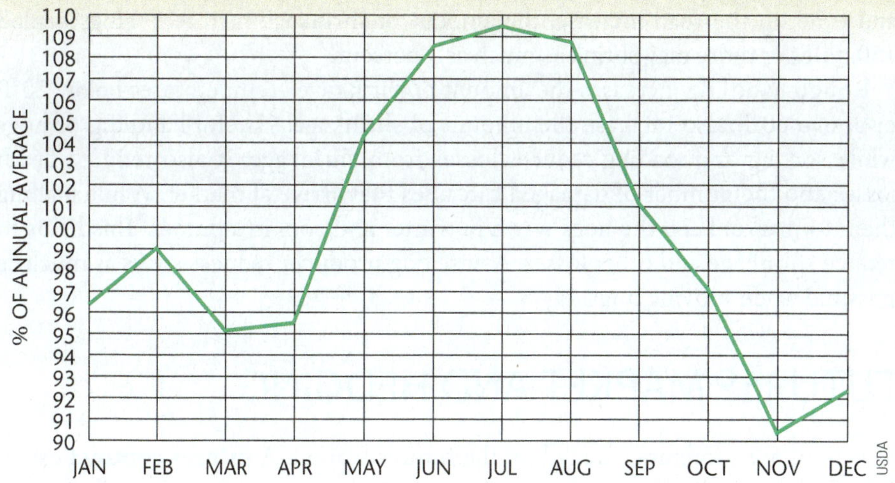

Figure 25-7 Seasonal trends in the price of slaughter barrows and gilts in the United States.

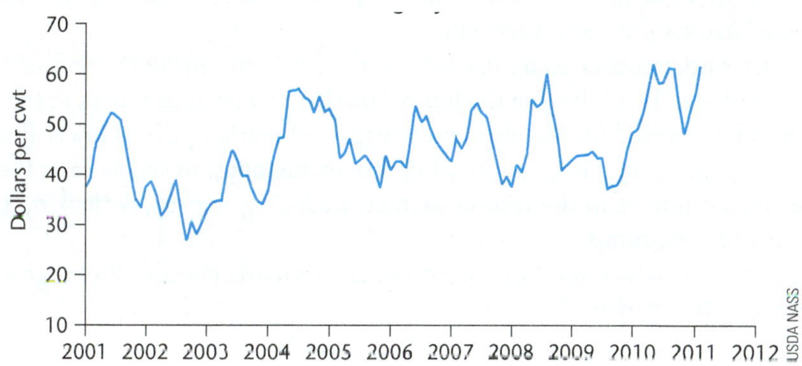

Figure 25-8 Prices received for hogs by month in the United States.

Slaughter hog prices tend to increase from April through July. Prices tend to move down from July to November and then show some recovery through February. They again decline from February to April.

When prices are on the increase, it may pay to feed hogs to heavier weights even though there will be some discount. However, when the price trend is downward, it generally will not pay to feed to heavier market weights. The hog producer should have the flexibility to feed to heavier weights part of the year and to lighter weights at other times of the year.

It must be remembered that the seasonal price trends are averages. Prices in any given year are affected by such factors as the number of hogs being marketed and the demand for pork (Figure 25-8). Hog prices for any given year do not always follow the long-term seasonal price patterns.

SHRINKAGE OF HOGS

Hogs lose weight as they are shipped to market. This weight loss is called **shrinkage**. The distance to market is one of the most important factors in determining the amount of shrinkage. Shrinkage of about 2 percent can be expected, regardless of how close to market the hogs are located. This is caused by the sorting, handling, loading, and hauling that takes place in the first few miles. As distance hauled and time on the road increase, the amount of shrinkage increases. Hogs hauled 150 miles or more may shrink as much as 4 percent.

Rough handling increases the amount of shrinkage. Temperatures below 20°F or above 60°F also increase the amount of shrinkage. Careful handling of hogs while sorting and loading reduces losses from shrinkage. It also reduces death losses and the number of damaged carcasses that arrive at market. While hauling them to market keep the hogs warm in winter and cool in summer. This helps to reduce shrinkage and other losses. A wise hog producer reduces stress as much as possible when moving hogs.

FUTURES MARKET AND HEDGING

Live hogs are sometimes traded on the futures market. A **futures contract** establishes a price for live hogs that are to be delivered at some future date. The unit of trading is 30,000 pounds. This is about 130 to 150 head of market hogs.

One objective of using the futures market is to obtain a higher price for hogs. Futures trading takes place when the futures price is higher than the expected market price of hogs at the time of delivery. The producer runs the risk that the expected market price is not accurate.

Another objective of using the futures market is to reduce the risk of loss if prices go down. By trading on the futures market, the producer locks in the price that will be received for the hogs. If the expected market price is lower than the costs of production, it does not pay to produce the hogs. In such a case, the producer would not be in the futures market. Reducing the risk by locking in the price is called **hedging**.

The producer who wants to hedge on the futures market needs a thorough understanding of the market. Detailed information on using the futures market may be obtained from brokers or from university specialists.

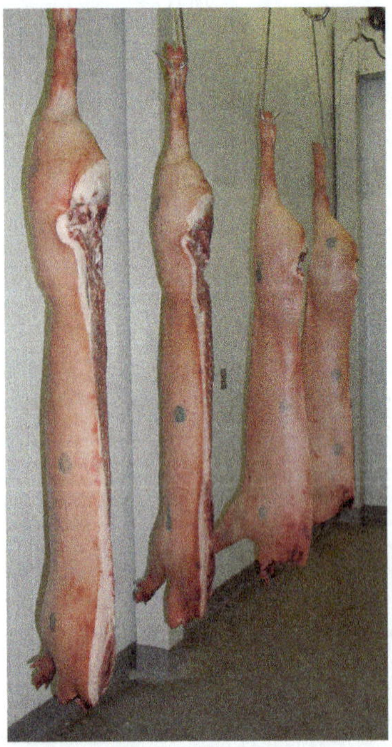

Figure 25-9 The current trend is for heavier market hogs. Packers want long hogs because more pork can be processed on the same amount of rail space. Note the length of the carcass on the left compared to the shorter carcasses to the right. Market weight greater than 260 pounds is common.

SUMMARY

How to market their hogs is one of the most important decisions that hog producers must make. Markets vary in the price offered, the services given, and the costs charged.

Most hogs in the United States are marketed by direct marketing. Smaller numbers are marketed through terminal markets and by auction market. Some group marketing systems have been established to help producers get higher prices.

Most hogs are sold on the basis of weight produced. Producers of quality meaty hogs can get more for their hogs by selling on a yield and grade basis.

Hogs are classified according to use, sex, weight, and quality. Butcher hogs are graded according to USDA official grading standards. Feeder pigs are also graded by USDA standards. The grades of market hogs are U.S. No. 1, U.S. No. 2, U.S. No. 3, U.S. No. 4, and U.S. Utility. Grading is based on the quality of the lean meat and the percent of four lean cuts that the carcass will produce.

Traditional recommendations have been to market hogs at 200 to 220 pounds. Top butcher prices are typically paid for hogs weighing 220 to 260 pounds (Figure 25-9). Meaty hogs can be fed to heavier market weights if the hog-feed price ratio is right, the discount for heavy hogs is not too great, and the seasonal price is rising. Hog prices tend to be lower in early spring and higher in the summer.

Shrinkage is increased as hogs are moved greater distances to market. Careful handling to reduce stress when marketing reduces shrinkage and other losses.

The futures market may be used to reduce the risk of loss. A producer using the futures market should consult with experts.

Student Learning Activities

1. Prepare a bulletin board display of charts showing price trends and local prices for hogs.
2. Take a field trip to a local hog market.
3. Ask a manager from a local packing plant to discuss damage to hog carcasses that results from mishandling while marketing.
4. Ask a hog buyer to speak to the class.
5. Ask a hog producer who uses the futures market to hedge to speak to the class on the use of hog futures.
6. Give an oral report on hog marketing.
7. Survey hog producers in the community to determine local marketing practices.
8. When planning and conducting a swine supervised experience program, follow good marketing practices as described in this chapter.

Discussion Questions

1. Compare direct marketing, terminal markets, and auction markets as methods of marketing hogs.
2. Explain how hogs are priced based on the total weight marketed.
3. Explain how hogs are priced on a yield and grade basis.
4. Describe the two uses of market hogs.
5. Describe the sex classes of market hogs.
6. How is the market grade of a hog determined?
7. Name the four lean cuts of the hog carcass.
8. Describe the five official USDA grades of market hogs.
9. Describe the six official USDA grades of feeder pigs.
10. Name and briefly explain the factors involved when deciding on the best weight to market hogs.
11. Describe the seasonal price trends for butcher hogs.
12. What factors affect the amount of shrinkage as hogs are moved to market?
13. How can the percent of shrinkage be kept low?
14. What are two reasons for a hog producer to use the futures market?

Review Questions

True/False

1. It is often hard to evaluate different markets on the basis of price.
2. Direct marketing accounts for the majority of hogs sold in the United States.
3. U.S. Cull is the lowest grade of feeder pigs.
4. Shrinkage decreases as hogs are moved greater distances to market.
5. The U.S. Utility feeder pigs show unthriftiness because of disease or poor care.

Multiple Choice

6. What percentage of market hogs sold in the United States is priced on the basis of carcass meat merit?

a. 50   c. 70
b. 60   d. 75

7. Most auction markets have hog sale days ______.
   a. monthly c. biweekly
   b. once or twice a week d. quarterly
8. The traditionally recommended weight for selling slaughter hogs has been ______.
   a. 200–220 pounds c. 230–250 pounds
   b. 175–200 pounds d. 150–175 pounds
9. Most market hogs today are sold on the basis of ______.
   a. fat-free lean percentage c. back fat
   b. loin eye area d. estimated fat cover

Completion

10. Making the right ______ decisions can mean the difference between profit and loss for the hog producer.
11. A ______ is an older female hog that has farrowed a litter of pigs.
12. The three official USDA swine muscle scores are ________, ________, and ________.

Short Answer

13. List three methods that may be used to sell hogs.
14. List two objectives of the futures market.
15. What are the three factors that are used to calculate fat-free lean percentage?

SECTION 6

Sheep and Goats

Chapter 26

Breeds and Selection of Sheep

Key Terms

banding instinct
band
staple
National Sheep Improvement Program (NSIP)
western ewe
native ewe
type
pelt

Objectives

After studying this chapter, the student should be able to

- describe major characteristics of the sheep enterprise.
- summarize the origin and development of the common breeds of sheep.
- identify the common breeds of sheep.
- select high-quality breeding stock using generally accepted criteria.

OVERVIEW OF THE SHEEP ENTERPRISE

Sheep are raised in every state in the United States (Figure 26-1). Range production is concentrated in 12 western states while native, or farm flock, production is found in the rest of the states. Small flocks of fewer than 100 head account for more than 90 percent of all sheep operations but only about 35 percent of the inventory in the United States (Figure 26-2). In recent years, the trend is toward smaller sheep operations. Range production accounts for more than 50 percent of the total sheep production in the United States; large flocks of 1,000 to 1,500 ewes are common in this area. The production of feeder lambs and wool is located mostly in the southern part of the range area, and market lamb production is more in the northern part of the area (Figure 26-3).

In the rest of the United States, farm flocks tend to be small. Sheep production on farms tends to be a secondary enterprise. About half of the sheep numbers in farm flocks are located in the Corn Belt states of Illinois, Indiana, Iowa, and Ohio and in parts of surrounding states. Both wool and market lambs are sold from these flocks. Purebred sheep production is also important in farm sheep enterprises.

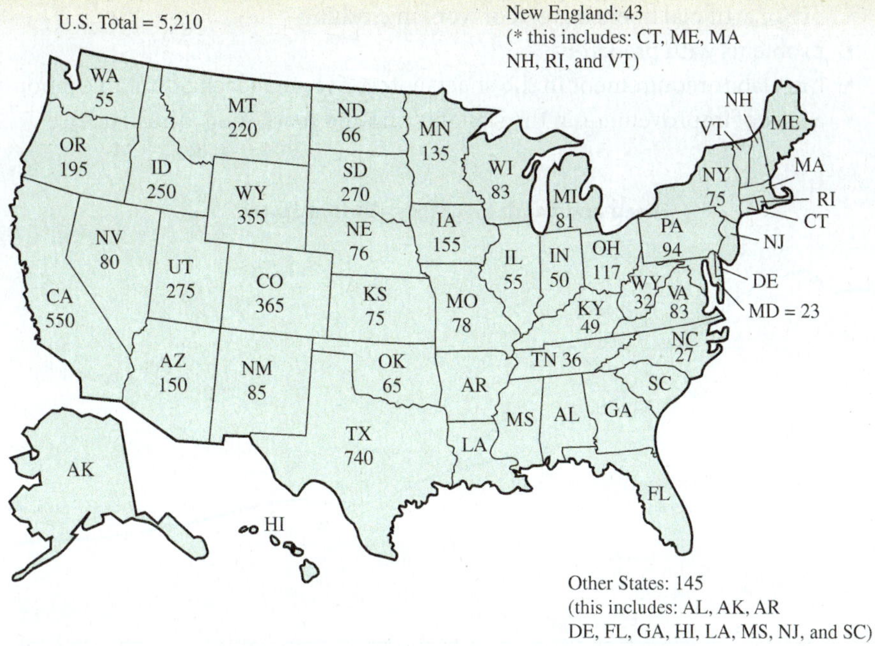

Figure 26-1 Sheep and lambs on farms—January 1, 2014 (1,000 head).
Source: USDA NASS, Sheep and Lamb Inventory by Class—States and United States: January 1, 2013–2014

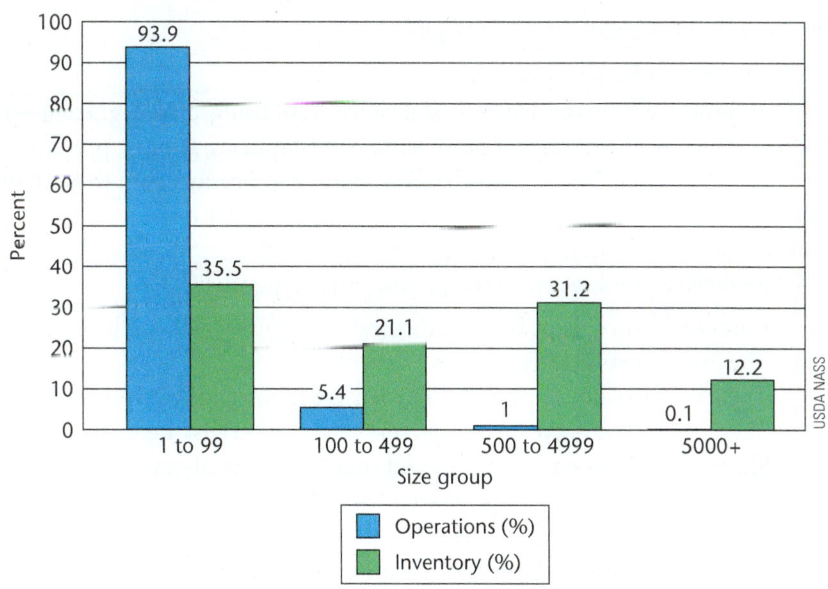

Figure 26-2 Breeding sheep: Percentage of operations and inventory, 2012.

The number of small sheep enterprises is declining. Most of the lamb feeding is concentrated in large commercial feedlots. The leading states in numbers of sheep and lambs on feed are Texas, California, Wyoming, and Colorado. Large commercial flocks of more than 500 head account for less than 2 percent of all producers but produce more than 40 percent of the total sheep and lambs in the United States.

Since 1990, sheep numbers in the United States have steadily declined (Figure 26-4). There are several factors that have contributed to the decline in the sheep industry in recent years:

- seasonal demand for lamb meat
- low per capita consumption
- low wool prices

USDA Agricultural Research Service; Photo by Stephen Ausmus

Figure 26-3 Sheep do an excellent job of converting roughage on marginal land into wool and meat.

- use of artificial fibers instead of wool in clothing
- problems with predators
- high labor requirement in the sheep enterprise and a lack of suitable labor
- a lack of improvement in the slaughtering and marketing infrastructure

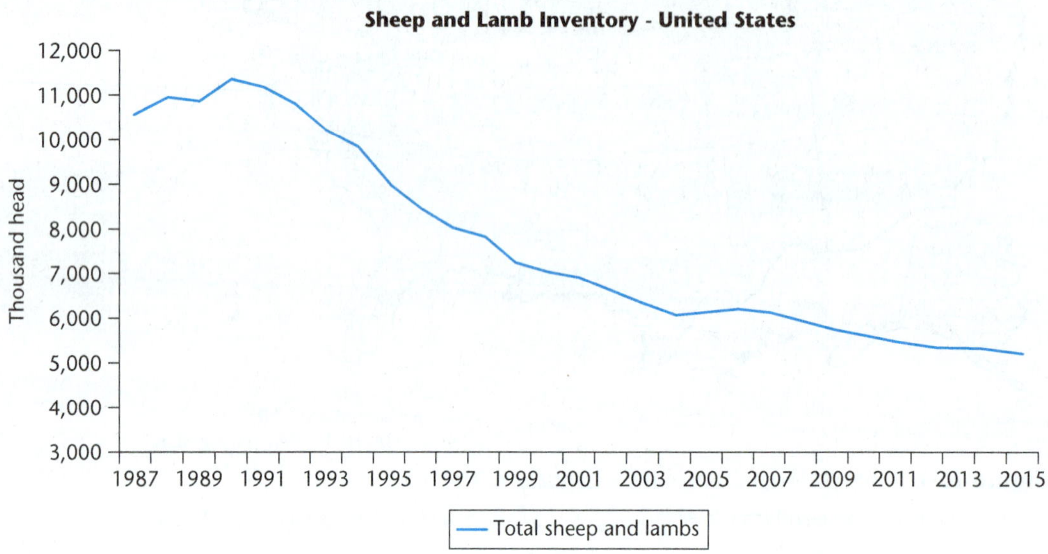

Figure 26-4 Since 1990, numbers of sheep and lambs have been steadily declining.
Source: USDA Overview of the United States Sheep and Goat Industry

Many producers in the western states are now using mixed grazing—that is, raising cattle and sheep on the same land. Mixed grazing is most prevalent in the Northern Plains and Texas. Two-thirds of the sheep producers in these states also raise cattle. The practice of mixed grazing provides more opportunity to increase livestock production than raising cattle or sheep alone.

Sheep make a good second enterprise on a farm because:

- Wool and lambs provide extra income.
- The initial costs are low.
- The enterprise does not require expensive housing or equipment.
- Sheep make use of pasture crops that might otherwise be wasted.
- Sheep can be fed on roughages and small amounts of grain.
- Market lamb returns compete well with other meat animal enterprises.
- Sheep can improve the pasture mix.
- Lamb, wool, and beef prices fluctuate independently.

There are some major disadvantages related to the sheep enterprise:

- Predatory animals such as coyotes and dogs prey on sheep.
- Sheep are susceptible to internal and external parasites.
- There are increased labor and management requirements.
- Wool prices are quite variable.

Figure 26-5 Wool samples. Long-hair wool on the left and short-hair wool on the right.

CLASSES OF SHEEP

There are a number of ways to classify sheep. The most commonly used classification is by type of wool (Figure 26-5). The wool type classifications are (1) fine wool, (2) medium wool, (3) long wool, (4) crossbred wool, (5) carpet wool, (6) fur sheep, and (7) hair sheep.

Fine Wool Breeds

The most common of the fine wool breeds are (1) Merino, (2) Rambouillet, and (3) Debouillet. These breeds were all developed from the Spanish Merino. They produce a fine wool fiber that has a heavy yolk, or oil content. Originally, these breeds did not produce good meat carcasses. However, through selection and breeding, the quality of the carcass has improved. Today, these breeds are still used primarily for wool production. They possess a strong flocking instinct, which allows herders in range areas to effectively look after and move large numbers of sheep. A high percentage of the sheep in the range areas are of the fine wool breeds. They have the ability to do well on poor-quality rangeland. Fine wool breeds will breed out of season and, thus, can produce lambs in the fall months.

Medium Wool Breeds

The most common medium wool breeds are (1) Cheviot, (2) Dorset, (3) Finnish Landrace, (4) Hampshire, (5) Montadale, (6) Oxford, (7) Shropshire, (8) Southdown, (9) Suffolk, and (10) Tunis. Medium wool breeds were originally bred mainly for meat, with wool production secondary. The fleece is medium in both fineness and length, although these breeds are typically raised mainly for meat production. Medium wool breeds are popular in both range and farm flock production.

Long Wool Breeds

The common long wool breeds are (1) Cotswold, (2) Leicester, (3) Lincoln, and (4) Romney. These breeds were developed in England and are larger than the other breeds of sheep. They produce long, coarse-fiber wool. They are hardy and prolific but tend to be late maturing. Depending on slaughter weight, the carcass quality may be poor, carrying too much fat. Excess fat may not be present at slaughter weights of less than 120 pounds. These breeds are used mainly in crossbreeding programs.

Crossbred Wool Breeds

Common crossbred wool breeds are (1) Columbia, (2) Corriedale, (3) Polypay, and (4) Targhee. Most crossbred breeds were developed by crossing long wool with fine wool breeds. Some of the newer breeds were developed by agricultural experiment stations involving different types of crosses.

Crossbred wool breeds were developed mainly to improve the carcass quality and the length of the wool fiber. These breeds have better banding (flocking) instinct than the long wool and medium wool breeds. **Banding** (or flocking) **instinct** refers to the tendency of sheep to stay together in a fairly tight group called a **band** or flock. This makes the flock easy to herd in large range areas. They are well adapted to the western range but are also popular in the Corn Belt states.

Carpet Wool Breed

There are several breeds of sheep used in countries outside of the United States for the purpose of producing carpet wool. The only breed used in the United States for this purpose is the Black-Faced Highland. The fleece is coarse, wiry, and tough. The length of the fiber varies from 1 to 13 inches.

Fur Sheep Breed

The only breed of sheep raised mainly for fur in the United States is the Karakul. The fur pelts are taken from the young lambs to make coats. The meat value of the breed is low.

Hair Sheep Breed

The hair sheep breeds include the (1) Barbados Blackbelly, (2) Dorper, (3) Katahdin, (4) Romanov, and (5) St. Croix. The breeds have coats of hair, not wool. The coat may thicken in the winter but will typically shed in warmer months. These sheep are primarily from the tropics and have generated a great deal of interest as a result of the decline in wool prices, the elimination of the need for shearing, and the breed's parasite resistance.

BREEDS OF SHEEP

Figure 26-6A Polled Merino ram.

Courtesy of The American and Delaine Merino Record Association

There are many breeds of sheep available to producers in the United States. Many of the breed associations register only small numbers with their purebred associations in any given year. In order of rank, Suffolk, Hampshire, Dorset, Dorper, Southdown, and Montdale have been the leading sheep breeds in the United States based on average numbers registered.

Fine Wool Breeds

Merino

Merino sheep originated in Spain and were first imported into the United States in 1793. There are at least 10 types of Merino sheep in the world. However, in the United States, the Delaine Merino sheep is the most common type. Delaine Merino sheep are practical for farms and are well adapted to range sheep production in western and southwestern states.

Figure 26-6B Horned Merino ram.

Courtesy of The American and Delaine Merino Record Association

The fleece of the Delaine Merino is white and grows about 2.5 to 4 inches per year and must be 21.5 microns or finer (spinning count 64 and higher). The rams may either be horned or polled and the ewes are polled (Figures 26-6A and 26-6B). Merinos are medium in size and have angular bodies. The Delaine Merino is considered to be the largest of the Merinos. Merinos have a strong banding instinct and are able to do well on poor grazing land and in all types of climate. Merino ewes are exceptional mothers, and their traits are inherited uniformly to offspring. Merino wool demands high prices around the world and, therefore, is called the "Golden Fleece."

Rambouillet

The Rambouillet originated in France, and was developed from the Spanish Merino. The first importations into the United States took place about 1840. About one-half or more of the crossbred sheep in the United States carry some Rambouillet blood. The breed is particularly popular in the western states and is the most popular of the fine wool breeds.

Rambouillets are white in color. They are large and have an angular, blocky body type. They produce a meatier carcass than the Merino, but it is not as good as the carcass of the breeds bred for meat production. There are both horned and polled rams. However, all ewes are polled. Their fleece grows about 3.5 inches per year.

Debouillet

The Debouillet breed originated in the United States. It was developed from crosses of Rambouillet and Delaine Merino. Development began in New Mexico. The Debouillet is adapted to the western range conditions.

The Debouillet is medium sized and has an angular body type. These animals are white faced and have wool on the legs. Rams may be horned or polled; ewes are polled. The body is smooth, and long-staple wool of 3 to 5 inches is produced.

Staple means the fibers of wool. Long-staple wool is the most valuable wool from the fleece.

Medium Wool Breeds

Cheviot

The Cheviot originated in the Cheviot Hills of northern England and southern Scotland in the 1800s. The first importations into the United States took place about 1838.

The Cheviot is small and has a blocky body type with a white face and legs and black nostrils. It carries its head erect and presents an alert appearance. The breed is polled. The fleece grows about 4 to 5 inches per year. The breed does not have a strong flocking instinct.

Dorset

The Dorset breed originated in southern England in the early 1800s. The first importations into the United States took place about 1887.

The Dorset is of medium size and has a blocky body type (Figure 26-7). The ears, nose, face, and legs are white. There are both polled and horned strains of Dorset. They produce a medium-coarse and light-weight fleece averaging 5 to 6 pounds. They produce a muscular carcass. The ewes will breed out of season, so fall lambs can be produced.

Sheep Breeder Magazine

Figure 26-7 Dorset ram.

Finnish Landrace (Finnsheep)

The Finnsheep originated in Finland and was first imported into the United States in 1968. The Finnsheep is small in size. The ears, nose, face, and legs are white (Figure 26-8). The breed is generally polled, although some rams may have horns. The Finnsheep produces a medium-coarse fleece that averages 4 to 5 pounds in weight. The carcass characteristics of the Finnsheep are not as desirable as those possessed by some of the other breeds of sheep.

The Finnsheep is noted for its high lambing rate. Its principal use in the United States is in crossbreeding programs to increase lambing rate. Mature crossbred (one-half Finnsheep) ewes will commonly produce a 200 to 250 percent lamb crop—that is, 200 to 250 lambs are produced for every 100 ewes. Finnsheep crossbred lambs are small at birth but have a high survival rate. When Finnsheep are used in crossbreeding, the lambs produce acceptable carcasses.

Photo by Fred Ward, USDA

Figure 26-8 Finnsheep.

Hampshire

The Hampshire breed originated in southern England during the late 1700s and early 1800s. It was first imported into the United States prior to 1840. All of the flocks from these importations disappeared during the Civil War. Later importations occurred around 1881.

Hampshires are large in size and have a blocky body type (Figure 26-9). The face, legs, ears, and nose are black with a wool cap on the poll. The breed is polled. Hampshires are medium in maturation rate, are good milkers, and produce lambs that are often ready for market at weaning. They produce a fleece of about 6 to 7 pounds of medium-fine wool. The lamb crop averages about 162 percent. Hampshires are one of the most popular of the medium wool breeds in the United States, especially in the Midwest. They cross well with fine wool or crossbred breeds to produce market lambs.

Courtesy of The American Hampshire Sheep Association

Figure 26-9 Hampshire ram. The 2006 National Champion.

Montadale

The Montadale breed originated in the United States in 1933. Mr. E. H. Mattingly, of St. Louis, Missouri, developed the breed by crossing Columbia ewes with Cheviot

Courtesy of Montadale Sheep Breeders Association

Figure 26-10 Montadale ram.

rams. The Montadale is a medium-large breed with a blocky body type (Figure 26-10). The face, ears, and legs are white, and there is no wool on the face and legs. The Montadale breed is polled and produces a good meat carcass. The fleece averages 10 to 12 pounds.

Oxford

The Oxford breed originated in south-central England. The breed was developed about 1836 from crosses of Hampshire, Cotswold, and Southdown sheep. The first importations into the United States occurred about 1846.

The Oxford is one of the largest of the medium wool breeds of sheep. It has a blocky body type. The face, ears, and legs are gray to brown. The breed is polled. Wool extends over the poll down to the eyes. The Oxford shears a heavy fleece, averaging 10 to 12 pounds. The ewes are prolific and are good milkers; the lambs grow quickly. The breed is mainly used in crossbreeding because of its large size.

Shropshire

The Shropshire breed originated in England around 1860. The first importations into the United States were about 1885. The Shropshire is one of the smaller of the medium wool breeds. It has a blocky body type and is polled. The face and legs are dark colored.

It has a heavy face covering of wool. Shropshires produce a small carcass. They become too fat when fed to heavy weights. The fleece shears about 9 pounds. The ewes are good mothers, and the lamb crop is often 150 percent. However, because of their small size, these sheep have lost popularity in the United States.

Southdown

The Southdown breed originated in southern England in the late 1700s. It is one of the oldest sheep breeds in the world and was used in the formation of most of the medium wool breeds. The first importations into the United States took place about 1803.

Courtesy of American Southdown Breeders Association, Fredonia, TX

Figure 26-11 Southdown.

The Southdowns are a moderate size sheep that are of gentle disposition and economical to feed and maintain (Figure 26-11). The face, ears, and legs are a gray or mouse brown. The breed is early maturing and polled. The ewes weigh 160 to 200 pounds when mature and are excellent mothers with a high number of multiple births. Rams weigh 225 to 250 pounds when mature and sire muscular, efficient lambs that yield a high-quality carcass. They shear a fleece of 5 to 7 pounds. Southdowns tend to become overfat at an early age. The breed is also very adaptable to all climates and environments.

Suffolk

The Suffolk breed originated in southern England and was developed in the early 1800s. The first importations into the United States took place about 1888.

Courtesy of American Southdown Breeders Association, Fredonia, TX

Figure 26-12 Suffolk ram.

The Suffolk is large and has a blocky, muscular body type (Figure 26-12). The face, ears, and legs are black. The breed is polled and has no wool on the head or legs. The Suffolk shears a fleece of 5 to 6 pounds. The lamb crop is often 150 percent or better. The lambs grow rapidly; deposit fat at a slower rate than other breeds; and produce lean, muscular carcasses with desirable yield grades. The Suffolk has gained popularity in the United States, especially in the production of market lambs.

Tunis

The Tunis breed, which originated in North Africa, is an ancient breed of sheep. It was imported into the United States about 1799. The breed is polled, and the color of the face is reddish brown to bright tan (Figure 26-13). The Tunis is of medium size and has an angular, blocky body. The wool is rather coarse, and there is none on the face.

Dale Huhnke of Hamlet, Indiana

Figure 26-13 Tunis sheep.

Long Wool Breeds

Cotswold

The Cotswold is an old breed of sheep that originated in England. Improvement of the breed occurred during the late 1700s. The first importations into the United States took place about 1832.

The Cotswold is large and has a blocky body type. The face and legs are white, although a grayish-white color is accepted. Dark spots may occur on the face and legs. The wool is coarse and grows in long, wavy curls. The Cotswold is polled and has a tuft of wool on the forehead. The Cotswold has been used in crossbreeding programs; however, it is not a popular breed in the United States.

Leicester

The Leicester originated in England. It was developed and imported into the United States in the late 1700s (Figure 26-14). The Leicester is large and has a blocky body type. The breed is white, polled, and has no wool on the face. The fleece does not give Leicester sheep good protection from the weather. They tend to become wet and chilled and are only moderately hardy. This breed is not a popular breed in the United States.

Photo by Ken Hammond, USDA

Figure 26-14 Leicester sheep.

Lincoln

The Lincoln originated in England. It is an old breed and was first imported into the United States in the late 1700s. The Lincoln is large and has a blocky body type. It is white, polled, and produces a long, coarse fleece. It is a late-maturing breed. The Lincoln produces 12 to 16 pounds of wool. This breed has not been popular in the United States and has been used mainly for crossbreeding.

American Romney Breeders Association

Figure 26-15 Romney sheep.

Romney

The Romney is also an old breed that originated in southern England. The first importations into the United States took place around 1904. The Romney is a large, hardy breed with a blocky body type that has been used in crossbreeding programs (Figure 26-15). The breed is white, polled, and produces a more compact, finer fleece than the other long wool breeds. The Romney is better adapted to wet, marshy areas than other long wool breeds of sheep.

Jay Daniel (Berry College)

Figure 26-16A Columbia ewe.

Crossbred Wool Breeds

Columbia

The Columbia originated in the United States in 1912. The breed was developed from a cross of Lincoln rams and Rambouillet ewes. The Bureau of Animal Industry did the primary work in developing the breed in Wyoming and Idaho. The purpose of the cross was to produce a breed that would be better adapted to the intermountain regions of the West.

The Columbia is a large, blocky breed and is the largest of the crossbred wool breeds (Figures 26-16A and 26-16B). The face, ears, and legs are white, and the breed is polled. There is no wool on the face. The Columbia is slightly longer legged than other breeds. It shears a 9- to 13-pound fleece, and produces a lean market lamb with acceptable leg scores. The Columbia is often used in the Midwest in crossbreeding with black-faced medium wool breeds to produce market lambs.

Columbia Sheep Breeders Association of America

Figure 26-16B Columbia ram.

Corriedale

The Corriedale originated in New Zealand and was developed between 1880 and 1910. It was a result of crossing Lincoln and Leicester rams with Merino ewes. The first importations into the United States took place in 1914. The animals were used in western range conditions.

The Corriedale is a medium-large, blocky breed. The face, ears, and legs are white, and the breed is polled. It produces good-quality wool and an acceptable carcass. Corriedales shear a 9- to 15-pound fleece.

Polypay

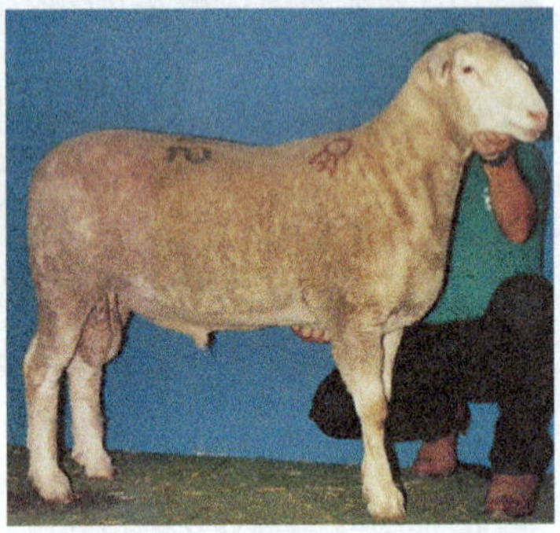

Sheep Breeder Magazine

Figure 26-17 Polypay ram.

The Polypay is a polled breed that was developed in the 1970s at the United States Department of Agriculture (USDA) sheep research station in Idaho (Figure 26-17). It was developed by crossing Finn, Dorset, Targhee, and Rambouillet sheep. The Polypay was developed with the goal of producing sheep that could produce large lamb crops, breed out of season, and produce a good fleece.

Targhee

The Targhee originated in the United States in 1927. It was developed in Idaho by the USDA from a cross of Rambouillet rams with Corriedale-Lincoln-Rambouillet ewes. A program of interbreeding was then followed based on production performance on the range. The Targhee is a medium-large, blocky breed with a white face. The breed is polled and has no wool on the face. It will shear about 10 to 12 pounds of wool.

Carpet Wool Breed

Black-Faced Highland

The Black-Faced Highland is an old breed, originating in Scotland. The first importations into the United States took place about 1861.

The breed is small in size. These sheep have a long, coarse outer coat of wool, with a finer undercoat. The length of the fiber varies from 1 to 13 inches. The face and legs are black. The rams have large, spirally curved horns, while the ewes have small, short, curved horns. The breed can live in rough land areas with little vegetation. The Black-Faced Highland breed is not common in the United States.

Fur Sheep Breed

Karakul

The Karakul originated in Asia and is an ancient breed. The first importations into the United States took place about 1909. The Karakul is a large, angular-bodied breed. The color of the face, ears, and legs is black or brown. The rams have horns, and the ewes are polled. The lamb pelts of this sheep at birth, and for a few days thereafter, are used as furs. The wool on the mature animal is 6 to 10 inches long. It is coarse, wiry, and of little value. The lambs produce poor carcasses. The Karakul has never been a popular breed in the United States.

Hair Sheep Breeds

Photo courtesy of the Barbados Blackbelly Sheep Association International, blackbellysheep.org

Figure 26-18 Barbados Blackbelly sheep.

Barbados Blackbelly

The Barbados Blackbelly originated on the island of Barbados in the Caribbean. The breed likely originated from crosses of African hair sheep with wool sheep from Europe. The Barbados Blackbelly sheep have a medium- to thick-hair coat that sheds in the spring. The coat can be brown, tan, or yellow, with black underparts and black points on the nose, forehead, and inside the ears (Figure 26-18). Rams have a thick covering of hair on the neck down to the brisket and may extend to the shoulder. Both rams and ewes are polled. Barbados Blackbelly sheep have an average lambing rate that ranges from 150 to 230 percent lamb crop. Lambs produce a light carcass. Barbados Blackbelly sheep have increased resistance to internal parasites and are more heat tolerant, which makes them popular in the southwestern United States.

USDA Agricultural Research Service; Photo by Kreg Leymaster

Figure 26-19 Dorper ram.

Dorper

The Dorper was developed in South Africa in the 1930s for arid regions (Figure 26-19). Dorper sheep have a white body with a white or black head. The coat is a mixture of hair and wool, which the sheep will shed in the spring. The breed is polled. Lambs are heavy muscled and grow rapidly. Lamb crop percentages range from 150 to 180 percent. Dorper sheep have the ability to breed out of season and produce fall lambs. In recent years, there has been a lot of interest in Dorper sheep in the United States.

Jay Daniel (Berry College)

Figure 26-20 Katahdin ram.

Katahdin

The Katahdin sheep (Figure 26-20) were developed in north central Maine by Michael Piel through crossing African hair sheep with various wool breeds. Katahdin sheep have an outer hair coat and wool undercoat that can be any color. The undercoat will thicken in the winter and shed in the spring. Lambs produce a well-muscled, lean carcass at 95 to 115 pounds live weight. Katahdin sheep have the ability to breed out of season and produce fall lambs. Lamb crop percentages are typically 200 percent. Katahdin sheep have increased resistance to internal parasites and have become very popular in the United States in recent years.

Romanov

The Romanov originated from the Volga Valley close to Moscow, Russia. Romanov sheep have a black head with white on the poll, black legs, and a mixture of wool and coarse black hair on the body. Although the Romanov is considered a hair sheep, shearing is still required. Romanov ewes have very high lamb crop percentages; quadruplets, quintuplets, and sextuplets are common. Seven live, healthy lambs in one litter is the North American record for a Romanov ewe. Half Romanov ewes will produce a 250 to 300 percent lamb crop. Lambs are also typically very vigorous, but produce light-muscled carcasses.

St. Croix

The St. Croix breed originated in the United States and British Virgin Islands. St. Croix have a solid white coat, with a mixture of hair and wool that sheds in the spring. Lamb crop percentages range from 150 to 200 percent. They have increased resistance to internal parasites.

SELECTION OF SHEEP

National Sheep Improvement Program

The **National Sheep Improvement Program (NSIP)** is a computerized, performance-based program for selection of sheep. The NSIP calculates expected progeny differences, or EPDs, for sheep of different breeds. The breeds for which the NSIP calculates EPDs include Columbia, Dorper, Dorset, Finsheep, Hampshire, Katahdin, Lincoln, Merino, Oxford, Polypay, Rambouillet, Shropshire, Suffolk, and Targhee. Producers of these breeds who choose to participate submit data and pay a fee based on the number of breeding animals. The NSIP then sends the data to Virginia Tech University to calculate expected progeny differences (EPDs). The EPDs are then included in breed-specific sire summaries and allow for comparison of rams and ewes within a breed. The traits for which EPDs are calculated vary some with breed but generally include maternal traits (number of lambs born per ewe lambing, maternal milk, and maternal growth), growth traits (weaning weight, post-weaning weight, and yearling weight), and wool traits (grease fleece weight, fiber diameter, and fiber length). Scientists are working to add carcass traits (fat thickness, ribeye area, and carcass value) and accelerated traits (date of first lambing and lambing interval) to the traits that have EPDs available for NSIP.

Ram Testing Stations

Ram testing stations provide information that allows the comparison of ram lambs from several different flocks. Producers looking for breeding stock can purchase performance-tested rams from ram testing stations.

Ram lambs are usually brought to a station when they are 7 to 12 weeks of age. They are tested for 8 to 12 weeks. They are weighed at regular intervals, usually every 2 or 3 weeks, and the average daily gain is determined. The gain of each ram lamb is compared to the average for its breed group. Feed efficiency data is collected at some stations for ram lambs from the same sire.

Ram lambs are evaluated for soundness at the end of the test period. Some stations also provide scrotal circumference measurements, fleece information, and ultrasound fat and loin eye measurements on the test lambs.

Genetic Testing

Various companies offer genetic testing to determine the specific genes an animal has for specific traits. The tests use a blood sample to determine the genes. Two traits that are commonly tested for include Spider Lamb Syndrome and scrapie susceptibility.

The spider genetic test determines if animals carry the spider gene. Spider Lamb Syndrome is a genetic disease that results in abnormally long limbs and death of the lamb either at birth or shortly after birth. Lambs must have two copies of the spider gene for the disease to be present. Sheep that have one copy of the spider gene appear normal and are called "carriers." The spider genetic test allows producers to determine if the sheep is a carrier. Carrier rams and ewes should not be used for breeding.

Scrapie is a disease of the central nervous system and is always fatal. More information on scrapie is available in the chapter on diseases and parasites in sheep and goats. Susceptibility of sheep to scrapie is influenced by variation in a gene called the prion gene. Two sections of this gene, called codons, are thought to be important in susceptibility to scrapie in the United States. These two codons are called codon 136 and codon 171. Sheep can have an A or V codon at 136 and a Q or R codon at 171. These codons only occur in certain combinations. Sheep that are AA RR are very resistant to scrapie, sheep that are AA QR are rarely susceptible, sheep that are AV QR are somewhat susceptible, and sheep that are QQ (AA QQ, AV QQ, or VV QQ) are susceptible to scrapie. Only genetic testing can determine which codons the sheep have.

Breed Selection

Personal preference is the main factor in selecting a breed of sheep. Other factors that should be considered are:

- how well the breed is adapted to the area
- the market for the product
- the availability of breeding stock
- the possibility of multiple births

Most of the breeds in the western areas carry a high percent of fine wool breeding. This is of value because of the banding instinct of these breeds. Sheep with a great deal of wool on the face may have more problems with wool blindness in areas where ice, snow, and some types of grasses are grown. Breeds that are more open faced do not have these problems. Some breeds are better suited to live and grow in range areas where the vegetation is not as lush and the ground is rougher.

A purebred breeder must consider the demand for the kind of breed that is produced. Some breeds produce meatier, leaner carcasses than others. If the main market is for meat, then breeds that produce a meaty carcass should be chosen. The market for the kind of wool produced varies from area to area. The breed selected must produce the kind of wool in demand.

The producer must consider what kind of breeding stock is available for crossbreeding programs. This depends on the popular purebred breeds in the area and the traits needed in the crossbreeding program. Costs increase if the producer must travel great distances to find the required breed.

Selecting Native or Western Ewes

Ewes that are produced in the western range area are referred to as **western ewes**. Those produced in other parts of the United States are called **native ewes**. Western ewes have a high percentage of fine wool breeding, while native ewes are generally of the medium, long, or crossbred wool types.

In the range area, western ewes are generally the desired type to buy. In the farm flock area, either type may be used. Native ewes generally have a larger lamb crop and produce a more muscular, leaner carcass. Costs are usually less, and these ewes are often better adapted to local conditions. Western ewes are less likely to have parasites. They are often more uniform in size and are generally hardier and have longer productive lives.

Selecting Breeding Sheep

All animals selected for the breeding flock must be healthy. Do not select sheep that have diseases or parasites. Some indications of poor health include dark, blue skins, paleness in the lining of the nose and eyelids, lameness, and lack of vigor.

Ewes

Check the udder for softness and soundness of the teats. Teeth, feet, legs, eyes, and breathing should all be normal. The incisor teeth should meet the dental pad.

The fleece should be heavy. Open-face ewes produce about 15 percent more wool than ewes with heavy wool covering on the face. Select ewes with uniform fleece and a reasonably long fiber length. A dense, compact fleece with bright luster is desired, and the yolk should be evenly distributed. Avoid open fleeces with dark-colored fibers. Large-framed ewes have less difficulty lambing. They are better milkers, are more prolific, and shear heavier fleeces.

Select uniform ewes of correct type. Correct type includes the following traits:

- generous length, depth, and width of body
- medium-short, thick neck
- wide, deep chest
- good spring of rib
- depth in the fore and rear flanks
- width and thickness through the loin
- strong, wide back; tight shoulder
- long, wide, and level over the rump
- straight legs with generous width between them

- strong pasterns, strong feet, and medium-sized toes
- feminine appearance
- purebred ewes show breed traits
- all parts of the body blend well together
- full, plump, muscular legs
- wide in the dock
- well-muscled, firm body

Well-grown yearling ewes are a better choice than older ewes. Medium, long, and crossbred wool breeds begin to decrease in productivity at 6 years of age. Fine wool breeds produce for about 1 year longer. Older ewes are worth less because they have fewer productive years remaining.

Up to the age of 4, the front teeth of the lower jaw are a good indication of age. Lambs have small, narrow teeth. At 1 year of age, the two center incisors are replaced by two broad, large permanent teeth. Each year thereafter, another set of permanent teeth appears until, at 4 years, the sheep has all its permanent teeth. At this age, the teeth begin to wear down. The sheep begins to lose some of its teeth at about 5 or 6 years of age. This condition is called *broken mouth.*

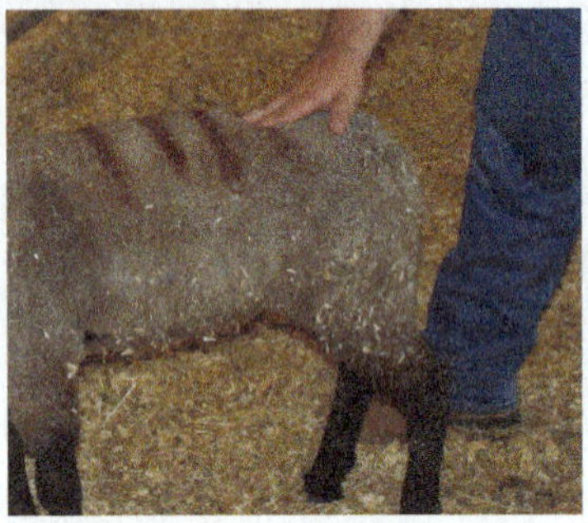

Courtesy of Frank B. Flanders

Figure 26-21 With the fingers close together and slightly cupped, check for finish over the back. Work from the rear to the front.

It is necessary to handle sheep to determine the exact conformation and muscling that is present because the wool hides some of these traits from visual inspection. When handling sheep, keep the fingers close together and slightly cupped, working from rear to front pressing firmly with the finger tips. Check for finish over the back and feel the shoulders for smoothness. Check length of loin, width and depth of loin, width of rump, and muscling in the leg. The length and quality of the fleece is also checked by handling. Figures 26-21 through 26-25 illustrate the handling of sheep to determine conformation and muscling.

Rams

Great care should be exercised in selecting the ram. Select a ram that will bring traits into the flock that are lacking in the ewes. The ram should be large, rugged, muscular, masculine, and have plenty of bone. If possible, determine the growth rate of the ram by checking the 90- to 120-day weight. Fast growth rate is 30 percent heritable and is a trait that is passed on to the lambs. Check for two well-developed, pliable testicles. The best indication of potential fertility in a young ram is the circumference of the scrotum. When the ram reaches puberty at 5 to 7 months of age, the average scrotal circumference is 12 to 13 inches. A ram with below-average scrotal circumference may be late maturing or have low fertility. Because these traits are inherited, do not select such rams for breeding stock. Select rams with above-average scrotal circumference at puberty. A good-quality ram that costs more to start with will usually be a better investment in the long

Courtesy of Frank B. Flanders

Figure 26-22 Check the length of the loin.

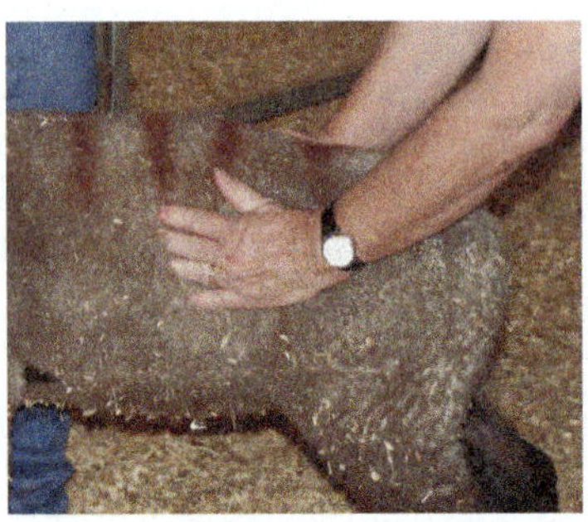

Courtesy of Frank B. Flanders

Figure 26-23 Check the finish over the ribs.

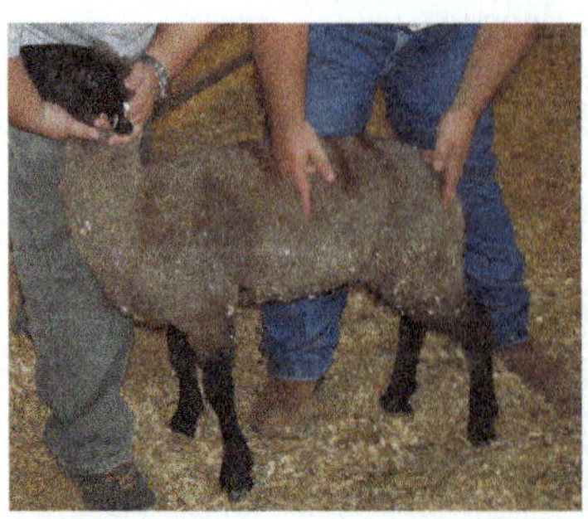

Courtesy of Frank B. Flanders

Figure 26-24 Check the width of the rump.

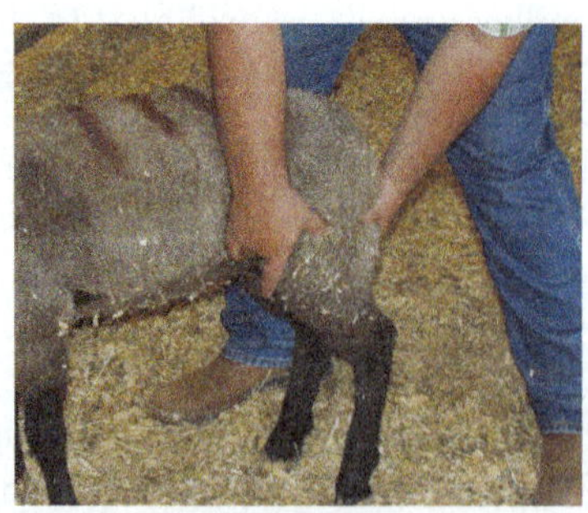

Courtesy of Frank B. Flanders

Figure 26-25 Check the muscling in the leg area.

run. Poor-quality, low performance rams will not improve the quality of the flock. Purchase rams with superior growth, muscle, soundness, and flesh quality.

JUDGING SHEEP

The general procedures for judging sheep are basically the same as those for judging beef cattle or hogs. However, sheep must be judged on two particular traits—meat and wool. Grooming sheep for the show ring helps accentuate the best qualities of the animal (Figure 26-26).

A judge must be familiar with the parts of the sheep. Figure 26-27 shows the parts of the sheep.

Courtesy of Frank B. Flanders

Figure 26-26 Students shearing and grooming their sheep before a show.

Market Lambs

Market lambs are judged on the basis of type, muscling, finish, carcass merit, yield, quality, balance, style, soundness, and smoothness. A number of these traits are described earlier in this chapter in the discussion of sheep selection.

Type is the general build of the sheep. Muscling is shown by the thickness and firmness of the leg, and the firmness of muscle over the top and in the shoulders (Figure 26-28). Meaty, heavily muscled lambs are more desirable than those that are overly fat or lightly muscled.

Excessive fat is undesirable. The back fat should measure 0.15 to 0.20 inch at the 12th rib. Finish is found by handling. Too much finish is shown by a soft, mellow touch. If the backbone and ribs can be felt too readily, the lamb is too thin. The lamb must produce a thick, meaty, correctly finished carcass.

The **pelt** consists of the skin and fleece. A pelt of 3/8 inch is desirable for show lambs. A lamb will have a high dressing percentage if it is heavily muscled and trim through the middle of the body with a 3/8-inch pelt. Too heavy a pelt will decrease dressing percentage. However, because the pelt is valuable, market lambs are more valuable with approximately 1 inch of fleece.

Quality, balance, style, and smoothness are defined and discussed in the section on judging beef cattle. The same definitions apply to sheep.

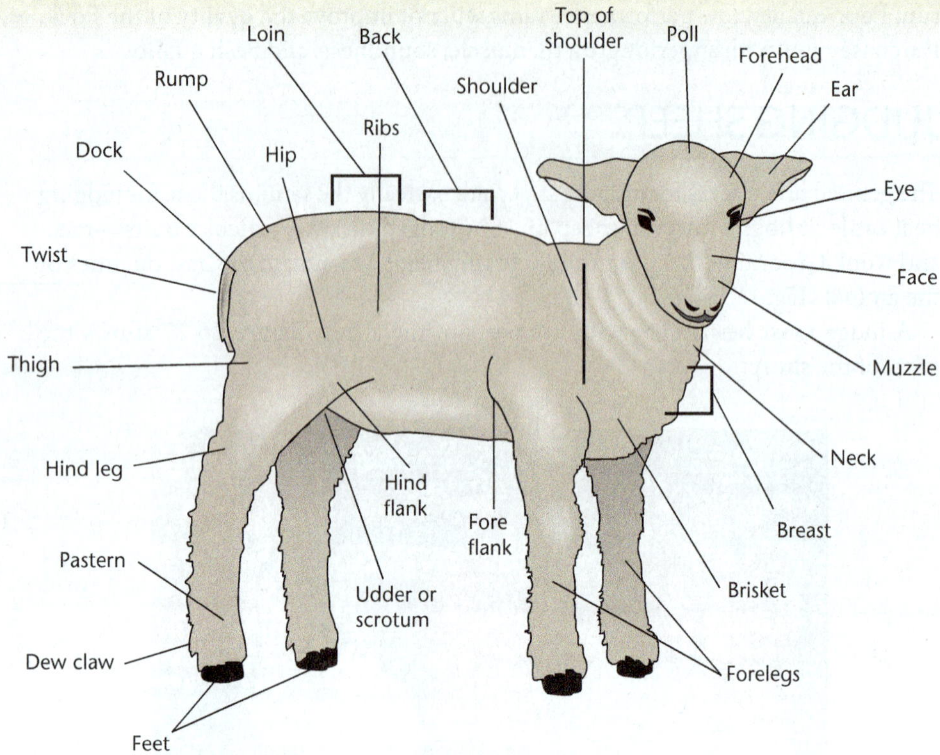

Figure 26-27 Parts of a sheep.

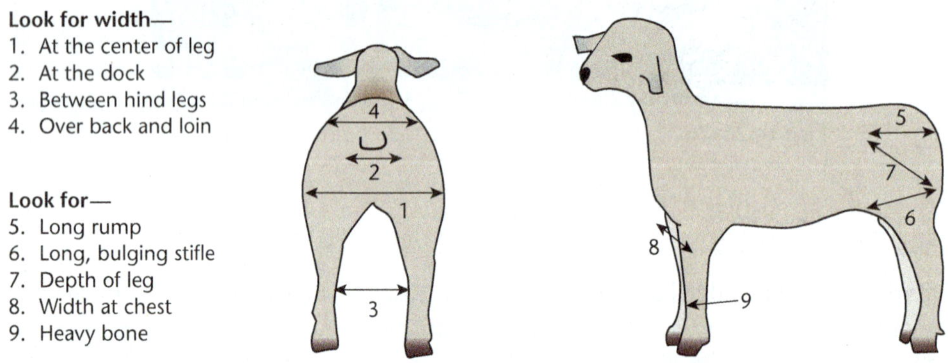

Figure 26-28 Rear and side view of a meaty lamb and what to look for when judging.

Breeding Sheep

The general traits to judge breeding sheep are discussed earlier in this chapter in the section on selection of breeding sheep. Breeding sheep are also judged on condition (finish in market lambs), size, feet, legs, bone, breed and sex character, and fleece.

Notes should be taken and oral reasons given, just as they are with beef and swine. The judging terms used are similar to those for beef cattle, only using the correct names for the parts of the sheep.

SUMMARY

Sheep are a major farm enterprise in the western range area of the United States. In the rest of the United States, farm flocks tend to be a secondary enterprise.

One way to classify sheep is by the type of wool. These types are fine wool, long wool, crossbred wool, medium wool, carpet wool, fur, and hair sheep. Fine wool breeding is more common in the western range areas. The medium wool and crossbred wool breeds are more common in the farm flock area. Long wool, carpet wool, and fur sheep are not common in the United States; however, hair sheep numbers are increasing.

Breed selection is mainly a matter of personal preference. However, adaptation of the breed to the area and purpose should also be considered. Sheep are selected on the basis of health, soundness, fleece, conformation, and age. It is necessary to handle sheep when selecting or judging to determine some of the traits present. Market lambs are judged on the additional trait of finish.

Student Learning Activities

1. Prepare a bulletin board display of pictures of the breeds of sheep.
2. Take a field trip to observe traits of the breeds found on local farms.
3. Survey the local area and prepare a list of purebred breeders that could serve as a source of breeding stock.
4. Name the parts of the sheep using a picture or live sheep.
5. Give an oral report on breeds, selection, or judging sheep.
6. Attend a purebred sheep sale and select sheep to buy using accepted criteria.
7. Practice good selection procedures when planning and conducting the selection of sheep for a supervised agricultural experience program.

Discussion Questions

1. What are the two types of sheep production in the United States, and where is each located?
2. Briefly describe the characteristics of each type of sheep production.
3. Discuss the advantages and disadvantages of sheep production.
4. Name the seven wool type classifications of sheep, list some typical breeds in each, and briefly describe each classification.
5. List and describe the breeds of sheep.
6. What factors should a producer consider when selecting a breed of sheep?
7. Compare native and western ewes.
8. Describe the characteristics that are considered desirable for breeding sheep.
9. Why are well-grown yearling ewes a better choice than older ewes when selecting breeding sheep?
10. Describe how the age of a sheep can be estimated by examining the teeth.
11. Describe how a sheep should be handled to determine conformation and muscling.
12. Describe the characteristics of a desirable breeding ram.
13. Draw a sketch showing the parts of a sheep.
14. Describe the characteristics of a desirable market lamb.

Review Questions

True/False

1. Many producers in the western states are now using mixed grazing, that is, grazing sheep and cattle on the same land.
2. A common way to classify sheep is by the type of wool.
3. Breed selection is not a matter of personal preference.
4. California is the leading state in number of sheep and lambs on feed.
5. Coyotes are a major predator of sheep.

Multiple Choice

6. Range production accounts for approximately __________ percent of the total sheep production in the United States.
   a. 10 c. 50
   b. 30 d. 70
7. Sheep are a major farm enterprise in the __________ area of the United States.
   a. southern range c. eastern range
   b. western range d. northern range
8. __________ is the general build of the sheep.
   a. Style c. Type
   b. Quality d. Balance
9. What is the only breed of sheep raised for fur in the United States?
   a. Karakul c. Merino
   b. Dorset d. Cheviot
10. The most commonly used method of classifying sheep is by __________.
    a. carcass type c. type of wool
    b. point of origin d. the number of young produced

Completion

11. Up until the age of 4, the front teeth of the lower jaw are good indicators of __________.
12. The __________ consists of the skin and fleece.
13. __________ refers to the tendency of sheep to stay together in a group.
14. __________ means the fibers of wool.
15. __________ wool demands high prices around the world, and is called "Golden Fleece."

Short Answer

16. List the leading states in numbers of sheep and lambs on feed.
17. How should a ram for breeding be selected?
18. What is the status of small sheep enterprises in the United States?
19. List three long wool breeds of sheep.
20. Where did the Merino sheep originate?

Chapter 27

Feeding, Management, and Housing of Sheep

Key Terms

hothouse lambs
accelerated lambing
drench
tagging
wether
prolapsed uterus
extension hurdle

Objectives

After studying this chapter, the student should be able to

- describe four systems of raising sheep.
- formulate a sheep feeding program.
- discuss management practices for sheep.
- describe housing and equipment needs for sheep.

SYSTEMS OF RAISING SHEEP

Sheep production may be broadly divided into purebred and commercial producers. A further classification of systems may be made based on the time of lambing as follows: (1) fall lambs, (2) early spring lambs, (3) late spring lambs, and (4) accelerated lambing. Commercial producers may also buy feeder lambs to finish for market.

Purebred producers produce breeding stock for commercial flocks and for other purebred flocks. A purebred producer must have experience in sheep production. The major goal of the purebred producer is breed improvement by providing commercial herds with superior breeding stock. Good production records are necessary.

Commercial producers maintain sheep flocks to produce meat and wool. The size of the flocks varies from small farm flocks to the large flocks found in the western range areas. In the farm flock areas, sheep tend to be a secondary enterprise. In the range areas, sheep are often the major livestock enterprise.

Fall Lambs

Lambs born before December 25 are generally referred to as fall lambs (Figure 27-1). Breeds of sheep that will breed out of season are necessary for this system. Some possible breeds are Rambouillet, Merino, Dorset,

Figure 27-1 Ewe with a lamb. The lambing seasons are referred to as spring and fall lambing.

Corriedale, and Tunis. Some crosses of these breeds, as well as Finnsheep and Romanov, will also breed out of season.

Fall lambs are marketed from early spring to June. The price for market lambs is usually higher at this time than during the rest of the year. In some areas, lambs are sold at 50 to 90 days of age. They weigh 35 to 60 pounds and are called **hothouse lambs**. These lambs go on a specialty market. Another specialty market is that for Easter lambs. Easter lambs are sold at weights of 20 to 40 pounds. This market exists in some large cities.

Some advantages of fall lambing include:

- more favorable weather
- better use of equipment
- lower feed and labor requirements
- generally lower feed costs
- better prices for lambs
- lower lamb mortality rate
- fits into an accelerated lambing program

Some disadvantages of fall lambing include:

- more grain is needed
- it may be difficult to breed the ewes
- lambs may pick up parasites from fall pasture
- lambing may occur at a busy time of the year
- lambs ready for market too early may not bring as good a price as those sold later in the year
- lambs have lighter birth weights
- lambing percentage is lower

Early Spring Lambs

Early spring lambs are born in January and February. Breeding must begin about August 1, and lambs are marketed before the end of June. Creep feeding of lambs is often necessary to prepare them for market on time. Lamb prices are usually better at this time than they are later in the year. There are fewer problems with parasites in the spring months. Lambing takes place when there is a lighter workload on the farm.

It may be more difficult to get ewes to breed in August. Weather conditions in some areas are more severe at lambing time. This system requires more grain and hay and does not make as much use of pasture. Better housing is needed, and more labor is required to save lambs.

Late Spring Lambs

Late spring lambs are born in March, April, and May. Breeding occurs later in the fall than for early spring lambs. This results in more ewes being settled in a shorter period of time. The lambing season is, therefore, shorter. Most of the feed comes from pasture and roughage. Little grain is required for this system. Producers who do not want to finish out the lambs may put them on the market as stockers and feeders.

Late spring lambing has several disadvantages. Parasite problems are worse with this system. The price for market lambs is not as good at the times these lambs reach market. If pasture is in short supply, additional feed is required to finish the lambs. Care of the lambs may take labor away from other farm tasks that must be done in the late spring and summer.

Accelerated Lambing

Accelerated lambing is a system that produces three lamb crops in 2 years. Ewes that will breed out of season must be used. It is better to use older ewes for this system. The major purpose is to increase production and, therefore, income, without greatly increasing production costs. To be successful, the producer must be an experienced sheep raiser. Better than average management ability is needed.

More labor is required for accelerated lambing. Early weaning of lambs must be practiced. The lambs are finished in the feedlot. Additional feeding is needed for both ewes and lambs.

Synchronized breeding is the forcing of ewes into heat with hormone treatments during a short 3- to 7-day period. When the hormones are withdrawn, most of the ewes come in heat in 1 to 3 days. Synchronized breeding is often used with accelerated lambing programs.

Feeder Lambs

Some farmers prefer to buy weaned lambs and finish them for market. This eliminates the need for a breeding herd. Careful selection of high-quality, healthy lambs is essential for success. This type of production involves a certain amount of risk. If the feeder pays too much, and the market goes down, the chances for profit are reduced. Lambs may be finished in a drylot or grazed on pasture. Good feeding practices are necessary for success.

FEEDING OF SHEEP

Gestation Feeding

Pasture or other roughage is the basic feed for the ewe flock during the gestation period (Figure 27-2). High-quality hay, silage, or haylage may be used. The hay may be legume, grass, or a mixture of legume and grass. Corn, grass, or legume silages may also be used. Haylage may be legume or grass, or a mixture of the two. Silage must be chopped finer than it is for beef cattle. Stubble and stalk fields may be used for pasture. Native range pastures, permanent pastures, rotation, and temporary pastures may all be used.

USDA

Figure 27-2 Pasture or other roughage is the basic feed for gestating ewes.

Rotating pastures increases the amount of feed that is available to the ewes. Rotation also helps to break the internal parasite cycle. Pastures should be rotated every 2 to 3 weeks, depending on the kind and quality of pasture available. They should not be overstocked. Sheep graze very close to the ground and may kill the vegetation. The stocking rate depends on the kind and quality of pasture used.

Salt, mineral mixes, clean water, and shade must be provided on pasture. Moving the salt and mineral from place to place will result in more even grazing of the pasture. Supplemental pastures may be needed with native range pastures.

If the pasture is poor, some feeding of hay may be needed. One to 2 pounds of legume hay will generally meet the ewes' needs. Silage may be substituted for hay at the rate of two to three parts of silage for each one part of hay.

During the last 6 weeks of the gestation period, the ewe should be fed some concentrate mixture. Corn, grain, sorghum, oats, barley, and bran are often used in concentrate mixes. A protein supplement may be needed, especially if the pasture or hay is of poor quality. Soybean meal, linseed meal, or commercial protein supplements are used. Urea may be used in sheep rations, but no more than one-third of the protein should come from urea. Gestating ewes may be self-fed. This practice reduces labor and increases the use of lower-quality roughage. The roughage and concentrate mix is ground and placed in a self-feeder. The main problem is to keep the ewes from becoming too fat. Controlling access to the self-feeder limits the feed intake. The rations used should contain a high percent of roughage.

Lactation Feeding

The amount of grain fed to the ewe should be reduced for the first 10 days after lambing. A mix of equal parts of oats and bran with hay fed free choice may be used. At about 10 days old, the lambs require more milk. Increasing the grain ration at that time will stimulate the ewe's production of milk. However, be sure the ewe has access to water.

For about 2 months after lambing, the ewe requires additional nutrients to produce the milk to feed the lamb. The amount of concentrate in the ration must be increased during this time. Ewes that are not on pasture need about 1.5 to 2.0 pounds of grain daily. Four to 6 pounds of high-quality legume hay should be fed per head per day. If the ewes are on good pasture at lambing, they usually will not need additional grain. Ewes that are nursing twins need more grain than those nursing singles. Also, heavier ewes need more feed than those of lighter breeds.

If low-quality hay or grass hay is used, increase the grain. Some protein and mineral supplement will also be needed with poor-quality legume hay or grass hay. After 2 months, the ration may be reduced to the amounts fed during the last 6 weeks of gestation. About a week before weaning, reduce the ewe's feed and water. This helps to decrease her milk production so that there will be fewer problems when the lambs are weaned. It also forces the lambs to eat more creep feed.

Feeding the Ram

Before the breeding season, the ram should only need pasture. During the breeding season, feed 1.0 to 1.5 pounds of corn or other grain mix if the ram is thin. Ram lambs require additional grain. During the winter, 3.5 to 5.0 pounds of hay should be fed. Heavier rams require the larger amounts. One pound of concentrate mix is enough for any weight of ram. Do not allow the ram to gain too much weight.

Flushing

Flushing is the practice of feeding a ration for 10 to 14 days before breeding and 2 weeks after breeding. This practice causes the ewe to gain rapidly. Flushing has the potential to increase the lamb crop by 10 to 20 percent. If ewes are already fat before breeding, do not flush them, or the lambing percentage may be reduced. Flushing may be achieved by putting the ewes on a better-quality pasture. Corn or oats, or a mixture of the two, fed at the rate of 0.5 to 0.75 pound per head per day is effective.

Feeding Lambs to Weaning

The lamb must receive the colostrum as soon as possible after it is born. Colostrum is the first milk produced by the ewe. It contains antibodies that help protect the lamb from infections and contains energy, protein, vitamins, and minerals needed by the lamb.

Lambs should be creep fed during the nursing period. A creep feeder is an area that allows only the lambs, and not the ewes, to enter. Lambs will start to eat grain at about 10 to 14 days of age. Feed small amounts of grain and clean out the trough each day. Coarsely grind or crush the grain until the lambs are 6 weeks of age. Pelleted feed may also be used.

Corn, oats, grain sorghum, and barley are good grains to use. A high-quality legume hay from a second or third cutting should be available to the lambs at all times. The creep ration should contain from 14 to 16 percent crude protein. Early weaned lambs should receive 18 percent crude protein in the ration. The addition of molasses increases the palatability of the ration. Soybean meal, cottonseed meal, linseed meal, or commercial supplements should be included. Urea should not be used for creep feeding lambs; however, antibiotics may be included. Lambs on good pasture probably will not benefit greatly from creep feeding.

Lambs are commonly weaned at about 3 months of age. Early weaning is being practiced by a number of producers. Lambs for early weaning should be at least 9 weeks old and weigh 40 to 50 pounds. The milk production of the ewe declines after 4 weeks. Lambs given high-quality feeds in the creep will gain faster and be ready for market sooner. Early weaning reduces the chances of lambs becoming infested with internal parasites. Early weaned lambs use feed more efficiently, requiring less feed per pound of gain.

Feeding Lambs from Weaning to Market

There is little difference in feeding practices for early weaned, late weaned, or feeder lambs. Use high-quality feeds and change rations slowly. Consult your veterinarian to determine the proper vaccination protocol for enterotoxemia (overeating disease) when starting on full feed to protect lambs on high concentrate rations.

Market lambs may be fed out in a drylot (Figure 27-3). Lambs that are eating well and weigh more than 40 pounds should not be put on pasture because that slows the rate of gain. Later lambs weaned after July 1 should be drenched before being put on pasture. A **drench** is a large dose of medicine mixed with liquid and ingested by the animal. Drenching is done to control internal parasites. Several different medicines may be used.

Feeder lambs may be put in the feedlot, or they may be fed on pasture for a while and then finished in a drylot. Reduce stress for feeder lambs when they arrive on the farm. Allow them to rest; provide grass or legume-grass hay and water. Isolate all sick lambs. Spray or dip to eliminate parasites if necessary. Lightweight lambs make better use of pasture than heavier lambs.

Jay Daniel (Berry College)

Figure 27-3 Self-feeding market lambs.

Lambs fed on pasture are grazed until about December 1, weather permitting. A supplemental ration is then fed at the rate of 1 pound per head per day. This ration is made up of grain and hay balanced to meet the lamb's needs. Around January 1, the grain portion of the ration is increased and the hay reduced. If the lambs are put in a drylot for finishing, then a protein supplement is added to the ration. If the lambs are fed out on good pasture, the protein supplement is not needed.

Drylot rations use more grains and less roughage. Faster weight gains will result if the grains make up about 65 percent of the ration and the roughage about 35 percent. Lightweight lambs can use more roughage in the ration. The ration should be about 15 percent protein.

Corn, grain sorghum, barley, wheat, and oats are all popular grain feeds for lambs. Legume or grass-legume hays are commonly used. The kind of feeds and hay used depends on the availability in the area. In some areas, cottonseed hulls and peanut hulls are used for roughage. Peanut hay may also be used where it is available. Soybean meal, cottonseed meal, linseed meal, peanut meal, or urea can be used for a protein supplement. Urea must not be used for more than one-third of the total protein in the ration. It is better to limit the use of urea to the last 25 pounds of feeding before market.

Water, salt, and minerals are needed. Always provide clean, fresh water. Salt and minerals may be fed free choice. The calcium-phosphorus ratio should be about 2:1. The copper content of the trace mineral mix should not be too high because sheep are sensitive to copper.

Lambs may be self-fed or hand-fed. It takes less labor to self-feed. Equipment and processing costs are slightly higher when self-feeding. Self-fed rations can be controlled better if ingredients are similar in particle size. Ground ratios should be avoided because feed will be too dusty and feed consumption will be decreased. Hand feeding should be done at regular times. It is better to hand feed three times a day rather than twice. Feed only as much as the lamb will eat in 30 minutes. The concentrate-roughage ratio in the ration should be changed every 7 to 10 days as the lambs become heavier. By the end of the feeding period, the lambs should be receiving 90 percent concentrate and 10 percent roughage.

Feeding Orphan Lambs

Because some ewes refuse to claim their lambs, there are usually orphaned lambs in sheep flocks at lambing time. Ewes that have lost a lamb may be used to care for orphaned lambs. Blindfolding the ewe is sometimes an effective way to encourage a ewe to accept an orphaned lamb. The ewe may accept the lamb if she is put into a headgate and the lamb is allowed to nurse. Rubbing orphan lambs with the afterbirth from lost lambs will often induce a ewe to accept an orphan. If all else fails, feed the lamb with a bottle and nipple.

The lamb must receive colostrum to have a chance to live. It may be necessary to use colostrum from another ewe. Some producers freeze extra colostrum so that it is on hand when needed.

Cow's milk may be used, but it does vary somewhat in composition from ewe's milk. Add 1 tablespoon of corn oil per pint of cow's milk. Warm the milk to body temperature but do not boil it, and keep utensils clean. Feed 3 ounces every 3 hours. By the end of a week, feed what the lamb will take in 5 minutes, 4 times per day. This will be about 0.5 to 0.75 pint.

A number of good commercial milk replacers are on the market for use in feeding orphan lambs. Follow the directions provided on the package.

Milk dispensers may be used to reduce the labor required. These may be homemade. Commercial milk dispensers are available. Keeping all equipment clean is essential for feeding lambs.

Protect the lambs from drafts and cold weather. Provide dry, well-bedded pens where heat lamps may be used in cold weather.

Start the lambs on a creep feed as soon as possible. If the lambs are eating well, they can be weaned from the milk at 4 to 6 weeks of age. They should be able to be weaned at no later than 8 weeks.

Feeding Replacement Ewes

Ewe lambs that are to be kept for flock replacement need to be well fed to be adequately developed at breeding time. Roughage and grain may be used for the ration. Half or more of the ration may be roughage, depending on the size and condition of the ewe. Rations must be adjusted to the growth and condition of the individual flock.

OTHER MANAGEMENT PRACTICES

Breeding

The fine wool breeds and the Dorset are nonseasonal breeders. Some sheep of other breeds will also breed out of season. Generally, the medium wool, long wool, and crossbred wool breeds are seasonal breeders. These breeds normally breed in the fall of the year. The length of the day appears to be the main controlling factor. Rams of all breeds are less seasonal in breeding habit than ewes. The medium wool breed rams are more often affected by periods of temporary sterility. Rams do affect the number of multiple births. This is an inherited trait, so it can be selected for when a ram is being chosen.

Most producers breed large, well-grown ewe lambs to produce lambs at about 1 year of age. Ewe lambs born early in the lambing season are more likely to breed as lambs. This practice requires good feeding to produce the growth and earlier sexual maturity that are necessary. Lambs do not have as long a breeding season as older ewes. There are breed differences in the ability to breed early. For example, Rambouillets are slow maturing. Production over the years will be greater if the ewe is bred as a lamb. Other producers breed to lamb at about 2 years of age. Since the gestation period is about 5 months, this means breeding at 18 to 19 months of age.

Check the ram for fertility before the breeding season. For best breeding results, shear the ram 6 to 8 weeks before the breeding season. Use a marking system to identify when ewes are bred. A marking harness on the ram may be used (Figure 27-4). It is also possible to use a marking pigment on the ram's brisket. This must be applied to the brisket every second or third day. Paint branding the ewes with an approved paint branding fluid aids in keeping better records.

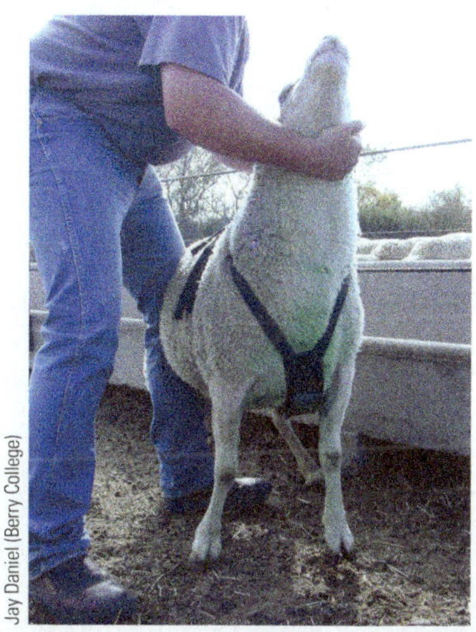
Jay Daniel (Berry College)

Figure 27-4 A marking harness on the ram will help identify bred ewes.

The number of ewes that a ram can breed depends on the age of the ram. When breeding on pasture, a ram lamb can breed about 15 ewes. A yearling ram can breed 25 to 35 ewes, and an aged ram can breed 35 to 45 ewes. A general rule of thumb is to keep 3 mature rams for each 100 ewes in the breeding flock.

When synchronized breeding is used, hand mating or limiting the number of matings per ewe is recommended. A vigorous ram can breed five to eight ewes per day when hand mating is used. Breed the ewe every 12 hours as long as she is in estrus.

Crossbreeding

Crossbreeding is recommended when producing market lambs. While several crossbreeding systems may be used, a three-breed rotational cross is the most desirable for producing market lambs.

Crossbred lambs have several advantages over straight-bred lambs. They:

- make more rapid gains
- are more hardy and vigorous
- have a lower mortality rate

The advantages of using crossbred ewes instead of straight-bred ewes include:

- greater fertility
- higher lamb survival rate
- higher lambing percentage
- better milk production

Characteristics to look for when selecting the breed of the ewe to use in a crossbreeding program include:

- early lambing ability
- high lambing rate
- greater ease of lambing
- better maternal instinct
- higher milk production
- greater longevity
- better wool quality and higher quantity of wool produced
- early sexual maturity
- greater potential for accelerated lambing
- good udder soundness

Characteristics to look for when selecting the breed of ram to use in a crossbreeding program include:

- rapid growth
- good carcass quality
- greater sexual aggressiveness
- above-average testicle size at puberty
- high fertility
- high survival rate of lamb offspring

Breeds considered particularly desirable for ewe selection include Rambouillet, Merino, Corriedale, Columbia, Targhee, and Polypay. Breeds considered particularly desirable for ram selection include Suffolk, Hampshire, Shropshire, Oxford, Southdown, and Texel. Commonly used ram breeds in crossbreeding programs are the Suffolk and Hampshire breeds.

Breeds that show desirable characteristics for selection of either ewes or rams for crossbreeding programs include Dorset and Montadale. Finnsheep and Romanov have high lambing rates but poorer fleece quality and quantity. When used in three-breed rotational crosses, Finnsheep and Romanov produce excellent market lambs. Other breeds not mentioned may be used in crossbreeding programs. The producer should give careful consideration to the desired characteristics of both ewe and ram breeds when making selections for breeding stock. Performance records are helpful when making selections for crossbreeding programs.

Gestation Management

Allow pregnant ewes plenty of exercise (Figure 27-5). Feeding away from the barn forces the ewes to exercise. Watch for signs of pregnancy disease (toxemia or ketosis). Ewes that are too fat or carrying more than one lamb are more likely to develop pregnancy disease.

USDA

Figure 27-5 Feeding ewes on pasture provides gestating ewes with needed exercise and, as with most ruminants, pasture is the most economical option for feeding.

The advantages of shearing ewes several weeks before lambing include the following:

- Conditions are more sanitary
- The lambs can nurse more easily
- Fleece contains less dirt and manure
- Less pen space is required
- Udder problems are easier to spot
- It is easier to see when the ewe is lambing

Care must be taken not to handle the ewes too roughly. Rough handling may cause premature lambing. Twice a year shearing is recommended in accelerated lambing systems.

If ewes are not sheared, they should be tagged or crutched out. This should be done at least 1 month before lambing. Tagging or crutching refers to shearing around the udder, between the legs, and around the dock. The wool around the vulva should be clipped. Ewes with a heavy face covering of wool should be clipped around the eyes.

Lambing Management

If possible, the ewes should not be disturbed during lambing. However, watch the flock carefully, and if a ewe is having difficulty lambing, give assistance. Pens used for lambing must be clean, warm, and dry. Pastures for lambing should be clean to help avoid the spread of parasites.

Be sure that the newborn lamb nurses. If it is weak, help it to nurse. Also check to be sure that the ewe is giving milk. During cold weather it may be necessary to use heat lamps to keep the lambs warm. The lambs must be dried off after birth to avoid chilling.

Some ewes do not claim their lambs. This problem is less common with older ewes than with ewes lambing for the first time. There is no sure way to persuade the ewe to claim the lamb. Tying the ewe until the lamb nurses may work, and rubbing the ewe's nose and the lamb with the ewe's milk may help. Blindfolding the ewe may also be effective.

Disinfect the lamb's navel shortly after it is born with tincture of iodine. Sore or irritated eyes should be treated immediately. Some lambs have eyelids that are turned under, which may cause blindness unless it is corrected. If the condition is not too bad, it can be corrected by working the eyelid outward several times a day. When it is severe, it may be necessary to stitch a fold of the eyelid to the skin. Use a sterile needle and silk or nylon thread. This is an inherited defect, so do not select breeding stock that has shown this trait.

Management from Lambing to Weaning

Docking, or cutting off part of the tail, is one of the first management practices performed after lambing. Dock lambs between 3 and 10 days of age. Docked lambs stay cleaner and, therefore, are less likely to get diseases or parasites.

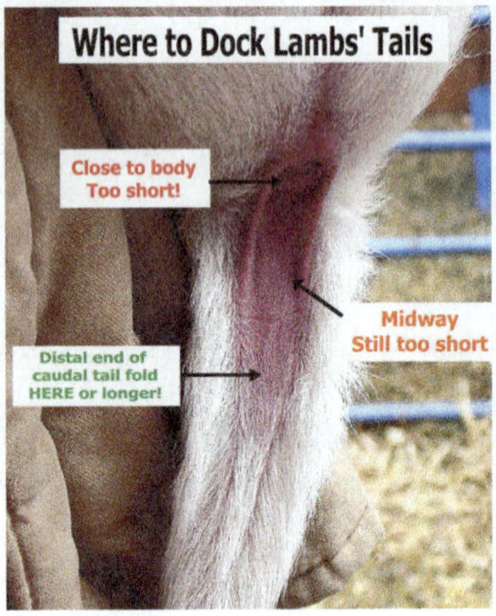

Figure 27-6 The American Veterinary Medical Association recommends the docking of lambs at the distal end of the caudal fold or longer to reduce the incidence of rectal prolapse.

The tail should be cut off at the distal end of the caudal fold or longer (Figure 27-6). Some people who show sheep prefer the tail to be docked extremely short. This does give a better appearance to the lamb, by giving the lamb a fuller and square rump. However, docking the tail too short can cause rectal prolapse. Many states have passed rules requiring longer docks on show lambs. Docking may be done with a knife, burdizzo, elastrator, emasculator, "all-in-one," electric docker, or hot docking iron. Follow good sanitation procedures. Clean and dip all instruments in a disinfectant before use.

Castrate ram lambs when they are young. Many producers dock and castrate lambs at the same time. Castration may be done with a knife, burdizzo, elastrator, "all-in-one," or emasculator. Sanitation procedures must be followed. If horses are or have been on the farm, vaccinate for tetanus.

Ear mark the lambs when they are docked and castrated. Ear marking makes it easier to tell the wethers from the ewes. A **wether** is a male lamb that has been castrated before reaching sexual maturity. If a different ear mark is used each year, the sheep can easily be identified as to age. Plastic tags may also be used for ear marking. The lambs should be vaccinated for sore mouth at 1 month of age or later.

Late lambs will do better in the summer if they are sheared. Shear at the time of weaning. A fly repellant is applied to any cuts that occur. In areas where spear and needlegrass are common, the faces, legs, bellies, or the entire lamb may have to be sheared.

Catching and Handling Sheep

While handling sheep, do not cause them to become excited. If they must be handled, confine them in a small area first. Do not handle sheep by the wool because this will cause bruises. Catch sheep around the neck or by the rear flank, but do not pinch the flank. Move sheep by placing the left hand under the jaw and the right hand under the dock. Guide the sheep with the left hand and urge them to move with the right hand.

Spraying and Drenching Sheep

In range areas, sheep are sprayed and drenched as they come out of the shearing pen. Spraying and drenching controls external and internal parasites. Always follow the current recommendations, which are available from extension specialists, universities, and veterinarians. Sprays and drenches should be changed from time to time to avoid a buildup of resistance in the parasites.

Foot Care

Trimming the feet of sheep helps to prevent problems with foot ailments. Feet should be trimmed twice a year. A sharp knife, pruning shears, or foot rot shears may be used for trimming. Do not trim into the quick. This will cause lameness and bleeding and may create a place for infection to start. Catch and examine any sheep that appear to be lame. Isolate all sheep that show signs of foot infection.

Culling Ewes

Ewes should be culled after the lambs are weaned. Cull ewes with bad udders, broken mouths, or that did not raise a lamb. Ewes that are fat at weaning time probably did not raise a lamb. Ewes with lumps in the udder should be culled along with ewes with prolapsed uteruses or ruptured abdomens. The condition referred to as **prolapsed uterus** occurs when the uterus protrudes from the vulva. Ewes that have difficulty in lambing may develop prolapsed uterus. An animal with this condition will die if not treated. A veterinarian should be called to administer treatment. Do not keep the ewe.

Records

Identification of individual sheep is necessary for a good system of record keeping. An ear tag designed especially for sheep and goats is commercially available. Birth and weaning weight records are needed, and fleece weight records should be kept. Weights at marketing and carcass evaluations are other important records. Forms are available from extension specialists and universities for keeping records on sheep flocks.

Shearing Sheep

The value of the fleece is lowered by the presence of dirt, manure, hay, straw, burrs, or other foreign matter. Follow management practices all year that will help keep the fleece clean.

Sheep are usually sheared early in the spring. In accelerated lambing systems, the sheep are sheared again in July or August. Most shearing is done by custom shearers. Shearing sheep is a specialized occupation that requires special skill.

Sheep should be sheared in a clean place. Shear only when the wool is dry. Take sheep off feed for a few hours before shearing. The fleece is removed in one piece. If possible, second cuts should not be made. Do not injure the sheep while shearing them.

Black-faced and black sheep are separated from the rest of the flock and sheared last. Yearlings are also sheared separately. Tender, coarse, black, and short fleeces are separated from the rest of the clip. If there is a large amount of foreign matter in the neck area, remove this and bag it separately.

Fleeces are packed in regulation wool sacks. Wool is stored in a clean, dry place until it is marketed.

PREDATOR LOSS

Losses from animal predators are a major problem for sheep producers in the United States, accounting for almost 40 percent of the total losses from all causes. The economic impact is severe, totaling more than $18 million annually in recent years. Nationally, coyotes account for the largest predator loss followed by dogs, both wild and domestic (Figure 27-7A and 27-7B). Other animal predators include

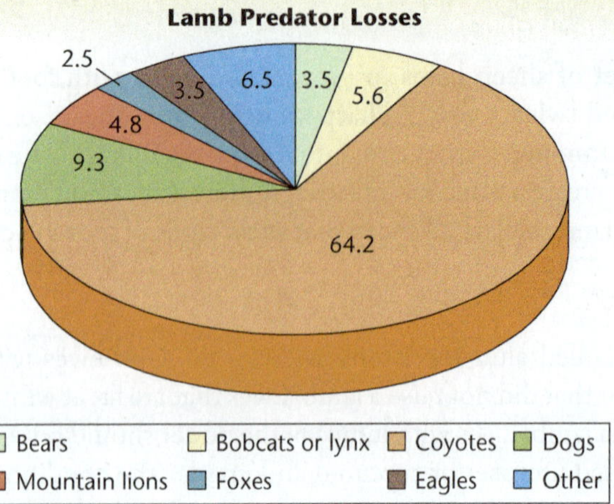

*Percentage may not sum to 100.0 due to rounding.

Figure 27-7A Lamb predator losses

Source: USDA, Sheep and Lamb Predator Death Loss in the United States

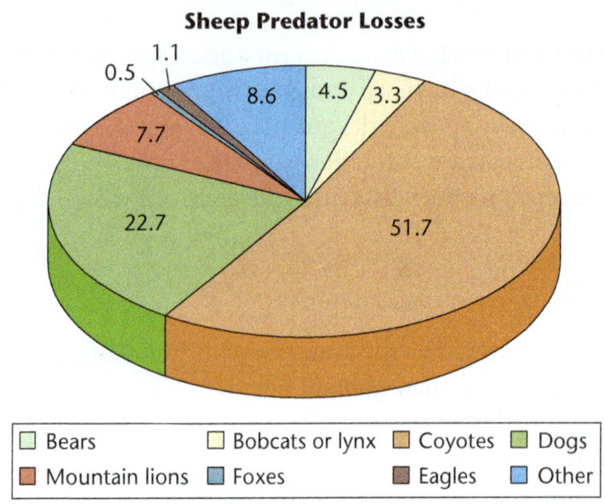

*Percentages may not sum to 100.0 due to rounding.

Figure 27-7B Sheep predator losses.

Source: USDA, Sheep and Lamb Predator Death Loss in the United States

mountain lions, bears, foxes, eagles, and bobcats. Losses from coyotes are highest in the western and mountain states, while dogs are the major source of losses in other parts of the United States.

Courtesy of Frank B. Flanders

Figure 27-8 Guard animals, such as donkeys, dogs, and llamas, are commonly used to protect sheep from predators.

Practical and effective methods of controlling predators depend on the geographic location of the producer. The use of properly constructed fences is a good method where the use of fences for pasture is common. Electric fences constructed of smooth, high tensile wire may be used. In areas where fences are not commonly used for pastures, the use of guard animals such as dogs, llamas, or donkeys may prove to be effective (Figure 27-8). Housing the sheep at night may be an effective predator control measure that can be used with small flocks. Sheep producers who are having animal predator problems should contact their local extension office or the USDA Animal Damage Control office for information about how to develop an effective control program.

HOUSING AND EQUIPMENT

It is not necessary or generally practical to provide expensive housing for sheep. Some use has been made of confinement housing systems and slotted floors for sheep. Producers who specialize in sheep production may want to consider such systems. Confinement facilities lend themselves to automated feeding systems. These systems are not economically viable for the average farm flock or for range production.

Housing for sheep should provide protection from winter weather. Sheep barns or sheds are often open to the south. The barn must be dry, draft free, and well ventilated. One square foot of window for each 20 square feet of floor area provides adequate light. Provide access to pens for cleaning. Loft floors must be tight to prevent chaff from falling through. Doors should be 8 feet wide so that sheep do not crowd each other as they go through. Locate feeding equipment for convenience.

Electricity may be needed in the lambing shed during cold weather. Night lighting, as well as heat lamps, is necessary in cold climates. Large flocks may justify a shepherd's room in the lambing barn. This can also be used to store other equipment.

Ewes require 12 to 15 square feet of floor space. An adjoining lot should provide 50 square feet of space per ewe. A hard-surfaced lot reduces the area needed by 40 percent. Provide 15 to 18 inches of space per ewe at hay racks. Two automatic watering cups are adequate for 30 ewes. Waterers and feeders should be 12 to 15 inches off the floor at the throat. This prevents the ewes from putting their feet in them. Twelve inches of feeder space per lamb is sufficient.

Hinged panels, from 4 to 10 feet in length, are useful for temporary pens. An extension hurdle can be used to crowd sheep into a corner. An **extension hurdle** or *gate* is a portable hurdle that can be lengthened or shortened as needed. It is usually about 3 feet high and 6 feet long. A sliding section inside the hurdle can be extended as needed when moving sheep. Maximum extension is about 6 feet. Mineral and salt feeders are also needed. Water may be provided in troughs or with automatic drinking cups.

A dipping tank is practical for larger flocks, while a smaller flock can be sprayed or dusted by hand. Commercial dipping tanks are available, or a homemade tank may be used.

Sheep must be fenced in. Fences that are 5 feet high are considered to be dog-proof, but most sheep can be held by a fence 3 to 4 feet high. In range areas, barbed wire fences will hold sheep. A cattle fence can be made sheep-proof by adding three or four barbs, 4 to 5 inches apart, on the lower section. Woven wire and electric fences can also be used.

Sorting and loading chutes may be used in larger flocks. Corrals are also useful for large flocks. Portable shelters and shades are used on pasture. A tilting squeeze, pregnancy testing cradle, blocking stand, weigh crate, and shipping crate are all useful pieces of equipment. A bag holder is needed for the shearing operation. Detailed plans for sheep equipment are available in the *Sheep Handbook Housing and Equipment* from the Midwest Plan Service.

Many items of miscellaneous equipment are necessary for sheep production. Examples include docking and castrating equipment, pruning shears, a balling gun, syringe, hand shears, tattooing, ear notching, and ear tagging equipment, scales, and record books.

SUMMARY

Purebred producers provide breeding stock for commercial and other purebred producers. Commercial producers raise sheep to market meat and wool. Lambs may be born in the fall, early spring, and late spring. Accelerated lambing systems produce three lamb crops in 2 years.

Roughage is the main feed for sheep. Pasture and good-quality legume or grass-legume hay is used. Silage may be used in place of hay. Corn, grain sorghum, oats, barley, and wheat are grains commonly used in sheep rations. Protein supplements are not needed if good-quality legume pasture or hay is available. Protein supplements to use, if needed, include soybean oil meal, cottonseed meal, linseed meal, and peanut meal.

Do not allow ewes and rams to become too fat. Some breeds of sheep will breed out of season, but others breed only in the fall. Prepare the ewe for breeding by clipping the wool from the vulva, udder, inside of the thigh, and around the eyes. Check the ram for fertility before the breeding season.

Creep feed lambs to get them off to a good start. Make sure lambs nurse and do not become chilled after lambing. Dock, castrate, mark, and vaccinate lambs. More concentrate than roughage is used to feed lambs to market weights.

Shear sheep in the spring, and provide a clean place to shear the sheep. Store the fleece in a dry place, properly bagged, until it is ready for market. Handle the fleece with care at shearing time to increase its value. Expensive housing is not necessary for sheep, but if needed, provide a dry, well-ventilated housing. Protect sheep from winter weather in cold areas. Use fencing to keep sheep in and predators out.

Student Learning Activities

1. Give oral reports on feeding, management, and housing of sheep.
2. Take field trips to observe good feeding, management, and housing for sheep.
3. Speak to a local sheep producer about the feeding, management, and housing for sheep.
4. Observe and practice activities such as docking, castrating, drenching, ear marking, and shearing sheep.
5. Do a community survey of sheep producers to determine feeding, management, and housing practices.
6. Prepare a display of equipment used for sheep production.
7. Use good feeding, management, and housing practices as described in this chapter when planning and conducting a sheep supervised experience program.

Discussion Questions

1. What are the functions of the purebred and commercial sheep producers?
2. Name and briefly describe the four systems of sheep production, based on lambing date.
3. Describe a good feeding program for the ewe flock during the gestation period.
4. Describe the feeding program for ewes for the first 10 days after lambing.
5. Describe a good feeding program for ewes during the lactation period, starting about 10 days after lambing.
6. Describe how ewes should be fed as weaning time approaches.
7. Describe a good feeding program for rams.
8. Describe the practice of flushing.

9. Describe a good feeding program for lambs from birth to weaning.
10. Describe a good feeding program for lambs from weaning to market.
11. How may a producer save more orphan lambs?
12. Describe feeding practices for replacement ewes.
13. What is meant by the term *seasonal breeder*?
14. At what age should ewes be bred for the first time?
15. Describe how ewes should be prepared for breeding.
16. What system is used to identify which ewes are bred during the breeding season?
17. What management practices should be followed with ewes during the gestation period?
18. What management practices should be followed during lambing?
19. Describe the management practices that should be followed with lambs from birth to weaning.
20. Describe how to catch and handle sheep.
21. Why are sheep sprayed and drenched?
22. Describe how to care for a sheep's feet.
23. Describe the kinds of ewes that should be culled from the breeding flock.
24. What kinds of records should be kept on the sheep enterprise?
25. Describe the management practices that produce a more valuable wool clip.
26. Briefly describe the kind of housing that is needed for sheep production.
27. What kind of fencing is used for sheep?
28. Briefly describe other kinds of equipment necessary for sheep production.

Review Questions

True/False

1. It may be more difficult to get ewes to breed in August.
2. For about 2 months after lambing, the ewe requires additional nutrients to produce milk for the lamb.
3. Gestating ewes should never be self-fed.
4. Lambs born before August 19 are generally referred to as fall lambs.
5. The lamb must receive the colostrum milk as soon as possible after it is born.

Multiple Choice

6. Accelerated lambing is a system that produces ______ lamb crops in 2 years.

   a. 1 c. 3
   b. 2 d. 4

7. Lambs are commonly weaned at about ______ months of age.

   a. 2 c. 4
   b. 3 d. 6

8. ______ is the practice of feeding a ration for 10 days to 2 weeks before breeding.

   a. Flushing c. Culling
   b. Waiting d. Rationing

9. A ______ is a large dose of medicine mixed with liquid ingested by the animal.

   a. wether c. ration
   b. drench d. parasite

10. Generally, ______ mature ram(s) breed 100 ewes in the breeding flock.

    a. 1 c. 4
    b. 10 d. 3

Completion

11. Early spring lambs are born during ______ and ______.
12. ______ ______ produce feeder lambs to finish for market.
13. ______ feed lambs to get them off to a good start.
14. ______ is the main feed for sheep.
15. A(n) ______ ______ is a portable hurdle or gate.

Short Answer

16. What is the basic feed for the ewe flock during the gestation period?
17. What milk helps to ensure the survival of lambs?
18. What are some advantages to shearing ewes weeks before lambing?
19. What are the main predators of sheep?
20. During breeding season, what feed is required for rams?

Chapter 28

Breeds, Selection, Feeding, and Management of Goats

Key Terms

buck
doe
doeling
kid
yearling
wether
niche market
dual-purpose breed
browse
wattle
mohair
cashmere
down
myotonic
kemp
milkstone
caping

Objectives

After studying this chapter, the student should be able to

- give a brief description of the goat enterprise.
- list and describe the common breeds of goats.
- select quality breeding stock using generally accepted criteria.
- discuss feeding and management of goats.
- describe housing and equipment required for goat production.

OVERVIEW OF THE GOAT ENTERPRISE

The goat industry in the United States once lacked in popularity and served a **niche market**, or a small segment of a larger market. However, in recent years, goat populations have increased throughout the United States, and producers sometimes cannot keep up with the demand. There is a national trend toward higher consumption of goat meat, milk, and other products. At one time, goat meat and products were only available in specialty or health-food stores. Goat milk, cheese, and butter can now be found in supermarkets, although goat meat is still rare in mainstream supermarkets (Figure 28-1).

Many factors have led to the growing popularity of the goat industry. The popular press, such as newspapers and news broadcasts, have published news stories relating to the health benefits and uses of goat products. Due to growing health concerns, goat meat, milk, and other products have increased in popularity. A similar-sized serving of goat meat has one-third fewer calories than beef, has a quarter fewer calories than chicken, and contains less fat. Goat shows and goat judging events have become increasingly popular, especially for youth groups such as 4-H and the

The following age and sex definitions are used for goats:

- **buck**: a reproductive male goat, any age
- **doe**: female goat, any age
- **doeling**: doeling: an unbred female goat under 2 years of age
- **kid**: young goat under 1 year of age, either sex
- **yearling**: goat over 1 year and under 2 years of age, either sex
- **wether**: castrated male goat

Goat Facts

- There are no official quality or yield grades for goat meat, but goats are covered under the Federal Meat Inspection Act. If harvested for sale, goat meat must be federally or state inspected.
- Goat meat is low in fat and is lower in cholesterol than chicken. It is lower in calories than the other common red meats such as beef, lamb, and pork.
- The per capita consumption of goat meat in the U.S. is very low. Goat meat is rarely seen in mainstream grocery stores due to low demand as well as a limited and inconsistent supply.
- Goats were one of the first animals used for milk by humans. Goat's milk is naturally homogenized, making it easier to digest and less allergenic than cow's milk.
- Goat production startup costs are low, making goats well suited for small farms. In recent years, increased demand for goat meat and milk has created new marketing opportunities for small farmers.
- Texas leads the U.S. in meat goat production with about 40 percent of all meat goat inventory. Tennessee, Oklahoma, California, and Missouri are also leading producers of meat goats.
- Goats are social animals and do not like to be alone. They are intelligent, agile, curious, and playful. They do not like to get wet and immediately seek shelter when it is raining.
- Like other ruminants, goats have a four compartment stomach that contains bacteria and protozoan that aid in breaking down roughages for digestion.

Figure 28-1 Facts and figures highlighting the U.S. goat industry.

National FFA Organization. As the ethnic population in the United States continues to increase, the future for continued demand and corresponding growth in the goat industry is promising. Other factors that have contributed to increased goat production in the United States include:

- the ability of goats to graze on poor land with browse plants
- an increase in small hobby farms
- increased research of goats and their uses
- goat being served as a delicacy in more upscale restaurants
- the ability of goats to control weeds and undesirable plants

CLASSES OF GOATS

There are a number of ways to classify goats. The simplest classification is by the products of the animal. The three major classifications are (1) dairy, (2) fiber, and (3) meat, with the two minor classifications being (4) pets or companions and (5) goatskin breeds. Some of the breeds are considered **dual-purpose breeds**. A dual-purpose breed is used to produce two or more products, such as meat and milk.

Dairy Goats

Although there are more than 70 breeds of goats used for dairy worldwide, only seven dairy goat breeds are common in the United States: (1) Alpine, (2) LaMancha, (3) Nigerian Dwarf, (4) Nubian, (5) Oberhasli, (6) Saanen/Sable, and (7) Toggenburg. The states with the largest number of dairy goats are Wisconsin, California, Iowa, Texas, and Minnesota; however, dairy goats are found in every state in the country.

There are only a few large dairy goat herds in the United States. Most dairy goats are kept in small numbers on farms for family milk production or for sale on a limited basis. Dairy goats require less space than dairy cows and are less expensive to maintain. In addition, dairy goats can also be used for meat.

Very little space is required for dairy goats. They are sometimes kept by people who live in nonrural, more urban areas; however, zoning restrictions often prohibit keeping goats and other livestock in some areas. A person considering keeping dairy or other types of goats should check the local zoning ordinances.

Fiber Goats

While there are about 20 breeds of goats used for fiber worldwide, only three are common in the United States: (1) Angora, (2) Cashmere, and (3) Miniature Silky Fainting Goat. Angora is the most popular fiber goat breed in the United States. More than 75 percent of the Angora goat population in the United States is found in Texas, which produces 90 percent of the total U.S. mohair. Fiber goats can also be used for meat and to help control brush and weeds. They do not compete with cattle for pasture because goats prefer browse plants while cattle prefer plants such as grasses and clovers. **Browse** is the shoots, twigs, and leaves of brush plants. Fiber goat flocks range in size from small farm flocks of 25 to 30 head to large range flocks of several thousand head.

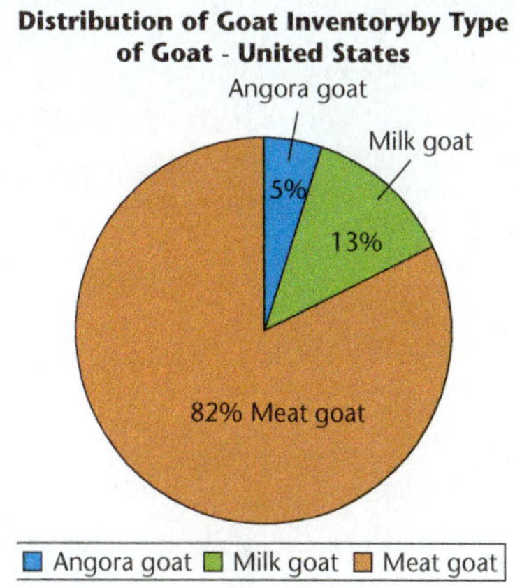

Figure 28-2 The most common type of goat in the United States is the meat goat. *Source*: USDA, NASS, Overview of the United States Sheep and Goat Industry

Meat Goats

A great portion of the increase in goat production is due to the increased popularity of meat goats (Figure 28-2). Goat meat was once a small segment of the total meat market, but consumption has been steadily increasing. Goats of all breeds may be harvested for meat production, with about 45 breeds that can be classified as meat type. The most common meat goat breeds in the United States are (1) Boer, (2) Kiko, (3) Kinder, (4) Myotonic, (5) Pygmy, (6) Savanna, and (7) Spanish.

BREEDS OF GOATS

Dairy Goats

Alpine

The Alpine breed (Figure 28-3) originated in France from Swiss foundation stock. It was first imported into the United States around 1922. The breed ranges in color from pure white to black, with a variety of color patterns. Color shades include fawn, brown, gray buff, and red. Alpines have erect ears and are short haired. They have no dewlap (a fold of flesh under the neck) and may be bearded or not bearded. Some are polled and some have horns. Bucks weigh 170 to 180 pounds, and does weigh 125 to 135 pounds. According to the American Dairy Goat Association (ADGA), the average milk production in a 1-year period for Alpines in 2013 was 2,559 pounds.

American Dairy Goat Association

Figure 28-3 French Alpine goat.

LaMancha

The American LaMancha (Figure 28-4) is a relatively new breed. It was developed by crossing a short-eared Spanish breed with several of the purebred breeds in the United States. The American LaMancha was formally recognized as a breed in 1958. LaMancha goats may be any color. They have straight faces with short hair. The breed has two different types of ears: gopher and elf. The gopher ear is 1 inch or less in length with little or no cartilage, and the end is turned up or down. The elf ear may be 2 inches long, cartilage shaping is allowed, and the end must be turned up or down. Bucks must have the gopher ear to be eligible for registration. According to the ADGA, the average milk production for LaMancha goats in 2013 was 2,171 pounds/year.

American Dairy Goat Association

Figure 28-4 LaMancha goat.

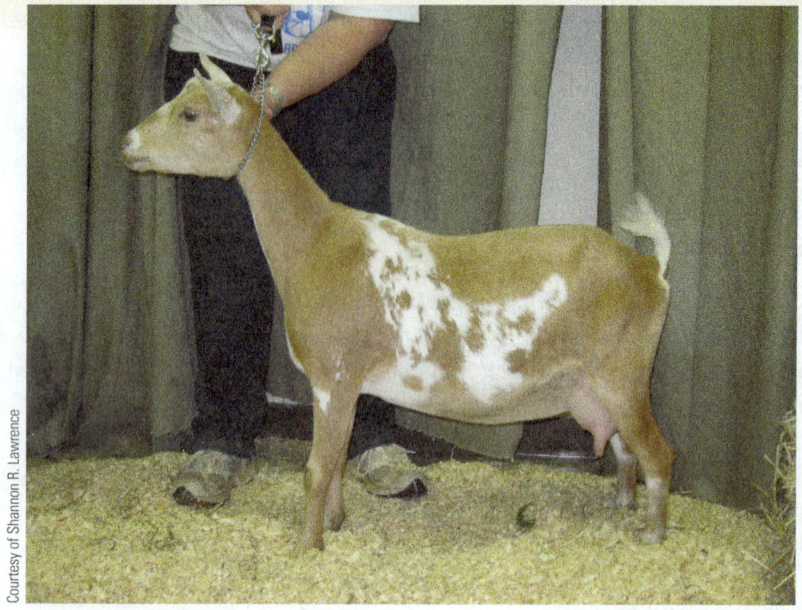

Figure 28-5 Nigerian Dwarf goat.

Nigerian Dwarf

The Nigerian Dwarf breed (Figure 28-5) originated in Western Africa. The first importations into the United States took place in the early 1900s. The breed is a miniature dairy goat breed but has the appearance of larger dairy goat breeds. Any color or color combination is acceptable for registration except Toggenberg markings. Nigerian Dwarfs can be horned or polled and have medium-length ears that are erect and alert. The hair is short and fine. Does stand at no more than 22.5 inches and bucks no more than 23.5 inches. The typical weight for this breed is around 75 pounds. The ADGA states that the 2013 average milk production for Nigerian Dwarfs was 719 pounds/year.

Nubian

The Nubian breed originated in Africa. The breed as it exists today in the United States was developed in England by crossing Nubian bucks with British dairy breeds. The breed was first imported into the United States around 1896. Nubians

Connection: Zoo Goats

Figure 28-6 Nigerian Dwarfs were once used as food for lions.

The story of how the Nigerian Dwarf came to the United States is unique. In the late 1800s and early 1900s, zoos were growing in popularity. Big cats such as lions and tigers were captured in Africa for zoo exhibits (Figure 28-6). Goats, including Nigerian Dwarfs, were often taken aboard ships as food for the big cats during the journey to the United States. The unique features of Nigerian Dwarf goats led some zoo owners to use them as show exhibits as well. Today, Nigerian Dwarfs are one of the seven common dairy goat breeds in the United States. They are the smallest of the dairy goat breeds.

may be any color or combination of colors except any Swiss colorations. Common shades include black, gray, cream, white, tan, and reddish-brown. They have short hair; long, pendulous ears; a Roman nose; and no fringe of long hair along the spine (Figure 28-7). Nubians can be polled or horned. The doe is beardless. Bucks usually weigh 175 to 180 pounds, and does weigh 130 to 135 pounds. Average milk production in 2013 according to the ADGA, was 1,964 pounds/year.

American Dairy Goat Association

Figure 28-7 Nubian goat.

Oberhasli

The Oberhasli goat breed (Figure 28-8) originated in Switzerland. These goats are medium in size. They are alert in appearance due to their forward-pointing ears. The breed has a coloration of bay ranging from light to a deep red bay, with black lining the head and a dorsal stripe to the tail, belly, knees, and hocks. Does can be solid black in color. Oberhaslis are short haired and have **wattles** (a projection of skin hanging from the chin) under their necks. Although all dairy goats can have wattles, they are more prominent in the Swiss breeds. Some are polled and some have horns. The U.S. standard for Oberhasli bucks is a minimum weight of 150 pounds and a minimum height of 30 inches. Oberhasli does have a minimum weight of 120 pounds and a minimum height of 28 inches. The ADGA lists the average milk production at 1,963 pounds/year for Oberhaslis in 2013.

American Dairy Goat Association

Figure 28-8 Oberhasli goat.

Saanen and Sable

The Saanen (Figure 28-9) originated in Switzerland. It was first imported into the United States about 1904. The color of this breed is white or light cream, with white being preferable. However, some Saanens are not all white. Traditionally, Saanens not completely white or cream were culled from the herd. Owners of non-white Saanens have created a breed of their own known as Sable. The hair for both breeds is short, and there is a fringe over the spine and thighs. The ears are erect, and bucks have a tuft of hair over the forehead. Saanens are both polled and horned. Both bucks and does are bearded. Bucks weigh 185 pounds or more, while does weigh 135 pounds or more. According to the ADGA, the average milk production is 2,613 pounds/year for Saanens and 2,222 pounds/year for Sables in 2013.

American Dairy Goat Association

Figure 28-9 Saanen goat.

Toggenburg

The Toggenburg originated in Switzerland. It was first imported into the United States around 1893. The body color varies from light fawn to dark chocolate. The markings for Toggenburgs are unique; ears are white with dark spots in the middle, there are two white/cream stripes down the face from the eyes to the muzzle (Figure 28-10), and the legs and from the tail down the backside of the rump are white or cream. Toggenburgs have short- to medium-length hair that lies flat, and the ears are erect. They may or may not have wattles. Toggenburgs are usually polled, although some may have horns. Bucks weigh 150 to 175 pounds , and does weigh 100 to 135 pounds. The ADGA lists the 2013 average milk production for Toggenburgs at 2,163 pounds/year.

American Dairy Goat Association

Figure 28-10 Toggenburg goat.

Fiber Goats

Angora

The Angora goat is an ancient breed. It originated in Turkey in the province of Angora, which is a mountainous region with a dry climate and great extremes in temperature. The first importation of this breed into the United States was in 1849. The Angora goat has been most popular in the southwestern United States.

Angoras are horned, although some polled individuals occur (Figure 28-11). They have long, thin, pendulous ears. Angoras are white, cream, red, fawn, and

Courtesy of Dr. Fred Speck

Figure 28-11 Angora goat.

varying shades of black and are open faced. Mature bucks weigh 125 to 175 pounds, and mature does weigh 80 to 90 pounds.

The fleece of the Angora goat is called **mohair**. Angoras are often classified based on the type of ringlet, or flat lock hair, in which the hair grows. The fleece grows at a rate of 6 to 12 inches per year, and the fleece is removed twice a year. Older goats produce a coarser fleece; therefore, kid fleece is considered more valuable. Average mohair production is 6 to 7 pounds per head per year. Selection programs have doubled this production in experimental groups.

Cashmere

Cashmere refers to the soft down or winter undercoat of fiber produced by most breeds of goats, except the Angora. **Down** is the soft furry or fine feathery coat of young birds and animals or the undercoat of adult birds and animals. Through selective breeding, several cashmere-producing types of goats have been developed. Breeding of goats to produce cashmere began in the Middle Eastern and Asian countries, including the Kashmir region, India, Pakistan, and China. The first cashmere goats were imported to the U.S. from Australia and New Zealand in the late 1980s. American breeders crossed cashmere goats with native stock to improve the cashmere-producing ability of the goats. Good results have been obtained by using Spanish meat goats as well as the Toggenburg, Saanen, and Nubian dairy breeds.

Most of the world supply of cashmere fiber comes from China, Afghanistan, Iran, Outer Mongolia, India, Australia, and New Zealand. The world demand for cashmere normally exceeds the supply. Cashmere fiber must be less than 19 microns in diameter; the usual range of diameter size is 16 to 19 microns. The yield of cashmere fiber is in the range of 30 to 40 percent of the total fleece weight. Solid-color goats are preferred for the production of cashmere fiber. The cashmere from these goats is white, brown, gray, or black. Cashmere from multicolored goats is not as desirable and is classed as white with color or mixed color.

Meat Goats

Boer

Courtesy of Frank B. Flanders

Figure 28-12 Boer goat buck.

The Boer goat breed originated in the Eastern Cape Province of South Africa in the early 1900s when farmers began selectively breeding for a meat-type goat using native goat breeds (Figure 28-12). During early development of the Boer goat, some crossbreeding occurred with Indian goats and European dairy goats. The name of the breed comes from the Dutch word "boer" meaning "farm." The term *Boer* was probably used to distinguish the breed from the Angora goats that were also raised in South Africa. Breeders selected for good conformation, rapid growth rate, high fertility, and short hair with red markings around the head and shoulders. Breed standards were established in 1959 with the formation of the Boer Goat Breeder's Association of South Africa.

According to current breed standards, the typical Boer goat has a white body with a red head, but no preference is given to any hair coloration or pattern. The breed is horned, has a Roman nose, and has pendulous ears. Polled individuals will occur occasionally, probably as a result of the European dairy breeds used in the early development of the breed. Mature males weigh 240 to 380 pounds, and mature females weigh 200 to 265 pounds.

Boer goats were introduced into the United States in 1993. Initially, the price of Boer breeding stock was artificially high due to high demand. The number of Boer goats in the United States has increased rapidly since their initial introduction, and prices for breeding stock have declined. The Boer goat has meat production characteristics that are superior to those of the Spanish goat. These characteristics make it valuable for producers interested in goat meat production.

Kiko

The Kiko goat breed (Figure 28-13) is a relatively new breed. The name "kiko" is Maori for "flesh" or "meat." The breed was developed in New Zealand in 1986 by crossing feral does with Nubian, Toggenburg, and Saanen bucks. The bucks and does that produced the best offspring under rugged conditions were chosen to produce future generations. The first Kikos were imported to the United States in 1995. The breed is generally white or cream; however, there is no color specification for registered Kiko goats. Kikos are often angular in appearance, have erect ears, and are horned. Generally, Kikos are good and capable mothers and are considered to be very hardy. They are popular because they seem to be more resistant to internal parasites than other breeds and often possess a good, lean carcass.

Kikogal Ranch, Coffeen, IL

Figure 28-13 Kiko buck and doe.

Kinder

The Kinder breed (Figure 28-14) originated in the state of Washington in 1985. The breed is crossbred between a Nubian doe and a Pygmy buck. The breed was developed for dual purposes of milk and meat production. Kinders can be any color, and all markings are acceptable. The coat is short and fine textured. The ears are long and wide and extend around the muzzle. Kinders are genetically horned. Wethers produce a desirable meat carcass. It is possible for a Kinder doe weighing 115 pounds to produce five kids, who in 14 months can weigh 80 pounds each and dress out at 50 pounds, thereby producing 250 pounds of meat in just over 1 year.

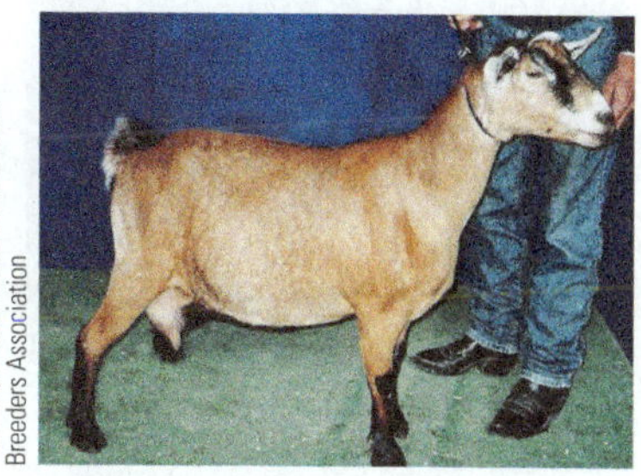

Courtesy of Sue Huston, Kinder Goat Breeders Association

Figure 28-14 Kinder doe.

Myotonic

The Myotonic breed (Figure 28-15) has its roots in Tennessee. The direct origin is unknown, but the breed was developed around 1885. Myotonic goats possess a strange characteristic that makes them unique. When they become excited or frightened, they often fall over or faint and lie stiff for a few seconds. The condition that causes this reaction is comparable to the human disease epilepsy. The word **myotonic** means spasm or temporary rigidity of the muscles. Because of this distinctive characteristic, Myotonic goats are often called wooden leg goats, stiff leg goats, or fainting goats. Myotonic goats are one of the few goat breeds developed in the United States.

Myotonic goats are most often black and white, but combinations of colors are common. They may be polled or horned. Myotonic goats vary in size. The smaller

Figure 28-15 Myotonic goats often faint or fall over when they become excited or frightened.

goats tend to be sought after as pets, while the medium to large goats are used for meat production. Generally, adult does weigh 80 to 110 pounds, but weight varies considerably by lines of the breed. Some bucks weigh close to 200 pounds. Myotonics are well muscled, which makes them popular for meat production.

Pygmy

The Pygmy goat breed originated in the French Cameroons area of Africa and was originally called the Cameroon Dwarf Goat. Early exports from Africa were sent to zoos in Sweden and Germany. From there, exports made their way to England, Canada, and the United States around 1959. Initially, Pygmy goats were primarily exhibited in zoos. Today, they are a dual-purpose breed used for meat and milk, as well as for FFA and 4-H projects. They are easily handled and make good pets because of their small size.

Courtesy of Linda Henwood

Figure 28-16 Pygmy goat doe.

Mature Pygmy goats (Figures 28-16 and 28-17) are 16 to 23 inches at the withers, with the legs and head being relatively short compared to the body length. Pygmy goats display a heavier, shorter bone pattern than the Nigerian Dwarf Dairy Goat. Genetically polled animals are not accepted for registry in the National Pygmy Goat Association (NPGA). Pygmy goats may be any color; preferred colors are white through gray and black in a grizzled (agouti) pattern. Breed-specific standards of the NPGA require that—except for solid black goats—the muzzle, forehead, eyes, and ears are accented in lighter tones than the body, and the front and rear hoofs, cannons, crown, dorsal stripe, and martingale are darker than the body color. Goats that are caramel colored must have light vertical stripes on the front sides of darker socks. These specific breed standards are not required for registry in the American Goat Society (AGS). Female goats may have no beard or one that is sparse or trimmed. Male goats should have a full, long, flowing beard.

Courtesy of Linda Henwood

Figure 28-17 Pygmy goat buck.

Savanna

The Savanna breed originated in South Africa and was introduced into the United States in the 1990s. The Savanna goat was developed by crossing a native multicolor, floppy ear doe and a large white buck. Through several generations of selective breeding and natural selection, a heat, parasite, and drought-resistant breed was established. The harsh climate of the savanna region of South Africa helped develop this hardy breed. Savannas have medium to large floppy ears with short, smooth hair. Savannas can produce quality cashmere. The coat is white to cream, while the skin is a black to brown color. Savannas have black horns that curve straight back from the crown of the head. The carcass is balanced and produces low-fat, tender meat.

Spanish

Photo by Larry Rana. USDA

Figure 28-18 Spanish goat.

The Spanish goat (Figure 28-18) is a mongrel descendant of the milk goat breeds. Around the 1500s, Spanish conquistadors traveled to the Americas and brought goats as a source of meat. The goats either escaped or were released, and through natural selection, evolved into today's Spanish goat. They are sometimes called brush goats or meat goats and are used for meat and milk production. However, the main use of Spanish goats is for meat. Spanish goats can live on brush and weeds on otherwise unproductive land. They are prolific and can survive with little care. Spanish goats are most common in the southern and southwestern United States. Size varies due to climate, terrain, and the line of breeding stock. Spanish goats are horned. All colors are acceptable, and the hair is usually short, although some have longer hair on the lower body and thighs. Some lines produce heavy cashmere coats.

SELECTION OF GOATS

Selection of Dairy Goats

For dairy goat production, breed choice is a matter of personal preference. Milk production is the main goal of most dairy goat owners. Owners should select individuals that give an indication of being good milk producers. It is necessary to be familiar with the parts of the animal in order to evaluate it properly. Figure 28-19 shows the parts of the dairy goat.

Scorecards have been developed by the American Dairy Goat Association for evaluation of type. The scorecard indicates the desirable traits to be used in selecting or judging dairy goats. Figure 28-20 shows the scorecard for evaluating bucks and does.

Young people who use dairy goats for 4-H and FFA projects may want to show their animals in dairy goat shows. The American Dairy Goat Association dairy goat showmanship scorecard (Figure 28-21) provides valuable suggestions for showing goats. All dairy goats must be registered to be shown at an ADGA show.

The front teeth on the lower front jaw of the goat may be used as a guide in determining the age of the animal (Figure 28-22). Teeth develop in the goat in the same manner as they do in sheep. The number of permanent teeth is an indication of the age of the goat.

Production records and pedigrees give additional information to be used in selection. Official production records from individuals close to the animal in the pedigree are of greater value than those of individuals beyond the fourth generation. Good dairy goats average 3 to 4 quarts of milk per day. The normal length of the lactation period is 10 months or 305 days.

Selection of Fiber Goats

Fiber goats are judged on the basis of body and fleece. These two factors are weighted equally. The body conformation is judged on the basis of breed type, conformation, amount of bone, constitution and vigor, and size and weight for age.

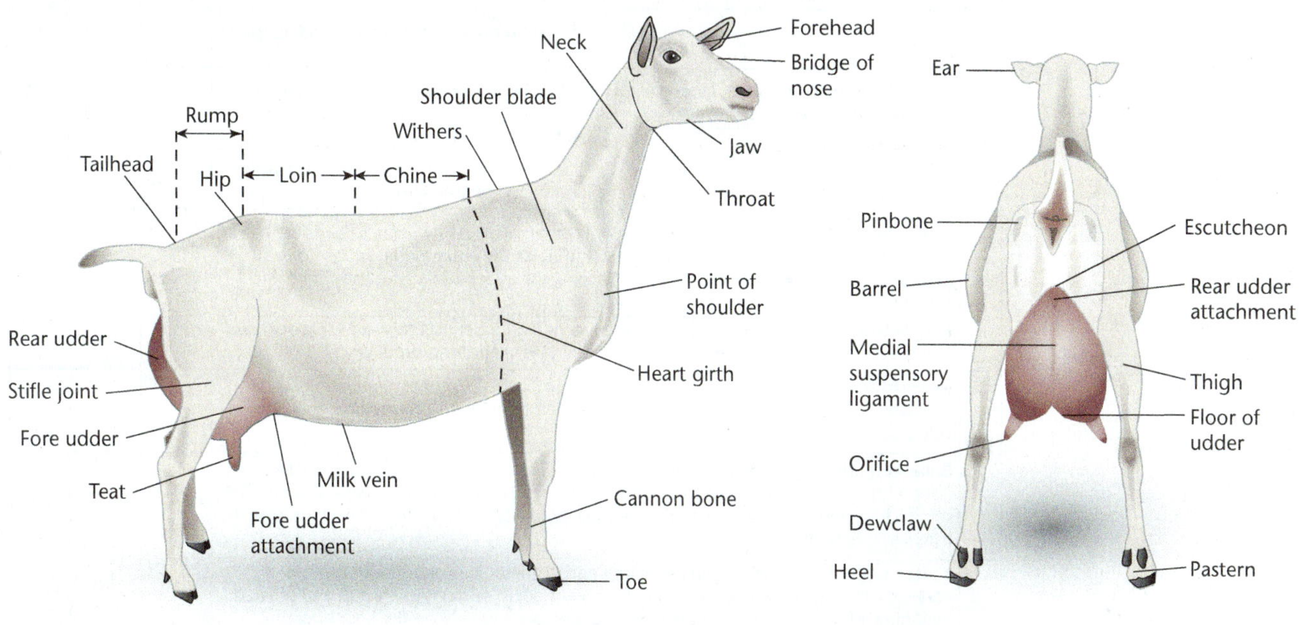

Figure 28-19 Parts of the dairy goat.

	POINTS	Senior Doe	Junior Doe	Buck
A.	**GENERAL APPEARANCE** An attractive framework with femininity (masculinity in bucks), strength, upstandingness, length, and smoothness of blending throughout that create an impressive style and graceful walk.	35	55	55
	Stature - slightly taller at withers than at hips with long bone pattern throughout.	2	2	2
	Head & Breed Characteristics - clean-cut and balanced in length, width, and depth; broad muzzle with full nostrils; well-sculpted, alert eyes; strong jaw with angular lean junction to throat; appropriate size, color, ears, and nose to meet breed standard.	5	10	8
	Front End Assembly - prominent withers arched to point of shoulder with shoulder blade, point of shoulder, and point of elbow set tightly and smoothly against the chest wall both while at rest and in motion; deep and wide into chest floor with moderate strength of brisket.	5	8	10
	Back - strong and straight with well-defined vertebrae throughout and slightly uphill to withers; level chine with full crops into a straight, wide loin; wide hips smoothly set and level with back; strong rump which is uniformly wide and nearly level from hips to pinbones and thurl to thurl; thurls set two-thirds of the distance from hips to pinbones; well defined and wide pinbones set slightly lower than the hips; tailhead slightly above and smoothly set between pinbones; tail symmetrical to body and free from coarseness; vulva normal in size and shape in females (normal sheath and testes in males).	8	12	10
	Legs, Pasterns & Feet - bone flat and strong throughout leading to smooth, free motion; front legs with clean knees, straight, wide apart and squarely placed; rear legs wide apart and straight from the rear and well angulated in side profile through the stifle to cleanly molded hocks, nearly perpendicular from hock to B, yet flexible pastern of medium length; strong feet with tight toes, pointed directly forward; deep heels with sole nearly uniform in depth from toe to heel.	15	23	25
B.	**DAIRY STRENGTH** Angularity and general openness with strong yet refined and clean bone structure, showing enough substance, but with freedom from coarseness and with evidence of milking ability giving due regard to stage of lactation (of breeding season in bucks). **Neck** - long, lean, and blending smoothly into the shoulders; clean-cut throat and brisket with adequate width of chest floor to support maintenance of body functions. **Withers** - prominent and wedge-shaped with the dorsal process arising slightly above the shoulder blades. **Ribs** - flat, flinty, wide apart, and long; lower rear ribs should angle to flank. **Flank** - deep, yet arched and free of excess tissue. **Thighs** - in side profile, moderately incurving from pinbone to stifle; from the rear, clean and wide apart, highly arched and out-curving into the escutcheon to provide ample room for the udder and its attachment. **Skin** - thin, loose, and pliable with soft, lustrous hair.	20	30	30
C.	**BODY CAPACITY** Relatively large in proportion in size, age, and period of lactation of animal (of breeding season for bucks), providing ample capacity, strength, and vigor.	10	15	15
	Chest - deep and wide, yet clean-cut, with well sprung foreribs, full in crops and at point of elbow.	4	7	7
	Barrel - strongly supported, long, deep, and wide; depth and spring of rib tending to increase into a deep yet refined flank	6	8	8
D.	**MAMMARY SYSTEM** Strongly attached, elastic, well-balanced with adequate capacity, quality, ease of milking, and indicating heavy milk production over a long period of usefulness.	35		
	Udder Support - strong medial suspensory ligament that clearly defines the udder halves, contributes to desirable shape and capacity, and holds the entire udder snugly to the body and well above the hocks. Fore, rear, and lateral attachments must be strong and smooth.	13		
	Fore Udder - wide and full to the side and extending moderately forward without excess non-lactating tissue and indicating capacity, desirable shape, and productivity.	5		
	Rear Udder - capacious, high, wide, and arched into the escutcheon; uniformity wide and deep to the floor; moderately curved in side profile without protruding beyond the vulva.	7		
	Balanced, Symmetry & Quality - in side profile, one-third of the capacity visible in front of the leg, one-third under the leg, and one-third behind the leg; well-rounded with soft, pliable, and elastic texture that is well collapsed after milking, free of scar tissue, with halves evenly balanced.	6		
	Teats - uniform size and of medium length and diameter in proportion to capacity of udder, cylindrical in shape, pointed nearly straight down or slightly forward, and situated two-thirds of the distance from the medial suspensory ligament on the floor of each udder-half to the side, indicating ease of milking.	4		
	TOTALS	**100**	**100**	**100**

Figure 28-20 ADGA Dairy Goat Scorecard for bucks and does.

		Points	Subtotal
E.	**SHOWMANSHIP**		
1.	**APPEARANCE OF ANIMAL**		40
	Condition and Thriftiness - showing normal growth neither too fat nor too thin.	10	
	Hair - clean and properly groomed. **Hoofs** - trimmed and shaped to enable animal to walk and stand naturally. **Neatly disbudded** if the animal is not naturally hornless.	10	
	Clipping - entire body if weather has permitted, showing allowance to get a neat coat of hair by show time; neatly trimmed tail and ears.	10	
	Cleanliness - as shown by a clean body as free from stains as possible, with special attention to legs, feet, tail area, nose, and ears.	10	
2.	**APPEARANCE OF EXHIBITOR**		
	Clothes and person neat and clean - white costume preferred.	10	10
3.	**SHOWING ANIMAL IN THE RING**		50
	Leading - enter, leading the animal at a normal walk around the ring in a clock-wise direction, walking on the left side, holding the collar with the right hand. Exhibitor should walk as normally and inconspicuously as possible. **Goat should lead readily** and respond quickly. **Lead equipment** should consist of a properly fitted collar or small link chain, which is inconspicuous, yet of sufficient strength to maintain proper control. **As the Judge studies the animal,** the preferred method of leading is to walk on the side away from the Judge. **Lead slowly** with the animal's head held high enough for impressive style, attractive carriage, and graceful walk.	10	
	Pose and show an animal so it is between the exhibitor and the Judge as much as possible. Avoid exaggerated positions, such as crossing behind the goat. **Stand or kneel** where both Judge and animal may be observed **Pose animal** with front feet squarely beneath and hind feet slightly spread. Where possible, face animal upgrade with her front feet on a slight incline. Neither crowd other exhibitors nor leave too much space when leading into a side-by-side position. **When Judge changes placing,** lead animal forward out of line, down or up to the place directed then back through the line, finally making a U-turn to get into position. When a Judge changes placing in a head to tail sequence, lead animal out of line and up or down the line on the side next to the Judge. It is the responsibility of another handler to accommodate changes by moving up or down on the side opposite the Judge. **To step animal ahead** - use slight pull on collar. If the animal steps badly out of place, return her to position by leading her forward and making a circle back through your position in the line. **When Judge is observing the animal,** if she moves out of position, replace her as quickly and inconspicuously as possible. **Be natural.** Overshowing, undue fussing, and maneuvering are objectionable.	15	
	Show animal to best advantage, recognizing the conformation faults of the animal you are leading and striving to help overcome them. Showmen may be questioned by the Judge on their knowledge of proper terminology for parts of a dairy goat, breed standards, evaluation of defects and ADGA Scorecards.	15	
	Poise, alertness, and courteous attitude are all desired in the show ring. Showmen should keep an eye on their animals and be aware of the position of the Judge at all times but should not stare at the Judge. Persons or things outside the ring should not distract the attention of the showmen. Respond rapidly to requests from judges or officials, and be courteous and sportsmanlike at all times, respecting the rights of other exhibitors. The best showmen will show their animals at all times not themselves and will continue exhibiting well until the entire class has been placed, the Judge has given his reasons, and he has dismissed the class.	10	
	TOTAL		**100**
Suggested Uniform: Long-sleeved white shirt, regulation white pants; 4-H or FFA necktie; 4-H or FFA cap (if applicable), with matching shoes and belt in either black, white, or brown.			

Figure 28-21 ADGA Dairy Goat Showmanship Scorecard.

The fleece is judged on the basis of fineness, uniformity and completeness of covering, oil content and luster, density, and character of fleece. Breeding animals are further selected on the basis of age and fertility.

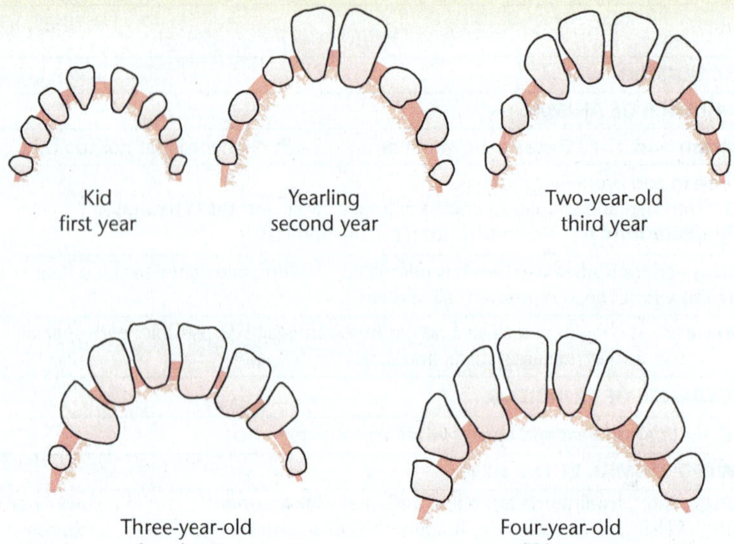

Figure 28-22 Determining the age of goats by examining the teeth.

Selection of good breeding stock increases production. Desirable traits are heritable, but a period of years is required for a selection program to significantly improve fiber production. Bucks must be carefully selected to improve the breeding flock. Good management practices must be followed if the selection program is to be effective.

Balance must be maintained between selection for body conformation and fleece. Too much emphasis on fleece tends to reduce the size and vigor of the goat. The yearling buck should weigh at least 80 pounds, and the yearling doe should weigh at least 60 pounds. The goat should have a wide, deep body with a good spring of ribs and wide loin. Strong feet and legs with adequate bone are important. An open face is considered better than too much wool covering the face.

A long, dense staple (fleece) is desirable for the Angora goat. The Angora should produce at least 1 inch of fiber growth each month. To determine density, part the fleece and observe how much skin area is exposed. Angora goats do not produce an undercoat of cashmere.

Cashmere down growth generally begins around the longest day of the year and stops on about the shortest day. If the down growth stops, it will shed naturally if not combed or sheared. Therefore, the best time to select Cashmere goats is toward the end of the growth stopping period so that the down is easier to assess. Part the guard hair to determine if down is available.

Kemp fibers are large, chalky white hairs. Do not select goats with large amounts of kemp in the fleece. The fleece should be bright and uniform in fineness and length; it should cover the body of the goat. A light covering under the jaws, throat, or belly is desirable. The mohair should feel soft to the touch. Also, the type of lock (ringlet or flat) is important to the purebred breeder, but it is of less importance to the commercial producer.

Breeding goats older than 8 years of age are not as productive as younger animals. The fertility of the doe is an important factor in selection. Does that do not breed regularly should be culled, while does that have multiple births should be kept. Records of fleece weight, length of staple, and percentage of kid crop help the producer do a better job of selection.

MEAT GOAT EVALUATION

Livestock Evaluation Events

In livestock evaluation events, classes with four animals each are provided for judging. The animals are designated by a number from 1 to 4. The contestants rank the animals in order from most desirable to least desirable. A sample class ranking may be 3-4-2-1, with 3 being the most desirable and 1 being the least desirable (Figure 28-23). Meat goat judging classes are generally either market goats or

Placing Classes

	Placing	Class 1	2	3	4	5	6	7	8	Placing	
1	1 2 3 4									1 2 3 4	1
2	1 2 4 3									1 2 4 3	2
3	1 3 2 4									1 3 2 4	3
4	1 3 4 2									1 3 4 2	4
5	1 4 2 3									1 4 2 3	5
6	1 4 3 2									1 4 3 2	6
7	2 1 3 4									2 1 3 4	7
8	2 1 4 3									2 1 4 3	8
9	2 3 1 4									2 3 1 4	9
10	2 3 4 1									2 3 4 1	10
11	2 4 1 3									2 4 1 3	11
12	2 4 3 1									2 4 3 1	12
13	3 1 2 4									3 1 2 4	13
14	3 1 4 2									3 1 4 2	14
15	3 2 1 4									3 2 1 4	15
16	3 2 4 1									3 2 4 1	16
17	3 4 1 2									3 4 1 2	17
18	3 4 2 1	●								3 4 2 1	18
19	4 1 2 3									4 1 2 3	19
20	4 1 3 2									4 1 3 2	20
21	4 2 1 3									4 2 1 3	21
22	4 2 3 1									4 2 3 1	22
23	4 3 1 2									4 3 1 2	23
24	4 3 2 1									4 3 2 1	24
		1	2	3	4	5	6	7	8		

Figure 28-23 In this example, the contestant ranked the animals in class order 3-4-2-1, indicating that animal 3 is the most desirable and 1 is the least desirable.

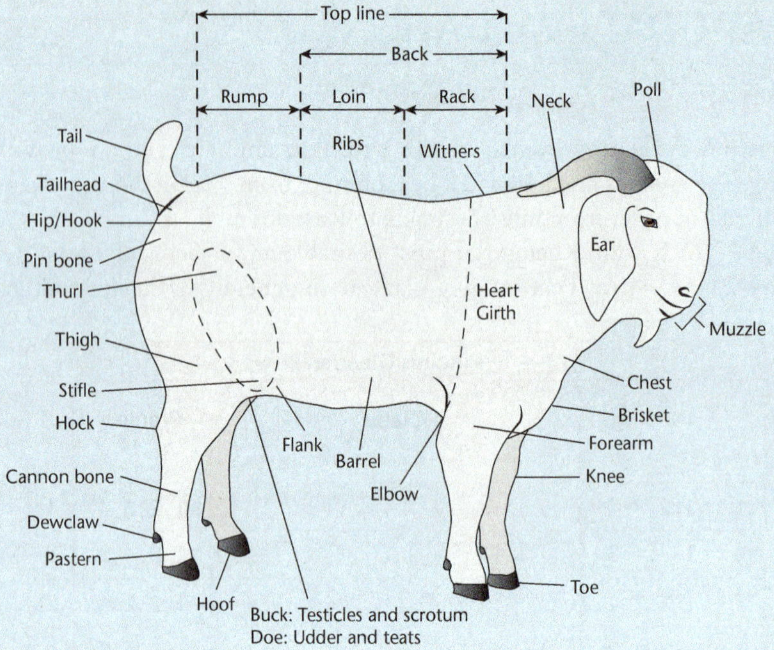

Figure 28-24 Parts of the meat goat.

breeding does. Competitions may include other components such as a written exam and oral reasons.

Beginners should follow some basic guidelines while learning to evaluate meat goats. First, it is important to learn the parts of a meat goat (Figure 28-24) and the desirable characteristics for marketing and breeding. Beginning competitors should study the characteristics of meat goats and establish a mental picture of the ideal animal that combines all the desirable traits of the species.

Some key guidelines for beginners are as follows:

- First, view each animal from about 25 feet away to get an overall impression of the animals. Evaluate meat goats from the ground up and then from the rump forward, identifying strengths and weaknesses. The evaluators should concentrate on moving their eyes from the hooves to the top and then from the rear of the animal to the head.
- View each animal from the rear, side, and front. Move closer to view the top. Be sure to view each animal from all angles.
- Evaluate each of the following characteristics of the animal. These are in order of importance:

Market Goats

1. muscling
2. soundness and structure
3. finish
4. capacity
5. balance and style

Breeding Does

1. soundness and structure
2. capacity
3. balance and style
4. muscling
5. condition

Muscling

The single most important trait in market goats is muscling. Market evaluation of muscling in meat goats is similar to judging muscle development in other species. First, look for thickness in the leg. The muscle should not taper off and should be complete all the way through the center of the leg. Next, look for width between the rear legs from the hooves to the top of the animal. A good indication of muscling is an animal that stands with its legs wide apart at the top and bottom (Figure 28-25). The best way to evaluate this trait is to observe the animal walking. The widest area should be the stifle. Evaluating the top may be difficult, but handling the animal helps determine the true degree of muscling. Many competition rules allow contestants to handle the animals to evaluate muscling. Check the rules prior to the competition to determine if handling the animals will be allowed. The top should be wide and flat, not steep and narrow. While handling the animal, feel the degree of muscling. The loin should feel firm and possess a bulge on both sides of the spine. The brisket provides another indicator of muscling. The brisket should possess extension and be broad, deep, and moderately firm.

Soundness and Structure

Soundness and structural correctness refer to bone and structure, which are the most important traits in breeding does. It is important to consider the hooves, pasterns, knees, hocks, and levelness of the top when evaluating this trait. The legs should be nearly straight with big feet and even toes, and the pasterns should be strong. The legs should be at the corners of the body, while having the front knees straight (Figure 28-26). The hocks should have some angulation, while not having too much set or being completely straight. The goat should be alert and move fluidly. The top should be strong and level, while not having profound angulation. The goat's head should be upright, and the neck should extend over the top of the shoulders.

Finish

Finish refers to the amount of fat cover on market animals. The term *condition* is used when referring to the amount of fat cover on breeding does. Goats deposit

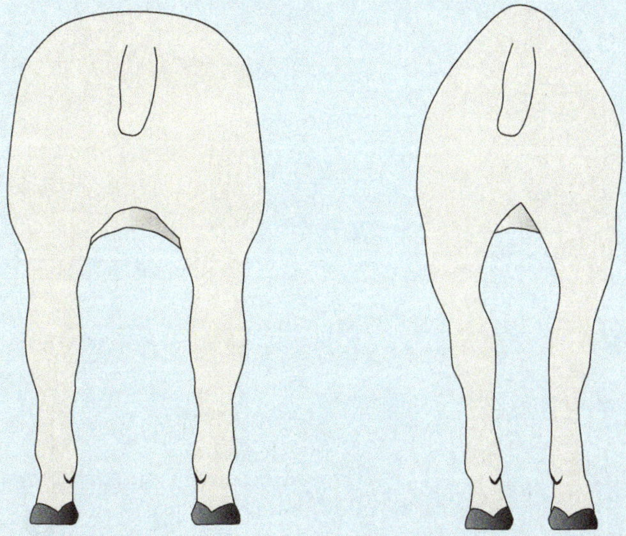

Figure 28-25 A good indication of muscling is an animal that stands with its legs wide apart at the top and bottom, while the top is wide and flat.

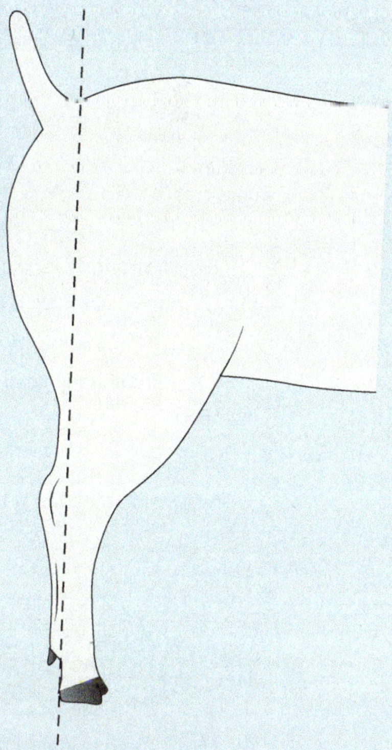

Figure 28-26 The rear legs should be angled at the hocks. If the hocks have too much angle, the legs extend too far underneath the body (sickle-hocked). If the legs are too straight (post-legged), the legs do not have enough flex.

fat on the interior of their bodies, first next to the organs and then on the outside of the body. Ideally, meat goats should have a thin, uniform cover of fat over the loin and rib area. There are no established standards for fat cover on goats; however, a measurement of 0.1 inch is a good basis for degree of finish on meat goats. Generally, a goat with a thick exterior fat cover is considered to be too fat. The brisket is an excellent place to evaluate the fat cover on meat goats. If the brisket possesses a significant amount of movement and feels flabby, generally the animal has too much fat cover. Handling the animal is the only reliable way to estimate fat cover.

Capacity

Capacity refers to overall body size and potential for muscling and growth. Evaluating capacity involves estimating the depth, length, and width of the goat's body. Meat goats should have a long, wide, and uniformly deep body (Figure 28-27). This is especially important for breeding does in reproduction. While viewing from the side, note the overall length of the animal and body depth and determine how they compare with other animals in the class. The topline should be strong. Also, note the length of the back and front halves of the body. A desirable meat goat should have a long back half. The back half should be at least the length of the front half; however, a longer back half is considered most desirable. A meat goat should have a good, wide base standing with their legs wide apart. Also, the ribs should be open and exhibit curvature. A wide chest is another indicator of capacity (Figure 28-28).

Balance and Style

Balance and style refer to how well the body parts blend together. Straightness of lines, distribution of weight, and size and shape of body parts are indicators of balance and style. All body parts should be well proportioned. An effective way to

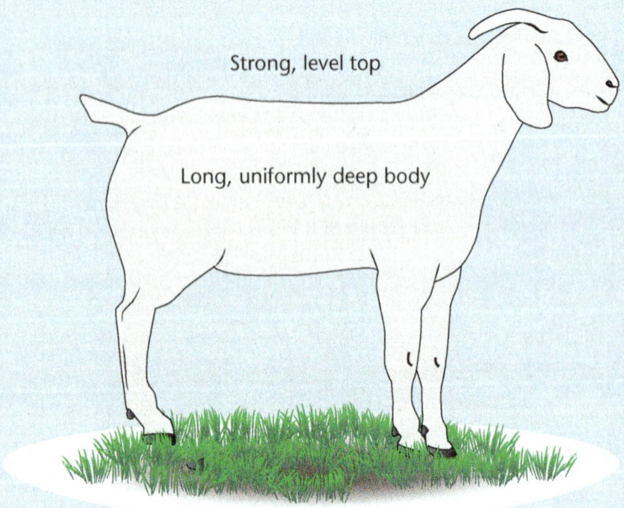

Figure 28-27 Desirable characteristics of meat goats include a long, uniformly deep body with a strong, level top. The rump should be long with a slight slope from the hooks to the pins. The legs should be strong with placement at the corners of the body. Meat goats should be heavily muscled and structurally correct.

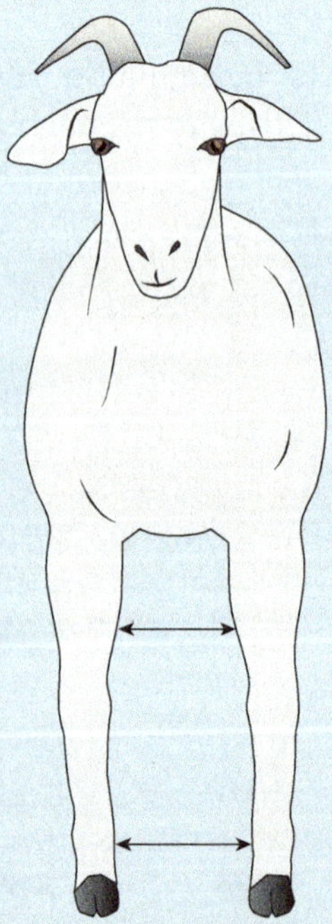

Figure 28-28 A wide chest is a good indicator of capacity.

evaluate meat goats is in thirds. Each third should flow from one to the other. The first third, the head and the neck, should flow to the next third, which is the start of the neck and the front half of the body. Lastly, the rack and pin bones should flow smoothly with the front half. The top of the body should be long and nearly level. A goat that shows profound curvature along the top is undesirable. The neck should not be ewe-necked but should have length and sit high on the shoulders. The shoulder should be well ordered and blend into the neck and fore rib. Generally, a goat that has good structural soundness is well balanced.

Sexual Characteristics

Youth evaluation events usually have classes of market goats and breeding does. Market goats are judged only for meat production, while does are judged on potential meat production of future offspring, capacity, and their reproduction potential. Does should possess a feminine appearance. On older does, the udder should possess a well-developed and functional udder with well-developed teats.

Courtesy of Rikki Griffith

Figure 28-29 Make written notes while judging characteristics.

Determining Your Placing

After the four animals in a class have been evaluated, they must be compared to each other and ranked from 1 to 4 (best to worst). Some general guidelines for placing include the following:

- Animals are evaluated for placing by considering three pairs: the top pair, the middle pair, and the bottom pair.
- Eliminate any easy top and bottom placings. Is there an obvious best and least desirable animal? If so, only the middle pair must be dealt with at that point.
- If there is a very good animal that is an obvious 1st place or a very poor animal that is obviously 4th place, then only the other three animals must be compared.
- If there are two very good animals and two less desirable animals, then concentrate on ranking the top two and then rank the bottom two animals.
- If all the animals are close in desirable characteristics and placing is difficult, look at each one and compare it to the ideal animal you have studied. Then rank them 1 to 4 based on the ideal. Base the placing on facts gathered through the evaluation, comparison to the ideal, and comparison to each other. Making written notes will help (Figure 28-29). The chapter on judging beef cattle discusses a suggested method for note taking.
- Keep in mind that a single undesirable trait on an otherwise desirable animal may be difficult to evaluate. A decision must be made on the degree of undesirability to determine where that animal will place compared to the other animals in the class.

Courtesy of Rikki Griffith

Figure 28-30 Oral reasons are given individually while standing before an official judge.

Reasons

In many judging competitions, contestants must defend their placings with oral reasons. Contestants explain why they ranked the animals as they did. Oral reasons are given individually while standing before an official reasons judge (Figure 28-30). Placing of the class is stated, and each animal or pair of animals is evaluated to justify the rankings. In giving oral reasons, the contestants should stand straight; direct their explanation to the judge; and speak clearly, confidently, and succinctly. It is important to know the common meat goat evaluation terminology in order to be successful. See the "Meat Goat Evaluation Terminology" sidebar for a list of common terminology.

MEAT GOAT EVALUATION TERMINOLOGY

Advantages

- larger framed
- larger, more growth oriented
- tracks wider and truer in front/rear
- more level in the rump
- more level topline
- heavier muscled
- more natural thickness
- thicker, squarer rump/hip
- wider tracking
- more style and balance
- nicer balanced
- more eye appealing
- longer, cleaner neck

Criticisms

- earlier maturing
- quick patterned
- steeped rump
- breaks in loin
- light muscled
- tapered into lower leg
- narrow tracking
- narrow chested
- poorly balanced
- weak topped
- coarse fronted
- shorter necked
- coarser made

Selection of Meat Goats

When selecting animals for breeding stock, producers must consider the purpose, functionality, and durability of the animal. Producers often examine the health, visual appearance, pedigree, and performance data of individual animals before selecting does and bucks for the meat goat herd. Many producers first learned to evaluate meat goats in youth competitions.

Meat goat projects for FFA and 4-H have become more popular with the rapid growth of the meat goat industry. The evaluation of meat goats is now a part of the National FFA Livestock Judging Career Development Event (CDE) and the National 4-H Livestock Judging Contest. Most states and districts conduct meat judging competitions to advance to the next level and eventually to the National Competition. The following is a basic guide for evaluating and showing meat goats.

In the Show Ring

Be sure to check for specific rules regarding shows. There are a few different types of shows. Generally, the categories of market goat shows are divided by weight, gender, and breed. Most goats shown are less than 1 year of age. To determine the age, the animal's teeth are checked prior to the show. Goats with milk teeth are usually less than 1 year of age. However, having proper records will ensure the age of the animal. Generally, goats are required to be castrated for showing unless it is a buck show. Always keep proper documentation on all show animals.

Showmanship is another important showing class. The purpose of showmanship is not to showcase the exhibitor, but to showcase the animal to the showman's best ability. In order to have a successful show, some basic guidelines are important:

- Know the show's rules prior to show time
- Lead the goat with a show chain (Check show guidelines)
- Always leave enough space between the animal and other exhibitors' animals
- Be aware of the surroundings, and keep an eye on the judge

Showmanship Tips

There are numerous things to do before and during the show. Below are a few tips on showing goats.

Before the Show:

- Be sure the goat is clipped and the hooves are trimmed prior to showing (Figure 28-31)
- Tip horns or dehorn the animal after birth
- Teach the animal to be calm and to lead in the show ring

The Day of the Show:

- Carefully clean the goat by washing off all dirt and debris
- Change into appropriate clothing before show time. Be neat and presentable
- Arrive early at the show site, and be on time for your class

In the Ring:

- Keep your eyes on the judge at all times. Smile!
- Keep the goat between the judge and yourself at all times
- Be a good sport. Congratulate winners, and shake the judge's hand after showing

Courtesy of James Hale

Figure 28-31 Be sure to clip the goat prior to showing.

- Set the goat up by keeping the animal's legs out
- Be prepared to brace the animal. Generally, bracing is not used, but be prepared in case the judge asks

FEEDING GOATS

Feeding Dairy Goats

Goats are ruminants and have digestive systems similar to those of cattle. Large amounts of roughage may be used as the basis for feeding goats. Grains are added to supplement low-quality roughages and to provide additional nutrients during growing and lactating periods.

Rations can be made up of grains, pasture, hay, silage, browse, and other feedstuffs (Figure 28-32). Commercial protein concentrates may be used as well as home-mixed protein feeds. Commercial dairy mixes will provide the protein needed by the dairy goat. Minerals, vitamins, salt, and water are also essential in the diet of the dairy goat.

USDA Agricultural Research Service, Photo by Scott Bauer

Figure 28-32 When grazing, goats select the young growing points first as they browse downward from the upper parts of a plant.

Feed good-quality roughages free choice to lactating does. Legumes, grasses, or grass-legume mixtures may be used. The quality is more important than the kind of hay being fed. More protein is needed in the concentrate if grass hays are fed. The crude protein content of the ration may range from 14 to 20 percent depending on needs. Lactating does are generally fed a 17 percent ration. No additional protein supplement is needed if good-quality legume hay or pasture is available for the dry does.

Although you can mix your own ration, most producers work with a local feed producer to produce a special blend or buy commercially available feed mixes. The amount of feed needed to produce quantity and quality of milk varies depending on environment, breed, stage of lactation, nursing offspring, and other management issues. Joining a local dairy goat club or speaking to others with dairy goats can help you figure out the amounts and feed types needed to maintain high-quality production. Trace minerals can be fed free choice to the herd. The mineral mixture should contain large amounts of copper. Goats, unlike sheep, require large amounts of copper in their daily diet.

Pasture and hay may be fed as the only feed for does during the gestation period. Keep the doe in good body condition. If additional feed is needed, use grain to maintain good body condition. Some breeders use bran before freshening and shortly after. Do not allow the doe to become too fat during the gestation period.

The buck can be supplied the needed nutrients with pasture and hay with a complement of free choice minerals. Concentrate may be added to the ration, depending on the body condition of the buck and the quality of the pasture and hay. Legume hays should not be fed to bucks. Do not allow the buck to become too fat. During the active breeding season, an additional 1 to 2 pounds of grain per day may be fed.

It is important that kids receive colostrum shortly after they are born. Some producers allow the kids to nurse from the does, while others prefer to take the kids away from the does and hand feed the colostrum. Train kids to drink from a pan or bucket as soon as possible.

Because the does are usually used for milk production, the kids should be fed away from the does. Begin feeding the kids 0.5 pint of milk four times per day. If the substitution is made gradually, cow's milk may be used in place of goat's milk. Milk replacers sold commercially may also be used. By the time the kid is 6 weeks old, the amount of milk being fed should be 5 pints per day. Kids begin to eat dry feed during the first 3 to 4 weeks. Provide fresh water and high-quality legume hay. Calf starter or rolled grains should be fed. Kids are usually weaned from milk at 8 to 12 weeks of age.

Growing kids are fed mainly on roughages. Low-quality roughages must be supplemented with a 12 to 14 percent protein mix. Grain may be fed at the rate of 0.5 to 1.5 pounds per head daily. Do not allow the kids to become too fat. Adjust the amount of concentrate feed based on body condition.

Feeding Fiber Goats

As long as fiber goats are on good range with a wide variety of brush, weeds, and grass, additional feeding is not necessary. Additional feed is needed in most parts of the U.S. during winter.

Fiber goats need green browse, grass, and weeds to produce good fibers. Guajillo and live oak are rated as excellent browse for fiber goats. Other common varieties of brush are rated from good to poor. Black persimmon, mesquite, and white brush are rated among the poorest.

Supplemental feeding may include pelleted feeds, 20 percent protein range cubes, shelled yellow corn, or other feeds. The amount to feed depends on the condition of the pasture and the goats. Pregnant does require more feed than dry goats. Protein may be provided by feeding 0.25 to 0.5 pound of cottonseed cake. Yellow corn may be fed at the rate of 0.5 to 1.0 pound per head per day. Goat cubes may be used in place of cottonseed cake and corn at the rate of 0.5 to 0.75 pound per head per day.

Goats may be self-fed on range, with salt added to the mixture to limit feed intake. Mix seven parts of concentrate to one part of salt if the feeders are located a mile or more from water. If the feeders are placed near water, mix three parts of concentrate to one part of salt to limit feed intake. A common concentrate mix is one part cottonseed meal, three parts ground grain, and one part salt. Alfalfa or other ground roughages may be used in the self-fed mixture. If ground roughage is used, lower the amount of salt in the mix. Moving the feeders from time to time results in better use of the pasture. Salt-controlled feeding should only be used if all other methods of feeding are impractical.

Roughages may be fed as hay. Commonly used hays include alfalfa, sorghum, peanut, Sudan, and Johnson grass. At least 1 pound of hay should be fed per head per day. Singed prickly pear and tasajillo may be fed free choice along with 0.25 pound of cottonseed cake per head per day.

Feeding Meat Goats

Feeding meat goats is generally the same as feeding fiber goats. They need basically the same amount of minerals and vitamins, but meat goats often require more protein and energy. A grain ration of 14 to 18 percent protein may be fed. The best way to reach these levels is to add a protein supplement, if needed. However, most meat goat producers use pasture or rangeland to meet the needs of their goats.

MANAGEMENT OF GOATS

Managing Dairy Goats

Standard-sized dairy goats are seasonal breeders, while Nigerian Dwarfs are polyestrous. The breeding season of standard dairy goats extends from September to March. Does are usually bred in September, October, or November and kid in February, March, or April. Depending on the breed, size, and body condition, yearling kids may be bred at 7 to 10 months of age. Body weight relative to the breed is more important than age and can affect lifetime performance. If breeding is postponed much beyond 10 months of age, the does may be less productive.

Keep the buck in a pen separate from the does during the milking season. The buck gives off a strong odor during this time, which can affect the odor and taste of the milk. It is generally a good practice to pen the buck separately throughout the year. A buck that does not receive enough exercise may become sterile. Do not overwork the buck during breeding. A year-old buck can breed about 25 does during the season, while a mature buck can breed about 50 does during one breeding season. Keep breeding records on every doe.

The gestation period for goats is 5 months (148 to 150 days). Watch the does for signs of kidding. The doe usually becomes more nervous a few hours before kidding, and the udder becomes swollen with milk. The doe appears to shrink in the belly, the flanks appear hollow, and the tail head seems to be higher than usual. These changes may occur 2 to 3 days before kidding. Do not help the doe with kidding unless she is obviously having difficulty. An experienced producer can provide help with difficult births. The inexperienced producer should obtain help from an experienced veterinarian.

Most does will want to be separate from the other goats in the herd after kidding. Does and their kids should be kept together and away from others for a period of at least 12 to 24 hours. This helps the doe bond with the kids and allows the producer to make sure the kids are being fed. If it is available, put kids on pasture after the first 24 hours. Provide a pen large enough for the kids to get plenty of exercise. Boxes, ramps, or other elevated objects in the yard provide the kids with items for running and jumping exercises.

A vaccination protocol should be set up for the kids as well as the rest of the herd prior to kidding season. Kids should be fully vaccinated before weaning from the bottle. Separate buck kids from doelings by the time they are 2 to 3 months of age. The buck kid usually becomes capable of breeding at the age of 3 months. Failure to separate the bucks from the does will result in does being bred too young.

Bucks that are not to be kept for breeding should be castrated at 8 to 12 weeks of age. Be sure that the testicles have descended into the scrotum before attempting to castrate. A knife, elastrator, or burdizzo may be used to castrate. Castrate before fly season if a knife or elastrator is to be used.

Dehorning, also known as disbudding, should be done when the kids are 3 to 5 days old. An electric dehorner may be used on young kids.

Unless goats are kept on hard surfaces, their hooves should be trimmed from time to time. Goats with untrimmed feet may go lame. Pruning shears may be used to trim the feet. A sharp jackknife or farrier's knife can also be used for trimming.

Goats may be marked for identification by tattooing, electronic identification, or ear tagging. Tattooing is done with an instrument, sold commercially, that is designed for this purpose. The tattoo is placed in the ear or in the soft tissue on one side of the tail. Electronic identification or microchips may be used to identify the goats. Because ear tags may tear out of the ear on a fence or in brush, ear tagging is not generally recommended. Identification is especially important for purebred producers.

Dairy goats are usually milked on a stand. The stand is 1.5 to 2.5 feet high. It has a stanchion to hold the doe while she is being milked. Milk absorbs odors from feed. Any strong-flavored feeds such as silage, onions, or cabbage should be fed only after the milking season.

Figure 28-33 Goats are milked either by hand or machine.

Does may be milked either by hand or by machine (Figure 28-33). The does should be clipped on the udder and flank area. This helps to keep dirt and hair out of the milk, especially if the doe is milked by hand.

Washing the udder with warm water and a sanitizer will stimulate milk letdown, as well as promote cleanliness. Hands and equipment should be sanitized between goats. Use a strip cup to check for mastitis. Milk the first two or three squirts of milk into the strip cup and examine it. Begin milking within 2 or 3 minutes after washing the udder. If you are milking by hand, use a hooded bucket to prevent dirt from getting into the milk.

Milking at regular intervals of 12 hours gives the maximum production. The same person should feed, handle, and milk the goat. Following a regular routine gives the most satisfactory results.

Strain the milk through a commercial strainer pad. Cool the milk as quickly as possible after milking. Commercial coolers designed for this purpose are available. Running water may also be used to cool the milk. After cooling, the milk must be kept refrigerated to prevent bacterial growth and spoilage.

Keep milking equipment clean and sanitized. Use a good dairy detergent for cleaning the equipment. Rinse the equipment with warm water as soon as the milking is done. Then scrub it in warm water using a dairy detergent. Brush all surfaces to be sure they are clean. Rinse again with an acidified rinse to prevent the buildup of **milkstone** (mineral deposits) on the equipment. Drain all rinse water and store the equipment bottom side up in a place free of dust and flies. Before using the equipment for the next milking, sanitize it with boiling water or rinse it with a dairy sanitizer. Follow directions on the package when using a dairy sanitizer.

Goat's milk is pure white. It has smaller fat globules and a softer curd than cow's milk. Because it is more easily digested, goat's milk is sometimes prescribed by doctors for people who have difficulty digesting cow's milk.

The cream in goat's milk rises more slowly than that in cow's milk because goat milk is naturally homogenized. Mechanical separation of cream is recommended. Both hard and soft cheese can be made from goat's milk. Butter made from the cream of goat's milk is white. The composition of goat's milk is similar to that of cow's milk. The fat content is generally higher, and the lactose (sugar) content is slightly lower in goat milk.

Managing Fiber Goats

Fiber goats breed in the fall and kid (give birth) in the spring. It is recommended that three or four bucks be run with each 100 does. Use a fresh pasture, or feed

0.25 to 0.33 pound of corn per head per day. Keep the does in good body condition during gestation to prevent abortion. It is recommended that bucks be at least 1 year old before being used for breeding. If the number of does bred is limited, a buck as young as 8 months old may be used.

Flushing does is a recommended practice. Flushing means to increase nutrient intake for several weeks before breeding. Flushing may increase the number of kids born, especially if the does are in thin body condition.

Two systems of kidding are commonly used. One system is to kid does in the pasture. A fresh pasture should be used and the does should be disturbed as little as possible. This system produces hardier goats with less labor.

The other system involves the use of stakes and kidding boxes. The kid is tied to a stake with 15 inches of rope. A box is provided for shelter. The doe is allowed to graze in a nearby pasture and is paired with the kid in the evening. Does and kids must be marked for identification with this system. In general, staking of the kids requires more labor. A modification of this system allows the kids to run free in a large pen. The does are paired with their kids in the evening. Allow kids to go to pasture with the does when they are about 1 month old. The stake or pen system is commonly used by purebred breeders because it facilitates record keeping.

Newborn kids should be handled as little as possible. Too much handling may cause the doe to disown them. Be sure that the kid nurses soon after it is born. If the doe refuses to claim the kid, put it in a small pen with other unclaimed kids for a few days. All unclaimed kids have to be fed colostrum soon after birth. If possible, milk the does that have just given birth. Feed the orphans as much of the colostrum as possible. Any remaining colostrum should be frozen in case it is needed at a later date. Raise orphan kids on a bottle, using cow's milk. Follow procedures outlined for feeding orphaned sheep.

Ear mark the kids and vaccinate them for sore mouth when they are about 1 to 1.5 months old. Ear marking is done only for identification. Because plastic tags may pull out in brush, most producers use a system of ear notching. Figure 28-34 shows the approved system for ear notching registered Angora goats.

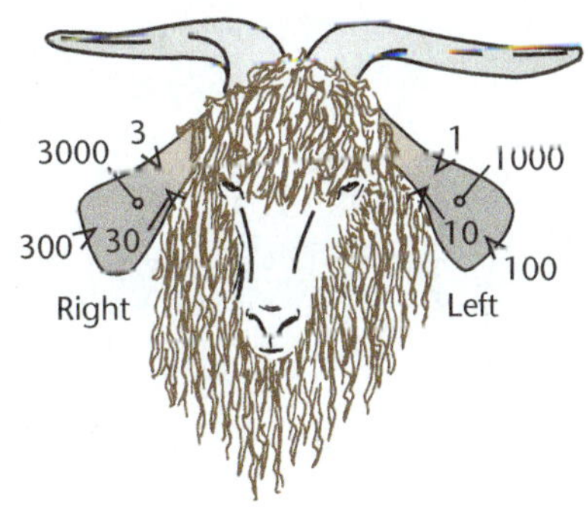

Figure 28-34 A system for identification of Angora goats by ear notching.

Castration is often done in November, December, or January following kidding in the spring. Castrating at the older age produces a heavier horn on the wethers. Heavier horns are preferred by buyers. A knife or burdizzo is used for castration. The knife is preferred for younger kids. The burdizzo is best for older animals.

Kids are usually weaned after the fall shearing (Figure 28-35). Move the does away from the kids when weaning. Leave the kids in the familiar pasture.

Courtesy of Dr. Fred Speck

Figure 28-35 Angora goat kids are usually weaned after the fall shearing.

Unlike sheep, Angora goats are generally sheared twice a year. They are sheared in spring before kidding and in fall prior to breeding. Spring shearing time is from February to April, depending on the region. Fall shearing time is mid-July to the end of September. Shearing procedures are similar to those for sheep.

The growth of cashmere down begins in late June and usually stops in late December. If the fiber is not harvested after growth stops, it will be shed naturally. The fleece shows two types of fiber: a very fine, crimpy down and the longer, outside coarse, straight guard hair. The fine cashmere fiber is separated by combing out or using a commercial dehairer on the sheared fleece.

The practice of **caping** is sometimes followed to protect the goats from the weather after shearing in the spring. In caping, a strip of mohair 3 to 4 inches wide is left down the goat's back. This should be sheared 1 month or 6 weeks after the first shearing. Special goat combs that leave 0.25 to 0.5 inch of stubble on the goats are sometimes used. This stubble provides about the same amount of protection from the weather as does caping. Using the special goat comb eliminates the problem of undesirable staple length that results from caping. Regular combs are generally used for shearing in the fall.

External and internal parasites are controlled by spraying and drenching. Most producers spray and drench goats as they come out of the shearing pen. Follow current recommendations for drenching substances. Always follow directions on the product being used.

Getting foreign material from pasture caught in the fleece is a serious problem in mohair production. Needlegrass, speargrass, grass burrs, cockle burrs, horehound, and other plant materials may contaminate the mohair. To minimize the problem, it is a good idea to move goats to pastures that contain few if any of these plants. Changes in shearing dates will sometimes help reduce plant contamination. Improving pastures by controlling noxious weeds is also recommended.

If oil is applied within 30 days after shearing to control external parasites, it will work out of the fleece before the next shearing. Some producers use oil before shearing to increase fleece weight. Fleeces are discounted by buyers when this practice is followed. The oil is difficult to remove from the fleece, the mohair is left dark and dingy, and the fiber luster is harmed. Artificial oils should not be used to increase fleece weight.

Managing Meat Goats

Meat goats are similar in many ways to fiber goats. Only the major differences are discussed here. Meat goats will breed year-round. Most producers follow a twice-a-year breeding system. Does are bred in February and March and again in September and October. Selection of breeding animals is based on certain characteristics. Size, body conformation, and rapid growth are important points to consider when making selections.

Many producers cross Spanish goats with Boer goats to increase meat production. The hybrid cross results in an animal that has some of the pest resistance of the Spanish goat and the meat-type body of the Boer goat. Another benefit is the ability of the Boer goat to have an extended breeding season. This makes it possible to have three kiddings every 2 years. Replacement doe kids are selected at weaning time. Doe kids are weaned in a drylot in order to teach the does to eat supplemental feed. The replacement does are returned to the breeding flock at about 1 year of age.

HOUSING AND EQUIPMENT

Housing and Equipment for Dairy Goats

The amount and kind of housing necessary for dairy goats depends on the number of goats owned, location, and the convenience desired by the producer. A small barn or shed is adequate for a farm with only a few goats. A producer who has a large flock may want more elaborate housing. Two types of housing are commonly used: loose pen and tie stalls. Combinations of housing are also used. The milking does are kept in tie stalls, and loose pens are used for the kids and yearlings.

When using loose housing, provide 15 square feet per goat. A shed, open to the south, may be used. Some provision for shutting the open side in cold weather may be needed in cold climates. More bedding is required for loose housing. Goats must be dehorned if loose housing is used. Loose housing does not cost as much as confinement housing.

Confinement housing is more expensive; however, less bedding is needed. Goats are kept in individual tie stalls or pens. The building may be insulated, and mechanical ventilation is needed. Bucks should be housed separately from the rest of the goats.

A separate milking area and milk room are recommended. If it is attached to the barn, use double doors to keep dirt, odors, and flies out of the milk area. Local Board of Health standards may regulate the design and construction of the milking area and milk room.

Woven wire fences are best for goats. The fence should be at least 4.5 to 5 feet high. Temporary fencing for dividing a pasture may be electric. Three wires are needed: at 10 inches, 20 inches, and 40 inches from the ground. Tethering is not generally recommended. If it must be done, move the location each day so the goat has feed. Provide shade on pasture if the goats are tethered. Feed racks, troughs, hurdles, and other equipment are about the same for goats as for sheep.

Milk Room Equipment:

- milk cooler
- sink for washing equipment
- hot and cold running water
- sanitation supplies
- place to store equipment
- milk pail, strainer, and brushes
- electric milking machine
- scale for weighing milk

Housing and Equipment for Fiber and Meat Goats

The housing and equipment needs for fiber and meat goats are very similar to those for sheep. Housing and equipment needs vary depending on the area of the country. Fiber and meat goats are produced with a minimum of housing and equipment. However, they need basic protection from the elements and weather extremes such as rain, wind, and cold. During the summer months, adequate shade and air circulation should be provided. Housing options vary by herd size, land availability, and operation type. Types of housing include open housing, confinement housing, kidding pens, and the use of existing buildings and structures. A dry, draft-free area is needed for kidding. Pasture kidding is most often used, but a covered area is needed during extreme conditions.

If animals are to be confined, each animal should be provided with a minimum of 20 square feet. Handling equipment such as corrals, chutes, and watering and feeding equipment make management much easier. Horned animals may require specialized equipment. On pasture, perimeter fencing should be adequate to keep predators out as well as the goats inside. Fencing for subdividing pastures is important. Subdivided pastures allow rotational grazing to help control internal parasites. Fiber and meat goats are often placed in pastures or on rangeland due to their ability to share land well with both sheep and cattle.

PROTECTION FROM PREDATORS

While good fencing and housing helps to protect goats, predators such as domestic dogs, wolves, coyotes, and wildcats are serious threats to the goat herd. Goats are easy prey for predators, and they require additional protection. Guardian animals such as dogs, donkeys, and llamas help protect the herd. The Great Pyrenees dog is one of the most widely used guardian animals.

SUMMARY

Goat populations are increasing rapidly in the United States. Reasons for the increased production of goats include a growth of the ethnic population; the ability of goats to utilize land unsuitable for other uses; an increase in demand for goat milk, cheese, and butter; an increase in small hobby farms; increased research on goat production; the rising popularity of chevon and cabrito as delicacies in more upscale restaurants; the increase in popularity as show animals, especially in youth organizations; and the use of goats as a green method of controlling weeds and other undesirable plants.

The simplest way to classify goats is by products made from the animal. The three major classifications are dairy, fiber, and meat, with the two minor classifications being pets/companions and for goatskin. Small herds of dairy goats are found in all parts of the United States.

Many small farms keep dairy goats to provide milk for the family and local markets. There are seven breeds of dairy goats commonly found in the United States: Alpine, LaMancha, Nigerian Dwarf, Nubian, Oberhasli, Saanen/Sable, and Toggenburg. Dairy goats are selected for milk production. Good dairy goats average from 3 to 4 quarts to 1 to 2 gallons of milk per day. The American Dairy Goat Association has developed scorecards that indicate the desirable traits to be used in selecting or judging dairy goats.

Fiber goats are raised for mohair and cashmere and are also used for meat production. There are about 20 breeds of goats used for fiber worldwide. Three fiber goat breeds are common in the United States: Angora, Cashmere, and Miniature Silky Fainting Goats. Fiber goats are judged on the basis of body and fleece. Balance must be maintained between selection for body conformation and fleece because too much emphasis on fleece tends to reduce the size and vigor of the goat. The body conformation is judged on the basis of breed type, conformation, amount of bone, constitution and vigor, and size and weight for age. The fleece is judged on the basis of fineness, uniformity and completeness of covering, oil content and luster, density, and character of fleece. Breeding animals are further selected on the basis of age and fertility.

Goats of all breeds may be harvested for meat, with about 45 breeds classified specifically as meat type. The most common meat type goat breeds in the United States are Boer, Kiko, Kinder, Myotonic, Pygmy, Savanna, and Spanish. When selecting animals for breeding stock, producers must consider the purpose, functionality, and durability of the animal. Producers often examine the health, visual appearance, pedigree, and performance data of individual animals before selecting does and bucks for the meat goat herd. Meat and dairy goat projects for FFA and 4-H have become more popular with the rapid growth of the goat industry.

There are several basic guidelines to follow when evaluating and showing meat goats. In livestock evaluation events, classes with four animals in each class are provided for judging. The contestants rank the animals in order from most to least desirable. Competitions may include other components such as a written exam and oral reasons. It is important to learn the parts of a meat goat and the desirable characteristics for marketing and breeding. The single most important trait in market goats is muscling. In breeding does, the most important trait is soundness and structural correctness. Finish, capacity, balance and style, and sexual characteristics are other categories that should be considered when placing a class of meat goats. In many judging competitions, contestants must defend their placing with oral reasons. Be sure to check the specific rules for the event.

Goats are ruminants and have digestive systems similar to those of cattle. Large amounts of roughage may be used as the basis for feeding goats. Grains are added to supplement low-quality roughages and to provide additional nutrients during growing and lactating periods. Dry or not reproducing dairy goats are fed mainly pasture and hay. Grain and protein supplements are fed to the milking does. Fiber goats are fed mainly on browse, with some supplemental feeding being added as needed. Feeding meat goats is generally the same as feeding fiber goats.

Fiber goats are seasonal breeders as are standard-sized dairy goats. The does are bred in the late fall and kid in the spring. Meat goats and Nigerian Dwarf goats will breed year-round and most producers use a twice-a-year breeding system. The gestation period for goats is 5 months (148 to 150 days). Body weight relative to the breed is more important than age when determining when to breed goats. Two systems of kidding are commonly used: kidding does in the pasture or using kidding boxes or pens. Goats may be marked for identification by tattooing, the use of microchips, or ear tagging.

Housing and equipment needs vary by herd size, location, land availability, and operation type. Care must be taken when milking dairy goats to keep the milk and equipment clean. Following a regular milking routine may help to increase production. The housing and equipment needs for fiber and meat goats are very similar, as well as for sheep. Fiber and meat goats are produced with a minimum of housing and equipment; however, they need basic protection from the elements and weather extremes. Good fencing and housing helps to protect goats from the elements and predators. Guardian animals, including dogs, donkeys, and llamas, are commonly used to help protect the herd from predators such as domestic dogs, wolves, coyotes, and wildcats.

Student Learning Activities

1. Prepare a PowerPoint™ presentation displaying pictures and characteristics of goat breeds.
2. Identify the parts of the goat on a picture or live animal.
3. Give an oral report on the selection, feeding, management, and housing and equipment requirements for one of the types of goats (dairy, meat, or fiber).
4. Judge a class of goats using the guidelines for judging provided in this chapter.
5. Practice selection and management procedures described in this chapter when conducting a supervised agricultural experience program on goats.

Discussion Questions

1. The goat industry in the United States has experienced an increase in production. Discuss at least three reasons for this increase.
2. Name and describe two goat breeds from the dairy, fiber, and meat goat types.
3. Describe the desirable characteristics of a dairy goat.
4. Discuss the factors to be considered when selecting fiber goats.
5. Describe the basic guidelines for evaluating meat goats.
6. What are the recommended feeding practices for dairy, fiber, and meat goats?
7. Discuss the recommended management practices for dairy, fiber, and meat goats.
8. Discuss the recommended practices for milking dairy goats.
9. Describe the practice and purpose of caping.
10. Describe the housing and equipment needs of dairy, fiber, and meat goats.

Review Questions

True/False

1. A buck is an unbred female goat.
2. There are three common meat goat breeds in the United States.
3. Cashmere refers to the soft down or winter undercoat of fibers produced by most breeds of goats, except the Angora.
4. Muscling is not an important trait in market goats.
5. Goats are ruminants and have digestive systems similar to those of cattle.

Multiple Choice

6. The fleece of the Angora goat is called_____.
   a. mohair c. wool
   b. cashmere d. fur
7. The shoots, twigs, and leaves of brush plants are called ________.
   a. cashmere b. wattles
   b. browse d. kemp
8. Texas produces ________ percent of the total U.S. mohair.
   a. 20 c. 60
   b. 40 d. 90
9. Which of the following is not used as a guardian animal to protect goat herds?
   a. Great Pyrenees dog c. wolves
   b. donkeys d. llamas
10. The most important trait to consider when evaluating breeding does is _____.
    a. capacity c. muscling
    b. finish d. soundness and structural correctness
11. Which of these is not matched to its correct definition?
    a. doeling: an unbred female goat
    b. kid: young goat under 1 year of age, either sex
    c. yearling: goat over 1 year and under 2 years of age, either sex
    d. wether: male goat of any age
12. Myotonic, as in myotonic goats, means:
    a. spasm or temporary rigidity of the muscles
    b. a mix of black and white coat color
    c. weak feet and legs
    d. large body capacity
13. If the four animals in a judging class are ranked 2-3-1-4:
    a. The most desirable animal is number 4.
    b. The most desirable animal is number 2.
    c. The most desirable animal cannot be determined from the information given.
    d. The top pair consists of animals number 1 and number 4.
14. In a judging event, the most important factors in selecting market meat goats are muscling, soundness and structure, finish, capacity, and ________.
    a. color c. line
    b. balance and style d. brisket thickness
15. Goats deposit fat on the ________ of their body first.
    a. interior c. front half
    b. outside d. back half

Completion

16. Some goat breeds are considered ________ purpose, that is, used to produce two or more products such as meat and milk.
17. A female goat of any age is known as a ________.
18. Fiber goats are judged on the basis of ________ and ________.

19. When selecting animals for breeding stock, meat goat producers must consider the ________, ________, and ________ of the animal.
20. The two types of housing commonly used for dairy goats are ________ and ________.

Short Answers

21. List the three major classifications and two minor classifications of goats.
22. What are the seven common dairy goat breeds and the three common fiber goat breeds in the United States?
23. Certain terminology is used when evaluating meat goats. Give five terms used for advantages and five terms for disadvantages.
24. What are the signs that a doe is about to kid?
25. List the equipment needed in the milk room.

Chapter 29

Diseases and Parasites of Sheep and Goats

Key Terms

virus
encephalitis
toxemia
convulsion
causative agent

Objectives

After studying this chapter, the student should be able to

- describe common diseases and parasites of sheep and goats.
- list prevention and control practices for common parasites of sheep and goats.
- outline a program to reduce losses from diseases and parasites.

GENERAL HEALTH RECOMMENDATIONS

A good health program for sheep and goats involves prevention, rather than treatment, of diseases and parasites. Sheep and goats do not respond well to treatment. When they become ill, they frequently do not recover. Sick sheep and goats are difficult to identify. Once a sick animal is identified, it often dies in spite of treatment because the illness has progressed too far before diagnosis. The help of a veterinarian may be needed to plan the use of drugs and to diagnose problems affecting the flock or herd.

Observing the vital signs (temperature, pulse rate, and respiration rate) in an animal can help in the early detection of health problems. Vital signs will vary with activity and environmental conditions. Normal vital signs in sheep and goats are:

- temperature (sheep): 100.9 to 103.8°F. The average is 102.3°F. Temperature is usually higher in the morning than in the afternoon; younger animals will show a wider range of temperature than mature animals
- temperature (goats): 101.7 to 105.3°F. The average is 103.8°F . Temperature is usually higher in the morning than in the afternoon; younger animals will show a wider range of temperature than mature animals

- pulse rate (sheep and goats): 70 to 80 heartbeats per minute
- respiration rate (sheep and goats): 12 to 20 breaths per minute

Body temperature is taken in the rectum using either a mercury thermometer or a battery-powered digital thermometer. Restrain the animal when inserting the thermometer into the rectum. The pulse rate is taken by finding the artery that runs down the inside of the hind leg. The pulse rate may also be taken by finding the artery that is high up on the inner surface of the thigh just where it emerges from the groin muscle. Respiration rate is determined by observing the number of times the animal breathes per minute.

Preventing Health Problems

1. Watch closely for signs of illness.
2. Use good feeding and management practices.
3. Handle animals with care; avoid stress.
4. Follow strict sanitation practices.
5. Treat wounds with disinfectants.
6. Select healthy animals for breeding.
7. Isolate newly purchased animals for 30 days.
8. Prevent contact with animals from other farms.
9. Control traffic and people in areas where animals are kept.
10. Isolate sick animals for treatment.
11. Prevent diseases by vaccinating.
12. Control parasites.
13. Rotate pastures to prevent parasite buildup.
14. Work with a veterinarian for prevention and treatment.

DISEASES OF SHEEP AND GOATS

A number of diseases are common to sheep, goats, and cattle. While there may be some differences in symptoms, the general description of many of these diseases is very similar. A discussion of the diseases listed in Table 29-1—which may affect sheep, goats, and cattle—is found in Chapter 17. Other diseases discussed in this section are more specific to sheep and goats.

Bluetongue (Sore Muzzle)

Bluetongue is a disease of sheep that is caused by a virus. A **virus** is a self-reproducing agent that is considerably smaller than a bacterium and can multiply only within the living cells of a suitable host. This disease occurs mainly in the western United States and is spread from sheep to sheep by a small gnat, a tiny biting midge (Figure 29-1). Bluetongue weakens a sheep's resistance to other diseases. Death loss (about 5 percent) is usually due to secondary infections such as pneumonia.

Sheep with the disease lose their appetite, become sluggish, and have a high fever. The ears, head, muzzle, and lips become swollen. The tissues inside the mouth become red and blue. The tongue develops ulcers and the sheep has difficulty eating. Lameness and swelling around the hoof occurs. There is a bad odor, and a discharge from the nose and eyes.

There is no treatment for bluetongue. Treat the secondary infections if the disease occurs. However, a vaccination is available that will prevent the disease. Vaccinate all ewes and rams at shearing time. Vaccinate replacement lambs at $3\frac{1}{2}$ months of age.

Caprine Arthritis Encephalopathy

Caprine arthritis encephalopathy (CAE) is a viral disease of goats that can cause encephalitis in kids and chronic joint disease in adults. There is no cure, so preventing spread of the disease is imperative. There are two kinds of CAE, each with distinct symptoms once the animal is infected. The first indication of CAE can be

USDA Agricultural Research Service; Scanning electron micrograph by Richard Nunamaker

Figure 29-1 The virus causing bluetongue in sheep and cattle is spread by a gnat, a tiny biting midge shown here.

TABLE 29-1 Diseases That May Affect Sheep, Goats, and Cattle

• actinobacillosis (wooden tongue)	• actinomycosis
• anthrax	• blackleg
• bloat	• brucellosis
• Johne's disease	• leptospirosis
• listeriosis	• malignant edema
• pinkeye	• shipping fever
	• white muscle disease

swollen joints, usually seen in the knees of goats. **Encephalitis** is an inflammation of the brain and may be caused by CAE. Encephalitic seizures usually kill infected kids. Goat herds should be tested every year by a veterinarian, and all incoming animals should be tested during the quarantine period and before entering the herd. This disease is passed from animal to offspring through the milk.

Caseous Lymphadenitis

Caseous lymphadenitis (CL) is a contagious disease caused by bacteria. It occurs in both sheep and goats. This disease is characterized by abscesses in or near the lymph nodes. Abscesses may burst open, exposing other animals to this disease. Once CL abscesses have burst, the bacteria can survive in the soil, on bedding, and on structures for several months. Any injury to the skin, including abrasions from shearing and castration and cuts from nails and other objects, provides an opportunity for infection. The best control for this disease is culling from the herd before the abscess bursts.

Enterotoxemia (Overeating Disease)

Enterotoxemia is a disease caused by a bacterium. **Toxemia** is a condition resulting from the spread of bacterial toxins through the bloodstream. This disease affects both sheep and goats, with lambs and kids being most often affected. The most common sign of the disease is finding dead lambs or kids. The dead animals have their heads drawn up in an arched and extended position as a result of the convulsions that occur before death. If an animal is having **convulsions** (involuntary muscular contractions), death will follow quickly. There is no treatment after the symptoms appear.

Enterotoxemia can be controlled through good management, proper feeding, and vaccination. Proper diagnosis of the disease is done through laboratory examination of tissue from animals that have died suddenly. Maintaining a steady intake of feed or milk is the best method of preventing this disease in lambs or kids. Gradually adjust rations for lambs/kids when increasing the concentrate level. Maintain a good source of clean drinking water. Chilling and stress can increase the incidence of enterotoxemia, especially in ewes/does that are milking heavily and nursing singles.

Ewes and does should be vaccinated 4 weeks and 2 weeks before lambing/kidding to prevent the disease in nursing young. In sheep, after ewes have been vaccinated, in succeeding years vaccinate them once about 2 to 3 weeks before lambing; lambs will secure protection from the disease through the colostrum milk. Vaccinate late-weaned lambs twice before weaning, once late in the nursing period and again 2 to 3 weeks later. When practicing early weaning (at 40 days), vaccinate lambs about 10 days before weaning and again about 10 days after weaning. Antibiotics in the feed and increasing the amount of roughage in the ration may help protect animals from enterotoxemia.

Kids should be vaccinated starting at 8 weeks old, if still receiving goat milk, and every 21 days after that for a total of three vaccinations. If the kids are being fed cow milk, their vaccination schedule is different. These kids should be given the first of three vaccinations at 5 to 6 weeks of age with the following vaccinations 21 and 42 days after the original vaccination. Dairy goats are usually vaccinated twice per year for enterotoxemia after the initial vaccination.

Foot Abscess (Bumblefoot)

Foot abscess is a disease that affects the soft tissue of the foot. It is infectious but, unlike foot rot, it is not contagious. It may occur in connection with foot rot when animals are in wet or muddy conditions, when the feet have been severely trimmed in wet weather, or when they are put on stubble pasture. Bacteria enter the foot through injuries, causing pus pockets or abscesses. There may be no visible swelling if the infection occurs in the toe or sole of the foot. If it occurs in the heel, there is generally a visible swelling. The disease may affect the joints and tendons. Foot abscess may cause permanent lameness in the animal. Treat the disease by draining the abscesses, applying a medicated dressing, and using systemic antibiotics. Isolate infected animals on soft, clean footing or slotted floors until healing is completed. Prevent foot abscess by correcting the conditions that can cause foot injuries.

Foot Rot

Foot rot is a disease caused by the presence of two different bacteria. Both types must be present for the disease to develop. Foot rot affects both sheep and goats. It is not the same disease as the foot rot that occurs in cattle and is caused by different bacteria.

Foot rot is extremely contagious and may affect the majority of the animals in the flock or herd when an outbreak occurs. It is more likely to occur when animals are on irrigated pastures, wet lowland pastures, or in areas with high rainfall. It is less likely to be a problem when animals are on sandy, well-drained soils and in areas of low rainfall. The disease will spread more rapidly if large numbers of animals are concentrated in a small area. The death loss from foot rot is low, but affected animals may lose weight. It can cause significant economic loss for sheep and goat producers.

Symptoms include lameness, loosening of the hoof wall, and a foul-smelling discharge. It does not form abscesses on the foot. Affected animals may move around on their knees.

Controlling foot rot involves regular inspection, proper trimming of the feet, keeping animals out of wet areas, keeping bedding dry, and the regular use of a footbath. Footbath solutions, as discussed in the next section on foot scald, may be used. Keeping the area around feed troughs dry and treating the area with disinfectants or drying agents may help prevent foot rot.

When an outbreak of foot rot occurs, separate the infected animals from the rest of the flock/herd and put them in a clean, dry area. Trim the feet if necessary. Treat the infected feet in a footbath solution. Severe cases may require the injection of antibiotics. Check the animals and treat them every 3 days for at least 4 treatments. Be sure the animals are completely free of foot rot before returning them to the uninfected flock/herd. A vaccination is available for contagious foot rot.

Foot Scald

Foot scald usually affects sheep during periods of extremely wet weather. It is believed to be caused by either the same bacteria or a different strain of the same bacteria that causes foot rot. It is not as severe as foot rot, although it resembles foot rot in its early stages. In the early stage of foot scald, the skin between the toes becomes inflamed and turns white. As the disease progresses, the rear part of the heel may separate from the hoof. Affected sheep become lame. The forefeet are more often affected than the rear feet. Move affected sheep to a dry area or onto slotted floors.

Treat the sheep with a footbath solution. A copper sulfate or zinc sulfate solution may be used to treat both foot scald and foot rot. Sheep that ingest copper sulfate may develop copper poisoning and copper sulfate solutions corrode metal; use care when handling a copper sulfate solution. Zinc sulfate solutions are nonirritating and less toxic than copper solutions; therefore, they are more commonly used.

Lamb Dysentery

Lamb dysentery affects sheep and is caused by bacteria. It affects mainly lambs 1 to 5 days of age and death losses can be high. Symptoms include loss of appetite, depression, diarrhea, and sudden death. Prevention of infection is the best control of dysentery. Follow strict sanitation practices and lamb in clean, dry housing and on clean pasture. Ewes may be vaccinated a few weeks before lambing. This will allow the ewe to develop antibodies that can be passed on to the newborn lamb in the colostrum.

Mastitis

Mastitis is a disease that affects sheep and goats, as well as cows, mares, and sows. This disease is caused by bacteria or injury to the udder. The udder becomes swollen, hard, hot, and sore. The affected animal may have a straddling walk. The milk is thick, yellow, and clotted. Often, the lamb or kid cannot nurse. Mastitis is treated with antibiotics. To aid in prevention of mastitis, remove any objects in the barn or pasture that could cause bruises, such as high door sills. Sanitation also helps to prevent the disease. Tag wool from the udder area on sheep and use proper milking procedures.

Navel-ill

Navel-ill affects sheep and goats and is caused by a bacterial infection of the navel. It affects mainly young animals within a few days after birth. Symptoms include fever, loss of appetite, depression, and swelling around the navel. The affected animal may suffer a paralysis of the legs or fall down and flail its legs. Death may occur quickly. A chronic form results in poor appetite, loss of weight, and lameness in the joints. Antibiotics are used to treat navel-ill. Prevention is achieved by good sanitation practices and the use of a disinfectant on the navel at birth.

Pneumonia

Pneumonia is an inflammation of the lungs that affects both sheep and goats. Exposure to cold, damp, and drafty conditions may cause pneumonia. It sometimes occurs as a secondary infection with another disease. Parasites in the lungs can also cause pneumonia. Symptoms include difficulty breathing, fever, coughing, loss of appetite, and diarrhea. Some animals die, some recover, and in others, the condition becomes chronic and extends for a period of several weeks. Treat by keeping the animal warm and dry and using antibiotics when the symptoms first appear. To prevent pneumonia, keep animals warm and dry, especially after shearing or dipping.

Scrapie

USDA Agricultural Research Service; Photo by Jack Dykinga

Figure 29-2 Sheep infected with scrapie. Scrapie affects the central nervous system of sheep and goats.

Scrapie (Figure 29-2) is a disease that affects the central nervous system of sheep and goats. The exact nature of the causative agent is not known. A **causative agent** is a biological pathogen that causes a disease; generally most are bacteria, fungi, or viruses. Scrapie is one of a class of brain diseases called transmissible spongiform encephalopathy (TSE). See the discussion of bovine spongiform encephalopathy (BSE) in Chapter 17 for more information about TSEs.

There is no cure or vaccine for scrapie, and it is always fatal. Scrapie is estimated to cost the U.S. sheep industry more than $20 million a year. Scrapie has a long incubation period, typically 2 to 5 years; therefore, the disease generally affects older animals. Early symptoms include a change in behavior and scratching or rubbing against fixed objects. The animal loses coordination and begins to walk abnormally, typically high stepping with the forelegs, hopping like a rabbit, and swaying the back end. Other symptoms include weight loss, wool loss, lip smacking, and biting at the feet and legs. Under stress, affected animals will tremble or go into a convulsive-like state.

Confirmation of the presence of scrapie in an animal is made after it dies by a microscopic examination of brain tissue. The presence of an abnormal prion protein in the brain tissue confirms the disease, but on a live animal, a biopsy may be taken from the lymphoid tissues on the inside of the third eyelid. There appear to be genetic variations that are related to susceptibility of scrapie and the length of the incubation period. The causative agent probably spreads from ewe to lamb by way of the placenta and placental fluids. A test may be used to determine an animal's genetic resistance to scrapie (Figure 29-3).

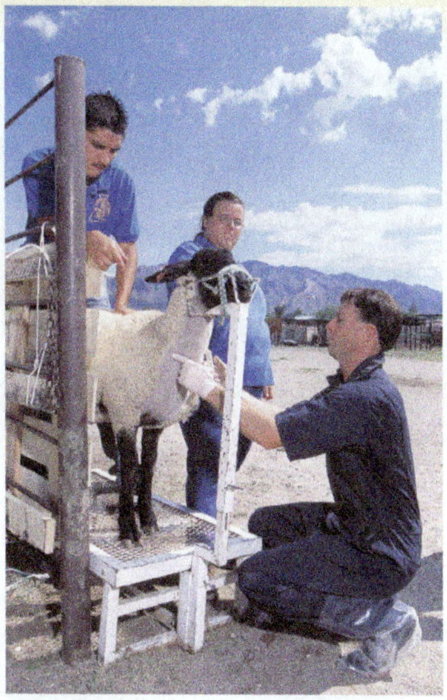
USDA Agricultural Research Service; Photo by Stephen Ausmus

Figure 29-3 Drawing blood to test for genetic resistance to scrapie.

In the United States, the U.S. Department of Agriculture's (USDA) Animal and Plant Health Inspection Service (APHIS) has used a voluntary program since 1992 in an effort to bring scrapie under control. Participants certify the origin of their flocks from scrapie-free flocks. This is a cooperative effort involving producers, veterinarians, state health officials, and APHIS. The interstate movement of sheep from scrapie-infected flocks is restricted.

The USDA initiated the National Scrapie Eradication Program (NSEP) in 2001. The goal of the NSEP is to identify and eliminate the last remaining scrapie cases in the United States by 2017. Furthermore, surveillance will be maintained at sufficient levels for 7 years so that by 2024, the United States can meet the World Organization for Animal Health's requirements for scrapie freedom. Since 2003, the program has reduced the prevalence of scrapie by more than 96 percent. As a part of the NSEP, certain classes of sheep and goats must be identified with USDA-approved tags and tattoos.

Sore Mouth

Sore mouth affects sheep and goats; it is caused by a virus and is more common in younger animals. Symptoms include blisters on the mouth, lips, and nose of the animal. The udder of an older animal affected with the disease will also show blisters. The blisters fill with pus, break, become open sores, and then scab over, turning a gray-brown color. It is difficult for the animals to nurse or eat, so weight is often lost. In areas where it occurs, sore mouth is prevented by vaccination. Treatment after the disease appears is not considered to be practical.

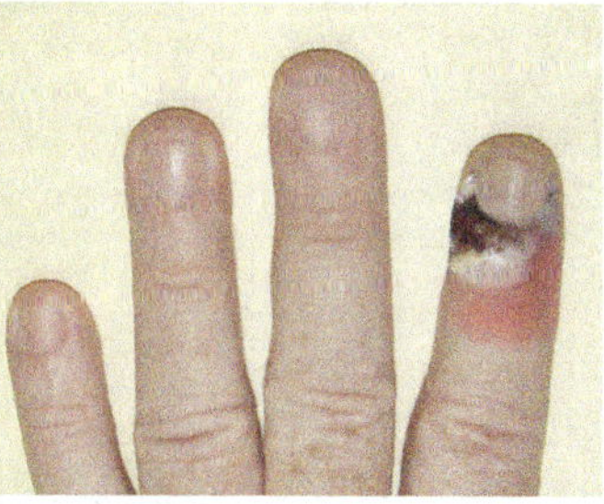
Courtesy of Frank B. Flanders

Figure 29-4 Be sure to wear protective gloves while handling an animal with sore mouth.

Sore mouth is a zoonotic disease, meaning that it is transmissible to humans. In humans it causes painful sores that may last for several weeks (Figure 29-4). People handling sheep or goats infected with sore mouth should wear rubber or plastic gloves. Exposed skin areas should be washed thoroughly. After washing, apply an antiseptic such as 70 percent isopropyl alcohol. The infective virus enters through small cuts or abrasions and mucus membranes (eyes, nose, and mouth). Keep small children away from infected sheep.

Tetanus

Tetanus affects sheep and goats; it is caused by bacteria that produce the tetanus toxin. These bacteria live in the soil and can enter the animal's body through wounds. Symptoms include stiffness, walking with a straddling gait, inability to eat, and rigid jaw and tail. The animal has muscle spasms that may lead to respiratory failure and death. There is no treatment after the symptoms appear. Tetanus is prevented by sterilization of docking, castrating, and shearing instruments and the use of disinfectants on open wounds. Vaccinations for the prevention of enterotoxemia will sometimes contain a vaccine for tetanus. Vaccination may be recommended for farms where the disease is known to exist.

Vibriosis (Campylobacteriosis)

Vibriosis or campylobacteriosis affects sheep and is caused by bacteria. Vibriosis in sheep is caused by a different bacterium than the one that causes vibriosis in cattle. The major symptom is abortion. One or two ewes out of the flock will abort, which is later followed by a rapid increase in the flock's abortion rates. Ewes will again breed normally after abortion. Isolate ewes that abort, and properly dispose of aborted fetuses. Move ewes that are not affected to a clean area. An annual vaccination is available for disease prevention, and antibiotics are used to help control an outbreak. Prevention is by annual vaccination.

The exact causes of abortions in sheep and goats are difficult to determine without laboratory analysis. To determine the cause of abortions, it is best to consult your local veterinarian. Many organisms that cause abortions in sheep and goats can also cause disease and abortions in humans. For this reason, pregnant women should not be in lambing or kidding areas, especially if abortions are being observed, and care should be taken in handling aborted tissues.

NUTRITIONAL PROBLEMS

Some nutritional problems are common to sheep, goats, and cattle. Bloat and white muscle disease, which are nutritional problems that may affect sheep, goats, or cattle, are discussed in Chapter 17. The following nutritional ailments pose particular problems for sheep and goats.

Constipation and Pinning

Constipation, or difficulty in passing feces, may affect sheep and goats. Pinning, which is the pasting of the tail to the anus by the feces, affects mainly very young animals. These problems may be caused by feeding coarse, dry, or indigestible feed. Overeating may also contribute to the problem. Constipation may be treated with a warm, soapy water enema or by feeding the animal 1 to 2 teaspoons of castor oil. Treat pinning by washing off the accumulated feces. Prevent these problems by proper feeding and by docking of the lamb's tail as soon as possible after birth.

Impaction

Impaction occurs when the rumen becomes filled with dry or indigestible feed. Sheep and goats are both subject to this problem. Causes include feeding too much dry feed, sudden changes in feed, overeating feeds when the ration is changed, lack of water, illness, or the presence of any disease that slows proper digestion. Symptoms include poor appetite, bad breath, no cud chewing, constipation, weakness, and, in extreme cases, death. Treat by providing plenty of water and taking the feed away from the animal for a few days. Feed a laxative feed for several days, and give the animal fish liver oil, mineral oil, or Epsom salts. Massage the flank to stimulate muscle contractions in the rumen. The problem can be prevented by proper feeding, good pasture, and exercise.

Milk Fever

Milk fever, a disease caused by a lack of calcium in the blood, occurs in both sheep and goats. Lambing ewes may be affected. In goats, it may occur shortly after kidding, or it may occur a month or so later in the lactation period. Symptoms include loss of appetite, restlessness, muscle tremors, and difficulty in standing. If milk fever is not treated, it can lead to a coma and death of the animal. A veterinarian should be called to treat animals with milk fever. A good ration and mineral supplement during gestation may help in the prevention of milk fever.

Night Blindness

Night blindness affects both sheep and goats and is caused by a lack of vitamin A in the diet. Symptoms include the inability to see at night, soreness of the eyes, loss of appetite, weakness, nervousness, and convulsions. Night blindness is treated by adding vitamin A to the diet. Fish liver oil, green pasture, and green, leafy hay are good sources of vitamin A. The disease can be prevented by feeding a ration and mineral supplement that includes sources of vitamin A.

Poisonous Plants

Certain plants are poisonous to both sheep and goats. However, more plants are poisonous to sheep than are poisonous to goats. Symptoms of plant poisoning vary with the type of plant involved. Treatment is often unsuccessful, but prevention includes the elimination of poisonous plants from the pasture. Many landscape plants are toxic to small ruminants. Animals generally will not eat poisonous plants unless they are forced to by lack of other feed. Do not overgraze pastures.

Pregnancy Toxemia (Ketosis)

Pregnancy toxemia affects sheep and goats, especially those carrying twins or triplets. It is a metabolic disorder that usually occurs during the last 6 weeks of gestation. During this time, the fetuses are growing rapidly and the ability of the ewe to take in an adequate amount of feed is reduced. Inadequate carbohydrate metabolism increases the ketone level in the blood. Symptoms include going off-feed, lagging behind the rest of the flock/herd, nervousness, unsteady gait, grinding the teeth, hard breathing, and frequent urination. If untreated, the animal may die.

To be effective, treatment must begin when the symptoms are first observed. Treatment should be under the supervision of a veterinarian. A concentrated source of energy is provided by an administration of oral propylene glycol. Do not use molasses or other sugar sources; that treatment may make the condition worse. The most effective treatment is to induce birth when the symptoms appear. Corticosteroids may be used to induce birth and also to increase carbohydrate metabolism.

Pregnancy toxemia may be prevented through proper nutrition by providing an adequate source of energy in the diet. Avoid stressful conditions and sudden feed changes during the last few weeks of gestation.

Urinary Calculi (Water Belly, Urolithiasis)

Urinary calculi is caused by the formation of small stones, called calculi, in the urinary tract. It generally affects male sheep and goats on high-concentrate rations. The calculi block the urethra, causing retention of the urine. If the condition is not treated quickly, the bladder may rupture, causing death. If the bladder does not rupture, the animal may die of uremic poisoning. While calculi may form in females, they usually do not block the urethra because it is larger than that of the male. Wethers castrated at a young age have a much smaller urethra and penis and are more likely to suffer from blockage of the urethra from smaller calculi. Breeding rams and bucks are sometimes affected by urinary calculi.

While there are several factors involved in the development of urinary calculi, the main cause is a narrow ratio of calcium to phosphorus in the diet. The ratio of calcium to phosphorus in the diet should be two parts calcium to one part phosphorus. This ratio may drop as low as 1:1 in high-concentrate diets; this contributes to the development of urinary calculi. Other factors that make the condition worse are cold weather, reduced salt intake, and reduced water intake.

Symptoms include standing with arched backs, depression, low feed intake, straining to urinate, calling out, and kicking at the belly. The abdomen may be swollen, especially if the bladder has ruptured; the penis may be swollen if the rupture has occurred in the urethra.

It is better to prevent urinary calculi than to try to treat it after it has occurred. Maintaining the proper ratio of calcium to phosphorus in the diet is the most effective method of prevention. It may be necessary to have a chemical analysis done on the feeds used in the ration to ensure that the calcium-to-phosphorus ratio is correct. Calcium carbonate (limestone) may be added to the ration to increase the calcium level in the diet. Adding ammonium chloride or ammonium sulfate to the diet may help to prevent the formation of calculi. Provide plenty of clean, fresh water at all times. In cold weather the water temperature should be maintained at 45 to 50°F to increase consumption. If a severe outbreak occurs, it may be necessary to add up to 4 percent salt in the ration. This will increase urine output and dilute the mineral content of the urine. If salt is added to the ration at this level, make sure a good supply of water is available.

If urinary calculi is diagnosed quickly, treatment can be effective. Surgery can be required to treat the condition. A veterinarian should be called immediately if the condition is suspected. Recovery is less likely if the bladder ruptures.

EXTERNAL PARASITES

Several external parasites attack sheep and/or goats. External parasites cause losses in the production of wool, mohair, meat, and milk. Sometimes they may cause the death of the animal. A combination of sanitation and the correct use of insecticides helps control most of the common external parasites of sheep and goats. Recommendations and regulations concerning the use of insecticides are constantly being updated. Producers should contact university entomologists, county extension personnel, or a local veterinarian for current information on permissible control measures.

USDA/APHIS

Figure 29-5 Flesh-eating screwworms once plagued southern states, but they have been eliminated from the United States by efforts of the USDA. The USDA also provides assistance to Mexico and other Central American countries to control the screwworm in order to help protect the United States from re-infestation.

Blowflies

Several species of blowflies attack sheep and goats. The screwworm, a maggot of one of these species, was a common pest of sheep and goats in the southwest United States (Figure 29-5). However, an eradication program has been under way for a period of years, and the screwworm has been eliminated from the United States. There has not been a case of screwworms in the United States since 1982. There is the possibility of re-infestations from other areas, especially Mexico and Central America, where it has not been eradicated. The United States is assisting these countries in eradication of the screwworm fly in their countries. Screwworms enter the animal's body through an open wound and feed on the living flesh in the wound. They breed only in the wounds.

Wool maggots are caused by another species of blowfly. These do not grow in live flesh but live in wool that is wet and is matted around wounds. Tagging sheep helps to prevent infestation. Shearing before warm weather and keeping sheep clean also helps to prevent trouble with blowflies. Sprays are available for treating infested animals.

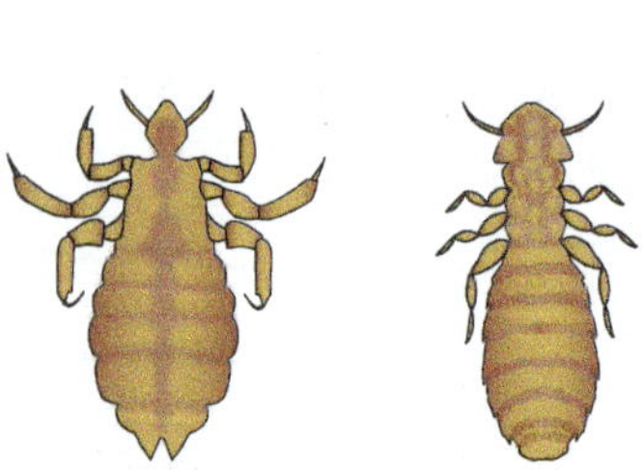

Figure 29-6 Two kinds of lice that attack sheep and goats: (left) a sucking louse; (right) a biting louse.

Lice

Lice (Figure 29-6) are tiny insects that live on animals. Several different kinds of lice attack sheep and goats. Some are blood sucking; others are biting or chewing lice. They spread rapidly from one animal to another. Lice irritate the animal's body and cause damage to wool. Animals that do a lot of rubbing may be infested with lice. Control is achieved by dipping, spraying, or dusting. Dipping is more practical for large flocks or herds and should be done in warm weather. Smaller flocks/herds should be dusted or sprayed. Dusting is the better method in cold weather.

Mange and Scab Mites

There are several species of mites that attack sheep and goats, causing a condition referred to as mange, scabies, or scab. Mites are tiny parasites that can hardly be seen with the naked eye. They burrow into the skin, causing irritation, and some cause scabs to form. The sheep scab mite is sometimes called wet mange. An eradication program against sheep scabies has been in operation in the United States since 1960. It is believed that it is now eradicated in the United States. If any cases appear, they must be reported to a veterinarian. The flock is quarantined, and control measures are carried out by trained personnel.

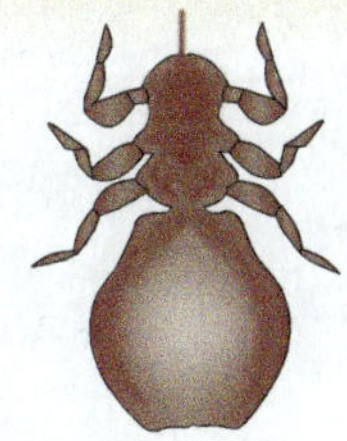

Figure 29-7 The adult sheep ked is a type of wingless fly. Keds are often mistaken for ticks, but note the ked (an insect) has six legs, while ticks have eight legs.

Sheep Bot Fly

The larvae of the sheep bot fly are found in the nasal cavities of sheep and goats. The fly is the adult stage of the insect. It is about $\frac{1}{2}$ inch long and looks like a honeybee. Infested animals shake their heads, sneeze, have difficulty breathing, and may hold their noses to the ground. They stamp their feet to try to keep the flies away.

Sheep Ked

The sheep ked (Figure 29-7) is sometimes called a sheep tick. However, it is not a true tick. It is a wingless fly that is about $\frac{1}{4}$ inch long and has six legs. It is found on both sheep and goats and is a bloodsucker. It is controlled by dipping, spraying, or dusting.

USDA Agricultural Research Service; Photo by Keith Weller

Figure 29-8 Several species of ticks attack sheep and goats. This is a black-legged tick that can carry Lyme disease.

Ticks

Several species of ticks attack sheep and goats. One type of tick is shown in Figure 29-8. The most serious of these is the spinose ear tick. The larvae of this tick lives in the ears of the animal. Ear infections may be caused by the irritation of the tick. The animals lose weight and become listless. Young animals may die from a heavy infestation.

INTERNAL PARASITES

Internal parasites are the most serious health problem of sheep and goats. Economic loss results from loss in weight, lower milk production, poor wool growth, wasted feed, and lower breeding efficiency. Death losses from internal parasites can be high at certain times of the year.

Good management is the key to controlling losses from internal parasites. Overgrazing pastures and failure to rotate pastures are two factors that contribute to problems with internal parasites. The eggs of parasites are present in the feces of infected animals and are deposited in pasture grasses. Rotating pastures allows time for the eggs of the parasites to die prior to consumption by grazing animals. A regular program of internal parasite tracking should be followed. Several anthelmintics are available for the control of internal parasites. Anthelmintics are chemical compounds used for deworming animals. Current regulations should be followed in the use of these substances. For up-to-date information on chemical control measures, contact county extension personnel, university specialists, or a local veterinarian.

Some of the symptoms of internal parasite infestation are rough hair coat, weight loss, slow gains, loss of appetite, diarrhea, and anemia. Swelling may be noticed beneath the lower jaw, and the animal may have a constant cough. Young animals are affected more severely by internal parasites than are older animals.

Coccidia

Coccidiosis is caused by the small protozoa called coccidia. These organisms live in the intestine of the sheep or goat. They cause the cell walls to rupture and the animal to bleed internally. A bloody diarrhea is one of the symptoms of coccidiosis. The sheep or goat picks up the protozoa organisms from contaminated feed and pastures. There are a number of species of protozoa that cause coccidiosis. Each kind of livestock is affected by its own type.

Drenching Procedure

CAUTION

When drenching, care must be taken not to injure the animal. If the animal's head is held too high, the material may enter the lungs, causing death. Read and follow the instructions on the label for the type of material being used.

Drenching is the oral administration of a liquid medication, generally for the control of internal parasites. It is usually given through the mouth of the animal. A syringe or an automatic drenching gun may be used to drench sheep or goats. The automatic drenching gun is more expensive, but it is more practical when a large number of animals need to be treated.

The animal's head must be kept in a level position during drenching. Hold the fingers under the muzzle and the thumb over the nose. Lift the upper lip with the thumb and carefully insert the nozzle along the side of the mouth. Administer the dose and release the animal so it can swallow.

During drenching of small flocks, the animals may be penned. The operator then moves among the sheep administering the drench. Each animal is marked as the operation is completed. With larger flocks, a chute should be used. The operator stands in the chute and catches each animal, holding it between the legs as he or she administers the drench.

Liver Fluke

The liver fluke lives in the liver of the infested sheep or goat. It causes bleeding in the liver. The eggs pass out in the manure. After hatching, the larvae enter a snail. They develop in the snail and then leave the snail and form a cyst, a protective covering, which is attached to vegetation. The sheep take in the cyst when eating the vegetation. Because the snail is a necessary part of the life cycle of flukes, the problem is more prevalent when animals have access to low, wet areas such as river bottoms or during times of excessive moisture.

Lungworms

Two species of lungworms infest sheep and goats: the thread lungworm and the hair lungworm. The adults of both types of worms live in the lungs. Eggs of the thread lungworm hatch in the intestine, and the larvae pass out in the feces. They are then picked up on the feed. The eggs of the hair lungworm hatch in the lungs. The larvae pass out in the feces and develop in land snails or slugs. The sheep or goats pick up the worms by eating the snails or slugs. The life cycle of the lungworm is shown in Figure 29-9.

Stomach and Intestinal Worms

Worms found in the stomach and intestines of sheep and goats include the (1) common stomach worm, (2) medium stomach worm, (3) bankrupt worm, (4) thread-necked strongyle, (5) nodular worm, (6) hookworm, (7) barber pole worm, and (8) tapeworm. The life cycles of all of these worms are similar. The life histories of three of these worms are shown in Figures 29-10, 29-11, and 29-12.

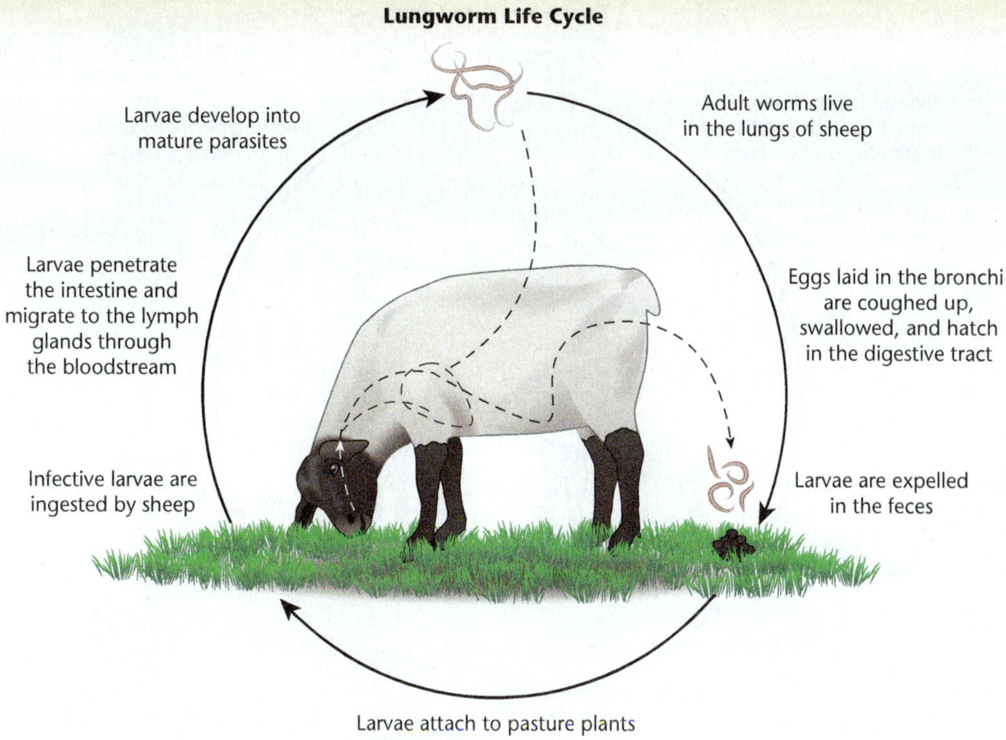

Figure 29-9 Life history of a lungworm of sheep.

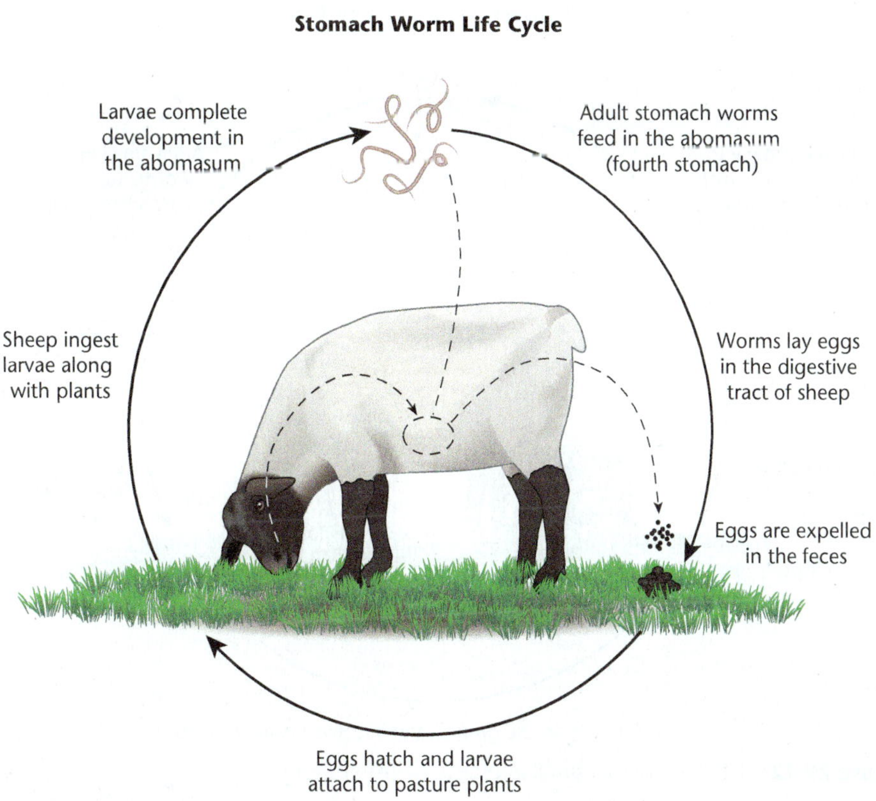

Figure 29-10 Life history of a common stomach worm of sheep.

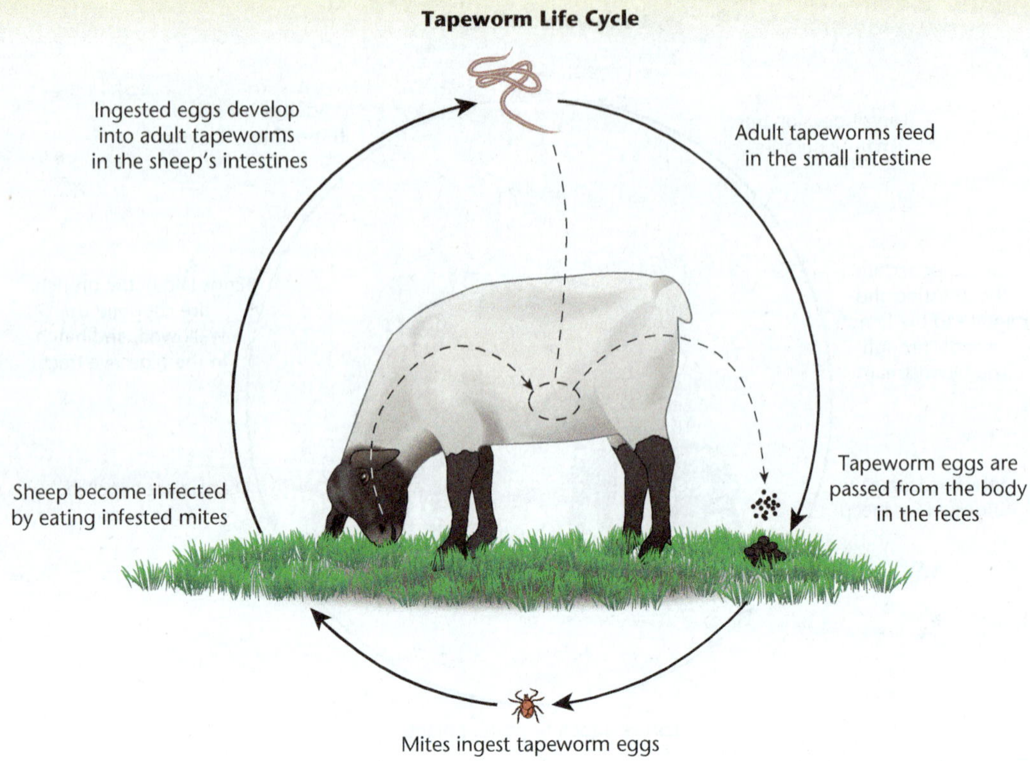

Figure 29-11 Life history of a tapeworm of sheep.

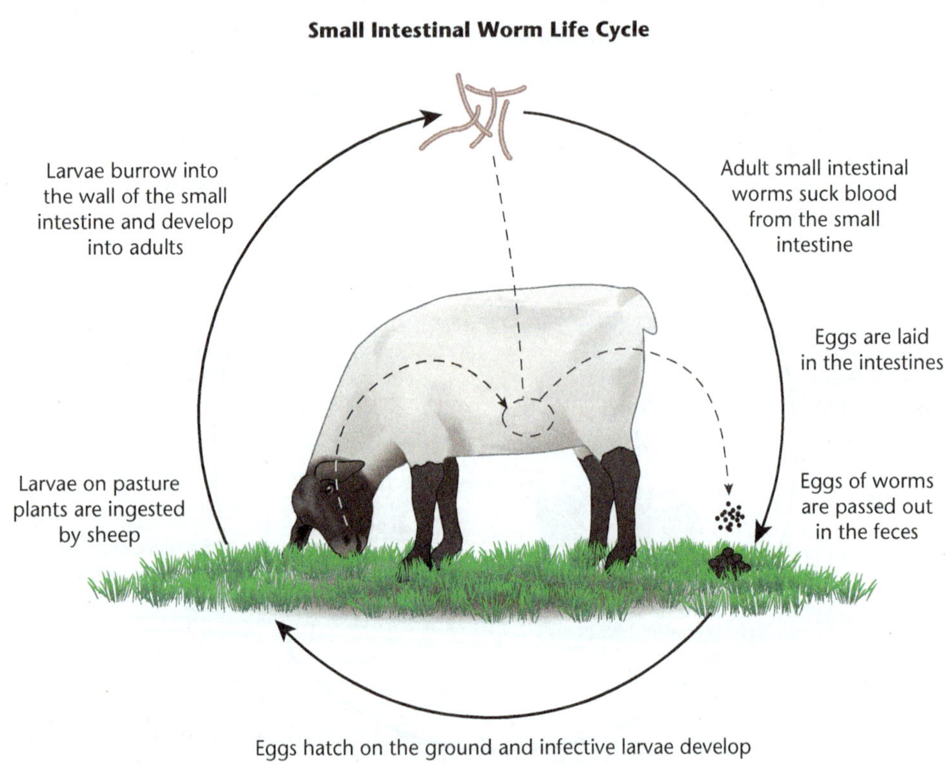

Figure 29-12 Life history of a small intestinal worm of sheep.

The following general cycle is found in each type of worm. The adult female deposits eggs that pass out of the animal in the manure. The eggs hatch on the ground. The larvae crawl up on blades of grass, which the sheep or goats eat. The larvae become adults inside the body of the animal. The cycle then begins again.

Female worms deposit from several hundred to several thousand eggs per day. The common stomach worm and the medium stomach worm are bloodsuckers. The other worms damage the wall of the intestine, making it more difficult for the animal to absorb nutrients. The common stomach worm is the most serious of all the internal parasites of sheep and goats.

SUMMARY

Several nutritional related problems, diseases, and parasites affect sheep and goats. Profits are reduced when diseases, parasites, and nutritional problems occur. Good management, feeding, vaccination, and sanitation programs help to prevent health problems. Planning prevention programs with the help of a veterinarian is recommended.

Sheep and goats have many diseases in common with cattle. Other diseases of sheep and goats include bluetongue, enterotoxemia, foot rot, foot abscess, foot scald, dysentery, navel-ill, pneumonia, scrapie, sore mouth, tetanus, and vibriosis. Some of these diseases are zoonotic. Many diseases are prevented by vaccination. Antibiotics are used to treat bacterial diseases. Isolating infected animals helps in control programs. The symptoms of many diseases are similar. The aid of a veterinarian is often needed to identify diseases, treat infected animals, prescribe a preventive program, and vaccinate healthy animals.

Stomach worms, intestinal worms, lungworms, and liver flukes are the most common internal worm parasites of sheep and goats. Coccidiosis is a parasite caused by protozoa called coccidia. Pasture rotation is important for the control of internal parasites in sheep and goats. The eggs of many internal parasites are spread over pasture grasses through animal feces. The larvae then infest the grass and are consumed by grazing animals where they develop into adults within the stomach or intestinal tract. Allowing pastures to rest between grazings provides time for larvae to die before they are consumed by grazing animals. Internal parasites can be treated by drenching. Care must be taken when drenching sheep and goats to avoid injuring the animals.

Flies, lice, mites, sheep keds, and ticks are the most common external parasites of sheep and goats. A combination of sanitation and insecticides are used to control most of these parasites.

A number of nutritional health problems affect sheep and goats, many of which can be prevented or cured with a good feeding management program. Common nutritional related problems include night blindness, milk fever, impaction, pregnancy toxemia, urinary calculi, constipation and pinning, and consuming poisonous plants.

Student Learning Activities

1. Prepare and give an oral report on one or more sheep and goat diseases and parasites.
2. Prepare a bulletin board display of pictures of sheep or goats with diseases and/or parasites.
3. Speak with a veterinarian about health problems of sheep or goats.

4. Conduct a community survey of health problems of sheep or goats. Develop recommendations that might improve the health of local herds or flocks.
5. Follow good health practices when planning and conducting a sheep or goat supervised experience program.

Discussion Questions

1. List the recommended practices that help prevent health problems of sheep and goats.
2. Under what conditions may foot rot spread rapidly through the flock or herd?
3. Describe the National Scrapie Eradication Program.
4. List the external parasites that affect sheep and goats, and describe the symptoms, prevention, and treatment of each.
5. How do external and internal parasites cause economic loss in sheep and goats?
6. Explain how sheep and goats may become infected with stomach worms while grazing.
7. Describe the typical symptoms of internal parasites in sheep and goats.
8. Describe the life cycle of tapeworms.
9. Describe how night blindness can be prevented and treated.
10. Describe the procedure for drenching sheep and goats.

Review Questions

True/False

1. Thick, yellow, and flaky milk may be an indication of mastitis.
2. A number of diseases are common to sheep, goats, and cattle.
3. The animal's head must be inclined while drenching.
4. Enterotoxemia does not affect sheep and goats.
5. Urinary calculi is a common problem in wethers castrated at a young age.

Multiple Choice

6. What type of parasites are the most serious health problem for sheep and goats?
   a. external
   b. internal
   c. ear
   d. feet
7. A lack of ________ in the blood causes milk fever in sheep and goats.
   a. vitamin K
   b. minerals
   c. calcium
   d. zinc
8. The oral administration of a liquid medication is called ________.
   a. drenching
   b. impaction
   c. convulsion
   d. lactation
9. Involuntary muscular contractions are called ________.
   a. parasites
   b. coccidiosis
   c. anthelmintics
   d. convulsions
10. Although it is a serious health risk for sheep, goats, and other animals, the United States has not had an outbreak of ________ since 1982.
    a. screwworms
    b. sheep bots
    c. night blindness
    d. stomach worms

Completion

11. ________ ________ is a disease that affects the tissue of the foot.
12. Liver fluke larvae live part of their life cycle in ________ before reinfecting sheep and goats.

13. ________ affects both sheep and goats and is caused by bacteria or injury to the udder.
14. The normal body temperature range for goats is ________ to ________, while the normal temperature range for sheep is ________ to ________.
15. A condition resulting from the spread of bacterial toxins through the bloodstream is known as ________.

Short Answer

16. Why should a person wear gloves when working with animals infected with sore mouth?
17. List three preventive methods for eliminating wool maggots, which are a species of blowflies.
18. What causes urinary calculi?
19. Under what conditions will animals sometimes eat poisonous plants?
20. What disease of sheep and goats affects the central nervous system and is related to BSE?

Chapter 30

Marketing Sheep, Goats, Wool, and Mohair

Key Terms

lamb
breeding sheep
shearer lamb
ewe
ram
cutability
feathering
grease
spinning count
wool top
hank
apparel wool
carpet wool
worsted
woolen
chevon
cabrito
clip

Objectives

After studying this chapter, the student should be able to

- discuss trends in the sheep and goat industries.
- describe the methods of marketing sheep, goats, wool, and mohair.
- define terminology used in marketing sheep, goats, wool, and mohair.
- list the grades of wool and mohair.
- explain methods of marketing meat, milk, and fiber products from sheep and goats.
- list products marketed from sheep and goats.
- explain the differences between lamb and mutton.
- describe how sheep and lamb products are promoted.
- compare and contrast the marketing of sheep and goats with respect to other types of livestock and poultry.
- define the classes and grades of sheep.

Markets for sheep, goats, and the products from these animals are not as well developed as markets for other livestock. In addition, the sheep and goat industries are small in comparison to animal industries such as swine, poultry, beef, and dairy. Finding a convenient market for sheep and goats is sometimes a challenge. While production of animals for meat and milk on contract is common for many farm animals, this is generally not the case for sheep and goats.

Connection: Memories of Mutton

Lamb and mutton have never been popular in the American diet. One reason can be traced to the food served to soldiers in World War II (Figure 30-1). During the war, American soldiers' rations often included mutton, the meat from sheep over a year old. Mutton is a tougher meat than lamb, which comes from younger animals. The pungent smell of cooking mutton and the overuse of this low-quality meat caused many returning soldiers to declare that they "would never eat mutton again."

In addition, several other factors led to the decline of the sheep industry, including the following:

- competition from imports
- foreign wool production subsidies
- losses from predator kills
- changes in consumer preferences
- competition from other meats and fibers
- changes in permits for grazing on public lands

Figure 30-1 Mutton was one of the least popular meals provided to troops during World War II, leading many to declare that they would never eat mutton again.

Sheep production for both meat and wool has been declining in popularity for several years for a number of reasons. Mutton and lamb have never been very popular with Americans. Wool production has declined more rapidly than lamb production because of competition from foreign production and the declining textile industry in the United States. However, the promotion of sheep and wool products, along with developing niche markets and improvements in production, has given reason for optimism about the future of the sheep and wool industries.

Goat products are becoming more popular. With the popularity of small farms and locally grown products, and an increasing ethnic population, goat production and the demand for goat products, including meat and milk, are increasing.

MARKETING SHEEP

Promotion of Sheep and Wool

Promotion of sheep and wool is provided by two organizations: the American Lamb Board and American Wool Trust. The American Lamb Board is charged with the promotion of lamb, and the American Wool Trust promotes wool.

The American Lamb Board was created by the U.S. Secretary of Agriculture and approved by producer referendum. The American Lamb Board consists of 13 volunteer members, including 6 producers, 3 feeders, 1 seedstock producer, and 3 first handlers. In a program similar to beef and swine checkoff programs, the American Lamb Board's efforts to promote lamb are funded by an assessment of 0.5 cent/pound on all sheep sold by producers, seedstock producers, feeders, and exporters in the United States. Unlike other livestock checkoff programs, an assessment of 30 cents per head is also paid by lamb first handlers (slaughter facilities). The American Lamb Board has been working to promote lamb consumption since July 2002 through advertising, public relations, culinary education, and retail promotions. The annual budget for 2013 was approximately $2 million, and 70 percent of funds were spent on promotions. The balance of funds was spent on administration, communication, oversight, and research activities. Because of limited funds, the American Lamb Board has focused on promoting lamb to chefs and targeted marketing in specific areas, as opposed to large-scale, nationwide marketing.

The American Wool Trust was established by the U.S. Congress in 2000. The American Wool Trust Fund provides funding aimed at increasing the competitiveness of American wool. A 14-member board oversees the spending of the American Wool Trust Fund, and funds are directed toward activities that include improvement of raw wool quality; development of new wool uses; adoption of new technologies in wool handling, evaluation, and storage; international marketing of wool; and market research.

Types of Markets for Sheep

Sheep may be marketed through terminal markets, local pools, sale barns, direct selling to the packer, or electronic marketing. Electronic marketing of lambs is a new development, and it is expected that an increasing number of animals will be marketed in this manner. Selling directly to the consumer has also increased. Direct selling to packers accounts for the largest volume of marketing. Auction markets, terminal markets, and local pools are still important ways to market sheep in some areas of the Midwest. The producer's choice of a market depends on the kinds of markets available locally, the prices paid, the numbers of lambs to be sold, and the type of available transportation.

Classes and Grades of Sheep

Sheep are classed according to age, use, sex, and grade (Figure 30-2). Age classes are lamb, yearling, and sheep. Lambs are further divided by age as hothouse lambs, spring lambs, and lambs. The classes by use are slaughter sheep, slaughter lambs, feeder sheep, feeder lambs, breeding sheep, and shearer lambs. The classes by sex are ewe, ram, and wether.

Photo courtesy of USDA NRCS

Figure 30-2 Sheep are classed according to age, use, sex, and grade.

Age

Hothouse lambs are under 3 months of age and usually weigh less than 60 pounds. They are sold in a specialty market between Christmas and Easter. Spring lambs are 3 to 7 months of age. They are finished at 110 to 130 pounds. Spring lambs are born in the fall and marketed in the spring and early summer. **Lambs** are 7 to 12 months of age. The most desirable market weight for lambs is 130 to 150 pounds. They are usually milk and grass fed. If they have been fed grain, they are called fed lambs. Small-framed lambs may be marketed at 95 to 100 pounds. Yearlings are between 1 and 2 years of age. The first pair of permanent incisor teeth is present, but the second pair is not. Sheep are considered to be more than 2 years old.

Use Classes

Slaughter sheep and lambs are marketed for immediate slaughter. Feeder sheep and lambs require more feeding to finish for slaughter. **Breeding sheep** are western ewes that are sold back to farms and ranches to be bred to produce more lambs. **Shearer lambs** are not finished enough for slaughter. They are shorn and fed to a higher level of finish before slaughter.

Sexes

A **ewe** is a female sheep or lamb. A **ram** is a male sheep or lamb that has not been castrated. A wether is a male lamb that was castrated when young. It does not show the sexual traits of the ram.

Grades

Live grades of sheep are based on quality and estimated yield. Quality grades are based on conformation and amount of finish. The four quality grades for lambs and yearlings are (1) Prime, (2) Choice, (3) Good, and (4) Utility. The four quality grades for slaughter sheep are (1) Choice, (2) Good, (3) Utility, and (4) Cull. For a description of body conformation and method of handling sheep to determine amount of finish, review the chapter on selection of sheep.

Estimated yield is based on five yield grades indicated by the numbers 1 through 5. Yield grade 1 is the highest yielding, while yield grade 5 is the poorest yielding. The thickness of fat over the rib eye determines the yield grade for all slaughter sheep. The relationship between fat thickness and yield grade is as follows:

Yield Grade	Back Fat
1	0.00 to 0.15 inch
2	0.16 to 0.25 inch
3	0.26 to 0.35 inch
4	0.36 to 0.45 inch
5	0.46 inch and greater

Yield grades identify differences in cutability among carcasses. **Cutability** refers to the yield of closely trimmed, boneless retail cuts that come from the major wholesale cuts. The major wholesale cuts are the leg, loin, rib, and shoulder (Figure 30-3). Figure 30-4 illustrates the retail cuts of lamb. The leg and loin generally yield one-half of the total carcass weight; they represent 75 percent of the value of the carcass. A lack of muscling in the leg or too much finish lowers the estimated yield grade. Grades for live animals are directly related to the quality and yield grades of the carcasses they will produce. The same grades that are used for live animals are used for carcasses.

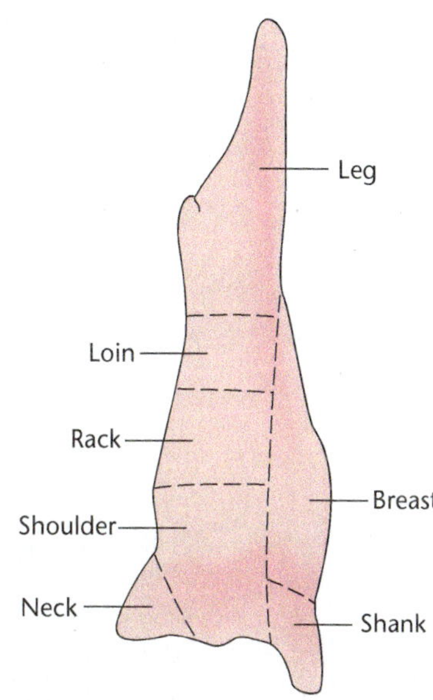

Figure 30-3 Wholesale cuts of lamb.

The carcass quality grades are based on conformation and quality of the lean. The amount of muscle development gives an indication of conformation. The quality of the lean is determined by the color, firmness of fat and lean, and amount of interior

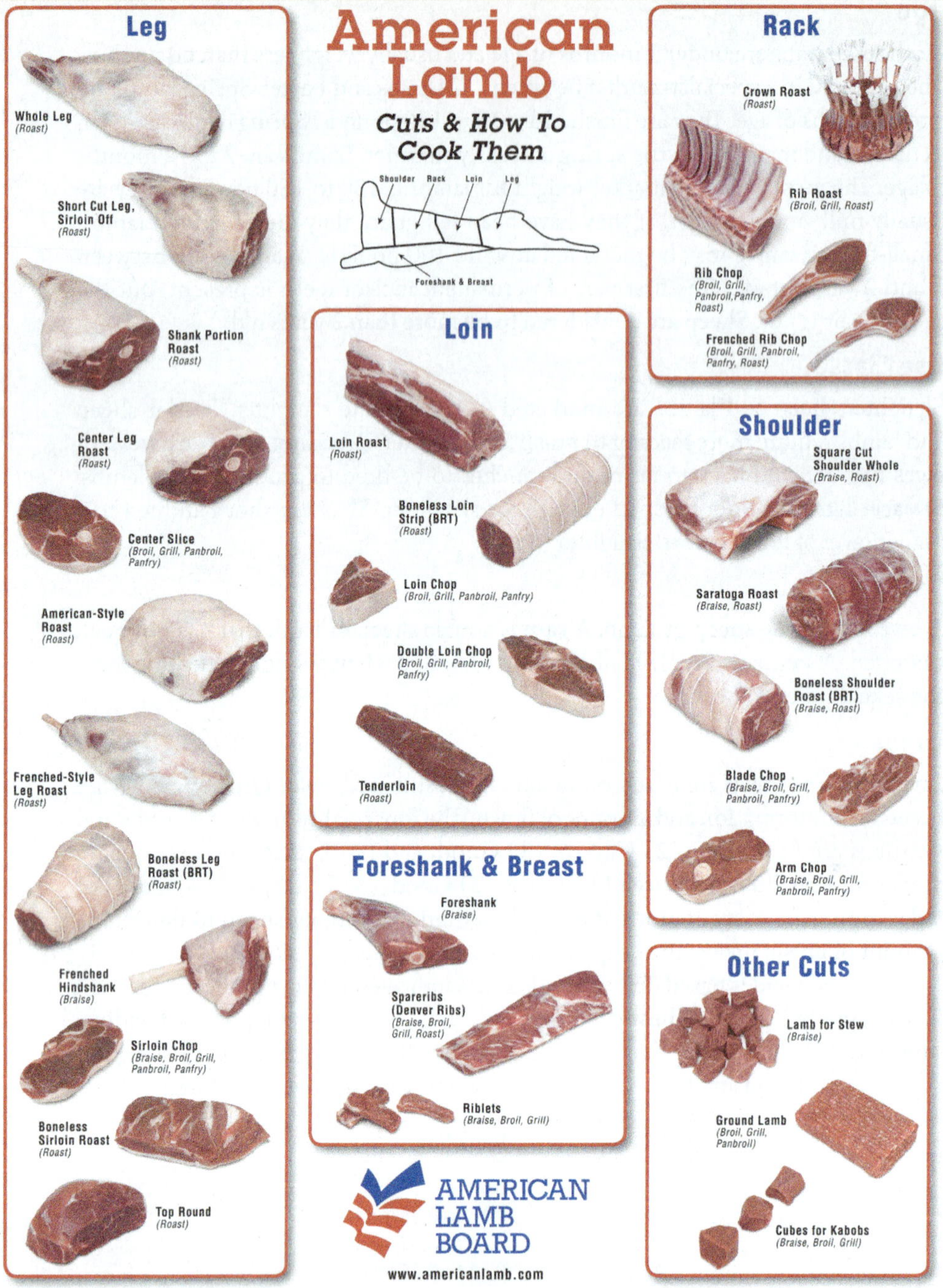

Courtesy of the American Lamb Board

Figure 30-4 Retail cuts of lamb—where they come from and how to cook them.

fat deposits. The maturity of the animal affects the standards for judging the quality of the lean.

The U.S. Department of Agriculture's standards for quality grades of lamb carcasses are briefly described in the following paragraphs. A complete description of standards is available from the USDA.

Minimum qualifications for Prime lamb carcasses are moderately wide and thick in relation to their length; moderately plump and full legs; moderately wide and thick backs; and moderately thick and full shoulders. Mature lambs have a moderate amount of feathering (intermingling of fat with lean) between the ribs and a modest amount of fat streaking within and upon the inside flank muscles, which are a light red color. The lean flesh and external finish are firm; the flanks are moderately full and firm.

Minimum qualifications for Choice lamb carcasses are slightly wide and thick in relation to their length; slightly plump and full legs; slightly wide and thick backs; slightly thick and full shoulders. Mature lambs have a small amount of feathering between the ribs; a slight amount of fat streaking within and upon the inside flank muscles, which are a moderately light red color; moderately firm lean flesh and external finish; and slightly full and firm flanks.

Minimum qualifications for Good lamb carcasses are moderately narrow in relation to their length; slightly thin, tapering legs; slightly narrow and thin backs and shoulders. Mature lambs have a slight amount of feathering between the ribs; traces of fat streaking within and upon the inside flank muscles, which are a slightly dark red color; moderately firm lean flesh and external finish; slightly full and firm flanks.

Utility lamb carcasses are those lambs with characteristics that are inferior to those specified as minimum for the Good grade.

Shrinkage

Sheep shrink as they are moved to market just like other classes of livestock. The amount of shrinkage depends mainly on the distance traveled and the time in transit. Most of the shrinkage occurs in the first few miles traveled. Shrinkage of fat lambs ranges from a little less than 2 percent to more than 8 percent. The average shrinkage is about 3 to 5 percent. Careful handling of sheep while they are being transported to market helps to reduce shrinkage and other losses.

Seasonal Prices

Lamb supplies vary from month to month throughout the year. This contributes to a seasonal variation in the price of lamb. Figure 30-5 shows an example of seasonal variation in lamb prices as a percentage of the yearly average. The variation shown covers prices received by farmers throughout the United States. Prices for any given year may not necessarily follow the long-term trend on a month-by-month basis.

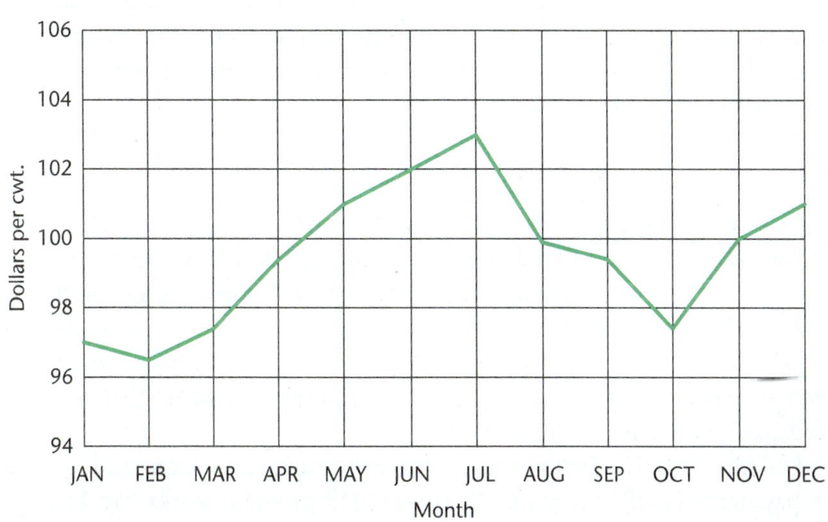

Figure 30-5 Seasonal variation in lamb prices received by farmers as dollars per cwt, United States.

Source: USDA, Annual Price Summary

Selling Purebred Sheep

Purebred sheep are sold by private sale or through auction sales sponsored by regional or state sheep associations. Purebred breeders often advertise through magazines that reach sheep producers and with signs at and near the farm.

Sheep from purebred flocks that do not meet the standards for breeding stock are sold for slaughter. The methods for marketing these sheep are the same as for any other commercial sheep producer.

MARKETING WOOL

Wool Markets

Wool is marketed through local buyers, wool pools, cooperatives, or warehouse operators, or by direct sale to wool mills. The choices open to the producer depend on the kinds of markets available locally. Local buyers are available in most wool-producing areas. Wool pools and marketing cooperatives have developed in many areas in an effort to obtain a better price for wool. Warehouse sales are common in the range states. Very little wool is marketed directly to wool mills.

Local buyers buy wool on a cash basis. They may or may not pay price differentials for different grades of wool. The producer pays no marketing costs. Prices paid are often lower than those that are paid by other types of markets.

Wool pools grade fleeces by fineness, staple length, color, and cleanness. The wool is sold on a sealed-bid basis by grade. Each producer is paid according to the weight and grade of the wool that was consigned. Because wool pools offer a greater volume of wool to prospective buyers, the prices received by the producer are often better.

Wool cooperatives act as marketing agents for producers. A cash advance or partial payment is made to the producer when the wool is delivered to the wool cooperative. The wool is fleece graded at the cooperative warehouse. Final settlement of price is made based on the price received for that grade of wool. Cooperatives also have the advantage of being able to offer larger volumes of wool to the buyer.

Some warehouses may buy the wool outright, while others market the wool on a commission basis. If the wool is handled on a commission basis, the producer does not receive payment until the wool is sold from the warehouse. Warehouse operators are familiar with the wool market and may be able to obtain better prices for the wool that is consigned to them.

Wool Prices

Wool prices vary seasonally. Prices are usually above the yearly average in March, April, May, June, July, and October, reaching their peak in May. They tend to be below the yearly average in January, February, August, September, November, and December. The lowest wool prices usually occur in November, December, and January. Figure 30-6 shows an example of the seasonal variation in wool prices received as a percent of the yearly average. Incentive payments that producers received are not included in the data.

The USDA currently offers producers marketing assistance loans and loan deficiency payments (LDP) for wool. The wool LDP program works like LDP programs for grain. Producers must enroll for the marketing assistance loan or LDP prior to selling their wool. Sheep producers can learn more about wool marketing assistance loans and LDP at their local USDA Farm Service Agency office.

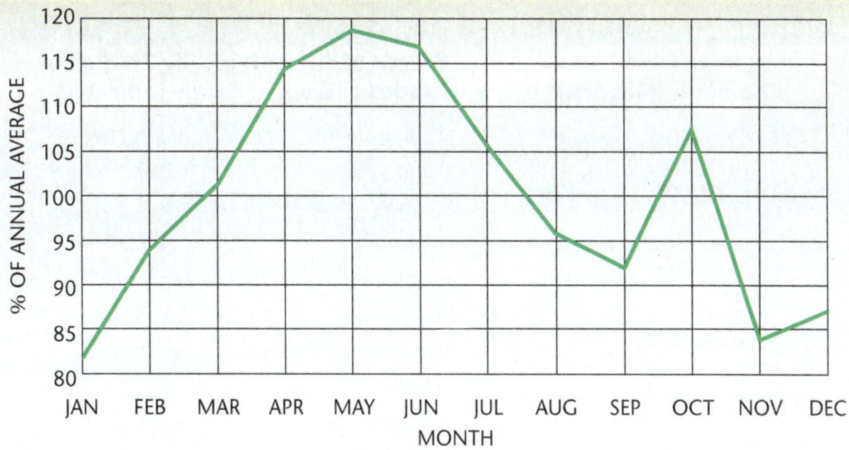

Figure 30-6 Seasonal variation in wool prices received by farmers as a percent of yearly average (not including incentive payments).
Source: USDA, Annual Price Summary

The value of the wool fleece is largely determined by the amount of clean wool that it produces. The term **grease** is used to indicate the presence of impurities in the fleece. Grease wool refers to the fleece before it is cleaned. Shrinkage is the amount of weight lost when the impurities are cleaned out of the fleece. Producing a fleece with less shrinkage increases the value of the fleece. The higher the shrinkage, the lower the grease value of the fleece. The length, density, and diameter of the wool fiber also affect the value of the wool by determining its grade.

The quality of wool produced in the United States is generally lower than the quality of imported wool. In the United States, wool is usually considered a by-product of the sheep industry, whereas foreign producers tend to take greater care to produce high-quality wool. Wool produced in the United States often contains residues from vegetable matter, urine stains, and dark hairs.

Wool Grades

Wool grades are based mainly on the diameter of the fiber (Figure 30-7). Density is the major factor in determining the weight of the fleece. Length of fiber is also considered when grading wool.

Three systems of grading wool are commonly used. These are the American (blood) system, the numerical count (USDA) system, and the micron system.

USDA Agricultural Research Service; Photo by Ken Hammond

Figure 30-7 Wool grades are based mainly on the diameter of the fiber.

The American System

The American system was originally based on the percent of Merino breeding in the sheep. A half-blood wool came from a sheep with half of the breeding from a fine wool breed and half from a coarse wool breed. Today, the system is used only to indicate the diameter of the wool fiber and does not indicate the amount of Merino breeding in the sheep. There are seven grades of wool in the American or blood system: (1) Fine, (2) $\frac{1}{2}$ blood, (3) $\frac{3}{8}$ blood, (4) $\frac{1}{4}$ blood, (5) Low $\frac{1}{4}$ blood, (6) Common, and (7) Braid. Fine has the smallest diameter fibers, and Braid has the largest.

The Numerical Count (USDA) System

The numerical system has 16 grades that reflect the spinning count. The official USDA grades range from finer than 80s to coarser than 36s. The **spinning count** is the number of hanks of yarn that can be spun from 1 pound of **wool top** (partially

TABLE 30-1 Comparison of Wool Grading Systems and the Grades of Wool Produced by Different Breeds

American Blood System	USDA Grades	Average Fiber Diameter (Microns)*
Fine	Finer than 80s	17.69 or less
	80s	17.70–19.14
	70s	19.15–20.59
	64s	20.60–22.04
$\frac{1}{2}$ Blood	62s	22.05–23.49
	60s	23.50–24.94
$\frac{3}{8}$ Blood	58s	24.95–26.39
	56s	26.40–27.84
$\frac{1}{4}$ Blood	54s	27.85–29.29
	50s	29.30–30.99
Low $\frac{1}{4}$ Blood	48s	31.00–32.69
	46s	32.70–34.39
Common	44s	34.40–36.19
Braid	40s	36.20–38.09
	36s	38.10–40.20
	Coarser than 36s	40.21 or more

Breeds of Sheep Typically Producing Grades of Wool in Range Indicated	
Delaine Merino	80s or finer
Rambouillet	64s–70s
Targhee	62s
Romeldale	58s–60s
Corriedale	60s
Southdown	56s–60s
Hampshire, Shropshire	56s–60s
Suffolk	54s–58s
Columbia	50s–58s
Dorset	50s–56s
Cheviot	48s–50s
Oxford	46s–50s
Tunis, Romney	44s–48s
Leicester	40s–46s
Lincoln, Cotswold	36s–40s
Highland	36s or coarser

*A micron is 1/25,400 of an inch.
Source: U.S. Department of Agriculture.

processed wool). A **hank** of yarn is 560 yards in length. For example, a 58s wool yields 32,480 yards of yarn (560 times 58). Finer than 80s has the smallest diameter fibers, while coarser than 36s has the largest diameter fibers.

The Micron System

The micron system of grading wool is the system that is used internationally. This system measures the actual diameter of the wool fiber. A micron is the unit of measurement used and is equal to 1/25,400 of an inch or 1/1,000,000 of a meter.

A given breed of sheep tends to produce more of one grade of wool than of another. Table 30-1 shows the relationship of breed to the grades produced in the American and numerical count system.

In addition to grades, wool is also classed according to use. The two major classes of wool are apparel wool and carpet wool. **Apparel wool** is used in the making of clothing. **Carpet wool** is used for products in the carpet industry. Carpet wool is coarser than apparel wool. The length and fineness of the wool influence the use for which it is suited. Finer grades can be shorter than coarser grades and still fall under the same use classification.

Apparel wool is further divided into **worsted** and **woolen** classes. Longer fiber wools are used in the worsted process. These make a relatively strong yarn. Shorter fibers are used in the woolen process, making a weaker yarn.

MARKETING GOATS FOR MEAT

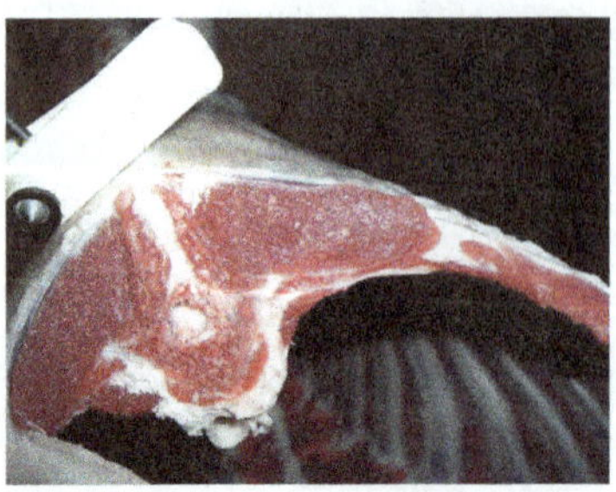

USDA

Figure 30-8 Goat meat is called chevon.

Meat goats are often marketed directly to the consumer. Goats may be marketed for meat from young kids to older adults. Generally, kids under nine months are preferred. Certain markets require graded animals. **Chevon** is the term used when marketing processed meat from goats (Figure 30-8). This naming convention is similar to that used for marketing meat from cattle; the term "beef" is more appealing to the consumer than "cow meat." Furthermore, marketing meat from goats is more appealing if the term "*chevon*" is used rather than "*goat meat*." The definition of chevon varies. According to the American Meat Goat Association, chevon may be from goats that are 48 to 60 pounds and 6 to 9 months of age. Bucks and does of older age are also used as chevon but they bring less per pound. Young kids that weigh 30 to 40 pounds are marketed as cabrito. **Cabrito** is Spanish for "little goat." Cabrito is considered a delicacy and is especially popular for barbecuing.

MARKETING GOAT MILK AND MILK PRODUCTS

Most goat milk is used on the farm where it is produced. In certain areas with high dairy goat populations, some milk is sold to consumers. The sale of goat milk must follow local regulations for the marketing of dairy products. Cleanliness and sanitation are essential for the production of a quality product. Some areas have evaporating and drying plants to handle goat milk. The milk is then marketed as an evaporated or dried product.

Goat cheese and goat butter are other products made from goat milk. Several kinds of cheeses are made from goat milk, but the most popular is feta (Figure 30-9). These products are marketed through retail outlets for cheese and butter.

Other goat milk products are bath and beauty products. Most of these products are sold directly to the consumer through co-operative markets or farmers' markets.

Photo by Shannon Lawrence

Figure 30-9 The most popular goat cheese is feta cheese.

MARKETING CASHMERE AND MOHAIR

The majority of the world's supply of cashmere is produced in Afghanistan, Iran, Mongolia, India, China, New Zealand, and Australia. Cashmere is considered a premium fiber because of its unique, soft, warm, and long-wearing nature. The demand for cashmere has typically exceeded the supply. The fibers must be separated by combing out the down or using a commercial dehairer on sheared fibers. Due to its high value, cashmere is often counterfeited. Product labels are required to reflect the true fiber content of an item. If fibers such as wool and silk are blended with cashmere, the product cannot be sold as 100 percent cashmere.

About 90 percent of the mohair in the United States is produced in Texas. Most of the mohair is marketed through warehouses. The procedure is the same as that for marketing wool through warehouses. Some of the warehouses grade the mohair. High-quality mohair clips will bring better prices when they are sold on a graded basis. A **clip** is the wool or mohair produced by a single shearing. Information gained from graded mohair can be used to improve the breeding program. Animals that produce higher grades of mohair should be kept for breeding purposes. This kind of selection will eventually improve the grade of mohair produced.

Mohair is graded on the basis of the fineness of the fibers. Kid hair is finer than adult hair and is the most valuable portion of the clip. The coarsest hair comes from bucks and old wethers. It is a good practice for the producer to pack these grades separately at shearing time. Grease mohair grades range from Finer than 40s to Coarser than 18s. These numbers refer to the number of hanks that can be spun from 1 pound of clean mohair. Average diameters range from less than 23.55 microns for the finest grade to more than 43.54 microns for the coarsest grade. It is not always profitable to grade mohair clips. The clip must contain more than 50 percent 26s to 28s and finer for it to be profitable to grade.

SUMMARY

Markets for sheep, goats, and the products from these animals are not as well developed as markets for other livestock. In addition, the sheep and goat industries are small in comparison with animal industries such as swine, poultry, beef, and dairy.

Sheep production for both meat and wool has been declining in popularity for several years for a number of reasons. However, the promotion of sheep and wool products, along with developing niche markets and improvements in production, has given reason for optimism about the future of the sheep and wool industries.

Promotion of sheep and wool is provided by the American Lamb Board and American Wool Trust. The American Lamb Board promotes lamb through advertising, public relations, culinary education, and retail promotions. The American Wool Trust Fund provides funding aimed at increasing the competitiveness of American wool.

Sheep may be marketed through terminal markets, local pools, sale barns, direct selling to the packer, or by electronic marketing. The producer's choice of a market depends on the kinds of markets available locally, prices paid, numbers of lambs to be sold, and the type of available transportation.

Sheep are classed according to age, use, sex, and grade. Age classes are lamb, yearling, and sheep. Lambs are further divided by age as hothouse lambs, spring lambs, and lambs. The classes by use are slaughter sheep, slaughter lambs, feeder sheep, feeder lambs, breeding sheep, and shearer lambs. The classes by sex are ewe, ram, and wether.

The U.S. Department of Agriculture sets standards for quality grades of lamb carcasses. Grades for live animals are directly related to the quality and yield grades of the carcasses they will produce. Live grades of sheep are based on quality and estimated yield. Quality grades are based on conformation and amount of finish.

Sheep shrink as they are moved to market just like other classes of livestock. Careful handling of sheep while they are being transported to market helps to reduce shrinkage and other losses. Lamb supplies vary from month to month throughout the year, which contributes to a seasonal variation in the price of lamb.

Purebred sheep are sold by private sale or through auction sales sponsored by regional or state sheep associations. Sheep from purebred flocks that do not meet the standards for breeding stock are sold for slaughter.

Wool is marketed through local buyers, wool pools, cooperatives, warehouse operators, or by direct sale to wool mills. The choices open to the producer depend on the kinds of markets available locally. Wool pools and marketing cooperatives have developed in many areas in an effort to obtain a better price for wool. The value of the wool fleece is largely determined by the amount of clean wool that it produces. The length, density, and diameter of the wool fiber also affect the value of the wool by determining its grade. In the United States, wool is usually considered a by-product of the sheep industry, whereas foreign producers tend to take greater care to produce high-quality wool.

Wool grades are based mainly on the diameter of the fiber. Density is the major factor in determining the weight of the fleece. Length of fiber is also considered when grading wool. The three systems of grading wool are the American (blood) system, numerical count (USDA) system, and micron system. The two major classes of wool are apparel wool and carpet wool.

Meat goats are often marketed directly to the consumer. Goats may be marketed for meat from young kids to older adults. Goat meat is called chevon. According to the American Meat Goat Association, chevon may be from goats that are 48 to 60 pounds and 6 to 9 months of age. Young kids that weigh 30 to 40 pounds are marketed as cabrito, which is considered a delicacy. Most goat milk is used on the farm where it is produced, but some goat milk is sold to consumers. Goat cheese and goat butter are other products made from goat milk.

The majority of the world supply of cashmere comes from Afghanistan, Iran, Mongolia, India, China, New Zealand, and Australia. Cashmere is considered a premium fiber because of its unique, soft, warm, and long-wearing nature. The demand for cashmere typically exceeds the supply.

About 90 percent of the mohair in the United States is produced in Texas. Most of the mohair is marketed through warehouses. Mohair is graded on the basis of the fineness of the fibers.

Student Learning Activities

1. Prepare a PowerPoint™ presentation that compares and contrasts the marketing of sheep and goats to other types of livestock and poultry.
2. Prepare a bulletin board of the various products marketed from sheep and goats.
3. Take a field trip to a local production unit or market for sheep, goats, wool, cashmere, or mohair.
4. Prepare a "bumper sticker" for the promotion of one or more sheep or goat products or for the industry as a whole. Be prepared to present your "bumper sticker" to the class. Remember that the objective is to promote the consumption or use of sheep and goat products.
5. Ask a producer of sheep, goats, wool, mohair, cashmere, or goat milk to speak to the class about his or her operation.

Discussion Questions

1. Discuss two factors that have led to the decline of the sheep industry.
2. How do the American Lamb Board and American Wool Trust work to promote sheep and wool?
3. Explain the factors that have contributed to the recent popularity of goats and goat products.
4. Discuss the seasonal variation in prices for lamb and wool.
5. Give two USDA standards for each of the quality grades of lamb carcasses.

Review Questions

True/False

1. A ewe is a male sheep or lamb that has not been castrated.
2. Promotion of sheep and wool is provided by the American Lamb Board and American Wool Trust.
3. The price of lamb varies by the season of the year.
4. Products made from goat milk include goat cheese and goat butter.
5. The production of sheep and sheep products has been declining in the United States for a number of years.

Multiple Choice

6. Meat from goats weighing more than 48 pounds and 6 to 9 months old is called ________.
   a. mutton c. beef
   b. veal d. chevon
7. What percent of the mohair in the United States is produced in Texas?
   a. 90% c. 10%
   b. 25% d. 50%
8. The classes by sex for sheep are ________, ________, and ________.
   a. cow, bull, steer c. ewe, ram, wether
   b. sow, boar, hog d. nanny, billy, wether
9. Long wool fibers make ________ yarn.
   a. strong c. long
   b. stiff d. short
10. The most popular cheese made from goat milk is ________.
    a. feta c. cheddar
    b. American d. blue

Completion

11. Apparel wool is further divided into _______ and _______ classes.
12. A term that refers to the yield of closely trimmed, boneless retail cuts that come from the wholesale cuts of sheep is _______.
13. Sheep are classed according to _______, _______, _______, and _______.
14. Goats are becoming more popular as a source of meat and _______.
15. The _______ _______ is the number of hanks of yarn that can be spun from 1 pound of wool top.

Short Answer

16. List the four quality grades for lambs.
17. What is the difference between mutton and lamb?
18. List the types of markets available for sheep.
19. What is the difference between chevon and cabrito?
20. What determines the yield grade of sheep?

SECTION 7

Horses

Chapter 31

Selection of Horses

Key Terms

foal
filly
colt
mare
stallion
gelding
gait
light horse
hand
pony
color breed
tobiano
tovero
overo
commensalism
donkey
ass
jack
jennet
mule
hinny
unsoundness
blemish
vice

Objectives

After studying this chapter, the student should be able to

- describe the characteristics of the horse industry.
- describe the common breeds of horses.
- describe the selection of a horse.

HORSES IN THE UNITED STATES

In the United States, horses are used primarily for recreational purposes. About 75 percent of all horses in the United States are owned for personal pleasure use. Ranching, racing, breeding, and commercial riding make up the other 25 percent.

The three main types of horse enterprises are breeding, training, and boarding stables. Horse breeding farms breed mares and sell the offspring. Some farms specialize in training horses for show or racing. Boarding stables feed and care for horses that are owned by people who do not have facilities for keeping horses. Good management skills, a high level of capital investment, and well-trained workers are required for success in horse enterprises. Breeding, feeding, caring for, and training horses all require a lot of labor. Today, it is estimated that the total horse population of the United States is about 9.2 million. Table 31-1 shows the top five leading states based on the number of horses in the United States. More than 4.6 million people are involved in the horse industry as owners, service providers, employees, and volunteers. Around 2 million people in the United States own horses. This means that 1 out of every 63 Americans is involved with horses.

The most popular breeds of horses in the United States, based on purebred registrations, are Quarter Horse, Thoroughbred, Paint, Appaloosa, and Arabian. These breeds account for more than 75 percent of all horses registered with breed associations in the United States. The most popular breeds of horses in the world, based on estimates of world numbers, are Thoroughbred, Quarter Horse, Arabian, and Appaloosa. These breeds account for more than 75 percent of all horse numbers worldwide.

The horse industry accounts for more than $39 billion of direct economic activity annually in the United States. Horse racing is a major spectator sport, and horse shows and rodeos draw large numbers of participants and spectators each year. It is estimated that more than 3.9 million horses are used for recreation and more than 2.7 million are used for showing.

Horses are beneficial in several ways. They contribute to the economic growth of the nation and provide people with an opportunity for physical exercise. Owning and caring for a horse contributes to an individual's sense of responsibility. Horse-related events provide an opportunity for members of a family to participate in activities together.

TABLE 31-1 Top States in Horse Numbers
1. Texas
2. California
3. Florida
4. Oklahoma
5. Kentucky

SELECTION OF HORSES

Definition of Terms

A **foal** is a young horse of either sex up to 1 year of age. A **filly** is a female less than 3 years of age; however, for Thoroughbreds, fillies include 4 year old females. A **colt** is a male less than 3 years of age, unless it is a Thoroughbred. In this case, colts include 4 year old males. A **mare** is a mature female, 4 years of age and older, except for Thoroughbred mares, which are 5 years of age or older. A **stallion** is a mature male 4 years of age or older; however, Thoroughbred stallions are 5 years of age or older. A **gelding** is a male that has been castrated.

Use of the Horse

The five general uses of horses are (1) pleasure, (2) breeding, (3) working stock, (4) show, and (5) sport. Generally, one horse cannot be used in all of these ways. The horse should be selected for its major intended use.

Sources of Horses

Horses can be purchased from breeders, private owners, auctions, and dealers. The most reliable source for obtaining a horse is from a breeder. Horses bought from breeders may be slightly more expensive, but the quality of the animal can be certified. Private owners may be a good source of horses, depending on the reasons they have for selling. Auctions and dealers are less reliable sources because it may be difficult to determine the soundness of the horses offered. The inexperienced person who desires to buy a horse should always obtain the aid of an experienced horse person when making the purchase.

Age of Horse to Buy

Horses from 5 to 12 years of age are in the prime of life. If the horse is for a young, inexperienced rider, a horse that is in this age range may be a wise choice. Older horses that are sound are also satisfactory for young riders. Young horses require additional training. The inexperienced rider may be unable to handle a young horse safely.

Sex of Horse to Buy

Stallions are often hard to manage and control. They are not suitable horses for the inexperienced rider. For pleasure riding, a gelding or mare is usually a better choice. A gelding is often very steady and dependable, while mares tend to be more excitable. Of course, a person who desires to raise a foal must purchase a young mare to breed.

Breed Selection

Certain breeds are better adapted to a particular use than others. Thus, the intended use of the horse may help to narrow the choice of breed. Personal preference is also a factor to consider when selecting a breed. A person who is not interested in breeding or extensive showing of horses may prefer to select a good grade or unregistered horse. These horses often make excellent mounts for trail and other pleasure riding. Purebred horses have a greater resale value and can be entered in more horse shows. The demand for a given breed should be considered if the owner's interest is in breeding and raising horses for resale.

Conformation

A person who is evaluating or judging horses must know the parts of the horse. The parts of the horse are shown in Figure 31-1. These terms are used extensively when working with horses and especially in equine competitions where oral reasons are required. See the section on unsoundness and blemishes for examples of how these terms are applied.

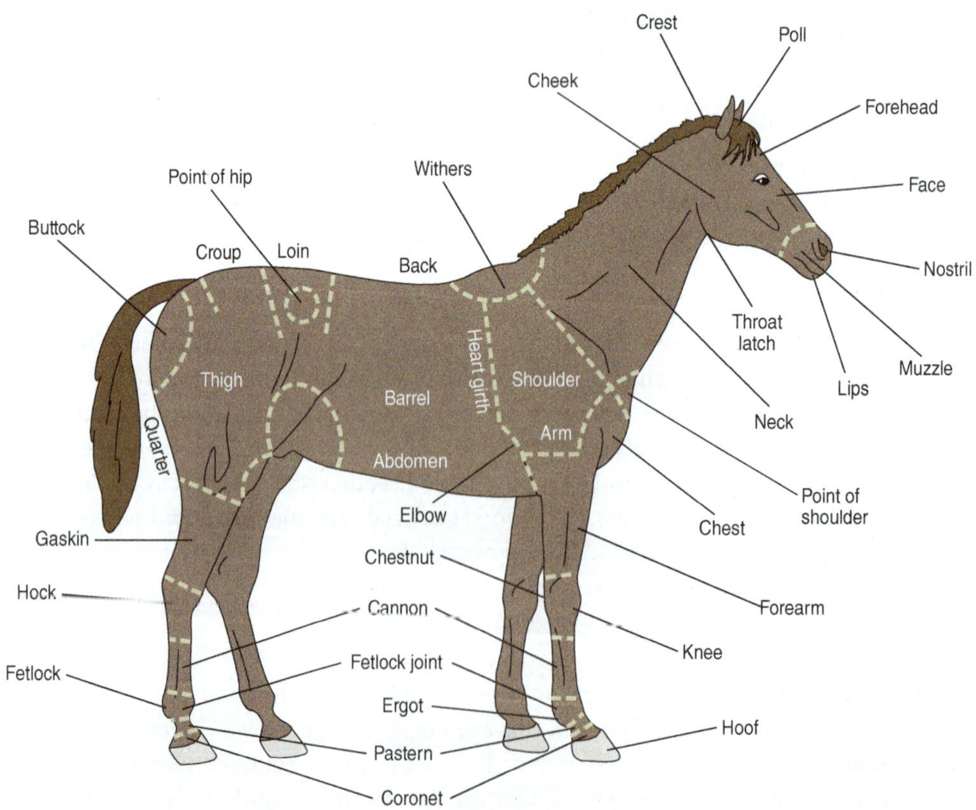

Figure 31-1 Parts of a horse.

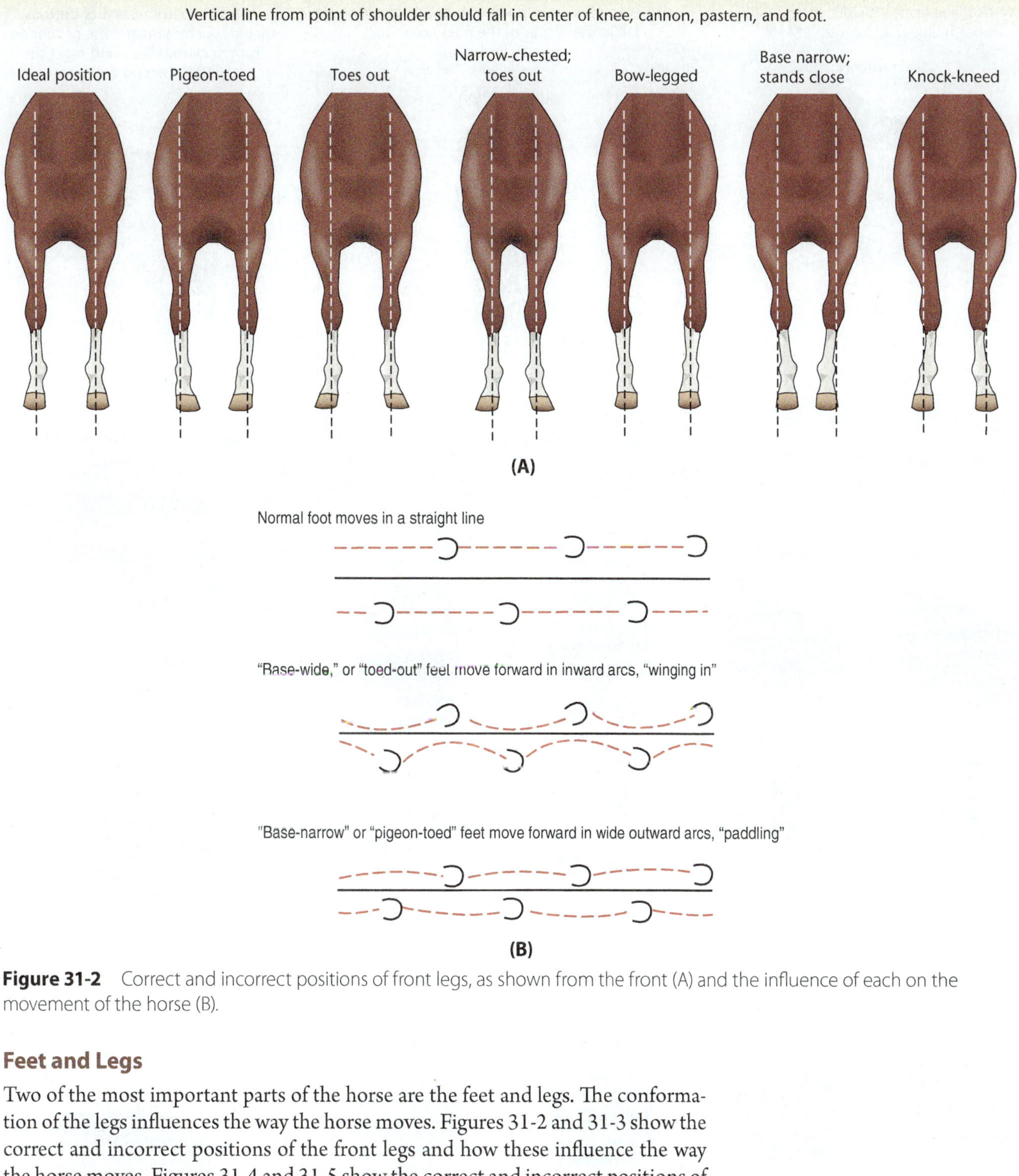

Figure 31-2 Correct and incorrect positions of front legs, as shown from the front (A) and the influence of each on the movement of the horse (B).

Feet and Legs

Two of the most important parts of the horse are the feet and legs. The conformation of the legs influences the way the horse moves. Figures 31-2 and 31-3 show the correct and incorrect positions of the front legs and how these influence the way the horse moves. Figures 31-4 and 31-5 show the correct and incorrect positions of the rear legs of the horse.

The feet are just as important as the legs of the horse. A detailed diagram of the hoof is shown in Figure 31-6. The weight of the horse is carried on the wall, bars, and frog. The sole normally does not touch the ground. The hoof must be properly trimmed to keep the horse standing squarely and moving straight. The hoof grows at the rate of about $\frac{3}{8}$ to $\frac{1}{2}$ inch per month. The hoof should be trimmed every 4 to 6 weeks. The hoof must be kept moist. A dry hoof will crack. If possible, it is helpful to allow the horse to stand occasionally in an area where the ground is moist. This

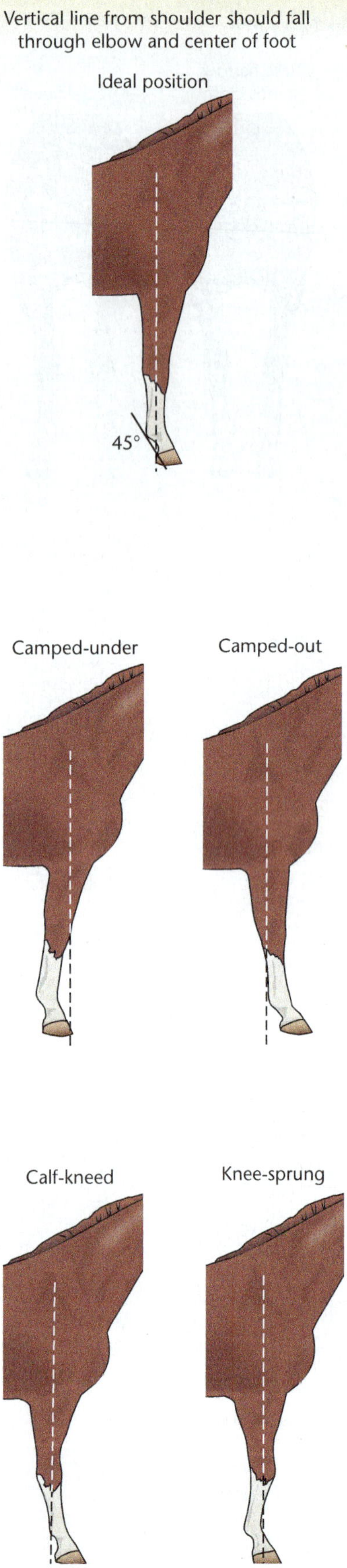

Figure 31-3 Correct and incorrect positions of the front legs, as shown from the side.

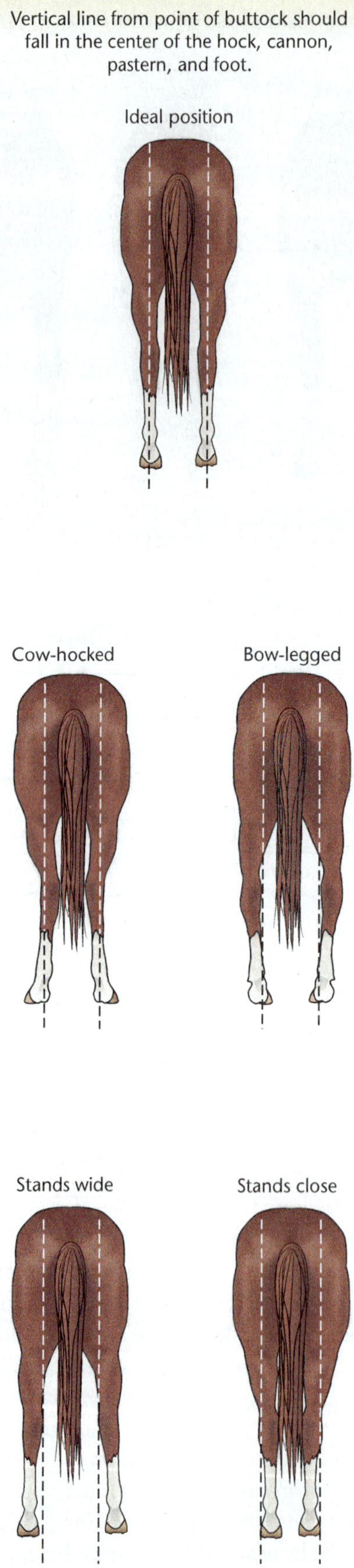

Figure 31-4 Correct and incorrect positions of the rear legs, as shown from the rear.

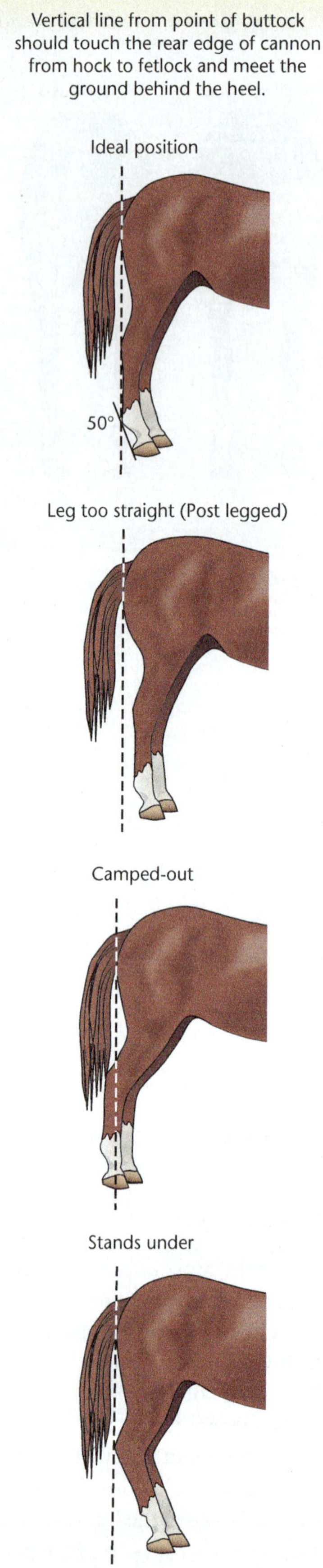

Figure 31-5 Correct and incorrect positions of the rear legs, as shown from the side.

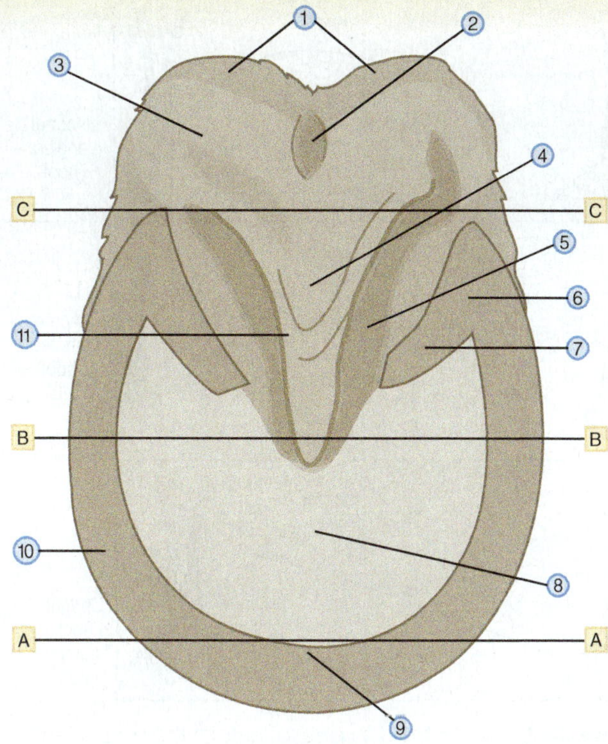

1. Horny bulb of the heel
2. Middle cleft of the frog
3. Branches of the frog
4. Body of the frog
5. Cleft of the frog
6. Buttress
7. Bars
8. Sole
9. White line
10. Wall
11. Commissure

A to A — Toe
A to B — Side wall
B to C — Quarter

Figure 31-6 The parts of the hoof.

helps to keep the hooves from drying out. A hoof dressing may also be applied to prevent drying.

The pastern and hoof should form a 50-degree angle with the ground. Figure 31-7 shows the correct and incorrect angles of the pastern and hoof. The effect of the angle on the arc of the hoof is also shown.

Body Colors

The four basic colors of horses are bay, black, brown, and chestnut. Bay varies from yellowish tan to bright mahogany. A dark bay is almost brown. A black horse has fine, black hair on the muzzle. A brown horse is very dark but will have tan or brown hair on the muzzle or flanks. The chestnut color is red with variations from a light yellow to a dark liver color. Brilliant red-gold and copper shades are also considered chestnut. The white horse is pure white and remains the same color all of its life.

The major variations in horse colors are dun, buckskin, white, gray, palomino, pinto, and roan. Dun is a yellow color. The variations are from a mouse color (grullo) to golden. Dun horses always have a stripe down their backs. Buckskins can resemble duns but generally have black manes and tails and black markings. Buckskins do not have a dorsal stripe like duns do. White horses are pure white and have pink skin. Gray is a mixture of black and white hair with black skin. The palomino color is golden with a light mane and tail. The color varies from light yellow to a bright copper. A Pinto horse is spotted with more than one color. Piebald is a white and black color combination, and skewbald is white with any other color except black. Roan horses have white hairs mingled with one or more

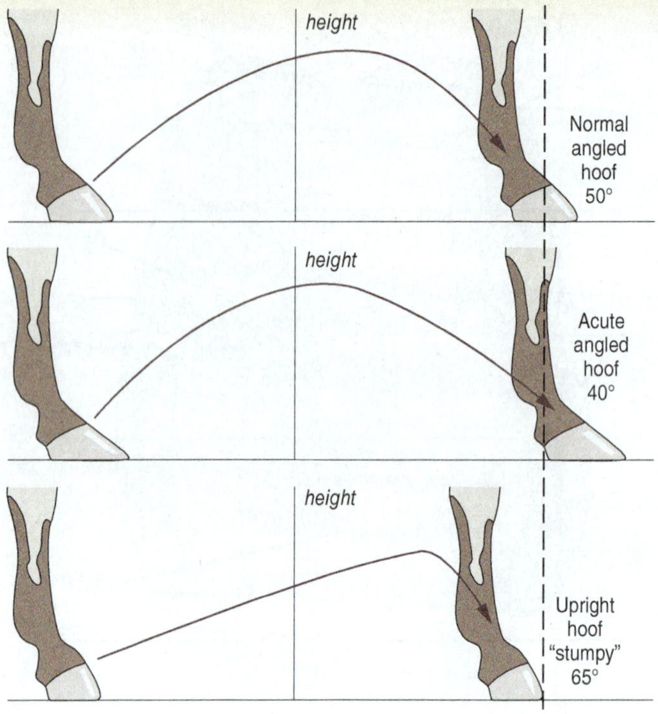

Figure 31-7 Correct and incorrect angle of the pastern and hoof and the arc of the hoof at each of the angles.

other hair colors. The blue roan is a mixture of black with white hair, while the bay roan is a mixture of bay with white hair. The strawberry roan is a mixture of chestnut with white hair.

Face and Leg Markings

Face and leg markings are often used, along with the color of the horse, to identify an individual horse. The common face markings are shown in Figure 31-8. The common leg markings are shown in Figure 31-9.

Figure 31-8 Common face markings found on horses.

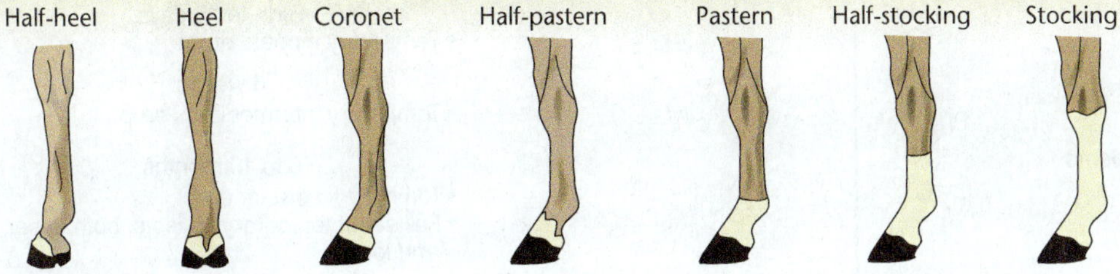

Figure 31-9 Common leg markings of horses.

Age of Horses

The approximate age of a horse can be determined by looking at the front teeth. This method is more accurate with younger horses than it is with older horses. A horse that has grazed sandy areas most of its life will appear older than it is because of the wear on the teeth. Figure 31-10 gives some guidelines to use in determining the age of horses by examining their teeth.

Gaits

The gait is the movement of the horse's feet and legs when the horse is in motion. The walk, trot, and gallop are the three natural gaits of the horse. Other gaits include the canter, stepping pace, running walk, fox-trot, amble, rack, and pace.

The walk is a slow, four-beat gait. Each foot leaves and strikes the ground separately from the other feet. It is the natural gait of the horse.

The trot is a fast, two-beat diagonal gait. Opposite front and hind feet leave and strike the ground at the same time.

The gallop is a fast, four-beat gait. Each foot strikes the ground separately. The feet strike the ground in the following order: (1) one hind foot, (2) the other hind foot, (3) the diagonal front foot, (4) the other front foot. For a brief moment, all four feet are off the ground. The lead changes when all four feet are off the ground. Most of the drive comes from the hindquarters. The extended gallop is called the run.

The canter is a slow, three-beat gait. The feet strike the ground in the following order: (1) one hind foot, (2) the other hind foot and the diagonal front foot, (3) the other front foot. The horse may lead from either the right or the left front foot. A Western adaptation of a very slow canter is called the lope.

The stepping pace is a slow, lateral, four-beat gait. The four feet strike the ground separately. The feet strike the ground in the following order: (1) right hind foot, (2) right front foot, (3) left hind foot, (4) left front foot.

The running walk is a slow, diagonal, four-beat gait. Each foot leaves and strikes the ground separately from the other feet. The front foot strikes the ground just ahead of the diagonal hind foot. This is a natural gait of the Tennessee Walking Horse.

The fox-trot is a slow, short, broken trot. The hind foot strikes the ground just ahead of the diagonal front foot.

The rack is a fast, even, four-beat gait. The time between each foot striking the ground is the same. The order of the feet striking the ground is the same as in the stepping pace.

The pace is a fast, two-beat gait. The front and hind feet on the same side leave and strike the ground at the same time. There is a brief moment when all four feet are off the ground at the same time.

The amble is a lateral movement of the horse. It is also called the traverse or side step. It is not a show gait. The horse moves to one side without going forward or backward.

Pedigree

Pedigree is of greatest importance when selecting race and show horses. It is good to remember that ancestors further back than the grandparents have contributed

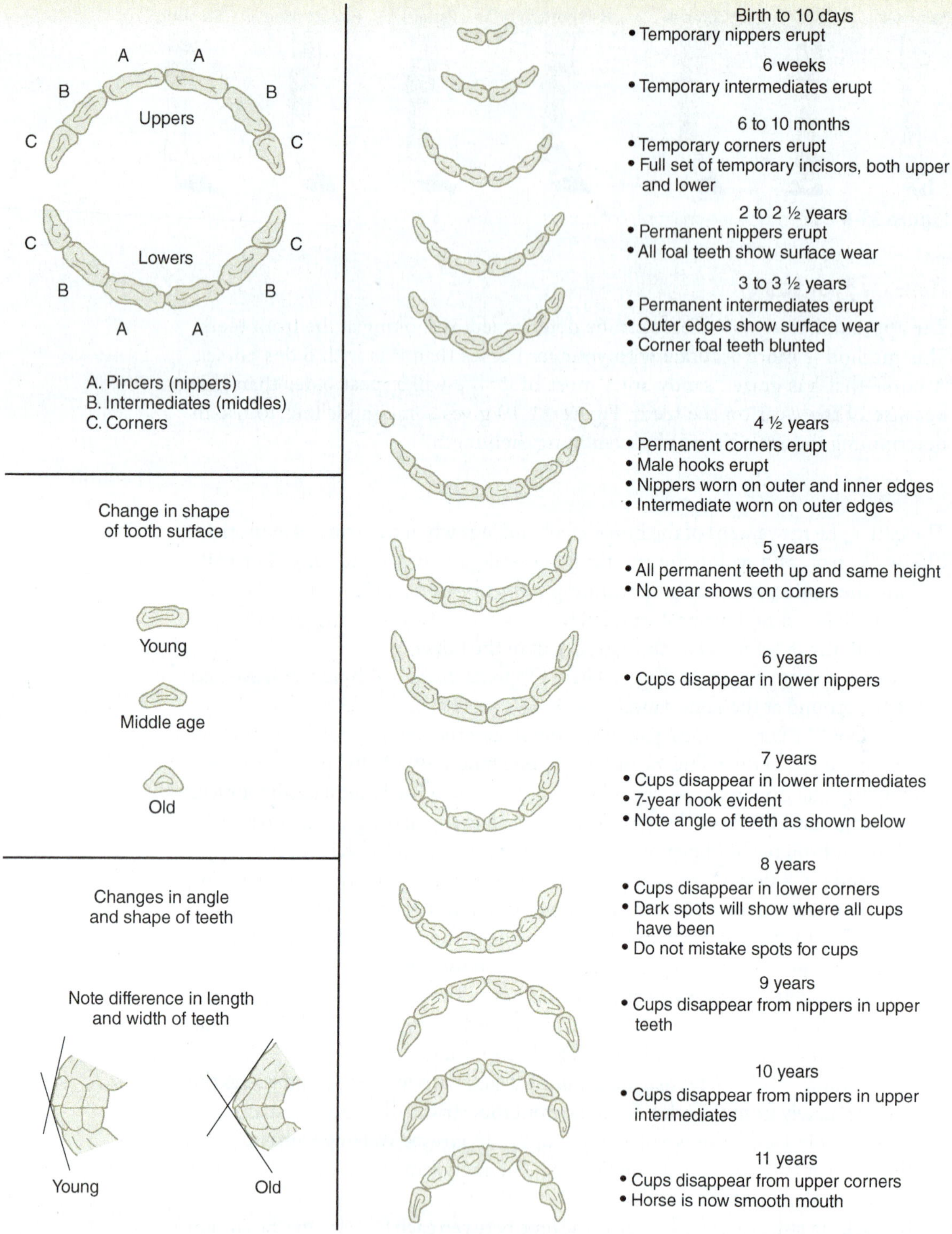

Figure 31-10 The number of permanent incisors may be used to determine age to 5 years. After that, use the shape of the biting surface and the angle of the incisors.

little to the genetic inheritance of the horse that is being considered for purchase. Other factors of selection such as conformation, soundness, training, and so forth, should be considered more important than pedigree for horses to be used for anything other than racing and show. When owners are searching for racehorses, one of the characteristics they look for are large nostrils (Figure 31-11). Large nostrils on racehorses allow them to take in more air when running.

Records

Breed associations have forms available for keeping records on registered horses. Breeding and production records are kept on both stallions and mares. More than one form may be used to keep records on horses. Identification records should be kept for both registered and nonregistered horses in case they are stolen or lost.

A record form is generally completed when a registered horse is sold to a new owner. Show and racehorses usually have performance records available.

Price

The purchase price of a horse may vary from a small amount to thousands of dollars. Mature, unregistered horses are generally lower in price. Registered horses with potential for showing are higher priced. Horses from outstanding breeding with racing potential will usually cost many thousands of dollars.

Prices of horses vary with the season. Prices are often lower during the fall and winter. The annual costs of keeping a horse must also be considered when thinking about price. Feed, housing, shoeing, and veterinary fees are annual expenses that add to the cost of having a horse. Tack and equipment prices should also be included when determining the cost of the horse to provide a more accurate picture of the cost of ownership. How the horse will be used will also influence the price of the horse as well as the other costs associated with horse ownership.

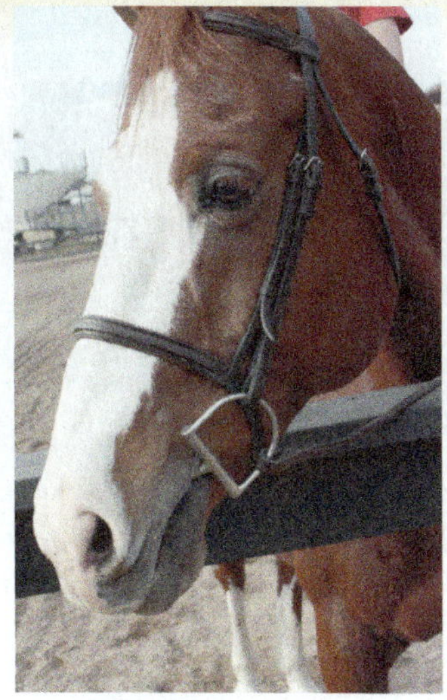

Courtesy of Frank B. Flanders

Figure 31-11 When selecting racehorses, an important criterion to consider is nostril size. Horses with large nostrils can take in more air when the animal is running.

Items Found on Breeding and Production Records

Stallions

- name and registration number
- identifying marks
- mare's owner when bred
- date of service
- mare's name
- mare's breed
- mare's registration number
- number of mares bred
- number of mares settled
- number of live foals
- performance records of foals

Mares

- name and registration number
- birth date
- owner
- breed
- sire and dam
- identifying marks
- number of times bred
- number of times settled
- number of live foals
- performance records of foals
- health records

BREEDS OF LIGHT HORSES AND PONIES

There are many breeds of light horses and ponies. More detailed information about these breeds may be secured by contacting the breed association. Breed associations register purebred horses of their respective breeds.

Light horses are used mainly for riding, driving, and racing. They measure 14-2 to 17 hands at the withers. A **hand** is 4 inches. A measure of 14-2 hands is 58 inches. In the measurement 14-2, the 14 represents the number of hands; the 2 indicates the number of inches. Light horses weigh from 900 to 1,400 pounds. **Ponies** measure under 14-2 hands and weigh from 300 to 900 pounds. They are used mainly for riding and driving.

American Bashkir Curly

The origin of the Curly horse in America is a mystery (Figure 31-12). There is much speculation as to how this breed arrived in the American West. Nearly all that is known about Curly horses today stems from the work of the Damele family of Nevada. Curly-haired horses were found in the wild herds inhabiting their ranch in 1898.

It was originally believed that the Russian Bashkir horse was also curly coated and the ancestor of the American Curly, hence the name American Bashkir Curly. With the introduction of new technology, we now know that the Russian Bashkir is unrelated to the American Bashkir Curly, and the name is maintained for public recognition only.

The American Bashkir Curly has very distinct characteristics that set the breed apart from other horse breeds. The curly hair is not typical horsehair but closely resembles mohair. The hair can actually be spun and woven into garments and is also considered hypoallergenic. Individuals who are allergic to straight-haired breeds will not usually have a reaction to the American Bashkir Curly. Curlys are intelligent and versatile. They are ridden in English, Western, Endurance, and Dressage and make excellent trail and driving horses. Therapeutic riding programs seek out Curlys for their quiet demeanor. The American Bashkir Curly Registry breed

Figure 31-12 American Bashkir Curly.

American Bashkir Curly "IM Mimosa" ABC3275 owned by Ingrid Fulcher of Curly Horse Ranch—Juniper Hills, CA

standard describes a Curly in part as being 14 to 16 hands and weighing 800 to 1,250 pounds. The head is of medium size with a well-defined jaw and throatlatch. The eyes are wide set with eyelashes that curl up. The neck is of medium length and deep at the base where it joins the shoulder. The back is noticeably short and deep through the girth. The legs are heavy boned with short cannon bones as compared to the forearm. They have a curly coat and display a gentle disposition.

American Creme

The American Creme originated in the United States. It is a color breed that may have the type of any of the breeds. A **color breed** is one that is registered on the basis of color. The American Creme is used for pleasure riding, exhibition, and stock horses.

Registration is in the American Creme Horse Registry. They are registered in one of four sections, according to the following classifications:

A—body ivory white, mane white (lighter than body), eyes blue, skin pink

B—body cream, mane darker than body, cinnamon buff to ridgeway, eyes dark

C—body and mane of the same color, pale cream, eyes blue, skin pink

D—body and mane of same color, sooty cream, eyes blue, skin pink

Combinations of these classifications are acceptable.

The American Creme was recognized as a separate breed for registration in 1970. From 1963 to 1970, Creme horses were registered in the National Recording Club. Creme horses so registered can now be transferred to the American Creme Section of the International American Albino Association.

Figure 31-13 American Paint horse.

Courtesy American Paint Horse Association © 2007

American Gotland Pony

The American Gotland originated in Sweden. The breed is small, ranging from 11 to 13 hands. Colors are bay, black, brown, chestnut, dun, palomino, or roan. Some have leopard or blanket markings. Gotlands were first imported into the United States in 1957. They are used for harness racing and as pleasure horses.

American Mustang

The American Mustang is a descendant of horses brought to the Americas by the Spanish. The breed originated in North Africa. Mustangs may be any color. They are used for show and pleasure riding, jumping, and endurance trail riding, and as stock horses. The breed association is the American Mustang Association Inc.

American Paint

The American Paint (Figure 31-13) originated in the United States from the descendants of horses brought to the Americas by the Spanish. The Paint horse was popular with the Plains Indians and the early American cowboys. Although the popularity of the breed faded for a time, there has been a renewed interest in the breed in recent years. The American Paint Stock Horse Association was formed in 1962. This later merged with the American Paint Quarter Horse Association to form the American Paint Horse Association.

Figure 31-14 American Paint horses are used as work or cow horses.

Courtesy American Paint Horse Association © 2007

The words *paint* and *pinto* both refer to spotted or two-tone horses with white and another color for body markings. The Pinto Horse Association of America Inc. registers horses of all breeds and types. The American Paint Horse Association registers on the basis of registered Paints, Quarter Horses, and Thoroughbreds. The Paint horse is used for pleasure, show, racing, and stock purposes (Figure 31-14).

Paint horses have three different color patterns: tobiano, tovero, and overo. The **tobiano** head is marked in the same way as that of a solid colored horse. The legs are white, at least below the knees and hocks, and there are regular spots on the body.

The **tovero** has dark pigmentation around the ears, which may expand to cover the forehead and/or eyes. One or both eyes are blue. Chest and flank spots vary in size and may accompany smaller spots, which extend forward across the barrel and up over the loin. The **overo** has variable color head markings. The white usually does not cross the back between the withers and the tail. One or more legs are dark colored. The body markings are irregular and scattered.

Sari Levin

Figure 31-15 American Saddlebred horse.

American Saddlebred Horse

The American Saddlebred horse (Figure 31-15) originated in Kentucky. The average height of the Saddlebred horse is 15 to 16 hands, and it ranges in weight from 1,000 to 1,200 pounds. American Saddlebred horses may be bay, chestnut, black, or gray. Roan, golden, or palomino animals appear occasionally. The horse may be three or five gaited. Three-gaited horses perform the walk, trot, and canter; five-gaited horses also have a slow gait (stepping pace, fox-trot, or running walk) plus the rack. They are used as pleasure, stock, and harness horses.

Saddlebred horses belonged to the former American Saddle Horse Breeders' Association, formed in Louisville, Kentucky in 1891. They now belong to and are registered in the American Saddlebred Horse Association.

American White

The American White Horse originated in the United States on the White Horse Ranch in Napier, Nebraska. It is a color breed, with many conformation types being represented. The color is snow white with pink skin, and dark eyes. American Whites are used as exhibition, pleasure, racing, and light draft horses.

American Whites are registered in the American Albino Association, which was established in 1937. Only Dominant White horses were registered until 1949. Classifications A, B, C, and D were added at that time to register Creme horses. The association now registers American White horses in one section and American Creme horses in another.

Notorio DW—Andalusian Stallion

Figure 31-16 Andalusian horse.

Andalusian

The Andalusian originated in Spain. There are only a few Andalusians in the United States because the Spanish did not permit export of the breed for many years (Figure 31-16). The colors are bay, white, gray, and occasionally black, roan, or chestnut. They are used for pleasure riding, jumping, exhibition, and bullfighting. The breed association is the International Andalusian and Lusitano Horse Association.

Appaloosa

The Appaloosa (Figure 31-17) originated in the United States from the descendants of horses brought to the Americas by the Spanish. The color patterns of the Appaloosa are variable. Most Appaloosas are white over the hips and loins with dark, round or egg-shaped spots. Occasionally, the entire body is mottled with spots. The eye is encircled with white. The hoofs are vertically striped with black and white.

The Nez Perce Indian tribe developed a selective breeding program using these spotted horses. After the Nez Perce Indians lost their war with the United States in 1877, the Appaloosa horses that belonged to them were scattered throughout the West. Interest in the Appaloosa horse increased in the late 1930s. The Appaloosa Horse Club Inc. was formed in 1938. Descendants of the Nez Perce horses were the foundation stock for the present breed. The Appaloosa is used for pleasure riding, showing, racing, parades, and as a stock horse.

Courtesy of Jillian Paige Photography

Figure 31-17 Appaloosa horse.

Courtesy of the Arabian Horse Association

Figure 31-18 Arabian horse.

The American Quarter Horse Journal

Figure 31-19 Dun-colored horses like the one pictured may be registered in the Buckskin registries.

Photo by David Field

Figure 31-20 Cleveland Bay horse.

Photo by David Field

Figure 31-21 Two Cleveland Bay horses (Pepper Pot Bay and Rosedale Atlantis).

Appaloosa Pony

The Appaloosa Pony has the same color traits as the Appaloosa Horse. The Pony of the Americas Club is the breed association that registers the pony. The ponies registered show the traits of the Appaloosa but do not meet the height requirements for horses.

Arabian

The Arabian horse originated in Arabia (Figure 31-18). It is small to medium in size, ranging from 850 to 1,100 pounds. The colors are mainly bay, gray, or chestnut, with a few being white or black. The skin is always dark, and the legs and head often have white markings. The Arabian is used for pleasure riding, racing, showing, and as a stock horse. The breed associations are the Arabian Horse Registry of America Inc. and the International Arabian Horse Association.

Buckskin

The Buckskin originated in the United States from horses of Spanish descent. The buckskin is a color breed, with many different types being registered. The colors of the Buckskin that are allowed for registration are buckskin, dun, red dun, or grulla (Figure 31-19). The American Buckskin is registered in the American Buckskin Registry Association Inc. The International Buckskin is registered in the International Buckskin Horse Association Inc. There are slight differences in the rules for registry in the two associations. The Buckskin is used for pleasure riding, as a stock horse, and for show.

Cleveland Bay

The Cleveland Bay originated in England and is Britain's oldest horse breed (Figures 31-20 and 31-21). The breed is solid bay with black points, such as black legs, black mane, and black tail. It is larger than many of the other light horse breeds, weighing 1,150 to 1,400 pounds and usually stands 16 to 16-2 hands high. The Cleveland Bay is a versatile breed with substantial stamina. It was originally used for pack and harness horses, but the breed declined with the coming of the automobiles and tractors. Today, they are still used for a variety of purposes, including pleasure riding and police horses. The breed association is the Cleveland Bay Horse Society of North America.

Connemara Pony

The Connemara Pony originated in Ireland (Figure 31-22). It was first imported into the United States in 1951. The animals may be bay, black, brown, cream, dun, or gray. Roan or chestnut appears occasionally. The Connemara is used for jumping and pleasure riding. The breed association is the American Connemara Pony Society.

Galiceno

The Galiceno (Figure 31-23) originated in Spain. It is a small breed, ranging from 12 to 13-2 hands high. The colors are bay, black, chestnut, dun, gray, brown, and palomino. The Galiceno is used mainly for pleasure riding. The breed association is the Galiceno Horse Breeders Association Inc.

Hackney

The Hackney originated in England. It varies in size from 12 to 16 hands and 800 to 1,200 pounds. The colors are bay, brown, or chestnut. White markings are common. Roan and black appear occasionally. The Hackney is used mainly as a harness or carriage horse. The breed association is the American Hackney Horse Society.

Missouri Fox Trotting Horse

The Missouri Fox Trotting horse originated in the Ozark Hills region of Missouri (Figure 31-24). The most common color is sorrel, usually with white markings. However, all colors can be found in this breed. The fox-trot gait is a distinguishing trait. Principal uses of the breed include pleasure riding and trail riding and as stock horses. The breed association is the Missouri Fox Trotting Horse Breed Association.

American Connemara Pony Soc

Figure 31-22 Connemara pony.

Morab

The Morab horse is a cross between the Morgan and the Arabian. The breed originated in California from foundation stock bred by Martha Doyle Fuller on her ranch in Clovis, California. This foundation stock had been developed since 1955. The Morab may be any color, but without spots on the body. The eyes and skin are black. The main uses of the breed are pleasure riding, showing, endurance riding and as stock horses.

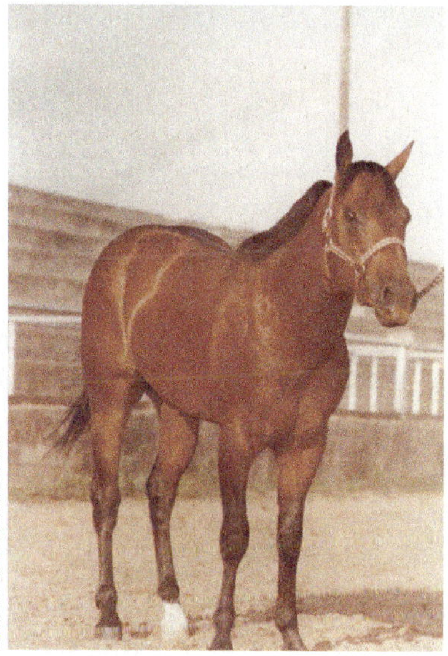

Galiceno Horse Breeders Assoc

Figure 31-23 Galiceno mare.

Morgan

The Morgan originated in the New England states. The breed is descended from a stallion named Justin Morgan that lived in the late 1700s (Figure 31-25). Bloodlines of the Morgan horse are found in the foundation stock of many of the light horse breeds in the United States. The color of the Morgan is bay, black, brown, or chestnut. It is used for pleasure riding and as a stock horse. The breed association is the American Morgan Horse Association Inc.

Morocco Spotted Horse

The Morocco Spotted horse originated in the United States. The breed carries bloodlines of Morocco Barb, Hackney, French Coach, and Saddle horse breeding. The Morocco Spotted horse is a color breed, with several breed types of light horses being recognized as eligible for registry. The main colors are black and white or bay and white. Other color combinations are chestnut and white, sorrel and white, blue and white, and palomino and white. Color patterns are either tobiano or overo, or a combination of the two. The tobiano pattern is more common. Stallions must have no less than 10 percent of the secondary color, not counting the face and legs. Mares and geldings must have no less than 3 percent of the secondary color, not counting face and legs. The Morocco Spotted horse is used for pleasure riding as a saddle horse, harness horse, and stock horse.

Missouri Fox Trotting Horse Breed Association

Figure 31-24 Missouri Fox Trotting horse. 2006 World Grand Champion.

Palomino

The Palomino (Figures 31-26 and 31-27) originated in the United States. Horses of palomino color were found among the descendants of horses brought to the Americas by the Spanish. The color of the breed is golden. It may be three shades lighter or darker than the color of a newly minted gold coin. The palomino color does not

Photo credit: Bentley Photography

Figure 31-25 Morgan mare.

The American Quarter Horse Journal

Figure 31-26 Palomino Quarter Horse.

Courtesy of Frank B. Flanders

Figure 31-27 Palomino mare.

breed true in crosses. Depending on the cross used, various ratios of palomino to other colors can result. The mane and tail are light colored. The Palomino is used for pleasure riding, and as a stock, harness, and parade horse. There are two breed associations that register Palominos: the Palomino Horse Association Inc. and the Palomino Horse Breeders of America.

Paso Fino

The Paso Fino originated in the Caribbean area. The colors of this breed are mainly solid bay, chestnut, or black with white markings. Palominos and pintos appear occasionally. The Paso Fino has a paso gait and is used for pleasure riding, showing, parades, and endurance riding. The paso gait is four-beat, with the legs on the same side moving together, and the hind foot striking the ground an instant before the front foot. There are two breed associations: the American Paso Fino Horse Association and the Paso Fino Horse Association Inc.

Peruvian Paso

The Peruvian Paso originated in Peru from horses that were brought to the Americas by the Spanish. The colors of the breed are bay, black, brown, sorrel, chestnut, dun, buckskin, red dun, grulla, palomino, gray, red roan, and blue roan. Solid colors are preferred, and the skin is dark. The horse is noted for the paso llano and sobreandando gaits. The paso llano is an equally spaced, four-beat gait. The feet strike the ground in the following order: (1) left hind foot, (2) left front foot, (3) right hind foot, and (4) right front foot. The sobreandando gait is usually faster and slightly more lateral than paso llano. The feet strike the ground in the same order as paso llano, but the left hind and left front feet strike closer together and the right hind and right front feet strike closer together. The Peruvian Paso is used mainly for pleasure, parade, and endurance riding. There are two breed associations: the American Association of Owners and Breeders of Peruvian Paso Horses and the Peruvian Paso Horse Registry of North America.

Pinto

Pinto Horse Association of America Inc., OK

Figure 31-28 Pinto horse.

The Pinto (Figure 31-28) originated in the United States. It is a color breed that may not have Appaloosa, draft, or mule breeding. The color patterns of the Pinto are tobiano and overo. Tobiano patterns usually have color on the head, chest, flanks, and some in the tail. The legs are usually white. The overo pattern usually has jagged-edged, white markings on the midsection of the body and neck area. The legs are usually colored rather than white. Only 4 square inches of white in the qualifying zone is required for horses. Three square inches are required for ponies, and 2 square inches are required for miniatures.

There are four types of Pinto horses: stock, pleasure, hunter, and saddle. The Pinto is used as a pleasure, show, stock, and parade horse. The breed association is the Pinto Horse Association of America Inc. The association also registers Pinto ponies and miniatures.

Pony of the Americas

Pony of the Americas Club Inc

Figure 31-29 Pony of the Americas.

The Pony of the Americas (POA) (Figure 31-29) originated in the United States and has Appaloosa coloring. Six color patterns are recognized: snowflake; frost; blanket; leopard; white with black spots on hindquarters; and red, blue, and marbleized roans. Other characteristics of a POA are mottled or partially colored skin, white sclera (area of the eye that encircles the iris), and striped hooves. The breed ranges in size from 11-2 to 14 hands. It is used for pleasure riding and showing by young people. The breed association is the Pony of the Americas Club Inc.

Quarter Horse

The Quarter Horse (Figure 31-30) originated in the United States. During the colonial era, horse racing was a common sport. Since the races seldom were longer than a quarter of a mile, the term *quarter miler* was used to describe these racehorses. Horses were bred that could run short distances faster than other breeds. The Quarter Horse was widely used during the westward expansion of the pioneers and on the western ranches.

The colors of the Quarter Horse are bay, black, brown, sorrel, chestnut, dun, buckskin, red dun, grullo, palomino, gray, red roan, and blue roan. Quarter Horses are used for pleasure riding, showing, racing, and as stock horses. The breed associations are the American Quarter Horse Association and the National Quarter Horse Registry Inc.

The American Quarter Horse Journal

Figure 31-30 American Quarter Horse (Little Peppy).

Racking Horse

The Racking Horse originated in the United States. It is small to medium in size. Average height is 15 hands, and average weight is 1,000 pounds. The colors of the breed include black, white, gray, chestnut, bay, brown, sorrel, roan, and yellow. White markings sometimes appear on the body. The Racking Horse is used for pleasure and show.

American Shetland Pony Stallion, Pan's Atomic

Figure 31-31 American Shetland Pony stallion (Pan's Atomic).

Rangerbred

The Rangerbred originated in the United States from foundation stallions imported from Turkey. These stallions were crossed on Mustang mares. Development of the breed took place in Colorado; therefore, the breed is also known as the Colorado Rangers. Rangerbreds can be any color. Spotted horses are common. The type is similar to the Appaloosa. They are used mainly as stock horses, although they are also shown in the ring. The breed association is the Colorado Ranger Horse Association Inc.

Shetland Pony

The Shetland Pony originated in the Shetland Islands. Shetland Ponies may be any horse color (Figure 31-31). Both broken and solid color patterns exist. The Shetland Pony is registered in two size classifications. (1) less than 10-3 hands and (2) 10-3 to 11-1 hands. The Shetland Pony is used for pleasure riding by children and for showing and racing. The breed association is the American Shetland Pony Club.

Courtesy of United States Trotting Assn

Figure 31-32 Standardbred pacer.

Spanish Barb

The Spanish Barb descended from horses that were brought to the Americas by the Spaniards. The Barb horse originally came from Africa to Spain. Many of the American light horse breeds carry Barb bloodlines. The most common colors are dun, grullo, sorrel, and roan. It is a small horse, standing 13-3 to 14-1 hands high, and it weighs between 800 and 975 pounds. Horses currently being registered with the breed association are mainly from three strains: the Romero-McKinley, Belsky, and Buckshot. The breed association is the Spanish-Barb Breeders Association.

Spanish Mustang

The Spanish Mustang originated in the United States from the descendants of horses brought to the Americas by the Spaniards. The Spanish Mustang may have any of the horse colors, with either solid or broken color patterns. The horse is small, standing 13 to 14-2 hands. The Spanish Mustang is used for pleasure riding and as a stock and packhorse. The breed association is the Southwest Spanish Mustang Association Inc.

Courtesy of United States Trotting Assn

Figure 31-33 Standardbred trotter.

© Vesty

Figure 31-34 Tennessee Walking Horse.

Standardbred

The Standardbred originated in the United States. The colors of the breed are mainly bay, black, brown, and chestnut. Other colors that may occur are gray, roan, and dun. The Standardbred was developed as a harness racing horse. Both trotters and pacers have been developed (Figures 31-32 and 31-33). The name "Standardbred" is derived from a "standard" that registered horses had to meet. Beginning in 1879, Standardbreds had to trot a mile in 2 minutes and 30 seconds or pace a mile in 2 minutes and 25 seconds. This standard is no longer required for registration. The breed registry is the United States Trotting Association. Standardbreds are also registered in the International Trotting and Pacing Association Inc.

Tennessee Walking Horse

The Tennessee Walking Horse (Figure 31-34) originated in the United States. Common colors include sorrel, chestnut, black, bay, roan, brown, white, gray, and golden. The feet and legs often have white markings. The breed is noted for its running walk gait. The horse is used mainly for pleasure riding and showing. The breed association is the Tennessee Walking Horse Breeders' and Exhibitors' Association.

Illinois Racing News

Figure 31-35 Thoroughbred horse.

Thoroughbred

The Thoroughbred originated in England (Figure 31-35). Development of the breed as a racehorse began in the seventeenth century. Common colors are bay, brown, black, and chestnut. Roan and gray occur occasionally. The face and legs often have white markings. The main use of the Thoroughbred is for racing. However, it is also used in crossbreeding programs. Thoroughbreds are registered with the Jockey Club. Horses with part Thoroughbred breeding may be registered with the American Remount Association Inc. and Thoroughbred Half-Bred Registry.

Connection: Got Your Goat?

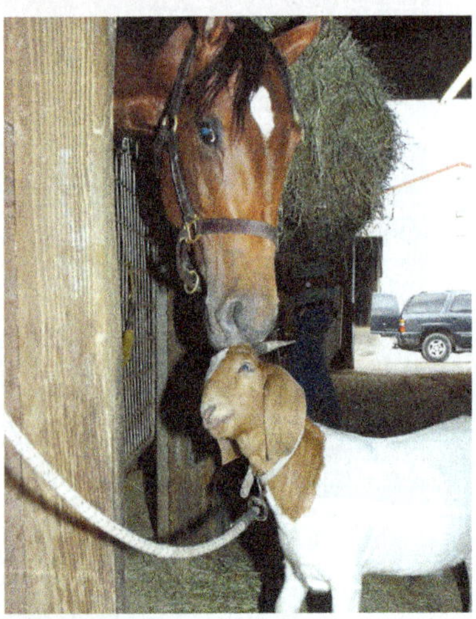

Mary Lou Coady/Coady Photography

Figure 31-36 Goats are sometimes put into the stall of a racehorse to soothe a skittish animal.

Have you ever heard the phrase "got your goat"? The saying began early in the horse racing industry. Goats were sometimes put into the stall of a racehorse to soothe a skittish animal. Since both are herd animals, they bonded naturally. The horse would develop strong emotional ties to the goat (Figure 31-36). If the goat was removed, the horse became upset. Sometimes, an unscrupulous horse owner would steal an opponent's goat the night before a race, knowing that his competitor's horse would be distressed at race time and would probably perform poorly.

The association between a goat and a horse is a type of commensalism. A commensal relationship is one in which one organism benefits while the other is neither harmed nor benefited. In this case, the horse benefits while the goat is not affected. Students interested in the behavior of animals may consider ethology as a career. *Ethology* is the study of animal behavior and is an important science in the production and use of animals.

Welsh Pony

The Welsh Pony originated in Wales. The Welsh Pony is a little larger than the Shetland Pony, but it is smaller than the light horse breeds. Two size classifications are registered: (1) ponies less than 12-2 hands and (2) ponies between 12-2 and 14 hands. The special use of the Welsh Pony is as a riding horse for children. Other uses include pleasure riding, trail riding, harness shows, hunting, and racing. Any of the horse colors, except piebald and skewbald, are accepted. The breed association is the Welsh Pony and Cob Society of America Inc.

Figure 31-37 American Walking Pony.

Walking Pony

The American Walking Pony (Figure 31-37) originated in the United States. It is a cross between the Welsh Pony and the Tennessee Walking Horse. The colors of either breed are accepted. The pony must be able to do the running walk gait. Its main use is for pleasure riding and showing. The breed association is the American Walking Pony Association.

Connection: Horsepower

What is horsepower? If you ask someone, he or she may guess that horsepower is the amount of power a horse can generate. The measurement of work in units of horsepower was developed by James Watt, inventor of the steam engine, which he patented in 1769. To convince potential buyers to purchase his steam engine, he created a way to measure its power based on the power of a horse. Watt determined that the average horse can perform 33,000 foot-pounds of work in one minute (Figure 31-38). Engines continue to be measured by horsepower. Automobile companies make good use of the horsepower measurement in automobile advertisements.

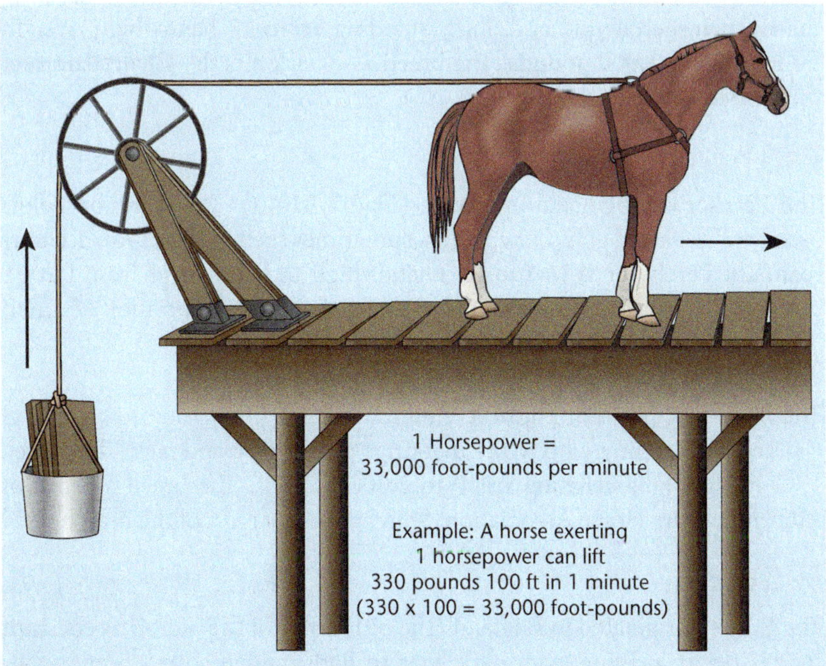

Figure 31-38 The measurement of work in units of horsepower was developed by James Watt.

The American Part-Blooded Horse Registry

Eligible horses that are not purebred may be registered in the American Part-Blooded Horse Registry. This registry was formed in 1939 in response to a demand for registering horses that were not purebred. Half bloods, grades, and crosses involving most of the light horse breeds are accepted for registry. Each group has its own registry book. The registry certificate authenticates age, breeding, color, and markings. Information and rules for registry may be obtained from the registrar of the association.

BREEDS OF DRAFT HORSES

At one time, the main source of power on the farm was the draft horse. With the mechanization of agriculture, the use and number of draft horses in the United States decreased. Draft horses are larger, heavier, and more muscular than the light horse breeds. They were selected and bred for the ability to pull heavy loads. The draft horse breeds originated in Europe. There are purebred associations for each of the draft horse breeds in the United States. Few draft horses are registered each year. More information on each breed of draft horse may be secured by contacting the breed association.

Belgian

The Belgian draft horse originated in Belgium. Common colors of the Belgian are bay, chestnut, and roan, and its average size is 15-2 to 17 hands high. Belgians range in weight from 1,900 to 2,200 pounds. The breed association is the Belgian Draft Horse Corporation of America.

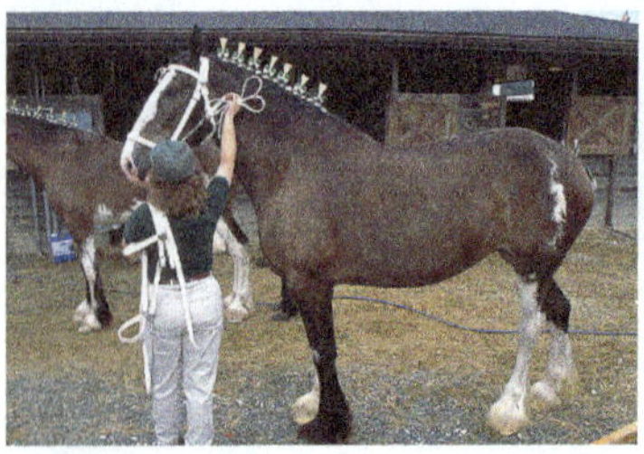

Figure 31-39 Clydesdale horse.

Clydesdale

The Clydesdale originated in Scotland. Common colors of the breed are bay and brown, both with white markings. Other colors such as black, chestnut, gray, and roan are sometimes seen (Figure 31-39). The Clydesdale is somewhat smaller than the Belgian, Percheron, and Shire, standing 16 to 17 hands high. Weights range from 1,700 to 1,900 pounds. The breed association is the Clydesdale Breeders of the United States.

Figure 31-40 Percheron horse.

Percheron

The Percheron originated in France (Figure 31-40). The common colors of the breed are black and gray. Other colors sometimes seen are bay, brown, chestnut, and roan. The Percheron is 16-1 to 16-3 hands high and may weigh from 1,900 to 2,100 pounds. The breed association is the Percheron Horse Association of America.

Shire

The Shire originated in England. Common colors are black, brown, bay, gray, and chestnut, frequently with white markings on the face and legs. The Shire is 16 to 17-2 hands high, weighing 1,900 to 2,000 pounds. The breed association is the American Shire Horse Association. Shires are the largest of the draft horse breeds.

Suffolk

The Suffolk originated in England. The only color of the Suffolk is chestnut. Seven shades occur, ranging from dark liver to light golden sorrel. Some white markings are found on the head and legs. Suffolks are the smallest of the draft horse breeds. They range in height from 15-2 to16-2 hands. The weight range is 1,600 to 1,900 pounds. The breed association is the American Suffolk Horse Association.

Donkeys and Mules

Photo by Shannon Lawrence

Figure 31-41 When a male donkey (left) is crossed with a female horse, the resulting offspring is called a mule (right).

Donkey is the common name for the ass. The **ass** is smaller than the horse, has longer ears, and a short, erect mane. The gestation period is 1 month longer than that of the horse. The male ass is called a **jack**, and the female ass is called a **jennet**. When a jack is crossed on a mare (female horse), the resulting offspring is called a **mule** (Figure 31-41). When a stallion (male horse) is crossed on a jennet, the resulting offspring is called a **hinny**. The hinny is smaller in size than the mule. Mules are usually sterile, meaning they will not reproduce. The ass and the mule are used mainly as work animals. Miniature donkeys are used as children's pets.

More information about donkeys and mules is available from breed associations: the American Donkey and Mule Society Inc., the Miniature Donkey Registry of the United States, the American Council of Spotted Asses, and the National Miniature Donkey Association.

UNSOUNDNESS AND BLEMISHES

Unsoundness is a defect that affects the usefulness of the horse. A **blemish** is an imperfection that does not affect the usefulness of the horse. The most serious unsoundness is one that affects the feet and legs. The following list defines some common unsoundnesses, blemishes, and faulty conformations. Unsoundnesses and blemishes are identified as follows: U = unsoundness, B = blemish.

Head

1. *Cataract* (U) is an opaqueness or cloudiness of the cornea of the eye.
2. *Blindness* (U) is the partial or complete loss of vision in the eye.
3. *Moon blindness (periodic ophthalmia)* (U) is a cloudiness or inflammation of the eye that occurs at repeated intervals. It may result in permanent blindness.
4. *Poll evil* (U) is an inflammation and swelling of the poll. It often results from bruising.
5. *Roman nose* is a faulty conformation.
6. *Parrot mouth* (U) is a condition where the upper jaw overshoots the lower jaw.
7. *Undershot jaw* (U) is a condition in which the upper jaw is shorter than the lower jaw.

Neck

8. *Ewe neck* is a faulty conformation of the neck where the topline of the neck is shorter than the bottom-line.

Withers and Shoulders

9. *Fistula of the withers* (U or B) is an inflammation and swelling in the area of the withers. It often results from bruising.
10. *Sweeney* (U) is a decrease in the size of a muscle or group of muscles, usually in the area of the shoulder.

Front Legs

11. *Shoe boil* or *capped elbow* (B) is a swelling at the point of the elbow.
12. *Knee sprung* (also called *buck-kneed* or *over in the knee)* is a bending forward of the knee. This is a faulty conformation.
13. *Calf-kneed* (also called *back at the knees*) is a condition in which the knees bend backward. This is a faulty conformation.

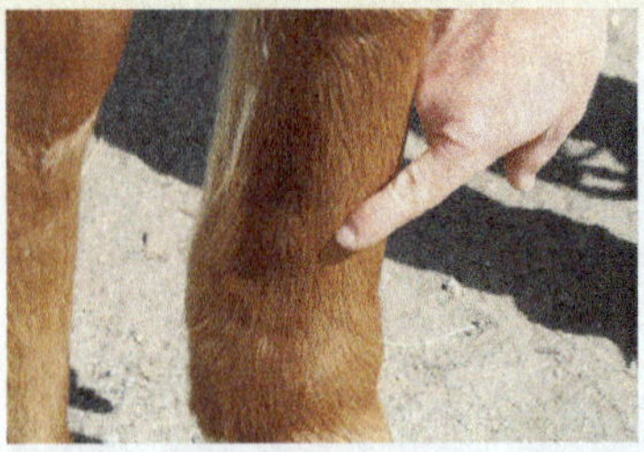

Figure 31-42 Horse with splints.

14. *Splints* (B) are deposits of bone that occur on the upper, inside part of the cannon bone. Figure 31-42 shows a horse with splints.
15. *Wind-puff (wind-gall, road-puff, road-gall)* (U) is a puffy swelling that occurs on either side of the tendons above the fetlock or knee.
16. *Bowed tendons* (U) are swellings of the tendons on the back of the leg. This may occur on either the front or back legs.

Feet

17. *Ringbone* (U) is a growth of bone on either or both of the bones of the pastern.
18. *Sidebone* (U) is a condition in which the lateral cartilage just above the hoof turns to bone.
19. *Quittor* (U) is a decay of the lateral cartilage resulting in an open sore.
20. *Quarter crack (sand crack)* (B) is a vertical split in the wall of the hoof.
21. *Navicular disease* (U) is an inflammation of the navicular bone and bursa inside the hoof.
22. *Founder* (U) is the inflammation of the sensitive lamina of the hoof. The *laminae* attach the hoof to the fleshy part of the foot.
23. *Contracted heel* (B) is a condition in which the heel draws in or contracts.
24. *Thrush* (B) is a disease of the frog of the foot. It is caused by filth and may result in lameness.
25. *Corns* (U) are reddish spots on the horny sole. They are usually caused by bruises or improper shoeing. Corns may result in swelling, a pus discharge, and lameness.
26. *Scratches (grease heel)* (U) is an inflammation of the back surfaces of the fetlocks. Scabs form on the area.

Hind Legs

27. *Stifled* (U) is a condition in which the patella (cap) of the stifle joint has been displaced.
28. *Stringhalt* (U) is a condition in which there is a sudden involuntary flexion of one or both hocks. The foot is jerked upward higher than normal.
29. *Thoroughpin* (U) is a puffy swelling just above the hock.
30. *Capped hock* (U or B) is a callus or firm swelling at the point of the hock.
31. *Bog spavin* (U) is a large, soft swelling on the inside and front of the hocks.
32. *Bone spavin (jack spavin)* (U) is a bony growth on the inside of the hock.
33. *Curb* (U) is a hard swelling on the back surface of the rear cannon, about 4 inches below the point of the hock.
34. *Cocked ankle* (U) is a condition in which the tendons become inflamed and shortened. The fetlock bends forward.
35. *Blood spavin* (B) is a swollen vein over the front and inside of the hock. It does not cause lameness.

Body

36. *Heaves (asthma, broken wind)* (U) is difficulty in breathing. It is caused by lung damage.
37. *Roaring* (U) is difficulty in breathing due to a paralysis of the nerve to the muscles of the larynx. There is a whistling or wheezing sound when the horse breathes in.
38. *Thick wind* (U) is difficulty in breathing in or out due to an obstruction in the respiratory tract.
39. *Rupture (hernia)* (U) is when any internal organ comes through the wall that contains it.

40. *Sway back* is a faulty conformation of the back.
41. *Knocked down hip* (U) is a fracture of the external angle of the hipbone. The point of the hip lowers as a result of the fracture.

VICES

Because of long periods of idleness or poor handling, horses sometimes develop bad habits that are called **vices**. The most common vices are cribbing, wind sucking, halter pulling, and kicking.

Cribbing (Figure 31-43) is a behavior in which a horse bites down on some part of a solid object. Wind sucking takes place when a horse presses the upper front teeth on some object and pulls back, at the same time sucking air into the stomach. Cribbing and wind sucking often occur together. Fitting a wide strap around the throat so that the larynx is compressed when pressure is put on the front teeth will help prevent or halt these two vices.

Figure 31-43 Stall panels worn down by cribbing horses.

Halter pulling occurs when the horse pulls back against the halter while tied. Using a heavy halter or heavy chain or rope around the neck when tying the horse will help to prevent this vice.

If hoof or shoe marks are on the walls of the stall, the horse may be a kicker. Capped hocks and scarred hind legs may also indicate a kicker. Padding the stall, hanging heavy chains from the ceiling, and hanging bags of straw from the ceiling are suggested methods of stopping this vice.

Pawing, another vice, refers to the horse using its front feet to paw or dig. Horses that paw may damage stalls and other facilities (Figure 31-44). Most commonly, a horse that paws will dig at the stall floor. Stall mats will discourage this behavior.

Figure 31-44 Pawing is a vice that is not only damaging to facilities, but dangerous to horses as well.

SUMMARY

About 75 percent of the horses in the United States are used for personal pleasure riding. Other uses include racing, ranching, breeding, and commercial riding. Today, it is estimated that the total horse population of the United States is about 9.2 million. More than 4.6 million people are involved in the horse industry as owners, service providers, employees, and volunteers. The horse industry is big business in the United States, accounting for more than $39 billion of direct economic activity annually.

Horses should be selected on the basis of conformation, use, age, sex, and soundness. Breed selection is a matter of use and personal preference. Reputable breeders are among the best sources to purchase a horse. They often keep the most accurate records and know the pedigree, temperament, and other details of the horses they sell. Other sources include private owners, auctions, and dealers.

The five basic colors of horses are bay, black, brown, chestnut, and white. Variations of color include dun, gray, palomino, pinto, and roan.

The gaits of the horse are the ways in which it moves. The common gaits are walk, trot, gallop, canter, stepping pace, running walk, fox-trot, rack, pace, and amble.

Pedigree and price must be considered when selecting a horse. Pedigree is more important for show and race horses. Prices may vary from a small amount to many thousands of dollars.

There are many breeds of light horses and ponies. Most of the horses in the United States are of the light breeds. There are few draft horses in the United States today. Ponies are smaller than light horses. Draft horses are larger than light horses.

Horses may have a variety of unsoundnesses and blemishes. Unsoundnesses are more serious than blemishes. The most serious unsoundnesses affect the feet and legs of the horse.

Student Learning Activities

1. Prepare a bulletin board display of pictures of the breeds of horses.
2. Present an oral report on the origin and traits of the breeds of horses.
3. Survey the community to determine the most popular breeds and numbers of horses locally.
4. Identify the parts of the horse on a picture or on a live horse.
5. On field trips to local farms or stables, observe conformation, unsoundnesses, blemishes, and vices of horses.
6. Determine the age of a live horse by examining the teeth.
7. Give an oral report on horse selection.
8. Follow the approved practices described in this chapter when selecting a horse while planning and conducting a horse supervised experience program.

Discussion Questions

1. Approximately how many horses are there in the United States?
2. Describe the economic impact of the horse industry in the United States.
3. List several ways in which horses are beneficial to people.
4. Compare the size and weight of light horses and ponies.
5. List and give a brief description of each of the breeds of light horses, ponies, and draft horses.
6. List the five general uses of horses.
7. Briefly explain the various sources from which horses may be secured.
8. A horse of what age is the best to buy? Why?
9. A horse of which sex is usually a better choice for pleasure riding?
10. What factors influence the choice of a horse breed to buy?
11. Draw a sketch of a horse and show the major parts.
12. Briefly describe the physical traits that are desirable in a light horse.
13. Explain the importance of a horse's feet and legs.
14. Name the five basic colors of horses.
15. Describe the five major variations in horse color.
16. Name and describe or sketch the common face markings found on horses.
17. Describe how the age of a horse can be estimated by looking at the teeth.
18. List and briefly describe some common unsoundnesses and blemishes of horses.
19. Name and describe some common vices a horse may have.
20. Name and describe the gaits of horses.
21. How important is a pedigree when selecting a horse?
22. What factors influence the price of a horse?
23. What costs must a horse owner be prepared to pay, in addition to the original cost of the horse?

Review Questions

True/False

1. The horse industry is a minor business in the United States.
2. Most of the horses in the United States are heavy breeds.

3. Pedigree and price must be considered when selecting a horse.
4. With today's technology, breeding, feeding, and caring for horses do not require a lot of labor.

Multiple Choice

5. What percentage of all horses in the United States are owned for personal pleasure use?
   a. 10% c. 60%
   b. 50% d. 75%
6. Light horses are used mainly for ___________.
   a. riding c. racing
   b. driving d. all of the above
7. What features of the horse can be used to determine age?
   a. eyes c. teeth
   b. legs d. hair coat

Completion

8. ___________ and ___________ are the two most important parts of the horse.
9. The ___________ is a lateral movement of the horse.

Short Answer

10. List the main types of horse enterprises.
11. What is the most serious unsoundness regarding selecting a good-quality horse?

Chapter 32

Feeding, Management, Housing, and Tack

Key Terms

meconium
imprinting
roached mane
float
farrier
tack
bridle
bit
halters
martingale
harness

Objectives

After studying this chapter, the student should be able to

- develop a feeding program for horses based on commonly accepted standards.
- describe good management practices for horses.
- describe housing and tack required for horses.

FEEDS FOR HORSES

Horses can utilize large amounts of roughage in the ration. Bacteria in the cecum break down roughages into a form that can be utilized by the horse. Mature, idle horses can be fed on roughage alone. Horses that are being ridden or otherwise worked require some concentrate added to the ration. Pregnant mares and growing foals also require concentrate in the ration in addition to roughage.

Pastures

The amount of pasture required per horse depends on the use of the pasture, how it is managed, the kind of pasture, and the amount of moisture available. One acre per horse will provide little more than an exercise area. The horse cannot obtain enough feed from 1 acre to meet its needs. Two acres (for the grazing season) are sufficient if the pasture is well managed and has enough moisture. In drier areas, a greater number of acres per horse is necessary. On native ranges in the western states, 2 to 10 acres per month per horse are required (Figure 32-1).

To produce the greatest amount of feed possible, pastures should be fertilized. Rotation of pastures increases their carrying capacity. In dryland

USDA

Figure 32-1 Large pastures provide grazing and areas to exercise for horses.

areas, irrigation can be used to improve the yield from the pasture. If only horses are grazed, the pasture should be clipped. Grazing cattle with horses reduces the need for clipping. Horses tend to graze unevenly on pastures, resulting in overgrazed and undergrazed areas (Figure 32-2). They also tear up the sod more than cattle do.

Kinds of Pasture

Kentucky bluegrass is considered the best all-around pasture for horses. It is palatable and provides the nutrients needed by horses. It maintains a tough turf that is less subject to damage.

Orchard grass is also an excellent horse pasture. It works well in combination with Kentucky bluegrass or a legume. It stands up well under close grazing.

Tall fescue is not as good as other grasses used for horse pasture. It is a good choice for areas of heavy horse traffic because it is not easily damaged. It is not as palatable as other grasses. Horses on tall fescue pasture should receive some grain in the ration. A legume should be seeded with fescue to make a better pasture.

Timothy is palatable to horses. However, it does not stand up well to close grazing and recovers more slowly than other pasture plants. Rotation grazing should be used with timothy pastures.

Bromegrass is a good pasture for horses. It requires good management practices such as rotation grazing, clipping, and fertilizing.

Reed canary grass is well adapted to areas with a high water table. It makes a good pasture for horses.

Bermuda grass is a good pasture for horses in the southern areas of the United States. It should be seeded with a legume.

Sorghum-Sudan grass crosses, Sudan grass, and Pearl millet are not recommended as horse pastures. In some areas of the United States, particularly the Southwest, problems with the disease cystitis syndrome have been reported on these pastures.

Crested wheatgrass, Intermediate wheatgrass, and Russian wild rye make good dryland pastures in the western United States. Crested wheatgrass is a good pasture for spring and late fall. Intermediate wheatgrass makes a good early summer pasture. Russian wild rye is a good pasture through the entire grazing season.

Native ranges have a low carrying capacity. To avoid killing out the desired species, rotation grazing is necessary. A given pasture should be left idle once every

Courtesy of Frank B. Flanders

Figure 32-2 Horses can damage pasture. They tend to graze unevenly, resulting in overgrazed and undergrazed areas.

3 years. Native rangelands are fragile. A range badly damaged by overgrazing may take many years to recover.

Legumes that make good horse pastures include alfalfa, white clover, ladino clover, red clover, alsike clover, lespedeza, and birds-foot trefoil. Any of the adapted legumes can be used in a given area. Horses do not bloat, so there is no danger in using legume pastures. The amount of legume in the mix should be about 35 to 40 percent.

Silage

Silage may be used to replace up to one-half of the hay in the horse ration. Corn silage is the best silage to use. Grain sorghum, grass, and grass-legume silages may also be used. Feed high-quality silage that is chopped fine and free of mold. Change the ration slowly when adding silage. Do not feed silage to foals and horses that are being worked hard because it is too bulky for these animals. Legume haylage can be used in place of the silage.

Legume Hay

Legume hay may be made from alfalfa, the clovers, and lespedeza. Legume hays are more palatable than grass hays. They also have a higher protein and mineral content. Alfalfa hay is the best of the legume hays. Red clover is a little lower in protein content than alfalfa hay. Lespedeza hay must be cut at an early stage to be of the best quality. Grass-legume mixtures are often used for horse hay and make an excellent combination for feeding horses.

Grass Hay

Common grass hays are timothy, bromegrass, orchard grass, Bermuda grass, prairie hay, and cereal hay. Generally, grass hays do not yield as much feed per acre as legume hays. They tend to be lower in protein, calcium, and vitamins. Timothy hay has long been considered the standard hay for feeding horses. However, it is low in protein and must be fed with a protein supplement or a high-protein grain, such as oats. It is better for mature horses than it is for younger, growing horses, stallions, or mares.

Bromegrass is most palatable when harvested in the bloom stage. Orchard grass is not quite as good horse hay as bromegrass. Several improved varieties of Bermuda grass hay are used in the southeastern United States. Prairie hay is not as palatable and is lower in protein than timothy hay. Oat, barley, wheat, and rye hays are often used in the Pacific Coast region of the United States. They should be cut in the soft to stiff dough stage for best quality.

Grain

Oats are considered to be the best grain for horse rations. They are palatable and bulky, which decreases digestive problems. They are higher in protein than corn but lower in energy value.

Figure 32-3 In addition to roughage, mixed rations of corn, oats, and pelleted feeds are often used for horses.

Corn is often a better buy on an energy basis than oats. It is especially good for thin horses and those that are worked hard. When corn is fed, care must be taken not to let the horse become too fat. Corn may cause colic, so it must be fed with care. A corn-oats mixture makes an excellent grain ration for horses. Figure 32-3 shows a mixture of corn, oats, and pellets.

Grain sorghum (milo) is best used in a grain mixture. Some varieties are not palatable to horses. Grain sorghum should be cracked or rolled for horse rations.

Barley should be rolled or crushed if fed to horses. It can be substituted for corn in the ration.

Wheat is usually too expensive to feed. It should not make up more than 50 percent of the grain mix. Wheat should be rolled or coarsely ground.

Wheat bran is a bulky, palatable feed for horses. It tends to be slightly laxative and is often fed to horses in stress conditions.

Cane molasses is used mainly to reduce the dustiness of the ration and to increase palatability. It should not make up more than 4 to 5 percent of the ration. If fed in excess, it acts as a laxative.

Protein Feed

Usually, little protein supplement is needed in horse rations. If at least one-half the roughage is legume, the protein needs of the horse, except in the case of milking mares, will be met. Protein is added to the ration of show horses to improve the hair coat and for additional energy. Protein supplement should be added to the ration if the quality of the roughage is poor.

Soybean meal is an excellent protein supplement for horses. It is high in protein and has a good balance of amino acids.

Cottonseed meal is not as palatable as soybean meal. It is used widely in the Southwest as a protein supplement for horses.

Linseed meal may be too laxative if fed with legume hay. Expeller-type linseed meal (a by-product of an oil extraction process) contains the fatty acid linoleic acid, which is often lacking in horse rations. Linseed meal puts a bloom on the hair coat of the horse.

Alfalfa meal, corn gluten meal, and meat meals may be used in horse rations, along with other protein supplements.

Commercial protein supplements are popular with producers who do not want to mix their own rations. Most commercial supplements are developed for a particular feeding program. The directions on the feed tag should be closely followed. Do not use commercial protein feeds that contain urea; horses do not utilize urea efficiently because the cecum is located too far down the digestive tract. In horses, non-protein nitrogen sources such as urea are converted to protein in the cecum. Too much urea can therefore be toxic to horses.

Minerals

Horses require salt, calcium, and phosphorus in the ration. The amount of salt needed by the working horse is large. Salt should be fed free choice. Calcium and phosphorus should be fed free choice, separate from the salt. Ground limestone, steamed bone meal, and defluorinated phosphate are common sources of calcium and phosphorus.

Feeding trace-mineralized, iodized salt free choice will supply the salt and trace minerals needed by horses. Check the tag on the trace-mineralized salt for the presence of selenium. Horses need some selenium in their diet, but not all trace-mineralized salt formulations contain this trace mineral. Mineral blocks like those shown in Figure 32-4 can be used to supply horses with necessary minerals.

Courtesy of Frank B. Flanders

Figure 32-4 Horses can receive many of the necessary nutrients from salt and mineral blocks specially designed for them.

Mares in early lactation need about twice as much calcium and phosphorus in their diet as do mares that are not nursing foals. Young horses should have a ratio of about 1.1 parts calcium to 1 part phosphorus in their diet. Older horses can tolerate a ratio of 6 parts calcium to 1 part phosphorus in their diet. Never feed more phosphorus than calcium in a horse's diet.

Vitamins

Horses on good pasture seldom need additional vitamin supplements in the ration. If the hay in the ration is of good quality and at least one-half legume, there probably is no need to add a vitamin supplement to the ration.

When horses are confined in barns, receiving poor-quality hay, or are on droughty pasture, the diet may not contain enough vitamin A, D, and E. In these cases, it may be wise to add a vitamin supplement to the diet. The supplement may be mixed in the feed, injected in the muscle, or provided in stabilized mineral blocks. Vitamin K and the B vitamins generally do not need to be added to the ration because horses can synthesize these in sufficient quantities in the body. However, when horses are under stress—including fast growth, intense training, heavy racing, or breeding—it may be wise to add a small amount of the B vitamins to the diet. Spent brewer's yeast is a good source of the B vitamins.

Excess vitamins in the diet are not usually toxic to horses. However, horses with liver or kidney problems may not be able to metabolize higher levels of vitamins. The fat-soluble vitamins (A, D, E, and K) are more likely to be toxic to horses with liver or kidney problems. Young horses and unborn foals may also have difficulty in metabolizing higher levels of vitamins. It is generally not wise to over feed vitamins under any conditions.

Water

Horses drink 10 to 12 gallons of water per day. Hard working horses drink more than this amount. Hot weather increases the need for water. A supply of fresh, clean water should be available at all times. Working horses must be watered at frequent intervals. However, horses that are hot should be cooled down and allowed to drink only small quantities of water at a time before being allowed to drink their fill. Be sure the water is not too cold when given to hard-working horses.

FEEDING HORSES

Horses are fed according to their size and weight, stage of development, body condition, and the amount of work they are performing. The major feedstuffs used for horses are grains, such as corn and oats, combined with roughages, such as alfalfa and Timothy pasture or hay. A diet of good-quality hay or pasture with salt and water supplemented with grains such as oats, corn, or barley will generally meet the minimum requirements of adult horses that are not pregnant, breeding, or involved in heavy work. Horses must be fed and watered regularly, at least twice each day.

The most accurate and in-depth information on nutritional requirements for horses can be found in the National Research Council (NRC) recommendations for horses. Nutritional requirements based on a horse's age, workload, and status are listed along with the nutritional value of different grains and forages. The NRC database can be accessed at http://nrc88.nas.edu/nrh/. This site allows you to specify the animal's age, weight, status and workload, and certain forages and other feedstuffs. From these specifications, the program determines how much of the horse's requirements are being met by a particular feed or combination of feedstuffs. It also calculates the specific nutritional needs for macronutrients, vitamins, and minerals.

Forages and Grains

The foundation of a good feeding program for horses is forage. Forage in the horse's diet is important for good digestive health and should comprise at least 50 percent of the ration. A general guideline for daily roughage intake is 1.0 percent of body weight. Forages may be supplied by pasture, hay, silage, and other roughages. Pasture is the easiest way to supply the horse's need for roughage. If horses have access to good pasture for several hours a day, it can be assumed that most of their roughage requirement is being met. When horses are housed in stalls, or when pasture is not available, horses should be fed good-quality hay. Compared with many other feeds, forage also provides horses with more chewing activity, which is an important part of normal behavior.

Nutrient needs not provided by roughage can then be satisfied by one or more grains. Grains are used to supplement roughages and should not provide more than 50 percent of the ration. Milo (grain sorghum), corn, barley, wheat, and oats are all used in horse feeds, with corn as the most common ingredient. In rations calling for corn, either milo, barley, or wheat can be substituted for corn on a pound-for-pound basis.

How Much to Feed

Most horses need the equivalent of 1.5 to 2.5 percent of their body weight in feed each day. Table 32-1 shows the expected feed consumption by horses based on body weight, age, and use. This can be forage only or, in most cases, a combination of forage and grains. However, highly active horses, work horses, lactating mares, and younger weaned horses may need up to 3 percent of their body weight in daily rations.

Most feeding recommendations for horses are based on the body weight of the horse. It is beneficial to know the approximate weight of the horse being fed for a more accurate estimate of the nutrient requirements and the amount of feed to provide. For example, if the recommendation for a horse is 1 percent of its body weight in roughage and the horse weighs 1200 pounds, then the horse should be fed 12 pounds of hay or other roughage, assuming that grazing is not provided.

It can be assumed that horses grazing on good pasture for several hours per day will consume their requirement for roughage and that hay is not required. Diets may still be supplemented with grain as necessary.

The ability to determine the body condition of horses is a useful skill in determining the amount to feed. In general, if the horse is too fat, reduce the amount of grain and increase the amount of hay. If the horse is too thin, increase the amount of grain in the diet.

TABLE 32-1 Expected Feed Consumption by Horses (Percent of Body Weight)[a]

	Forage	Concentrate	Total
Mature horses			
Maintenance	1.5–2.0	0–0.5	1.5–2.0
Mares, late gestation	1.0–1.5	0.5–1.0	1.5–2.0
Mares, early lactation	1.0–2.0	1.0–2.0	2.0–3.0
Mares, late lactation	1.0–2.0	0.5–1.5	2.0–2.5
Working horses			
Light work	1.0–2.0	0.5–1.0	1.5–2.5
Moderate work	1.0–2.0	0.75–1.5	1.75–2.5
Intense work	0.75–1.5	1.0–2.0	2.0–3.0
Young horses			
Nursing foal, 3 months	0	1.0–2.0	2.5–3.5
Weanling foal, 6 months	0.5–1.0	1.5–3.0	2.0–3.5
Yearling foal, 12 months	1.0–1.5	1.0–2.0	2.0–3.0
Long yearling, 18 months	1.0–1.5	1.0–1.5	2.0–2.5
2 year old (24 months)	1.0–1.5	1.0–1.5	1.75–2.5

Source: Adapted from *Nutrient Requirements of Horses*, National Academy of Sciences

[a]Air-dry feed (about 90% DM).

Figure 32-5 Pelleted feeds are a popular and convenient choice for horse feeds.

Pelleted and Other Processed Feeds

The use of pelleted rations and other processed feeds for horses is gaining in popularity (Figure 32-5). These rations are carefully balanced by the manufacturer. Convenience is one of their strong points, and there is often less waste when pelleted rations are used. Pelleted rations are generally more expensive than roughage and grain rations. Horses tend to eat bedding and chew on wood more when fed pelleted feeds without hay.

Feed companies manufacture a wide variety of pelleted and other processed feeds specifically for various types of horses. For example, a feed company might manufacture one feed for broodmares and another feed for performance horses. Consumers should read the feed tag and other information on the bag to choose the best feed for their horse. The best results are obtained when manufactured feeds are used for their intended use. Poor results are often observed when a feed formulated for one type of horse is used for another type horse. The law requires feed companies to list the amount of crude protein, crude fat, crude fiber, and ingredients on a guaranteed analysis tag.

Commercial feed mixes are usually available as a sweet feed or in pelleted form. Horses prefer sweet feeds over other forms. Sweet feeds may contain a number of ingredients, including whole or slightly processed grains, a supplemental protein, vitamins, minerals, and molasses. Pelleted feeds may contain the same ingredients and nutrient composition as sweet feeds, but the materials are ground and pressed into a pellet. With sweet feeds and other nonpelleted feeds, horses often sort the ingredients, eating their choice and wasting other ingredients.

Complete Feeds

Most processed feeds are used to supplement roughages. However, most companies market complete feeds for horses that are well balanced and ready-to-feed. A complete feed is a fortified grain/forage mix that is formulated with high-quality fiber sources to raise the total fiber percentage in the feed so that little or no additional forage is needed. Some fiber sources in complete feeds include alfalfa, beet pulp, and soy hulls. These are all good digestible fiber ingredients for horses.

It is important to read and follow the feeding recommendations when buying a complete feed; the recommended feeding amount both with and without hay should be listed on the tag. As the amount of hay is decreased, the amount of complete feed must be increased.

Although complete feeds contain the recommended amount of roughage, they may not supply enough of the long-stem roughages, such as hay, to provide horses with their normal chewing behavior. This can lead to excessive chewing on wood stalls and fences because a lack of enough roughage in the diet is a major factor leading to wood chewing.

Feeding Precautions

Do not feed moldy or dusty feed or allow a horse that is heated to drink too much water. Water the horse before feeding if water is not available free choice. Hay should be fed before grain. A tired horse should be fed only half of the grain ration, and the rest should be fed 1 hour later. Do not work a horse right after a full feed of grain.

Horses should not be allowed to become too fat or too thin. Horses in proper body condition can perform the work and pleasure activities desired by their owners. Horses that are too fat or too thin cannot perform properly. Horses should have a strong and healthy appearance. Their hair should be smooth and glossy, and the mane and tail should be glossy.

Feeding Breeding Stock

Special attention should be given to feeding breeding horses. Stallions and mares that are too fat or too thin may have reproductive problems.

Pregnant mares that are healthy, and not too thin or too fat, usually do not need extra grain in their diet if they are fed good-quality hay. Pregnant mares that are older or thin may benefit from the addition of some grain in the diet. If additional grain is added to the ration, do not overfeed the mare or she will become too fat.

Mares that are too fat may have more trouble at foaling time. Feed a small amount of grain along with the hay for the first 7 to 10 days after foaling. Feeding too much grain can increase the amount of milk produced; this may cause the foal to scour. After seven 7 to 10 days, slowly increase the amount of grain in the ration.

The body condition of stallions should be monitored and adjusted as needed going into the breeding season. It is important that the stallion not be too fat because being overweight can decrease the stallion's libido and fertility. The weight the stallion places on his hind legs during breeding must be considered because excessive weight can lead to joint soreness and damage.

CARE OF BROODMARE AND FOAL

Breeding the Mare

Horses have a low conception rate. The national average is 50 to 60 percent. However, application of the basic principles of reproduction can help to increase the conception rate.

The mare is more likely to conceive if bred in the months of April, May, or June. This also means that the foal will be dropped the following spring, which is a desirable time of the year for foaling. In the spring, foaling can take place on pasture, which reduces problems with diseases and parasites.

The best age to breed mares for the first time is 3 years old, so they will foal at about 4 years old. A well-grown filly can be bred as a 2-year-old but requires extra care in feeding so that her own body, as well as the developing fetus, will grow properly.

Pasture breeding requires less labor. However, the conception rate is likely to be lower. Hand breeding increases the conception rate.

Broodmares (mares used for reproduction) will usually produce foals until they are 14 to 16 years old. Mares may be rebred during foaling heat, which occurs about 9 days after foaling. The practice of rebreeding during the first heat period after foaling heat (about 25 to 30 days after foaling) is becoming more popular. Mares that do not conceive will continue to come into heat about every 21 days.

Mares that are too thin or too fat have lower conception rates. Mares must be given enough exercise and the proper rations before breeding.

Care of Pregnant Mares

The pregnant mare requires a great deal of exercise. Ride or drive the broodmare for a short period of time each day. Provide pasture in which the mare may run. Do not confine the mare to a stall where she will not get exercise.

Care at Foaling Time

The mare's udder swells about 2 to 6 weeks before foaling time. A week or 10 days before foaling, the muscles over the buttocks appear to shrink and the abdomen drops. The teats fill out to the ends about 4 to 6 days before foaling. Wax appears on the ends of the nipples about 12 to 24 hours before foaling. Just before foaling, the mare becomes nervous and restless. Other indications that foaling is near

Figure 32-6 Clean stalls and pasture help in parasite and disease control.

include pawing, lying down and getting up frequently, sweating, lifting the tail, and frequent urinating in small amounts.

Foaling on Pasture In warm weather, a clean pasture away from other livestock is the best place for the mare to foal. However, it is more difficult to assist the mare with foaling on pasture if help is needed.

Foaling in a Box Stall A box stall for foaling should be at least 16 3 16 feet. A smooth, well-packed clay floor is preferred. No obstructions, such as feed mangers, should be present in the stall. Isolate the mare from other livestock as far as possible. Stable the mare in the stall for several days ahead of foaling to allow her to become used to the area. Keep the stall clean and provide large amounts of fresh bedding (Figure 32-6).

Someone should be nearby the mare when foaling begins, although it is best to remain out of the mare's sight unless she needs help. Mares that are too thin, too fat, or are foaling for the first time are more likely to need help at foaling time. If the mare appears to be having difficulty, call a veterinarian for help.

Mares usually foal in about 15 to 30 minutes. A normal presentation of the foal is front feet first with heels down, followed by nose and head, shoulders, middle, hips, and hind legs and feet. If the presentation is not normal, or if the mare is taking too much time to foal, call a veterinarian for assistance.

After foaling, be sure the foal is breathing. Remove any mucus from its nose. The navel cord usually breaks 2 to 4 inches from the belly. If it does not break, cut it with clean, dull scissors or scrape it apart with a knife. Dip the navel cord in tincture of iodine to prevent infection.

The foal must nurse shortly after birth to obtain the colostrum milk. Colostrum milk is higher in nutrients and contains antibodies the foal needs.

The foal should have a bowel movement within 4 to 12 hours after it is born. The feces that are impacted in the bowels during prenatal growth are called **meconium**. If they are not eliminated shortly after birth, the foal could die. Feeding colostrum milk usually results in the foal having a bowel movement. If it does not, give the foal an enema. If the foal scours, reduce its milk intake.

Caring for the Suckling Foal

The foal will begin to eat some grain and hay at 10 days to 3 weeks of age. A creep feeder may be used on pasture. In confinement, provide a low grain box for the foal. When well fed during this period, foals reach about one-half their normal mature weight during the first year.

Imprinting

Proper handling of a newborn foal makes training a horse later in life much easier. **Imprinting** is a behavior-shaping process that takes place within the first 24 hours of a foal's life. This process helps the foal bond to humans and teaches them immediate submission. Imprinting can begin by toweling off the foal and making it lie down. The foal will struggle at first and needs to be gently and calmly restrained. Then its face, mouth, and entire body is touched until the foal quits moving and accepts whatever the handler is doing. Handling the foal's feet, ears, and muzzle prepares the young animal for clipping and trimming later in life, while applying pressure to the girth area will prepare it for saddling. If not done within the first 24 hours of life, the process will not be as effective because foals are precocial animals. This means that they are fairly independent at birth and quickly learn what to do and what not to do. Imprinting will greatly affect the foal's future as a racing, pleasure, or work animal and will make handling the animal more enjoyable.

Training the Foal

Training the foal should begin when it is 10 to 14 days old (Figure 32-7). Begin by putting a well-fitted halter on the foal. After a few days, tie the foal in the stall next to the mare. Be sure the foal does not become tangled in the rope. Keep the foal tied for only 30 to 60 minutes each day. Handle and groom the foal during this time. Pick up its feet to allow it to become used to having its feet handled. Lead the foal with the mare for a few days. Later, lead it by itself at both the walk and the trot. Train the foal to respond to commands to stop and go as they are made. Stand the foal in show position when stopping. It should stand squarely on all four legs with its head up. Use care and patience when working with the foal. Be firm, but do not mistreat it.

USDA Agricultural Research Service; Photo by J. Kim Kaplan

Figure 32-7 Training for foals should begin when they are 10 to 14 days old.

Weaning the Foal

Foals can be weaned at 4 to 6 months of age. Earlier weaning may be desired if the mare has been rebred during foal heat, is being worked hard, or if either the foal or mare is not doing well. A foal that has had a good creep ration will suffer little setback when weaned. Decreasing the ration of the mare a few days before weaning will help reduce problems when drying her up.

The foal should be left in the quarters (stall or pasture) and the mare moved to another area at weaning. Do not allow the foal to see the mare for several weeks. Foals may be allowed to run on pasture after several days. Several foals running together may injure each other. To prevent this, only a few should be run together at one time. Make sure that fences are in good repair so that the foals do not become caught in them. Separate timid foals from the rest of the animals, and do not run weaned foals with older horses on pasture.

Castration

Any colt not intended for breeding purposes should be castrated by a veterinarian. Castrating the colt at about 1 year of age will result in better development of the foreparts of the horse. Castration should be done in the spring before hot weather and flies become a problem.

Breaking

A foal that has been trained at an early age will not need breaking because it is used to being handled. Saddling and harnessing may be done during the winter before the foal is 2 years old.

Orphan Foals

An orphan foal may be put on another mare for nursing, or it may be fed milk replacer or cow's milk formula. Feed the foal about 0.5 pint each hour. Larger foals need about 1 pint. The feeding interval can be increased to every 4 hours after 4 or 5 days. The quantity may be increased after about 1 week. Be sure that the foal receives some colostrum milk. Colostrum milk may be frozen and stored for use when needed. Use a bottle and rubber nipple to feed the foal at first and after 2 weeks switch to a bucket. Keep all utensils clean. Start the foal on dry feed as soon as possible.

Drying Up the Mare

When drying up the mare, an oil preparation, such as camphorated oil or a mixture of lard and spirits of camphor, may be rubbed on the udder. The mare should be put on grass hay or low-quality pasture during this time. At first, the udder fills up and becomes tight. Do not milk out the udder at this time. After about 5 to 7 days, the udder becomes soft and flabby. At this time, milk out whatever small secretion remains in the udder.

Figure 32-8 Horse grooming equipment (left to right): mane/ear clippers, mane/tail comb, metal currycomb, soft brush, stiff brush, half moon sweat scraper, hoof pick, and sweat scraper.

Grooming

Grooming improves the appearance of the horse. It cleans the hair and skin and reduces the chances for skin diseases and parasites. Blood circulation is also improved during grooming. Equipment commonly used for grooming is shown in Figure 32-8.

Follow a pattern when grooming the horse. The near side is the horse's left side; the off side is the horse's right side. Begin on the near side at the head and work toward the rear. Then follow the same procedure on the off side. Follow good safety procedures when working around the horse.

Use the currycomb on areas caked with dirt and sweat, but do not use the currycomb around the head or below the knees and hocks. A wet sponge or soft brush is used to remove dirt in those areas. Use the currycomb in a circular motion. Finish cleaning the horse with a brush. Always brush in the same direction in which the hair lies (Figure 32-9) with short brisk strokes. The dirt is thrown out of the hair by flicking the brush up at the end of each stroke.

A grooming cloth is used to remove any dust remaining on the horse. Cloths made of soft linen, terrycloth, or jersey make good grooming cloths.

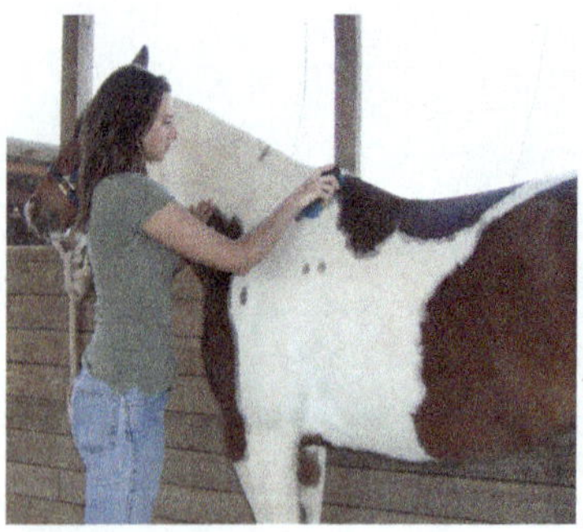

Figure 32-9 Always brush in the same direction in which the hair lies.

The mane and tail comb is used to remove dirt and other foreign matter from the mane and tail. The mane is lifted to comb the underside. A **roached** mane is trimmed down to the crest of the neck, leaving the foretop and a wisp of mane at the withers. Other styles of clipping the mane are used depending on the type and breed of horse being shown. Remove tangles in the mane and tail using the fingers and a coarse-toothed comb. Trying to pull the tangles out makes them tighter.

The hair may be trimmed from some parts of the body to improve grooming. The mane hair just behind the ears is often trimmed close so that the bridle and halter will fit better (Figure 32-10). Hairs outside the ears may be trimmed to help keep the ears cleaner. Occasionally, the feather is trimmed from the fetlock and pastern. If a shorter tail is desired, hairs of the tail should be pulled, rather than trimmed. Pull mainly from the underside of the tail. The tail of the three-gaited saddle horse is trimmed closely for a distance of 6 to 8 inches from the base.

CARE OF TEETH

Problems with a horse's teeth can sometimes cause the horse to go off feed even though a good-quality ration is being fed. The horse will often develop an unthrifty appearance. The teeth often wear unevenly and develop sharp edges because the upper molars and premolars normally overlap the bottom teeth. These sharp edges may cut into the horse's cheek while chewing, causing pain. Indications of this problem include weight loss, slobbering, dribbling of feed while eating, and a reluctance to eat. The teeth should be filed with a float. A **float** is a long-handled rasp with a guard to prevent injury to the horse during treatment. Milk teeth sometimes remain in too long and need to be removed. If they are not removed, crooked permanent teeth may result. Regular examination and care of the horse's teeth is recommended.

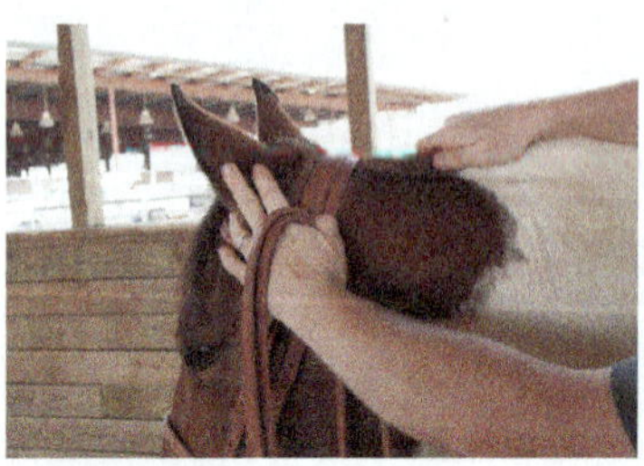

Figure 32-10 The mane hair right behind the ears may be trimmed close so that the headstall will fit better.

FOOT CARE

Each foot must be picked up for proper cleaning and inspection. The feet should be inspected every day. Follow a pattern when working with the feet. Start with the near front foot, then the near hind foot, next the off front foot, and finally the off hind foot. Figures 32-11A through 32-11D show the proper way to pick up a horse's foot.

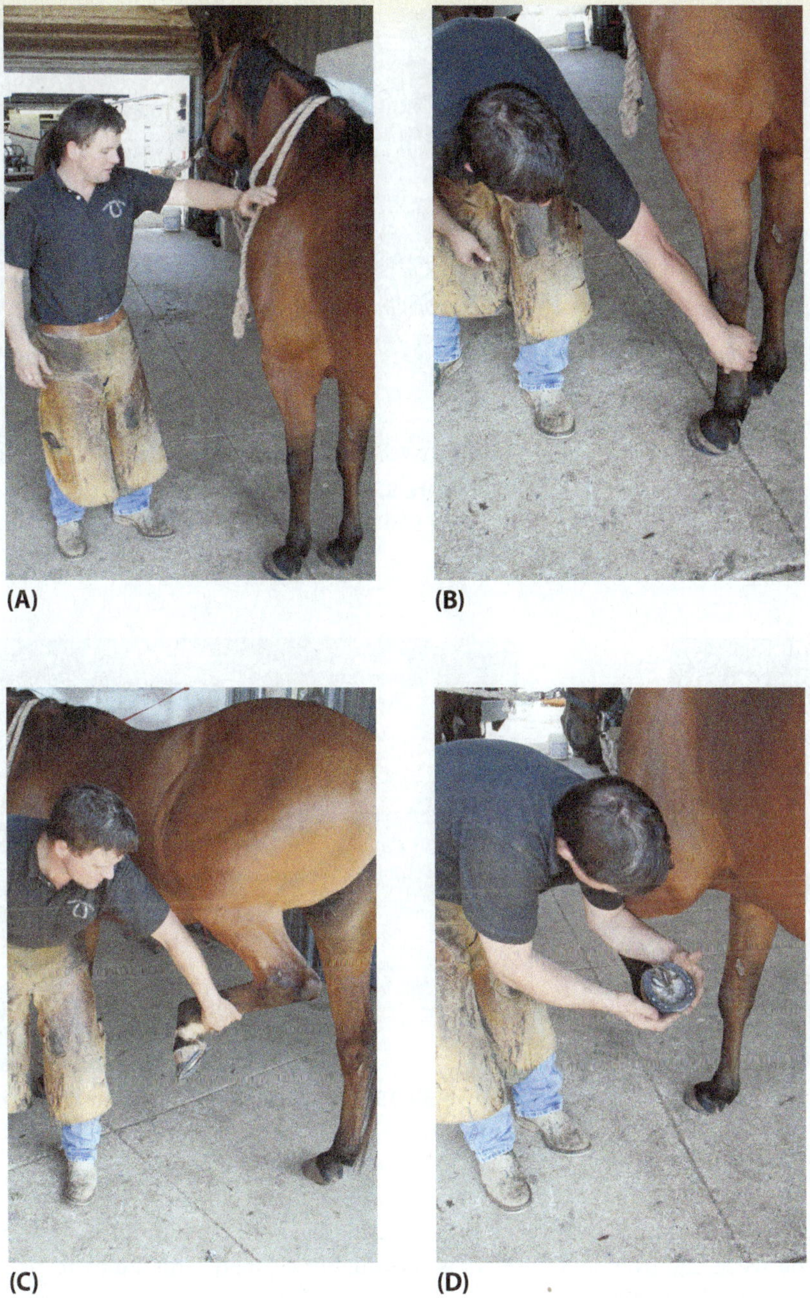

Courtesy of Wayne Nunn Farrier Service. Photo by Frank B Flanders

Figure 32-11 The proper way to pick up a horse's feet. (A) Always introduce yourself to a horse and become acquainted with the animal before trying to pick up its feet. Remember, some horses startle easily, so it is best to stay calm and gentle. (B) Near front foot: slide your hand nearest to the horse down the cannon to the fetlock. While applying pressure to the cannon, you can place your shoulder against the horse's shoulder. With this, the horse should shift its weight to the other side so you can lift its foot. (C) Near hind foot: with your back to the horse's side, run your nearest hand over the hindquarter and down the horse's leg. Grasp the back of the cannon just above the fetlock and lift the foot forward. (D) Once the horse has picked up its foot, support the horse's leg by holding the hoof and not the cannon.

Courtesy of Wayne Nunn Farrier Service. Photo by Frank B. Flanders

Figure 32-12 Tools used by a farrier (left to right): hoof testers, hoof nippers, rasp, hoof knife, hoof gauge, turning hammer, pulloffs, driving hammer, clinch pan, clinch cutter, crease nail puller, and clinchers.

Courtesy of Wayne Nunn Farrier Service. Photo by Frank B. Flanders

Figure 32-13 A farrier uses a forge to heat the shoe and make it easier to shape.

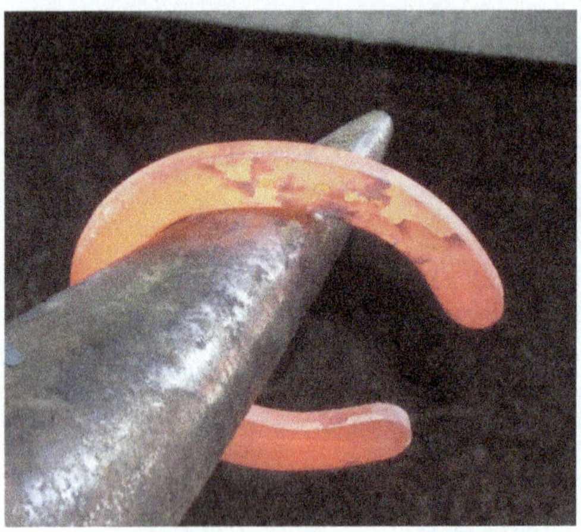

Courtesy of Wayne Nunn Farrier Service. Photo by Frank B. Flanders

Figure 32-14 Horseshoes heated in a forge are moved quickly to an anvil for shaping.

Courtesy of Wayne Nunn Farrier Service. Photo by Frank B. Flanders

Figure 32-15 Nippers are used to trim the outer surface of the hoof to a proper angle and length to fit the conformation of the animal.

Clean the hoof from the heel to the toe, using the hoof pick. Be sure the depression between the frog and the bars is thoroughly cleaned. This helps to prevent thrush and other infections of the foot. Check for stones, nails, injuries, or loose shoes when cleaning the foot.

A **farrier** is a person who works on horses' feet. The tools used by a farrier are shown in Figures 32-12 and 32-13. Notice the bright glow of the horseshoe in Figure 32-14 immediately after removing it from the forge. Figures 32-15 through 32-18 show a farrier preparing a horse's hoof for shoeing.

The hoof should be trimmed every 4 to 6 weeks. If the horse is shod, the shoes should be replaced every 4 to 6 weeks. Failure to provide proper care for the feet and hooves may result in lameness.

Horses are shod to protect the hooves from wearing too much on hard surfaces. Shoes also help to correct defects in stance and gait. They provide better traction on ice and in mud. Poor hoof structure and growth may be corrected with shoes. Shoes also help to protect the hoof from cracks, corns, and contractions.

Always fit the shoe to the foot. If shoes are left on too long without refitting, the hooves will grow out of proportion, which throws the horse off balance.

Courtesy of Wayne Nunn Farrier Service. Photo by Frank B. Flanders

Figure 32-16 The hoof knife is used in conjunction with the nippers to pare the dead and flaky tissue from the sole of the hoof.

Courtesy of Wayne Nunn Farrier Service. Photo by Frank B. Flanders

Figure 32-17 A rasp is used to prepare a ground-bearing surface by eliminating jagged and sharp corners on the bottom and around the edge of the hoof.

Courtesy of Wayne Nunn Farrier Service. Photo by Frank B. Flanders

Figure 32-18 A farrier will often take a hot shoe from the forge and press it to the bottom of the horse's hoof. The slight burning not only shows the farrier where to make alterations to the shoe and smooths any uneven areas on the horse's foot, but it also helps kill bacteria that cause white line disease. This procedure does not cause any pain to the horse.

Courtesy of Victoria Churchill

Figure 32-19 A small horse barn with space for feed, bedding, tack storage, and a runout paddock for each horse. Notice the hole in the boards where horses have chewed because of boredom.

BUILDINGS FOR HORSES

Horses do not require elaborate or expensive barns. They need a shelter that will protect them from the cold, storms, sun, and wind. A horse barn or shelter should be located in an area with good drainage. A structure with good ventilation, one without drafts, is important. A space should be provided for storage of feed, bedding, and tack. Figure 32-19 shows a small horse barn that has these features. Plans for horse barns are available from many state cooperative extension services. Figure 32-20 shows a four-stall horse barn. A nine-stall horse barn is shown

Jennifer Osborne

Figure 32-20 A horse barn with 12 × 12 feet box stalls and a covered way in front of the stalls.

Source: Image redrawn based on plan perspective originally created by the University of Tennessee Cooperative Extension in conjunction with the USDA. Horsebarn Indoor Exercise Area. 6128

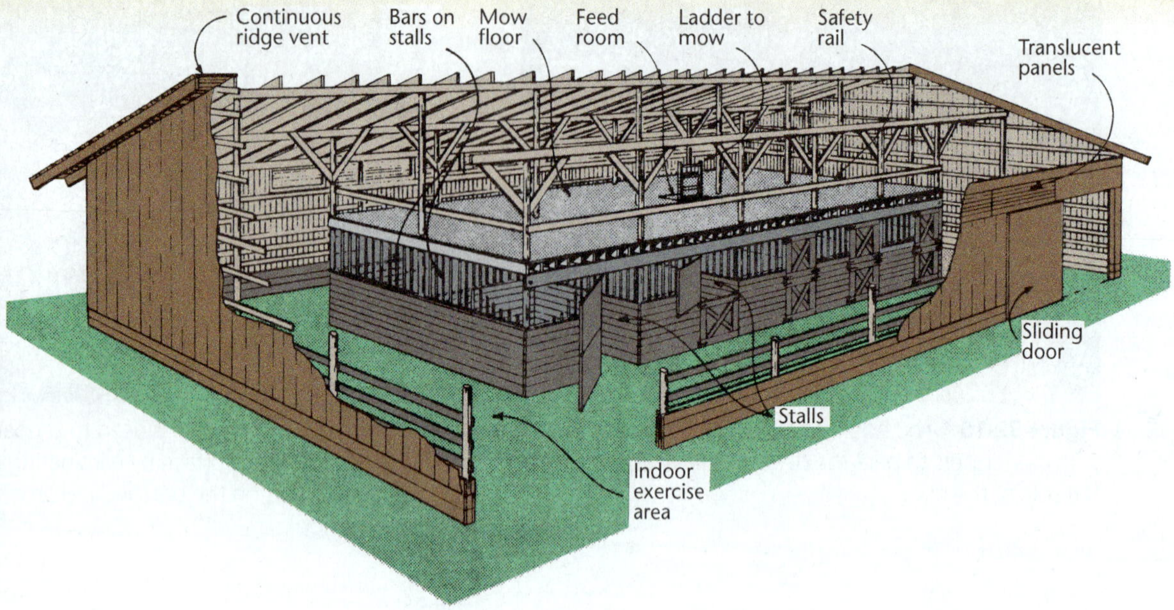

Figure 32-21 A pole-type, nine-stall horse barn with an indoor exercise area.

TABLE 32-2 Box and Tie Stalls for Horses

Horse	Box Stall	Tie Stall
Small horse	10 × 10 to 10 × 12 feet	3 × 6 feet
Medium horse	10 × 12 to 12 × 12 feet	5 × 9 feet
Large horse	12 × 12 to 16 × 16 feet	5 × 12 feet
Mare at foaling	16 × 16 to 16 × 20 feet	N/A

in Figure 32-21. A barn for a commercial boarding and riding stable is shown in Figure 32-22. The dimensions for box and tie stalls for various sizes of horses and a mare at foaling time are shown in Table 32-2.

Floors made of clay, sand, soil, gravel, or wood are better for horses than floors made of concrete. If concrete floors are used, it is important to utilize stall mats and bedding on top of the concrete floors to help protect the horses hocks, joints, and feet. Stall mats are also useful when covering dirt or sand floors because they facilitate the cleaning of the stall without removing soil. Straw or

Jennifer Osborne

Figure 32-22 A barn for a commercial boarding and riding stable.

wood shavings make good bedding materials for horses. Specialized bedding such as recycled cardboard shavings can be used for horses with allergies. When planning the location of the horse barn, make provisions for cleaning the manure from the barn. Removal of manure from the barn and premises helps reduce the fly population and smell. Some barns completely remove manure from the property (Figure 32-23).

Courtesy of Frank B. Flanders

Figure 32-23 Some farm managers have manure hauled away from the farm for sanitation purposes and to reduce fly and other insect problems.

FENCES AND CORRALS

Fences built of wood or poles are the best choice for horses. Woven wire may be used, although the mesh must be small enough to prevent the horse from catching its hoof in it. Barbed wire is not a good material for horse fences. It can cause injury to the horses. A smooth wire or a board may be fastened along the top of a wire fence to help prevent injuries.

It is convenient to have a corral or paddock near the horse barn. This provides a space for the horses to exercise. Heavy lumber is the best fencing material for a corral or paddock. Place the boards or fencing on the inside of the corral rather than on the outside. If chewing becomes a problem, cover the top board with $\frac{1}{4}$ inch chicken mesh wire.

EQUIPMENT FOR FEEDING AND WATERING

A hayrack or manger reduces waste when feeding horses. A grain feeder may be attached to the wall of a stall or placed on a wooden shelf. It should be easily removable for cleaning. Provision should be made for supplying fresh, clean water. A water tank or trough may be used. Measures, such as covering the trough, must be taken to prevent the water from freezing in cold climates. Wooden salt and mineral feeders are also needed. Two separate containers, or one container with two compartments to keep the salt and mineral separate, may be used.

HORSE TRAILERS

Two important considerations in the selection of horse trailers are the safety and protection of the horse. Standard safety equipment includes brakes, brake lights, taillights, and a safety chain on the hitch. The law in many states requires these items.

A padded chest bar helps to prevent injury to the horse. The floor of a horse trailer should have a rough surface to provide good footing. Rough cut lumber for flooring is ideal. Flooring that has a slick finish is a hazard to the animals and to people, especially when water or excrement are present. Trailer floor mats may also be used for traction. Mats should have ridges on both sides, providing drainage and air movement on the underside and traction for animals on the upperside. Protect the horse from drafts in the trailer and close all drop-down windows when traveling. Escape doors located in the front of straight-load trailers allow a person to leave the trailer after leading a horse inside. Figure 32-24 shows a horse being led into a slant-load trailer, which provides more comfortable transportation for the horse and allows the handler to exit from the rear of the trailer.

Courtesy of Frank B. Flanders

Figure 32-24 Loading a horse into a slant-load trailer.

HORSE TACK

Tack is the equipment used for riding and showing horses. Tack that is made of good-quality materials and that fits the horse properly should be selected. A good fit is the most important factor to consider in selection of tack. Basic tack consists of

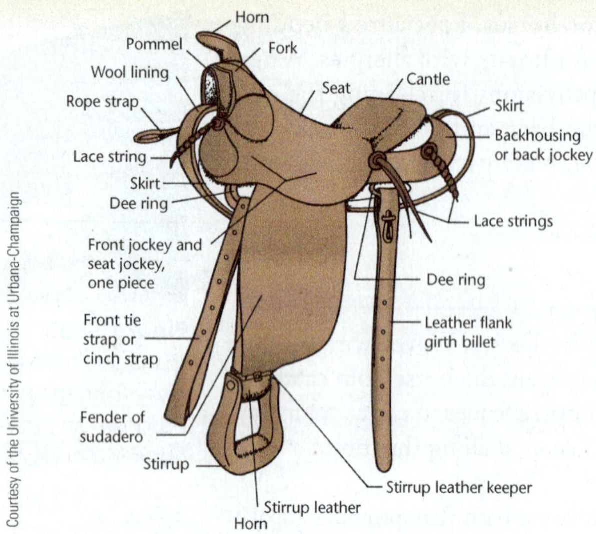

Figure 32-25 The parts of the Western saddle.

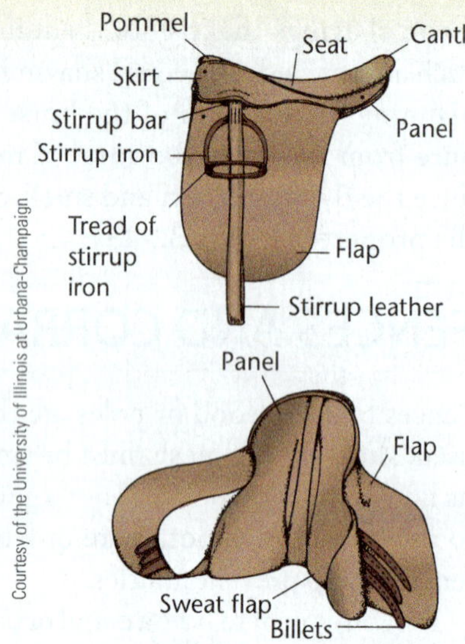

Figure 32-26 Parts of the English saddle.

the saddle, saddle pad or blanket, bridle, halter, and lead rope. Other kinds of tack required depend on the kind of riding being done.

Tack is expensive and should be given proper care. Leather should be kept clean and oiled. Use saddle soap to clean leather. Several kinds of oils are available for use on leather; select one that does not rub off on clothing. Use only enough oil to keep the leather soft and pliable. Repair or replace any parts of the tack that become worn or broken.

Saddles

There are many types and styles of saddles. The two most common types are the Western saddle and the English saddle.

Western saddles are large and heavy, but they are more durable than other types of saddles. Western saddles give a comfortable ride and are designed to be used for work with cattle on Western ranges and are used for Western-style riding. A saddle pad or blanket is usually used with the Western saddle. Parts of the Western saddle are shown in Figure 32-25.

English saddles are lighter than Western saddles and have a flat seat (Figure 32-26). There are many styles, depending on the type of riding being done. English saddles are used for pleasure riding, training, racing, jumping, and polo.

Selecting a saddle

The type of saddle selected depends on the style of riding to be done. The saddle must fit both the horse and the rider. It is especially important that the saddle fit the horse at the withers and along the spine. A poorly fitted saddle will often result in injury to the horse. Stores that sell tack have a wide variety of saddles available (Figure 32-27).

Figure 32-27 A wide variety of saddles are available for different styles of riding.

Bridles and Bits

There are many styles of bridles and bits. The style selected depends on the type of riding to be done. The bit must be matched to the horse's needs.

The main purpose of the **bridle** is to hold the bit in the horse's mouth. Common types of bridles are shown in Figure 32-28. The Weymouth (A) is a double bridle. A double bridle has two bits and two sets of reins. The double bridle is usually

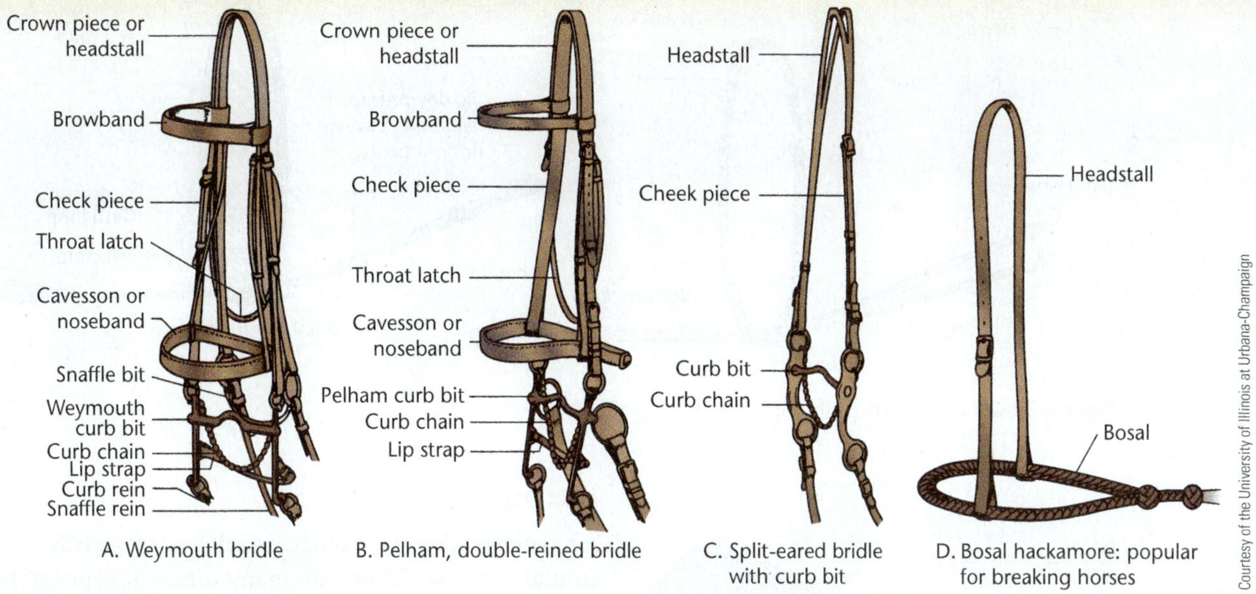

Figure 32-28 Common types of bridles.

used when showing three and five-gaited horses. The Pelham bridle (B) is used for pleasure riding, polo, and hunting. The split-eared bridle (C) may be used on working stock horses. The hackamore (D) is similar to a bridle except it has no bit.

The purpose of the **bit** is to help control the horse. Care must be taken not to injure the horse's mouth by improper use of the bit. The bit must be the right length to fit the horse's mouth. The bridle should be adjusted in such a way that the bit just raises the corners of the mouth.

Common English riding bits include the Weymouth curb bit, Pelham curb bit, Walking horse bit, snaffle bit, and Dee race bit. The curb bit is used in Western show classes. The snaffle bit is the most commonly used bit.

Common Western riding bits used for working stock and roping horses include the hackamore bit, roper curbed-cheek bit, and spade mouth bit. Common driving bits used with harness horses include the Liverpool bit, bar bit, and half-cheek snaffle bit.

Halters

Halters are used for tying or leading horses. They are made of rope, nylon, or leather. When putting a horse on pasture, always remove the halter to prevent it from becoming caught on some object.

Martingales

A **martingale** is a device that prevents the horse from lifting its head too high. The two types of martingales are the standing and the running (Figure 32-29). The standing martingale is attached to the horse's head. The running martingale is attached to the reins by two rings.

Riding Clothes

Many different types and styles of riding clothes are designed for various kinds of riding. A good riding boot should be used when pleasure riding. Tennis shoes are not safe for riding or for working around horses because they provide neither security in the stirrup nor protection from being stepped on by the horse. If a riding boot is not worn, a good leather shoe with a reinforced toe is the next best choice.

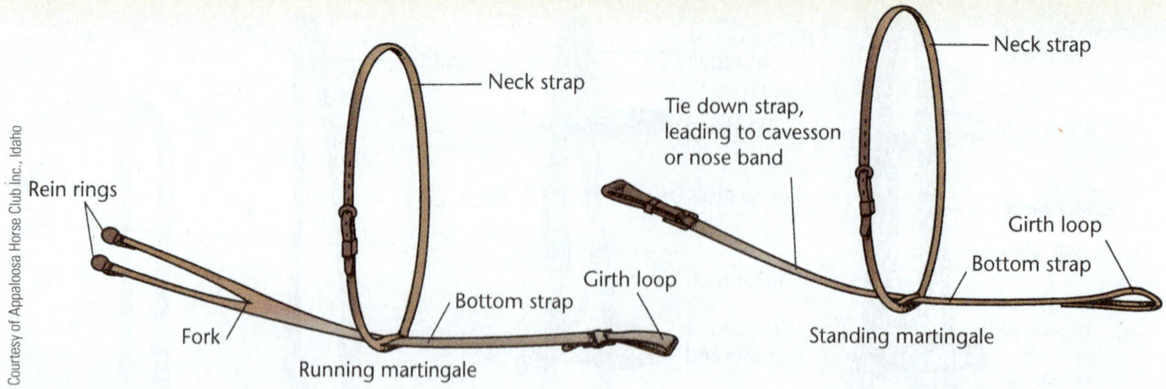

Figure 32-29 Two types of martingales.

Harness

A **harness** is the gear used to attach a horse or other draft animal to a load. There are many different types of harnesses used with horses, including carriage, show, racing, and wagon harnesses. Harness racing is a type of horse racing in which the horses race at a certain gait, trotting or pacing. A racing harness is shown in Figure 32-30. Figure 32-31 shows some of the parts of a carriage harness.

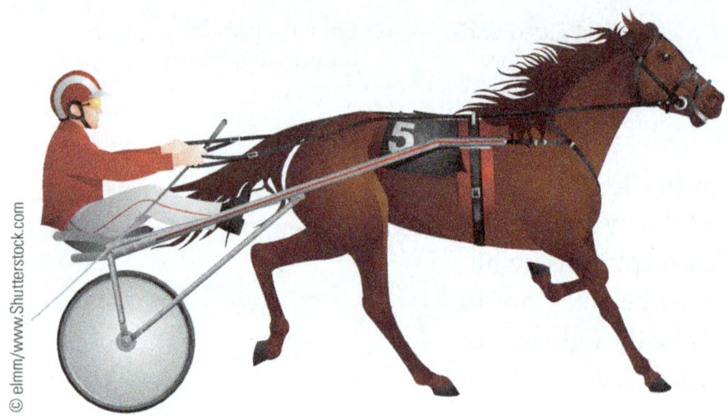

Figure 32-30 Harness racing is a type of horse racing in which the horses race at a particular gait, trotting or pacing.

Knots

Horse owners use many different types of knots. All knots should be tied so that they can be quickly released in case of an emergency. The most common types of knots are shown in Figure 32-32.

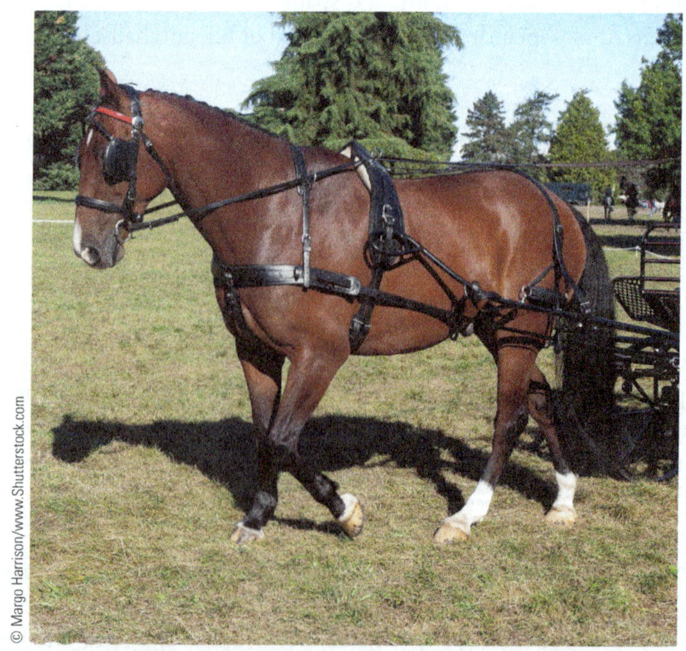

Figure 32-31 A harness is the gear used to attach a horse or other draft animal to a load.

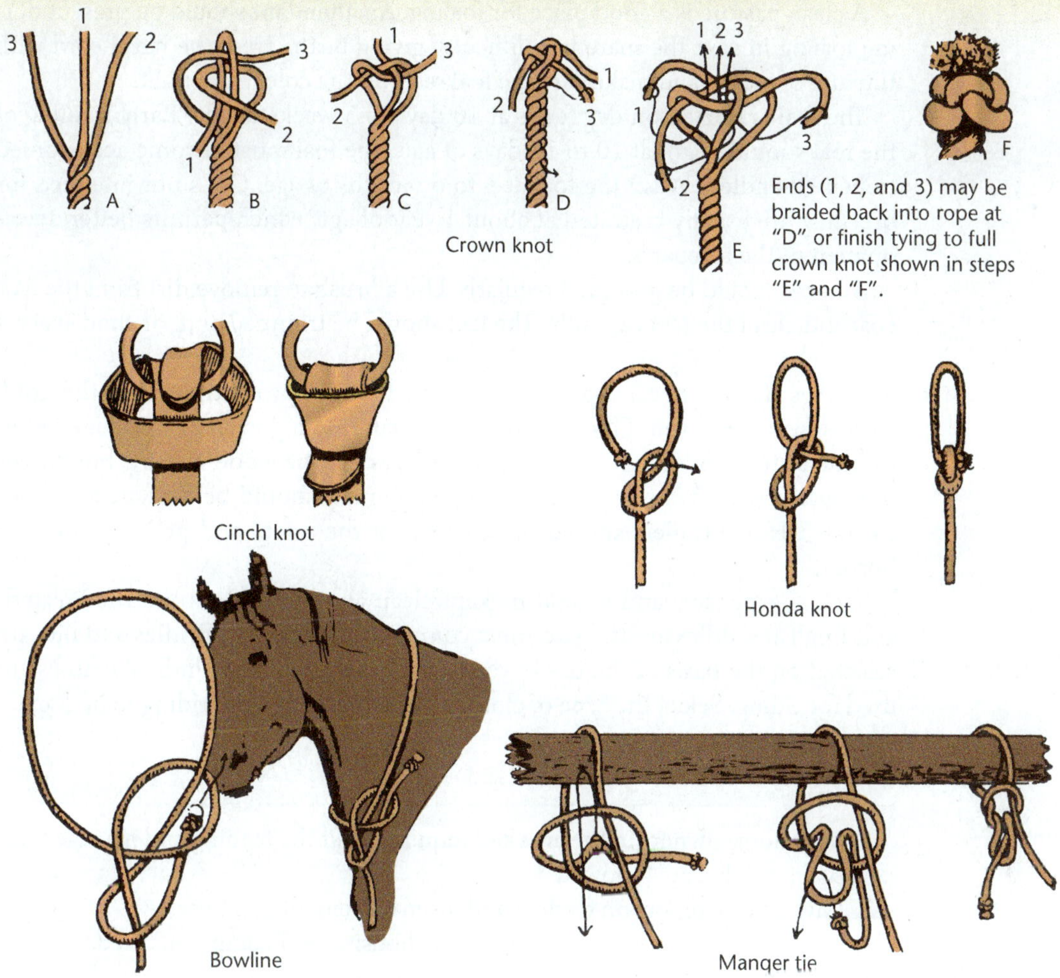

Courtesy of Appaloosa Horse Club Inc., Idaho

Figure 32-32 Knots commonly used by horse owners.

SUMMARY

Mature idle horses can be fed on a ration composed solely of roughage. Horses that are working, pregnant mares, and growing foals need some concentrate in the ration. Grass and legume pastures can provide much of the roughage required by horses. Legume and grass hays are also good sources of roughage for horses. Silage can be used for part of the roughage in the diet.

Oats are the preferred grain for horses. Other grains such as corn, milo, barley, and wheat can also be used in the ration. Usually, little protein supplement is needed in the ration. Soybean meal is an excellent supplement to use if one is needed. Commercial protein supplements and complete pelleted rations may be used for feeding horses.

Horses require salt, calcium, and phosphorus in the ration. Additional vitamin supplements are seldom necessary. A clean, fresh supply of water should be available for horses.

Horses are fed according to their size, stage of growth, condition, and amount of work they perform. Regular feeding and watering is important.

A mare is more likely to conceive if bred in the spring. Most horse owners prefer to breed mares at 3 years old so that they will foal when they are 4 years old. Pregnant mares need exercise and the proper rations.

A clean pasture is a good place for foaling. An attendant should be present during foaling in case the mare has difficulty giving birth. Treat the navel cord with tincture of iodine and make sure the foal receives the colostrum milk.

The foal is started on dry feeds at 10 days to 3 weeks of age. Early training of the foal should begin at 10 to 14 days of age. The foal must become accustomed to being handled. Wean the foal at 4 to 6 months of age. Colts not intended for breeding are usually castrated at about 1 year of age, which permits better development of the foreparts.

A horse should be groomed regularly. Use a brush to remove dirt from the hair coat and clean the feet carefully. The feet should be trimmed and/or shod every 4 to 6 weeks.

Horses do not need expensive shelters. Provide protection from the cold, storms, wind, and sun. Floors made of clay, sand, soil, gravel, or wood are better for horses than those made of concrete. Fences may be wood or wire, but do not use barbed wire. Watering and feeding equipment should be designed for easy cleaning. Horse trailers should be selected for the safety and protection of the horse.

Tack is expensive and should be kept clean and in good repair. The Western and English saddles are the two most common types in use. Bridles and bits are selected on the basis of the use of the horse. Many different kinds of clothes are used for riding. Select the type of clothes based on the kind of riding to be done.

Student Learning Activities

1. Visit horse owners in the area and inquire about the feeding and management practices that are followed.
2. Give an oral report on feeding and/or management practices.
3. Visit local horse owners to observe the housing and equipment used.
4. Prepare a bulletin board display of pictures of tack used with horses.
5. Give an oral report on housing and equipment for horses.
6. Identify the parts of saddles, bridles, and bits.
7. Practice tying knots used by horse owners.
8. Apply the feeding, management, and housing principles described in this chapter when planning and conducting a horse supervised experience program.

Discussion Questions

1. What determines how much pasture is needed per horse?
2. How should pastures be maintained to improve their value for horses?
3. List and briefly describe some common pastures used for horses.
4. Describe the place of silage in the feeding of horses.
5. Compare legume and grass hays as feeds for horses.
6. List and briefly explain the grains used for feeding horses.
7. Briefly explain the use of protein feeds for horses.
8. Briefly explain some advantages and disadvantages of using pelleted feeds for horses.
9. What minerals should be included in horse rations?
10. Briefly explain the use of vitamin supplements in horse rations.
11. How important is water in horse feeding?
12. What practices should be followed in caring for the pregnant mare?
13. What are the signs that indicate the mare is about to foal?
14. What help should be given to the mare and foal at foaling time?

15. Describe some possible steps to follow when beginning to train the foal.
16. Describe the practices to follow when weaning the foal.
17. At what age and in which season should a colt be castrated?
18. Describe the practices that help to save orphan foals.
19. Describe the procedure to follow when grooming a horse.
20. Describe the care of a horse's feet.
21. Describe the kind of building used for horses.
22. What type of fencing is needed for horses?
23. What equipment is required for feeding and watering horses?
24. Describe some characteristics of a good horse trailer.
25. Define tack and describe how to care for it.
26. Name and describe the two most common kinds of saddles used for riding horses.
27. Name and describe the common kinds of bridles.
28. What are the uses of the halter?
29. What is the use of the martingale?
30. Name the common kinds of knots used by horse owners.

Review Questions

True/False

1. Pregnant mares and growing foals do not require concentrate in the ration in addition to roughage.
2. Pastures never need to be fertilized.
3. Orchard grass is an excellent horse pasture.
4. Bromegrass is most palatable when harvested during the blooming stage.

Multiple Choice

5. What kind of pasture is considered the best all-round for horses?
   a. timothy c. tall fescue
   b. crested wheatgrass d. Kentucky bluegrass
6. What is the average conception percentage rate for horses?
   a. 10–20 c. 50–60
   b. 30–40 d. 80–90
7. Horses drink ______ to ______ gallons of water a day.
   a. 1; 2 c. 8; 10
   b. 4; 6 d. 10; 12

Completion

8. A ______ is a person who works on horses' feet.
9. ______ is the equipment used for riding and showing horses.
10. ______ are used for tying and leading horses.

Short Answer

11. The training of a foal should start at what age?
12. What minerals do horses require in their rations?
13. How is a bit used with a horse?

Chapter 33

Diseases and Parasites of Horses

Key Terms

anhydrosis
anthrax
azoturia
colic
Cushing's syndrome
distemper
encephalomyelitis
equine infectious anemia
founder
heaves
tetanus
ascarid
bot
Pinworms
strongyle

Objectives

After studying this chapter, the student should be able to

- identify common diseases and parasites of horses.
- describe prevention measures for diseases and parasites of horses.

PRINCIPLES OF HEALTH CARE

Diseases and parasites cost horse owners hundreds of thousands of dollars every year. A sick horse loses its usefulness to the horse owner. However, inexperienced horse owners may not always recognize horse health problems when they occur. Therefore, some general principles of preventive health care can be followed to help reduce losses from diseases and parasites.

Proper feeding and management help to prevent losses from diseases and parasites. A horse that is properly fed has a higher resistance to diseases and parasites. Dusty or moldy feed should never be used. The mycotoxin *fumonisin* is sometimes found in moldy corn. It is toxic to animals and may cause brain or liver disorders in horses. The amount of water a horse drinks after it has been worked hard should be carefully controlled.

Cleanliness and sanitation also help to prevent horse health problems. Disease organisms and parasites grow in organic waste material. Stalls, barns, and exercise areas must be kept clean, thereby removing or reducing many sources of infection. The use of clean pastures helps to break the life cycle of many parasites that affect horses (Figure 33-1). Do not keep horses in barns that are too warm and humid.

Planning and practicing an immunization and parasite control program prevents many problems before they occur. Planning should be done with

Figure 33-1 Clean pastures help break the life cycle of many of the parasites that affect horses.

the advice of a veterinarian who is familiar with local disease and parasite problems. Horses can be vaccinated against some diseases. However, to be effective, vaccination must be done before the horse becomes infected. Periodic examination of the feces by a veterinarian will detect the presence of parasites. Successful treatment depends on the use of the right drug at the right time. The correct diagnosis of the problem is important to determine the proper treatment to be used. A veterinarian is best qualified to make such a diagnosis. The services of a veterinarian can save the horse owner money that might otherwise be spent on remedies that will not cure the problem.

Proper exercise and good grooming help to keep horses in good health. Horses must have regular exercise to maintain good muscle tone and to prevent stiffness. Good grooming keeps the horse clean and provides an opportunity for the owner to observe any problems that may be developing. Early treatment of horse health problems results in less loss of time and money for the horse owner.

When a disease does occur, isolate the affected horse to prevent the possible spread of the disease. Water and feed containers for the sick horse should be kept separate from other horses. Watering troughs that are provided for horses at fairs and shows should not be used. Draw water from the tap into a pail rather than dipping it out of a trough.

Do not allow the horse to come in contact with horses that are known to be sick. Access to the area in which the horse is kept should be controlled to prevent the possible spread of diseases by infected horses. Stress lowers the horse's resistance to disease. Therefore, care should be taken to avoid chilling and other causes of stress in horses.

Observing the vital signs (temperature, pulse rate, and respiration rate) in an animal can help in the early detection of health problems. Vital signs will vary with activity and environmental conditions. Normal vital signs in horses are:

- temperature (normal range): 99 to 100.8°F. The average is 100.5°F. Temperature is usually higher in the morning than in the afternoon; younger animals will show a wider range of temperature than mature animals
- pulse rate (normal range): 32 to 44 heartbeats per minute
- respiration rate (normal range): 8 to 16 breaths per minute

Body temperature is taken in the rectum using either a mercury thermometer or a battery-powered digital thermometer. Restrain the animal when inserting the thermometer into the rectum. The pulse rate is taken by finding the artery on the lower edge of the jaw. It may also be taken by finding the artery along the inside of the forearm where it travels down the bone. Respiration rate is determined by observing the number of times the animal breathes per minute.

DISEASES AND DISORDERS

Anhydrosis

Anhydrosis is a condition in which horses do not sweat normally. The ability of a horse to sweat is an integral part of body temperature regulation. Some horses have sweat glands that do not function at all and others that function at a less than normal level. Anhydrosis should be suspected when a horse does not sweat normally during exercise.

Management practices such as riding or working the horse only when it is cool and keeping the horse out of the sun may be necessary. Using fans or air conditioning in the barn will help keep the horse cool. Using a higher-fat diet helps because less heat is generated in the digestion of fat as compared to other nutrients. Consultation with a veterinarian may provide some other possible solutions for a specific set of circumstances. One treatment that has shown success is using a thyroid medication. The proper dosage must be used because an excessive amount of thyroid medication may cause other health problems.

Anthrax

Anthrax is a serious bacterial disease that is zoonotic. The spores formed by the bacteria can live for many years in the soil. Symptoms include high fever, blood in the feces, rapid breathing, swellings on the body (especially around the neck), and in the later stages, depression. The horse may bleed from all body openings. The death rate is usually high.

The disease can be prevented by vaccination. Anthrax can occur anywhere; however, it is most common in temperate agricultural regions. Outbreaks tend to occur after periods of drought followed by heavy rains, but natural incidences in the United States tend to be low. A local veterinarian should be consulted for information on vaccination.

Anthrax can be transmitted to people. Never open the carcass of a horse that has died of anthrax. Use care in disposing of the carcass. Burn or bury the carcass and cover it with quicklime. To combat anthrax, isolate sick horses and vaccinate healthy ones. Change pastures, quarantine the area, and practice strict sanitation measures.

Azoturia (Monday-Morning Sickness)

Azoturia is a nutritional disorder. It develops when a horse is put to work following a period of idleness. The horse becomes stiff, sweats, and has dark-colored urine. The muscles become swollen, tense, and paralyzed.

Azoturia can be prevented by decreasing the amount of grain fed when the horse is idle. The horse should exercise during idle periods and be started back to work slowly. When symptoms appear, stop the horse from working and do not allow it to move. Use blankets to keep the horse warm and dry. Call a veterinarian for treatment.

Bruises and Swellings

Horses have many chances of becoming bruised. Bleeding may occur under the skin. Apply cold compresses until the bleeding and swelling stop. Then apply heat and liniments to the affected area.

Colic

Colic is not a specific disease, but rather a disease complex encompassing a wide range of conditions that affect the horse's digestive tract. It is usually caused by some type of obstruction that blocks the flow of feed through the intestine, resulting in abdominal pain. The pain occurs when the intestine is distended by an accumulation of gas, fluid, or feed. When it occurs, colic must be treated immediately.

While there are a number of things that can cause colic in horses, the major cause (about 90 percent of the time) is spasmodic colic, which is usually caused by overexcitement or sudden feed changes. Colic may be caused by nutritional factors, such as a ration that is too high in energy, sudden changes in feed, feeding a high-fiber ration, or feeding a poor-quality feed. Teeth problems or other mouth disorders that result in improper chewing of the feed may cause colic. If the horse is allowed to drink excessive quantities of cold water before being cooled down after heavy exercise, colic may result. Diseases that cause high fever and reduced intake of feed and water may also cause colic. Feeding an excessive amount of grain or twisting of the intestine can result in colic.

Symptoms of colic include severe abdominal pain, uneasiness or restlessness, looking at the flank region, getting up and down, kicking at the belly, sweating, and shifting weight. As the problem continues, the horse may lie down and roll; have an increased pulse and respiration rate; have congested mucous membranes (gums); or experience strain, sweat, and bloat.

It is better to follow good management practices to prevent colic than to have to treat the condition after it occurs. A major preventive measure is to follow a good deworming program to control worms, especially the large strongyles. Check horses' teeth at least twice a year to eliminate chewing problems and other mouth disorders. Feed high-quality rations on a definite schedule. Feeding small amounts of the ration two or three times a day is better than feeding a large amount less frequently. Make changes in the ration slowly, over a 7- to 10-day period. Make sure there is a good supply of fresh, clean water, and do not allow a horse to drink too much cold water before cooling down immediately after heavy exercise. Make sure there are enough good-quality hay, salt, and minerals for the horses.

If a horse develops colic, call the veterinarian immediately. Colic can be treated satisfactorily if the proper treatment is started quickly. Before the veterinarian arrives, do not allow the horse to eat or lie down and roll; keep horses walking if they want to lie down and roll. If the horse is not trying to lie down and roll, let it stand quietly. The horse may be allowed to drink water. The severe pain may cause the horse to become violent, so the handler must be careful not to get injured. Upon arrival, the veterinarian will proceed with the proper treatment. In some cases, surgery may be necessary.

Cushing's Syndrome

Cushing's syndrome, also referred to as hyperadrenocorticism, is a disease caused by a small benign tumor in the pituitary gland. The disease is usually seen in older horses, but it can occur in animals of all ages. Symptoms of the disease include excessive drinking and urination, abnormal hair growth and shedding, development of a swayback stance and potbelly, deposition of fat above the eye, loss of muscle over the topline, and chronic laminitis. A blood test is used to monitor hormone levels to detect the disease. Once diagnosed, a horse can be given medications to treat the symptoms; however, there is no cure for Cushing's syndrome.

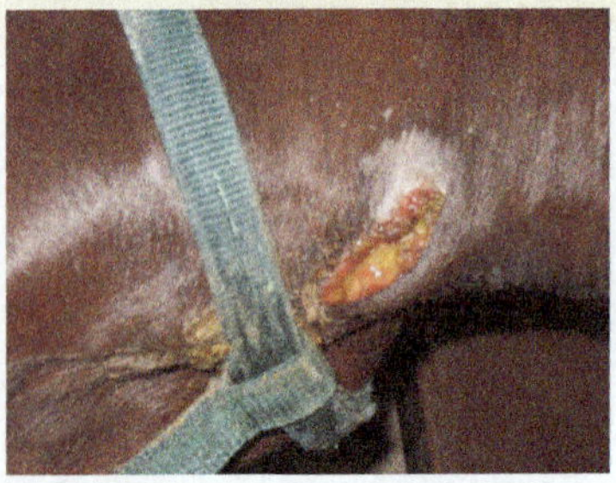

Jennifer Osborne

Figure 33-2 A horse with strangles whose abscesses have ruptured.

Distemper (Strangles)

Distemper is caused by a bacterium. It spreads quickly from horse to horse, especially where large numbers are together in one place. Contaminated feed, watering troughs, and tack—or direct contact—will spread the disease. Young horses are more likely to get distemper than are older horses.

Symptoms include high fever, loss of appetite, and depression. There is a pus-like discharge from the nose. The lymph nodes in the lower jaw and the throat swell.

Isolation of newly arrived animals for 2 to 4 weeks and vaccination are methods of prevention. However, because vaccines may cause some problems, vaccination is usually done only on farms with a history of distemper. Consult a veterinarian concerning vaccination.

Antibiotics are used to treat distemper. Abscesses may be drained surgically, if necessary (Figure 33-2). Give the horse complete rest. Protect it from cold and drafts. Call a veterinarian for diagnosis and treatment.

Encephalomyelitis (Sleeping Sickness)

Encephalomyelitis, a disease that affects the brain, may be caused by any one of several different viruses. The most common forms of the disease are known as Eastern and Western. Outbreaks of Venezuelan encephalomyelitis have occurred in the United States. The viruses are carried by mosquitoes.

The symptoms of the various forms of the disease are similar. High fever, depression, lack of coordination, lack of appetite, drowsiness, drooping ears, and circling are signs of the disease. The horse may die, or it may recover. Death rate may be as high as 90 percent from the Eastern and Venezuelan types. The Western type has a death rate of about 20 to 30 percent.

Eastern and Western encephalomyelitis are prevented by vaccination. Two injections are given about 1 to 2 weeks apart. The vaccination is repeated each year. Venezuelan encephalomyelitis vaccination is given once each year. Controlling mosquitoes helps in preventing the disease. There are no effective treatments for encephalomyelitis. Call a veterinarian if the disease is suspected.

Equine Abortion

Abortion in mares may be caused by bacteria, viruses, or fungi. Other causes include hormone deficiencies, carrying twins, genetic defects, or other miscellaneous factors. Abortion may occur at various times during pregnancy, depending on the cause.

Virus abortions (rhinopneumonitis and equine arteritis) may be prevented by vaccinations. Bacterial abortions (caused by several different types of bacteria) are best prevented by strict sanitation measures at breeding time. Isolate horses that have aborted. The bedding and aborted fetus should be burned or buried. The area should be disinfected. There are no vaccines for fungi-caused abortions.

Equine Infectious Anemia (Swamp Fever)

Equine infectious anemia is caused by a virus, and it is carried from horse to horse by bloodsucking insects. Symptoms include fever, depression, weight loss, weakness, and swelling of the legs. Death often occurs within 2 to 4 weeks. Chronic forms cause recurring attacks. Pregnant mares infected with the disease may abort. Horses with the chronic form become carriers of the disease.

Infected horses are put down and their carcasses carefully disposed of. Only buy horses that have been tested and found free of the disease. Control all bloodsucking insects. Practice sanitation and sterilize all instruments used on horses after use on each horse. There is no vaccine or treatment for the disease.

Equine Influenza (Flu)

Influenza is caused by viruses. It spreads quickly where large numbers of horses are brought together. Symptoms include a high temperature, lack of appetite, and a watery nasal discharge. Younger horses are more likely to become infected.

Isolate newly arrived horses and those that have the disease. A vaccine may be used each year to prevent the disease. However, because there are several strains of viruses involved, the horse may be infected by a strain of the disease that was not protected against in the vaccine used. For treatment, use antibiotics and allow the animal to rest. A veterinarian should be consulted.

Equine Protozoal Myeloencephalitis

Equine protozoal myeloencephalitis (EPM) is caused by the protozoan *Sarocystis neurona*. It is a neurological disease of horses that has traditionally been difficult to diagnose. Symptoms can appear at any age, but it occurs most often in horses under 4 years of age. EPM is spread by the opossum. The opossum excretes sporocysts of the protozoan in its feces, which may be accidentally ingested by horses in feed or while grazing. The disease cannot be spread from horse to horse.

Symptoms of EPM may be exhibited in a wide variety of neurological signs, depending on where in the horse's central nervous system the organism localizes. The most common symptoms are stumbling and incoordination, head tilts, and in some cases muscle wasting.

Diagnosing EPM is a matter of matching up the history and symptoms of a horse, and then ruling out other possible causes. The introduction of the West Nile virus in 1999, another neurological disease, has increased the difficulty of diagnosing EPM. Diagnostic tests have recently been developed to aid in the diagnosis of EPM.

Fescue Toxicity

There are three syndromes associated with tall fescue infected with the fungus *Acremonium coenophialum* growing within the plant (endophyte). Fescue foot, bovine fat necrosis, and fescue toxicity are all caused by the endophyte. Of these three, fescue toxicity is the most important, especially in horses. The fungus produces toxins that inhibit prolactin, a hormone that is essential in the last months of gestation for udder development and colostrum formation. The toxin can also cause lameness; sloughing off the end of the tail; poor weight gain; and an increase in temperature, pulse, and respiration rate.

There is no treatment for fescue toxicity. When symptoms are observed, the animals must be removed from fescue pasture immediately. Do not feed hay made from tall fescue to pregnant mares or use it for bedding. It is recommended that pregnant mares be removed from infected tall fescue pasture at least 3 months before foaling. A laboratory test is used to determine the presence of the fungus in tall fescue. There are fungus-free varieties of tall fescue available for pasture use (Figure 33-3).

Courtesy of Frank B. Flafnders

Figure 33-3 Fescue toxicity can cause a variety of problems in horses and is especially dangerous to pregnant mares. The mare and foal can be affected when the mare eats endophyte-infected fescue. Existing fescue may be replaced with endophyte-free fescue varieties.

Founder (Laminitis)

Founder is a nutritional disorder. Common causes are overeating of concentrates, sudden changes in feed, drinking too much water, or standing in a stall for long periods of time. Founder may occur in an acute or chronic form.

Symptoms of the acute form include swelling of the sensitive laminae on one or more feet, lameness, fever, and sweating. Distortion of the hoof is common with the chronic form.

Care in feeding and management help to prevent founder. Use cold applications to treat the acute form. Call a veterinarian for additional treatment. Chronic cases are treated by trimming the hoof and shoeing the horse.

Fractures

Fractures are broken bones. Strain or injury may cause a bone to break. The horse with a fracture usually shows signs of extreme pain when it tries to move. Care in handling horses helps to reduce the chances of fracture. Treatments are seldom practical or feasible, except in the case of very valuable animals. Always call a veterinarian for treatment.

Heaves (Broken Wind, Asthma)

Heaves is a nutritional disorder that affects the respiratory system. It often occurs when moldy or dusty hay is fed. It is more common in horses over 5 years of age.

An affected horse has difficulty in breathing. The air must be forced from the lungs by the abdominal muscles. Other symptoms include a dry cough, nasal discharge, and loss of weight.

Care in selecting feed is the best preventive measure. Never feed moldy or dusty hay. Changing to a pelleted ration may help if the disease has not progressed too far. Putting the horse on pasture may result in some improvement. In advanced cases, there is no effective treatment.

Lameness

Lameness in horses may occur from many different causes. Much of the unsoundness of the feet and legs results in lameness. A veterinarian should be called for the diagnosis and treatment of lameness.

Navel-ill (Joint-ill, Actinobacillosis)

Navel-ill is a condition caused by bacteria. It affects newborn foals. The foal refuses to nurse and shows swelling and stiffness in the joints. The animal may have a fever. The foal does not move around. In older foals, there is a loss of appetite and weight loss. Sanitation and dipping the navel in tincture of iodine at birth help to prevent navel-ill. Antibiotics are used in treatment.

Periodic Ophthalmia (Moon Blindness)

The exact cause of moon blindness is not known, but it is believed to be some type of infection. The disease affects older horses more often than younger ones. One or both eyes become swollen and the horse keeps its eyes closed. A watery discharge may be secreted from the eye, and the cornea may become cloudy. The attack usually clears up in a week to 10 days. The horse may become blind, or the eye may be only slightly affected. Attacks recur at periodic intervals. Keep the horse at rest in a partially darkened stall. Call a veterinarian for treatment. If left untreated, permanent blindness usually results.

Pneumonia

Pneumonia is caused by bacteria and viruses. Stress, such as chilling, increases the chances of infection. Inhaling dust, smoke, or liquids can also cause pneumonia. It sometimes occurs as a complication of other diseases.

Symptoms include fever, rapid breathing, loss of appetite, and chest pains. Sanitation and prevention of stress will help prevent the disease. There are no vaccines. Prompt treatment by a veterinarian is necessary.

Poisonous Plants

Certain plants found in pastures and hay may be poisonous to horses. Symptoms of poisoning vary with the type of plant involved. A few of the plants that are poisonous to horses are bracken, goldenrod, horsetail, locoweed, oleander, ragwort, and tarweed (see Figure 33-4). The best way to control plant poisoning is to remove the plants from pastures and hay fields. Treatment after poisoning occurs is often unsatisfactory.

Figure 33-4 Goldenrod is a toxic plant that affects horses, typically in summer and late fall. Goldenrod causes cardiac and nervous system problems.

Rabies

Rabies is a disease caused by a virus. The virus enters the horse's body when it is bitten by an infected dog or wild animal. An affected horse may become quite violent, sometimes attacking other animals or people. The affected animal will usually drool. The horse eventually becomes paralyzed and dies. Rabies can be prevented by vaccinating horses with an approved equine rabies vaccine. Horses as young as 3 months can be vaccinated; a yearly booster is required. The disease can also be controlled by vaccinating dogs and controlling wild animals that may carry the disease. Always call a veterinarian if rabies is suspected.

Tetanus (Lockjaw)

Tetanus is caused by bacteria. The bacteria usually enter the animal's body through a puncture wound but may enter through other types of wounds. The horse becomes nervous and stiff. Muscle spasms and paralysis follow. Death usually occurs in untreated cases. About 30 percent of all horses that are infected recover.

Tetanus is prevented by vaccination. Annual booster shots are given. An unvaccinated horse is given tetanus antitoxin serum if it is injured. Call a veterinarian for treatment of tetanus.

Vesicular Stomatitis

Vesicular stomatitis is caused by a virus. The horse drools and blisters form in the mouth. Provide the infected horse with water and soft feed. There is no vaccination for this disease.

West Nile Virus

West Nile is a type of virus that causes encephalitis, the inflammation of the brain. The virus is most commonly spread by mosquitoes that have acquired the disease from infected birds. Because mosquitoes transmit the disease, it has the potential to affect humans, livestock, and poultry (Figure 33-5). The first documented case of the West Nile virus in the United States was reported in September 1999. The incidence rate of West Nile virus in horses has been declining in recent years, dropping from a high of 15,257 cases in 2002 to 377 cases in 2013. Even though the number of reported cases has declined over the last decade, West Nile virus is still a very dangerous disease for horses as well as humans. In 2013, there were 119 human deaths from the West Nile virus in the United States.

Symptoms of West Nile virus typically develop between 3 and 14 days after being bitten by the affected mosquito. Symptoms include ataxia, or stumbling and incoordination; depression or apprehension; weakness of limbs; partial paralysis; the inability to stand; and sometimes death. Horses may be infected without showing any clinical signs.

Protecting your animals from mosquito bites is the best method of prevention. Eliminating any potential mosquito breeding site is essential. Mosquitoes tend to breed in sources of still water. These can include discarded tires, bird baths, clogged

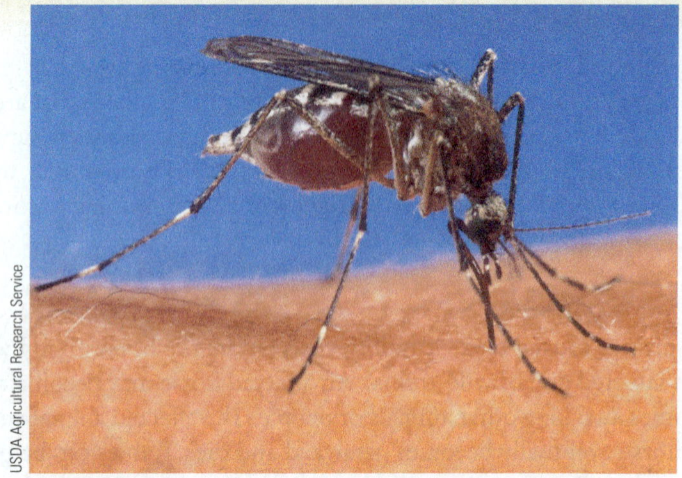

Figure 33-5 Mosquitoes can acquire the West Nile virus from infected birds and transmit the disease to humans, livestock, and poultry.

roof gutters, outdoor water holding devices, any puddle that lasts for more than 4 days, and water that may have collected on top of a swimming pool cover. It is generally recommended that horses be vaccinated for West Nile virus. The vaccines have been shown to be highly effective in preventing the disease.

EXTERNAL PARASITES

Common external parasites of horses are (1) flies, (2) lice, (3) mites, (4) ringworm, and (5) ticks. Most of these external parasites also attack beef cattle and other livestock. A discussion of general control measures for external parasites is found in the chapter concerning diseases and parasites of beef cattle. These control measures can be applied to the control of external parasites on horses.

Flies

Biting Flies

Biting flies that attack horses include black flies, deerflies, horn flies, horseflies, mosquitoes, and stable flies. Descriptions of these flies are found in the chapter concerning diseases and parasites of beef cattle.

Horse Botflies

Horse botflies produce larvae that are parasites of horses. There are three species of botflies. The common botfly and the throat botfly are found throughout the United States. The nose botfly is common in the Midwest and Northwest.

The common botfly is about the size of a honeybee. It has mottled wings and black and yellow hair on the body. The throat botfly is smaller than the common botfly and has no markings on its wings. The nose botfly is the smallest of the three types. The wings have no markings, and the body hair is dark.

Botflies lay their eggs on the hairs of the horse. The common botfly lays eggs on the horse's front legs. The throat botfly lays eggs on hairs under the chin, around the throat, and around the lips. The nose botfly lays eggs on the hairs of the mouth and lips. In all cases, the larvae enter the horse through the mouth and migrate to the stomach. After developing for 10 to 11 months, the larvae pass out of the horse in the feces. The larvae pupate in the soil for about 30 days. Adult flies then emerge and reproduce. The adult flies do not eat. They live from a few days to 3 weeks. Botfly season in the northern states is from the middle of June until freezing weather. In the southern states, it is from March to December.

Damage by botflies is both direct and indirect. The direct damage is the inflammation caused by the larvae in the stomach. This interferes with digestion. The horse may also be affected with colic from the digestive disturbance caused by the larvae.

The flies cause indirect damage when laying their eggs by frightening and annoying the horses. Horses spend time and effort fighting the flies. They may run to get away from them. Weight loss may result from this excessive activity. Horses that are attacked by the flies while being ridden may run away, possibly endangering the rider. When burrowing in, larvae cause indirect damage by irritating the tongue, gums, and lips. Horses rub against objects to try to relieve the itching and may injure their lips and noses.

Control is accomplished by treating horses with drugs that kill the larvae in the stomach. A veterinarian should be consulted for the current recommended drug treatments. Treatments should be given several times during the fly season and at least once during the winter. In areas where botflies are a problem, community control programs have been effective.

Face Flies, Houseflies, and Screwworm Flies

Face flies, houseflies, and screwworm flies attack horses as well as other kinds of animals. Descriptions of these flies are found in the chapter concerning diseases and parasites of beef cattle.

Lice, Mites, Ringworm, and Ticks

Lice, mites, ringworm, and ticks may also attack horses. Descriptions of these external parasites are found in the chapter concerning diseases and parasites of beef cattle. Common ticks are shown in Figure 33-6.

Courtesy of Frank B. Flanders

Figure 33-6 Ticks can spread a number of diseases, including Lyme disease.

INTERNAL PARASITES

More than 75 different species of internal parasites affect horses. The most important of these are ascarids, bots, pinworms, and strongyles. **Ascarids** are a type of parasitic roundworms that infest the intestines of animals. **Bots** are the larvae of the botfly and are discussed earlier in this chapter, in the section on external parasites. **Pinworms** may cause severe itching. **Strongyles** are a type of parasitic, bloodsucking worms that attack the organs and tissues of animals.

Internal parasites are so widespread that all horses are affected by them. Heavy infestations of internal parasites lead to poor physical condition. In extreme cases, infestation may cause death. While all ages of horses are affected, young horses are more seriously affected.

Symptoms

Some general indications of internal parasites in horses are:

- weight loss
- listlessness and poor performance
- dry, rough hair coat
- poor appetite
- bowel problems and colic
- periodic lameness
- breathing problems and coughing
- anemia
- foals that do not grow well and develop pot bellies

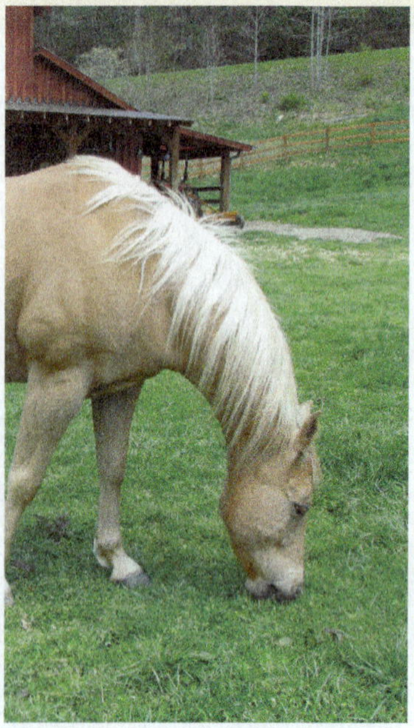

Courtesy of Frank B. Flanders

Figure 33-7 The life cycles of ascarids, pinworms, and strongyles are similar. Adult parasites lay eggs that are passed out of the animal's body in fecal matter. Eggs hatch and the larvae attach to vegetation. Horses can pick up larvae and eggs from pasture grasses. Other sources include water and contaminated feed.

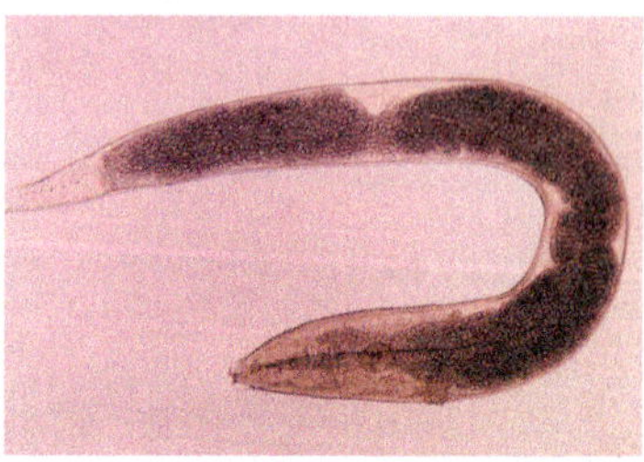

© Dr. Robert Calentine/Visuals Unlimited/Getty Images

Figure 33-8 Pinworms cause irritation in the anal region of the horse.

Diagnosis and Treatment

The only sure diagnosis of internal parasites is by an examination of the horse and the feces by a veterinarian. Worm eggs in the feces will reveal what kind of parasite is affecting the horse. Treatment with the proper drug depends on identification of the parasite.

Treatment for internal parasites is based on the use of drugs. No one drug is effective against all of the different kinds of parasites. Worm medications can be purchased in several different forms and are administered in different ways. A variety of types of wormers should be used to prevent resistance. Wormers can be used to kill parasites. Always follow the manufacturer's directions when using any medication.

Life Cycles

The life cycles of ascarids, pinworms, and strongyles are very similar (Figure 33-7). They follow this general pattern:

1. Eggs are passed out in the feces.
2. Eggs develop to infective stage on vegetation or in litter, or eggs hatch and larvae attach to vegetation.
3. The horse picks up infective eggs or larvae from vegetation or contaminated litter or water.
4. Eggs hatch, and larvae migrate through tissues of the horse's body.
5. Larvae develop into mature worms and lay eggs.

There are differences in the life cycles of these parasites, especially in the parts of the body to which the larvae migrate.

Ascarids

Ascarids migrate to the liver and the lungs. Later they are coughed up, reswallowed, and go to the small intestine. They are not bloodsuckers. They are the largest of the worms that affect horses. They may rupture the wall of the small intestine and cause death.

Pinworms

Pinworms travel to the large intestine and do not migrate through other tissues of the body. They cause irritation in the anal region of the horse. The horse rubs the rear quarters to relieve the itching, resulting in the loss of hair from the tail. A pinworm is shown in Figure 33-8.

Large Strongyles

Larvae of the large strongyles migrate to the arteries, liver, and gut wall. Adult large strongyles are bloodsuckers. Blood clots may form in the arteries, resulting in complete blockage and death. Large strongyles are considered the most serious of the internal parasites.

Small Strongyles

Larvae of the small strongyles migrate to the intestine. There they cause digestive problems. They are not as serious a problem as large strongyles.

Prevention

Sanitation and good management practices are the basis for any parasite prevention program. The eggs of the worms pass out in the feces of the horse. Therefore, proper handling of manure helps in controlling these parasites. Manure should not be spread on horse pastures. In confined areas, pick up the manure at least twice a week. On pastures, use a chain drag to spread horse droppings so the sunlight can destroy eggs and larvae.

Do not overstock pastures. Rotation grazing helps break the life cycle of the worms. Do not graze young horses with older horses. It is best to alternate horses with cattle or sheep on pasture.

Place hay and grain in mangers or bunks rather than on the ground. Be sure that the water supply is clean. Keep stalls clean and replace soiled bedding with clean bedding as often as possible. On earth floors, remove 10 to 12 inches of the surface each year and replace with clean soil.

SUMMARY

The effects of diseases and parasites are very costly for horse owners. Proper feeding and good management help to reduce losses from diseases and parasites. Cleanliness and sanitation are the basis of disease and parasite prevention programs. Proper exercise and grooming are necessary to help keep horses in good health.

The most serious diseases of horses are distemper, encephalomyelitis, equine infectious anemia, and equine influenza. Vaccinations are available for some of the diseases that affect horses. A veterinarian should be consulted concerning vaccination programs.

The common external parasites of horses are flies, lice, mites, ringworm, and ticks. The most serious internal parasites are ascarids, bots, pinworms, and strongyles. Insecticides are used to control external parasites. Good management practices designed to break the life cycle of the parasite are used to control internal parasites. A regular worming program using specific drugs should be followed to treat internal parasites.

Student Learning Activities

1. Prepare an oral report on horse health problems.
2. Ask a veterinarian to speak to the class on horse health problems.
3. Survey horse owners in the community concerning common horse health problems and treatments.
4. Establish a good horse health program to follow when planning and conducting a horse supervised experience program.

Discussion Questions

1. List the recommended practices for keeping horses in good health.
2. List the diseases of horses, and describe the symptoms of each. Also describe the prevention and treatment of each disease.
3. Describe the life cycle of horse botflies, the damage they cause, and the control measures used for these parasites.
4. Name the four most common internal parasites that affect horses.
5. List the general symptoms of internal parasites of horses.
6. Describe how internal parasites are diagnosed and treated.
7. Describe the life cycles of the internal parasites of horses.
8. List several means of preventing internal parasites in horses.

Review Questions

True/False

1. A virus causes anthrax.
2. Encephalomyelitis is a disease that affects the brain and is caused by several different bacteria.
3. Bacteria, viruses, or fungi may cause abortion in mares.

Multiple Choice

4. A disease caused by a virus that enters the body from the bite of an infected dog or wild animal is ______.
   a. lockjaw
   b. rabies
   c. pneumonia
   d. tetanus
5. ______ is a condition in which horses do not sweat normally.
   a. Pneumonia
   b. Colic
   c. Distemper
   d. Anhydrosis

Completion

6. ______ is a nutritional disorder that affects the respiratory system.
7. A disease caused by bacteria that enter the animal's body through a puncture wound is called ______ .

Short Answer

8. List three plants that are poisonous to horses.
9. Name five external parasites of horses.
10. Why are flies an indirect problem when managing horses?
11. List some symptoms of distemper.

Chapter 34

Training and Horsemanship

Key Terms

longeing
sacking out
horsemanship
neck-reining
riata
posting
gymkhana

Objectives

After studying this chapter, the student should be able to

- explain the basic principles of training a horse.
- describe basic horsemanship procedures.
- list basic procedures for showing a horse.

UNDERSTANDING THE BEHAVIOR OF HORSES

The behavior of horses is based on survival instincts that have developed over a period of several million years. The ancestors of the horse survived by being alert, by hiding or running from danger, and by adapting to changing conditions.

Horses' eyes see independently. Horses can see to the front, side, and rear at the same time. The retina of the eye is arranged so that part of it is closer to the lens than other parts. Horses must raise their head to see objects at a distance and lower their heads to bring close objects into focus. It is difficult for horses to judge height and distance. Horses' eyes adjust to changing light conditions slower than human eyes.

While many believe that horses are color-blind, there is no scientific proof of this. Horses react quickly to sudden movement. This is why a horse will sometimes shy and throw a rider at the sudden movement of an object along a trail.

Horses have good memories. Therefore, they can be trained and will remember what they learn. Horses remember the rewards and punishments they are given. Reward or punishment must be given immediately following the desired or undesired behavior so that the horse will associate the two.

Horses band together. This is a survival instinct that originated when horses lived only in the wild. When horses ran wild, the horse in the center of the band was safer from attack. The banding instinct also means that horses imitate the behavior of other horses. This has application when riding in a group.

Horses generally have good hearing. They are able to hear tones higher than the human ear can hear. A sound that the rider does not hear may frighten the horse.

Horses have a good sense of smell. A stallion can detect a mare in heat from a great distance if he is downwind of the mare. A young horse should be allowed to smell the saddle and blanket the first few times it is saddled. This helps to reassure the horse that these are not dangerous objects.

Horses have several areas on their bodies that are particularly sensitive. These include the mouth, feet, flanks, neck, and shoulders. These sensitive areas are used in training and controlling the horse.

The rider communicates to the horse through the voice, hands, legs, and weight. The horse can learn and remember voice commands. The sensitive mouth responds to the control of the bit. The legs of the rider can direct the horse by applying pressure to sensitive areas of the skin. The horse is able to sense the security or lack of security of the rider. An insecure rider cannot obtain the best performance from a horse.

TRAINING THE YOUNG HORSE

Training a horse requires skill, patience, and careful handling. Handling a foal while it is young makes the job easier (Figure 34-1). The foal should be handled each day for a short period of time. This helps the foal to overcome its fear.

Haltering

A foal may be halter-broken when it is only a few weeks old. Crowd the foal into a corner of the pen and gently place the halter on its head. If the mare is gentle, she may be used to help crowd the foal into the corner.

Let the foal become used to the feel of the halter by leaving it on for a short period of time. Petting the foal and giving it a small amount of grain will help it to associate the halter with a pleasant experience. Repeat this procedure for a week or two. After the foal has learned to accept the halter, it can be taught to lead.

Leading

To teach the foal to lead, put a loop of rope over the foal's rump. Fasten a lead rope to the halter. Have one end of the rump rope passed through the halter. Pull on the halter rope and the rump rope. This encourages the foal to move forward. The foal may jump forward when the rump rope is pulled. Be alert to avoid being stepped on by the foal. Working with the foal for about 30 minutes a day for several days will teach it to lead.

Jennifer Osborne

Figure 34-1 Handling foals often when they are young makes training them later much easier.

Working with the Feet

The foal should be taught to allow its feet to be picked up and handled. This can be done after teaching it to lead. With the foal tied, pick up each foot and put it down. Be gentle with the foal, petting it to keep it calm. Work with the front feet first and then the hind feet. Keep working with the feet until the foal learns to yield its feet without struggling.

Longeing

Longeing is training the horse at the end of a 25- to 30-foot line. The horse is worked in a circle. This training may be started when the horse is a yearling, but many horse owners do not begin this training until the horse is 2½ years old. This is because excessive stress and torque on a yearling's joints can cause growth problems.

One end of the line is fastened to the halter. The trainer attempts to make the horse move in a circle. Training begins with a small circle, and as the horse learns to respond to commands, the circle is enlarged. A light whip may be used to get the horse moving. Never hit the horse hard with the whip; a light touch on the hindquarters is all that should be used.

Teach the horse to circle at a walk. After it has learned this, advance to the trot and slow canter. Work the horse in both directions equally so the horse learns skill in moving both ways in a circle. The faster gaits should be taught only in a large circle.

The horse can be taught to respond to voice commands. Always use the same commands so the horse learns to associate the command with the action. The longe line can be used to train and exercise the horse (Figure 34-2). It is a good way to exercise a horse at a show.

Figure 34-2 A longe line is used to warm up a horse before riding or a show or just for exercise.

Saddling

The horse must be prepared for saddling several days before it is first saddled. This is often accomplished by using a desensitizing process known as **sacking out**, where various objects, such as blankets, water bottles, plastic bags, rags, etc., are introduced to the horse before the actual saddle. It is best if one person holds the horse with a lead rope while another person introduces the different objects to it. A trainer can tie a horse if necessary, but horses often tend to sit back on a lead rope if they cannot get away from an object that scares them. Begin sacking out by rubbing a sack or saddle pad over the head, neck, back, rump, and legs. The sack is then flipped over and about the body and legs. To further rid the horse of any fear of movement, a soft cotton rope can be placed over the back and pulled back and forth around the body and legs. Walk around the horse slowly twirling a rope. These actions need to be repeated until the horse is comfortable with every object used and will stand still when around these items. If the horse acts terrified, remove the object, let it calm down, and try again later. Repeating this process for several days will help to prepare the horse for saddling.

When it is time to saddle the horse, first let it see and smell the saddle. Slide the saddle blanket on and off the horse several times until it becomes used to it. Before placing the saddle on the horse, the girths and right stirrup are laid over the seat of the saddle. Lift the saddle gently into place. Lower the girth and stirrup on the off side by hand. Do not push them off, causing them to hit the side of the horse.

Reach under the horse with the left hand and bring the girth up to the latigo. Slip the latigo through the ring in the girth. Run the latigo up through the Dee ring on the saddle and back through the girth. Draw the girth snug (Figure 34-3). Walk the horse a few steps and then draw the girth up again. Repeat the operation with the back girth. Do not let the back girth hang too loose or the horse will catch a foot in it. There should be room for a hand between the back girth and the horse. Leave the saddle on for a time to let the horse become used to its feel. The saddling procedure should be repeated for several days before attempting to ride the horse.

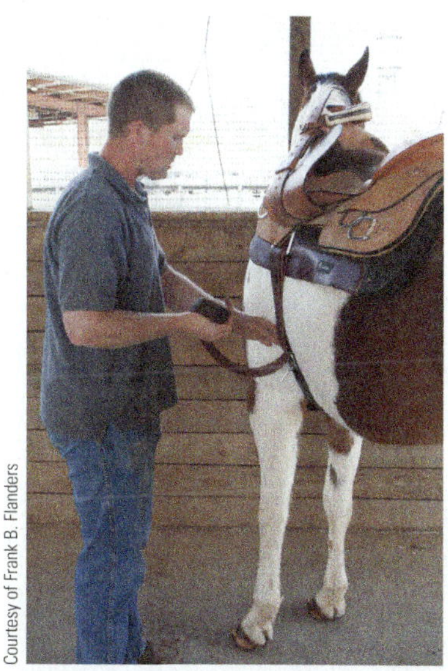

Figure 34-3 Make sure the girth is pulled tight while saddling a horse.

Use of Hackamore and Bridle

Some horse owners use a hackamore when starting to ride young horses. The hackamore prevents injury to the horse's mouth. The horse learns to respond to pressure that indicates the direction in which to go.

A bridle with a snaffle bit may also be used to train young horses. The halter is left on and the bridle is slipped over it. Be sure that the reins are even in length and tie them to the saddle horn. The bit should be in contact with the mouth when the head is held naturally. Lead the horse around for a few minutes. Do this for several days to allow the horse to become used to the bit.

Driving

Some trainers ground-drive the horse before mounting and riding. A halter, hackamore, or bridle may be used. Driving lines should be about 25 feet long. They are passed through the stirrups and attached to the halter, hackamore, or bit. A lead rope with a helper may be needed the first few times this is done. With the helper leading, the horse is led forward. Use voice commands such as "whoa" and "back" to fit the action. Start and stop frequently. Ground-driving helps the young horse relax.

Mounting and Riding

When mounting, turn the left stirrup to receive the foot. Hold the reins in the left hand. The left hand may hold a lock of mane, the bridle, or be placed on the horse's neck. The right hand is placed on the saddle horn. Step up into the left stirrup. Do this several times to get the horse used to the weight. When the horse accepts the weight, lie across the saddle and have another person lead the horse a couple of steps. Repeat this step until the horse is comfortable walking in a circle. Next, step into the stirrup, bring the right foot slowly over the saddle, and ease into the saddle. Do not allow the right foot or leg to drag over the horse's rump. Take hold of the reins with both hands. Be careful that the horse does not buck. If a horse throws its rider the first time, it will be more difficult to ride. Dismounting is the reverse of mounting.

Allow the horse to walk on its own for a short time. Stop the horse with the voice command "whoa" and a slight pull on the reins. Start and stop the horse for 15 or 20 minutes. Repeat for several days until the horse is used to the rider. At that time, proceed with training the horse to obey other commands. Remember that control of the horse is through voice, hand, leg pressure, and weight. Move from the walk to the trot. Teach the horse to turn by using the reins. Move from the trot to the lope.

After the horse has had several weeks of this training in a small area, it is time to move into more open spaces. Start at a slow gait and move into a lope. Vary the gaits during each training session. As the horse becomes used to the riding, the time spent each day may be increased.

The horse learns best at a slow gait. After it has learned the basic commands, it can be trained to the proper leads. Work the horse in a circle at the lope. Change directions frequently. Moving to the right trains the horse to pick up its right lead, and moving to the left trains it to pick up its left lead. When the horse has learned leads properly, it may be trained for various show events. The help of a professional may be needed for advanced training.

HORSEMANSHIP

Horsemanship is the art of riding a horse. It is also called *equitation*. The two general types of equitation are Western and English. The basic principles of horsemanship are similar for both styles of riding. There are differences in tack and clothing.

The American Horse Shows Association, breed associations, and individual shows establish rules for judging equitation classes. A person entering a show must become familiar with the rules that govern that particular show.

It is not possible to learn horsemanship by reading about it. The person who wishes to become a competent rider must learn from an instructor. Classes in horsemanship are offered in many localities. Youth groups such as 4-H clubs and FFA chapters may offer instruction and practice in horsemanship for interested members.

In any type of equitation, the rider must master the basic skills of mounting, correct seat, and dismounting. The basic principles are the same for any type of riding.

Courtesy of Frank B. Flanders

Figure 34-4 The horse is mounted from the near (left) side. A mounting block may be used for taller horses.

Mounting

A horse is always mounted from the near side (Figure 34-4). Two body positions are used. When mounting a strange horse, stand by the horse's left shoulder and face a quarter turn to the rear. When mounting a gentle horse, stand by the stirrup fender and face directly across the saddle.

Beginning from either position, the left foot is placed in the stirrup (Figure 34-5). The left hand holds the reins and is placed on the neck of the horse just in front of the withers. The right hand is placed on the saddle horn in Western equitation. In English equitation, the right hand grasps the off side of the cantle. The rider takes one or two hops on the right foot to gain momentum and springs up. The right leg is swung over the horse, taking care not to drag across the rump of the horse (Figure 34-6). The weight is shifted to the left leg to maintain balance. The right foot is slipped into the stirrup before the body weight is settled into the saddle.

Courtesy of Frank B. Flanders

Figure 34-5 Begin mounting by placing the left foot in the stirrup.

Seat Position

The Western rider sits erect and relaxed in the center of the saddle. Stirrup length should be such that the heels are lower than the toes (Figure 34-7). The balls of the feet are placed on the treads of the stirrups. Keep the toes pointed in the direction of travel, without being too far in or out. Maintain contact with the saddle with the calves, knees, and thighs. The elbows are kept close to the body. The rein hand is held just above and in front of the saddle. The free hand is held relaxed.

As the horse moves, the rider moves to stay in balance with the horse. It is important to coordinate body rhythm with the movement of the horse. Practice will help the rider learn how to do this.

Dismounting

Dismounting is the reverse of mounting. The left hand holding the reins is placed on the withers. The right hand is placed on the saddle horn or pommel (depending on whether Western or English equitation is used). The rider rises in the stirrups and slips the right foot free (Figure 34-8). The right foot is swung over the horse's back, keeping clear of the rump (Figure 34-9). In English equitation, the right hand is moved to the cantle of the saddle in preparation for stepping down. It is correct for the rider to either step down or slide down after the right leg is swung over the horse's back. Figures 34-10 and 34-11 show how to dismount a horse without stepping down first. The technique used depends on the size of the horse and the rider. The right leg is kept in close to the near side of the horse. The rider faces slightly forward when the right foot touches the ground. The left foot is removed from the stirrup by pushing down on the heel to slip the foot out of the stirrup. Never roll the left foot on its side to slip it out of the stirrup.

Courtesy of Frank B. Flanders

Figure 34-6 When mounting, the rider pushes off the left foot and swings the right leg over while being careful not to drag it across the horse's rump.

Courtesy of Frank B. Flanders

Figure 34-7 When in proper seat position, the rider sits relaxed in the center of the saddle. Notice that the heels remain lower than the toes.

Courtesy of Frank B. Flanders

Figure 34-8 During dismounting, the left hand holds the reins in place while the right hand is placed on the horn or pommel. The right foot is slipped out of the stirrup.

Courtesy of Frank B. Flanders

Figure 34-9 The right foot is swung over the horse's back, remaining clear of the rump during dismounting.

Courtesy of Frank B. Flanders

Figure 34-10 While the rider places most of the weight on the saddle, the left foot is slipped out of the stirrup during dismount.

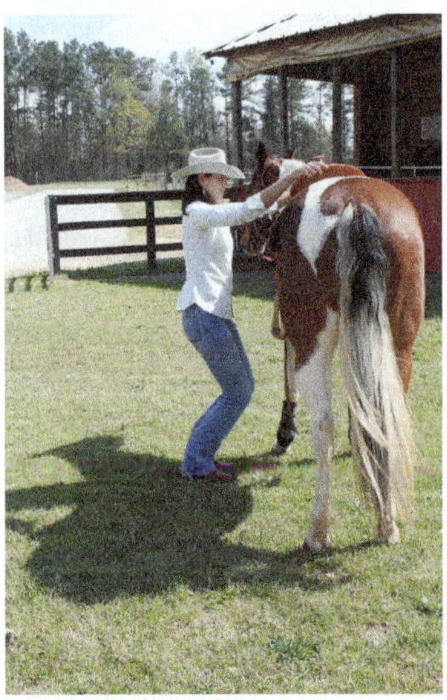

Courtesy of Frank B. Flanders

Figure 34-11 The rider safely slides down to complete the dismount.

Controlling the Horse

The basic aids used in controlling the horse are the voice, hands, legs, and weight. A well-trained horse will respond to these aids when they are properly used. After training, a horse will respond to light applications of these aids.

When the same words are used consistently during training, the horse learns to associate the words with the action. The horse should be spoken to in a soft, quiet, and firm voice. Yelling at the horse will only make it nervous.

The rider's hands are used to help guide the horse. A light hand is recommended for controlling the horse. The reins are held with a small amount of slack. The correct manner of holding the reins depends on the style of riding. The rider must learn the rules for the style of riding being done.

The rider must maintain body balance to avoid pulling on the reins for bracing. Using light pulls and slacking on the reins signal the horse. The reins should never be pulled steadily with great force. This will ruin the horse's mouth. **Neck-reining** is controlling the response of the horse by the weight of the rein against the neck. It takes patience and training to teach a horse to respond to neck-reining.

The rider's legs are used to control forward speed and the movement of the hindquarters. Pressure from the calves and heels is a signal to the horse to move forward. Pressure is also used to change gaits, to help stop the horse, and in preparation for backing. Pressure by one leg or the other will get the horse to respond by swinging the hindquarters away from the pressure.

A well-trained horse will respond to slight shifts in the body weight of the rider. The horse shifts its body to balance the weight of the rider. This causes the legs of the horse to be held in place by the weight of the rider or be left free to move because of the absence of weight.

The aids are used to start and stop, move into the gaits, take the correct leads, turn, and back the horse. Instruction and practice are necessary for the rider to learn how to use the aids correctly.

SHOWING AT HALTER

Horses shown at halter are judged for conformation, soundness, and action. The horse must be properly groomed and trained before the show. When showing, the rider must be properly dressed for the class. Clothes must be neat and clean. Fairness and courtesy are always appropriate in the show ring.

The horse is led into the show ring at a brisk walk. The ring steward indicates the direction in which to move. When the horses are lined up, enter the line from the

Figure 34-12 Always turn to face the judge during evaluation.

rear. Stand the horse in position evenly with the other horses in the class. Some breeds are shown in a stretched position and others are not. Learn the correct position for the breed of horse being shown.

When in line, stand on the near side of the horse and even with the horse's head. Always leave room on either side of the horse so the judge can see the horses. This also prevents possible injury from other horses in line. Always turn to face the judge as the class is judged. When the judge is inspecting the near side, move to the front of the horse (Figure 34-12).

The horse is shown at the walk and the trot. Each horse is asked to perform individually by the ring steward. The horse is led to the indicated line and stopped about one horse length from the person standing at the end of the line. The horse is walked in a brisk manner to the other end of the line. Always lead from the near side and never in front of the horse. At the other end of the line, stop about one horse length from the person standing there. When given the signal to return, turn the horse to the right. Pivot the horse on its hindquarters. Trot the horse to the other end of the line. When given the signal, return to the line of horses waiting to be shown. Again, enter the line from the rear.

Various shows and arenas have different classes and patterns of showing. Study the rules for the show being entered. Always watch the ring steward for signals. After entering the ring, continue showing the horse until the class is placed and the judge has given reasons. Leave the ring in the same brisk manner used when entering. Exhibitors should display good sportsmanship at all times.

EQUITATION CLASSES

Horses can be shown in Western or English equitation. In these classes, the rider is being judged for ability to ride and control the horse. The horse is not being judged as it is in halter classes. The right kind of horse should be used for the class in which it is performing. It must be well trained to perform the maneuvers required in the class.

The rider is judged on a number of points, including position in the saddle, use of the hands, proper tack and dress, and performance of the various gaits appropriate for the class. The rider may be asked to mount and dismount. There are a number of individual tests the rider may be asked to demonstrate.

There are many performance classes in the various styles of riding. A person desiring to enter this type of competition should secure a copy of the official rulebook of the American Horse Shows Association. Any given show may also have rules with which the exhibitor must be familiar.

Western Equitation

The rider in Western equitation classes should wear a Western hat and cowboy boots. Spurs may be worn but are not necessary. Chaps are permitted in some shows but not in all; therefore, rules for the show being entered should be followed.

Use a Western saddle in Western equitation shows. Snaffle, curb, spade, and other similar bits are usually allowed in Western shows. Hackamores, martingales, or tie-downs are generally not allowed. Chin straps or curb chains must be at least a ½ inch wide and lie flat against the jaw. Wire curbs are not allowed. A coiled rope or **riata** may be attached to the saddle.

The rider maintains an erect position, properly balanced in the center of the saddle and facing the front. The horse is gripped with the inside of the thighs, knees, and calves of the legs of the rider. The rider's legs should be nearly parallel to the side of the horse. The upper body of the rider is relaxed but erect. The arms are held with the elbows close to the rider's side. Elbows should not be allowed to flap with the movement of the horse.

The reins may be carried in either hand, but the hand is not changed during the show. Hold the rein hand about 3 inches above and in front of the saddle horn. Place the other hand on the thigh or carry it in front of the waist, keeping the elbow close to the body.

The reins are usually held between the forefinger and thumb. Some shows permit a finger to be placed between the reins unless a romal is used or the ends of the split reins are held in the hand not used for reining. If the reins are held in the right hand, carry the bight (end of reins) on the right side; carry the bight on the left side if the reins are held in the left hand. Control of the horse is done mainly by neck-reining with a semi-loose rein. Do not put strong, constant pressure on the bit.

During Western equitation, the rider maintains a firm seat in the saddle. Do not stand in the stirrups at a trot or lope. Rising and posting the trot are not required in Western equitation as they are in English equitation.

In Western equitation, the horses enter the ring at a walk. During the show, the horse is expected to demonstrate a walk, trot (jog), and lope (canter) both clockwise and counterclockwise.

Voice commands should not be heard, nor should leg, feet, and hand cues be noticeable during the competition. Do not slap the horse with the reins. A well-trained horse will maintain good form during the walk. Judges discount horses that dance, prance, toss the head, and chafe and pull at the bit during the walk.

The trot is a diagonal two-beat gait, which should be performed at moderate speed. If the horse appears to be racing during the trot, it is going too fast; if it appears sluggish, it is going too slow. The lope or canter is a three-beat gait. It should be performed reasonably slowly, at a restrained gallop. The rider must maintain a firm seat and lean slightly forward with the horse while performing this gait.

Important points that are emphasized in the canter are the ability to execute the correct lead and to change leads easily and quickly. The correct lead is with the front leg toward the inside of the ring. The front and hind legs move in unison during the canter. Begin the canter by lifting the horse slightly with the reins and giving a gentle boot on the fence side of the ring. Do not jerk on the reins, as this may cause the horse to bounce or stop.

During the show, stay on the rail and do not bunch up in a group of riders. Always show courtesy to other riders. Do not hesitate to pass when necessary. Passing is done on the inside of the ring; do not pass on the rail side.

When the judge requests the riders to reverse direction, the turn may be made either toward or away from the rail. In some shows, however, the rules require the turn to be made away from the rail. Riders must be sure to determine the rule for the show in which they are competing. Turning to the inside is more difficult and provides a test of the rider's control of the horse. Be sure to have a firm grip on the reins when executing the turn.

After the gaits are completed, the riders are usually asked to line up the horses in the center of the ring. The horse should stand quietly when lined up. Riders may be asked to back their horses from this line. The horse should back in a straight line when the rider gives a slight pull on the bit. The horse should not pull up or turn its head when backing. Points are lost if the horse swings its head or wrings its tail while backing.

Sometimes riders are asked to dismount and remount their horses. Upon dismounting, the rider should hold the reins in the right hand and stand facing the horse. Riders are sometimes asked to individually demonstrate the figure eight, square stops, and quarter and half turns. The judge will be looking for balance, correct leads, head and tail carriage of the horse, and good form. Additional maneuvers sometimes requested include lope and stop, roll backs, 360-degree turns, riding a serpentine course showing flying lead changes, and riding without stirrups. Judges may ask riders questions about equitation, tack, or conformation.

English Equitation

Informal clothing—which includes a solid-colored saddle suit (navy, black, brown, beige, or gray), a shirt (white or pastel), a tie (four-in-hand, either solid-colored or striped), jodhpur boots (black or brown), and a soft or hard dark derby—is worn during English equitation. Gloves (white, brown, or black) may be worn but are not required. An alternative to a saddle suit is jodhpurs (without flare or cuff), black or brown jodhpur boots, a shirt, a tie, and a sweater or vest.

Formal attire is required for evening saddle seat equitation and three-gaited classes. This consists of a tuxedo saddle suit, a white stiff-front shirt with a bow tie, and a top hat. Formal jodhpur pants have satin stripes down each leg, and formal jodhpur boots are patent leather. Formal attire is never worn in five-gaited classes.

For hunt seat equitation, the rider should wear a dark-colored, three-button coat; this may be mid-hand or fingertip length with one vent in mid-back. The rider should also wear tan, fawn, black, or brown breeches or flare-top jodhpurs with cuffs. The rider's tie may be plaid, striped, or solid-colored and may be stock, ratcatcher, or four-in-hand. Wear black or brown high boots or jodhpur boots and a black hunt cap for hunt seat equitation (Figure 34-13). Black, brown, or tan gloves may also be worn if desired.

Do not wear fancy hats, jockey caps, sleeveless coats, or brightly colored suits in English equitation. Riders with long hair should tie it up or wear a hair net because contestant numbers are worn on the back and must be visible to the judge.

Figure 34-13 English equestrian riders often wear breeches, coat, white shirt, and helmet.

During English equitation, the rider should maintain an upright balanced position that is comfortable; do not lean forward or backward. Look straight ahead between the ears of the horse. Stirrups should be adjusted so the rider's feet hang naturally when mounted. The stirrup should touch the anklebone.

The rider sits close to the front of the saddle. The ball of the foot is kept evenly on the stirrup bar, with the heels down and toes tilted slightly upward. This position helps the rider maintain a balanced and centered position.

The rider keeps the hands just above the pommel of the saddle at about the waistline. The back of the hands is kept up, with the thumbs close together and the knuckles nearly vertical. The upper arms are kept parallel to the body with the elbows in so the arms do not flap.

In English equitation, the reins are always carried in both hands. If a single-rein bridle is used, the rein comes through the palm and is held between the forefinger and the thumb; when a double-reined bridle is used, the snaffle reins are outside the little fingers and the curb reins are between the little and fourth fingers. An alternative that is acceptable is to hold the curb rein between the middle finger and the fourth finger with two fingers between the snaffle and curb. The reins come out of the palm over the first finger and are held in place by the thumb. If a crop is carried, place it in the hand away from the rail.

When jumping in hunt seat equitation, the rider should lean slightly forward in the seat to help the horse over the jump. The rider's hands are carried closer to the horse's neck when jumping because the horse has a lower head set than the regular saddle horse.

If the show includes hunt seat equitation without jumping, the gaits requested are usually a walk, a trot, and a canter, in both directions in the ring. Riders may be asked to hand gallop and sit (but not post) a slow trot. The body should be kept in a comfortable vertical position for the walk and slow trot; the rider should lean slightly forward at the posting trot, canter, and gallop.

In English equitation, the rider needs a high level of skill because horses used in this type of equitation are often more high-strung and animated than those used in Western equitation. The rider signals the proper gait to the horse by posture, hand signal through the bit, and pressure from the legs. The walk begins by gathering the reins in the hands above the pommel and urging the horse forward using slight pressure from both legs; the trot begins by shortening the reins slightly and signaling the horse with a definite pressure from the legs.

During the trot, the rider is expected to post on the correct diagonal (position of the horse's front legs). **Posting** is when the rider rises and sits in the saddle. The rider rises slightly out of the saddle when the horse's shoulder on the fence side rises and sits down when that shoulder falls. Be careful not to rise too far out of the saddle or to sit with a bump.

The canter is usually begun from a walk or from a standing position. In some forms of hunt seat equitation, the canter may be entered from the trot. The canter is a three-beat gait, with the fore and hind legs moving together. It begins by shortening the reins, turning the horse's head slightly toward the rail, and signaling the horse with the heel behind the girth on the rail side. The correct lead is accomplished when the horse takes the first step with the foreleg toward the center of the ring. If the horse takes the wrong lead, stop and begin again. If the reins are held too tightly, the horse may hop and pound with its front legs; if the reins are held too loosely, the horse may lunge into the canter.

During the competition, it is permissible to pass other horses on the inside away from the rail. Do not get bunched up in a group of horses. If this happens, the rider should turn the horse across the ring to the opposite rail. When the judge requests that the horses be reversed, stop and turn the horse away from the rail.

When the horses are lined up, the horse should be positioned squarely with some stretch. Arabians and jumpers may not be required to stretch. When backing the horse, take one or two steps forward and then back the horse in a straight line for about four steps. When backing, the rider's legs guide the horse. Riders are usually not required to dismount in English equitation.

Riders may be asked to perform some individual work, such as demonstrating change of diagonals at the trot or showing leads in making a figure eight. The judge may also request that the rider drop the reins so the routine used for picking them up may be checked.

GYMKHANA

Gymkhana is the term used for games on horseback. Many horse shows have gymkhana events (Figure 34-14). Rules, regulations, and diagrams for gymkhana events are available from the American Horse Shows Association, the American Quarter Horse Association, the Appaloosa Horse Club Inc., and the sponsors of local horse shows.

Some typical events in gymkhana include the following:

- pole bending
- clover-leaf barrel race
- rescue race
- sack race
- wheelbarrow race
- keyhole race
- saddling race
- team baton race

Appaloosa Horse Club

Figure 34-14 Games at horse shows are popular events.

Other original events may be included in a local show. Many of these events are timed. The rider and horse are working against the clock and the fastest time wins.

RODEOS

Rodeos are popular horse events in all parts of the United States. A horse must be well trained to participate in rodeo events. Rodeo events require a high degree of skill and experience on the part of the person entering.

Typical rodeo events include the following:

- saddle bronc riding
- bareback bronc riding
- calf and steer roping
- bulldogging (steer wrestling)
- chuck wagon races
- wild horse races
- reining contests
- cutting horse contests

Other events and contests may be included in any given rodeo.

TRAIL RIDING

One of the most popular activities involving the use of horses is trail riding. There are many clubs and groups that organize trail rides. A trail ride may be a 1-day event or an overnight event.

Both the rider and the horse should prepare for trail riding. The rider must develop the endurance needed for trail riding. Short rides each day before the trail ride will help both the rider and the horse prepare for the trail ride. The horse must be properly fed and conditioned and must become used to the kinds of obstacles that will be found on trail rides.

The trail to be ridden must be carefully selected. Permission must be obtained in advance for crossing private property. A variety of conditions should be provided for on the trail ride. Busy highways should be avoided if at all possible. Provision should be made for rest stops. If the trail ride is to last overnight, arrangements for camping out must be made. Planning must include measures for handling accidents and emergencies.

Planning for the ride includes determining who will be in charge during the ride. Appropriate equipment and supplies must be secured. Rules and procedures for each rider to follow must be agreed upon before the ride begins. Rules and procedures are based on safety requirements and consideration of others on the trail ride.

SUMMARY

Understanding the reasons for horse behavior is helpful when training and riding horses. Horses have to move their heads to bring objects into focus. This is one reason horses may shy at sudden movements. Horses have good memories, making it possible to train them to respond to commands. Sensitive areas on the body of the horse are used to help control the animal.

Training should begin when the horse is very young. First lessons include haltering, leading, and yielding the feet. When the horse is older, training can begin on the longe line. The horse learns to respond to spoken commands. Prepare a horse for saddling by sacking out. Lift the saddle into place carefully. Be sure that it is properly fastened. Many owners prefer to use a hackamore when beginning to teach a horse to be ridden.

A bridle may be used with a snaffle bit. Ground-driving may be used to further train the horse before beginning to ride. Use care when mounting a horse for the first time. Make sure the horse will stand for the weight on its back. Teach the horse at slow gaits first. Later it may be trained at faster gaits.

The two general types of equitation are Western and English. The basic principles are the same for each. The main differences are in the tack and clothes used. Mounting, seat position, and dismounting are basic skills that every rider must have.

The voice, hands, legs, and weight of the rider control the horse. A well-trained horse will respond quickly to the signals of the rider. Horses shown at halter are judged on conformation, soundness, and action. Lead the horse from the near side and always face the judge when showing at halter. The horse is shown at the walk and the trot in halter classes.

Equitation classes are judged on the performance of the rider. The rider must have skill in riding and controlling the horse. Position in the saddle, use of the hands, proper tack and dress, and the performance of the various gaits are the main points judged in equitation classes.

Other popular horse events are gymkhana, rodeos, and trail riding. Participation in these events adds to the pleasure of owning a horse.

Student Learning Activities

1. Observe a demonstration on training a young horse.
2. Observe demonstrations on haltering, bridling, saddling, mounting, dismounting, and riding horses.

3. Attend horse shows to observe correct procedures.
4. Prepare an oral report and/or demonstrations on horsemanship and showing horses.
5. Follow good practices when planning and conducting horse training and riding as a part of a horse supervised experience program.

Discussion Questions

1. What are the survival instincts of horses that have developed over the years that help to explain their behavior?
2. Describe the eyesight of horses.
3. How is a horse's memory used to help train the horse?
4. Name the most sensitive areas on a horse's body.
5. Name four ways a rider communicates commands to a horse.
6. Describe the proper procedures for (a) haltering, (b) leading, (c) working with the feet, (d) longeing, and (e) saddling the horse.
7. Describe how the hackamore and the bridle are used.
8. Describe how ground-driving is used when training a horse.
9. Describe the proper way to mount and ride a horse during training.
10. Define the term *horsemanship*.
11. How may a rider learn horsemanship?
12. Describe the proper way to mount a horse in Western and English equitation.
13. Describe the proper seat position in the saddle.
14. Describe the proper way to dismount from a horse.
15. How does the rider use the basic aids in controlling a horse?
16. Explain how a horse should be shown at halter.
17. Describe the showing of a horse in equitation classes.
18. Name some of the events that are often included in gymkhana.
19. Name some typical events included in a rodeo.
20. List the preparations that are necessary for trail riding.

Review Questions

True/False

1. Leaving the halter on the foal for a short period of time will help it get used to wearing it.
2. It is generally believed that horses are color-blind.
3. The foal should not be handled each day in order to help it overcome fear of the handler.
4. The horse is able to sense security or the lack of security in the rider.

Multiple Choice

5. A process called sacking out is used for several days before a horse is to be ______.

   a. vaccinated c. haltered
   b. saddled d. mounted

6. Longeing is training a horse at the end of a ______ to ______ foot line.

   a. 5; 10 c. 25; 30
   b. 15; 20 d. 35; 40

Completion

7. ______ is the reverse of mounting.
8. The ______ prevents injury to the horse's mouth.
9. When mounting, turn the ______ stirrup to receive the foot.

Short Answer

10. Name four basic aids used in controlling a horse.
11. On which side should a horse be mounted?
12. Horses shown at halter are judged on what three things?
13. List the two general types of equitation.

SECTION 8

Poultry

Chapter 35

Selection of Poultry

Key Terms

vertical integration
breed
variety
type
class
cross-mating
breed crossing
inbreeding
hybrid
dual-purpose breed
comb
bantam
hatchability
pinfeather
drake
straight-run chick
pullet
cockerel
sexed chick
culling
molting
axial feather
trap nest

Objectives

After studying this chapter, the student should be able to

- describe the nature of the poultry industry.
- identify common breeds of poultry.
- explain the selection of poultry for production.

THE POULTRY INDUSTRY

Types of Enterprises

The size of poultry enterprises range from small farm flocks to large commercial operations. Most of the poultry raised in the United States is produced in large commercial operations. Regardless of the size of the enterprise, there are three important factors for poultry success: proper feeding, good management, and sanitation.

The three general types of chicken enterprises are egg production, broiler production, and raising replacement pullets. Most of the turkeys, ducks, and geese in the United States are raised for meat production. Except for hatching purposes, there is little market for turkey, duck, and goose eggs.

In egg production operations, laying hens are kept to produce eggs. Laying hens may be confined in cages or the farmer may use a floor-pen system. Cleaning, grading, and packaging of eggs usually occurs on the farm. When the production cycle is completed, the hens are sold for meat.

Broiler production operations involve raising chickens for meat. High-quality rations are fed to secure rapid, efficient gains. Generally, several flocks of birds are fed and marketed each year. Replacement pullet

TABLE 35-1 Rate of Lay and Per Capita Consumption, Projected through 2015

Year	Rate of Lay: Eggs Per Layer (number)	Per Capita Consumption of eggs (number)
1970	218	310
1975	232	276
1980	242	272
1985	247	256
1990	251	236
1995	254	236
2000	257	252
2005	262	255
2010	269	248
2015	275	257

Source: USDA

operations involve raising chicks, which are then sold to egg production or broiler production operations, depending on the type of bird raised.

Vertical integration is common in the poultry industry. **Vertical integration** means that two or more steps of production, marketing, and processing are linked together. Vertical integration is usually set up by feed manufacturers or poultry processors. They provide the financing needed and have most control of the management decisions that are made in the production process. Most of the broilers and turkeys in the United States are produced under some type of vertical integration arrangement. More than one-half of the eggs are produced in this manner. Vertical integration has resulted in much larger and more efficient operations in the poultry industry.

Numbers and Trends in Production and Consumption

The average number of laying hens on farms in the United States decreased until the late 1980s but has been steadily increasing since then. The number of farms with laying hens has decreased, while the average size of flock per farm has increased. Much of the egg production in the United States is concentrated in large commercial operations, many of which produce eggs under some type of vertical integration arrangement.

Annual egg production has increased significantly since 1970. This is partly due to an increase in the number of layers but is mainly due to a significant increase in the rate of lay per hen (Table 35-1).

As shown in Table 35-1, the consumption of eggs per person has declined since 1970. In recent years, however, consumption has begun to increase. The long-term decline in consumption was probably related to health concerns about the cholesterol level in eggs.

The five leading states in numbers of laying hens and egg production are Iowa, Ohio, Pennsylvania, Indiana, and California, which are responsible for almost 50 percent of all U.S. layers and egg production (Figure 35-1). The rank order in numbers of laying hens and egg production will vary from year to year; however, these five states are generally the major egg-producing states in the United States.

The production of broilers in the United States has increased about five times since 1970. The per person consumption of poultry meat has also increased significantly since 1970 (Table 35-2). A major increase in the per capita consumption of

TABLE 35-2 Per Capita Consumption of Broilers, Projected through 2020

Year	Per Capita Consumption (Pounds)
1970	36.6
1975	36.3
1980	45.8
1985	51.0
1990	59.5
1995	68.0
2000	76.9
2005	85.8
2010	82.4
2015	86.3
2020	91.5

Source: USDA

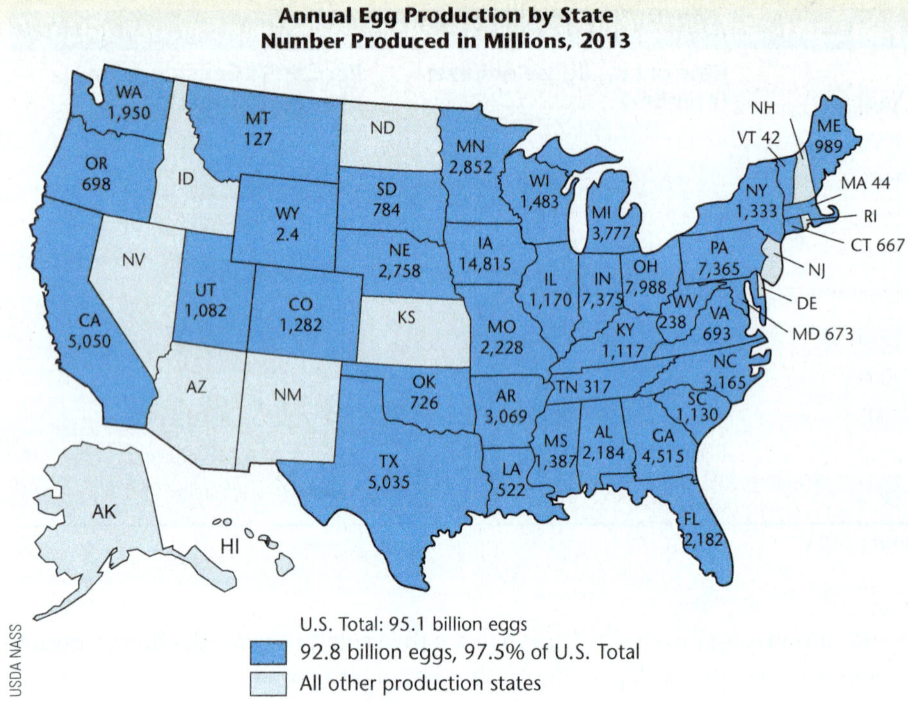

Figure 35-1 Annual egg production by state.

poultry meat began in the late 1970s as consumers became increasingly concerned about the level of cholesterol in their diets. Poultry meat is perceived by the consumer as having a lower level of cholesterol than beef or pork.

The five leading states in broiler production are Georgia, Arkansas, Alabama, Mississippi, and North Carolina (Figure 35-2). The rank order of states in numbers of broilers produced will vary from year to year. These five states are usually the major broiler-producing states, together producing more than 50 percent of the total broiler production in the United States.

Since 1970, the overall trend in per capita consumption of turkey has been upward (Table 35-3). As with broilers, this may partially reflect consumer concern regarding cholesterol in the diet. Turkey meat is also perceived by consumers as having less cholesterol than beef or pork.

Turkey production tends not to be concentrated in any one geographical area of the United States (Figure 35-3). The six leading states in turkey production are Minnesota, North Carolina, Arkansas, Missouri, Virginia, and Indiana. The rank order of turkey-producing states will vary from year to year; however, these six states tend to be the major turkey-producing states in the United States. These states are responsible for more than 60 percent of the United States' turkey production.

Between 25 and 35 million ducks are raised commercially on an annual basis in the United States. Although many breeds are raised for game purposes, the White Pekin breed is most often used for meat. Duck consumption is generally about one-third to one-half pound per person per year. Most ducks for meat are produced in the Midwest, with a high percentage produced in Indiana.

Consumption of goose per person per year is even less than that of duck, with annual production only being about 300,000 to 350,000 geese raised per year in the United States. The leading states in production are Texas, Minnesota, and South Dakota.

TABLE 35-3 Per Capita Consumption of Turkey, Projected through 2020

Year	Per Capita Consumption (Pounds)
1970	8.1
1975	8.3
1980	10.3
1985	11.6
1990	17.5
1995	17.7
2000	17.4
2005	16.7
2010	16.4
2015	16.1
2020	16.5

Source: USDA

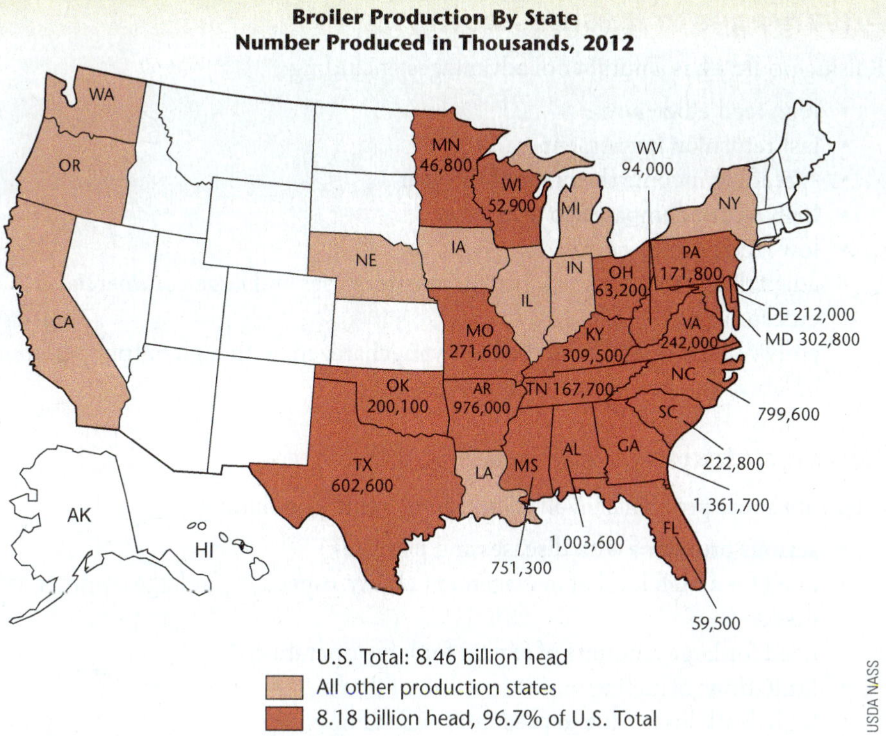

Figure 35-2 Broiler production by state.

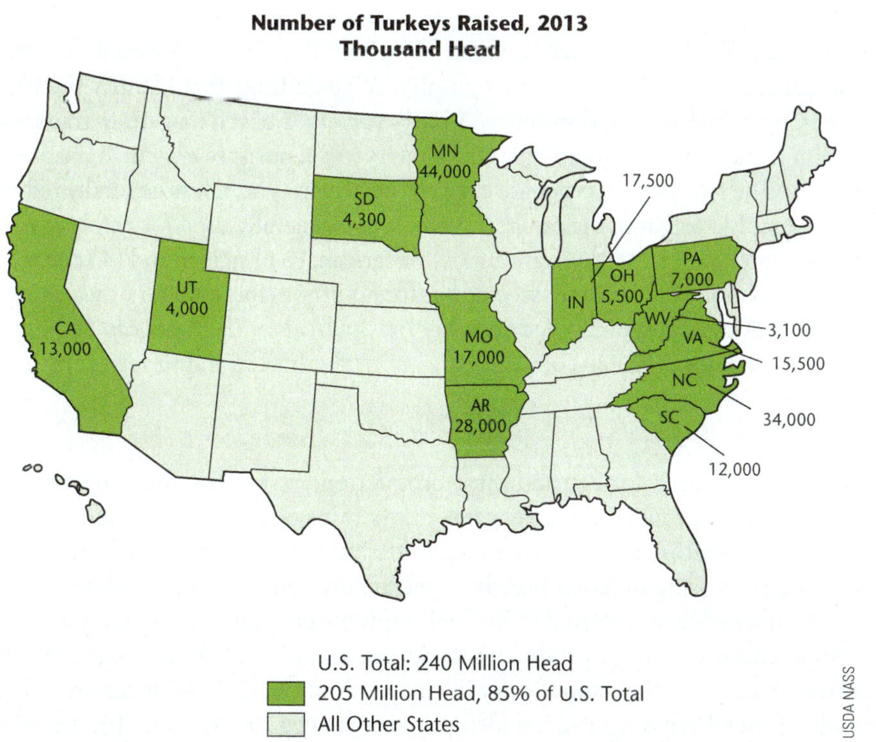

Figure 35-3 Turkey production in major turkey-producing states.

Advantages of Raising Poultry

Raising poultry has a number of advantages, including:

- high feed efficiency
- fast return on investment
- spreading income throughout the year
- high return compared to feed costs
- low land requirements
- adaptability to both small part-time enterprises and large commercial enterprises
- ability of the operation to be highly mechanized, with high output per hour of labor

Disadvantages of Raising Poultry

There are some problems involved in raising poultry, including:

- serious problems with diseases and parasites
- need for a high level of management ability, especially for large commercial flocks
- need for large amounts of capital for large operations
- limitations of zoning on the location of flocks
- high death losses due to predators and stampeding
- need for careful control of product quality
- need for careful marketing
- need for high volume for an economical enterprise
- problems with waste disposal and odor

Photo by Andrea Glover

Figure 35-4 Rhode Island Red.

Courtesy of Frank B. Flanders

Figure 35-5 The Barred Plymouth Rock was one of the foundation breeds for the broiler industry in the 1920s. The breed is popular among poultry fanciers.

BREEDS OF POULTRY

Poultry are divided into breeds, varieties, types, and classes. A **breed** is a group of related fowls that breed true to a number of given traits that identify the breed. Breeds are subdivided into varieties. A bird's **variety** is based on certain traits, such as color of plumage and comb type. **Type** refers to the purpose for which the poultry are bred. The two general types are egg type and meat type. **Class** generally refers to the geographic origin of the poultry. Four classes generally used for chickens in the United States are (1) Mediterranean, (2) American, (3) English, and (4) Asiatic.

There are no breed registry associations for poultry as there are for other farm animals. The American Poultry Association, Inc., publishes *The American Standard of Perfection;* however, not all commercially important breeds and varieties are listed.

Chickens

Most chickens used for egg and meat production are produced by cross-mating, breed crossing, or inbreeding. Very few pure strains of chickens are used commercially. **Cross-mating** is crossing two or more strains within the same breed. **Breed crossing** is crossing different breeds to get the desired traits. **Inbreeding** refers to the breeding of closely related individuals within a breed to get the desired traits.

Most commercial egg-producing flocks are made up of White Leghorn strain crosses because of their superiority in egg production. **Hybrid** flocks are a direct result of breed crossing and are also popular for egg production. The Leghorns and the Hybrids are white egg layers. Brown egg layers are Rhode Island Reds (Figure 35-4), New Hampshires, and Barred Plymouth Rocks (Figure 35-5).

These are larger in size than the Leghorn strains and, therefore, have higher feed costs per dozen eggs produced. They are not as popular for large commercial flocks. Smaller farm flocks sometimes use these breeds because they produce more meat for home use and are considered to be dual-purpose breeds. **Dual-purpose breeds** are used for egg and meat production, although they are not as efficient as breeds that are bred for one purpose.

Commercial crosses that are bred for meat production are generally used for broiler production operations. A common cross is White Plymouth Rock females (Figure 35-6) mated to Cornish males. These crosses do not have a high rate of egg lay but are highly efficient as meat producers.

Dr. Charles Wabeck

Figure 35-6 White Plymouth Rock hen. The White Plymouth Rock is used in some broiler breeding lines.

Characteristics of some typical breeds of chickens in four classes are shown in Table 35-4. There are a number of different varieties of most chickens. Color of plumage, type of comb, and size identify the different varieties. Refer to Henderson's Handy Dandy Chicken Chart for more information about other breeds.

There are a number of different types of **combs** found in chickens. Some of these are (1) single, (2) rose, (3) pea, (4) cushion, (5) buttercup, (6) strawberry, (7) V-shaped, and (8) walnut. The single comb is the most common type. The other comb types mentioned are mutations. Figure 35-7 shows some of the comb types of chickens.

Bantams are small chickens, usually weighing from 16 to 30 ounces as adults. There are nearly 350 varieties of bantams. Most of them are of the same breeds as

TABLE 35-4 Characteristics of Some Typical Breeds of Chickens

Class and Breed	Eggs	Skin	Comb	Eyes	Earlobes	Shanks	Plumage	Comments
Mediterranean								
Leghorn (white)	White	Yellow	Single	Reddish bay	White	Yellow	White	All three are small in size and are used mainly for egg production; Leghorn is the most popular.
Minorca (Black)	White	White	Single	Brown	White	Dark slate	Black	
Andalusian (Blue)	White	White	Single	Reddish bay	White	Dark slate blue	Slate blue	
American								
Plymouth Rock (Barred)	Brown	Yellow	Single	Reddish bay	Red	Yellow	Barred (sex-linked)	Dual purpose; used in crosses for sexing chicks at hatching.
Plymouth Rock (White)	Brown	Yellow	Single	Reddish bay	Red	Yellow	White	Primary use—Broiler.
New Hampshire	Brown	Yellow	Single	Reddish bay	Red	Yellow	Red	Primary use—Broiler.
Rhode Island Red	Brown	Yellow	Single	Reddish bay	Red	Yellow	Dark red	Dual purpose.
English								
Cornish (White)	Brown	Yellow	Pea	Pearl	Red	Yellow	White	Used in development of male lines for crossbreeding.
Australorp	Tinted	White	Single	Brown	Red	Dark slate; bottom feet white	Black	Used in production of crossbreeds.
Asiatic								
Brahma (Light)	Brown	Yellow	Pea	Reddish bay	Red	Yellow; feathered	Columbian (white and black)	Used in crossbreeding for meat production.

Source: U.S. Department of Agriculture, National Agricultural Statistics Services, various years

Figure 35-7 Comb types found in chickens.

the larger chickens. Bantams are generally used for show and as pets. They are not generally bred for egg or meat production.

Turkeys

The American Standard of Perfection recognizes only one breed of turkeys. However, a number of varieties of turkeys exist, including the Broad Breasted Bronze, Broad Breasted Large White, Beltsville Small White, Narragansett, Black, Slate, and Bourbon Red. The commercially important varieties are the Broad Breasted Bronze, Broad Breasted Large White, and Beltsville Small White. The Broad Breasted Large White is the most commonly raised meat turkey.

The Broad Breasted Bronze

The Broad Breasted Bronze has black plumage and dark-colored pinfeathers. The females have white tips on black breast feathers. The beard is black. Females normally do not have beards. The shanks and feet are black on young turkeys and change to a pinkish color on the adults. The beak is light at the tip and dark at the base. The Broad Breasted Bronze is the largest of the turkey varieties.

The reproductive ability of the Broad Breasted turkey varieties is not as good as the Beltsville Small White. Fewer eggs with lower fertility and **hatchability** (number of young) are produced. Artificial insemination is generally used with all varieties of Broad Breasted turkeys because the heavy males are not good breeders.

The Broad Breasted Large White

The Broad Breasted Large White was developed from crosses of the Broad Breasted Bronze and the White Holland. The color of the plumage is white. The males have a black beard. Some females have small beards. The shanks, feet, and beak are white to pinkish white, and the throat wattle is red. Because the pinfeathers are white, they are not easily seen and, therefore, do not lower the market grade of the carcass. A **pinfeather** is a feather that is not fully developed. The term generally refers to feathers that are just coming through the skin. Pinfeathers are not easily removed from the carcass. Dark pinfeathers are seen more easily than white pinfeathers. This detracts from the appearance of the carcass and lowers its value. Because market grade is partially associated with value, presence of dark pinfeathers also lowers the market grade of poultry. The Broad Breasted Large White has the body conformation of the Broad Breasted Bronze but is slightly smaller when fully grown.

The Beltsville Small White

The Beltsville Small White (Figure 35-8) was developed by the U.S. Department of Agriculture (USDA). It is similar to the Broad Breasted Large White in color and body type. Hens average about 10 pounds, while toms average about 17 pounds.

Figure 35-8 Beltsville Small White turkey.

Figure 35-9 The White Pekin is the breed generally used in commercial meat production.

Ducks

The best breeds of ducks for meat production are the White Pekin, Aylesbury, and Muscovy. Other meat-producing breeds include the Rouen, Cayuga, Swedish, and Call. The best egg-laying breeds are the Khaki Campbells and Indian Runners.

The White Pekin

The White Pekin (Figure 35-9) is the breed generally used in commercial meat production. It typically reaches a market weight of 7 pounds in 8 weeks. The adult duck weighs about 8 pounds. White Pekins originated in China and were brought to the United States in the late 1870s. They have white feathers, orange-yellow bills, reddish-yellow shanks and feet, and yellow skin. The eggs are tinted white. White Pekins are not good setters (will not sit on a nest of eggs) and will seldom raise a brood. White Pekins are nervous and must be handled with care.

The Aylesbury

The Aylesbury originated in England, where it is still the most popular meat duck. They are the same size as the White Pekin. They have white feathers, white skin, flesh-colored bills, and light-orange legs and feet. Their eggs are tinted white. They are not as nervous as the White Pekin. However, they do not set well and, like the White Pekin, seldom raise a brood.

The Muscovy

The Muscovy is not related to the other duck breeds. It originated in South America and includes a number of varieties. The White Muscovy is the best meat duck of the Muscovy varieties. The meat is of excellent quality if ducks are marketed before 17 weeks of age. The Muscovy has white feathers and white skin. The adult **drakes** (males) average 10 pounds, and the adult ducks average 7 pounds. They are good setters and will raise a brood. A cross between the female Muscovy and a drake of the Mallard-type produces a sterile hybrid. The sterile hybrid produces a good meat yield.

Figure 35-10 Khaki Campbell ducks may lay close to an average of 365 eggs per duck per year compared to an average of about 250 eggs per year for chickens.

The Khaki Campbell

The Khaki Campbell (Figure 35-10) originated in England. Ducks of this breed are excellent egg producers, with some strains averaging close to 365 eggs per duck per year. The drake is khaki colored with a brownish-bronze lower back, tail covert, head, and neck. The bill is green, and the legs and toes are dark orange. The females are khaki colored with seal-brown heads and necks. The female's bill is greenish-black, and the legs and toes are brown. At 8 weeks of age, these ducks average 3.5 to 4 pounds. Adult drakes and ducks average 4.5 pounds.

The Indian Runner

The Indian Runner originated in the East Indies. The three varieties of this breed are White, Penciled, and Fawn and White. The feet and shanks of the three varieties are orange to reddish-orange. The Indian Runner stands erect and is about the same weight as the Khaki Campbell. It does not produce as many eggs as the Khaki Campbell.

Geese

The five common breeds of geese found in the United States are the Toulouse, Embden, Chinese, Pilgrim, and African.

The Toulouse

The Toulouse originated in France. It is dark gray with a white abdomen. It has a fold of skin (dewlap) that hangs down from the throat at the upper end of the neck. The Toulouse has a pale orange bill, deep reddish-orange shanks and toes, and dark brown or hazel eyes. The adult gander weighs about 26 pounds. The adult goose weighs about 20 pounds.

The Embden

The Embden is a white breed that originated in Germany. Weights of the geese in this breed are similar to those of the Toulouse.

The Chinese

The Chinese breed exists in two varieties: white and brown. It originated in China. The Chinese has a knob on its beak. Adult ganders weigh about 12 pounds. The adult goose weighs about 10 pounds. The Chinese is popular as an exhibition and ornamental breed. The White Chinese is also popular in some areas where it is used for weeding crops such as strawberries, asparagus, sugar beets, mint, and orchards.

The Pilgrim

The Pilgrim male is white, and the female is gray and white. The adult gander weighs about 14 pounds, and the adult goose weighs about 13 pounds.

The African

The African is gray with a brown shade. It has a knob on its beak and dewlap. The knob and bill are black. The head is light brown and the eyes are dark brown. The adult gander weighs about 20 pounds, and the adult goose weighs about 18 pounds.

SOURCES OF CHICKENS

Commercial hatcheries are responsible for almost all of the chicken production in the United States. Commercial hatcheries obtain hatching eggs from foundation breeders who raise parent flocks. Foundation breeders usually produce hatching eggs under some type of contract with the hatchery. Large commercial egg and meat producers buy chicks or started pullets from the hatchery.

Small farm flock owners may obtain chickens from (1) a hatchery as day-old chicks, (2) a grower who produces ready-to-lay pullets, or (3) as second-year layers from a commercial flock. It is possible, although not too practical, to hatch chicks on the farm.

Day-old chicks should be bought from a reputable hatchery that is U.S. Pullorum-Typhoid Clean. The chicks should be secured from a hatchery that is as close to home as possible. The less distance the chicks have to travel, the lower the loss.

Day-old chicks require brooding facilities, which increases the costs of raising them to production age.

Day-old chicks can be bought as straight-run or sexed chicks. **Straight-run chicks** cost less than sexed chicks. Straight-run chicks are about one-half **pullets** (female) and one-half **cockerels** (male). **Sexed chicks** are sorted into pullets and cockerels. Sex-linked characteristics may be used to sort chicks by sex at the hatchery. Sexing chicks by visual inspection is a specialized occupation requiring special training and practice. The decision as to which to buy depends on how the chickens are to be used. Straight-run chicks will provide some birds for meat. If birds are wanted only for egg production, then pullets should be purchased. If only meat production is desired, cockerels should be purchased.

Started chickens may be available from a grower who specializes in this type of operation. These chickens require less equipment and care than day-old chicks. Started pullets are generally sold at about 6 to 8 weeks of age. Ready-to-lay pullets are sold at about 16 to 20 weeks of age. Started chickens cost more than other chickens because of the feed, care, culling, and management that has already been invested in them.

Most commercial flocks of laying hens are replaced after 12 to 15 months. They go through a molting period and will then lay again. A small farm flock owner could get cheaper chickens with a bred-in potential for a high rate of lay from this source.

Farm flocks used for meat production are usually started from day-old chicks and should use meat-type breeds. Straight-run chicks are usually bought.

Orders for chicks from hatcheries should be placed several weeks before the birds are to be delivered. Chickens that are hatched in February and March will be in production when egg prices are higher.

CULLING CHICKENS

Culling is the process of removing undesirable chickens from the flock. Culling is generally not done in large commercial egg-laying flocks. These birds have been bred to lay and usually maintain high rates of lay.

Small farm flocks can be improved by a regular culling program. When the chickens arrive on the farm, remove any deformed, weak, or diseased chickens. After laying begins, the flock should be culled regularly to remove those birds that are not laying or are laying at a low rate. Heavy culling should not be necessary during the first 8 to 9 months of laying.

If the flock is to be kept for a second year of production, culling should take place. However, keeping hens for a second year of production is a questionable practice. Hens usually lay about 20 to 25 percent fewer eggs in the second year of production.

Three things are considered when culling birds for egg production: (1) present production, (2) past production, and (3) rate of production. Culling for these traits is based on the appearance and condition of the body. To judge, select, or cull chickens, a person must know the names of the parts of the chicken (Figure 35-11).

The hen should be examined properly when decisions are being made on culling. Most chickens are naturally nervous and must be handled gently. Rest the bird on the palm of one hand with the legs between the fingers. Use the other hand on the back and wings of the hen to steady it while you are picking it up. During examination, the hen should be held with its head toward the examiner. With the hen being held on the palm of one hand, use the other hand to examine the body. Keep the hen as quiet as possible during examination. Make sure the hen's feet are on the floor before releasing it after the examination.

Figure 35-11 Parts of a male chicken.

To determine whether or not the hen is laying, examine the condition of the comb, wattles, eyes, beak, pubic bones, abdomen, and vent. A laying hen has large, bright red, soft and waxy comb and wattles. The eyes are bright and prominent. The beak is bleaching or bleached. The eyelids and eye ring are bleached. The pubic bones are flexible and spread wide enough for two to four fingers to be placed between them. The abdomen is full, soft, and pliable. The vent is moist, bleached, and enlarged.

A nonlaying hen has a small, pale, scaly, and shrunken comb. The beak is yellow, and the eyes are dull and sunken. The eyelids and eye rings are also yellow. The pelvic bones are stiff and close together with room for less than two fingers between them. The abdomen is full and hard. The vent is dry, puckered, and yellow.

Past production is indicated by the amount of yellow pigment left in the body and the time of molt. **Molting** is the process of losing the feathers from the body and wings. Producing a large number of eggs bleaches the yellow pigment from the hen's body. The beak, eye ring, earlobes, and shanks are bleached white. The feathers are worn and soiled. Molting will occur in late September or October.

The pigment bleaches from the hen's body in a definite order as laying progresses. The pigment leaves the vent first. It becomes fully bleached after about 1 week of laying. The eye ring is next to bleach, requiring about 1 week to 10 days to fully bleach. The ear lobe is the third part to bleach and requires about 1 week to 10 days to become fully bleached. The beak bleaches next, starting at the base and progressing to the tip. This takes 4 to 6 weeks. The shanks are the last to bleach. The shanks bleach in this order: (1) front of shanks, (2) rear of shanks, (3) tops of toes, and (4) hock joint. It requires 4 to 6 months for the shanks to fully bleach.

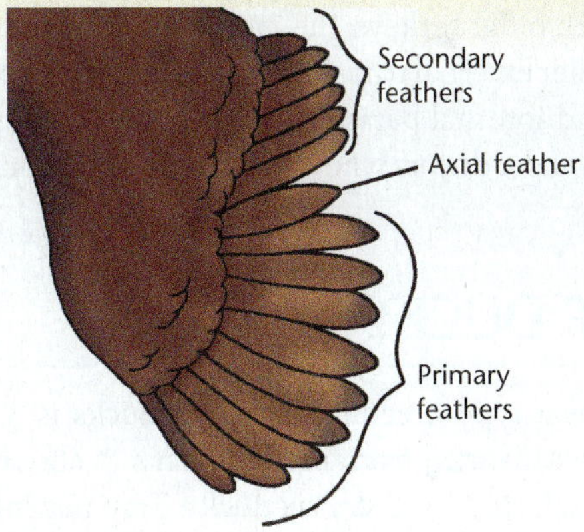

Figure 35-12 Location of primary, axial, and secondary feathers on the chicken wing.

When the hen stops laying, the pigment returns to the body parts in the same order in which it left. Return of the color takes about one-half of the time required by bleaching.

A hen that has been a poor layer will show yellow pigment in the body parts mentioned above. The molt will come before September. Molting is a normal process after the hen has been in production. Poor producers usually molt early and may take as many as 20 weeks to complete the molt. High-producing hens molt late and complete the molt rapidly, some as quickly as 6 to 8 weeks. The hen does not lay eggs during the molt.

Molting occurs in a definite order. The neck feathers fall out first. Next the feathers fall from the back, wings, and then body. There are 10 primary feathers on the outer part of the wing. These are separated from the secondary feathers on the inner part of the wing by the **axial feather**. Figure 35-12 shows the location of these feathers. The molt begins with the primary feather next to the axial feather. A slow-molting hen sheds one primary feather about every 2 weeks. A fast-molting hen may shed several primary feathers at the same time. It takes about 6 weeks for each new feather to become fully grown.

A high rate of lay is indicated by the shape and refinement of the head, the width and depth of the body, the abdominal capacity, the softness and pliability of the abdomen, the thinness of the pubic bones, the thinness and pliability of the skin, and the shape of the shanks.

A hen with a high rate of lay has a moderately deep and broad head. The face is free of wrinkles, and the comb and wattles are fine and smooth textured. The hen has a large body capacity, the abdomen is soft and pliable, and the pubic bones are flexible and thin. The skin is soft and pliable. The shanks are flat or wedge shaped.

A hen with a poor rate of lay has a long, shallow head and a back that is narrow and tapering. The body capacity is poor and the abdomen is hard. The pubic bones are narrow and thick, the skin is tight and thick, and the shanks are round and rough.

SOURCES OF TURKEYS

Turkeys should come from flocks that are tested for and found free of the Arizona and mycoplasma infections, fowl typhoid, paratyphoid, and pullorum disease. Turkey flocks can be started with hatching eggs, day-old poults, or started poults. If the flock is started with older turkeys, the birds should be examined for lice, northern fowl mites, and other external parasites. Isolate the flock for 3 weeks to make sure it is free of disease and internal parasites. Most turkey growers buy started poults 7 to 8 weeks of age from producers who specialize in the business of breeding flocks and hatching.

SOURCES OF DUCKS

The production of meat-type breeding stock for ducks is a specialized business. Potential breeders are separated from market flocks at about 6 to 7 weeks of age. One drake should be selected for every six ducks. Trap nesting and progeny testing can be used to select breeding stock that have high fertility, hatchability, and egg production. A **trap nest** is a nest equipped with a door that automatically closes when the bird enters the nest. The bird is held in the nest until released by the farmer. The purpose is to identify the egg laid by a particular bird. Body weight, conformation, and feathering are other important factors in the selection of market ducks.

Eggs must be carefully handled to maintain hatchability. Hatching eggs are stored at a temperature of 55°F and a relative humidity of 75 percent. Hatching eggs may be stored for 2 weeks. After the first week they must be turned daily. Eggs must be stored with the small end down.

CAUTION

Do not inhale the formaldehyde fumes used for fumigation. The room containing the incubator must be well ventilated.

In commercial flocks, artificial incubators are used to hatch the eggs. Muscovy eggs require 35 days of incubation. Other domestic duck breeds require 28 days in the incubator. Fumigation of the eggs in the incubator will prevent bacteria from entering the egg and killing the developing embryo.

Eggs in the incubator are candled after 7 or 8 days of incubation. Infertile eggs and those with dead embryos are removed. Candling is often repeated after the eggs have been in the incubator for 25 days (32 days for Muscovy eggs).

Ducklings must be protected from chilling after they hatch. Large commercial producers usually take the ducklings from the incubator as they hatch. Small producers may prefer to leave the ducklings in the incubator until the hatch is completed.

A person who is raising only a few ducks may use natural incubation. Muscovy ducks will set on the nest and hatch a brood with little or no trouble. Other breeds do not set eggs very well. Eggs from these breeds may be hatched by setting them under a broody hen. Keep feed and water close to the nest.

SOURCES OF GEESE

Geese may be secured as day-old goslings from a hatchery. They are incubated by artificial means in much the same manner as ducks. Geese, like ducks, can be hatched using natural incubation. Geese eggs require 28 days for incubation.

SUMMARY

Most of the poultry raised in the United States is produced in large commercial flocks. Chicken enterprises include egg production, meat production, and raising pullets for replacement. Most of the turkeys and geese are raised for meat. Ducks may be raised for game or meat. Much of the commercial poultry industry is vertically integrated.

The production of eggs in the United States has increased since 1970, while per capita consumption has varied over the years. The five leading states in numbers of laying hens and egg production are Iowa, Ohio, Pennsylvania, Indiana, and California. There has been a major increase in broiler production and per capita consumption of poultry since 1970. Broiler production tends to be concentrated in the southeastern states, with the five leading states being Georgia, Arkansas, Alabama, Mississippi, and North Carolina.

Since 1970, the overall trend in per capita consumption of turkey has been upward. Turkey production is generally not concentrated in any one part of the United States. The six leading states in turkey production are Minnesota, North Carolina, Arkansas, Missouri, Virginia, and Indiana.

Most of the ducks raised in the United States are raised in the Midwest. The leading state in duck production is Indiana. Geese are raised all over the United States, with most production in Texas, Minnesota, and South Dakota.

Most chickens used for egg production are Leghorn strain crosses. Crossbred chickens are generally used for broiler production. In both cases, the chickens are bred for the type of production desired.

Most of the turkeys raised in the United States are Broad Breasted Large Whites. The White Pekin duck is most commonly used for meat production. The most popular meat breeds of geese are the Toulouse and the Embden.

Most of the poultry is supplied by commercial hatcheries. Poultry may be purchased as day-old, started birds, or ready-to-lay chickens.

Hens are culled to remove unproductive birds from the flock. Culling is based on appearance and condition of the body. Pigmentation and molt are indicators of egg production.

Student Learning Activities

1. Prepare and give an oral report on the nature of the poultry industry.
2. Give an oral report on common breeds of poultry.
3. Prepare a bulletin board display of pictures and descriptions of common poultry breeds.
4. Survey the local community to determine the common breeds of poultry and the sources of poultry used by local producers.
5. Demonstrate the proper method of culling hens for indications of production.
6. Follow the approved practices described in this chapter when planning and conducting the selection of poultry for a supervised agricultural experience program.

Discussion Questions

1. What is meant by the term *vertical integration*? How important is it in the poultry industry?
2. Name the five leading states in the egg production.
3. Name the five leading states in broiler production.
4. Name the six leading states in turkey production.
5. Where in the United States is the largest number of ducks raised?
6. Name the leading states in geese production.
7. List the advantages and disadvantages of raising poultry.
8. Name the four classes of chickens generally raised in the United States.
9. What breeding methods are commonly used to produce most of the chickens used for egg and meat production in the United States?
10. What strain crosses are popular for commercial egg producing flocks? Why?

11. What common commercial cross is often used for meat-producing chickens? Why?
12. Name and describe the commercially important varieties of turkeys.
13. Name and describe the common breeds of geese.
14. Compare day-old chicks with started chicks as sources for the producer.
15. What three things are considered when culling chickens for egg production?
16. Describe the characteristics of a laying and nonlaying hen.
17. Name the factors a producer should consider when selecting turkey poults.
18. Name the factors a producer should consider when selecting ducks.

Review Questions

True/False

1. Most poultry raised in the United States is produced in large commercial operations.
2. Vertical integration, at this time, is not common in the poultry industry.
3. Despite the fluctuation in the total number of laying hens, annual egg production has increased since 1990.
4. About 60 percent of the ducks raised in the United States are raised on Long Island, New York.
5. A hen with a high rate of lay has a moderately deep and broad head.

Multiple Choice

6. About ______ geese are raised annually in the United States.
   a. 3,000
   b. 30,000
   c. 300,000
   d. 3,000,000
7. The process of removing undesirable chickens from the flock is called ______.
   a. culling
   b. separating
   c. sifting
   d. pulling

Completion

8. ______ is the process of losing the feathers from the body and wings.
9. ______ chicks are about one-half pullets and one-half cockerels.

Short Answer

10. Name the three general types of chicken enterprises.
11. What turkey breed is most commonly raised for meat?
12. Name the three best breeds of ducks for meat production.
13. What three factors are needed for success with poultry?
14. List the order in which molting occurs.

Chapter 36

Feeding, Management, Housing, and Equipment

Key Terms

ad libitum
grit
phase feeding
feeding efficiency
poult
brooder
hover guard
beak trimming
foot-candle
negative pressure system
positive pressure system
tunnel ventilation
dry cleaning
capon
desnooding
broodiness
spraddled legs

Objectives

After studying this chapter, the student should be able to

- describe accepted feeding practices for different kinds of poultry.
- describe approved management practices for different kinds of poultry.
- list housing and equipment required for various kinds of poultry enterprises.

FEEDING POULTRY

Chickens

The cost of feed is about two-thirds of the total cost of chicken egg and meat production. All management practices affect profits in chicken production. Feeding practices must meet the needs of the chickens and still allow room for profit.

Rations for the chickens must supply the protein, carbohydrates, minerals, vitamins, and water that poultry require. Most of the energy in poultry diets is supplied by grains, grain by-products, and animal and vegetable fats and oils. Poultry only have a limited ability to use high-fiber feeds such as roughages because of their relative inability to digest the fiber. Grains make up 50 to 80 percent of the total ration for chickens. Corn is the most commonly used grain in poultry rations. Other grains such as oats, wheat, grain sorghum (milo), and proso millet may be substituted for part of the corn in poultry rations.

When poultry are fed **ad libitum** (given all they will eat), they tend to eat enough to meet their energy requirements. Chickens fed low-energy diets will eat more feed than those fed high-energy diets. Therefore, the amount of required nutrients in a poultry ration must be adjusted in relation to the energy level in the ration in order to ensure that the birds consume the right amount of the needed nutrients. The concentration of nutrients must be increased in high-energy diets because the birds will eat less of the ration per day. The concentration of nutrients should be reduced in low-energy rations because the birds will eat more of these rations per day. High-energy rations usually result in higher efficiency in converting feed to meat or eggs as compared to low-energy rations.

Pelleted feeds contain more nutrients per volume of feed; this results in a higher intake of nutrients with the consumption of a given amount of feed as compared to nonpelleted feeds. Adding fat to the ration increases the amount of energy in the diet.

When substituting other grains for corn in poultry diets, care must be taken to ensure that the level of required nutrients in the ration is adjusted to compensate for the difference in energy level and feed intake. Corn has a higher energy level than other grains that can be used in poultry rations. Generally, other grains should be substituted for only part of the corn in the ration because of their lower energy level and higher fiber level. For example, oats or barley should be limited to no more than 10 to 15 percent of the ration.

When poultry feeds are prepared, corn should be coarsely ground. Oats, barley, and millet should be finely ground. Wheat and grain sorghum should be processed through a roller mill. If wheat is ground, care must be taken not to grind it too fine.

Fats may be used to increase the energy level of low-energy rations. Animal and vegetable fats should be limited to no more than 5 to 10 percent of the diet. Fats will increase the palatability of the diet, decrease dustiness, and improve the texture of the feed. Fats are used more often in broiler rations to increase the energy level of the diet. In hot weather, feed containing added fat may become rancid unless it has been properly stabilized.

Proteins supply the essential amino acids needed by chickens. The amino acids most commonly deficient in chicken rations are arginine, glycine, lysine, methionine, and tryptophan. Common sources of protein for chickens include soybean meal, canola meal, and animal by-products (blood meal and fish meal). Balancing the ration on the amino acid needs rather than crude protein content ensures that the nutritional needs of the chickens are met and may also result in a lower-cost ration.

From 55 to 75 percent of the egg and the chicken's body weight is water. Younger chickens have the highest water content. Chickens need a fresh, clean supply of water at all times.

The daily water consumption of chickens varies with the type of chicken, the temperature, type of diet fed, rate of growth, egg production, and type of watering equipment used. Table 36-1 shows the daily water consumption of chickens at moderate temperatures (68 to 77°F). Water consumption will increase at higher

TABLE 36-1 Daily Water Consumption of Chickens at Various Ages per 1,000 head

	Broilers	White Leghorn Hens	Brown Hens
Age In Weeks	**Gallon/1,000 Head/Day**	**Gallons/1,000 Head/Day**	**Gallons/1,000 Head/Day**
1	8.5	7.5	7.5
4	37.7	18.9	26.4
8	75.5	30.2	34.0
12		37.7	41.5
16		45.3	45.3
20		60.4	56.6

Source: National Academy of Sciences

temperatures. On very hot days, chickens may consume as much as three times the water used on cooler days (Figure 36-1).

The mineral requirements of chickens include calcium, phosphorus, magnesium, manganese, iron, copper, zinc, and cobalt. Calcium can be supplied by the addition of oyster shell or limestone. Current recommendations for calcium are that it makes up 3.4 percent of the ration for laying hens. It is especially important for laying hens because the formation of the shell of the egg requires a great deal of calcium. Commercial mineral mixes can be used to supply the other minerals needed in the ration.

Vitamins required by chickens include A, D3, E, K, thiamine, riboflavin, B12, niacin, pantothenic acid, folic acid, biotin, and choline. Grains, protein sources, and green feeds will supply some of the vitamin needs. However, it is common practice to use a vitamin premix when preparing chicken rations. This ensures that the necessary vitamins are present in the ration.

Figure 36-1 Chicks drinking from an automated watering system.

The use of additives in poultry rations is discussed in Chapter 7. Proper management practices must be followed whenever feed additives are used in the ration.

Some substances not readily identified are found in dried whey, marine and packing house by-products, distillers' solubles, green forages, soybeans, corn, and other natural materials. These substances have been found to be beneficial to chickens. These factors stimulate growth, increase reproduction, improve egg quality, or reduce the toxicity of some minerals. Many poultry rations for young birds and breeders contain some of these feeds to add these factors to the diet.

Chickens and other poultry do not have teeth. They need **grit** in the ration to help grind their food. Most grinding of the feed takes place in the gizzard, a thick-walled, muscular part of the stomach of birds. Grit is usually small particles of granite and comes in small, medium, and large sizes for use with chicks, growing chickens, and adult chickens, respectively. Limestone and oyster and clam shells may also be used as a source of grit in diets. The more the feed is ground, the greater available surface area there is for digestion and, therefore, the greater absorption of nutrients. Grit is often used when hard, coarse, or fibrous feeds are fed in order to increase the available surface area for grinding. If mash or finely ground feeds are used, the value of grit is lowered. Because modern poultry production diets are composed of ground ingredients, grit is not required.

The different kinds of chicken feeds are all available as commercially manufactured feeds. Commercial feeds are available in the form of mash, pellets, or crumbles. Chickens may produce slightly better with the pelleted or crumble forms; however, they tend to cost more than mash. Pellet and crumble feeds reduce feed waste, and the chickens may eat them a little better. The manufacturer's directions for use must be followed closely. Always follow feed tag directions closely when using feeds containing additives.

There are four systems that may be used for preparing rations:

- using a complete commercial feed
- using a commercial protein concentrate and mix with local or homegrown grains
- using a commercial vitamin-mineral premix and soybean meal and mix with local or homegrown grain
- buying all the individual ingredients and mix the ration

The choice of system depends on the age of the chickens being fed, the relative costs, the size of the enterprise, and the equipment available for grinding and mixing feeds. Farmers with small flocks generally find it more economical to use a complete commercial feed. Large commercial poultry enterprises generally use some system of mixing either at a local mill or on the farm. For large operations, mixing the feeds usually results in a lower feed cost per ton.

Phase feeding is a system of making specific feeds to be used to meet the changing nutritional requirements of chickens. A number of factors—such as rate of lay, egg size, body maintenance needs, and temperature—are used to determine the feed content. The purpose is to lower feed cost. There are some problems in the use of phase feeding. The procedure for taking into account all of the factors when making up a ration is a complicated one. When more than one age group of chickens is being fed, more feed storage bins and feeders are needed. It requires a high level of management ability on the part of the producer. Phase feeding is more practical for large commercial operations than it is for small farm flocks.

Chickens should be fed only fresh feed. Overfilling feeders results in wasted feed. Hanging feeders should be filled only three-fourths full. Trough feeders should be filled only two-thirds full. Fill feeders in the early morning and refill during the day as needed. Keep the feed clean and never supply moldy or dirty feed. Feeders should be cleaned periodically.

Figure 36-2 Feed bins used for poultry feed storage.

Feed should be stored in a dry place where rats or mice cannot get to it. Feed should be handled and stored in bulk for large operations (Figure 36-2).

A good manager watches the feed consumption of the flock closely. A drop in feed intake is often the first sign of trouble. Stress, disease outbreak, molting, or other management problems are often first indicated by a drop in feed consumption.

Feed intake is influenced by the energy level of the ration, level of protein intake, environmental temperature, body size, amount of feather coverage, and rate of growth or egg production. Feed intake is lower at higher environmental temperatures and higher at lower environmental temperatures. Feed intake will change about 1.5 percent for each 1.8°F above or below 68 to 70°F. While temperature is the major factor that affects the level of feed intake, a low intake of protein may cause a higher total feed intake.

Feeding for Egg Production

Chicks must be given feed and water as soon as they are put into the brooder house. Starter rations should contain 18 percent protein. Complete starter feeds should be used for chicks and they should be fed without grain until they are 6 weeks old. About 40 pounds of starter mash per 100 chicks are required for the first 2 weeks. From 2 to 6 weeks, 250 to 300 pounds of starter mash are needed per 100 chicks. Chick-size grit should be provided when the birds are young and a medium size given as they become older.

For chicks up to 2 weeks of age, 100 linear inches of feeder space are required for 100 chicks. Two hundred linear inches of feeder space are needed per 100 chicks from 3 to 6 weeks of age. Up to 2 weeks of age, 25 linear inches of waterer space are needed per 100 chicks. From 3 to 6 weeks of age, 50 linear inches of waterer space are required per 100 chicks.

For chicks at 6 weeks of age, the feed should be changed to a growing ration. From 6 to 14 weeks, the ration should contain 15 percent protein. The ration should contain 12 percent protein from 14 to 20 weeks. Grain may be fed from 6 weeks on. Grain and mash should be fed in separate feeders. Begin with 10 pounds of grain to 100 pounds of mash. Gradually increase the amount of grain until the ration is one-half grain and one-half mash. For chicks at 18 to 20 weeks of age, change the ration to a laying mash. The change should be made over a 2-week period of time.

From 7 to 12 weeks of age, 250 linear inches of feeder space and 50 linear inches of waterer space are needed per 100 birds. From 13 to 20 weeks, 300 linear inches of feeder space and 100 linear inches of waterer space are needed per 100 birds.

A laying ration should contain 14.5 percent protein. The cost per dozen eggs produced should be used in deciding which is the most economical ration to use. Egg production, body size, health, and temperature all affect the amount of feed a

Figure 36-3 The make-up of a chicken's feed ration depends on whether it is used as a broiler (left) or as a layer (right).

laying hen eats (Figure 36-3). Light breeds eat about 24 pounds per 100 hens per day. A hen of the light-weight breeds eats a total of 85 to 90 pounds of feed per year. A heavy-breed hen eats 95 to 115 pounds per year. About 2 to 5 pounds of oyster shell and 1 pound of grit per year per hen are required.

Light-weight breeds require 300 linear inches of feeder space and 50 linear inches of waterer space. Heavy-weight breeds need 400 linear inches of feeder space and 100 linear inches of waterer space per 100 birds.

The ration for the laying flock should not be changed suddenly. This creates stress in the hens that will result in a drop in egg production. Even changing from a coarse-ground feed to a fine-ground feed may create stress and a drop in production.

Three systems of feeding are commonly used for laying hens: (1) all mash, (2) mash and grain, and (3) cafeteria.

The all-mash system is the use of a complete ground feed. This system is well adapted to use with mechanical feeding systems. It is often used by commercial egg producers.

In the mash and grain system, grain and mash are fed separately. A 20 to 26 percent protein supplement is fed with a light grain feeding in the morning. Corn is fed in the evening. Some of the grain may be fed in the litter. This causes the hens to stir up the litter and helps keep it drier.

The cafeteria system allows the birds to balance their own rations. Grain is fed in one feeder, and a 26 to 32 percent protein supplement is fed in another feeder. Feed is kept in the feeders at all times. Older hens may tend to eat too much grain and not enough protein supplement when this system is used. About one-fourth of the feeders should contain a protein supplement and three-fourths of the feeders should contain grain.

Feeding efficiency is the number of pounds of feed required to produce a dozen eggs. Records must be kept of the amount of feed used and the egg production in order to calculate feed efficiency. Divide the total pounds of feed fed by the number of dozen eggs produced to calculate how many pounds of feed it took to produce one dozen eggs. Feed efficiency should be about 4.3 pounds of feed per dozen eggs. Wasted feed, low rate of lay, health problems, or other management problems lower feed efficiency.

Feeding Broilers

For chicks from day-old to 6 weeks, use a broiler starter ration with at least a 23 percent protein level. Replacement chick starters, designed for egg production-type birds, should not be used for broiler starter rations. Broiler starter rations should contain 3 percent added fat and a coccidiostat. The protein level is adjusted to the energy level of the ration. Use chick-sized grit scattered on top of the mash or feed in separate hoppers. At 3 to 6 weeks of age, broilers should be fed a ration containing 20 percent protein.

Broilers need 100 linear inches of feeder space and 25 linear inches of waterer space per 100 birds up to 2 weeks of age. Provide 300 linear inches of feeder space and 50 linear inches of waterer space for birds from 3 to 6 weeks of age.

Feed a finishing ration of 18 percent protein to birds from 6 weeks old to market age (usually about 8 to 9 weeks of age). The finishing ration should contain 3 percent or more of added fat. Broiler rations are fed as complete mixed feeds and never as separate grain and protein supplement, as rations for laying hens might be.

Provide 350 linear inches of feeder space and 75 linear inches of waterer space per 100 birds from 6 weeks to market age. Broilers are full fed at all times to obtain the fastest possible gains.

Capons and roasters are fed to heavier weights than broilers. They are fed the same ration as broilers up to 6 weeks of age. After 6 weeks, grain is added to the ration in addition to the finishing mash until it is being fed as one-half of the ration.

Vitamins and trace minerals are included in rations in small amounts. Broiler rations must be mixed properly and the quality carefully controlled. This is the responsibility of the feed mill where the mixing is done. Inventory control, assaying of mixed feeds (analysis for content), and assaying of ingredients are methods of ensuring proper mixing and quality control.

Feed conversion in broilers refers to the amount of feed needed to produce 1 pound of live weight. It is found by dividing the total weight of feed consumed by the total weight of broilers marketed. Feed conversion after condemnations is found by subtracting the weight of the condemned meat from the weight of the birds marketed before dividing into the total weight of the feed used.

Feed conversion is affected by a number of factors, including:

Feed conversion is calculated by dividing the weight of feed consumed by the weight of broilers marketed.

- genetic background of the strain being fed
- type of feed used
- temperature
- amount of feed wasted
- additives used in the feed
- general management of the operation

Other factors also affect feed conversion. Current feed conversion in the broiler industry averages about 1.85 pounds of feed per 1 pound of gain for males and 1.95 pounds of feed per 1 pound of gain for females.

Feeding the Breeding Flock

A special mash is made for use with breeder flocks. It contains 14.5 percent protein and is fortified with extra vitamin A, D, B_{12}, riboflavin, pantothenic acid, niacin, and manganese. This mash is more expensive than laying-hen mash. The breeding flock should be started on breeder mash about 1 month before hatching eggs are to be kept. Breeding flocks may be fed a ration that is one-half mash and one-half grain. The breeding flock is fed in much the same way as is the laying flock.

Replacement pullets for breeding flocks of meat-type chickens tend to grow too fast and develop sexual maturity too early. This results in an increased number of small eggs that are not good for hatching purposes. Egg production tends to be at a low level during what should be its peak.

Several methods may be used to slow down the rate of growth and sexual maturity. These include limiting feed intake, skip-a-day feeding, low-protein diets, and low-lysine diets.

A limited-feed intake program is based on feeding at the rate of about 70 percent of the normal feed level. This program is begun when the birds are from 7 to 9 weeks of age and is continued until they are 23 weeks old. They are then full fed on a laying mesh.

A skip-a-day program involves full feeding every other day. On alternate days, only 2 pounds of grain are fed per 100 birds. This feeding program is followed from the 7th or 9th week to the 23rd week.

A low-protein diet is one in which the protein level of the ration is 10 percent. This program is also followed from the 7th or 9th week to the 23rd week.

A low-lysine diet will also slow down rate of growth and sexual maturity because of the amino acid imbalance. The diet contains 0.4 to 0.45 percent lysine and 0.6 to 0.7 percent arginine. It is followed from the 7th or 9th week until the 23rd week. The protein level of this diet is 12.5 to 13 percent.

Turkeys

The general principles of feeding turkeys are similar to those for feeding chickens. Major differences are in the protein levels required and the importance of the vitamins biotin and pyridoxine in the turkey diet.

Feeding Turkey Poults

Poults (young turkeys) must be fed and watered as soon as possible after hatching. When feeding and watering is delayed beyond 36 hours, the poults have difficulty learning to eat. It may be necessary to force feed them to get them started. Start poults on a turkey prestarter or starter ration. The protein content should be 28 percent. Prestarters are generally used only for the first week. They contain higher levels of antibiotics, vitamins, amino acids, and energy than do starter feeds. Prestarters are especially useful if the poults are under unusual stress conditions. From 4 to 8 weeks of age, the starter should contain 26 percent protein and a higher energy level.

Feeding Growing Turkeys

Turkeys may be moved to range or fed in confinement (Figure 36-4) at 8 to 10 weeks of age. Confinement feeding results in faster gains. However, range feeding may result in as much as 10 percent feed savings. A good range contains forage or grain crops. The stocking rate for range varies from 100 to 250 turkeys per acre.

Growing turkeys should be separated by sex because toms have a higher protein requirement than hens. Turkeys generally give better feed conversions when fed complete mixed rations. Pelleting the ration gives the best results. As the turkeys become older, the energy levels of the ration are increased and the protein level is decreased. The tables in the Appendix show the amounts needed at various age levels.

The daily water consumption of turkeys is shown in Table 36-2. As with chickens, water consumption will increase in hot weather. For male turkeys at 8 to 12 weeks

Figure 36-4 Turkey poults raised in confinement.

TABLE 36-2 Water Consumption of Turkeys at Moderate Temperatures (68–77°F)

	Large White Turkeys	
	Males	Females
Age (weeks)	Gallons/1,000 Head/Day	Gallons/1,000 Head/Day
1	14.5	14.5
4	62.3	48.1
8	151.7	120.0
12	234.7	175.9
16	261.2	178.9

of age and females at 8 to 11 weeks of age, the ration should have a protein content of 22 percent. Rations for male turkeys at 12 to 16 weeks of age and female turkeys at 11 to 14 weeks of age should have a protein content of 19 percent. Rations for male turkeys at 16 to 20 weeks of age and female turkeys at 14 to 17 weeks of age should have a protein content of 16.5 percent. Rations for male turkeys at 20 to 24 weeks of age and female turkeys at 17 to 20 weeks of age should have a protein content of 14 percent. Male turkeys are usually marketed at about 24 weeks of age and female turkeys at 20 weeks of age.

Feeding Breeding Turkeys

Turkeys should be selected at about 16 weeks of age for the breeding flock. A holding diet containing 12 percent protein and of medium energy level is usually fed to female breeding turkeys from 16 to 30 weeks of age and males from 16 to 26 weeks of age. A breeding ration containing 14 percent protein is then fed. Males in the breeding flock are fed a breeding ration beginning at about 26 weeks of age. To control the weight of males in the breeding flock, they may be fed a limited diet. Females are normally fed all they will eat.

Feeding Ducks and Geese

Commercial feeds in mash, pelleted, or crumble form are available for ducks and geese. Pelleted or crumble forms are recommended. If a commercial feed for ducks or geese is not available, a chicken feed may be used. However, be sure the chicken feed does not contain a coccidiostat if it is used for ducks or geese because coccidiostats are toxic to these birds.

A starter ration containing 22 percent protein is used for the first 2 weeks. Feed all that the birds will eat. Geese should be allowed out on pasture during the day where they can eat grass and bugs. During the first several days, the feed should be placed on rough paper, paper plates, or egg fill flats. Do not place the feed on a smooth surface that could cause the young birds to slip and fall as this may cause leg injuries. Supply an insoluble grit in addition to the feed.

Geese may be allowed to forage for feed after they are 2 weeks old. Under this feeding program, they are fed grain for the last 2 or 3 weeks and marketed at about 18 weeks of age. Geese may also be fed a grower ration while they are allowed to forage for feed. Under this feeding program they may be switched to a high-energy finishing diet and marketed at about 14 weeks of age. Another feeding method is to feed the geese in full confinement and market them at about 10 weeks of age. Geese marketed under this feeding program are called "junior" or "green geese."

Ducks may be raised either in total confinement or in a growing house; this should open onto an exercise area that includes water for swimming or wading. Pelleted feeds are better than mash feeds for ducks (Figure 36-5). They are usually put on a starter diet containing 22 percent protein for 2 weeks. Then, they are fed a grower diet containing 18 percent protein and finished on a diet containing 16 percent protein. Ducks are usually ready for market in 7 to 8 weeks.

Figure 36-5 Feed can be provided to ducks in self-feeders.

Provide a supply of fresh water that is easily accessible at all times. Ducks and geese consume large amounts of water. The waterers should be designed in such a way that the birds cannot get into them to swim. Water for swimming is not necessary.

Pasture is especially important for geese. They will start to eat grass when they are only a few days old. Geese can live entirely on pasture after they are 5 to 6 weeks of age. However, a growing ration is recommended. Timothy, bluegrass, ladino, white clover, and bromegrass are good pastures. Barley, wheat, and rye make good fall pastures. Geese do not eat alfalfa, sweet clover, or lespedeza.

If the pasture is not of good quality, some additional grain such as cracked corn, wheat, or milo should be fed. Pastures should be rotated. Young goslings should be protected from rain for the first several weeks. Provide shade in hot weather. One acre of pasture will feed about 20 birds. Ducks will eat some green feed, but are not as good at foraging as geese. Pasture is not necessary for ducks. However, farm flocks are usually not confined.

A breeder-developer ration is fed to birds being kept as breeders. These birds are given a breeder diet when they are older. Breeder diets contain less energy than grower diets. The breeder diet should be used starting about 1 month before eggs are to be kept for hatching. Breeder flocks require more calcium, which can be provided by feeding oyster shell.

MANAGING POULTRY

Chickens

Brooding Chickens

Small farm flocks are usually brooded on the floor using hover-type **brooders** (Figure 36-6) or infrared heat lamps. Large commercial flocks are usually brooded in battery brooder units or in wire cages in houses with controlled heat and ventilation. The brooder house must be prepared for the chicks. Be sure that it is in good repair. Clean and disinfect the house before the chicks arrive. Use clean litter for brooding chicks. Commercial litters such as peat moss or sugarcane pulp may be used. Other materials such as ground corncobs, chopped straw, wood shavings, or sawdust may be used. Battery brooders and wire cages do not use litter.

Photo by Joe Valbuena, USDA

Figure 36-6 Broiler chicks being brooded under hover-type brooders in a commercial broiler operation.

Three to 4 inches of litter is placed on the floor. A deep litter system is commonly used. This means that fresh litter is added as needed. It may reach a depth of 8 to 10 inches. Stir the litter to keep it from packing.

Brooders should be put into operation 2 days before the chicks arrive. This gives the producer a chance to check the equipment and make sure it is working properly. The temperature under the hover must be 90 to 95°F for day-old chicks. Brooder temperatures are measured about 3 inches off the floor and about 3 inches inside the outer edge of the hover. Brooder temperatures are reduced about 5°F each week until the temperature reaches 70 to 75°F. Brooder temperatures must be checked at least twice a day.

The behavior of the chicks gives some indication of proper brooder temperature. If the chicks crowd close together and cheep, they are too cold. If they move out from under the brooder, pant, and hold their wings away from their bodies, they are too hot.

A **hover guard** (chick guard or brooder guard) is used for the first week of brooding to prevent the chicks from wandering away from the heat and becoming chilled. This is a corrugated cardboard or cloth-covered wire about 12 inches high. It is placed about 2 feet from the hover edge and all the way around it after the chicks are placed underneath. Move it back a short distance each day. Use a wire guard if the chicks are beak trimmed. When using infrared lamps for brooding, make a circle with the chick guard about 8 feet in diameter around the heated area to keep the chicks confined to the area under the lamp and prevent chilling. With any type of brooder, remove the chick guard after 1 to 2 weeks. Blocking the corners of the brooder house with cardboard will prevent the chicks from crowding into the corners and smothering. Chicks may crowd into the corners when they are frightened or if they are cold and have not yet learned to return to the heated area. Chicks need heat until they are well feathered. This will be at 4 to 8 weeks of age, depending on the season.

Seven to 10 square inches of space are needed under the hover for each chick. More space is required per chick with electric brooders than with other types. The house should provide a total of 0.75 to 1.0 square feet of floor space for broiler chicks. Leghorn-type pullet chicks require 1.5 to 2 square feet of floor space. Heavier breeds need 2.5 to 3 square feet of floor space.

Roosts are not used for broiler chickens. Using roosts for broiler chicks causes crooked breast bones and breast blisters. Laying hens may use roosts. If they do, then roosts should be provided for the young chickens when they are about 4 to 6 weeks of age. About 4 to 6 inches of roost space are needed per bird.

A small amount of light should be provided under the brooder. Gas brooders often provide enough light and no additional light is needed. A 7.5- to 10-watt bulb should be used under an electric brooder. Room lights are often used for the first 2 or 3 days. After that, no room light is needed at night. Commercial broiler producers often use lights 24 hours a day for 4 to 6 weeks. Sometimes they are used until the broilers reach market weight.

A brooder house must provide fresh air. Be sure that there are no drafts on the floor. If the litter is wet and there is a strong odor, the ventilation needs to be improved. This will help prevent respiratory diseases.

Small, trough-type feeders may be used when the chicks are young. As they become older, the feeders should be larger. Mechanical and hanging feeders may be used. Round or trough-type waterers may be used with young chicks, but automatic waterers may be used when the chicks get older.

Heat lamps (infrared type) may be used for brooding small numbers of chicks. Lamps of this type will brood from 50 to 75 chicks. A chick guard must be installed to prevent drafts and confine the chicks to the heated area. Hang the lamps about 18 inches from the litter. The lamps are raised as the chicks grow.

Large commercial operations often use battery or wire cage brooding. Battery brooders are heated with electric heaters. The entire house is heated when wire cages are used. Houses must be well insulated and ventilated.

Day-old chicks in battery brooders require 10 square inches of space per chick. Space should be increased by 10 to 15 square inches every 2 weeks. Using plastic or plastic-coated wire floors reduces breast blisters in broilers.

Wire cages are commonly used when brooding commercial-size laying flocks. Leghorn-type chickens need 20 to 30 square inches of floor space for the first 7 to 8 weeks. They require 45 to 55 square inches from 8 to 18 weeks. Heavier breeds require about 25 percent more floor space per chicken.

Figure 36-7 Cannibalism in poultry involves the pecking and consumption of skin, tissues, or organs of flock mates. Good management and selection of genetic stock that is not prone to cannibalism are the best ways to prevent cannibalism. However, beak trimming may also be necessary.

Beak trimming (Figure 36-7) is done to control cannibalism (picking). The causes of cannibalism are not well understood, but overcrowding, nutritional deficiencies or changes in rations, vent protrusions, and environmental factors are found to be common contributors to its occurrence. Beak trimming is the cutting off of one-third to one-half of the upper beak and one-fourth of the lower beak and can be done at any age. However, if done too young, it may not be permanent. Beak trimming at 9 to 12 weeks of age is usually permanent. Birds may be ordered from the hatchery already with trimmed beaks. Beak trimming does not affect the health or growth of the chickens.

Beak trimming is recommended in layers and was once practiced on broilers as well. However, many broilers are no longer beak trimmed for a number of reasons. Broilers produced commercially reach market weight at about 6 weeks of age, reducing time for cannibalism to become a problem. Also, broiler houses are kept in low light conditions, reducing aggressive behavior and activity of the birds. Methods used to minimize the problem of cannibalism in layers include beak trimming and the selection of genetic stock less prone to this behavior.

Raising Laying Pullets

Replacement pullets for laying flocks are raised in open housing confinement (Figure 36-8).

Pullets may be raised in the same house that is used for brooding, or they may be moved to a growing house. Pullets should be culled when they are moved to the growing house. They may be grown in cages, using one of two systems: partial cage growing or complete cage growing. Partial cage growing is floor brooding for the first 6 to 10 weeks and then moving the pullets to cages. Complete cage growing is brooding and growing entirely in cages.

Light affects the age of sexual maturity and the rate of egg production. The length of the day is the controlling factor. Thus, increasing day length speeds up sexual maturity, and decreasing day length slows down sexual maturity. If pullets reach sexual maturity too early, they will lay small eggs for several months.

Several systems of controlling light during the growing period are used. The basic principle involved is not to grow pullets between 12 and 22 weeks of age under conditions of increasing light. Dark, windowless houses are used in some systems to permit total control of light. Some systems reduce the light gradually during the growing period. Others maintain a constant day length during the growing period. Day length is increased when the pullets are moved into the laying house.

Less feed is required for growing pullets on range. However, more labor is required, and there are greater problems with disease and predators. Clean range must be used and rotated. One acre can carry from 300 to 500 birds, but local agricultural zoning laws vary. Birds of different ages should not be mixed on range. Range shelters and shade are needed. Range growing of pullets is better adapted to small flocks than large commercial operations.

Courtesy of Chore-Time Poultry Production Systems, Milford, Indiana, USA

Figure 36-8 Pullets being raised for layer production in an open house. They will be moved to battery cages in a layer house when they reach the proper age.

Laying Flock Management

Laying flocks may be housed in open-floor systems or caged systems. Smaller flocks often use open-floor systems. The trend in large commercial flocks is toward using some system of caged layers (Figure 36-9).

Courtesy of Chore-Time Poultry Production Systems, Milford, Indiana, USA

Figure 36-9 Eggs roll to the front of the layer cages and are moved on an automated system to the packing house.

Laying houses must be cleaned and disinfected before pullets are moved in. The laying house should remain empty for at least 1 week after cleaning before moving the pullets in. The house and equipment should be in good repair and new litter put in the house.

Handle pullets gently when moving them to the laying house. They are easily frightened. Moving them in the dark will excite them less. Watch the pullets for a few days to make sure they do not pile together and suffocate. Nests should be cleaned regularly, and new nesting material should be added when needed. Keep waterers and feeders clean. Keep the litter stirred, and add new litter when needed. Do not allow the litter to become wet and caked over.

Several types of cages are used for cage laying systems, including single cages, multiple-bird cages, and colony cages.

Single cages allow room for one bird. Single cages may be arranged in single rows, back-to-back, stair-step, or double-deck. Single row is two rows of cages separated by an aisle. This system is generally used in narrow houses. The back-to-back system uses two rows of single cages placed back-to-back. This permits a sharing of common water and feed troughs. The capacity of the house is increased by the use of this system. A stair-step system is created by offsetting one tier of single cages below another (Figure 36-10). Manure drops to the floor rather than on dropping boards. Double-deck systems place one row of cages directly on top

Courtesy of Chore-Time Poultry Production Systems, Milford, Indiana, USA

Figure 36-10 The cages used to house these laying hens are arranged in a stair-step system.

of another. Dropping boards are used to catch the manure. Problems with this system include the difficulty of removing the manure and of providing enough light to the chickens in the lower cages.

Most multiple-bird cages house 2 to 10 birds per cage. Cost per bird is lower than with single cages. Cage arrangements are the same as with single-bird cages. The stair-step arrangement is considered to be the most profitable.

Colony cages contain 11 to 60 birds. Advantages of colony cages include lower housing and labor costs. Problems include higher death loss, lower production, and more dirty and broken eggs. As more hens are placed in the colony cage, the production level decreases.

Feeding in cages may be done by hand or automatically. Watering may be done by the use of continuous flow troughs, drip nipples, or automatic cups.

Laying hens require a minimum of 14 hours of light per day. Many controlled lighting systems use a maximum of 17 hours of light per day. Artificial lighting must be used as the length of available natural light decreases with the season. Several types of lighting systems are used. These include all-artificial lights, morning light, evening light, a combination of morning and evening light, and all-night lighting.

Light is increased to 12 hours at 20 weeks of age. It is then increased gradually (15 to 20 minutes per week) until the maximum desired light is reached. Lights are generally controlled by automatic time clocks.

All-artificial lighting is used in windowless houses. Morning lighting systems turn on lights 2 to 5 hours before sunrise and turn off at daybreak. Evening lighting systems turn on in the evening. Dimmers are used to prevent a sudden change from light to dark. Morning–evening systems use lights at each end of the day. All-night lighting is generally used only for second-year laying flocks. Low-wattage bulbs are used in this system.

Regardless of the system used, lights should provide 1 foot-candle of light per bird. A foot-candle is a unit of illuminance on a surface that is everywhere one foot from a point source of a candle or light. Keep the bulbs clean to maintain efficiency of light output. The general principle involved in providing light for laying hens is never to decrease the amount of light while the flock is in production.

Reducing Heat Stress

Extreme heat is a problem that may affect the growth or egg-laying ability of poultry. Heat stress can also increase the mortality rate in a flock. Birds do not have sweat glands. They depend on evaporation, mainly by panting, to control body temperature. Excessive panting increases the pH of the blood, which reduces the amount of calcium and bicarbonate in the blood. This results in thin-shelled eggs that are more likely to break when handled. Heat stress also reduces appetite, which reduces rate of gain, lowering feed efficiency.

Buildings need to be designed and managed so that a temperature that is as comfortable as possible for poultry is maintained. Poultry are generally comfortable in a range of 55 to 75°F. Good ventilation is necessary to help keep the interior temperature of the poultry house in the comfort zone for the birds. Circulation fans, fogging nozzles, and evaporative cooling pads (Figure 36-11) are often used to help control the temperature in the poultry house.

Some systems combine natural air flow with circulating fans, while other systems depend entirely on circulating fans for air movement. If the fans exhaust the air from the building, it is called a negative pressure system. If the fans bring fresh air into the building, it is called a positive pressure system. **Tunnel ventilation** uses air inlets at one end of the building and exhaust fans at the other end to move air

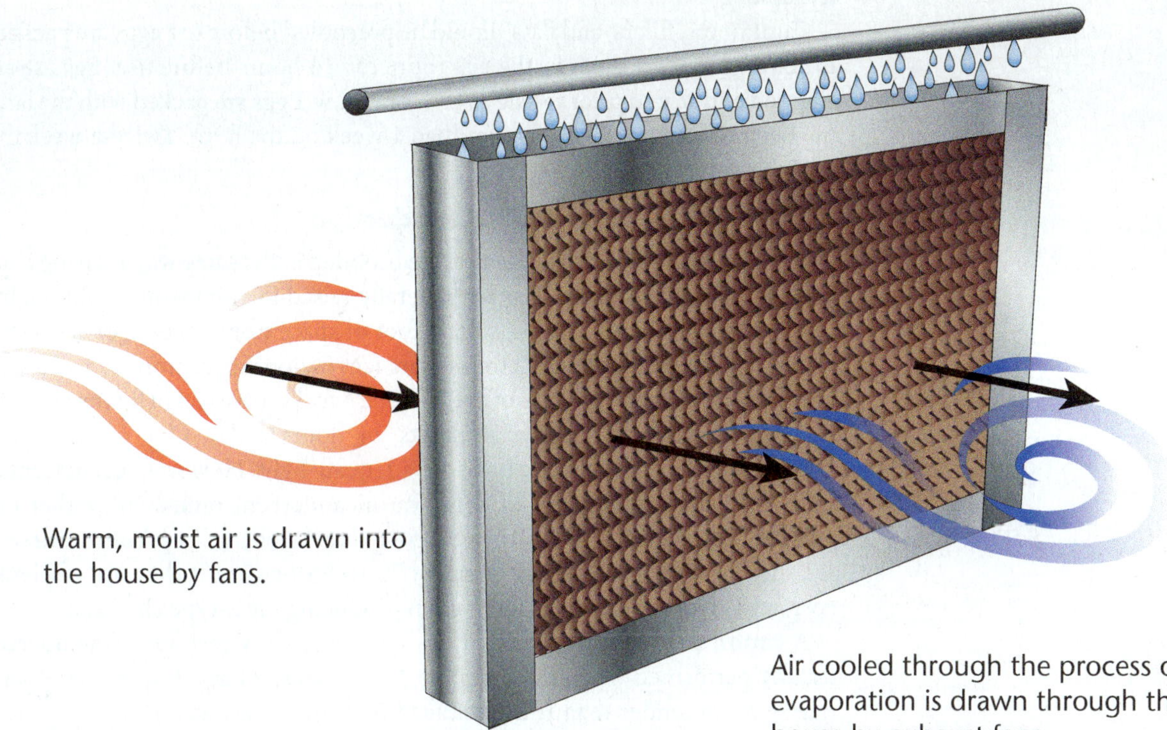

Figure 36-11 Evaporative cooling systems are often used to help control the temperature in poultry houses.

through the building at a minimum velocity of about 350 feet per minute. This system does not use a pressure difference between the inside and the outside of the building to move the air. The use of fogging nozzles and evaporative cooling pads are additional options that may be used with circulating fans and tunnel ventilation. Fogging nozzles and evaporative cooling pads are both methods that decrease temperature through water evaporation.

Care and Handling of Eggs

Eggs lose quality rapidly if they are not handled carefully. Quality is maintained by keeping eggs cool and humid. All equipment used in handling eggs must be kept clean. Open plastic or rubber-coated wire baskets will let the eggs cool more rapidly than closed containers. Egg room temperature should be kept at 55 to 60°F. The humidity should be between 75 and 80 percent. Eggs absorb odors from the environment where they are stored. Never store eggs near anything with a strong odor, such as onions, paint, or kerosene.

In floor systems, eggs should be gathered at least three times a day. Caged-layer systems may use some type of automatic egg gathering system such as a conveyor belt. Eggs may also be gathered using mechanized carts.

Most of the eggs will be clean if good management practices are followed. If a high number of eggs are dirty, management practices should be reviewed. Dirty eggs must be cleaned. The method of cleaning is often specified by the egg dealer. Eggs may be dry cleaned or washed. **Dry cleaning** is done by buffing the eggs with fine sandpaper, emery cloth, or steel wool. However, this may increase the number of cracked eggs and requires a great deal of labor.

Egg-washing machines are available and provide the fastest way to clean eggs. Only dirty eggs should be washed. Always follow the equipment manufacturer's directions for the use of the egg washer. Eggs that are washed incorrectly lose quality rapidly.

Containers, fillers, and flats should be precooled before the eggs are packed. This is done by storing them in the egg room for 24 hours before use. Eggs should be shipped and stored prior to sale at 45°F or below. Eggs are packed with the large end up. Eggs should not be held more than 1 week on the farm. Twice-a-week marketing is preferred.

Managing Chickens for Meat Production

Chickens raised for meat production are brooded in the same way as laying chickens. Chickens for meat production are generally raised in confinement, although range can be used. Broilers require 1 square foot of space from 2 weeks of age to market. Capons and roasters require 2 to 3 square feet of floor space from 10 to 20 weeks of age. After they reach 20 weeks of age, provide 4 to 5 square feet of floor space per bird.

Broiler producers usually use lights 24 hours a day. One 60-watt bulb per 200 square feet of floor space provides adequate light. However, recent trends have supported the replacement of common incandescent bulbs with either energy-efficient cold cathode (CC) or compact fluorescent (CFL) bulbs. These bulbs are shown to reduce lighting costs by 70 to 85 percent if they are implemented correctly. No nests or roosts are used in producing meat-type chickens.

A **capon** is a male chicken that has been surgically castrated. The operation is usually performed when the birds are 3 to 5 weeks of age. Capons produce meat that is more tender than that produced by broilers. However, they require more labor and investment to produce.

Breeding Flock Management

Hatching eggs are produced under agreements with a hatchery. The hatchery generally provides the farmer with a breeding stock containing special bloodlines that the hatchery wants produced.

Ten to 14 cockerels (males) per 100 pullets should be started. These are later culled down to 8 to 12 cockerels at 10 weeks of age. Leghorn breeds require one cockerel for each 15 to 17 hens at mating time. Heavier breeds require one cockerel for each 12 to 15 hens.

Brooding and management practices are done in the same way as for laying flocks. Breeding stock is often revaccinated for Newcastle disease and bronchitis about 1 month before breeding begins. This gives the newly hatched chicks a temporary immunity to these diseases.

Mating begins at least 2 weeks before the eggs are to be delivered to the hatchery. Do not put too many birds in one mating pen because this will reduce the matings.

Hatching eggs should be gathered three or four times per day. Gather more often if the weather is very hot or cold. Handle hatching eggs in the same careful way that other eggs are handled. Store hatching eggs at 55 to 65°F and at 75 percent humidity. Always pack hatching eggs with the large end up. If eggs have to be held more than 1 week, tip the crate sharply to prevent the yolk from sticking to the shell.

Hatching eggs must be fumigated within 2 hours after they are gathered. This will help prevent the spread of egg-borne diseases such as pullorum, fowl typhoid, paratyphoid, paracolon, and navel infection. A spray or dip may be used instead of fumigation.

Records

Good records are the key to good management. Records are used to determine the amount of money made from the operation. They are used as the basis for making management decisions and for selection of breeding stock in breeding programs.

Laying flock records must include information on egg production, death losses, amount of feed used, cost of feed, amount of money received from egg sales, and costs of buildings, repairs, and equipment. Meat production operations require records of feed and pounds of birds sold, as well as other production cost figures. From these types of records, the producer can determine profit or loss, and what management changes should be made.

Turkeys

Brooding Turkeys

The brooder house must be cleaned and disinfected before each flock of poults is placed in it. Old litter is removed and the house is allowed to dry out. Fresh litter is spread on the floor. Shavings, wheat and beardless barley straw, peat moss, shredded cane, rice hulls, processed flax straw, and cedar tow may be used for litter. The litter should be covered with a strong, rough-surfaced paper for the first few days. Enclose the area with a poult guard about 1.5 to 3 feet from the edge of the brooder hover. The paper cover is taken up after 5 to 6 days. Add fresh litter as needed after removing the cover. Keep the litter dry. The poult guard should be 16 to 18 inches high. It may be made of wire, heavy corrugated paper, or lightweight aluminum.

Temperature under the brooder for the first 2 weeks should be 95°F for dark poults and 100°F for light poults. The brooder temperature should be lowered 5°F each week until heat is no longer needed.

Provide ventilation without drafts. Lights are kept on for the first 2 weeks of brooding. After 2 weeks, houses with windows need only dim light at night. Windowless houses need about 1 foot-candle of light for 16 hours and half of a foot-candle during the other 8 hours. Roosts are seldom used when brooding turkeys. Up to 8 weeks of age, a minimum of 1 square foot of floor space is needed per poult. Larger-type poults require slightly more floor space. Poults in force-ventilated houses may be given slightly less floor space.

Turkeys that are to be raised in confinement should be debeaked at 3 to 5 weeks of age, but turkeys to be raised on range should not be debeaked unless picking or fighting occurs. **Desnooding** is the removal of the tubular fleshy appendage on the top of the head and helps to prevent head injuries from picking or fighting.

Small or medium-type turkeys that are to be raised on range should have the flight feathers of one wing clipped to prevent flying. It is not necessary to wing clip confinement-raised turkeys. Toe clipping is done on day-old poults at the hatchery because it prevents scratched and torn backs.

Trough-type feeders and waterers can be used for turkeys. Mechanical feeding and watering equipment is available (Figure 36-12). Feeders and waterers must be cleaned regularly. Moving feeders and waterers helps to prevent wet litter.

Range-Growing Turkeys

Turkeys can be raised on range (Figure 36-13), although almost all commercial production has moved to confinement production. Feed costs are somewhat lower on range. However, losses from diseases, predators, and weather are higher.

Poults can be moved to range at 8 weeks of age. Poults should be moved to range only when the weather is good. The number of turkeys per acre will vary depending on the type of range. On average, 1 acre of clay or clay-loam soil can carry from 250 to 300 birds. On very sandy soil, up to 1,000 turkeys may be put on an acre. One system of range raising involves moving the turkeys to a clean area every 1 to 2 weeks. This helps to prevent diseases and parasites. Range should be rotated with other crops. A 3- or 4-year rotation is used. Legumes or grass, or a mixture of the two, may be used for turkey ranges.

Courtesy of Chore-Time Poultry Production Systems, Milford, Indiana, USA

Figure 36-12 Young turkeys (poults) learn to eat from an automatic feeder.

Figure 36-13 Turkeys being raised on range. Most turkeys are raised in indoor facilities, similar to broiler houses.

Turkeys raised on range require shelter and protection from predators, such as dogs and wild animals. Double-strand electric fences provide protection against many predators. Six-foot-high, heavy-gauge poultry fencing provides protection from dogs, coyotes, and foxes. Confinement of the turkeys at night also protects against predators.

Confinement-Growing Turkeys

Most turkeys produced commercially are raised in confinement housing. Production in the turkey industry is becoming more like commercial broiler production. While some range production may still be used, the vast majority of turkeys are raised in housing ranging from pole-barn type structures to the latest modern tunnel ventilation units similar to those used for broilers.

Poults may be moved to confinement houses when they are 6 to 8 weeks of age. Check the poults the first few nights to be sure they have settled down. Use lights for the first few nights. Be sure that the house is properly ventilated.

Large-type toms require 5.5 square feet of floor space per bird. Hens need 3.5 square feet, and mixed flocks need 4.5 square feet of floor space per bird. Softwood shavings or wheat straw are often used for litter.

Managing Breeding Turkeys

Breeding stock should be blood tested for pullorum disease, fowl typhoid, paratyphoid, infectious sinusitis, and Arizona and *Mycoplasma meleagridis* infections. Vaccinations are based on the requirements of the hatching egg buyer. More information about diseases of poultry can be found in Chapter 37.

Low fertility is a problem when natural mating is used. Because of this, artificial insemination is widely practiced with turkey breeding. The first insemination is given when egg production begins. A second insemination is given 1 week later. This is followed with insemination at 2-week intervals.

Breeding flocks may be kept on limited range or in confinement. Confinement allows better disease control and management. The labor cost is lower in confinement. Breeding flocks require more floor space than market turkeys.

Toms should receive artificial light 2 to 3 weeks before they are used with the hens. Hens should receive light 30 days before the eggs are needed. A 14-hour day is used for normal spring lighting.

Broodiness of hens is a problem with breeding flocks. **Broodiness** is when a hen stops laying eggs and wants to sit on a nest of eggs to hatch them. Remove broody hens from the flock and confine them in a separate pen. Use slatted floors in the pens, and keep the pens well lighted.

Eggs should be gathered at least every 2 hours. Hold eggs at room temperature for 24 hours. Store eggs at 55 to 65°F until they go into the incubator. Store eggs with the large ends up or place them on their sides in trays. Eggs stored more than 1 week must be turned. Dirty eggs lower hatchability. Keep eggs clean by keeping nests clean and gathering eggs often. Eggs that are too dirty can be washed. Eggs should be delivered to the hatchery at least twice a week.

Ducks and Geese

Brooding

Ducklings and goslings can be brooded naturally or artificially. Broody chicken hens or females of ducks or geese can be used for natural brooding. Young birds not hatched by the female should be placed under her at night. Twelve eggs may be placed under a duck, 9 to 10 eggs under a goose, and 4 to 6 goose eggs or 10 to 11 duck eggs under a chicken hen. Goose eggs set under a hen should be turned twice a day. A dry shelter is needed for natural brooding.

Brooders used for chickens can be used for artificial brooding of ducks and geese. Ducks and geese grow faster than chickens and require heat for a shorter period of time. Five to 6 weeks of brooding is usually long enough.

An infrared heat lamp can be used for a small number of eggs. Hang the lamp about 18 inches above the floor. One lamp will brood about 30 ducklings or 25 goslings.

Other types of chicken brooders have a capacity of about one-half chick capacity for ducklings and one-third chick capacity for goslings. The hover needs to be 3 to 4 inches higher than for chicks. Set the brooder at a starting temperature of 90°F. Reduce the temperature by 5 to 10°F each week until a temperature of 70°F is reached.

Slats or litter flooring may be used for ducks. Three-quarter-inch mesh wire, plastic, or wooden slats placed 4 inches above the floor is used. Slatted flooring keeps the ducklings away from manure and moisture. Chopped straw, wood shavings, or peat moss can be used for litter. Litter must be dry and free of mold.

Chopped straw, sawdust, wood shavings, crushed corn cobs, or peat moss make good litters for goslings. Cover the litter with a rough paper for the first 4 or 5 days. This prevents the goslings from eating the litter. Smooth paper should not be used. It causes *spraddled legs,* a condition in which the leg is dislocated from the socket where it is attached to the body. Spraddle-legged goslings do not recover. The litter must be kept dry, wet spots should be removed, and fresh litter should be added as needed.

Rearing

Feeding, watering, and pasture for ducks and geese were discussed earlier in this chapter. Ducks and geese should be moved outside as soon as the weather permits. Young ducklings and goslings should not be allowed out in the rain until they are well feathered.

Pasture for ducks and geese should be fenced. Young birds should be confined with a 2-inch mesh poultry netting. After the birds are 4 to 6 weeks of age, an ordinary woven wire fence will hold them. The fence need not be high because the birds seldom fly. Electric fencing has been used successfully with waterfowl.

Birds confined in yards cause a buildup of manure. This must be cleaned periodically at intervals determined by the number of birds in the yard.

HOUSING AND EQUIPMENT

The trend in poultry production has been toward large commercial flocks. With this trend has come an increase in confinement housing for poultry. Initially, pole-type construction with curtain sides was commonly used for confinement operations. This type of construction was relatively economical to build, maintain, and clean. Today, most confinement poultry houses are clear span structures with insulated ceilings and walls. It is easier to provide proper ventilation in this type of housing (Figures 36-14 and 36-15). There is also a trend toward housing layers in structures that are light-tight. The control of lighting significantly influences egg production. The companies that provide the birds in vertically integrated systems usually specify the type of housing to be used. State colleges of agriculture can assist in designing poultry buildings for specific geographic areas.

Courtesy of Chore-Time Poultry Production Systems, Milford, Indiana, USA

Figure 36-14 In tunnel-ventilated houses, cool cells (wet pads) are attached to one end of the house. Air pulled into the house by fans is cooled through the process of evaporation.

Courtesy of Chore-Time Poultry Production Systems, Milford, Indiana, USA

Figure 36-15 Exhaust fans attached to one end of a tunnel-ventilated poultry house pull air through cooling pads down the length of the facility.

Some general characteristics are common to all good poultry houses. Houses should be easy to clean and disinfect. Mechanical cleaning equipment is used in some types of houses. Dampness is a problem in poultry houses. All types of housing must be properly ventilated to help prevent dampness. Insulation should be used in areas where it is needed. Water and electricity are essential for modern poultry operations. Egg, feed, and equipment storage rooms are often included in the house. Bulk feed storage is generally used for large operations.

Automatic and semiautomatic feeding, watering, and cleaning equipment are common in poultry enterprises. The use of this type of equipment saves time and labor. It also makes it possible for large numbers of birds to be handled in the operation.

Poultry raised on range need shelters and shade. Portable feeders and waterers are used on range.

SUMMARY

The greatest single cost of poultry enterprises is feed. Commercially prepared feeds are available for all kinds of poultry. Smaller flock owners generally find it easier to use a complete commercial feed. Large commercial flock owners usually use some system of mixing ingredients to make a complete feed. Poultry require carbohydrates, protein, minerals, vitamins, and water in their rations. Grit is used to help the bird grind the feed in the gizzard. Oyster shell provides calcium for egg production.

Poultry may be brooded on the floor or in cages. Small flocks generally use floor brooding. Large operations may use battery brooder units or wire cages. Proper temperature control is necessary for successful brooding.

Light affects the sexual maturity and egg production rates of poultry. Controlled lighting is generally used in poultry enterprises. Laying flocks may be handled using floor systems or in cages. Large operations are more likely to use cage systems. Eggs must be carefully handled to maintain quality. Chickens raised for meat production are usually raised in confinement. Turkeys are raised on range or in confinement, with the vast majority of producers moving to confinement production similar to broiler production.

Ducks and geese can be brooded naturally or artificially. They are easy to raise and do not require expensive housing or equipment.

Automatic or semiautomatic feeding, watering, and cleaning equipment are frequently used in the poultry industry.

Student Learning Activities

1. Give an oral report on feeding, management, housing, or equipment for different kinds of poultry.
2. Prepare a bulletin board display of pictures of housing and equipment used for different kinds of poultry.
3. Visit poultry production enterprises to observe feeding, management, and housing.
4. Survey poultry producers in the local community concerning feeding, management, and housing practices. Prepare a summary of the results.
5. Follow the practices described in this chapter when planning and conducting the feeding, managing, housing, and equipping of a poultry supervised experience program.

Discussion Questions

1. How important is water in the chicken ration?
2. List the vitamins and minerals commonly required in a chicken ration.
3. What level of calcium should be fed in a chicken ration?
4. Briefly explain the use of additives in chicken rations.
5. Why is grit used in the chicken ration?
6. List three forms in which commercially manufactured chicken feeds may be purchased.
7. List four systems that may be used for preparing chicken rations.
8. Briefly explain the use of phase feeding for chickens.
9. List the good management practices to follow when feeding chickens.
10. Why should the poultry producer keep a close watch on the feed consumption of the flock?
11. How much feeder and waterer space is needed for light and heavy breeds of laying hens?
12. Why should the producer avoid sudden changes in the ration for laying hens?
13. How is feed efficiency determined for laying hens?
14. Describe the kind of starter and finishing rations that should be used for broilers.
15. How is feed conversion determined for broilers?
16. Describe a good feeding program for the breeding flock, turkeys, ducks, and geese.
17. List the recommended practices for brooding chicks, turkeys, ducks, and geese.
18. Describe the principles of using artificial light when raising poultry.
19. Compare and contrast the recommended practices for managing the laying flock and the breeding flock.
20. List the recommended practices for egg handling.
21. List the kinds of records that should be kept by a poultry producer.
22. Describe the growing of turkeys on range.
23. Describe confinement raising of turkeys.
24. Describe the practices that should be followed for rearing ducks and geese.
25. Describe the kinds of housing and equipment used for poultry.

Review Questions

True/False

1. Fats added to the ration increase palatability, decrease dustiness, and improve the texture of the feed.
2. Feed intake is also increased by the energy level of the ration.
3. The ration can be altered immediately if conditions change without creating stress on the laying flock.

Multiple Choice

4. The most commonly used grain in poultry rations is ______.
   a. oats　　c. milo
   b. wheat　　d. corn
5. Essential amino acids needed by chickens are supplied by ______.
   a. minerals　　c. proteins
   b. carbohydrates　　d. calcium

Completion

6. A ______ is a male chicken that has been surgically castrated.
7. ______ is done by buffing the eggs with fine sandpaper.
8. ______ is done to control cannibalism in poultry.
9. A ______ ______ is used for the first week of brooding to prevent the chicks from wandering away from the heat and becoming chilled.

Short Answer

10. What kind of house is used for chicks and provides heat through infrared heat lamps?
11. Name five nutrients that poultry rations must supply. What part of the ration supplies these nutrients?
12. Name three types of cages used in cage-laying systems.

Chapter 37

Diseases and Parasites of Poultry

Key Terms

intranasal vaccination
intraocular vaccination
wing web vaccination
synovia
bacterin
virulence

Objectives

After studying this chapter, the student should be able to

- establish a disease and parasite control program for poultry.
- identify the symptoms of common poultry diseases.
- identify the common parasites of poultry.
- describe methods of preventing disease and parasite infestation of poultry.

MAINTAINING POULTRY HEALTH

A program of preventive management is the best way to control poultry diseases and parasites. Such a program involves sanitation, good management, vaccination, and control of disease and parasite outbreaks (Figure 37-1).

Sanitation

Poultry houses should be completely cleaned and disinfected before new birds are moved in. The following steps are used to achieve a clean poultry house.

1. Take all movable equipment outside of the house. Clean off the manure, and wash and disinfect the equipment. Expose the equipment to sunlight.
2. Remove all manure and litter from the house.
3. Sweep down the walls and ceilings.
4. Scrape and brush the floor.
5. Scrub and hose the inside and outside of the house using high pressure. A portable steam cleaner can be used for this operation.

6. Spray the inside of the house with an approved disinfectant. Spray should be applied to all surfaces, including the walls, ceiling, and floor. Do not spray the disinfectant into waterers or feeders.
7. Use new, clean, dry, non-dusty litter on the floor.
8. Leave the house empty for 2 weeks to break disease cycles.
9. Lock the door to prevent people from entering and contaminating the clean house.

USDA Agricultural Research Service; Photo by Keith Weller

Figure 37-1 Chick receiving vaccine for coccidia.

Whenever possible, do not allow visitors to enter poultry houses, pens, and yards. Clean coveralls and disinfected rubber footwear should be worn by anyone who must enter the poultry area. Place a foot pan with disinfectant in it at each door to be used before entering. Replace the disinfectant in the pan frequently.

Use only clean and disinfected equipment. Be cautious about allowing used poultry crates, egg cases, and feed bags to be brought onto the farm. These can spread diseases.

All dead birds must be disposed of promptly and properly. Use of incinerators, composting, or deep burying are recommended for disposal of dead birds. Be sure that disposal methods meet Environmental Protection Agency (EPA) regulations.

Dispose of manure by spreading it thinly on land that is not used for poultry. Eliminate places for pests, such as flies, to breed. Do not pile up manure outside of the poultry house. Control lice and mites inside of the house by using approved chemicals.

Other Health Management Practices

In addition to sanitation, a number of other management practices contribute to good flock health. Among these practices are the following:

1. Buy poultry replacement stock from a reliable, disease-free source.
2. Purchase day-old chicks and poults. Older birds brought to the farm are more likely to bring disease and parasites.
3. If possible, keep birds of only one age on the farm. Use an all-in/all-out program. Bring all the birds onto the farm at one time and remove them all at one time.
4. If it is necessary to keep birds of different ages, separate the flocks by at least 40 feet.
5. Separate chickens and turkeys. It is best to have only one or the other on the farm.
6. Separate breeder flocks from other poultry. It is best not to have any other poultry on the farm if a breeder flock is kept.
7. Keep pets and flying birds out of the poultry house. Flying birds can be kept out by screening the windows.
8. Provide proper ventilation in the poultry house.
9. Control rats and mice. Use baits and traps as necessary. Make feed bins and storage rooms rat-proof. Eliminate places for rats and mice to breed by cleaning up trash and junk.
10. Feed balanced rations to prevent nutritional diseases (Figure 37-2). Make sure the feed is mixed properly. Provide plenty of fresh, clean feed and water.
11. Maintain good health records. Records should be kept of vaccinations, disease problems, and medicines used.
12. If outside labor is used, those workers should be discouraged from keeping poultry on their property to lessen the transmission of disease and parasites.

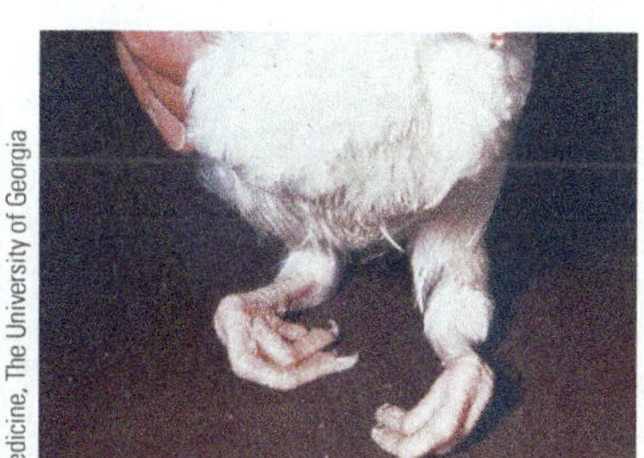
Dr. Jean Sander, Department of Avian Medicine, The University of Georgia

Figure 37-2 Deformity caused by a deficiency of riboflavin.

Vaccination

Vaccination is not a substitute for good flock health management, but it is helpful in controlling certain diseases. Vaccines are available for Newcastle disease, Marek's disease, infectious bronchitis, fowl pox, epidemic tremors, fowl cholera, laryngotracheitis, infectious bursal disease, erysipelas, reovirus, salmonella, chicken anemia virus, aspergillosis, infectious coryza, chicken herpes virus, turkey herpes virus, encephalomyelitis, and viral hepatitis. Vaccinations should be used only in areas where the disease is known to exist. A vaccination program should be planned for each flock. Some vaccines for certain diseases can only be used with the permission of the state veterinarian. When planning a vaccination program, obtain help from a veterinarian, the Cooperative Extension Service, a hatchery, or feed dealer.

Vaccinations cause stress in poultry. Only vaccinate healthy birds. Read and follow all directions on the vaccine.

Several methods may be used to vaccinate poultry. Individual bird vaccinations are given by injection, intranasally, intraocularly, or through the wing web. **Intranasal vaccination** is placement of the vaccine directly into the nose opening. **Intraocular vaccination** is placement of the vaccine directly into the eye. **Wing web vaccination** is the process of injecting the vaccine into the skin on the underside of the wing web at the elbow. A grooved, double needle instrument is used for wing web vaccination.

Flock treatments are given in the water, by spray, or through dust. The method used depends on the disease to be controlled. Individual vaccination causes more stress on the birds than flock treatments. Vaccinations for some diseases can be given in more than one way.

Controlling Disease Outbreaks

It is better to prevent a disease outbreak than to try to control it once it has occurred. Following the sanitation, management, and vaccination suggestions discussed earlier in this chapter will help the poultry producer prevent disease outbreaks from occurring.

The poultry flock should be checked daily for signs of disease. A sudden drop in feed and water consumption is often a sign of health problems. Watch the birds to see how they are eating and drinking. If more than 1 percent of the flock is sick, a disease is probably present. Death rate is another sign of disease. During the first 3 weeks, the normal death rate for chicks is about 2 percent. For turkeys, it is about 3 percent. After 3 weeks of age, the death rate should not be more than 1 percent per month. A sudden increase in the death rate is an indication of disease.

Most diseases can be accurately diagnosed only in a laboratory. The producer should use the services of a veterinarian or the state diagnostic laboratory to determine which disease is causing the problem. The procedure for collecting needed information and specimens is specified by the laboratory. This procedure should be carefully followed. The recommendations of the veterinarian or laboratory for control of the disease must also be followed for best results.

DISEASES AND DISORDERS

The list of diseases and disorders that affect poultry is extensive. Brief descriptions of the most common are included in this chapter. Positive diagnosis and recommendations for treatment should be obtained from a veterinarian or diagnostic laboratory.

Amyloidosis

Amyloidosis, also called "wooden liver disease," is a common disease of adult birds. Brown layer chickens are particularly susceptible. The liver becomes hard and fluids accumulate within the body cavity. Death rate may be as high as 10 percent. There is no known cure.

Aortic Rupture/Dissecting Aneurysm

Aortic rupture/dissecting aneurysm affects mainly male turkeys between the ages of 8 and 20 weeks. The exact cause is not known, but the disease seems to be related to a high intake of energy feeds. An artery ruptures and the bird bleeds to death internally. Affected birds which appear to be in good condition die suddenly. Aortic rupture is prevented by feeding a lower-energy diet and by using a continuous feeding of tranquilizers at a low level.

Arthritis/Synovitis

The arthritis/synovitis syndrome is an inflammation of the joints and synovial membranes. The synovial membrane is a thin, pliable structure surrounding a joint. The membrane secretes **synovia**, which is a transparent lubricating fluid. The causes include injury, diet, and infection. The three most common forms are infectious synovitis, staphylococci arthritis, and viral arthritis.

Infectious synovitis is caused by a bacterium. It affects chickens and turkeys. The birds become lame, lose weight, and usually have diarrhea. Chickens show swelling of the foot pads and shanks. Turkeys show swelling of the feet and hocks. Death loss from this disease is low. Prevention is through good sanitation, and treatment involves the use of antibiotics.

Staphylococci arthritis is also caused by a bacterium. Affected birds have swollen joints and diarrhea, and show signs of depression. The acute form causes death. Good management helps to prevent the disease. Antibiotics may be used to treat this condition.

Viral arthritis is caused by a virus. The birds become lame and show a slow rate of gain. The disease affects only chickens.

Aspergillosis (Brooder Pneumonia)

Aspergillosis is caused by a fungus or mold. It affects chickens and turkeys primarily, but it can also affect ducks and wild and pet birds. The acute form occurs in young birds and results in a high death rate. A chronic form affects older birds. The symptoms in young birds include loss of appetite, gasping, sleepiness, convulsions, and death. Older birds show a loss of appetite, gasping, and loss of weight. The disease is prevented by using litter that is free of mold. There is no treatment but the disease can be prevented with a vaccine.

Avian Influenza (Fowl Plague)

Avian influenza (AI), sometimes called bird flu, is a contagious disease of birds. AI is caused by a virus which occurs commonly in wild birds and can be transmitted to turkeys and chickens. Influenza viruses vary widely in their ability to cause disease. Chickens and turkeys are often infected with strains of influenza causing low to moderately severe symptoms. Strains of AI with extremely aggressive and deadly results are not common. However, a major outbreak of a highly pathogenic strain of avian influenza could be disastrous to the U.S. poultry industry.

Clinical signs of avian influenza vary widely. Symptoms include coughing, sneezing, weight loss, depression and droopiness, sudden drop in egg production,

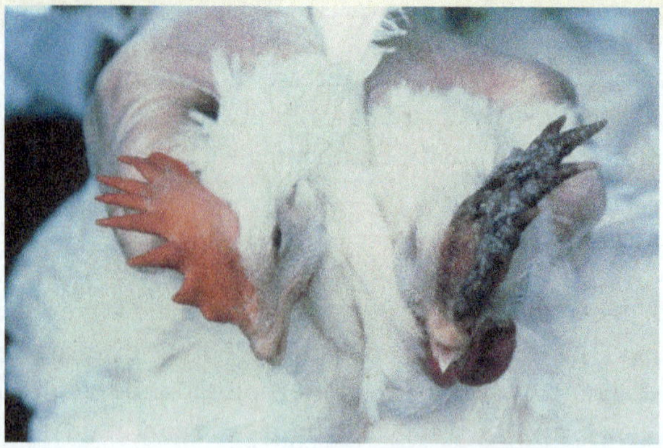

USDA/APHIS

Figure 37-3 Avian influenza causes purple discoloration of the comb. The chicken on the right is infected with avian influenza.

loss of appetite, purplish-blue coloring of the wattles and comb (Figure 37-3), diarrhea, bloody nasal discharge, loss of ability to stand, and death losses in a flock.

An unusually aggressive subtype of avian influenza is caused by the H5N1 virus. It is highly contagious, and can be deadly. Bird flu and human flu are both caused by viruses, but each virus generally affects either birds or people, not both. The H5N1 virus is one strain that has spread to humans. This virus strain has killed many humans in Asia since 1997. Unlike seasonal flu, in which infection usually causes mild to moderate respiratory symptoms most people, H5N1 infections may be severe, with rapid deterioration and a high fatality rate. People cannot become infected with bird flu from properly handled and cooked poultry and eggs. Proper cooking will kill the virus in the unlikely event that poultry and eggs are contaminated. Most cases of avian influenza infection in humans have resulted from direct contact with infected poultry.

Photo courtesy of Dr. Jean Sander, College of Veterinary Medicine, The University of Georgia

Figure 37-4 Avian pox is caused by a virus that can affect turkeys, chickens, and other birds.

Avian Leukosis

Avian leukosis is a complex of diseases caused by viruses. The most common forms are lymphoid leukosis and Marek's disease.

Lymphoid leukosis usually affects both turkeys and chickens. Sometimes the birds die without showing any symptoms of disease. Usual symptoms are loss of weight, loss of appetite, and diarrhea. The bone form of the disease affects younger birds and is a serious problem in the broiler industry. The affected birds show lameness. Prevention involves the use of resistant strains of birds, sanitation in the hatchery, and keeping the flock isolated. There is no treatment for the disease.

Marek's disease affects mainly young chickens, although it may occur in older birds. It is caused by a virus different from the one that causes lymphoid leukosis. The main symptom is lameness or paralysis of one or both legs. The disease is prevented by vaccination at hatching time. There is no treatment.

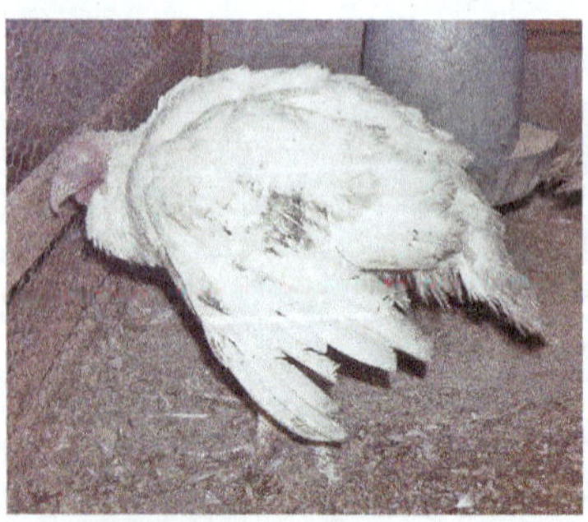

Photo courtesy of Dr. Jean Sander, College of Veterinary Medicine, The University of Georgia

Figure 37-5 A turkey suffering from fowl pox. The growth rate is slowed and egg production is reduced.

Avian Pox (Fowl Pox)

Avian pox is caused by a virus and affects turkeys, chickens, and other birds. Symptoms include wart-like scabs around the head and comb (Figures 37-4 and 37-5), yellow cankers in the mouth and eyes, and, in turkeys, yellowish-white cankers in the throat. Young birds have a slower growth rate. Egg production in layers is reduced. The disease is prevented by vaccination. There is no treatment.

Avian Vibrionic Hepatitis

Avian vibrionic hepatitis is caused by bacteria. It affects chickens in either an acute or chronic form. The acute form causes a higher death loss. The birds show loss of weight, diarrhea, listlessness, shrunken comb, and a drop in egg production. Good management and sanitation help in prevention. Antibiotics are available for treatment of avian vibrionic hepatitis.

Blackhead

Blackhead is caused by a protozoan parasite. It affects chickens, turkeys, and other birds, but it is more prevalent in turkeys. The disease affects mainly young birds between 6 and 16 weeks of age. Symptoms include droopiness, loss of appetite, darkening of the head, and a watery, yellow diarrhea. Death losses are usually low but may be as high as 90 percent if the disease is not controlled. Diseased chickens often show no signs of the disease.

Brooding on wire or slatted floors helps in prevention. Never house chicken and turkey flocks together. Rotate pastures to help prevent the disease if pastures are used in production. It may be necessary to continuously feed a preventive medication at low levels in the drinking water.

Bluecomb (Transmissible Enteritis; Mud Fever)

Bluecomb in turkeys is caused by a virus and affects all ages. Symptoms in poults include droopiness, dehydration, and gaseous, watery diarrhea. Death loss in poults can be as much as 100 percent. Older birds show a loss of appetite, loss in body weight, diarrhea, and a blueness of the head. Death loss in older birds is not high. Birds that recover may be carriers of the disease.

Sanitation practices help to prevent the disease. If an outbreak has occurred, empty the house and allow it to remain vacant for at least 30 days. Antibiotics may be used in treating bluecomb.

Bluecomb (Pullet Disease)

Bluecomb in chickens is not the same disease as bluecomb in turkeys. The cause is associated with hot weather, water deprivation, toxin, and/or a virus. Symptoms include a drop in egg production, loss of appetite, loss of weight, diarrhea, and a darkening of the comb. The crop may become compacted with feed. Sanitation and good management help to prevent the disease. Treatment is with a broad-spectrum antibiotic.

Bumblefoot (Staphylococcosis)

Bumblefoot is a common disease of poultry that affects the feet. Bumblefoot results from injury or abrasion to the lower surface of the foot, which allows for the introduction of staphylococcus bacteria. Symptoms include swelling, reluctance to walk, and a swelling in the center of the foot. Bumblefoot can cause lameness and slow growth rate. Birds on muddy, dirty range or litter are more likely to get bumblefoot. Management that eliminates these conditions helps in prevention. Bumblefoot infections can be successfully treated with antibiotics. Administer the antibiotic according to label directions.

Botulism (Limberneck; Food Poisoning)

Botulism is the result of the bird eating decayed material that contains a toxin (poison) produced by a bacterium. Symptoms include weakness, trembling, paralysis, loose feathers, and dull, closed eyes. Affected birds usually die. Prevention is achieved by eliminating sources of decayed material that might be consumed by the birds,

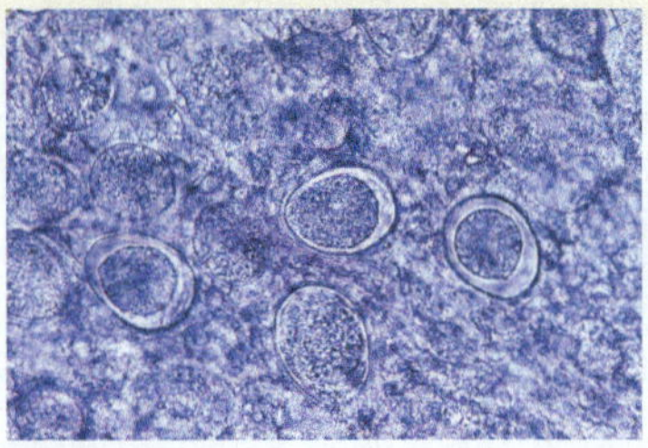

Dr. Jean Sander, College of Veterinary Medicine, The University of Georgia

Figure 37-6 Coccidiosis is caused by protozoan parasites called coccidia.

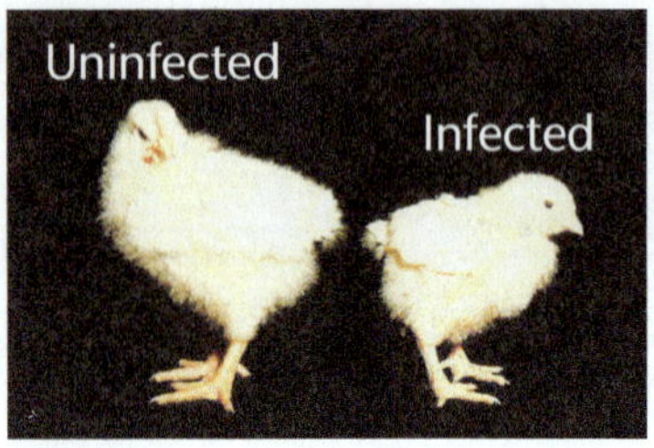

USDA

Figure 37-7 A common effect of coccidiosis is poor growth.

especially in hot weather. Contaminated water is often a source of the toxin. Be sure that the water supply is clean. Isolate sick birds. An antitoxin may be used to save valuable individual birds.

Cage Layer Fatigue

Cage layer fatigue appears to be a nutritional problem involving minerals. It commonly affects young pullets that are producing eggs at a high rate. The disease is not common if the diet is properly balanced. The main symptom is paralysis. Proper diet prevents the disease. Treatment is usually not practical.

Coccidiosis

Coccidiosis is caused by a number of protozoan parasites called coccidia (Figure 37-6). Different species of coccidia affect different kinds of poultry. Birds 3 to 8 weeks of age are most frequently affected by this disease. Symptoms include droopiness, huddling of birds, loss of appetite, loss of weight, and diarrhea, which may be bloody (Figure 37-7). Death loss may be high. The best prevention is the feeding of a coccidiostat in the ration of chicks and poults. Do not feed ducks or geese a chicken feed containing a coccidiostat. Brooding on wire or slat floors will also help to prevent the disease. Medications may be used to treat affected birds.

Coliform Infections (Colibacillosis)

Coliform infections represent a variety of diseases caused by various strains of the bacteria Escherichia coli. The diseases range from severe to mild. Young birds are more commonly affected, although adults may be affected too. The symptoms vary with the type of infection. Death losses may be high in young birds. Some of the symptoms of coliform infections include fever, ruffled feathers, listlessness, difficulty in breathing, and diarrhea. Prevention includes sanitation, fumigation of hatching eggs, and reduction of stress. Medications for treatment are not very effective but may reduce losses.

Duck Virus Enteritis (Duck Plague)

Duck virus enteritis is caused by a virus. It affects ducks, geese, and other waterfowl. Symptoms include watery diarrhea, nasal discharge, dehydration, tremor, decreased egg production, and droopiness. Death loss is high. Sanitation and isolation of birds helps to prevent the disease. There is no treatment.

Degenerative Joint Disease

Degenerative joint disease is seen in chickens and turkeys. It is caused by physical damage, developmental effects, or rapid growth. Symptoms include lameness and reduced breeding performance. To prevent the disease, growth-control programs may be practiced, especially during a period of rapid growth. Birds often need to be segregated as soon as symptoms are demonstrated for treatment purposes. Birds may need to be culled if they do not improve.

Degenerative Myopathy (Green Muscle Disease)

The exact cause of degenerative myopathy is not known, but it may be an inherited condition. The inner large breast muscle degenerates and becomes a greenish color. There is a dent in the muscle of one or both sides of the breast. The disease affects mainly turkey breeding hens. Breeding stock that shows signs of the condition should not be used.

Dietary Disorders

A number of diet-related disorders affect poultry. Feeding a well-balanced ration containing the minerals and vitamins needed will prevent most of these disorders.

Epidemic Tremor (Avian Encephalomyelitis)

Epidemic tremor is caused by a virus and affects all ages of chickens, pheasants, turkeys, and quail. The symptoms usually appear only in young birds. They become paralyzed and tremble. Death loss is high. Prevention is achieved by vaccinating breeding stock. There is no treatment.

Erysipelas

Erysipelas is caused by a bacterium. In poultry, it is most common in turkeys between 4 and 7 months of age. An increased risk occurs when the housing or land has been previously used for pigs or sheep. The first sign of disease is usually several dead birds. Sick birds show weakness, lack of appetite, chronic scabby skin, and yellowish or greenish diarrhea. Death loss can be high. Turkeys can be vaccinated for erysipelas. It can be treated with antibiotics or vaccination with a bacterin. A bacterin is a vaccine made with weakened bacteria. Vaccination should be done only if the disease is present in the area.

Fatty Liver Syndrome

The cause of fatty liver syndrome is unknown. It affects mainly caged birds on high-energy diets. Symptoms include a drop in egg production, pale comb and wattles, and obesity. Dead birds may be found when no other symptoms have been noticed. Treatment and control measures involve dietary adjustment. A nutrition specialist should be consulted.

Fowl Cholera, Endemic Fowl Cholera (Pasteurellosis)

Fowl cholera is caused by bacteria. It may occur in acute or chronic form. Symptoms of the acute form include purple color of the comb, difficult breathing, watery diarrhea, weakness, droopiness, loss of appetite, loss of weight, lameness, and sudden death.

Endemic fowl cholera is a chronic condition that affects mainly chickens, but it can affect a range of poultry species. Good management has almost eliminated it from commercial flocks. Symptoms include a nasal discharge, inflammation of the eye, and swelling of the sinus. Prevention is achieved by isolating infected birds and using stock from sources that are free of the disease. If an outbreak occurs, dispose of all birds and clean and disinfect the house. Vaccinate with bacterins in areas where the disease is present. Antibiotics are used for treatment.

Hemorrhagic Anemia Syndrome

Hemorrhagic anemia syndrome affects chickens of all ages, but mainly those between 4 and 12 weeks of age. Its cause is associated with medication toxicities, mycotoxins, and viral infections. Symptoms include loss of weight, diarrhea, weakness, and anemia. Death losses may be high with the acute form. Sulfa medications and antibiotics may make the condition worse. Addition of liver solubles to the ration has resulted in some improvement.

Hemorrhagic Enteritis

Hemorrhagic enteritis can occur in turkeys from 3 weeks to 6 months of age but is more common in older birds. It is caused by a virus. Sudden death of several birds on range is often the main symptom. A change in the ration and moving birds to new range may help in treatment. A serum made from the blood of birds that have recovered can be injected to treat the flock.

Hexamitiasis

Hexamitiasis is caused by a protozoan parasite. It does not affect chickens. Turkeys and some game birds are affected up to 10 weeks of age. Poults huddle together and become listless. There is rapid weight loss with a watery, foamy, yellowish diarrhea. Sanitation and brooding on wire or slat floors help to prevent the disease. Broad-spectrum antibiotics have been used with some success in treatment.

Inclusion Body Hepatitis

Inclusion body hepatitis is caused by a virus. The disease is common in commercial chicken flocks. The sudden death of chickens in the flock is often the only symptom. Diseased chickens show symptoms of anemia, yellowing of the skin, depression, and weakness. Death follows these symptoms fairly quickly. There is no vaccine or treatment.

Infectious Bronchitis

Infectious bronchitis is caused by a virus and affects only chickens. Symptoms include difficult breathing, sneezing, gasping, and nasal discharge. Egg production of layers is reduced, as well as the growth rate of broilers. Death rate of young chicks is high. Older birds also show loss of appetite and slower growth. Prevention is by vaccination, isolation of flocks, and sanitation. Antibiotics are used where permitted to control secondary colibacillosis.

Infectious Bursal Disease (Gumboro)

Infectious bursal disease is caused by a virus. It affects young chickens. Symptoms include fever (in the early stages), ruffled feathers, loss of appetite, difficulty in defecation, slight tremors, and dehydration. Birds do not want to move and are unsteady when walking. Vent picking often occurs. Death rate is high. Prevention is by vaccination and proper biosecurity measures. There is no treatment.

Infectious Coryza

Infectious coryza is caused by a bacterium. It affects mainly older chickens. Symptoms include swelling around the eyes and wattles, nasal discharge, and swollen sinuses. Egg production in layers is reduced. Good sanitation and management will help in prevention. Vaccines are used in areas of historically high concentration. Medications are used for treatment.

Laryngotracheitis

Laryngotracheitis is caused by a virus. It affects mainly older chickens and turkeys. Symptoms include difficult breathing, coughing, and sneezing. Death rate is high. Egg production in layers is reduced. Vaccination is used in areas where the disease exists. The use of the vaccine is restricted. Infected birds should be quarantined. There is no treatment.

Leucocytozoonosis

Leucocytozoonosis is caused by a protozoan parasite. It affects mainly turkey poults and ducklings. Symptoms include lack of appetite, droopiness, weakness, thirst, and difficult breathing. The death rate is high. Sanitation, isolation of brooding, and good management help in prevention. There is no treatment.

Mycoplasmosis

Three species of the bacterium Mycoplasma cause respiratory diseases in poultry. These diseases include chronic respiratory disease/air sac syndrome in chickens, infectious sinusitis in turkeys, airsacculitis in turkeys, and infectious synovitis in chickens and turkeys. Infectious synovitis is discussed earlier in this chapter.

Symptoms of chronic respiratory disease include nasal discharge, coughing, sneezing, swelling below the eyes, loss of weight, and drop in egg production. Isolation, sanitation, good management, and avoiding stress will help prevent the disease. Treatment is with antibiotics.

Symptoms of infectious sinusitis include watery eyes, nasal discharge, swollen sinuses, coughing, and difficult breathing. Prevention and treatment are the same as for chronic respiratory disease.

While there are a number of causes of airsacculitis in turkeys, the bacterium Mycoplasma meleagridis is one of the main ones. Breeding flocks as well as young poults may be affected. Yellow deposits are found on the air sacs and sometimes on the lungs of infected birds. Prevention is achieved by treating hatching eggs with antibiotic solutions. Treatment is with antibiotics.

Mycotoxicosis

Fungi or mold growing on feed or litter produce toxins. If these toxins are taken in by the bird, mycotoxicosis can result. A variety of symptoms may indicate the presence of the disease. Growth rate may be slower, egg production may be reduced, the hatchability of eggs may be lowered, or the bird may die. The condition is best prevented by storing feeds in such a way that the mold does not grow. Do not feed moldy feed or use moldy litter.

Necrotic Enteritis

Necrotic enteritis appears to be caused by toxins produced by bacteria. Sudden death of healthy birds is one of the first indications of the disease. Death rate in a flock may be high. Birds that do not die have slower growth rates and poor feed conversion. Medications are used in prevention and treatment. Controlling coccidiosis helps to prevent the disease.

Newcastle Disease

Newcastle disease is a major threat to the poultry industry. This highly contagious disease occurs in several forms, each caused by a different virus. The strains of the disease vary greatly in virulence, which is the capacity of a microorganism to cause disease. Young birds show both breathing problems and signs of nervousness, including difficult breathing, sneezing, gasping, paralysis, early molt, and tremors.

Adults show breathing problems, and egg production of layers is reduced. Sanitation and vaccination are the methods of prevention. There is no treatment.

New Duck Disease (Infectious Serositis)

New duck disease is caused by bacteria. It is one of the most serious diseases affecting ducklings. Symptoms include sneezing, eye discharge, diarrhea, twisted neck, and loss of balance. The mortality rate is high. Proper management and vaccination are used for prevention. Antibiotics are used in treatment.

Omphalitis

Omphalitis, also known as "mushy chick" disease and "navel-ill," appears to be caused by several bacteria. It affects young birds. Symptoms include discoloration around the navel, loss of appetite, diarrhea, and drowsiness. The death rate is high. Good management and sanitation in the hatchery are necessary for prevention. Broad-spectrum antibiotics are used for treatment.

Ornithosis

Ornithosis is caused by a bacterium. In humans it is called parrot fever. The disease is not a major problem in the poultry industry, but workers in turkey-processing plants have contracted the disease from infected birds. Symptoms include coughing, sneezing, reduced weight gain, and decreased egg production. Isolation and vaccination of flocks may help in prevention. Antibiotics are used in treating the disease. Affected flocks are quarantined and treated under supervision. Flocks that have had the disease should not be kept for breeding purposes.

Poisoning

Birds may be poisoned by eating or drinking any toxic materials. Sprays, disinfectants, nitrates, pesticides, and poisonous plants are a few of the sources of poisoning. Convulsions are a common symptom of poisoning. Birds raised on range are more likely to be poisoned than those raised in confinement. Prevent birds from eating poisonous substances, and allow proper time for disinfectants between flocks to ensure bird safety.

Round Heart Disease

The exact cause of round heart disease is unknown. It may be the result of a virus, heredity, or both. It affects mainly young male chickens and turkeys. Stress conditions appear to increase the chances of the disease appearing. The heart becomes enlarged, rounded, and flabby. The liver may be yellow in color. The condition can be prevented by making sure there is adequate heat during brooding and that only disease-free stock is purchased. There is no treatment.

Salmonella and Paracolon Infections

Salmonella and paracolon infections are caused by a number of bacteria. Four diseases of poultry that are caused by these bacteria include pullorum disease, fowl typhoid, paratyphoid, and paracolon infections.

Symptoms of pullorum include white diarrhea, droopiness, and chilling. It affects mainly young birds, and death losses may be very high. Prevent the disease by getting stock from pullorum-free hatcheries. Treatment is not very effective, and the recovered birds become carriers.

Fowl typhoid affects mainly birds 12 weeks of age and older. Symptoms include loss of appetite, thirst, pale comb and wattles, listlessness, yellow or green diarrhea, and sudden death. The death rate may be high. Prevent the disease by securing stock from disease-free flocks and practicing good sanitation on the farm.

Paratyphoid affects mainly young birds. The symptoms are similar to pullorum, as is the death rate. Sanitation in the hatchery and on the farm helps to prevent the disease. Medications may be used in treatment to lower the death loss.

Paracolon infections are generally known as Arizona infections. A large number of bacteria are involved. The symptoms, control, and treatment are similar to paratyphoid. Laboratory diagnosis is needed to identify the disease.

Trichomoniasis

Trichomoniasis is caused by a protozoan parasite. Turkeys, chickens, and other birds are affected. Losses are highest among young and growing birds. Symptoms include loss of appetite, loss of weight, and droopiness. Sanitation is the best way to prevent the disease. Affected birds may be treated with copper sulfate added to the drinking water.

Ulcerative Enteritis

The cause of ulcerative enteritis is a bacterium. Chickens and quail are affected. The acute form may cause sudden death. Symptoms of the chronic form include ruffled feathers, humped up posture, anemia, and listlessness. A white diarrhea may be present. Antibiotics are used for the prevention and treatment of the disease.

Photo courtesy of Frank B. Flanders

Figure 37-8 External parasites are rarely found in large commercial poultry operations. However, these pests are common in home poultry flocks.

EXTERNAL PARASITES

A number of external parasites are problems in poultry production. Included are chicken mites, northern fowl mites, scaly leg mites, several species of lice, fowl ticks, fleas, bedbugs, and flies. The most serious of these parasites are mites and lice.

External parasites have all but been eliminated in large commercial poultry operations. This is due to the birds' limited contact with wild birds and other sources of parasites and diligent parasite control measures. In recent years, external parasites have been mainly a problem of small backyard flocks. With the current popularity of locally produced food and organic and free-range poultry, the number of small poultry producers is increasing (Figure 37-8). These smaller operations are much more susceptible to parasites than larger poultry operations that adhere to strict biosecurity practices.

Mites

Poultry mites are about 1/40 of an inch long and are barely visible without magnification. Mites are arachnids, the same family as spiders, and have eight legs. They will bite humans as well as chickens, causing small red skin lesions and intense itching. People who handle birds, gather eggs, or are in or near poultry housing infested with mites are often attacked (Figure 37-9).

Photo courtesy of Frank B. Flanders

Figure 37-9 Mites are bloodsucking pests and will readily attack people who come into contact with infested birds or their housing.

Common Chicken Mites

The common chicken mite is also called the red mite or the roost mite. The chicken mite sucks blood from the birds, reducing egg production and slowing growth. Young chickens may die due to diseases carried by the chicken mite. Common chicken mites feed on birds during the night and hide in cracks and crevices of the chicken house during the day. The life cycle of the common chicken mite is shown in Figure 37-10.

Northern Fowl Mite

The northern fowl mite is a bloodsucker occurring mostly in cooler climates. It spends all of its life cycle on the bird. Wild birds such as sparrows carry these mites and should be kept out of the poultry house. The northern fowl mite causes lower egg production, slow growth, irritation, and may cause death.

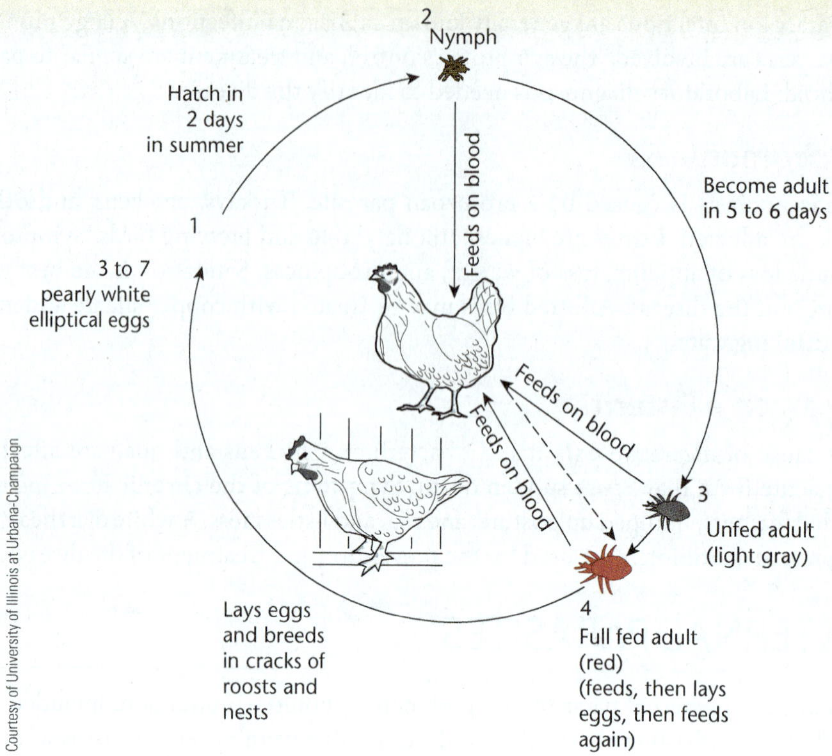

Figure 37-10 Life cycle of the common chicken mite.

Scaly Leg Mite

This mite causes scaly leg. It lives under the scales on the feet and legs of the bird. Heavy infestations cause a rough appearance and enlargement of the legs.

Poultry Lice

Lice, unlike mites, are insects and have six legs. The most common type of poultry louse is the body louse. Other lice that attack poultry are the shaft louse, fluff louse, wing louse, and the head louse. Lice may be up to 1/8 inch long, depending on species. All poultry lice have chewing mouth parts. They feed on skin, scab tissue, and feather parts. They do not suck blood but will feed on blood from damaged tissue. Poultry lice are species-specific and will not feed on humans. Infected birds lose weight, egg production drops, and young birds may die due to anemia.

Fowl Tick

The fowl tick is a dark-colored bloodsucker. It hides during the day in cracks and crevices. Fowl ticks can potentially cause paralysis due to toxins in their saliva.

Fleas and Bedbugs

Fleas and bedbugs are bloodsuckers, and may become a problem in the poultry house. Fleas stay on the birds. Bedbugs feed on the bird at night and hide in cracks and crevices during the day.

Flies

Flies do not directly attack poultry. However, they may become a nuisance around the poultry house. Heavy fly infestations may cause problems with neighbors.

Control of External Parasites

The basis of a control program is the sanitation procedures outlined earlier in this chapter. Insecticides may be used to control these parasites. Because recommendations and regulations on the use of insecticides change from time to time, no specifics are listed here. Check with poultry specialists, veterinarians, or the Cooperative Extension Service for current recommendations on the use of insecticides.

INTERNAL PARASITES

Internal parasites that affect poultry include large roundworms, cropworms (capillaria), cecal worms, tapeworms, flukes, gapeworms, and gizzard worms. The life cycles of some of these parasites are illustrated in Figures 37-11 through 37-15. Worm infestations generally cause slow growth and lower production. Medications are available for the treatment of some of these worms.

Control of worms is best achieved by strict sanitation. In range production, rotation of pastures helps prevent infestation. Controlling insects and keeping wild birds out of poultry houses is important in control. Birds raised in confinement and on wire or slat floors seldom have serious infestations of worms. Good management indicates that the flocks should be checked periodically for the presence of worms. Current recommendations on the use of medications for worm control may be secured from poultry specialists, veterinarians, or the Cooperative Extension Service.

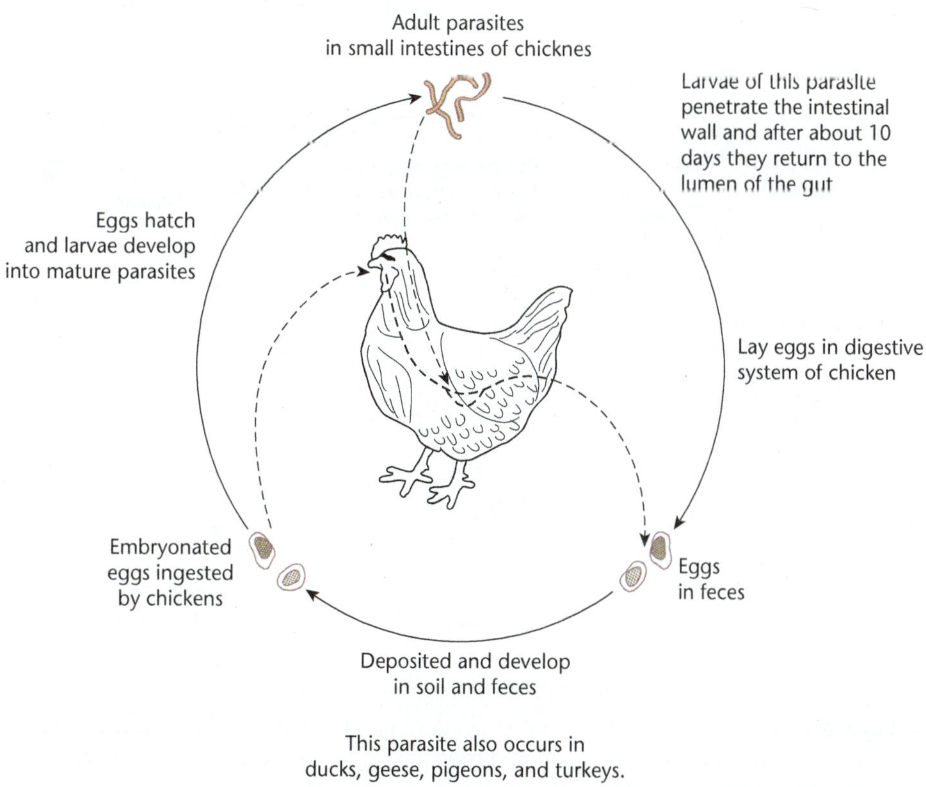

Courtesy of University of Illinois at Urbana-Champaign

Figure 37-11 Life cycle of large roundworms in poultry. Large roundworms spend their entire life cycle in the small intestine. The eggs must pass through a period of development before they can infect another fowl.

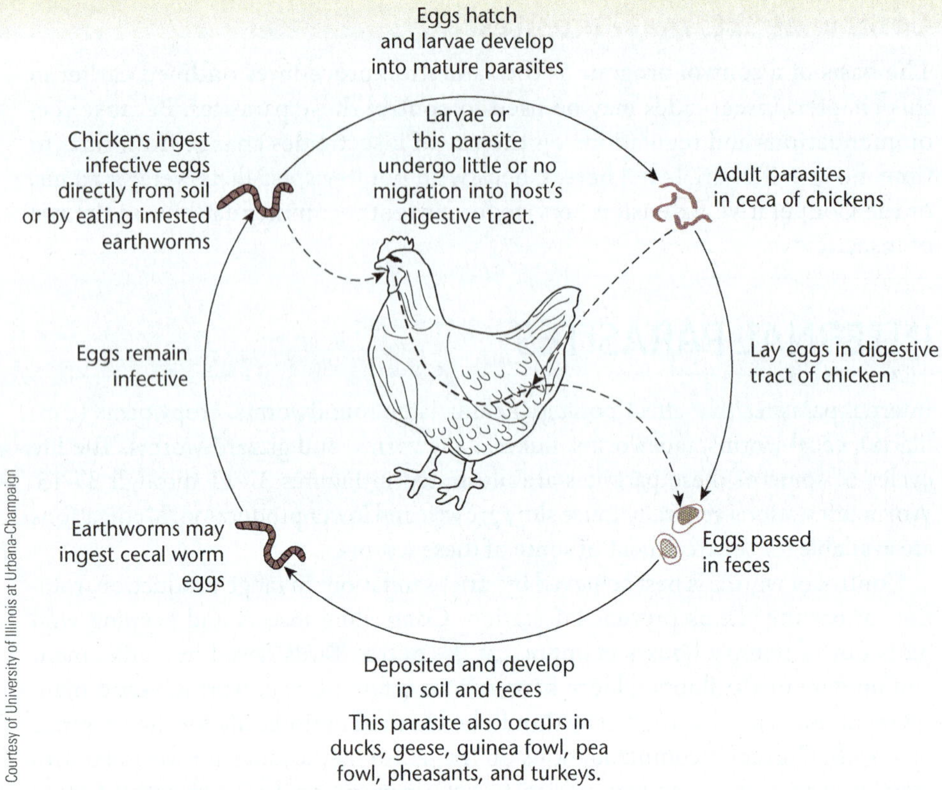

Figure 37-12 Life cycle of cecal worms in poultry. Cecal worms are found only in the ceca of chickens. Eggs must pass through a period of 7 to 12 days before infesting another fowl.

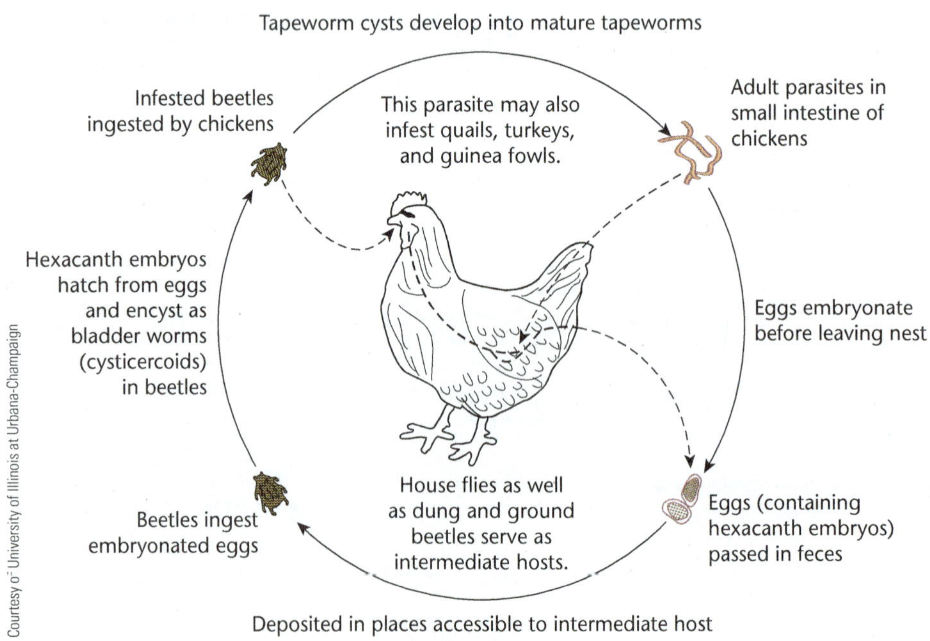

Figure 37-13 Life cycle of tapeworms in poultry. Tapeworms have an indirect life cycle; the eggs must develop in an intermediate host such as flies, beetles, slugs, earthworms, or grasshoppers.

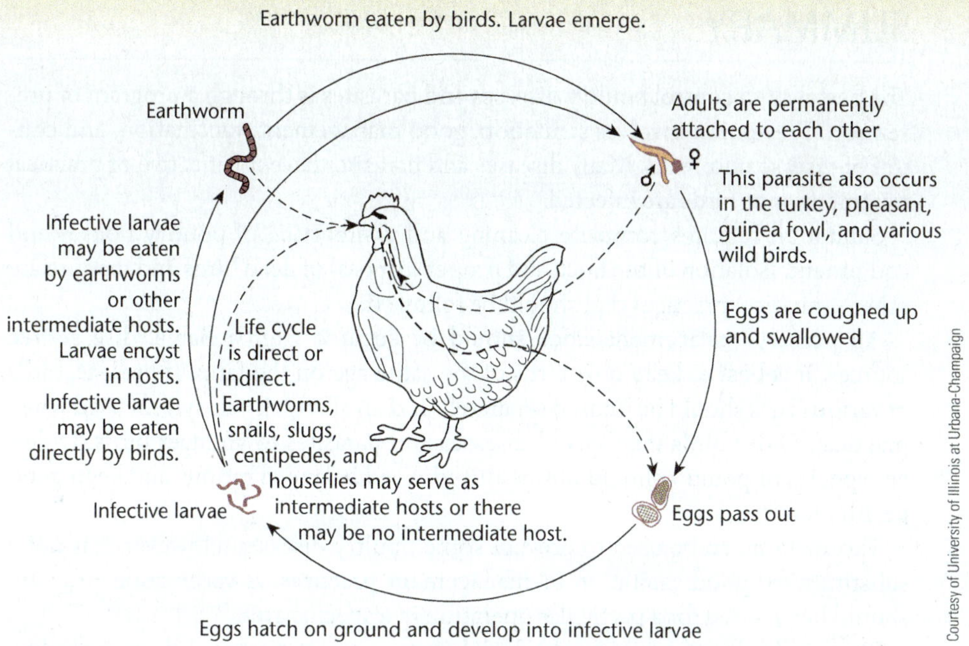

Figure 37-14 Life cycle of gapeworms in poultry. Gapeworms attach themselves to the lining of the trachea. Eggs are present in discharges coughed up by the bird or in the feces.

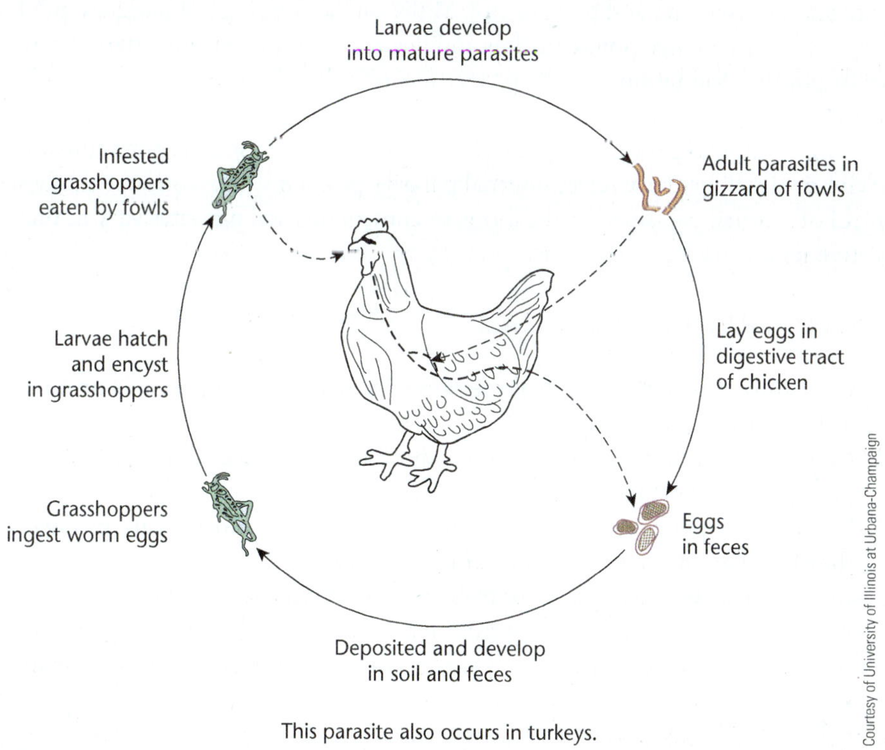

Figure 37-15 Life cycle of gizzard worms in poultry. Gizzard worms have an indirect life cycle. That is, the eggs must develop in an intermediate host such as grasshoppers, weevils, and certain species of beetles.

SUMMARY

The best way to control poultry diseases and parasites is through a program of prevention. Prevention involves sanitation, good management, vaccination, and control of disease outbreaks. Many diseases and parasites have no effective or practical treatment once birds are infected.

Sanitation requires complete cleaning and disinfecting of poultry houses and equipment. Isolation of the flock and proper disposal of dead birds and manure are other sanitation practices that should be followed.

All poultry replacement stock should be secured from reliable, disease-free sources. It is best to keep only birds of the same age on the farm. If present, birds of various ages should be housed separately and an all-in/all-out system used when practical. Older birds may spread diseases and parasites to younger birds. Different species of poultry should not be mixed. Feed balanced rations and keep good health records.

Vaccination can be used to prevent some poultry diseases. However, it is not a substitute for good sanitation or management practices. A vaccination program should be planned for a particular operation in a specific area.

The poultry flock should be checked daily for signs of disease. A drop in production or an increase in the death rate may indicate the presence of disease. Accurate diagnosis of poultry diseases is best done in a laboratory. Many diseases affect poultry. A large number have similar symptoms and effects. Some affect all kinds of poultry, while others affect only some species.

External parasites are rarely a problem in large commercial poultry operations due to strict biosecurity and pest control. Mites and lice often infest backyard poultry flocks and free-range poultry. Mites and lice are the most serious external parasites of poultry. Sanitation and the proper use of pesticides can keep these under control.

The main internal parasites of poultry are worms. Birds raised in confinement or on wire or slat floors have fewer internal parasite problems. Prevention is the best control of internal parasites. Medications to control internal parasites vary in their effectiveness.

Student Learning Activities

1. Give an oral report on control measures for diseases and parasites or on an individual disease or parasite.
2. Interview a veterinarian on the control of poultry diseases and parasites of poultry and report to the class.
3. Survey local poultry producers to determine disease and parasite problems they may have had and the control measures used.
4. Practice disease and parasite control measures on the home farm.
5. Use the disease and parasite control measures described in this chapter when planning and conducting a poultry supervised agricultural experience program.

Discussion Questions

1. Outline the steps for a good sanitation program for poultry houses.
2. List important health management practices for poultry.
3. Briefly explain why the vaccination of poultry is part of a good health program.
4. Describe how to control disease outbreaks in a poultry flock.
5. List and briefly describe diseases that affect poultry.

6. List and briefly explain the common external parasites of poultry.
7. List and briefly explain the control of internal parasites of poultry.

Review Questions

True/False

1. Visitors may enter poultry houses for short periods of time without special precautions or posing biosecurity issues.
2. If a good vaccination program is in place, a poultry owner can be lax on some of the disease prevention systems.
3. The disposal of manure may be handled by spreading it thinly on land that is not used for poultry.
4. Most poultry diseases can be accurately diagnosed without the use of a laboratory.
5. The poultry body louse is the most common type of external poultry parasite.

Multiple Choice

6. Parasites can be controlled by the use of __________.
   a. vitamins c. pesticides
   b. molluscicides d. antibiotics
7. A vaccine made from weakened bacteria is a(n) __________.
   a. fungicide c. herbicide
   b. bacterin d. pesticide

Completion

8. __________ __________ is caused by a virus and affects turkeys, chickens, and other birds.
9. The best way to control poultry diseases and parasites is through a program of __________.

Short Answer

10. List four steps involved in preventive management of diseases and parasites in poultry.
11. Name four microorganisms that cause poultry diseases.
12. Name two signs of common poultry diseases.
13. What does an all-in/all-out program mean in poultry production?
14. What is the cause of cage fatigue in chickens?
15. List five ways a poultry flock can be contaminated with disease.

Chapter 38

Marketing Poultry and Eggs

Key Terms

viscera
green geese
candling

Objectives

After studying this chapter, the student should be able to

- summarize the production and price trends in poultry and eggs.
- describe the methods of marketing poultry and eggs.

PRICE AND PRODUCTION TRENDS IN POULTRY AND EGGS

Long-Term Production and Price Trends

Since 1970, the growth of the poultry industry has increased rapidly. Broiler meat produced in the United States is five times greater today than in 1970 (Figure 38-1). Figure 38-2 illustrates the recent trends in broiler production and prices. Generally, both price and production of broilers have increased over the past several years.

Figure 38-3 shows the recent trends in egg prices and production. Egg production and prices have shown variation during this period. Production has been generally trending upward, and since 2000, egg production has steadily increased.

Figure 38-4 reflects the recent trends in turkey per capita consumption in the United States. While there has been some variation from year to year, the general trend has been upward.

Production and prices tend to follow the laws of supply and demand. It is probable that vertical integration in the poultry industry has also had an effect on the long-term trends in production and prices.

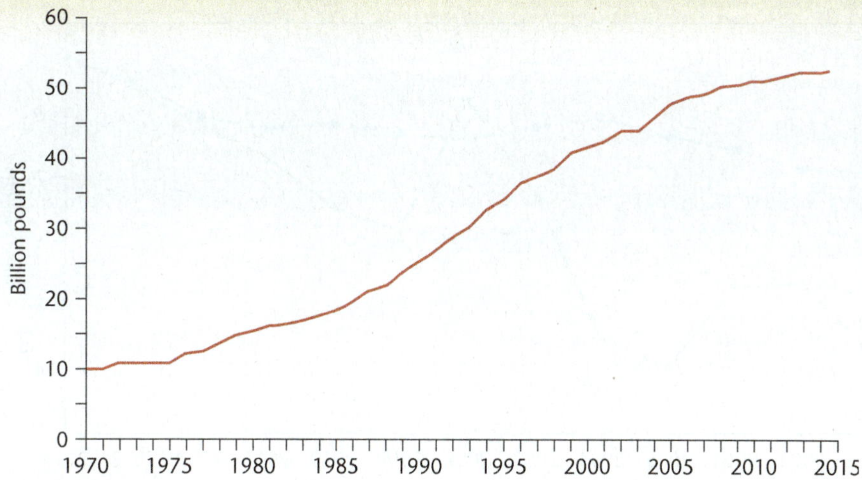

Figure 38-1 Pounds of Broilers Produced in the United States.
Source: USDA NASS

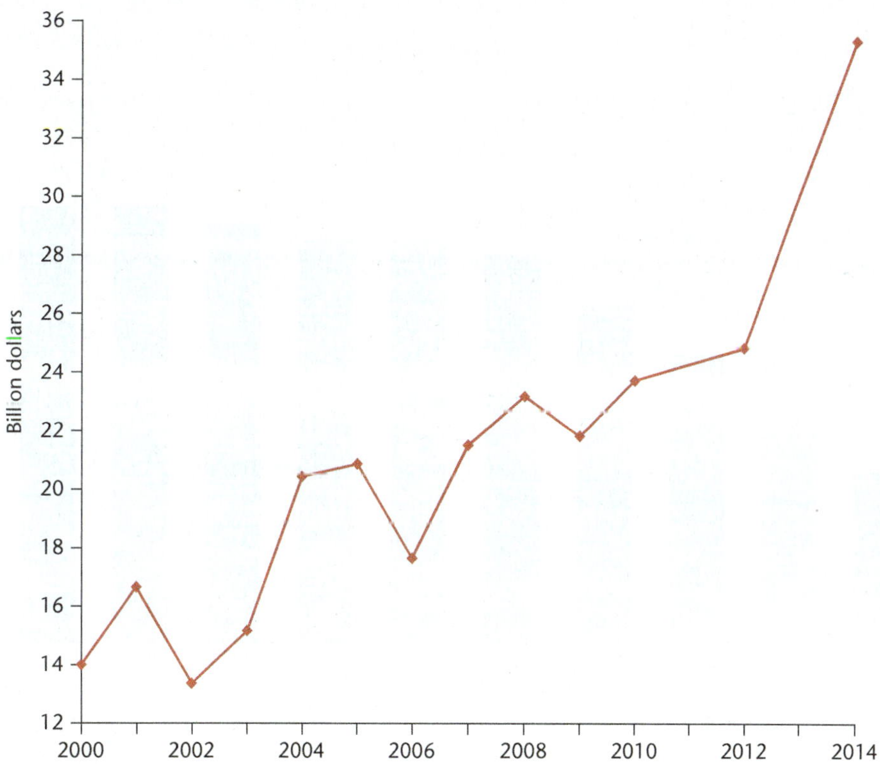

Figure 38-2 Value of Broiler Production, United States.
Source: USDA NASS

MARKETING POULTRY

Broilers and Other Chickens

Vertical Integration Contract

Approximately 99 percent of the broilers raised in the United States are grown under some type of vertical integration contract. As a result of vertical integration, there are few actual sales of live birds. The birds are owned by the integrator rather than the farmer who feeds them. The contract the farmer has with the integrator spells out the terms under which payment is made for growing the birds.

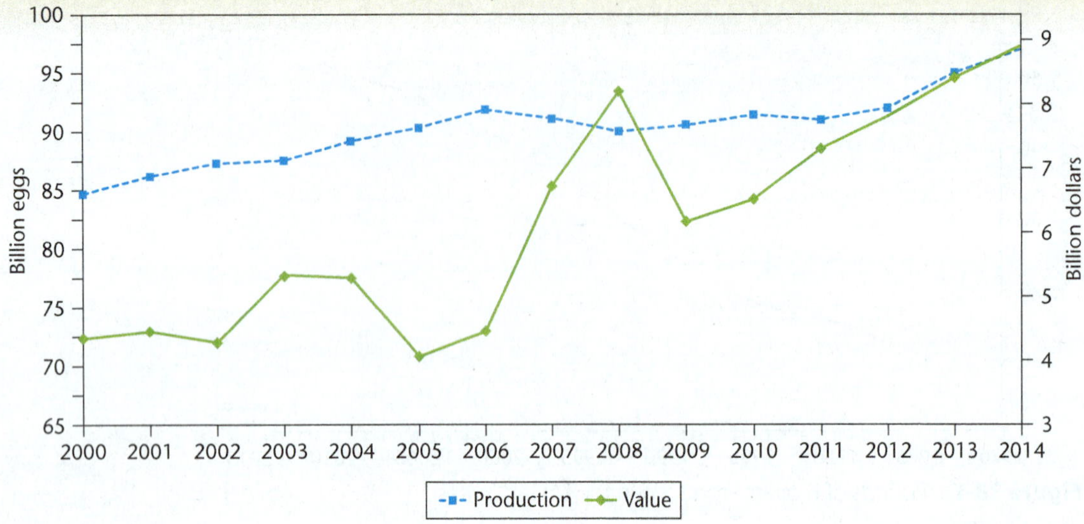

Figure 38-3 Egg Production and Value, United States.
Source: USDA NASS

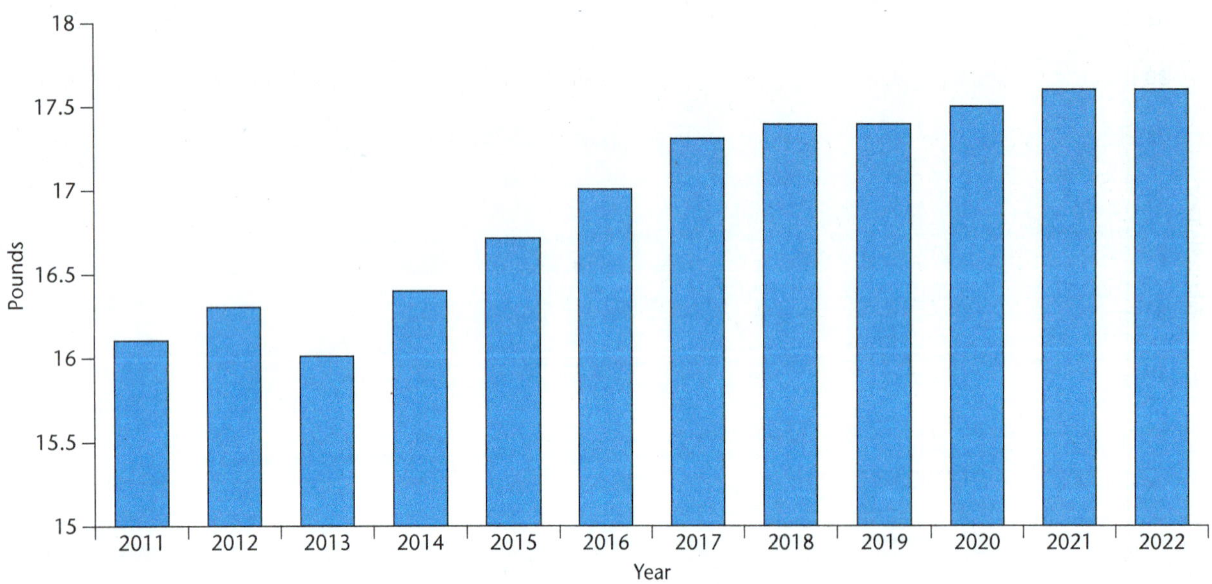

Figure 38-4 Consumption of Turkey Per Capita.
Source: USDA NASS

A contract usually gives a guaranteed minimum payment for the birds. In addition, there is often a bonus clause that provides for payment of an additional amount if the feed conversion is better than the average. Payments are made on the basis of live weight at the time the birds are marketed. Contracts are regulated by the Packers and Stockyards Act, which requires written contracts.

Poultry Not Under Contract

Poultry not produced under contract may be sold live, dressed, or ready-to-cook. Sales may be made to the buyer at the farm, at auction, to dealers, to brokers, or to processing plants. If the poultry is processed on the farm, it may be sold to wholesale dealers or direct to the consumer. However, very little poultry in the United States is sold in this manner.

Preparing Birds for Market

Chickens should not be fed for 12 hours before slaughter but should be provided with water. Handle birds carefully to prevent injuries and bruises. Avoid crowding birds in delivery coops. It is best to move birds in the early morning if possible.

Turkeys

About 75 percent of the young turkeys produced in the United States are marketed between August and December. Fryer-roaster turkeys are marketed throughout the year. Most turkeys are marketed through integrated firms, processors, or cooperatives.

Large broad-breasted male turkeys are usually ready for market at 24 weeks, while females are ready at 20 weeks. Small- and medium-type turkeys are ready at 22 weeks. Fryer-roasters are ready to market at 12 to 13 weeks and light roasters at 17 to 20 weeks.

Preparing Turkeys for Market

Turkeys are usually not taken off of feed and water before being moved to market for slaughter. Care must be used when handling turkeys to prevent injuries and bruises that will cause a lowering of the market grade. Turkeys may be handled with less confusion and injury to the birds if they are confined to a small area, such as a chute or pen, when loading them for market. In confinement systems, a small, darkened room works well for catching turkeys.

Ducks

Meat ducks of the Pekin type are usually ready for market at 7 to 8 weeks of age. Muscovy ducks are ready in 10 to 17 weeks. Ducks are marketed when they have a good finish and are free of pin feathers.

Ducks are taken off of feed 8 to 10 hours before they are slaughtered. They are provided with water until slaughter. Use care when catching, handling, and moving ducks to prevent injuries or bruises. Small numbers of ducks may be processed on the farm for direct sale to consumers. Larger producers sell to processing plants that also sell the feathers, feet, heads, and **viscera** (internal organs) as by-products. Small producers who process their own ducks usually do not find a market for these by-products.

Geese

Geese are usually sold alive off the farm to live-poultry buyers or to processing plants. Most geese are marketed in the fall and winter. The best prices are generally obtained in large cities around Thanksgiving and Christmas. Usually, geese are marketed when they weigh 11 to 15 pounds. Geese may be full fed for fast growth. Geese fed in this way are called **green geese** or junior geese. They are marketed at about 10 weeks when they weigh 10 to 12 pounds.

Goose feathers are valuable in the bedding and clothing industries. Buyers of goose feathers are found in most large cities. The feathers may be sold to a feather processing plant. One pound of dry feathers can be produced from three geese.

CLASSES OF READY-TO-COOK POULTRY

Chickens

The USDA sets standards for six classes of ready-to-cook chickens (Figure 38-5), as follows:

1. Rock Cornish game hen or Cornish game hen. A young immature chicken of either sex, usually less than 5 weeks of age, with a ready-to-cook weight of not more than 2 pounds.

Photo by Joe Valbuena, USDA

Figure 38-5 Ready-to-cook poultry.

2. Broiler or fryer. A young chicken of either sex, usually less than 10 weeks of age, that is tender-meated with soft, pliable, smooth-textured skin and flexible breastbone cartilage.
3. Roaster or roasting chicken. A young chicken, usually less than 12 weeks of age, of either sex, that is tender-meated with soft, pliable, smooth-textured skin and flexible breastbone cartilage.
4. Capon. A surgically neutered male chicken, usually less than 4 months of age, that is tender-meated with soft, pliable, smooth-textured skin.
5. Hen, fowl, baking chicken, or stewing chicken. An adult female chicken, usually more than 10 months of age, with meat less tender than that of a roaster or roasting chicken and with a nonflexible breastbone tip.
6. Cock or rooster. An adult male chicken with coarse skin, toughened and darkened meat, and nonflexible breastbone tip.

Turkeys

The USDA sets standards for four classes of ready-to-cook turkeys, as follows:

1. Fryer-roaster turkey. A young, immature turkey of either sex, usually less than 12 weeks of age, that is tender-meated and has soft, pliable, smooth-textured skin and flexible breastbone cartilage. This class may also be labeled "young turkey."
2. Young turkey. A turkey generally under 6 months of age, of either sex, with tender meat and soft, pliable, smooth-textured skin. The breastbone cartilage is a little less flexible than in a fryer-roaster turkey.
3. Yearling turkey. A fully matured turkey usually under 15 months of age, of either sex, that is reasonably tender-meated. It has reasonably smooth-textured skin.
4. Mature or old (hen or tom) turkey. An adult turkey, usually more than 15 months of age, of either sex, with coarse skin and toughened flesh. Sex designation is optional.

Ducks

The USDA sets standards for three classes of ready-to-cook ducks, as follows:

1. Duckling. A young duck, usually less than 8 weeks of age, of either sex, that is tender-meated and has a soft bill and soft windpipe.
2. Roaster duck. A young duck, usually less than 16 weeks of age, of either sex, that is tender-meated and has a bill that is not completely hardened and a windpipe that is easily dented.
3. Mature duck or old duck. An adult duck, usually more than 6 months of age, of either sex, with toughened flesh, a hardened bill, and a hardened windpipe.

Geese

The USDA sets standards for two classes of ready-to-cook geese, as follows:

1. Young goose. An immature goose of either sex that is tender-meated and has a windpipe that is easily dented.
2. Mature or old goose. An adult goose of either sex that has toughened flesh and a hardened windpipe.

GRADES OF READY-TO-COOK POULTRY

All poultry slaughtered for human food in the United States must be processed, handled, packaged, and labeled in accordance with federal law. The applicable laws are the Poultry Products Inspection Act of 1957 and the Wholesome Poultry Products

Act of 1968. Before poultry is graded for quality, it must be inspected for wholesomeness and fitness for human food (Figure 38-6). Mandated federal inspection is paid for by the government. Quality grading is voluntary and is paid for by the processor. Unsound or unwholesome poultry are not eligible for quality grading.

The U.S. quality grades of poultry are grades A, B, and C. These grades are applied both to the ready-to-cook carcass and to the parts of the carcass. Parts include such things as poultry halves, breast, leg, thigh, drumstick, wing, and tenderloin. No grade standards exist for giblets, detached necks and tails, wing tips, and skin.

USDA

Figure 38-6 A U.S. Department of Agriculture poultry inspector checks broilers for wholesomeness as they make their way through a processing plant.

The quality grade is determined by an evaluation of the following factors: (1) conformation, (2) fleshing, (3) fat covering, (4) defeathering, (5) exposed flesh, (6) discolorations, (7) disjointed and broken bones, and (8) freezing defects. Grade A is the only grade of poultry that is generally sold at retail. Occasionally, grades B and C are sold at retail, but it is more common for these grades to be used in processed poultry products, where they are cut up, chopped, or ground. Figures 38-7 through 38-13 illustrate representative carcasses in the various poultry grades. A given lot of poultry may occasionally contain a small percentage of a quality lower than the grade specified because some defects are permitted. This is an unavoidable necessity due to today's production-type processing methods.

The standards for grading ready-to-cook poultry are found in the *USDA Poultry Grading Manual.* These grades apply to all kinds and classes of poultry. The *Poultry Grading Manual* regulations govern the grading of poultry, and standards, grades, and weight classes for shell eggs, which can be found on the USDA's Agricultural Marketing Services website.

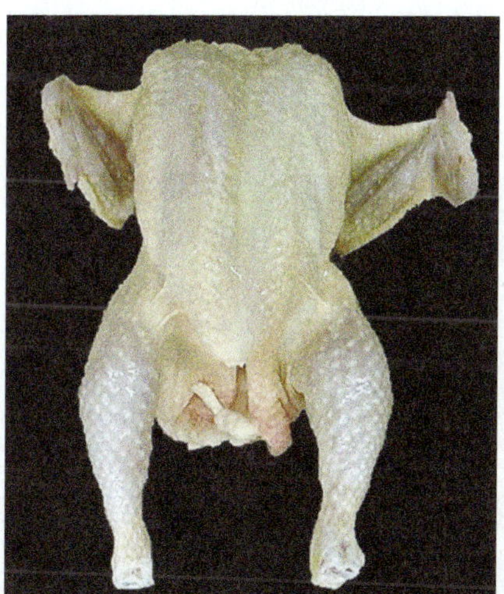

USDA

Figure 38-7 Grade A chicken carcass.

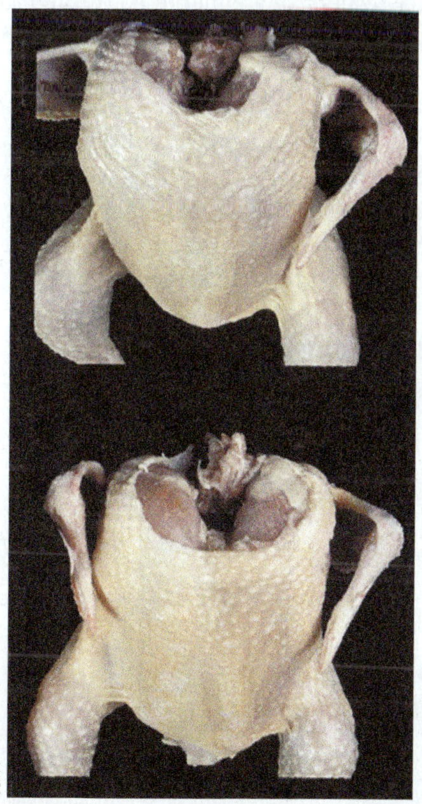

USDA

Figure 38-8 Grade A chicken carcass (top). The carcass shown at the bottom is Grade B due to excess neck skin removed during processing.

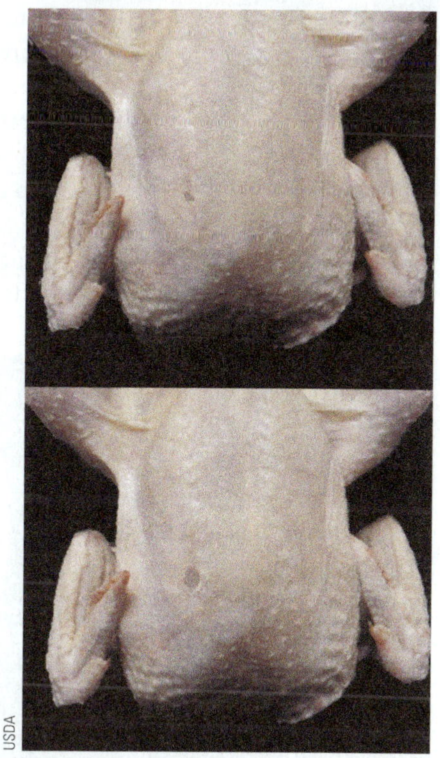

USDA

Figure 38-9 Chicken carcasses. Grade A (top)—the exposed flesh does not exceed $\frac{1}{4}$-inch tolerance for breast and legs. Grade B (bottom)—exposed flesh exceeds $\frac{1}{4}$-inch tolerance for breast and legs.

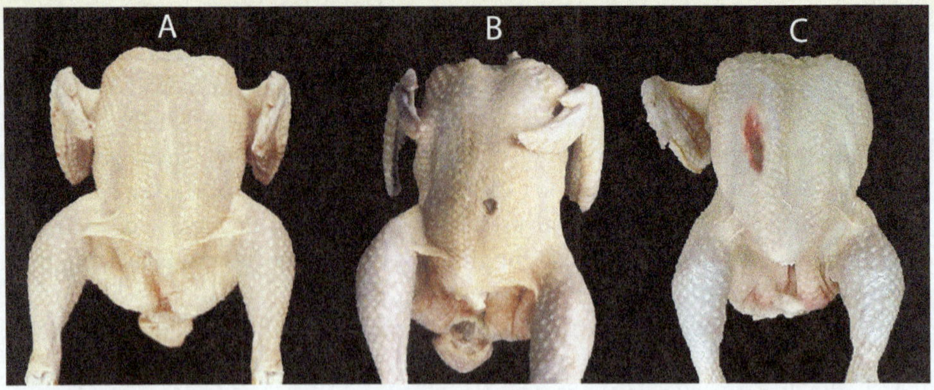

Figure 38-10 Chicken carcasses, Grades A, B, and C. The carcass in the middle has exposed flesh greater than ¼ inch. Grade C, on the right, has an entire wing missing and discoloration greater than 2 inches and more than moderate in color.

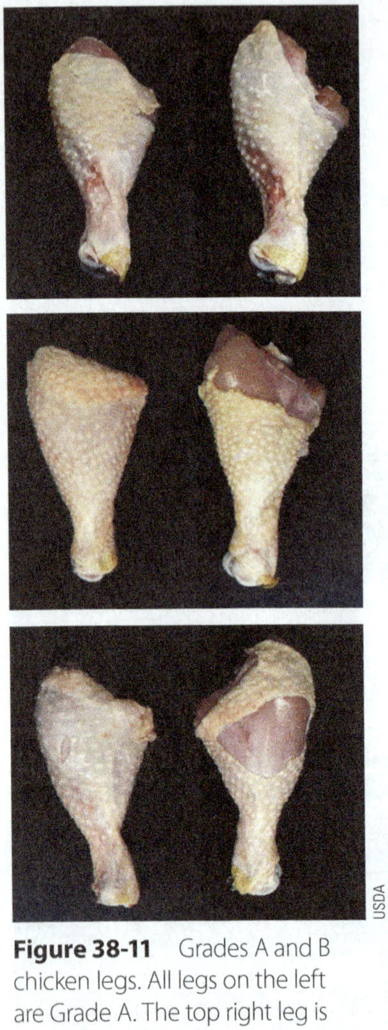

Figure 38-11 Grades A and B chicken legs. All legs on the left are Grade A. The top right leg is Grade B due to discoloration. The top left leg has some discoloration but within tolerance. The right leg of the middle pair has too much flesh exposed and is Grade B. The bottom pair, Grade A on the left and Grade B on the right; the Grade B leg has too much flesh exposed. Grade A has some flesh exposed but is within tolerance.

MARKETING EGGS

Vertical integration is not as widespread in egg production as it is in broiler production. However, more than half of the eggs produced in the United States are produced under some type of vertical integration in the egg industry. Eggs produced under a vertical integration contract are marketed in much the same manner as broilers. The grower does not own the birds, and the eggs are paid for under the terms of the contract.

Eggs that are not produced under contract are sold to local buyers, are sold to produce dealers, or are marketed through cooperatives. Some producers may sell directly to consumers at the farm or at roadside stands. These types of marketing practices require more labor than selling to produce dealers or through cooperatives. An increasing interest in organic foods has created a good market for locally grown organic eggs sold directly to the consumer or local supermarkets and restaurants.

Eggs are highly perishable. They must be handled carefully to prevent spoiling. Egg quality is best preserved by keeping the eggs cool and at the correct humidity after they are laid and until they are used.

Figure 38-12 Grade A turkey carcass.

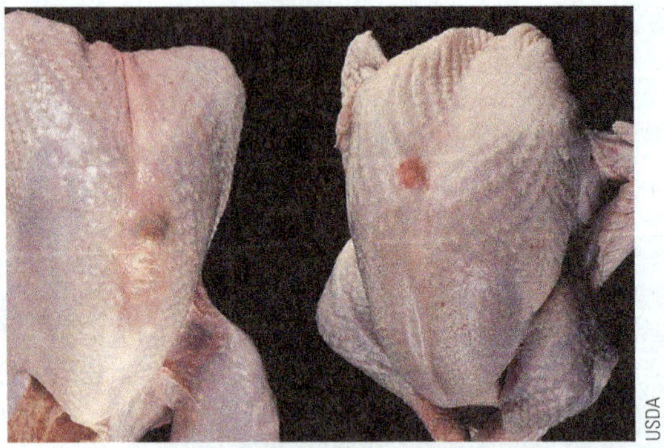

Figure 38-13 Grade B turkey carcasses.

CLASSES AND GRADES OF EGGS

The USDA sets standards for weight classes and grades of shell eggs. Grades of eggs are based on these four factors: (1) shell, (2) air cell, (3) white, and (4) yolk. Table 38-1 is a summary of the U.S. standards for quality of shell eggs. Examples of various grades of eggs are shown in Figures 38-14 through 38-19.

Classes of eggs are based on weight shown as ounces per dozen. The weight classes for U.S. Consumer Grades of shell eggs are shown in Table 38-2 and illustrated in Figure 38-20.

TABLE 38-1 Summary of U.S. Standards for Quality of Individual Shell Eggs (Based on Candled Appearance)

	Specifications for Each Quality Factor		
Quality Factor	**AA Quality**	**A Quality**	**B Quality**
Shell	Clean Unbroken Practically normal	Clean Unbroken Practically normal	Clean to slightly stained* Unbroken Abnormal
Air cell	1/8 inch or less in depth Unlimited movement and free or bubbly	3/16 inch or less in depth Unlimited movement and free or bubbly	Over 3/16 inch in depth Unlimited movement and free or bubbly
White	Clear Firm	Clear Reasonably firm	Weak and watery Small blood and meat spots present**
Yolk	Outline: slightly defined Practically free from defects	Outline: fairly well defined Practically free from defects	Outline: plainly visible Enlarged and flattened Clearly visible germ development but no blood Other serious defects

Standards of Quality for Eggs with Dirty or Broken Shells

Dirty	Check
Unbroken. Adhering dirt or foreign material, prominent stains, moderate stained areas in excess of B quality	Broken or cracked shell but membranes intact, not leaking***

*Moderately stained areas permitted (1/32 of surface if localized, or 1/16 if scattered).
**If they are small (aggregating not more than 1/8 inch in diameter).
***Leaker has broken or cracked shell and membranes, and contents leaking or free to leak.
Source: USDA

TABLE 38-2 U.S. Weight Classes for Consumer Grades for Shell Eggs

Size or Weight Class	Minimum Net Weight per Dozen	Minimum Net Weight per 30 Dozen	Minimum Weight for Individual Eggs at Rate Per Dozen
	Ounces	**Pounds**	**Ounces**
Jumbo	30	56	29
Extra large	27	50.5	26
Large	24	45	23
Medium	21	39.5	20
Small	18	34	17
Peewee	15	28	—

Source: U.S. Standards Grades, and Weight Classes for Shell Eggs, Agricultural Marketing Service, Poultry Division, U.S. Department of Agriculture, April 6, 1995

USDA

Figure 38-14 Grade AA, or Fancy Fresh, egg content covers small area; white is thick and stands high, with small amount of thin white; yolk is firm, round, and stands high.

USDA

Figure 38-16 Grade A egg content covers a moderate area; white is reasonably thick and stands fairly high; yolk is firm and high.

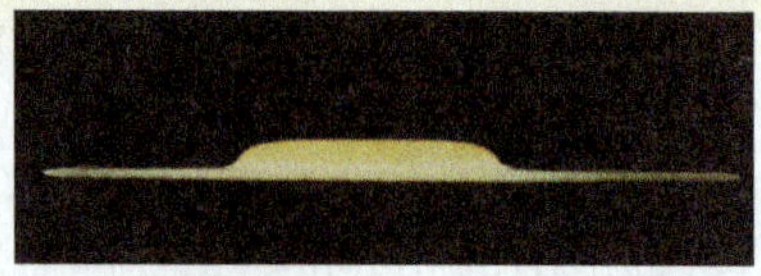

USDA

Figure 38-18 Grade B egg content covers a very wide area; white is weak and watery; it has no thick white; a large amount of thick white is thinly spread. Yolk is enlarged and flattened.

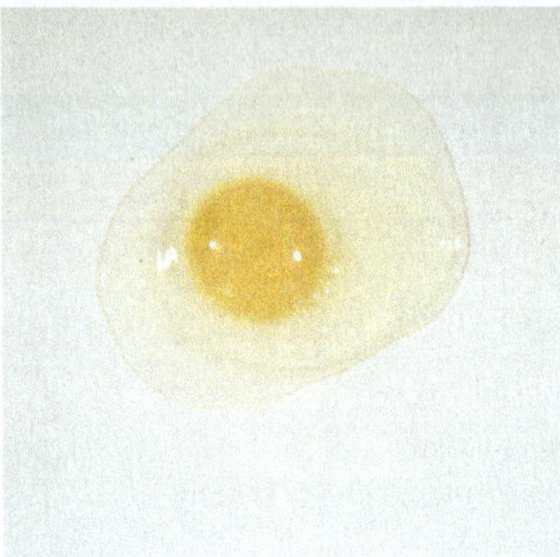

USDA

Figure 38-15 Grade AA egg, top view. The egg covers a small area.

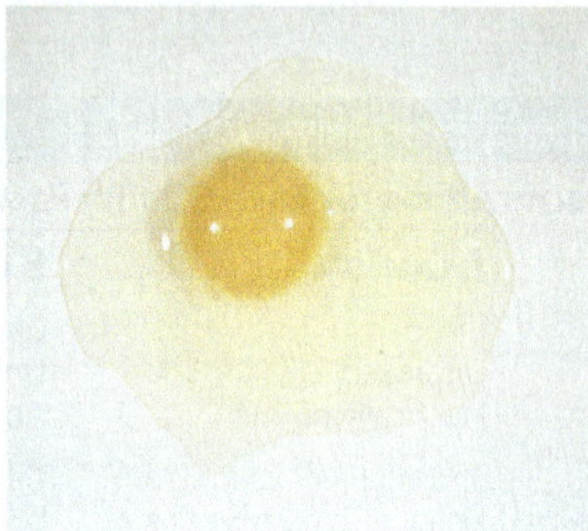

USDA

Figure 38-17 Grade A egg, top view. Egg content covers a moderate area.

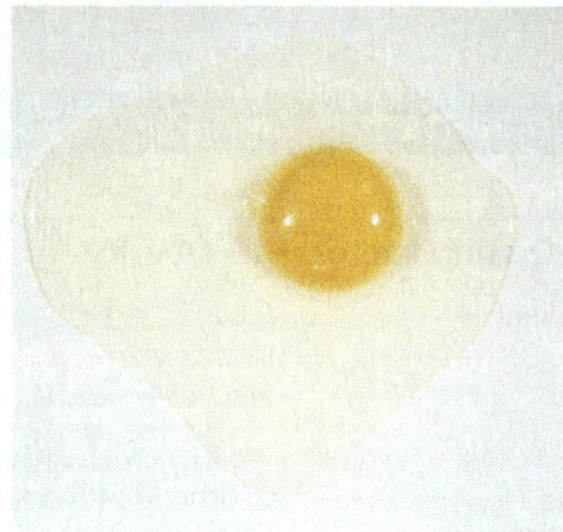

USDA

Figure 38-19 Grade B egg, top view. Egg content covers a very large area.

USDA Agricultural Research Service; Photo by Keith Weller

Figure 38-21 Hand candling of eggs using a high-intensity light.

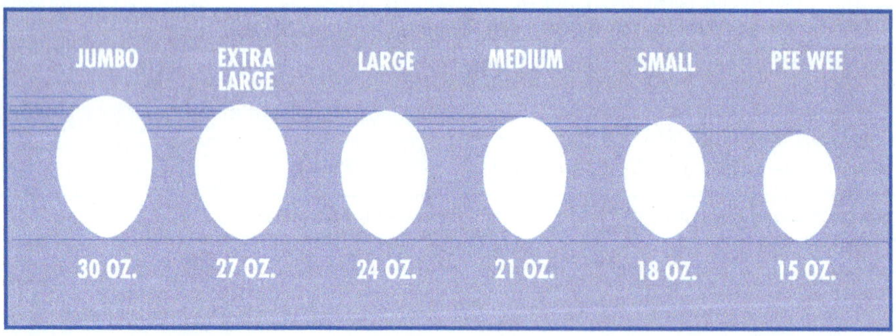

USDA

Figure 38-20 U.S. weight classes for eggs. Minimum weight per dozen eggs.

© iStockPhoto.com/Chalabala

Figure 38-22 Mass candling of eggs.

The interior quality of the egg is determined by a process called candling. The eggs are examined by using a high-intensity light. Figure 38-21 shows eggs being hand candled. Figure 38-22 shows the use of a conveyor belt to mass candle the eggs. This is the most common method of candling.

U.S. standards for grades of eggs do not consider the color of the egg. The color of the egg does not affect its nutritive value. However, some markets prefer white eggs, whereas others prefer brown eggs. Therefore, eggs are sometimes sorted by color for the market in which they are to be sold.

Consumer, wholesale, and procurement grades of eggs are different classifications used in the marketing of eggs. A complete description of these classifications is found in the *U.S. Standards, Grades, and Weight Classes for Shell Eggs*, which is available from the USDA.

SUMMARY

Since 1970, the growth of the poultry industry has increased rapidly. The long-term trend in production of broilers has been upward. Egg production and prices have shown variation; however, production has been generally trending upward. While there has been some variation from year to year in turkey per capita consumption, the general trend has also been upward. Long-term trends in prices show variations resulting from supply and demand and the influence of vertical integration on the market.

About 99 percent of the broilers raised in the United States are produced under some type of vertical integration contract. Most of the turkeys in the United States are marketed through integrated firms, processors, or cooperatives. Poultry must be carefully handled during marketing to avoid injuries and bruises.

Ducks and geese may be processed on the farm for direct sales to consumers. Larger producers generally sell to processing plants. Goose feathers are valuable in the bedding and clothing industries.

The USDA has established six classes of ready-to-cook chickens, four classes of ready-to-cook turkeys, three classes of ready-to-cook ducks, and two classes of ready-to-cook geese. Classes are based on age, tenderness of meat, smoothness of skin, and hardness of breastbone cartilage.

The three grades of ready-to-cook poultry are known as A, B, and C. Standards for grading poultry are found in the *USDA Poultry Grading Manual.*

More than one-half of the eggs produced in the United States are marketed under some type of vertical integration contract. Other eggs are sold to local buyers, produce dealers, cooperatives, or direct to consumers. Eggs must be handled carefully to maintain high quality.

The standards for weight classes and grades of eggs are set by the USDA. Grades are based on (1) shell, (2) air cell, (3) white, and (4) yolk. Weight classes are based on the ounces per dozen eggs.

Student Learning Activities

1. Prepare a display of eggs showing the different grades and weight classes.
2. Talk with a poultry producer about marketing poultry products and give a report to the class about what you learn.
3. Take a field trip to a poultry processing plant to observe inspection and grading procedures.
4. Talk with a poultry grader or egg candler and give a report to the class about what you learn.
5. Prepare a bulletin board display of the grades of poultry.
6. Follow the practices described in this chapter when planning and conducting the marketing of meat or eggs from a poultry supervised experience program.

Discussion Questions

1. Describe the long-term production and price trends for broilers, eggs, and turkeys.
2. Describe the marketing of broilers under contract.
3. List other methods besides contract that are used to market broilers.
4. Describe how broilers should be handled just before and at marketing.
5. List the common methods of marketing turkeys.
6. At what age are turkeys usually marketed?

7. At what age are ducks usually marketed?
8. Describe the marketing of geese.
9. Name and describe the six USDA classes of ready-to-cook chickens.
10. Name and describe the four USDA classes of ready-to-cook turkeys.
11. Name the market classes of ready-to-cook ducks and geese.
12. List the factors that are used for quality grading of ready-to-cook poultry.
13. Describe the grading of ready-to-cook poultry.
14. List and briefly describe the methods of marketing eggs.
15. Briefly describe the U.S. standards for quality of shell eggs.
16. List the weight classes for U.S. Consumer Grades of shell eggs.
17. How is the interior quality of shell eggs determined?
18. What effect does the color of an egg have on its nutritive value?

Review Questions

True/False

1. The quantity of broiler meat produced in the United States has been decreasing steadily since 1970.
2. Approximately 99 percent of the broilers raised in the United States are grown under vertical integration contracts.
3. Meat ducks of the Pekin type are usually ready for market in 12 weeks.
4. Eggs are highly perishable and must be stored carefully.
5. Chickens should not be fed for 12 hours before slaughter.

Multiple Choice

6. The long-term trend in the production of broilers and turkeys has been ______.
   a. unchanged
   b. upward
   c. downward
   d. in cycles
7. Classes of eggs are based on weight shown as ______ per dozen.
   a. ounces
   b. pounds
   c. grams
   d. kilograms

Completion

8. Ducks are taken off feed ______ to ______ hours before they are slaughtered.
9. Geese are marketed when they weigh ______ to ______ pounds.

Short Answer

10. During which months are most young turkeys marketed in the United States?
11. About how many weeks are required to produce a large broad-breasted turkey for market?
12. List three USDA grades for poultry.
13. List four quality factors used to determine the grade of eggs.
14. Name four of the size and weight classes for eggs.

SECTION 9

Dairy Cattle

Chapter 39

Breeds of Dairy Cattle

Key Terms

Grade A milk
Grade B milk
registered

Objectives

After studying this chapter, the student should be able to

- describe the characteristics of the dairy enterprise.
- identify the major breeds of dairy cattle.
- describe the breeds of dairy cattle, giving their origin and breed characteristics.

CHARACTERISTICS OF THE DAIRY ENTERPRISE

Dairy cows must be milked at regular intervals, two times per day, 7 days per week. Some herds are milked three times per day. The dairy cow is a creature of habit (Figure 39-1). A change in routine, such as a wide variation in milking time or strangers in the milking area, can reduce milk production. The modern dairy farmer uses mechanical equipment to milk, feed, and care for the dairy herd. Milk is moved by pipeline from the milking machine to the bulk tank where it is stored until it is picked up for market by the milk transport. Few modern dairy farm workers handle milk in cans.

In addition to the regular milking chores, most dairy farmers raise the crops needed to feed the dairy herd. Corn, small grains, hay, and pasture are raised on the farm to feed the cows. Some dairies, especially larger operations, buy most or all of the needed feed.

The dairy barn and milking parlor must be kept clean. In stanchion barns, the manure must be removed daily during the winter when the cows are kept inside most of the time. Milking parlors become dirty as the cows pass through for milking. These parlors must be kept washed and clean for the collection of milk. Insects and rodents must be controlled in and around the dairy buildings.

USDA Agricultural Research Service; Photo by Peggy Greb

Figure 39-1 The dairy cow is a creature of habit. Milk production can be reduced by small changes in routine, such as strangers in the milking parlor.

Courtesy of Harveststore-Engineered Storage Products Co

Figure 39-2 Dairy farms require high capital investments in land, cattle, facilities, and equipment.

Advantages of the dairy enterprise:

- Dairy cattle use feed roughages that might otherwise be wasted
- Dairying provides a steady income throughout the year
- Labor is utilized throughout the year
- With good management, death losses in the dairy herd are usually low

Disadvantages of the dairy enterprise:

- A high capital investment is needed
- The labor requirement is high
- The operator is confined to a regular schedule of milking
- Training and experience are needed before entering into the dairy business
- It takes a relatively long time to develop a high-producing dairy herd

The dairy farmer must keep good records to run an efficient operation. Milk production records are kept to help in culling cows and in the breeding program. Records of breeding and calving are kept. Crop, animal health, and business records are also necessary.

A successful dairy farmer must have patience and be willing to work long hours. The capital investment needed to get into dairy farming is high. The average investment is $6,000 to $7,000 per cow. A person must be able to get credit to borrow needed money. Experience on a dairy farm is needed. Experience may be gained by working for a dairy farmer before trying to start a new operation (Figure 39-2).

The dollar return on the money invested in the dairy farm may be from 6 to 9 percent. The average labor per cow per year is about 27 hours. The dollar return per hour of labor varies depending on feed and other costs. Good management is needed to get the best returns per hour of labor invested in the dairy enterprise.

TRENDS IN DAIRY PRODUCTION

The number of dairy cows in the United States has shown a steady decline for many years (Figure 39-3). There was a slight increase in numbers of dairy cows during the early 1980s, with the numbers again showing a decline since 1985.

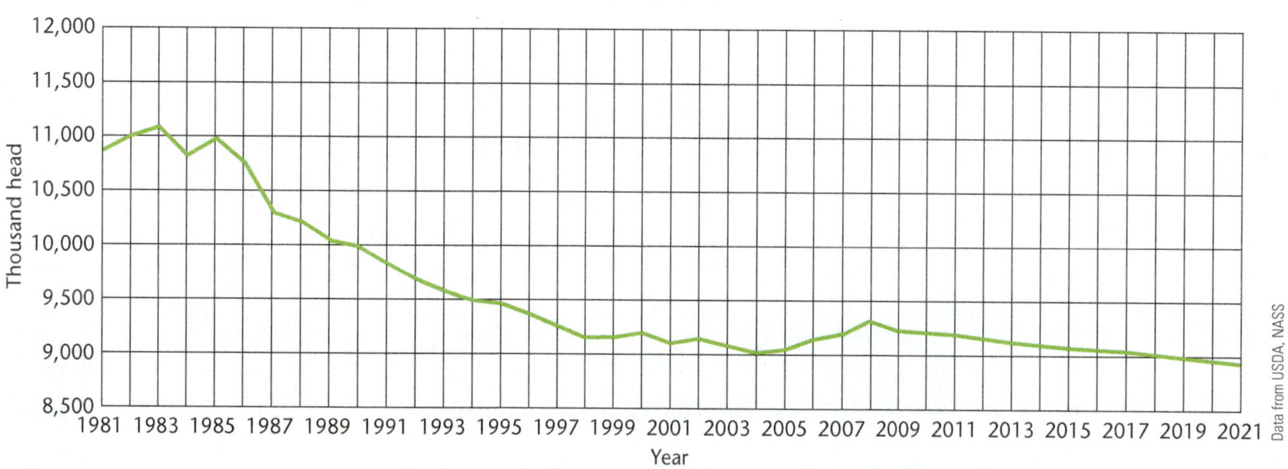

Figure 39-3 Trend in milk cow numbers, 1981–2021 (projected).

Periodically, dairy termination programs or dairy buy-out programs have been administered by the USDA and the dairy industry. The programs encourage dairy farmers to sell entire herds for slaughter and are designed to reduce an oversupply of milk in the marketplace by reducing the number of dairy farms. The effectiveness of the programs has been modified by an increase in the amount of milk produced per cow in the herds that remained in production.

There has been an ongoing trend in the United States toward fewer dairy farms, with the remaining herds becoming larger. In 1954, there were 2,167,000 farms in the United States that reported one or more dairy cows; by 2001, there were only 97,560 farms reporting one or more dairy cows, and by 2012, there were only 51,481 farms, a decline of 47 percent since 2001. The average size of dairy herds per farm in 1954 was 10 head of cattle; by 2001, the average size of dairy herds per farm had increased to 94.2 head of cattle, and by 2012, the average size had increased to 179 head of cattle per farm.

The total production of milk in the United States, while varying from year to year, has shown a general upward trend (Figure 39-4). This has occurred despite a general downward trend in the number of dairy cows in the United States because of an increase in milk production per cow (Figure 39-5). This increase in production per cow is the result of better feeding, breeding, and management of the dairy herds. Dairy cattle production facts are shown in Figure 39-6.

Grade A milk is produced on farms that have been certified to meet certain minimum standards. These standards deal with the buildings and conditions under

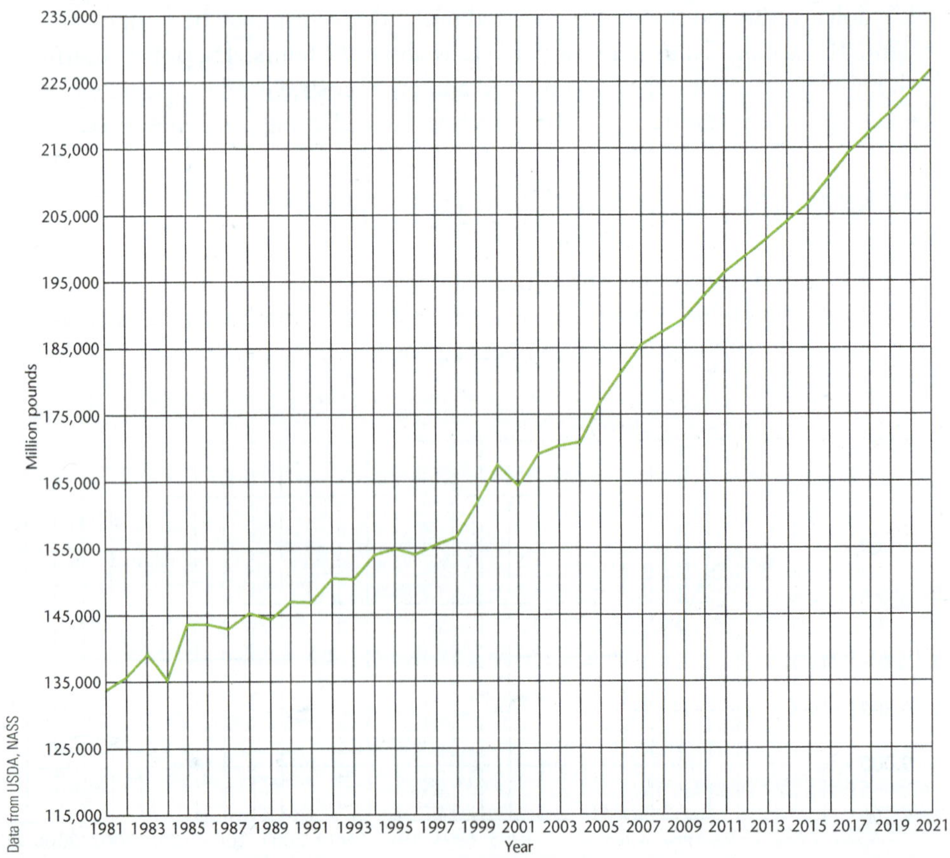

Figure 39-4 Total milk production, 1981–2021 (projected).

which the milk is produced. The purpose of the standards is to ensure that the milk is pure enough to be used for fluid milk consumption.

Grade B milk (manufacturing milk) is produced under conditions that are less controlled than those for producing Grade A milk. Grade B milk is used for processing into dairy products such as cheese, butter, and powdered milk. The processing removes any contamination in the milk and makes the product safe for human use.

There has been a trend in the United States toward the production of more Grade A milk and less Grade B milk. In the late 1940s, 46 percent of the milk

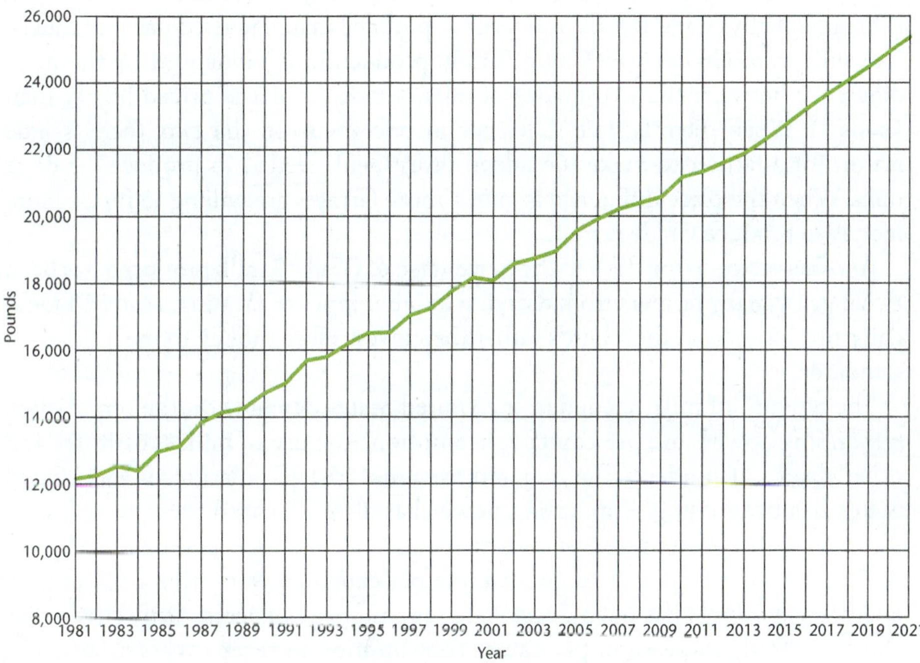

Data from USDA, NASS, Milk Production per Cow by Year, US

Figure 39-5 Milk production per cow, 1981–2021 (projected).

Dairy Facts

- About 97% of American dairy farms are family owned.
- The average size of a dairy herd in the United States is about 179 cows.
- California is currently the number one milk producing state, accounting for 21% of the U.S. total.
- Typically, a dairy cow produces 6.5 gallons of milk daily.
- The record amount of milk produced by a single cow in a year is 72,170 pounds (8,392 gallons).
- Dairy cows are typically milked twice a day during lactation.
- An average mature dairy cow weighs about 1,400 pounds.
- A dairy cow consumes an average of 50 pounds of feed each day.
- The total number of dairy farms and dairy cows in the United States is decreasing, but total milk production in the United States is increasing.

Figure 39-6 Facts about dairy production in the United States.

produced in the United States was eligible for fluid use. "Milk eligible for fluid use" may be interpreted as meaning Grade A milk. Sources of statistical data do not identify Grade A and Grade B milk but refer instead to "milk eligible for fluid use" and "manufacturing milk." However, "milk eligible for fluid use" may be considered to be Grade A milk because it is assumed that it is either produced under Grade A inspection or would qualify for Grade A if it were subject to Grade A inspection. "Manufacturing milk" is considered to be Grade B milk.

Approximately 98 percent of the milk currently produced in the United States is eligible for fluid use; that is, it has been produced under conditions that meet Grade A requirements in most states. Not all Grade A milk is used for fluid consumption. There is generally an oversupply of milk eligible for fluid use. This milk is blended with Grade B milk and used to produce manufactured dairy products. The shift from Grade A to Grade B milk production is influenced by the price difference between the two grades of milk. Grade A milk is priced higher than Grade B. When there is little difference in price between the two, there is little reason for a farmer to make the added investment needed to produce Grade A milk. When the price difference is larger, more farmers are willing to invest more money to produce Grade A milk.

Another factor in the drop in the percentage of Grade B milk produced has been a change by many farmers from dairying to other types of farming. Many Grade B dairy producers sold their herds rather than invest more money to meet Grade A standards.

The amount of milk needed in the United States depends on the population, purchasing power, and per capita consumption. The use of milk substitutes and other drinks also affects the demand for milk. The population of the United States has been rising. Purchasing power has also increased, despite inflation. The per capita consumption of fluid milk and dairy products is shown in Figures 39-7A, 39-7B, and 39-7C. In general, the per capita consumption of fluid milk and dairy products has been decreasing for a number of years. Only cheese has shown a steady increase in per capita consumption in recent years. Many substitute products (such as margarine) have come on the market and compete with dairy products. Many times, these substitutes are lower priced than dairy products and gain more acceptance from consumers. Substitutes for dairy products may be expected to continue to compete for the consumer's dollar in the marketplace.

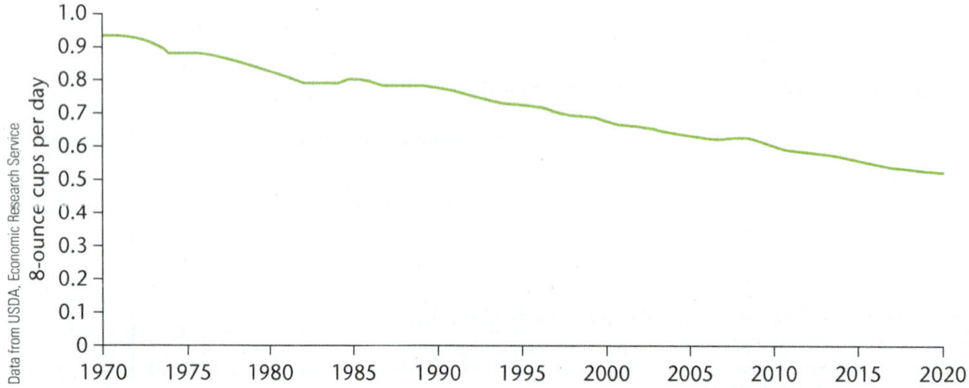

Figure 39-7A Daily per capita consumption of fluid milk by year. The daily consumption of milk per person is projected to decline to about one-half cup by 2020.

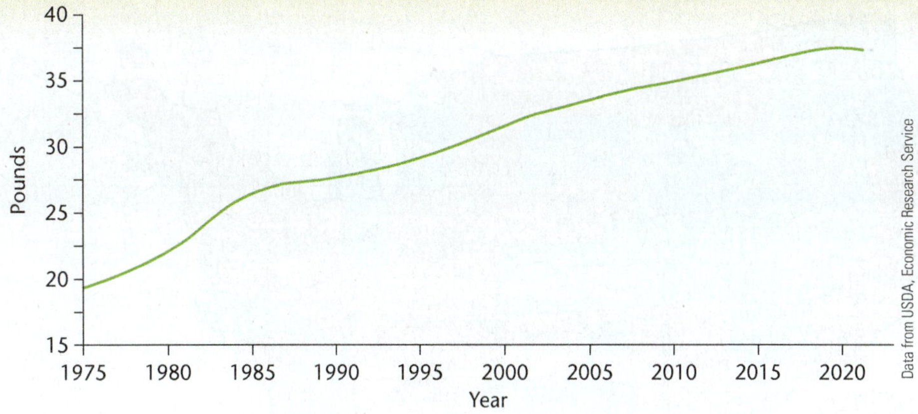

Figure 39-7B Consumption of cheese per capita by year. The yearly per capita consumption of cheese is projected to increase to 36.5 pounds by 2020.

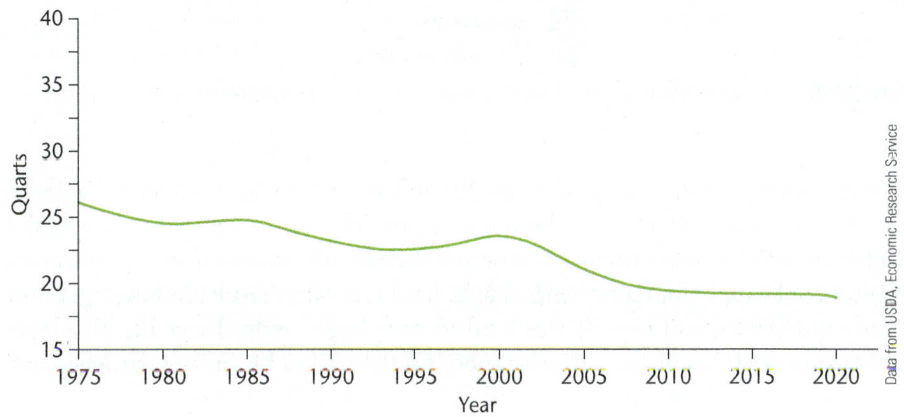

Figure 39-7C Consumption of ice cream per capita by year. The yearly per capita consumption of ice cream is projected to decline to about 18 quarts by 2020.

Dairy cows are found in every state in the United States (Figure 39-8). The leading states in dairy cow numbers are California, Wisconsin, New York, Idaho, Pennsylvania, Texas, and Minnesota. The rank order of states that lead in dairy cow numbers varies from year to year; however, the states listed here are usually among the top 10.

The production of milk in the United States has been shifting from the midwestern states to the western and southwestern states since the middle 1960s. The rate of this geographic shift has increased since the 1980s. California now produces more milk than Wisconsin. The Midwest produced about 50 percent of the nation's milk supply in 1965; it now produces only about 34 percent. The West and Southwest now produce about 34 percent of the nation's milk supply. The Northeast has shown a smaller decline in milk production, producing about 21 percent of the total supply in 1965 and about 18 percent in recent years.

The increase in milk production in the West and Southwest is attributed to five main factors:

- a long-term population shift to these areas
- more favorable climate
- economies of scale
- higher milk production per cow
- a more favorable attitude toward economic growth

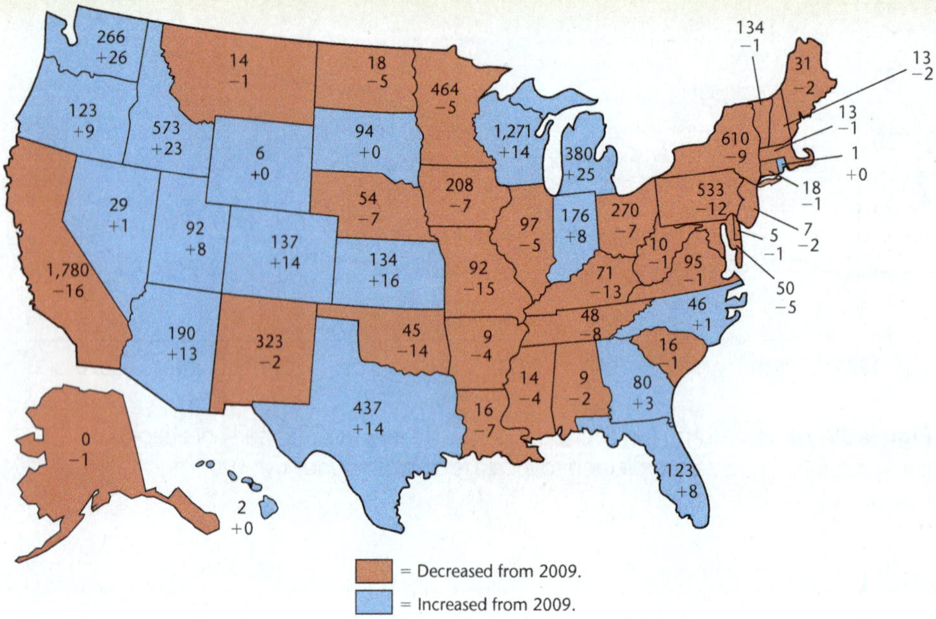

Figure 39-8 Milk cow inventory—United States: 2013 and changes from 2009 (1,000 head).
Source: USDA, NASS

Population growth creates a demand for milk and other dairy products. Facilities are cheaper, and there are fewer herd health problems where the climate is warmer and dryer. High-quality forages can be produced with irrigated farming, ensuring a good feed supply for dairy herds. Large herd size often results in lower per cow production costs, and new dairies tend to have large herds. There has also been active support for economic development on the part of businesses, bankers, and dairy farmers in the western and southwestern states. This has contributed to the expansion of dairy enterprises in those states.

Dairy herds that exceed 500 head of cows in size account for 3.1 percent of the total number of dairy operations in the United States; however, they account for 48 percent of the total inventory of dairy cows (Figures 39-9A and B). Small herds

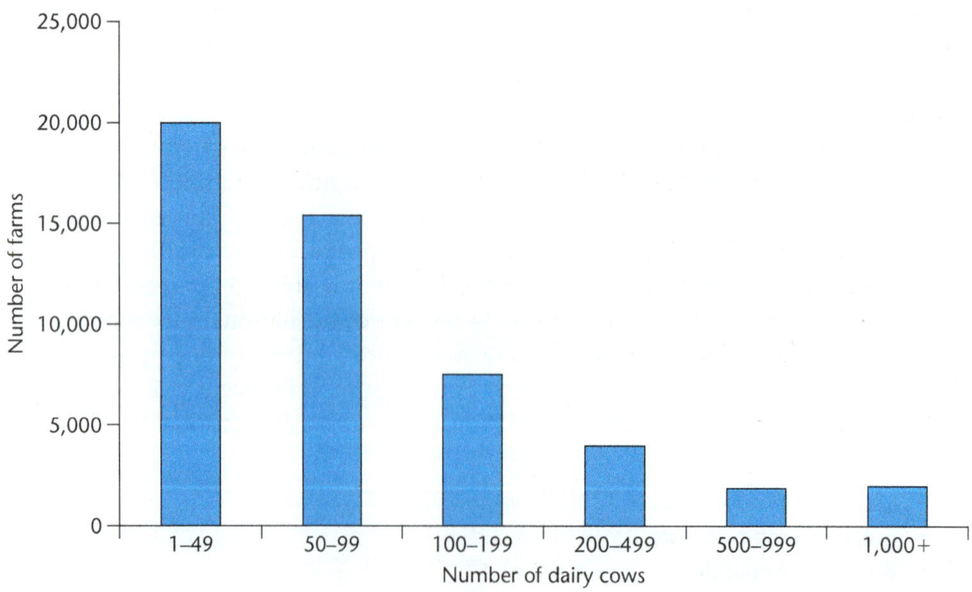

Figure 39-9A Number of dairy farms in the United States by size group.
Source: USDA, NASS

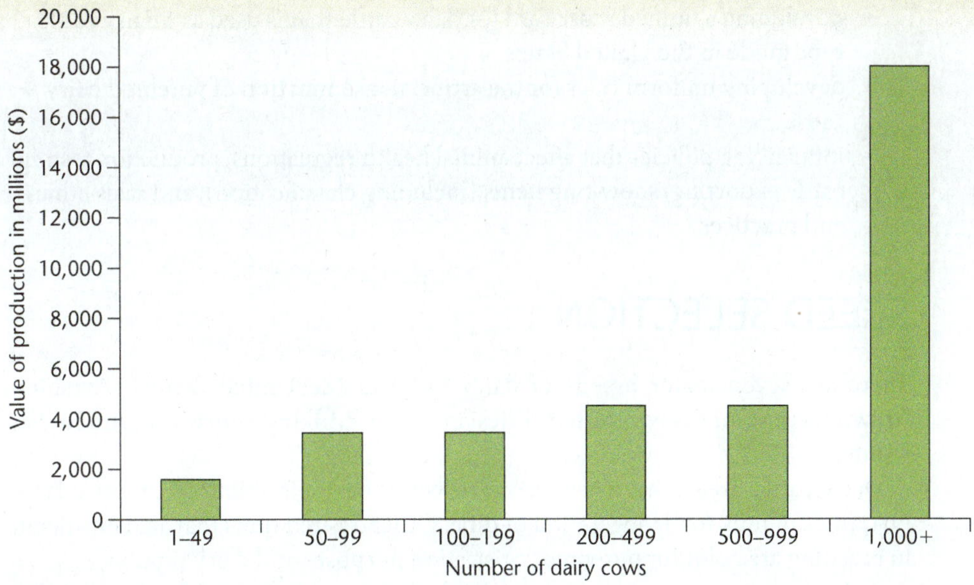

Figure 39-9B Value of milk production in the United States by size group.
Source: USDA, NASS

of less than 49 cows make up 11.4 percent of the total inventory but account for almost 68 percent of the total number of operations. More than 79 percent of all the dairy cows in the United States are found in herds of 100 head or more, making up 17.3 percent of the total number of dairy operations. These statistics reflect a long-term trend in the United States toward larger dairy herds and fewer dairy farms.

It is becoming more difficult in many areas to meet environmental regulations, especially when the dairy herd is large. The control of manure runoff is a major concern and has led to a slowdown in the expansion of dairy operations into larger herds. Manure disposal requires large parcels of land, and this has limited expansion of some dairy herds in states with high population densities, such as California.

GOVERNMENT INFLUENCE IN DAIRYING

Government at national, state, and local levels influences the dairy enterprise. Examples are the Dairy Herd Improvement Program, milk marketing orders, support prices, import quotas, taxes, zoning ordinances, local health regulations, and animal waste disposal regulations.

THE PUREBRED DAIRY CATTLE ASSOCIATION

The Purebred Dairy Cattle Association (PDCA) was organized in 1940 by the five major dairy breed associations. The PDCA currently includes representatives from the Ayrshire, Brown Swiss, Guernsey, Holstein, Jersey, Red & White, and Milking Shorthorn breed associations. The World Dairy Expo—located in Madison, Wisconsin—currently provides management services for the PDCA.

The PDCA works to improve the dairy industry. Some of the areas in which the PDCA provides leadership include:

- developing uniform rules for official testing in national production testing programs

- developing a unified score card for dairy cattle that is used as a dairy cattle type guide in the United States
- developing uniform rules for the artificial insemination of purebred dairy cattle
- influencing policies that affect animal health regulations, production testing, cattle exporting, show-ring items (including classification), and sales ethics and practices

BREED SELECTION

There are seven major breeds of dairy cattle in the United States—Ayrshire, Brown Swiss, Guernsey, Holstein-Friesian, Jersey, Milking Shorthorn, and Red & White.

Dairy cattle breeds have been selected based on their ability to produce large quantities of milk for a long period of time. Other desired traits that are considered in breeding are color for purebred registration purposes and dairy type.

Animals that meet the requirements of the breed association and are recorded in the herdbook of the association are called **registered**. Grade animals are those that are the result of mating a registered bull with a nonregistered cow or one of mixed breeding. Offspring of animals that are not registered but that have near ancestors that are registered are also called grade.

The name of a breed may be used for either registered or grade animals. The terms *registered* or *grade* should be used in front of the breed name, for example, registered Ayrshire or grade Ayrshire. In the case of grade animals, the breed name used indicates the visible traits of that breed can be easily seen in the animal.

An animal is eligible for registration when the sire and dam are both registered in the association herdbook. The animal must also meet any other qualifications described by the association, such as color. Each breed association sets its own rules for qualifications for registering animals in the association.

Each dairy breed association also has a register of performance. This provides systems of production testing for owners of registered cattle. The rules for performance testing for each breed may be obtained from the respective breed associations.

Several dairy breed associations have established programs to allow owners of grade animals to register their cattle in an official herdbook. The associations benefit from an increased membership base. Cattle owners benefit from the increased value of their cattle. Other benefits include more available records for progeny testing bulls, Dairy Herd Information Association (DHIA) records may be used in sire summaries, more identified superior bulls become available for use in artificial insemination programs, and crossbred cattle can eventually be upgraded to a registered status. The procedures for these programs vary among associations. Dairy cattle owners interested in these programs should contact the appropriate breed association for more information and current regulations.

It is good practice for a dairy to select a breed that best suits its needs and will thrive in that geographic location. It is more important to select individual cows within a selected breed that are high producers than to put too much emphasis on which breed to select. There are some general breed differences that are discussed in more detail as each breed is described.

Some general guidelines for the selection of a breed include:

- selecting a breed that is common in the area
- personal preference
- market requirements for the product

Selecting a breed common in the area increases availability of breeding stock. There will also be a better market for surplus animals.

Personal preference is a matter of individual likes and dislikes. Breeds differ in the percent of milkfat produced. Some markets may prefer a lower-testing product while others prefer a higher test. Since market demand may shift, it is not wise to put too much emphasis on a current market demand.

CHARACTERISTICS OF THE BREEDS OF DAIRY CATTLE

Ayrshire

History

The Ayrshire breed originated in the county of Ayr in the southwestern part of Scotland (Figure 39-10). The Ayrshire breed was developed during the last part of the eighteenth century. No known direct importations were used in the development of the breed. Animals were selected that had the desired traits and were mated to produce the color and type wanted by the breeders.

The first importation of Ayrshires into the United States was in 1822. Some importations have taken place since then, mainly from Scotland and Canada. The greatest numbers of Ayrshires are found in the northeastern United States.

The breed association is the Ayrshire Breeders' Association organized in 1875.

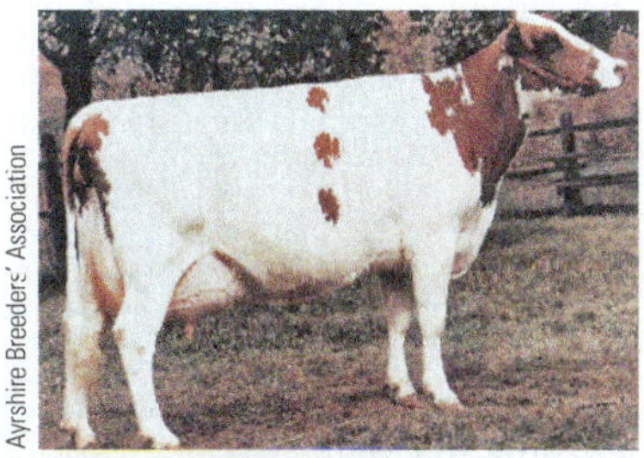

Ayrshire Breeders' Association

Figure 39-10 Ayrshire cow.

Description

The Ayrshire may be any shade of cherry red. Other colors are mahogany, brown, or white. White may be mixed with red, mahogany, or brown. Each color should be clearly defined. The preferred color is a distinctive red and white. Objectionable colors are black or brindle.

The horns curve up and out. They are of medium length, small at the base, and tapered toward the tips.

Ayrshires have straight lines and well-balanced udders. The udders are attached high behind and extend forward. The teats are medium in size.

Ayrshires are vigorous and strong. They have excellent grazing ability. Mature cows weigh about 1,200 pounds. Mature bulls weigh about 1,800 pounds.

The average milk production per Ayrshire cow is 17,230 pounds per year, and, they average about 4 percent milkfat.

Brown Swiss

History

The Brown Swiss breed originated in Switzerland (Figure 39-11). It is thought to be one of the oldest of the dairy breeds. It was developed in the Alps. It is believed that no outside breeding was used in the development of the breed after records were kept. Brown Swiss were first imported into the United States in 1869, but only a small number have ever actually been imported into the United States. Total importations are believed to be about 25 bulls and 130 cows. All of the approximately 820,000 registered Brown Swiss in the United States are descendants of those importations.

The breed association is the Brown Swiss Cattle Breeders' Association of the U.S.A., organized in 1880.

Brown Swiss Cattle Breeders' Association of the U.S.A

Figure 39-11 Brown Swiss cow.

Description

Brown Swiss are solid brown, ranging from light to dark. White and off-color spots are objectionable colors. The nose and tongue are black.

The horns incline forward and slightly upward. They are of medium length and taper toward black tips.

The Brown Swiss are large-framed cattle. Mature cows weigh about 1,500 pounds, and bulls weigh about 2,000 pounds. The heifers mature more slowly than other dairy breeds. Brown Swiss have a quiet, docile temperament. They are considered to be good grazers.

Brown Swiss are the longest lived of the dairy breeds. They have a high heat tolerance. Bulls of this breed have been used recently in beef crossbreeding programs.

Brown Swiss average milk production per cow is 22,252 pounds per year, and they average about 4 percent milkfat.

Guernsey

History

Agri Graphics Ltd

Figure 39-12 Guernsey cow.

The Guernsey breed originated on the Isle of Guernsey, which is located in the English Channel off the Coast of France (Figure 39-12). The development of the breed began about 1,000 years ago. Monks on the Isle of Guernsey crossed two breeds of French dairy cattle. These were the Fromond du Leon from Brittany and the Norman Brindle from Normandy. The Guernsey breed was developed through selection of desired traits and elimination of crossbreeding.

Guernseys were first imported into the United States in 1831. Major importations occurred after 1870. Total importations over the years were about 13,000 cattle.

The breed association is the American Guernsey Association, organized in 1877. Guernseys rank second in total number of dairy cattle registered in the United States.

Description

The Guernsey may be any shade of fawn with white markings. Black and brindle colors are objectionable. The skin is yellow. A clear or buff muzzle is preferred over smoky or black.

The horns curve outward and to the front. They are medium in length, small and yellow at the base, and taper toward the tips.

The Guernsey is an early-maturing breed. These cows are adaptable and have a gentle behavior. Mature cows weigh about 1,100 pounds, and bulls weigh about 1,800 pounds.

Guernsey average milk production per cow is 16,000 pounds per year, and they average about 4.5 percent milkfat. Guernseys produce milk that is golden in color.

Holstein-Friesian

History

The Holstein-Friesian breed originated in the Netherlands (Figure 39-13). Development occurred in the northern province of Friesland and in nearby northern Germany. It is not known when it became a distinct breed. It is believed that selection and breeding that resulted in the Holstein-Friesian breed started about 2,000 years ago. The breed is commonly called Holstein in the United States and Canada. Elsewhere in the world it is called Friesian.

Ideal Mature Holstein cow. Courtesy of Holstein Association USA Inc

Figure 39-13 Holstein cow.

The first importations of Holsteins into the United States were by early Dutch settlers between 1621 and 1664. The first significant importations into the United States were in 1857, 1859, and 1861. Most present-day Holsteins trace their ancestry to importations between 1877 and 1905. There have been no importations since 1905.

The first breed association was called the Association of Breeders of Thoroughbred Holstein Cattle and was organized in 1871. This association published nine volumes of the Holstein Herdbook. In 1877, the Dutch Friesian Cattle Breeders' Association was organized. They published four volumes of the Dutch Friesian Herdbook. These two associations merged in 1885 to form the Holstein-Friesian Association of America. The herdbook is now called the Holstein-Friesian Herdbook.

About 90 percent of all dairy cattle in the United States are of Holstein breeding, making it the most popular breed of dairy cattle. There are about 1,500,000 registered Holsteins in the United States.

Description

Holsteins are black and white. A recessive gene occasionally causes a red and white color to appear. The switch (tail) has white on it. Solid black or solid white animals are not registered. Off-colors include black on the switch, solid black belly, one or more legs encircled with black that touches the hoof at any point, and black and white intermixed to give gray spots.

The horns incline forward and curve inward. They are of medium length and taper toward the tips.

Although these cows are referred to as Holsteins, the official name is Holstein-Friesian. Holsteins are the largest of the dairy breeds. Mature cows weigh about 1,500 pounds, and bulls weigh about 2,200 pounds. Cows have large udders. Holsteins have excellent grazing ability and a large feed capacity. The cows are generally quiet, but the bulls can be mean and dangerous. Holsteins are adaptable to a wide range of conditions.

Holsteins rank first among the dairy breeds in average milk production per cow at an average of 23,151 pounds per year, and they average about 2.5 to 3.6 percent milkfat.

Jersey

History

The Jersey breed originated on the Isle of Jersey, which is located in the English Channel off the coast of France (Figure 39-14). It is not known what cattle were the source of the breeding stock that was developed into the Jersey breed. One theory is that the breed developed from early cattle from Normandy and Brittany in France. A law was passed on the Isle of Jersey in 1763 that prohibited the importation of cattle from France. Selection for desired traits developed the breed without further outside breeding after that time.

The earliest importation of Jersey cattle into the United States occurred in about 1815. The major importations occurred between 1870 and 1890. The first Jersey cattle registered by the American Jersey Cattle Club were imported in 1850.

The breed association is the American Jersey Cattle Association, organized in 1868.

Description

Jerseys are cream to light fawn to almost black in color. Some animals have white markings. The muzzle is black. The switch and tongue may be black or white.

The Jersey is the smallest of the dairy breeds. Mature cows weigh about 1,000 pounds, and bulls weigh about 1,600 pounds.

The horns curve inward and are inclined forward. They are of medium length and taper toward the tips.

Jersey cattle have excellent udders that are well attached. They are adaptable and efficient users of feed. Jerseys have excellent grazing ability even on poor pastures.

Ideal-type Jersey, (c) American Jersey Cattle Association, Reynoldsburg, Ohio

Figure 39-14 Jersey cow.

The cows may be somewhat nervous, and the bulls can be mean and very aggressive. Jerseys are early maturing and have excellent dairy type.

Jerseys average milk production per cow is 16,431 pounds per year, and they average about 4.9 percent milkfat. They rank first among the dairy breeds in average milkfat produced per cow.

Photo by Cybil Fisher. Reprinted with permission of the American Milking Shorthorn Society

Figure 39-15 Milking Shorthorn.

The Milking Shorthorn

History

The Milking Shorthorn originated in the Valley of the Tees River, which is located in northeastern England (Figure 39-15). In 1783, the first Milking Shorthorns came to the United States, arriving in Virginia as the "Milk Breed" Shorthorns, but they were often referred to as "Durhams." They were most used for meat, milk, and power. In 1919, the American Milking Shorthorn Society (AMSS) was incorporated and took care of the registration and promotion of Milking Shorthorns. By 1969, they became an official dairy breed, and in 1972, the AMSS became members of the Purebred Dairy Cattle Association.

Description

Milking Shorthorns are red, red and white, white, or roan. Roan is a very close mixture of red and white, and this color combination is not seen in any other breed of cattle. The Milking Shorthorn is what is considered a dual-purpose breed, meaning it is used for both beef and milk production. This docile breed has a long productive life and is most known for its versatility and efficiency in converting feed to milk. Milking Shorthorns average milk production per cow is 15,000 pounds per year, and they average about 3.8 percent milkfat.

Red and White

History

Red & White Dairy Cattle Association

Figure 39-16 Red and White.

The Red and White dairy cattle breed is derived from the Holstein-Friesian breed (Figure 39-16). The red and white colors, as compared to black and white colors of Holsteins, appear occasionally in calves born in Holstein herds. Eventually, some dairy farmers began selecting for the red and white color, and the new breed evolved. The Red and White Dairy Cattle Association (RWDCA) was established in 1964.

Description

The Red and Whites are colored just as the name suggests, red and white. A single recessive gene controls the red coloring in this breed, and any other color on the cow is strictly prohibited. This breed is known for its rugged, feminine qualities and its adequate size and vigor. The Red and Whites are the same as Holsteins and should weigh about 1,400 pounds when mature and in milk. Production and other characteristics of Red and White dairy cattle compare closely with the Holstein breed. Average milk production and milkfat are similar to the Holstein-Friesian.

SUMMARY

Dairying is an enterprise that requires patience and a willingness to work long hours. Dairy cows must be milked regularly 7 days per week. The dairy enterprise requires a large investment of capital. Dairying provides a steady income year-round. A person needs training and experience to be a successful dairy farm operator.

The number of dairy farms in the United States has been declining in recent years, but the dairy cattle numbers have recently begun to increase slightly. The average size of the dairy herds has increased as well as the average production per cow. There has been a trend toward the production of more Grade A milk and less Grade B milk. The per capita consumption of dairy products and fluid milk has been declining.

The seven major breeds of dairy cattle in the United States are Ayrshire, Brown Swiss, Guernsey, Holstein-Friesian, Jersey, Milking Shorthorn, and Red and White. Dairy cattle breeds have been developed by selecting for quantity and persistence of milk production. Dairy breed associations register purebred dairy animals. They also keep registers of performance.

More emphasis should be put on selecting good individuals for the herd rather than on which breed to select. There are some differences among dairy breeds with regard to total production per cow and percentage of milkfat produced. Holsteins rank first in average production of milk per cow per year, and Jerseys rank first in average percentage of milkfat.

Student Learning Activities

1. Present an oral report on the characteristics of the dairy enterprise.
2. On field trips in the community, observe different breeds of dairy cattle and their characteristics.
3. Conduct a survey of local dairy producers to determine what dairy breeds are most common in the community.
4. Prepare a bulletin board display of pictures of the dairy breeds.
5. Present an oral report to the class on the history and characteristics of a dairy breed.
6. When planning and conducting a dairy supervised experience program, take into consideration information given in this chapter regarding dairy breeds and characteristics of a dairy enterprise.
7. A high-producing Holstein may produce up to 28,000 pounds of milk per year. How many 8-ounce glasses of milk does she produce?

Discussion Questions

1. Describe briefly some of the activities that a dairy farm operator must perform.
2. Discuss the advantages and disadvantages of a dairy enterprise.
3. List and briefly describe some of the important trends in dairy farming.
4. What is Grade A milk? What is manufacturing milk?
5. Name the states that lead in dairy cow numbers.
6. Name some of the influences of government in the dairy enterprise.
7. Discuss characteristics of the seven major breeds of dairy cattle in the United States.
8. What is the difference between registered and grade dairy animals?

Review Questions

True/False

1. Dairy cattle must be milked at regular intervals, two times a day, 6 days a week.
2. The number of milk cows on farms in the United States has shown a general decline since 1981.
3. There are seven major breeds of dairy cattle in the United States.

4. Holsteins are the smallest of the dairy cattle breeds.
5. Jerseys rank first among dairy breeds in average milkfat production.

Multiple Choice

6. What is the breed of cattle that ranks first in average milk production per cow per year?
   a. Jersey
   b. Holstein
   c. Guernsey
   d. Brown Swiss
7. The average dollar return on money invested in the dairy farm ranges from ______.
   a. 6 to 9%
   b. 10 to 14%
   c. 15 to 20%
   d. 25 to 39%

Completion

8. Holsteins are ______ and ______ in color.
9. The ______ breed of dairy cattle originated in the county of Ayr in Scotland.

Short Answer

10. Describe the Jersey cattle breed.
11. Where did the Brown Swiss breed originate?
12. Who probably imported the first Holsteins to the United States?
13. What is the advantage of selecting a common dairy cattle breed?
14. List three reasons for the increase in milk production in the West and Southwest.
15. Which dairy breed is the largest in physical size and in numbers used for dairy production in the U.S.?

Chapter 40

Selecting and Judging Dairy Cattle

Key Terms

type
pedigree
linear classification
meaty

Objectives

After studying this chapter, the student should be able to

- select desirable breeding and production dairy cattle.
- identify the parts of a dairy cow.
- evaluate dairy cattle.

SELECTING DESIRABLE BREEDING AND PRODUCTION ANIMALS

The selection of desirable dairy animals for breeding and production is based on the animal's physical appearance, health, milk production records, and pedigree.

The physical appearance (conformation and some aspects of dairy strength) of the animal is referred to as **type**. Type and milk production are closely related. Cows with good dairy type usually produce more milk for a longer period of time. It takes a great deal of practice and study to learn the ideal dairy animal type.

Health records include a history of vaccinations and the general health of the herd from which the animal comes. The apparent health of the animal being considered is also important.

Milk production records that show past performance may or may not be available for the individual animal. If the cow has been in production and in a herd where DHIA records were kept, the record should be available. Such records give some indication of the possible production of the offspring of the cow. Production records for young cows may be used as the basis for predicting future performance. Younger animals that have not been in production will not have production records. Bulls are evaluated on the basis of the production records of their daughters. Young bulls may not have such records available.

Production records should show the pounds of milk produced, the pounds of milkfat produced, and the percentage of milkfat. To properly evaluate production records, more information is needed. This includes the number of times milked per day, age, feed and care received, and the length of time the cow carried a calf during each lactation.

Production records can only give an estimate of the ability of the cow to transmit high production ability to her daughters. The best indication of a cow's transmitting ability is in the record of her offspring. The best foundation stock is a cow with high production records that also has daughters with high production records.

The **pedigree** is the record of the animal's ancestors. Pedigrees that are most valuable as a basis for selection give the name, registration number, type rating, production record, and show-ring winnings of each ancestor for three or four generations. Such a record gives a more complete picture of the possible inheritance of type and production than information on only the sire and dam. Pedigrees must be studied carefully because they sometimes contain misleading information.

JUDGING DAIRY ANIMALS

The Dairy Cow Unified Scorecard

Judging dairy animals is a process of comparing the individuals being judged with an ideal dairy type. The ideal dairy type is described in the Dairy Cow Unified Scorecard (Figure 40-1).

This scorecard was developed by the Purebred Dairy Cattle Association (PDCA) to describe the general traits of a good dairy cow. These general traits are the same for all breeds of dairy cattle. Specific breed characteristics are included in the scorecard. The scorecard also shows the parts of the dairy cow. A good judge must learn the parts of the dairy animals in order to be able to use the right terms when judging the animals. This scorecard is used to learn to judge dairy animals and for type classification.

Breed associations use the scorecard for type classification. The individual's physical appearance is compared to the ideal, and a type classification score, also known as a number score, is assigned to the animal. Type classification is completed by a representative of the breed association at the request of the animal's owner. The official scores for Holsteins are as follows:

Excellent	90–100 points
Very Good	85–89 points
Good Plus	80–84 points
Good	75–79 points
Fair	65–74 points
Poor	50–64 points

The first part of the scorecard contains four main divisions with the points for a perfect score listed for each. The dairy judge must first become familiar with these divisions and the points for each.

The second part of the scorecard lists the breed characteristics. The dairy judge must learn these to be able to determine the amount of breed character being shown by the animal being judged.

The third part of the scorecard is the evaluation of defects. This part helps the dairy judge determine the importance of any defects that the animal being judged may have.

Linear classification is a modification of type classification that utilizes a computer program to score dairy cattle on a number of individual traits. The computer

DAIRY COW UNIFIED SCORECARD

Breed characteristics should be considered in the application of this scorecard.

MAJOR TRAIT DESCRIPTIONS

Perfect Score

There are four major breakdowns on which to base a cow's evaluation. Each trait is broken down into body parts to be considered and ranked.

1) Frame - 15% — 15

The skeletal parts of the cow, with the exception of rear feet and legs. Listed in priority order, the descriptions of the traits to be considered are as follows:

Rump (5 points): Should be long and wide throughout. Pin bones should be slightly lower than hip bones with adequate width between the pins. Thurls should be wide apart. Vulva should be nearly vertical and the anus should not be recessed. Tail head should set slightly above and neatly between pin bones with freedom from coarseness. **Front End (5 points):** Adequate constitution with front legs straight, wide apart, and squarely placed. Shoulder blades and elbows set firmly against the chest wall. The crops should have adequate fullness blending into the shoulders. **Back/Loin (2 points):** Back should be straight and strong, with loin broad, strong, and nearly level. **Stature (2 points):** Height including length in the leg bones with a long bone pattern throughout the body structure. Height at withers and hips should be relatively proportionate. Age and breed stature recommendations are to be considered. **Breed Characteristics (1 point):** Exhibiting overall style and balance. Head should be feminine, clean-cut, slightly dished with broad muzzle, large open nostrils and strong jaw.

2) Dairy Strength - 25% — 25

A combination of dairyness and strength that supports sustained production and longevity. Major consideration is given to general openness and angularity while maintaining strength, width of chest, spring of fore rib, and substance of bone without coarseness. Body condition should be appropriate for stage of lactation. Listed in priority order, the descriptions of the traits to be considered are as follows:

Ribs (8 points): Wide apart. Rib bones wide, flat, deep, and slanted towards the rear. Well sprung, expressing fullness and extending outside the point of elbows. **Chest (6 points):** Deep and wide floor showing capacity for vital organs, with well-sprung fore ribs. **Barrel (4 points):** Long, with adequate depth and width, increasing toward the rear with a deep flank. **Thighs (2 points):** Lean, incurving to flat and wide apart from the rear. **Neck (2 points):** Long, lean, and blending smoothly into shoulders; clean-cut throat, dewlap, and brisket. **Withers (2 points):** Sharp with chine prominent. **Skin (1 point):** Thin, loose, and pliable.

3) Rear Feet and Legs - 20% — 20

Feet and rear legs are evaluated. Evidence of mobility is given major consideration. Listed in priority order, the descriptions of the traits to be considered are as follows:

Movement (5 points): The use of feet and rear legs, including length and direction of step. When walking naturally, the stride should be long and fluid with the rear feet nearly replacing the front feet. **Rear Legs-Side View (3 points):** Moderate set (angle) to the hock. **Rear Legs-Rear View (3 points):** Straight, wide apart with feet squarely placed. **Feet (3 points):** Steep angle and deep heel with short, well-rounded closed toes. **Thurl Position (2 points):** Near central placement between the hip and pin bones. **Hocks (2 points):** Adequate flexibility with freedom from swelling. **Bone (1 point):** Flat and clean with adequate substance. **Pasterns (1 point):** Short and strong with some flexibility, having a moderate, upright angle.

4) Udder - 40% — 40

The udder traits are evaluated. Major consideration is given to the traits that contribute to high milk yield and a long productive life. Listed in priority order, the descriptions of the traits to be considered are as follows:

Udder Depth (10 points): Moderate depth relative to the hock with adequate capacity and clearance. Consideration is given to lactation number and age. **Rear Udder (9 points):** Wide and high, firmly attached with uniform width from top to bottom and slightly rounded to udder floor. **Teat Placement (5 points):** Squarely placed under each quarter, plumb and properly spaced. **Udder Cleft (5 points):** Evidence of a strong suspensory ligament indicated by clearly defined halving. **Fore Udder (5 points):** Firmly attached with moderate length and ample capacity. **Teats (3 points):** Cylindrical shape; uniform size with medium length and diameter; neither short nor long is desirable. **Udder Balance and Texture (3 points):** Udder floor level as viewed from the side. Quarters evenly balanced; soft, pliable, and well collapsed after milking. **(Note: In the Holstein breed, an equal emphasis is placed on fore and rear udder (7 points each). All other traits are the same as listed above.)**

TOTAL — 100

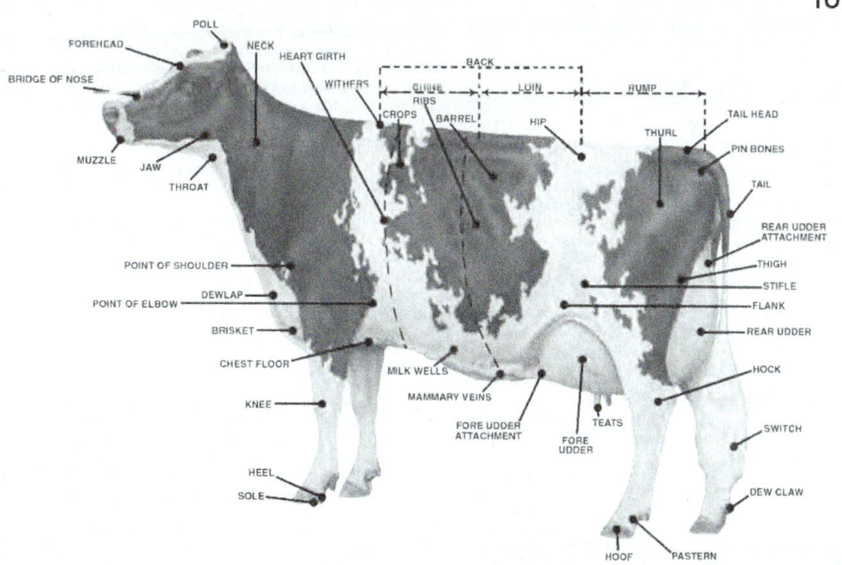

Figure 40-1 The Dairy Cow Unified Scorecard.

THE SEVEN BREEDS

BREED CHARACTERISTICS

Except for differences in color, size, and head character, all breeds are judged on the same standards as outlined in the Unified Score Card. If any animal is registered by one of the dairy breed associations, no discrimination against color or color pattern is to be made.

Ayrshire Strong and robust, showing constitution and vigor, symmetry, style and balance throughout, and characterized by strongly attached, evenly balanced, well-shaped udder.
HEAD- clean cut, proportionate to body; broad muzzle with large, open nostrils; strong jaw; large, bright eyes; forehead, broad and moderately dished; bridge of nose straight; ears medium size and alertly carried.
COLOR- light to deep cherry red, mahogany, brown, or a combination of any of these colors with white, or white alone, distinctive red and white markings preferred.
SIZE- a mature cow in milk should weigh at least 1200 lbs.

Brown Swiss Strong and vigorous, but not coarse. Adequate size with dairy quality. Frailness undesirable.
HEAD- clean cut, proportionate to body; broad muzzle with large, open nostrils; strong jaw; large, bright eyes; forehead, broad and slightly dished; bridge of nose straight; ears medium size and alertly carried.
COLOR- body and switch solid brown varying from very light to dark; muzzle has black nose encircled by a white ring; tongue and hooves are dark brown to black.
SIZE- a mature cow in milk should weigh at least 1400 lbs.

Guernsey Strenth and balance, with quality and character desired.
HEAD- clean cut, proportionate to body; broad muzzle with large, open nostrils; strong jaw; large, bright eyes; forehead, broad and slightly dished; bridge of nose straight; ears medium size and alertly carried.
COLOR- shade of fawn and white markings throughout clearly defined.
SIZE- a mature cow in milk should weigh 1200-1300 lbs.; Guernsey does not discriminate for lack of size.

Red & White Rugged, feminine qualities in an alert cow possessing adequate size and vigor.
HEAD- clean cut, proportionate to body; broad muzzle with large, open nostrils; strong jaw; large, bright eyes; forehead, broad and slightly dished; bridge of nose straight; ears medium size and alertly carried.
COLOR- must be clearly defined red and white; black-red and brindle is strictly prohibited.
SIZE- a mature cow in milk should weigh at least 1400 lbs. and be well balanced.

Holstein Rugged, feminine qualities in an alert cow possessing Holstein size and vigor.
HEAD- clean cut, proportionate to body; broad muzzle with large, open nostrils; strong jaw; large, bright eyes; forehead, broad and moderately dished; bridge of nose straight; ears medium size and alertly carried.
COLOR- black and white or red and white markings clearly defined
SIZE- a mature cow in milk should weigh at least 1400 lbs.
UDDER- equal emphasis is placed on fore and rear udder (7 points each), all other traits are the same as listed on the PDCA scorecard.

Jersey Sharpness with strength indicating productive efficiency.
HEAD- proportionate to stature showing refinement and well chiseled bone structure. Face slightly dished with dark eyes that are well set.
COLOR- some shade of fawn with or without white markings; muzzle is black encircled by a light colored ring; switch may be either black or white.
SIZE- a mature cow in milk should weigh at least 1000 lbs.

Milking Shorthorn Strong and vigorous, but not coarse.
HEAD- clean cut, proportionate to body; broad muzzle with large, open nostrils; strong jaw; large, bright eyes; forehead, broad and slightly dished; bridge of nose straight; ears medium size and alertly carried.
COLOR- red or white or any combination (no black markings allowed).
SIZE- a mature cow in milk should weigh 1400 lbs.

FACTORS TO BE EVALUATED

The degree of discrimination assigned to each defect is related to its function and heredity. The evaluation of the defect shall be determined by the breeder, the classifier or judge, based on the guide for discrimination and disqualifications given below.

HORNS
No discrimination for horns.

EYES
1. Blindness in one eye: *Slight discrimination.*
2. Cross or bulging eyes: *Slight discrimination.*
3. Evidence of blindness: *Slight to serious discrimination.*
4. Total blindness: *Disqualification.*

WRY FACE
Slight to serious discrimination.

CROPPED EARS
Slight discrimination.

PARROT JAW
Slight to serious discrimination.

SHOULDERS
Winged: *Slight to serious discrimination.*

CAPPED HIP
No discrimination unless affects mobility.

TAIL SETTING
Wry tail or other abnormal tail settings: *Slight to serious discrimination.*

LEGS AND FEET
1. Lameness- apparently permanent and interfering with normal function: *Disqualification.* Lameness- apparently temporary and not affecting normal function: *Slight discrimination.*
2. Evidence of crampy hind legs: *Serious discrimination.*
3. Evidence of fluid in hocks: *Slight discrimination.*
4. Weak pastern: *Slight to serious discrimination.*
5. Toe out: *Slight discrimination.*

UDDER
1. Lack of defined halving: *Slight to serious discrimination.*
2. Udder definitely broken away in attachment: *Serious discrimination.*
3. A weak udder attachment: *Slight to serious discrimination.*
4. Blind quarter: *Disqualification.*
5. One or more light quarters, hard spots in udder, obstruction in teat (spider): *Slight to serious discrimination.*
6. Side leak: *Slight discrimination.*
7. Abnormal milk (bloody, clotted, watery): *Possible discrimination.*

LACK OF ADEQUATE SIZE
Slight to serious discrimination. (Note: Guernsey does not discriminate for lack of size.)

EVIDENCE OF SHARP PRACTICE
(Refer to PDCA Code of Ethics)
1. Animals showing signs of having been tampered with to conceal faults in conformation and to misrepresent the animal's soundness: *Disqualification.*
2. Uncalved heifers showing evidence of having been milked: *Slight to serious discrimination.*

TEMPORARY OR MINOR INJURIES
Blemishes or injuries of a temporary character not affecting animal's usefulness: *Slight to serious discrimination.*

OVERCONDITIONED
Slight to serious discrimination.

FREEMARTIN HEIFERS
Disqualification.

Figure 40-1 *(Continued)*

program uses a wider range of scoring for individual traits than is found on the Dairy Cow Unified Scorecard and develops an overall type classification score for each animal.

GENERAL DESCRIPTION OF THE SCORECARD

Dairy character as described on the scorecard has the highest positive correlation to milk production. Other characteristics on the scorecard are also important when visually evaluating dairy animals for their ability to efficiently produce milk over a long period of time.

Frame (15 Percent)

The frame is defined as the skeletal structure of the cow, except the feet and legs. In priority order, the areas considered when evaluating the cow's frame are rump, stature, front end, back/loin, and breed characteristics. The rump is the highest priority because it is closely related to reproductive efficiency and the support and placement of the udder. The width of the pelvic region affects the ease of calving. The animal should be properly proportioned throughout with a strong and straight topline.

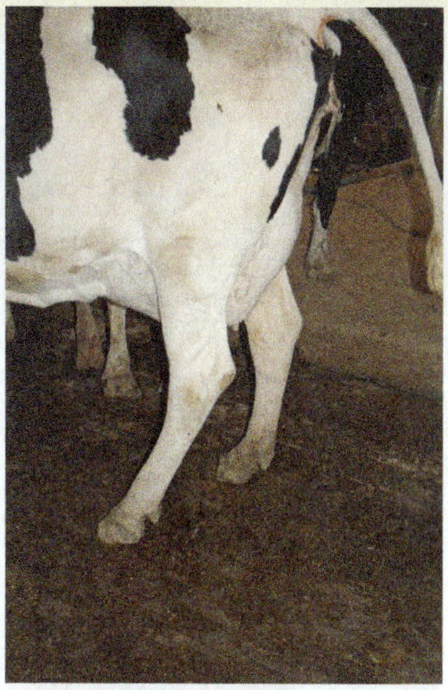

Courtesy of Frank B. Flanders

Figure 40-2 This back leg has too much set (sickle hocked).

Dairy Strength (25 Percent)

Dairy strength indicates the milking ability of the animal, along with strength that supports good milk production. This category is a combination of dairyness (production capacity) and body strength that will help the animal to be productive over several years. The priority order for evaluating dairy strength is ribs, chest, barrel, thighs, neck, withers, and skin. The ribs should be wide as well as the chest. The chest should also be deep, and the barrel should be long and show good capacity, which is needed so the animal can consume the amount of feed needed for high milk production. When viewed from the rear, the thighs should be wide enough to provide plenty of space for the udder attachment. The age of the animal is also taken into consideration when evaluating the length, depth, and width of the body. When evaluating an animal for dairy strength, the body condition score appropriate for the stage of lactation should be considered.

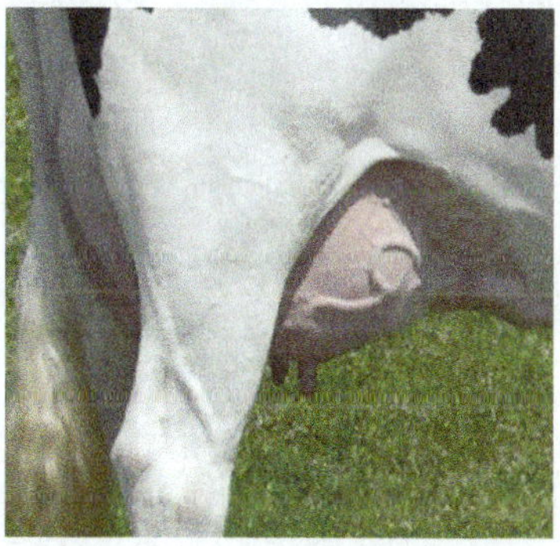

Courtesy of Dr. Dennis Sheppard

Figure 40-3 The main purpose of the dairy cow is to produce milk. The udder is the most important part of the dairy cow's body.

Feet and Legs (20 Percent)

Feet have a little higher priority than rear legs when evaluating feet and legs. The ability of an animal to reproduce efficiently over a long period of time is closely related to the structure and strength of its feet and legs. Proper placement of the legs improves the ability of the animal to move about with ease. Width between the rear legs provides room for a large udder.

Correct set to the hocks affects the ability of the animal to stand and walk on concrete surfaces over a long period of time. Too much set at the hocks (sickle hock) will cause the legs to weaken as the animal becomes older (Figure 40-2). Legs that are too straight place too much stress on the hocks.

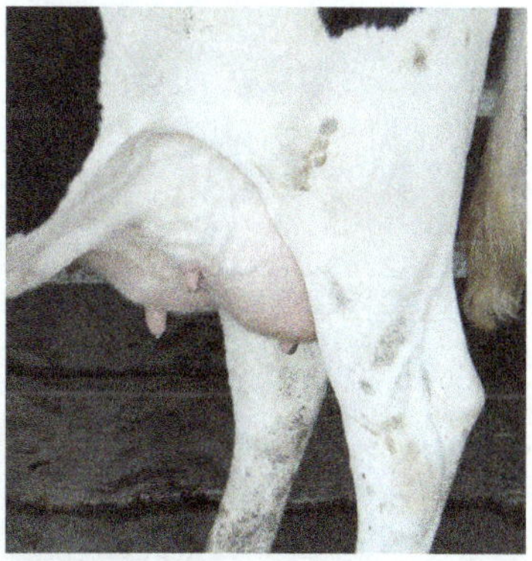

Courtesy of Frank B. Flanders

Figure 40-4 A blind quarter does not secrete milk and is a disqualification on the Dairy Cow Unified Scorecard.

Udder (40 Percent)

The priority order for evaluating the characteristics of the udder is udder depth, teat placement, rear udder, udder cleft, fore udder, teats, and balance and texture.

The main purpose of the dairy cow is to produce milk. The udder is the most important part of the dairy cow's body (Figure 40-3). Udders that have poor conformation, are weakly attached, and are poorly balanced do not stand up well under the stress of high production. A blind quarter, for example, is a serious defect and is a disqualification on the Dairy Cow Unified Scorecard (Figure 40-4).

Courtesy of Frank B. Flanders

Figure 40-5 From the side, look at the topline, rump, barrel, heartgirth, shoulders, udder, teats, flank, legs, and neck.

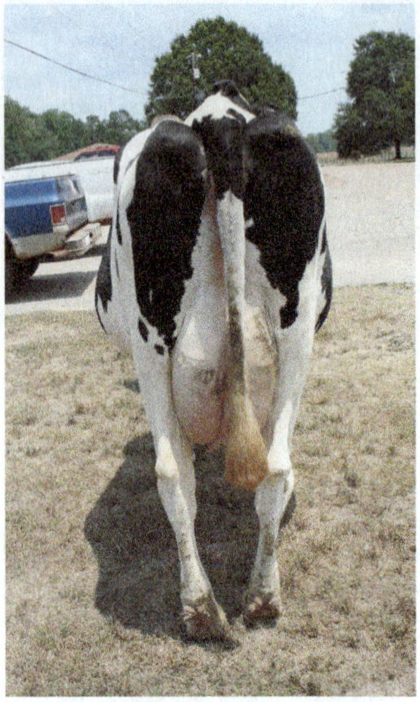

Courtesy of Frank B. Flanders

Figure 40-6 From the rear, look at the barrel, pinbones, tailhead, udder, and legs.

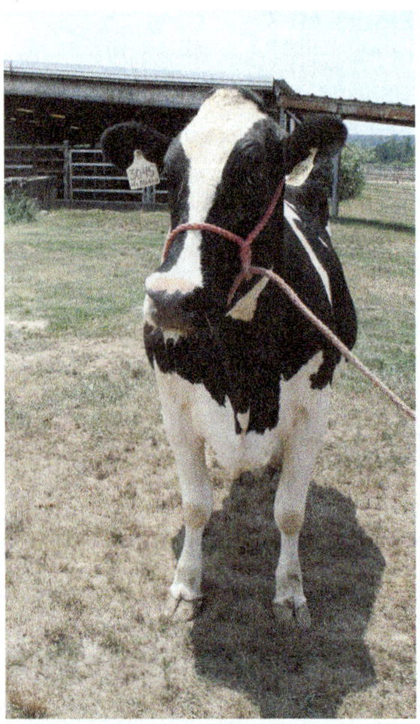

Courtesy of Frank B. Flanders

Figure 40-7 From the front, look at the head, chest, and front legs.

The size of the udder is generally related to milk-producing capacity. Cows with small udders are usually not high-producing cows. The udder should be soft, pliable, and elastic. If it is still quite firm and large after milking, it is probably full of fibrous or scar tissue. An udder that is still firm after milking is referred to as **meaty**.

The size and placement of the teats is important for ease of machine milking. Teats that are uneven in size or poorly placed make it more difficult to use the milking machine. Teats should be 1.5 to 2.5 inches long. When the udder is full, the teats should hang straight down.

The mammary veins are blood vessels that circulate blood to the udder. The size of these mammary veins indicates the amount of blood circulation to the udder. Large mammary veins are desirable.

The capacity of the mammary system is reduced by a small udder, deep cuts between the quarters or halves, meaty texture, and small mammary veins. A cow with a poor mammary system is not a good foundation cow for the dairy farmer who wants a high-producing herd.

Steps in Judging

First, view the animal or group of animals from a distance of 20 to 30 feet. Look at the side, front, and rear of the animal. The animal should be standing level or with the front feet slightly higher than the rear.

From the side view, look at the topline, rump, barrel, heartgirth, shoulders, udder, teats, flank, legs, and neck (Figure 40-5). From the rear view, look at the barrel, loin, hips, pinbones, tailhead, udder, and legs (Figure 40-6). From the front view, look at the head, chest, and front legs (Figure 40-7).

Make mental or written notes on each animal as it is viewed on this first inspection. First impressions are usually more accurate and should be carefully noted. Compare each animal with the ideal described in the Dairy Cow Unified Scorecard. If there is an easy top or bottom animal or top and bottom pair in the class, make note of it.

The class of animals being judged is usually walked slowly around the judge in a clockwise direction. Look at style, carriage, straightness of legs and topline, udder attachment, and blending of parts while the cows are walking.

After observing the animals from a distance, move in for a closer inspection. Observe the shape of the withers, quality of hair, mellowness of hide, texture of udder, and development of the mammary veins in this close inspection.

A class of four animals is compared by pairs, that is, the top pair, middle pair, and bottom pair. When you are judging dairy cattle, a procedure for note taking and giving reasons should be followed.

When a pair is close, make the decision on the mammary system. If the mammary system is equal in the pair, use dairy character to determine the placing. If, in a close pair, one cow is dry and the other milking, place the milking cow over the dry cow.

Look for any defects that individual animals may have. Defects are described in the Dairy Cow Unified Scorecard. The influence of the defect on the placing ranges from disqualification to slight discrimination depending on the degree of seriousness. For example, permanent lameness will put a cow at the bottom of the class regardless of other traits she may have.

Question Class (Type Analysis Questionnaire—TAQ)

In some judging contests, a different method is being used for judging. This involves asking a series of about 5 to 10 questions about the class. During or after judging, the contestant places beside each question the number of the cow that most closely corresponds to the correct answer. The official judge determines the correct answers. Scoring is done by giving 5 points for the correct answer, 3 points for the second most correct answer, 1 point for the third most correct answer, and 0 points for the wrong answer. If 10 questions are used, there are a total of 50 points for the class.

Judging Dairy Heifers

Dairy heifers are judged on much the same points as dairy cows. Heifers do not have as much development as mature cows. Therefore, the judge must visualize how the heifer will develop as she matures. This is especially true for the mammary system. When examining the udder, place emphasis on uniformity of quarters, placement and size of teats, length and width, and rear and fore attachments. A heifer does not have the depth of barrel and dairy temperament of a mature cow.

Springing (pregnant) heifers often carry some surplus fat. This may make them appear to be coarse over the withers. This accumulated fat disappears when the heifer comes into production.

DAIRY JUDGING TERMS

Comparative terms are used when giving reasons for the placement of a dairy class. The following lists contain terms that may be used to compare one animal with another.

General Appearance

- more feminine head
- more breed character in head and neck
- neck blends more smoothly with the shoulders
- blends smoother at the point of shoulders
- shoulders that blend more smoothly with the body
- tighter through the shoulders
- fuller through the crops
- stronger through the front
- stronger over the back and loins
- stronger and straighter in the loin and chine
- more strength in the loin
- weaker in the loin
- longer and more open in the rib
- stronger through the heart
- more power and width in the chest
- wider and more nearly level rump
- smoother over the rump

- higher at the thurls
- wider in the thurls
- more nearly level from hips to pins
- longer from hips to pins
- wider and more open at the pins
- more smoothly blended in at the tail setting
- standing more squarely on the legs
- standing straighter on rear legs
- straighter on rear legs
- stands straighter on legs as she walks
- crooked rear legs
- sickle hocked
- cow hocked
- straighter on the pasterns
- shorter and stronger pasterns
- weaker on the pasterns
- deeper in the heel
- shallow in the heel
- toes out in front
- showing evidence of lameness
- a larger cow with more size and scale
- more strength and substance
- more breed type and character
- lacks the breed character
- more style and balance
- shows more style when on the move
- smoother throughout

Dairy Strength

- has a more milky appearance
- more dairy-like
- cleaner cut head and neck
- longer in the neck
- cleaner at the throat
- longer and thinner in the neck and cleaner in the throat
- sharper over the withers
- more open ribbed
- more spread to the ribs
- flatter and more open of rib
- sharper in hips and pins
- neater and more refined tailhead
- cleaner over the pinbones
- thinner and cleaner in the thighs
- flatter in the thighs
- cleaner and more refined
- more angular throughout
- cut in behind the shoulders
- smoother behind the shoulders
- fuller in the crops
- narrow chested
- wider on the chest floor
- deeper through the heart
- pinched at the heart
- fuller in the heartgirth
- fuller in the fore ribs
- deeper in the fore and rear ribs
- a deeper-ribbed cow
- deeper and more open ribbed
- more spring of fore rib
- more spring of rear rib
- more open in the rib
- showing greater arch to the ribs
- showing more spring to her rear ribs
- greater spring of rib and depth of body
- narrow body, lacking spring of rib
- more capacity and spring in the barrel
- longer and deeper in the barrel
- greater body capacity
- greater feed capacity
- greater depth of body
- longer bodied
- longer in the body
- deeper bodied
- shows more stretch
- deeper in the flank

Mammary System

- more evenly balanced udder
- more balanced udder
- udder shows more balance from front to rear
- unbalanced udder
- a more uniformly balanced udder that is more level on the floor
- not as badly quartered (or halved)
- udder shows more defined cleavage
- udder is badly quartered (or halved) and hangs forward
- cut up between the quarters
- a more symmetrical udder with more uniform-sized quarters
- udder shows more desirable halving and quartering
- more nearly level on the floor
- showing more quality and texture
- a softer, more pliable udder
- more milky udder with numerous veining
- carries out fuller in the rear quarters
- more balance of the fore quarters
- more balance of rear quarters
- carries udder closer to the body
- fore udder attachment blends more smoothly in with the floor of the barrel
- carries her udder closer to the hocks
- udder is more strongly attached both fore and rear
- udder is more firmly attached both fore and rear

- weaker attachment
- stronger fore udder attachment
- is more firmly attached in the front
- smoother (or stronger) in the fore udder attachment
- carrying farther forward
- loose in front attachment
- a larger, more capacious udder
- more quality of udder
- more capacity and bloom
- more symmetry
- weaker fore udder attachment
- higher and wider rear udder attachment
- rear attachment is broken down, allowing the quarter to point out
- very low and narrow in the rear attachment
- weaker rear udder attachment
- tilted udder
- short front udder
- stronger center support in the udder
- teats more squarely placed
- teats more evenly spaced
- more uniformly placed teats
- teats too close together
- teats strut
- rear teats too close together
- more width between rear teats
- teats of more desirable size, shape, and placement
- teats of more uniform size
- more desirable in size and shape of teat
- teats hang more nearly plumb
- weaker quarter
- lighter quarter
- underdeveloped rear quarters

SUMMARY

Physical appearance, health, milk production records, and pedigree are important factors to consider when selecting dairy animals for breeding and production. Dairy animals are judged by comparing the animal being judged with an ideal dairy type. The Dairy Cow Unified Scorecard describes the general traits of a good dairy cow. The scorecard is also used for type classification of dairy cows.

The desirable traits described on the scorecard are related to the ability of the cow to produce milk over a long period of time. The mammary system is one of the most important parts of the dairy cow and carries the most weight when scoring and evaluating the cow. Follow a definite routine when judging dairy cows. Look at the animal from the side, rear, and front. First impressions are usually more accurate. Use dairy judging terms when giving oral or written reasons.

Student Learning Activities

1. Using pictures or a live animal, name the parts of the dairy cow.
2. Judge a class of dairy cows using pictures or live animals. Give oral reasons.
3. Give an oral report to the class on the selection of desirable breeding cows.
4. Arrange and take a field trip to a local dairy farm to observe type classification of the herd.
5. Use the information given in this chapter regarding selecting and judging dairy cattle when planning and conducting a dairy supervised experience program.

Discussion Questions

1. Name the four things that are considered when selecting dairy animals for breeding and production.
2. Briefly describe each of the four things named in question 1.
3. What is the Dairy Cow Unified Scorecard?
4. On an outline of a dairy cow provided by the instructor, name the parts of the dairy cow.
5. Name the seven type classifications for dairy cows.
6. Briefly describe the breed characteristics for each breed of dairy cow.
7. Name and briefly describe common defects of the dairy cow.

8. Briefly describe how each of the following is related to the milk producing ability of the dairy cow.
   a. general appearance
   b. dairy strength
   c. mammary system
9. Describe the steps to follow when judging a dairy cow.
10. Describe how a question class is used when judging dairy cows.
11. How does judging dairy heifers differ from judging dairy cows?

Review Questions

True/False

1. Judging dairy animals is a process of comparing the individuals being judged with an ideal dairy type.
2. Dairy character is an indication of milking ability.
3. The size of the udder is generally unrelated to milk producing capacity.
4. Dairy heifers are not judged on the same points as dairy cattle.
5. When a judging pair is close, make the decision based on the mammary system.

Multiple Choice

6. The main purpose of dairy cattle is to produce ______.
   a. leather
   b. meat
   c. milk
   d. all of the above
7. Which of the following is **not** one of the major classification traits as described on the Dairy Cow Unified Scorecard?
   a. frame
   b. differential
   c. udder
   d. dairy strength

Completion

8. The pedigree is the record of the animal's ______.
9. The physical appearance of an animal is referred to as ______.
10. Pregnant heifers are referred to as ______.
11. The class of animals being judged is usually walked slowly around the judge in a ______ direction.
12. ______ ______ is a modification of type classification that utilizes a computer program to score dairy cattle on a number of individual traits.

Short Answer

13. What is included in the health records of dairy cattle?
14. How are dairy animals judged?
15. Which classification trait on the Dairy Cow Unified Scorecard carries the highest number of points?

Chapter 41

Feeding Dairy Cattle

Key Terms

challenge (lead) feeding
transponder
double cropping
body condition score
ionophore
NEL

Objectives

After studying this chapter, the student should be able to

- select appropriate feeds for dairy animals.
- select rations for dairy cows for maximum production.
- select rations for replacement animals for fast, economical growth.
- select an appropriate feeding method.

Feed costs are about 40 to 50 percent of the total cost of producing milk. Cows need to be fed balanced rations to give the most profitable level of production. Milk production of the individual cow is limited by heredity. Differences in milk production among cows are due to about 25 percent heredity and 75 percent environment. Feeding has the most influence on the amount of milk any cow produces. Proper feeding and care allows the cow to produce closer to her potential ability.

Nutrients are used by the dairy animal for the following needs:

- growth of the immature animal
- pregnancy; needs are small during the first 6 months and large during the last 2 or 3 months of the gestation period
- fattening or the regaining of normal body weight lost in lactation and during the dry period
- maintenance of the mature animal; needs vary according to the size of the animal
- milk production; needs vary with the pounds of milk produced and its composition

Maintenance and milk production needs are the two most important to consider when balancing rations for dairy cows. Cows fed correctly use about one-half of the feed for maintenance and one-half for milk production. Nutrients are used to meet other needs before being used for milk production. The cow will draw on body reserves for her maintenance needs if not fed enough feed. Milk production will then go down.

Milk production records are needed to develop a good feeding program. High- and low-producing cows are identified by good records. The feeding program is then adjusted for individual cows or groups of cows. If the cow is not fed enough, milk production goes down. If fed too much, the cow gets too fat and production costs increase.

Dairy cattle are ruminants. The basis for dairy rations is roughage. Dairy cows can use large amounts of roughage such as pasture, hay, haylage, or silage. When these are raised on the dairy farm, they are the cheapest source of nutrients needed by the cow. Feed good-quality roughages and supplement the ration with concentrates (grains and protein feeds).

Moldy feeds may produce toxins that interfere with metabolism, lower the nutritive value of the feed, and cause reproductive problems. *Fusarium* mold, aflatoxins, and mycotoxins are sometimes found in feeds. If moldy feed must be used, dilute it with feed that does not contain mold; this will reduce the toxic effect of the mold. Proper drying and storing of feeds will reduce mold problems.

METHODS OF FEEDING DAIRY COWS

Traditional

At some farms concentrates are fed to cows individually according to milk production. The roughages and concentrates are generally fed free choice but are fed separately. Roughages are usually fed free choice in feed bunks to the entire herd or in mangers in stanchion barns (Figure 41-1). Traditionally, concentrate mix is fed individually in mangers in stanchion barns or in the milking parlor during milking. Modern farms are getting away from this and are feeding free choice, which saves milking time and labor.

Disadvantages of traditional feeding include the following:

- It is hard to measure the amount of forage (roughage) each cow eats; therefore, it is hard to balance the ration with the right amount of concentrate for each cow

USDA Agricultural Research Service; Photo by Scott Bauer

Figure 41-1 Roughages are usually fed free choice.

- Low-producing cows are often overfed concentrates
- High-producing cows are often underfed concentrates
- Grain-feeding facilities are required in the milking parlor
- The level of dust in the milking parlor increases
- Cleanup of uneaten grain in the milking parlor is required
- Milking in the parlor may be delayed while waiting for the cow to finish eating the grain mix
- Feeding in the parlor slows down the milking
- Cows do not stand as quietly and defecate more during milking
- More labor is necessary
- There is less control over the total feeding program
- Cost of equipment is higher
- Careful records of individual production and continual adjustment of concentrate feeding to match production is required

Advantages of traditional feeding include the following:

- Less specialized equipment is needed
- Theoretically, each cow feeds according to individual needs based on production
- It permits adjusting concentrate feeding to the stage of lactation
- It permits challenge feeding of each cow

Challenge or Lead Feeding

The practice of feeding higher levels of concentrate to challenge the cow to reach her maximum potential milk production is called **challenge (lead) feeding**. Challenge feeding gives more concentrates to the cow early in the lactation period and less later in lactation. Dry cows in good condition should get about $\frac{1}{2}$ pound of concentrate per 100 pounds of body weight. Two to 3 weeks before calving, increase the amount of concentrate gradually, about 1 pound per day, until the cow is eating from 1 to $1\frac{1}{2}$ pounds per 100 pounds of body weight at calving. About 3 days after calving, increase the concentrate by 1 to 2 pounds per day for 2 to 3 weeks after calving. At this time, the cow should be getting about 2 pounds of concentrate per 100 pounds of body weight. Continue to increase the level of concentrate feeding until milk production levels off. At maximum milk production, hold the level of concentrate feeding constant. This practice is best used on small, high-producing farms. It would require immense labor and time to perform this on large farms.

Keep accurate production records. When production drops, decrease the level of concentrate feeding. Decrease by 1 pound of concentrate for each 3 pounds of milk production drop. This practice requires excellent record keeping and daily milk weighing.

If the cow loses body weight early in the lactation period, she needs more concentrate. Cows that do not respond to challenge feeding should be culled.

Feeding Total Mixed Rations

A method of feeding that is becoming more popular with modern dairy producers is feeding a total mixed ration (Figure 41-2). A total mixed ration is one that has all or almost all of the ingredients blended together. This ration is then fed free choice to all the cows in a group. In large herds, the cows are usually divided into several groups.

The total mixed ration contains roughages and concentrates combined to meet the energy, protein, mineral, vitamin, and crude fiber needs of the cows. When feeding a total mixed ration free choice, no concentrates are fed in the milking parlor.

Figure 41-2 A method of feeding that is becoming more popular with modern dairy producers is feeding a total mixed ration.

Advantages of feeding a total mixed ration include:

- Each cow receives a balanced ration
- Each cow is challenged to produce to her maximum genetic potential
- Feeds are used more efficiently
- NPN (nonprotein nitrogen) is utilized more effectively
- Fewer cows have digestive problems or go off-feed
- It is not necessary to feed minerals free choice separately from the rest of the ration
- Rations can be changed more easily without affecting consumption
- Less labor is required for feeding
- Problems with low milkfat test are reduced
- Costs of cow housing and feeding facilities are lowered
- Cows that gain weight early in the lactation period are quickly identified and may be culled
- Weight gains usually come later in the lactation period
- It is possible to substitute cheaper grains and urea in the ration because silage tends to mask the taste and dustiness of feed ingredients
- There are several advantages of not feeding concentrates in the milking parlor:
  a. Cost of parlor construction and maintenance of feeding equipment is less
  b. Cows stand more quietly in the parlor during milking
  c. There is less dust and manure in the parlor
  d. No concentrate cleanup is necessary, and there is no clogging of the drain system by concentrates in the parlor
  e. It takes less time to milk the cows
  f. There is less wasted concentrate

Disadvantages of feeding a total mixed ration include:

- Special equipment for weighing and mixing the ration is necessary
- It may be more difficult at first to get cows to enter the milking parlor when they are not fed there
- It is hard to include hay in the total mixed ration
- If hay is fed separately, the ration may not be balanced for some cows
- Low-producing cows tend to get too fat
- Cows need to be divided into groups for most efficient feeding

Special Equipment for Feeding Total Mixed Rations

The dairy farmer must have two items of special equipment for successful use of total mixed rations. These are a mixer-blender unit, which may be mobile or stationary, and a weighing device. The mixer-blender unit is necessary in order to get the ration completely and uniformly mixed. If the ration is not mixed for a long enough period of time, the blend will not be uniform. If it is mixed for too long a period of time, the blend may be pulverized or shredded. The mixer-blender must be able to handle haylage. Some units cannot handle large amounts of haylage, resulting in equipment failure. Haylage needs to be chopped fine for proper handling in the mixer-blender unit.

The various feeds being used in the ration must be carefully weighed to ensure a balanced ration. Weighing devices may be stationary scales or electronic scales on the mixer-blender unit.

Feed Analysis

Analysis of feeds being used in dairy rations is recommended regardless of the type of feeding system used. However, it is especially important for the successful use of total mixed rations. Feed analysis is needed to properly balance the ration and reduce feed costs.

The most important feeds to be analyzed are the roughages used in the ration. Roughages have a greater variation in nutrient content than do grains. However, grains may be analyzed for nutrient content if desired.

Feed testing services are available from both private and state university laboratories in many states. Some feed companies provide feed testing services for their customers. The dairy farmer should contact the feed testing laboratory to get the procedures for sampling and packaging feed samples before sending in samples to be tested.

Grouping Cows to Feed Total Mixed Rations

Part of the success of feeding total mixed rations depends on dividing the cow herd into groups. This is done to more nearly match the ration to the needs of the different groups.

Several factors may be considered when deciding on groups for total mixed ration feeding. These are herd size, facilities, and time the cow must wait in the holding area for the milking parlor. Large herds should be divided into several groups, but it may be practical to have only one group in a small herd. The kind of facilities available may limit the number of different groups that may be used. If a cow has to be in a holding area for more than 2 hours before milking, the herd should be divided into smaller groups to reduce this waiting time. Cows should not be kept away from feed and rest for more than 2 hours.

The most common way to group cows is according to production level. All dry cows are kept in a separate group. Some dairy farmers prefer to keep first calf heifers in a separate group. Cows with mastitis or other health problems may also be kept in a separate group or milked separately.

In larger dairy herds, at least three groups should be used. These are high-, medium-, and low-producing cows. Higher overall milk production and increased profits will usually result from this type of grouping.

High-producing cows need more concentrate in their ration than do low-producing cows. They are more efficient in converting feed into milk and are the most profitable cows in the herd. Low-producing cows tend to get too fat when fed at the same level as high-producing cows. Dry cows should be in a separate group from lactating cows.

Computerized or magnetic feeders can be used in small herds where two or three groups are not practical. This creates another group without physically dividing the cows and permits high-producing cows equipped with the computerized or magnetic tag to get access to additional feed.

Move fresh cows into the high-producing group about 3 days after calving. Leave them in this group for at least 2 months to challenge them to reach their maximum potential of production. At the end of 2 months, move any cow whose production is not high enough to a lower-producing group.

Move cows from one group to another at no more than monthly intervals. This shift is based on Dairy Heard Information Association (DHIA) production records for individual cows. Shift cows in small groups rather than one cow at a time. Other factors should be considered when shifting cows from one group to another. These include physical condition of the cow, age, stage of lactation, stage of gestation, and individual cow temperament. A short drop in production should be expected after the movement of the cow. Changes in environment or procedure will result in the cow producing less milk until she becomes comfortable with her new surroundings.

The ration for each group is determined by average production, size of cows, and the average milkfat test for the group. Rations must be adjusted when the quality of the forage being used changes. Analyze forages at least once a month.

Make sure plenty of water is available at all times for each group. If hay is not included in the total mixed ration, it may be fed to the group in a separate manger.

Make sure there is enough bunk space so cows may eat free choice whenever they want to. Keep feed in the manger at all times, especially for the high-producing group.

Several problems may occur when grouping cows. It takes more labor to change groups at milking time and to sort cows when moving them from one group to another. Proper facility design reduces this problem.

Automatic Concentrate Feeders

There are three types of automatic concentrate feeders currently in use on dairy farms: magnetic, electronic, and transponder. These systems automatically control access to concentrate feed by individual cows. These systems require large amounts of initial investment. Therefore, they are used mainly by large farms.

A magnetic or electronic device is attached to each cow and allows access to feed. The **transponder** controls the amount of feed each cow receives. The purpose is to allow high-producing cows to have more concentrate while limiting the concentrate intake of low-producing cows. Good management is needed to successfully use these feeding systems.

ROUGHAGES FOR DAIRY CATTLE

From 60 to 80 percent of the dry matter in the dairy ration should come from roughages. High-quality roughages can lower the cost of feed for the dairy herd. Roughages used for dairy cattle rations are hay, silage, and pasture.

There can be a wide variation in the nutrient value of forages depending on the variety, stage of maturity when harvested, moisture content at harvest, growing conditions, method of harvesting, and method of storing. Evaluation of nutrient content should be made for:

- dry matter (DM), to determine dry matter intake and proper storage
- crude protein (CP) and adjusted crude protein (ACP). These will be different if the forage has been heat damaged. Heat damage reduces available protein

- acid detergent fiber (ADF). As ADF increases, the energy content of the forage is reduced
- neutral detergent fiber (NDF). The level of NDF is a good indicator of forage intake by the animal. Lactating cows should get at least 21 percent of their daily dry matter intake from the NDF in the forage
- calcium (Ca) and phosphorus (P) to ensure proper mineral balance in the diet

Legume and legume-grass mixtures can be compared using a relative feed value (RFV) figure based on the quality of the forage as determined by its ADF and NDF content. An ADF content of 41 percent and an NDF content of 53 percent are considered equal to an RFV value of 100. The equation for the relative feed value is

$$\text{RFV} = (\text{Digestible dry matter} \times \text{Daily dry matter intake}) + 1.29$$

Forages should have a RFV of 120 or higher for high-producing cows.

Representative samples of the feed are necessary for accurate analysis. Follow the instructions from the testing laboratory for securing, handling, and sending representative samples.

Hay

The feeding value of any hay depends on the kind of hay, stage of maturity when cut, and the harvesting method. Second and third cuttings of hay are generally higher in nutrient value than first cuttings. Early-cut hay has a higher feed value than hay that is more mature when harvested.

A high percent of the nutrient value of hay is in the leaves. Harvesting methods that retain more of the leaves result in a higher nutritional value. Windrowing hay when it is too dry causes a high leaf loss. Hay should be windrowed at about 35 to 40 percent moisture content. The use of hay crimpers or conditioners reduces leaf loss during harvesting. The moisture content of the hay should be 20 percent or less when it is baled. This will reduce spoilage and helps maintain nutrient value. Hay that is rain damaged has a lower feeding value compared to similar hay that is not rain damaged. Legume hays contain more protein than grass hays.

Hay has traditionally been handled in a baled form. This requires a lot of hand labor. Some dairy farmers use large, round bales as a method of harvesting hay. This saves labor but usually results in a greater loss of nutritional value. Handling hay in a cubed or wafered form reduces the labor needed. However, machine costs are higher.

Dairy cattle eat much higher quantities of high-quality hay compared to low-quality hay. More concentrate is needed to balance the ration when low-quality hays are used.

Alfalfa is the best hay for dairy cattle use (Figure 41-3). It yields more protein and total digestible nutrients (TDN) per acre than other hay crops. Alfalfa hay should be cut at the 1/10 bloom stage for highest feeding value.

Red clover hay is lower in protein and less palatable compared to alfalfa hay. Sweet clover hay tends to be coarse and stemmy. Sweet clover is better as a pasture rather than a hay for dairy cattle. Lespedeza hay is similar to alfalfa hay in feeding value.

Nonlegume hays such as timothy, millet, oats, prairie grass, Johnsongrass, and Coastal Bermuda grass are generally lower in protein and TDN than alfalfa hay. A high level of fertilization can improve the feeding value of these grass hays.

Mixtures of legume-grass hay are popular in some areas. Alfalfa-bromegrass, alfalfa-timothy, and alfalfa-orchard grass mixtures are high in nutritive value if harvested at the time recommended for alfalfa.

Figure 41-3 Alfalfa field partially cut.

Green Chop

Some dairy farmers harvest forage by chopping it daily and feeding it in bunks. This harvesting method reduces field losses. A major problem is the need to chop the forage each day, especially in bad weather or during times of high labor requirements for other farm operations.

The moisture content of green chop forage varies considerably. As the crop matures, cows generally eat less of the green chop forage. This reduces the amount of energy received from the forage part of the diet.

Silage

Almost any crop can be made into silage. Hay crops, corn, and small grains are common kinds of silage used in dairy rations. Silage has a higher moisture content than dry roughages. Silage fits well with the feeding of total mixed rations. It is in a form that is easily handled by automated equipment. Legume-grass silage may be put into a bunker silo at 60 to 65 percent moisture content.

Courtesy of Frank B. Flanders

Figure 41-4 Silos are used to store grain or silage.

Haylage

Haylage is a hay crop made into silage. The first cutting of hay is often harvested as haylage. Both grasses and legumes may be used for haylage. Haylage may be put into a conventional silo at 50 to 60 percent moisture and in an oxygen-limiting silo at 40 to 50 percent moisture (Figure 41-4). When stored in bunkers, it may be harvested at 75 to 80 percent moisture. This saves the maximum nutrients from standing crop to feeding. Legumes should be cut at 1/10 bloom and grasses at the early heading stage. Harvest Sudan grass and sorghum-Sudan hybrids before the heads emerge from the boot. Harvest small grains for silage in the boot to early milk stage. Soybeans should be cut for silage when the beans are forming in the pod. As maturity at harvest increases beyond the recommended stage, the palatability and digestibility of the feed decreases.

Grass and legume silages are lower in energy and higher in protein content than corn silage. The amount of dry matter in haylage varies with percentage of moisture at the time it is put in the silo. Cows have difficulty in eating enough high-moisture haylage to maintain high milk production. However, when haylage is put in the silo at 50 to 65 percent moisture content, little difference has been shown experimentally in milk production when compared to cows fed dry hay. Cows apparently use the nutrients in haylage more efficiently.

Putting haylage in the silo at too low a moisture content may cause heat damage. This occurs during the fermentation process in the silo. Heat-damaged haylage provides less energy and protein in the ration. It has a darker color (golden brown to black), caramel odor, and high palatability. Cows like to eat heat-damaged haylage but do not maintain high milk production on it.

If the crop is too dry, water may be added at the silo. Propionic acid may be added to reduce heat damage.

CAUTION

Propionic acid is corrosive and will cause severe burns. Always use goggles, wear rubber gloves, wash skin immediately upon contact, and clean all equipment right after use.

Some dairy producers prefer to put the forage crop directly into the silo without letting it wilt. Dry ground material, such as grain, or chemical preservatives are added to reduce moisture content and nutrient loss. Materials such as ground grain, dried beet or citrus pulp, dried brewer's grains, hominy, or soybean flakes may be used as preservatives in forage silage. When the forage is direct cut, 100 to 200 pounds of this material may be added per ton of silage. Preservatives, such as propionic acid, may be added to high-moisture forage silage. No additive or preservative will need to be used if the crop has been wilted to the recommended level before it is put into the silo.

Good management during ensiling helps preserve the best possible nutritive value of the haylage. Cutter knives should be set at 3/8 inch. Fill the silo as fast as possible and get a good pack on the material. Distribute the material evenly in the silo. Make sure silo walls and doors are tight to keep air out. If possible, put wetter material on top. Wait 3 to 5 weeks after filling the silo before starting to feed out of it.

Corn Silage

Corn silage is a popular roughage in dairy cattle rations (Figure 41-5). Some dairy farmers use corn silage as the only roughage for the dairy herd.

Corn silage yields more energy per acre than other forages. It is a highly palatable feed that is easily stored and handled. Corn silage requires less labor to harvest and feed than many other roughages. The quality of corn silage is usually more consistent than other roughages. Good corn silage contains about 50 percent grain on a dry matter basis.

Corn silage contains about 8 percent protein on a dry matter basis. When corn silage is used as the main forage in a dairy ration, it must be supplemented with protein. This can be done by feeding a concentrate mix that is higher in protein. Urea or other NPN sources may be added to the corn silage to increase the protein content. A protein supplement, like soybean oil meal, may be added to the ration.

Compared with other forages, corn silage is low in minerals. Additional minerals must be fed to balance the ration.

Harvest corn silage when the kernels have reached the dent stage. Dry matter should be about 30 to 35 percent. Set cutter knives on the field chopper at 1/4 to 3/8 inch.

Courtesy of Harveststore-Engineered Storage Products Co

Figure 41-5 Silage is often used as a roughage in dairy cow rations.

Sorghum Silage

Sorghum silage may be made from both forage and grain sorghums. The nutrient content is about the same as corn silage. The crude protein content and energy level are a little lower than corn silage. Sorghum silage is usually a little lower in digestibility than corn silage. Forage sorghum yields about the same per acre as corn silage. Grain sorghum yields per acre are usually a little lower than corn silage.

Dairy cows produce about the same when fed either grain sorghum silage or corn silage. However, cows must eat more grain sorghum silage than corn silage to maintain the same level of production.

Intermediate-type sorghum silage yields more per acre than corn silage. It is taller than combine-type sorghum. It is grown in some areas that are not well adapted to corn. Milk production is lower when intermediate-type sorghum silage is fed.

Coastal Bermuda Grass Silage

Coastal bermuda grass silage has a lower feeding value than corn silage. Storage loss in upright silos is high. Harvesting coastal Bermuda grass as silage is not recommended.

Small-Grain Silage

Small-grain silage may be made from small grains such as oats, rye, barley, and wheat. The nutritive value of small-grain silages is less than corn silage.

Barley and wheat silage will maintain higher levels of milk production than oat silage. In some areas, the use of small-grain silage will permit **double cropping** (growing two crops on the same ground in the same year).

Harvest barley, wheat, and oats at the boot to early head stage in order to produce the best yield of energy and protein per acre. These small grains may be harvested by direct cut without wilting.

Harvest rye for silage at the boot stage of maturity. Cut and wilt it before putting it into the silo.

Growing Austrian winter peas with barley for silage will increase the yield per acre. The protein content of the silage is also increased. This mixture may be direct cut and put in a conventional upright silo without wilting. If the crop is to be stored in a sealed (air tight) silo, wilt it to 50 to 55 percent dry matter before ensiling.

USDA Agricultural Research Service, Photo by Wally Wilhelm

Figure 41-6 Baled corn stalks may be used for dairy cows during certain stages, but the product is of low quality.

Straw

Straw from small grains (oats, wheat, etc.) is not recommended in rations for cows that are milking. These straws are low in energy, protein, minerals, and vitamins. They mainly add fiber to the ration. Some straw in the ration can be used for dry cows and older heifers. Supplement must be added if straw is used in the ration.

Corn Stover

Corn stover is not recommended in rations for cows that are milking. It may be used in rations for dry cows and heifers. Corn stover is low in protein and carotene (Figure 41-6). Supplements must be added if corn stover is used in the ration.

Pasture

The effective use of pasture for the dairy herd requires good management. The use of pasture reduces the labor needed for feed and manure handling. Pastures are an excellent source of roughage for dry cows and growing heifers.

Several problems may occur when pastures are used for lactating cows. These include:

- drop in milk production after cows are on pasture for a time
- drop in milkfat test
- off flavors in the milk
- bloating of cows
- reduced grain intake
- watery feces
- difficulty in getting cows to come into the milking parlor

It is hard for lactating cows to get enough dry matter in the ration if pasture is used as the main source of roughage. Pasture has a high moisture content and generally cannot meet the energy needs of high-producing cows. Several management practices may be used to help reduce the problems with lactating cows on pasture.

Limit the grazing time on pasture for lactating cows to 1 to 2 hours per day. Feed dry forage before letting the cows out to pasture. Bring the cows back to the barn several hours before milking. Feed silage and/or hay again at this time. Continue feeding the concentrate mixture at the same rate as when the cows are not on pasture. Protein content of the concentrate mix may be reduced some if the pasture is of good quality.

Rotational grazing may be used. Divide the pasture with a temporary fence such as an electric fence. The high-producing cows are allowed to graze on the fresh pasture. Lower-producing cows, dry cows, and bred heifers are put into the pasture that has been grazed over by the high-producing cows. Clip the pasture regularly to keep weeds down, and allow fresh regrowth of the pasture. Pastures need to be well fertilized to provide high-quality forage.

Use pasture as a supplementary roughage. Make sure the lactating cows have access to silage and/or hay to make up for nutritional shortages as pasture quality changes through the grazing season.

Grasses or grass-legume mixtures make good pastures for dairy cattle. Timothy, bromegrass, orchard grass, coastal Bermuda grass, alfalfa, sweet clover, and ladino clover are examples of crops used for dairy pastures.

Temporary pasture for dairy cattle may be provided by Sudan grass, rye, oats, hybrid sorghum, and pearl millet. Prussic acid poisoning may occur on Sudan grass if it is grazed before it is 18 inches high. Do not pasture Sudan grass after a freeze as the new growth may cause prussic acid poisoning. Rye may cause off flavors in milk. Removing lactating cows from rye pasture several hours before milking will help prevent this problem.

GRAINS FOR DAIRY COWS

Grains are included in the dairy ration mainly for their energy content. Usually, the most limiting factor in milk production is a shortage of energy in the ration. It is important that lactating dairy cows get enough energy in the ration. Grains contain about 70 to 80 percent TDN.

Grains that are processed before feeding are more digestible. Methods of processing grains include grinding, rolling, crimping, pelleting, or cracking. Grains that are ground too fine may lower digestibility and percent of milkfat. Rumen acidosis may also result from feeding grain that is too finely ground.

Corn

Corn is the most commonly used grain in dairy cattle rations. It is high in energy value and is palatable. Home-grown corn is usually a cheaper energy source than other grains.

Corn and cob meal contains about 90 percent of the TDN and is higher in fat than shelled corn. The higher fiber content helps keep the percentage of milkfat higher. Cows stay on feed better when corn and cob meal is used in the ration.

High-Moisture Corn

Ear or shelled corn may be stored at a relatively higher moisture content in a silo. High-moisture ground ear corn is stored at 28 to 32 percent moisture and high-moisture shelled corn at 25 to 30 percent moisture.

An organic acid (such as propionic or acetic) may be used as a preservative for high-moisture corn. More preservative is needed for high-moisture ear corn than for high-moisture shelled corn.

Ear corn should be ground before being put in the silo. Shelled corn may or may not be ground before ensiling. It should be ground before feeding.

High-moisture corn is a good feed for dairy cattle. On a dry matter basis, it is equal to dry corn in feeding value. More high-moisture corn must be fed to get the same nutrient intake as dry corn.

Oats

Oats are an excellent feed for dairy cows. They are lower in energy value and the protein content is higher than corn. Oats add fiber and bulk to the grain mix and are a little lower in digestibility than corn. Oats should replace no more than one-half of the corn in the ration.

Barley

Barley has about the same overall energy value for dairy cattle as corn. It is a little higher in protein compared to corn. Barley that is rolled is more palatable than when it is ground. Finely ground barley should make up no more than 50 percent of the grain ration.

Wheat

Wheat is high in energy and protein. Dairy cattle like wheat. Because of price, it is not usually used in dairy cattle rations. Wheat should make up no more than 50 percent of the grain ration.

PROTEIN SUPPLEMENTS FOR DAIRY COWS

A protein supplement is added to the ration to make up the difference between what the cow needs and what is supplied by the rest of the ration. The amount and quality of the forage in the ration has a great effect on the amount of protein needed in the concentrate mix. Protein supplements are usually the most expensive part of the ration. Therefore, care must be taken to carefully balance the ration to hold the feed cost as low as possible. The quantity of protein is more important than the quality of the protein. Purchase protein for dairy cows on the basis of cost per pound of protein supplied in the supplement.

Corn Gluten Meal

Corn gluten meal is a by-product from the wet milling of corn for starch and syrup. It may have 40 percent or 60 percent crude protein content; the 60 percent meal is the most common. The energy content of corn gluten meal is slightly lower than corn grain. Because it has a lower palatability, limit the amount of corn gluten meal to no more than 5 pounds per head per day.

Distillers' Dried Grains

Distillers' dried grains are a by-product of grain fermentation for alcohol production. The nutrient value of distillers' dried grains varies depending on the grain used as its source. They contain 23 to 30 percent protein.

Soybean Meal

Soybean meal is the by-product left when oil is extracted from soybeans. Soybean meal is a commonly used protein supplement. It is economical and an excellent source of protein. Soybean meal is found in many commercial dairy protein supplements. It is palatable and highly digestible.

Soybeans

Ground, unprocessed soybeans may be used in grain mixes for dairy cows as an added protein source (Figure 41-7). If used, they should not make up more than 20 percent of the grain mix. Do not use urea in the ration if soybeans are included.

© iStockphoto/Ana Simin

Figure 41-7 Soybeans.

Processing by roasting or cooking can increase the palatability and stability of the soybeans. Soybeans are usually too high-priced to be used in dairy cow rations.

Sunflower Meal

Sunflower meal supplements range from 28 to 45 percent protein. Sunflower meal is a good source of protein and phosphorus. Because it is less palatable, it should not be topdressed on the feed.

Linseed Meal

Linseed meal is the by-product left when oil is extracted from flax seed. It is a good protein supplement. Linseed meal is usually higher in cost than some other protein supplements. It is sometimes used for fitting show or sale cattle because it adds a shine to the hair coat. Linseed meal is palatable and slightly laxative.

Cottonseed Meal

Cottonseed meal is made from hulled cotton seeds after the oil has been extracted. It is high in protein and may be used as a protein supplement in dairy rations. Cottonseed meal is palatable and may cause constipation.

Urea and Other Nonprotein Nitrogen

Urea may be used in dairy rations to supply some of the protein needed by cows. The use of urea usually lowers the cost of the ration. The nitrogen content of urea is 46 percent. It supplies a protein equivalent of 287 percent (46×6.25).

Urea has a bitter taste and must be mixed completely in the ration for cattle to eat it. Too much urea in the ration can be toxic to cattle. Feed no more than 0.4 pound of urea per head per day. The concentrate mix should be no more than 1 percent urea. If urea is added to corn silage at the maximum recommended rate of 10 pounds per ton (0.05 percent), then the concentrate mix should contain no more than 0.5 percent urea.

It takes the rumen bacteria time to adapt to urea in the ration. It is recommended that the amount of urea in the ration be gradually increased over a period of 7 to 10 days to reach the desired amount in the ration. The amount of urea fed should be limited to 0.2 pound per day during early lactation.

Other NPN products may be used as protein substitutes in the ration. Monoammonium phosphate (11 percent nitrogen, 68.75 percent crude protein equivalent) and diammonium phosphate (18 percent nitrogen, 112.5 percent crude protein equivalent) are two examples of other NPN sources for dairy cattle.

Liquid protein supplements are made by putting an NPN source (usually urea) in a liquid carrier. Molasses is the most common carrier. Other carriers such as fermentation liquors or distillers' solubles may be used. Molasses provides an energy source needed for the utilization of the NPN. Phosphoric acid is often added to provide phosphorus and help stabilize the nitrogen.

Most liquid protein supplements contain 32 to 33 percent crude protein equivalent. Liquid protein supplements can be used to provide part of the protein needed in the ration. They may be fed free choice, topdressed (usually on the forage), or blended into a total mixed ration.

There is generally no economic advantage to using liquid protein supplements. The decision to use a dry or a liquid protein supplement depends on the available labor and equipment. Convenience of use is also a factor. The use of liquid protein supplement may be of most value for low-producing cows, dry cows, yearlings, and older heifers when corn silage is the major forage in the ration.

BY-PRODUCTS AND OTHER PROCESSED FEEDS FOR DAIRY CATTLE

Figure 41-8 shows different types of processed feed ingredients for dairy cattle.

Alfalfa Meal

Alfalfa meal is made by grinding alfalfa hay. High-quality hay produces a high-quality meal. It may be pelleted. Because of the grinding, the fiber content will not help in maintaining milkfat test. Alfalfa meal is lower in digestibility than alfalfa hay.

Courtesy of Frank B. Flanders

Figure 41-8 Top, left to right: citrus pulp, whole cottonseed, soybean hulls. Bottom, left to right: corn meal, distillers' grain, soybean meal.

USDA Agricultural Research Service; Photo by Doug Wilson

Figure 41-9 Alfalfa field.

Alfalfa Leaf Meal

Alfalfa leaf meal is made by grinding the leaves of the alfalfa plant (Figure 41-9). It is higher in protein than alfalfa meal.

Beet Pulp

Beet pulp may be plain or mixed with molasses (Figure 41-10). It is palatable, bulky, and slightly laxative. Beet pulp is low in protein and high in energy. Up to 30 percent of the ration dry matter may be composed of beet pulp.

Courtesy of Frank B. Flanders

Figure 41-10 Beet pulp.

Brewer's Grain

Brewer's grain is a by-product of the brewing industry. It may be wet or dry. Wet brewer's grain contains about 80 percent water and, therefore, requires the feeding of a large amount to get the needed dry matter intake. On a dry matter basis, brewer's grain is higher in protein and lower in net energy compared to corn.

The ration must be gradually adjusted if brewer's grain is to be included. Wet brewer's grain requires special handling and storage.

Citrus Pulp

Citrus pulp has about the same feeding value as beet pulp. It is lower in protein than beet pulp. Citrus pulp is sometimes fed when the price is competitive.

Cottonseed, Whole

Whole cottonseed is high in fat, fiber, and energy but only medium in protein content. The whole cottonseed is white and fuzzy; delinted cottonseed is black and smooth. Do not feed more than 7 pounds per cow per day.

Corn Gluten Feed

Corn gluten feed is a by-product of the corn wet milling industry. It is medium in energy and protein content but high in fiber. It may be purchased as either a wet or dry feed. Limit the amount of corn gluten feed to not more than 25 percent of the total ration dry matter.

Hominy Feed

Hominy feed is a mixture of the starch part of the corn kernel, the bran, and the corn germ. It is a little higher than ground corn in feeding value. It is palatable and high in energy. Hominy feed may be used in place of ground corn in dairy rations.

Malt Sprouts

Malt sprouts are bitter and need to be mixed with other feeds. They have a medium protein level and only a moderate amount of energy. No more than 20 percent of the grain mix should be composed of malt sprouts.

Molasses

Both cane and beet molasses may be used in dairy rations. They are used mainly to increase the palatability of the ration. Molasses should be no more than 5 to 7 percent of the grain mix. Cane molasses is more commonly used than beet molasses (Figure 41-11). Molasses supplies energy in the ration and is low in protein.

Figure 41-11 Sugar cane.

Potatoes

Cull potatoes, potato meal, and potato pulp may be used in dairy rations. Potato meal has a high fat content. Potato pulp is palatable and about equal to hominy feed when it is not more than 20 percent of the concentrate mix. Cull potatoes are similar to corn silage in feeding value. Cull potatoes should be chopped when fed. Limit maximum intake to 30 pounds per head per day.

Soybean Hulls

Soybean hulls, soyhulls, and soybean flakes are all by-product feeds from soybean processing. Their fiber content is highly digestible, and they may replace starch in the diet. They do not replace forage fiber. These feeds should be limited to no more than 45 percent of the grain ration.

Wheat Bran

Wheat bran adds bulk and fiber to the concentrate mix. It is palatable and slightly laxative. Wheat bran is low in energy and medium in protein content. It should not make up more than 20 to 25 percent of the concentrate mixture.

Wheat Middlings

Wheat middlings, also known as wheat midds, are by-products of wheat milling for flour. They are fine particles of wheat bran, wheat shorts, wheat germ, and other wheat products. They contain a moderate amount of energy and protein. The crude fiber content cannot exceed 9.5 percent. If used at too high a level in dairy cow diets, they may cause a reduction in milk production. Limit the use of wheat middlings to no more than 20 percent of the grain mix. Wheat midds can be purchased as a loose meal or pellets. Feed manufacturers often include wheat midds as an ingredient in commercial feeds and feed supplements.

Whey

Both dried and liquid whey may be fed to dairy cows (Figure 41-12). Up to 10 percent of the concentrate mix may be dried whey. Dried whey may be added to silage when the silo is being filled. Use 20 to 100 pounds per ton of wet silage.

Liquid whey may be fed free choice. Because of the high water content (90 percent) cows must consume 15 to 25 gallons per day to get enough dry matter intake.

Figure 41-12 Whey protein is a by-product of cheese making. It is often used as supplementary protein for body builders.

One hundred pounds of liquid whey contains about the same amount of energy as 7.5 pounds of most dairy concentrates.

Liquid whey feeding requires adaptation by the cattle. Water must be limited for about 4 weeks to get the cattle drinking the desired amount of liquid whey. A large amount of liquid in the gutter may be a problem if liquid whey is fed in stanchion barns. A constant supply of fresh liquid whey is necessary. Flies may become a problem when liquid whey is fed.

MINERALS FOR DAIRY CATTLE

Minerals that are needed by dairy cows include calcium, phosphorus, magnesium, potassium, sodium, chloride, sulfur, iodine, iron, copper, cobalt, manganese, zinc, selenium, and molybdenum (Table 41-1).

Well-balanced rations that include mineral supplements will meet the mineral needs of dairy cows. Steamed bonemeal, dicalcium phosphate, and limestone may be used to supply calcium. Dicalcium phosphate, monocalcium phosphate, and monosodium phosphate may be used to supply phosphorus. Salt supplies sodium and chloride. The use of trace-mineralized salt will supply many of the minerals needed in small amounts. Commercial mineral mixes fed at the recommended level will normally supply the needed minerals. The ratio of calcium to phosphorus in the ration of lactating cows should be 1.2:1 to 2:1.

Do not feed too much phosphorus in the ration. Doing so increases costs and the amount of pollution caused by manure. The amount of phosphorus in the manure can be reduced by 25 to 30 percent by limiting the amount of supplemental phosphorus fed to not more than 0.38 percent of the diet for cows during the lactation period. Reproduction problems from a lack of phosphorus in the diet do not occur until the level drops to 0.15 to 0.2 percent of the ration.

Zinc methionine may be added to the diet at the rate of 4.5 grams per cow per day for cows that have feet and leg problems. Research and field responses to this treatment have varied, depending on the forages fed, feeding system used, and herd health. The economic benefit of the treatment must be balanced against its cost.

TABLE 41-1 Vitamins and Minerals Required and Commonly Supplied to Dairy Cattle

Vitamins	Minerals
Vitamin A	Calcium
Vitamin D	Chloride
Vitamin E	Cobalt
	Copper
	Iodine
	Iron
	Manganese
	Magnesium
	Molybdenum
	Phosphorus
	Potassium
	Selenium
	Sodium
	Sulfur
	Zinc

VITAMIN NEEDS OF DAIRY CATTLE

Nutrient requirement tables give the vitamin A and D requirements for dairy cattle. Some supplementation of vitamins A, D, and E may be needed in dairy rations (Table 41-1). Usually there are enough vitamins in the feeds used in dairy rations. Some conditions such as poor-quality forage, high levels of grain feeding, or lack of sunshine may cause a need for additional vitamins.

A commercial vitamin premix may be added to the concentrate mix to supply needed vitamins. Massive doses of vitamins taken over too long a period of time may be toxic.

The addition of 6 grams of niacin per cow per day to the diet has resulted in increased milk production and feed intake and a decrease in problems with ketosis. The niacin improves rumen digestion and protein synthesis and controls fatty liver. It may be added to the diet beginning 2 weeks before calving and continued for the first 10 to 12 weeks of lactation. Major benefit is achieved when niacin is added to the diet of cows that are high-producing, are prone to ketosis problems, or are too fat.

WATER NEEDS OF DAIRY CATTLE

Milk is 85 to 87 percent water. Lactating dairy cows require more water in relation to their size than any other farm animal. Dairy cattle suffer more quickly from a lack of water than from a lack of any other nutrient. As the air temperature increases, the need for water also increases (Table 41-2). Always keep plenty of fresh water available for dairy cattle.

Keep the water supply free of bacteria and high levels of nitrates and sulfates. A high level of blue-green algae in the water may be toxic to cattle. Do not allow cattle to drink from ponds or lakes that contain a heavy growth of algae.

BODY CONDITION SCORE

Body condition score in dairy cattle refers to the amount of fat the animal is carrying. Scores range from 1 (very thin) to 5 (excessive fat). Body condition scoring is done by observing the amount of depression around the tailhead, the amount of fat covering the pin and pelvic bones, and the amount of fat covering the loin area. The amount of fat covering is determined by feeling with the hands and making as objective a judgment as possible about the amount of fat covering that is present.

Body condition score 1 is a thin animal showing a deep depression around the tailhead and no fat covering over the rump and loin. Body condition score 2 shows a shallow cavity around the tailhead and a small amount of fat covering the rump and loin areas. Body condition score 3 shows no cavity around the tailhead and fatty tissue over the whole rump and loin area (Figure 41-13). Body condition score 4 shows folds of fatty tissue over the tailhead and patches of fat over the rump with fairly heavy fat covering over the loin. Body condition score 5 shows the tailhead buried in fatty tissue and heavy fat covering over the rump and loin areas.

For maximum efficiency in milk production, dairy cows must not be too thin or too fat. Using body condition scoring to evaluate the condition of the herd will help the dairy farmer improve feed efficiency and herd health. Thin cows have more health problems; cows that are too fat have more difficulty calving and a higher risk of fatty liver syndrome. Heifers being raised for breeding should also be checked for

TABLE 41-2 Water Intake Guideline for Dairy Cattle

Type of Cattle by Weight (lb)	Milk (lb)	Air Temperature		
		40°F or Less	60°F	80°F
Heifers		(gal water/day)	(gal water/day)	(gal water/day)
200	N/A	2.0	2.5	3.3
400	N/A	3.7	4.6	6.1
800	N/A	6.3	7.9	10.6
1,200	N/A	8.7	10.8	14.5
Dry Cows				
1,400	N/A	9.7	12.0	16.2
1,600	N/A	10.4	12.8	17.3
Lactating Cows				
1,400	60	22.0	26.1	24.7
	80	27.0	31.9	38.7
	100	32.0	37.7	45.7

USDA

Figure 41-13 This dairy cow has a body condition score (BCS) of 3.

body condition score. Those that are too fat will have more difficulty calving and may have poorer mammary development, which will reduce their lifetime production potential. Dairy farmers planning to use bovine somatotropin (bST) should check their herd for body condition. The use of bST is not recommended for cows that are too thin or too fat.

Cows should be checked for body condition at the following times:

1. **Shortly after calving so feed adjustments may be made if necessary.** At the time of calving, a cow should have a body condition score of 3.0 to 3.5. It is normal for the cow to lose some weight during the early part of the lactation period, but the body condition score should not drop below 3.0 to 2.5. A very high-producing cow may drop as low as body condition score 2.0 during this time.
2. **Early in the lactation period, between 1 to 4 months.** Body condition score during this part of the lactation period should be around 2.5 to 3.0. Increase the energy level of the ration for cows that are too thin, and make sure they have sufficient feed intake to maintain the proper body condition. Cows should reach their peak milk production during this part of the lactation period. If the cow is maintaining a body condition score in the range of 3.0 to 3.5 but does not produce the expected amount of milk, check the ration for protein level, calcium, potassium, phosphorus, and magnesium intake. Make sure there is an adequate water intake. Deficiencies in these nutrients may prevent the cow from reaching her maximum milk production while still maintaining adequate body condition score.
3. **Around the middle of the lactation period or at about 4 months.** The cow should have a body condition score of about 3.0 during this period. If it appears the animal is becoming too fat (body condition score of 3.5 to 4.0) at this time, energy intake should be reduced. If the animal is becoming too thin, increase the energy intake. Problems often begin earlier in the lactation period, so if a number of animals in the herd are out of body condition at mid-lactation, it is wise to do an earlier check on body condition during the next lactation period.
4. **Check for body condition score again near the end of the lactation period (after 8 months).** The cow should be building body reserves at this time in preparation for the next lactation period. The body condition score at this time should be about 3.5. Adjust the ration for energy level if the animal is too thin or too fat. Do not allow the cow to become too fat before drying off in preparation for the next lactation period.

Adjustments are made in the energy, protein, fiber, and acid detergent fiber levels of the ration to correct problems with body condition scores during the lactation and dry periods of the production cycle. The next several sections of this chapter discuss various feeding recommendations that will help the dairy farmer better manage the body condition score of the animals in the herd.

FEEDING LACTATING DAIRY COWS

Some general guidelines may be followed when feeding the dairy herd. The total ration should contain from 18 to 19 percent crude protein on a dry matter basis during early lactation. Reduce the crude protein to 13 percent on a dry matter basis late in the lactation period. About 75 to 80 percent of the crude protein is digestible and available for use by the cow.

Protein degradability must be considered when feeding high-producing cows. Some protein is not digested in the rumen but is passed on to the small intestine before it is digested or degraded. This is called "bypass" or "escape" protein. Sources of bypass protein include heat-treated whole soybeans or soybean meal, distillers' grains, feather meal, blood meal, and meat and bonemeal.

A shortage of energy is usually the most limiting factor in milk production. The total ration should contain 60 to 70 percent TDN. This is equivalent to 0.60 to 0.80 megacalorie of net energy per pound of feed. A megacalorie is 1,000,000 calories. During early lactation, provide a minimum of 0.78 megacalorie per pound of dry matter as the net energy level for lactation. During the later stage of lactation, this may be reduced to 0.70 megacalorie per pound. During the dry period, provide 0.60 megacalorie per pound of dry matter.

The basis of any dairy feeding program is forage. Use high-quality legume forages during early lactation. Feed the cow 1.5 to 2.8 pounds of forage dry matter per 100 pounds of body weight.

Fiber in the ration is needed to maintain milkfat percent in the milk. There should be a minimum of 15 percent crude fiber in the ration. During early lactation, provide a minimum of 18 percent acid detergent fiber (ADF) in the dry matter. During late lactation, increase this to 21 percent or more ADF in the dry matter. The dry matter in the diet should contain a minimum of 21 percent neutral detergent fiber (NDF) from forages.

Rations low in "fiber" reduce the amount of time cows spend chewing. When cows spend less time chewing, less saliva is produced; this reduces the pH level and affects the ratio of acetate to propionate in the rumen. Lower saliva production also affects the milkfat percentage and the rate of passage of material through the rumen. Care must be taken to provide sufficient levels of "effective fiber" in the dairy ration in addition to the recommended minimum of 15 percent crude fiber.

An increase in milk production and higher feed efficiency can be achieved by properly balancing the amount of NDF and soluble carbohydrates (sugars and starches) in the dairy ration. Too high a level of soluble carbohydrates in the ration may result in rumen acidosis, lower milkfat test, cows going off-feed, and laminitis. Feed intake and the energy level of the ration are lowered when the level of neutral detergent fiber is too high. The recommended level of soluble carbohydrates in the ration is 30 to 35 percent, and the recommended level of NDF in the ration is a minimum of 28 percent. An ADF level of 19 to 21 percent is recommended. The proper amount of fiber improves rumen digestion and helps maintain the right pH level in the rumen. Soluble carbohydrates are needed for proper microbial growth in the rumen.

TABLE 41-3 Four Feeding Phases for Dairy Cows

Phase One	• Occurs during the first 70 days of lactation • Highest milk production
Phase Two	• 70 to 140 days after calving • Decreasing milk production • Highest dry matter intake
Phase Three	• 140 to 305 days after calving • Continual decrease of milk production
Phase Four	• Dry period of 45 to 60 days before the next calving • Beginning of new lactation period

The proper balance may be accomplished by combining high soluble carbohydrate sources with lower soluble carbohydrate sources in the ration. Shelled corn has a high soluble carbohydrate level. This may be balanced with beet pulp, oats, or soyhulls, which have lower soluble carbohydrate levels to prevent milkfat depression. Milkfat depression may occur with a ration that is too high in shelled corn. It is better to use alfalfa forage (lower soluble carbohydrate) rather than corn silage (higher soluble carbohydrate) with high-moisture shelled corn (higher soluble carbohydrate). High-moisture ear corn works well with corn silage in dairy rations because of a better balance of soluble carbohydrates between these two ingredients.

Both major and trace minerals are needed by the dairy cow. The concentrate mix should contain 0.5 to 1 percent mineralized salt. One percent of the concentrate mix should be a calcium-phosphorus supplement.

Grain and protein supplements are usually the most expensive part of the ration. Buy feeds on the basis of the least cost per unit of nutrient supplied. Generally, home-grown grains will lower the cost of the ration.

Feed requirements vary with the stage of lactation and gestation. Four feeding phases (Table 41-3) are identified based on milk production, fat test, dry matter intake, and changes in body weight during lactation. Phase one occurs during the first 70 days of lactation. Milk production is highest during this phase. Phase two is from 70 to 140 days after calving. Milk production is decreasing and dry matter intake is at its highest during this phase. Phase three is from 140 to 305 days after calving. Milk production continues to decrease during this phase. Phase four is the dry period of 45 to 60 days before the next calving and the beginning of a new lactation period. Tables 41-4 and 41-5 show some sample rations for these different feeding periods.

The most critical feeding period is during phase one. Milk production is increasing rapidly at this time. Maximum milk production is reached 4 to 6 weeks after calving. Increase the amount of grain in the ration by 1 to 1.5 pounds per day during the first 10 weeks of lactation. Keep the fiber level in the ration above 15 percent to keep the rumen working properly. Keep the grain level no higher than 50 to 55 percent of the total dry matter in the ration. Do not increase the rate of grain feeding too fast or feed more grain than is needed. The cows may go off-feed or have problems with acidosis or displaced abomasums if this is done. Too much grain in the diet can also decrease the milkfat percentage. During phase one, the diet must contain 19 percent or more crude protein. Feeding 1 pound of soybean meal or an equivalent commercial supplement for each 10 pounds of milk produced over 50 pounds is recommended. Problems with low peak production and ketosis develop if the nutrient needs of the cow are not met.

Some cows lose weight during early lactation. Adding 1 to 1.5 pounds of fat or oil to the diet during this period can increase milk yield, improve fat test, improve reproductive ability, and improve the general health of the cow. Fats from animal

TABLE 41-4 Example Rations for Lactating Cows (1,350-Pound Cow, 3.8 Percent Fat Test)

Item	Phase 1 (lb)	Phase 2 (lb)	Phase 3 (lb)
Milk/day	90	80	50
DM intake/day[a]	49	51	38
Ration 1			
Alfalfa hay (88% DM)—140 RFV, 20% CP	28.00	34.00	27.00
Corn	17.85	20.40	13.60
Oats	3.15	3.60	2.40
Soybean meal—44%	5.00		
Dicalcium phosphate—18% P	0.50	0.45	0.30
Salt, vitamins, trace minerals	0.30	0.25	0.25
Ration 2 (hay limit fed)[b]			
Alfalfa grass hay—113 RFV, 16% CP	10.00	10.00	10.00
Corn silage	41.00	70.00	57.00
Corn	13.60	9.35	5.10
Oats	2.40	1.65	0.90
Soybean meal—44%	11.50	8.20	4.50
Dicalcium phosphate—18% P	0.40	0.30	0.25
Limestone	0.40	0.30	0.15
Salt, vitamins, trace minerals	0.30	0.25	0,25

[a] Estimated average intake during the phase.
[b] Feed amounts may have to be limited during phases 2 and 3 to avoid overconditioning.
Source: J. G. Linn, M. F. Hutjens, W. T. Howard, L. H. Kilmer, and D. E. Otterby, Feeding the Dairy Herd, North Central Regional Extension Publication 346, Agricultural Extension Service, University of Minnesota, Revised 1989.

TABLE 41-5 Example Dry Cow Rations (1,400-Pound Cow)

Ration	lb
Ration 1 (grass forage)	
Orchard grass hay—12% CP	25.00
Corn	3.00
Soybean meal	0.50
Limestone	0.15
Trace mineral salt and vitamins	0.10
Ration 2 (limited legume forage)[a]	
Alfalfa hay—RFV 140, 20% CP	12.00
Corn silage	43.00
Monosodium phosphate	0.10
Trace mineral salt and vitamins	0.10

[a] Ration contains excess energy as formulated and may overcondition cows in some situations.
Source: J. G. Linn, M. F. Hutjens, W. T. Howard, L. H. Kilmer, and D. E. Otterby, *Feeding the Dairy Herd*, North Central Regional Extension Publication 346, Agricultural Extension Service, University of Minnesota, Revised 1989

sources are better than unsaturated vegetable fats. Commercial products are available to provide added fats and oils to the ration of dairy cows. The cost of these products as compared to the benefits resulting from their use must be carefully considered. Economical sources of oil include raw soybeans, heat-treated soybeans, raw whole cottonseed, and sunflower seeds.

During phase two, feed the cow to maintain peak milk production as long as possible. The cow reaches her maximum feed intake during this phase and should be maintaining or slightly gaining body weight. Maximum grain intake should be 2.3 percent of body weight, and minimum forage intake should be 1.5 percent of body weight. Calculate intake amounts on a dry matter basis. The recommended level of feed intake will maintain rumen function and normal milkfat test. To keep the rumen functioning at optimum level, use feeds that are high in digestible fiber when grain, on a dry matter basis, makes up 55 to 60 percent of the ration. For best results, feed forages and grain several times a day. Problems that may occur during phase two include lower milk production, low-fat test, silent heat, and ketosis.

During the last part of the lactation period (140 to 305 days after calving) milk production is usually dropping. The nutrient needs of the cow are less. To avoid waste, match grain intake to milk production. Cows that are too thin during this stage of lactation should be fed extra grain. It takes less feed to restore desirable body condition during this period than when the cow is dry.

Younger cows (2- and 3-year-olds) need extra amounts of feed nutrients for continued growth during lactation. If their nutrient needs for growth are not met, they often will not reach their full milk-producing potential. Two-year-olds should get 20 percent more and 3-year-olds 10 percent more nutrients.

The feeding sequence, number and size of feedings, and nutrient balance in each feeding can affect total milk production, milkfat test, and general herd health. The feeding sequence is especially important when total mixed rations are not used. Feeding 4 to 8 pounds of forage dry matter approximately 40 to 80 minutes before the grain is fed will increase saliva production and slow the rate at which the grain moves through the rumen, thereby improving digestion. Total mixed rations should be fed at least three to four times per day to maintain an acceptable level of feed intake. Increasing the number of feedings of a total mixed ration above four times daily has resulted in increased milk production for some dairy farmers. Limit the amount of grain fed per cow per feeding to a maximum of 5 to 7 pounds. Electronic grain feeders that are currently available can provide from 6 to 13 feedings per day. If forages are fed separately, feed them several times per day so they stay fresh and palatable and to increase feed intake; clean uneaten feed out of the feed bunks to prevent an accumulation of moldy, unpalatable feed. Nutrients should also be balanced at each feeding to increase milk production. For example, if a feeding consists of a low-energy, high-fiber feed such as grass haylage, then add a high-energy feed with a protein source. To function properly, the rumen microbes need a balance of nutrients at each feeding.

FEEDING DRY COWS

The nutrient needs of dry cows are generally not as high as during lactation. Nutrients are needed for the developing calf and to replace losses in body weight that occurred during lactation. Care must be taken not to overfeed the cow. Overfeeding will result in cows getting too fat during the dry period. Limit the amount of corn silage fed because of its high energy content.

Maintain a total dry matter intake during this period at about 2 percent of body weight. Roughage intake must be at least 1 percent of body weight. The amount of grain needed depends on the quality and type of forage fed. Feed no more than 1 percent of body weight daily; 0.5 percent is usually enough.

If no grain was fed early in the dry period, some should be fed during the last 2 weeks before calving. This will prepare the rumen for digesting grain during the lactation period. Begin grain feeding with about 4 pounds, and slowly increase the amount fed daily.

If retained placentas or metabolic disorders such as milk fever, ketosis, hypocalcemia, and displaced abomasum are a problem in the herd, then feeding a transition ration may be necessary. In addition to adding grain to the diet, the protein level should be raised from 12 percent to 14 or 15 percent. Dry matter intake drops considerably about 5 days before calving. The ration needs to be adjusted to compensate for the reduced feed intake. Some other additives that may help prevent metabolic disorders are niacin (6 grams/day), anionic salts, and yeast (10 to 113 grams/day). Anionic salts are not palatable and must be mixed in the feed. These additives are expensive and should only be used if problems are noted in the herd and then only during the last 7 to 10 days before calving.

Do not feed an excessive amount of calcium and phosphorus. Calcium intake of 1.7 to 2.8 ounces and phosphorus intake of 1 to 1.4 ounces per cow per day is usually enough. The calcium-phosphorus ratio should be about 2:1. Milk fever problems generally increase when the ration contains more than 0.6 percent calcium and 0.4 percent phosphorus on a dry matter basis.

If the feed sources are of poor quality, add vitamins A and D to the ration. This will increase the calf survival rate and reduce problems with retained placenta and milk fever.

Include trace minerals in the ration. Iodine and cobalt are especially important during the dry period. When urea is fed in the milking ration, begin feeding it about 2 weeks before calving.

FEEDING HERD REPLACEMENTS

Feeding Calves from Birth to Weaning

Proper feeding of the calf from birth to weaning is critical if death losses are to be held to a minimum. Calves are born with little immunity to disease. Colostrum milk is the first milk secreted by the cow after calving. Colostrum milk, as compared to regular milk, is high in fat, nonfatty solids, total protein, and antibodies that protect against disease. It is critical that the calf receives colostrum milk within a few hours after being born. The ability of the calf to directly absorb antibodies drops sharply after 24 hours. It is best if the calf gets its first feeding of colostrum milk within 30 minutes after being born. The amount of this first feeding should be equal to 4 to 5 percent of the calf's body weight. The calf should receive an amount of colostrum during the first 24 hours equal to 12 to 15 percent of its body weight. Good-quality colostrum milk is thick and creamy. Colostrum milk that is thin and watery should not be fed to calves because it does not contain the nutrients and antibodies they need. Do not feed newborn calves colostrum milk that is bloody or from cows infected with mastitis. The second milking after calving usually contains about 60 to 70 percent of the antibodies found in the first colostrum milk.

Wash the cow's udder and teats before the calf nurses. The calf is less likely to pick up dirt and germs when the udder and teats are clean.

Colostrum milk may be fed from a bottle or a nipple pail. It is more work to feed the calf this way. However, it is easier to keep track of how much the calf drinks.

Feeding too much colostrum milk may cause scours (diarrhea) and increase stress on the calf.

Extra colostrum milk may be frozen and used in case there is none available when other calves are born. After giving birth, cows may die, have milk fever, not let down the milk, or not produce colostrum milk. Freeze the colostrum milk in a container that will not break or split when frozen. Half-gallon cardboard milk cartons may be used for freezing colostrum milk and provide about the right amount for a newborn calf.

The extra colostrum milk may be stored in large plastic containers at room temperature and allowed to ferment, producing fermented or sour colostrum. The fermentation process produces an acid that prevents the milk from spoiling. Stir the fermented colostrum milk each day. Dilute it half and half with warm water for feeding.

Courtesy of Frank B. Flanders

Figure 41-14 Mixing milk replacer for a dairy calf.

Extra colostrum milk may be fed to older calves, but it must be diluted half and half with warm water. Failure to dilute the colostrum milk when feeding it to older calves may result in scours (diarrhea) or calves going off-feed.

Colostrum from several different cows can be mixed together for feeding. This makes a wider range of antibodies available to the calves being fed. This practice may be especially useful if calf diseases are a major problem on the farm.

After 3 days, the calf may be fed whole milk or milk replacer. Feed about 5 pounds to small-breed calves. Large-breed calves are fed about 7 pounds. Feed this amount for 3 weeks on an early weaning program. Feed for 4 to 5 weeks on a liberal milk feeding program. Do not force the young calf to drink more than it will take in 3 to 5 minutes. Overfeeding a liquid diet may cause scours (diarrhea). If the calf develops scours, cut back on the amount of milk or milk replacer fed. When the calf stops scouring, gradually increase feeding to the recommended level (Figure 41-14).

Use milk replacer of high quality. High-quality milk replacers are made from dairy products rather than plant products. Milk replacers that have the following protein sources are preferred:

- skim milk powder
- buttermilk powder
- dried whole whey
- delactosed whey
- casein
- milk albumin

Plant protein sources are not as digestible as milk protein sources. The following protein sources in milk replacers are considered to be of poorer quality:

- meat solubles
- fish protein concentrate
- soy flour
- distillers' dried soluble
- brewer's dried yeast
- oat flour
- wheat flour

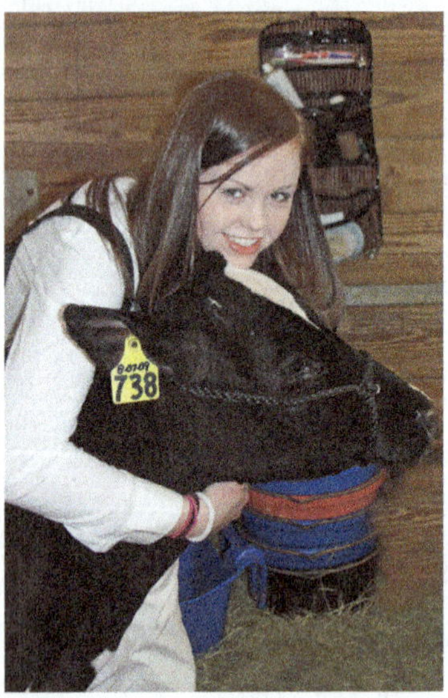

Courtesy of Frank B. Flanders

Figure 41-15 Replacement dairy heifers are often loaned out to high school students by dairies for use as Supervised Agricultural Experience (SAE) projects. Many calves are still on a milk or milk replacer diet when received by the student. They are gradually weaned and put on a roughage and grain diet. The student in this photo is about to enter the show ring with her heifer.

A milk replacer should contain at least 20 to 22 percent crude protein if all protein sources are from dairy products. If poorer quality (plant) protein sources are used, they should contain at least 22 to 24 percent crude protein.

A crude fat level of 10 to 20 percent is recommended. Fat reduces scours and provides needed energy. Animal fat sources are better than plant fat sources. Homogenized soy lecithin is a good fat source in milk replacers. Adding fat to either whole milk or a milk replacer during the first 4 weeks of feeding has shown to result in a faster rate of gain for calves kept in hutches during cold weather (Figure 41-15).

Lactose (milk sugar) and dextrose are good carbohydrate sources in milk replacers. Starch and sucrose (table sugar) are not good carbohydrate sources for use in milk replacers.

Follow the instructions on the feed tag for mixing the milk replacers with water. A bottle or nipple pail may be used to feed milk or milk replacer. Make sure the pail is clean for each feeding. A dirty pail may increase the chances of the calf getting scours.

Calves are usually fed twice a day in two equal feedings. If the calf is weak, it may be advisable to feed it three times a day. Divide the milk or milk replacer into three equal amounts.

Calves are usually weaned between 4 and 8 weeks of age. With good management and proper feeding, they may be weaned as early as 3 weeks of age. However, calves weaned this early may show a reduced growth rate for 7 to 10 days. By 12 weeks of age, they usually weigh as much as calves weaned at an older age. Calves weaned later than 8 weeks may become too fat. Calves should be eating at least 1 pound per 100 pounds of body weight per day of a good-quality calf starter at the time they are weaned.

Catrina Kennedy Pollard

Figure 41-16 Bottle feeding a dairy calf.

Calf starters should be fed when the calf is about 3 days old and no later than 10 to 12 days of age. A good-quality, palatable starter should contain 75 to 80 percent total digestible nutrients and 16 to 20 percent crude protein.

Teach the calf to eat starter by rubbing some on its muzzle or putting a small amount in the milk pail after each milk feeding. The calf will soon start to eat the starter (Figure 41-16).

Grain starters are fed with forage; complete starters include the forage. Using a complete starter provides better control over nutrient intake and is recommended. Feed a complete starter on a free choice basis, adding some forage when the calves reach 3 months of age. Feed a grain starter on a free choice basis until daily intake reaches 4 to 5 pounds. Start feeding a good-quality forage with the grain starter about 1 week before the calves are weaned. Calves younger than 3 months should not be fed corn silage or be allowed to graze from the pasture. The high moisture content of these feeds may reduce feed intake and growth. Low-moisture haylage may be fed if it is kept fresh. Both complete and grain starters are fed until the calves are about 4 months of age.

Commercial calf starters are usually pelleted; home-mixed starters may be used and should be coarsely ground, rolled, or crushed to improve palatability. Nutrients needed in only small amounts are more uniformly mixed in pelleted form. Commercial calf starters usually have vitamins A, D, and E along with minerals and antibiotics added. If the milk replacer contains antibiotics, do not use a starter with antibiotics. Always follow label instructions when using milk replacers or calf starters that contain antibiotics.

Preventing calf scours will reduce death losses. Some of the causes of calf scours and preventative measures are listed here:

Causes	Preventive Measures
Overcrowding	Provide 24–28 square feet of bedded area per calf.
Poor ventilation	No direct drafts; minimum of 4 air changes per hour in winter; 15 in summer.
Calves getting wet	Dry bedding; good ventilation; do not spray calves with water.
Overfeeding liquid diet	Follow recommendations for feeding milk or milk replacer.
Low resistance to disease	Provide vitamins A, D, and E orally or by injection right after birth.
Not getting colostrum milk	Make sure the calf nurses right after birth.
Dirty feeding pails	Clean completely right after feeding. Store upside down to drain.

Figure 41-17 Dairy replacement heifers must be fed properly to become high yielding dairy cows.

Baby calves that get scours dehydrate (lose water from the body) rapidly. This often results in death. The calf becomes weak and refuses to drink. A commercial solution may be bought for feeding, or a home-mixed solution may be used. Examples of home-mixed solutions are available from state extension services. The calf may be fed by carefully putting a tube down the esophagus. Tube feeding equipment can be purchased at feed and livestock supply stores. When inserting the tube, make sure it is going down the esophagus and not the trachea. If the tube is allowed to go down the trachea, the liquid will be deposited into the lungs and will cause the calf to asphyxiate. The tube can be felt going down the esophagus by placing a hand on the left side of the calf's neck. After the tube is placed, hold the bag containing the solution above the calf's head. This allows the liquid to drain into the calf's stomach.

Feeding the Calf from Weaning to 1 Year

Replacement heifers must be fed properly if they are to be ready for breeding at the right time. Heifers that have not been fed to grow properly will not produce as well when they are in the milking herd (Figure 41-17).

A ration made up of forage fed free choice and limited grain may be fed. The amount of protein supplement needed depends on the amount and quality of the forage fed. Pasture may be used for some of the forage.

Feeding 4 to 5 pounds of grain per day is recommended. Do not allow replacement heifers to become too fat. Reduce the grain intake if the heifer is getting too fat.

Energy intake levels need to be carefully controlled as heifers grow from 3 months of age to puberty (9 to 11 months of age). Allowing too much energy intake during this time will result in depressed growth hormone levels, decreased mammary tissue formation, increased fat deposits in the immature mammary gland, and lowered milk yield potential when the heifer comes into production. Research has shown that the effects of overfeeding energy during this period are permanent; that is, the heifer will produce less milk not only during her first lactation, but also through all later lactations. Too little energy intake, however, will result in slower growth and delayed breeding. The effects of a limited underfeeding of energy during this period can be overcome by feeding higher levels of energy during pregnancy and the first lactation. Large-breed heifers (Holsteins and Brown Swiss) should gain an average of 1.7 pounds per day from 3 months of age until 2 months before calving. Guernsey and Ayrshire heifers should gain an average of 1.5 pounds, and Jersey heifers should gain an average of 1.3 pounds per day during this period.

Ionophores are antimicrobials that target bacteria in the rumen and result in increased carbon and nitrogen retention by the animal, which increases production efficiency. Monensin is a commonly used ionophore in dairy heifers. Ionophores may be feed to dairy heifers until calving but may not be fed to milking animals. Ionophores specifically:

- manipulate rumen fermentation.
- improve efficiency of feed utilization.
- increase weight gain of cattle.

Rate of gain has been shown to generally improve by 0.1 to 0.2 pound per day. Do not use ionophores if more than 50 percent of the diet is a high-energy feed such as corn silage. Nonprotein nitrogen sources, such as urea, must be limited to no more than 15 percent of the crude protein equivalent when an ionophore is fed. Follow label directions for use and feeding amounts of ionophores.

Feeding Heifers 1 to 2 Years of Age

A good-quality forage is the basis for feeding heifers during this period. Some grain may be needed for proper growth. A gain of 1.6 to 1.8 pounds per day is desirable. A good pasture requires no added forage or grain. Heifers on mature or heavily grazed pasture will need additional feed. Be sure the ration meets the mineral and vitamin needs of the heifer.

Feeding during the last 2 months of gestation is especially important because the condition of the heifer at calving will strongly influence the amount of milk produced during the first lactation. The heifer should be growing more rapidly (up to 2 lb per day) during the last 2 or 3 months of gestation. Begin feeding grain about 6 weeks before calving at the rate of 1 percent of body weight. The exact amount of grain needed depends on the size and condition of the heifer and the quality of the forage being fed. Be sure the heifer has a balanced ration. Limit the amount of salt in the diet during the last 2 weeks of gestation. Excess salt can cause udder edema. Heifers that are too fat or too thin have more problems at calving time.

BALANCING RATIONS FOR DAIRY CATTLE

The steps used in balancing rations for dairy cattle are the same as for any other kind of farm animal. A general discussion of balancing rations is given in Chapter 8. Nutrient requirements and feed composition tables for dairy cattle are available from university cooperative extension services.

Example Ration Balancing Problem

Balance a ration for a group of cows with an average body weight of 1,300 pounds and an average milk production of 50 pounds per day testing 3.0 percent fat.

Step 1: Find daily requirements.

Daily Requirements	Crude Protein (lb)	NE (Mcal)	Ca (lb)	P (lb)
Maintenance	1.06	9.57	0.046	0.037
Milk production	3.85	14.5	0.12	0.008
Total daily requirements	4.91	24.07	0.166	0.117

The requirements for milk production are found by multiplying the average daily production by the requirement per pound of 3.0 percent milk found in a table of nutrient requirements for dairy cattle. The requirement per pound of 3.0 percent milk for crude protein is 0.077. Therefore, 50 times 0.077 equals 3.85 pounds of crude protein needed. The same type of calculation is carried out for the other requirements.

Step 2: Calculate nutrients provided by forage fed. Daily forage intake on a dry matter basis is 1.5 to 2.0 percent of body weight. A 1,300-pound cow will eat from 19.5 to 26 pounds (dry matter basis) of forage per day ($1,300 \times 0.015 = 19.5$; $1,300 \times 0.02 = 26$).

Forage intake may need to be limited for high-producing cows so they will eat enough concentrate to meet their needs. If forage intake is limited to less than 1 percent of body weight, milkfat test may be lowered.

This example will use 1.5 percent of body weight for forage intake. The total amount of forage fed is, therefore, 19.5 pounds. Assume one-half the forage is first-cutting alfalfa-bromegrass hay and the other one-half is corn silage. Assume further that the hay is one-half alfalfa and one-half bromegrass.

All calculations of nutrient content in this example are made on a dry matter basis (Table 41-6). The final ration is then converted to an as-fed basis.

TABLE 41-6 Nutrient Content of Feed

Feed	lb Fed	Crude Protein (lb)	NE_L (Mcal)	Ca (lb)	P (lb)
Hay					
Alfalfa	4.875	0.84	2.8745	0.0609	0.0112
Bromegrass	4.875	0.51	3.0956	0.0146	0.0170
Corn silage	9.75	0.78	7.0317	0.0263	0.0195
Total nutrients from forage	19.5	2.13	13.0018	0.1018	0.0477

TABLE 41-7 Calculating Nutrients for Concentrate Mix

	Crude Protein (lb)	NE_L (Mcal)	Ca (lb)	P (lb)
Total nutrients needed	4.91	24.0700	0.1660	0.1170
From forage	−2.13	−13.0018	−0.1018	−0.0477
To be supplied by concentrate mix	2.78	11.0682	0.0642	0.0693

Step 3: Nutrients needed in the concentrate mix. Subtract the nutrients supplied by the forage from the total needed by the animals (Table 41-7).

Step 4: Amount of concentrate mix required. An estimate of the pounds of concentrate mix needed is found by dividing the Mcal of NEL needed by the Mcal supplied by the concentrate mix. Some references use therms for this calculation. One therm equals one megacalorie (Mcal). The average value of 0.84 Mcal per pound is used for calculating on a dry matter basis. Use 0.76 Mcal per pound for calculating on an as-fed basis.

The calculation for this example is $11.0682 \div 0.84 = 13.1764$ pounds of concentrate mix needed to balance the ration.

Step 5: Percent of protein needed. The percent of protein needed in the concentrate mix is found by dividing the amount of protein needed by the pounds of concentrate mix needed and multiplying by 100. Therefore, $2.78 \div 13.1764 \times 100 = 21.09$ percent.

Step 6: Use of the Pearson square method to mix grain and supplement. A simple way to determine how much protein supplement to mix with the grain is to use the Pearson square. The use of the Pearson square is explained in Chapter 8.

Assume that ground ear corn is to be used with soybean oil meal for the concentrate mix. Usually, rations are mixed with more than two ingredients. This example uses only two ingredients to simplify the illustration. Methods of calculating a ration using more than one grain or protein supplement are given in Chapter 8. The Pearson square is set up as follows:

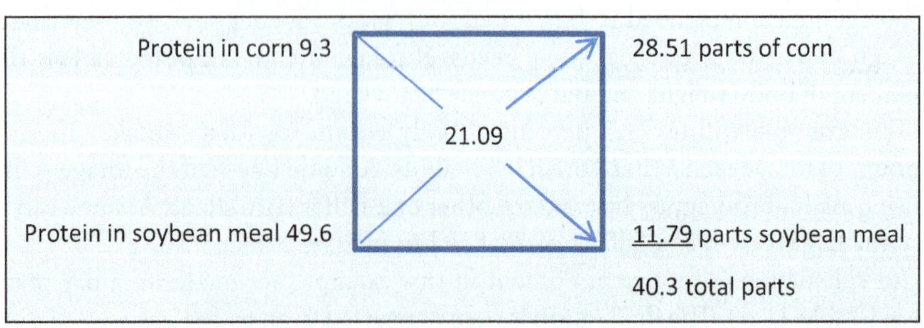

The percent of ground ear corn is 28.51 ÷ 40.3 × 100 = 70.74.

The percent of soybean oil meal is 11.79 ÷ 40.3 × 100 = 29.26.

To convert the percentage on a 100 percent dry matter basis to an as-fed basis, follow these steps:

1. Divide the percentage of each feed on a dry matter basis by the percentage of dry matter in that feed. Ground ear corn is 87% dry matter and soybean oil meal is 89% dry matter.
2. Add the results to get a total.
3. Divide each result in step 1 by the total in step 2 and multiply by 100. The result is the percentage of each feed to include when preparing the concentrate mix.

A ton of this mix on an as-fed basis contains 1,424 pounds (2,000 × 0.712) ground ear corn and 576 pounds (2,000 × 0.288) soybean oil meal.

The total dry matter intake of cows weighing 1,300 pounds and producing 50 pounds of milk per day is about 2.7 percent of body weight. This is about 35 pounds (1,300 × 0.027) for the cows in this example. The forage dry matter intake is 19.5 pounds. Therefore, the cows will eat 15.5 pounds of the concentrate mix per day.

Step 7: Compare nutrients provided to requirements. The nutrients provided by this ration are compared to the requirements, as shown in Table 41-8.

The ration provides a little more protein and energy than required. Feeding a little less of the concentrate mix per day will correct this. It is slightly deficient in calcium and phosphorus. Adding a mineral supplement to the mix will provide the needed amounts. It is recommended that 2,000 international units (IU) of vitamin A per pound of mix be added to the ration. This will ensure that the cow gets the needed amount of vitamin A.

Step 8: Determine amount needed on an as-fed basis. The amount of the ration on an as-fed basis is found by dividing the amount of feed on a dry matter basis by the percent of dry matter in each feed. The calculations are as follows:

Alfalfa-bromegrass hay (average dry matter 89%)

9.75 ÷ 0.89 = 10.96 pounds

Corn silage

9.75 ÷ 0.35 = 27.86 pounds

Concentrate mix (average dry matter 87.6%)

15.5 ÷ 0.876 = 17.7 pounds

TABLE 41-8 Nutrients Provided as Compared to Requirements

	lb Fed (Dry Matter) (lb)	Crude Protein (lb)	NE (Mcal)	Ca (lb)	P (lb)
Forage	19.5	2.13	13.00	0.1018	0.0477
Ground ear corn (15.5 × 0.7074)	10.96	1.02	9.15	0.0055	0.0285
Soybean oil meal (15.5 × 0.2926)	4.54	2.25	3.83	0.0163	0.0340
Nutrients from ration	35.00	5.5	25.98	0.1236	0.1102
Requirements	35.00	4.9	24.07	0.166	0.117
Surplus (deficit)		0.5	1.91	(0.0424)	(0.0492)

FEEDING AND REPRODUCTION

Underfeeding or overfeeding energy feeds to developing heifers can lead to reproduction problems. Underfeeding delays the time the heifer reaches first heat. This causes a delay in breeding and shortens the productive life of the cow.

Overfeeding causes first heat to be reached earlier. However, there is a greater chance of the cow having breeding problems later in life. Overfed cows become sterile at a higher rate than cows fed proper rations. This also shortens the productive life of the cow. Bulls that are too fat have problems with sperm production.

A protein shortage in the ration may cause silent heats or discontinued heats. Protein shortage is more common when corn silage is the main forage fed. Care must be taken to properly balance the ration for protein needs when most of the forage fed is corn silage.

Vitamin A shortage can also cause breeding problems. The addition of vitamin A to the ration will ensure that the cow's needs are met.

FEED INVENTORY

A feed inventory is useful when planning current and future feeding programs. A list of the kinds and amounts of feed available and feed needs for the dairy herd is made.

The following management decisions can be made from the feed inventory.

- Sell extra feed
- Buy additional feed needed
- Put more feed (forage or corn) in the silo
- Make adjustments in the feeding program to fit feeds available
- Plan future cropping program

Worksheets and information on capacities of feed storage structures are available from the Cooperative Extension Service in many states.

ON-FARM FEED PROCESSING

Feed processing may be done by commercial mills or on the farm. The amount of feed used per year is a major factor in deciding which alternative to use.

On-farm feed processing requires a large investment in equipment. Additional time is also needed. Care must be taken to be sure trace ingredients are completely mixed in the ration.

The current cost of commercial mixing must be compared to the cost for on-farm feed processing. On-farm feed processing begins to be cheaper than commercial feed processing at about 200 to 250 tons of feed processed per year.

SUMMARY

Feed costs are about 40 to 50 percent of the total cost of producing milk. Feeding balanced rations permits the cow to produce closer to her maximum genetic potential. Dairy rations should be made up of good-quality roughages supplemented with grain and protein supplement.

Some farmers have fed cattle on an individual basis based on individual milk production. There is a trend toward grouping dairy cows for feeding total mixed rations.

Hay, pasture, green chop, silage, and haylage are generally used for roughage in dairy rations. Alfalfa hay and corn silage are two of the most common roughages used.

Corn is the most commonly used grain for dairy rations. Oats are also an excellent dairy feed. Barley and wheat are less commonly used.

Soybean oil meal, linseed meal, and cottonseed are often used as protein supplements. Urea may be used as a nonprotein nitrogen source in dairy rations.

A variety of by-product and other processed feeds may be used for dairy cattle feeding. Alfalfa meal, beet pulp, and brewer's grain are examples of by-product and other processed feeds used in dairy rations.

Calcium and phosphorus are two of the most important minerals needed when balancing dairy rations. Salt and trace minerals are also needed.

Vitamins A, D, and E often need to be added to dairy rations. A commercial vitamin premix added to the concentrate will supply needed vitamins.

Lactating dairy cows need more water for their size than any other farm animal. A good supply of fresh water is vital for maximum production.

Lactating cows need to be fed according to the stage of milk production they are in. The first 10 weeks after calving is the most critical time for meeting nutritional needs of dairy cows. The most limiting factor in milk production is usually a shortage of energy in the ration.

Newborn calves should get colostrum milk within 30 minutes of birth and no later than 12 hours after birth. Colostrum is the first milk produced by the fresh cow. It contains antibodies that help protect the calf from disease.

Calves should be started on solid feed as early as they will eat it. Proper feeding of replacement heifers to get good growth is essential for high milk production. Poor feeding during the growth period often causes reproductive problems.

Student Learning Activities

1. Prepare a classroom display of roughages and concentrates used locally in dairy rations.
2. Observe and compare methods used locally for harvesting and storing forages for dairy cattle. Give an oral report to the class.
3. Visit local dairy farms that use traditional and total mixed ration methods of feeding. Give an oral report to the class.
4. Calculate a balanced ration for the milking herd on the home farm or for a local dairy farm.
5. Select an appropriate feeding program and balance the ration when planning and conducting a dairy supervised experience program.

Discussion Questions

1. Why is it important to feed dairy cows properly?
2. Describe the importance of roughage in dairy rations.
3. List the advantages and disadvantages of traditional dairy feeding methods.
4. List the advantages and disadvantages of total mixed rations for dairy cattle.
5. Why is it especially important to analyze the roughages used in dairy rations?
6. Describe how cows may be grouped for feeding total mixed rations.
7. What problems may occur when grouping cows for feeding total mixed rations?
8. Name three types of automatic concentrate feeders used for dairy cattle and describe their purpose.
9. Describe the use of hay as a roughage for dairy cattle.

10. Name and compare the different kinds of silage often used in dairy rations.
11. Briefly explain the use of pastures for dairy cattle.
12. Name and compare grains commonly used in dairy rations.
13. Name and describe the common protein supplements often used in dairy rations.
14. Briefly explain the use of urea as a nonprotein nitrogen source for dairy cattle.
15. Briefly explain the importance of minerals for dairy cattle.
16. Briefly explain the use of vitamins for dairy cattle.
17. Why is water important for dairy cattle?
18. Why is the body condition score of dairy cows and heifers an important concept for dairy farmers to understand and use?
19. When during the lactation cycle should cows be checked for body condition score, and what are the appropriate scores for each period?
20. What general guidelines need to be followed when feeding lactating dairy cows?
21. Briefly explain feed requirements for the various stages of lactation for the dairy cow.
22. What are the recommended practices for feeding dry cows?
23. Why is colostrum milk important for newborn calves?
24. List recommended practices to prevent calf scours.
25. Balance a ration (using local feeds) for a dairy cow weighing 1,400 pounds and producing 60 pounds of 3.5 percent milk daily.
26. Balance a ration (using local feeds) for a mature dry cow weighing 600 kilograms.
27. Balance a ration (using local feeds) for a small-breed, 300-kilogram growing dairy heifer that is 83 weeks of age and gaining 600 grams per day.
28. Briefly explain the relationship between feeding and reproductive problems in dairy cattle.
29. Briefly explain on-farm feed processing.

Review Questions

True/False

1. Feed costs are about 25 percent of the total cost of producing milk.
2. Dairy cattle are not ruminants; therefore, roughage is not the basis for their feed.
3. High-producing cows need more concentrates in their rations than low-producing cows.
4. Large round bales of hay are less labor demanding than the traditional small bales.
5. It is hard for lactating cows to get enough dry matter in the ration if pasture is the main source of roughage.

Multiple Choice

6. To provide high-quality forage, pastures need to be well ______.

a. aerated   c. watered
b. clipped   d. fertilized

7. The most commonly used grain in dairy cattle rations is ______.

a. wheat   c. oats
b. corn   d. beans

8. The calf should receive colostrum during the first 24 hours equal to ______ percent of its body weight.

   a. 8–11 c. 14–17
   b. 18–20 d. 12–15

Completion

9. The moisture content of hay should be ______ percent or less when it is baled.
10. Grains are included in the dairy ration mainly for their ______ content.
11. Putting haylage in the silo at too low a moisture content may cause ________.

Short Answer

12. Why is a protein supplement used in the cattle ration?
13. Why is corn the most commonly used grain in dairy cattle ration?
14. Why must urea be mixed completely in the ration in order for cows to eat it?
15. List four by-products or other process feeds used in dairy cattle rations.
16. Why is it necessary for dairy cattle to have adequate water in their rations?
17. Why is it undesirable to overfeed heifers?

Chapter 42

Management of the Dairy Herd

Key Terms

net income
gross income
Dairy Herd Improvement Association (DHIA)
estimated relative producing ability (ERPA)
predicted transmitting ability (PTA)
commercial heat detector

Objectives

After studying this chapter, the student should be able to

- describe the use of records in managing the dairy herd.
- determine which cows should be culled from the dairy herd.
- set goals for the dairy farm.
- manage dairy cows during all phases of production.
- raise dairy replacement heifers.
- describe approved dairy management practices.

The data presented in previous chapters show a trend toward larger dairy herds and fewer dairy farms. A larger dairy herd does not always mean more net income from dairying. **Net income** is income (profit) left after expenses have been deducted from gross income. **Gross income** is the total of all income received. The dairy farmer must follow good management practices to be successful. Selection, feeding, facilities, milking practices, herd health, and marketing are discussed in other chapters in this section. This chapter describes other management practices that are necessary for successful dairy farming.

RECORDS

Good records are needed to serve as the basis for sound management decisions. The dairy farmer must have records to analyze the business. An analysis of the business is needed to make improvements that will increase net income.

The dairy farmer should keep the following kinds of records:

- production records on individual cows and on the herd
- feed use records
- breeding and calving records
- health records
- cow identification records
- financial records (entire farm and enterprise)
- inventory

DHIA Testing Programs

- Dairy Herd Improvement (DHI)
- Dairy Herd Improvement Registry (DHIR)
- AM-PM testing programs
- Owner-Sampler (O-S)
- Weigh-a-Day-a-Month (WADAM)
- Basic Production and Management (BPM)

DAIRY HERD IMPROVEMENT ASSOCIATIONS (DHIA)

Dairy herd testing is available through the Cooperative Extension Service working with local dairy herd improvement associations. A **Dairy Herd Improvement Association (DHIA)** is formed by a group of dairy farmers in the local area to provide production testing services. The DHIA hires a supervisor. Some testing plans require the supervisor to travel to each dairy farm in the association. This person takes and weighs milk samples from each cow in the herd. The milk is tested for fat content, and a written report is provided to the dairy farmer.

Other testing plans require the dairy farmer to weigh and take the milk samples. Herds of all types and sizes, both registered and grade, are included in DHIA programs.

The use of DHIA testing has increased over the years. Net income usually increases on farms that use DHIA testing programs compared to those that do not.

The type of testing program the dairy farmer decides to use depends on time required, cost, and use of information. These factors vary with the type of testing program. The lowest-cost plans are O-S, WADAM, and BPM. The BPM plan requires the least amount of time. All plans provide information for herd management. However, DHI and DHIR are official tests that provide information that may be used in the sale of purebred or surplus cattle.

Dairy Herd Improvement

The Dairy Herd Improvement (DHI) testing program requires the DHIA supervisor to visit the farm once a month. The supervisor weighs and samples the milk from each cow. Both evening and morning milkings are included. The supervisor also records information on feeding, breeding, calving, and management of the herd.

The records are official because they are made from information collected by the DHIA supervisor. Costs vary from one area to another. DHI records are recognized by the USDA and breed associations.

DHI records may be used to:

- identify low-producing cows for culling
- improve feeding programs
- identify problem breeders
- identify cows with chronic mastitis problems
- develop yearly progress records
- provide records for the sale of breeding stock
- establish values if cattle are lost in a disaster
- give recognition to outstanding dairy operations
- provide information for research
- provide information for pedigrees and sale publicity in purebred herds
- provide information for sire and cow evaluation

Dairy Herd Improvement Registry

The Dairy Herd Improvement Registry (DHIR) testing program is conducted in the same way as DHI, with some added requirements. These involve enrollment in the program, cow identification, and surprise tests. The dairy farmer must make arrangements with the appropriate breed association to start or stop a DHIR testing program. The records of DHIR are recognized by the USDA and the breed associations.

Surprise tests are unannounced visits by the test supervisor to test the herd. Surprise tests are done when requested by the breed association or when production levels exceed a preset standard.

AM-PM Testing

The AM-PM testing program is conducted by the DHIA supervisor. Only one milking is weighed and sampled on the test day. Alternate morning and evening milkings are weighed and sampled in consecutive test periods. The test is official if the entire herd is milked twice daily and the milking times are recorded. The DHIA supervisor records the beginning and ending times of milking on the test day as well as the two previous milkings.

This testing program is very accurate as compared to DHI testing. The cost is less because the supervisor does not have to return for the second milking on the test day.

Owner-Sampler (O-S)

The O-S testing program requires the dairy farmer to weigh and sample the milk from each cow. This is done once a month. The equipment and forms are supplied by the DHIA.

This testing program provides the same information as DHI, DHIR, and AM-PM testing programs. The results are not official. They are not used in USDA or breed association reports. Owner-sampler testing provides information needed for on-farm herd management. The cost is less than DHI, DHIR, and AM-PM testing.

Weigh-a-Day-a-Month (WADAM)

The Weigh-a-Day-a-Month program involves the dairy farmer taking weights on milk given by each cow. This is done once a month at two consecutive milkings. No milk sample for testing is taken. The breed average or the milk plant fat test is used for the herd.

The cost of WADAM testing is low. Less individual information on each cow is produced by this testing program. The results are not official for USDA or breed association purposes.

Basic Production and Management (BPM)

The BPM testing program requires the dairy farmer to gather data and report it. Weighing and (if desired) samples are taken at two consecutive milkings once a month. The amount of information from this testing program depends on what the dairy farmer wants. It may be only production with no fat test, or samples may be taken and fat tests made. Breeding and feeding records may be included if desired.

No individual body weight, ear tag, or feed information is required for this program. The cost depends on the amount of information the dairy farmer wants. The records are not official.

IDENTIFICATION OF DAIRY ANIMALS

Keeping individual records on dairy cows requires a system of permanent identification. Calves need to be permanently identified shortly after birth. Methods of identification include ear tags, ear badges, neck chains, tattoos, freeze brands, photographs, and ink sketches.

It is recommended that ear tags and neck chains be used together. Neck chains are easier to read in elevated milking parlors. It is easier to read ear tags when the cows are eating (Figure 42-1). Tattoos provide a permanent mark if the ear tag and/or neck chain is lost. The tattoo is placed in the ear of cattle. Ear tattoos can only be read from a close proximity. They become very hard to read under certain circumstances, such as being at a distance, dirt in the ear, or if hair has grown over the edge of the ear.

A tag inventory record book is recommended. This written record includes the permanent tattoo number, vaccination tag, ear tag, and neck chain number for each cow. Large herd owners may find it useful to keep a location book. This will show by the number the groups and pens to which each cow is assigned. This book saves time when it is necessary to locate a given cow.

USDA

Figure 42-1 A Holstein cow identified with an ear tag.

COMPUTERS FOR COW IDENTIFICATION AND MANAGEMENT

Computer systems that tie together electronic identification of individual cows with various routine management tasks have been developed and are continuing to be improved. Several methods of electronically identifying individual animals include ear tags, neck tags, leg straps, and implanted devices. Ear tags and neck tags are currently the most commonly used methods; most of the devices currently available are too large to be used as implants. Implanting also reduces the range of the device.

Electronic identification units are typically powered by radio signals from a stationary transmitter located at the point where the information is used. This may be at a feed dispensing station, in the milking parlor, or at some other location, depending on the management task to be done. The devices may also be battery powered. The identification unit transmits a signal that identifies the specific animal to which it is attached. The range of the transmitted signal varies from as little as 2 inches to as much as 3 feet, with a typical range being 6 inches.

The most common use of these systems is for computer-controlled feed-dispensing stations. The systems permit control over the feeding of concentrates on an individual basis to cows kept in groups. Upon identifying the individual cow in the feeding station, the computer determines the proper amount of concentrate to be fed, based on the cow's production record that is kept in the computer database. Some programs even determine the appropriate percentage of protein to be included in the concentrate for the individual cow. The computer program keeps track of the amount of feed each cow has received during the day and dispenses the proper amount of feed to avoid feeding too much concentrate at one time. Computer programs currently available typically have the capability to dispense the feed from 6 to 20 times per day. More complex programs are available that will take into account the amount and quality of forage the animal is receiving and adjust the concentrate mix accordingly, based on the cow's weight, milk production, and milkfat test.

Research is being conducted on the utilization of electronic identification devices and computer programs for other management tasks. These include estrus detection, mastitis detection, milk production, milkfat test, maximum and average milk flow, milking time, and internal biological changes in the cow's body. Some of these capabilities are already becoming available in commercial systems.

When individual cows are identified in the milking parlor, some systems are capable of accessing a database that provides the operator with information regarding health needs, time to observe for heat detection, cows that are off-feed, and dry cows.

The increased use of electronic identification with computer programs can be expected to improve the management capabilities of dairy farmers. Records collected on the farm can be transmitted electronically to computers for DHIA, breed association, artificial insemination association, veterinary clinic, or university analysis.

STANDARDIZING LACTATION RECORDS

In order to compare production records of dairy cows, it is necessary to standardize records to the same basis. The following are usually considered when standardizing production records:

- length of lactation
- number of times milked per day
- age at calving
- month of year when calving

The standard length of lactation used for comparing production records is 305 days. Tables of factors for each dairy breed have been developed for adjusting production to the 305-day standard. Factors for both milk and percent milkfat are used.

The standard for number of times milked per day is twice daily (2×). Some dairy farmers milk their cows three times per day (3×). A table of factors is used to convert 3× records to 2× .

A first calf heifer will not produce as much milk as she will when mature. A table of factors has been developed to predict mature equivalent (ME) production for the cow. Age at calving adjustment factors have been developed for different regions of the United States. These are based on DHIA records.

The month of the year during which the cow calves also affects production. Cows calving in the summer months generally produce less milk as compared to cows calving at other times of the year. The adjustment factors for age at calving and month of calving are combined on a regional basis into one table. The factors are different for each breed of dairy cows.

All of the tables of factors have been developed by the USDA from DHIA records. These tables are available from the DHIA, the Cooperative Extension Service, and breed associations.

A production record that is reported as 305-2×-ME means that the record has been adjusted to 305 days lactation, twice daily milking, and it is a mature equivalent record. If a 305-2× -ME record is used in promoting the sale of breeding stock, the cow's actual production should also be given.

Another adjustment in production records is often made to account for variations in early and late test periods. A cow usually increases milk yield and decreases fat yield during the first two test periods of the lactation. Milk yield usually decreases and fat yield increases during the last test period of the lactation. Tables of factors are used to adjust for these normal changes between tests. Making this adjustment results in a more accurate measure of production.

CULLING

Culling is permanently removing cows from the herd. The most common reason for culling is low milk production. Cows are also culled because of reproduction problems, diseases, udder problems, or other miscellaneous reasons (Table 42-1).

Cows that are poor producers are generally not profitable. Good production records serve as a basis for culling for low production.

The **estimated relative producing ability (ERPA)** is a prediction of 305-2× -ME production compared to other cows in the herd. This information is available for herds on DHI testing.

The ERPA may be used to help cull low-producing cows from the herd. Cows with low ERPAs will probably be relatively poor producers in future lactations.

Suggestions for deciding which cows to cull include:

- those with the lowest ERPA
- first calf heifers producing 30 percent or more below herd average
- other cows producing 20 percent or more below herd average
- those with lowest USDA cow index
- all calves from cows that are low producers (bottom 15 to 20 percent of the herd)
- those that are still not bred 150 days after calving
- those that have repeated health problems
- those that show chronic mastitis problems
- those with poor udders and/or feet and leg problems
- nervous and/or hard to handle cows

The dairy farmer must decide when, during the lactation period, a given cow is to be culled. An analysis of production and financial records will show an economic break-even point for the herd. Both variable and fixed costs are included in finding the break-even point. The break-even point is that production level at which the costs of producing 100 pounds of milk equal the price received for that milk. A guide that may be used is to cull the cow when her production drops to the break-even point.

The cost, availability, and estimated productivity of replacement cows must be considered when deciding how many cows to cull. The number of cows that can be handled by the facilities on the farm must also be considered. If the facilities can handle more cows than are available as replacements, then the break-even point should be calculated on only variable costs. The total fixed costs remain the same regardless of how many cows are in the herd.

TABLE 42-1 Why Cows Are Removed from the Milking Herd

Reason	Percentage
Low production	32.5
Reproduction problems	26.6
Mastitis	10.4
Disease	7.7
Teat or udder injury	7.2
Poor udder conformation	5.0
Accidents and injury	4.0
Poor feet and legs	2.0
Other poor conformation	1.2
Hard to milk or leaks milk	1.9
Poor disposition	0.8
Other miscellaneous reasons	0.7

GOALS FOR THE DAIRY FARM

A good set of records will tell the dairy farmer what has happened in the business. When planning management changes, it is wise to have goals to work toward. DHIA records provide information that may be used when setting herd goals. Goals may be set for production, feeding, reproduction, management, and culling.

BUDGETS FOR THE DAIRY FARM

Because of the high cost of operating a dairy farm, wise financial planning is essential. A budget is an estimate of income and expenses for a period of time, typically 1 year. The budget may be compared to current typical figures for dairy farms in the area. Summaries of income and cost figures are available from the Cooperative Extension Service.

Expenses that are much higher or lower than typical figures indicate areas where the dairy farmer needs to carefully examine the business. These wide variations may indicate management problems that need to be corrected.

A budget is usually required by a lender when the dairy farmer is borrowing money for operating expenses or capital investment. A typical dairy budget may include the items listed here.

Receipts

- milk sales
- cull cow sales
- calf sales

Expenses—Cash Variable

- feed purchased
- veterinary and medicine
- breeding fees
- supplies (cleansers, sanitizers, paper towels, inflations, etc.)
- DHIA
- milk hauling and marketing
- bedding purchased
- utilities and fuel (dairy share)
- hired labor (dairy share)
- machinery and equipment repair
- building repair
- taxes (dairy share)
- interest on borrowed capital

Expenses—Cash Fixed

- property taxes (dairy share)
- insurance (dairy share)
- miscellaneous (dues, magazines, travel, accounting, legal fees, etc.)

Expenses—Noncash Variable

- home-grown feeds (market value)

Expenses—Noncash Fixed

- depreciation—dairy share (machinery, equipment, buildings, purchased animals)
- interest on capital investment—dairy share (equity in buildings, equipment, land, cows, milk base)

The dairy farmer who is faced with making decisions about additional investment in the business must calculate the debt repayment capacity of the business. Information from the dairy budget is useful in making this calculation.

Debt repayment ability is calculated as follows:

Total business cash receipts
Minus: Business cash operating expenses
Minus: Family living expenses
Minus: Interest and principal payments on existing debt
Balance remaining: Cash available for new investment

A lender will consider the equity that the farmer has in the business when determining the risk involved in loaning money for capital investment. Equity is the percent of the business the farmer actually owns. Equity of less than 50 percent is considered very risky. Most lenders prefer that the borrower have at least 60 percent equity. The management ability of the farmer is also considered by the lender when making the decision to lend money.

The dairy farmer with limited capital to invest must often decide which of several alternate investments will be made. The decision must be based on the expected benefits, timing of benefits, and safety of each investment.

The expected benefits are often measured in dollars. The net return from each alternate investment is found by estimating expenses and income produced. Non-cash benefits, such as labor saved, are harder to calculate. These are evaluated on an individual preference basis.

It is usually preferable to have the benefits produced early rather than late. Total benefits are usually higher if some net gain is produced shortly after the investment rather than much later.

Evaluating risk is often hard to do. The more difficult it is to calculate the cost of the investment or the benefits received, the higher the risk.

It is important to maintain enough cash flow (available money) to pay operating costs. If the additional investment reduces cash flow to too low a level, it may not be a wise investment. Lenders will generally want to evaluate the cash flow situation when deciding about making a loan for capital investment purposes.

Computer programs are available that will do budget problems and provide information for wise decision making on alternate investments in the farm business. Information on these computer programs is available through the Cooperative Extension Service.

BUY OR LEASE DAIRY COWS

An alternative to investing additional capital to buy dairy cows is leasing the cows. It is necessary to carefully prepare budgets on these alternatives before making such a decision.

Things to be considered include available capital, current price of cows, terms of the leasing arrangement, costs of production, and individual goals.

Each situation is different. No general recommendation to buy or to lease can be made. The dairy farmer who considers leasing dairy cows needs to carefully read all the terms of the lease. Make sure there is no misunderstanding about any of the terms. Deal only with reputable cow leasing firms. Ask others who have leased cows about their experience.

ADJUSTING TO CHANGES IN THE ECONOMY

Production costs and prices received change as the economy moves up or down. The long-range goals of the dairy farmer cannot be changed each time the economy changes. The dairy farmer who is following good management practices will not need to make major changes.

Some minor adjustments may be made to keep costs as low as possible and still maintain high production. For example, the ration may be changed to take advantage of a comparable feed with a lower per nutrient cost as feed prices vary. The use of a computer to calculate least cost rations is helpful.

Careful management of the breeding program is always needed regardless of the economy. Timely breeding and high conception rates are always goals of the efficient manager. It is false economy to use semen from inferior bulls to temporarily lower breeding costs. The dairy farmer is breeding for the future of the herd. Temporary changes in the economy do not change this.

When milk prices drop or feed costs increase, it may be wise to cull inferior cows sooner.

When the economy is moving down, less cash may be available. This can cause cash flow problems. Capital investments may need to be delayed or more carefully planned. Place emphasis on reducing risk and getting maximum benefits sooner.

LABOR MANAGEMENT

It takes about 27 hours of labor per year to take care of one dairy cow and her replacement. In larger dairy herds, labor efficiency slightly reduces the labor required per cow. One person can provide about 3,000 hours of labor per year. Therefore, one person can effectively take care of a herd of 100 to 110 cows. Larger herds need additional labor. In some cases, this can be supplied by other members of the family. If family labor is not available, then additional labor must be hired. Large dairy farmers may have several employees, with some labor specialization taking place. The dairy farmer who hires labor needs to have skills in managing people. Employees who are dissatisfied can quickly ruin a dairy business.

Some suggestions for effective labor management include:

- recruiting good help by stressing job benefits, interviewing prospective employees, and maintaining a reputation as a good place to work.
- being as competitive as possible with the industry in terms of wages, benefits, hours worked, vacations, etc.
- putting a dollar value on the benefits offered.
- maintaining good personal working relationships with employees. Recognize their needs for satisfactory interpersonal relationships. Communicate effectively with employees.
- maintaining good working conditions. For example, consider straight shifts instead of split shifts for milking in large herds of 200 or more cows.
- training new employees.
- using incentives for excellence on the job. Be sure the incentive program is clearly spelled out in writing. It should be based on criteria that can be measured by records.

MANAGING DRY COWS

Dry cows are those that are not producing milk. Most cows need a dry period between lactations. Good management during the dry period increases total herd profits.

Accurate breeding records are needed to determine when the cow is due to calve. The lactation period begins with calving. The average gestation (length of pregnancy) for dairy cows is 283 days. A 10-day variation in gestation is considered normal. The cow should be dry for 45 to 50 days. The date to begin the dry period is calculated back from the projected date of calving.

Conditioning for the dry period is done during the last few weeks of lactation. USDA research shows that body fat is replaced more efficiently during late lactation than during the dry period. Cows should not be too fat or too thin at the end of the lactation. Weight is controlled by adjusting the grain-to-roughage ratio. Give thin cows a higher percentage of grain than fat cows.

There are three ways to dry off the cow:

- Stop milking her
- Do not milk her out completely the last few days
- Milk her every other day for several days

The first method is recommended in most cases. Milk left in the udder causes pressure that stops milk secretion. This helps the drying-off process. Do not feed grain or silage for 2 or 3 days. Reduce water and forage intake for 1 or 2 days. After the feed has been reduced for the recommended time, stop milking the cow.

Cows producing less than 35 pounds of milk per day can be dried off by stopping grain feeding and milking. High-producing cows need to have their production reduced before drying off.

Routine treatment for mastitis at drying off is recommended. Treat the cow for mastitis at the last milking. Use an approved dry cow mastitis treatment and a teat dip. Watch the udder for abnormal swelling for 2 or 3 weeks after drying off.

Separate dry cows from the milking herd. They may be grouped with the bred heifers. Allow dry cows to get plenty of exercise. Watch for any health problems that might develop.

Do not overfeed dry cows. Feed mainly good-quality roughage up until the last 2 or 3 weeks before calving. Limit body weight gain to no more than 100 pounds from late in the lactation to the next calving. The developing calf will gain about 40 pounds during the last 8 weeks of the gestation period.

When dry cows get too fat, there are more problems with ketosis, depressed appetite, milk fever, displaced abomasum, and cows that are down (downer cows). Cows that are too fat have more problems at calving time.

Research shows that it is very beneficial to treat dry cows for internal parasites. Follow the advice of a veterinarian on a treatment program.

During the last few days of the dry period, watch the cow closely for signs of calving. The muscles around the tailhead relax, giving a sunken appearance. The vulva swells, and the udder becomes larger and the teats distend. There may be some leaking of milk from the teats.

Check the udder carefully for signs of mastitis. Quarters with lumpy material, watery fluid, or blood may indicate the presence of mastitis. If necessary, treat the quarter for mastitis.

During warm weather, the pasture is the best place for calving. A clean, well-bedded stall is needed for cold or inclement weather. The stall surface should provide good footing. A slippery floor may cause the cow to slip during calving and suffer muscle injury.

A 12 × 12 feet calving pen is large enough. Make sure calving pens are clean, disinfected, and well bedded with fresh, clean bedding.

During calving, watch the cow but do not disturb her. Give help at calving only if it becomes apparent that the cow is having unusual difficulty. Young cows and heifers are more likely to need assistance than older cows, and cows giving birth to twins or bull calves are more likely to need assistance at calving.

A cow about to give birth is restless, lying down and getting up frequently. When the amniotic sac (the membranes surrounding the fetus) breaks, the calf should move into the vagina within 2 hours. This process may take as long as 4 hours in heifers. Normal birth occurs about $\frac{1}{2}$ to 1 hour after the calf moves into the vagina. Heifers may take about 2 hours to give birth. Observe the cow carefully to make sure the presentation of the calf is normal. Normally, the front feet appear first with the head placed on both front legs (anterior presentation). Posterior presentation sometimes occurs and is normal if the back legs and tail appear first. If the calf appears in any other presentation, its position must be corrected for the birth to proceed without serious problems. A veterinarian should be called immediately to provide assistance if the presentation of the calf is not normal. Failure to provide proper assistance may result in the death of the calf and the cow.

The placenta (afterbirth) should be expelled within 8 hours after the calf is born. Retained placentas occur more often when the birth is difficult, when infection is present, or when the cow is too fat. The normal rate of retained placentas in a dairy herd is 5 to 10 percent. A retained placenta rate higher than 10 percent is an indication of poor management, inadequate feeding, or poor sanitation. If the placenta is not expelled within 12 hours after the calf is born, call a veterinarian for assistance.

After calving, watch the cow for signs of milk fever, ketosis, or other health problems. Special health problems of dairy cows are discussed in the chapter on dairy herd health.

Provide fresh water and hay for the cow right after calving. It takes several days to get a cow on full feed after calving. Gradually increase grain feeding. Do not allow too much loss in body weight.

TABLE 42-2 Heritability of Dairy Traits

Trait	Heritability (%)
Fat, solids not fat, protein	25
Stature	50
Teat placement	20–31
General appearance	25–29
Milk or fat production	25
Final type classification score	20–30
Mastitis resistance	20–30
Milking qualities	20–30
Rump	25
Feed efficiency	25
Back	23
Fore udder	21
Rear udder	21
Udder support	21
Body capacity (type score)	15–25
Mammary system	15–25
Dairy character	19
Hind legs	15
Longevity	0–15
Front end	12
Feet	11
Head	10
Rate of maturity	0–10
Breeding efficiency	0–10
Disposition	0–10
Udder quality	0

RAISING DAIRY REPLACEMENTS

An important part of successful dairy management is raising replacements for the milking herd. About 30 percent of the average milking herd must be replaced each year. The dairy farmer must either raise or purchase replacements if the herd size is to be maintained or increased.

Approximately 50 percent of the calves born each year are heifers. Calf death losses must be held low. A goal for death loss is under 5 percent of the calves born. Replacements need to be selected from those animals with the highest potential for milk production.

The advantages of raising dairy herd replacements include:

- less cost than buying replacements
- greater control of genetic improvement
- less chance of bringing disease into the herd
- use of labor, feed, and facilities that might otherwise not be used
- provision of herd replacements when needed
- increased income from sale of extra calves
- personal satisfaction from herd improvement

Herd Breeding Program

Improving the herd by raising replacements begins with a well-managed herd breeding program. Selecting good sires and dams is the key to genetic improvement. The traits selected in a dairy herd breeding program have different levels of heritability (Table 42-2).

The lower the heritability, the slower the genetic progress in improving that trait. Genetic progress is very slow when heritability is less than 10 to 15 percent. Genetic improvement is slower when selection is made for more than one trait at a time. A breeding program must place emphasis on selection for the most economically valuable traits. Commercial dairy farmers should place emphasis on selecting for milk production.

Genetic evaluation of dairy bulls and cows is based on a USDA-DHIA Animal Model method. The evaluation values for both bulls and cows are known as **predicted transmitting ability (PTA)**, which is the average genetic value for a certain trait that an animal transmits to its offspring. The PTA value replaces the predicted difference value for bulls that was previously used.

Select semen from proven sires with a high PTA for use in the herd breeding program. Younger bulls that have not been completely evaluated may be a good choice if both their sire and dam have high PTA values. Semen from younger bulls is usually not as expensive. Breed no more than 20 to 30 percent of the herd to a young sire that is the offspring of parents with high PTA values. This will give an opportunity to develop records that will prove the genetic value of these younger sires for future breeding programs. Use a proven bull with a high PTA on the rest of the herd.

Due to the high cost of purchasing and maintaining a bull, natural breeding is not recommended. If a herd owner chooses to use natural breeding, always select a bull whose sire and dam both have high PTA values. Less genetic progress in improving desirable traits will be made in a herd where natural breeding is used. More genetic improvement can be gained at a lower cost with the use of artificial insemination.

Keeping good records is an essential part of herd improvement. Records of feeding, reproduction, health, production, and sires used should be kept on every cow in the herd. It is recommended that herds be in a dairy herd improvement program. Breed all heifers to good dairy bulls for their first calving.

The effective herd breeding program is based on the goals of the individual dairy farmer. Some goals that the dairy farmer may wish to achieve include:

- high milk production
- calving interval of 12 to 13 months
- reduction of health problems
- improvement in dairy type
- cows that are easy to handle (good disposition)
- ease of milking
- longevity in the herd
- production of quality breeding stock

Inbreeding is the mating of close relatives. Avoid inbreeding when managing the dairy herd breeding program. Inbreeding usually results in lower milk production.

Research indicates that it is difficult to predict the influence of a sire on calving difficulty. It is not recommended that calving difficulty scores be considered when breeding milking cows. Sires that are rated below average in calving difficulty should not be used on heifers.

Artificial Insemination

Most dairy herds in the United States use artificial insemination (AI). The use of AI allows the dairy farmer a wide choice of genetically superior bulls. The risk of disease is less with AI.

The amount of money a dairy farmer should spend for a unit of semen depends on the goals set for the herd. The predicted transmitting ability and the fertility rate of the bull are important for all dairy farmers. The value of the offspring as breeding stock is of importance mainly to the purebred breeder. The commercial dairy farmer can afford to pay more for semen from bulls that will improve the production level of the herd.

For dairy farmers, heifer calves are desirable over bull calves because of their value as additions to the milking herd or to sell to other producers. The value of a heifer calf over a bull calf can be substantial in the dairy business. Sex-selected semen, also known as gender-selected and sex-sorted semen, is available to assure dairy farmers a high percentage of heifer calves. Approximately 90 percent of the calves born through artificial insemination with sex-sorted semen can be expected to be of the gender selected.

Heat Detection

The major problem with AI on the farm is detecting cows in heat. The key to detecting cows in heat is careful and frequent observation of the herd. A good record-keeping system will help by indicating when to expect a cow to come in heat.

The average cow comes in heat every 21 days. Observe the cow at least twice a day for signs of heat from day 17 to day 25 after the last heat period. More frequent daily observation will reduce the number of missed heats. Three times per day for 20 minutes each time is recommended.

It is difficult to observe standing heat when cows are confined to stanchions. They must be turned out to pasture or an exercise area to be observed for signs of standing heat. Observe cows for signs of heat at times when feeding or other routine work with the herd is not being done. Signs of heat include:

- Attempts to mount other cows indicates the cow is coming into heat
- Restlessness, bawling, excessive walking, drop in feed intake, and decrease in milk production all indicate the cow is coming into heat
- Slight swelling and reddening of the vulva indicates the cow is coming into heat
- Pale white or opaque mucus discharge from the vulva indicates the cow is coming into heat
- The cow stands when mounted by other cows. This is the best indication of heat in the dairy cow.
- An increase of swelling and reddening of the vulva indicates the cow is in "standing heat"
- A thin stream of clear mucus discharge from the vulva indicates the cow is in heat

Dairy cows stay in standing heat an average of 15 to 18 hours. The range in length of heat is 6 to 36 hours. A bloody discharge from the vulva indicates the heat period is over. Record the date and watch the cow for signs of heat again in 14 to 18 days.

A small percentage of dairy cows have quiet heat periods, meaning they do not show the usual signs of heat. Watching for the next heat on the basis of the bloody discharge from the vulva will help in detecting heat in these cows.

Commercial heat detectors are available to help detect heat in a herd of cows. An example of a heat mount detector is a device that uses a pad and dye system. This device is glued to the tailhead of each cow. It contains a red dye capsule that releases the dye when the cow is mounted by another animal. It requires 4 to 5 seconds of continual pressure to release enough dye to stain the pad red. If the cow is mounted when she is not in heat, she will usually not stand long enough for the dye to be released. A device such as this works best when the herd has access to open areas. It has less value when animals are crowded in holding areas or around feed bunks. The device may be accidentally activated in these areas. Low-hanging tree branches or other low objects may also accidentally activate the device.

Another type of commercial heat detector is the chin-ball marking device. It is a special halter with a cone-shaped unit containing a stainless steel ball bearing attached to the underside of the halter. It contains a marking fluid that is available in several colors. When the cow is mounted, the fluid leaves a visible color on the cow's back. This method requires isolating the cow in question along with several others from the herd for detection.

An implanted computer chip can be used to help determine when a cow is in heat. The computer chip is enclosed in a case that is pressure sensitive and is implanted beneath the skin on the tailhead of the cow. The device records when the cow is mounted. A reader device that may be hand-held or mounted. The reader device reads the computer chip. The information is fed into a computer that provides a printout showing which cows are in heat. The implant lasts for the lifetime of the cow. Implanting is done by a veterinarian.

A bull (marker bull) that has been surgically modified so it cannot breed the cow may be used with a heat detection device. Surgical modification may involve a vasectomy or removal of the penis. A bull that has had a vasectomy may still spread disease from cow to cow. Bulls with the penis removed will not spread disease. The use of a bull for heat detection means extra feed expense. Also, bulls can be dangerous.

Spayed heifers, steers, or estrogen-treated cows may be used with these commercial devices for heat detection. There is less expense and danger when using these animals.

The success of these methods still depends on good records, proper animal identification, and careful observation of the herd by the dairy farmer. The use of heat detectors, combined with careful observation, can increase the heat detection rate in the herd. In a recent study, the use of a heat mount detector device resulted in an 87 percent heat detection rate compared to only 50 percent when the cows were observed only at milking time.

Estrus or heat synchronization products are available for use with dairy herds. Depending on the product used, the animal will come into heat within a specified number of hours after the product is administered or the animal is bred at a specified time after the product is removed. These products work well if properly used. However, they will not make up for poor management.

One of the most common products for estrus synchronization is the naturally occurring hormone prostaglandin. One or more injections of prostaglandin is used to make the cow come into heat. Other estrus synchronization products are available.

The use of estrus synchronization products requires careful planning, proper facilities, and adequate labor. While conception rates with these products are generally satisfactory, research indicates that they are slightly lower than those achieved by breeding with observed estrus.

Time of Breeding

Inseminate the cow from the middle to the last half of the standing heat period. The egg is released from the ovary (ovulation) about 10 to 14 hours after the end of standing heat. The range in time of ovulation is 3 to 18 hours after heat. Conception is highest when insemination is done 12 to 18 hours before ovulation. Breed cows in the afternoon when they are observed in standing heat that morning. When standing heat is observed in the afternoon, breed the cow the next morning. It is better to breed late (up to 6 hours after standing heat) than to breed in the first half of the heat period.

About 5 percent of dairy cows will show signs of heat 2 to 3 weeks after breeding even though they are pregnant. If the cow is rebred at this time, do not insert the inseminating tube completely through the cervix. Doing so may cause the cow to abort. Insert the tube no further than mid-cervix when rebreeding cows that may be pregnant.

After the cow is bred, watch her carefully for the next heat period. She may not have settled on the previous breeding. If no more heat periods are seen, have a veterinarian check the cow about 2 months after the last service to confirm pregnancy.

A herd conception rate of 1.5 to 1.8 services per conception is a desirable goal for the dairy farmer. A herd average of more than two services per conception is an indicator of serious reproductive problems in the herd or improper breeding techniques.

Calving Interval

A 12 to 13 month calving interval is the most profitable. Healthy cows may be bred about 40 days after calving. However, it is recommended that most cows be bred 50 to 60 days after calving. This allows time for the reproductive organs to return to normal after pregnancy and calving. A delay in breeding may be needed for cows that had calving or post-calving problems. Do not breed if any infection is seen in the reproductive tract.

Most cows show signs of heat 34 to 35 days after calving. Cows that have not shown a heat period by 45 days after calving should be checked by a veterinarian. Breeding the cow at the second heat after calving will keep her on schedule for the calving interval if she settles. The desirable calving interval can still be maintained if she has to be bred one more time to settle. If additional services are needed to settle the cow, the calving interval will be longer than recommended for maximum profit.

Breeding Heifers

It is recommended that heifers be bred according to size rather than age (Table 42-3). With proper nutrition, heifers should reach the right size for breeding at about 14 to 15 months of age. Heifers should weigh about 60 percent of their mature weight at the time of breeding. There is a significant increase in lifetime milk production for heifers that calve at 22 to 24 months of age compared to those that calve when they are older.

Freshening Date and Milk Base

The milk base is the amount of milk that may be sold at the Class I price from a farm. A dairy farmer's milk base is established during the late summer and early fall months. There is an economic advantage in a high milk base. Therefore, it is a good management practice to breed cows and heifers so they will freshen during this time. Freshening in the fall and winter results in more total milk production from each cow.

Some cows fail to conceive the first few times they are bred. Heat periods are sometimes missed. Therefore, there is a tendency for cows that originally freshened in the fall to, over a period of several years, freshen in the spring.

Breed heifers to freshen in the fall. This will help keep the milk base higher. It is easier to breed heifers for fall freshening than to try to move the freshening date of older cows back to the fall. However, do not delay breeding a heifer for more than 2 months just to help establish a high milk base.

Care of the Newborn Calf

Check the calf as soon as it is born to be sure it can breathe. Wipe any mucus or fetal membrane from its nose. Give artificial respiration by pressing on and releasing the chest wall if the calf is not breathing.

TABLE 42-3 Recommended Age and Weight for Heifers at Time of Breeding by Breed

Breed	Minimum Weight (lb)	Minimum Age (months)
Holstein, Brown Swiss, and Red and White	825–875	14–15
Ayrshire and Guernsey	680–700	14–15
Jersey	580–600	14–15
Milking Shorthorn	680–780	14–15

Usually, the cow will lick the calf clean right after birth (Figure 42-2). If not, dry the calf with a cloth, towel, or clean burlap sack. Dip the navel cord in a 7 percent tincture of iodine solution to prevent infection. Sometimes, there will be bleeding from the navel cord. If so, tie it off with a sterile cotton or linen cord.

A healthy calf will be on its feet within 15 to 20 minutes after it is born (Figure 42-3). It will be nursing within 30 minutes. A weak calf must be helped to its feet and be held so it can nurse. The newborn calf must receive colostrum milk if it is going to live. It may be necessary to hand feed a weak calf colostrum milk using a clean nipple bottle. If the calf is too weak to nurse, use a stomach tube to feed the colostrum milk.

Anemia may be prevented by giving the calf an injection of 150 mg of iron dextrin solution within a few hours of birth. An injection of vitamins A, D, and E is also recommended.

Mark the calf for permanent identification before it is removed from its dam. A permanent identification may be made by a freeze brand, tattoo, photograph, or sketch. Some breed associations require a photograph or sketch for identification. Others require a tattoo inside the ear. Registration forms and permanent identification requirements may be secured from the appropriate breed association. Permanent identification in all herds is important for managing the breeding program. Date of birth, sire, and dam are recorded. Sire identification is especially important to aid in developing sire performance information in DHI and DHIR programs.

The adjustment period for both the cow and the calf is shortened by removing the calf from the cow within a few hours after it is born. Continue feeding colostrum from the dam for 2 or 3 days. Use a nipple pail or bottle feeder to feed the calf. This will help prevent some digestive problems caused by the calf gulping its milk too fast from an open bucket. It also saves time because it eliminates having to teach the calf to drink from a pail.

Keep all feeding equipment clean and sterile to prevent disease. Wash and sanitize feeding equipment with the same sanitizer strength used to sanitize milking equipment. Clean the equipment after each feeding.

Dehorning

Horns on dairy cattle do not have any useful purpose. Cows without horns are easier and safer to handle. Horns can cause serious injury to other cows and to people. It is recommended that all calves be dehorned at 1 to 2 weeks of age. It is easier and less dangerous to dehorn young calves rather than waiting until they are older.

The horn button feels like a hard lump under the skin. An electric dehorner is the best way to destroy the horn growing tissue. It may be necessary to clip the hair around the horn if the horn button is hard to find.

Allow the iron to reach a cherry red heat before using it. Hold the animal tightly and touch the hot iron to the horn button. Hold it in place a few seconds. The skin should show a continuous copper-colored ring around the horn button. If it does not, apply the iron again. The horn button will drop off after several weeks.

A dehorning tube may be used on animals up to 4 months of age. Several sizes are available. The dehorning tube removes the horn button by cutting it out. Because a small open wound is left, a fly repellent must be used during fly season. The tube must be disinfected between each calf.

Horns may be removed chemically. Caustic potash is used. This is not as neat a method as electric dehorning. Care must be taken not to allow any of the chemical to run down the face or side of the animal's head.

Clip the hair around the horn button. Apply petroleum jelly or grease around the clipped area to keep the chemical on the horn area. Rub the horn button with the

CAUTION

Some cows will attack a person coming near the calf after giving birth. If the cow shows signs of this aggressive behavior, secure or separate her before going near the calf.

Figure 42-2 A cow will lick the calf clean right after birth.

Figure 42-3 A healthy calf will be on its feet soon after it is born.

CAUTION

Caustic potash must not get on the hands or fingers of the person doing the dehorning. It may cause serious burns. Wrap the caustic stick in paper or cloth to hold it. Follow manufacturer's directions for the use of the caustic stick.

stick of caustic potash until a slight bleeding occurs. The horn button will drop off after several weeks.

Keep the calf tied for at least 1 day after applying the chemical. This will keep the chemical from being rubbed off onto other calves in the area.

Older animals may be dehorned with various kinds of mechanical dehorners and saws. Refer to the chapter on the management of the cow-calf herd for illustrations and further discussions of dehorning cattle.

Removing Extra Teats

Extra teats have no useful purpose. They may interfere with milking. Sometimes they leak milk, develop mastitis, or become abscessed. Extra teats are best removed when the calf is 2 to 6 weeks of age.

If it is not obvious which are the extra teats, have the calf examined by a veterinarian. Have a veterinarian remove extra teats from older animals.

Lay the calf on the floor and hold it firmly when removing extra teats. Wash and disinfect the area around the teat to be removed. Stretch the teat slightly and cut it off close to the udder with sharp scissors or a knife. Be sure the scissors or knife is clean and sterilized. Use an antiseptic such as iodine on the cut to prevent infection.

A Burdizzo may be used to remove extra teats. This is a type of clamp which, when applied, crushes the blood vessels. The teat will fall off after a period of time.

Raising Young Bulls

In general, follow the same practices and feeding for young bulls as for young heifers. Place a ring in the nose when the bull shows signs of becoming dangerous. As the bull grows, the ring needs to be replaced periodically with larger, heavier rings. A bull is more easily controlled with a staff attached to the ring than by a halter.

Limited use may be made of a well-grown bull at 1 year of age. The bull should be at least 2 years old before being used heavily. Heavy service is breeding 40 to 50 cows per year.

Bull calves that are not being raised for breeding stock may be sold for veal or fed out for beef. See the beef section of this text for information on raising beef feeders.

Introducing the Heifer to the Milking Herd

About 2 weeks before calving, introduce the heifer to the milking herd routine. The heifer is easier to manage if she becomes used to the milking parlor or stanchion barn before she freshens. This also allows the heifer to adjust to the feeding program of the milking herd. Handle heifers gently. Heifers that are handled roughly are more difficult to manage in the milking herd.

CAUTION

Use care when trimming the animal's hind feet. If the animal feels unsafe, it can be prone to kicking. Being kicked by the animal can result in serious injury or death.

HOOF TRIMMING

Care of the feet of heifers and cows is often overlooked. Hooves may grow too long and need trimming. If they are not trimmed, they will crack and break off. Cows with poor feet often have lower milk production. Bulls with poor feet may have low-quality semen and refuse to breed cows. Some conformation problems in younger animals may be corrected with proper hoof trimming.

Tools needed for hoof trimming include a hard rubber mallet, a straight wood chisel, a T-handle chisel, rasp, hoof nippers, and a pair of hoof knives.

Stand the animal on a 4 × 8 foot, $\frac{3}{4}$-inch sheet of plywood. Confine the animal in a stanchion or tie it securely with a halter.

Trim the front feet first so the animal gets used to the operation. Using the chisel, shorten the toes to the desired length. Do not remove too much because this will cause lameness. Use the rasp to smooth and shape the outer surface.

To trim the sole of the foot, place the animal's knee on a box or a bale of hay or straw. Clean and trim the inner sole with a hoof knife. Use a sharp chisel to level the sole. Do not cut too deeply into the sole.

Check for hoof rot when trimming the feet. Clean out cracks with a hoof knife. Soak cracks with pine tar. If corns are found between the toes, call a veterinarian to treat them.

MAINTAINING MILK PRODUCTION IN HOT WEATHER

Heat stress caused by high temperatures and high humidity will lower milk production and conception rates in dairy cattle. Cows produce the most milk when the temperature ranges from 25 to 65°F. Milk production begins to drop above 80°F. At 90°F, the drop in milk production ranges from 3 to 20 percent. The combination of heat and humidity increases stress. When the temperature reaches 100°F and the humidity is 20 percent, heat stress begins to be serious, and some type of cooling activity needs to be used. At 100°F and 50 percent humidity, the stress on dairy cattle begins to reach a dangerous level. A combination of 100°F and 80 percent humidity can be fatal for dairy cows.

Problems with heat stress are reduced by keeping cows cool. Cows that are outside need access to some type of shade. This can be accomplished by trees in the pasture or artificial shades made from metal (such as aluminum) sheets. The effectiveness of artificial shade can be improved by painting the top white. Other types of artificial shade may be constructed by suspending snow fence, canvas, or woven plastic fiber from poles.

Provide 6 to 8 feet of height under the shade. Make sure that ventilation is adequate. About 40 square feet is needed for each cow. Use fans inside buildings to provide ventilation and reduce the humidity.

Pasture quality usually declines in hot weather. Feed more hay and silage as the pasture quality goes down. Do not put the cows on extremely poor-quality pasture. The cows' energy is used in looking for feed instead of for milk production.

Use the best quality hay or silage available. Low-quality, high-fiber roughage reduces milk production. Maintain high grain rations to help maintain production.

Lush green pasture may lower milkfat percentage. Free choice feeding of hay helps maintain fat test.

Provide plenty of fresh water. The need for water may increase up to 50 percent in hot weather. Feeding dry feed increases the need for water. High-producing cows drink more water than low-producing cows.

Controlling flies helps to prevent a drop in milk production during hot weather. Follow a good fly control program.

Do not force cattle to go long distances to pasture in the hot sun. Avoid forcing cattle to stand in unshaded areas in the hot sun for long periods of time while waiting to be milked. Do not crowd cattle too closely in pens and lots. Water-sprinkling systems in the barnyard or free stall area will help keep cows cool.

It is important to reduce stress as much as possible during hot weather. Too much stress greatly reduces milk production.

SUMMARY

Records are an important management tool on the dairy farm. Dairy Herd Improvement Associations provide many records that help the dairy farmer increase net profits. Several different DHIA programs are available to the dairy farmer. Dairy production records are standardized to a 305-day, twice daily, mature equivalent basis for comparison purposes.

A regular culling program needs to be followed to increase net profits from the dairy herd. Records help determine which cows should be culled from the herd.

A dairy farmer needs to set goals. A budget will help determine how well the goals are being met as well as serve as a basis for management planning. Budgets are needed when the dairy farmer wants to borrow money.

The dairy farmer must manage labor efficiently. Large dairy farms often hire additional labor. Good labor management skills are necessary for success.

Proper feeding and management of dry cows is an important part of dairy farming. Do not allow dry cows to become too fat or too thin. Provide clean quarters for calving.

It is best to raise the replacements needed for the dairy herd. Select for economically important traits when breeding cows. Dairy sires and cows are evaluated for genetic value on predicted transmitting ability. Base a breeding program on the goals for the dairy farm. Dairy farmers make extensive use of artificial insemination. Better sires are available when artificial insemination is used.

Observe cows carefully for standing heat. Breed during the latter part of the standing heat period. Dairy cows should calve at 12 to 13 month intervals. Breed cows 50 to 60 days after calving. Breed heifers to calve at 22 to 24 months of age.

Make sure the newborn calf receives colostrum milk. Mark the calf for permanent identification. Dehorn calves and remove extra teats.

Milk production often drops in hot weather. Proper feeding and keeping cows cool will help maintain production.

Student Learning Activities

1. Talk to dairy farmers in the local area who are enrolled in DHIA programs. Prepare and present an oral report to the class on DHIA programs.
2. Prepare a bulletin board display of various methods of identifying dairy animals.
3. Discuss culling dairy cows with local dairy farmers. Present an oral report to the class.
4. On a field trip to a local dairy farm, observe dehorning and removal of extra teats from dairy calves.
5. On a field trip to a local dairy farm, observe hoof trimming on dairy animals.
6. Prepare and present an oral report to the class on any aspect of dairy herd management.
7. Follow the management practices described in this chapter when planning and conducting a dairy supervised experience program.

Discussion Questions

1. What kinds of records should a dairy farmer keep?
2. What is the value of DHIA programs to the dairy farmer?
3. What methods may be used to identify dairy animals?
4. Why and how are lactation records standardized?
5. What factors are considered when deciding which cows to cull from the milking herd?

6. What items should be included in a budget?
7. How is debt repayment ability calculated?
8. Compare buying to leasing dairy cows.
9. How should the dairy farmer adjust to changes in the economy?
10. How long should a cow be dry?
11. What are the signs of calving in dairy cows?
12. What are the advantages to the dairy farmer of raising herd replacements?
13. Briefly explain the heritability of traits in dairy cattle.
14. Describe how sires are evaluated for breeding purposes.
15. How are cows evaluated for breeding purposes?
16. Why is artificial insemination widely used on dairy farms?
17. What are the signs of heat in dairy cows?
18. When is the best time to breed a dairy cow during the heat period?
19. What is the most profitable calving interval in dairy cows?
20. How can the proper calving interval be maintained in the dairy herd?

Review Questions

True/False

1. There is a trend toward larger dairy herds and fewer dairy farms.
2. Computers have not been found to be useful in identification and management of cattle.
3. Identification of dairy animals includes ear tags, ear badges, nose rings, neck chains, and tattoos.
4. The DHIR testing program has added requirements over the DHI program.
5. Some dairy herd improvement associations hire supervisors to travel to each farm to test milk samples.
6. In order to compare production records of dairy cows, it is not necessary to standardize records to the same basis.

Multiple Choice

7. How many days should a dairy cow be dry between lactation periods?
   a. 25–35 c. 45–50
   b. 35–45 d. 50–60
8. A single person can effectively handle a dairy herd of about ______ cows.
   a. 100 c. 200
   b. 150 d. 250
9. The average cow comes in heat every _____ days.
   a. 31 c. 21
   b. 25 d. 18

Completion

10. Permanently removing animals from the herd is called ______ .
11. Equity in a dairy farm of less than ______ percent is considered risky.

Short Answer

12. Why is a budget necessary for a dairy farmer?
13. What is one alternative to investing additional capital to buy dairy cattle?
14. Why is it necessary for a dairy farmer to have skills in managing people?
15. Why is it important to manage dry cows effectively?
16. Why is it important to avoid heat stress in cows?
17. What is a marker bull?
18. What is heat detection?

Chapter 43

Milking Management

Key Terms

alkaline cleaners
acid cleaners
off flavors

Objectives

After studying this chapter, the student should be able to

- describe the function of the cow's udder.
- describe recommended milking practices.
- describe methods of maintaining milk quality.
- describe the cleaning and sanitizing of milking equipment.
- list and describe off flavors in milk.

FUNCTION OF THE UDDER

The cow's udder is made up of four glands called quarters (Figure 43-1). The udder is attached to the lower abdominal wall by ligaments. Each quarter has a teat that provides an outlet for the milk. A circular muscle (sphincter muscle) at the end of the teat controls the flow of milk.

The udder contains alveoli with specialized cells that synthesize the milk. The raw material for the making of the milk is carried to the alveoli by the blood. The alveoli contain milk cavities. A tubule leads from each alveolus to small ducts that lead to large milk ducts. The large milk ducts empty into a gland cistern, which holds about 16 ounces of milk. Milk passes from the gland cistern through the teat cistern and then through the streak canal at the end of the teat.

A narrow streak canal and a strong sphincter muscle make the cow harder to milk. A wider streak canal and a weaker sphincter muscle make the cow easier to milk. If the sphincter muscle is too weak, the cow will leak milk when the udder is full.

Between milkings, milk is stored mainly in the milk cavities, tubules, and small ducts. The growth and function of the udder is controlled by hormones.

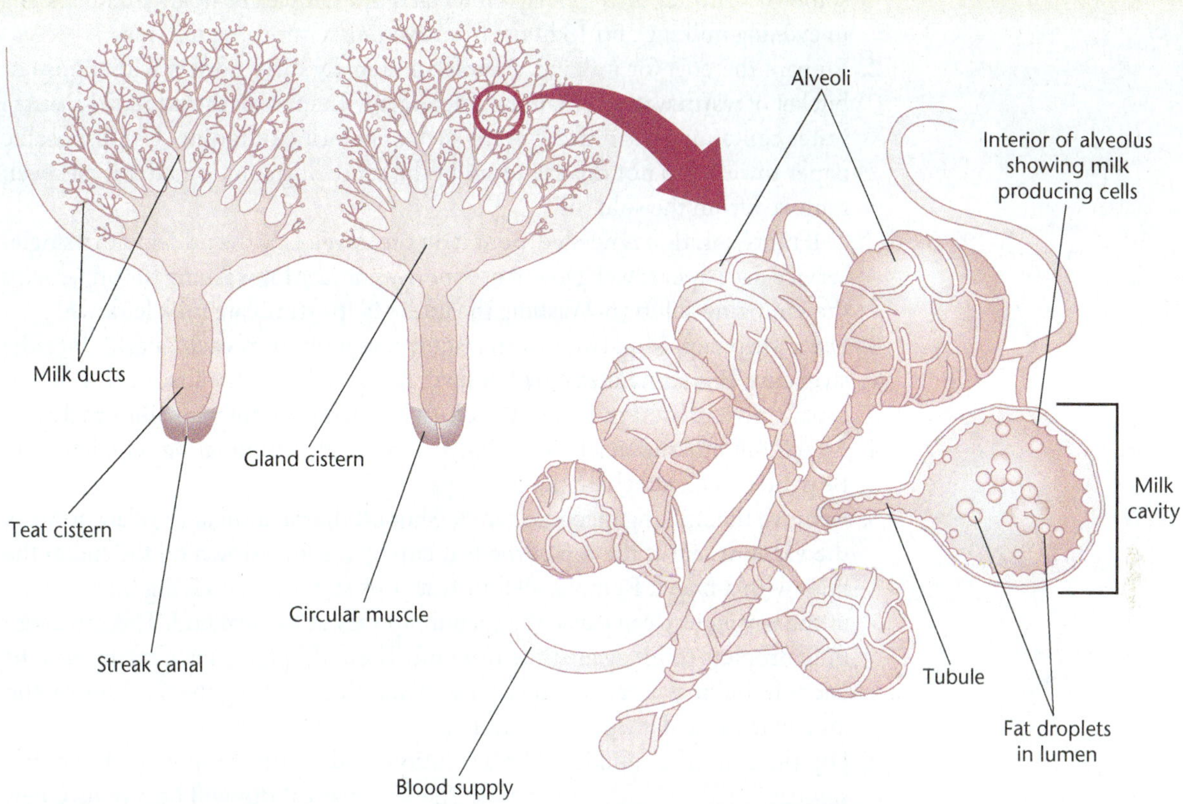

Figure 43-1 A side view of the cow's udder showing detail of the alveoli and alveolus.

Milk Letdown

Milk letdown occurs when the cow responds by a conditioned reflex to sensory stimuli such as the washing of the udder. A hormone (oxytocin) is released by the pituitary gland. This hormone is carried by the blood to the udder. It causes muscle-like fibers to contract around the alveoli and ducts. This forces the milk into the larger ducts and gland cistern. Milk letdown must occur before the milk can be effectively removed from the udder. The effect of the oxytocin lasts about 5 to 7 minutes. Milking must be completed in that time to remove most of the milk from the udder.

The udder produces milk all the time. However, as pressure from produced milk increases, the rate of production slows down. Regular milking is important to maintain high production. High-producing cows will stop increasing the amount of milk in the udder after about 8 to 10 hours. In lower-producing cows, this takes about 16 to 20 hours.

About 80 percent of the milk is removed from the udder at each milking. If too much milk is left in the udder, the pressure builds up quicker. This causes the cow to dry up sooner.

Gentle washing of the udder helps stimulate milk letdown. Attach the milking machine gently to the teats within 1 minute of washing (Figure 43-2). Frightening or hitting the cow will create an emotional disturbance that releases the hormone adrenalin, causing it to interfere with the milk letdown process.

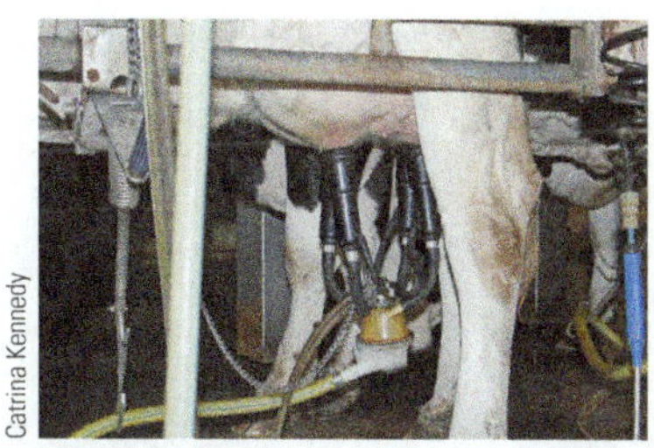

Catrina Kennedy

Figure 43-2 Attach the milking machine within 1 minute after stimulating milk letdown.

RECOMMENDED MILKING PRACTICES

1. Follow a regular routine. Dairy cows respond with higher production when milked regularly at about the same times each day. Milking interval (time between milkings) should be about equal. The daytime interval can be

somewhat shorter than the night interval. For example: 11 hours from morning to evening milking and 13 hours from evening to morning milking.

2. Prepare the cow for milking. Prewash extremely dirty udders with a hose or bucket of warm water containing a detergent. Then wash the udder with warm water containing a sanitizing agent such as chlorine or iodine. Use disposable paper towels. Do not dip the towel in the sanitizing solution after it has been removed from the solution.

   If more solution is needed, pour it on the towel. Dry the udder with a single-service dry paper towel. Do not use sponges or rags for washing the udder with the sanitizing solution. Washing the udder helps stimulate milk letdown.
3. Use a strip cup. Milk two or three squirts of milk from each quarter into the strip cup. This stimulates milk letdown. It also removes the first milk, which is usually high in bacteria count. Watch each quarter carefully for abnormal milk.
4. Attach the milking machine within 1 minute after stimulating milk letdown. Be gentle when attaching the teat cups.
5. Remove the milking machine gently. Shut off the vacuum at the claw or break the vacuum seal at the top of the teat cup by pushing down on the top of the liner with a finger. Remove all four teat cups at once. Removing the machine in the wrong way can cause the vacuum to vary at the teat end. This can cause milk droplets to hit against the teat end. These droplets may contain mastitis bacteria and may infect the quarter with mastitis. Leaving the machine on too long may cause damage to the udder.
6. Dip the teats after milking. Use a solution made for teat dipping. Do not use sanitizer made for other purposes. The use of a teat dip will help reduce new mastitis infections, but it is not a cure for existing infections. Dip at least two-thirds of the teat in the solution. Do this as soon after removing the milking machine as possible. Follow directions on the label of any commercial preparation for teat dipping.
7. Milker's hands must be kept clean. Wash before starting to milk and after handling any infected cow.
8. Do not try to operate too many units. This will result in overmilking and can cause udder damage. Two units per person in a stanchion barn are recommended. Three units per person in a milking parlor are usually enough. A good milker may be able to handle four units in a parlor. A maximum of six units per person is recommended in parlors with automatic milking units.
9. Milking order is important. Milk heifers, cows in early lactation, and normal cows first. Cows with udder infections are milked last or into a separate tank for disposal.

Machine Stripping

Machine stripping is the practice of trying to get all of the milk out of the udder before removing the milking machine. This has been a common and recommended practice for many years. About 0.15 to 1.2 pounds of milk are left in the udder when machine stripping is not done. However, research shows that this does not have any significant effect on somatic cell count, milk composition, yield of milk, or bacteria count. The main benefit of not machine stripping cows in large herds is the time saving. Teat damage is also reduced when machine stripping is not used.

Three-Times-per-Day Milking

Milking cows three times per day will increase total milk production from 6 to 20 percent as compared to two times per day milking. Milkfat percentage will decrease slightly. Some costs will increase with three-times-per-day milking.

There appears to be little effect on herd health. Various studies indicate that an increase of 8 to 10 percent in milk production is needed for three-times-per-day milking to be profitable. A dairy farmer who is considering three-times-per-day milking will need to carefully evaluate the effect on the individual operation before making a final decision.

A major cost when milking three times per day is labor. Some dairy farmers milk three times per day in the winter and two times per day when field work increases. Research shows that greater increases in production from three-times-per-day milking come during the later stages of lactation. Therefore, it is generally recommended that three-times-per-day milking be continued for the entire lactation period.

MILK QUALITY

The production of high-quality milk requires good management practices. The characteristics of high-quality milk include:

- being free of dirt and other sediment
- having low bacteria count
- having no chemical contamination
- having a low somatic cell count
- having no water added
- having a good flavor

Management practices that will help to produce high-quality milk include:

- cleanliness of cows, lots, barns, milking parlors, milk houses, and milking equipment
- keeping cows healthy
- proper cooling of milk
- using correct cleaning and sanitizing methods on equipment
- keeping chemicals out of milk
- preventing off flavors in milk

Keeping Sediment Out of Milk

Keep the cows clean. Cows that are kept in clean facilities will stay cleaner than cows kept in facilities that are dirty. Keep manure cleaned out and use enough clean, dry bedding in barns and pens.

Make sure there is enough ventilation to reduce humidity and odor problems. Keep loafing and holding areas dry. Paved holding areas are easier to keep clean and dry.

Clipping cows will reduce dirt and bacteria in the milk. It also helps control parasites such as lice and will save time when preparing the cow for milking. Clip cows in the fall and again in late winter. Areas to be clipped include the switch, udder, hind legs, underline, flanks, thighs, thurls, and topline to the poll.

Wash the cow's udder before each milking. This helps keep the milk clean and stimulates milk letdown.

Keep the milking area free of dust. Do not sweep the barn just before milking. Dry feed increases the amount of dust in the air. If the dust from dry feed is a problem, feed at least 1 hour before milking or after completing milking.

Keep the milking machine units off the floor and out of the manure. If a unit falls to the floor, clean it off before putting it on the cow.

Strain or filter all milk. Strain milk into cans in the milk house, where there is less dust in the air. Do not bang the strainer to force milk through. Strainers or filters that frequently clog indicate a problem with cleanliness or mastitis. Find and correct the problem. Produce clean milk at the start and do not depend on the strainer or filter to clean dirty milk.

Keeping Bacteria Count Low

Bacteria are single-celled organisms. High bacteria counts lower milk quality. Several types of bacteria may be found in milk. Some grow in warm or hot milk, while others grow in cold milk.

Problems caused by bacteria in milk include:

- souring of milk
- reduced shelf life of milk
- off flavors in milk
- ropiness in milk
- disease transmission through milk

Bacteria counts are lowered by following good management practices, which include:

- cleanliness in the milking area
- keeping the milking area dry
- reducing dust in the milking area
- clipping cows
- proper cleaning and sanitizing of equipment
- washing udders with sanitizing solutions
- proper milking procedures
- rapid cooling of milk
- rinsing and washing equipment properly after milking
- keeping cows healthy and free of mastitis

CLEANING AND SANITIZING MILKING EQUIPMENT

Two kinds of deposits may be found on milking equipment. Organic deposits come from the fat, protein, and sugar found in milk. Mineral deposits come from inorganic salts—such as calcium, magnesium, and iron—that are in the water and milk.

Milkstone is a combination of the organic and inorganic materials. When equipment is not properly cleaned, milkstone will build up to a point where it can be seen. Deposits on milking equipment provide a place for bacteria to grow and multiply.

Alkaline, chlorinated alkaline, and acid cleaners may be used to remove deposits from milking equipment. Cleaners do not sanitize the equipment. Alkaline and chlorinated **alkaline cleaners** remove organic deposits. **Acid cleaners** remove inorganic deposits. Both alkaline and acid cleaners are used in order to remove all types of deposits.

Do not use household soaps or detergents to clean dairy equipment. Soaps leave a greasy film on the equipment. Some household cleaners may cause odors or flavors in the milk.

Cleaners must be used at the proper concentration and temperature to be effective. Cleaning solutions need to be kept above 110°F to be effective. Follow the manufacturer's directions for the proper use of cleaners.

Even after proper cleaning, there are still some bacteria on the equipment. A sanitizing solution is used just before milking to kill these bacteria. Chlorine, iodine, or quaternary ammonium compounds may be used to sanitize the equipment. Follow the manufacturer's directions when using any sanitizer.

CAUTION

Do not mix chlorinated alkaline cleaners with acid cleaners. Poisonous chlorine gas is released when these two types of chemicals are mixed.

The method of cleaning and sanitizing varies with the type of equipment. General procedures to follow are given in this text. Follow the manufacturer's directions for the type of cleaner and sanitizer used.

Metal Parts of the Milker

Rinse equipment in lukewarm water (100 to 120°F) as soon as the milking is done. Milk solids that are allowed to dry are hard to remove later. Take the equipment apart. Soak the parts for 5 minutes in a dairy cleaning solution (120 to 130°F). Use a hard-bristled brush or plastic sponge to wash the metal parts. Do not use metal or stainless steel sponges because these will scratch the surface. Bacteria grow and multiply on scratched surfaces, and it is hard to kill these bacteria.

Put the metal parts in an acid rinse (100 to 120°F). This is especially important if the water is hard. The acid rinse helps prevent a buildup of milkstone. If there is a buildup of milkstone on the metal parts, soak them in an acid cleaner before washing them.

Drain and store the parts upside down. The parts will dry better when stored upside down and bacteria do not multiply as easily on a dry surface.

Sanitize the parts in a dairy sanitizer just before the next milking. Drain well but do not wash before starting to milk.

Inflations and Rubber Parts

Daily cleaning of these parts begins with a rinse in lukewarm water as soon as milking is done. Use a commercial cleaner to wash the rubber parts. Rinse them in an acid solution. Store them dry or in an acid solution until the next milking.

Inflations last longer if one set is used 1 week and a second set the following week. Store the inflations when they are not in use in a lye solution or a commercial cleaner made for this purpose. Inflations may be stored dry. Remove them from the lye or cleaner solution after soaking for 2 hours. Rinse and wash them and let them air dry. Store them in a clean place. Rubber inflations absorb milkfat. This practice removes the milkfat. A lye solution may be made by mixing 1/2 pound of lye in 5 gallons of water.

Inspect inflations regularly. Replace any damaged, stretched, soft, spongy, or rough inflations.

CAUTION

Lye solutions are caustic. Handle with care. Store in a crock, plastic pail, or stainless steel pail. Keep all lye solutions out of the reach of children. Wear rubber gloves to protect the hands.

Vacuum Lines

Clean vacuum lines on a regular schedule. Vacuum lines can become clogged with dirt, rust, insects, and milk. This reduces the airflow through the line. A reduced airflow causes milking and mastitis problems.

Use a lye solution or a nonfoaming cleaner in hot water. The lye solution is prepared by mixing 2 to 3 ounces of lye per 1 gallon of hot water. Do not put more solution into the system than the trap will hold. If the system has a vacuum reserve tank, use no more solution than one-half the volume of the tank. These precautions are necessary to prevent getting any of the solution into the vacuum pump.

Begin with the stall cock nearest the vacuum pump and allow 1 quart of solution to be drawn into the line. Use a flexible tube from the stall cock to the pail of solution. Allow air to enter the line before moving to the next stall cock. Repeat the procedure at each stall cock until the last one in the line is reached.

Drain the trap and throw the solution away. If the line is extremely dirty, repeat the procedure.

Rinse the line with hot water. Then wash the line with a hot solution of acid cleaner. Use 3 ounces of a nonfoaming cleaner per gallon of water for this solution. Start with the stall cock farthest from the vacuum pump and draw some solution through each stall cock.

chinahbzyg/www.Shutterstock.com

Figure 43-3 Milk flows down the pipeline into a holding tank, then into a bulk tank for cooling and storage.

Rinse the acid cleaner from the line with hot water. Start at the stall cock farthest from the vacuum pump.

Drain the trap and open all the stall cocks. Run the vacuum pump for 5 minutes to dry the line.

Pipeline Milkers

Pipeline milkers are cleaned in place (CIP) (Figure 43-3), meaning they are not taken apart for cleaning. The cleaning solution is circulated through the system. Methods used include vacuum circulating, pressure or pump circulating, vacuum gravity circulating, and reserve vacuum circulating.

Manufacturer's instructions must be carefully followed for proper cleaning of pipeline milkers. Cleaners made for pipelines must be used. Cleaning requirements are different than those for hand cleaning of milking machines. Stronger solutions that are low foaming are used. Label directions for use must be followed. A general procedure for cleaning pipeline milkers is described here.

As soon as milking is completed, flush the system using a large volume of lukewarm water. Water that is too hot will set the milk film and cold water will not remove the fat.

Following directions for the specific cleaner used, prepare the cleaning solution. Use hot (160°F) water when preparing the solution. Keep the water above 120°F during circulation of the solution through the pipeline. A booster heater may be needed to maintain solution temperature.

Circulate the solution as recommended, usually 10 to 20 minutes. Use a lot of clean water to rinse the system, especially if the water is hard. Drain the system and allow it to air dry. Use forced hot air to dry plastic tubing that is greater than 10 feet long. Store plastic tubes in a drying position.

Just before the next milking, circulate a nonfoaming sanitizer through the system. Drain the sanitizer from the system. Do not wash the sanitizer from the system.

Bulk Tank

Bulk tanks are cleaned by hand or by an automatic mechanical system. Always rinse the tank as soon as the milk is pumped out. This is usually done by the milk hauler. Use lukewarm (100 to 120°F) water to rinse the tank.

Hand cleaning is done using a hard-bristled, long-handled brush. All inside surfaces are scrubbed with the washing solution. Mix a chlorinated alkaline or other bulk tank cleaner in hot (120 to 160°F) water according to the manufacturer's directions. Use a plastic pail that can be placed inside the tank when scrubbing the tank. Take the outlet valve apart and soak it in the wash solution. Rinse the tank using an acid rinse. Just before the next milking, sanitize the tank with an approved bulk tank sanitizing solution. Drain the sanitizer from the tank. Do not rinse the tank after draining the sanitizer.

Mechanical washing is done by spraying the solution with a spray device in the tank. Use a nonfoaming cleaner in mechanical systems.

Follow the manufacturer's directions when preparing the washing solution. Use hot water to mix the solution. Keep the solution above 120°F during the wash cycle.

The solution is sprayed for about 10 to 15 minutes on the inside of the tank. Take apart the manhole cover, outlet valve, and gaskets before starting the wash cycle.

Drain the tank after the wash cycle. Brush by hand any parts not contacted by the wash solution. Be sure the outlet valve, tank cover, agitator, calibration rod, and support bridge for the agitator rotor are cleaned. Wash the outside of the tank with the solution drained from the tank.

Rinse the tank with lukewarm water. Then rinse with a foamless acid rinse. Drain the tank completely.

Just before the next milking, rinse the tank with an approved tank sanitizer. Drain the sanitizer from the tank. Do not rinse the tank after draining the sanitizer.

COOLING MILK

Bacteria grow and multiply rapidly in warm milk. Cool milk to 60°F within 20 minutes and to 40°F within 90 minutes after it is drawn from the cow. Do not allow milk to warm up after it is cool. A bulk tank that fails to cool milk properly must be checked for malfunction.

PREVENTING CHEMICAL CONTAMINATION

Many chemicals and drugs are used by dairy farmers. Improper handling or use of these products can result in milk contamination. Drug and chemical residues are not permitted in milk. Contaminated milk cannot be sold.

Prevent contamination by proper use of chemicals and drugs. Follow label directions for discarding milk when cows are treated for mastitis. Use only sprays and dusts that are approved for dairy barns and cows. Be sure feed is free of pesticide residues. Use only recommended pesticides on forage crops that are to be fed to dairy cows. Always read and follow label directions on any chemical or drug used on the farm.

PREVENTING OFF FLAVORS IN MILK

Milk has a clean, pleasant, and slightly sweet flavor. Off flavors cause dissatisfied customers, and milk sales go down when customers do not like the taste of the milk. The dairy farmer must follow good management practices to prevent off flavors in milk.

Off flavors in milk include:

- feed flavors
- rancid
- foreign flavors
- sanitizer, medicinal
- salty
- flat
- unclean, cowy, or barny
- high acid (sour), malty
- oxidized

The most common off flavor in raw milk is caused by feed (Figure 43-4). Any feed with a strong smell can cause off flavors in the milk. Silage, weeds, grasses, and moldy feeds all cause off flavors. Remove cows from pasture 2 to 4 hours before milking. Feed silage in the barn only after milking instead of before. Be sure there is plenty of ventilation in the barn.

Figure 43-4 The most common off flavor in milk is caused by feed.

Rancid flavor is strong and bitter. Improper handling of milk can cause rancid flavor. Too much agitation or cooling, warming, and recooling milk should be avoided to prevent rancid flavor. Milk is agitated by foaming that is caused by air bubbles in the line. Keep equipment in good operating condition to prevent air entering the line.

A flat flavor in milk is caused by adding water. Be sure equipment is drained before adding milk and do not add water to milk.

Poor barn ventilation causes unclean, cowy, or barny flavors. Provide good ventilation in the barn and milking area. Make sure all milk-handling equipment is clean.

Kerosene, creosote, paint, and fly spray all cause foreign flavors in milk. Never store these materials in or around the barn or milk house.

Sanitizer and medicinal flavors are caused by improper use of these materials. Follow label directions carefully when using sanitizers and medications. Apply udder medication after milking and wash udders before the next milking. Use teat-dips only after milking.

A salty flavor is caused by cows in late lactation or cows with mastitis. Always milk cows with mastitis separately from the rest of the herd. Do not milk cows more than 10 or 12 months in one lactation.

High acid (sour) and malty flavors are caused by high bacteria counts. Keep equipment, cows, and facilities clean and sanitized. Cool milk rapidly to prevent bacteria growth.

Chemical reactions in the milk can cause an oxidized flavor where the milk tastes like wet cardboard. The most common cause is contamination of the milk by small amounts of copper or iron. Lack of green feed in the ration and exposure of the milk to sunlight are other common causes. Do not allow milk to come in contact with copper or iron. The use of stainless steel on all milk contact surfaces reduces the problem. Also, keep milk away from sunlight.

SUMMARY

The cow's udder is made up of four glands called quarters. Alveoli in the udder contain specialized cells that synthesize milk. Udder function is controlled by hormones. Milk letdown occurs when the cow responds by a conditioned reflex to sensory stimuli. Gentle washing of the udder stimulates milk letdown.

Follow a regular routine when milking cows. Handle cows gently with as little stress as possible. Do not over milk the cows.

Produce high-quality milk by keeping cows and facilities clean, rapidly cooling the milk, and properly cleaning the equipment. Bacteria lower the quality of the milk. Good management practices help to reduce bacteria in the milk.

Mineral and milkstone deposits may be found on milking equipment. Alkaline, chlorinated alkaline, and acid cleaners are used to remove deposits from equipment. Use caution and follow the manufacturer's directions for the use of cleaners and sanitizers.

Off flavors reduce milk quality. There are many causes of off flavors in milk. Good management practices help reduce off flavors.

Student Learning Activities

1. Prepare and present an oral report to the class on one aspect of milking management.
2. On a field trip to a dairy farm, observe proper methods of cleaning and sanitizing milking equipment.
3. Talk with a dairy plant field person about cleaning milking equipment, keeping bacteria count low, and preventing off flavors in milk. Give an oral report to the class.
4. Talk with a local dairy farmer about proper milking techniques. Give an oral report to the class.
5. When planning and conducting a dairy supervised experience program, follow the recommended milking practices and other milking management practices discussed in this chapter.

Discussion Questions

1. Describe how the udder of the dairy cow functions.
2. Describe milk letdown in dairy cows.

3. Outline recommended milking procedures for the dairy herd.
4. Briefly explain three-times-per-day milking.
5. What are the characteristics of high-quality milk?
6. What practices help produce high-quality milk?
7. Describe ways to keep sediment out of the milk.
8. List problems caused by high bacteria count in milk.
9. What practices help keep the bacteria count low in milk?
10. What kinds of deposits are found on milking equipment?
11. What kinds of cleaners are used to remove deposits from milk equipment?
12. What kinds of sanitizing solutions are used to sanitize milk equipment?
13. Why must milk equipment be both cleaned and sanitized?
14. Describe the cleaning and sanitizing procedure for each kind of milking equipment.
15. How may chemical contamination of milk be prevented?
16. List and describe the common off flavors in milk.

Review Questions

True/False

1. A narrow streak canal and strong sphincter muscle make a cow harder to milk.
2. Gentle washing of the cow's udder helps stimulate milk letdown.
3. Milk letdown is not necessary to effectively remove milk from the udder.
4. The cow's udder is made up of two glands called halves.

Multiple Choice

5. What is a major additional cost of milking three times a day compared to two times per day?

   a. labor
   b. equipment
   c. barns
   d. medicine

6. Milking cows three times per day will increase total milk production from 6 percent to ______ percent when compared to milking two times per day.

   a. 10
   b. 20
   c. 30
   d. 40

Completion

7. The udder contains ______ with specialized cells that synthesize the milk.
8. ______ is a combination of organic and inorganic material.

Short Answer

9. What is the major cost of increasing the frequency of milkings per day?
10. What is machine stripping?
11. What is the best way to keep sediment out of milk?
12. List two kinds of deposits found on milking equipment.
13. Why is it important to store milking machine parts upside down after washing?
14. What causes the most common off flavor in raw milk?

Chapter 44

Dairy Herd Health

Key Terms

leukocytes
somatic cells
placenta

Objectives

After studying this chapter, the student should be able to

- outline procedures for maintaining herd health.
- describe the proper use of drugs for treating herd health problems.
- describe dairy herd health problems and treatments.

Dairy cattle are subject to most of the same disease and parasite problems that affect beef cattle. See Chapter 17 for a discussion of cattle diseases and parasites. There are some health considerations that need to be emphasized concerning dairy cattle, which are discussed in this section.

HERD HEALTH PLAN

The dairy farmer should develop an overall plan for maintaining the health of the dairy herd. An effective plan puts emphasis on the prevention of problems. A veterinarian should be consulted to prepare the overall herd health plan (Figure 44-1).

A planned program of regular vaccination and herd testing should be implemented. Diseases of major concern are brucellosis, infectious bovine rhinotracheitis (IBR), bovine virus diarrhea (BVD), and parainfluenza-3 (PI_3). Symptoms, control, and prevention of these diseases are discussed in Chapter 17. Control of other diseases is needed in herds or areas where they are a problem.

It is important to keep health records on all animals in the herd and make individual physical examinations as needed. It is not only important to keep health records, but to keep accurate reproduction records as well. Have a veterinarian examine any cows with breeding or calving problems, and make routine pregnancy examinations. Follow a planned program of calf health care. Prevent disease by vaccination and other management practices. To prevent loss, treat calf health problems, such as scours, quickly. Use the services of a veterinarian whenever cows show health problems. Diagnosis and treatment of health problems is a specialized skill. Losses can be high with improper diagnosis and treatment. Mastitis, which is one of the major causes of economic losses in dairy herds, should have its own specific control program to be followed.

Herd health problems are reduced by the following management practices:

- proper feeding of the herd
- good facilities that are ventilated properly
- using clean, dry bedding
- proper cleaning and sanitation of facilities and equipment
- controlling disease carriers, such as flies, birds, and rodents
- raising the replacements needed for the herd
- requiring the health records of replacement animals and isolating them from the rest of the herd for 30 days
- isolating all sick animals from the herd
- using a veterinarian for quick, accurate diagnosis and treatment of health problems
- controlling access to the dairy facilities by posting a sign at the farm entrance informing people that entry is limited and restricted to certain areas to control diseases
- requiring visitors to use some type of protective footwear covering, such as plastic boots, or use rubber boots and a disinfectant foot dip at the entrance to the facilities
- not allowing visitors to have unlimited access to areas where the cattle are kept
- having bulk milk pickup and feed delivery points as far from the cows as possible and restricting access by these vehicles to only those designated areas

Photo by Peggy Greb. USDA

Figure 44-1 The services of a veterinarian should be used on a regular basis to maintain herd health.

Dairy Quality Assurance Program

The Dairy Quality Assurance Program is designed to help dairy farmers produce high-quality milk (Figure 44-2). Participation is on a voluntary basis. The program was developed by the National Milk Producers Federation and the American Veterinary Medical Association. A producer, in cooperation with a veterinarian, goes through a 10-point checklist of management practices to become certified.

USDA

Figure 44-2 A healthy herd is of the utmost importance for maximum milk production.

The program identifies several critical control points that help the herd owner produce a quality product. Emphasis is placed on following a preventive health program, including vaccination, housing, nutrition, and sanitation. Producers must use the correct drugs, store them properly, and follow required withdrawal times. The use of drug screening tests and good record keeping are other important aspects of the program. Other management practices, including sanitation, feeding, and milking equipment maintenance, are also reviewed. Producers who complete the certification and follow the management guidelines can reduce the risk of drug violations in the milk by more than 50 percent.

Use of Drugs for Treatment of Dairy Cows and Calves

CAUTION

Label directions for period of use and withholding of milk from the market must be followed; failure to do so may result in severe economic loss due to condemnation of entire tanks of milk. Kits are available for testing individual samples of milk before adding them to the bulk tank if there is any doubt about the presence of drug residues.

- Read drug labels carefully. Hundreds of changes are made in drug labels each year, and many of these changes affect the way drugs are used in treating animals
- Use drugs only in the animal species indicated on the label. Drugs meant for one kind of animal can cause adverse drug reactions or illegal drug residues in another species
- Always make sure to give the proper amount of drug for the kind and size of animal being treated. Overdosing can cause drug residue violations
- Make sure to calculate preslaughter drug withdrawal and milk discard times accurately. Remember, withdrawal and discard times begin with the last drug administration
- Always use the correct route of drug administration. Giving oral drugs by injection can cause loss of drug effectiveness. Giving injectable drugs incorrectly can lead to adverse reactions, reduced effectiveness, illegal drug residues, and possibly the death of an animal
- Avoid "double-dosing" the animals. Using the same drug in the feed supply and then by injection can cause illegal residues
- Keep an accurate record of the drugs used, and identify the animals receiving the drugs. Sending an animal to market too soon after it has been treated or shipping a treated animal because it was not properly identified can be a costly mistake
- Good drug use records also help when professional animal health care is needed. Veterinarians need to know how much and what kinds of drugs have been given before they can treat animals effectively and safely
- When injecting animals, select needles and injection sites with care. Depending on the animal, and sometimes the drug, the wrong needle size, spacing or number of injection sites, or drug amount per site can result in tissue damage, reduced drug effectiveness, or illegal drug residues
- Remember, feed containing drugs also can cause illegal residues. Make sure there is a reliable source of drug-free feed for the animals to eat during withdrawal periods and that storage bins and feed troughs are cleaned thoroughly before the withdrawal feed is put in them
- For a complete explanation of all the precautions needed when using any particular drug or feed medication, first consult the drug label or feed tag. If there are any questions about the proper use of any drug, consult the feed dealer or a veterinarian

Mastitis Control

Mastitis is a serious economic problem for dairy farmers. Figure 44-3 shows the effect one type of mastitis has on milk.

The presence of mastitis in the dairy herd causes losses by:

- lowering milk production from infected cows
- increasing the cull rate in the herd
- the cost of treatment
- loss of infected milk that must be thrown away
- increased labor cost to treat infected cows
- possible loss of permit to sell milk if infection becomes serious enough

Mastitis is usually caused by bacteria that get into the udder through the teat opening, but can also enter through an injury to the teat.

Mastitis may be acute or chronic. The symptoms of the acute form include:

- inflamed udder
- swollen, hot, hard, tender quarter
- drop in milk production
- abnormal milk (lumpy, stringy, straw-colored, contains blood, yellow clots)
- the cow going off-feed, showing depression, dull eyes, rough hair coat, chills or fever, constipation
- death

The symptoms of the chronic form of mastitis include:

- abnormal milk (clots, flakes, watery)
- slight swelling and hardness of udder that comes and goes
- sudden decrease in milk production

Chronic mastitis is a more serious economic problem than the acute form. However, either form may cause permanent damage to the udder.

Leukocytes are white blood cells that fight infection. The presence of mastitis causes an increase of leukocytes in the milk. **Somatic cells** are leukocytes and body cells. While all normal milk contains some somatic cells, the goal for the dairy herd should be an average of no more than 150,000 to 200,000 cells per milliliter. Ninety percent of the herd should be below 200,000 cells per milliliter. The somatic cell count can vary considerably from month to month in cows that have a mastitis infection. Daily per head milk losses increase as the somatic cell count increases, ranging from 1.5 pounds at 72,000 cells per milliliter to 6.0 pounds at somatic cell counts of more than 1 million per milliliter. Somatic cell counts above 500,000 usually indicate a bacterial infection, a cow in late lactation, udder injury, or an old cow. Problem cows may need to be culled from the herd.

Current regulations state that bulk Grade A milk picked up at the farm cannot have a somatic cell count exceeding 750,000 cells per milliliter. A violation can result in the loss of the Grade A permit. This can result in serious economic loss for the dairy farmer.

Several tests are used to detect high somatic cell counts. The California Mastitis Test (CMT) is a common test to screen the herd for mastitis. It gives an estimate of the somatic cell count and should be used at least once a month.

A small paddle with four cup compartments is used. Milk about 1 teaspoon of the first milk from a quarter into a cup. Check each quarter separately. A chemical that reacts with the milk is placed in the cups. The presence of leukocytes is shown by the reaction. A slight precipitation shows a low leukocyte count. The development of a heavy gel and a purple color shows a high leukocyte count.

Tests for somatic cell count can be done in a laboratory. Some tests are chemical. Electronic instruments have been developed for giving more accurate somatic cell counts in milk. A direct microscope count may also be made.

A carefully planned and followed mastitis control program is essential for the dairy farm. The control program must be designed to both reduce the number of

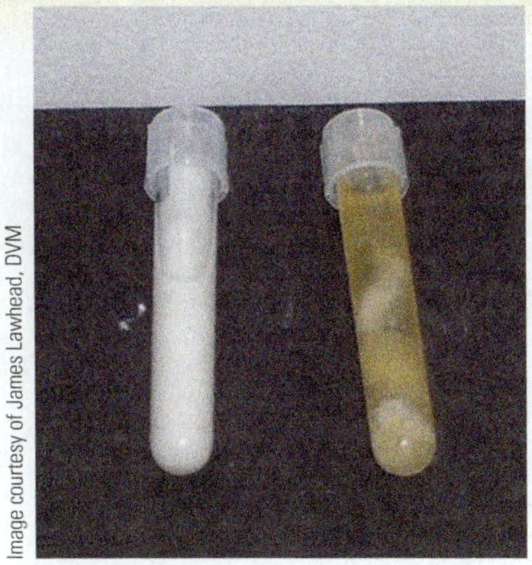

Image courtesy of James Lawhead, DVM

Figure 44-3 Normal milk is shown on the left and an abnormal secretion from a cow with coliform mastitis is shown on the right.

new infections and effectively treat existing infections in the dairy herd. An additional objective of the control program must be the avoidance of drug residues in the milk.

The following practices should be followed in an effective mastitis control program:

1. Maintain milking equipment in proper operating condition.
2. Practice proper milking procedures.
3. Identify the bacteria causing the infection and determine the extent of the infection in the herd.
4. Promptly treat identified cases of mastitis.
5. Treat all quarters of cows at drying off time.
6. Cull cows with chronic mastitis problems that do not respond to treatment.

Improperly maintained and operating milking equipment can contribute to a mastitis problem in the herd. The milking vacuum should be relatively stable at 11 to 12 inches of mercury at the claw. The pulsator should maintain 45 to 60 pulsations per minute with a milk/rest (pulsation) ratio of 50/50 to 60/40. Specific brands of equipment may be designed to operate at different specifications; if so, be sure the equipment is operating within the design specifications. Check the system while all units are operating. The end of the teat may be damaged if the pulsation ratio and/or rate are too high. This can allow bacteria to enter the teat. Vacuum fluctuation, liner slip, and liner flooding may allow bacteria to enter the teat through the teat canal.

Bacteria can also be transmitted mechanically from cow to cow by the milker claw; this problem can be reduced by properly sanitizing the liners and claw between cows and by milking cows with high somatic cell counts or known infections of mastitis after the other cows in the herd are milked. Care must be taken to keep the sanitizing solution from becoming contaminated when treating the liners and claw between cows. Contaminated sanitizing solution will not prevent the spread of mastitis.

Large herds with a serious mastitis problem may justify the expense of solid-state backflushing units, which are now available for use in milking parlors. These units backflush the equipment with a sanitizing solution between each cow. The flush cycle takes from 1 to 3 minutes and uses up to 1.5 gallons per cow. This can help reduce the spread of the infection by the equipment.

Proper milking procedure includes washing the udder before applying the milking claw and dipping the teats in a sanitizing solution after milking is completed. Procedures are discussed in detail in Chapter 43.

Identifying the bacteria that are causing the mastitis problem is essential for proper treatment. A determination can then be made of the proper antibiotic to use in treating the infection. Bacterial culture tests and somatic cell counts taken from four or five daily bulk tank samples will give an indication of the extent of mastitis infection in the herd.

Cows showing symptoms of acute mastitis infection should be treated with intravenous or intramuscular injections of the proper antibiotic. This treatment should always be done under the supervision of a veterinarian. Cases that are not quite as severe may be treated with mastitis infusion tubes in the affected quarters.

Treat all quarters of cows being dried off with an approved dry cow mastitis treatment. This practice will help prevent infection during the dry period. If infection is present, the cure rate is higher than if treatment is done during lactation. Treatment also helps damaged tissue regenerate, reduces infection at freshening time, and avoids the problem of drug residues in marketable milk.

Cows with chronic mastitis infection that does not respond to treatment should be culled from the herd; this will eliminate a possible source of infection for other cows.

Displaced Abomasum

Displaced abomasum (DA) is a condition in which the abomasum moves out of place in the abdominal cavity. It is more common in dairy cattle than in beef cattle. Treatment generally requires surgical correction and, in most cases, results in rapid recovery and return to production (Figure 44-4). Nonsurgical treatments such as rolling the cow over or forcing her up a sharp incline have sometimes proven successful. The majority of the cases occur shortly after calving.

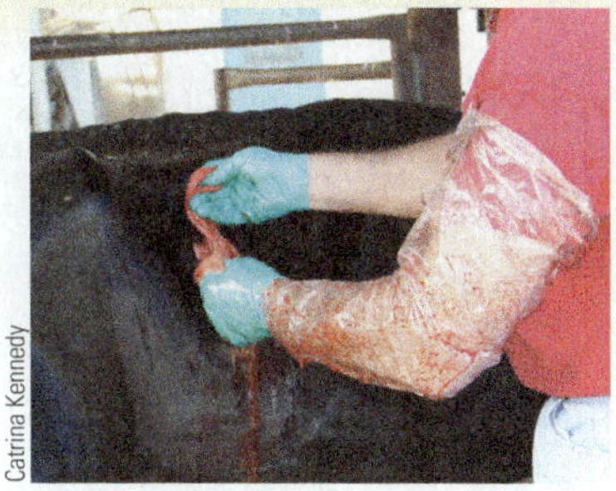

Catrina Kennedy

Figure 44-4 A displaced abomasum in this Holstein cow required surgery. The veterinarian practices normal surgical procedures of anesthesia, cleansing, etc., on the farm.

The symptoms of displaced abomasum include:

- poor appetite
- reduced fecal discharge
- soft or pasty feces
- diarrhea
- drop in milk production
- dull, listless, thin appearance

The kind of ration fed appears to be involved in causing displaced abomasum. Too rapid an increase in grain feeding just before calving increases the chances of DA. Poor-quality, moldy roughage or too much silage in the ration also increases DA. Do not overfeed silage and concentrates to dry cows but rather increase the amount of concentrate in the ration slowly at calving time.

Retained Placenta

Retained placenta is a condition in which the **placenta** (afterbirth) is not discharged within 12 to 24 hours after calving. It is normal for 10 to 12 percent of dairy cows to have retained placenta. A higher rate indicates a problem that needs attention.

A number of causes may be involved in retained placenta, including:

- infection in the reproductive tract during pregnancy
- deficiencies of vitamin A or E, iodine, and selenium
- an out-of-balance calcium-to-phosphorus ratio in the diet
- the cow being too fat (fed too much carbohydrate feeds)
- stress at calving
- breeding the cow too soon after calving

Good management that prevents the listed causes will help reduce retained placenta. Call a veterinarian to treat a cow with retained placenta.

Ketosis

Ketosis is a nutritional disorder in dairy cows in which their blood sugar drops to a low level. It is caused by not feeding enough energy feeds to meet the cow's needs for high milk production. Ketosis usually occurs in the first 6 to 8 weeks after calving.

Symptoms of ketosis include:

- the cow going off-feed shortly after calving
- a drop in milk production
- loss in body weight
- the cow becoming dull and listless
- an odor of acetone in breath, urine, and milk

Feeding a properly balanced ration will help prevent ketosis. A veterinarian should be called to determine the proper treatment if a cow develops ketosis. Common treatments include injection of glucose into the bloodstream, injection of hormones (cortisone or adrenocorticotrophic hormone), or oral feeding of propylene glycol or sodium propionate. Feeding molasses will not cure ketosis.

Metritis

Metritis is an infection in the uterus. It usually affects a cow within 1 to 10 days after calving. A higher rate of metritis is seen in cows that are too fat at calving time.

Symptoms of metritis include:

- loss of appetite
- fever
- drop in milk production
- abnormal (thick, cloudy, gray, foul odor) discharge from the vulva
- standing with back arched
- rapid death in severe cases

Feeding a properly balanced ration to dry cows helps prevent metritis. Keep the calving area clean and sanitary. Metritis is treated with intrauterine antibiotic drugs. Consult a veterinarian for proper treatment.

Milk Fever

Milk fever (parturient paresis) is caused by a shortage of calcium salts in the blood. It is more common in older, high-producing cows and usually occurs within a few days after calving.

Symptoms of milk fever include:

- loss of appetite
- reduction in quantity of feces passed
- the cow becoming excited in the early stage
- staggering
- the cow becoming depressed
- cold skin, dry muzzle
- paralysis
- lying on brisket with head turned back toward side
- in later stages, lying on side with head stretched out
- bloating
- death if not treated

Feed a balanced ration to dry cows with the correct calcium-to-phosphorus ratio. Milk fever is treated by intravenous injection of calcium. Call a veterinarian for treatment.

Internal Parasites

Common internal parasites of cattle are discussed in Chapter 17. A regular program of treatment for internal parasites should be followed for the dairy herd. Consult a veterinarian to set up a regular treatment program.

All mature dairy cows should be treated for worms after each lactation. Worm replacement heifers near the end of their pregnancy.

External Parasites

Common external parasites of cattle are discussed in Chapter 17. Care must be taken when using insecticides on dairy farms. Insecticides in milk are illegal and very small amounts can be detected. Use only insecticides approved for dairy animals and facilities. Follow label directions carefully to avoid illegal residues in the milk.

SUMMARY

An effective herd health plan emphasizes prevention of problems. Regular vaccination and herd testing is important. Good management practices help prevent health problems.

Mastitis is the most serious disease that affects dairy cattle. Careful management and proper treatment is needed to keep losses low. Be careful when using drugs to avoid illegal residues in the milk.

A good herd health plan, set up with the assistance of a veterinarian, will help the operation run more efficiently and can lead to an increase in net profits.

Student Learning Activities

1. Speak with a veterinarian about mastitis control in dairy herds.
2. Plan a herd health program for a local dairy farm.
3. Prepare and present an oral report to the class on one aspect of dairy herd health.
4. In order to maintain good herd health, use the information in this chapter when planning and conducting a dairy supervised experience program.

Discussion Questions

1. Describe management practices that will help maintain good herd health.
2. What care should be taken when using drugs for treatment of dairy cows and calves?
3. What losses are caused by mastitis?
4. What are the symptoms of acute and chronic mastitis?
5. Describe the California Mastitis Test.
6. How is mastitis treated?
7. Discuss the symptoms, prevention, and treatment of each of the following:
   a. displaced abomasum
   b. retained placenta
   c. ketosis
   d. metritis
   e. milk fever
8. Briefly explain control of internal and external parasites in the dairy herd.

Review Questions

True/False

1. Dairy cattle are not subject to the same diseases and parasites that affect beef cattle.
2. The Dairy Quality Assurance Program is designed to help farmers produce high-quality milk.
3. Mastitis is one of the major causes of economic losses in dairy herds.
4. All mature dairy cattle should be treated for worms during each lactation.
5. Metritis is an infection of the uterus in cattle.

Multiple Choice

6. Ketosis usually occurs during the first ______ to ______ weeks after calving.
   a. 2; 4
   b. 4; 6
   c. 6; 8
   d. 8; 10

7. One objective of a mastitis control program is the avoidance of _____ residue in the milk.

   a. feed
   b. drug
   c. protein
   d. bacteria

Completion

8. _____ _____ is using the same drug in the feed supply and then by injection.
9. Mastitis may be _____ or _____.

Short Answer

10. What does an effective plan for herd health emphasize?
11. How does mastitis cause economic problems for dairy farmers?
12. Why is it important to sanitize the milker claw and liners?
13. What is a displaced abomasum?
14. What are leukocytes?

Chapter 45

Dairy Housing and Equipment

Key Terms

inflation
pulsator

Objectives

After studying this chapter, the student should be able to

- explain housing needs of dairy cattle.
- discuss the economics of dairy cattle housing and equipment.
- describe milking equipment.
- describe milk handling systems.
- describe manure handling systems.

PLANNING THE DAIRY FACILITY

The modern dairy farm requires a high investment in buildings and equipment (Figure 45-1). Careful planning is necessary before building new facilities or remodeling old facilities. Dairy farming is a high labor requirement enterprise. Mechanization can reduce labor requirements, but the cost of mechanization is high. Sources of ideas for dairy facilities include farm publications, state universities, the Cooperative Extension Service, other dairy farmers, and manufacturers of dairy equipment. Efficiency, economy, and convenience are important considerations when planning dairy facilities.

Important areas to be considered when planning dairy facilities include the following:

- location of the facility
- size of the planned herd
- laws and regulations that apply to dairy farms
- source and amount of money available
- type of milk market available
- amount of labor available

Courtesy of Frank B. Flanders

Figure 45-1 A modern dairy operation with a free stall barn and feed mill.

- kind of housing system to be used
- kind of milking system to be used
- feed handling system to be used
- manure handling system to be used
- what future expansion might be desired

Location of the Facility

Location of the dairy is a major decision because once the facility is built, it cannot be moved. Sometimes, the decision is a matter of choosing between remodeling and building on a new location. Each farm presents a different set of circumstances. While no general recommendation is made, it is often better to build completely new facilities rather than try to remodel old ones. Review all the planning considerations previously listed before making a final decision.

A well-drained location is needed. Easy access for vehicles to handle feed, milk, livestock, and manure should be provided. Make sure there is a good water supply. Provide room for expansion of the facility. Consider prevailing wind direction. Locate the dairy facility downwind from the farmhouse. Try to avoid a location that will cause odor problems for close neighbors.

Electrical Service

A dairy farm must have a continuous supply of electricity. Make sure the power supply is adequate to meet the needs of the operation. A standby generator is essential in case of a power failure.

HOUSING FOR THE MILKING HERD

Many types of housing systems are used for the milking herd. The two most common types are stall (stanchion) barns and free stall barns. Stall barns are more common when the herd size is 80 cows or fewer. Free stall barns are more commonly used with herds larger than 80 cows. Both systems can be mechanized to reduce labor requirements.

Stall Barns (Stanchions)

A stall barn is a traditional-style dairy housing and is currently only used on small-scale family farms. Each cow is confined to an individual stall in a stall barn and is held in a stanchion or a tie stall.

Stanchions have metal yokes that are fastened at the top and bottom that can pivot from side to side. In tie stalls, the cow is fastened by a neck chain or strap to the front of the stall. There is a pipe across the front of the stall to prevent the cow from walking forward through the stall. Tie stalls give greater comfort to the cow, but more labor is needed to fasten and release each cow from a tie stall.

Advantages of the stall barn include:

- allows more individual attention to each cow
- easier to observe and treat each cow
- better display of breeding stock

Disadvantages of the stall barn include:

- more labor and time needed to fasten and release cows
- harder to use with a milking parlor
- harder to keep the area clean
- more bedding needed
- more labor needed for feeding
- stooping required when milking in stall barns
- moisture problem is greater, especially in older barns

Figure 45-2 It is recommended that cows face out in a stanchion barn.

Stalls must be long enough for cows to lie down with their udders on the platform. There must be enough width for the cow and for the person milking the cow. Teat and udder injuries are more likely to occur when the stall is too small. Stalls that are too long or too short make it hard to keep the cows clean.

Stalls may be arranged with two rows facing either in or out. The recommended arrangement is to have the cows facing out (Figure 45-2). This makes it easier to milk the cows and to install a gutter cleaner. The dairy farmer spends more time working behind the cows as compared to working in front of the cows.

The use of deep bedding (typically oat straw) has been the traditional way of keeping cows clean in stall barns. Bedding is expensive, takes a lot of labor, is getting harder to find, and does not work well with liquid manure systems.

A bedding chopper is a machine powered by a small gas engine. It chops a bale of bedding material as it is moved down the alley behind the cows. It saves time and labor when using bedding. Poor-quality or rained-on hay may be used for bedding when put through a bedding chopper.

Stall mats made of rubber have been developed as an alternate to deep bedding. Mats that are held in place on top of the concrete tend to get manure, urine, and bedding underneath them. This causes an odor problem.

It is hard to clean the area under the mats. Rubber mats can be placed in the fresh concrete during construction. This reduces some of the problems. However, these mats will eventually come loose. Cleaning under them is harder than cleaning under those placed on top of the concrete.

There is no completely satisfactory system for the cow platform surface that will keep cows clean. Properly adjusted cow trainers will help force the cow to deposit manure and urine in the gutter. The cow trainer is an electrical device that contacts the cow when she arches her body to defecate or urinate. The electrical charge causes the cow to move back toward the gutter.

Use a finely chopped bedding material on cow mats. In tie stalls, use metal tie chains to complete the electric ground for the cow trainer. The rubber mat insulates the cow from the charge. Stanchion stalls provide enough ground through the metal yoke of the stanchion.

Gutter grates are steel bars placed over the gutter. These help keep the cow's tail out of the gutter when she lies down. They also increase the length of the cow bed. The steel bars are installed parallel to the gutter.

Feed carts are used in stall barns to reduce the labor needed to feed the cows. Carts may be homemade or purchased from commercial suppliers. Self-propelled, self-unloading carts are available for use in large stall barns. A separate cart equipped with a scale is often used for feeding grain.

Water cups are used to provide water for the cows in the stalls. There is always some spillage from water cups. Place the cups ahead of the stall curb to prevent wet stall platforms. Provide drainage in the feed manger area to take care of spilled water.

Too many windows in a stall barn make it harder to keep it warm in the winter and cool in the summer. They are expensive to install and maintain. Do not depend on windows as part of the ventilation system.

A minimum of 10 foot candles of light is required by Grade A regulations in all work areas during milking. Place a 100-watt bulb every 10 linear feet in the area behind the cows to meet this requirement. Fewer bulbs may be used in the feed alley area. Place alternate lights on different circuits to reduce energy use during times when the cows are not being milked.

Stall barns must be properly insulated and ventilated. Health and moisture problems are common in poorly insulated and poorly ventilated barns.

Ventilation designs must also conform to local climatic conditions. A common design is to provide enough ventilation capacity for a minimum of four air changes per hour in the winter. For summer capacity, plan for a minimum of 30 air changes per hour. Fans controlled by thermostats are used for ventilation capacity above the minimum.

When a manure pit is located beneath the barn, it must be ventilated continuously. Failure to do so will create a serious health hazard.

Fresh air must be brought into the building without causing drafts. Provide many small inlets for fresh air. Slot and ceiling intake systems are the most commonly used. Some moisture condensation is normal at the point of fresh air entry.

Provision must be made for electric power failure. Automatic warning systems and a standby generator are recommended.

Remodeling

Older barns may be remodeled to improve stall sizes and reduce labor requirements. Many old barns are not worth remodeling. Whether or not to remodel is an individual decision that must be evaluated in each case. Follow recommendations for stall and gutter widths. Other dimensions may be changed to fit the existing building. Increased herd size may be taken care of by adding to the length of the barn. It is usually not practical to try to widen an existing barn. Be sure the remodeled barn is properly ventilated. Older barns seldom have the proper ventilation.

Free Stall Barns

A free stall barn is widely used and is the most common among commercial dairy farmers. It is a loose housing system in which stalls are provided for the cows (Figure 45-3). The cows are not fastened in the stalls. They may enter and leave the stalls whenever they want to. The system usually has a resting and a feeding area. The stalls are located in the resting area.

Courtesy Brenda Hastings, Hastings Dairy in Burton, OH

Figure 45-3 A free stall barn is a loose housing system in which stalls are provided for the cows to enter as they choose. Sand provides comfortable bedding in the stalls.

Advantages of the free stall barn include the following:

- it requires less bedding than other loose housing systems
- cows stay cleaner compared to other loose housing
- there are fewer injuries to teats and udders
- it requires less space than for other loose housing systems
- it is easier to use with a milking parlor
- it is easier to use automatic feeding equipment
- cow disposition is better
- there is less disturbance from boss cows and cows in heat

Disadvantages of free stall barns include the following:

- manure is usually fluid, making it harder to handle
- some cows will not use the stalls and must be trained to enter them
- some systems require more daily labor for manure handling

A number of different designs are used for free stall housing systems. An enclosed warm system is fully enclosed, insulated, and mechanically ventilated. An enclosed cold system is fully enclosed, not insulated, and uses natural ventilation. Some designs use an open front with a roof, no insulation, and natural ventilation. An additional paved area may be provided outside of the roofed area. Another design provides a building only for resting. The feeding area, which may or may not be roofed, is outside and paved.

A stall is provided for each cow. Current practice is to allow about 10 percent more cows in the area than there are stalls. Not all the cows lie down at the same time.

To force the cow to move back when standing up, place a neck board across the top of the stall about 2 feet from the front of the stall. This causes more of the droppings to be deposited in the alley. The cows will stay cleaner.

Floors of stalls may be earth or concrete. Compacted clay is preferred to other types of earth fill. Deep holes may develop in earth floors. These can cause injury to the cow's hips and legs. Dirt from an earth floor may get into the liquid manure system and cause problems. Concrete adds to the cost but eliminates some of the problems of earth floors. Rubber mats may be used on concrete floors. Mattresses made from a 22-oz, woven, nylon polyester material are durable and may be used in free stall housing. The mattresses are stuffed with shredded tires and do not shift under a cow's weight.

Several different kinds of materials may be used for bedding in free stalls. Sawdust and wood shavings make excellent bedding materials. Other materials such as chopped straw, chopped hay, shredded corn stalks, ground corncobs, and peanut hulls are used for bedding. Manure must be removed from the stalls each day. Use a rake to move the manure into the alley behind the stall.

Free stall barns usually have two, three, or four alleys. The width of the barn depends on the number of rows of stalls and the width of the alleys. A variety of plans and arrangements for free stall barns are available from the Cooperative Extension Service and agriculture engineers at state universities.

Slat floors may be used in alleys as part of the manure handling system. The manure is worked through the slats by cow traffic. The manure is held in a pit beneath the floor until it is removed. Continuous ventilation must be provided in enclosed barns to prevent serious health hazards.

Scrapers or small tractors with blades are used to remove manure from solid floor alleys. The manure may be moved into a holding area and later hauled to the field. Manure may also be disposed of in lagoons.

MILKING IN STALL BARNS

Three types of milking equipment are commonly used in stall barns—pail milkers, suspension milkers, and pipeline milkers.

Pail Milkers

Pail milkers are set on the floor beside the cow. The milker claw is attached to the udder and draws the milk into the pail. The pail must be emptied by hand after each cow is milked. The machine is operated by a vacuum pump with a vacuum line installed along the top of the stalls.

Suspension Milkers

Suspension milkers are hung from a strap, called a surcingle, which is placed over the back of the cow. This type of milker is also operated by a vacuum pump and has a vacuum line along the top of the stalls.

Pipeline Milkers

Pipeline milkers consist of a vacuum line and a milk line that are installed along the top of the stalls. The milker claw is attached to the udder of the cow. An inlet for the milker claw is usually located between each set of two stalls. The milk is drawn through the milk line into the bulk tank located in the milk house.

Pipeline milkers are more expensive than pail or suspension milkers. More vacuum is needed to operate the system. However, less labor is needed in handling the milk. Milk can be kept cleaner when a pipeline milker is used.

MILKING PARLORS

A milking parlor consists of a separate area in which the cows are milked. Usually, there is a pit in which the operator works. This eliminates the stooping that is necessary when milking in stall barns. Milking parlors are more commonly used with free stall housing. However, they may also be used when cows are kept in stall barns.

Advantages of milking parlors include the following:

- They reduce the labor and stooping needed to milk the cows
- A lower vacuum is needed than in pipeline milkers in stall barns
- Less stainless steel pipe is needed as compared to pipeline milkers in stall barns
- They can handle herd expansion easily
- Total cost of the system may be less because free stall housing is usually cheaper to build than stall barns
- They require less milk handling than pail or suspension milkers

Disadvantages of milking parlors include the following:

- There is a high investment cost involved in building the milking parlor. However, total cost of the system may be less than stall barns
- When grain is fed in the milking parlor, high-producing cows may not have enough time to eat their ration of grain while being milked
- Less individual attention can be given to each cow in the herd

There are four types of milking parlors:

- herringbone
- side-opening
- rotary or carousel
- polygon

Herringbone

The most common milking parlor in current use is the herringbone (Figure 45-4). Common sizes range from the double-4 to the double-10. In the herringbone parlor, the cows enter and leave in groups. Cows stand at an angle to the operator

Catrina Kennedy Pollard

Figure 45-4 Currently, the most common milking parlor is the herringbone.

pit, which is usually about 30 inches below the level at which the cows stand. A double-4 means that four cows stand on each side of the operator pit. Operator travel is reduced by having the cows stand at an angle to the operator pit.

Gates that allow the cows to enter and leave the parlor are controlled by the operator. Cows are kept in a holding area while waiting to be milked. A crowding gate is often used in the holding area to force the cows into the milking parlor. Holding areas should be paved to help keep the cows clean.

The pipeline milker may be along only one side of the parlor or along both sides. It is more expensive to have the pipeline along both sides. However, milking can be done faster with more units.

Cows may be fed grain in the milking parlor. The operator controls the amount of grain each cow receives. With the increased use of group feeding, it is becoming less common to feed grain in the milking parlor.

One disadvantage of the herringbone is that all the cows on one side must enter and leave at the same time. This means that a slow-milking cow will hold up an entire group. Separating slow-milking cows from the rest of the herd will help solve this problem.

Courtesy of Frank B. Flanders

Figure 45-5 A side-opening milking parlor.

Side-Opening

Side-opening milking parlors are arranged with the cows standing parallel to the operator pit (Figure 45-5). Each cow enters and leaves the parlor individually. Most common sizes are two, three, or four milking stalls on each side of the operator's pit. Walking distances between udders are 8 to 10 feet as compared to 3 to 4 feet for herringbone parlors. One advantage of side-opening parlors is that slow-milking cows do not hold up an entire group of cows during milking.

Rotary or Carousel

Rotary milking parlors are arranged so that cows enter onto a turning platform that rotates slowly (Figure 45-6). Cows may be arranged to stand parallel to the operator pit or in a herringbone pattern next to the operator pit. A large number of cows can be milked in a small space with the rotary parlor. It is a high-cost type of milking parlor. Two or more operators are usually needed to milk cows in a rotary parlor.

Each cow enters and leaves the rotary parlor by itself. The milker is attached and the cow is milked while the platform rotates. If the cow is not done milking by the time she reaches the exit, she must go all the way around again.

Polygon

The polygon milking parlor consists of a herringbone arrangement on four sides of the operator pit. Common sizes have 4, 5, 6, 8, or 10 cows on each side. The capacity of the polygon parlor, in terms of cows milked per hour, is about 25 percent greater than in herringbone parlors.

Generally, the polygon parlor is arranged in a diamond shape with corners not exceeding a 60-degree angle. Other shapes are possible, but the diamond shape is the most common.

The recommended procedure is for the operator to move around the side of the pit preparing cows, attaching milkers, and removing milkers.

It is not recommended that the operator cross from side to side in the pit to perform the milking operation.

© polat/www.Shutterstock.com

Figure 45-6 A rotary milking parlor rotates slowly as cows are milked.

MECHANIZATION IN THE MILKING PARLOR

There are a number of ways to increase mechanization in milking parlors. These include crowding gates, power gates, feed gates, prep stalls, and automatic detaching units.

Crowding gates are used to move cows forward in the holding pen area. Do not use electrically charged gates.

Power gates are used for entrance and exit of cows to and from the parlor. Pneumatic (air-operated) gates are commonly used. Controls are placed in several locations in the operator pit to reduce walking.

Feed gates are used to cover feed in the milking parlor. The feed is automatically covered when the cows are released. The purpose is to keep cows from stopping at the first milking stall as they come into the parlor. They also prevent cows from stopping to eat from stalls as they move out of the parlors. This helps control cow movement and cuts down on the amount of operator time needed to chase cows that do not want to leave the area.

Prep stalls are located ahead of the milking stall. A timed wash spray is used to preclean the cow and stimulate milk let down.

Automatic detaching units are used with the milk claw or suspension cup milker. These units detect a decreased flow of milk and automatically shut off the vacuum. These units speed the milking and prevent overmilking.

HOLDING AREA

A paved holding area is recommended for confining the cows before they enter the milking parlor. A long, narrow area is best because this allows the use of a crowding gate to move the cows forward to the milking parlor. Provide 12 to 15 square feet for each cow. Do not leave cows in the holding area more than 2 hours. In high-temperature climates, mister systems are used for evaporative cooling. Fans also are installed for use when temperatures rise to cool the cows as well as to draw heat and fumes away from the milking parlor. Slope the holding area to provide good drainage. Clean the area after each milking.

MILK HOUSE

The milk house contains the equipment for filtering, cooling, and storing milk. Dairy utensils are also cleaned and stored in the milk house. The design and construction of the milk house is subject to approval by public health agencies. Be sure all plans are approved before beginning construction.

The size of the milk house depends on the size of the herd and the type of equipment to be used. Consider the possibility of herd expansion when planning the size of the milk house.

Locate the milk house on the side of the barn away from the cow pens or holding areas. Provide easy access for milk trucks to pick up milk.

Heat the milk house to prevent freezing and to keep the floors dry. Heat may come from the compressor, water heater, and a furnace. Insulate the milk house for the type of climate found locally.

Use a fan to ventilate the milk house. A filter is used to help keep dust out of the area. Screened air inlets are used in the milk house. Use screens on any milk house windows that may be opened.

Provide floor drains to remove water. Do not locate floor drains under the bulk tank or the milk outlet valve. Check local and state building codes for proper drain location.

Use concrete for the floor of the milk house. Slope the floor $\frac{1}{4}$ inch per foot toward the drain. Extend the footing about 6 inches above the floor to protect the wall from excess water.

Use smooth, tight construction for the interior walls and ceiling. The material used must permit washing of the walls and ceiling.

Provide plenty of light in the milk house. Moisture-proof lighting fixtures should be used.

Use solid, tight-fitting, and self-closing doors. Direct openings between the milk house and the barn or milking parlor must have doors. Use screen doors in addition to the solid doors on any outside openings.

Water under pressure must be available in the milk house. An automatic hot water heater is also needed. Clean-in-place pipelines may require extra-capacity hot water heaters. Properly vent any gas water heaters used. Two-compartment sanitary wash and rinse vats are needed. Provide facilities for hand washing.

Office facilities are included in some milk house designs. Toilet facilities are also sometimes included. The toilet facility may not open directly into the milk room. Make sure there is proper waste disposal from any toilet facilities. Do not run waste disposal from toilet facilities into the waste disposal system for the milk room and milking parlor. Milk room and milking parlor wastes may be disposed of through a common system. Because there is usually a large amount of fibrous material in the system, use a settling tank that can be easily cleaned.

BULK TANKS

Milk is cooled and stored in a stainless steel tank (Figure 45-7). A refrigerant is compressed by a compressor. Water or air may be used to cool the refrigerant. When water is used, the heat that is removed is used to heat water for use in the milk house. Direct expansion bulk tanks have the milk contact surface cooled directly by the refrigerant. Some bulk tanks use an ice bank that is located in the wall of the tank. The refrigerant circulates through the evaporator coils. This forms the ice bank. The milk is cooled by coming in contact with the walls of the tank in which the ice bank is located. Direct expansion systems require a larger compressor than ice bank systems. However, ice bank systems use more total energy.

Atmospheric tanks are not under pressure. They may be opened at any time. It is easy to reach all surfaces for cleaning. Vacuum tanks are under pressure. They are typically round with curved end plates. It is harder to clean vacuum tanks.

The size of the bulk tank needed depends on the amount of milk produced each day and the frequency of milk pickup. For everyday pickup, buy a tank large enough to hold three milkings plus 10 percent. For every-other-day pickup, use a tank large enough to hold five milkings plus 10 percent.

Consider future herd expansion when buying a bulk tank. It may be wise to buy a tank larger than those recommended in the previous paragraph if herd expansion is probable in the near future.

Bulk tanks must be capable of quickly cooling milk to meet local requirements for Grade A production. Pre-coolers may be located on the milk line before it reaches the bulk tank to remove some of the heat from the milk.

Locate the bulk tank at least 2 feet from the wall on the side and rear of the tank. Provide at least 3 feet between the wall and the outlet end of the tank.

Courtesy of Frank B. Flanders

Figure 45-7 A vacuum bulk milk tank.

STRAY VOLTAGE

Stray electrical current, also called stray voltage, sometimes causes a problem in dairy facilities. A voltage difference as small as $\frac{1}{2}$ volt can make cows nervous and cause them to withhold milk. Stray electrical current may originate off the farm, or it may originate on the farm. The dairy farmer needs to work with the local utility

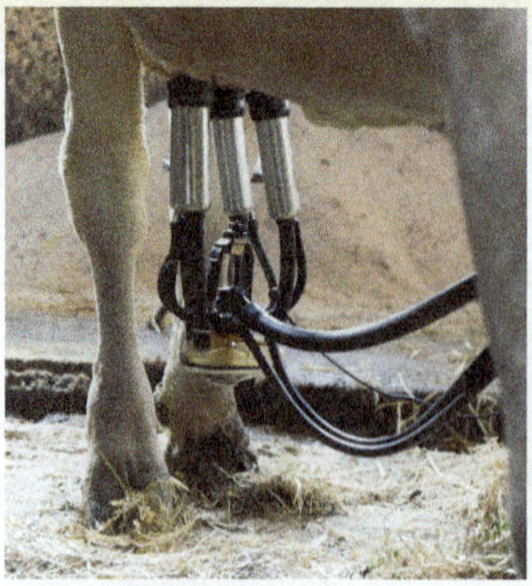

Figure 45-8 A milking unit attached to the udder.

company to identify the source of the stray current and take steps to correct the problem. Properly bonding all the metal parts of the facility that cows may contact will prevent stray currents because no voltage difference will exist between different metal objects. Permanent bonds between metal objects can be made by welding connections between them. A grounding wire (minimum size #8) should be connected between the metal stalls and the grounding bar in the electrical service panel. Make sure all connections meet electrical code requirements. It may also be necessary, especially in an older facility, to cover the floor with a nonconductive material. Asphalt, concrete, or epoxy materials may be used.

MILKING EQUIPMENT

The four parts of a milking system are the milking unit, the pulsation system, the vacuum supply system, and the milk flow system.

Milking Unit

The milking unit is attached to the udder (Figure 45-8). The parts of the milking unit include the teat-cup assembly, the claw or suspension cup, and the connecting air and milk tubes. The teat-cup assembly consists of a steel shell with a liner. This fits over the teats. The liner is called an **inflation**. The inflation squeezes and relaxes on the teat as the pulsator operates. This causes the milk to flow into the system.

Many different types of inflations are on the market. They may be molded, one-piece stretch, or ring-type stretch. Both wide- and narrow-bore inflations are available. Inflations may be made of natural rubber, synthetic rubber, or a combination of the two.

Molded inflations are easier to use and usually last longer than other types. Narrow-bore (less than $\frac{3}{4}$ inch) inflations cause less mastitis problems than wide-bore inflations. Narrow-bore shells and liners usually milk the cow faster than wide-bore shells and liners. Synthetic rubber inflations do not absorb milk particles as much as natural rubber inflations. They are easier to keep clean and do not tear as easily. The best molded inflations are made from synthetic rubber. Straight liners that are tightly stretched should be made from natural rubber.

The dairy farmer should try different types of inflations to see which work best for a specific herd. Make sure that the inflations match the type of shells in the system. Check the warranty on the machine to see if the use of a different type of inflation voids the warranty.

Narrow-bore inflations may be used for about 1,200 to 1,500 individual cow milkings. After this much use, they tend to stretch and bulge. Natural rubber inflations will last for about 800 individual cow milkings. The use of alternate sets of inflations each week make the inflations last longer. Always replace an entire set of four inflations at the same time rather than just one at a time. This makes it easier to keep track of the number of individual cow milkings per inflation.

Teat-cup shells are made of stainless steel. Several lengths and styles are manufactured. Make sure the inflation matches the type and style of shell used.

There are two types of milk receiving units used in the milking unit. One is the claw and the other is the suspension cup. The claw type is simpler than the suspension cup type. The suspension cup has a milk-receiving container on the bottom. It requires a support arm or surcingle.

An air bleeder vent in the claw increases the rate of milk flow away from the end of the teat. Make sure these air bleeder vents are kept clean.

A plugged vent causes slower milking and problems with the teat cup dropping off the udder. A plugged vent can also cause milk flooding in the inflation, claw, or short hose leading from the teat cup. Milk flooding can cause a wide variation in the vacuum level at the end of the teat and may also increase mastitis problems. The milk may contain bacteria that will enter the end of the teat.

Some newer types of inflations have an air inlet just below the shell. This reduces the problem of inflation flooding. More vacuum reserve is needed because this type lets more air into the system.

Air vents that are too large can cause too much foaming in the milk line and receiver jar. This can cause rancid milk.

The use of clear plastic in the construction of the claw or suspension cup permits the operator to see the milk flow from each quarter. This makes it possible to see when the quarter is milked out.

Synthetic rubber or clear plastic is used for milk hoses. It is easier to see the milk flow in clear plastic hoses. Rubber has more flexibility and lasts longer. The plastic hoses tend to crack at the end that is placed on the pipeline nipples. Do not use hoses that are too long. Looping long hoses increases vacuum changes in the system.

Air hoses are made of rubber. The air hose should be about the same length as the milk hose. A good seal around the pipe nipple is necessary for proper functioning.

Pulsation System

The pulsation system controls the action of the inflation. This is done with a **pulsator** that alternately allows vacuum and atmospheric pressure into the space between the inflation and the metal teat-cup shell. Parts of the system include the pulsator, pipe, and hoses.

Pulsators may be vacuum or electrically operated. Air is used to move a plunger or slide valve in the vacuum-operated pulsator. This opens and closes the air passage to give the pulsation action. A needle valve controls the rate of pulsation. Temperature changes affect the operation of the vacuum-controlled pulsator. This type of pulsator must be kept at normal temperatures for best pulsation action. The electrically controlled pulsator is operated by a low-voltage electric current. The pulsation rate is constant. Temperature changes do not affect the operation of this type of pulsator.

The inside of the inflation is always under vacuum. When air is let into the space between the inflation and the shell, the inflation collapses against the teat. This is called the rest or massage phase of operation. The milk does not flow at this time. The teat is massaged in an upward direction. This lets the blood circulate out of the teat.

When a vacuum is created between the inflation and the shell, the inflation opens. This is called the milk phase of operation. The milk flows out of the teat and into the milking system.

The pulsator ratio is the ratio of time between the open and closed phases in the complete cycle. The range of ratios is 50:50 to 70:30 (milking time to resting time). A wider ratio gives faster milking. However, too wide a ratio may cause teat damage because of overmilking.

The pulsator rate is the number of times the inflation opens and closes per minute. Normal rates range from 45 to 70 per minute. The speed of milking is about the same within these rates. Too slow a pulsation rate slows milking. Too fast a pulsation rate allows too much air into the system, wears out inflations faster, does not give enough massage to the teat, and may slow the rate of milking. Follow the manufacturer's directions for the rate of pulsator operation.

A unit pulsator operates only one milking unit. A master pulsator operates two or more milking units at the same time. The unit pulsator is more commonly used.

Some milking systems use an alternating pulsation. That is, two quarters are milked while the other two are rested. Most machines milk all four quarters at the same time. There does not appear to be any advantage of one system over the other.

A pipe carries air from the pulsator to the vacuum pump. Follow the manufacturer's recommendations for the proper pipe size for the type of system used. When the pipe is installed, slope it down toward the vacuum pump.

A flexible rubber hose connects from the pipe to the milking unit. Sometimes, the pulsator is at the stall cock, where the hose connects to the pipeline. Having the pulsator on the milking claw reduces the amount of pulsated hose.

Vacuum Supply System

The vacuum supply system includes the vacuum pump, vacuum lines, vacuum tank, vacuum controllers, and gauges. The functions of the vacuum supply system are to:

- cause the milk to flow from the end of the teat
- massage the teat
- cause the inflations to flex
- move milk through the system

The vacuum produced in the system is about one-half atmosphere of pressure. It is important to maintain a steady rate of vacuum. The pump must be able to constantly remove the air at a rate equal to the amount of air entering the system.

Several types of vacuum pumps are used in milking systems. These include piston-type, rotary-vane, and centrifugal water displacement pumps.

The piston-type pumps stand up well under use. They operate at low to moderate speeds and are dependable.

Rotary-vane pumps may be high (1750 rpm or higher) or low (400 to 800 rpm) speed. High-speed pumps are noisier but have a higher capacity for moving air. Low-speed pumps are quieter, use less oil, and usually require less maintenance. Rotary pumps move more air at the same horsepower than do piston pumps.

Centrifugal water displacement pumps are quiet, reliable, and stand up well under use. They require a method of disposing of large volumes of water. Some recycle the water. This works well except where the water is quite hard.

The vacuum pump must be able to move the amount of air needed by the milking system. Vacuum pumps are rated in cfm (cubic feet per minute) at a given vacuum level and rpm (revolutions per minute). Use care when comparing pump ratings. A wide variation exists in pump ratings.

Two methods are used to rate the cfm of vacuum pumps. These are the ASME (American Society of Mechanical Engineers) Standard and the New Zealand Standard (NZ). The two rating systems use different pressure conditions to measure the volume of air moved. The ASME system uses normal atmospheric pressure. The NZ system uses 15 inches of mercury vacuum. Fifteen inches of vacuum is one-half normal atmospheric pressure. This means that the NZ Standard shows twice the volume of air as compared to the ASME Standard (1 cfm ASME = 2 cfm NZ).

Pump size needed for a particular dairy operation depends on:

- the number of milking units in the system
- the size and length of pulsating lines
- the type of pulsator used
- whether the system is a bucket or pipeline type
- the amount and type of other vacuum-operated equipment in the system
- air leakage in the system

The pump needs to have enough capacity to prevent a drop of more than $\frac{1}{2}$ inch of mercury under normal operating conditions. The vacuum level should be restored within 5 seconds.

Locate the vacuum pump as close as is practical to the milking area. Pipe the exhaust from the pump to the outside of the building. Use an exhaust pipe at least as large as the connection on the pump. The exhaust usually contains some oil fumes. Direct the exhaust pipe down and away from the building to keep oil off the side of the building.

Maintain the oil at the proper level in the pump. Keep the belt aligned and at the proper tension. Follow the manufacturer's directions for routine servicing.

Vacuum lines that are too small or partly clogged reduce the flow of air through the system. Follow the manufacturer's recommendations for proper size of vacuum lines to fit the system being used. Regular routine maintenance includes cleaning the vacuum lines.

Vacuum tanks are usually included in the system unless the vacuum lines are oversized. The vacuum tank prevents excessive vacuum fluctuation in the system. It is important to keep the vacuum level as even as possible.

A vacuum regulator is included in the system to maintain the right level of vacuum. Several types are used, including ball, poppet, and sliding sleeve valves. When the vacuum level reaches a preset amount, the vacuum regulator opens to let air into the system. The vacuum regulator is often located on the vacuum supply pipe just ahead of the sanitary trap or the vacuum tank. Make sure the regulator is kept clean and is functioning properly. Do not attempt to change the setting on the vacuum regulator. If the vacuum level is not normal, look for the cause and correct it. This applies to systems in which the milk goes directly from the cow into the milk pipeline.

A vacuum gauge is used to show the vacuum level in the system. Locate the gauge near the sanitary trap. Several gauges may be installed in the system to give an indication of the operating vacuum levels throughout the system.

A sanitary trap is used to separate the air line from the milk line in the system. This keeps liquid from moving from one part of the system to the other. Slope the connecting pipe from the milk receiver jar downward toward the sanitary trap. This prevents a reverse flow of liquid from the trap to the milk receiver jar.

Milk Flow System

The milk flow system includes the sanitary milk line, milk inlet valves, filters, milk receiver jar (Figure 45-9) and releaser, and milk pump. The bulk tank is usually not considered to be a part of the milk flow system. When no receiver jar is used and a vacuum bulk tank is used, it is a part of the system.

The purpose of the milk flow system is to move the milk from the milking unit to the bulk tank. A stable vacuum needs to be maintained from the sanitary trap to each milking unit.

Stainless steel or glass is used for the sanitary milk line. Stainless steel will not break and can be welded in place. Glass allows the operator to watch the flow of milk through the system. Sanitary couplings are used to connect sections of line that are not welded.

The milk line should not be more than 6.5 feet above the cow platform. Low-level milk lines are installed at or below the level of the cow's udder. Low-level installations prevent vacuum fluctuation in the line.

Slope the milk line toward the receiver jar. Do not use any vertical risers in the line. Have as few sharp bends in the line as possible. Use the proper size of line for the number of units on the system.

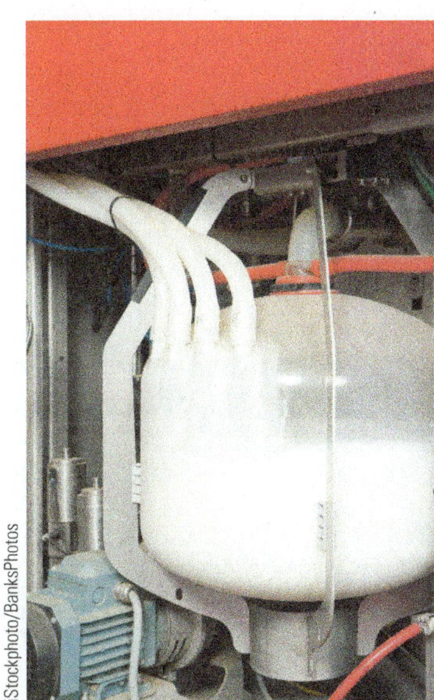

Figure 45-9 A receiver jar is used between the milking unit and the bulk tank.

Locate milk inlet valves in the top half of the milk line. This prevents a reverse flow of milk in the milk hose when it is attached to the milk line. It also helps maintain the vacuum between the milk line and the milking unit.

A milk filter is used to remove sediment from the milk before it goes into the bulk tank. Milk must be filtered before it is cooled. Do not depend on the milk filter to detect mastitis in individual cows.

There are four types of milk filters:

1. Unit filters are used in the suspension cup or the milk hose between the claw and the milk line. This type clogs easily and must be watched closely. A clogged filter reduces the vacuum and slows the milking.
2. In-line suction filters are commonly used with vacuum bulk tanks. They are placed in the line just ahead of the tank and must be watched closely to prevent clogging.
3. In-line pressure filters are used with a milk pump. They do not affect the vacuum in the milk line. If they become clogged, they may break and allow sediment to go through to the bulk tank. This is the most common type of filter used in closed milking systems that are under pressure.
4. Gravity filters are not under pressure. They do not affect the milking vacuum and are less likely to break down.

Milk filters may be made of woven or nonwoven material. Woven material can stand higher vacuum pressures. However, they do not filter as well as nonwoven material filters. Nonwoven material filters do a better job of taking small particles out of the milk.

The milk receiver–releaser collects the milk from the pipeline. It is made of either stainless steel or glass. Stainless steel will not break and gives more flexibility for placement of inlets. Stainless steel milk receivers are available in larger sizes than glass milk receivers. The glass receiver provides visibility of the milk flow.

The vacuum line and the milk line are connected to the milk receiver. The fittings are at or near the top of the receiver to maintain a stable vacuum. Electrodes are used to start and stop the milk pump.

The milk pump moves the milk from the milk receiver to the bulk tank. Most pumps operate under constant vacuum. Differential vacuum milk releasers are sometimes used. However, they create a demand for more vacuum in the system and give more trouble in operation.

Several methods are used to weigh the milk on test days. With modern computerized systems, most testing is done with the computer. Older systems include the use of milk meters, weigh jars, and buckets. Some milk meters are hard to clean in place. They also require additional vacuum to operate. Weigh jars slow the milking slightly. A regular milking machine bucket may be used to catch and weigh each cow's milk. They do not affect the vacuum in the system. However, the use of a bucket slows the milking operation.

HOUSING FOR DAIRY HERD REPLACEMENTS

Young calves up to 2 months of age may be housed in individual portable hutches (Figure 45-10) or in confinement calf barns. Either system works well with good management. Keep any type of housing clean, dry, well ventilated, and free from direct drafts.

Calf hutches are used successfully even in areas with cold winters. Hutches vary in size. A typical size is 4 feet $\times$ 8 feet for the shelter. Other sizes may be used depending on individual preference. The hutch is usually about 4 feet high. An exercise area of similar size is provided in front of the shelter. Exterior grade plywood

Figure 45-10 Young calves up to 2 months of age may be housed in individual portable hutches.

may be used to construct the hutch. Burlap, canvas, or a plastic drop cloth may be used to close the hutch at night in severe weather.

The hutch may be bedded with wood shavings, sawdust, or straw. A manure pack will add warmth in the winter. The hutch is easily moved for cleaning. In wet areas, use a raised board floor to keep bedding dry.

Place the calf in the hutch as soon as the calf is dry after it is born. There is less shock to the calf than if it were to become used to warm conditions before being moved to the hutch.

Advantages of calf hutches include the following:

- low cost
- easy to design and build
- easy to clean and disinfect
- fewer disease problems
- easy to move from contaminated areas
- gets the calf used to existing weather conditions, therefore there is less shock when it is moved to minimal housing facility
- natural ventilation
- calves grow as well in hutches as those housed in permanent barns

Disadvantages of calf hutches include the following:

- takes more labor to feed and care for calves
- takes more bedding
- need another building to store feed
- more discomfort for person taking care of calves during bad weather

Confinement calf barns may be cold or warm. A cold calf barn is not heated or insulated and uses natural ventilation. Warm housing is heated, insulated, and uses a forced air ventilation system. Cold housing systems cost less than warm housing. Warm housing requires less space per calf, is easier to mechanize, is less likely to have its water freeze, is easier to use for treating sick animals, and provides greater operator comfort.

Use individual pens for young calves. This keeps the calves from sucking on each other after feeding. When the calves are weaned, they may be kept in group pens (Figure 45-11).

Figure 45-11 When calves are weaned, they may be kept in group pens.

Elevated pens or floor-level pens may be used for individual calves. Elevated pens do not require bedding but are more uncomfortable for the calves in cold weather. Individual elevated pens may be 2 × 4 feet. Floor-level pens may be 4 × 4 feet in warm housing and 4 × 6 feet in cold housing.

In group pens, provide 20 to 25 square feet of floor area for each calf up to 8 months of age. Provide 25 square feet of bedded area per calf in open front loose housing. Do not put more than five to seven calves in a group pen.

Pole barn construction may be used for calf housing. Use concrete floors with gutters along the rear of the pens. Slope the floor toward the gutter to make manure handling easier.

A typical calf barn might have a row of individual pens along each side of a feed alley. Provide side alleys for moving calves in and out of the pens.

It is convenient to have a feed and supply storage room in one end of the calf barn. Include a hot water heater and a sink in the storage room.

Cold calf housing may be as simple as a row of pens open to the south. Each pen may be 4 × 8 feet. Provide an 8-foot exercise area in front of each pen. A service alley with hay and bedding storage is provided on the interior back of the building.

Include a feed manger and watering facility for each pen, regardless of the type of housing used.

Heifers from 2 Months to Calving

Cold housing may be used for heifers after they are weaned. Heifers may be grouped in pens, or free stall housing may be used. During warm weather, heifers may be kept on pasture. Provide shade and shelter for the animals on pasture.

Heifers in confinement housing should be grouped in pens by age and size. For heifers up to 6 months of age, provide 25 square feet per animal. From 6 months to 9 months of age, provide 30 square feet per animal. From 9 months to 15 months of age, provide 35 square feet per animal. From 15 months to calving provide 40 square feet per animal.

Solid floors may be used in pens or free stalls. A manure pack will provide added heat in cold weather. However, the manure pack system uses a lot of bedding. Less bedding is needed in free stalls, but almost daily removal of the manure is required.

Slat floors may be used in free stall housing. Bedding needs are less and daily manure handling is eliminated.

Several stanchions should be included for holding heifers for artificial insemination. A head gate and squeeze chute will hold animals for treatment when required.

Do not feed and water animals in the resting area. Provide an outside lot for this purpose. Put pavement around feed bunks, waterers, and along the building. This will keep the animals out of the mud and make it easier to clean the area.

Cold housing is generally open on one side. Face the open side to the south or east. Make the building 30 feet deep in moderate climates and 40 feet deep in cold climates. Allow enough height for a 3- to 4 -foot manure pack. Use a door opening height of at least 10 feet. The floor of the building can be packed earth. Slope it toward the open side. Keep the floor about 1 foot above the outside grade to prevent water and manure from running into the resting area.

Do not use a manure pack in the summer. If the building is used in the summer, a paved floor is recommended. Dry cows may be kept in open housing similar to that described for heifers. It is generally not good management to run the dry cows with the milking herd. Feed needs are different for dry cows.

MANURE HANDLING

Manure must be handled properly to maintain clean conditions for milk production. The best use for manure is as a fertilizer for crop production. The manure handling system must meet local health and milk market requirements. Handle manure in such a way that odors, runoff, seepage, and insects are controlled.

Manure may be handled in either a solid or liquid form. Liquid manure usually contains less than 15 percent solid material. A solid material content greater than 20 percent requires that the manure be handled as a solid.

Manure may be hauled every day as a solid and spread on crop ground. Another method is to haul the manure from storage as either a solid or liquid and spread it on crop ground. This is done when weather conditions are right, the ground is available, and the labor is available.

Manure mixed with bedding is handled as a solid. If the moisture content drains away or evaporates, the manure is also handled as a solid.

Barn cleaners located in the gutter are used in stall barns. A tractor scraper and a front-end loader may be used in free stall housing. A scrape-off ramp located at one end of the barn will permit the direct loading of the manure into the spreader. Numerous methods of removing manure from barns are used (Figure 45-12).

The manure may be hauled daily or stored for later hauling. Daily hauling is not always convenient in bad weather or when cropland is not available on which to spread the manure (Figure 45-13). The cost is relatively low and odor problems are reduced with daily hauling.

Some type of stacking equipment is needed for storage of solid manure for a long period of time. Provide enough room for storage for up to 6 months. A manure loader on a tractor is needed to load the manure for hauling to the field.

Liquid manure may be stored in several ways, including:

- under the barn with slat floors
- a below-ground storage tank outside the barn
- an outside basin with sloped earthen sides
- above-ground silo storage

The size of the storage area depends on the number of cows in the herd and the length of time the manure is to be held. It is better to plan for extra capacity than to have too little storage volume. For example, Holsteins need about 2 cubic feet per cow per day.

Manure may be moved to the storage area by pumps or by scraping. In slotted-floor facilities with a tank below, it drops directly into the storage area.

Automatic scrapers may be used. Tractors with scraper blades are sometimes used. The manure may be scraped through grates directly into a storage tank or into a holding tank to be pumped into the storage area.

Several types of pumps are used to move manure. Centrifugal pumps have a high capacity. Piston, auger, and diaphragm pumps have lower capacity. Large, hollow piston pumps work well with manure that has a low solid content.

The liquid manure in storage must be agitated just before unloading the storage tank. Centrifugal pumps with a bypass may be used. Recirculating pumps, paddle wheels, and inclined augers are also used for agitation in the storage tank.

Liquid manure is hauled to the field in a tank holding up to 5,000 gallons. Tanks may be loaded by gravity or with pumps. High-capacity centrifugal or vacuum pumps are used. The tank wagon is unloaded on the field by pressure or gravity.

Liquid runoff from the barnyard must be controlled to prevent contamination of rivers, lakes, or ponds. Check state Environmental Pollution Agency regulations for control of barnyard runoff.

Courtesy of Frank B. Flanders

Figure 45-12 Gray water (wastewater from milking, cleaning, etc.) is sometimes collected and used to flush free stall barns. Note the metal gate on this collection tank.

Courtesy of Frank B. Flenders

Figure 45-13 Manure spreader.

CAUTION

Toxic gases are produced by liquid manure in storage tanks, especially during agitation. Make sure there is enough ventilation, especially in storage areas under the barn.

FEED HANDLING

Many of the same kinds of feeding facilities are used for dairy cattle as are used for beef cattle. See the chapter on beef cattle housing and equipment for a discussion of silos, feed bunks, waterers, and other equipment.

SUMMARY

A high investment is required for buildings and equipment on the modern dairy farm. Dairy facilities must be planned for efficiency, economy, and convenience.

Two common types of housing used for the milking herd are stall (stanchion) barns and free stall barns. Large herds are more commonly housed in free stall barns.

Pail milkers, suspension milkers, and pipeline milkers are used in stall barns. Milking parlors with pipeline milkers are used with free stall housing. They may also be used for stall barn housing.

The most common type of milking parlor is the herringbone. Other types are side-opening, rotary or carousel, and polygon.

The use of crowding gates, power gates, feed gates, stimulating wash sprays, and automatic detaching units increase mechanization in the milking parlor.

Cows are held in a paved holding area while they are waiting to enter the milking parlor.

A milk house contains the equipment for filtering, cooling, and storing the milk. The milk house must be heated, insulated, ventilated, and kept clean.

Milk is stored in bulk tanks. These are large refrigeration units. Dairying requires the use of large amounts of energy. Careful management can save 10 to 20 percent of the energy normally used on a dairy farm.

A milking system is made up of the milking unit, the pulsation system, the vacuum supply system, and the milk flow system. The milking unit is attached to the udder of the cow. The pulsation system controls the action of the milking unit. The vacuum supply system produces and distributes the vacuum necessary to operate the milking unit. The milk flow system moves the milk from the cow to the bulk tank.

Young calves may be housed in individual portable hutches or confinement barns. Either system works well with good management. Housing must be clean, dry, well ventilated, and free from drafts.

Older heifers may be grouped in pens or kept in free stall barns. Divide animals by age and size.

Manure may be handled as a solid or a liquid. Manure handling on the modern dairy farm is highly mechanized.

Student Learning Activities

1. Take field trips to local dairy farms in the area. Observe and report on the kinds of facilities used.
2. Prepare visuals on housing facilities and/or feed-handling facilities for a dairy farm and present an oral report to the class.
3. Prepare a bulletin board display of pictures of dairy housing and equipment.
4. Use the information in this chapter to ensure adequate housing and equipment when planning and conducting a dairy supervised experience program.

Discussion Questions

1. List the factors that should be considered when planning facilities for dairy cattle.
2. What are the two most common types of housing systems used for the milking herd?
3. Briefly describe pail and suspension milkers.
4. What are the advantages and disadvantages of milking parlors?
5. Name the four types of milking parlors.
6. Briefly describe each type of milking parlor.
7. Describe some ways that mechanization in the milking parlor can be increased.
8. Describe the holding area needed for cows waiting to enter the milking parlor.
9. List the characteristics of a good milk house.
10. Describe the types of bulk tanks used to cool milk.
11. How much energy does it take to operate the average dairy farm in the United States?
12. Describe the parts of the milking unit.
13. Describe how the pulsation system works.
14. What are the functions of the vacuum supply system?
15. Describe the kinds of pumps used in the vacuum supply system.
16. List and briefly describe the other parts of the vacuum supply system.
17. Briefly describe the parts of the milk flow system.
18. Compare the use of individual hutches with confinement barns for housing dairy herd replacements.
19. Describe housing requirements for older heifers.
20. Compare handling manure as a solid and as a liquid.

Review Questions

True/False

1. The size of the milk house depends on the size of the herd and the type of equipment used.
2. A standby generator is not essential on a dairy farm because the work can be done manually.
3. Stall barns require less labor to fasten and release cows than other types of barns.

Multiple Choice

4. Stall barns are more commonly used for herd sizes up to ______ cows.

   a. 120 c. 90
   b. 100 d. 80

5. A barn width of ______ feet is recommended for larger breeds.

   a. 36 c. 25
   b. 30 d. 20

Completion

6. A ______ ______ is used to remove sediment from the milk before it goes into the bulk tank.
7. The ______ ______ moves the milk from the milk receiver to the bulk tank.

8. The ______ ______ ______ includes the sanitary milk lines, milk inlet valves, filters, milk receiver jar and releaser, and milk pump.
9. Types of vacuum pumps used in milking systems include ______, ______, and ______.
10. ______ milking parlors are arranged so that cows enter onto a turning platform that rotates slowly.
11. The most common milking parlor in current use is ______.
12. The two most common types of housing systems are ______ and ______.
13. The modern dairy farm requires high investment in ______ and ______.

Short Answer

14. A bedding chopper is a machine powered by what?
15. How is water removed from the milk house?

Chapter 46

Marketing Milk

Key Terms

Grade A milk
Grade B milk
Agricultural Marketing Service (AMS)
federal milk marketing order (FMMO)
Commodity Credit Corporation (CCC)
Class I milk
Class II milk
Class III milk
Class IV milk
National Agricultural Statistics Service (NASS)

Objectives

After studying this chapter, the student should be able to

- describe price, supply, and demand trends for milk and dairy products.
- discuss markets for milk.
- describe the grades of milk.
- explain the pricing of milk.

INTRODUCTION

The dairy farmer produces milk to sell at a profit. Careful management can help reduce the costs of producing the milk. A knowledge of the marketing structure for milk can help the dairy farmer make wise management decisions. Management decisions are influenced by the following factors:

- price, supply, and demand trends for milk and dairy products
- markets available for milk
- pricing structure and regulation of milk marketing

PRICE OF MILK

Milk-to-Feed Price Ratio

The average annual price received by farmers for all milk marketed in the United States fluctuates by month and year. Figure 46-1 shows the typical variation in milk prices by month. The milk-to-feed price ratio is derived by dividing the price of 100 pounds of milk by the price of 100 pounds of a 16 percent complete dairy feed. This ratio shows a fairly wide variance by month and year (Figure 46-2). High milk prices and lower feed costs

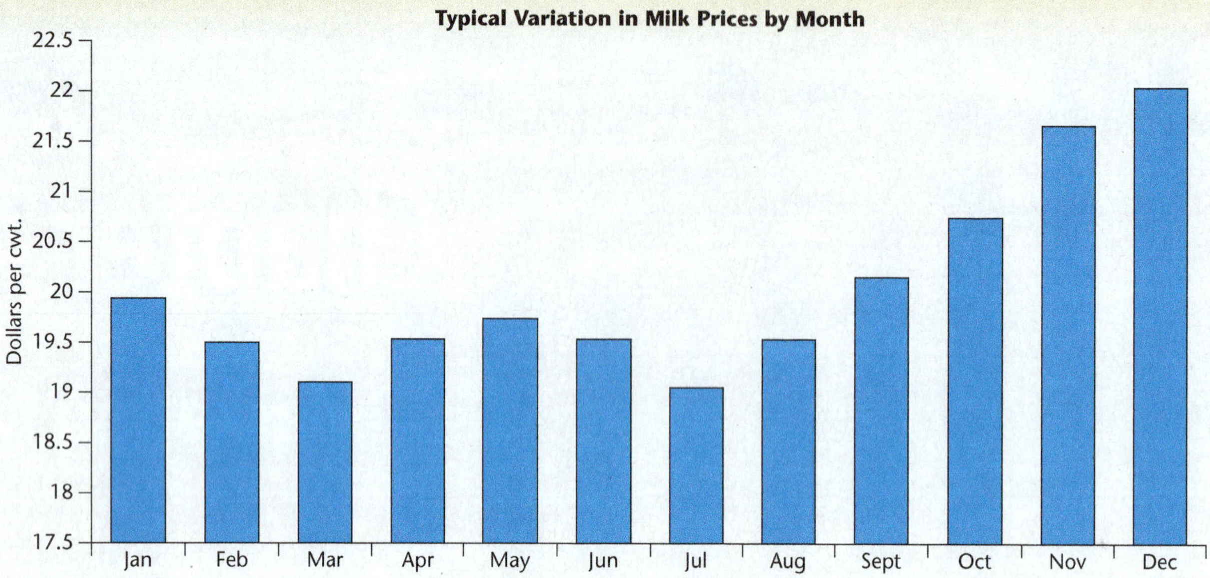

Figure 46-1 The prices the producer receives for milk vary widely by month and year.
Source: USDA, NASS

Milk-Feed Price Ratio

Market Year	Sep	Oct	Nov	Dec	Jan	Feb	Mar	Apr	May	Jun	Jul	Aug	Avg
2008/09	1.90	2.02	2.01	1.92	1.60	1.51	1.55	1.58	1.49	1.47	1.59	1.80	1.70
2009/10	2.00	2.11	2.26	2.42	2.33	2.34	2.18	2.19	2.18	2.26	2.30	2.36	2.24
2010/11	2.36	2.40	2.23	1.98	1.96	2.01	2.12	1.81	1.71	1.86	1.88	1.83	2.01
2011/12	1.84	1.82	1.89	1.81	1.72	1.56	1.48	1.41	1.34	1.38	1.34	1.36	1.58
2012/13	1.58	1.73	1.73	1.65	1.58	1.52	1.48	1.54	1.53	1.52	1.53	1.68	1.59
2013/14	1.88	2.13	2.27	2.3	2.46	2.6	2.56	2.43	2.24	2.2	2.36	2.61	2.33

Figure 46-2 Milk-feed price ratio.

Note: National average prices are used in the calculation. Prices of commercial feeds are based on current U.S. prices received for corn, soybeans, alfalfa hay, and all wheat.

Source: USDA, NASS, Agricultural Prices

result in a higher milk-to-feed price ratio. The higher the milk-to-feed price ratio, the higher the potential a producer has for making a profit. However, many other factors affect the profitability of a dairy.

Seasonal Trends

The amount of milk produced in the United States varies by season (Figure 46-3). The seasonal variation in production is less now than it used to be. Currently, national average production is highest in the spring and lowest in the winter.

The price for milk also varies seasonally (Figure 46-4). Highest prices are generally paid in October, November, and December.

DEMAND FOR DAIRY PRODUCTS

Products Made from Milk

The three major uses of milk are as fluid product, cheese, and butter. These three types of product use about 95 percent of the milk processed in the United States.

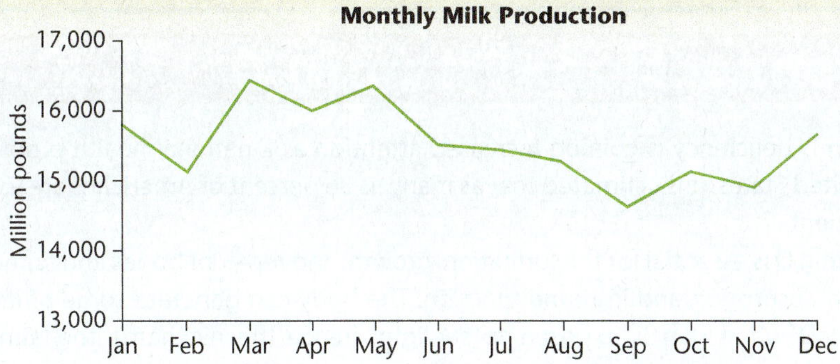

Figure 46-3 U.S. Monthly Milk Production.

Source: USDA, NASS

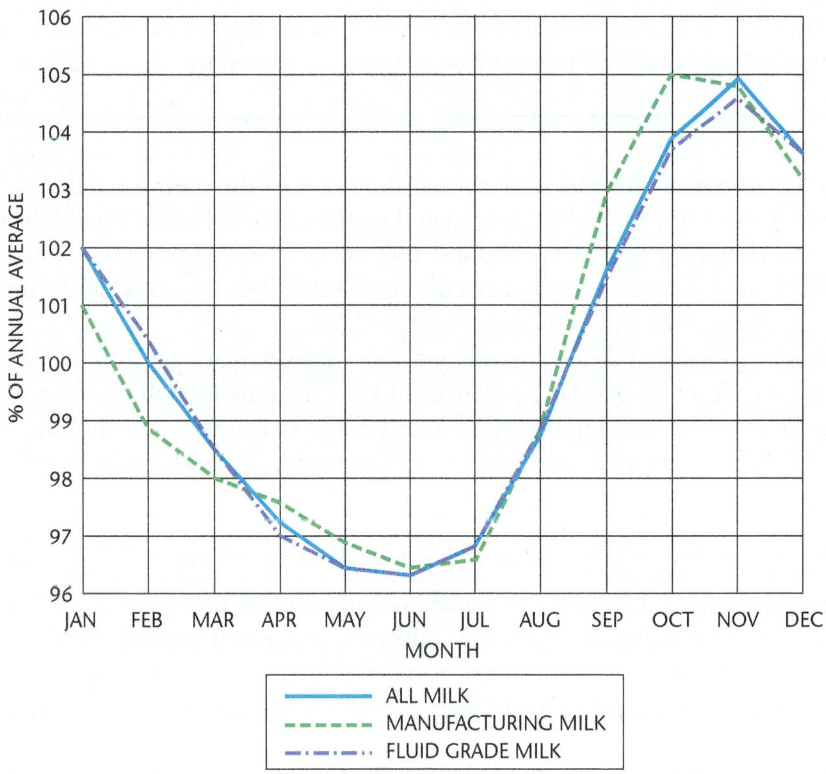

Figure 46-4 Monthly variation in pricing received by farmers for all milk, manufacturing milk, and fluid grade milk as a percent of annual average.

Source: Annual Price Summary, USDA

Fluid milk products are mainly whole milk (3.25 percent milkfat), low fat (from 0.5 to 2.0 percent milkfat), and skim milk (less than 0.5 milkfat). Coffee cream, whipping cream, half and half, and sour cream are also fluid milk products. About 25 percent of the milk produced in the United States is used for fluid milk products.

Many types of cheeses are produced from milk. About 50 percent of the milk produced in the United States is used for making cheese. About 67 percent of the cheese made in the United States is American-type, mainly cheddar and colby. The main Italian-type cheese made is mozzarella for pizza. Italian-type cheese makes up about 25 percent of the total cheese produced in the United States. Swiss cheese accounts for about 6 percent of the total cheese produced. Other types of cheese make up the rest of the cheese produced in the United States.

Did You Know?

Twenty-five gallons (215 pounds) of milk can make 18 gallons of ice cream, 21 pounds of cheese, or 10 pounds of butter.

Connection: Vitamin D

Vitamin D deficiency is gaining increased attention as a national health problem in the United States. It is estimated that as many as 36 percent of Americans are vitamin D deficient.

Vitamin D is essential for the formation, growth, and repair of bones and for normal calcium absorption and immune function. The body can generate some of its own vitamin D when skin is exposed to sunlight, hence the nickname the "sunshine vitamin." It can also be obtained from some foods and dietary supplements.

Vitamin D is important to help reduce risk of diseases such as osteoporosis, high blood pressure, some cancers, diabetes, and other disease. Deficiency in vitamin D can cause rickets in children, which leads to deformities.

Only a few foods are naturally high in vitamin D. Many food products, such as milk, are fortified with vitamin D to meet the recommended amounts.

The production of butter uses about 20 percent of the milk produced in the United States. Only the milkfat is used in the production of butter. The solids-not-fat (SNF) is used to produce nonfat dry milk and condensed skim milk. More butter and nonfat dry milk are produced during the months when milk production is highest. The surplus milk is used for these products.

About 5 percent of the milk produced in the United States goes for making ice cream, yogurt, and other dairy products. Most of the milk used in this category is used for ice cream and yogurt. Other dairy products include evaporated and condensed milk, evaporated and condensed buttermilk, dry buttermilk, dry whole milk, dry skim milk, dry cream, dry whey, and lactose.

Trends in the Consumption of Dairy Products

Figure 46-5 shows the change in per capita consumption of yogurt in the United States. Cheese and yogurt are the only milk products with increased consumption from 1970 to 2009. The consumption of other products—including fluid milk and cream, butter, ice cream, and evaporated and condensed milk—has decreased since 1970.

Increased competition from imitation and substitute dairy products has caused a lower demand for real dairy products. Some of these products are margarine, imitation milk and cheese, non-dairy creamers, and non-dairy whipped products. Many of these products are sold at lower prices than real dairy products.

Did You Know?
Ninety-nine percent of all U.S. households purchase milk. The average American consumes almost 25 gallons of milk a year—that's 400 glasses of milk!

Research and development activities are needed to improve dairy products and develop new ones. The National Dairy Council and the United Dairy Industry Association have research and development programs for dairy products. These efforts must be maintained and expanded if recent trends toward a decrease in demand for dairy products are to be reversed.

Projections of demand for dairy products show an expected increase in use of fluid low-fat milk, fresh cream, cheese, ice cream, and butter. The trend toward lower demand for fluid whole milk, cottage cheese, and ice milk is expected to continue.

The demand for dairy products is affected by population, price of dairy products, price of competing products, purchasing power of consumers, and promotional advertising for dairy products.

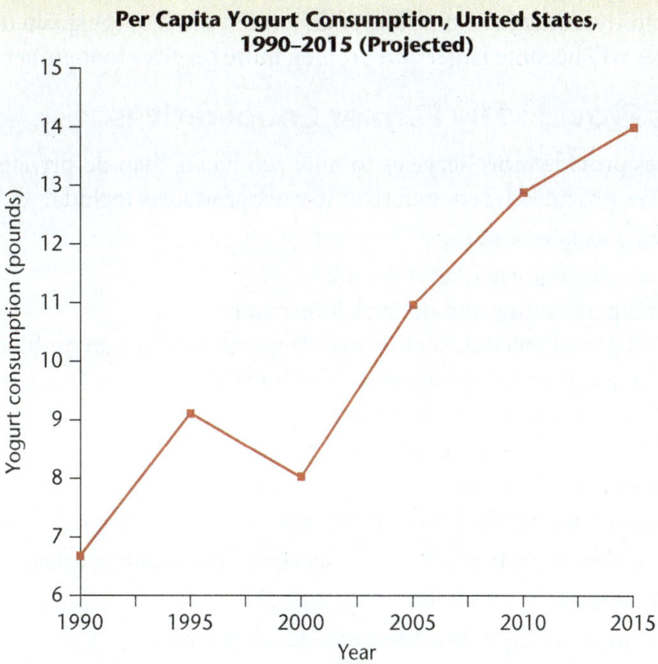

Figure 46-5 Per capita yogurt consumption in the United States.
Source: USDA/Economic Research Service

Advertising Dairy Products

Before 1983, the generic or non-brand-specific promotion of dairy products was funded primarily by voluntary contributions by dairy farmers through local and state dairy organizations. Congress established a mechanism for a dairy checkoff program with the passage of the Dairy and Tobacco Adjustment Act of 1983. This established the National Dairy Promotion and Research Board for the purpose of developing and administering promotion, research, and nutrition programs for the dairy industry.

Dairy farmers approved the continuation of the dairy checkoff program through a referendum held in 1993. In 1995, an entity called Dairy Management Inc. (DMI) was organized to coordinate local and national dairy promotion programs. This organization established the U.S. Dairy Export Council to help promote and market U.S. dairy products in international markets.

The promotion of fluid milk and cheese receives the highest priority for the use of the checkoff money. It is estimated that about 200 billion more pounds of milk have been sold than were projected by the USDA since the checkoff program began in 1984. Most of the surplus of dairy products that existed in 1984 has been sold, mainly because of promotion activities funded by the dairy checkoff. Promotion programs are funded by dairy farmers and do not use tax money. The USDA does exercise an oversight function in relation to all producer checkoff programs.

MARKETS FOR MILK

More than $21 billion worth of milk is produced each year on dairy farms in the United States. About 86 percent of this is sold through farmer milk marketing cooperatives. The rest is sold to private firms, used on the farm, and sold directly to consumers.

Looking to the future, the number of dairy cooperatives is expected to decline. Cooperatives will become larger and provide more services to members.

Services Provided by Farmer Cooperatives

Cooperatives provide more services to milk producers than do private firms. The major services provided by cooperatives to milk producers include:

- checking weights and tests
- guaranteeing daily markets for milk
- providing marketing and outlook information
- providing field services, such as assisting with production problems
- collecting and ensuring payment from buyers
- assisting with inspection problems
- providing insurance programs
- negotiating hauling rates
- selling milking supplies and equipment

Cooperatives also provide marketwide services. The major marketwide services provided by cooperatives include:

- maintaining quality control and related lab services
- direct farm-to-market movement of milk
- handling milk in excess of Class I use
- participating in federal order hearings
- paying milk haulers
- negotiating Class I prices and service charges
- maintaining a full supply of milk
- balancing milk supplies among processors to reduce reserve requirements
- making out-of-market raw milk sales

HAULING MILK TO MARKET

Most of the milk marketed in the United States is hauled from the farm to the plant in bulk trucks (Figure 46-6). Fewer dairy farms, larger herds, and fewer dairy plants have made it necessary to haul milk farther. Larger bulk trucks are being used now. Fuel and labor costs have increased sharply in recent years.

All of these factors have combined to increase the costs of hauling milk from the farm to the dairy plant.

A charge per hundredweight for hauling is usually deducted from the price the farmer receives for milk. Several studies have shown that the amount deducted does not cover the total cost of hauling. The amount not covered is absorbed by the plant buying the milk. This has the effect of lowering the price the plant is able to pay for milk.

Because of competition for milk suppliers, some plants use various methods of calculating the charge for hauling. In some cases, the cost is based partially on the volume of milk picked up at the farm. In other cases, a per stop charge is made in addition to the charge per hundredweight. It is expected that milk hauling charges will continue to increase in the future.

USDA

Figure 46-6 Bulk milk must be picked up from the farm and delivered to the processing plant on a regular basis.

MILK GRADES

In 1924, the U.S. Public Health Service developed the Standard Milk Ordinance to help state and local governments prevent diseases that are spread through milk. The adoption of this ordinance is voluntary; however, it is widely used as the model for regulating the production and processing of **Grade A milk** or fluid milk. The ordinance has been revised many times since 1924 and is now titled the Grade "A" Pasteurized Milk Ordinance. This ordinance is recognized by public health agencies and the milk industry as the national standard for milk sanitation. A copy of the complete ordinance may be obtained from the Public Health Service/Food and Drug Administration or may be downloaded from the Internet off of the FDA website.

The ordinance sets forth standards for cleanliness of facilities, temperature for storing milk, bacterial count, somatic cell count, and chemical residue in milk, as well as other factors relating to the production and processing of milk for human consumption (Figure 46-7). Grade A raw milk must be cooled to 45°F or less within 2 hours after milking. The blend temperature after the first and subsequent milkings cannot exceed 50°F. Each farmer's milk cannot exceed 100,000 bacterial count per milliliter before it is mixed with other producers' milk. The bacterial count of milk from several producers mixed together cannot exceed 300,000 bacterial count per milliliter before it is pasteurized. Each farmer's milk cannot exceed 750,000 somatic cells per milliliter. There can be no detectable antibiotics in the milk.

Courtesy of Frank B. Flanders

Figure 46-7 The Grade "A" Pasteurized Milk Ordinance sets standards for cleanliness, temperature storage, bacterial count, somatic cell count, chemical residue in milk, and other factors relating to milk for human consumption.

Inspections of dairy farms are made regularly to make sure there are no violations of the standards. The permit to sell Grade A milk may be suspended if violations are found that are not corrected.

Grade B milk is produced under standards that allow it to be used for manufacturing dairy products but not to be used for fluid milk consumption. The requirements for producing Grade B milk are not as strict as those for Grade A milk. However, the standards became stricter beginning July 1, 1980, for dairy plants that are inspected and approved by the USDA.

The standards are found in "Milk for Manufacturing Purposes and Its Production and Processing, Recommended Requirements," which became effective November 12, 1996. A complete copy of the standards may be obtained from the USDA's **Agricultural Marketing Service (AMS)** Dairy Programs or downloaded from the Internet. Many states have adopted these standards as the standard for Grade B milk sold to plants within the state. The adoption of the standard is voluntary. However, plants that do not meet the standards cannot sell products to the Commodity Credit Corporation under milk price support programs.

The Grade B milk standard addresses many of the same issues as the Grade A milk standard. If the bacterial count exceeds 500,000 per milliliter, the producer is given a warning. If two of the last four consecutive bacterial counts exceed 500,000 per milliliter, a written warning is given to the producer and the regulatory authority is notified. Another sample must be taken within 21 days after the written notice is issued. If the bacteria count still exceeds 500,000 per milliliter, the milk is not permitted on the market. The producer can be given a temporary permit to ship milk if a subsequent test shows a bacteria count below 500,000 per milliliter. Full reinstatement comes only after three out of four consecutive tests show bacteria counts below 500,000 per milliliter.

A somatic cell count above 750,000 per milliliter is considered excessive. The same testing procedures and time limits as described for bacteria counts also apply to somatic cell counts. No drug residues are permitted in the milk.

FEDERAL MILK MARKETING ORDER PROGRAM

The **federal milk marketing order (FMMO)** program began with the passage of the Agricultural Marketing Agreement Act of 1937. Since then, the U.S. Congress has passed a series of laws establishing price support programs for dairy products.

These price support programs have been implemented through the purchase of surplus butter, cheese, and nonfat dry milk products by the **Commodity Credit Corporation (CCC)**. The CCC has not removed much product from the market in recent years. It is a part of the industry, but does not affect the industry like it did in the past. Table 46-1 shows measures of growth in federal milk orders.

The Federal Agriculture Improvement and Reform Act of 1996 required the consolidation of the 31 federal milk marketing orders into no fewer than 10 and no more than 14 orders. This act also eliminated the dairy price support operations of the CCC after December 31, 1999. It permitted the Secretary of Agriculture to review and revise related issues such as the method for determining the price of milk in the market orders.

Federal milk marketing orders are generally established at the request of the producers or through producer organizations. Two-thirds of the producers affected by the order must approve it by referendum. The order regulates the first buyers of milk and not the producers. The purposes of a federal order market are to:

- establish and maintain orderly market conditions
- establish prices that are reasonable
- ensure a sufficient quantity of pure and wholesome milk
- provide safeguards to protect producers from unfair and abusive trade practices

A milk order may be suspended or terminated by the Secretary of Agriculture. It may also be terminated at the request of more than 50 percent of the producers who supply more than one-half of the milk in the order. Federal milk market orders are administered by the USDA through a local administrator appointed by the Secretary of Agriculture. Federal milk marketing order areas are shown in Figure 46-8.

In federal order markets, Grade A milk is divided into classes, based on the final use of the milk, for the purpose of pricing. The federal order price is a minimum price; market conditions may cause the price paid for milk to be above the federal order price.

TABLE 46-1 Measures of Growth in Federal Milk Orders

Year	Number of Markets	Number of Pooled Producers	Total Milk Produced (million pounds)	Producer Milk Used in Class I (million pounds)	Percent Used as Class I (%)
1950	39	156,584	18,660	11,000	58.9
1955	63	188,611	28,948	18,032	62.3
1960	80	189,816	44,812	28,758	64.2
1965	73	158,077	54,444	34,561	63.5
1970	62	143,411	65,104	40,063	61.5
1975	56	123,855	69,249	40,106	57.9
1980	47	117,490	83,998	41,034	48.9
1985	44	116,765	97,762	42,201	43.2
1990	42	100,397	102,396	43,783	42.8
1995	33	88,717	108,548	45,044	41.5
2000	11	69,590	116,920	45,989	39.3
2005	10	53,036	114,682	44,570	38.9
2010	10	45,918	126,909	44,970	35.4
2012	10	40,750	122,388	43,492	35.5

Source: USDA, Agricultural Marketing Service

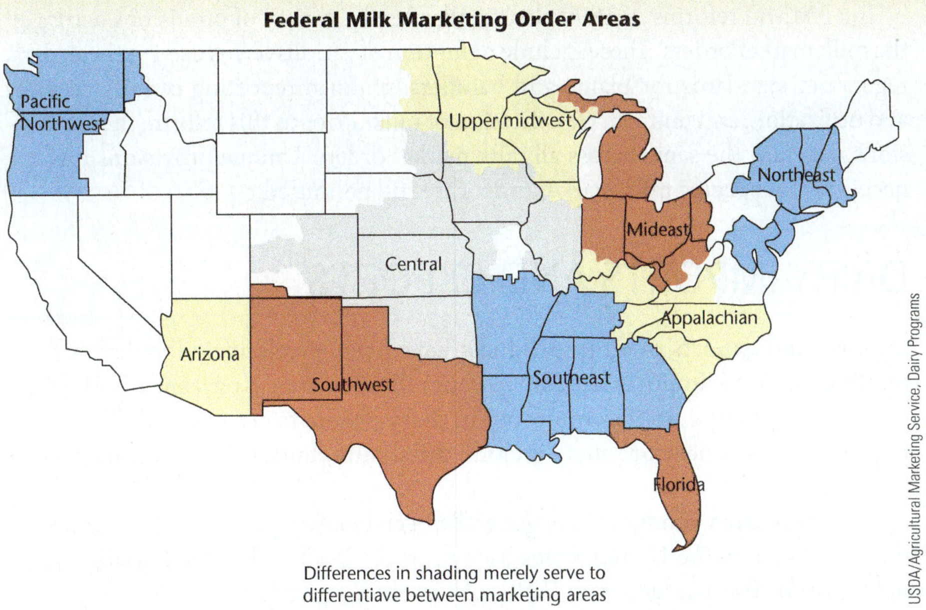

Figure 46-8 Federal milk marketing order areas.

More than 95 percent of the milk sold to plants and dealers is Grade A. **Class I milk** is used for whole milk, skim milk, buttermilk, eggnog, and flavored milk drinks. **Class II milk** is used for ice cream, yogurt, cottage cheese, and cream. **Class III milk** is used for most cheeses. **Class IV milk** is used for butter and any milk product in dried form. Class I milk has the highest value and Class IV milk the lowest.

Prior to the reform of the federal milk market order system in 1999, the price of milk was determined by a basic formula price (BFP). This price was based on the price of Grade B milk in selected areas of Minnesota and Wisconsin. When a significant volume of milk sold was Grade B, it was believed that this represented the true competitive market value of milk. Grade A milk prices were then determined by setting differentials based on the class of milk. Because less than 5 percent of the milk currently sold to plants and dealers is Grade B, the price paid probably no longer represents a true competitive market value for milk. The FMMO reform of 1999 addressed this problem by establishing a different method for determining the price of Grade A milk.

This system of pricing uses multiple-component pricing, with differentials for various classes of milk. The base price for Class III and Class IV milk is determined by the value of protein, butterfat, and other nonfat solids used in manufactured dairy products. Monthly average commodity prices are used to determine the component values. Commodity prices tracked include block cheese, barrel cheese, butter, dry whey, and nonfat dry milk. Estimated manufacturing costs are deducted from the commodity price in order to arrive at a price based only on the component values. The USDA's **National Agricultural Statistics Service (NASS)** surveys and reports on these prices. Class I milk in a given milk market order is then priced based on a differential for that market order. The Class II price of milk is set at about 70 cents over the price of Class IV milk in all market orders.

Each producer in a milk market order receives a blend price for milk. This is a weighted average price determined from proceeds received for all the milk sold in all the classes in the milk market order. By using a blend price, all producers share both the higher price for Class I milk and the lower prices paid for milk used for manufactured products.

Figure 46-9 Imports and exports of dairy products have been generally less than 2 percent of the total U.S. production.

The FMMO reforms of 1999 also standardized a variety of provisions across all the milk market orders. These include definition of various terms used in the orders, regulatory standards for plants and handlers, uniform reporting of milk receipts and utilization, and uniform classification of milk. Prior to this reform, these provisions were not the same across all milk market orders. Unique provisions that are necessary for specific milk market orders are still permitted.

DAIRY IMPORTS AND EXPORTS

Imports and exports of dairy products have been at relatively low levels for a number of years. Imports have been generally less than 2 percent of total U.S. production (Figure 46-9). A number of dairy products are covered by specific import quotas. These products include dried milk, butter, and several types of cheeses.

Exports of dairy products have generally been below 1 percent of total U.S. production. Prices in the United States have generally been higher than in other parts of the world. This has tended to limit exports of dairy products.

SUMMARY

The average annual price received by farmers for all milk marketed in the United States fluctuates by month and year. Milk production and prices vary seasonally.

About 95 percent of the milk processed in the United States is used for fluid product, cheese, and butter. Per capita consumption of cheese and yogurt has increased from 1970 to 2009. The consumption of other products—including fluid milk and cream, butter, ice cream, and evaporated and condensed milk—has decreased since 1970.

About 86 percent of the milk produced in the United States is marketed through cooperatives. Cooperatives provide many services to milk producers as well as to the entire market.

Most milk is hauled to market in bulk trucks. The cost of milk hauling has increased a great deal in recent years. A charge for hauling milk is deducted from the dairy farmer's milk check.

Milk is divided into classes based on use for pricing purposes. Class I milk has the highest price.

Federal milk marketing orders are established in many areas. Minimum prices for Grade A milk are set in these areas. The actual price paid is often above the minimum set in the federal order market.

Dairy imports and exports have generally been at a low level for a number of years.

Student Learning Activities

1. Prepare a chart that shows the price and production trends for milk.
2. Prepare and give an oral report to the class on any aspect of milk marketing.
3. Visit a local milk marketing cooperative and prepare a report for the class.
4. Have a representative from a milk marketing cooperative speak to the class.
5. When planning and conducting a dairy supervised experience program, use the information in this chapter to become more efficient in marketing your milk production.

Discussion Questions

1. Describe the long-term trends in the price of milk.
2. Briefly explain the seasonal trends in the price of milk.
3. What kinds of products are made from milk?
4. Describe trends in the consumption of dairy products.
5. What methods are being used to try to increase the demand for dairy products?
6. How is most of the milk in the United States marketed?
7. List the marketwide services provided by marketing cooperatives.
8. How is most of the milk in the United States hauled to market?
9. How is the hauling paid for?
10. What is the difference between Grade A and Grade B milk?
11. Describe the classes of milk.
12. Briefly explain federal milk marketing orders.
13. How is milk priced in federal milk marketing orders?

Review Questions

True/False

1. The amount of milk produced and the price for milk received varies seasonally.
2. The trend toward a lower demand for fluid whole milk, cottage cheese, and ice milk is expected to continue.
3. Increased competition from imitation and substitute dairy products has caused a lower demand for real dairy products.
4. The production of butter uses about 10 percent of the milk produced in the United States.
5. Grade B milk is produced under standards that allow it to be used for drinking.

Multiple Choice

6. Which of the following is not considered one of the four major uses of milk?

   a. fluid product　　c. yogurt
   b. butter　　d. cheese

7. What is Grade A milk in Class III used for?

   a. cottage cheese　　c. yogurt
   b. butter　　d. ice cream

Completion

8. The highest milk production occurs during the __________ months.
9. Dairy farmers in the United States produce more than ______ billion dollars worth of milk each year.

Short Answer

10. How has hauling milk to market been affected by changes in dairy farming?
11. List three standards set forth by pasteurized milk ordinances.
12. Who can suspend or terminate a milk order?
13. List two services provided by farmer milk cooperatives.
14. What has been the limiting factor in milk exports?

SECTION 10

Alternative Animals

Chapter 47

Rabbits

Key Terms

herbivorous
coprophagy
kindling
pseudopregnancy
palpate
hutch

Objectives

After studying this chapter, the student should be able to

- describe the common breeds of rabbits.
- discuss nutrient requirements and feeding practices for rabbits.
- discuss reproduction and breeding of rabbits.
- discuss management of the rabbit herd.
- describe common diseases and prevention of disease problems with rabbits.
- discuss housing and equipment for rabbits.
- discuss marketing of rabbits.

CLASSIFICATION OF RABBITS

Rabbits belong to the order Lagomorpha, which is divided into two families: Leporidae, containing rabbits and hares, and Ochotonidae, containing the rock rabbit called pika. The three genera of Lagomorpha are *Lepus* (hares), *Sylvilagus* (American cottontail), and *Oryctolagus* (wild European and domestic rabbits). The wild European and domestic rabbits belong to the species *Oryctolagus cuniculus*. The various species of rabbits do not interbreed.

The ancestor of the domestic rabbit in the United States is the European wild rabbit. It is believed that the European wild rabbit developed in the area around the Mediterranean Sea. The modern breeds of domestic rabbit that are raised in the United States have been developed since the eighteenth century.

USES OF RABBITS

The primary use of rabbits in the United States is for meat production. Rabbits are also raised as pets, used in research laboratories, and used for wool production. There are about 200,000 producers of meat rabbits in the United States. Approximately 8 to 10 million pounds of rabbit meat are consumed each year in the United States. About 6 to 10 million rabbits are raised each year in the United States for all purposes. The annual world consumption of angora wool produced from rabbits is about 20 million pounds.

Rabbits produce a white meat that is palatable and nutritious; high in protein; and low in calories, fat, and cholesterol. Operations raising rabbits for meat range from small producers marketing just a few rabbits locally per year to large commercial farms that market many thousands of rabbits annually.

Figure 47-1 Checkered Giant.

The skins from meat rabbits have some commercial value; however, to be commercially successful, a large volume is necessary. Skins are used for fur garments, in slipper and glove linings, in toy making, and as felt. Remnants of flesh from dried skins are used in making glue.

Rabbits are also used for research purposes in medical schools, laboratories, and hospitals. Research relating to venereal disease, cardiac surgery, hypertension, virology, infectious diseases, toxins and antitoxins, and immunology is conducted utilizing rabbits. About 300,000 rabbits are used annually for research purposes. Persons living near research facilities may find a market for rabbits raised for research purposes.

BREEDS

Forty-seven breeds of rabbits are listed in the American Rabbit Breeders Association (ARBA) Standard of Perfection. Breed standards are based on color, type, shape, weight, fur, wool, and hair. Persons desiring complete information on these breeds of rabbits should contact the American Rabbit Breeders Association. Information regarding rabbit raising, shows, judging, registration requirements, and sources of breeds is available from the ARBA.

The selection of a breed should be based on its expected use. Medium- and heavy-weight breeds are best adapted for meat production. Some breeds are used primarily for show and fur purposes. Persons wishing to market rabbits for research purposes should contact potential buyers in the area to determine the type, age, and size of rabbits preferred. Table 47-1 lists some of the common breeds of rabbits and their primary uses. Figures 47-1 through 47-14 show some typical breeds of rabbits.

Figure 47-2 Dutch.

Figure 47-3 English Spot.

Figure 47-4 Californian.

TABLE 47-1 Some Representative Breeds of Rabbits

Breed	Color	Mature Weight (pounds)	Major Uses
Californian	Body white; nose, ears, feet, and tail colored.	8–10.5	Meat and show
Champagne d'Argent	Under fur dark slate blue; surface fur blue, white, or silver; liberal sprinkling of long black guard hairs.	9–12	Meat and show
Checkered Giant	Body white; black spots on cheeks, sides of body, and hindquarters. Wide spine stripe. Ears and nose black with black circles around the eye.	11 and over	Show and fur
Dutch	Body black, blue, chocolate, tortoise, steel gray, or gray. White saddle or band over the shoulder, under the neck, and over front legs and hind feet.	3.5–5.5	Show and research
English Spot	Body white with black, blue, chocolate, tortoise, lilac, gray, or steel gray spots. Spots on nose, ears, cheeks; circles around eyes. Spine stripe from base of ears to end of tail. Side spots from base of ears to middle of hindquarters.	5–8	Meat, show, and research
Flemish Giant	Body steel gray, light gray, sandy, blue, black, white, or fawn.	13 and over	Meat and show
Himalayan	Body white; nose, ears, feet, and tail colored.	2.5–5	Show and research
New Zealand	Body white, red, black, or broken.	9–12	Meat, show, and research

At maturity small breeds weigh 3 to 4 pounds, medium breeds weigh 9 to 12 pounds, and large breeds weigh 14 to 16 pounds.

Breeding stock should be purchased from reliable breeders. Rabbits purchased for breeding purposes should be evaluated on the basis of their health, vigor, longevity, reproduction ability, and desirable type and conformation.

Figure 47-5 New Zealand White.

FEEDING

Rabbits are simple-stomached animals and are herbivorous. **Herbivorous** means their diet comes mainly from plant sources. Rabbits have an enlarged cecum and therefore can use more forage in their diet than can other simple-stomached animals such as swine and poultry. The relative cost of feeds and local availability are two important factors to consider when selecting rabbit feeds.

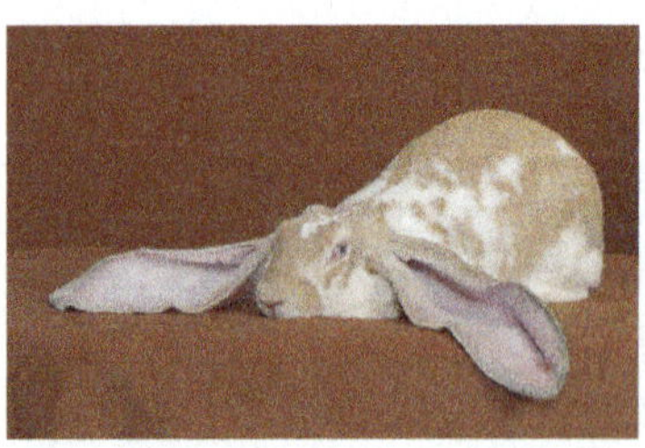

Figure 47-6 English Lop.

Nutrient Requirements

The nutrient requirements for rabbits are available from cooperative extension services. Rabbits fed well-balanced rations are relatively efficient converters of feed into meat, having a conversion ratio of about 3:1 (3 pounds of feed for each pound of meat produced).

Energy

It is believed that rabbits, like many other animals, adjust their energy intake to meet their needs.

Rabbits are efficient users of starch, which is found in cereal grains. Rabbits show a preference for barley or wheat over corn when given a choice of cereal grains. Diets that are based on corn have produced poorer growth rates as compared to barley or oat-based diets. Oat-based diets appear to give the best results for lactation rations. Because the energy levels of oats, barley, and wheat are lower than corn, it appears that other factors such as palatability of the ration influence the intake level and thus the growth rate in rabbits.

The palatability of the diet may be improved with the addition of fat. Research has shown that rabbits prefer a diet with some fat added. Vegetable oils are a good source of fat in rabbit diets.

Figure 47-7 Jersey Wooly.

Figure 47-8 English Angora.

Figure 47-9 Tan.

Figure 47-10 Rhinelander.

The energy requirement of rabbits is influenced by the environmental temperature. As temperatures decrease, the rabbit requires more energy to maintain normal body temperature. To compensate for this increased energy need, either the intake level of feed must be increased or the energy content of the ration must be increased. If the rabbit is already consuming feed at its maximum rate, then the energy content of the diet must be increased. The addition of fat to the ration is probably the best way to increase its energy level. Care must be taken not to have too high an energy level in the diet because excess energy levels may increase the incidence of diarrhea and enteritis.

Fiber

Research indicates that rabbits do not make as efficient use of plant fiber in the diet as do cattle, horses, and swine. However, the data also indicate that plant fiber is necessary in rabbit diets for the normal functioning of the digestive tract. Crude fiber levels lower than 6 to 12 percent of the ration have been shown to increase the incidence of diarrhea and enteritis in rabbits.

High fiber levels in the ration lower the energy level of rabbits; this may reduce feed efficiency. Lowering the fiber level during cold weather will increase the energy level of the diet and help provide the needed energy intake for maintenance of body temperature.

Protein

Essential amino acids need to be included in the ration for rabbits. Lysine and methionine are usually the amino acids that are found to be deficient in rabbit rations. While there is some bacterial protein synthesis in the cecum, it is not enough to meet the essential amino acid requirements of rabbits. Nonprotein nitrogen sources such as urea should not be used in rabbit diets.

Research has shown that soybean meal or fish meal promotes better growth rates than other protein supplements when the alternative supplements do not have essential amino acids added. When essential amino acids were added to protein supplements such as cottonseed meal, rapeseed meal, horsebeans, and peas, growth rates similar to those achieved with soybean and fish meals were attained. The amino acid composition of the protein supplement has the greatest influence on its value when feeding rabbits.

The relative cost of protein supplements must be considered when formulating rations. Animal and fish meal supplements are generally more expensive than legume and oil seed supplements.

Rabbits do make efficient use of the protein in plants. This means that large amounts of alfalfa can be used in rabbit diets. Research has shown that adequate

Figure 47-11 Netherland Dwarf.

Figure 47-12 Palomino.

Figure 47-13 Harlequin.

American Rabbit Breeders Association, Inc. Bloomington, IL, www.arba.net

Figure 47-14 American Fuzzy Lop.

growth rates can be maintained when alfalfa protein concentrate is substituted for soybean meal in the ration.

Coprophagy

Coprophagy refers to the ingestion of fecal matter. Rabbits produce two kinds of feces, one hard and one soft; the soft feces is ingested directly from the anus as it is excreted, usually when the rabbit is unobserved. The practice of coprophagy is similar to rumination practiced by ruminants in the sense that it provides a method of passing feedstuffs through the digestive tract a second time. It is sometimes referred to as pseudorumination. It is believed that coprophagy permits rabbits to synthesize the necessary B vitamins, contributes to better utilization of plant proteins, and provides some additional digestion of other nutrients. Rabbits begin the practice of coprophagy when they begin eating solid feed at about 3 to 4 weeks of age. The practice is normal in rabbits and does not indicate any nutritional deficiencies.

Minerals

A calcium-to-phosphorus ratio of 1:1 in rabbit diets will meet the need for these mineral elements. Rabbits can tolerate high levels of calcium in the diet without adverse effects. Levels of phosphorus above 1 percent of the diet reduce the palatability of the ration and may lower feed intake. Alfalfa and other legumes are good sources of calcium, while cereal grains are good sources of phosphorus; a combination of alfalfa and grain will generally supply the calcium and phosphorus needed in rabbit diets. Complete pelleted feeds will also supply the needed amounts of these minerals.

The use of iodized salt at the rate of 0.5 percent of the diet will supply the needed sodium, chlorine, and iodine for rabbits. Salt spools may be used, but they are more expensive, require more labor, and are corrosive to wire cages. Commercial rations for rabbits usually contain enough of these minerals; the use of salt spools, therefore, is generally not required.

Other required trace minerals in rabbit diets are usually provided through properly formulated commercial feeds. Deficiencies of trace minerals are rarely observed with rabbits.

Vitamins

While it is known that rabbits need fat-soluble vitamins in their diet, there is little available research information relating to quantities needed. Commercial feeds for rabbits are normally fortified with adequate amounts of the fat-soluble vitamins. Additionally, alfalfa is a good source of vitamin A. Vitamin K is synthesized in the intestinal tract of the rabbit; therefore, deficiencies rarely occur.

Several of the water-soluble vitamins, especially riboflavin, pantothenic acid, biotin, folic acid, and vitamin B12, are synthesized by rabbits in amounts sufficient to meet their needs. No additional supplementation of the diet is usually needed for these vitamins. Some fortification of the ration with choline and niacin is suggested. Additional pyridoxine is usually not needed because this vitamin is found in both cereal grains and forages and is synthesized in the digestive tract as well. No additional vitamin C is needed in the ration.

Water

Rabbits need a good supply of clean, fresh water at all times. A commercial rabbitry should use an automated water system to meet the water needs of rabbits. Rabbits usually consume 2.5 to 3 times more water than dry matter. Water consumption increases with both hot and cold air temperatures. If the water supply is limited, feed intake is reduced. A meat doe, weighing 10 to 12 pounds, and an 8-week-old litter of seven will consume about 1 gallon of water in each 24-hour period. Water intake is increased when the diet contains higher levels of protein and fiber.

Additives

The use of antibiotics on a regular basis in rabbit rations is not generally recommended. Because of the withdrawal period required before marketing, the use of antibiotics in the feed would require the use of a different feed formulation during the withdrawal period. Because there is little research to indicate that the regular use of antibiotics in rabbit feeding significantly improves rate of gain or feed efficiency, it is not considered to be economical to use them on a continuous basis.

Water treatment with drugs is recommended when it is required for the treatment of an outbreak of disease. Manufacturer's recommendations should be followed when using any drug treatment. Current regulations regarding drug treatments are available from the U.S. Food and Drug Administration and must be followed.

Courtesy of Frank B. Flanders

Figure 47-15 Dry forages, such as hay, usually make up between 40 and 80 percent of a rabbit's diet.

FEEDS

Hay

Dry forages, such as hay, usually make up between 40 and 80 percent of a rabbit's diet (Figure 47-15). Legume hays, especially alfalfa, make good roughage feeds for rabbits. Grass hays are not as palatable for rabbits as legume hays but may be used. The protein level of grass hays is lower than that of legume hays; therefore, protein supplementation of the ration is needed if they are used. Grass hays harvested before they are in bloom are more desirable than those cut after bloom. Hay provides bulk and fiber in the diet and its use reduces the incidence of fur chewing in rabbits.

Select hays that are leafy, fine stemmed, green, and well cured. Never feed hay containing mold or mildew. Coarse hay should be cut into 3- or 4-inch lengths to reduce waste.

Green Feeds/Root Crops

Fresh green feeds such as grasses, palatable weeds, cereal grains, and leafy vegetable crops may be fed. These feeds are high in vitamins, minerals, and proteins. They are of special value when feeding breeding animals (Figure 47-16).

Root crops such as carrots, sweet potatoes, turnips, mangels, beets, and Jerusalem artichokes may be fed to rabbits as substitutions for green feeds during the winter when the green feeds are not available.

Green feeds and root crops are supplements to the concentrate in the diet. They will produce choice carcasses when fed to meat rabbits and may be used for maintenance for mature animals that are not in production.

Do not overfeed green feeds or root crops to rabbits that are not accustomed to them. Do not use the feed if it is spoiled or contaminated. Put the feed in a manger rather than on the floor, and remove any that has not been eaten before feeding any additional amounts.

Commercial producers generally do not use fresh green feeds or root crops because of their high cost per unit of nutrient, the amount of labor required for their use, and their lower energy value; the low energy value reduces both the efficiency of meat production and their value for lactating females.

Courtesy of Frank B. Flanders

Figure 47-16 Legumes such as alfalfa, shown here, and clover make good roughage feeds for rabbits.

Grains

Oats, barley, wheat, buckwheat, grain sorghum, rye, and soft varieties of corn may be fed, either whole or milled (Figure 47-17). To reduce waste, flinty varieties of

USDA

Figure 47-17 Soft varieties of corn may be fed to rabbits, either whole or milled.

corn should be processed (ground or cracked) before feeding. One cereal grain may be substituted for another in the ration, although rabbits do show a preference for barley, wheat, or oats over corn.

Bran, middlings, shorts, and other cereal by-products may be used in mash mixtures or pellets for rabbits.

Protein Supplements

Soybean, linseed, sesame, and cottonseed oil meals are good protein supplements for use in mash or pelleted feeds. They should not be mixed with grains because they will settle out and be wasted. When feeding whole grains, feed the protein supplement in cake, flake, or pelleted form. Soybean oil meal is the most widely used protein supplement for rabbit rations.

Cottonseed meal must be treated to remove gossypol, which is toxic to rabbits. Cottonseed meal should be limited to no more than 5 to 7 percent of the total diet.

Animal protein supplements are generally not used in rabbit production because of their higher cost. Rabbits can meet their protein needs from plant protein sources.

PELLETING FEEDS

Courtesy of Frank B. Flanders

Figure 47-18 Both the rate of gain and feed efficiency of rabbits are improved when pelleted rations are used. Commercial producers use complete pelleted rations.

Rabbits prefer a pelleted ration to one in mash form. Both rate of gain and feed efficiency are improved when pelleted rations are used (Figure 47-18). Commercial producers of meat rabbits usually use complete pelleted rations. In some areas, the only commercial rabbit feeds available are in pelleted form.

There are two types of pelleted feeds available: all-grain pellets, which are designed to be used with hay, and complete pellets (green pellets). Complete pellets generally contain all the nutrients needed by rabbits.

The recommended pellet size is $\frac{1}{8}$ to $\frac{3}{16}$ inch in diameter and $\frac{1}{8}$ to $\frac{1}{4}$ inch in length. Rabbits will bite off only part of the pellet if it is too big, and the rest of the pellet is dropped; the dropped portion is usually wasted.

STORING FEEDS

Feeds stored for more than 4 weeks lose feed value and palatability. Keep rodents and insects out of stored feeds to avoid contamination of the feed. Do not allow cats or dogs in feed storage areas; cat or dog droppings may contain tapeworm eggs that could infest rabbits.

Store pellets in a dry area; excess humidity causes the pellets to soften and decreases the palatability of the feed.

FEEDING METHODS

Rabbits are generally hand-fed in small rabbitries. Feed is placed by hand in feed crocks or troughs. Regular feeding is more important than the number of times per day feed is provided. Rabbits generally eat more feed at night than during the day.

Rabbits may be self-fed or full-fed by means of a hopper or self-feeder. Rabbits on full feed usually make more efficient use of feed; full feed results in more pounds of gain per pound of feed consumed. Gains are also generally faster on full feed. Commercial meat producers usually utilize full feeding with automated feeding equipment.

DIETS FOR COMMERCIAL PRODUCTION

Table 47-2 gives some examples of diets that are adequate for feeding rabbits in commercial production.

TABLE 47-2 Examples of Adequate Diets for Commercial Production

Kind of Animal	Ingredients	Percent of Total Diet*
Growth, 1 pound to 9 pounds	Alfalfa hay	50
	Corn, grain	23.5
	Barley, grain	11
	Wheat bran	5
	Soybean meal	19
	Salt	0.5
Maintenance, does, avg. 10 pounds	Clover hay	70
	Oats, grain	29.5
	Salt	0.5
Pregnant does, avg. 10 pounds	Alfalfa hay	50
	Oats, grain	45.5
	Soybean meal	4
	Salt	0.5
Lactating does, avg. 10 pounds	Alfalfa hay	40
	Wheat, grain	25
	Sorghum, grain	22.5
	Soybean meal	12
	Salt	0.5

*Composition given on an as-fed basis.

Source: Reprinted from *Nutrient Requirements of Rabbits*, (Second Revised Edition), 1977, with permission of the National Academy Press, Washington, D.C.

MAINTENANCE FEEDING

Junior does (females) and bucks (males), mature dry does, and herd bucks that are not in service can be fed a maintenance ration. A fine-stemmed, leafy, legume hay will provide the nutrients needed for maintenance. Adding all-grain pellets or complete pellets to the maintenance ration is recommended if coarse legume hay or grass hay is fed. Medium-weight breeds should be fed about 2 ounces of pellets several times per week. Light-weight breeds should be fed less, and heavier-weight breeds should be fed more pellets along with the coarse legume or grass hay. Observe the rabbits closely to make sure they do not become too fat.

FEEDING HERD BUCKS IN SERVICE

Allow herd bucks free access to high-quality hay during the breeding season. Feed 4 to 6 ounces of a complete pellet if hay is not included in the ration. Do not allow bucks to become too fat, but feed them enough so they stay in good condition while in service.

FEEDING GROWING JUNIOR DOES AND BUCKS

A daily allowance of 2 to 4 ounces of all-grain or grain-protein pellets and free access to good-quality hay will provide the proper nutritional level for rabbits of the medium-weight breeds. A daily ration of 4 to 6 ounces of a complete pelleted feed is sufficient for proper growth of medium-weight breeds. Careful observation of the growing juniors is required to make sure they do not become too fat. Decrease the amount fed to light-weight breeds and increase the amount fed to heavier-weight breeds.

Alfalfa pellets containing 99 percent no. 2 or better-grade leafy alfalfa meal and 1 percent salt may be fed as the only ration to growing juniors from weaning until breeding. The pellets should contain 15 percent protein. A coarse alfalfa crumble or turkey alfalfa crumble may be substituted for alfalfa pellets if the latter are not locally available.

FEEDING PREGNANT AND NURSING DOES

After does are bred, continue feeding them the maintenance ration until it has been determined that they are pregnant. Pregnancy may be determined about 12 to 14 days after breeding by palpating, which is described further in the section on rabbit breeding. Good-quality hay or hay pellets may be fed. Limit the amount of feed when feeding an all-pellet ration to keep the does from becoming too fat.

After pregnancy is confirmed, does may be full-fed a complete pelleted feed. A ration of good-quality hay with a full feed of concentrates may also be fed. Grain plus protein pellets or all-grain pellets may be fed as the concentrate in the diet. Commercial rabbitries usually feed a complete pelleted ration. Pregnant females will normally eat 6 to 8 ounces of feed per day.

Do not make sudden changes in the diet or the doe may go off-feed; any change in the diet must be made gradually. Feed $\frac{1}{4}$ of the new ration and $\frac{3}{4}$ of the old ration for 3 or 4 days; then feed 50 percent of each for 3 or 4 days; complete the change by feeding $\frac{3}{4}$ of the new ration and $\frac{1}{4}$ of the old ration for another 3 or 4 days.

On the day of **kindling** (giving birth), the does should be fed about one-half the normal daily ration. After kindling, gradually increase the amount of daily feed until the does are back on full feed at the end of 1 week.

Nursing does may be fed the same ration as pregnant does. Full-feed the does until the litter is weaned at about 2 months of age. Satisfactory diets include hay with an all-grain pellet, a grain-protein mixture, or a complete pelleted feed. Nursing does will eat 6 to 8 ounces of feed per day until the litter is 3 weeks old. During the nursing period of 3 to 8 weeks, does will eat from 1 to 2 pounds of feed per day.

RABBIT BREEDING

Does remain in heat for long periods of time during the breeding season. They do not show regular estrous cycles. Normally, ovulation occurs in the female about 10 hours after she is bred to the male; the sperm fertilize the eggs shortly after ovulation. The gestation period for rabbits is normally 30 to 32 days.

Litter size will vary with breeds and strains of rabbits. More prolific breeds will average about eight young per litter. Poor nutrition will lower the litter size.

The light-weight breeds of rabbits become sexually mature at an earlier age than the medium- and heavy-weight breeds. They may be bred when they are 4 to 5 months of age. The medium-weight breeds may be bred when they are 5 to 6 months of age, and the heavy-weight breeds may be bred at 8 to 10 months of age. Does generally reach sexual maturity earlier than bucks. Commercial rabbitries often allow bucks to grow a month longer than does before using them for breeding for the first time.

Prolific does that are in good physical condition can be rebred 6 weeks after kindling, even though they are still nursing their young. This makes it theoretically possible for a doe to produce five litters per year (assuming that there are no conception failures). Does that are rebred after weaning their litters at 8 weeks of age can produce four litters per year. Commercial rabbitries often use breeding intervals of 21, 28, or 35 days to increase production. Does that are maintained in good physical condition will generally produce litters in commercial rabbitries for 2.5 to 3 years.

Does that are full-fed a properly balanced ration can be rebred before weaning the current litter. Do not rebreed does that are in poor physical condition. Feed a ration that improves their physical condition before attempting to rebreed them. If the litter dies at birth or is small and the doe is in good physical condition, she may be rebred about 3 or 4 days after kindling.

In commercial rabbitries, a regular breeding schedule should be followed, regardless of whether or not the doe shows signs of being ready for mating. These signs include restlessness and nervousness, rubbing the chin on feeding and watering equipment, and attempting to join with rabbits in other cages.

The doe is taken to the buck for mating. Mating generally occurs within a few minutes of placing the doe with the buck. Return the doe to her cage as soon as mating is completed.

Records of matings that include the date and identification of the doe and buck may be kept. One buck should be kept for each 10 breeding does. Mature, healthy, vigorous bucks may be mated several times per day for short periods of time.

Does sometimes exhibit **pseudopregnancy** (false pregnancy); this may result from an infertile mating or one doe's riding another during a period of sexual excitement. The pseudopregnancy lasts for 17 days and the doe cannot be bred during this period. Separating the does 18 days before mating will allow them to pass through any pseudopregnancy before breeding.

The conception rate for rabbits is higher during the spring. Research has shown that the conception rate in March and April may be as high as 85 percent, while it may be as low as 50 percent during September and October. There is a high level of individual variation in fertility among does and bucks; this factor should be considered when selecting breeding stock.

Artificial insemination is not commonly practiced in commercial rabbitries. It is sometimes used with rabbits used in research.

Pregnancy should be confirmed by **palpating** the doe 12 to 14 days after mating. Restrain the doe by holding the ears and a fold of skin over the shoulders in one hand. With the other hand, reach under the shoulder to the area between the hind legs and in front of the pelvis. Place the thumb on the right side and the fingers on the left side of the two uteri to palpate the fetuses. Move the hand gently back and forth, exerting a slight pressure. If the doe is pregnant, the fetuses may be felt as small marble-shaped forms, which will slip between the thumb and fingers. Do not exert too much pressure. Use caution when palpating to avoid bruising or tearing the tissue, which may cause abortion.

KINDLING

Prepare for kindling by providing a nesting box in the cage 27 days after the doe is bred. The nesting box should contain bedding materials. Suitable materials include clean straw or wood shavings. The doe will prepare a nest in the nesting box by pulling fur from her body to line the nest.

Feed intake is often reduced just before kindling. Adding some green feed to the diet may have a beneficial effect on the digestive system.

Kindling usually occurs at night. Does in good physical condition seldom have problems kindling. They are often nervous after kindling and should not be disturbed until they have quieted down.

If the doe fails to pull enough fur to cover the litter (or kindlings) on the hutch floor, warm the litter and add fur to keep them warm. Fur may be pulled from the doe's body for this purpose, or extra fur may be kept on hand to use.

The litter should be inspected no later than the day after kindling and any dead, deformed, or undersized young should be removed from the nest box. Take care not to disturb the doe when inspecting the litter. This inspection will not cause the doe to disown the litter. Nervous does may be quieted by placing feed in the hutch immediately after the inspection.

The litters of breeds used for meat production should contain seven to nine young. If the litter has more than nine young, the extra rabbits may be transferred to a smaller litter; make this transfer in the first 3 or 4 days after kindling. Place the young rabbits with a litter of approximately the same age. More uniform development and finish at weaning will result from keeping the litters at the desirable number.

Young rabbits usually open their eyes about 10 or 11 days after birth. If an infection prevents the eyes from opening normally, they should be treated promptly. Wash the eyes with warm water. After the tissue softens, the eyes may be opened using a gentle pressure. An antibiotic eye ointment should be used to treat the eyes if pus is present.

LOSSES AT KINDLING

Does sometimes kindle on the hutch floor rather than in the nest box. If this happens, the young may die of exposure unless they are warmed and placed in the nest box.

Frightened does may kindle prematurely. If a doe is frightened after kindling, she may jump into the nest box and injure or kill the young rabbits by stamping with her back feet. Take care to prevent does from becoming frightened just prior to or at kindling time.

If the doe fails to produce milk, the young will die in 2 or 3 days. Make sure the doe is producing milk and is feeding the young. If the doe is not producing milk, transfer the young rabbits to litters where they can nurse and be properly cared for.

While it is not common, a doe may eat her young. This may occur if the diet is not adequate to meet her needs. Does that are disturbed, nervous, or frightened after kindling may also eat their young. Providing a good diet and using care when handling the doe will usually prevent this problem. Does that continue to eat their young despite proper feeding and care should not be kept.

WEANING

Young rabbits will start to eat solid feed at about 19 to 20 days of age. They will come out of the nest box at this time for feeding.

Does usually nurse their young at night. If some of the young are out of the nest box and others are not, she will nurse only one group or the other, but not both. The doe will not carry the young rabbits back into the nest box to nurse; make sure all the young are nursing properly or losses will occur.

Rabbits may be weaned at 8 weeks of age. Meat rabbits should weigh about 4 pounds and be ready for market at that time. Leaving the rabbits with the doe for 9 or 10 weeks will produce fryers weighing about 4.5 to 5.5 pounds.

Producers who use an accelerated breeding program, which breeds does at less than 35 days after kindling, should wean the young at 5 to 7 weeks; this allows the doe to be in better physical condition for the next litter. It is a good practice to allow the doe to have a few days between weaning a litter and the birth of the next litter.

HANDLING RABBITS

Do not handle rabbits unless it is necessary. Young rabbits that are to be kept for breeding purposes may be handled occasionally to get them used to the moving done during breeding.

Do not lift rabbits by their ears or legs. To properly lift a rabbit, grasp the loose skin over the shoulders with one hand and support the weight of the rabbit with the other hand under the rump. Rabbits that weigh less than 4 pounds may be lifted and carried by the loin region. When carrying the rabbit in this manner, keep the heel of the hand toward the tail of the rabbit. Take care not to bruise or damage the skin of the animal.

SEXING RABBITS

Rabbits that are to be kept for breeding stock should be separated by sex at weaning. Hold the rabbit on its back, restraining it by holding the front legs up along the head. With the other hand, depress the tail back and down. Use the thumb to gently depress the area in front of the sex organs to expose the reddish mucous membrane. The organ protrudes as a rounded tip in the buck; it protrudes as a slit with a depression at the end next to the anus in the doe.

IDENTIFICATION OF BREEDING RABBITS

Tattoo breeding rabbits in the ear to mark them for identification. This provides a permanent mark that does not disfigure the ears. Do not use ear tags or clips for marking rabbits; these often tear out of the ear. An adjustable box may be used to restrain the rabbit while tattooing the ears. This makes it possible for one person to perform the operation. Biological and livestock supply houses have tattooing equipment available.

CASTRATION

Rabbits raised for meat purposes are generally not castrated. There is limited research to indicate that castrating young bucks will improve feed conversion, rate of growth, or carcass quality.

CARE DURING HOT WEATHER

In hot weather, rabbits need to be kept cool. Good ventilation without drafts is necessary in the rabbitry. Pregnant does and newborn litters are especially susceptible to hot weather. Young rabbits become restless when the temperature is too high. Heat stress in older rabbits is indicated by rapid respiration, excessive moisture around the mouth, and sometimes slight bleeding around the nostrils. Rabbits that show symptoms of heat stress should be moved to a cooler area.

Automatic sprinkling equipment may be used in hot, dry areas to help cool rabbitries. Evaporative coolers also may be used to help keep the rabbitry cool. In areas where the humidity is high, use cooling fans in the rabbitry.

CARE DURING COLD WEATHER

Rabbits should be protected from direct exposure to rain, sleet, snow, and wind. Provide adequate ventilation in enclosed buildings to reduce excess moisture. Respiratory diseases are more common in rabbits exposed to drafts, high humidity, and cold. Mature rabbits are less susceptible to cold weather than younger rabbits if they are kept free of drafts. Make sure young rabbits are kept warm during cold weather.

Commercial rabbitries are increasingly turning to the use of controlled-environment housing for rabbits as a means of eliminating extremes in temperature. Seasonal variations in production are also reduced.

MAINTAINING HEALTHY RABBITS

Sanitation and Disease Control

Good sanitation in the rabbitry is the key to disease control. Remove manure and bedding on a regular basis. Protect water and feed from urine and feces contamination. Contaminated feed must be removed daily. Store feed in rodent-proof areas. Clean watering and feeding equipment frequently by washing it in hot, soapy water. Rinse the equipment thoroughly and dry it in the sun or use rinse water with a disinfectant added and then rinse with clear water.

CAUTION

Dry heat from a flame may be used to disinfect equipment. However, the use of a flame in the rabbitry is a fire hazard and should be used cautiously.

Do not overcrowd animals in the rabbitry; overcrowding makes it more difficult to follow good sanitation practices. When rabbits are overcrowded, their resistance to disease is lowered and diseases spread more rapidly through the herd.

After an outbreak of disease or parasites in the rabbitry, clean and disinfect hutches and equipment. Commercial rabbitries often use steam under pressure to clean facilities and equipment. Soak caked fecal matter before using a steam cleaner to facilitate removal of the feces. Make sure hutches and equipment are dry before putting rabbits in them. Nest boxes should be cleaned and disinfected after each use.

Movable equipment that has been thoroughly cleaned may be placed in the sun for a period of time to disinfect it. Dry heat from a flame may be used to disinfect equipment.

Maintaining Health

Good sanitation, proper management, adequate diet, and plenty of clean, fresh water will help the grower maintain healthy rabbits. Observe the herd closely for signs of disease outbreak. Isolate any animals that appear to be sick and keep them isolated for at least 2 weeks. Treatment and caring for sick animals should be done only after completing work with the healthy rabbits. Practice good personal sanitation by washing hands and disinfecting boots after working with sick animals. Send some of the sick animals to a diagnostic laboratory if the cause of illness is not readily apparent. Sick animals that cannot be successfully treated should be put down and properly disposed of. Dead animals should be burned or buried. Do not use open pits for disposal of dead animals.

Preventing Disease Transmission

The most common health problems of rabbits are pasteurellosis, ear mange, and coccidiosis; these are often transmitted by contact with infected rabbits or by mechanical carriers. Table 47-3 describes some other health problems of rabbits.

Adult rabbits that have had a disease may appear healthy but can still spread the disease; infectious organisms may be present in the feces, urine, or moisture droplets exhaled with the breath of these animals. Pasteurellosis and liver coccidiosis are often spread by contact carriers. Isolate new animals or those that have returned from shows from the rest of herd for a period long enough to determine that they are free of disease. A minimum of 2 weeks of isolation is recommended. Tests to determine if the rabbit is a carrier of disease are often too expensive to be practical.

Sometimes the only solution to a disease problem in a rabbitry is to remove all the animals and clean and sanitize the facility. After the facility has been cleaned, use only disease-free animals to repopulate the rabbitry.

Disease organisms may be mechanically transmitted from one rabbitry to another. People such as feed salesmen, service workers, buyers, and visitors who travel from place to place may carry disease organisms; growers who check or treat sick animals and then work with the healthy animals may carry an infection

TABLE 47-3 Common Health Problems in Rabbits

Health Problem	Symptoms	Cause	Treatment/Control
Caked breasts	Breasts are swollen and firm; hard knots form at sides of nipples. May occur after kindling, when litter is small, when litter dies, or after weaning.	Excess supply of milk compared to rate of removal.	Reduce amount of concentrate in diet for several days. Reduce amount of feed before kindling, weaning, or if litter dies.
Coccidiosis, intestinal	Diarrhea containing blood. Slow or no weight gain, or loss of weight. May be pot bellied after recovery.	Coccidia—protozoan parasite infects intestinal tract (*Eimeria magna*).	Treat with coccidiostat. Practice good sanitation. Prevent contamination of feed and water by feces.
Coccidiosis, liver	Diarrhea, loss of appetite, weight loss, pot belly may develop. Death may occur. White nodules on the liver.	Coccidia—protozoan parasite infects liver (*Eimeria steidae*).	Same as coccidiosis, intestinal.
Ear mange (canker)	Scratching ears, shaking head. Scabs at base of inner ear.	Rabbit, goat, and cat ear mite.	Swab ear with medication. Isolation. Treat entire herd with medication once a month.
Fur block	Loss of appetite, loss of weight. Sometimes diarrhea. Fur is rough. Pneumonia may develop.	Small intestine blocked by fur. Blocks flow of feed through digestive system. Not enough fiber or roughage in diet.	Increase roughage in diet. Give mineral oil to affected animal.
Heat prostration	Respiration rate becomes rapid. Moisture around mouth and nose may contain blood. Ears and mouth become blue-tinged. Prostration followed by death.	High temperature often accompanied by high humidity.	Put affected animals in cool place. Keep animals cool with proper ventilation in buildings.
Mastitis (Blue breast)	Breasts become swollen, feverish, turn black and purple. Abscesses may form. Animal has fever.	Infection of breasts by bacteria (usually *Staphylococcus* or *Streptococcus*).	Treat with antibiotic. Sanitation and less concentrate in the diet helps prevent.
Paralyzed hindquarters	Drags hind legs, cannot stand, no control of bladder or bowel.	Injury that breaks back, displaces disk in spinal cord, or damages nerves in spinal cord.	No treatment: dispose of affected animal. Prevent by careful handling. Protect from disturbances.
Pinworm	Mild infestation: no symptoms. Severe infestation: slow growth, weight loss, reduced disease resistance. Worms found in cecum and large intestine.	Pinworms (*Passalurus ambiguus*).	Treat with worm medicine. Prevent with good sanitation.
Ringworm (favus)	Hair loss in circular patches, scaly skin with red raised crust. Matted fur. May occur on any part of the body.	Fungus (*Trichophyton* and *Microsporum*). May be transmitted to humans: Use caution when handling animals.	Individual treatment with medication applied to skin. Herd treatment with oral or systemic medication. Disinfect cages and equipment.
Skin mange	Dry, scaly, irritated skin, scratching, loss of fur on head, ears, and neck.	Mange mites (rabbit fur mite and scabies or itch mite).	Consult veterinarian. Treat with miticide. Isolate herd, control rodents.
Sore hocks	Infected, inflamed, bruised, or abscessed bare areas on hind legs. Animal shifts weight to front legs. May affect front legs in severe cases. Secondary infections may occur.	Wet floors in hutch, unsanitary conditions, irritation from wire floors. Nervous stompers more likely to be affected.	Place animals on dry, solid surface. Good sanitation. Cull severe cases. Select breeding stock with well-furred feet.
Tapeworm larvae	Mild cases: no symptoms. Severe cases: diarrhea and loss of weight.	Larva from dog or cat tapeworm. Infestation occurs when rabbit ingests feed or water contaminated by eggs or tapeworm segments.	No practical treatment. Prevent by keeping dogs and cats out of rabbitry.
Urine-hutch burn	External genitals and anus become inflamed. Crusts may form, bleeding may occur. Pus may develop in severe cases.	Condition starts with wet, dirty hutches. Bacteria infect skin membranes.	Treat with antibiotic cream. Prevent with good sanitation practices.
Vent disease (Spirochetosis)	Blisters, lesions, scabs around genital organs. Lesions may occur around nose, ears, or mouth. (Sometimes confused with symptoms of urine-hutch burn.)	Spirochete (*Treponema cuniculi*). Usually transmitted by mating.	Treat with antibiotic. Do not use infected animals for breeding. Cull infected animals. Do not loan bucks to other breeders.

to the healthy rabbits. Keep dogs, cats, birds, and rodents out of the rabbitry; these animals may carry disease organisms. Control insects in the rabbitry to help prevent the spread of disease.

COMMON HEALTH PROBLEMS

Pasteurellosis

Pasteurellosis is the most serious health problem of rabbits. It appears in a number of forms, including snuffles, pneumonia, pyometra, orchitis, otitis media, conjunctivitis, subcutaneous abscesses, and septicemia. All of these diseases are associated with the bacteria *Pasteurella multocida.*

Snuffles, or cold, is caused by a bacterial infection of the nasal sinuses. Symptoms include sneezing, rubbing the nose, and a nasal discharge. Snuffles is usually a chronic condition that may develop into pneumonia. Stress contributes to the development of this disease. Antibiotics are used in treatment.

Pneumonia occurs when the bacterial infection develops in the lungs. Symptoms include labored breathing, depression, nasal discharge, elevated body temperature, and a bluish eye and ear color. This disease is the major single cause of death in mature rabbits. Antibiotics are used in treatment.

Pyometra is pus in the uterus of female rabbits. It occurs when *Pasteurella* bacteria get into the uterus during mating with a buck that has infected testicles. There is no treatment.

Orchitis is an infection of the testicles of the buck by the *Pasteurella* bacteria. Symptoms include enlarged testicles, which contain pus. Generally, this condition is not treated. If the infection occurs in the membranes of the penis, swelling, reddening, and pus may be observed; antibiotics may be used to treat this condition.

Otitis media is an infection of the middle ear. If the infection spreads to the inner ear, the equilibrium of the rabbit is affected, and the animal will tilt its head. This condition is referred to as wry neck. Treatment of otitis media is usually not attempted because of the difficulty of getting antibiotics to the site of the infection. Wry neck may also be caused by some other bacterial infection, parasitic infestation, or injury.

Conjunctivitis occurs when the *Pasteurella* bacteria infect the membranes covering the eye and the inner part of the eyelid. The condition is sometimes called weepy eye. Symptoms include inflammation of the eyelids, discharge from the eye, and wet, matted fur around the eye. Antibiotic eye ointment is used in treatment.

Subcutaneous abscesses are soft swellings under the skin of the rabbit. The abscess is opened and drained and an antibiotic ointment is applied.

Septicemia, which is caused by *Pasteurella* bacteria, is an infection of the blood stream. It is usually of short duration and shows no visible symptoms. Because of the lack of symptoms and its short duration, treatment is usually not attempted.

Enteritis

Enteritis is a disease complex associated with bloating and diarrhea. This disease complex can be a serious health problem in rabbits. Death loss is often high. Enteritis takes several forms.

Mucoid enteritis is most commonly found in meat rabbits. Symptoms include loss of appetite, bloated appearance, grinding of teeth, and high death rate. The cause of this condition is not known. Reducing stress and feeding an adequate amount of fiber in the diet helps prevent the condition. Antibiotics may be used to treat secondary infections.

Tyzzer's disease is caused by the organism *Bacillus piliformis*. Symptoms include diarrhea, listlessness, loss of appetite, and dehydration. Affected animals usually die within 72 hours. Stress combined with exposure to fecal-contaminated feed and bedding contributes to the transmission of this disease. Antibiotic treatment of the entire herd has shown some success in reducing the effect of Tyzzer's disease.

Enterotoxemia is characterized by an acute diarrhea coupled with dehydration and death within 24 hours. No specific cause has been identified. The disease appears to be associated with a diet that is high in energy and low in fiber. Feeding a diet lower in energy and higher in fiber (18 percent) may help prevent this disease. Treating the drinking water with antibiotics may also help.

KEEPING RECORDS

An important part of commercial rabbit raising is keeping good records. Records can provide information for culling unproductive animals from the herd, selecting breeding stock, determining feed efficiency, and determining costs and returns from the enterprise. Good records are also essential for tax purposes.

The doe performance record card should include identification data, date bred, buck used, date kindled, number of young born alive and dead, number of young retained, litter number, date weaned, number weaned, age at weaning, and weight at weaning. The buck performance card should include identification data, does bred, date of breeding, number of young kindled alive and dead, number weaned, and weaning weights. Annual summaries of the individual performance records may be made. Various record card forms are available from commercial firms that supply rabbitries.

Financial records should show investment in capital equipment, production expenses, and income records. The cost of production and returns over expenses may be determined from good records. Financial records also provide the information needed for income tax purposes.

Computers may be used for keeping the necessary records of the rabbitry. Commercial database and general accounting programs may be used for keeping records. There are some software packages available specifically for the rabbit producer. These include pedigree, doe and buck performance record, and feed programs. There are a number of farm accounting programs available for computers that can be adapted for use in rabbit production.

FACILITIES AND EQUIPMENT

Buildings

Climatic conditions, local building codes, and available capital are factors to consider when determining the kind of housing for rabbit hutches. Buildings of simple design that protect the rabbits from the weather and provide adequate ventilation are desirable.

In mild climates, hutches may be outside if shade is provided. Trees or a lath superstructure may be used for shade.

Where the weather is hot part of the year, some method of cooling the building is necessary. Overhead sprinklers or foggers inside the building may be used for cooling. In hot, humid areas, evaporation coolers with fans may be used to provide cooling in the building. High-volume commercial operations often use automated ventilation and cooling equipment. In areas where the weather is cold part of the year, the building must be designed to provide more protection.

Figure 47-19 Hutches are pens or cages for rabbits.

Hutches

Hutches are provided for individual rabbits (Figure 47-19). These may be made of wood or wire construction. Hutches should be about 2.5 feet deep and 2 feet high. Length varies with breed: small breeds need 3 feet, medium breeds need 3 to 4 feet, and large breeds need 4 to 6 feet.

Hutches may be arranged in one-, two-, or three-tier configurations. If enough room is available, the one-tier arrangement is the most convenient for caring for the rabbits. Commercial rabbitries usually hang all-wire cages from eye hooks in the ceiling of the building.

Wire cages of various designs are available from commercial firms. Self-cleaning wire cages require no bedding and are easy to maintain. Wire mesh floors are usually used in self-cleaning hutches. Commercial rabbitries use self-cleaning cages because of the high labor requirement of solid-floor cages. Scraper blades are used to remove the manure from the building.

Feeding and Watering Equipment

Feeding equipment should be designed to prevent feed waste and contamination. Crocks, hoppers, and hay mangers with troughs may be used for feeding rabbits. Commercial rabbitries may find it desirable to use electric feed carts to reduce labor costs.

Water crocks may be used in the hutch to provide fresh, clean water for rabbits. However, the labor requirement is high when large numbers of rabbits are raised. Commercial rabbitries usually use automated watering systems to reduce labor costs. In cold climates, automated watering systems must be protected against freezing; this may be done by heating the building or wrapping the pipes with heating cables.

Nest Boxes

Nest boxes are used to provide seclusion for the doe when kindling and to provide protection for the litter. The type used should be simple to clean.

Several types of nest boxes are available for use in rabbitries. Be sure the type used is large enough to prevent crowding but small enough to keep the young rabbits together.

Nest boxes may be made of wood or metal; apple packing boxes or nail kegs may be used. Metal nest boxes may be cold, with water condensation being a problem in cold climates. Counterset nest boxes are recessed below the hutch floor; this type provides a more natural environment for the young rabbits. They are also easier to keep clean. Make sure nest boxes are well drained and ventilated. In cold weather, a well-insulated nest box will provide protection for the young rabbits at temperatures as low as –15 to –20°F.

MARKETING RABBITS

Commercial rabbitries usually sell fryers to processors who slaughter the rabbits and market the meat. Processors usually use trucks to pick up the live animals at the rabbitry. Rabbits with colored pelts bring lower prices; therefore, commercial operations usually raise only white-pelted rabbits.

The USDA defines a fryer as "a young domestic rabbit carcass weighing not less than 1.5 pounds and rarely more than 3.5 pounds processed from a rabbit usually less than 12 weeks of age." The rabbits are marketed when they reach fryer weight, which is a live weight of 3 to 6 pounds. Live weights of 4 to 4.75 pounds produce the best carcasses, with a dressing percent ranging from 50 to 59 percent. Ready-to-cook rabbits are graded based on three quality grades: A, B, and C. The following criteria are considered when quality grading rabbits:

- conformation
- fleshing
- cuts and tears
- disjointed and broken bones, and missing parts
- discolorations
- freezing defects

For further information on these specifications consult the USDA Agricultural Marketing Service (AMS).

The USDA defines a roaster as "a mature or old domestic rabbit carcass of any weight, but usually over 4 pounds processed from a rabbit usually 8 months of age or older." Roasters have a dressing percent of 55 to 65 percent. Culls from the breeding herd may be fattened and sold as roasters. It is usually not profitable to feed younger rabbits to roaster weights because of the higher feed requirements and the possibility of additional death loss.

Some small-volume producers slaughter, package, and market their own rabbits. Strict sanitation must be practiced when doing this. Producers should determine local slaughtering regulations and restrictions when planning to market their rabbits in this manner.

SUMMARY

Modern breeds of domestic rabbits in the United States are descendants of the European wild rabbit and have been developed since the eighteenth century. The major use of domestic rabbits is for meat production. Rabbits produce a white meat that is high in protein and low in calories, fat, and cholesterol. Medium- and heavy weight breeds are best suited for meat production. The most common breed used for this purpose is the New Zealand White.

Rabbits are simple-stomached, herbivorous animals. The major part of their diet comes from plants and plant products. Most commercial rabbitries use complete pelleted feeds for rabbit rations. A continuous supply of fresh, clean water should be available for rabbits. Feed additives are usually not used on a regular basis in rabbit feeding.

Maintenance rations may be fed to junior does and bucks, mature dry does, and herd bucks not in service. A fine-stemmed, leafy, legume hay will provide the nutrients needed for maintenance. Herd bucks in service should be fed 4 to 6 ounces of a complete pellet feed if hay is not included in the ration. Growing junior does and bucks may be fed 4 to 6 ounces of a complete pelleted ration. Pregnant does may be fed 6 to 8 ounces of a complete pelleted feed, while nursing does will eat 6 to 8 ounces during the first 3 weeks of lactation and 1 to 2 pounds of feed per day for the remainder of the nursing period.

The gestation period for rabbits is 30 to 32 days. Does in good physical condition can be rebred 6 weeks after kindling. The litter is normally weaned at 8 weeks of age. On an accelerated breeding program, a doe can produce five litters per year, although four litters per year is more common. Meat rabbits are usually sold for market as fryers when they are weaned at 8 weeks of age.

Good sanitation in the rabbitry is the key to disease control. Keep manure and soiled bedding cleaned out, isolate sick animals and new breeding stock coming into the herd, and prevent visitors from coming into contact with the rabbits. Keep dogs, cats, birds, rodents, and insects out the rabbitry because they sometimes carry diseases or parasites that may be transmitted to the rabbits.

Rabbit producers need to keep good records. Performance records and financial records are important. A computer may be used to keep the necessary records for the rabbitry.

Buildings for rabbits need to protect the animals from weather conditions. Commercial rabbitries generally use wire cages and utilize automated feeding and watering equipment. Nest boxes are provided for kindling and to protect the young rabbits during their first few weeks of life.

Commercial rabbitries usually sell live fryers to processors. The live weight of fryers ranges from 3 to 6 pounds; the best carcasses dress out at 50 to 59 percent. Roasters are older rabbits that have been culled from the herd.

Student Learning Activities

1. Talk with a commercial rabbit raiser. Give an oral report to the class and describe the business.
2. Present an oral report on commercial rabbit raising.
3. Prepare a bulletin board display on the breeds of rabbits.
4. When planning and conducting a rabbit supervised experience program, use the information discussed in this chapter to be as successful as possible.

Discussion Questions

1. List the uses of rabbits in the United States.
2. Describe and list the major uses of each of the following breeds of rabbits: (a) Californian, (b) Dutch, (c) New Zealand.
3. What are the typical mature weights of (a) small breeds, (b) medium breeds, (c) large breeds?
4. Why can rabbits use more forage in their diet than other simple-stomached animals, such as swine and poultry?
5. How may the palatability of rabbits' rations be improved?
6. What effect does temperature have on the energy requirements of rabbits?
7. What percentage of the ration for rabbits should consist of fiber? What effects do higher or lower levels of fiber have on rabbit health and growth?
8. What is coprophagy? How does coprophagy affect rabbit nutrition?
9. Briefly explain the need for vitamins in rabbit diets.
10. Briefly explain the need for water in rabbit diets.
11. Briefly explain the use of feed additives in rabbit rations.
12. Why are legume hays preferred over grass hays in rabbit diets?
13. Which legume hay is considered the best for rabbits?
14. Briefly explain the use of green feeds and root crops for rabbits.
15. Which protein supplement is most commonly used in rabbit feeds?
16. Why do commercial rabbitries usually full-feed rabbits for meat production?
17. Briefly explain maintenance feeding for junior does and bucks, mature dry does, and herd bucks that are not in service.
18. What kind of a ration should be fed to growing junior does and bucks?
19. Briefly explain the feeding of pregnant does.
20. Briefly explain the feeding of nursing does.
21. What are some common breeding intervals used by commercial rabbitries?
22. Describe the method of determining pregnancy by palpating the doe.

23. List several causes of losses at kindling and describe how to prevent these losses.
24. Describe the method of determining the sex of young rabbits.
25. List and describe several good management practices in a disease control program in rabbitries.
26. List and describe several good management practices to prevent the transmission of disease in rabbitries.
27. Briefly explain the enteritis disease complex.
28. Briefly explain the kind of buildings needed for rabbits.
29. What kind of cage is most commonly used in commercial rabbitries?
30. What is the purpose of the nest box? Describe a nest box.
31. Define (a) fryers, (b) roasters.

Review Questions

True/False

1. The ancestor of the domestic rabbit in the United States is a wild rabbit found in the southwestern part of the country.
2. Rabbits are herbivorous.
3. The primary use of rabbits in the United States is for wool production.
4. The gestation period for rabbits is 30 to 32 days.

Multiple Choice

5. Rabbits consume how much more water than dry matter?

   a. 7 to 8 times   c. 1 to 2 times
   b. 6 times   d. 2.5 to 3 times

6. The term for the rabbit house is a ______.

   a. shed   c. coop
   b. hutch   d. pen

Completion

7. In the marketing of rabbits, those with ______ pelts bring lower prices.
8. The environmental temperature influences the ______ requirement of rabbits.

Short Answer

9. In rabbit raising, what is kindling?
10. List four rabbit types that are raised in the United States.
11. What are the three most common health problems of rabbits?
12. Why is it important not to store rabbit feed for more than 4 weeks?

Chapter 48

Bison, Ratites, Llamas, Alpacas, and Elk

Key Terms

bloodtyping
ratite
cria
sire
dam
velvet

Objectives

After studying this chapter, the student should be able to

- describe the origin, history, and general characteristics of bison.
- describe the characteristics of the bison industry.
- discuss the management, feeding practices, health maintenance, facilities and equipment, and marketing of bison.
- discuss the characteristics and origin of ratites.
- discuss management practices used when caring for ratites.
- discuss facilities needed for ratites.
- discuss maintaining the health of ratites.
- discuss getting started with ratites.
- discuss the marketing of ratites and their products.
- discuss the history of and getting started in business with llamas and alpacas.
- describe the characteristics of llamas and alpacas.
- discuss the feeding and management of llamas and alpacas.
- discuss the uses, production, and management of elk.

BISON

Origin

The American bison is a member of the Bovidae family, genus *Bison*, species *bison*. The Bovidae family includes cattle, sheep, and goats. It is believed that bison crossed the Bering Strait land bridge from Asia to North America approximately 20,000 to 30,000 years ago. The American bison is related to the European bison (*Bison bonasus*), also known as the wisent, which is a species that is almost extinct and is mainly found in parks and zoos. The name "buffalo" is often used when referring to the bison; however, this is not a correct use of the word. "Buffalo" refers to the Cape buffalo or water buffalo found in other parts of the world but not in the United States.

History

Bison were an important source of food, skins, bones, and fuel (dried dung) for the Plains Indians before European explorers reached the North American continent. It is estimated that there were between 30 and 60 million head of bison ranging the western part of the continent at that time. European settlers and explorers engaged in a methodical slaughter of bison during the nineteenth century. By the end of that century, there were probably no more than 300 bison left on the North American continent. The bison had been hunted almost to extinction.

Through the efforts of a few conservationists and ranchers, the few remaining animals were saved from extinction. A slow rebuilding of bison numbers continued throughout the twentieth century. A 1929 census reported 3,385 bison. By the late 1980s, the number of bison in North America had grown to more than 80,000. Legal protection of the bison in Yellowstone Park, the establishment of preserves like the National Bison Refuge in Montana, along with individuals raising bison on their own land, have helped restore the bison population. Census of Agriculture estimates place the size of the U.S. bison herd at 220,000 animals, with most of the production (about 200,000 head) on private ranches. Today, bison producers can be found in all 50 states, every Canadian province, and several countries overseas. The Canadian herd is estimated at 250,000 head. This is an amazing comeback for a species that hovered on the brink of extinction in 1900. The American bison is no longer on the endangered species list (Figure 48-1).

Figure 48-1 The American bison is no longer on the endangered species list.

Description and Characteristics

Bison have a hump over the front shoulders. The bull has larger horns than the cow; the horns curve outward and up from the head. The head is large, and the body narrows down toward the hindquarters. Long, dark hair covers the head and forequarters; the hindquarters are covered with shorter hair. The bull has a black beard that is about 12 inches long. A mature bull may weigh from 1,500 to 2,000 pounds and stands about 6.5 feet at the hump; the body ranges from 9 to 12 feet in length. The mature cow is smaller than the bull, weighing about 1,000 pounds. Bison have 14 pairs of ribs instead of the 13 pairs found in cattle.

With proper care and management, bison flourish in a wide range of environments. They survived on grass and the open plains of America for 10,000 years without extra hay, feed, shelter, or other care provided by people. Their hair coat thickens in cold weather, providing them with protection from low temperatures. Bison can find grass under heavy snow cover and will eat snow to get the water they need. They can survive equally well in warm climates.

Bison are territorial animals; that is, they will defend their territory from intruders. They can successfully fight off most predators. People working with bison must exercise care. While generally not aggressive, they are wild animals; cows with calves and bulls in rutting (breeding) season can be dangerous.

CHARACTERISTICS OF THE BISON INDUSTRY

The bison industry in the United States is small compared with other livestock enterprises on farms and ranches. Bison is a small, niche livestock market. Bison require much less care, handling, and shelter than beef cattle. People interested in becoming involved in this industry have several options:

- absentee ownership
- hobby
- small producer
- medium-size ranches
- large ranches

Types of Bison Ownership

Absentee Ownership

People who want to invest in bison but do not have the land or knowledge to raise them might choose to contract with an existing bison ranch to own one or more animals. The extent of control the absentee owner has over the bison owned depends on the terms of the contract. Absentee ownership offers investment and tax shelter opportunities. Because of rising costs and other problems associated with this option, there are few absentee ownership programs currently being offered in the bison industry.

Hobby

A person with an interest in bison may keep only a few (usually fewer than 25) head as a hobby or attraction. This kind of ownership may be just for personal pleasure and interest or may be to serve as an attraction to some other type of existing business, such as a restaurant, museum, recreational area, or tourist attraction. Care must be taken to protect the bison and the public in such operations.

Small Producer

A small producer is a person with a small herd of bison (usually 25 to 100 head) who gets some income from the herd, usually from the sale of breeding stock. The small producer may have a goal of expanding into a larger operation.

Medium-Sized Ranches

A person with a medium-sized ranch has a herd of 100 to 250 bison. Sales from this kind of operation usually include breeding stock and excess bulls for a meat market.

Large Ranches

The large producer is one with a herd of bison ranging in size from several hundred to several thousand head. The bison are the major source of income in this operation. Income comes from the sale of breeding stock, and excess bulls sold or fed out for a meat market.

Breeding Stock

Because of a growing interest in raising bison, the demand for breeding stock has increased in recent years. There are a limited number of bison available for breeding purposes; therefore, the price of breeding animals has risen. There is a continuing strong demand for bison breeding stock. In addition to private owners, surplus bison are being offered for sale at auctions from some public herds.

Demand for Meat

Excess bulls and animals not suitable for breeding stock are sold for meat. Bison produce a healthy, lean, red meat that is low in fat, calories, and cholesterol (Figure 48-2). It has a high protein content without a gamey or wild flavor. Hormones and subtherapeutic antibiotics are rarely used in the production of bison, making the meat desirable for those who have concerns about the presence of these substances in the food supply. The average American eats less than a pound of bison meat a year compared with about 66 pounds of beef. However, demand has been growing recently for bison. One factor in the increase in demand is that some national chain restaurants now carry bison on the menu. In addition, bison meat is readily available on many Internet websites with delivery directly to the home anywhere in the United States. The bison meat presents an exotic dining experience that many people seek, especially when eating out and traveling. The end result has been that demand often outstrips supply. As America's appetite for bison grows, ranchers have steadily been increasing the size of their herds.

© iStockphoto/Debbi Smirnoff

Figure 48-2 Bison meat is a lean, red meat that is low in fat, calories, and cholesterol.

MANAGEMENT OF BISON

Selecting Bison for the Herd

When starting out in the business of raising bison, it is usually best to begin by buying heifer calves. Calves adjust to a new environment more easily than older animals (Figure 48-3). Yearlings from a good herd are also a good choice. If buying 2- or 3-year-old cows, be sure they are bred and tested for pregnancy.

Photo by Alaska Interior Game Ranch Inc

Figure 48-3 Bison calf with herd.

To develop a good breeding herd, select animals from several herds that have developed superior breeding animals. The prospective buyer should learn as much as possible about the health program, the source of stock, and the calving rate of the herd from which the purchase is made. Look at the entire herd to get an idea of the overall condition of the animals. Determine the vaccination and worming schedule followed; ask about any other health problems. Inbreeding is a serious problem with bison because the current stock began with so few animals. Generally, buy from herds that avoid inbreeding. Good calving rates (90 to 95 percent) indicate a vigorous culling program with generally sound herd health practices.

When selecting animals, look for large, lengthy animals with well-developed hindquarters. An animal with good conformation has a flat back with a slight slope toward the tail. The top and back show a rounded, angular development. A large-frame body with adequate width and depth over its length is desirable. The parts of the bison are shown in Figure 48-4.

Avoid animals with a severe sloping of the back downward toward the tail and a severe taper from the hind legs toward the tail. This is referred to as a pencil-shaped rear end. Do not select females that show masculine traits or males that show feminine traits. Both conditions appear to correlate with sterility. Crooked legs and deformed body parts are also undesirable.

The presence or absence of horns, hair coat color, and length of hair on the forehead are generally matters of personal preference. They have little overall effect on the desirability of the animals as breeding stock.

Herd Management

Handling Bison

Always use caution when working with bison. They are wild animals and have great strength. Bison require different handling procedures than cattle. Many bison

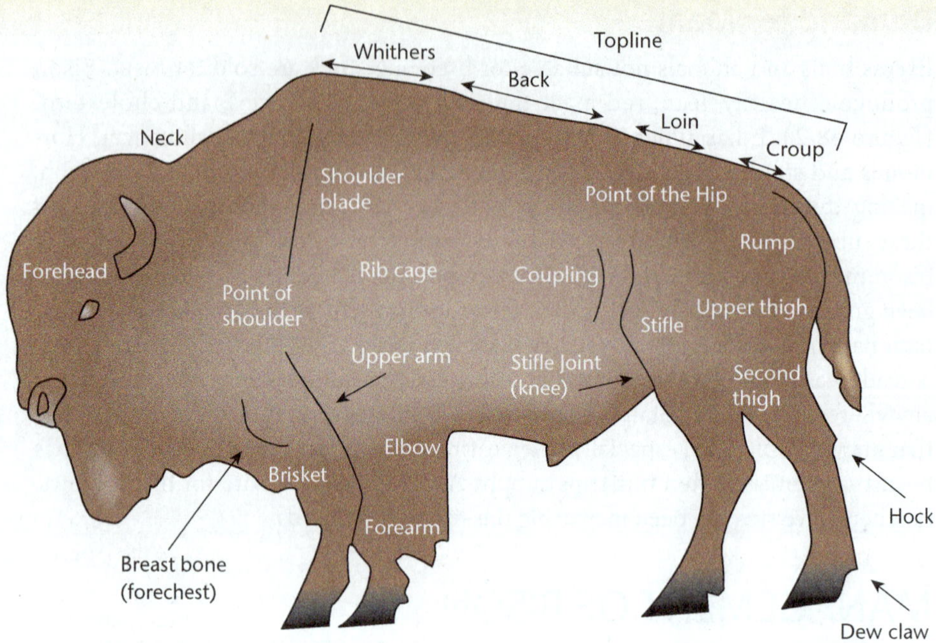

Figure 48-4 Parts of the bison.

producers agree with the saying, "You can get a bison to go anywhere he wants to go." Serious injury to workers may result from carelessness around the bison herd.

Do not attempt to work with bison during the breeding season and when cows are calving. Bulls become aggressive and dangerous during the breeding season. Cows that are calving will aggressively protect their calves, charging workers to drive them away. Young calves are not as dangerous as adult animals but should still be handled with care.

When working with bison, plan ahead and move slowly. Do not attempt to force the animals. Bison generally follow the herd leaders; therefore, if the herd leaders can be moved to the desired area, the rest of the herd will tend to follow. Using feed may help to get the bison to move to another area.

Use a closed truck or trailer with a top to transport bison. A stock trailer is recommended for use in hauling bison. Get the bison used to being confined by holding them in a closed corral for a time before transporting them. To prevent injuries in the trailer, separate the young animals from the older animals. Keep bulls in a compartment separate from other bison.

Bringing New Animals into the Herd

New animals being brought into a herd should be isolated in a separate corral or holding area for a period of 2 to 4 weeks before mixing them with the rest of the animals. This allows them to become acclimated to the new area and gives them time to settle down. There will be some pushing and shoving when the new animal is moved into the herd as a new dominance order is established.

Breeding

Bison reach maturity about 1 year later than cattle. Generally, the females can be bred when they are 2 years old. The gestation period is 275 days. Bison generally breed in July and August, with the calves being born the following spring. It is usually a good practice to keep yearling heifers separated from the breeding herd during the breeding season because some may become pregnant at this age. Breeding bulls can be put into service when they are 2 years old. Breeding bulls may be semen tested to ensure that they are not sterile. A normal ratio is 1 mature bull to every 10 cows or 2 younger bulls with 10 cows. Some producers let the bulls run with the herd all year; others keep them separate except during the breeding season. Artificial insemination is not generally used with bison herds.

Inbreeding should be avoided; replace herd bulls about every 2 years to avoid breeding back to mothers or sisters in the herd. Inbreeding has the following undesirable effects:

- Fertility is decreased in bulls that are the product of inbreeding
- Feed efficiency is lower; it may take as much as 2 years longer to produce a desirable carcass
- Dished in, shortened faces can appear in the offspring
- Unsoundness in the legs, a condition called rabbit-legged bison, appears

Outbreeding (mating animals that are not closely related) develops individuals with more vigor, size, and hardiness and reduces the effects of recessive genes. Outbreeding is recommended in bison herds.

Weaning

Some producers regularly wean calves and others do not. Bison calves may be weaned when they are 8 to 9 months old. Weaned calves need to be started on supplements by creep feeding them for 1 to 2 months before weaning. Worm, ear-tag, and vaccinate calves at weaning time. Vaccinate heifer calves for brucellosis. The weaned calves should be kept in a shelter with plenty of feed and water. The major reason for weaning calves is to reduce stress on the cows; they also appear to grow faster.

Culling

Bison herds should be culled to remove old, unproductive, or undesirable individuals. Cull animals that do not meet the criteria for productivity, health, vigor, temperament, and physical characteristics. Usually, culling is not necessary until several years after the herd has been established. Keep only the best-quality bulls for breeding purposes. The remainder should be removed, generally as 2-year-olds, and sold for meat. This will maintain herd size at a desirable level.

Branding and Identification

Most states do not require bison to be branded for identification purposes because they are classified as wild animals. A herd owner needs to check state requirements. Some owners prefer not to brand bison because branding reduces the value of the hides. Most owners use ear tags as a method of identifying animals. A disadvantage of ear tags is that they may be lost.

Dehorning

The advantages of dehorning bison are as follows:

- animals, especially cows, are not as dangerous
- animals are more docile
- less damage to hides and meat in the feedlot
- reduces the space requirements in the feedlot
- animals require less feed bunk space

Those who oppose dehorning state the following reasons:

- detracts from the natural appearance of bison
- proper handling can minimize the danger and economic losses without changing the animals' appearance

Owners who keep bison primarily for an attraction and do not handle the herd extensively probably will not want to dehorn the animals. Large herds that are kept for meat, where extensive handling is necessary, probably should consider dehorning at least the females and the bulls that do not appear to have good trophy heads.

To prevent maggot infestation, dehorn at least 6 weeks before fly season begins or after a killing frost in the fall. Commercial dehorning instruments are available. Remove horns below the hairline. Cauterizing the wound with a hot iron helps prevent bleeding and the growth of spurs from missed horn cells.

Castration

Bison bulls are usually not castrated. The bulls are sold for meat before they reach breeding age; therefore, there is no undesirable flavor in the meat from uncastrated animals. Castrated bison appear to grow more slowly than those that are not castrated.

Immobilization

Occasionally it may be necessary to immobilize an animal for treatment. There are several drugs that are used for this purpose. A veterinarian should always be present when it is necessary to immobilize an animal. Generally, do not use drugs to immobilize a bison unless it is absolutely necessary.

Orphan Calves

In general, do not remove calves from their mothers unless it is necessary. Circumstances that may make it necessary to remove the calf from the mother include if the cow:

- dies
- is unhealthy and not able to care for the calf
- is unable to nurse the calf
- is too wild to handle and must be removed from the herd

When it is necessary to raise an orphan calf, it is important that the calf receive colostrum milk in the first few hours of life. Bison colostrum milk is usually not commercially available; dairy cow colostrum milk will work as well and is more often available. After the calf has received colostrum milk, goat's milk or a mixture of evaporated milk and water may be used to bottle feed the calf. Lamb milk replacer is a better choice than beef milk replacer if goat's milk or evaporated milk is not available. Lamb milk replacer is higher in protein and fat; its composition is closer to bison milk than cattle milk replacer.

The calf should be fed 0.5 to 1 pint of milk four to six times per day for about 2 weeks. Gradually increase the amount of milk per feeding until the calf is being fed 1.5 to 2 quarts of milk replacer twice a day. Watch carefully for signs of scours. Overfeeding or too rich a mixture may cause scours. Treat calves with scours the same way a beef calf is treated.

Provide access to water, hay, and grain free choice. The calf will start eating at a fairly young age. Provide a pen with shelter from the weather and an area for exercise. Wean calves from the bottle when they are about 6 to 7 months old. Hay supplemented with grain can be fed after the calf is weaned.

Registering Animals and Bloodtyping

The National Bison Association founded the North American Bison Registry in 1980. **Bloodtyping** is used to identify individual animals and to determine parentage, which is possible if the sire and dam are also bloodtyped. The use of bloodtyping can determine that the animal is a purebred bison and not a crossbred animal. It is a good management tool for selecting breeding stock, especially herd bulls. Detailed information on registering animals and bloodtyping is available from the National Bison Association.

FEEDING AND NUTRITION OF BISON

Feeding the Bison Herd

The nutrient requirements of bison are similar to those for beef cattle. Bison herds can use a variety of grasslands to meet their basic nutritional requirements (Figure 48-5). When the size of the rangeland available is limited in relation to the size of the herd, it may be wise to use a rotational grazing system. One method of rotational grazing is to move the herd to different pastures during the grazing season. Another method is to fence the pasture into several smaller areas and move the herd from one area to another. When rotational grazing is used, the pasture should be allowed about 30 days for regrowth before moving the herd back onto it. Rotational grazing is recommended when the amount of pasture available is limited. It does require additional investment in fence building and time.

Photo by Ron Nichols, USDA

Figure 48-5 Bison can use a variety of grasslands to meet their basic nutritional requirements.

When the size of the pasture is large in relation to the size of the herd, it is possible to allow the bison to range freely over the entire area. If enough space is available, bison tend to move from one area to another each day. This method of grazing works best on large rangeland areas where it can be used most easily and efficiently.

During the winter, when pasture is not available, the herd requires supplemental feeding. Hay or other forages may be used along with protein supplement, salt, and minerals. Range cubes are commercially available and are often used for supplemental winter feeding. The protein level of the ration should be from 11 to 13 percent. The amount of protein supplement needed depends on the quality of the hay or other forage fed.

Be sure there is an adequate supply of fresh, clean water for the herd. Mineral requirements for bison are similar to those for beef cattle. Free choice feeding of an iodized salt mix along with a good commercial mineral mix will generally meet the requirements of the bison. Salt blocks and mineral blocks are commercially available for use on pasture.

Finishing in the Feedlot

Extra young bulls from the herd are usually finished in the feedlot to be sold for meat. After weaning, bull calves that are to be fed out may be fed a roughage ration in drylot or carried on pasture until the next year when they are moved into the feedlot for finishing. The animals are then fed a finishing ration until they reach a live weight of 1,000 to 1,100 pounds, at which point they are marketed as meat. Bison will generally yield about 62 percent dressed meat.

Bison may be successfully started on feed with a ration of prairie and alfalfa hay, supplemented with some corn. After they become accustomed to eating corn, a free choice ration of alfalfa hay and corn may be used. The ration should contain 10 to 12 percent protein. Soybean oil meal may be used to add protein to the ration if necessary. A complete cubed ration containing roughage, grain, and supplement may be used. There is no apparent advantage to grinding, cracking, or rolling the corn in the ration when it is not fed in a complete ration. Consumption is about 10 pounds of alfalfa and 7 to 10 pounds of corn per head per day. A mineral supplement should be available free choice. Make sure there is plenty of water at all times.

Bison do best in the feedlot if all are about the same age and size. Dehorned animals require less pen and feed bunk space. Self-feeders for the concentrate mix are recommended.

Average daily gain for well-managed bison in the feedlot can be as much as 2 to 3 pounds. Because of genetic differences, some bison do not gain well in the feedlot. If they do not appear to be gaining adequately after 3 to 4 months on feed, they should be considered for slaughter because they will probably never make adequate gains no matter how long they are fed.

BISON HERD HEALTH

Preventing health problems in the bison herd is better than trying to treat the problem after it occurs. Proper feeding and good sanitation will help maintain a herd in a healthy condition. Keep feedlots, feed bunks, and waterers clean. Bring only healthy animals into the herd. Worming and vaccinating new animals is generally recommended. Work with a veterinarian to determine the proper treatment of new breeding stock as well as the existing herd.

Bison are subject to many of the same health problems that affect other ruminants such as cattle and sheep. Some diseases that may affect bison include the following:

- anaplasmosis—bison are rarely affected but can be carriers; testing is recommended
- anthrax—vaccinate if a problem exists
- blackleg—vaccination is recommended
- brucellosis—vaccinate female calves at 3 to 10 months to keep the herd clean
- coccidiosis—bloody scours
- enterotoxemia—vaccination is recommended
- hoof rot—foot rot
- leptospirosis—vaccination is recommended
- malignant edema—vaccination is recommended
- shipping fever—use preventive treatment

Other diseases that may affect bison include malignant catarrhal fever and tuberculosis. Malignant catarrhal fever is caused by a virus. Symptoms include a high fever and death. Bison may be infected by coming into contact with sheep that carry the virus but do not develop symptoms. There is no treatment. Prevent infection by keeping bison away from contact with sheep.

There are three forms of tuberculosis: human, bovine, and avian. The bacteria that cause tuberculosis, mycobacterium, only multiply in the host. The disease affects the lungs, causing a chronic weakening condition. In some animals it becomes acute and may cause death. Infection is spread from animal to animal by bodily discharges that get on the feed, in the water, or in the manure. Removal and disposal of infected animals is the only way to control this disease.

Using Medications

Medications that are approved for cattle are usually safe for use with bison. Bison are generally subject to the same external and internal parasites that affect cattle. A general discussion of these parasites is found in Chapter 17. Medications may be administered through water or feed, by spraying or pouring on the animal, and by injection. Always follow the advice on the label, including withdrawal time before slaughter, when administering medications. It is usually a good management practice to consult a veterinarian when health problems arise.

Poisonous Plants

Bison are affected by the same poisonous plants that affect cattle and sheep. In some cases, the bison appear to be less susceptible to some poisonous plants than cattle. It is a good management practice to avoid exposing animals to poisonous plants and moldy feeds whenever possible.

FACILITIES AND EQUIPMENT FOR BISON

Fences

Bison need higher fences than cattle (Figure 48-6). Fences should be at least 6 feet high. Bison can jump higher than some horses; if they can get their nose over a fence they may try to jump over it. Fences surrounding the pasture area need to be strong and high enough to keep the animals confined.

Barbed wire can be used to fence bison; however, because of bisons have thick hair and hide, these fences are not as effective as other types. Double-strand (one roll on top of another) woven wire fences will confine bison. Woven wire fences are more expensive and require more labor to install. High-tensile wire fencing, which is a smooth wire with high tensile strength, works well with bison. High-tensile wire is less costly than woven wire and about the same cost as barbed wire fencing. An electric fence may be used if the bison are trained to respect it. Pipe or a combination of pipe and cable is especially useful for small pastures, feedlots, and corrals.

Photo by Alaska Interior Game Ranch Inc

Figure 48-6 This game fence in Alaska will be 8 feet high when completed.

Temporary fencing may be used when moving bison from one area to another. Portable cattle panels may work for temporary fencing if they are tall enough. Black plastic has been used successfully to cover a weak area in an existing fence or to make a temporary holding pen. Use care when working with bison if temporary fencing is used. The animals may soon discover that the fencing is not substantial and will go through it.

Small pastures need higher and stronger fences than larger pastures. Corral-type fencing should be used in small areas where bison are confined for long periods of time.

Gates

Pasture gates may be made of the same material as the fence. Locate pasture gates away from corners because bison do not like to be driven into a corner area. Corral gates should be constructed of heavier material than pasture gates. Solid gates are better for working with bison in the corral. Use rope-activated, spring-loaded gate latches if possible. These are safer and easier to use than other types of gate latches.

Cattle Guards

Ordinary cattle guards do not work well with bison. They have a tendency to jump over them. A cattle guard for use with bison should be at least 12 to 16 feet long and 8 feet wide. Keeping the cattle guard hole cleaned out or using a light-colored material in the bottom of the hole may deter bison from attempting to jump over it.

Handling Facilities

Handling facilities must be built stronger than pasture facilities because bison tend to become excitable in confined areas. Bison are extremely large and strong animals; the safety of the workers must always be considered when designing handling facilities.

Catch Pen

A catch pen is used to move bison from the pasture to the corral. Use a funnel or triangular shape with an exit gate in the corner leading to the holding pen.

Holding Pen

A holding pen is used to sort animals before working with them. The pen should be 6 to 7 feet in height.

Figure 48-7 A running alley is used to move bison from the catch pens to the crowding area. Note the sturdy construction, solid panels, and 8-foot-high walls.

Figure 48-8 The squeeze chute and drop-in gates as seen from the alleyway.

Figure 48-9 The squeeze chute is used to hold bison for testing, vaccination, or other work. Note the heavy-duty nature of bison handling equipment compared to cattle equipment.

Running Alley

The running alley is used to move the animals from the catch pen or holding pen into the crowding area (Figure 48-7).

Crowding Pen

The two basic types of crowding pens are funnel and circle or half-circle. Bison remain calmer in solid-sided pens. Catwalks built along the sides of the pen make it easier for workers to handle gates with greater safety. Gates in the crowding pen help move the bison into the working area. The alleyway leading from the crowding pen to the working area should be equipped with a series of gates to prevent the animals from turning back. The animals should be moved single file through this alleyway to the working area. An adjustable alleyway permits working with different sizes of animals more easily. A scale might be included in the area just before the squeeze chute.

Squeeze Chute

The squeeze chute is used to hold the animal for testing, vaccination, administering medications, or other work that may need to be done when the animal is confined (Figures 48-8 and 48-9). The animal is held in a head catch, and the sides can be squeezed together to provide further confinement. An ordinary cattle squeeze chute is generally not suitable for use because it is not tall or strong enough for a mature bison. Squeeze chutes designed for bison are commercially available.

MARKETING BISON

Bison are marketed as breeding stock, for meat, and for by-products. Herd owners may choose to market their bison in any or all of these categories.

Selling Breeding Stock

Bison sold as breeding stock may be marketed by sale to private individuals or at auction. Personal reputation and advertising are keys to successful marketing of breeding stock. Certificates of health are generally needed, especially if the buyer is from a different state. Breeding stock should be guaranteed free of brucellosis and tuberculosis. Testing is required for these two diseases. Some states require additional testing before bison can be brought into the state.

Careful selection of the best stock for sale as breeding animals will help maintain the reputation of the seller and improve the bison industry. Poorer-quality animals should only be marketed for meat.

A general livestock auction is not a good choice for selling bison. Well-advertised bison auctions will attract the kind and number of buyers the seller wants. These auctions are a better choice for the seller. Video auctions have been held for bison; these provide exposure of animals to a large number of potential buyers without having to transport the bison.

Selling to Feedlot Operators

Feedlot operators need a consistent source of animals. The bison herd owner who does not want to feed bison for meat can sell the surplus animals to a feedlot operator.

Meat Marketing

Bison herd owners who want to market some of their animals as meat have two general types of markets to consider. The traditional bison market consists of consumers who want a low-fat red meat. Animals sold to this market are generally grass fed with little or no finishing in the feedlot. The gourmet market consists of consumers who want grain-fed meat. Animals sold to this market are finished on grain in a feedlot before being marketed.

Several channels are open to the producer for marketing bison meat:

- distributors and meat brokers
- restaurants
- diet and health food stores
- direct to consumers

There are an increasing number of meat distributors and meat brokers that purchase bison meat for resale. When selling to restaurants, the producer must be able to provide delivery of the kinds of cuts the restaurant needs. Diet and health food stores are looking for wholesome products that are packaged to their specifications. Direct sales to consumers may involve selling entire carcasses, halves, or quarters. Some consumers will want only a small number of cuts packaged for their use. Direct sales may also be done through mail order. All customers want a quality product with good service.

Marketing By-Products

There are a number of by-products from bison that the producer may want to market. These include.

- heads
- robes
- skulls
- leather goods
- wool
- bones and horns

Heads, robes, and skulls are often used for decoration. Bison bones and horns are used in many decorative ways. Some heads are mounted with the hump area; others do not include the hump. Winter robes are hides that have been tanned while covered with the winter hair coat. These may be used for wall hangings, furniture throws, rugs, pillow covers, seat cushion covers, and clothing. Bison skulls may be bleached or used without bleaching. They are sometimes used by artists for paintings or are decorated with stones. Leather garments are in high demand. Bison shed their hair in the spring of the year and this may be gathered and carded like lamb's wool. It can then be spun into yarn. By-products from bison are a specialty market that can add profit to the enterprise.

RATITES

Introduction

There is an interest in the United States in raising ratites as an alternative to other livestock production. Commercial production of ostriches, primarily for feathers, began in the 1800s in South Africa. After World War II, a market developed for the meat and leather from ostriches. South Africa has retained a virtual monopoly on the commercial production of ostriches until recently.

Most of the ratite industry in the United States involves ostriches, emus, and rheas. The care and management of ostriches, emus, and rheas are similar.

Ratites are a group of flightless birds that include the ostrich, emu, rhea, cassowary, and kiwi. The breastbone of ratites is flat, lacking the keel that is characteristic of most birds that fly. The shape of the breastbone resembles a raft; the name ratite comes from the Latin words *ratis* (raft) and *ratitus* (shape of a raft). The ostrich at one time was found in the Middle East and Africa; currently, its natural range is only in Africa. The natural range of the only surviving species of emu is in Australia. The rhea is from South America.

Description

Ostrich

Figure 48-10 Male and female ostrich with young ostriches.

The ostrich is the largest of the ratites and is also the largest bird in existence. Mature ostriches are 7 to 9 feet in height and weigh from 200 to 350 pounds; males are generally larger than females (Figure 48-10). The ostrich is capable of running at speeds of up to 30 to 50 miles per hour for up to 30 minutes. The length of its stride is from 15 to more than 20 feet, depending on its size. It has only two toes on its feet; the other ratites have three. The male ostrich has black feathers with white feathers on the wings and tail; the female is more drab and has light gray to grayish brown feathers. The feathers of the male become brighter during the mating season. Ostriches have long necks and small heads with large eyes and a short, flat bill.

The domestic ostrich is a result of crossbreeding several subspecies, all of which belong to the species *Struthio camelus.* The initial purpose of the crossbreeding was to improve feather quality. The emphasis on breeding today is selection for meat production.

Emu

Figure 48-11 The emu is the second largest bird in the world, with the ostrich being the largest.

The emu is the second largest bird in existence; mature adults are 5 to 6 feet high and weigh from 125 to 150 pounds (Figure 48-11). The female is dominant and the male is slightly smaller than the female. Emus are fast runners, being capable of speeds up to 30 miles per hour. The head and neck of mature birds are colored a grayish blue. There are black feathers on the head and mottled brown feathers on the rest of the body. The plumage is coarse and hair-like. The emu has three toes on its feet.

Rhea

The rhea (Figure 48-12) is similar in appearance to the ostrich but is smaller and has three toes on its feet instead of two. The mature rhea is about 5.50 feet high and weighs 44 to 55 pounds. Like the ostrich, the rhea can run very rapidly. The feathers of the rhea are pale gray to brown; the head and neck are completely feathered and there are no tail feathers. The long body feathers droop over the posterior of the bird.

Breeding and Egg Production

Ostrich

The male ostrich matures at about $2\frac{1}{2}$ years of age, the female at about 2 years of age. Ostriches will mate under both monogamous and polygamous conditions; breeders in the United States typically mate bonded pairs. They engage in elaborate courtship behavior; the male vocalizes with a loud booming noise, inflates his neck area, sits on his hocks, and moves his wings up and down and his head from side to side. The female indicates receptive behavior by bowing, ruffling her wings, sitting, and allowing the male to mount her from the back.

Figure 48-12 An adult rhea.

In the United States, mating and egg laying generally begin in March and may extend to September; under some conditions, the breeding season may be shorter or longer. Breeding pairs should be established about 2 months before the breeding season begins. The female begins to lay eggs about 5 to 10 days after the first mating. The male digs a shallow depression in the ground, forming a nest where the female lays the eggs. One egg is usually laid every other day; the total number of eggs laid during the season varies widely. When the eggs are left in the nest after they are laid, a total of about 15 to 20 eggs will be laid. More eggs will be laid if each egg is removed from the nest after it is laid. While estimates of total egg production range up to 90, the average number is closer to 40 to 50 eggs per hen. Ostrich eggs weigh from 3 to 5 pounds. The difference between an ostrich egg and a chicken egg is shown in Figure 48-13. A diet that does not meet the needs of the hen will reduce the number of eggs laid. The male aggressively defends the female and the nest. Both the male and female sit on the nest. The male usually sits on the nest at night and the female sits on it during the day.

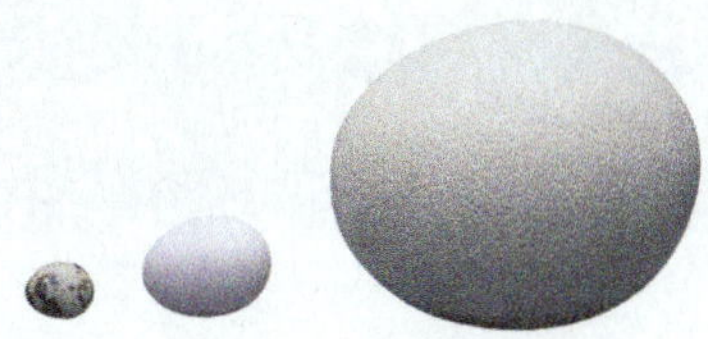

Figure 48-13 An ostrich egg (right) can weigh up to about 5 pounds. This is about 40 times more than one large chicken egg. A quail, chicken, and ostrich egg are shown from left to right.

Emu

Most breeders pair emus in monogamous relationships; however, it is possible to group them in polygamous groups of one male to more than one female. Emu females are the dominant member of the pair; they choose the male for mating, choose the nesting site, and defend their territory. Sexual maturity is reached in $1\frac{1}{2}$ to 3 years. The mating season usually is from November to March. Vocalization occurs during courtship, with the male making a grunting sound and the female making a drumming sound. They both strut and display their neck feathers. Eggs are laid in a shallow depression in the ground. On average, 30 eggs are laid by one female during the egg laying season. Typically, one egg is laid about every 3 to 4 days (Figure 48-14). Emu eggs weigh about 1.5 pounds. The male sits on the nest to incubate the eggs if they are not removed for artificial incubation.

Figure 48-14 Emu checking its egg.

Rhea

In their natural habitat, rheas are polygamous; a group usually consists of one male and six or more females. The females lay their eggs in the same nest, and the male sits on the nest to incubate the eggs. The management of rhea breeding in domestic production is similar to practices followed for the ostrich.

Incubation and Hatching of Eggs

Most ratite producers in the United States use artificial rather than natural incubation for eggs. The incubation time for ostrich eggs ranges from 39 to 59 days, with an average time of 42 days; for emu eggs, the range is 46 to 56 days, with an average time of 50 days. Commercial incubators and hatchers are available for use by ratite producers.

Collect the eggs from the nests each day and store them in a cool place at 55 to 65°F at a humidity of 75 percent. Before placing the eggs in the incubator, store them with the large end up or on their side and turn them three times each day. To achieve maximum hatchability, the eggs should be placed in the incubator within 7 days. Eggs that are dirty should be cleaned by sanding or washing in water containing a sanitizing agent that is 10 degrees warmer than the egg temperature. Do not use water that is too hot. The shell of the egg is porous, and care must be taken not to allow water and bacteria to penetrate the egg.

Use care when handling eggs to prevent contamination. Some producers wear disposable latex gloves when collecting and handling the eggs. Contamination may also be prevented by thoroughly washing the hands before handling eggs.

The incubator temperature for ratite eggs is usually set in a range of 97.5 to 99°F. The producer should follow the directions for the specific brand of incubator being used. Humidity in the incubator is adjusted to achieve the desired moisture loss from the eggs. Typical relative humidity settings for ostrich eggs are in a range of 35 to 40 percent and 24 to 35 percent for emu eggs. Follow the manufacturers' directions for relative humidity setting. Eggs are weighed weekly to determine the amount of moisture loss that is occurring. This information is used to set the appropriate relative humidity in the incubator. The moisture loss should be about 12 to 16 percent over the incubation period.

Eggs need to be positioned properly so the air cell is at the top and turned while in the incubator. Ostrich eggs can be candled prior to placement in the incubator to determine the position of the air cell. Emu and other ratite eggs that are opaque should be placed on their side. Hand turn the eggs through a 90-degree angle several times daily. Some incubators automatically turn the eggs 12 to 24 times during a 24-hour period.

One to 2 days before hatching, the chick will break through the internal shell membrane into the air cell. At this time the egg is moved to the hatcher. Twelve to 18 hours after internal piping occurs, the chick will begin to break through the shell of the egg. The chick will normally hatch within 12 hours after first breaking the egg shell. They are moved to the brooding area after they are dry.

Not all eggs set in the incubator will hatch. On average, about 70 percent of ostrich eggs set in the incubator will hatch; the average for emu eggs ranges from 50 to 80 percent. To determine the number of fertile eggs use this formula:

$$\frac{\text{Number of fetile eggs}}{\text{Number of total eggs set}} \times 100 = \text{Percentage of fertility}$$

Fertile eggs are any that showed indications of fertility even if an embryo did not develop. Hatchability of fertile eggs is determined by this formula:

$$\frac{\text{Number of chicks hatched}}{\text{Number of fertile eggs set}} \times 100 = \text{Percentage hatch of fertile eggs}$$

If the percentage of fertile eggs is low, then a problem with the breeding program is indicated. If the percentage of hatched eggs is low, then a problem probably exists in the incubator.

It is important to inspect the material left in the hatcher tray after the birds have hatched. This material will give an indication of possible problems. Common problems found in the hatch waste breakout are indicated in Table 48-1.

TABLE 48-1 Common Symptoms of Problems Found in Hatch Waste Breakout

Symptom	Check for
Smelly, brown-colored albumen	Contamination of eggs
Missing or small lower jaws	High/low selenium, low calcium
Chicks with head away from air cell	Proper positioning in incubator tray
Extremely small embryo or chick, no feathering seen (early death)	High incubator temperatures during beginning period of incubation
Little formation of air cell, failure to pip by embryo	High humidity in incubator
Low water loss within eggs, edema seen in chicks	High humidity in incubator, vitamin E/calcium/selenium deficiencies
Chicks with dry membranes stretched and surrounding chick	Low humidity in incubator and possibly within the hatching cabinet
Liquid seeping from eggs	Contamination of eggs
Chicks hatched with large amounts of yolk exposed	Late high temperatures in incubator and hatcher, low water loss during incubation
Spraddled legs, problems standing	Slick hatcher trays, giving improper footing. Low calcium in some cases
No embryo seen within egg, undeveloped germinal disk	Infertility, lack of breeding activity

Source: G. P. Martin, R. C. Fanguy, and J. S. Jeffrey, *The Incubation of Ratite Eggs* (College Station, TX: Texas A&M University, 2012)

Brooding

Ratite chicks need supplemental heat for several weeks after they hatch. The brooder area at floor level should be kept at about 90°F for 10 days to 2 weeks. After that, the temperature can be reduced about 5 degrees every 2 weeks until supplemental heat is no longer needed. Do not allow chicks to become chilled or overheated. Provide enough space so they can move away from the heat source if necessary.

The brooding area needs to be clean, dry, and well ventilated. Provide protection from inclement weather if necessary. A variety of materials can be used for the floor area, including wood shavings, straw, rice hulls, or clean sand. The material used should be easy to clean. Do not put any slick material such as newspaper, cardboard, or plastic on the floor. On a slick floor chicks may develop "spraddle leg," which may be fatal. Ratite chicks tend to eat the litter, which may result in impaction in the digestive tract. This may be prevented by covering the litter with burlap for the first 2 weeks. Stir the litter periodically to keep it dry and prevent caking.

Do not overcrowd the chicks in the brooding area because this may result in their piling up together and suffocating. As they get older, they may be moved to larger pens. It is a good idea to provide outside exercise areas if the weather permits. This area should be free of trash, small rocks, or other material the birds may eat.

Chicks should be weighed and marked with numbered leg bands at the time they are placed in the brooder. Ratite chicks are not fed for 1 to several days after they have hatched so they will absorb the nutrients from the yolk sac. Feeding too soon can prevent the absorption of the yolk nutrients.

Sexing Young Birds

It is difficult to determine the sex of young ratite birds by observation alone. An examination of the sex organs is generally required. This is most easily done when the birds are 1 to 2 months of age. Because the sexual organs are small in immature birds, it requires some practice to accurately determine the sex of the individual bird. Both the penis and the clitoris are located in the ventral cloaca. To determine the sex of the bird, apply pressure on the cloaca so that it inverts, revealing the penis or clitoris. The penis is larger, curved, and has more cartilage than the clitoris.

Feeding

Only a limited amount of research has been done on the nutritional requirements of ratites. Commercial ratite starter, grower, and breeder rations in mash or pellet form are available from a number of feed companies. Most of these are based on nutritional research with chickens, turkeys, and game birds. The protein content of commercial feeds for ratites ranges from 16 to 22 percent.

After ratite chicks have absorbed the yolk nutrients (usually 1 to several days after hatching) they may be fed a starter ration. Commercial ratite, turkey, or game bird starter rations may be used. Do not use a ration that is too high in fiber because this may cause intestinal blockage. The fiber content of the ration should range from 5 to 15 percent. Some growers feed a vitamin/mineral supplement in addition to the starter ration. Never feed moldy feed.

Limited feeding for ratite chicks is generally recommended for a period of up to 3 months. Typical feeding patterns are two or three times per day for about 1 hour for each feeding period. Limiting the feed intake during this time reduces problems with bacterial enteritis that may occur if continuous feeding is done.

A continuous supply of fresh water should be available. Some growers provide supplemental feed such as cut alfalfa, clover, or alfalfa pellets during this time. Grit may need to be supplied if fresh roughage is provided.

Impaction of the digestive system is sometimes a problem with young ratites. They have a habit of eating an excessive amount of sand, small stones, or fiber if these materials are available. Some growers add mineral oil to the ration about once a week to reduce the problem of impaction. Impaction problems are less severe with older birds, probably because of the larger size of their digestive system.

The birds may be switched to a grower ration at 2 to 3 months of age; the grower ration is fed until the birds reach maturity. Some growers continue to limit feed the birds at this time; others begin free choice unlimited feeding at 10 to 12 weeks of age. Either system appears to be satisfactory. Supplemental feeding of alfalfa pellets or green forage may be done. Birds that have access to green forage should be provided with grit.

A breeder ration is fed when the birds are mature and beginning egg production. The ration needs to provide sufficient calcium and phosphorus for egg production. Forages such as alfalfa, oats, Bermuda grass, or wheat may be used as part of the diet; a complete breeder ration should also be fed to birds that have access to forage.

While birds are not laying eggs, a maintenance ration may be used. This ration should be sufficient to prevent weight loss but not allow the birds to gain weight. About 2 months before the laying season begins, switch to a breeder ration that provides more protein.

Some sample starter, grower, and breeder ration mixes for ostriches are shown in Table 48-2.

Handling and Transporting

Mature ratites can be dangerous to handle; they can kick with sufficient force to inflict serious injury to a human. They become nervous, jump about, and flail their wings when they are frightened or threatened. It is necessary to restrain a mature bird when handling it for medical treatment such as vaccination or examination.

Young birds can be picked up by the legs or by supporting the chest with one hand while holding the other hand on the back. Wrapping the young bird in a large towel makes it easier to handle and transport.

TABLE 48-2 Ostrich Ration Mixes

Ostrich Starter		Ostrich Grower		Ostrich Breeder	
Ingredient Name	**Amount (Pounds)**	**Ingredient Name**	**Amount (Pounds)**	**Ingredient Name**	**Amount (Pounds)**
Soybean meal	26.50	Soybean meal	26.00	Soybean meal	27.40
Corn grain, yellow	25.20	Corn grain, yellow	22.90	Corn grain, yellow	21.00
Alfalfa (dehydrated)	18.60	Alfalfa (dehydrated)	15.00	Corn screenings	10.20
Peanut hulls	9.53	Sorghum	15.00	Sorghum	10.00
Sorghum	8.50	Wheat midds	6.50	Alfalfa (dehydrated)	9.80
Corn screening	5.50	Fish meal (menhaden)	5.00	Meat and bone meal	8.18
Calcium carbonate	1.84	Peanut hulls	4.80	Calcium carbonate	4.90
Dicalcium phosphate	1.84	Calcium carbonate	1.47	Peanut hulls	4.80
Nutra blend vitamins	1.25	Nutra blend vitamins	1.25	Nutra blend vitamins	1.25
Fat (animal)	0.80	Dicalcium phosphate	1.23	Dicalcium phosphate	1.20
Salt	0.60	Fat (animal)	0.50	Fat (animal)	0.60
DL methionine	0.07	Salt	0.30	Salt	0.60
		DL methionine	0.04	DL methionine	0.06
Total weight	100.00	Total weight	100.00	Total weight	100.00

Source: J. Berry, *Ostrich Production*, OSU Extension Facts No. 3988 (Stillwater, OK: Oklahoma State University, 2007).

It may take several people to handle a mature bird. Because birds kick to the front or side, it is safer to approach them from the rear when preparing to restrain them. An opaque hood is often placed over the bird's head when handling is necessary. Usually, the bird will calm down when hooded. Restrain the bird by grasping its wings and pushing down. An adult bird that is used to being handled can be moved by guiding it with one hand holding a wing and the other hand holding the tail. Another method of moving a bird involves two people: one behind the bird holding it by the wings and lifting on the rump, and the other holding it by the neck or beak and guiding it in the desired direction.

It is easier to move birds if alleys are constructed between pens. A solid or semisolid panel may be used to push the bird down the alley. Birds are easier to handle under limited light conditions.

Transport ratites in enclosed trailers that are well ventilated and have a nonslip floor. Old rugs or carpets may be used to provide a nonslip surface on the floor of the trailer. Make sure there are no sharp projections in the trailer that might injure the birds. Separate birds of different sizes when transporting them in a trailer.

Identification

It is recommended that plastic leg bands be placed on the chicks when they hatch. Larger leg bands that can be adjusted in size may be used when the birds get older; these help identify the birds at a distance without having to restrain them for microchip reading. Microchips can be implanted, usually in the piping muscle just behind the head, at hatching time for identification purposes. Electronic identification is widely used in the ratite industry. Some insurance companies require microchip identification for insurance purposes. The microchip is implanted in young birds and can be read with a microchip reader to facilitate record keeping. Some breeders implant a second microchip in the tail head of older birds because it is easier to read when the bird is drinking or eating. Some companies offer DNA identification services. Tattooing has also been used for identification of birds.

Electronic identification provides proof of ownership, helps in breeding programs by keeping track of genetic bloodlines, and makes it easier to trace and recover missing birds. Most birds offered for sale have the microchip electronic identification devices already implanted.

Facilities

Young ratite chicks can be kept in small pens with adequate shade and protection from inclement weather. Generally, they should be moved inside at night and kept in a heated room until they are 2 to 3 months of age. Provide larger pens for older birds; shelter will be needed if the temperature of the region is cold. Ground cover of alfalfa, clover, or grass is recommended.

The size of pen for breeding pairs of ratites may be from 0.25 to 3 acres. One-third to one-half acre per pair is adequate if space is limited. Do not mix other animals in the same area because this may have a negative effect on their nesting and breeding. Provide an adequate area for exercise along with housing and shade. Woven wire, chain link, or smooth wire fencing may be used. Fences should be 5 feet high and flush with the ground to keep other animals out of the pen.

Alleys built between the pens make it easier to move birds from one area to another. Restraint facilities are useful for treating sick or injured birds. An isolation pen for sick birds is recommended. High solid walls should be used in crowding areas and chutes because this makes it easier to handle the birds.

Figure 48-15 Young ratite chicks, such as these young emus, may suffer from impaction after ingesting sticks, fabric, sand, glass, etc.

Health

Ratites are subject to many of the same illnesses and parasites as other poultry. It is wise to establish a good working relationship with a veterinarian who is knowledgeable about ratites to plan a good health maintenance program. Always consult a veterinarian when treatment for an illness or injury is necessary. Birds should be examined by a qualified veterinarian before they are purchased.

Birds under 6 months of age have the highest mortality rate. Emu chicks appear to have a lower mortality rate than ostrich chicks. Yolk sac and umbilical infections are two common problems with young chicks. Careful management at hatching will help reduce these problems. Sometimes chicks become listless, stop eating and drinking, and die; this condition is called malabsorption syndrome or fading chick syndrome. Another health problem of young chicks is impaction caused by ingesting foreign items such as sticks, fabric, sand, and grass (Figure 48-15). Fecal impaction has been treated successfully with surgery, by adding mineral oil to the diet, or with the use of a laxative. Impaction is generally not a major health problem with older birds. Respiratory diseases are more common in young birds as compared to older birds. Antibiotics may be used to treat respiratory diseases.

Ratites are affected by both internal and external parasites, including nematodes, ticks, lice, and mites. Several anthelmintic drugs and dusting powders may be used for treatment. Consultation with a veterinarian to determine the appropriate treatment is recommended.

A good disease prevention program is based on the premise of reducing the opportunity for infectious organisms to enter the facilities where birds are kept. The three major components of a good program are isolation, traffic control, and sanitation. Isolation means to keep the birds in a facility that does not allow other animals to enter, keeping birds separated by age group, and isolating new birds until their health status is determined (2 to 4 weeks). Traffic control refers to not allowing people uncontrolled access to the facilities and birds. Sanitation refers to making sure all equipment is clean and disinfected, and that the facilities are properly cleaned and disinfected at regular intervals.

Getting Started in Business

Serious interest in raising ratites in the United States began to develop in the mid-1980s. During its first 10 years, the ratite industry in the United States was primarily in a breeder phase; that is, birds were being raised to supply breeding stock to increase the total numbers. Prices for breeding stock were very high during this breeder phase. By the mid-1990s, the ratite industry began to move from a breeder phase to a slaughter and product production phase.

There are several ways to get started in ratite production. Buying eggs and hatching them requires the lowest initial investment; however, there are several disadvantages to this method. The mortality rate is highest, incubation and brooding equipment is necessary, eggs are generally not insurable, too many chicks of one sex may be produced, and it is 2 to 3 years before any return on investment is realized. Generally, only people with experience in raising ratites should purchase eggs. Emu eggs are seldom sold because they are opaque and cannot be candled to determine fertility. Purchase of sexed chicks from 1 day to 6 months of age is another way to get started. Advantages of this method of getting started include knowing the sex of the birds, the purchase of an incubator can be delayed, conformation can be judged, and the initial investment is less than buying adults. The mortality rate is still relatively high, and it is still 2 to 3 years before any return on investment is realized.

Yearling pairs of male and female birds may be purchased. The advantages of buying yearling birds include a lower mortality rate, birds are insurable, and conformation can be judged. Birds of this age are still 1 to 2 years from producing a return on investment.

Getting started in ratite production by buying 2-year-old breeding pairs is closer to producing a return on investment; other advantages are the same as for buying yearling pairs. The major advantages of buying a proven breeding pair are that their reproductive potential is known and a quicker return on investment can be realized.

People interested in ratite production need to carefully research the costs, the requirements for facilities, their ability to care for these birds, and the risks that are involved. They need to visit current operations and talk with those who are in the business. Ratites are currently being raised in the United States primarily for meat, hide, oil, and feathers. Some early investors in the industry have lost money, and many have gotten out of the business. One of the major problems currently facing producers is the lack of facilities for slaughtering and processing ratites. The market for products from ratites is limited and retail prices are very high.

Purchases of breeding birds should be made from reputable breeders who are willing to guarantee the fertility and health of the birds. It is also important to purchase unrelated breeding stock. It is not possible to get unrelated breeding stock from a seller who has only one pair of breeding birds.

Before purchasing, be sure birds are healthy (get a health certificate from a licensed veterinarian), have been implanted with microchip identification or are DNA identified, and are guaranteed as to sex. Microchip numbers should be verified and checked against a current list of stolen birds. Obtain written records on the birds that include origin, family bloodline, and vaccinations. Get all warranties or guarantees in writing. Contact other customers of the breeder from whom the birds are being purchased, the Better Business Bureau, and the State Attorney General's Office to determine if there have been any problems with this breeder.

Products

Ratite products include meat, leather, oil, feathers, and other by-products. The meat is promoted as a low-fat, low-cholesterol red meat. Research shows that an ostrich has a dressing percentage of about 57 percent, and an emu has a dressing percentage of about 54 percent. While not currently generally available in the United States, except in some restaurants, ratite meat is popular in many parts of Europe.

Ratite hides are used to produce leather products, including shoes, boots, coats, belts, purses, wallets, and briefcases. Emu leather is thinner and finer textured than ostrich leather.

Emu oil is produced from the fat and is used in cosmetic and pharmaceutical products. One emu will produce 1.3 to 1.6 gallons of oil.

Ratite feathers are used in fashion, costuming, and feather dusters. Feathers can be harvested once a year. Eggshells and toenails are used in arts and crafts.

LLAMAS AND ALPACAS

Introduction and History

Forty million years ago, the Camelidae family originated on the North American continent. About 3 million years ago, some members of this family migrated to the Asian and African continents and others migrated to South America. The Camelidae family became extinct on the North American continent about 10,000 to 12,000 years ago. The Dromedary camel and the Bactrian camel are

Photo by Bill Tarpenning, USDA

Figure 48-16 Llamas.

the present-day descendants of those that migrated to Asia and Africa. The llama, alpaca, guanaco, and vicuna are the present-day descendants of those that migrated to South America.

The llama and alpaca were domesticated from the guanacos in Peru about 4,000 to 5,000 years ago. In South America, llamas have long been used as pack animals and for their meat, milk, hides, and wool; alpacas are not usually used as pack animals but are valued for their wool. There are currently about 3.5 million llamas and 3.5 million alpacas in South America and more than 70,000 llamas and more than 5,000 alpacas in the United States and Canada.

Llamas were first imported into the United States in the late 1800s to be exhibited in zoos. Interest in raising llamas and alpacas increased in the 1970s, with a primary emphasis on raising and selling breeding stock. In addition to breeding stock, llamas are used as pack animals, for fiber production, as guard animals for sheep herds, as show animals, and for pets. Alpacas are also used for breeding stock, fiber production, investments, and as pets; they are generally not used as pack animals or guard animals.

Initial investment costs for both llama and alpaca breeding stock vary widely in different parts of the United States. Prices for llama breeding stock can range from several hundred to several thousand dollars each. Generally, alpaca prices are in the same range as llama prices. Many producers advertise their animals for sale on their web pages.

Photo by Bill Tarpenning, USDA

Figure 48-17 Young llama.

Most llamas and alpacas in the United States are registered with the International Llama Registry (ILR). This is a closed registry, which means that both parents of an animal must be registered before the offspring can be registered. Bloodtyping with the ILR is commonly used to maintain and track genealogy records.

Characteristics

Llama wool color is usually white but may be black; sometimes, there are shades of beige, brown, red, or roan. The wool coat may be marked in a variety of patterns from solid to spotted (Figures 48-16 and 48-17). The average weight of mature llamas is 280 to 450 pounds. Llamas have a long neck, and they stand 5.5 to 6 feet at the head and approximately 4 feet at the shoulder. The life span of the llama is about 20 to 25 years.

Alpacas have 22 distinct natural colors, including black, white, roan, pinto, brown, red, fawn, rose, and gray (Figure 48-18). There are two types of alpacas based on the type of fleece they produce. Huacaya fiber has a dense, crimped fiber; Suri fiber is usually white and is not crimped but hangs straight down and is curly. An adult alpaca produces about 4 pounds of fleece per year. The average weight of mature alpacas is 100 to 175 pounds; overall, alpacas are about one-half the size of llamas. Alpacas are about 4.5 feet at the head and approximately 3 feet at the shoulders. The life span of the alpaca is about 20 to 25 years.

Llamas and alpacas have six incisor teeth on the front bottom jaw and a hard dental pad (no teeth) on top. The back of their jaw contains five molar teeth on the top and bottom on both sides. Their upper lip is split in a manner that allows them to use it in grasping forage when grazing.

USDA

Figure 48-18 Alpacas.

Their feet are two toed and have a broad, leathery pad on the bottom that helps give them excellent footing on many types of terrain. They have a scent gland located on the side of each rear leg and one between the toes.

Feeding

Llamas and alpacas are modified ruminants; the stomach has three compartments instead of four. They chew their cud just as other ruminants do, and the stomach functions in essentially the same manner. They make more efficient use of feed than do cattle or sheep, converting more of it to the energy and protein they need to meet their requirements. Care must be taken not to overfeed these animals; it will cause them to become too fat. The feeding program is based on roughage, with concentrates as a supplement to the diet. The total daily dry matter intake should be in the range of 1.8 to 2.0 percent of body weight. The crude fiber content of the diet should be about 25 percent.

Llamas will eat both forage and browse; alpacas are less likely to eat browse. Browse is the shoots, twigs, and leaves of brush plants found growing on rangeland. A moderate-producing pasture will carry three to five adult animals per acre. If a good-quality forage is available, grain supplementation of the diet is usually not needed except during the last 2.5 to 3 months of pregnancy and during the first 3 months of lactation if the animal is not too fat. Working pack animals may also need some grain supplementation.

The protein requirement is generally low for llamas and alpacas. A maintenance ration should contain 8 to 10 percent crude protein; a 10 to 12 percent protein level is recommended for growing animals and for pregnant and nursing females. Feed for an early-weaned cria should contain 16 percent protein, with the amount being gradually reduced to 12 percent by the time it is 6 months of age. A **cria** (pronounced cree-ah) is the name for a baby llama, alpaca, or other camelidae.

Make sure the newborn cria receives the colostrum milk. In an emergency, cow or goat colostrum milk may be used. While it may be necessary to bottle-feed a premature, weak, or orphaned cria to save its life, this practice should be held to a minimum. Cria that are bottle-fed have a tendency to imprint on and become aggressive toward humans. If bottle-feeding of colostrum milk is necessary, feed a total of about 10 percent of the cria's weight in two to three feedings during the first day. After the first day, llama or goat milk, or a milk replacer, may be fed at a rate of 10 percent of the cria's body weight per day. Follow the manufacturer's directions for preparing and using the milk replacer.

Cria will begin eating a small amount of forage when they are 2 or 3 weeks of age. They may be creep-fed a limited amount of a 16 percent grain and forage mix starting at about 3 months of age and continuing until they are weaned. Feeding a free choice mineral supplement with the proper calcium-to-phosphorus ratio is recommended when creep feeding cria.

While llamas and alpacas have less need for water than most other farm animals, it is recommended that a fresh, clean supply of water be available at all times. Provide loose, trace-mineralized salt in containers to protect it from the weather. A calcium-to-phosphorus ratio of 1.2:1 to 2.0:1 is recommended. A free choice vitamin-mineral mix is used to provide the necessary vitamins and minerals in the diet.

Good-quality hay that is not moldy and that has been analyzed for nutrient content should form the basis for any llama or alpaca feeding program. Care must be taken when feeding alfalfa hay because it may be too high in protein content. Bromegrass, timothy, fescue, orchard grass, oat hay, and pea hay are good choices for feeding.

Many different kinds of pasture, including bromegrass, timothy, orchard grass, bluegrass, white clover, and alfalfa, are suitable for llamas and alpacas. It may be necessary to limit access to pasture to prevent excessive dry matter intake.

Commercially prepared complete feeds in pelleted form are available for llamas and alpacas. While this is a convenient method of feeding, the diet may not contain enough fiber unless the animals are on pasture or being fed hay with the pelleted feed. Using a pelleted feed is more expensive than other methods of feeding. When using a pelleted feed, care must be taken not to overfeed the animals; also, some animals may choke on the pelleted feed.

Management

Llamas and alpacas are herd animals that desire the companionship of other animals. When kept in isolation, they do not thrive. If a person wants only one, it can be kept with other domesticated animals such as sheep. Some sheep ranchers use llamas as guard animals to protect the flock from predators.

When llamas or alpacas are kept on a breeding farm, there are several groupings that may be used.

- All the females and their young cria: Generally, it is not wise to keep a stud with this group. Doing so makes it difficult to maintain good breeding records and makes it harder to work with the group because of the territorial behavior of the male. Some males experience a decline in libido when kept with a group of females.
- Pregnant females: Keeping pregnant females in a separate group makes it easier to meet their nutritional needs and to more closely monitor them as they approach parturition. Do not keep any males with this group; they sometimes attempt to initiate breeding behavior as the females approach parturition.
- Weanlings: Females up to 2 years of age and males up to 9 months of age may be kept in this group. It is easier to meet the nutritional needs of this age group when they are kept in a separate group. Males are removed from the group by the time they are 9 months of age to prevent breeding problems.
- Young males from 9 to 18 months of age: Keeping this group separate prevents breeding problems with females. They should not be kept with older males because of problems with fighting.
- Castrated (gelded) males 18 months of age and older: Many breeders castrate all the males they do not plan to use as studs.
- Individual stud male: When males that have not been castrated are kept together they tend to fight. Some breeders do keep uncastrated males together successfully if they are not near a group of females.
- New purchases in a quarantine area: Quarantining all animals brought into the herd for 2 weeks reduces disease problems. The quarantine area must be located so contact with other animals is not possible.

Good record keeping is essential when managing breeding herds. The following kinds of records should be kept:

- identification of **sire** (the male parent of an animal) and **dam** (the female parent of an animal)
- breeding dates
- parturition dates
- transfers among groups
- medical problems, treatments, and vaccinations

Llamas and alpacas respond well to training and are relatively easy to handle; however, they do not like to be touched and do not respond well to physical punishment. They are intelligent, naturally curious about their surroundings, and have a strong herd instinct, tending to move together as a group. They may be moved in the pasture by utilizing the flight zone concept; approach them from the direction opposite to the direction of desired movement. A large group may require two or more handlers to keep the group moving in the desired direction. After they have been moved from pasture to a smaller space, an individual animal may be caught, when necessary, by moving it into a corner or smaller space. Approach the animal slowly to put a halter on its head. Feeding in a small pen makes it easier to catch individual animals; just shut the gate after they have entered for feeding, and then proceed to isolate and catch the individual.

Occasionally, the toenails of llamas and alpacas need to be trimmed. This is usually not necessary if they are kept on, or have intermittent access to, stony ground, roughened concrete, or sandy lots. If the nails grow too long, the animal may become lame. Trimming is done with the same side-cutting nippers that are used on sheep. The animal will probably need to be restrained for nail trimming. Care must be taken not to cut into the live flesh around the nail.

At about 2 years of age, fighting teeth (canines) erupt in males. These teeth have sharp points and cutting edges and should be removed to prevent injury if the animals fight. The owner needs to recheck these teeth periodically because they will continue to erupt until the animal is 4 to 5 years of age. These teeth should be cut off when they are about 0.25 inch in length. A flexible braided cutting wire called an obstetrical wire is used for this procedure. The animal needs to be restrained when these teeth are removed. The wire is hooked behind the tooth and drawn back and forth in a sawing motion to cut the tooth. Soft tissue is not normally damaged by the obstetrical wire. Fighting teeth are also present in females, but they are small and erupt when the animal is about 5 to 6 years of age; normally they are not removed.

In areas where temperatures go above 90°F, several methods of keeping the animals cool may be used. These include:

- providing shade in an area with good air circulation.
- shearing once a year.
- providing an area filled with sand or pea gravel that can be wetted down.
- using a sprinkler system or providing a shallow wading pool.
- scheduling breeding so they do not give birth during hot weather.
- using fans or evaporative coolers in barns.
- limiting the feeding of forages and high-protein supplements. The digestion of these feeds produces a higher heat increment than does the digestion of concentrates. However, be sure the crude fiber content of the diet remains at about 25 percent.
- feeding in the evening so digestion takes place during the cooler night hours.

Reproduction

Llamas and alpacas do not have estrus cycles; they are induced ovulators and can be bred at any time during the year. The act of mating stimulates the release of an egg for fertilization. Ovulation occurs 24 to 36 hours after copulation. Females can be bred at 1 year of age; however, the recommended age for breeding is $1\frac{1}{2}$ to 2 years of age. The gestation period is about 11 to 11.5 months. Parturition normally occurs during daylight hours; assistance at birth is rarely needed. The birth weight of llama crias is about 20 to 35 pounds and about 10 to 20 pounds for alpaca crias. Crias are weaned at about 6 months of age. Llama males become capable of breeding at about 7 to 9 months of age; however, it is recommended they not be used for breeding until they are approximately 3 years old. Male alpacas are used for breeding at about 2 to 3 years of age.

Health

There are few disease problems with llamas and alpacas. It is generally wise to have a good working relationship with a veterinarian who is knowledgeable about them.

A good preventive medicine program, including proper nutrition, vaccinations, parasite control, and sanitation of facilities should be followed. Vaccines that are currently used for other domestic animals are used for llamas and alpacas; there are no vaccines currently available that have been developed specifically for these animals. Animals should be examined by a veterinarian before they are purchased and should be accompanied by a certificate of health.

The same internal and external parasites that affect cattle and sheep also affect llamas and alpacas. Control these parasites with the same methods used for cattle and sheep. The animals should be dewormed once or twice a year if internal parasites are a problem.

Digestive disorders are rare in llamas and alpacas. They rarely bloat on feed; however, overeating grain can cause severe, sometimes fatal, illness. Poisonous plants should be eliminated from pasture areas to prevent ingestion.

Housing and Fencing

Llamas and alpacas require little housing, especially in a mild climate. In areas where temperatures may be extremely cold or hot, some shelter should be provided. A closed barn will provide adequate shelter in cold weather; a three-sided shelter is usually sufficient in a milder climate. Orient the shelter to provide maximum protection from the prevailing wind. Good ventilation must be provided in closed barns. Many breeders provide an area where sick animals or a new mother with her cria can be kept for treatment. Heat lamps may be used to provide warmth for the cria in cold weather.

Providing hay mangers or feed bunks will prevent animals from picking up parasites from the ground. Feed can be protected by placing the mangers or bunks in a sheltered area.

Catch corrals make it easier to manage such activities as vaccinating, nail trimming, shearing, brushing, and training. A holding chute may be built into a corner of the catch corral. Holding pens, alleyways, and sorting gates help control the animals when it is necessary to work with them.

Fencing requirements vary with herd size, groupings used, pasture size, and management practices. Four-foot fences are usually adequate if there is little pressure for the animals to want to leave the enclosure. Llamas, however, can jump a fence of this height. A fence height of 5.5 to 6 feet may be necessary if the pressure to leave the enclosure is high. Higher pressure against a fence can be caused by separating cria from mothers at weaning time, keeping studs away from unbred females, and keeping stud males in separate enclosures. In some areas, it may be necessary to erect fences sufficiently secure to keep out wild and domestic predators.

Fences may be constructed from a variety of suitable materials, including woven wire, chain link, smooth wire strand, cattle wire panels, wooden rails, or boards. Cost and durability of fencing are factors to be considered when constructing fencing. Electric fencing is effective once the animals learn what it is. Barbed wire should never be used because it may cause injury. Standard farm livestock gates work well and allow machinery access to pasture areas if necessary.

USDA Agricultural Research Service; photo by Keith Weller

Figure 48-19 Elk are members of the deer family (Cervidae).

ELK

American elk are members of the deer family (Figure 48-19). Elk are primarily raised for their **velvet** but are also bred for meat, by-products, and breeding stock. A mature bull elk produces 30 to 40 pounds of velvet annually.

Commercial deer farming is a steadily increasing market due to the increasing demand for deer and elk products and minimal acreage requirements. The labor requirements for elk production are minimal in comparison to a beef cow-calf operation, and the profit potential can make it a viable enterprise. Elk have a high fertility rate and a long reproductive life. Elk calve easily and wean their calves early. Elk heifers (young females) are able to reproduce at 16 months of age, and bulls are reproductively mature at 24 to 30 months of age. Elk have a calm disposition and can tolerate cold winters and hot summers and have a very low susceptibility to disease.

SUMMARY

The American bison once numbered in the millions on the North American continent and were an important resource for the Plains Indians. Hunted almost to extinction, their numbers are now on the increase. A small but growing number of people are raising bison for sale as breeding stock and for meat. Bison are large, strong animals and must be handled with care. Bison require much less care, handling, and shelter than beef cattle; however, the nutrient requirements are similar to those of beef cattle. Breeding herds can generally be maintained on good grassland pasture. Animals being fed for meat are finished on roughage and concentrate rations.

There is some interest in the raising of ratites as an alternative to other livestock in the United States. Ratites are flightless birds and include ostrich, emu, rhea, cassowary, and kiwi. The ratite industry in the United States is mainly concerned with ostriches, emus, and rheas. The ostrich is the largest of the birds still in existence, and the emu is the second largest. The ostrich hen will usually produce 40 to 50 eggs per year and the emu hen will produce about 30 eggs each year. Producers typically use artificial incubation to hatch the eggs. Only a limited amount of research has been done on the nutritional requirements of ratites. Some feed companies produce commercial ratite feeds. Mature ratites can be dangerous to handle. Electronic identification is widely used in the ratite industry in the United States. Producers need to follow a good disease prevention program. The meat of ratites is promoted as a low-fat, low-cholesterol red meat.

Llamas and alpacas were domesticated in Peru about 4,000 to 5,000 years ago. Llamas are used as pack animals, as well as for meat, hides, and wool. Alpacas are not usually used as pack animals. There has been a growing interest in the United States in raising llamas and alpacas. The initial investment required for breeding animals is high. Llamas and alpacas are modified ruminants, having three stomachs. The feeding program is based on the use of roughages. Commercially prepared complete pelleted feeds are available for llamas and alpacas. These animals respond well to training and are relatively easy to handle. They do not have estrus cycles; they are induced ovulators and can be bred at any time of the year. There are relatively few health problems but a producer should follow good health management practices. Housing needs for llamas and alpacas are generally simple.

American elk are members of the deer family. They are primarily raised for their velvet but are also bred for meat, by-products, and breeding stock.

Student Learning Activities

1. Visit an operation where bison are being raised.
2. Talk with someone who raises bison. Give an oral report to the class.
3. Prepare and give an oral report to the class on any aspect of raising bison.
4. Visit a facility where ratites are being raised.
5. Talk with someone who raises ratites. Give an oral report to the class.

6. Prepare and give an oral report to the class on any aspect of raising ratites.
7. Visit a facility where llamas or alpacas are being raised.
8. Talk to someone who raises llamas or alpacas. Give an oral report to the class.
9. Prepare and give an oral report to the class on any aspect of raising llamas or alpacas.
10. Use the information discussed in this chapter when planning and conducting an alternative animal supervised experience program.

Discussion Questions

1. Briefly explain the five types of bison ownership.
2. What are the desirable and undesirable physical characteristics to look for when selecting bison for the breeding herd?
3. At what age are bison heifers usually bred?
4. What months of the year are considered to be the breeding season for bison?
5. Why should the bison breeder avoid inbreeding?
6. Briefly explain procedures to be followed when weaning bison calves.
7. Briefly explain the advantages and disadvantages of dehorning bison.
8. Briefly explain the recommended practices for feeding the bison herd on pasture.
9. Briefly explain the recommended practices for finishing bison for meat.
10. Name and describe the kinds of fencing that might be used for bison.
11. List and briefly explain the methods of marketing bison.
12. Describe the most common species of ratites used in the ratite industry in the United States.
13. Discuss the breeding of ratites.
14. Discuss the brooding of ratite chicks.
15. Discuss the feeding of ratite chicks and birds.
16. Describe the proper method of handling and transporting mature ratites.
17. What methods may be used for identification of individual ratites, and why are they used?
18. Describe facilities that might be used for raising ratites.
19. What health problems may occur with ratites?
20. How might a person get started in the ratite business, and what precautions should be taken?
21. What products are produced in a ratite business?
22. Describe the characteristics of llamas and alpacas.
23. Discuss the feeding of llamas and alpacas.
24. Discuss the principles of managing llamas and alpacas.
25. Discuss reproduction of llamas and alpacas.
26. Discuss health management of llamas and alpacas.
27. What kind of housing and fencing are needed for llamas and alpacas?

Review Questions

True/False

1. Bison were never an important source of food, skins, or bones for any people in North America.
2. The American bison is not related to the European bison.
3. There is a growing interest in raising bison.
4. A sire is the female and a dam is the male parent of an animal.

Multiple Choice

5. What is the average daily gain for well-managed bison in the feedlot?

   a. 3 to 5 pounds c. 4 pounds
   b. 2 to 3 pounds d. 0.5 to 1 pound

6. What is a serious problem affecting bison because the current stock began with so few animals?

   a. inbreeding c. outcrossing
   b. crossbreeding d. linebreeding

Completion

7. Using a ration too high in fiber can result in ______________ blockage in ratite chicks.
8. ______________ oil is produced from fat and is used in cosmetic and pharmaceutical products.

Short Answer

9. Describe the ostrich in terms of its height, weight, and speed.
10. When is the mating season for emus?
11. When were llamas first imported to the United States?
12. Describe the meat produced by bison. Why are people interested in bison meat?
13. List the products from elk.

Appendix

TABLE 1 Age Computing Chart[a]

Day	Jan.	Feb.	March	April	May	June	July	Aug.	Sept.	Oct.	Nov.	Dec.
1	1	32	60	91	121	152	182	213	244	274	305	335
2	2	33	61	92	122	153	183	214	245	275	306	336
3	3	34	62	93	123	154	184	215	246	276	307	337
4	4	35	63	94	124	155	185	216	247	277	308	338
5	5	36	64	95	125	156	186	217	248	278	309	339
6	6	37	65	96	126	157	187	218	249	279	310	340
7	7	38	66	97	127	158	188	219	250	280	311	341
8	8	39	67	98	128	159	189	220	251	281	312	342
9	9	40	68	99	129	160	190	221	252	282	313	343
10	10	41	69	100	130	161	191	222	253	283	314	344
11	11	42	70	101	131	162	192	223	254	284	315	345
12	12	43	71	102	132	163	193	224	255	285	316	346
13	13	44	72	103	133	164	194	225	256	286	317	347
14	14	45	73	104	134	165	195	226	257	287	318	348
15	15	46	74	105	135	166	196	227	258	288	319	349
16	16	47	75	106	136	167	197	228	259	289	320	350
17	17	48	76	107	137	168	198	229	260	290	321	351
18	18	49	77	108	138	169	199	230	261	291	322	352
19	19	50	78	109	139	170	200	231	262	292	323	353
20	20	51	79	110	140	171	201	232	263	293	324	354
21	21	52	80	111	141	172	202	233	264	294	325	355
22	22	53	81	112	142	173	203	234	265	295	326	356
23	23	54	82	113	143	174	204	235	266	296	327	357
24	24	55	83	114	144	175	205	236	267	297	328	358
25	25	56	84	115	145	176	206	237	268	298	329	359
26	26	57	85	116	146	177	207	238	269	299	330	360
27	27	58	86	117	147	178	208	239	270	300	331	361
28	28	59	87	118	148	179	209	240	271	301	332	362
29	29		88	119	149	180	210	241	272	302	333	363
30	30		89	120	150	181	211	242	273	303	334	364
31	31		90		151		212	243		304		365

[a]The numbers under the months represent the day of the year. In a leap year, one (1) must be added to the numbers beginning on February 29. To determine the age of an animal: Follow down the column under the month until the birth date is found in the Day column; this will be the day of the year. Next, find the current date, using the same method. Subtract the smaller number from the larger number to find the age of the animal. For example, an animal is born on February 17 (day of the year is 48); the current date is August 25 (day of the year is 237). Subtract 48 from 237 = 189. The animal is 189 days old.

TABLE 2 Selected Feed Ingredients

Feed Name	International Reference Number	Dry Matter (%)	Crude Fiber[a] (%)	Crude Protein[a] (%)	Acid Detergent Fiber[a] (%)
ALFALFA *Medicago sativa*					
1. fresh	2-00-196	24	26.2	19.7	35
2. fresh, mid-bloom	2-00-185	24	28.0	18.3	35
3. hay, sun-cured, mid-bloom	1-00-063	90	26.0	17.0	35
4. meal, dehydrated, 17% protein	1-00-023	92	26.2	18.9	35
5. silage, late vegetative	3-00-204	21	34.6	20.0	47
6. silage, full bloom	3-00-207	26	34.9	17.5	38
7. silage	3-00-212	41	33.0	17.8	—
8. silage wilted, mid-bloom	3-00-217	38	30.0	15.5	35
BAHIA GRASS *Paspalum notatum*					
9. fresh	2-00-464	30	30.4	8.9	38
10. hay, sun-cured	1-00-462	91	32.0	8.2	41
BARLEY *Hordeum vulgare*					
12. grain	4-00-549	88	5.7	13.5	7
13. grain, Pacific coast	4-07-939	89	7.1	10.8	9
14. hay, sun-cured	1-00-495	87	27.5	8.7	—
15. silage	3-00-512	31	30.0	10.3	—
BERMUDA GRASS, COASTAL *Cynodon dactylon*					
16. fresh	2-00-719	29	28.4	15.0	—
17. hay, sun-cured	1-00-716	90	30.7	6.0	38
BLOOD					
18. meal	5-00-380	92	1.1	87.2	—
19. meal spray, dehydrated	5-00-381	93	0.6	88.9	—
BLUEGRASS, KENTUCKY *Poa pratensis*					
20. fresh, early vegetative	2-00-777	31	25.3	17.4	29
21. fresh, mature	2-00-784	42	32.2	9.5	40
22. hay, sun-cured	1-00-776	89	31.0	13.0	—
BONEMEAL					
23. feeding (more than 10% P)	6-00-397	94	—	24.5	—
BONEMEAL					
24. steamed	6-00-400	97	2.0	13.2	—
BREWERS					
25. grains, dehydrated	5-02-141	92	14.9	25.4	24
26. grains, wet	5-02-142	21	14.9	25.4	23
BROME *Bromus* spp.					
27. fresh, early vegetative	2-00-892	34	24.0	18.0	31
28. hay, sun-cured, late bloom	1-00-888	89	37.0	10.0	43
29. hay, sun-cured	1-00-890	91	33.3	9.7	—
BROME, SMOOTH *Bromus inermis*					
30. fresh, early vegetative	2-00-956	30	22.8	21.3	27
31. fresh, mature	2-08-364	55	34.8	6.0	—
32. hay, sun-cured, mid-bloom	1-05-633	90	31.8	14.6	37
CANARY GRASS, REED *Phalaris arundinacea*					
33. fresh	2-01-113	27	29.5	11.6	28
CITRUS *Citrus* spp.					
37. pulp, silage	3-01-234	21	15.6	7.3	—
38. pulp, dried	4-01-237	91	12.7	6.7	22
CLOVER, CRIMSON *Trifolium incarnatum*					
41. fresh, early vegetative	2-20-890	18	28.0	17.0	—
42. hay, sun-cured	1-01-328	87	30.1	16.0	—
CLOVER, LADINO *Trifolium repens*					
43. fresh, early vegetative	2-01-380	19	14.0	24.7	—
44. hay, sun-cured	1-01-378	90	21.2	22.0	32
CLOVER, RED *Trifolium pratense*					
45. fresh, early bloom	2-01-428	20	23.2	19.4	31
46. hay, sun-cured	1-01-415	89	28.8	16.0	36

(continues)

TABLE 2 Selected Feed Ingredients *(continued)*

Feed Name	International Reference Number	Dry Matter (%)	Crude Fiber[a] (%)	Crude Protein[a] (%)	Acid Detergent Fiber[a] (%)
CORN, DENT YELLOW *Zea mays indentata*					
47. stover (straw)	1-28-233	85	34.4	5.9	39
48. cobs, ground	1-02-782	90	35.0	2.8	35
49. cobs, ground	1-28-234	90	36.2	3.2	35
50. distillers grains, dehydrated	5-02-842	92	13.0	29.5	—
51. distillers grains, dehydrated	5-28-235	94	12.1	23.0	17
52. distillers grains w/solubles, dehydrated	5-02-843	92	10.0	29.8	—
53. distillers grains w/solubles, dehydrated	5-28-236	92	9.9	25.0	18
54. distillers solubles, dehydrated	5-28-237	93	5.0	29.7	7
55. ears, ground (ground ear corn)	4-02-849	87	9.0	9.3	—
56. ears, ground (corn and cob meal)	4-28-238	87	9.4	9.0	11
57. ears with husks, silage	4-28-239	44	11.6	8.9	—
58. gluten, meal	5-28-241	91	4.8	46.8	9
59. gluten, meal 60% protein	5-28-242	90	2.2	67.2	5
60. gluten, meal 41%	5-12-354	91	3.8	42.1	—
61. gluten, w/bran (corn gluten feed)	5-28-243	90	9.7	25.6	12
62. grain, opaque 2 (high lysine)	4-28-253	90	3.7	11.3	—
63. grain	4-02-879	87	2.4	10.6	—
64. grain, grade 2, 69.5 kg/hl	4-02-931	88	2.2	10.1	—
65. grain, dent yellow	4-02-935	89	2.9	10.9	3
66. grain	4-02-985	88	2.0	10.9	—
67. grain, cracked, dent yellow	4-20-698	89	2.6	10	3
68. grain, ground, dent yellow	4-26-023	89	2.6	10	3
69. grain, flaked	4-28-244	89	2.6	10	3
70. grain, high moisture	4-20-770	77	2.6	10	3
71. grits by-product (hominy feed)	4-03-011	90	6.7	11.5	13
72. silage (stalkage) (stover)	3-28-251	31	31.3	5.9	43
73. silage, well eared	3-28-250	33	23.7	8.1	28
74. silage (ears with husks)	3-28-239	44	11.6	8.9	14
75. silage, dough stage	3-02-819	26	24.5	7.8	31
COTTON *Gossypium* spp.					
76. seeds, meal solvent extracted, 41% protein	5-01-621	91	13.3	45.2	17
77. seeds, meal solvent extracted	5-01-619	92	11.7	45.3	—
78. seeds, meal prepressed solvent extracted, 41% protein	5-07-872	91	14.1	45.6	19
CUP GRASS, TEXAS *Eriochloa sericea*					
79. fresh, late vegetative	2-29-996	40	—	7.8	—
80. fresh, mature	2-30-059	83	—	5.0	—
CURLY MESQUITE *Hilaria belangeri*					
81. browse, fresh, early vegetative	2-01-723	23	25.6	17.2	—
DALLIS GRASS *Paspalum dilatatum*					
82. fresh, early vegetative	2-01-738	26	30.1	23.2	—
83. fresh, full bloom	2-01-739	30	32.2	7.1	—
84. dicalcium phosphate	6-01-080	96	—	—	—
fats and oils					
85. fat, animal, hydrolyzed	4-00-376	99	—	—	—
86. fat, animal-poultry	4-00-409	99	—	—	—
87. oil, vegetable	4-05-077	100	—	—	—
FESCUE *Festuca* spp.					
88. hay, sun-cured, early vegetative	1-06-132	91	26.0	12.4	32

(continues)

TABLE 2 Selected Feed Ingredients *(continued)*

Feed Name	International Reference Number	Dry Matter (%)	Crude Fiber[a] (%)	Crude Protein[a] (%)	Acid Detergent Fiber[a] (%)
FESCUE, KENTUCKY 31 *Festuca arundinacea*					
89. hay, sun-cured	1-20-800	92	24.9	18.2	—
FESCUE, MEADOW *Festuca elatior*					
90. grazed	2-01-920	27	29	11.5	—
91. hay, sun-cured	1-01-912	88	33	10.5	43
FLAX, COMMON *Linum usitatissimum*					
94. seeds, meal solvent extracted (linseed meal)	5-02-048	90	10.1	38.3	19
GRASS-LEGUME					
95. silage	3-02-303	29	31.8	11.3	—
KUDZU *Pueraria*					
99. hay, sun-cured	1-02-478	91	39.1	14.3	—
LESPEDEZA, COMMON—LESPEDEZA, KOREAN *Lespedeza striata—Lespedeza stipulacea*					
100. fresh, early bloom	2-20-885	28	32	16.4	—
101. fresh, late vegetative	2-07-093	25	32	16.0	—
102. grazed	2-02-568	31	38	14.9	—
103. hay, sun-cured	1-08-591	91	32	13.9	—
104. hay, sun-cured, mid-bloom	1-26-026	93	30	14.5	—
105. hay, sun-cured, mid-bloom	1-02-554	92	28.8	15.1	—
106. hay, sun-cured, mid-bloom	1-21-021	93	30	14.5	—
107. hay, sun-cured, full bloom	1-20-887	89	30.7	14.3	—
MEAT					
112. meal rendered	5-00-385	94	2.8	54.8	—
113. with bone, meal rendered	5-00-388	93	2.4	54.1	—
114. meat and bonemeal, 50%	5-09-322	94	2.4	50.9	—
115. meat meal, 55%	5-09-323	93	2.3	55.6	—
MESQUITE, HONEY *Prosopis glandulosa*					
116. browse, fresh late vegetative	2-29-999	48	—	16.2	—
MESQUITE, SPICIGERA *Prosopis spicigera*					
117. hay, sun-cured	1-30-160	86	22.1	14.2	—
MILK *Bos taurus*					
118. fresh (cattle)	5-01-168	12	—	26.7	—
119. skimmed, fresh (cattle)	5-01-170	10	0.2	31.2	—
120. skimmed, dehydrated	5-01-175	94	0.2	35.8	—
MILLET, PEARL *Pennisetum glaucum*					
122. grain	4-03-118	91	4.7	—	—
MILLET, PROSO *Panicum miliaceum*					
123. grain	4-03-120	90	6.8	12.9	17
MOLASSES AND SYRUP *Beta vulgaris altissim*					
125. beet, sugar, molasses, more than 48% invert sugar, more than 79.5 degrees brix	4-00-668	78	—	8.5	—
MOLASSES AND SYRUP *Saccharum officinarum*					
126. sugar cane, molasses, more than 46% invert sugars, more than 79.5 degrees brix (black strap)	4-04-696	75	—	5.8	—
OATS *Avena sativa*					
130. grain	4-03-309	89	12.1	13.3	16
131. grain, Pacific coast	4-07-999	91	12.3	10.0	—
132. hay, sun-cured	1-03-280	91	30.4	9.3	36
133. silage, late vegetative	3-20-898	23	29.9	12.8	—
134. silage	3-03-298	31	31.5	9.6	—

(continues)

TABLE 2 Selected Feed Ingredients *(continued)*

Feed Name	International Reference Number	Dry Matter (%)	Crude Fiber[a] (%)	Crude Protein[a] (%)	Acid Detergent Fiber[a] (%)
ORCHARD GRASS *Dactylis glomerata*					
135. fresh, early vegetative	2-03-439	23	24.7	18.4	31
136. fresh, early bloom	2-03-442	25	30.0	16.0	—
137. fresh, mid-bloom	2-03-443	31	33.5	11.0	41
138. hay, sun-cured, early bloom	1-03-425	89	31.0	15.0	34
139. hay, sun-cured	1-03-438	91	35.1	11.2	—
PANGOLA GRASS *Digitaria decumbens*					
141. fresh	2-03-493	21	30.5	10.3	38
142. hay, sun-cured	1-26-214	91	36.0	7.1	43
PEANUT *Arachis hypogaea*					
143. hay, sun-cured	1-03-619	91	33.2	10.8	—
144. kernels, meal mechanical extracted (peanut meal)	5-03-649	93	7.5	52.0	6
145. kernels, meal solvent extracted (peanut meal)	5-03-650	92	10.8	52.3	—
PRAIRIE PLANTS, MIDWEST					
151. hay, sun-cured	1-03-191	92	34.0	5.8	—
152. hay, sun-cured, mid-bloom	1-07-956	95	32.1	7.0	—
PRICKLY PEAR *Opuntia* spp.					
153. fresh	2-01-061	17	13.5	4.8	—
154. fresh, mature	2-01-059	21	13.7	3.1	—
155. fruit, fresh, immature	4-30-020	26	—	6.8	—
RAPE *Brassica napus*					
156. grazed, early vegetative	2-03-865	18	13	16.4	—
157. fresh, early bloom	2-03-866	11	15.8	23.5	—
158. seeds, meal solvent extracted	5-03-871	91	13.2	40.6	—
REDTOP *Agrostis alba*					
159. fresh, mid-bloom	2-03-890	39	29.0	7.4	—
160. fresh, full bloom	2-03-891	26	30.0	8.1	—
161. fresh	2-03-897	29	26.7	11.6	—
162. hay, sun-cured, mid-bloom	1-03-886	94	30.7	11.7	—
RYE *Secale cereale*					
163. distillers grains, dehydrated	5-04-023	92	13.4	23.5	—
164. grain	4-04-047	88	2.5	13.8	—
165. hay, sun-cured	1-04-004	93	33.3	8.5	—
RYEGRASS, ITALIAN *Lolium multiflorum*					
166. fresh	2-04-073	25	23.8	14.5	—
167. hay, sun-cured, early vegetative	1-04-064	89	19.7	15.2	38
168. hay, sun-cured, late vegetative	1-04-065	86	23.8	10.3	42
169. hay, sun-cured, early bloom	1-04-066	83	36.3	5.5	45
RYEGRASS, PERENNIAL *Lolium perenne*					
170. fresh	2-04-086	27	23.2	10.4	—
171. hay, sun-cured	1-04-077	86	24.6	8.6	30
SAGE, BLACK *Salvia mellifera*					
172. browse, fresh, stem-cured	2-05-564	65	—	8.5	—
SAGEBRUSH, BIG *Artemisia tridentata*					
173. browse, fresh, stem-cured	2-07-992	65	—	9.3	30
SAGEBRUSH, BUD *Artemisia spinescens*					
174. browse, fresh, early vegetative	2-07-991	23	—	17.3	—
175. browse, fresh, late vegetative	2-04-124	32	22.7	17.5	—
SAGEBRUSH, FRINGED *Artemisia frigida*					
176. browse, fresh, mid-bloom	2-04-129	43	33.2	9.4	—
177. browse, fresh, mature	2-04-130	60	31.8	7.1	35
SAGEBRUSH, MEXICAN *Artemisia ludoviciana albula*					
178. browse, fresh, mature	2-30-052	44	—	10.2	—

(continues)

TABLE 2 Selected Feed Ingredients *(continued)*

Feed Name	International Reference Number	Dry Matter (%)	Crude Fiber[a] (%)	Crude Protein[a] (%)	Acid Detergent Fiber[a] (%)
SAGEBRUSH, SAND *Artemisia filifolia*					
179. browse, fresh, early vegetative	2-04-133	29	22.6	12.2	—
180. browse, fresh, mature	2-04-135	45	31.7	7.2	—
SALTBUSH, NUTTALL *Atriplex nuttallii*					
181. browse, fresh, stem-cured	2-07-993	55	—	7.2	—
SORGHUM *Sorghum bicolor*					
184. aerial part, sun-cured, full bloom	1-04-371	90	23.8	6.4	0.62
185. distillers grains, dehydrated	5-04-374	94	12.7	34.4	—
186. grain	4-04-383	90	2.6	12.4	0.04
187. grain, 8–10% protein	4-20-893	87	2.0	9.7	9
188. grain, more than 10% protein	4-20-894	88	2.0	13.0	—
189. silage	3-04-323	30	27.9	7.5	38
SORGHUM, JOHNSONGRASS *Sorghum halepense*					
190. hay, sun-cured	1-04-407	89	33.5	9.5	—
SORGHUM, KAFIR *Sorghum bicolor caffrorum*					
191. grain	4-04-428	89	2.3	12.3	—
SORGHUM, MILO *Sorghum bicolor subglabrescens*					
192. grain	4-04-444	89	2.5	11.3	5
SORGHUM, SORGO *Sorghum bicolor saccharatum*					
193. silage, mature	3-04-467	27	27.6	6.6	—
194. silage	3-04-468	27	28.3	6.2	—
SORGHUM, SUDAN GRASS *Sorghum bicolor sudanense*					
195. fresh, early vegetative	2-04-484	18	23.0	16.8	29
196. hay, sun-cured, late vegetative	1-04-474	88	31.9	13.9	33
197. hay, sun-cured	1-04-480	91	36.0	8.0	42
198. silage	3-04-499	28	33.1	10.8	42
SOYBEAN *Glycine max*					
199. hay, sun-cured, mid-bloom	1-04-538	94	29.8	17.8	40
200. hay, sun-cured	1-04-558	89	33.7	16.0	—
201. seeds, meal solvent extracted, 44% protein	5-20-637	89	7.0	49.9	—
202. seeds, meal solvent extracted	5-04-604	90	6.5	49.9	—
203. seeds without hulls, meal solvent extracted	5-04-612	90	3.7	55.1	6
SUGAR CANE *Saccharum officinarum*					
204. bagasse, dehydrated	1-04-686	91	49.0	1.5	61
SWEETCLOVER, YELLOW *Melilotus officinalis*					
206. hay, sun-cured	1-04-754	87	33.4	15.7	—
TIMOTHY *Phleum pratense*					
207. fresh, mid-bloom	2-04-905	29	33.5	9.1	37
208. hay, sun-cured, late vegetative	1-04-881	89	27.0	17.0	29
209. hay, sun-cured, mid-bloom	1-04-883	89	31.0	9.1	36
TREFOIL, BIRDSFOOT *Lotus corniculatus*					
210. fresh	2-20-786	24	24.7	21.0	—
211. hay, sun-cured	1-05-044	92	30.7	16.3	36
212. hay, sun-cured, mid-bloom	1-20-790	91	—	14.5	38
TRITICALE *Triticale hexaploide*					
213. grain	4-20-362	90	4.4	17.6	8
VETCH *Vicia* spp.					
215. hay, sun-cured	1-05-106	89	30.6	20.8	33

(continues)

TABLE 2 Selected Feed Ingredients *(continued)*

Feed Name	International Reference Number	Dry Matter (%)	Crude Fiber[a] (%)	Crude Protein[a] (%)	Acid Detergent Fiber[a] (%)
WHEAT *Triticum aestivum*					
216. bran	4-05-190	89	11.3	17.1	15
217. grain	4-05-211	89	2.9	16.0	8
218. grain, hard red spring	4-05-258	88	2.9	17.2	13
219. grain, hard winter	4-05-268	88	2.8	14.4	4
220. grain, soft red winter	4-05-294	88	2.4	13.0	—
221. grain, soft white winter	4-05-337	89	2.6	11.3	4
222. flour by product, less than 9.5% fiber (wheat middlings)	4-05-205	89	8.2	18.4	10
223. flour by-product, less than 7% fiber (wheat shorts)	4-05-201	88	7.7	18.6	—
WHEAT, DURUM *Triticum durum*					
224. grain	4-05-224	88	2.5	15.9	—
WHEATGRASS, CRESTED *Agropyron desertorum*					
225. fresh, early vegetative	2-05-420	28	22.2	21.5	—
226. hay, sun-cured	1-05-418	93	32.9	12.4	36
WILDRYE, RUSSIAN *Elymus junceus*					
227. fresh	2-05-469	33	22.4	14.1	—
YEAST, BREWERS *Saccharomyces cerevisiae*					
231. dehydrated	7-05-527	93	3.1	46.9	—
YEAST, PRIMARY *Saccharomyces cerevisiae*					
233. dehydrated	7-05-533	93	3.3	51.8	—
YEAST, TORULA *Torulopsis utilis*					
234. dehydrated	7-05-534	93	2.4	52.7	—
YUCCA *Yucca* spp.					
235. flowers, fresh	2-05-536	15	13.3	19.7	—
236. leaves, fresh, immature	2-30-050	41	—	7.3	—

[a] 100 percent dry matter basis.

Glossary/Glosario

A

abomasum the fourth compartment, or true stomach, of the ruminant animal.

abomaso el cuarto compartimento (o estómago verdadero) del animal rumiante.

abscess a localized swelling of body tissues with a pus-filled cavity.

absceso inflamación localizada de tejidos corporales con una cavidad llena de pus.

absorption the process by which digested nutrients are taken into the bloodstream.

absorción proceso a través del cual los nutrientes digeridos pasan a la corriente sanguínea.

accelerated lambing a system that produces three lamb crops in 2 years.

paridera acelerada sistema que produce tres partos de ovejas en 2 años.

acid cleaner a cleaning compound made up of a combination of mild acids and wetting agents used to clean milkstone from milking equipment.

limpiador ácido compuesto limpiador que comprende una combinación de ácidos suaves y agentes humidificadores que se usa para limpiar la piedra de leche en los equipos de ordeñe.

acid detergent fiber (ADF) that portion of a feed that contains mainly cellulose, lignin, and some silica.

fibra con detergente ácido (ADF) parte de un alimento que contiene principalmente celulosa, lignina y algo de sílice.

ad libitum a system of feeding livestock in which no limit is placed on feed intake.

ad libitum sistema para alimentar ganado en el cual no se pone límite a la ingesta de alimento.

additive gene effects when many different genes are involved in the expression of a trait.

efectos de los genes aditivos cuando muchos genes diferentes participan en la expresión de un rasgo.

adenine a purine base found in DNA and RNA; it normally pairs with thymine in DNA.

adenina una base púrica que se encuentra en el ADN y el ARN; normalmente se aparea con la timina en el ADN.

aflatoxin a compound produced by a mold, especially *Aspergillus flavus*, that can contaminate stored animal feed.

aflatoxina compuesto producido por un moho, especialmente el *Aspergillus flavus*, que puede contaminar alimento para animales almacenado.

afterbirth the membranes expelled after the birth of the fetus.

placenta membranas que se expelen después del nacimiento del feto.

agricultural biotechnology the science of altering genetic and reproductive processes in animals and plants.

biotecnología agrícola la ciencia de alterar procesos genéticos y reproductivos en animales y plantas.

agricultural marketing service (AMS) division of the USDA that administers programs that facilitate the efficient, fair marketing of U.S. agricultural products and provides various services to ensure the quality and availability of food for consumers.

servicio de comercialización agrícola (AMS) división del USDA que administra programas para facilitar la comercialización eficiente y equitativa de productos agrícolas de EE. UU. y que ofrece diversos servicios para garantizar la calidad y la disponibilidad de alimentos para los consumidores.

agronomic nitrogen rate the amount of available nitrogen per unit of yield necessary to produce a given crop.

índice agronómico de nitrógeno la cantidad de nitrógeno disponible por unidad de rendimiento que se requiere para producir un cultivo determinado.

agrosecurity the use of all possible means and procedures to guard against deliberate or incidental harm to the food production system.

agroseguridad el uso de todos los medios y procedimientos posibles para proteger contra el daño deliberado o incidental al sistema de producción de alimentos.

agroterrorism the deliberate use of biological or chemical weapons to bring harm to agricultural enterprises.

agroterrorismo el uso deliberado de armas biológicas o químicas para dañar emprendimientos agrícolas.

albino an animal with pink eyes and white skin and hair.

albino animal con ojos rosados y piel y pelo blancos.

alimentary canal in poultry, the food-carrying passage that begins at the mouth and ends at the vent.

canal alimentario en las aves de corral, el pasaje para transporte del alimento que comienza en la boca y finaliza en la zona excretora.

alkaline cleaners group of general cleaners used to clean milking equipment; includes such substances as caustic soda (lye), soda ash, baking soda, and metasilicate of soda.

limpiadores alcalinos grupo de limpiadores generales que se usan para limpiar equipos de ordeñe; incluyen sustancias como soda cáustica (lejía), ceniza de soda y metasilicato de soda.

allele one member of a pair or several genes that are located on a specific position on homologous chromosomes.

alelo un miembro de un par o de varios genes ubicados en una posición específica en cromosomas homólogos.

all-in/all-out a method of raising hogs that moves pigs as a group from nursery, through growing and finishing, and to market.

todos dentro/todos fuera método de cría de porcinos que traslada a los cerdos en grupo desde el criadero, durante el crecimiento el acabado y al mercado.

alveoli in dairy animals, the part of the udder that manufactures milk; in the respiratory system, the place where the exchange of gases occurs by diffusion.

alvéolos en animales lecheros, la parte de la ubre que fabrica leche; en el sistema respiratorio, el lugar donde se produce el intercambio de gases por difusión.

amino acid a compound that contains carbon, hydrogen, oxygen, and nitrogen; certain amino acids are essential for growth and maintenance of cells.

aminoácido compuesto que contiene carbono, oxígeno, hidrógeno y nitrógeno;

ciertos aminoácidos son esenciales para el crecimiento y el mantenimiento de las células.

amphiarthroses joint a joint that consists of discs of a fibrous cartilage that separate and cushion the vertebrae, allowing very limited movement.

articulación anfiartrosis una articulación con discos de cartílago fibroso que separan y acolchan las vértebras, lo cual permite un movimiento muy limitado.

ampicillin a type of penicillin effective against gram-negative and gram-positive bacteria; used to treat infections of the intestinal, urinary, and respiratory tracts.

ampicilina tipo de penicilina efectiva contra bacterias grampositivas y gramnegativas; se usa para tratar infecciones de los tractos intestinal, urinario y respiratorio.

anabolism the formation and repair of body tissues.

anabolismo la formación y la reparación de tejidos corporales.

anaphase one of the stages of mitosis and meiosis; at this stage the chromosomes move to opposite ends of the nuclear spindle.

anafase una de las etapas de la mitosis y la meiosis; en esta etapa, los cromosomas se trasladan a los extremos opuestos del huso del núcleo.

Anas boschas wild mallard duck believed to be the ancestor of all domestic breeds of ducks.

Anas boschas ánade salvaje que se cree que es el ancestro de todas las razas domésticas de patos.

androsterone a steroid hormone that intensifies masculine characteristics.

androsterona hormona esteroide que intensifica los rasgos masculinos.

anestrus a period of sexual dormancy between two estrus periods.

anestro período de latencia sexual entre dos períodos de estro.

animal protein a protein supplement that comes from animals or animal byproducts.

proteína animal suplemento proteico que proviene de animales o de derivados de animales.

anydrosis a condition in which animals do not sweat normally.

anhidrosis condición en la cual los animales no transpiran de manera normal.

animal protein protein supplements that come from animals or animal byproducts.

proteína animal suplementos proteicos que provienen de animales o de derivados de animales.

animists ancient Greeks who believed that humans and animals shared and exchanged souls.

animistas antiguos griegos que creían que los humanos y los animales compartían e intercambiaban almas.

anthelmintic a chemical compound used for treating internal worms in animals.

antihelmíntico compuesto químico usado para tratar gusanos internos en los animales.

anthrax a serious bacterial disease that is zoonotic.

ántrax enfermedad bacterial grave que es zoonótica.

antibiotic a chemical agent that prevents the growth of a germ or bacteria.

antibiótico agente químico que previene el crecimiento de un germen o una bacteria.

antibodies substances that protect an animal from infections and poisons.

anticuerpos sustancias que protegen a un animal de infecciones y venenos.

antimicrobial a substance that can destroy or inhibit the growth of microorganisms, i.e., an antimicrobial drug.

antimicrobiano sustancia que puede destruir o inhibir el crecimiento de microorganismos, es decir, un fármaco antimicrobiano.

anus the opening at the end of the large intestine that is the termination of the digestive system and through which feces pass out of the body.

ano apertura en el extremo del intestino grueso que es la terminación del sistema digestivo y a través de la cual las heces se eliminan del cuerpo.

apparel wool a fine wool used in making clothing.

lana para ropa lana fina usada para hacer ropa.

arsenical a drug containing arsenic.

arsenical fármaco que contiene arsénico.

arterioles smaller branches of arteries that connect to the capillary bed.

arteriolas ramas menores de las arterias que se conectan con el lecho capilar.

artificial insemination the placing of sperm in the female reproductive tract by other than natural means.

inseminación artificial la colocación de espermatozoides en el tracto reproductivo femenino por un medio que no es natural.

ascarid a type of parasitic roundworms that infest the intestines of animals.

ascárido tipo de gusanos redondos parásitos que infestan los intestinos de los animales.

asexual reproduction a method of reproduction that does not involve gametes, such as simple cell division in bacteria.

reproducción asexual método de reproducción que no involucra gametos, como la división celular simple en las bacterias.

as-fed basis data on feed composition or nutrient requirements calculated on the basis of the average amount of moisture found in the feed as it is used on the farm; sometimes referred to as *air-dry basis*.

según se administra datos sobre los requisitos de composición o de nutrientes del alimento calculados a partir de la cantidad promedio de humedad que hay en el alimento según se usa en el establecimiento agrícola; a veces se denomina *con base en secado al aire*.

Asiatic urial a breed of wild sheep believed to be ancestors of some present-day domestic breeds.

urial asiático raza de carneros salvajes que se cree que fueron los ancestros de algunas razas domésticas actuales.

ass a four-footed animal of the genus *Equus* that is smaller than a horse, has longer ears, and a short, erect mane.

asno animal cuadrúpedo del género *Equus* más pequeño que un caballo, con orejas más grandes y crin corta y erecta.

associates degree a post-secondary college level degree awarded after the completion of a course of study typically lasting 2 years.

título de asociado título de nivel terciario postsecundario que se otorga tras completar un curso de estudio que típicamente dura 2 años.

atrophy the wasting away or decreasing in size of muscle, fat, or any tissue or organ.

atrofia el desgaste o la disminución del tamaño de músculos, grasa o cualquier tejido u órgano.

auction market a market in which cattle are sold by public bidding, with the animals going to the highest bidder; also called *local sale barn* or *community auction*.

mercado de subasta mercado donde se vende ganado por subasta pública y el mejor postor se lleva los animales; también se denomina *establo de venta local* o *subasta comunitaria*.

axial feather in poultry, the feather that separates the primary feathers on the outer part of the wing from the secondary feathers on the inner part of the wing.

pluma axial en las aves de corral, la pluma que separa las plumas principales en la parte externa del ala de las secundarias en la parte interna del ala.

azoturia a nutritional disorder that develops when a horse is put to work following a period of idleness; also called *Monday-Morning Sickness*.

azoturia trastorno nutricional que se desarrolla cuando un caballo se pone a trabajar después de un período de inactividad; también se denomina *enfermedad del lunes por la mañana*.

B

bachelors degree a post-secondary college degree awarded after the completion of a course of study typically lasting 4 years.

título de licenciatura título terciario postsecundario que se otorga tras completar un curso de estudio que típicamente dura 4 años.

backgrounding the growing and feeding of calves from weaning until they are ready to enter the feedlot.

recría crecimiento y alimentación de terneros desde el destete hasta que están listos para ingresar al pastadero.

bacteria one-celled microorganisms.

bacterias microorganismos unicelulares.

bacterin a liquid containing dead or weakened bacteria that causes the animal to build up antibodies that fight diseases caused by bacteria.

bacterina líquido que contiene bacterias muertas o atenuadas y que causa que el animal desarrolle anticuerpos que combaten enfermedades causadas por bacterias.

balanced ration feed allowance for an animal during a 24-hour period that has all the nutrients the animal needs in the right proportions and amounts.

ración balanceada provisión de alimento para un animal durante un período de 24 horas que tiene todos los nutrientes que el animal necesita en las proporciones y las cantidades correctas.

band a group of sheep; also called a *flock*.

rebaño grupo de ovejas; también se denomina *manada*.

banding instinct tendency of sheep to stay together in a group; also called *flocking instinct*.

instinto de rebaño tendencia de las ovejas a mantenerse juntas en un grupo; también se denomina *instinto de manada*.

bantam small chicken, usually weighing from 16 to 30 ounces as an adult.

bantam pollos pequeños que suelen pesar entre 450 y 850 g (entre 16 y 30 oz) en la adultez.

barrow a male swine castrated while young, in which physical traits of the boar have not developed.

capón cerdo macho castrado cuando es joven en el cual no se han desarrollado los rasgos físicos del padrillo.

basal metabolism the processes that produce energy for the functioning of the heart, for breathing, and for other vital body processes.

metabolismo basal procesos que producen energía para el funcionamiento del corazón, para respirar y para otros procesos corporales vitales.

beak trimming cutting off part of the upper and lower beak of poultry.

corte de pico amputación parcial de la parte superior e inferior del pico de aves de corral.

bile a fluid produced by the liver that aids in the digestion of fats and fatty acids.

bilis líquido producido por el hígado que ayuda a digerir grasas y ácidos grasos.

binding quality a measure of how well feed particles stick together in the pellet.

capacidad de unión medida del grado en que las partículas de alimento se adhieren juntas en la pastilla.

biosecurity protection from biological harm to living things from diseases, pests, and bioterrorism.

bioseguridad protección a seres vivientes contra daños biológicos por enfermedades, pestes y bioterrorismo.

biotechnology technology concerning the application of biological and engineering techniques to microorganisms, plants, and animals.

biotecnología tecnología referida a la aplicación de técnicas biológicas y de ingeniería a microorganismos, plantas y animales.

bioterrorism the deliberate use of biological or chemical weapons; in agriculture, it is referred to as *agroterrorism*.

bioterrorismo el uso deliberado de armas biológicas o químicas; en agricultura, se denomina *agroterrorismo*.

bit the metal part of a bridle that goes in the horse's mouth.

bocado la parte metálica de una brida que va en la boca del caballo.

bladder storage compartment for liquid waste in the animal's body.

vejiga compartimento de almacenamiento para desechos líquidos en el cuerpo de un animal.

blemish an imperfection that does not affect the usefulness of a horse.

tacha imperfección que no afecta la utilidad de un caballo.

bloodtyping a process whereby an individual's blood group is identified by a serological test on a sample of blood.

determinación del grupo y tipo sanguíneo proceso por el cual se identifica el grupo sanguíneo de un individuo mediante el análisis serológico de una muestra de sangre.

boar a male swine that has not been castrated.

padrillo cerdo macho que no ha sido castrado.

boar taint refers to the odor of boar meat.

olor sexual se refiere al olor de la carne de padrillo.

body condition score a system of estimating body energy (relative body fat) in animals using a numeric score.

puntaje del estado corporal sistema para estimar la energía corporal (grasa corporal relativa) en animales que usa un puntaje numérico.

bomb calorimeter an apparatus used to measure the amount of heat given off by any combustible substance; used for determining the calorie content of feeds.

calorímetro de bomba aparato que se usa para medir la cantidad de calor emitida por cualquier sustancia combustible; se usa para determinar el contenido calórico de los alimentos.

boot stage the point in growth at which the inflorescence (the flowering part of the plant) expands the boot (the upper sheath of the plant).
etapa de encañado punto del crecimiento en el cual la inflorescencia (la parte de la planta que florece) expande la caña (la vaina superior de la planta).

Bos indicus humped cattle originating in tropical countries and common in modern cattle production in the United States.
Bos indicus ganado con joroba originario de países tropicales y común en la producción moderna de ganado en EE. UU.

Bos taurus domestic cattle originating from either the Aurochs or the Celtic Shorthorn.
Bos taurus ganado doméstico descendiente del uro o del shorthorn celta.

bot the larvae of the botfly.
bot larvas de moscardón.

bovine somatotropin (bST) a hormone produced naturally by the pituitary gland of the cow. (See *somatotropin*.)
somatotropina bovina (bST) hormona producida naturalmente por la glándula pituitaria de la vaca. (Ver *somatotropina*.)

bovine spongiform encephalopathy (BSE) degenerative disease that affects the central nervous system of cattle.
encefalopatía espongiforme bovina (EEB) enfermedad degenerativa que afecta el sistema nervioso central del ganado.

breed a group of related animals that breed true to a number of given traits that identify the breed.
raza a grupo de animales relacionados que se reproducen repitiendo un número de rasgos determinados que identifican a la raza.

breed crossing in poultry, the crossing of different breeds to get the desired traits.
cruza de razas en las aves, la cruza de distintas razas para obtener los rasgos deseados.

breeding rack equipment used for hand mating.
caja de reproducción equipo utilizado para apareamiento manual.

breeding sheep ewes that are sold back to farms and ranches to be bred to produce more lambs.
ovejas reproductoras ovejas que se vendenz nuevamente a establecimientos agrícolas y ranchos para que se apareen y produzcan más carneros.

breeding value the value of the sire or dam as a parent; based on differences that exist between a large number of offspring and the average performance of a trait within a population.
valor reproductivo el valor de un macho o una hembra para producir descendencia; se basa en las diferencias que existen entre un gran número de crías y el rendimiento promedio de un rasgo dentro de una población.

bridle headgear used to control a horse.
brida arnés para la cabeza que se usa para controlar a un caballo.

bronchi the two airways that lead into an animal's lungs (singular bronchus).
bronquios las dos vías aéreas que se dirigen al interior de los pulmones de un animal.

bronchiole smaller branch of the bronchus.
bronquiolo rama menor del bronquio.

brood refers to an animal that is kept for breeding purposes.
camada se refiere a un animal que se conserva para fines de reproducción.

brooder the enclosure in which young animals are kept, or are allowed free access, which furnishes heat until they adapt to outside temperatures.
incubadora el recinto donde se permite el libre acceso de animales pequeños, o donde se los conserva, y que les proporciona calor hasta que se adaptan a las temperaturas externas.

broodiness in poultry, when a hen stops laying eggs and wants to sit on a nest of eggs to hatch them.
cloquera entre las aves de corral, cuando una gallina deja de poner huevos y quiere sentarse en un nido de huevos para empollarlos.

browse the shoots, twigs, and leaves of brush plants found growing on rangeland.
follaje los brotes, las ramas y las hojas de arbustos que crecen en pastizales.

buck a male of such species as goats, deer, rabbits, etc.
macho animal masculino de especies como las cabras, los renos, los conejos, etc.

bull a male bovine of any age that has not been castrated.
toro bovino macho de cualquier edad que no ha sido castrado.

bullock a male beef animal, usually under 24 months of age, which may be castrated or uncastrated and does show some of the characteristics of a bull.
novillo animal vacuno macho, generalmente de menos de 24 meses, que puede estar castrado o no y que muestra algunas de las características de un toro.

burdizzo a pincers-like instrument with wide jaws, used for bloodless castration of animals with a pendant scrotum; also used for bloodless killing of poultry and docking lamb tails.
burdizzo instrumento similar a una pinza con mordazas amplias que se usa para la castración sin sangre de animales con escroto colgante; también se usa para el sacrificio sin sangre de aves y para acoplar las colas de las ovejas.

by-product a product of significantly less value than the major product.
derivado un producto de valor significativamente inferior al producto principal.

C

cabrito the meat from young kids slaughtered at approximately 30 to 40 pounds.
cabrito la carne de crías jóvenes sacrificadas cuando pesan aproximadamente 13 a 18 kg (30 a 40 libras).

calf a beef animal under 1 year of age.
ternero animal vacuno de menos de 1 año.

cancellous bone material spongy material usually found in the ends of long bones.
material óseo esponjoso material esponjoso que suele encontrarse en los extremos de los huesos largos.

candling a process that determines the interior quality of an egg with the use of a high-intensity light.
ovoscopio proceso que determina la calidad interior de un huevo con el uso de una luz de alta intensidad.

caping the practice of leaving a strip of mohair down the Angora goat's back to protect it from the weather.
capeado práctica de dejar una tira de mohair a lo largo del lomo de la cabra de Angora para protegerla de la intemperie.

capon a male chicken that has been surgically castrated.
capón pollo macho castrado quirúrgicamente.

carbohydrate organic compound containing carbon, hydrogen, and oxygen.
carbohidrato compuesto orgánico que contiene carbono, hidrógeno y oxígeno.

cardiac muscle striated involuntary muscle found only in the muscular wall of the heart.
músculo cardíaco músculo estriado involuntario que solo se encuentra en la pared muscular del corazón.

carpet wool a coarse wool used for making carpets.
lana para alfombras lana rústica usada para hacer alfombras.

carrying capacity the number of animals that can be grazed on a pasture during the grazing season.
capacidad de transporte el número de animales que pueden pastar en una pastura durante la temporada de pastoreo.

cashmere the soft down or winter undercoat of fiber produced by most breeds of goats, except the Angora.
cachemira capa blanda inferior o invernal de fibra producida por la mayoría de las razas de cabra, excepto la de Angora.

castration the removal of the testicles of a male animal.
castración extirpación de los testículos de un animal macho.

catabolism the breakdown of body tissues into simpler substances and waste products.
catabolismo descomposición de tejidos corporales en sustancias más simples y productos de desecho.

catarrhal inflammation of mucous membranes, especially of the nose and throat.
catarral inflamación de membranas mucosas, especialmente de la nariz y la garganta.

cattle beef animals over 1 year of age.
reses animales vacunos de más de 1 año.

cattle guard a gate that allows equipment to be driven into an area without the operator having to stop and open a gate.
guardaganado puerta que permite ingresar equipo a un área sin que el operador tenga que detenerse para abrir una puerta.

causative agent a biological pathogen that causes a disease; generally most are bacteria, fungi, or viruses.
agente causal patógeno biológico que causa una enfermedad; generalmente la mayoría son bacterias, hongos o virus.

cecum a blind pouch located at the point where the small intestine joins the large intestine.
ciego bolsa ciega ubicada en el punto donde el intestino delgado se une con el grueso.

cell a microscopic mass of protoplasm that is the structural and functional unit of a plant or animal organism.
célula masa microscópica de protoplasma que es la unidad estructural y funcional de una planta o un organismo animal.

cell membrane a semipermeable membrane that surrounds the cell nucleus and cytoplasm.
membrana celular membrana semipermeable que rodea al núcleo y al citoplasma celular.

centriole a structure located just outside of the cell nucleus that organizes the spindle during mitosis and meiosis.
centriolo estructura ubicada justo fuera del núcleo de la célula y que organiza al huso durante la mitosis y la meiosis.

centromere that part of the chromosome where the spindle fiber attaches during mitosis.
centrómero parte del cromosoma donde se fija la fibra del huso durante la mitosis.

cervix in the female reproductive system, the neck of the uterus, which separates the uterus from the vagina.
cérvix en el sistema reproductivo femenino, el cuello del útero, que separa el útero de la vagina.

challenge feeding the practice of feeding higher levels of concentrate to challenge the cow to reach her maximum potential milk production; also known as lead feeding.
alimentación de provocación la práctica de proporcionar niveles mayores de concentrado en la alimentación para provocar a la vaca para que alcance su máximo potencial de producción de leche; también se *denomina alimentación adelantada.*

chemobiotic a combination of an antibiotic and an antibacterial formed to combat a problem that is not susceptible to either one individually.
quimiobiótico combinación de un antibiótico y un antibacterial formada para combatir un problema que no responde a ninguno de esos dos elementos por separado.

chevon meat from goats.
chevon carne de la cabra.

cholesterol a steroid that occurs mainly in the bile, gallstones, brain, and blood cells. It is the most common of the animal sterols and is a precursor of a form of vitamin D.
colesterol esteroide que se presenta principalmente en la bilis, los cálculos biliares, el cerebro y las células de la sangre. Es la forma más común de esteroles animales y precursor de una forma de vitamina D.

chromatid one of the two daughter strands of a duplicated chromosome, joined by a single centromere, that separate during cell division to form individual chromosomes.
cromátida una de las dos cadenas hijas de un cromosoma duplicado, unidas por un solo centrómero y que se separan durante la división celular para formar cromosomas individuales.

chromosome small rod-shaped structure found in the nucleus of a cell that contains the genes.
cromosoma estructura pequeña en forma de barra que se encuentra en el núcleo de una célula y contiene los genes.

chyme partially digested feed that moves from the stomach to the small intestine.
quimo alimento parcialmente digerido que pasa del estómago al intestino delgado.

citizenship the status of a person recognized as a member of a government entity such as a city, state, or country who has certain defined rights and responsibilities. Rights may include the right to vote, work, and own property. Responsibilities may include a duty to obey laws, pay taxes, serve in the military, and respect the rights of others.
ciudadanía estatus de una persona reconocida como miembro de una entidad de gobierno, como una ciudad, un estado o un país, con ciertos derechos y responsabilidades definidos. Los derechos pueden incluir el derecho a votar, a trabajar y a poseer propiedad. Las responsabilidades pueden incluir la obligación de obedecer leyes, de pagar impuestos, de hacer el servicio militar y de respetar los derechos de otras personas.

class in poultry, a term that generally refers to geographic origin; in plant and animal classification, a division lower than a phylum and higher than an order.

clase en las aves de corral, término que generalmente se refiere al origen geográfico; en la clasificación de plantas y animales, una división inferior a la afiliación y superior al orden.

Class I milk Grade A milk used in all beverage milks such as whole milk, skim milk, buttermilk, etc.

leche de clase I Leche de categoría A que se usa en todas las leches para beber, como leche entera, leche desnatada, suero de leche, etc.

Class II milk Grade A milk used in fluid cream products, yogurts, or perishable manufactured products such as ice cream, cottage cheese, etc.

leche de clase II Leche de categoría A que se usa en productos con crema líquida, yogures o productos elaborados perecederos, como helado, queso cottage, etc.

Class III milk Grade A milk used to produce cream cheese and hard manufactured cheeses.

leche de clase III Leche de categoría A que se usa para producir queso crema y quesos elaborados duros.

Class IV milk Grade A milk used to produce butter and any milk in dried form.

leche de clase IV Leche de categoría A que se usa para producir mantequilla y todas las leches deshidratadas.

clip the wool or mohair produced by a single shearing.

recorte la lana o el mohair producidos mediante una sola esquila.

clitoris the sensory and erectile organ of the female reproductive system.

clítoris órgano sensorio y eréctil del sistema reproductivo femenino.

cloaca in poultry, an enlarged part of the digestive tract where the large intestine joins the vent.

cloaca en las aves, una parte agrandada del tracto digestivo donde el intestino grueso se une con la zona excretora.

clone one of two or more cells or organisms that are genetically identical to each other.

clon una o más células u organismos que son genéticamente idénticos entre sí.

closebreeding an intensive form of inbreeding in which the animals being mated are very closely related and have ancestry that can be traced back to more than one common ancestor.

endogamia forma intensiva de apareamiento en la cual los animales apareados están muy íntimamente emparentados y tienen una ascendencia que puede rastrearse a más de un ancestro en común.

coccidiostat any of a group of chemical agents mixed in feed or drinking water to control *coccidiosis*, a growth-retarding and occasionally fatal disease of poultry.

coccidiostético cualquiera ejemplar de un grupo de agentes químicos mezclado en alimentos o en agua para beber para controlar la *coccidiosis*, una enfermedad que retarda el crecimiento y, ocasionalmente, es mortal en las aves.

cockerel a young male chicken.

gallito pollo macho joven.

codominance in heterozygotes the situation in which both alleles of a gene pair are fully expressed; neither being dominant or recessive to the other.

codominancia en heterocigotas, la situación en la cual ambos alelos de un par de genes se expresan plenamente; ninguno es dominante o recesivo con respecto al otro.

codon a set of three nucleotides that make up the genetic code that specifies the specific amino acid to be inserted at a particular position in a polypeptide chain during protein synthesis.

codón conjunto de tres nucleótidos que forman el código genético que especifica el aminoácido específico que se debe insertar en una posición en particular en una cadena de polipéptidos durante la síntesis de proteínas.

cold confinement house a hog shelter whose temperature is only slightly warmer than the outside temperature.

alojamiento de encierro frío refugio para cerdos cuya temperatura es solo un poco mayor que la temperatura exterior.

colic a disease complex encompassing a wide range of conditions that affect the digestive tract; usually caused by some type of obstruction that blocks the flow of feed through the intestine, resulting in abdominal pain.

cólico complejo de enfermedades que abarca una amplia variedad de condiciones que afectan el tracto digestivo; suele estar causado por algún tipo de obstrucción que bloquea el flujo del alimento a través del intestino, lo cual causa dolor abdominal.

color breed in horses, a breed that registers horses on the basis of color; horses in the breed may have a variety of conformation types.

reproducción por color en los caballos, tipo de reproducción que registra los animales con base en el color; los caballos de la reproducción pueden tener diversos tipos de conformación.

colostrum the milk produced in the first few days after parturition.

calostro la leche producida en los primeros días después de la parición.

colt generally, a male horse under 3 years of age; the Thoroughbred breed includes males that are 4 years old.

potrillo generalmente, un caballo macho de menos de 3 años; la raza purasangre incluye machos de 4 años.

comb in poultry, a fleshy crest on the top of the head.

peine en las aves, una cresta carnosa en la parte superior de la cabeza.

comfort zone the range of effective ambient temperatures within which an animal does not have to increase normal metabolic heat production to offset heat loss to the environment; also called *thermoneutral zone.*

zona de bienestar rango de temperaturas ambiente efectivas dentro del cual un animal no necesita aumentar la producción metabólica normal de calor para compensar la pérdida de calor al medio ambiente; también se denomina *zona termoneutral.*

commensalism a relationship in which one organism benefits while the other is neither harmed nor benefited.

comensalismo relación en la cual un organismo se beneficia y el otro no se perjudica ni se beneficia.

commercial feed tag the label attached to a bag of feed purchased at a grain elevator or feed store.

etiqueta de alimento comercial etiqueta que se fija a una bolsa de alimento que se compra en un elevador de granos o en una tienda de alimentos.

commercial heat detector a device that helps detect heat in a herd of cows.

detector de calor comercial dispositivo que ayuda a detectar calor en un hato de vacas.

commercial protein supplement mixes of animal and vegetable protein feeds made by commercial feed companies that are usually made for one class of animals.

suplemento proteico comercial alimentos con mezclas de proteínas vegetales y animales fabricados por compañías de alimentos comerciales y que habitualmente se producen para una clase de animales.

commission the fee charged by commission firms for their services in marketing a product.

comisión tarifa que cobran las firmas a comisión por sus servicios en la comercialización de un producto.

Commodity Credit Corporation (CCC) a part of the USDA Farm Service Agency that was created to stabilize, support, and protect farm income and prices; also helps maintain balanced and adequate supplies of agricultural commodities and aids in their orderly distribution.

La Corporación de Créditos para Productos Básicos (CCC) una parte del organismo de servicios a establecimientos agrícolas del Departamento de Agricultura de EE. UU. (USDA) que se creó para estabilizar, brindar apoyo y proteger los ingresos y los precios de los establecimientos agrícolas, también ayuda a mantener suministros equilibrados y adecuados de los productos básicos agrícolas y ayuda a su distribución ordenada.

compact bone material material composed of Haversian systems that is usually found on the outside of bones and surrounding the bone marrow.

material óseo compacto material compuesto de sistemas haversianos y que suele encontrarse en la parte exterior de los huesos y alrededor de la médula ósea.

complete mixed feed a ration in which all ingredients are mixed together.

comida mixta completa una ración en la que todos los ingredientes se mezclan juntos.

concentrate feed containing less than 18 percent crude fiber when dry; grains and protein supplements are concentrates.

concentrado alimento que contiene menos del 18 por ciento de fibra cruda cuando está seco; los granos y los suplementos proteicos son concentrados.

conception rate the percentage of a group of animals that becomes pregnant when bred.

tasa de concepción porcentaje de un grupo de animales que queda preñado cuando se los aparea.

conformation the type, form, and shape of a live animal, usually with reference to some performance characteristic.

conformación el tipo, la formación y la forma de un animal vivo, generalmente con respecto a alguna característica de rendimiento.

contemporary group a group of animals that have been raised under the same conditions, have the same genetic background, and are of the same sex.

grupo contemporáneo grupo de animales criados en las mismas condiciones, con los mismos antecedentes genéticos y del mismo sexo.

convulsion involuntary muscular contractions.

convulsión contracciones musculares involuntarias.

coprophagy the practice of ingesting feces.

coprofagia la práctica de ingerir heces.

copulation the mating of a male and a female.

copulación el apareamiento de un macho y una hembra.

corpus luteum a reddish-yellow mass that forms in a ruptured follicle in the ovary of mammals; the hormone progesterone is released by the corpus luteum.

cuerpo lúteo masa amarillo rojizo que se forma en un folículo roto del ovario de los mamíferos; la hormona progesterona es liberada por el cuerpo lúteo.

corticotropin a hormone that is produced and secreted by the pituitary gland; may be used for the treatment of some breathing problems, severe allergies, mononucleosis, and leukemia.

corticotropina hormona que produce y segrega la glándula pituitaria; se puede usar para el tratamiento de algunos problemas respiratorios, alergias graves, mononucleosis y leucemia.

cortisone a hormone that comes from the adrenal glands of an animal;may be used for the treatment of rheumatoid arthritis, adrenal insufficiency, some allergies, diseases of the connective tissues, and gout.

cortisona hormona que proviene de las glándulas suprarrenales de un animal; puede usarse para el tratamiento de la artritis reumatoide, insuficiencia suprarrenal, algunas alergias, enfermedades de tejidos conectivos y gota.

counteractant a substance that attempts to neutralize the odor of wastes so no odor remains.

contrarrestante sustancia que intenta neutralizar el olor de desechos para que no quede olor.

cow a female bovine that has had one or more calves; or an older female that has not had a calf but has matured.

vaca bovino hembra que ha tenido uno o más terneros; o una hembra mayor que no ha tenido terneros pero ha madurado.

cow-calf system a system of beef production that involves keeping a herd of beef cows that are bred each year to produce calves, which are sold to cattle feeders.

sistema vaca-ternero sistema de producción de carne vacuna que implica conservar un hato de vacas para carne que se reproducen todos los años y así producir terneros, que se venden a alimentadores de ganado.

Cowper's gland an accessory gland in the male reproductive system that produces a fluid which moves ahead of the seminal fluid, cleaning and neutralizing the urethra.

glándula de Cowper glándula accesoria del sistema reproductivo masculino que produce un líquido que se desplaza antes del líquido seminal y limpia y neutraliza la uretra.

creep feeding a system of feeding young domestic animals by placing a special fence around special feed for the young; the fence excludes mature animals but permits the young to enter.

alimentación diferencial sistema para alimentar a animales domésticos jóvenes mediante la colocación de una cerca especial alrededor del alimento especial para los jóvenes; la cerca excluye a los animales maduros pero permite el ingreso de los jóvenes.

cria a baby llama, alpaca, or other camelidae.

cría llama, alpaca u otro camélido bebé.

crimping passing a feed crop through a set of corrugated rollers that are set close together.

prensado paso de un alimento a través de un grupo de rodillos corrugados ajustados muy cerca entre sí.

crisscrossing mating crossbred females to a sire belonging to one of the parent breeds of the female; also called *backcrossing*.

cruzamiento alterno apareamiento de hembras de razas cruzadas con un macho perteneciente a una de las razas de origen de la hembra; también se denomina *retrocruzamiento*.

critical control points (CCPs) points at which identified hazards may be controlled. (See *Hazard Analysis and Critical Control Point [HACCP]*.)

puntos de control críticos (CCP) puntos en los cuales se pueden controlar riesgos identificados. (Ver *Análisis de riesgos y punto de control crítico [HACCP].*)

crop in poultry, an enlargement of the gullet that serves as a storage area for feed.

buche en las aves, un agrandamiento del esófago que sirve como área de almacenamiento para el alimento.

crossbreeding the mating of two animals from different breeds.

cruza el apareamiento de dos animales de distintas razas.

cross-mating in poultry, the crossing of two or more strains within the same breed.

apareamiento cruzado en las aves, la cruza de dos o más estirpes dentro de la misma raza.

crossover the formation of new chromosomes resulting from the splitting and rejoining of the original chromosomes.

cruce la formación de nuevos cromosomas que resulta de la división y la reunión de los cromosomas originales.

crude protein the amount of ammoniacal nitrogen in a feed multiplied by 6.25.

proteína cruda la cantidad de nitrógeno amoniacal en un alimento multiplicada por 6,25.

cryptorchidism an inherited trait that causes one or both testicles to be held in the body cavity.

criptorquidismo rasgo heredado que causa que uno o ambos testículos se retengan en la cavidad del cuerpo.

cubing processing a feed by grinding and then forming it into a hard form called a cube; cubes are larger than pellets.

cubeteo procesamiento de un alimento mediante molienda y, seguidamente, formación de cuerpos rígidos llamados cubos; los cubos son más grandes que las pastillas.

cud in ruminants, a ball-like mass of feed that is brought up from the stomach to be rechewed.

contenido ruminal en los rumiantes, una masa tipo bola de alimento que asciende del estómago para volver a ser masticada.

culling removing animals from a herd that are not as productive or desirable as the others in the herd.

selección controlada eliminación de animales de una manada que no son tan productivos o deseables como el resto.

Cushing's syndrome a disease caused by a small benign tumor in the pituitary gland; also called *hyperadrenocorticism.*

síndrome de Cushing enfermedad causada por un tumor benigno pequeño en la glándula pituitaria; también se denomina *hiperadrenocorticismo.*

cutability the yield of closely trimmed, boneless retail cuts that come from the major wholesale cuts of an animal carcass.

proporción de cortes el rendimiento de cortes al menudeo bien recortados y sin hueso que se obtienen de los cortes mayoristas principales de la carcasa de un animal.

cyst a swelling containing a fluid or semi-solid substance.

quiste inflamación que contiene un líquido o una sustancia semisólida.

cytoplasm material surrounding the nucleus of a cell.

citoplasma material que rodea el núcleo de una célula.

cytosine a pyrimidine base that is a part of DNA and RNA; pairs with guanine in DNA.

citosina una base pirimidina que forma parte del ADN y el ARN; se aparea con la guanina en el ADN.

D

Dairy Herd Improvement Association (DHIA) a group of dairy farmers in a local area that is formed to provide production testing services.

Asociación para la Mejora de Hatos Lecheros (DHIA) grupo de agricultores lecheros en un área local formado que proporcionan servicios de pruebas de producción.

dam the female parent of an animal.

madre la hembra madre de un animal.

dark cutting beef a condition in which the lean meat is darker than normal in color. It usually has a gummy or sticky texture; also called *black cutting beef.*

carne vacuna oscura condición en la cual la carne magra es más oscura que lo normal. Suele tener una textura gomosa o pegajosa; también se denomina *carne vacuna negra.*

debris basin a structure that is used to catch runoff from animal pens.

vaciadero estructura que se usa para atrapar los escurrimientos de los corrales de animales.

deficiency a lack of a certain element in an animal's ration.

deficiencia falta de un elemento determinado en la ración de un animal.

demand the amount of a product that buyers will purchase at a given time for a given price.

demanda cantidad de un producto que comprarán los compradores en un tiempo determinado por un precio determinado.

deodorant a chemical that kills the bacteria that cause an odor.

desodorante sustancia química que extermina las bacterias que causan un olor.

deoxyribonucleic acid (DNA) a compound composed of deoxyribose (a sugar), phosphoric acid, and nitrogen. The DNA molecule contains genes on strands in the form of a double helix.

ácido desoxirribonucleico (ADN) a compuesto formado por desoxirribosa (un azúcar), ácido fosfórico y nitrógeno. La molécula de ADN contiene genes en cadenas con forma de doble hélice.

deoxyribose the five-carbon sugar found in DNA.

desoxirribosa azúcar de cinco carbonos que se encuentra en el ADN.

desnooding the removal of the tubular fleshy appendage on the top of the head of some types of poultry.

descaperuzado extirpación del apéndice carnoso tubular en la parte superior de la cabeza de algunos tipos de aves de corral.

dewlap a hanging fold of skin under the neck of animals, especially some breeds of cattle and goats.

papada pliegue de piel colgante debajo del cuello de los animales, especialmente algunas razas de ganado y cabras.

diarthroses joints joints that allow free movement and have a fluid-filled cavity; also called *synovial joints.*

diartrosis articulaciones móviles que permiten el movimiento libre y tienen una cavidad llena de líquido; también se denominan *articulaciones sinoviales.*

diet the type and amount of food and drink habitually ingested by a person or an animal; ration without reference to a specific time period.

dieta tipo y cantidad de alimento y bebida que habitualmente ingiere una persona o un animal; ración sin referencia a un período específico.

diffusion process in which food, oxygen, and wastes pass through a semipermeable membrane.

difusión proceso en el cual el alimento, el oxígeno y los desechos atraviesan una membrana semipermeable.

digestible energy (DE) gross energy of a feed minus the energy remaining in the feces of the animal after the feed is digested.

energía digerible (ED) energía bruta de un alimento menos la energía restante en las heces del animal una vez digerido el alimento.

digestible protein (DP) that portion of the crude protein in a feed that can be utilized by an animal.

proteína digerible (PD) parte de la proteína cruda en un alimento que puede ser utilizada por un animal.

digestion the process of breaking down feed into simple substances that can be absorbed by the body.

digestión proceso de descomposición del alimento en sustancias simples que pueden ser absorbidas por el cuerpo.

digestive deodorant bacteria that create a digestive process that eliminates an odor.

desodorante digestivo bacterias que crean un proceso digestivo que elimina un olor.

digestive system the parts of the body involved in chewing and digesting feed; also called the *digestive tract*.

sistema digestivo las partes del cuerpo que participan en la masticación y la digestión del alimento; también se denomina *tracto digestivo*.

diploid an individual or cell having two sets of chromosomes.

diploide individuo o célula que tiene dos conjuntos de cromosomas.

dipping vat a pit filled with a liquid containing insecticides, ovicides, repellents, etc., through which animals are forced to pass for disinfestation.

baño de inmersión una fosa llena de un líquido que contiene insecticidas, ovicidas, repelentes, etc., a través de la cual se obliga a los animales que pasen para desinfestarlos.

disposal the final step in controlling runoff from feedlots, in which water collected in holding ponds is used for irrigation or allowed to evaporate.

disposición el paso final en el control de los escurrimientos de los pastaderos en el cual el agua recolectada en fuentes de retención se usa para riego o se deja evaporar.

distemper (equine) in horses, a disease caused by a bacterium that results in symptoms such as high fever, pus-like discharge from the nose, and swollen lymph nodes; also called *strangles*.

moquillo (equino) en los caballos, una enfermedad causada por una bacteria que causa síntomas como fiebre alta, secreción tipo pus de la nariz y ganglios linfáticos inflamados; también se denomina *adenitis equina*.

distiller's grains a by-product of grain alcohol fermentation used as animal feed.

granos de destilador derivado de la fermentación de alcohol de grano utilizado como alimento animal.

diversion preventing surface water from outside the feedlot from getting onto the feedlot.

desvío impedir que el agua superficial del exterior del pastadero ingrese al pastadero.

diversion terrace a structure that forces water to go around the feedlot.

terraza de desvío estructura que obliga al agua a pasar alrededor del pastadero.

doctoral degree generally the highest degree awarded in the education system; may be an academic or professional degree that qualifies the holder to teach at the university level or to work in a specific profession.

título de doctorado generalmente el título más alto que se otorga en el sistema educativo; puede ser un título académico o profesional que califica al titular para enseñar en el ámbito universitario o para trabajar en una profesión específica.

doe a female of those species in which the male is called a buck; for example: goats, deer, rabbits, etc.

hembra animal femenino de las especies en las cuales el animal masculino se denomina macho; por ejemplo: cabras, renos, conejos, etc.

doeling an unbred female goat under 2 years of age.

cabrita la cabra hembra no apareada de menos de 2 años.

domesticate to adapt the behavior of an animal to fit the needs of people.

domesticar adaptar el comportamiento de un animal para adecuarlo a las necesidades de las personas.

dominant gene one of a pair of genes that hides the effect of the other gene in the pair.

gen dominante gen dentro de un par que oculta el efecto del otro gen del par.

donkey the common name for the ass.

burro nombre común del asno.

double cropping the practice of growing two crops on the same ground in the same year.

doble cultivo la práctica de realizar dos cultivos en el mismo terreno el mismo año.

down the soft furry or fine feathery coat of young birds and animals or the undercoat of adult birds and animals.

pelusa capa peluda suave o plumosa fina de las aves y los animales jóvenes o la capa inferior de las aves o los animales adultos.

draft animal an animal used for pulling loads.

animal de tiro animal que se usa para arrastrar cargas.

drainage channel pipe for runoff water from the feedlot to be diverted into.

canal de drenaje tubo dentro del cual se debe desviar el agua de escurrimiento del pastadero.

drake a male duck.

pato pato macho.

drench a large dose of medicine in a liquid form administered to the animal through the mouth.

drench dosis grande de medicamento en forma líquida que se administra al animal por la boca.

dry cleaning a method of cleaning eggs in which the eggs are buffed with fine sandpaper, emery cloth, or steel wool.

limpieza en seco método para limpiar huevos en el cual los huevos se pulen con lija fina, tela esmeril o lana de acero.

dry weight the weight of a product or material when the moisture content is removed.

peso seco el peso de un producto o un material cuando se ha eliminado el contenido de humedad.

dual-purpose breed a breed used to produce two or more products such as meat and milk.

raza de doble propósito raza que se usa para producir dos o más productos, como carne y leche.

dystocia painful or slow delivery or birth.
distocia parto o nacimiento doloroso o lento.

E

effective ambient temperature (EAT) the combined effect of air temperature, humidity, precipitation, wind, and heat radiation on the efficiency of energy use by farm animals.
temperatura ambiente efectiva (TAE) efecto combinado de temperatura del aire, humedad, precipitación, viento e irradiación de calor sobre la eficiencia del uso de energía por parte de los animales de granja.

efficiency in gain refers to the amount of feed needed for each pound of gain.
eficiencia en aumento de peso se refiere a la cantidad de alimento necesaria para cada 0,454 kg (1 lb) de aumento de peso.

elastrator band a tight rubber band placed by a special instrument around the scrotum above the testicles that cuts off the blood supply to the testicles, causing them to waste away.
banda elastradora banda de caucho apretada que se coloca mediante un instrumento especial alrededor del escroto por encima de los testículos y que corta el suministro de sangre a los testículos, causando su muerte.

electrolyte a solution containing salts and energy sources used to feed young animals suffering from scours (diarrhea).
electrolito solución que contiene sales y fuentes de energía que se usan para alimentar a animales jóvenes que sufren de disentería (diarrea).

embryo the early stage in the development of the fetus.
embrión la primera etapa en el desarrollo del feto.

embryo transfer moving an embryo from one animal to another.
transferencia de embrión trasladar un embrión de un animal a otro.

encephalitis an inflammation of the brain that occurs in goats and may be caused by caprine arthritis encephalopathy.
encefalitis una inflamación cerebral que se produce en las cabras y puede ser causada por una artritis encefalitis caprina.

encephalomyelitis a disease that affects the brain, may be caused by any one of several different viruses; also known as *sleeping sickness*.
encefalomielitis enfermedad que afecta el cerebro; puede ser causada por cualquiera de varios virus; también se denomina *enfermedad del sueño*.

endoskeleton a skeletal system found inside the body.
endoesqueleto sistema esquelético que se encuentra en el interior del cuerpo.

endotoxin poisonous substances produced by certain bacteria.
endotoxina sustancias venenosas producidas por ciertas bacterias.

energy feed livestock feed containing less than 20 percent crude protein; most grains are energy feeds.
alimento energético alimento para ganado que contiene menos del 20 por ciento de proteína cruda; la mayoría de los granos son alimentos energéticos.

environment the total of the external conditions and influences that affect the life and development of living organisms.
medio ambiente el total de las condiciones y las influencias externas que afectan la vida y el desarrollo de los organismos vivos.

enzootic a disease that affects or is prevalent among animals in a specific geographic area.
enzoótica enfermedad que afecta o prevalece entre animales de un área geográfica específica.

enzyme an organic catalyst that speeds up the digestive process without being used up in the process.
enzima catalizador orgánico que acelera el proceso digestivo sin que se agote en el proceso.

Eohippus tiny (about 1-foot high), four-toed ancestor of today's horse; native to the North American continent about 58 million years ago.
Eohippus ancestro diminuto (aprox. de 30 cm o 1 pie de altura), con cuatro dedos en las patas, del caballo actual; nativo del continente de América del Norte hace aproximadamente 58 millones de años.

epididymis a long, coiled tube that is connected to each testicle.
epidídimo tubo largo y enroscado conectado a cada testículo.

epinephrine a hormone, also known as adrenaline, that is produced by the medulla of the adrenal glands in mammals.
epinefrina hormona también denominada adrenalina, que es producida por la médula de las glándulas suprarrenales en los mamíferos.

epithelium in poultry, a thick, horny membrane-like material that lines the muscular stomach or gizzard.
epitelio en las aves, un material grueso, calloso tipo membrana que reviste el estómago muscular o molleja.

equine infectious anemia disease that is caused by a virus and is carried from horse to horse by bloodsucking insects; also called *swamp fever*.
anemia infecciosa equina enfermedad causada por un virus y transmitida de un caballo a otro por insectos que chupan la sangre; también se denomina fiebre de *los pantanos*.

ergot a fungus that, if ingested, will cause abortion in pregnant sows.
cornezuelo hongo que, si se ingiere, causa el aborto en las cerdas preñadas.

esophagus the tube like passage from the mouth to the stomach; sometimes called the *gullet*.
esófago pasaje con forma de tubo de la boca al estómago; a veces se denomina *gola*.

estimated breeding value (EBV) a method of combining performance data on a given trait about an individual animal and its close relatives to determine the individual's ability to transmit that performance trait to its offspring.
valor de reproducción estimado (EBV) método de combinar datos de rendimiento sobre un rasgo determinado acerca de un animal individual y sus parientes cercanos para determinar la capacidad del individuo de transmitir ese rasgo de rendimiento a su descendencia.

estimated relative producing ability (ERPA) in dairy cattle, a predication of 305-day, two-times-per-day milking, mature equivalent (305-2×-ME) production compared to other cows in the herd.
capacidad productiva relativa estimada (ERPA) en el ganado lechero, un predicamento de 305 días, dos veces al día, de producción equivalente madura (305-2×-ME) de ordeñe en comparación con otras vacas del hato.

estray a domestic animal of unknown ownership that is running at large.
errante animal doméstico de propiedad desconocida que en general está corriendo.

estrogen a hormone produced by the ovaries.

estrógeno hormona producida por los ovarios.

estrus the time during which the female will accept the male for copulation; also referred to as being "in heat."

estro tiempo durante el cual la hembra acepta al macho para copular; también se denomina estar "en celo".

ewe a female sheep or lamb.

oveja carnero o cordero hembra.

exoskeleton typically a hard shell on the outside of the body.

exoesqueleto típicamente una cubierta dura en el exterior del cuerpo.

expected progeny difference (EPD) an estimate of the genetic value of an animal in passing genetic traits to its offspring.

diferencia esperada de progenie (EPD) estimado del valor genético de un animal para transmitir rasgos genéticos a sus descendientes.

extension hurdle a portable hurdle (gate) that can be lengthened or shortened as needed.

valla de extensión valla (puerta) portátil que puede alargarse o acortarse según sea necesario.

F

fagopyrin a photosensitizing agent found in buckwheat that can cause rashes and itching when white pigs that ingest it are exposed to sunlight.

fagopirina agente fotosensibilizante que se encuentra en el trigo sarraceno y que puede causar erupciones y picazón cuando los cerdos blancos que lo ingieren se exponen a la luz solar.

farrier a person who works on horses' feet.

herrador persona que trabaja con las patas de los caballos.

fat organic compound composed of carbon, hydrogen, and oxygen that is solid at body temperature; fats contain more carbon and hydrogen than do carbohydrates and are mainly glyceryl esters of certain acids that are soluble in ether but not in water.

grasa compuesto orgánico formado por carbono, hidrógeno y oxígeno, sólido a temperatura corporal; las grasas contienen más carbono e hidrógeno que los carbohidratos y son principalmente ésteres de glicerilo de ciertos ácidos solubles en éter pero no en agua.

feathering the streaks of fat visible on the ribs of a carcass.

difuminado bandas de grasa visibles sobre las costillas de una carcasa.

feces undigested material that is passed out of the digestive system through the anus.

heces material no digerido que se elimina del sistema digestivo a través del ano.

federal milk marketing order (FMMO) establishes rules under which dairy processors purchase fresh milk from dairy farmers supplying a marketing area.

orden federal de comercialización de la leche (FMMO) establece las reglas según las cuales los procesadores lácteos compran leche fresca a los tamberos que suministran una zona de comercialización.

feed additive a material added to livestock feed, usually an antibiotic, that is not a nutrient but enhances the growth efficiency of the animal.

aditivo de alimento material que se agrega al alimento de ganado, habitualmente un antibiótico, que no es un nutriente pero que mejora la eficiencia de crecimiento del animal.

Feed Additive Compendium a publication that lists feed additives in current use and the regulations for their use.

Compendio de aditivos para alimentos publicación que enumera los aditivos para alimentos de uso actual y las regulaciones para su uso.

feed composition table a table showing the nutrients found in feeds.

tabla de composición de alimentos tabla que muestra los nutrientes que se encuentran en los alimentos.

feeder calf a weaned calf that is under 1 year of age and is sold to be fed for more growth.

ternero para alimentador ternero destetado de menos de 1 año que se vende para ser alimentado y que crezca más.

feeding efficiency the ratio of units of feed needed per one unit of production, also called *feed conversion*.

eficiencia de alimentación relación de unidades de alimento necesarias por cada unidad de producción; también se denomina *conversión de alimento*.

feeding standard a table of nutrient requirements for an animal.

estándar de alimentación tabla de requisitos de nutrientes para un animal.

fertilization the union of a sperm cell with an egg cell.

fertilización unión de un espermatozoide con un óvulo.

fetus unborn animal.

feto animal nonato.

fiber complex carbohydrates such as cellulose and lignin.

fibra carbohidratos complejos como celulosa y lignina.

fibrous joints joints that are joined by fibrous cartilage tissue and generally do not permit any type of movement; also called *synarthroses joints*.

articulaciones fibrosas articulaciones unidas por tejido cartilaginoso fibroso y que generalmente no permiten ningún tipo de movimiento; también se denominan *articulaciones sinartrosis*.

fillback period the time during which cattle are fed.

período de llenado el tiempo durante el cual el ganado se alimenta.

filly generally, a female horse under 3 years of age; the Thoroughbred breed includes females that are 4 years old.

potra generalmente, un caballo hembra de menos de 3 años; la raza purasangre incluye hembras de 4 años.

first cross mating a sire from one breed to a female from another breed; also called a *two-breed cross*.

primera cruza apareamiento de un macho de una raza con una hembra de otra; también se denomina *cruza de dos razas*.

float a long-handled rasp with a guard to prevent injury to the horse during treatment.

flotador raspador de mango largo con una protección para evitar lesiones al caballo durante el tratamiento.

flushing increasing the amount of feed fed to an animal for a short period of time, usually just prior to breeding.

descarga aumentar la cantidad de alimento que se proporciona a un animal durante un período breve, habitualmente justo antes del apareamiento.

foal a young horse of either sex up to 1 year of age.

potrillo caballo joven de cualquier sexo de hasta 1 ño.

food irradiation the treatment of food with radioactive isotopes to kill bacteria, insects, and molds that are present in the food.

irradiación de alimento tratamiento del alimento con isótopos radiactivos para exterminar bacterias, insectos y moho presentes en el alimento.

foot-candle a unit of illuminance on a surface that is everywhere one foot from a point source of a candle or light.

pie candela unidad de iluminancia sobre una superficie que es de un pie hacia todos lados desde un punto de origen de una candela o una luz.

founder a nutritional disorder that is commonly caused by overeating of concentrates, sudden changes in feed, drinking too much water, or standing in a stall for long periods of time.

laminitis trastorno nutricional comúnmente causado por exceso de ingesta de concentrados, cambios de alimento, exceso de agua o permanecer de pie en un establo durante períodos prolongados.

frame score in cattle, a score based on subjective evaluation of height or actual measurement of hip height when calves are 205 days old.

puntaje marco en el ganado, puntaje basado en la evaluación subjetiva de la altura o la medición real de la altura de la cadera cuando los terneros tienen 205 días.

fraternally related refers to animals who are born at the same time but come from different egg cells.

emparentados fraternalmente se refiere a animales nacidos al mismo tiempo pero de distintos óvulos.

freckles black pigmented spots on swine; considered an undesirable trait for the Yorkshire breed.

pecas manchas pigmentadas de color negro en los cerdos; se consideran un rasgo no deseable en la raza Yorkshire.

free choice making feed available to the animal at all times.

libre elección poner alimento a disposición del animal en todo momento.

full feed giving an animal all it wants to eat.

alimento pleno darle a un animal todo lo que desee comer.

fullblood an animal of unmixed breed.

pura sangre animal de raza sin mezclas.

funnel the part of the oviduct of the chicken that receives the yolk from the ovary and stores the sperm cells received from the rooster.

embudo parte del oviducto de la gallina que recibe la yema del ovario y almacena los espermatozoides recibidos del gallo.

futures contract a document that establishes a price for live hogs that are to be delivered at some future date.

contrato a futuro documento que establece un precio por porcinos vivos que se deben entregar en alguna fecha futura.

G

gait the movement of the horse's feet and legs when the horse is in motion.

paso movimiento de las pezuñas y las patas del caballo cuando el animal está en movimiento.

Gallus gallus a wild jungle fowl found in India and believed to be an early ancestor of most tame chickens.

Gallus gallus ave selvática silvestre encontrada en India que se considera un antiguo ancestro de la mayoría de los pollos domesticados.

gamete a mature germ cell (sperm or egg) that is capable of initiating the formation of a new individual by fusion with another gamete.

gameto célula germinal madura (espermatozoide u óvulo) capaz de iniciar la formación de un nuevo individuo por fusión con otro gameto.

gander a male goose.

ganso ganso macho.

gastric juice a fluid secreted by glands in the wall of the stomach, containing hydrochloric acid and the enzymes gastric lipase, pepsin, and rennin.

jugo gástrico líquido segregado por glándulas de la pared del estómago que contiene ácido clorhídrico y las enzimas lipasa gástrica, pepsina y renina.

gelding a male horse that has been castrated.

caballo capón caballo macho que ha sido castrado.

gene a complex molecule located on the chromosomes that is involved in the transmission of inherited traits.

gen molécula compleja ubicada en los cromosomas que participa en la transmisión de rasgos heredados.

genetic engineering a technology involving the removal, modification, or addition of genes to a DNA molecule; also known as *recombinant DNA (rDNA) technology*.

ingeniería genética tecnología que involucra la extracción, la modificación o el agregado de genes a una molécula de ADN; también se denomina *tecnología del ADN recombinante (rDNA)*.

genetics the study of heredity, or the way in which traits of parents are passed on to offspring.

genética el estudio de la herencia o la manera en que los rasgos se transmiten a la descendencia.

genotype the kinds of gene pairs possessed by the animal.

genotipo las clases de pares de genes que posee el animal.

gestation the time during which the animal is pregnant.

gestación tiempo durante el cual el animal está preñado.

gilt a young female swine that has not farrowed and is not showing any signs of pregnancy.

cerda joven cerda hembra joven que no ha parido y que no muestra ningún signo de preñez.

gizzard in poultry, the muscular stomach that crushes and grinds the feed and mixes it with digestive juices.

molleja en las aves, el estómago muscular que quiebra y muele el alimento y lo mezcla con jugos digestivos.

gossypol a material found in some cottonseed meal that is toxic to swine and certain other simple-stomached animals.

gosipol material que se encuentra en cierta harina de semilla de algodón y que es tóxico para los cerdos y para algunos otros animales de estómago simple.

grade refers to any animal not eligible for registry; result of mating a registered male with a native female or one of mixed breeding.

mejora genética se aplica a un animal no elegible para registro; es el resultado de aparear un macho registrado con una hembra nativa o una de raza mestiza.

Grade A milk milk produced under high standards that make it acceptable for fluid use.

leche de categoría A leche producida según estándares altos que la hacen aceptable para el uso como líquido.

Grade B milk milk produced under standards that allow it to be used for manufacturing dairy products but not for fluid consumption.

leche de categoría B leche producida según estándares que permiten que se use para fabricar productos lácteos pero no el consumo como líquido.

grading up the mating of purebred sires to grade females.

mejorar genéticamente apareamiento de machos puros para mejorar genéticamente a las hembras.

grass fed a process by which an animal is fed grass for its lifetime. The animal cannot be fed grain or grain by-products and must have continuous access to pasture during the growing season. Roughages that do not contain grain (hay, haylage, etc.) are also acceptable feed sources. In addition, routine minerals and vitamins supplementation may also be included in the animal's diet.

alimentado con pasto proceso por el cual un animal es alimentado con pasto durante toda la vida. No se puede alimentar al animal con granos ni derivados de granos y se le debe permitir el acceso permanente a pasturas durante la temporada de crecimiento. El forraje que no contiene granos (heno, henolaje, etc.) también es una fuente de alimento aceptable. Además, se pueden incluir suplementos minerales y vitamínicos de rutina en la dieta del animal.

grease impurities present in a fleece.

grasa impurezas presentes en un vellón.

Grecian ibex species of wild goat believed to be ancestors of some of today's domestic breeds.

cabra montesa griega especie de cabra salvaje que se cree que fue el ancestro de algunas de las razas domésticas actuales.

green geese geese full fed for fast growth and marketed at 10 to 13 weeks of age when they weigh 10 to 12 pounds; also called *junior geese*.

ganso de pradera gansos alimentados a pleno para un crecimiento rápido y comercializados a las 10 ó 13 semanas de edad, cuando pesan entre 4,5 y 5,5 kg (entre 10 y 12 lb); también se denominan *gansos junior*.

grinding processing a feed by breaking it up into smaller particles.

molienda procesamiento de un alimento deshaciéndolo en partículas más pequeñas.

grit small particles of granite used in poultry rations to help in grinding the feed in the gizzard.

arenilla partículas pequeñas de granito que se usan en las raciones para aves con el objeto de ayudar a moler el alimento en la molleja.

gross energy total amount of heat released by completely burning a feed in a bomb calorimeter.

energía bruta cantidad total de calor liberado al quemar por completo un alimento en un calorímetro de bomba.

gross income the total of all income received.

ingresos brutos el total de todos los ingresos recibidos.

growing period the period from weaning to about 100 pounds.

período de crecimiento período desde el destete hasta que pesan aprox. 45,5 kg (100 lb).

growth promotant pelleted synthetic or natural hormones or hormone-like compounds placed under the skin or in the muscle of the animal that are used to lower production costs by improving both the rate and efficiency of gain; known as *hormone implants*.

promotor del crecimiento hormonas naturales o sintéticas o compuestos tipo hormona en pastillas que se colocan debajo de la piel o en el músculo del animal y que se usan para reducir los costos de producción al mejorar tanto el índice como la eficiencia del aumento de peso; se denominan *implantes hormonales*.

growthy a term used to describe an animal that is large and well developed for its age. Refers to an animal that is balanced with adequate muscle and has sufficient frame and bone structure to indicate good growth potential.

desarrollable término usado para describir a un animal grande y bien desarrollado para su edad. Se aplica a un animal equilibrado, con musculatura adecuada y marco y estructura ósea suficiente como para indicar un buen potencial de crecimiento.

guanine a purine base that occurs in both RNA and DNA; pairs with cytosine in DNA.

guanina base de purina presente en el ARN y en el ADN; se aparea con la citosina en el ADN.

gymkhana games on horseback.

gincana juegos ecuestres.

H

halter a rope or leather headgear used to tie or lead an animal.

cabestro arnés para la cabeza de cuerda o de cuero que se usa para amarrar o dirigir a un animal.

hand a unit of measurement used to describe the height of horses; 1 hand is 4 inches.

palmo unidad de medida usada para describir la altura de los caballos; 1 palmo mide 10,16 cm (4 pulg.).

hank a measure of length for yarn; a hank of worsted yarn contains 560 yards.

madeja medida de longitud para hebras; una madeja de estambres contiene 512 m (560 yardas).

haploid a cell that has only a single set of chromosomes.

haploide célula que tiene un solo conjunto de cromosomas.

harness the gear used to attach a horse or other draft animal to a load.

arnés equipo usado para fijar un caballo u otro animal de tiro a una carga.

hatchability the percentage of fertile eggs that hatch.

viabilidad porcentaje de huevos fértiles que eclosionan.

haylage low-moisture grass silage.

henolaje silaje de pasto con bajo contenido de humedad.

Hazard Analysis and Critical Control Point (HACCP) a system used to monitor the production of food (developed for NASA in 1971); currently used on a voluntary basis at many meat and poultry processing plants.

Análisis de riesgos y punto de control crítico (HACCP) sistema usado para controlar la producción de alimentos (desarrollado por la NASA en 1971); actualmente se utiliza de forma voluntaria en muchas plantas procesadoras de carnes y aves.

headgate a device for restraining livestock.

compuerta de toma dispositivo para contener al ganado.

heaves a nutritional disorder that affects the respiratory system that often occurs when moldy or dusty hay is fed.

huélfago trastorno nutricional que afecta el sistema respiratorio y que suele presentarse cuando se proporciona alimento que contiene moho o polvo.

hedging the practice of reducing risk in a futures market by locking in a price.

cobertura de riesgo la práctica de reducir el riesgo en un mercado a futuro mediante el bloqueo de un precio.

heifer a female bovine that has not had a calf or has not matured as a cow.

vaquilla bovino hembra que no ha tenido terneros o que no ha madurado como vaca.

hemoglobin the red pigment in the red blood cells of people and animals which carries oxygen from the lungs to other parts of the body; it is a complex chemical compound made up of iron, carbon, hydrogen, and oxygen, and is essential to life in red-blooded animals.

hemoglobina pigmento rojo de los glóbulos rojos que transporta oxígeno de los pulmones a otras partes del cuerpo; es un compuesto químico complejo formado por hierro, carbono, hidrógeno y oxígeno y es esencial para la vida en los animales de sangre roja.

hen a female of domestic poultry.

gallina hembra de las aves domésticas.

heparin a material that comes from the lungs of animals and is used to prevent blood clotting during operations; it also prevents heart attacks.

heparina material que proviene de los pulmones de los animales y que se usa para prevenir la coagulación de la sangre durante las operaciones; también previene ataques cardíacos.

herbivorous the practice of eating plants as the main part of the diet.

herbívoro la práctica de alimentarse con plantas como parte principal de la dieta.

heritability the amount of the difference between animals that is passed from the parent to the offspring.

heredabilidad cantidad de la diferencia entre animales que se transmite de padres a hijos.

heritability estimate the likelihood of a trait being passed on from parent to offspring.

estimado de heredabilidad probabilidad de que un rasgo se transmita de padres a hijos.

heterosis improvement in the offspring resulting from favorable combinations of gene pairs; sometimes called *hybrid vigor*.

heterosis mejora de la descendencia que resulta de combinaciones favorables de pares de genes; a veces se denomina *vigor híbrido*.

heterozygous gene pair a gene pair that carries two unlike genes for a trait.

par de genes heterocigotas par de genes que transmite dos genes diferentes para un rasgo.

high-moisture storage harvesting a feed crop when it has a high moisture content and storing it in a silo.

almacenamiento de humedad alta cosechar un cultivo para alimento cuando tiene alto contenido de humedad y almacenarlo en un silo.

hinny the offspring of a cross between a stallion and a jennet.

burdégano la cría de una cruza entre un caballo macho y una burra.

holding pond a temporary storage area for runoff water.

fuente de retención área de almacenamiento temporal de agua de escurrimiento.

homeothermic refers to warm-blooded animals that are able to maintain a fairly constant body temperature.

homeotérmico se refiere a animales de sangre caliente que pueden mantener una temperatura corporal casi constante.

homologous pairs of chromosomes that are the same length, that have their centrioles in the same position, and that pair up during synapsis in meiosis.

homólogos pares de cromosomas que tienen la misma longitud, tienen los centriolos en la misma posición y se aparean durante la sinapsis en meiosis.

homozygous gene pair a gene pair that carriers two like genes for a trait.

par de genes homocigotas par de genes que transportan dos genes similares para un rasgo.

hormone an organic material given off (secreted) by a body gland that helps to regulate body functions.

hormona material orgánico segregado por una glándula del cuerpo que ayuda a regular funciones corporales.

horsemanship the art of riding a horse; also called *equitation*.

destreza ecuestre el arte de montar a caballo; también se denomina *equitación*.

hot carcass weight the unchilled weight of a carcass after harvest and the removal of the head, hide, intestinal tract, and internal organs.

peso muerto en caliente el peso sin enfriar de un cadáver de animal después del sacrificio y la remoción de la cabeza, la piel, el tracto intestinal y los órganos internos.

hothouse lambs young sheep sold at 50 to 90 days of age, weighing 35 to 60 pounds.

carneros de invernáculo ovejas jóvenes que se venden con 50 a 90 días de edad y con un peso de 15,9 a 27,2 kg (35 a 60 lb).

hover guard a corrugated cardboard or cloth-covered wire about 12 inches high used during the first week of brooding to prevent chicks from wandering away from the heat and becoming chilled.

protección flotante alambre cubierto de cartón corrugado o de tela de aprox. 30,5 cm (12 pulg.) que se usa durante la primera semana de incubadora para evitar que los polluelos se alejen del calor y se enfríen.

husklage the material that comes from the corn combine that can be used as a feed.

chalaje material que proviene de la segadora trilladora de maíz que puede usarse como alimento.

hutch a pen or cage for rabbits.

conejera corral o jaula para conejos.

hybrid an animal produced from the crossing or mating of two animals of different breeds.

híbrido animal producido mediante la cruza o el apareamiento de dos animales de distintas razas.

I

identical refers to animals that come from the same egg cell.

idéntico se refiere a animales que provienen de la misma célula huevo.

imprinting a behavior-shaping process that takes place within the first 24 hours of an animal's life.

impronta proceso de formación de comportamientos que tiene lugar durante las primeras 24 horas de vida de un animal.

inbreeding the mating of related animals.

endogamia apareamiento de animales emparentados.

incomplete dominance a situation in which one gene does not completely hide or mask the effect of the other gene in a gene pair.

dominancia incompleta situación en la cual un gen no oculta o enmascara completamente el efecto del otro gen en un par de genes.

incubation keeping eggs at the right temperature and humidity for hatching.

incubación conservación de huevos a la temperatura y con la humedad correctas para que eclosionen.

inflation in reference to milking equipment, the liner used in the teat cup.
inflación con respecto a equipos de ordeñe, el revestimiento usado en la taza de la tetina.

infundibulum in the female reproductive system, the funnel-shaped end of the oviduct that is close to the ovary.
infundíbulo en el sistema reproductivo femenino, extremo en forma de embudo del oviducto que está cerca del ovario.

insulin the hormone from a part of the pancreas that promotes the utilization of sugar in the organism and prevents its accumulation in the blood.
insulina hormona de una parte del páncreas que promueve la utilización de azúcar en el organismo y previene su acumulación en la sangre.

integrated resource management (IRM) a management tool available to beef producers that uses a team approach to improve the competitiveness, efficiency, and profitability of their beef business.
gestión integrada de recursos (GIR) herramienta de gestión disponible para los productores de ganado vacuno que usa un enfoque en equipo para mejorar la competitividad, la eficiencia y la rentabilidad de su negocio ganadero.

interest things that hold a person's attention.
interés cosas que retienen la atención de una persona.

intestinal juice fluid produced by the walls of the small intestine that contains peptidase, sucrase, maltase, and lactase.
jugo intestinal líquido producido por las paredes del intestino delgado que contiene peptidasa, sacarosa, maltasa y lactasa.

intestine, large the tube from the small intestine to the anus; shorter and larger in diameter than the small intestine.
intestino, grueso tubo desde el intestino delgado hasta el ano; más corto y de mayor diámetro que el intestino delgado.

intestine, small the long, folded tube attached to the lower end of the stomach.
intestino, delgado tubo largo, plegado, conectado al extremo inferior del estómago.

intranasal vaccination in poultry, the placement of a vaccine directly into the nose opening.
vacunación intranasal en las aves, colocación de una vacuna directamente en el orificio de la nariz.

intraocular vaccination in poultry, the placement of a vaccine directly into the eye.
vacuna intraocular en las aves, colocación de una vacuna directamente en el ojo.

ionophore antimicrobials that target bacteria in the rumen and result in increased carbon and nitrogen retention by the animal, which increases production efficiency.
ionóforo antimicrobianos específicos para bacterias del rumen y que dan como resultado el aumento de la retención de carbono y de nitrógeno por parte del animal, lo cual mejora la eficiencia de la producción.

isthmus in poultry, that part of the oviduct where the two shell membranes are added to the egg.
istmo en las aves, parte del oviducto donde las dos membranas de la cáscara se agregan al huevo.

J

jack a male ass.
borrico asno macho.

jennet a female ass.
burra asno hembra.

K

kemp large, chalky white hairs found in the fleece of some breeds of goats, especially the Angora.
pelo muerto pelos grandes, calcáreos, que se encuentran en el vellón de las cabras, especialmente las de Angora.

kid a goat of either sex that is under 1 year of age.
chivato cabra de cualquier sexo que tiene menos de 1 año.

kindling giving birth; generally used when referring to rabbits.
parición dar a luz; se usa en general para referirse a los conejos.

L

lamb a young sheep.
cordero oveja joven.

larynx the upper portion of the windpipe that contains the vocal cords.
laringe porción superior de las vías aéreas que contiene las cuerdas vocales.

legume a plant of the family Leguminosae that carries its seeds in a pod that splits along its seams; many legumes have nitrogen fixing bacteria in nodules on the roots that can transform nitrogen in the air into a form (NH_3) that can be used by the plant; peanuts, soybeans, clovers, and alfalfa are common legumes used in agriculture.
legumbre planta de la familia Leguminosae que guarda las semillas en una vaina que se parte a lo largo de sus costuras; muchas legumbres tienen bacterias que fijan nitrógeno en nódulos en las raíces y pueden transformar el nitrógeno del aire en una forma (NH_3) que puede ser utilizada por la planta; el maní, la soja, el trébol y la alfalfa son legumbres de uso frecuente en la agricultura.

leukocytes white blood cells that prevent infection.
leucocitos glóbulos blancos de la sangre que previenen las infecciones.

light horse a horse that is used mainly for riding, driving, and racing.
caballo liviano caballo que se usa principalmente para montar, conducir y correr carreras.

limit-fed a method of feeding in which the amount of feed given the animal is controlled or limited to less than the animal would eat if given free access to the feed.
alimentación limitada método de alimentación en el cual la cantidad de alimento que se proporciona al animal está controlada o se limita a menos que lo que el animal comería si tuviera libre acceso al alimento.

linear classification a modification of type classification that utilizes a computer program to score dairy cattle on a number of individual traits.
clasificación lineal modificación de la clasificación por tipos que utiliza un programa de computadora para puntuar ganado lechero con respecto a varios rasgos individuales.

linebreeding a form of inbreeding in which the animals being mated are more distantly related than in closebreeding and their ancestry can be traced back to one common ancestor.
consanguinidad indirecta forma de endogamia en la cual los animales que se aparean tienen un parentesco más distante que en la endogamia intensiva y la ascendencia puede rastrearse hasta un ancestro en común.

linecrossing mating animals from two different lines of breeding within a breed.

cruza de líneas apareamiento de animales de dos líneas de razas diferentes dentro de una raza.

linkage the tendency for genes that are located close together on the chromosome to stay together.

vínculo tendencia de genes ubicados cercanos en el cromosoma a mantenerse juntos.

locus the location of a given gene on a chromosome.

locus ubicación de un gen determinado en un cromosoma.

longeing training a horse at the end of a 25- to 30-foot line.

trabajo a la cuerda entrenamiento de un caballo en el extremo de una línea de 7,6 a 9,2 m (25 a 30 pies).

lower critical temperature (LCT) the temperature at which animals will show symptoms of cold stress; feed intake and metabolic heat production increase.

temperatura crítica inferior (TCI) temperatura en la cual los animales muestran síntomas de estrés por frío, aumento de la ingesta de alimento y de la producción de calor metabólico.

lysosome that part of the cytoplasm in most cells that contain various hydrolytic enzymes involved in digestion.

lisosoma parte del citoplasma de la mayoría de las células que contiene diversas enzimas hidrolíticas que participan en la digestión.

M

magnum in poultry, that part of the oviduct where the thick white of the egg is secreted.

magno en las aves, parte del oviducto donde se segrega la clara del huevo.

marbling the presence and distribution of fat and lean in a cut of meat.

marmolado la presencia y la distribución de grasa y carne magra en un corte de carne.

mare generally, a mature female horse, 4 years of age or older; in the Thoroughbred breed, 5 years of age or older.

yegua generalmente, un caballo hembra maduro, de 4 años o más; en la raza purasangre, de 5 años o más.

Markhors species of wild goat believed to be ancestors of some of today's domestic breeds.

markhors especie de cabras salvajes que se consideran ancestros de algunas de las razas domésticas actuales.

martingale a strap attached to a horse's girth, passed through the forelegs, and fastened to the bit (standing martingale) or the reins (running martingale); the purpose is to hold the horse's head down so it will not rear up.

martingala banda que se fija en el perímetro torácico de un caballo, se pasa a través de las patas delanteras y se sujeta al bocado (martingala de pie) o a las riendas (martingala de correr); el propósito es mantener la cabeza del caballo hacia abajo para que no vaya hacia atrás.

masking agent a substance that covers up the odor of wastes with the introduction of another odor.

agente enmascarador sustancia que cubre el olor de los desechos con la introducción de otro olor.

masters degree an academic degree given by a college or university to recognize mastery of a specific field that requires at least 1 year of additional study after a bachelors degree is earned.

título de maestría título académico otorgado por una institución terciaria o una universidad para reconocer el dominio de un campo específico y que requiere al menos 1 año de estudios adicionales posteriores a recibir un título de licenciado.

maternal breeding value (MBV) a prediction of how the daughters of a bull will milk based on weaning weight information.

valor reproductivo materno (MBV) predicción de la producción de leche de las hijas de un toro con base a la información del peso al momento del destete.

meaty refers to an udder that is still firm after milking.

carnosa se refiere a una ubre que sigue firme tras el ordeñe.

mechanists ancient Greeks who believed that neither humans nor animals had souls.

mecanicistas antiguos griegos que creían que ni los seres humanos ni los animales tenían alma.

meconium the feces that are impacted in the bowels during prenatal growth.

meconio las heces impactadas en los intestinos durante el crecimiento prenatal.

medicated early weaning (MEW) a system in which pigs are weaned and isolated in a clean facility for finishing as in SEW, but the pigs are medicated as well with broad-spectrum antibiotics and vaccines.

destete precoz medicado (DPM) sistema en el cual los cerdos se destetan y se aíslan en instalaciones limpias para la terminación como en el DPS, pero los cerdos también son medicados con antibióticos de amplio espectro y vacunas.

meiosis division of gamete in which chromosome pairs split, each individual chromosome going to one of the new gametes.

meiosis división de gameto en la cual los pares de cromosomas se dividen y cada cromosoma individual va a uno de los nuevos gametos.

melengestrol acetate (MGA) a synthetic hormone similar to progesterone that suppresses estrus.

acetato de melengestrol (AMG) hormona sintética similar a la progesterona que suprime el estro.

metabolism the sum of the processes, both chemical and physical, that are used by living organisms and cells to handle nutrients after they have been absorbed from the digestive system.

metabolismo la suma de los procesos tanto químicos como físicos que usan los organismos y las células vivas para manejar los nutrientes una vez que han sido absorbidos del sistema digestivo.

metabolizable energy (ME) for ruminants, the gross energy in the feed eaten minus the energy found in the feces, the energy in the gaseous products of digestion, and the energy in the urine; for poultry and simple-stomached animals, the energy in the gaseous products of digestion is not considered when determining metabolizable energy.

energía metabolizable (EM) para los rumiantes, la energía bruta en el alimento comido menos la energía encontrada en las heces, la energía en los productos gaseosos de la digestión y la energía en la orina; para las aves y los animales con estómago simple, la energía en los productos gaseosos de la digestión no se considera cuando se determina la energía metabolizable.

metaphase in mitosis and meiosis, the stage between prophase and anaphase when the chromosomes are aligned along the metaphase plate.

metafase en la mitosis y la meiosis, etapa entre la profase y la anafase cuando los cromosomas se alinean a lo largo de la placa de metafase.

micronutrients feed ingredients, such as minerals and vitamins, that are used in small amounts in the ration.
micronutrientes ingredientes del alimento, como minerales y vitaminas, que se usan en cantidades pequeñas en la ración.

middling coarsely ground wheat mixed with bran.
mediocre trigo molido grueso mezclado con afrecho.

milkstone a combination of organic and inorganic materials that accumulate on improperly cleaned milking equipment.
piedra de leche combinación de materiales orgánicos e inorgánicos que se acumula en equipos de ordeñe limpiados incorrectamente.

mineral inorganic substance needed in small amounts for proper nutrition.
mineral sustancia inorgánica necesaria en pequeñas cantidades para una nutrición adecuada.

mitosis cell division that increases the number of total cells and results in growth.
mitosis división celular que aumenta el número total de células y da como resultado el crecimiento.

mohair the fleece of the Angora goat.
mohair el vellón de la cabra de Angora.

molting in poultry, the process of losing the feathers from the body and wings.
muda en las aves, proceso de perder las plumas del cuerpo y de las alas.

Mouflons a breed of wild sheep believed to be ancestors of some present-day domestic breeds.
muflones raza de ovejas salvajes que se cree fueron los ancestros de algunas razas domésticas actuales.

mouth that part of the digestive system through which feed enters the animal's body.
boca parte del sistema digestivo a través de la cual el alimento ingresa al cuerpo del animal.

mule the offspring of a cross between a jack and a mare; usually sterile.
mulas crías de una cruza entre un burro y una yegua; suelen ser estériles.

multiple farrowing arranging the breeding program so that groups of sows farrow at regular intervals throughout the year.
partos múltiples organización del programa de reproducción de modo que grupos de cerdas tengan cría a intervalos regulares a lo largo de todo el año.

muscle tone the state of tension that is maintained by the paired muscles balancing the contraction of each with the other.
tono muscular estado de tensión que mantienen los músculos emparejados al equilibrar la contracción de cada uno con el otro.

mutation the appearance of a new trait in the offspring that did not exist in the genetic makeup of the parents.
mutación aparición de un nuevo rasgo en la descendencia que no existía en la configuración genética de los padres.

mytonic refers to the spasm or temporary rigidity of the muscles.
mitónico se refiere al espasmo o la rigidez temporal de los músculos.

N

National Agricultural Statistic Service (NASS) division of the USDA that conducts surveys and prepares reports covering almost every aspect of U.S. agriculture, including production and supplies of food and fiber, prices paid and received by farmers, farm labor and wages, etc.
Servicio Nacional de Estadísticas Agrícolas (NASS) división del Departamento de Agricultura de EE. UU. (USDA) que realiza encuestas y prepara informes que cubren casi todos los aspectos de la agricultura de EE. UU., incluida la producción y los suministros de alimentos y fibra, los precios pagados y recibidos por los agricultores, la mano de obra y los salarios agrícolas, etc.

National Sheep Improvement Program (NSIP) a computerized, performance-based program for selection of sheep.
Programa Nacional de Mejora de Ovejas (NSIP) programa computarizado, basado en el rendimiento, para la selección de ovejas.

native ewe a ewe that is not produced in the western United States.
oveja nativa oveja no producida en el oeste de EE.UU.

NEg net energy used for animal growth.
ENc energía neta usada para el crecimiento del animal.

NEL net energy used for milk production (lactation).
ENL energía neta usada para la producción de leche (lactación)

NEm net energy used for animal maintenance.
ENm energía neta usada para el mantenimiento del animal.

neck-reining a method of controlling the response of a horse by the weight of the rein against its neck.
riendas al cuello método para controlar la respuesta de un caballo mediante el peso de la rienda sobre su cuello.

negative pressure system a system in which fans exhaust the air from the building.
sistema de presión negativa sistema en el cual se extrae el aire del edificio mediante ventiladores.

net energy (NE) metabolizable energy minus the heat increment; used for maintenance, for production, or both.
energía neta (EN) energía metabolizable menos el incremento de calor; se usa para mantenimiento, para producción o para ambos.

net income income (profit) left after expenses have been deducted from gross income.
ingresos netos ingresos (utilidad) que queda una vez deducidos los gastos de los ingresos brutos.

neutral detergent fiber (NDF) the portion of a feed that is of lower digestibility, consisting of the more insoluble material found in the cell wall; mainly cellulose, lignin, silica, hemicellulose, and some protein.
fibra detergente neutra (FDN) porción de un alimento que es de menor digestibilidad, que comprende el material más insoluble que se encuentra en la pared celular; principalmente celulosa, lignina, sílice, hemicelulosa y algunas proteínas.

niche market a small segment of a larger market.
nicho del mercado pequeño segmento de un mercado mayor.

nitrogen-free extract (NFE) simple carbohydrates such as sugar and starches that are easily digested.
extracto libre de nitrógeno (ELN) carbohidratos simples como el azúcar y los almidones, que se digieren fácilmente.

nonadditive gene effect controls traits by determining how gene pairs act in different combinations with one another.

efecto genético no aditivo controla los rasgos mediante la determinación del modo en que actúan los pares de genes en distintas combinaciones unos con otros.

non-ruminant an animal that has a simple, one-compartment stomach; for example: pigs, horses, and poultry.

no rumiante animal que tiene un estómago simple, de un compartimento; por ejemplo: cerdos, caballos y aves.

nucleotide a combination of one of the nitrogenous bases, one phosphate, and one deoxyribose; forms the basic constituent of DNA and RNA.

nucleótido combinación de una de las bases nitrogenosas, un fosfato y una desoxirribosa; es el constituyente básico del ADN y del ARN.

nucleus the center of a cell that contains the hereditary material of the cell and controls the cell's metabolism, growth, and reproduction.

núcleo el centro de una célula que contiene el material hereditario de la célula y controla el metabolismo, el crecimiento y la reproducción de la célula.

nutrient a chemical element or compound that aids in the support of life.

nutriente elemento o compuesto químico que ayuda a sostener la vida.

nutrient sparing a substance that allows animals to use available nutrients more effectively.

ahorrador de nutrientes sustancia que permite que los animales usen nutrientes disponibles de manera más efectiva.

O

off flavor any undesirable odor or flavor in milk.

sabor anormal cualquier olor o sabor no deseado en la leche.

oil organic compound composed of carbon, hydrogen, and oxygen that is liquid at body temperature.

aceite compuesto orgánico formado por carbono, hidrógeno y oxígeno que es líquido a la temperatura corporal.

omasum the third compartment of the ruminant stomach.

omaso el tercer compartimento del estómago del rumiante.

one hundred percent (100%) dry matter basis data on feed composition or nutrient requirement calculated on the basis of all the moisture being removed from the feed.

cien por ciento (100%) base de materia seca datos sobre la composición del alimento o requisito de nutrientes calculada en base a toda la humedad extraída del alimento.

oocyte one of the intermediate cells in the process of oogenesis; produces a secondary oocyte and a polar body.

oocito una de las células intermediarias en el proceso de oogénesis; produce un oocito secundario y un cuerpo polar.

oogenesis formation, development, and maturation of an ovum.

oogénesis formación, desarrollo y maduración de un huevo.

ootid one of the intermediate cells in the process of oogenesis; develops into the ovum.

oótido una de las células intermediarias en el proceso de oogénesis; se desarrolla en el huevo.

organ a distinct part of an animal which carries on one or more particular functions.

órgano parte diferenciada de un animal que lleva a cabo una o más funciones particulares.

ossein gelatin-like protein that is part of the composition of bones.

oseína proteína tipo gelatina que es parte de la composición de los huesos.

osteocyte a bone cell, particularly one encased in hard bone.

osteocito célula ósea, especialmente una que forma parte de un hueso duro.

outcrossing the mating of animals of different families within the same breed.

cruzamiento exogámico apareamiento de animales de familias diferentes dentro de la misma raza.

ovary organ in the female reproductive system that produces eggs and the two female sex hormones.

ovario órgano del sistema reproductivo femenino que produce óvulos y las dos hormonas sexuales femeninas.

overo a color pattern in horses in which the head has variable color markings; the white usually does not cross the back between the withers and tail; one or more legs are dark colored; and the body markings are irregular and scattered.

overo patrón de color en los caballos en el cual la cabeza tiene marcas de colores variables; el blanco habitualmente no cruza la espalda entre la cruz y la cola; una o más patas son oscuras y las marcas del cuerpo son irregulares y dispersas.

oviduct the tube in the female reproductive system that carries the eggs from the ovaries to the uterus; also called *Fallopian tubes.*

oviducto tubo del sistema reproductivo femenino que transporta los óvulos de los ovarios al útero; también se denomina *trompa de Falopio.*

ovulation the release of the egg cell from the ovary.

ovulación liberación del óvulo desde el ovario.

ovum female gamete or reproductive cell; also called *egg.*

óvulo gameto o célula reproductiva femenina; también se denomina *huevo.*

oxidation any chemical change which involves the addition of oxygen or its chemical equivalent.

oxidación cualquier cambio químico que implica el agregado de oxígeno o su equivalente químico.

P

palatable good tasting.

palatabilidad de buen sabor.

palpate the process used to examine a doe to determine pregnancy.

palpar proceso usado para examinar a una hembra y así determinar la preñez.

pancreatic juice fluid secreted by the pancreas that contains the enzymes trypsin, pancreatic amylase, pancreatic lipase, and maltase.

jugo pancreático líquido segregado por el páncreas que contiene las enzimas tripsina, amilasa pancreática, lipasa pancreática y maltasa.

papilla in poultry, the organ in the wall of the cloaca that deposits the sperm cells into the hen's reproductive system.

papila en las aves, órgano de la pared de la cloaca que deposita los espermatozoides en el sistema reproductivo de la gallina.

parturition the act of giving birth.

parición acto de dar a luz.

Pasang species of wild goat believed to be ancestors of some of today's domestic breeds.

pasang especie de cabra salvaje que se considera el ancestro de algunas de las razas domésticas actuales.

pedigree the record of the ancestors of an animal.

pedigree registro de los ancestros de un animal.

pelleting grinding a feed into small particles and then forming it into a small, hard form called a pellet.

empastillado moler un alimento en partículas pequeñas y, después, darle una forma pequeña y dura denominada pastilla.

pelt the natural, whole skin covering, including the hair, wool, or fur of smaller animals, such as sheep, foxes, etc.

cuero cobertura de piel natural, completa, incluido el pelo, la lana o la piel de animales más pequeños, como ovejas, zorros, etc.

penis male reproductive organ that contains the urethra and deposits the sperm into the female reproductive tract.

pene órgano reproductivo masculino que contiene la uretra y deposita los espermatozoides en el tracto reproductivo femenino.

performance stimulants term sometimes applied to feed additives and hormone implants.

estimulantes del rendimiento término que a veces se aplica a aditivos del alimento y a implantes hormonales.

performance testing a method of collecting records on an animal herd to be used for selecting the most productive animals.

pruebas de rendimiento método para reunir registros sobre un hato de animales que se usarán para seleccionar a los ejemplares más productivos.

pharynx the cavity that connects the mouth and nasal cavity to the throat; a passage common to the digestive and respiratory tracts.

faringe cavidad que conecta la boca y la cavidad nasal con la garganta; pasaje que comparten los tractos digestivo y respiratorio.

phase feeding changes in an animal's diet to adjust for age and stage of production, to adjust for season of the year and for temperature change, to account for differences in body weight and nutrient requirements of different strains of animals, or to adjust one or more nutrients as other nutrients are changed for economic or availability reasons.

alimentación por fases cambios en la dieta de un animal para hacer ajustes según la edad y la etapa de producción, la estación del año y cambios de temperatura, para contemplar diferencias en el peso corporal y los requisitos de nutrientes de distintas cepas de animales, o bien, para hacer ajustes en uno o más nutrientes a medida que se cambian nutrientes por motivos económicos o de disponibilidad.

phenotype the physical appearance of an animal.

fenotipo la apariencia física de un animal.

pinfeather in poultry, a feather that is not fully developed.

pluma inmadura en las aves, pluma que no está totalmente desarrollada.

placenta in mammals, the structure by which the fetus is nourished in the uterus.

placenta en los mamíferos, la estructura a través de la cual se nutre el feto en el útero.

plasma the liquid portion of blood or lymph.

plasma parte líquida de la sangre o linfa.

point source a place from which pollutants may originate; for example, a large feedlot.

origen puntual lugar de donde pueden originarse contaminantes; por ejemplo, un pastadero extenso.

polled not having horns.

mocho que no tiene cuernos.

pollutants impurities in the environment, such as odors or waste from a feedlot.

contaminantes impurezas en el medio ambiente, como olores o desechos de un pastadero.

polynucleotide a linear sequence of a number of nucleotides.

polinucleótido secuencia lineal de varios nucleótidos.

pony a small-type horse, under 14-2 hands at the withers.

pony caballo de tipo pequeño, de menos de (14-2 palmos) en la cruz.

porcine somatotropin a protein produced naturally by the pituitary gland of the pig.

somatotropina porcina proteína producida naturalmente por la glándula pituitaria del cerdo.

porcine stress syndrome (PSS) a condition in swine characterized by extreme muscling, nervousness, tail tremors, skin blotching, and sudden death.

síndrome de estrés porcino (SEP) afección de los cerdos que se caracteriza por movimientos musculares extremos, nerviosismo, temblores en la cola, manchas en la piel y muerte súbita.

positive pressure system a system in which fans bring fresh air into the building.

sistema de presión positiva sistema en el cual se lleva aire fresco dentro del edificio mediante ventiladores.

posting in horseback riding, the practice of rising and sitting in the saddle.

trote levantado en equitación, la práctica de elevarse y sentarse en la montura.

poult a young turkey whose sex cannot yet be distinguished.

pollo de pavo pavo joven cuyo sexo todavía no se puede distinguir.

PQA Plus (Pork Quality Assurance Plus) a management education program with major emphasis on swine herd health and animal well-being.

Aseguramiento de la Calidad de los Cerdos Plus (PQA Plus) programa de educación en gestión con énfasis principal en la salud de las piaras de cerdos y el bienestar de los animales.

preconditioning the process of preparing calves for the stress of being moved into the feedlot.

acondicionamiento previo proceso de preparación de los terneros para el estrés de ser trasladados al pastadero.

predicated transmitting ability (PTA) in dairy cattle, the average genetic value for a certain trait that an animal transmits to its offspring.

capacidad de transmisión predicada (PTA) en el ganado lechero, el valor genético promedio de determinado rasgo que un animal le transmite a su descendencia.

probiotics compounds such as yeasts and lactobacilli that change the bacterial population in the digestive tract to a more desirable type.

probióticos compuestos como levaduras y lactobacilos que modifican la población bacteriana del tracto digestivo a un tipo más deseable.

production testing measuring a brood female's production by the performance of her offspring.

pruebas de producción medición de la producción de crías hembra a través del rendimiento de su descendencia.

progeny testing evaluation of a male by the performance of a number of his offspring.

pruebas de progenie evaluación de un macho por el rendimiento de varios de sus descendientes.

progesterone hormone produced by the ovaries that maintains pregnancy in the animal.

progesterona hormona producida por los ovarios que mantiene la preñez en el animal.

prolapsed uterus a condition in which the uterus protrudes from the vulva.

útero prolapsado una afección en la cual el útero sobresale por la vulva.

prophase the first stages of mitosis and meiosis.

profase las primeras etapas de mitosis y meiosis.

prostate gland an accessory gland of the male reproductive system that produces a fluid that is mixed with the seminal fluid.

glándula prostática glándula accesoria del sistema reproductivo masculino que produce un líquido que se mezcla con el líquido seminal.

protein an organic compound made up of amino acids and containing carbon, hydrogen, oxygen, and nitrogen.

proteína compuesto orgánico formado por aminoácidos y que contiene carbono, hidrógeno, oxígeno y nitrógeno.

protein supplement livestock feed that contains 20 percent or more protein.

suplemento proteico alimento para ganado que contiene un 20 por ciento o más de proteínas.

protoplasm the material that constitutes the living matter of cells; it includes the nucleus and cytoplasm.

protoplasma material que constituye la materia viva de las células; incluye el núcleo y el citoplasma.

protozoa one-celled animals.

protozoarios animales unicelulares.

pseudopregnancy false pregnancy.

seudopreñez falsa preñez.

puberty the age at which sexual maturity is reached.

pubertad edad en la cual se alcanza la madurez sexual.

pullet a young female chicken.

polla pollo hembra joven.

pulmonary circulation system the body system that carries blood through the lungs of a human or animal.

sistema de circulación pulmonar sistema corporal que transporta la sangre a través de los pulmones de un ser humano o un animal.

pulsator a control unit used with milking equipment to control the action of the inflation in the teat cup shell.

pulsador unidad de control que se usa con los equipos de ordeñe para controlar la acción de la inflación en la cubierta de la taza de la tetina.

purebred designating an animal belonging to one of the recognized breeds of livestock; such animals are registered or eligible for registry in the official herdbook of the breed.

pura raza designa a un animal que pertenece a una de las razas de ganado reconocidas; esos animales están registrados o pueden estarlo en el registro oficial de la raza.

Q

quadruped an animal that walks on four legs.

cuadrúpedo animal que camina en cuatro patas.

quality grade grade given to a beef carcass; closely related to marbling, age of the animal, and color of the lean. The most common quality grades are prime, choice, select, and standard.

grado de calidad grado asignado al cadáver de un vacuno; está estrechamente relacionado con el marmolado, la edad del animal y el color de la carne magra. Los grados de calidad más comunes son: primera, elección, selección y estándar.

R

rabbitry a place where rabbits are kept.

conejera lugar donde se conservan los conejos.

radura the international symbol that must be placed on all foods treated by irradiation.

radura símbolo internacional que debe colocarse en todos los alimentos tratados por radiación.

ram a male sheep or lamb that has not been castrated.

carnero oveja o cordero macho no castrado.

ration the total amount of feed that an animal is allowed during a 24-hour period.

ración cantidad total de alimento que se proporciona a un animal durante un período de 24 horas.

ratite a group of flightless birds that include ostrich, emu, rhea, cassowary, and kiwi.

ratite grupo de aves que no vuelan; incluye avestruces, emús, ñandúes, casuarios y kiwis.

recessive gene one of a pair of genes, the effect of which is hidden by the other gene in the pair.

gen recesivo gen de un par cuyo efecto queda oculto por el otro gen del par.

recombinant bovine somatotropin (rbST) bovine somatotropin manufactured by genetic engineering. (*See genetic engineering, bovine somatotropin, and somatotropin.*)

somatotropina bovina recombinante (rbST) somatotropina bovina fabricada mediante ingeniería genética. (*Ver ingeniería genética, somatotropina bovina y somatotropina.*)

rectum the last part of the large intestine.

recto la última parte del intestino grueso.

registered refers to any animal that meets the requirements of a breed association and is recorded in the herdbook of the association.

registrado se refiere a cualquier animal que cumple con los requisitos de una asociación de raza y está registrado en el registro de la asociación.

relative feed value (RFV) a number used to compare the quality of various legume and legume-grass forages as determined by their acid detergent fiber and neutral detergent fiber content.

valor relativo del alimento (RFV) número que se usa para comparar la calidad de diversos forrajes de legumbres y de legumbres y pasto, según lo determina su contenido de fibra detergente ácida y de fibra detergente neutra.

rennet the salted or dried stomach of a suckling calf that is used as a milk coagulant in cheese making; it also helps babies digest milk.

rennet estómago salado o seco de un ternero lechal que se usa como coagulante de la leche en la fabricación de quesos; también ayuda a los bebés a digerir la leche.

repeatability in dairy cattle, a measure of the confidence that can be placed on the predicted transmitting ability being a true measure of a bull's ability to transmit genetic characteristics.

repetibilidad en el ganado lechero, una medida de la confianza que se puede tener en que la capacidad de transmisión predicha sea una medida verdadera de la capacidad de un toro de transmitir características genéticas.

reproduction the production of offspring.

reproducción producción de descendencia.

restricted feeding feeding 75 to 80 percent of full feed; also referred to as limited feeding.

alimentación restringida alimentación que comprende entre el 75 y el 80 por ciento de la alimentación plena; también se denomina alimentación limitada.

reticulum the second compartment of the ruminant stomach.

retículo el segundo compartimento del estómago de un rumiante.

retractor muscle part of the male reproductive system that helps extend the penis from the sheath and draws it back after copulation.

músculo retractor parte del sistema reproductivo masculino que ayuda a extender el pene desde la vaina y lo retrae después de la copulación.

riata rawhide, horsehair, or hemp rope, generally 35 to 45 feet long and arranged with a large loop on one end.

lazo lonja de cuero, crin o cáñamo, generalmente de (35 a 45 pies) de longitud con un gran bucle en un extremo.

ribonucleic acid (RNA) regulates protein synthesis in animals; primary function is carrying the genetic message from DNA for the building of the polypeptide chains that begin the process of protein synthesis; contains the bases adenine, guanine, cytosine, and uracil bonded to the ribose.

ácido ribonucleico (ARN) regula la síntesis de las proteínas en los animales; la función principal es transmitir el mensaje genético del ADN para la construcción de las cadenas de polipéptidos que inician el proceso de síntesis de proteínas; contiene las bases adenina, guanina, citosina y uracilo unidas a la ribosa.

ribosome a small particle composed of RNA and protein that is in the cytoplasm of cells and is active in protein synthesis.

ribosoma partícula pequeña compuesta por ARN y proteína que está en el citoplasma de las células y es activa en la síntesis de proteínas.

ridgeling a male in which one or both testicles are held in the body cavity; also called ridgel.

con testículos no descendidos macho en el cual uno o ambos testículos están retenidos en la cavidad del cuerpo; también se denomina medio castrado.

roached mane refers to a mane that is trimmed down to the crest of the neck, leaving the foretop and a wisp of mane at the withers.

crin recortada se refiere a la crin recortada hasta la cresta del cuello, dejando la parte más alta y un mechón de crin en la cruz.

rolling processing grain through a set of smooth rollers that are set close together; sometimes called flaking.

arrollado procesamiento del grano a través de un conjunto de rodillos lisos ajustados muy cerca entre sí; a veces se denomina escamado.

rotation grazing grazing forage plants on well-managed pastures in such a manner as to allow for a definite recovery period following each grazing period; this includes alternate use of two or more pastures at regular intervals or the use of temporary fences within pastures to prevent overgrazing.

pastoreo de rotación plantas de forraje para pastoreo en pasturas bien administradas de manera que se permita un período de recuperación definido después de cada período de pastoreo; incluye el uso alternado de dos o más pasturas a intervalos regulares o el uso de cercas temporales para evitar el pastoreo excesivo.

rotavirus wheel-shaped viruses that cause gastroenteritis, especially in newborn animals.

rotavirus virus con forma de rueda que causa gastroenteritis, especialmente en animales recién nacidos.

roughage a feed containing more than 18 percent crude fiber when dry; examples: hay, silage, and pasture.

concentrado alimento que contiene más del 18 por ciento de fibra cruda cuando está seco; ejemplos: heno, silaje y pastura.

rumen the first and largest compartment of the ruminant stomach.

rumen el primer y mayor compartimento del estómago del rumiante.

rumen organisms bacteria found in the rumen of cattle and other ruminant animals.

organismos del rumen bacterias que se encuentran en el rumen de las reses y de otros animales rumiantes.

ruminant an animal that has a stomach divided into several compartments; for example, cattle, sheep, goats.

rumiante animal que tiene un estómago dividido en varios compartimentos; por ejemplo: reses, ovejas, cabras.

rumination in ruminants, the process of chewing the cud.

rumiación en los rumiantes, proceso de masticar el contenido ruminal.

S

sacking out a process that introduces various objects such as blankets, water bottles, plastic bags, rags, and others to the horse before the actual saddle.

enmantado proceso que introduce diversos objetos como mantas, botellas de agua, bolsas de plástico, trapos y otros sobre el caballo antes de colocarle la montura

saliva fluid secreted into the mouth by glands and containing enzymes that aid in digestion.

saliva líquido segregado dentro de la boca por glándulas; contiene enzimas que ayudan a la digestión.

sanitizing making a surface sanitary by using a sanitizing agent such as chlorine or quaternary ammonium compound.

desinfección sanitizar una superficie mediante el uso de un agente desinfectante, como un compuesto con cloro o amonio cuaternario.

scab a disease that attacks barley and is poisonous to hogs.

fusariosis de la espiga enfermedad que ataca la cebada y es venenosa para los porcinos.

scrotum the sac-like part of the male reproductive system outside the body cavity that contains the testicles and the epididymis.

escroto parte con forma de saco del sistema reproductivo masculino afuera de la cavidad del cuerpo que contiene los testículos y el epidídimo.

segregated early weaning (SEW) a system in which pigs are weaned from the sow at 9 to 20 days and placed in an isolated, clean facility.

destete temprano segregado (DTS) sistema en el cual los cerdos se destetan de la madre entre los 9 y los 20 días y se colocan en instalaciones aisladas y limpias.

selection identification and use for breeding purposes of those animals with traits that are considered by the breeder to be desirable.

selección identificación y uso para reproducción de los animales y los rasgos que el criador considera deseables.

self-fed a method of feeding in which the animal is given free access to all the feed it will eat.

autoalimentación método de alimentación en el cual el animal tiene libre acceso a todo el alimento que desee comer.

semen the mixture of the seminal and prostate fluid and the sperm.

semen mezcla del líquido seminal y el prostático con los espermatozoides.

semiconservative replication a process that results in a new DNA double helix molecule that consists of one old strand and one new strand.

duplicación semiconservadora proceso que da como resultado una nueva molécula de ADN con doble hélice que comprende una cadena antigua y una nueva.

seminal fluid fluid in the seminal vesicles that mixes with the sperm cells.

líquido seminal líquido en las vesículas seminales que se mezcla con los espermatozoides.

seminal vesicles an accessory gland in the male reproductive system that produces a fluid that protects and transports the sperm.

vesículas seminales glándula accesoria en el sistema reproductivo masculino que produce un líquido que protege y transporta los espermatozoides.

sexed chicks young chickens sorted into groups of male and female.

pollos por sexo pollos jóvenes separados en grupos de machos y hembras.

sex-influenced gene gene that is dominant in one sex but recessive in the other.

gen influenciado por el sexo gen que es dominante en un sexo y recesivo en el otro.

sex-limited gene the phenotypic expression of some genes is determined by the presence or absence of one of the sex hormones; its expression is limited to one sex.

gen limitado por el sexo la expresión fenotípica de algunos genes está determinada por la presencia o la ausencia de una de las hormonas sexuales; su expresión está limitada a un sexo.

sex-linked genes genes that are carried only on the sex chromosomes.

genes vinculados al sexo genes que solo se transportan en los cromosomas sexuales.

sexual reproduction a method of reproduction that involves the union of a sperm and an egg.

reproducción sexual método de reproducción que implica la unión de un espermatozoide y un óvulo.

shearer lamb a lamb that is shorn and fed to a higher level of finish before slaughter.

oveja legadora oveja esquilada que se alimenta para llevarla a un nivel de acabado más alto antes del sacrificio.

sheath in mammal reproductive systems, a tubular fold of skin that covers the penis.

vaina en el sistema reproductivo de los mamíferos, pliegue tubular de piel que cubre el pene.

shigella rod-shaped bacteria of the genus *Shigella*, including some species that cause dysentery.

shigellas bacterias con forma de barra del género *Shigella*, que incluye algunas especies que causan disentería.

shrinkage the weight lost by an animal while it is being shipped to market.

achicadura peso perdido por un animal mientras se está enviando al mercado.

sigmoid flexure the part of the male reproductive system that helps to extend the penis from the sheath for the purpose of copulation.

flexura sigmoidea parte del sistema reproductivo masculino que ayuda a extender el pene desde la vaina a los fines de la copulación.

sire the male parent of an animal.

progenitor el macho padre de un animal.

sire summary a record that provides information on traits that are economically important to cattle producers.

resumen de progenitores registro que brinda información sobre los rasgos económicamente importantes para los productores de ganado.

skatole a white crystalline organic compound, C_9H_9N, which has a strong fecal odor and is found in feces.

escatol compuesto orgánico blanco cristalino, C_9H_9N, de fuerte olor fecal y que se encuentra en las heces.

slaughter calf a calf that is between 3 months and 1 year of age at the time of slaughter.

ternero de sacrificio ternero entre 3 meses y 1 año de edad al momento del sacrificio.

smooth involuntary muscle unstriated muscle that surrounds the hollow internal organs of the body, such as the blood vessels, stomach, intestines, and bladder.

músculo liso involuntario músculo no estriado que rodea los órganos internos huecos del cuerpo, como los vasos sanguíneos, el estómago, los intestinos y la vejiga.

somatic cells leukocytes and body cells.

células somáticas leucocitos y células del cuerpo.

somatotropin a polypeptide hormone secreted by the anterior lobe of the pituitary gland that promotes growth of the body and influences the metabolism of proteins, carbohydrates, and lipids; also called growth hormone or somatotropic hormone.

somatotropina hormona polipeptídica segregada por el lóbulo anterior de la glándula pituitaria que promueve el crecimiento del cuerpo e influencia el metabolismo de proteínas, carbohidratos y lípidos; también se denomina hormona del crecimiento u hormona somatotrópica.

soring the prohibited practice of using chemical or mechanical irritants on the forelegs of the horse.

ulceración práctica prohibida de usar irritantes químicos o mecánicos en las patas delanteras del caballo.

sow an older female swine that has farrowed or is showing signs of pregnancy.

cerda porcino hembra mayor que ha parido o que muestra signos de preñez.

specific pathogen free (SPF) swine breeding stock that are surgically removed from the sow under antiseptic conditions and raised in disease-free conditions.

libre de patógenos específicos (LPE) plantel de cerdos reproductores que se extraen quirúrgicamente de la cerda en condiciones antisépticas y se crían en condiciones libres de enfermedades.

sperm male gamete or reproductive cell.

espermatozoide gameto o célula reproductiva masculina.

spermatic cord protective sheath around the vas deferens.

cordón espermático vaina protectora alrededor del vaso deferente.

spermatid any one of the four haploid cells that form during meiosis in the male that develop into spermatozoa.

espermátida cualquiera de las cuatro células haploides que se forman durante la meiosis en el macho y que se desarrollan en espermatozoides.

spermatocyte diploid cells that divide by meiosis to produce four spermatids.

espermatocito células diploides que se dividen por meiosis para producir cuatro espermátidas.

spermatogenesis formation and development of spermatozoa by meiosis.

espermatogénesis formación y desarrollo de espermatozoides por meiosis.

spermatogonia cells in male gonads that are progenitors of spermatocytes.

espermatogonia célula en las gónadas masculinas que son los progenitores de los espermatocitos.

spinning count the number of hanks of yarn that can be spun from 1 pound of wool top.

recuento de giros número de madejas de hilado que pueden enroscarse a partir de 0,45 kg (1 lb) de estambre de lana.

split-sex feeding feeding barrows and gilts separately to improve performance.

alimentación por sexos separados alimentar a capones y cerdas por separado para mejorar el rendimiento.

spraddled legs a condition in which the leg is dislocated from the socket where it is attached to the body.

patas abiertas afección en la cual la pata está dislocada del receptáculo donde se fija al cuerpo.

spray-dried porcine plasma (SDPP) a by-product of blood from pork slaughter plants; also called plasma protein.

plasma porcino secado por aspersión (PPSA) derivado de la sangre de mataderos de cerdos; también se denomina proteína plasmática.

spring lambs young sheep that are 3 to 7 months of age and usually weigh from 70 to 90 pounds.

ovejas de primavera ovejas jóvenes de 3 a 7 meses de edad, generalmente de 31,8 a 40,8 kg (70 a 90 lb).

stag a male animal, usually beef or swine, which was castrated after reaching sexual maturity and shows the physical traits of the uncastrated male.

torete animal macho, habitualmente bovino o porcino, castrado tras alcanzar la madurez sexual y que muestra los rasgos físicos del macho no castrado.

staggy having the appearance of a mature male.

torete que tiene la apariencia de un macho maduro.

stallion generally, a mature male horse 4 years of age or older; in the Thoroughbred breed, 5 years or older.

garañón generalmente, un caballo macho maduro de 4 años o más; en la raza purasangre, de 5 años o más.

staple a fiber of materials such as wool, cotton, or flax, either in its natural state or after it has been carded or combed; also refers to the length, fineness, condition, or grade of the fiber.

fibra fibra de materiales como lana, algodón o lino, ya sea en estado natural o una vez cardado o peinado; también se refiere a la longitud, la finura, la condición o el grado de la fibra.

steer a male bovine animal that was castrated before reaching sexual maturity.

novillo animal bovino macho castrado antes de alcanzar la madurez sexual.

sterile unable to reproduce.

estéril que no se puede reproducir.

stillage a by-product of grain alcohol fermentation; also called distillers grains.

residuo de destilación derivado de la fermentación del alcohol de grano; también se denomina granos de destilador.

stockmanship the knowledgeable and skillful care and handling of livestock, using a holistic approach to meet the needs of animals. It includes providing for the proper health, nutrition, housing and general welfare and handling of animals. Also known as *animal husbandry*.

ganadería el cuidado y la manipulación informados y diestros de ganado, mediante un enfoque holístico para cubrir las necesidades de los animales. Incluye tomar medidas para la salud, la nutrición, el alojamiento y el bienestar general y la manipulación de los animales. También se denomina *zootecnia*.

stockyard a series of pens or yards where market animals are collected for sale.

corral serie de rediles o patios donde se reúnen animales de mercado para la venta.

stomach the organ in the digestive system that receives the feed and adds chemicals that help in the digestive process.

estómago órgano del sistema digestivo que recibe el alimento y agrega sustancias químicas que ayudan en el proceso digestivo.

straightbreeding mating animals of the same breed.

reproducción directa apareamiento de animales de la misma raza.

straight-run chicks young chickens not sorted by sex.

pollos no sexados pollitos jóvenes no separados por sexo.

striated voluntary muscle muscle that has dark bands that cross each muscle fiber; "meat" of farm animals; also called skeletal muscle.

músculo voluntario estriado músculo que tiene bandas oscuras cruzando cada fibra muscular; "carne" de los animales de granja; también se denomina músculo esquelético.

strongyle a type of parasitic, bloodsucking worms that attack the organs and tissues of animals.

estróngilo tipo de gusano parásito hemíptero que ataca los órganos y los tejidos de los animales.

subclinical disease a disease that is present in an animal's body at levels too low to cause visible effects.

enfermedad subclínica enfermedad presente en el cuerpo de un animal en niveles demasiado bajos como para causar efectos visibles.

subtherapeutic feeding an antimicrobial drug below the dosage level used to treat diseases; for example, the subtherapeutic feeding of penicillin to livestock.

subterapéutico administración de un fármaco antimicrobiano por debajo del nivel de dosificación usado para tratar enfermedades; por ejemplo, la administración subterapéutica de penicilina al ganado.

superovulation the stimulation of more than the usual number of ovulations during a single estrus cycle due to the injection of certain hormones.

superovulación estimulación de un número de ovulaciones superior al habitual durante un solo ciclo de estro debido a la inyección de ciertas hormonas.

supply the amount of a product that producers will offer for sale at a given price at a given time.

suministro cantidad de un producto que los productores pondrán a la venta a un precio determinado en un momento dado.

Sus scrofa European wild boar.
Sus scrofa padrillo salvaje europeo.

Sus vittatus East Indian wild pig.
Sus vittatus cerdo salvaje de india oriental.

swirl hair growing in a circular pattern from the roots.
remolino pelo que crece de forma circular desde las raíces.

synchronization of estrus the use of various compounds (usually hormones) to cause all of the females in a herd to come into heat within a short period of time.
sincronización del estro uso de diversos compuestos (generalmente hormonas) para que todas las hembras de un hato entren en celo en un período breve.

synovia a transparent lubricating fluid secreted by the synovial membrane.
sinovia líquido lubricante transparente segregado por membrana sinovial.

synovial membrane a thin, pliable structure surrounding a joint.
membrana sinovial estructura delgada, flexible que rodea una articulación.

syrinx the structure located at the lower end of the trachea that enables birds to make sounds.
siringe estructura ubicada en el extremo inferior de la tráquea que permite que las aves produzcan sonidos.

system a group of organs that carries out a major function.
sistema grupo de órganos que cumple una función importante.

systemic circulation system the body system that carries blood through the rest of the body and eventually returns the deoxygenated blood to the heart.
sistema circulatorio sistémico sistema corporal que transporta la sangre a través del resto del cuerpo y finalmente devuelve la sangre desoxigenada al corazón.

systemic insecticide an insecticide capable of absorption into plant sap or animal blood and lethal to insects feeding on or within the treated host.
insecticida sistémico insecticida que se puede absorber en la savia de una planta o la sangre de un animal y es mortal para los insectos que se alimenten con o dentro del huésped tratado.

T

tack the equipment used for riding and showing horses.
arreo equipo usado para montar y exhibir caballos.

tagging shearing a ewe around the udder, between the legs, and around the dock; also called crutching.
etiquetado esquilar una oveja alrededor de la ubre, entre las patas y alrededor del rabo; también se denomina crutching.

Tahrs species of wild goat believed to be the ancestor of some of today's domestic goats.
tahrs especie de cabra salvaje que se considera el ancestro de algunas de las razas domésticas actuales.

tail docking cutting off part of an animal's tail.
formar el rabo cortar parte de la cola de un animal.

talent a natural aptitude a person possesses for performing an activity particularly well.
talento aptitud natural que posee una persona para realizar una actividad especialmente bien.

tankage animal tissues and bones from animal slaughter houses and rendering plants that are cooked, dried, and ground and used as a protein supplement.
tancage tejidos y huesos animales de mataderos y de plantas de fusión de grasas que se cocinan, se secan y se muelen y se usan como suplemento proteico.

teat the outlet for milk produced in the udder.
teta la salida para la leche que produce la ubre.

teat cup shell the metal part of the milking unit that contains the inflation and is attached to the teat during the milking operation.
cubierta de la taza de la tetina la parte metálica de la unidad ordeñadora que contiene la inflación y que se fija a la teta durante la operación de ordeñe.

telophase final stage of mitosis or meiosis; in this stage the chromosomes are grouped in new nuclei.
telofase etapa final de mitosis o meiosis; en esta etapa los cromosomas se agrupan en nuevos núcleos.

testicles male organs that produce the sperm cells.
testículos órganos masculinos que producen los espermatozoides.

testosterone male hormone that controls the traits of the male animal.
testosterona hormona masculina que controla los rasgos del animal macho.

tetanus a condition caused by bacteria that results in nervousness, muscle stiffness and spasms, paralysis, and sometimes death. The bacteria usually enter the animal's body through a puncture wound but may enter through other types of wounds.
tétano afección causada por bacterias que produce nerviosismo, rigidez espasmos musculares, parálisis y, a veces, la muerte. Generalmente, las bacterias ingresan al cuerpo del animal a través de una herida punzante, pero pueden ingresar a través de otros tipos de heridas.

three-breed rotation cross mating crossbred females with a sire of a third breed.
cruza de rotación de tres razas apareamiento de hembras cruzadas con un macho de una tercera raza.

thrombin a material that comes from the blood of animals and is a coagulant used in surgery to help make blood clot.
trombina material que proviene de la sangre de animales y es un coagulante que se usa en cirugía para ayudar a que la sangre coagule.

thymine pyrimidine base that is a part of DNA; normally pairs with adenine in DNA.
timina base pirimidina que forma parte del ADN; normalmente se aparea con la adenina en el ADN.

tilting table a device used for restraining an animal in a horizontal position.
mesa inclinable dispositivo que se usa para inmovilizar a un animal en posición horizontal.

tobiano a color pattern in horses in which the head is a solid color; the legs are white, at least below the knees and hocks; and there are regular spots on the body.
tobiano patrón de color en los caballos en el cual la cabeza es de color liso, las patas son blancas, al menos debajo de las rodillas y los codillos, y hay manchas regulares en el cuerpo.

tom a male turkey.
pavo pavo macho.

total digestible nutrients (TDN) the total of the digestible protein, digestible nitrogen-free extract, digestible crude fiber, and 2.25 times the digestible fat.
nutrientes digeribles totales (NDT) el total de la proteína digerible, el extracto libre de hidrógeno digerible, la fibra cruda digerible y 2,25 veces la grasa digerible.

tovero a color pattern in horses in which there is dark pigmentation around the ears, which may expand to cover the forehead and/or eyes; one or both eyes are blue; and chest and flank spots vary in size and may accompany smaller spots, which extend forward across the barrel and up over the loin.

tovero patrón de color en los caballos en el cual hay una pigmentación oscura alrededor de las orejas que puede extenderse hasta cubrir la frente y/o los ojos; uno o ambos ojos son azules y las manchas del pecho y los flancos varían de tamaño y pueden acompañar manchas más pequeñas que se extienden hacia adelante por la barriga y hacia arriba sobre el lomo.

toxemia a condition resulting from the spread of bacterial toxins through the bloodstream.

toxemia afección que resulta de la difusión de toxinas bacterianas por la corriente sanguínea.

toxin a poisonous substance produced by the metabolism of plant or animal organisms.

toxina sustancia venenosa producida por el metabolismo de organismos vegetales o animales.

trace organic compound any of certain chemical compounds necessary in minute quantities for optimum growth and development of plants and animals.

compuesto orgánico de oligo elementos cualquiera de ciertos compuestos químicos necesarios en cantidades mínimas para un crecimiento y desarrollo óptimos de plantas y animales.

trachea the passage through which air passes to and from the lungs (windpipe).

tráquea canal a través del cual el aire pasa hacia y desde los pulmones (vía aérea).

transgenic refers to genetically engineered plants and animals.

transgénico se refiere a plantas y animales producidos mediante ingeniería genética.

transponder a radio receiver activated for transmission by the reception of a predetermined signal.

transpondedor receptor de radio que se activa para transmitir por la recepción de una señal predeterminada.

trap nest a nest equipped with a door that automatically closes when the bird enters the nest.

nido trampa nido equipado con una puerta que se cierra automáticamente cuando el ave ingresa al nido.

trenbolone acetate a synthetic male hormone.

acetato de trembolona hormona masculina sintética.

tunnel ventilation a system that uses air inlets at one end of the building and exhaust fans at the other end to move air through the building at a minimum velocity of 350 feet per minute.

ventilación en túnel sistema que usa tomas de aire en un extremo del edificio y ventiladores extractores en el otro para mover aire a través del edificio a una velocidad mínima de 107 m (350 pies) por minuto.

turbinate bone a thin, bony plate on the wall of the nose.

hueso cornete placa ósea delgada sobre la pared de la nariz.

type the general build of an animal; or, the purpose for which an animal is bred.

tipo complexión general de un animal, o el propósito para el cual se reproduce un animal.

U

udder the milk-producing gland of mammals such as cows.

ubre glándula productora de leche en mamíferos como las vacas.

ultrasonics the use of high frequency sound waves to measure fat thickness and loin-eye area.

ultrasonido uso de ondas sonoras de alta frecuencia para medir el espesor de la grasa y el área del ojo de lomo.

umbilical cord the part of the fetal membranes that connects to the navel of the fetus and carries the blood vessels that, before birth, supply nutrients and oxygen to and carry off waste products from the placenta.

cordón umbilical parte de las membranas fetales que se conecta con el ombligo del feto y tiene los vasos sanguíneos que, antes del nacimiento, suministran nutrientes y oxígeno y retiran los productos de desecho de la placenta.

unsoundness a defect that affects the usefulness of a horse.

falsedad defecto que afecta la utilidad de un caballo.

upper critical temperature (UCT) the temperature at which animals will show symptoms of heat stress; feed intake is generally lower.

temperatura crítica superior (TCS) temperatura en la cual los animales mostrarán síntomas de estrés por calor; la ingesta de alimento suele ser menor.

uracil pyrimidine base that is a part of RNA.

uracilo base pirimidina que forma parte del ARN.

urea a synthetic nitrogen source that is manufactured from air, water, and carbon.

urea fuente de nitrógeno sintética que se fabrica con aire, agua y carbono.

urethra the tube that carries urine from the bladder.

uretra tubo que lleva la orina hasta la vejiga.

urine liquid waste collected in the bladder.

orina desecho líquido que se recolecta en la vejiga.

uterus part of the female reproductive system where the fetus grows; also called the *womb*.

útero parte del sistema reproductivo femenino donde crece el feto; también se denomina *vientre*.

V

vagina in the female reproductive system, the passage between the cervix and the vulva.

vagina en el sistema reproductivo femenino, pasaje entre el cuello del útero y la vulva.

variety a subdivision of bird breeds based on certain traits, such as color of plumage and comb type.

variedad subdivisión de razas de pájaros con base en ciertos rasgos, como el color del plumaje y el tipo de peine.

vas deferens the tube that connects the epididymis with the urethra, providing a passageway for the sperm cells.

vaso deferente tubo que conecta el epidídimo con la uretra y ofrece una vía para los espermatozoides.

veal meat from calves.

carne carne de ternera.

vealer a calf less than 3 months old that is grown for veal.

ternero ternero de menos de 3 meses que se cría para carne.

vegetable protein protein supplement that comes from plant sources.

proteína vegetal suplemento proteico que proviene de plantas.

velvet the soft, furry covering on the developing antlers of elk.

terciopelo cubierta blanda y peluda en la cornamenta en desarrollo de los alces.

vent the external opening of the lower end of the digestive system in poultry.

ventilación apertura externa del extremo inferior del sistema digestivo de las aves de corral.

vertical integration the linking together of two or more steps of production, marketing, and processing.

integración vertical enlace de dos o más pasos de producción, comercialización y procesamiento.

vice any seriously undesirable habit of an animal.

vicio cualquier hábito seriamente indeseable de un animal.

villi small finger-like projections that line the walls of the small intestine; they increase the absorption area of the small intestine.

vellosidades proyecciones pequeñas con forma de dedos que recubren las paredes del intestino delgado; aumentan la superficie de absorción del intestino delgado.

virulence the capacity of a microorganism to cause disease.

virulencia capacidad de un microorganismo de causar enfermedad.

virus a self-reproducing agent that is considerably smaller than a bacterium and can multiply only within the living cells of a suitable host.

virus agente que se autorreproduce, considerablemente más pequeño que una bacteria y que solo puede multiplicarse dentro de las células vivas de un huésped adecuado.

viscera the internal organs of an animal.

vísceras órganos internos de un animal.

vitalists ancient Greeks who believed that animals had souls but were not as advanced as humans.

vitalistas antiguos griegos que creían que los animales tenían alma pero no eran tan avanzados como los seres humanos.

vitamin an organic compound needed in small amounts for nutrition.

vitamina compuesto orgánico necesario en pequeñas cantidades para la nutrición.

vulva the external opening of the female reproductive and urinary systems.

vulva apertura externa de los sistemas reproductivo y urinario femeninos.

W

warm confinement house a hog shelter that maintains a desired temperature regardless of the outside temperature.

alojamiento de encierro caliente refugio para cerdos que mantiene la temperatura deseada independientemente de la temperatura externa.

wattle an projection of skin hanging from the chin or throat, especially in poultry and some breeds of goats.

barbillón proyección de piel que cuelga de la barbilla o la garganta, especialmente en las aves de corral y en algunas razas de cabra.

western ewe an ewe that is produced in the western range area of the United States.

oveja occidental una oveja que se produce en la zona del cordón occidental de EE. UU.

wether a male sheep or goat that has been castrated before reaching sexual maturity.

carnero castrado oveja o cabra macho que ha sido castrado antes de alcanzar la madurez sexual.

wind chill index the measure of the combined effect of air temperature and speed of air movement.

índice de enfriamiento por el viento medida del efecto combinado de la temperatura del aire y la velocidad del movimiento del aire.

wing web vaccination in poultry, the process of injecting a vaccine into the skin on the underside of the wing web at the elbow.

vacunación intraalar en las aves, proceso de inyectar una vacuna en la piel debajo del patagio en el codo.

withdrawal period the length of time a feed additive must not be fed to an animal prior to slaughter.

período de supresión tiempo durante el cual no se debe administrar un aditivo de alimento antes del sacrificio.

wool top partially processed wool.

estambre lana parcialmente procesada.

woolen a division of apparel wool that uses shorter fibers, making a weaker yarn.

lanoso división de la lana para ropa que usa fibras más cortas, lo cual hace que la hebra sea más débil.

worsted a division of apparel wool that uses longer fiber wools, making a relatively strong yarn.

estambre división de la lana para ropa que usa lanas de fibras más largas, lo cual hace que la hebra sea relativamente fuerte.

Y

yardage the fee charged for the use of stockyard facilities.

yardaje tarifa que se cobra por el uso de instalaciones de corral.

yearling an animal between 1 and 2 years of age.

tusón animal entre 1 y 2 años de edad.

yearling feeders cattle that are 1 to 2 years of age and are sold to be fed to finish for slaughter.

alimentadores de tusones ganado de 1 a 2 años de edad que se vende para ser alimentado para lograr el acabado para el sacrificio.

yield the dressing percent (weight of the chilled carcass compared to the live weight) of the animal.

rendimiento porcentaje de agregados (peso de la carcasa enfriada comparado con el peso vivo) del animal.

yield grade a numerical score (with 1 being the highest yielding and 5 being the lowest yielding) given to beef carcasses; this score is based on the estimated carcass weight in boneless, closely trimmed retail cuts from the round, loin, rib, and chuck.

grado de rendimiento puntuación numérica (donde 1 es el rendimiento máximo y 5 el mínimo) que se aplica a los cadáveres de bovinos; esta puntuación se basa en el peso estimado del cadáver en cortes al menudeo sin hueso, bien recortados, de redondo, lomo, costilla y cogote.

Z

zoning laws used by counties, cities, and other local governments to define how property may be used.

leyes de zonificación usadas por condados, ciudades y otros gobiernos locales para definir cómo puede usarse la propiedad.

zoonoses diseases and parasites that may be transmitted between humans and animals.

zoonosis enfermedades y parásitos que pueden transmitirse entre seres humanos y animales.

zygote fertilized egg cell formed by the union of two gametes.

cigota célula huevo fertilizada formada por la unión de dos gametos.

Index

Note: Page numbers followed by *f* and *t* represent figures and tables respectively.

C

D

E

H

Q

R

S

X

Y

Z